W9-DGG-344

Award-winning teacher Michael D. Johnson demystifies how the human body works and teaches you how to be a smarter consumer of health and science information.

12
Sensory Mechanisms

Colorized SEM of the surface of the tongue, showing papillae (pointing toward the back of the tongue).

Current Issue

DWD: Driving While Distracted

Linda Doyle died on September 3, 2008, when her Toyota Rav4 was struck broadside by a Ford pickup truck in an intersection in Oklahoma City. The 20-year-old driver of the truck, a college student named Christopher Hill, admitted to police that he ran a red light at 45 miles per hour because he had been distracted by a cell phone call. Asked what color the light was when he went through it, he said, "I never saw it."

Your brain receives sensory inputs from many different sources at once. It copes with all this sensory input by focusing on input considered most important at the moment, sometimes at the expense of other input. Anything that distracts you while driving is likely to lower your ability to drive safely. Distractions may be visual (taking your eyes off the road), cognitive (thinking about something else), or manual (taking your hands off the wheel). Typical distractions include texting or talking on the phone, putting on makeup, eating, reading, or using a navigation system. Some distractions, such as texting, are especially risky because they involve all three types of distractions at once (visual, cognitive, and manual).

According to the National Highway Traffic Safety Administration (NHTSA), more than half a million injuries and nearly 6,000 deaths a year are due to car crashes involving distracted drivers. The highest proportion of distracted drivers involved in fatal crashes are under 20 years old.

Driving while on the phone is the most common distraction. The NHTSA recommends that drivers not use their cell phones while driving except in an emergency, but it doesn't appear that anyone is listening. According to the NHTSA, at any given moment during daylight hours nearly 800,000 drivers are on their cell phones while driving

Reacting to the statistics and to high-profile deaths like Linda Doyle's, some states are beginning to take action. Six states (California, Oregon, Washington, New Jersey, New York, and Connecticut) now prohibit all drivers from using handheld cell phones while driving. Nineteen states prohibit text messaging. Seventeen states and the District of Columbia have placed special restrictions on cell phone use by drivers with learner's permits or drivers under a certain age. Other states are likely to follow suit.

The use of handheld cell phones is not illegal yet in Oklahoma but Linda Doyle's daughter, Jennifer Smith, continues to press lawmakers for change. She's also suing Samsung, maker of the phone used by the driver, and the Sprint Nextel

Christopher Hill

service that provided the phone coverage in the area. The suit alleges that the companies failed to warn consumers about the dangers of driving while using their product. Sprint Nextel says that it includes adequate safety messages in its packaging, user manuals, advertising, and even on its Web site.

According to legal experts, Ms. Smith's lawsuit faces an uphill battle because crashes are caused by drivers, not cell phones, and because deep down inside, most drivers understand that talking on the phone while driving is a distraction but they continue to do it anyway. Lawyers for Samsung and Sprint Nextel are likely to argue that the companies had no responsibility to Linda Doyle because she did not use their phone or phone service; their only responsibility is to the young man who was driving the other vehicle, and he's not suing them.

Nevertheless, momentum is building to do something about the number of accidents being attributed to distracted drivers. "Every single time someone takes their eyes or their focus off the road—even for just a few seconds—they put their lives and the lives of others in danger," said Transportation Secretary Ray LaHood at a national summit on distracted driving. "Distracted driving is unsafe, irresponsible and in a split second, its consequences can be devastating."

Christopher Hill, the driver of the pickup truck that killed Linda Doyle,

and Jennifer Smith, her daughter, would undoubtedly agree. Christopher Hill plead guilty to negligent homicide (a misdemeanor) and was sentenced to five years of probation plus 240 hours of community service. He deeply regrets the accident and took responsibility for his actions, saying it was his choice to talk on the phone while driving. Linda Doyle's family says they have forgiven him. These days he talks to schools and community groups about the dangers of driving while distracted. Jennifer Smith, for her part, is now a spokesperson for FocusDriven, the first national nonprofit organization for increasing awareness about the dangers of driving while distracted.

Both Christopher Hill and Jennifer Smith say they no longer use their cell phones while driving.

Jennifer Smith

The facts...
- More than 6,000 deaths and 500,000 injuries per year are attributed to driving while distracted.
- At any given moment during daylight hours, more than 800,000 drivers are talking on their cell phones.
- The highest proportion of distracted drivers is under 20 years old.
- Laws against driving while distracted (texting, talking on a handheld phone) vary from state to state, from complete bans to no ban at all. There is no national standard.

Questions to consider

1 Do you think that texting while driving should be illegal nationwide? What about talking on a handheld phone?

2 When you cross a state line, how do you know whether texting or using a phone is illegal in the state you are entering? Should you be told?

3 In several states (Utah and New Hampshire), using a handheld phone is a violation only if the driver commits another moving offense while on the phone. Is that a reasonable solution?

275

■ **NEW!** Now located at the start of each chapter, popular **Current Issue** essays draw you into the subject with interesting science and health news items, connecting human biology to real-world issues. Each essay provides contrasting views on the featured hot topic. New topics include the controversies surrounding driving while distracted, the promise and perils of genetic testing, the effectiveness of antioxidants, and the implications of the black market in human tissues and organs.

Each essay includes a summary of **The Facts** behind the topic, providing context for you to think critically about the issue presented.

Questions to Consider at the end of each essay ask you to formulate your own opinions on the featured issue.

Join the conversation
on the latest in human biology-related issues

MJ's Human Biology Blog

"MJ's Human Biology blog" highlights recent items in the news or in scientific journals that cover new findings and hot topics relevant to the teaching of human biology. The host is Dr. Michael Johnson, author of Johnson's Human Biology: Concepts and Current Issues. The postings have recently been opened to comments so that Human Biology instructors and students can share their own views on the topics.

MONDAY, OCTOBER 11, 2010

Cell Phones Can't Cause Cancer?

One of the limitations of the scientific method is that it is virtually impossible to prove that something doesn't exist or could never happen based on empirical data alone. And if that "something" has the potential for great harm, then common sense would dictate that we should err on the side of caution and continue to assume that it could exist, or might happen. It's what's known as the precautionary principle.

Take cell phones and cancer, for example. Despite tens of millions of dollars spent on research, no one has ever proven that mobile phones cause brain cancer. And yet, researchers (and the public) are still

Contributors
- administrator
- Michael Johnson

Topics
- blood (6)
- cancer (15)
- cells (1)
- cells and tissues (8)

NEW! MJ's Human Biology Blog offers you an opportunity to read and respond to the author's thoughts on human biology-related issues. The author's blog is frequently updated with relevant news items on the latest developments in science and health research. Access the blog through the Human Biology Place at **www.humanbiology.com**

Share   Report Abuse   `Next Blog»`   Create Blog   Sign In

MJ's Human Biology Blog

Sensing Danger in the Air

Scientists have long known that some animals can sense danger in the air, but little was known about how they did it. Now they have a clue. According to a report in *Science*, specialized neural cells in the olfactory system of mice can detect unidentified alarm pheromones given off by other mice under stress. The scientists collected air from around stressed mice and then exposed other mice to it. Normal mice froze (a typical danger reaction in mice) when exposed to the alarm-pheromone scented air, whereas mice whose special olfactory neural cells had been destroyed did not respond to alarm-pheromone scented air at all.

It would be interesting to know whether humans also give off alarm pheromones. Could this be an explanation for why people seem to internalize the stress of others around them? ■

Reference: Brechbuhl, Julien, *et al.* Grueneberg Ganglion Cells Mediate Alarm Pheromone Detection in Mice. *Science* 321: 1092–1095, 2008.

NEW! Thoughtfully selected **blog entries appear alongside relevant content** throughout each chapter so you can connect what you learn to science and health news and research.

Interact online
with media that brings human biology to life

Engaging media make studying easier and interactive.

NEW! With a Pearson **eText** you can study online whenever and wherever you want. The *Human Biology* eText is an online copy of the book, powered with interactive and customization functions, including a new whiteboard-style annotation tool.

NEW! Media callouts throughout the book spotlight **Web Animations** of key biology concepts that you can access online at the Human Biology Place.

Human Biology Place | www.humanbiology.com

Improved and updated, the **Human Biology Place** open-access website helps you get the most from your book and the course.

Loaded with useful resources, this site features:

- An online chapter guide
- Animated Web activities with accompanying quizzes
- Text-specific self tests and essay questions that encourage critical thinking
- Crossword puzzles
- Flashcards and a glossary
- Answers to textbook questions
- Links to human biology-related news stories updated every day
- Science Daily RSS feeds

Move beyond memorizing

Critical thinking questions and comprehensive chapter review help you truly understand the concepts and issues of human biology.

■ NEW! **Key Concepts** at the beginning of each chapter focus your attention on the main ideas to take away from the chapter.

Key concepts

>> **Sensory mechanisms provide information** about the world around us and also about conditions within our own bodies.

>> **Various types of receptors convert different types of sensory stimuli (physical touch, odors, light, etc.) into action potentials.** The brain interprets incoming action potentials correctly by where they go in the brain.

>> **Some receptors adapt quickly,** which is why you stop noticing some stimuli after a while or they may seem less intense.

>> **Receptors are located throughout the body.** They provide us with information about body position, touch, temperature, vibration, pressure, and especially pain.

>> **The five special senses (taste, smell, hearing, balance, and vision) originate from special areas of the body.** Four of them (taste, smell, hearing, and vision) provide us with detailed information about the external world.

✓ Quick Check Will the immune system of an Rh *positive* woman attack blood cells from an Rh *negative* baby? Why or why not? ■

a) When an Rh positive man fathers a child by an Rh negative woman, the fetus may inherit the Rh positive antigen.

b) During pregnancy or more commonly at childbirth, a small amount of fetal blood enters the mother's circulation.

c) Over the next several weeks the woman develops antibodies and an immune memory against the Rh antigen.

d) When the woman becomes pregnant with her second Rh positive child, her immune system quickly produces antibodies that attack the fetus' red blood cells.

Figure 7.13 **How Rh factor incompatibility can affect a fetus.**

✓ *In which population—Caucasian, African American, or Native American—would it be least risky to do an emergency blood transfusion without blood typing either the donor or the recipient?*

■ NEW! **Quick Check** critical thinking questions at the end of challenging text sections and accompanying selected figures help you better understand the material.

Get the most from your course

Health & Wellness

Cholesterol and Atherosclerosis

As noted in Chapter 2, cholesterol is a key component of all cell membranes and the precursor molecule for several hormones. All cells require a certain amount of it for normal functioning. However, too much cholesterol in the blood can lead to a condition called *atherosclerosis*—a thickening of an arterial vessel wall due to the buildup of fatty materials containing cholesterol. Left untreated, atherosclerosis contributes to heart attacks, strokes, aneurysms, and peripheral vascular disease.

Most of the cholesterol in the blood is bound to certain carrier proteins. Together, the cholesterol and the protein are called a *lipoprotein*. There are two types of lipoproteins, based on their densities. One of them, called *low-density lipoprotein (LDL)*, is considered "bad" in terms of atherosclerosis. When present in normal amounts, LDL transports cholesterol throughout the body and makes it available to cells. However, when there is too much LDL it begins to attach to the cells lining the arterial blood vessel wall and then makes its way into the cells. Once inside the cell, LDL triggers an inflammatory response that ultimately results in the buildup of fatty deposits called atherosclerotic *plaques* within the blood vessel wall. Eventually these plaques may rupture, causing blood clots to form that can occlude arteries and cause heart attacks and strokes.

High-density lipoproteins (HDLs), on the other hand, are considered "good" because they target cholesterol for removal. HDLs pick up free cholesterol and carry it to the liver, where it is detached from the protein, mixed with bile, and secreted into the small intestine. Some of the cholesterol in bile is excreted from the body with the feces, although some is reabsorbed, to be used again.

Risk factors for atherosclerosis include factors that raise blood cholesterol (obesity, sedentary lifestyle, and a high-fat diet), smoking, diabetes, hypertension, and a family history of atherosclerosis. Before age 45 men have a 10 times greater risk than women; however, women's risk rises after menopause.

According to the American Heart Association, a total cholesterol of under 200 mg/dl is considered desirable. Ideally, HDL should be greater than 60 mg/dl and LDL should be less than 100 mg/dl. A total cholesterol of greater than 240 mg/dl along with a high LDL and/or low HDL would be cause for concern.

If you're having trouble remembering which lipoprotein is bad for you and which is good, just remember that cholesterol is a lipid, and lipids are less dense than protein or water. So "low-density means more cholesterol," and therefore low-density lipoprotein (LDL) is the "bad" one.

Some degree of atherosclerosis is common with advancing age. However, lifestyle can make a big difference in how rapidly atherosclerosis develops and whether it becomes severe. At the end of this chapter we look at what you can do to lower your risk of atherosclerosis and other cardiovascular conditions. ■

a) Cross-section of an artery narrowed by atherosclerotic plaque.

b) The same photo with the atherosclerotic plaque removed, showing how a normal artery would look.

Atherosclerosis.

■ Streamlined and updated **Health & Wellness** boxes reflect new advances in health and medical research and provide you with practical consumer information, such as the causes and risks of carbon monoxide poisoning and the prevalence and consequences of Viagra abuse.

■ Enhanced! **Test Yourself** end-of-chapter quizzes help you prepare for tests using a new multiple-choice question format that includes questions for Bloom's Taxonomy levels 2 and above (understand, apply, analyze, evaluate, and create). Answers for chapter quizzes are available at the back of the book so you can check your work.

■ Each chapter ends with Test Yourself multiple-choice questions, and **Apply What You Know** critical-thinking questions so you can review concepts, test yourself on content, and apply biology to your everyday life.

Chapter 9 *The Immune System and Mechanisms of Defense* **217**

8. Describe how cells that belong to a particular individual are identified so that the individual's immune system doesn't attack them.
9. Explain how cytotoxic T cells kill target cells.
10. Describe how a vaccine produces immunity from a specific disease.

Test Yourself

Answers can be found in Appendix A.

1. In which of the following ways are bacterial cells similar to human cells?
   a. Bacterial cells have cell walls.
   b. Bacterial cells have a single, circular chromosome.
   c. Bacterial cells use ATP to fuel cellular activities.
   d. Bacterial cells lack mitochondria.

2. Which of the following statements about viruses is true?
   a. Viruses require a host cell in which to reproduce.
   b. Viruses are very small bacteria.
   c. Viral infections can generally be controlled with antibiotics.
   d. Viruses are composed of protein only.

3. Which of the following pathogenic agents causes a self-propagating misfolding of proteins in nerve cells?
   a. bacteria
   b. prions
   c. viruses
   d. helminths (worms)

4. Consider the following group of diseases: hepatitis, chicken pox, warts, and measles. What do these diseases have in common?
   a. They are all caused by bacteria.
   b. They are readily treated with antibiotics.
   c. They are all caused by viruses.
   d. They are very common in patients infected with HIV.

5. Which of the following is true regarding prion diseases?
   a. They are caused by a deadly type of virus.
   b. They can readily be treated with antibiotics.
   c. They can be prevented by vaccinations.
   d. They cause accumulation of misfolded proteins in brain cells.

6. Which of the following is a benefit of resident bacteria?
   a. Resident bacteria cause the stomach to be acidic.
   b. Resident bacteria produce antiviral compounds that prevent viral infections.
   c. Resident bacteria can out-compete harmful bacteria and lower the incidence of infection.
   d. Resident bacteria digest cellulose within the human digestive tract.

7. DiGeorge syndrome is a congenital disease that results in a poorly developed, non-functioning thymus gland. Which of the following would be a likely problem experienced by a baby with DiGeorge syndrome?
   a. lack of B cells
   b. lack of antibodies
   c. lack of T cells
   d. lack of macrophages

8. The following are steps in phagocytosis: (1) Bacterium is digested by lysosomal enzymes, (2) phagocyte approaches bacterium, (3) phagocytic vesicle fuses with lysosome, and (4) phagocyte engulfs bacterium, forming a phagocytic vesicle. In which order do these steps occur?
   a. 4-2-3-1
   b. 2-4-3-1
   c. 2-3-4-1
   d. 4-1-3-2

9. In which of the following choices is the cell correctly matched with its function?
   a. eosinophil: produces antibodies
   b. B lymphocyte: directly attacks foreign cells

   c. basophil: secretes histamine
   d. T lymphocyte: phagocytizes bacteria

10. Each of the following processes helps combat infection except:
    a. inflammation
    b. fever
    c. autoimmunity
    d. antibody production

11. The primary immune response is:
    a. faster than the secondary immune response
    b. longer lasting than the secondary immune response
    c. less effective than the secondary immune response
    d. due to the presence of memory cells

12. Compared to active immunization, passive immunization:
    a. provides immediate protection
    b. is longer lasting
    c. creates a large number of memory cells
    d. may occasionally cause the disease it is intended to prevent

13. Which of the following increases the likelihood of successful organ transplant?
    a. matching the ABO blood group antigens
    b. matching the MHC tissue antigens
    c. administration of immunosuppressive drugs
    d. all of these choices

14. Which of the following does not belong with the others?
    a. lupus erythematosus
    b. rheumatoid arthritis
    c. anaphylactic shock
    d. Type I diabetes mellitus

15. Which of the following statements about HIV is true?
    a. Latex condoms are 100% effective in blocking transmission of HIV.
    b. Most individuals who progress to Phase II of HIV infection remain in Phase II and never progress to Phase III (AIDS).
    c. HIV specifically impairs the cell-mediated immune response.
    d. Anti-HIV medications such as AZT and maraviroc can cure HIV infection.

Apply What You Know

Answers can be found at the Human Biology Place.
www.humanbiology.com

1. When you get a minor infection in a small cut in the skin, sometimes soaking it in hot water speeds the healing process. How might the heat help?
2. Explain why antibiotics don't work against viruses.
3. Everyone knows that bacteria can cause disease. Suppose that we could actually remove all bacteria from our bodies all at once. Would that be a good thing or a bad thing?
4. In 1918 a pandemic of a deadly flu strain killed upward of 30 million people. If that same flu strain were to come around again, would we be better off or worse off than those who were alive in 1918?
5. Researchers have been working on an effective vaccine for gonorrheal infections for some years. One promising vaccine is delivered via a spray into the nasal cavity. This is clearly not the site of gonococcal infection. Why would one administer a vaccine meant to protect the reproductive tract into the nasal cavity?
6. Why is it that most people get the chicken pox only once, but they can get a cold or the flu over and over again throughout a lifetime?
7. The immune system is supposed to defend us from harmful microorganisms. Why doesn't it always work? In other words, why do some people still get sick and die?

Tap into a wealth of resources
to make teaching efficient and learning fun

For Instructors

Instructor Resource DVD
978-0-321-74252-0 • 0-321-74252-4

Augment your lectures and increase student comprehension with a range of ready-to-use media supplements.

- An image library of all the art (a selection of which has customizable labels and stepped out art), tables, and photos from the book
- Editable PowerPoint® lecture presentation slides with embedded links to *ABC News* Videos and Web Activities
- Clicker questions
- Animated Web Activities
- BLAST! Animations
- BioFlix® 3D animations with labeled PowerPoint slides

- 28 *ABC News* video clips
- PowerPoint slides for *Scientific American: Concepts and Current Issues in Biology*, Volumes 1–6
- *Interactive Physiology® for Human Biology* slides, worksheets, and answer sheets
- Interactive Quiz Show games
- Microsoft® Word files for the Instructor Manual and Test Bank
- An all-electronic set of ready-to-print Transparency Masters of all textbook images

Instructor Guide
978-0-321-74253-7 • 0-321-74253-2

Contains Chapter Objectives, a Suggested Lecture Outline, Lecture Hints, a Media Guide, Class Demonstrations and Student Activities, Suggestions for Possible Assignments, MJ's Human Biology Blog, Integrated Case In Point Sections, and additional resources.

TestGen® Computerized Test Bank
978-0-321-74350-3 • 0-321-74350-4

Includes 75-100 questions per chapter, consisting of Multiple-Choice, True/False, Matching, Short Answer, and Essay questions. Instructors are able to build, edit, print, and administer tests using a computerized bank of questions developed to cover all text objectives.

Course Management

Blackboard Premium
www.pearsonhighered.com/elearning

For Students

Human Biology Place Companion Website
www.humanbiology.com

See page W3 for more information.

Study Guide
978-0-321-74251-3 • 0-321-74251-6

Contains a chapter summary and key concepts, exercises (including matching, short answer, labeling, table completion, fill-in-the-blank, paragraph completion, word choice, true/false, crossword puzzle and new looking back and looking forward questions), a chapter test, and an answer key.

Laboratory Manual
978-0-321-75028-0 • 0-321-75028-4
Explores fundamental concepts of human biology in a laboratory setting through 23 full-color laboratory exercises, each followed by critical thinking questions.

Interactive Physiology for Human Biology CD-ROM
978-0-321-59539-3 • 0-321-59539-4

Provides outstanding animations, activities, self testing, and additional resources. *Can be packaged with the bound textbook at no additional charge.*

Scientific American: Current Issues in Biology, Volumes 1-6

Volume 1: 978-0-8053-7507-7 | 0-8053-7507-4
Volume 2: 978-0-8053-7108-6 | 0-8053-7108-7
Volume 3: 978-0-8053-7527-5 | 0-8053-7527-9
Volume 4: 978-0-8053-3566-8 | 0-8053-3566-8
Volume 5: 978-0-321-54187-1 | 0-321-54187-1
Volume 6: 978-0-321-59849-3 | 0-321-59849-0

Give your students the best of both worlds—accessible, dynamic, relevant articles from *Scientific American* magazine that present key issues in biology, paired with the authority, reliability, and clarity of Benjamin Cummings' biology texts. Articles include questions to help students check comprehension and make connections to science and society.

This resource is available at no additional charge when packaged with a new text.

Human Biology

Concepts and Current Issues

SIXTH EDITION

Michael D. Johnson

Weill Cornell Medical College in Qatar

Benjamin Cummings

Boston   Columbus   Indianapolis   New York   San Francisco   Upper Saddle River
Amsterdam   Cape Town   Dubai   London   Madrid   Milan   Munich   Paris   Montréal   Toronto
Delhi   Mexico City   São Paulo   Sydney   Hong Kong   Seoul   Singapore   Taipei   Tokyo

Editor-in-Chief: Beth Wilbur

Executive Director of Development: Deborah Gale

Acquisitions Editor: Becky Ruden

Senior Development Editor: Susan Teahan

Associate Editor: Brady Golden

Editorial Assistant: Leslie Allen

Executive Marketing Manager: Lauren Harp

Senior Media Producer: Jonathan Ballard

Executive Managing Editor: Erin Gregg

Managing Editor: Michael Early

Senior Production Project Manager: Shannon Tozier

Production Service: PreMedia Global

Illustrations: Imagineering

Text and Cover Design: tt eye, Tani Hasegawa

Manufacturing Buyer: Michael Penne

Senior Photo Editor: Donna Kalal

Photo Research: Kristin Piljay

Cover Printer: Lehigh-Phoenix Color/Hagerstown

Printer and Binder: R.R. Donnelley/Willard

Cover Image: Zap Art/Getty Images

Credits and acknowledgments borrowed from other sources and reproduced, with permission, in this textbook appear on [p. C-1].

Library of Congress Cataloging-in-Publication Data

Johnson, Michael D., 1948–
   Human biology : concepts and current issues / Michael D. Johnson.—6th ed.
     p. cm.
   Includes bibliographical references and index.
   ISBN-13: 978-0-321-70167-1 (alk. paper)
   ISBN-10: 0-321-70167-4 (alk. paper)
   1. Human biology—Textbooks.   I. Title.
   QP34.5.J645 2012
   612—dc22

                  2010046361

ISBN 10: 0-321-70167-4; ISBN 13: 978-0-321-70167-1 (Student edition)

ISBN 10: 0-321-74347-4; ISBN 13: 978-0-321-74347-3 (Exam copy)

ISBN 10: 0-321-75035-7; ISBN 13: 978-0-321-75035-8 (Books a la Carte)

About the Author

Dr. Michael D. Johnson spent most of his youth in the fields and forests of rural Washington, observing nature. He earned his B.S. degree in Zoology from Washington State University and then moved East to earn a Ph.D. in physiology from the University of Michigan. After completing a Postdoctoral Research Fellowship at Harvard Medical School he joined the faculty of West Virginia University, where he remained for most of his career.

In 2008 Dr. Johnson moved to Qatar, on the Arabian Peninsula, to take the position of Associate Dean for Premedical Education at Weill Cornell Medical College in Qatar. In his new position he directs the premedical education of students from more than 25 countries.

Dr. Johnson received several teaching awards during his career, including the West Virginia University Foundation Outstanding Teacher award and the Distinguished Teacher Award of the School of Medicine. He is a member of the American Physiological Society, the Human Anatomy and Physiology Society, the National Association of Biology Teachers, and the American Association for the Advancement of Science.

Whether teaching undergraduates or medical students, Dr. Johnson has a keen interest in instilling in students an appreciation of science. He seeks to show students how the advancement of scientific knowledge sometimes raises unforeseen ethical, political, economic, and social issues for all of us to discuss and solve. Through his teaching and this book, he encourages students to become scientifically literate so that they will feel comfortable making responsible choices as consumers of science.

You can contact Dr. Johnson at mdj5412@gmail.com with comments or questions about this book.

Contents

1 Human Biology, Science, and Society 2

Current Issue
Mandatory Childhood Vaccinations 2

1.1 The characteristics of life 4

1.2 How humans fit into the natural world 7
Living things are grouped according to their characteristics 7

MJ's Human Biology Blog
Thimerosal and Autism 8

The defining features of humans 8
Human biology can be studied on any level of biological organization 9

1.3 Science is both a body of knowledge and a process 11
The scientific method is a process for testing ideas 11
Making the findings known 14

Health & Wellness
The Growing Threat of Antibiotic-Resistant Bacteria 15

A well-tested hypothesis becomes a theory 15

1.4 Sources of scientific information vary in style and quality 15

1.5 Learning to be a critical thinker 16
Become a skeptic 16
Appreciate the value of statistics 16
Learn how to read graphs 17
Distinguish anecdotes from scientific evidence 17
Separate facts from conclusions 18
Understand the difference between correlation and causation 18

MJ's Human Biology Blog
Correlation versus Causation 18

1.6 The role of science in society 19
Science improves technology and the human physical condition 19
Science has limits 20

MJ's Human Biology Blog
Scientific Uncertainty and Shared Responsibility 20

The importance of making informed choices 20

2 The Chemistry of Living Things 24

Current Issue
Functional Foods and Dietary Supplements—Safe and Effective? 24

2.1 All matter consists of elements 26
Atoms are the smallest functional units of an element 26
Isotopes have a different number of neutrons 27

2.2 Atoms combine to form molecules 28
Energy fuels life's activities 28
Chemical bonds link atoms to form molecules 29
Living organisms contain only certain elements 31

2.3 Life depends on water 32
Water is the biological solvent 32
Water helps regulate body temperature 33

2.4 The importance of hydrogen ions 33

MJ's Human Biology Blog
I Don't Hear You... 34

Acids donate hydrogen ions, bases accept them 34
The pH scale expresses hydrogen ion concentration 34
Buffers minimize changes in pH 35

MJ's Human Biology Blog
Getting That Caffeine Buzz 35

2.5 The organic molecules of living organisms 35
Carbon is the common building block of organic molecules 36
Macromolecules are synthesized and broken down within the cell 36

2.6 Carbohydrates: Used for energy and structural support 37
Monosaccharides are simple sugars 37
Oligosaccharides: More than one monosaccharide linked together 38
Polysaccharides store energy 38

2.7 Lipids: Insoluble in water 39
Triglycerides are energy-storage molecules 39
Phospholipids are the primary component of cell membranes 39
Steroids are composed of four rings 40

2.8 Proteins: Complex structures constructed of amino acids 40
Protein function depends on structure 42
Enzymes facilitate biochemical reactions 43

2.9 Nucleic acids store genetic information   **44**

2.10 ATP carries energy   **46**

3 Structure and Function of Cells   50

Current Issue
The Use of Human Stem Cells   50

3.1 Cells are classified according to their internal organization   **52**
Eukaryotes have a nucleus, cytoplasm, and organelles   52
Prokaryotes lack a nucleus and organelles   52

3.2 Cell structure reflects cell function   **53**
Cells remain small to stay efficient   54

MJ's Human Biology Blog
Stem Cell Therapy for Parkinson's?   55

3.3 A plasma membrane surrounds the cell   **55**
The plasma membrane is a lipid bilayer   56

3.4 Molecules cross the plasma membrane in several ways   **57**
Passive transport: Principles of diffusion and osmosis   57

MJ's Human Biology Blog
Birth Dating Human Cells   57

Passive transport moves with the concentration gradient   58
Active transport requires energy   59
Endocytosis and exocytosis move materials in bulk   60
Information can be transferred across the plasma membrane   61
The sodium-potassium pump helps maintain cell volume   61
Isotonic extracellular fluid also maintains cell volume   63

3.5 Internal structures carry out specific functions   **63**
The nucleus controls the cell   63
Ribosomes are responsible for protein synthesis   63
The endoplasmic reticulum is the manufacturing center   65
The Golgi apparatus refines, packages, and ships   66
Vesicles: Membrane-bound storage and shipping containers   66
Mitochondria provide energy   67
Fat and glycogen: Sources of energy   67

3.6 Cells have structures for support and movement   **68**
The cytoskeleton supports the cell   68
Cilia and flagella are specialized for movement   69
Centrioles are involved in cell division   69

3.7 Cells use and transform matter and energy   **69**
Glucose provides the cell with energy   70
Fats and proteins are additional energy sources   75
Anaerobic pathways make energy available without oxygen   76

4 From Cells to Organ Systems   80

Current Issue
Can Lipodissolve Melt Away Fat?   80

4.1 Tissues are groups of cells with a common function   **82**

4.2 Epithelial tissues cover body surfaces and cavities   **82**
Epithelial tissues are classified according to cell shape   83
The basement membrane provides structural support   84

MJ's Human Biology Blog
My Mother's Cells Within Me   85

4.3 Connective tissue supports and connects body parts   **85**
Fibrous connective tissues provide strength and elasticity   85
Specialized connective tissues serve special functions   87

MJ's Human Biology Blog
Fat Cells Are Replaced Throughout Life   88

4.4 Muscle tissues contract to produce movement   **88**
Skeletal muscles move body parts   88
Cardiac muscle cells activate each other   89
Smooth muscle surrounds hollow structures   89

4.5 Nervous tissue transmits impulses   **89**

4.6 Organs and organ systems perform complex functions   **89**
The human body is organized by organ systems   89
Tissue membranes line body cavities   92
Describing body position or direction   93

4.7 The skin as an organ system   **93**
Skin has many functions   93
Skin consists of epidermis and dermis   93

Health & Wellness
Suntans, Smoking, and Your Skin   96

4.8 Multicellular organisms must maintain homeostasis   **97**
Homeostasis is maintained by negative feedback   97
Negative feedback helps maintain core body temperature   98
Positive feedback amplifies events   99

5   The Skeletal System   102

Current Issue
A Black Market in Human Bones?   102

5.1   The skeletal system consists of connective tissue   104
Bones are the hard elements of the skeleton   104
Bone contains living cells   104
Ligaments hold bones together   106
Cartilage lends support   106

5.2   Bone development begins in the embryo   106
MJ's Human Biology Blog
A Really Costly Drug   107

5.3   Mature bone undergoes remodeling and repair   108
Bones can change in shape, size, and strength   108
Bone cells are regulated by hormones   109
Bones undergo repair   109

5.4   The skeleton protects, supports, and permits movement   109
The axial skeleton forms the midline of the body   109
The appendicular skeleton: Pectoral girdle, pelvic girdle, and limbs   112

5.5   Joints form connections between bones   114
Joints vary from immovable to freely movable   114
Ligaments, tendons, and muscles strengthen and stabilize joints   115
MJ's Human Biology Blog
Is Running Hard on Knees?   117

5.6   Diseases and disorders of the skeletal system   117
Sprains mean damage to ligaments   117
Bursitis and tendinitis are caused by inflammation   117
Arthritis is inflammation of joints   117
Health & Wellness
Treating a Sprained Ankle   118
Osteoporosis is caused by excessive bone loss   118
MJ's Human Biology Blog
Treating "Pre-osteoporosis"   119

6   The Muscular System   122

Current Issue
Drug Abuse Among Athletes   122

6.1   Muscles produce movement or generate tension   124
The fundamental activity of muscle is contraction   124
Skeletal muscles cause bones to move   124
A muscle is composed of many muscle cells   126
The contractile unit is a sarcomere   127

6.2   Individual muscle cells contract and relax   128
Nerves activate skeletal muscles   128
Activation releases calcium   128
Calcium initiates the sliding filament mechanism   128
When nerve activation ends, contraction ends   130
MJ's Human Biology Blog
New Drug Test for Athletes   131
Muscles require energy to contract and to relax   131
Health & Wellness
Delayed Onset Muscle Soreness   133

6.3   The activity of muscles can vary   133
Isotonic versus isometric contractions: Movement versus static position   133
The degree of nerve activation influences force   133
Slow-twitch versus fast-twitch fibers: Endurance versus strength   135
Exercise training improves muscle mass, strength, and endurance   136
MJ's Human Biology Blog
Beating the Testosterone Doping Test   136
MJ's Human Biology Blog
Stretching and Sports Injuries   137

6.4   Cardiac and smooth muscles have special features   137
How cardiac and smooth muscles are activated   137
Speed and sustainability of contraction   138
Arrangement of myosin and actin filaments   138

6.5   Diseases and disorders of the muscular system   139
Muscular dystrophy   139
Tetanus   139
Muscle cramps   140
Pulled muscles   140
Fasciitis   140

7   Blood   142

Current Issue
Should You Bank Your Baby's Cord Blood?   142

7.1   The components and functions of blood   144
Plasma consists of water and dissolved solutes   146
Red blood cells transport oxygen and carbon dioxide   146
Hematocrit and hemoglobin reflect oxygen-carrying capacity   147

All blood cells and platelets originate from stem cells 147
RBCs have a short life span 148
RBC production is regulated by a hormone 149
White blood cells defend the body 149

MJ's Human Biology Blog
The Spleen Stores Monocytes 150

Platelets are essential for blood clotting 151

7.2 Hemostasis: Stopping blood loss 151
Vascular spasms constrict blood vessels to reduce blood flow 151
Platelets stick together to seal a ruptured vessel 152

MJ's Human Biology Blog
Platelet-Rich Plasma Therapy Revisited 152

A blood clot forms around the platelet plug 152

Health & Wellness
Donating Blood 153

7.3 Human blood types 154
ABO blood typing is based on A and B antigens 155
Rh blood typing is based on Rh factor 156
Blood typing and cross-matching ensure blood compatibility 157

7.4 Blood disorders 157
Blood poisoning: Infection of blood plasma 157
Mononucleosis: Contagious viral infection of lymphocytes 157

MJ's Human Biology Blog
Cleansing Blood with Magnets 158

Anemia: Reduction in blood's oxygen-carrying capacity 158
Leukemia: Uncontrolled production of white blood cells 158
Multiple myeloma: Uncontrolled production of plasma cells 159
Thrombocytopenia: Reduction in platelet number 159

8 **Heart and Blood Vessels 162**

Current Issue
Comparative Effectiveness Research 162

8.1 Blood vessels transport blood 164
Arteries transport blood away from the heart 164
Arterioles and precapillary sphincters regulate blood flow 166
Capillaries: Where blood exchanges substances with tissues 166
Lymphatic system helps maintain blood volume 167
Veins return blood to the heart 168

8.2 The heart pumps blood through the vessels 169
The heart is mostly muscle 169

MJ's Human Biology Blog
Boosting Cardiac Repair Mechanisms 169

The heart has four chambers and four valves 170
The pulmonary circuit provides for gas exchange 171
The systemic circuit serves the rest of the body 172

Health & Wellness
Cholesterol and Atherosclerosis 173

MJ's Human Biology Blog
A Beating Heart is Created in the Laboratory 174

The cardiac cycle: The heart contracts and relaxes 174
Heart sounds reflect closing heart valves 175
Cardiac conduction system coordinates contraction 176
Electrocardiogram records the heart's electrical activity 176

8.3 Blood exerts pressure against vessel walls 177
Measuring Blood Pressure 178
Hypertension: High blood pressure can be dangerous 179
Hypotension: When blood pressure is too low 180

8.4 How the cardiovascular system is regulated 180
Baroreceptors maintain arterial blood pressure 180
Nerves and hormones adjust cardiac output 180
Local requirements dictate local blood flows 181
Exercise: Increased blood flow and cardiac output 182

8.5 Cardiovascular disorders: A major health issue 182
Angina: Chest pain warns of impaired blood flow 182
Heart attack: Permanent damage to heart tissue 183
Heart failure: The heart becomes less efficient 183
Embolism: Blockage of a blood vessel 184
Stroke: Damage to blood vessels in the brain 184

8.6 Reducing your risk of cardiovascular disease 184

MJ's Human Biology Blog
Stress Reduction and Heart Attacks 185

9 **The Immune System and Mechanisms of Defense 188**

Current Issue
AIDS: A Crisis in Africa, a Challenge for the World 188

9.1 Pathogens cause disease 190
Bacteria: Single-celled living organisms 190
Viruses: Tiny infectious agents 191
Prions: Infectious proteins 192

MJ's Human Biology Blog
Prion-like Activity in Neurodegenerative Disorders 192

Transmissibility, mode of transmission, and virulence determine health risk 192

9.2   The lymphatic system defends the body   193

Lymphatic vessels transport lymph   193

MJ's Human Biology Blog
A Way to Cure HIV Infection?   193

Lymph nodes cleanse the lymph   195
The spleen cleanses blood   195
Thymus gland hormones cause T lymphocytes to mature   195
Tonsils protect the throat   195

9.3   Keeping pathogens out: The first line of defense   196

Skin: An effective deterrent   196
Impeding pathogen entry in areas not covered by skin   196

9.4   Nonspecific defenses: The second line of defense   197

Phagocytes engulf foreign cells   197
Inflammation: Redness, warmth, swelling, and pain   198
Natural killer cells target tumors and virus-infected cells   199
The complement system assists other defense mechanisms   199
Interferons interfere with viral reproduction   200
Fever raises body temperature   200

9.5   Specific defense mechanisms: The third line of defense   201

The immune system targets antigens   201
Lymphocytes are central to specific defenses   201
B cells: Antibody-mediated immunity   201
The five classes of antibodies   203
Antibodies' structure enables them to bind to specific antigens   203
T cells: Cell-mediated immunity   203

9.6   Immune memory creates immunity   206

9.7   Medical assistance in the war against pathogens   207

Active immunization: An effective weapon against pathogens   207
Passive immunization can help against existing or anticipated infections   208
Monoclonal antibodies: Laboratory-created for commercial use   208
Antibiotics combat bacteria   209

9.8   Tissue rejection: A medical challenge   209

9.9   Inappropriate immune system activity causes problems   209

Allergies: A hypersensitive immune system   209
Autoimmune disorders: Defective recognition of "self"   210

9.10   Immune deficiency: The special case of AIDS   211

HIV targets helper T cells of the immune system   212
HIV is transmitted in body fluids   212
AIDS develops slowly   213
The AIDS epidemic: A global health issue   214
Risky behaviors increase your chances of getting AIDS   214
Sex can be safer   215
New treatments offer hope   215

10   The Respiratory System: Exchange of Gases   218

Current Issue
Limiting Exposure to Secondhand Smoke   218

10.1   Respiration takes place throughout the body   220

10.2   The respiratory system consists of upper and lower respiratory tracts   220

The upper respiratory tract filters, warms, and humidifies air   222
The lower respiratory tract exchanges gases   222

MJ's Human Biology Blog
Snus—Smokeless Tobacco Made Easy   223

10.3   The process of breathing involves a pressure gradient   227

Inspiration brings in air, expiration expels it   227

MJ's Human Biology Blog
Smoking and Breast Sagging   227

Lung volumes and vital capacity measure lung function   228

10.4   Gas exchange and transport occur passively   229

Gases diffuse according to their partial pressures   229
External respiration: The exchange of gases between air and blood   230
Internal respiration: The exchange of gases with tissue fluids   230
Hemoglobin transports most oxygen molecules   230
Most CO_2 is transported in plasma as bicarbonate   232

10.5   The nervous system regulates breathing   233

A respiratory center establishes rhythm of breathing   233
Chemical receptors monitor CO_2, H^+, and O_2 levels   233

Health & Wellness
Carbon Monoxide: An Invisible, Odorless Killer   235

We can exert some conscious control   235

10.6   Disorders of the respiratory system   236

Reduced air flow or gas exchange impedes respiratory function   236
Microorganisms can cause respiratory disorders   236

MJ's Human Biology Blog
Living with Cystic Fibrosis   237

Lung cancer is caused by proliferation
of abnormal cells   238
Pneumothorax and atelectasis: A failure of gas
exchange   238
Congestive heart failure impairs lung function   238

11 The Nervous System: Integration and Control   242

Current Issue
Medically Induced Coma   242

11.1   The nervous system has two principal parts   244

**11.2   Neurons are the communication cells
of the nervous system   245**

11.3   Neurons initiate action potentials   246
Sodium-potassium pump maintains resting
potential   246
Graded potentials alter the resting potential   247
An action potential is a sudden reversal of
membrane voltage   247
Action potentials are all-or-none
and self-propagating   249

11.4   Neuroglial cells support and protect neurons   249

**11.5   Information is transferred from a neuron to its
target   251**
Neurotransmitter is released   251
Neurotransmitters exert excitatory or inhibitory
effects   252
Postsynaptic neurons integrate and process
information   253

**11.6   The PNS relays information between tissues
and the CNS   253**
Nerves carry signals to and from the CNS   253
Sensory neurons provide information to the CNS   254
The somatic division controls skeletal muscles   254
The autonomic division controls automatic body
functions   255
The sympathetic and parasympathetic divisions oppose
each other   255

11.7   The brain and spinal cord constitute the CNS   257
Bone, meninges, and the blood-brain barrier protect the
CNS   258
The spinal cord relays information   259

11.8   The brain processes and acts on information   260
The hindbrain: Movement and automatic
functions   260
The midbrain: Vision and hearing   261
The forebrain: Emotions and conscious thought   261

11.9   Brain activity continues during sleep   263

MJ's Human Biology Blog
Young Adults Turn to Sleeping Pills   264

**11.10   The limbic system is the site of emotions
and basic behaviors   265**

**11.11   Memory involves storing and retrieving
information   265**

**11.12   Psychoactive drugs affect higher brain
functions   266**

11.13   Disorders of the nervous system   267
Trauma   267
Infections   267

Health & Wellness
Repairing Spinal Cord Injuries   268

Brain tumors: Abnormal growths   268

MJ's Human Biology Blog
Your Cell Phone's Radiation Emission   269

Disorders of neural and synaptic transmission   269

MJ's Human Biology Blog
Ginkgo Doesn't Prevent Dementia   270

12 Sensory Mechanisms   272

Current Issue
DWD: Driving While Distracted   274

**12.1   Receptors receive and convert
stimuli   276**
Receptors are classified according to stimulus   276

MJ's Human Biology Blog
Sensing Danger in the Air   276

The CNS interprets nerve impulses based on origin
and frequency   277
Some receptors adapt to continuing stimuli   278
Somatic sensations and special senses provide sensory
information   278

**12.2   Somatic sensations arise from receptors
throughout the body   278**

MJ's Human Biology Blog
Can You Taste Bitter Foods?   278

Mechanoreceptors detect touch, pressure, and
vibration   279
Mechanoreceptors indicate limb position, muscle length,
and tension   279
Thermoreceptors detect temperature   280
Pain receptors signal discomfort   280

12.3   Taste and smell depend on chemoreceptors   282

Taste: Chemoreceptors bind with dissolved substances   282

Smell: Chemoreceptors bind with odorants   282

12.4   Hearing: Mechanoreceptors detect sound waves   284

The outer ear channels sound waves   284

The middle ear amplifies sound   284

The inner ear sorts and converts sounds   285

12.5   The inner ear plays an essential role in balance   287

Sensing rotational movement   288

Sensing head position and acceleration   288

12.6   Vision: Detecting and interpreting visual stimuli   289

Structure of the eye   289

Regulating the amount of light and focusing the image   290

Eyeball shape affects focus   291

MJ's Human Biology Blog
Myopia Is on the Rise   292

Light is converted into action potentials   292

Health & Wellness
LASIK to Correct Vision Problems   293

Rods and cones respond to light   293

Rods provide vision in dim light   293

Cones provide color vision and accurate images   294

Visual receptors adapt   294

12.7   Disorders of sensory mechanisms   294

Disorders of the ears   294

Disorders of the eyes   295

The Endocrine System   300

Current Issue
Dealing with Type 2 Diabetes   300

13.1   The endocrine system produces hormones   302

13.2   Hormones are classified as steroid or nonsteroid   304

Steroid hormones enter target cells   304

Nonsteroid hormones bind to receptors on target cell membranes   304

Hormones participate in negative feedback loops   305

MJ's Human Biology Blog
Athlete Caught Doping with GH   306

13.3   The hypothalamus and the pituitary gland   306

The posterior pituitary stores ADH and oxytocin   306

The anterior pituitary produces six key hormones   308

Pituitary disorders: Hypersecretion or hyposecretion   310

13.4   The pancreas secretes glucagon, insulin, and somatostatin   311

MJ's Human Biology Blog
Glucose Monitoring Devices Are Inaccurate   312

13.5   The adrenal glands comprise the cortex and medulla   312

The adrenal cortex: Glucocorticoids and mineralocorticoids   312

The adrenal medulla: Epinephrine and norepinephrine   313

13.6   Thyroid and parathyroid glands   313

The thyroid gland: Thyroxine speeds cellular metabolism   314

Parathyroid hormone (PTH) controls blood calcium levels   315

13.7   Testes and ovaries produce sex hormones   317

Testes produce testosterone   317

Ovaries produce estrogen and progesterone   317

13.8   Other glands and organs also secrete hormones   317

Thymus gland hormones aid the immune system   317

The pineal gland secretes melatonin   317

Endocrine functions of the heart, the digestive system, and the kidneys   318

13.9   Other chemical messengers   319

Histamine is important in inflammation   319

Prostaglandins: Local control of blood flow   319

Nitric oxide has multiple functions   319

Growth factors regulate tissue growth   319

13.10   Disorders of the endocrine system   319

Diabetes mellitus: Inadequate control of blood sugar   319

MJ's Human Biology Blog
Inhaled Insulin (Who Cares?)   320

Hypothyroidism: Underactive thyroid gland   320

Hyperthyroidism: Overactive thyroid gland   320

Addison's disease: Too little cortisol and aldosterone   320

Cushing's syndrome: Too much cortisol   321

The Digestive System and Nutrition   324

Current Issue
Is "Overweight" Overstated?   324

MJ's Human Biology Blog
Is Being Overweight a Health Risk?   326

14.1 **The digestive system brings nutrients into the body   327**

The walls of the GI tract are composed of four layers   328

Five basic processes accomplish digestive system function   328

Two types of motility aid digestive processes   328

14.2 **The mouth processes food for swallowing   329**

MJ's Human Biology Blog

Obesity in Close Mutual Friends   329

Teeth bite and chew food   330

The tongue positions and tastes food   330

Saliva begins the process of digestion   330

14.3 **The pharynx and esophagus deliver food to the stomach   331**

14.4 **The stomach stores food, digests protein, and regulates delivery   332**

Gastric juice breaks down proteins   332

Stomach contractions mix food and push it forward   333

14.5 **The small intestine digests food and absorbs nutrients and water   333**

14.6 **Accessory organs aid digestion and absorption   334**

The pancreas secretes enzymes and NaHCO$_3$ 334

MJ's Human Biology Blog

California Bans *Trans* Fats   335

The liver produces bile and performs many other functions   336

The gallbladder stores bile until needed   336

14.7 **The large intestine absorbs nutrients and eliminates wastes   337**

14.8 **How nutrients are absorbed   337**

Proteins and carbohydrates are absorbed by active transport   337

Lipids are broken down, then reassembled   338

Water is absorbed by osmosis   338

Vitamins and minerals follow a variety of paths   338

14.9 **Endocrine and nervous systems regulate digestion   339**

Regulation depends on volume and content of food   339

Nutrients are used or stored until needed   339

14.10 **Nutrition: You are what you eat   340**

MyPyramid plan offers a personalized approach   340

Carbohydrates: A major energy source   341

Lipids: Essential cell components and energy sources   342

Complete proteins contain every amino acid   342

Vitamins are essential for normal function   343

Minerals: Elements essential for body processes   343

Fiber benefits the colon   343

14.11 **Weight control: Energy consumed versus energy spent   345**

BMR: Determining how many Calories we need   345

Energy balance and body weight   346

Physical activity: An efficient way to use Calories   346

Healthy weight improves overall health   346

14.12 **Disorders of the digestive system   347**

Disorders of the GI tract   347

Disorders of the accessory organs   348

Malnutrition: Too many or too few nutrients   348

Obesity: A worldwide epidemic?   348

14.13 **Eating disorders: Anorexia nervosa and bulimia   349**

15 **The Urinary System   354**

Current Issue
How Should We Allocate Scarce Kidneys?   354

15.1 **The urinary system contributes to homeostasis   356**

The kidneys regulate water levels   357

The kidneys regulate nitrogenous wastes and other solutes   357

MJ's Human Biology Blog

Using Urine to Diagnose Disease   357

15.2 **Organs of the urinary system   358**

Kidneys: The principal urinary organs   358

Ureters transport urine to the bladder   358

Urinary bladder stores urine   358

Urethra carries urine from the body   359

MJ's Human Biology Blog

Buying/Selling Kidneys   359

15.3 **Nephrons produce urine   360**

The tubule filters fluid and reabsorbs substances   360

Special blood vessels supply the tubule   360

15.4 **Formation of urine: Filtration, reabsorption, and secretion   362**

Glomerular filtration filters fluid from capillaries   362

Tubular reabsorption returns filtered water and solutes to blood   364

Tubular secretion removes other substances from blood   365

MJ's Human Biology Blog

Encouraging Organ Donations   365

15.5 **The kidneys can produce dilute or concentrated urine   366**

Producing dilute urine: Excreting excess water   366

Producing concentrated urine: Conserving water   366

15.6   Urination depends on a reflex   367

15.7   The kidneys maintain homeostasis in many ways   368
ADH regulates water balance   368
Aldosterone regulates salt balance   369
The renin-angiotensin system controls blood volume and blood pressure   369
Atrial natriuretic hormone protects against blood volume excess   370
Kidneys help maintain acid-base balance and blood pH   371
Erythropoietin stimulates production of red blood cells   371
Kidneys activate vitamin D   372

15.8   Disorders of the urinary system   372
Kidney stones can block urine flow   372
Urinary tract infections are often caused by bacteria   372
Acute and chronic renal failure impair kidney function   372
Dialysis cleanses the blood artificially   373
Kidney transplants are a permanent solution to renal failure   373

16   Reproductive Systems   376

Current Issue
Would You Like a Boy or a Girl?   376

MJ's Human Biology Blog
The Dark Side of Gender Preference   378

16.1   The male reproductive system delivers sperm   378
Testes produce sperm   379
Accessory glands help sperm survive   379
Sperm production requires several cell divisions   380

MJ's Human Biology Blog
Infertility Patients Favor Stem Cell Research   380

Testosterone affects male reproductive capacity   381

16.2   The female reproductive system produces eggs and supports pregnancy   382
Ovaries release oocytes and secrete hormones   382
The uterus nurtures the developing embryo   382
The vagina: Organ of sexual intercourse and birth canal   383
Mammary glands nourish the infant   384

16.3   Menstrual cycle consists of ovarian and uterine cycles   384
The ovarian cycle: Oocytes mature and are released   384

MJ's Human Biology Blog
Is She a Woman?   386

The uterine cycle prepares the uterus for pregnancy   386
Cyclic changes in hormone levels produce the menstrual cycle   388

Health & Wellness
Erectile Dysfunction and Viagra Abuse   389

16.4   Human sexual response, intercourse, and fertilization   389
The male sexual response   389
The female sexual response   390
Fertilization: One sperm penetrates the egg   390

16.5   Birth control methods: Controlling fertility   390
Abstinence: Not having intercourse   391
Surgical sterilization: Vasectomy and tubal ligation   391
Hormonal methods: Pills, injections, patches, and rings   391
IUDs are inserted into the uterus   392

MJ's Human Biology Blog
Birth Control Method Failures   393

Diaphragms and cervical caps block the cervix   393
Chemical spermicides kill sperm   393
Condoms trap ejaculated sperm   393
Withdrawal and periodic abstinence   393
Pills that can be used after intercourse   393
Elective abortion   394
The future in birth control   394

16.6   Infertility: Inability to conceive   394
Infertility can have many causes   394
Enhancing fertility   395

16.7   Sexually transmitted diseases   396
Bacterial STDs: Gonorrhea, syphilis, and chlamydia   398
Viral STDs: HIV, hepatitis B, genital herpes, and HPV   399
Other STDs: Yeast infections, trichomoniasis, and pubic lice   400
Protecting yourself against STDs   400

17   Cell Reproduction and Differentiation   404

Current Issue
Should We Clone Humans?   404

17.1   The cell cycle creates new cells   406

17.2   Replication, transcription, and translation: An overview   407
Replication: Copying DNA before cell division   408
Mutations are alterations in DNA   409
Mechanisms of DNA repair   410
Transcription: Converting a gene's code into mRNA   410

MJ's Human Biology Blog
DNA Mutations Between Generations   410

Translation: Making a protein from RNA   412

17.3   Cell reproduction: One cell becomes two   413
Mitosis: Daughter cells are identical to the parent cell   413
Cytokinesis divides one cell into two identical cells   415
Mitosis produces diploid cells, and meiosis produces haploid cells   415
Meiosis: Preparing for sexual reproduction   415
Sex differences in meiosis: Four sperm versus one egg   417

17.4   How cell reproduction is regulated   418

17.5   Environmental factors influence cell differentiation   419
Differentiation during early development   419
Differentiation later in development   420

17.6   Cloning an organism requires an undifferentiated cell   420
Embryo splitting: Producing identical offspring   420
Somatic Cell Nuclear Transfer: Cloning an Adult   421

17.7   Therapeutic cloning: Creating tissues and organs   422

MJ's Human Biology Blog
Re-creating Undifferentiated Cells   422

18   Cancer: Uncontrolled Cell Division and Differentiation   426

Current Issue
Voluntary Breast and Ovary Removal   426

18.1   Tumors can be benign or cancerous   428

18.2   Cancerous cells lose control of their functions and structures   429

18.3   How cancer develops   430
Mutant forms of proto-oncogenes, tumor suppressor genes, and mutator genes contribute to cancer   430

MJ's Human Biology Blog
Human Gene Patents Invalidated   431

A variety of factors can lead to cancer   431
The immune system plays an important role in cancer prevention   433

18.4   Advances in diagnosis enable early detection   434
Tumor imaging: X-rays, PET, and MRI   434
Genetic testing can identify mutated genes   435
Enzyme tests may detect cancer markers   435

18.5   Cancer treatments   436
Conventional cancer treatments: Surgery, radiation, and chemotherapy   436
Magnetism and photodynamic therapy target malignant cells   436
Immunotherapy promotes immune response   436
"Starving" cancer by inhibiting angiogenesis   437
Molecular treatments target defective genes   437

18.6   The 10 most common cancers   437
Skin cancer: Look for changes in your skin   438
Lung cancer: Smoking is leading risk factor   438
Breast cancer: Early detection pays off   439

Health & Wellness
Breast Self-Examination and Routine Mammograms   440

Prostate cancer: Most common after age 50   440
Cancers of colon and rectum: Tests can detect them early   440

MJ's Human Biology Blog
The PSA Test for Prostate Cancer   441

Lymphoma: Cancers of lymphoid tissues   441
Urinary bladder cancer: Surgery is often successful if done early   441
Kidney cancer: Detected during examination for a renal-related problem   442
Cancer of the uterus: Unusual uterine bleeding is major symptom   442

MJ's Human Biology Blog
A DNA Test for Cervical Cancer   442

Leukemia: Chemotherapy is often effective   442

18.7   Most cancers can be prevented   443

19   Genetics and Inheritance   446

Current Issue
The Promises and Perils of Genetic Testing   446

19.1   Your genotype is the genetic basis of your phenotype   448

19.2   Genetic inheritance follows certain patterns   449
Punnett square analysis predicts patterns of inheritance   449
Mendel established the basic principles of genetics   449
Dominant alleles are expressed over recessive alleles   451

Health & Wellness
Cystic Fibrosis   452

Two-trait crosses: Independent assortment of genes for different traits   452

19.3   Other dominance patterns   454
Incomplete dominance: Heterozygotes have an intermediate phenotype   454
Codominance: Both gene products are equally expressed   454

19.4   Other factors influence inheritance patterns and phenotype   456
Polygenic inheritance: Phenotype is influenced by many genes   456
Both genotype and the environment affect phenotype   457
Linked alleles may or may not be inherited together   458

MJ's Human Biology Blog
Do Identical Twins Have Identical DNA?   458

19.5   Sex-linked inheritance: X and Y chromosomes carry different genes   458
Sex-linked inheritance depends on genes located on sex chromosomes   459
Sex-influenced traits are affected by actions of sex genes   460

19.6   Chromosomes may be altered in number or structure   461
Down syndrome: Three copies of chromosome 21   461
Alterations of the number of sex chromosomes   462
Deletions and translocations alter chromosome structure   463

19.7   Many inherited genetic disorders involve recessive alleles   463
Phenylketonuria is caused by a missing enzyme   463
Tay-Sachs disease leads to brain dysfunction   463
Huntington disease is caused by a dominant-lethal allele   464

MJ's Human Biology Blog
Genetic Screening Tests   464

19.8   Genes code for proteins, not for specific behaviors   465

20   DNA Technology and Genetic Engineering   468
Current Issue
Genetically Engineered Plants   468

20.1   DNA sequencing reveals structure of DNA   470

20.2   DNA can be cloned in the laboratory   471
Recombinant DNA technology: Isolating and cloning genes   471

MJ's Human Biology Blog
Whatever Happened to Golden Rice?   472

Cloning DNA fragments: The polymerase chain reaction   473
Identifying the source of DNA: DNA fingerprinting   474

20.3   Genetic engineering creates transgenic organisms   475
Transgenic bacteria have many uses   475
Transgenic plants: More vitamins and better pest resistance   476
Transgenic animals: A bigger challenge   477

20.4   Gene therapy: The hope of the future?   478
Gene therapy must overcome many obstacles   478

MJ's Human Biology Blog
FDA Approves a Genetically Engineered Drug   478

Vectors transfer genes into human cells   479

MJ's Human Biology Blog
That's One Small Step for Gene Therapy...   479

Success with SCID gives hope   480
Research targets cystic fibrosis and cancer   480

21   Development and Aging   484
Current Issue
Who Should Make Life and Death Decisions for You?   484

21.1   Fertilization begins when sperm and egg unite   486
The journeys of egg and sperm   486
One sperm fertilizes the egg   486
Twins may be fraternal or identical   488

Health & Wellness
Prenatal Diagnostic Techniques   489

21.2   Development: Cleavage, morphogenesis, differentiation, and growth   489

21.3   Pre-embryonic development: The first two weeks   490

21.4   Embryonic development: Weeks three to eight   491
Extra-embryonic membranes   491
The placenta and umbilical cord   493
The embryo develops rapidly   493

21.5   Gender development begins at six weeks   494

21.6   Fetal development: Nine weeks to birth   496
Months three and four   496
Months five and six   496
Months seven through nine   496

21.7 Birth and the early postnatal period 496
Labor ends in delivery 496
Cesarean delivery: Surgical delivery of a baby 498
The transition from fetus to newborn 498
Lactation produces milk to nourish the newborn 499

21.8 From birth to adulthood 500
The neonatal period: A helpless time 500
Infancy: Rapid development and maturation of organ systems 500
Childhood: Continued development and growth 501
Adolescence: The transition to adulthood 501

21.9 Aging takes place over time 501
What causes aging? 502

MJ's Human Biology Blog
Caloric Restriction and Longevity 503

Body systems age at different rates 503

MJ's Human Biology Blog
Hormone Replacement Therapy Revisited 504

Aging well 504

21.10 Death is the final transition 505

22 Evolution and the Origins of Life 508

Current Issue
Who Were the Flores People? 508

22.1 Evidence for evolution comes from many sources 510
The fossil record: Incomplete but valuable 510
Comparative anatomy and embryology provide more evidence 512
Comparative biochemistry examines similarities between molecules 513
Biogeography: The impact of geographic barriers and continental drift on evolutionary processes 513

22.2 Natural selection contributes to evolution 514
Random mutations underlie evolution 514
Natural selection encourages changes in the gene pool 515
Genetic drift and gene flow alter populations 515
Mass extinctions eliminate many species 515
Evolutionary trees trace relationships between species 516

22.3 The young Earth was too hot for life 516

22.4 The first cells were able to live without oxygen 516
Organic molecules formed from atmospheric gases 516
Self-replicating RNA and DNA formed 517
The first living cells were anaerobic 517

MJ's Human Biology Blog
Creating Synthetic Life 518

22.5 Photosynthetic organisms altered the course of evolution 518
Photosynthesis increased oxygen in the atmosphere 518
Aerobic organisms evolved 518
The rise of animals and our human ancestors 518

22.6 Modern humans came from Africa 519
Humans are primates 519
Evolution of *Homo sapiens* 520

MJ's Human Biology Blog
Walking Like a Modern Human 522

Differences within the human species 522

23 Ecosystems and Populations 526

Current Issue
Waiting for the Next "Big One" 526

23.1 Ecosystems: Communities interact with their environment 528

23.2 Populations: The dynamics of one species in an ecosystem 528
Where a species lives: Habitat and range 528
Population growth rate tends toward biotic potential 529
Environmental resistance limits biotic potential 529

23.3 Human population growth 530
Zero population growth has not yet been achieved 531
Population age structure is linked to economic development 531

23.4 Communities: Different species living together 532
Overlapping niches foster competition 532
Succession leads toward a climax community 532
Ecosystems: Communities and their physical environment 533

23.5 Energy flows through ecosystems 533
Producers capture and convert energy, consumers rely on stored energy 533

MJ's Human Biology Blog
Carbon Dioxide and Forest Growth 535

A food web: Interactions among producers and consumers 535
The lower levels of an ecological pyramid support consumer populations 536
Human activities disrupt ecological pyramids 537

23.6   Chemical cycles recycle molecules in ecosystems   538

The water cycle is essential to other biogeochemical cycles   538

MJ's Human Biology Blog
Dwindling Phosphorus Supplies   538

The carbon cycle: Organisms exchange CO_2 with the atmosphere   539
Nitrogen: An essential component of nucleic acids and proteins   540
Phosphorus: A sedimentary cycle   540

24   Human Impacts, Biodiversity, and Environmental Issues   544

Current Issue
Global Warming   544

24.1   Pollutants impair air quality   546

Excessive greenhouse gases lead to global warming   546

MJ's Human Biology Blog
Dwindling Arctic Sea Ice   547

CFCs deplete the ozone layer   547
Pollutants produce acid precipitation   548
Smog blankets industrial areas   548

24.2   Pollution jeopardizes scarce water supplies   549

Water is scarce and unequally distributed   549
Urbanization increases storm water runoff   549

Human activities pollute freshwater   550

MJ's Human Biology Blog
China's Future Water Shortage   550

Groundwater pollution may impair human health   551
Oil pollution damages oceans and shorelines   551

24.3   Pollution and overuse damage the land   552

24.4   Energy: Many options, many choices   553

24.5   Humans are creating a biodiversity crisis   554

MJ's Human Biology Blog
Energy Sustainability in 20 Years?   555

Humans alter and destroy habitats   555
Urbanization is a major force for environmental change   556
Biodiversity is healthy for humans too   556

24.6   Sustainable development supports future economic needs   556

Measuring sustainability and quality of life   556
Strategies to support sustainable development   556

Glossary   G-1

Answers to Figure Check, Quick Check, and Test Yourself Questions   A-1

Credits   C-1

Index   I-1

Preface

In 1997 scientists copied (cloned) an adult animal for the first time. In 2003 the entire sequence of a complete set of human DNA (the human genome) was first published. In 2010 scientists synthesized a DNA molecule of over a million base pairs, and then used it to create the first living cell from non-living components. Technically, humans can now create life.

Those of us who find these discoveries fascinating—and yes, even exciting!—have an obligation to help our students understand science and the impact science has on their lives. Science is too much fun and far too important to be left to scientists.

The Focus Is on the Student

This book was written for students who do not yet have a strong background in science so that they, too, might share in the joy and wonder of science. Every effort was made to make the book accurate and up-to-date while keeping it inviting, accessible, and easy to read. The look and feel of the text is intentionally like that of a news magazine, peppered with short features likely to be of interest to the student and with a strong visual appeal.

Each chapter begins with a Current Issue that highlights a recent controversy or ethical/social/political issue related to human biology. The main narrative of each chapter begins with Key Concepts that summarize the most important points within the chapter. Quick Check questions throughout the narrative or linked to select figures allow the students to check their understanding as they go along. Finally, each chapter ends with 15 new multiple-choice questions so that students can check their progress.

Students are naturally curious about how their own bodies work and about human diseases. We capitalize on this curiosity with Health & Wellness boxes that highlight timely health topics. In addition, organ system chapters generally conclude with a section covering the more common human diseases and disorders.

A new feature of this edition is the inclusion of MJ's Human Biology Blog, 2-4 blog entries per chapter, taken from the blog Website developed specifically for this text, http://www.humanbiologyblog.blogspot.com. Blog entries highlight recent discoveries or news items relevant to the subject of each chapter. We hope that MJ's Human Biology Blog entries and the blog Website will encourage curious students to dig a little deeper into specific topics that interest them.

Unifying Themes Tie the Subjects Together

Several unifying themes in biology hold the chapters together. Homeostasis, the state of dynamic equilibrium in which the internal environment of an organism is maintained fairly constant, is one of those recurrent themes. The concept of homeostasis ties in with another recurrent theme: that structure and function are related. Structure/function relationships are the very core of the study of anatomy and physiology, and both of these fields in turn rely on the most unifying concept in all of biology: evolution. Only in the context of evolution can anatomy and physiology be fully understood; without the concept of evolution, very little in biology makes sense.

A predominant theme of this book is that each of us has choices to make—choices that will affect ourselves, other humans, and the entire planet. Should all children be vaccinated against childhood diseases? Should we spend time and money preparing for a pandemic that may never occur? Will we be willing and able to slow the rate of global warming? Is it important that we save other species from extinction, and if so, how would we go about it? Students are encouraged to formulate their own views on these and other topics so that they will feel comfortable with the choices they make.

The Organization Fits the Course

This book was designed to accommodate the fairly standard format for college courses in human biology. There are chapters that introduce science and chemistry, chapters that cover basic human biology from cells through the human organ systems, and finally chapters on evolution, ecosystems and populations, and human impacts on the environment.

With such broad coverage, however, there is never enough time to teach all that is interesting, exciting, and relevant about human biology in one semester. Fortunately, because each chapter was written to stand on its own, this book allows for a certain degree of flexibility. Instructors wishing to emphasize the basics of human anatomy and physiology or focus on the medical aspects of human biology could omit or deemphasize the last two chapters. Instructors should feel free to present the organ system chapters in a different order if they feel more comfortable doing so. Within chapters, diseases and disorders sections could be omitted or considered optional. Those interested in a more cellular or molecular approach might want to give greater emphasis to Chapters 4 and 17–21 and move quickly through the organ systems chapters. Those more interested in the broader picture of where humans came from and how humans fit into the world order may want to allow sufficient time for the last three chapters, even if it means that they must move quickly or selectively through the organ system chapters. All of these approaches are equally valid.

However much you cover, dig in and enjoy your course!

Michael D. Johnson

Acknowledgments

The Sixth Edition of *Human Biology: Concepts and Current Issues* is once again the product of the continued hard work and dedication of the people at Benjamin Cummings Publishing, led by Editor-in-Chief Beth Wilbur, Executive Director of Development Deborah Gale, and Acquisitions Editor Becky Ruden. Becky directs a team that functions as smoothly and professionally as any in the business.

On a day-to-day basis, I depended on Senior Developmental Editor Susan Teahan. I could always pass an idea on to Susan and know that it would be given careful thought and consideration. Her influence on the Sixth Edition may be subtle, but it's there. I am blessed to have had her advice and support.

Changes to the art and photos in the Sixth Edition are the result of the hard work of artists at Imagineering and photo coordinator Donna Kalal. Once again, photo researcher Kristin Piljay found the new photos you see in this edition.

Several new contributors had a significant impact on this edition. Kathleen Hunt wrote the Quick Check questions and the accompanying answers, and Suzanne Long contributed the Test Yourself multiple-choice questions (and answers) at the end of each chapter.

Accuracy and clarity have been checked and rechecked by the hundreds of insightful faculty members around the country over the past ten years. Reviewers specific to this edition are listed below.

Thanks go to the outstanding administrative support team at Pearson Education, BenjaminCummings. It includes Assistant Editor Leslie Allen, Senior Media Producer Jonathan Ballard, Associate Editor Brady Golden, Production Supervisor Shannon Tozier, Copyeditors Kevin Gleason and Cheryl Smith, and Proofreader Chris Sabooni.

Once again, the textbook is supported by a wonderful set of ancillary materials. Thanks go to Robert Sullivan of Marist College, who wrote the Instructor Guide, to Suzanne Long of Monroe Community College, who developed the PowerPoint Lecture Slides, to David Rivers of Loyola College in Maryland, who revised the Test Bank, to Judith Stewart at the Community College of Southern Nevada, whose Study Guide will certainly help those students who choose to use it, and to Bert Atsma of Union County College, whose Laboratory Manual continues to complement the best human biology courses. In addition, Tom Farruggella of Marist College supplied the Web Publishing System (WPS) Multiple Choice Quizzes, Sue Scambis of Valencia Community College wrote the WPS True-False Quizzes, and Erwin Bautista of UC Davis was responsible for the Instructor Quiz Shows. Nancy O'Keefe of Purdue University Calumet supplied the Web Tutorials, Web Links, Biology in the News Essay, Weighing the Issues, and Crosswords.

Last but not least, I'd like to thank my wife, Pamela, for her whole-hearted support and understanding over the years.

Reviewers

Shazia Ahmed,
Texas Woman's University

Barbie Baker,
Florida Community College, Jacksonville

Erwin Bautista,
UC Davis

Mike Boms,
SUNY New Platz

Thomas J. Farruggella,
Marist College

Susan Keys,
Springfield College

Lee Lee,
Montclair State University

Katherine Lockwood,
University of New Hampshire

Suzanne Long,
Monroe Community College

Dr. Tony Morris,
Fairmont State University

James Murphy,
Monroe Community College

Nancy O'Keefe,
Purdue University Calumet

Louise Petroka,
Gateway Community Technical College

Dr. Barbara Pleasants,
Iowa State University

Alison Shakarian, PhD,
Salve Regina University

Susan Skambis,
Valencia Community College

Ed Spicka,
SUNY Geneseo

New to This Edition

Improved Pedagogical Framework for Chapter Openers. Instructors often tell us that one of the most effective ways to make human biology relevant to their students is to bring current issues into the classroom. For the 6th edition, each chapter opens with the popular *Current Issue* essay to engage students from the beginning and set the context for the science that follows.

Inclusion of Key Concepts. A concise summary of 4-6 *Key Concepts* is incorporated into the beginning of each chapter. These focus students' attention on ideas that are fundamental to building a conceptual framework for understanding human biology.

Addition of the Author's Blog. To rouse students' interest in the science they encounter in their everyday lives, we've incorporated entries from the author's own blog, *MJ's Human Biology Blog* (http://www.humanbiologyblog.blogspot.com). Each chapter includes 2-4 blog entries that share the author's insights into current biology news.

Integrated Critical Thinking and Problem Solving Questions. Each chapter now includes 7-10 new *Quick Check* questions at the end of select sections and process diagrams. These higher-order thinking questions are aimed at Bloom's Taxonomy level 2 and above to give students practice developing their reasoning skills. At the end of each chapter, a revised *Test Yourself* quiz now includes 15 multiple-choice questions—also aimed at Bloom's Taxonomy level 2 and above—to give students practice answering questions they are likely to encounter on exams.

Connections to Online Components. Because human biology courses are rapidly moving online or being taught as an online/on-campus hybrid, and to intertwine media more closely with the book, we now offer two new electronic, learning components. *Web Animations* with an icon and URL references are now threaded throughout the chapter, directing students to a convenient, online tool that shows human biology concepts in action. Now available is Pearson's *eText*, an online interactive version of the book that features highlighting, annotation, and bookmarking capabilities available to students 24/7.

Refreshed Visual Content. To revitalize the visual content, more than 75 new photos replace images from the previous edition, and 6 figures are new. To help students' with their understanding of anatomical and process illustrations, orientation diagrams are included for 7 more figures. So that students can understand figures quickly, we've moved figure legends into part labels wherever possible.

Chapter-specific Changes:

Chapter 1 Human Biology, Science, and Society

- New *Current Issue* on "Mandatory Childhood Vacci-nations" highlights parental concerns that vaccinations may cause autism

- Revised *Health Watch* (renamed *Health & Wellness)* "The Growing Threat of Antibiotic-Resistant Bacteria"
- Three new *MJ's Human Biology Blogs*:
  - "Thimerosal and Autism"
  - "Correlation versus Causation"
  - "Scientific Uncertainty and Shared Responsibility"

Chapter 2 The Chemistry of Living Things

- Two new *MJ's Human Biology Blogs*:
  - "I Don't Hear You..."
  - "Getting That Caffeine Buzz"

Chapter 3 Structure and Function of Cells

- Revised and updated *Current Issue* "The Use of Human Stem Cells"
- Revised explanation of the structural and functional properties of cholesterol
- Two new *MJ's Human Biology Blogs*:
  - "Stem Cell Therapy for Parkinson's?"
  - "Birth Dating Human Cells"
- New orientation diagram of parts of the cell—Golgi apparatus, lysosomes, and peroxisomes—for Figure 3.18
- New orientation diagram of a mitochondrion for Figure 3.24

Chapter 4 From Cells to Organ Systems

- Two new *MJ's Human Biology Blogs*:
  - "My Mother's Cells Within Me"
  - "Fat Cells Are Replaced Throughout Life"

Chapter 5 The Skeletal System

- Revised and expanded the *Current Issue* "A Black Market in Human Bones"
- Revised coverage on osteoporosis in women after menopause
- Three new *MJ's Human Biology Blogs*:
  - "A Really Costly Drug"
  - "Is Running Hard on Knees?"
  - "Treating "Pre-osteoporosis"
- New orientation diagram of disks of the spine for Figure 5.8

Chapter 6 The Muscular System

- Three new *MJ's Human Biology Blogs*:
  - "New Drug Test for Athletes"
  - "Beating the Testosterone Doping Test"
  - "Stretching and Sports Injuries"
- New orientation diagram of muscles in the upper arm for Figure 6.2
- New orientation diagram of myofibril for Figure 6.7
- New orientation diagrams of myofibril and thick and thin filaments for Figure 6.8

Chapter 7 Blood

- Revised *Current Issue* "Should You Bank Your Baby's Cord Blood?"
- Three new *MJ's Human Biology Blogs*:
  - "The Spleen Stores Monocytes"
  - "Platelet-Rich Therapy Revisited"
  - "Cleansing Blood with Magnets"
- New Figure 7.13 rendering of how Rh factor incompatibility can affect a fetus

Chapter 8 Heart and Blood Vessels

- New *Current Issue* "Comparative Effectiveness Research"
- Revised *Health Watch* (renamed *Health & Wellness*) "Cholesterol and Atherosclerosis"
- Incorporated coverage of heart attack symptoms in women
- Three new *MJ's Human Biology Blogs*:
  - "Boosting Cardiac Repair Mechanisms"
  - "A Beating Heart is Created in the Laboratory"
  - "Stress Reduction and Heart Attacks"
- New Figure 8.11 rendering of the cardiac cycle

Chapter 9 The Immune System and Mechanisms of Defense

- Revised and updated the *Current Issue* "AIDS: A Crisis in Africa, a Challenge for the World"
- Significantly revised section on prions
- Updated coverage on the AIDS epidemic
- Updated Table 9.3 on AIDS cases in adults and adolescence by sex and exposure category
- Two new *MJ's Human Biology Blogs*:
  - "Prion-like Activity in Neurodegenerative Disorders"
  - "A Way to Cure HIV Infection?"
- Updated data in Figure 9.23 graphs on number of deaths due to AIDS and people living with HIV/AIDS

Chapter 10 The Respiratory System: Exchange of Gases

- Revised *Current Issue* "Limiting Exposure to Second Hand Smoke"
- New section on colds and the flu.
- New section on pneumothorax and atalectasis
- Three new *MJ's Human Biology Blogs*
  - "Snus—Smokeless Tobacco Made Easy"
  - "Smoking and Breast Sagging"
  - "Living with Cystic Fibrosis"

Chapter 11 The Nervous System: Integration and Control

- Revised the *Current Issue* "Medically Induced Coma" to include the rationale behind this controversial procedure
- Three new *MJ's Human Biology Blogs*:
  - "Your Cell Phone's Radiation Emission"
  - "Young Adults Turn to Sleeping Pills"
  - "Ginkgo Doesn't Prevent Dementia"
- New orientation diagram of meninges for Figure 11.13

Chapter 12 Sensory Mechanisms

- New *Current Issue* "DWD: Driving While Distracted"
- New *Health & Wellness* "LASIK to Correct Vision Problems"

- New section on age-related macular degeneration
- Three new *MJ's Human Biology Blogs*:
  - "Sensing Danger in the Air"
  - "Can You Taste Bitter Foods?"
  - "Myopia Is on the Rise"
- New Figure 12.14 rendering of the structures of the human eye

Chapter 13 The Endocrine System

- Focused the *Current Issue* solely on Type 2 diabetes
- Clarified the reason for inclusion of renin, an enzyme, in discussion of hormones
- Expanded coverage of Type 1 diabetes for mention of suspected causes: autoimmune, genetic, and environmental
- Three new *MJ's Human Biology Blogs*:
  - "Athlete Caught Doping with GH"
  - "Glucose Monitoring Devices Are Inaccurate"
  - "Inhaled Insulin (Who Cares?)"
- Updated data in graph on the number of Americans with diabetes (in *Current Issue*)

Chapter 14 The Digestive System and Nutrition

- Inserted passage describing challenges by critics of specific recommendations made in USDA's MyPyramid
- Three new *MJ's Human Biology Blogs*:
  - "Obesity in Close Mutual Friends"
  - "Is Being Overweight a Health Risk?"
  - "California Bans Trans Fats"
- Revised Figure 14.16 of USDA dietary guideline pyramid

Chapter 15 The Urinary System

- Revised *Current Issue* "How Should We Allocate Scarce Kidneys?"
- Clarified the role of antidiuretic hormone (ADH) in water balance
- Separated last edition's section on aldosterone, renin, and atrial natriuretic hormone (ANH) into three sections, taking on the hormones one at a time and expanding coverage on each
- Discussed whether it's safe to donate a kidney in the section on kidney transplants
- Three new *MJ's Human Biology Blogs*:
  - "Using Urine to Diagnose Disease"
  - "Buying/Selling Kidneys"
  - "Encouraging Organ Donations"

Chapter 16 Reproductive Systems

- Revised *Current Issue* "Would You Like a Boy or a Girl?"
- Refocused the *in vitro* fertilization section to cover the broader topic of artificial reproductive technologies (ART)
- Updated data in Table 16.3 on failure rates in various contraceptive methods
- Four new *MJ's Human Biology Blogs*:
  - "Infertility Patients Favor Stem Cell Research"
  - "The Dark Side of Gender Preference"

- "Is She a Woman?"
- "Birth Control Method Failures"

Chapter 17 Cell Reproduction and Differentiation

- Revised *Current Issue* "Should We Clone Humans?"
- Moved material from last edition's *Current Issue* to main narrative for two added sections, one on cloning organisms and one on therapeutic cloning
- Two new *MJ's Human Biology Blogs*:
  - "DNA Mutations Between Generations"
  - "Re-creating Undifferentiated Cells"
- Revised Figure 17.7 of genetic code of mRNA

Chapter 18 Cancer: Uncontrolled Cell Division and Differentiation

- Expanded *Current Issue* "Voluntary Breast and Ovary Removal"
- New *Health & Wellness* "Breast Self-Examination and Routine Mammograms"
- Expanded section on angiogenesis for inclusion of recent anti-angiogenic drug research
- Updated data in Table 18.3 on top ten cancers ranked by estimated incidence
- Updated data in Table 18.4 on recommendations for cancer screening
- Three new *MJ's Human Biology Blogs*:
  - "Human Gene Patents Invalidated"
  - "The PSA Test for Prostate Cancer."
  - "A DNA Test for Cervical Cancer"
- New Figure 18.6 rendering of the development of cancer cells

Chapter 19 Genetics and Inheritance

- Revised *Current Issue* "The Promises and Perils of Genetic Testing"
- Created two new main sections; one on dominance patterns and the other on factors that influence inheritance and phenotype
- Two new *MJ's Human Biology Blogs*:
  - "Do Identical Twins Have Identical DNA?"
  - "Genetic Screening Tests"
- New Figure 19.1 rendering of a pair of autosomes

Chapter 20 DNA Technology and Genetic Engineering

- Revised *Current Issue* "Genetically Engineered Plants"
- Three new *MJ's Human Biology Blogs*:
  - "Whatever Happened to Golden Rice?"
  - "FDA Approves a Genetically Engineered Drug"
  - "That's One Small Step for Gene Therapy..."

Chapter 21 Development and Aging

- Expanded *Current Issue* "Who Should Make Life and Death Decisions for You?" to include the options available to make one's wishes known
- Inserted brief discussion on age at which the brain reaches full maturity.
- Two new *MJ's Human Biology Blogs*:
  - "Caloric Restriction and Longevity"
  - "Hormone Replacement Therapy Revisited"

Chapter 22 Evolution and the Origins of Life

- Revised *Current Issue* "Who Were the Flores People?"
- Included the new hominid *Ardipithecus ramidus*" in the section on human evolution
- Revised Table 22.1 on the taxonomic classification as it applies to modern humans to include time line data (e.g., *Chordata*, 550 mya)
- Two new *MJ's Human Biology Blogs*:
  - "Creating Synthetic Life"
  - "Walking Like a Modern Human"
- New Figure 22.13, artist's rendering of *Ardipithecus ramidus*

Chapter 23 Ecosystems and Populations

- Refocused last edition's *Current Issue* on bird flu for a broader scope, including other pandemics such as the Black Plague and swine flu
- Two new *MJ's Human Biology Blogs*:
  - "Carbon Dioxide and Forest Growth"
  - "Dwindling Phosphate Supplies"

Chapter 24 Human Impacts, Biodiversity, and Environmental Issues

- Re-focused the *Current Issue* on "Global Warming" to introduce the newer concept of tipping points
- Revised section on chlorofluorocarbons (CFCs)
- New coverage on Gulf of Mexico oil spill in 2010
- Expanded coverage of energy options and choices
- Revised the section on how humans alter and destroy habitats
- Three new *MJ's Human Biology Blogs*:
  - "Dwindling Arctic Sea Ice"
  - "China's Future Water Shortage"
  - "Energy Sustainability in 20 Years?"
- Updated data in Figure 24.4 graph on rising temperature and atmospheric carbon dioxide concentration

1

Human Biology, Science, and Society

Crew of the space shuttle Atlantis, November 20, 2007.

Mandatory Childhood Vaccinations

On January 1, 2009, New Jersey became the first state to require flu shots for children who attend licensed day care and pre-school programs. New Jersey now requires immunization (vaccinations) for 13 vaccine-preventable communicable diseases, more than any other state. New Jersey is not alone, however, in requiring vaccinations—all 50 states currently have some kind of school immunization requirement. (All 50 states also permit exemptions under certain conditions.)

Childhood Vaccinations Save Lives

The states' rationale is clear: childhood vaccines introduced since the 1950s have all but wiped out many communicable diseases in the United States, including measles, mumps, whooping cough (pertussis), polio, and diphtheria. In the 1940s and '50s, before vaccines against these diseases were available, the five diseases combined caused an estimated 900,000 cases of disease and 7,700 deaths per year. By 2004 there were only 27 deaths from all five diseases combined—a 99.6% reduction. The

number of cases of measles dropped from more than 500,000 per year before the measles vaccine was available to only 62 cases per year in recent years.

Recently, however, public health officials have noticed an uptick in the number of cases of measles and whooping cough, two diseases that are highly sensitive to vaccination rates. In the first seven months of 2008 there were more cases of measles than at any time since 1996. Most of the measles victims in 2008 had not been vaccinated, even though they were eligible for the vaccine (children under 12 months of age are not yet eligible).

Parents Resist Mandatory Vaccination

The rise in measles and whooping cough coincides with more than a doubling of exemptions from school immunization programs granted for "philosophical or personal beliefs" between 1991 and 2004. Why are parents increasingly refusing to have their children vaccinated when the evidence is so overwhelming that vaccinations prevent communicable diseases? Their reasons tend to fall into two categories: 1) a belief that the vaccines (or something in them) may be contributing to what they view as an epidemic of childhood chronic diseases, including especially autism, and 2) a dislike of government intervention into personal decisions.

Compared to parents who vaccinate their children, parents who choose not to vaccinate their children tend to believe that the risk of their child getting the disease is low and that the disease itself is not very severe. The latter view is understandable, because most parents today have not lived through a major outbreak of any communicable disease. Today's parents were born after the scourge of polio, for example. Polio killed nearly 10% of its victims and crippled countless others for life before the polio vaccine became available in 1955.

Public health officials are watching these developments with concern. Not all people in a community can be vaccinated, and so the prevention of widespread outbreaks of vaccine-preventable diseases in communities depends in part on "herd immunity." The concept is that when most people in a community (or herd) have been vaccinated, the disease has a much harder time spreading from individual to individual. In other words, high vaccination rates benefit the community overall (especially young children), in addition to protecting the individual who has been vaccinated. Says Dr. Anne Schuchat, director of the National Center for Immunization and Respiratory Diseases, "The vaccine against measles is highly effective in preventing infections, and high immunization levels in the community are effective at preventing or drastically reducing the size of outbreaks."[1]

A Link Between Vaccinations and Autism?

Parents who oppose mandatory vaccinations for safety reasons often point to cases of children who developed autism shortly after receiving a vaccine. Their celebrity spokesperson is actress and former *Playboy* model Jenny McCarthy, whose son is autistic. Ms. McCarthy is on the board of Generation Rescue, a nonprofit organization that claims to be able to treat autism effectively with a special diet.

Actress Jenny McCarthy has appeared on the Oprah Winfrey Show opposing mandatory vaccinations.

Medical professionals and research scientists continue to point out that the available scientific evidence does not support the argument that vaccination can cause childhood diseases, including autism. But for many parents, scientific studies are not as convincing as an appearance by Ms. McCarthy on the Oprah Winfrey Show with an emotional story about ill children.

Some parents oppose mandatory childhood vaccinations because they are philosophically opposed to government intervention into what they see as a personal choice. Says Barbara Loe Fisher, a mother and the cofounder of the National Vaccine Information Center, representing parents against forced vaccinations, "... If the State can tag, track down and force citizens against their will to be injected with biologicals of unknown toxicity today, there will be no limit on which individual freedoms the State can take away in the name of the greater good tomorrow."[2]

Parents in favor of vaccines are mounting lobbying campaigns as well. Their celebrity advocate is actress Amanda Peet, now a spokesperson for Every Child By Two, a vaccine-advocacy group founded by former first lady Rosalynn Carter. Ms. Peet once called anti-vaccine parents "parasites" for relying on other children's immunity to protect their own. She later apologized for the word, and suggested that parents should get their advice from doctors, not celebrities like herself (and presumably Ms. McCarthy).

It would be a shame if vaccines became such a hot-button issue that preventable diseases such as polio returned. We need to find a way to address parents' concerns about vaccine safety and about the role of government in our lives, while at the same time protecting the public from preventable, communicable diseases. How we choose to do that is up to all of us.

Questions to consider

1 What should medical professionals, politicians, or even just concerned citizens do, if anything, to help parents understand the risks and benefits of vaccines?

2 Will you vaccinate your children? Why or why not? What would you like to know in order to make an informed decision?

[1] www.cdc.gov/media/pressrel/2008/r080821.htm
[2] www.vaccineawakening.blogspot.com

The facts . . .

- Childhood vaccination programs have been effective in all but eliminating certain communicable diseases.
- All 50 states have childhood vaccination (immunization) programs as a requirement for school attendance—all states also allow for certain exemptions.
- Exemptions from vaccination (and communicable diseases) are on the rise. Many parents object to mandatory vaccination programs out of concern that the vaccines may cause autism or certain other chronic childhood diseases.
- The available scientific evidence does not support the argument that vaccinations can cause childhood diseases, including autism.

>> **Living things have certain characteristics** that make them different from nonliving things. Living things harness energy and use energy to create unique chemical compounds, grow, and ultimately reproduce.

>> **Humans are just one of several million different life-forms** on Earth. Our closest relatives are the other primates (including monkeys and apes). Features that *taken together* define humans as unique are bipedalism, opposable thumbs, a large brain, and a capacity for complex language.

>> **Science is a process for studying the natural world.** It is based on observable, quantifiable data obtained by repeatedly questioning, observing, and drawing conclusions.

>> **Science helps us understand what *is*, not what *should* be.** It does not provide us with "right" answers or give meaning to our lives.

>> **We make choices** about how to use scientific knowledge every day whether we are consciously aware of it or not. We owe it to ourselves to make informed choices.

You were born into exciting times, when scientific discoveries are happening more rapidly than at any other time in human history. Like the Industrial Revolution of the nineteenth century and the discovery of DNA in the twentieth, today's scientific innovations will change the human condition forever.

In your lifetime people may be able to select or modify their children's features before they are born. People may even be able to have clones (copies) made of themselves. At the very least, certain diseases that threaten us now will become curable. Perhaps your grandchildren will not even know what AIDS is because the disease will have disappeared.

What you are witnessing is the power of science. **Science** is the study of the *natural world,* which includes all matter and all energy. Because all living organisms are also made of matter and energy, they are part of the natural world (**Figure 1.1**). Biology is one of many branches of science. More specifically, **biology** (from the Greek words *bios,* life, and *logos,* word or thought) is the study of living organisms and life's processes. It is the study of life. Other branches of science are chemistry, physics, geology, astronomy, and related fields such as medicine.

This text is specifically about *human* biology. We will explore what it means to be alive. We will see how the molecules that make up our bodies are created from molecules in the air and in our food and drink. We will learn how our cells grow and divide, and how we evolved from single celled organisms that arose from nonliving chemical elements nearly 3.5 billion years ago. We will explore how our bodies function, why we get diseases, and how we manage to survive them. We will look at how we develop into adults, reproduce, and influence the destinies of other organisms on Earth.

With the power of science comes an awesome responsibility. All of us, individually and collectively, must choose how to use the knowledge that science gives us. Will human cloning be acceptable? Can we prevent global warming? Should your insurance company be able to reject you for coverage because genetic testing shows that you may develop cancer 40 years from now? Should you be required to vaccinate your children against certain infectious childhood diseases? (See the Current Issue feature, *Mandatory Childhood Vaccinations.*)

We all have to make responsible decisions concerning not only our own health and well-being but also the long-term well-being of our species. This book considers many aspects of human interaction with the natural world. We'll contemplate human functioning within the environment and the impact of humans on the environment. Along the way we'll confront a variety of social and personal issues and discuss the choices we might make about them. Because biology is the study of life, we begin by defining life itself.

↩ **Recap** Science is the study of the natural world, which consists of all matter and energy. Biology is the study of living organisms. ▪

1.1 The characteristics of life

What is life? On one hand, this question seems easy and on the other so abstract that it is more like a riddle. We all think we can recognize life even if we can't define it easily. Children learn early to distinguish between living and nonliving things. Remember that childhood game "animal, vegetable, or mineral"? In it, children distinguish what is alive (animals and plants) from what is not (minerals).

Most biologists accept the following criteria as signs of life:

▪ *Living things have a different molecular composition than nonliving things do.* Everything in the natural world, both living and nonliving, is composed of the same set of approximately 100 different chemical *elements.* However, only a few elements are present in any abundance in living organisms. In addition, living organisms can combine elements in unique ways, creating

Photo taken from the Hubble Space Telescope showing a tiny portion of the universe. Studies by astronomers have shown that all matter on Earth originated inside stars or with the Big Bang.

Studying unusual species such as this deep sea glass squid allows biologists to understand the processes by which a species successfully survives. Many different environments exist in the world, but the same physical and chemical laws govern them all.

The natural world comprises all matter and energy. An erupting volcano spewing liquid rock and heat is the result of energy that still remains from the creation of Earth nearly 4.6 billion years ago.

Jane Goodall has dedicated her life to studying the needs and behaviors of chimpanzees. The DNA of humans and chimps is almost the same, yet important physical and behavioral differences are obvious. Evolution examines how these differences arose.

Figure 1.1 **Studies of the natural world.**

certain *molecules* (combinations of elements) that nonliving things cannot create. These molecules of life (proteins, carbohydrates, lipids, and nucleic acids) are found in all living organisms and often persist in the remains of dead organisms. Variations in these molecules in different life-forms account for the diversity of life.

■ *Living things require energy and raw materials.* The creation of the molecules of life doesn't happen by accident, at least under present conditions on Earth. The transformation of molecules from one form to another requires energy. The term **metabolism** refers to the physical and chemical processes involved in transforming energy and molecules so that life can be maintained. All living things take in raw materials

and energy from the environment and metabolize them into the molecules and energy that they need to survive. Plants use the energy of sunlight and chemicals obtained from soil, water, and air. Animals and all other forms of life ultimately obtain their energy and raw materials from water, air, plants, or other animals.

■ *Living things are composed of cells.* A **cell** is the smallest unit that exhibits all the characteristics of life (**Figure 1.2**). All cells come only from existing cells. There is always at least one cell in any living thing, and some organisms (called *unicellular* organisms) are *only* one cell. *Multicellular* organisms are composed of many cells or many different types of cells.

a) Several *Staphylococcus aureus*, the bacterium that causes food poisoning (SEM × 50,000).

b) Some of the many cells that line the inner surface of the human stomach (SEM × 500).

Figure 1.2 Cells are the smallest units of life. Some organisms consist of just one cell (unicellular), whereas others contain many cells (multicellular).

- *Living things maintain homeostasis.* All living organisms must maintain an internal environment compatible with life, and the range of chemical and physical conditions compatible with life is very narrow. The maintenance of a relatively constant internal environment is called **homeostasis.** Living things have developed remarkable ways of regulating their internal environment despite sometimes dramatic changes in the external environment. Single cells and unicellular organisms are surrounded by a membrane that allows the cell (or organism) to maintain internal homeostasis by providing a selective barrier to the entry and exit of various substances. In multicellular organisms the tissues, organs, and organ systems work together to maintain homeostasis of the fluid that surrounds all cells. We discuss the importance of homeostasis further in Chapter 4.
- *Living things respond to their external environment.* Stay out in the cold too long and you are likely to respond by moving to a warm room. Plants respond to their environment by turning their leaves toward light or by growing roots toward sources of nutrients and water. Even bacteria respond to their environment by moving toward nutrients (and away from noxious stimuli) and by increasing their growth rate.
- *Living things grow and reproduce.* Living organisms have the capacity to grow and ultimately produce more living organisms like themselves (Figure 1.3). The ability to grow and reproduce is determined by the genetic material in cells, called deoxyribonucleic acid (DNA). Some nonliving things can get larger, of course; examples are glaciers and volcanic mountains. However, they cannot create copies of themselves.

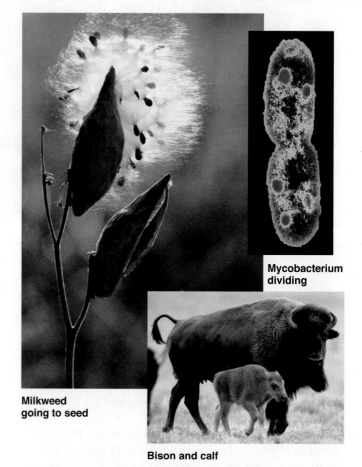

Mycobacterium dividing

Milkweed going to seed

Bison and calf

Figure 1.3 Living things grow and reproduce.

■ *Populations of living things evolve.* The various forms of life may change over many generations in a process known as evolution. Evolution explains why there are so many different forms of life on Earth today.

✅ **Quick Check** Imagine that you are studying a deep-sea hot vent, and you notice a peculiar structure near one of the vents that seems to grow from week to week. How could you determine whether it is a living organism? ■

🌐 **Web Animation** *Signs of Life* at www.humanbiology.com.

Although all these characteristics are necessary to describe life fully, not all of them apply to every living thing all the time. Individual organisms do not evolve, nor do they necessarily reproduce or always respond to their surroundings. However, populations of similar organisms have the *capacity* to perform these functions.

↩ **Recap** All living things are composed of cells. Living things require energy and raw materials, maintain homeostasis, respond to their external environments, grow and reproduce, and evolve over many generations. ■

1.2 How humans fit into the natural world

Living things are grouped according to their characteristics

To find order in the diversity of life, biologists have long sought ways to categorize living things. In 1969 a classification system of *five kingdoms* was proposed. In this system the fundamental criteria for classifying organisms are the presence or absence of a nucleus, the number of cells, and the type of metabolism. One kingdom comprises organisms lacking a nucleus (the prokaryotes), and the remaining four kingdoms comprise organisms with cell nuclei (the eukaryotes).

The five-kingdom classification system (**Figure 1.4a**) places all single celled prokaryotic organisms in the kingdom **Monera,** which comprises the bacteria, some of the oldest, smallest, simplest, and most successful organisms on Earth. Of the four kingdoms of eukaryotes, three of them (**Animalia, Plantae,** and **Fungi**) consist of multicellular organisms. The criteria for classifying animals, plants, and fungi are based largely on the organism's life cycle, structure, and mode of nutrition. Plants, for example,

Answers to ✅ questions can be found in Appendix A.

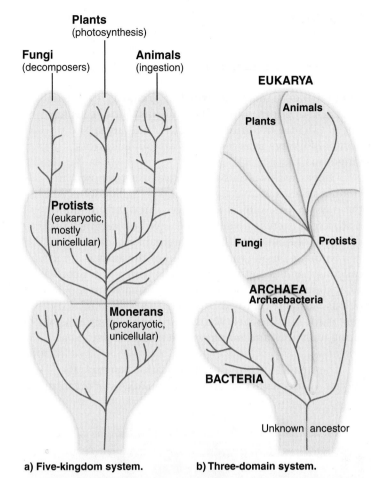

a) **Five-kingdom system.**   b) **Three-domain system.**

Figure 1.4 Categorizing organisms. a) The five-kingdom system classifies organisms based on the presence or absence of a cell nucleus, cellular structure (multicellular or unicellular), and certain characteristics of structure, life cycle, and mode of nutrition. b) The three-domain system places prokaryotes in one of two domains based on their biochemical and molecular characteristics, and puts all eukaryotes in a single domain.

contain a green pigment called chlorophyll that allows them to capture the energy of sunlight, which they convert for their own use in a process called **photosynthesis.** Animals get the energy they need by eating plants or other animals, which requires structures specialized to digest and absorb food. Most animals can move about to obtain food. Fungi (yeasts, molds, and mushrooms) are *decomposers,* meaning that they obtain their energy from decaying material. The fourth group of eukaryotes, the kingdom **Protista,** comprises unicellular and relatively simple multicellular eukaryotes such as protozoa, algae, and slime molds. All animals, plants, and fungi are thought to have evolved from single celled protistan-type organisms.

In recent years new techniques in molecular biology and biochemistry have distinguished two fundamentally different types of prokaryotes. For this reason, many biologists now advocate a classification system that begins with *domains,* a higher classification level that encompasses kingdoms. In the three-domain system (see Figure 1.4b) the organisms of the kingdom Monera are distributed across two domains (**Bacteria** and **Archaea**), each comprising a kingdom, while all organisms whose cells have nuclei fall into the third domain, the **Eukarya.** The domain Eukarya thus comprises all four eukaryotic kingdoms.

Other classification systems have been proposed, including systems with five, seven, and eight kingdoms. All these systems are subject to change as new information is discovered. There is no debate, however, concerning the classification of humans within the animal kingdom.

Humans belong to a subgroup of the animal kingdom called *vertebrates,* defined as animals with a nerve cord and a backbone. Within the vertebrates, humans are subclassified as *mammals,* defined as vertebrates with mammary glands for nursing their young. Among the mammals, we, along with apes and monkeys, are further classified as *primates.*

The smallest unit of any classification system is the species. A **species** is one or more populations of organisms with similar physical and functional characteristics that interbreed and produce fertile offspring under natural conditions. All living humans belong to the same **genus** (the second smallest unit of classification) and species, called *Homo sapiens.* We share features that make us different from any other species on Earth, and we can interbreed.

No one knows how many species of living organisms exist on Earth. Estimates range from about 3 million to 30 million, but only about 2 million species have been identified so far.

✅ **Quick Check** While studying a drop of pond water under a microscope, you notice two tiny single celled organisms, one with a nucleus and one without. Are they prokaryotes or eukaryotes, which one is more closely related to humans, and do you have enough information to determine what domain and kingdom they are in? Explain. ▪

The defining features of humans

Humans are not the largest animal, nor the fastest or strongest. Our eyesight and hearing are not the best. We can't fly, we swim poorly, and we don't dig holes in the ground very well with our hands. Nevertheless we possess several features that, taken together, define how we are different from other organisms and explain how we have managed to survive for so long:

- *Bipedalism.* Humans are the only mammals that prefer to stand upright and walk on two legs. Bipedalism (from the Latin *bi-,* two, and *pes,* foot) frees our hands and forearms for carrying items ranging from weapons to infants. Birds walk upright, too, of course, but they do not have the advantage of being able to carry things with their forelimbs.
- *Opposable thumbs.* Humans and several other primates have thumbs that can be moved into position to oppose the tips of the fingers. However, only humans have the well-developed muscles that enable us to exert a certain type of precise control over the thumb and fingers. For instance, we tend to pick up and manipulate small objects between the tip of the thumb and the tip of either the index or second finger (**Figure 1.5**). In contrast, chimpanzees more naturally grasp objects between the thumb and the side of the index finger. Threading a needle or suturing a wound would be difficult for a chimpanzee.
- *Large brain.* Humans have a large brain mass relative to body size. The evolution of a large brain seems to have coincided with the advent of stone tools, leading some scientists to suggest that a large brain was required for the complex motions associated with tool use. Other scientists believe that a large brain was necessary for language, and that language developed as

MJ's Human Biology Blog

Thimerosal and Autism

The results of a study by researchers at the Department of Public Health in California do not support the hypothesis that autism is caused by thimerosal, a preservative containing ethylmercury that was once used in childhood vaccines. The researchers found that the incidence of autism rose steadily in California from 1995 to 2007, even though thimerosal was removed from most childhood vaccines in 2001. If thimerosal were responsible for causing autism there should have been a sharp decline in new cases of autism after 2004. The new findings and a commentary about them are published in *The Archives of General Psychiatry.*

It is doubtful, however, that the study will do much to reassure parents, some of whom continue to believe passionately that their child's autism was a direct consequence of childhood vaccinations despite research findings to the contrary. ▪

Reference: Fombonne, Eric. Thimerosal Disappears but Autism Remains. *Archives of General Psychiatry* 65: 15–16, 2008.

Figure 1.6 **Language.** Human societies throughout the world have developed complex written languages.

Human   Chimpanzee

Figure 1.5 **How humans and chimpanzees hold small objects.** Although the hands of humans and chimpanzees seem similar, only humans tend to hold objects between the tip of a thumb and the tips of the fingers.

social interactions among humans became more important.

- *Capacity for complex language* Many animals vocalize (produce sounds) to warn, threaten, or identify other members of their species, and a few (such as dolphins) have developed fairly complex forms of communication. However, humans have developed both complex vocal language and a system of signs, symbols, and gestures for communicating concepts and emotions. Throughout the world, every group of human beings has developed a complex spoken language. Humans have also placed their languages into written form, permitting communication over great distances and spans of time (Figure 1.6).

It should be stressed that none of these features necessarily make us any better than any other species, only different.

✓ **Quick Check** Suppose you are a paleontologist who has just uncovered a complete fossil skeleton of some kind of human-like primate. Which areas of the skeleton could you study to determine whether the fossil is closely related to humans? Explain your reasoning. ∎

Human biology can be studied on any level of biological organization

Figure 1.7 shows how humans fit into the grand scheme of things in the natural world. The figure also shows how humans—or any living thing, for that matter—can be studied on any level of biological organization, from the level of the atom to the level of the biosphere. This text examines human biology on progressively larger scales.

Our study of human biology begins with the smallest units of life. Like rocks, water, and air, humans are composed of small units of natural elements called *atoms* and *molecules.* We'll introduce the basic chemistry of living things in Chapter 2. In Chapter 3 we'll see how the atoms and molecules of living things are arranged into the smallest of living units called *cells.* Next (Chapter 4) we'll learn how, in multicellular organisms, groups of similar cells become *tissues,* how groups of tissues that carry out a specific function constitute an *organ,* and how organs may work together in an *organ system* to carry out a more general function. The structures and functions of specific human organs and organ systems are the subjects of Chapters 5–16. For example, we'll learn why—and how—more blood flows through the lungs than through any other organ and what happens to your dinner as it makes its way through your digestive system. Chapters 17–21 consider humans as complete *organisms,* including how cells reproduce, how we inherit traits from our parents, and how we develop, age, and die. Finally, we'll discuss how *communities* of living organisms evolved and how life began (Chapter 22), and how humans fit into and alter the *ecosystems* in which we live and the entire *biosphere* of the natural world (Chapters 23 and 24).

Figure 1.7 Levels of organization in human biology.

Table 1.1 lists some current issues, controversies, and "hot topics" that relate to human biology. Many of these issues and controversies also concern fields outside the sphere of science, such as economics, law, politics, and ethics. What you learn in this course will help you make informed decisions about these and future issues that will come up in your lifetime.

Your ability to make good judgments in the future and to feel comfortable with your decisions will depend on your critical thinking skills. We turn now to a discussion of what science is, how we can use the methodology of science to improve our critical thinking skills, and how science influences our lives.

Table 1.1 **Examples of issues and controversies associated with the levels of human biology**

Level of organization	Definition	Issues and controversies
Atom and molecule	**Atom:** Smallest unit of an element of matter **Molecule:** More than one atom in a stable association	Disposing of radioactive wastes Role of unstable molecules (free radicals) in cancer and aging
Cell	Smallest unit of life	Cloning adult animals, plants, and humans from a single cell
Tissue	An association of cells with the same general structure and function	Using human fetal tissues in research
Organ	An association of several tissue types that carry out a specific function	Increasing the supply of human organs for transplantation Transplanting animal organs into humans
Organ system	Two or more organs that work together to carry out a general function, such as digestion or movement	Enhancing human performance with drugs or by genetic engineering
Organism	An individual living being composed of several organs or organ systems	Testing for heritable diseases for which there are no cures Abortion Deciding who should pay for human behavior-related illnesses such as those caused by smoking
Population	A group of individuals of the same species living in the same area	Rationing medical care Determining who gets the scarce human organs available for transplantation
Community	Several populations of different species who inhabit the same area and interact with each other	Impact of humans on the well-being and survival of other species Genetic engineering of plants and animals for human purposes Using animals in medical research and cosmetics testing
Ecosystem	All of the organisms in a given area plus all of the nonliving matter and energy	Environmental pollution Destruction of ecosystems due to overuse by humans
Biosphere	All ecosystems combined. The portion of Earth occupied by living organisms, plus those organisms.	Global warming Destruction of the ozone layer

Recap Classification systems place living things into groups. The most inclusive group is a domain, and the smallest is a species. Humans belong to the kingdom Animalia within the domain Eukarya. Our genus and species is *Homo sapiens*. Humans walk on two legs (are bipedal) and can grasp small objects between the tips of the thumb and first finger. Humans also have large brains relative to body mass, and the capacity for complex spoken and written languages. Biology can be studied at any level, from atom to biosphere. ∎

1.3 Science is both a body of knowledge and a process

We have already said that science is the study of the natural world. More explicitly, science is two things: *knowledge* (organized, reliable information) about the natural world, and the *process* we use to get that knowledge. *Scientific knowledge* refers to information about the natural world. The process of science, or the way scientific knowledge is acquired, is generally called the **scientific method,** although in practice this term encompasses a variety of methods. Throughout this book you will be presented with scientific knowledge, but it's good to remember that this information was obtained slowly over time by the scientific method.

Scientific knowledge enables us to describe and predict the natural world. Scientific knowledge is also empirical, meaning that it relies on observation and experimentation. Through the scientific method, scientists strive to accumulate information that is as free as possible of bias, embellishment, or interpretation.

The scientific method is a process for testing ideas

Although there is more than one way to gather information about the natural world, the scientific method is a systematic

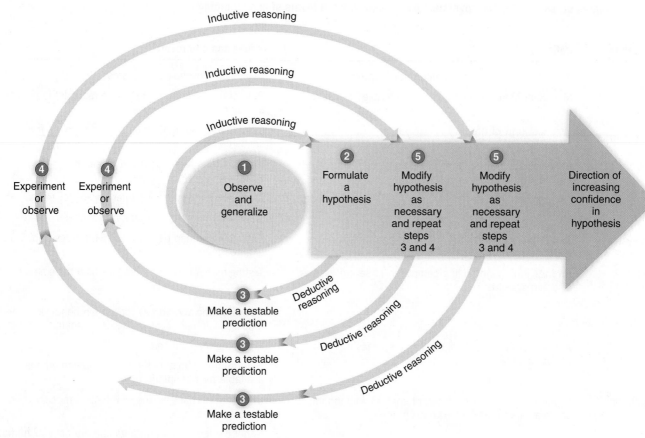

Figure 1.8 The scientific method. Observations and generalizations lead to the formulation of a hypothesis. From the hypothesis, specific predictions are made that can be tested by experimentation or observation. The results either support the hypothesis or require that it be modified to fit the new facts. The cycle is repeated. Ultimately the scientific method moves in the direction of increased confidence in the modified hypothesis.

✓ *Explain in your own words the difference between inductive and deductive reasoning, and explain why the scientific method needs both.*

process for developing and testing predictions (**Figure 1.8**). You probably already use the scientific method, or at least elements of it, in your own everyday problem solving. Sometimes we go through the steps without thinking about them consciously, but we follow them nevertheless.

Step 1: Observe and generalize When we observe the world around us and make generalizations from what we learn, we are employing *inductive reasoning* (extrapolating from the specific to the general case). Usually we don't even think about it and don't bother to put our observations and generalizations into any kind of formal language, but we do it just the same. For example, you are probably convinced that it will *always* be colder in winter than in summer (a generalization) because you have observed that every winter in the past was colder than the preceding summer (specific observation). The difference between common experience and good science is that science uses generalization to make a prediction that can be tested.

Taking an example from biological research, let's start with two observations and a generalization:

> *Observation 1:* Rats given a particular drug (call it Drug X) have lower blood pressures than rats not fed the drug.
>
> *Observation 2:* Independently, researchers in Canada showed that Drug X lowers blood pressures in dogs and cats.
>
> *Generalization:* Drug X lowers blood pressure in all mammals.

Step 2: Formulate a hypothesis Observations and generalizations are used to develop a *hypothesis*. A **hypothesis** is a tentative statement about the natural world. Importantly, it is a statement that can lead to testable deductions.

> *Hypothesis:* Drug X would be a safe and effective treatment for high blood pressure in humans.

✅ **Quick Check** Consider the following statement: "Sasquatches are giant apes that live in the forests of the Pacific Northwest, but they avoid people and leave no evidence of their existence." Does this statement qualify as a scientific hypothesis? ■

Step 3: Make a testable prediction Hypotheses that cannot be tested are idle speculation, so much hot air. But many hypotheses are so sweeping and comprehensive that ways must be found to test them under a variety of conditions. For example, you probably would not be convinced that Drug X is safe and effective for all people under all conditions until you had at least tested it in quite a few people under many different conditions. To have confidence in your hypothesis, you must make testable predictions (also called *working hypotheses*) based on the hypothesis and then test them one at a time. Predictions employ *deductive reasoning* (applying the general case to the specific). Often they are put in the form of an "if … then" statement, in which the "if" part of the statement is the hypothesis. For example:

> *Prediction: If* Drug X is a safe and effective treatment for high blood pressure in humans, *then* 10 mg/day of Drug X will lower blood pressure in people with high blood pressure within one month.

Notice that the prediction is very specific. In this example the prediction specifies the dose of drug, the medical condition of the persons on whom it will be tested, the expected effect of the drug if the prediction is correct, and a specified time period for the test. Its specificity makes it testable—yes or no, true or false.

✅ **Quick Check** Suppose you are designing a research study to test the hypothesis that regular exercise helps people sleep better. Develop a specific, testable prediction from this hypothesis. ■

Step 4: Experiment or observe The truth or falsehood of your prediction is determined by observation or by experimentation. An **experiment** is a carefully planned and executed manipulation of the natural world to test your prediction. The experiment that you conduct (or the observations you make) will depend on the specific nature of the prediction.

When testing a prediction, scientists try to design experiments that can be conducted under strictly controlled conditions. Experiments conducted under specific, controlled conditions are called *controlled experiments* because they have the distinct advantage of accounting for all possible **variables** (factors that might vary during the course of the experiment) except for the one variable of interest, called the *controlled variable*. In this case the controlled variable is blood pressure. Therefore, in the example of Drug X, you could follow these steps in your controlled experiment (**Figure 1.9**):

■ Select a large group of human subjects. In this case you would specifically choose people with high blood pressure rather than people with normal blood pressure.

Select a large number of appropriate subjects.

Randomly divide the subjects into two groups.

Group 1   Group 2

Treat the groups equally in all ways but one.

| Experimental group: receives treatment | Control group: receives placebo |

Observe or make measurements.

Are blood pressures lower in the experimental group?

Compare results.   **Yes**   **No**

Hypothesis received support.   Modify hypothesis to fit the new findings.

Figure 1.9 The steps in a controlled experiment. In this example the experimental variable is blood pressure. Because of the way the experiment is designed, the only difference between the two groups is the presence or absence of the experimental treatment itself.

 Suppose that instead of randomly assigning subjects to the two groups, subjects with high blood pressure are more likely to be put in the experimental group. Would this be a better or a worse experiment, or would it make no difference? Explain.

■ *Randomly* divide the larger pool of subjects into two groups. Designate one the **experimental group** and the other the **control group.** The importance of random assignment to the two groups is that all other factors that might affect the outcome (such as age or gender differences in responsiveness to the drug, or other previous health problems) are automatically equalized between the two groups. In effect, the control group accounts for all unknown factors. If you treat the experimental and control groups identically at this point, the probability is that the average blood pressures of the groups will be equal, as long as the groups are large enough.

■ Treat subjects in the two groups *exactly* the same except that only the experimental group gets the drug. Treat the experimental group with 10 mg/day of Drug X for one month and the control group with a **placebo,** or "false treatment." If you deliver Drug X in a pill, the placebo should be an identical-looking pill with no drug in it. If you administer Drug X as an injection in a saline solution, the placebo should be an injection of the same volume of saline.

When working with human subjects, scientists must take the power of suggestion into account. It is important that no subjects in either group know which group has received the drug. If members of the experimental group find out that they are getting a drug to lower their blood pressure, this knowledge may accidentally influence the outcome of the experiment. To eliminate the power of suggestion as a variable, researchers conduct experiments "blind," meaning that the subjects don't know whether they are getting the placebo or the drug. Sometimes experiments are done "double-blind," so that even the person administering the drugs and placebos does not know which is which until the experiment is over.

■ Measure blood pressures in both groups at the end of one month and compare them using the appropriate statistical (mathematical) tests. If the experimental group's blood pressures are statistically lower, then your prediction is verified and your hypothesis receives support.

✓ **Quick Check** A researcher tries to test the hypothesis that "exercise helps people sleep better" with the following study: He tells four friends about the hypothesis, ask them to jog for an hour on a given day, and then calls them the next day to ask how well they slept. What are some problems with this experiment? ■

Step 5: Modify the hypothesis as necessary and repeat steps 3 and 4 If your prediction turns out to be false, you will have to modify your hypothesis to fit the new findings and repeat steps 3 and 4. For example, perhaps the drug would lower blood pressure if you increased the dose or gave it for a longer period of time.

Even if your prediction turns out to be true, you're still not done. This is because you've tested only one small part of the hypothesis (its effectiveness under one specific set of conditions), not all the infinite possibilities. How do we know the drug is safe, and how safe is "safe," anyway? Does it cause a dangerous drop in blood pressure in people with normal blood pressures? Does it cause birth defects if taken by pregnant women? Does long-term use lead to kidney failure? Does the drug's effectiveness diminish over time? Specific predictions may have to be stated and tested for each of these questions before the drug is allowed on the market.

Only after many scientists have tried repeatedly (and failed) to disprove a hypothesis do they begin to have more confidence in it. A hypothesis cannot be proved true; it can only be supported or disproved. As this example shows, the scientific method is a process of elimination that may be limited by our approach and even by our preconceived notions about what to test. We move toward the best explanation for the moment, with the understanding that it may change in the future.

 Web Animation *The Scientific Method* at www.humanbiology. com.

Making the findings known

New information is not of much use if hardly anybody knows of it. For that reason, scientists need to let others know of their findings. Often they publish the details in scientific journals to announce their findings to the world. Articles in *peer-reviewed* journals are subjected to the scrutiny of several experts (the scientists' peers) who must approve the article before it can be published. Peer-reviewed journals often contain the most accurate scientific information.

An unspoken assumption in any conclusion is that the results are valid only for the conditions under which the experiment was done. This is why scientific articles go into such detail about exactly how the experiment was performed. Complete documentation allows other scientists to repeat the experiments themselves or to develop and test their own predictions based on the findings of others.

Try to apply the scientific method to a hypothesis dealing with some aspect of evolution or to a global problem such as cancer or AIDS, and you begin to appreciate how scientists can spend a lifetime of discovery in science (and enjoy every minute!). At times the process seems like three steps forward and two steps back. You can bet that at least some of what you learn in this book will not be considered accurate 10 years from now. Nevertheless, even through our mistakes we make important new observations, some of which may lead to rapid advances in science and technology. Just in the last 100 years we've developed antibiotics, sent people to the moon, and put computers on millions of desks.

Health & Wellness

The Growing Threat of Antibiotic-Resistant Bacteria

When antibiotics—drugs that kill bacteria—became available in the 1940s, they were hailed as a breakthrough. Indeed, many people owe their lives to them. But there is a downside: Indiscriminate use of antibiotics leads to antibiotic resistance in some strains of bacteria. For example, in recent years a new strain of the *Staphylococcus aureus* bacterium has appeared that is resistant to all penicillin-type antibiotics. This frightening superbug, called MRSA (methicillin-resistant *Staphylococcus aureus*) can cause serious skin infections and even penetrate bones and lungs, killing approximately 20% of those with severe infections. Over 25% of all *Staphylococcus aureus* ear/nose/throat infections in children are now MRSA.

The rise of MRSA actually makes biological sense. Whenever an antibiotic kills most but not all of a population of bacteria, the surviving bacteria are the ones that are naturally most resistant to the

antibiotic. With the total population of bacteria now decreased, these antibiotic-resistant survivors multiply. So the more we use antibiotics, the more we encourage the rise of resistant strains of bacteria.

Because antibiotics have been so effective in the past, we have grown to rely on them and we tend to use them indiscriminately. Over 50 million pounds of antibiotics are produced each year in the United States, about half of which are fed to livestock or sprayed on fruit trees. Researchers estimate that fully one-third of

Skin lesions caused by MRSA.

all antibiotic prescriptions given to nonhospitalized patients in the United States are not needed.

The key to preserving the effectiveness of antibiotics is to use them only when necessary. Things we can do:

- Don't ask your doctor for antibiotics for a viral illness such as a cold or flu. Antibiotics kill bacteria but have no effect on viruses.
- Take antibiotics only when needed and as prescribed. Complete the full course of treatment.
- Reduce your use of antibacterial hand creams, soaps, and laundry detergents.
- Support farmers' efforts to reduce their use of antibiotics in cattle feed and on fruit trees.
- Support research efforts to find new antibiotics.

Used properly and judiciously, antibiotics will remain in our antibacterial arsenal for many years.

Many people associate science with certainty, whereas in reality scientists are constantly dealing with uncertainty. That is why we find scientists who don't agree or who change their minds. Building and testing hypotheses is slow, messy work, requiring that scientists constantly question and verify each other.

A well-tested hypothesis becomes a theory

Many people think that a *theory* is a form of idle speculation or a guess. Scientists use the word quite differently. To scientists, a **theory** is a broad hypothesis that has been extensively tested and supported over time and that explains a broad range of scientific facts with a high degree of reliability.

A theory is the highest status that any hypothesis can achieve. Even theories, however, may be modified over time as new and better information emerges. Only a few

hypotheses have been elevated to the status of theories in biology. Among them are the theory of evolution and the cell theory of life.

Recap The scientific method is a systematic process of observation, hypothesis building, and hypothesis testing. A theory is a broad hypothesis that has withstood numerous tests.

1.4 Sources of scientific information vary in style and quality

We are constantly bombarded by scientific information, some of it accurate and some not. What can you believe when the facts seem to change so quickly? All of us need to know how to find good information and evaluate it critically. Different sources of scientific information may have very different goals, so look for those that can best inform you at your own level of understanding and interest.

Some scientific knowledge is highly technical. As a result, scientists have a tendency to speak on a technical level and primarily to each other. As we've already mentioned, scientists often communicate by means of articles in specialized peer-reviewed journals such as *Nature* and *Science*. Articles in peer-reviewed journals are concise, accurate, and documented so thoroughly that another scientist ought to be able to duplicate the work after reading the article. Generally they refer extensively to previous literature on the subject. Articles in peer-reviewed journals make for laborious reading, and they are usually as dry as toast. But bear in mind that their purpose is primarily to inform other experts.

Other helpful print sources are science magazines and nonfiction books meant for the well-educated public. The goal is to inform the interested reader who may have only a limited background in science. The authors are usually science writers or experts who translate the finer scientific points into language that we can all understand. The information is generally accurate and readable, although the reader may not understand some of the details. Generally these articles and books tell readers who want to delve more deeply into the subject where to find more information.

General interest news magazines and daily newspapers also report on selected hot topics in science. Their goal is to get the information out as quickly as possible to a wide audience. Coverage is timely but less in-depth than in science magazines, and may not include the details you need to check the validity of the statements. A decided plus is that magazines and newspapers often discuss social, political, economic, legal, or ethical ramifications of the scientific findings, something generally lacking in the previous sources. Although the scientific information is usually accurate, the reporter may not understand the subject fully and may not provide adequate context. The best articles point readers to the original sources. Television (for instance, the Discovery Channel, *Nova*) also presents science-related topics to the public.

Since the 1980s, scientists and researchers have used the Internet to communicate and share ideas. The recent expansion of the World Wide Web has made the Internet accessible to the general public, opening exciting new sources of scientific information. Nearly all universities now have Web pages; the site addresses end in ".edu" (for "educational") rather than ".com" (for "commercial"). A number of scientific and professional organizations have created Web sites that offer helpful information for both scientists and consumers. Examples of organizations with Web sites include the National Institutes of Health, the American Cancer Society, and the American Heart Association. The Web addresses of government agencies and non-profit organizations generally end in ".gov" and ".org," respectively.

Be aware that the Internet can also be a source of misinformation. At present the Internet is less closely regulated than print and broadcast media, so it can be difficult to tell the difference between objective reports and advertisements. In addition, participants in online chat rooms and special interest groups may promote their own opinions as proven truths. It pays to be skeptical.

 Recap The best sources of scientific information translate difficult or complex information accurately into understandable terms and have enough references that you can check the information if you wish. ∎

1.5 Learning to be a critical thinker

Many scientists are motivated by strong curiosity or a sense of wonder and awe about how the natural world works. Exploring the frontiers of knowledge requires a great deal of creativity and imagination. Like many people, however, scientists may leap to conclusions or resist new ideas. A few may be driven by self-interest. To combat these natural human tendencies, good scientists try to use certain tools of critical thinking. You too can learn to use these tools, regardless of whether you choose a career in science. The sections below describe some of the simple tools that anyone can use to improve their critical thinking skills.

Become a skeptic

Good scientists combine creativity and imagination with **skepticism**, a questioning attitude. If you've ever bought something based on claims about how well it works and then been disappointed, you know the value of skepticism. Question everything and dig a little deeper before believing something you read and hear. Here are some questions you might ask yourself:

- Who says that a particular statement is true?
- What evidence is presented?
- Are the persons speaking on a subject qualified by training or skill to speak authoritatively about it?
- Are they being paid, and if so, how might that affect what they have to say?
- Where's the evidence to back up a claim?

Skepticism is particularly important for claims that are new, startling, and not yet verified by other scientists. Listen carefully to the debate between scientists in the public arena. A new scientific claim may take several years to be checked out adequately.

Appreciate the value of statistics

Statistics is the mathematics of organizing and interpreting numerical information, or **data.** Scientists use statistics to determine how much confidence they should place in information. Most scientists would be willing to accept experimental results with confidence if (according to statistical tests) they would get the same outcome 19 of every 20 times they repeat the experiment, or 95% of the time. When you see numerical averages followed by a smaller "+/−" number, the smaller number represents an expression of confidence in the certainty of the results, called the "standard error." In graphs, the standard errors are represented as small lines that extend above and below the average number.

Statistics are important in many disciplines. During elections, we may hear pollsters report, for example, that "52% of the respondents said that they will vote for the president. The poll has a margin for error of +/−3%." This tells you the pollsters are relatively certain that the actual percentage who will vote for the president is somewhere between 49 and 55%, still too close to call.

Learn how to read graphs

Just like a picture, a graph is worth a thousand words. Graphs display data obtained from observations and experimental results in a way that is economical and easy to grasp. Graphs can also be used to clarify the meaning of experimental results.

Most graphs are plotted on two lines, or axes (singular: axis). The horizontal axis at the bottom is called the *abscissa* (from math you may know this as the *x*-axis), and the vertical axis is called the *ordinate* (*y*-axis). By convention the *independent variable*, such as time, distance, age, or another category that defines groups, is generally plotted on the abscissa. The *dependent variable*, so called because its variation may depend on the independent variable, is plotted on the ordinate.

Graphs can take a variety of forms, from plots of individual data points to lines or bars of average values (Figure 1.10). When reading a graph, first check the scales and the legends on the abscissa and the ordinate to determine what the graph is about. Be careful to look for a "split axis," in which the scale changes. An example is shown in Figure 1.11. A split axis is sometimes a convenient way of representing data that cover a wide range on one axis, but it can also be used to deliberately mislead people unfamiliar with reading graphs.

Distinguish anecdotes from scientific evidence

Anecdotal evidence takes the form of a testimonial or short unverified report. Although an anecdote may be true as stated, it in no way implies scientific or statistical certainty. It cannot be generalized to the larger population because it is not based on empirical evidence. Advertising agencies sometimes use anecdotes to influence you. The actor on television who looks sincerely into the camera and says "Drug X worked for me" may be telling the truth—the drug may work for him. But this does not prove the drug will work for everyone, or even for 10% of the population.

Nonscientists (and even scientists) often say things like "My grandmother swears by this remedy." Again, the statement may be true, but it is not scientific evidence. Listen carefully to how the evidence for a statement is presented.

 Quick Check Suppose one brand of cold medication has personal testimonials on its Web site from three different people who all say that the medication helped them get over colds faster, while another brand has similar stories from *30* people. Does this prove that the second medication is better than the first? Why or why not? ■

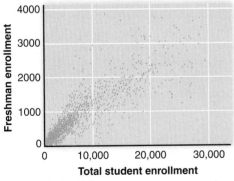

a) A scatter plot showing enrollment at each individual college. Each data point is known as an observation.

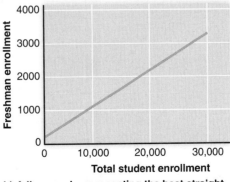

b) A line graph representing the best straight line fit of the data in a).

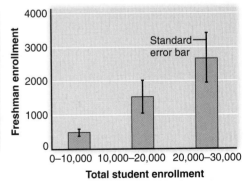

c) A bar graph in which university enrollments are lumped together in three class sizes and the freshman enrollments are then averaged. Standard error bars indicate that the data have been analyzed statistically.

Figure 1.10 Types of graphs. Each of these graphs reports the relationship between freshman enrollment and total student enrollment at approximately 1,500 U.S. colleges and universities.

 In the bar graph, why are the standard error bars in the third bar much higher than in the first bar? Put another way, what do standard error bars actually tell us? (Look back at the first graph for a hint.)

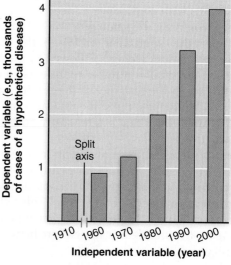

a) Graph with regular axis.

b) Graph with split axis.

Figure 1.11 How a split axis affects a graph. The graph in b) is redrawn from the data in a) by splitting the abscissa and omitting the data for the years 1920–1950. The effect is a consolidated graph that fits in less space, but it might mislead you into thinking that the number of cases of the disease has been rising steadily since 1910, instead of only since 1960.

Separate facts from conclusions

A *fact* is a verifiable piece of information, whereas a *conclusion* is a judgment based on the facts. The news media often mix facts with conclusions without indicating which is which. Almost every evening on the business news we hear statements like "The Dow Jones Industrial Average declined 50 points today on renewed concern over the consumer price index." The first half of the sentence (about the decline) is a verifiable fact. The second half is conjecture on the part of the reporter.

Or consider the following: "The average global temperature was 0.1° C higher this year than last year. The rise in temperature proves that global warming is occurring." Again, fact is followed by conclusion. The conclusion may not be warranted if temperature fluctuations up and down of 0.3° C are normal from year to year.

Understand the difference between correlation and causation

A close pattern or relationship (a correlation) between two variables does not necessarily mean that one causes the other. The catch-phrase is *"correlation does not imply causation."* A good example of a correlation without causation is the close correlation between ice cream sales and drownings—when ice cream sales are up in the summer months, so are drownings. Does that mean that eating ice cream causes people to drown? Hardly. Ice cream sales and drownings also correlate with (and are most likely caused by) a third

Correlation versus Causation

According to a study of the nearly 140,000 women who were enrolled in the Women's Health Initiative, women who breast-fed their babies had a lower risk of developing heart disease and diabetes later in life than women who did not breast-feed their babies. The headline of the New York *Times* article about the research read "Breast-Feeding Benefits Mothers, Study Finds." Is this a correct summary of the research?

The answer is "NO"! This is a classic case of the common misunderstanding about the relationship between correlation and causation. Yes, there is a clear correlation between breast-feeding and a lower risk of diabetes and heart disease later in life, according to the study. But that is not proof that the act of breast-feeding is what reduces the risk. What if women who breast-fed their children are just more health conscious overall throughout life? What if they exercise more often, or have healthier diets?

A more correct headline would be "Breast-Feeding May Benefit Mothers, Study Suggests." Indeed, the article itself goes on to say that some experts are cautioning that an association (between breast-feeding and health benefits) does not prove a causal relationship, and that more research would be needed to determine the exact cause of the effect (lower risk). ■

factor not mentioned in the original correlation—warmer temperatures during the summer.

If the above example seems too obvious, try this one: In 1999 a study at a major university found that children who slept with a light on were more likely to develop nearsightedness (myopia) later in life. But does this mean that sleeping with a light on *causes* nearsightedness? In fact, a follow-up study in 2000 found no direct causal relationship between sleeping with a light on and the development of nearsightedness. The follow-up study showed that children who develop nearsightedness are more likely to have parents who are nearsighted, suggesting (but not proving) a genetic cause. It also showed that parents who are nearsighted are just more likely to leave the light on!

In the above example, the original scientific observation was stated correctly (lights-on *correlates* with nearsightedness). But anyone who became convinced that sleeping with a light on *causes* nearsightedness would have been wrong. Be skeptical of causal statements that are based only on a good correlation, for the true cause may not be obvious at first. (See MJ's Human Biology Blog, Correlation Versus Causation.)

Of course, a close correlation is likely whenever a true causal relationship does exist. So although a correlation does not necessarily prove causation, it can be a strong hint that you *may* have found the true cause, or at least that the true cause is nearby and may even be linked to both of the variables you're observing.

Recap Healthy skepticism, a basic understanding of statistics, and an ability to read graphs are important tools for critical thinking. Know anecdotal evidence when you see it, and appreciate the differences between fact and conclusion and between correlation and causation. ■

1.6 The role of science in society

How do we place science in its proper perspective in our society? Why do we bother spending billions of dollars on scientific research when there are people starving in the streets? These are vital questions for all of us, so let's look at why we study the natural world in the first place.

Science improves technology and the human physical condition

Science gives us information about the natural world upon which we can base our societal decisions. Throughout history some of the greatest benefits of science have been derived from the *application* of science, called **technology,** for the betterment of humankind (**Figure 1.12**). Time and time again, scientific knowledge has led to technological

a) Modern farmers grow food crops much more efficiently than in the past, thanks to advances in such diverse fields as genetics, chemistry, and even the aerospace industry. Global positioning satellites and computers allow farmers to administer fertilizers precisely where needed, thereby eliminating waste, reducing environmental degradation, and improving yields.

b) This scientist is collecting insects from a tree-top in a tropical rainforest. Studies such as this improve our understanding of the interrelationships of organisms in an ecosystem and have yielded rare natural chemical compounds that may prove useful in human and animal medicine.

Sep 22 2004

c) A satellite map documenting depletion of the ozone layer over Antarctica in 2004. The area of greatest depletion appears dark blue. Studies such as this one allow scientists to document when and where ozone depletion occurs so that they can better understand its causes and cures.

Figure 1.12 Benefits of science. These photos show typical scenes of scientists and the application of science (technology) for the betterment of the human condition.

advances that have increased the productivity and hence prosperity of both industries and nations. Science has given us larger crop yields, more consistent weather predictions, better construction materials, better health care, and more efficient and cleaner sources of power, to name just a few benefits. It has made global transportation and communication possible.

Many people are concerned that overuse of our technological capabilities may lead to problems in the future. Science can help here, too, by helping us identify problems early on. We can see the early warning role of science in the "Health & Wellness" discussion of bacterial resistance to antibiotics (on page 15). Only by understanding a problem can we learn how to solve it. Science helps us correct our mistakes.

Science has limits

Scientific knowledge is limited to physical explanations for observable events in the natural world. It cannot prove or disprove the existence, or importance to us, of things that fall outside the realm of the natural world, such as faith or spiritual experiences. Many scientists have a strong faith or belief that cannot be tested by science, because faith does not depend on logical proof or material evidence. They believe that the search for meaning and the search for knowledge are complementary, not contradictory.

In addition, science alone cannot provide us with the "right" answers to political, economic, social, legal, or ethical dilemmas. Humans have minds, a moral sense, and a sense of history and the future. How we use scientific knowledge is up to all of us, not just scientists. For example, given the current state of knowledge about how cells grow and divide, scientists may eventually be able to clone an adult human being. Whether or not we should permit cloning, and under what circumstances, are important topics of public debate. It's not for scientists to decide alone.

This does not mean scientists are without moral obligation. As experts in their fields, they are in a unique position to advise us about the application of scientific knowledge, even if the choices ultimately rest with all of us.

A practical limitation of science is that some information, including data that may be useful in improving human health, cannot be obtained by observation or experimentation. Our society places a very high value on human life, and therefore we don't experiment on humans unless the experiment is likely to be of direct benefit to the subject (the use of experimental cancer drugs falls into this category). This is why it is hard to investigate the danger of street drugs like cocaine or anabolic steroids. No good scientist would ever deliberately give healthy humans a drug that might cause injury or death, even if the resulting information could save lives in the future. Our society currently does permit experiments on animals as

MJ's Human Biology Blog

Scientific Uncertainty and Shared Responsibility

Here's an interesting dilemma for you: An industrial chemical is known to cause a particular type of cancer in rats. A lifelong employee at the plant that makes the chemical is diagnosed with the cancer, so he sues the company for millions of dollars. Ultimately the employee loses the case and receives nothing, because the company's lawyers argue (correctly) that the worker could have gotten the cancer from some other source, or that the worker had a genetic predisposition to cancer. In other words, the worker cannot prove scientifically beyond a reasonable doubt that *his* cancer was caused by exposure to the chemical while working in the company's plant. Having won the case, the company continues to expose its current workers to the chemical.

Do we consider this to be fair, equitable, and just? Should the entire burden of uncertain science be borne by just one side?

Some lawyers are now arguing for the concept of "shared responsibility" when the scientific evidence leaves room for uncertainty. Perhaps the company should be asked to pay at least a small amount to its workers who develop the cancer as acknowledgment that their chemical might have caused the workers' cancers—not enough for the workers to have "won the lottery" or to bankrupt the company, but at least enough to help defray the workers' medical expenses. Over time that might cause the company to reconsider continuing to expose its current workers.

What do you think? ■

substitutes for human subjects, however, provided that federal guidelines are strictly followed (**Figure 1.13**).

The importance of making informed choices

You live in a science-oriented society. Throughout this book we present a common theme: Every day you make decisions about how you and society choose to use the knowledge that science gives us. Whether or not you are conscious of it, whether or not you deliberately take action, you make these choices daily.

Should Olympic athletes be allowed to use bodybuilding drugs? Do you think the use of pesticides is justified in order to feed more people? How do you feel about the cloning of human beings? Are you willing to eat a proper

Figure 1.13 Animals in research. In this society we allow the use of animals for research in certain circumstances. This researcher is using a noninvasive technique for measuring blood flow in the skin as part of a study of vascular diseases.

diet to stay healthy—and by the way, what *is* a proper diet? Who should pay for health care for the poor?

Our knowledge has advanced rapidly. By the end of this book you will know more about genetics and evolution than did the scientists who originally developed the theories about them. With knowledge comes the responsibility for making choices. From global warming to genetic engineering to personal health, each of us must deal with issues that concern our well-being and the future of the biological world in which we live. We owe it to ourselves, as individuals and as a society, to acquire the knowledge and skill we need to make intelligent decisions. Your choices can make a difference.

Recap Science and technology have improved the human condition. Science cannot, however, resolve moral dilemmas. Scientists can advise us on issues of science, but we as a society must decide how to put this scientific knowledge to use. ∎

Chapter Summary

The characteristics of life *p. 4*

- All living things acquire both matter and energy from their environment, transforming them for their own purposes.
- The basic unit of life is a single cell.
- Living things maintain homeostasis, respond to their external environments, and reproduce.

How humans fit into the natural world *p. 7*

- The biological world can be organized into kingdoms. A common scheme classifies life into five kingdoms: Monera, Protista, Fungi, Plantae, and Animalia.
- Features that define humans are bipedalism, well-developed opposable thumbs, a large brain, and the capacity for complex language.
- Humans are part of communities of different organisms living together in various ecosystems.

Science is both a body of knowledge and a process *p. 11*

- Scientific knowledge allows us to describe and make predictions about the natural world.
- The *scientific method* is a way of thinking, a way of testing statements about the natural world (*hypotheses*) by trying to prove them false.
- A *theory* is a hypothesis that has been extensively tested and that explains a broad range of scientific facts with a high degree of reliability.

Sources of scientific information vary in style and quality *p. 15*

- Articles in peer-reviewed scientific journals are generally written for other scientists. They may be difficult to read but are very accurate.
- Journals, books, and television shows on popular science present scientific knowledge efficiently to the general public.
- Web sites vary widely in the quality and accuracy of the information they present.

Learning to be a critical thinker *p. 16*

- Skepticism is a questioning attitude ("prove it to me"). Critical thinking requires skepticism.
- Knowing how to read graphs and understand basic statistics can help you evaluate numerical data.
- Being able to recognize an anecdote, tell fact from conclusion, and distinguish between correlation and causation can help you evaluate the truth of a claim.

The role of science in society *p. 19*

- The application of science is called technology.
- Science is limited to physical explanations of observable events.
- How to use science is up to us.

Terms You Should Know

biology, *4*	hypothesis, *12*
control group, *14*	science, *4*
data, *16*	scientific method, *11*
experiment, *13*	theory, *15*
experimental group, *14*	variable, *13*
homeostasis, *6*	

Concept Review

Answers can be found at the Human Biology Place.
www.humanbiology.com

1. Living things have a different molecular composition from non-living things. What makes this possible?
2. Explain the meaning of the term *homeostasis*.
3. Name four features that together contribute to our uniqueness and define us as human.
4. Describe the difference between a hypothesis and a prediction (or working hypothesis).
5. Discuss the role of scientists in helping us solve economic, social, and ethical dilemmas.

Test Yourself

Answers can be found in Appendix A.

1. To which of the following domains of life do humans belong?
   a. prokarya
   b. eukarya
   c. animalia
   d. mammalia
2. To which of the following domains do unicellular organisms which lack nuclei belong?
   a. Eukarya
   b. Archaea
   c. Bacteria
   d. both Archaea and Bacteria
3. New scientific knowledge is gained through a multistep process known as
   a. the scientific method
   b. hypothesis development
   c. variable testing
   d. observation testing
4. An experiment designed and conducted under strictly managed conditions is a:
   a. replicated experiment
   b. controlled experiment
   c. "blind" experiment
   d. peer-reviewed experiment
5. A broad hypothesis that has been supported by repeated experimentation is known as:
   a. a proven hypothesis
   b. a supported hypothesis
   c. a theory
   d. a dogma
6. Which of the following is used when developing a hypothesis?
   a. observations
   b. inductive reasoning
   c. controlled experiments
   d. both observations and inductive reasoning

7. The smallest unit of life that demonstrates all the properties of life is:
   a. an organism
   b. an organ system
   c. a molecule
   d. a cell
8. Consider all of the organisms (human as well as nonhuman) that occupy your college campus. From a biological standpoint, this would be:
   a. a community
   b. an ecosystem
   c. a biome
   d. a population
9. Which of the following lists the steps of the scientific method in order?
   a. observation—prediction—experimentation—hypothesis development
   b. hypothesis development—observation—experimentation—prediction
   c. prediction—hypothesis development—experimentation—observation
   d. observation—hypothesis development—prediction—experimentation
10. In graphs, which of the following is usually plotted on the abscissa (*x* axis)?
    a. controlled variable
    b. independent variable
    c. dependant variable
    d. placebo
11. An acceptable scientific hypothesis:
    a. can be tested
    b. can be proven true
    c. can be proven false
    d. both (a) and (c)
12. Drug A is being tested for its effectiveness in shortening the duration and severity of influenza in humans. In designing an experiment to test Drug A, which of the following would be an important consideration?
    a. Participants can choose whether to be in the experimental or control group.
    b. The experimental group will contain only males and the control group will contain only females.
    c. The experimental group should contain 1,000 subjects, but the control group should include 100 subjects.
    d. The experimental group will receive Drug A and the control group will receive a placebo.
13. Jenna has been telling her friends about how successful she was at losing 10 pounds by using a dietary supplement she purchased at a health food store. This is an example of:
    a. a proven hypothesis
    b. anecdotal evidence
    c. a controlled experiment
    d. a scientific theory
14. The maintenance of a relatively stable internal environment is:
    a. metabolism
    b. evolution
    c. constancy
    d. homeostasis

15. All of the following are features that collectively distinguish humans from other animals except:
    a. bipedalism
    b. large brain
    c. ability to evolve as a species
    d. capacity for complex language

Apply What You Know

Answers can be found at the Human Biology Place.
www.humanbiology.com

1. A magician has a coin that he says (hypothesizes) has heads on both sides, but he's unwilling to show you both sides. To convince you, he flips it three times and gets heads each time. Do you believe that the coin has two heads? What if he gets heads 10 times in a row? 100 times? What would it take (by coin flip) to prove that the coin does *not* have two heads? With this example, explain the difference between having relative confidence in the truth of a hypothesis, proving it to be true, and proving it to be false.

2. Your roommate is writing a paper on the subject of cocaine and birth defects in humans and wonders why there don't seem to be published reports of controlled experiments in humans on the subject; all the studies are on rats! Describe to her how such a controlled experiment would have to be designed and conducted, and convince her that it would never be permitted by any responsible regulatory agency.

3. You have a friend who truly believes in the existence of ghosts and says he has scientific evidence; he and his two roommates have all seen them. Explain to your friend what it means to have scientific evidence. Think about what data are, how they are gathered, and why personal experiences do not meet the criteria to be considered scientific evidence.

4. An episode of an old TV show was about doctors living in a tropical environment where the heat is unusually oppressive. An orderly comes to seek relief from the heat, and one doctor gives a supply of sugar pills to the orderly and tells him they are an experimental drug designed to keep humans cool in hot weather. The gullible orderly takes them, and while others are sweating, he claims to suffer no effects from the heat to the point of not even sweating. He later finds out the drug is fake, and immediately complains of being overheated.

    This is a fictitious demonstration of the "placebo effect." Explain how the placebo effect can be avoided when testing new drugs.

5. You are trying to convince your friend who smokes cigarettes that he should quit. You explain to him that smoking and the incidence of lung cancer are strongly correlated. Your friend says that that does not prove smoking causes lung cancer. Is your friend correct? If so, explain why he is correct. What would you say to him?

6. On the radio, you hear an interview with a climatologist discussing global warming. The interviewer asks the scientist what proof she has that humans are to blame for global warming and the resulting rise in sea levels. The scientist responds that while we cannot prove humans are to blame for global warming, there is much evidence that human activities are responsible for the rising temperatures. The interviewer says, "Aha, so you have no proof that humans are to blame, and this is all nothing more than a theory, and isn't a theory nothing more than an opinion?" What differentiates a scientific theory from an opinion?

7. Explain why religious explanations cannot disprove a scientific theory, and conversely, why science cannot prove or disprove a religious belief.

2
The Chemistry of Living Things

Crystals of lactose, a carbohydrate. Human milk contains 6% lactose.

Functional Foods and Dietary Supplements—Safe and Effective?

Red Bull energy drink to boost our energy; glucosamine and chondroitin for our aching joints; extracts of *Ginkgo biloba* to improve our memories: We seem to have an appetite for functional foods and dietary supplements that promise to improve our health or make us feel better.

Functional foods, also sometimes called "nutraceuticals," are food or drink products that are said to have benefits beyond basic nutrition. Some are natural products; others are fortified foods or completely artificially created products. Red Bull is a functional food because of the manufacturer's claim that it "boosts

energy levels." Whole-wheat bread is a functional food, too, if the claim is made that the insoluble fiber in wheat bran "contributes to the maintenance of a healthy digestive tract."

Dietary supplements are products that are not normally part of the diet, but that you choose to take to improve your health or well-being. They include your daily multi-vitamin, as well as any supplemental minerals, amino acids, body-building products, plant extracts, and hormones you take by choice. Extracts of *Echinacea purpurea* taken to fight infections and extracts of *Ginkgo biloba* for improved memory are both dietary supplements.

Do you know what's in this can?

Some natural-product dietary supplements, such as herbal remedies, have been around for hundreds or even thousands of years. Many of the ingredients in them probably do have specific health benefits. However, their effectiveness and safety in most cases have never been tested scientifically. How did this state of affairs come about, and are we comfortable with it? To understand this issue, it will help you to understand how and why functional foods and dietary supplements are regulated differently than pharmaceutical drugs (and why drugs cost so much!).

Regulatory Issues

The U.S. Food and Drug Administration (FDA) is responsible for overseeing the safety and efficacy of pharmaceutical drugs (drugs created specifically for the treatment or prevention of disease). By law, a pharmaceutical company must prove beyond a reasonable doubt that a new drug is safe and effective in humans before it can be sold to the public. On average it takes 12–15 years and costs about $800 million to bring one new drug to market. Companies can afford it only because they can patent the drug, giving them exclusive ownership and marketing rights for a certain number of years. Although patients sometimes complain about the high cost of prescription medications, pharmaceutical companies respond that the price reflects their steep development costs.

Because the ingredients in functional foods and dietary supplements occur in nature, they cannot be patented. Anyone can purify and package them. But without the assurance of patent protection, manufacturers and producers cannot afford to spend what it would cost to test their safety and effectiveness. Recognizing this, back in 1994 the dietary supplement manufacturers asked for (and were granted) an exemption from the FDA drug approval process. Under the Dietary Supplements Health and Education Act of 1994, dietary supplements and functional foods can be produced and sold until they are *proven unsafe*. And why would any manufacturer choose to spend time and money to prove its own product unsafe? The FDA does not even need to be notified of "adverse events"; the agency must rely on voluntary information supplied by consumers and health professionals. As a result, the FDA reported only about 500 adverse events per year associated with dietary supplements over a five-year period. In contrast, the American Association of Poison Control Centers received nearly 7,000 reports involving dietary supplements in one year alone.

Producers and distributors of functional foods and dietary supplements also have considerable latitude in advertising their products; the only restriction is that they are not allowed to claim that their product prevents or treats specific medical conditions or diseases. For example, producers of cranberry juice products are free to say that cranberry juice "helps maintain urinary tract health" (a rather vague health claim), but they cannot claim that cranberry juice "prevents the recurrence of urinary tract infections" because that would represent a specific medical claim. Nevertheless, many consumers do use cranberry juice to treat urinary tract infections or to prevent their recurrence, simply because they believe that it works. And indeed it may; it's just that it has never been scientifically tested to the standards of a pharmaceutical drug.

With all that latitude in producing and marketing their products, it's not surprising that the functional foods and dietary supplements industries and the advertising industry that supports them have grown rapidly. U.S. sales of dietary supplements now top $18 billion a year.

Questions of Safety and Efficacy

Proponents of functional foods and dietary supplements argue that because many of these products have been in use for a long time, any adverse effects should have shown up by now. Critics argue that many of the ingredients can now be synthesized chemically, and thus used at much higher concentrations and in different combinations than ever occur in nature. The active ingredients in Red Bull, for example, are all synthetically produced.

How will you determine if the supplement you're taking is safe and effective?

Other concerns include inaccurate product labeling and improper manufacturing processes. Manufacturers are not required to report quality control information to the FDA, so there is no assurance that the product actually contains what the manufacturer says it does. For example, independent tests found that products labeled as containing the same dosage of ginseng actually varied by a factor of 10 (some contained none at all). And California investigators found that nearly a third of all imported Asian herbal remedies they tested contained lead, arsenic, mercury, or drugs not mentioned on the label at all.

Consumers want to be assured that the dietary supplements and natural and fortified food products they use are safe. They'd like to know that the health claims about these products are true. How to achieve that goal and still ensure that the products remain available is an ongoing issue. In the meantime, it's up to you to know what is in the products you choose to put in your body.

Questions to consider

1 Who do you think should be responsible for ensuring that dietary supplements and functional foods are safe? Would you be willing to accept more regulation if it meant fewer products would be available? Explain your position.

2 What dietary supplements or functional foods do you use? Do you know what's in them and do you understand why you're using them?

The facts...

- Functional foods and dietary supplements are popular with consumers. Americans consume more than $18 billion worth of dietary supplements every year.

- Functional foods and dietary supplements don't pass through the same rigorous approval procedure that is required of all new pharmaceutical drugs.

- Some products may not be effective; others may not be safe. However, increased regulation to ensure efficacy and safety would mean fewer products would be available.

» **The natural world consists of matter and energy.** The smallest functional unit of matter is an atom.

» **Chemical bonds link atoms together to form molecules.** These bonds form naturally because the molecules are more stable than the atoms that comprise them. One of the most important naturally occurring stable molecules is water.

» **Water is the universal biological solvent.** Water comprises most of the fluid within cells and also surrounds all cells in multicellular organisms. Most of life's chemical reactions take place in it.

» **Living things harness energy and use it to make complex molecules not otherwise found in nature.** The four classes of "organic" molecules made by living organisms are proteins, carbohydrates, lipids, and nucleic acids.

» **Carbon is the common building block of all four classes of organic molecules** because of the many ways it can form chemical bonds with other atoms.

that exist in the natural world. Every substance is made from the same basic units of matter and the same types of energy (Figure 2.1). Everything in the natural world, including you and me, is governed by the laws of chemistry.

Yet there is something very special about the chemistry of living things. Living organisms have the ability to grow and reproduce, unlike inanimate objects. Living things have evolved to take advantage of the rules of chemistry that govern the natural world. Living organisms create special combinations of matter not generally found in nonliving things. They also have developed the ability to store energy so that they may later turn it to their own purposes.

In this chapter we consider how the laws of chemistry serve life. We begin with an introduction to basic chemistry.

2.1 All matter consists of elements

Matter is anything that has mass and occupies space. All matter is composed of elements. An **element** is a fundamental (pure) form of matter that cannot be broken down to a simpler form. Aluminum and iron are elements, and so are oxygen and hydrogen. There are just over 100 known elements, and together they account for all matter on Earth. The *periodic table of elements* arranges all known elements into groups according to their similar properties (Figure 2.2).

Atoms are the smallest functional units of an element

Elements are made up of particles called *atoms*. An **atom** is the smallest unit of any element that still retains the physical and chemical properties of that element. Although we now know that atoms can be split apart under unusual circumstances (such as a nuclear reaction), atoms are the smallest units of matter that can take part in chemical reactions. So, for all practical purposes, atoms are the smallest functional units of matter.

The central core of an atom is called the *nucleus*. The nucleus is made of positively charged particles called **protons** and a nearly equal number of neutral particles called **neutrons,** all tightly bound together. An exception is the smallest atom, hydrogen, whose nucleus consists of only a single proton. Smaller negatively charged particles called **electrons** orbit the nucleus. Because electrons are constantly moving, their precise position at any one time is unknown. You may think of electrons as occupying one or more spherical clouds of negative charge around the nucleus called *shells*. Each shell can accommodate only a certain number of electrons. The first shell, the one closest to the nucleus, can

As mentioned in Chapter 1, the natural world consists of matter and energy. *Chemistry* is the study of matter and the energy that causes matter to combine, break apart, and recombine into all the substances, both living and nonliving,

a) A magnified view (× 15,000) of a portion of a skeletal muscle cell.

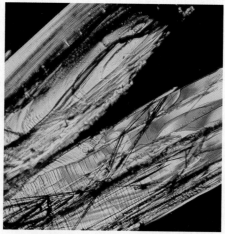

b) A magnified view (× 30) of a crystal of aspirin.

Figure 2.1 All matter is made of atoms. The three most common atoms in both muscle and aspirin are carbon, hydrogen, and oxygen.

Figure 2.2 **The periodic table shows all known elements in order of increasing atomic number.**
For example, nitrogen (N) has 7 protons, so its atomic number is 7. The table also organizes elements into groups based on the number of electrons in their outer shells (elements in group 1 have one electron, those in group 3 have three electrons, and so on). Scientists are interested in the number of electrons in the outer shell because these particles affect how an atom interacts with other elements.

hold two electrons, the second can accommodate up to eight, and the third shell (if there is one) also accommodates eight. Each type of atom has a unique number of electrons. Under most circumstances the number of electrons equals the number of protons, and, as a result, the entire atom is electrically neutral (Figure 2.3).

Protons and neutrons have about the same mass and both have much more mass than electrons. (*Mass is measured chemically and is not dependent on gravity. For the purpose of this text, however, mass and weight are about the same.*) The protons and neutrons in the atom's nucleus account for over 99.9% of the atom's mass.

In the periodic table and in chemical equations, atoms are designated by one- or two-letter symbols taken from English or Latin. For example, oxygen is designated by the letter O, nitrogen by N, sodium by Na (from the Latin word for sodium, *natrium*), and potassium by K (Latin *kalium*). A subscript numeral following the symbol indicates the numbers of atoms of that element. For example, the chemical formula O_2 represents two atoms of oxygen linked together, the most stable form of elemental oxygen.

In addition to a symbol, atoms have an *atomic number* representing the characteristic number of protons in the

nucleus and an *atomic mass* (or mass number), which is generally fairly close to the total number of neutrons and protons.

 Quick Check Nitrogen's atomic number is 7. Just from the atomic number, can you determine how many protons, neutrons, and electrons a nitrogen atom has, and how many electrons are in its first and second electron shells? (Assume the atom is electrically neutral.) ∎

Isotopes have a different number of neutrons

Although all the atoms of a particular element have the same number of protons, the number of neutrons can vary slightly. Atoms with either more or fewer neutrons than the usual number for that element are called **isotopes.** Isotopes of an element have the same atomic number as the more common atoms but a different atomic mass. For example, elemental carbon typically consists of atoms with six protons and six neutrons, for an atomic mass of 12. The isotope

Answers to **questions can be found in Appendix A.**

a) Hydrogen
1 proton

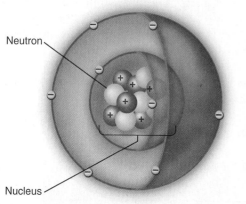

b) Oxygen
8 protons
8 neutrons
8 electrons
in 2 shells

c) Sodium
11 protons
11 neutrons
11 electrons
in 3 shells

Figure 2.3 The structure of atoms. Atoms consist of a nucleus, comprising positively charged protons and neutral neutrons, surrounded by spherical shells of negatively charged electrons.

of carbon known as carbon-14 has an atomic mass of 14 because it has two extra neutrons.

Isotopes are always identified by a superscript mass number preceding the symbol. For instance, the carbon-14 isotope is designated ^{14}C. The superscript mass number of the more common elemental form of carbon is generally omitted because it is understood to be 12.

Many isotopes are unstable. Such isotopes are called *radioisotopes* because they tend to give off energy (in the form

of radiation) and particles until they reach a more stable state. The radiation emitted by radioisotopes can be dangerous to living organisms because the energy can damage tissues.

Certain radioisotopes have a number of important scientific and medical uses. Because the rate of decay to more stable energy states is known for each radioisotope, scientists can determine when rocks and fossils were formed by measuring the amount of radioisotope still present. The carbon-14 isotope is commonly used for this purpose. In medicine, radioisotopes are used to "tag" molecules so that radiation sensors can track their location in the body. For example, physicians use radioisotopes to locate areas of damaged tissue in a patient's heart after a heart attack. Radioisotopes are also used to target and kill certain kinds of cancer. Certain radioisotopes that emit energy for long periods of time are used as a power supply in heart pacemakers.

⟲ Recap Atoms are made up of protons, neutrons, and electrons. Radioisotopes are unstable atoms with an unusual number of neutrons that give off energy and particles as they decay to a more stable state. ▪

2.2 Atoms combine to form molecules

A **molecule** consists of a stable association between two or more atoms. For example, a molecule of water is two atoms of hydrogen plus one atom of oxygen (written H_2O). A molecule of ordinary table salt (written NaCl) is one atom of sodium (Na) plus one atom of chlorine (Cl). A molecule of hydrogen gas (written H_2) is two atoms of hydrogen. To understand *why* atoms join together to form molecules, we need to know more about energy.

Energy fuels life's activities

Energy is the capacity to do "work," the capacity to cause some change in matter. Joining atoms is one type of work, and breaking up molecules is another—and both require energy. Stored energy that is not actually performing any work at the moment is called **potential energy** because it has the *potential* to make things happen. Energy that is actually *doing* work—that is, energy in motion—is called **kinetic energy** (Figure 2.4).

You can visualize the difference between potential energy and kinetic energy in the water held behind a dam: there is tremendous potential energy in the water held in reserve. When the water is released, potential energy is converted into kinetic energy: rushing water that can be put to work turning turbines. Similarly, the spark of a match converts the potential energy in firewood to kinetic energy in the form of heat and light.

Potential energy is stored in the bonds that hold atoms together in all matter, both living and nonliving. Living organisms take advantage of this general principle of chemistry by using certain molecules to store energy for their own

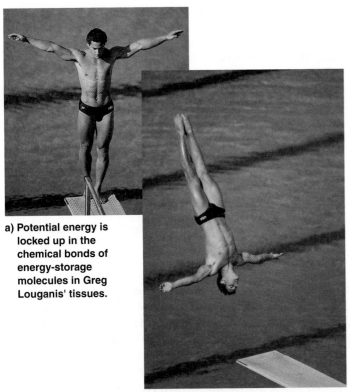

a) Potential energy is locked up in the chemical bonds of energy-storage molecules in Greg Louganis' tissues.

b) Kinetic energy is energy in motion.

Figure 2.4 **Energy.**

use. When the chemical bonds of these energy-storage molecules are broken, potential energy becomes kinetic energy. We rely on this energy to do biological work, such as breathing, moving, digesting food, and many other tasks.

Recall that electrons carry a negative charge, whereas protons within the nucleus have a positive charge. Electrons are attracted to the positively charged nucleus and repelled by each other. As a result of these opposing attractive and repulsive forces, each electron occupies a specific shell around the nucleus. Each shell corresponds to a specific level of electron potential energy, and each shell farther out represents a higher potential energy level than the preceding one closer to the nucleus. When an electron moves to a shell closer to the nucleus, it loses energy. To move to a shell that is farther from the nucleus, the electron must absorb energy.

Chemical bonds link atoms to form molecules

A key concept in chemistry is that atoms are most stable when their outermost electron shell is completely filled with the maximum number of electrons that it can accommodate. An atom whose outermost electron shell is not normally completely filled tends to interact with one or more other atoms in a way that fills its outermost shell. Such interactions generally cause the atoms to be bound to each other by attractive forces called **chemical bonds.** The three principal types of chemical bonds are called covalent, ionic, and hydrogen bonds.

Covalent bonds involve sharing electrons. One way that an atom can fill its outermost shell is by sharing a pair of electrons with another atom. An electron-sharing bond between atoms is called a **covalent bond** (Figure 2.5). Covalent bonds between atoms are among the strongest chemical bonds in nature, so strong that they rarely break apart. In structural formulas, a covalent bond is depicted as a line drawn between two atoms. For example, the structural formula for hydrogen is H—H.

Hydrogen gas offers an example of how a covalent (electron-sharing) bond fills the outermost shells of two atoms. Each of the two hydrogen atoms has just one electron in the first shell, which could accommodate two electrons. When joined together by a covalent bond (forming H_2, a gas), each atom has, in effect, a "full" first shell of two electrons. As a result, H_2 gas is more stable than the same two hydrogen atoms by themselves. The sharing of one pair of electrons, as in H_2, is called a *single* bond.

Oxygen gas is another example of covalent bonding. An oxygen atom has eight electrons: two of these fill the first

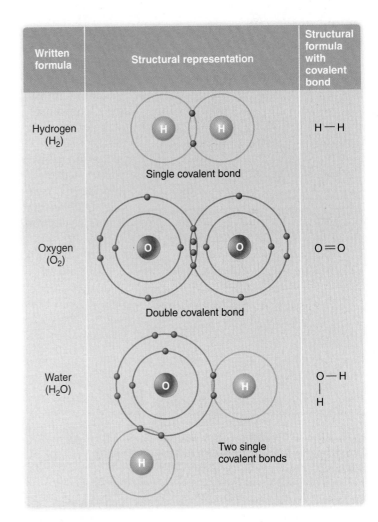

Figure 2.5 **Covalent bonds.** Sharing pairs of electrons is a way for atoms to fill their outermost shell.

☑ Draw the structural formula of CH_4. *Hint: Carbon has 2 electrons in its inner shell and 4 in its second shell.*

electron shell, and the remaining six occupy the second electron shell (which can accommodate eight). Two oxygen atoms may join to form a molecule of oxygen gas by sharing two pairs of electrons, thus completing the outer shells of both atoms. When two pairs of electrons are shared, the bond is called a *double bond*. In structural formulas, double bonds are indicated by two parallel lines. For example, the structural formula for oxygen is O=O.

A molecule of water forms from one oxygen and two hydrogen atoms because this combination completely fills the outermost shells of both hydrogen and oxygen. The prevalence of water on Earth follows from the simple rule that matter is most stable when it contains the least potential energy. That is, both hydrogen and oxygen are more stable when bonded together (as H_2O) than as independent atoms.

✓ **Quick Check** Can you determine the number of bonds between atoms in a single molecule of carbon dioxide, CO_2? Hint: Oxygen has eight electrons and carbon has six electrons. ▪

Ionic bonds occur between oppositely charged ions. A second way that atoms can fill their outer shell of electrons is to give up electrons completely (if they have only one or two electrons in their outermost shell) or to take electrons from other atoms (if they need one or two to fill their outermost shell). Such a loss or gain of electrons gives the atom a net charge, because now there are fewer (or more) electrons than protons in the nucleus. The net charge is positive (+) for each electron lost and negative (−) for each electron gained.

An electrically charged atom or molecule is called an **ion.** Examples of ions are sodium (Na^+), chloride (Cl^-), calcium (Ca^{2+}), and hydrogen phosphate (HPO_4^-). Notice that ions can have a shortage or surplus of more than one electron. (Ca^{2+} has lost two electrons.)

Ever heard the expression "opposites attract"? It should come as no surprise that oppositely charged ions are attracted to each other. When two oppositely charged ions

come together, an **ionic bond** is formed (Figure 2.6). In aqueous (watery) solutions, ionic bonds are much weaker than covalent bonds, and so ionic bonds tend to break rather easily. In the human body, for example, almost all of the sodium is in the form of Na^+, and most of the chlorine is in its ionized form, called *chloride* (Cl^-). Very little exists as NaCl.

Ions in aqueous solutions are sometimes called *electrolytes* because solutions of water containing ions are good conductors of electricity. As you will see, cells can control the movement of certain ions, creating electrical forces essential to the functioning of nerves, muscles, and other living tissues.

Weak hydrogen bonds form between polar molecules. A third type of attraction occurs between molecules that do not have a net charge. Glance back at the water molecule in Figure 2.5 and note that the two hydrogen atoms are found not at opposite ends of the water molecule, but fairly close together. Although the oxygen and the two hydrogen atoms share electrons, the sharing is unequal. The shared electrons in a water molecule actually spend slightly more of their time near the oxygen atom than near the hydrogen atoms because the oxygen atom attracts electrons more strongly than do the hydrogen atoms. The uneven sharing gives the oxygen region of a water molecule a partial negative charge and the two hydrogen regions a partial positive charge, even though the water molecule as a whole is electrically neutral.

Molecules such as water that are electrically neutral overall but still have partially charged regions, or *poles,* are called *polar* molecules. According to the principle that opposites attract, polar molecules arrange themselves so that the partial negative pole of one molecule is oriented toward (attracted by) the partial positive pole of another molecule. The weak attractive force between oppositely charged regions of polar molecules that contain covalently bonded hydrogen is called a **hydrogen bond.** Hydrogen bonds between water molecules are so weak that they continually break and re-form, allowing liquid water

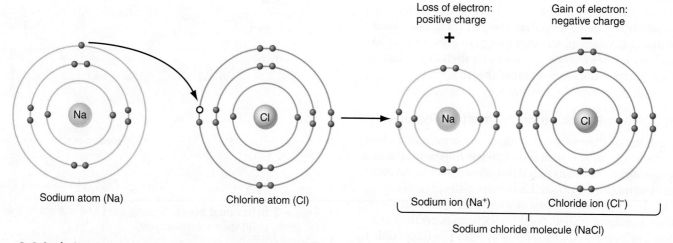

Loss of electron: positive charge **+**

Gain of electron: negative charge **−**

Sodium atom (Na) Chlorine atom (Cl) Sodium ion (Na^+) Chloride ion (Cl^-)

Sodium chloride molecule (NaCl)

Figure 2.6 Ionic bonds. Electrically charged ions form when atoms give up or gain electrons. The oppositely charged ions are attracted to each other, forming an ionic bond.

to flow (**Figure 2.7**). When water gets cold enough to freeze, each water molecule forms four stable, unchanging hydrogen bonds with its neighbors. When water is vaporized (becomes a gas), the hydrogen bonds are broken and stay broken as long as the water is in the gas phase.

Hydrogen bonds are important in biological molecules. They're what give proteins their three-dimensional shape, and they keep the two strands of the DNA molecule together. The structures of both proteins and DNA are described later in this chapter.

Table 2.1 summarizes covalent, ionic, and hydrogen bonds.

HBP **Web Animation** *Atoms, Ions, and Bonding* at www.humanbiology.com

Living organisms contain only certain elements

Although there are nearly 100 different elements in nature, living organisms are constructed from a limited number of them. In fact, about 99% of your body weight consists of just six elements: oxygen, carbon, hydrogen, nitrogen, calcium, and phosphorus (**Table 2.2**). However, the less common elements are still important, and life on Earth as we know it would not be possible without them.

Even the largest atoms are small compared to the structures in living organisms. To appreciate the vast size differences between the atoms, cells, and organs in your body, imagine that a sodium atom is the size of a penny. On this scale, one of your red blood cells would be 1/2 mile in diameter, and your heart would be larger than the entire Earth!

Next, let's look at some of the most important matter of living systems: water, hydrogen ions, and a host of molecules that contain a backbone of carbon atoms.

↩ **Recap** Electrons farthest from the nucleus have more potential energy than electrons close to the nucleus. Strong covalent bonds form between atoms when they share pairs of electrons, ionic bonds form between oppositely charged ions, and weak hydrogen bonds occur between oppositely charged regions of polar molecules. ∎

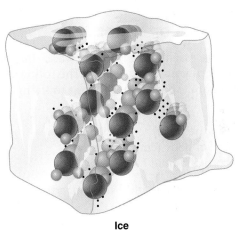

Molecule —

Oxygen (O) —

Hydrogen (H) —

Water

Ice

Figure 2.7 **Hydrogen bonds.** In water, weak hydrogen bonds continually form, break, and re-form between hydrogen and oxygen atoms of adjacent water molecules. Ice is a solid because stable hydrogen bonds form between each water molecule and four of its neighbors.

Table 2.1 **Summary of the three types of chemical bonds**

Type	Strength	Description	Examples
Covalent bond	Strong	A bond in which the sharing of electrons between atoms results in each atom having a maximally filled outermost shell of electrons	The bonds between hydrogen and oxygen in a molecule of water
Ionic bond	Moderate	The bond between two oppositely charged ions (atoms or molecules that were formed by the permanent transfer of one or more electrons)	The bond between Na^+ and Cl^- in salt
Hydrogen bond	Weak	The bond between oppositely charged regions of molecules that contain covalently bonded hydrogen atoms	The bonds between molecules of water

Table 2.2 The most common and important elements in living organisms*

Element	Atomic symbol	Atomic number	Atomic mass	% of Human weight	Functions in life
Oxygen	O	8	16.0	65	Part of water and most organic molecules; also molecular oxygen
Carbon	C	6	12.0	18	The backbone of all organic molecules
Hydrogen	H	1	1.0	10	Part of all organic molecules and of water
Nitrogen	N	7	14.0	3	Component of proteins and nucleic acids
Calcium	Ca	20	40.1	2	Constituent of bone; also essential for the action of nerves and muscles
Phosphorus	P	15	31.0	1	Part of cell membranes and of energy storage molecules; also a constituent of bone
Potassium	K	19	39.1	0.3	Important in nerve action
Sulfur	S	16	32.1	0.2	Structural component of most proteins
Sodium	Na	11	23.0	0.1	The primary ion in body fluids; also important for nerve action
Chlorine	Cl	17	35.5	0.1	Component of digestive acid; also a major ion in body fluids
Magnesium	Mg	12	24.3	Trace	Important for the action of certain enzymes and for muscle contraction
Iron	Fe	26	55.8	Trace	A constituent of hemoglobin, the oxygen-carrying molecule

*The elements are listed in descending order of their contribution to total body weight. Atomic number represents the number of protons in the nucleus. Atomic mass is roughly equivalent to the total number of protons and neutrons because electrons have very little mass. Note that 99% of your body weight is accounted for by just six elements.

2.3 Life depends on water

No molecule is more essential to life than water. Indeed, it accounts for 60% of your body weight. The following properties of water are especially important to living organisms:

- Water molecules are polar.
- Water is a liquid at body temperature.
- Water can absorb and hold heat energy.

These properties make water an ideal solvent and an important factor in temperature regulation.

Water is the biological solvent

A **solvent** is a liquid in which other substances dissolve, and a **solute** is any dissolved substance. Water is the ideal solvent in living organisms specifically because it is a polar liquid at body temperature. As the solvent of life, water is the substance in which the many chemical reactions of living organisms take place. Let's look at a simple example of a solute dissolving in water to better understand how the polar nature of water facilitates the reaction.

Consider a common and important solid, crystals of sodium chloride (NaCl), or table salt. Crystals of table salt consist of a regular, repeating pattern of sodium and chloride ions held together by ionic bonds (**Figure 2.8**). When salt is placed in water, individual ions of Na^+ and Cl^- at the surface of the crystal are pulled away from the crystal and are immediately surrounded by the polar water molecules. The water molecules form such a tight cluster

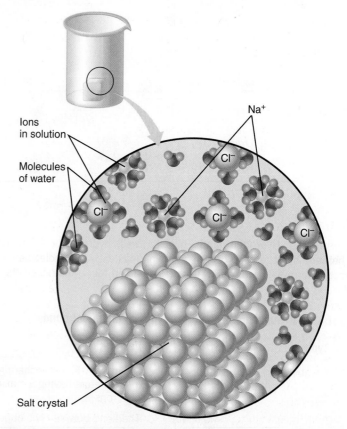

Ions in solution

Molecules of water

Na^+

Cl^-

Cl^-

Cl^-

Cl^-

Cl^-

Salt crystal

Figure 2.8 How water keeps ions in solution. The slightly negative ends of polar water molecules are attracted to positive ions, whereas the slightly positive ends of water molecules are attracted to negative ions. The water molecules pull the ions away from the crystal and prevent them from reassociating with each other.

around each ion that they are prevented from reassociating back into the crystalline form. In other words, water keeps the ions dissolved. Note that the water molecules are oriented around ions according to the principle that opposite charges attract.

Polar molecules that are attracted to water and interact with it easily are called **hydrophilic** molecules (Greek, meaning "water-loving"). Nonpolar, neutral molecules such as cooking oils do not interact easily with water and generally won't dissolve in it. They are said to be **hydrophobic** (Greek, meaning "water-fearing"). When water and oil are mixed, the water molecules tend to form hydrogen bonds with each other, excluding the oil from regions occupied by water. Over time the oil is forced together into larger and larger drops until it is separated from the water completely.

Because water is a liquid at body temperature, it can flow freely. This makes it an excellent medium for transporting solutes from one place to another. Indeed, the blood in our cardiovascular system is over 90% water. As a liquid, water also occupies space. It fills our cells (the intracellular space) and the spaces between cells (the intercellular space).

 Quick Check You may have noticed when making salad dressings that vinegar and olive oil don't mix together easily. Knowing that oils are nonpolar and hydrophobic, what can you say about vinegar? Explain your reasoning. ▪

Water helps regulate body temperature

An important property of water is that it can absorb and hold a large amount of heat energy with only a modest increase in temperature. In fact, it absorbs heat better than most other liquids. Water thus may prevent large increases in body temperature when excess heat is produced. Water also holds heat well when there is a danger of too much heat loss (for instance, when you go outdoors wearing shorts on a cool day). The ability of water to absorb and hold heat helps prevent rapid changes in body temperature when changes occur in metabolism or in the environment.

Our bodies generate heat during metabolism. We usually generate more heat than we need to maintain a constant body temperature of 98.6° Fahrenheit (37° Celsius), so losing heat is generally more of a priority than conserving it. One way we can lose heat quickly is by evaporation of water. When water is in contact with air, hydrogen bonds between some of the water molecules at the surface of the water are broken, and water molecules escape into the air as water vapor. It takes energy to break all those hydrogen bonds, and that energy comes from heat generated by the body and transported to the skin by the blood. Evaporation of sweat is just one of the mechanisms for the removal of heat from the body (**Figure 2.9**). How the body regulates body temperature is discussed in more detail in Chapter 4.

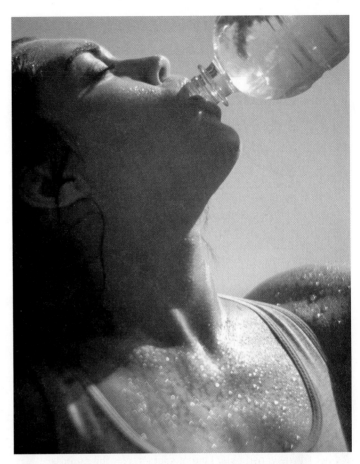

Figure 2.9 Water contributes to the regulation of body temperature.

You can demonstrate the cooling power of evaporation for yourself. The next time you perspire heavily, notice that your exposed skin may actually feel cool to the touch.

 Recap Most biological molecules dissolve readily in water because water is a polar molecule. The liquid nature of water facilitates the transport of biological molecules. Water absorbs and holds heat and can lower body temperature through evaporation. ▪

 Web Animation *Water and Chemistry* at www.humanbiology.com

2.4 The importance of hydrogen ions

One of the most important ions in the body is the hydrogen ion (a single proton without an electron). In this section we will see how hydrogen ions are created and why it is so important to maintain an appropriate concentration of them.

MJ's Human Biology Blog

I Don't Hear You...

How does the public respond when a published scientific report shows that a dietary supplement is ineffective, or even worse, potentially harmful? To find out, scientists at the National Institutes of Health examined the sales trends of five different dietary supplements before and after the publication of negative research results. There were no significant declines in sales for four of the five supplements (saw palmetto, Echinacea, glucosamine, and St. John's wort) after published reports that the supplement was ineffective. But sales of the fifth supplement (Vitamin E) declined about 33% after a report suggested that high doses of Vitamin E might actually be harmful.

Why did consumers ignore the reports that supplements just didn't work, but responded to a report of potential harm? Researchers speculate that reports of harm might have higher impact because of greater news coverage, or that some supplements (such as Vitamin E) might be recommended more often by physicians who are more likely to read and understand scientific reports, or even that it depends on the type of person who takes a particular kind of supplement, the purpose of the supplement, and the availability of alternatives.

Still, it must be discouraging for public health officials to learn that consumers aren't getting the message, don't believe the message, or just don't care whether their supplements work or not. ■

Reference: Tilburt, Jon, *et al.* Does the Evidence Make a Difference in Consumer Behavior? Sales of Supplements Before and After Publication of Negative Research Results. *J. Gen. Intern. Med.* 23: 1495–1498, 2008.

Acids donate hydrogen ions, bases accept them

Although the covalent bonds between hydrogen and oxygen in water are strong and thus rarely broken, it can happen. When it does, the electron from one hydrogen atom is transferred to the oxygen atom completely, and the water molecule breaks into two ions—a *hydrogen ion* (H^+) and a *hydroxide ion* (OH^-).

In pure water, only a very few molecules of water are dissociated (broken apart) into H^+ and OH^- at any one time. However, there are other sources of hydrogen ions in aqueous solutions. An **acid** is any molecule that can donate (give up) an H^+ ion. When added to pure water, acids produce an *acidic* solution, one with a higher H^+ concentration than that of pure water. (By definition, an aqueous solution with the

same concentration of H^+ as that of pure water is a *neutral* solution.) Common acidic solutions are vinegar, carbonated beverages, black coffee, and orange juice. Conversely, a **base** is any molecule that can accept (combine with) an H^+ ion. When added to pure water, bases produce a basic or *alkaline* solution, one with a lower H^+ concentration than that of pure water. Common alkaline solutions include baking soda in water, detergents, and drain cleaner.

Because acids and bases have opposite effects on the H^+ concentration of solutions, they are said to neutralize each other. You have probably heard that a spoonful of baking soda in water is a time-honored way to counteract an "acid stomach." Now you know that this home remedy is based on sound chemical principles.

The pH scale expresses hydrogen ion concentration

Scientists use the pH scale to indicate the acidity or alkalinity of a solution. The **pH scale** is a measure of the hydrogen ion concentration of a solution. The scale ranges from 0 to 14, with pure water having a pH of 7.0, the neutral point. A pH of 7 corresponds to a hydrogen ion concentration of 10^{-7} moles/liter (a *mole* is a term used by chemists to indicate a certain number of atoms, ions, or molecules). An *acidic* solution has a pH of *less* than 7, whereas a *basic* solution has a pH of *greater* than 7. Each whole number change in pH represents a 10-fold change in the hydrogen ion concentration in the opposite direction. For example, an acidic solution with a pH of 5 has an H^+ concentration of 10^{-5} moles/liter (100 times greater than pure water), whereas an alkaline solution with a pH of 9 has an H^+ concentration of 10^{-9} moles/liter (1/100 that of water). Figure 2.10 shows the pH scale and indicates the pH values of some common substances and body fluids.

The pH of blood is 7.4, just slightly more alkaline than neutral water. The hydrogen ion concentration of blood

pH values	Concentrations of H^+ (moles/liter)	
14		Drain opener
13	10^{-13}	Bleach
12		
11	10^{-11}	Ammonia cleanser
10		Soapy water
9	10^{-9}	Baking soda
8		
7	10^{-7}	Human blood, tears
6		Saliva, urine
5	10^{-5}	Black coffee
4		Tomatoes
3	10^{-3}	Vinegar, cola
2		Lemon juice
1	10^{-1}	Hydrochloric acid
0		Concentrated nitric acid

More alkaline

Neutral pH

More acidic

Figure 2.10 The pH scale. The pH scale is an indication of the H^+ concentration of a solution.

plasma is low relative to the concentration of other ions. (The hydrogen ion concentration of blood plasma is less than one-*millionth* that of sodium ions, for example.) It is important to maintain homeostasis of this low concentration of hydrogen ions in the body because hydrogen ions are small, mobile, positively charged, and highly reactive. Hydrogen ions tend to displace other positive ions in molecules, and when they do they alter molecular structures and change the ability of the molecule to function properly.

Changes in the pH of body fluids can affect how molecules are transported across the cell membrane and how rapidly certain chemical reactions occur. They may even alter the shapes of proteins that are structural elements of the cell. In other words, a change in the hydrogen ion concentration can be dangerous because it threatens homeostasis.

 Quick Check A chemist has a solution that has a pH of 3. She adds a chemical to it, and shortly afterwards the solution has a pH of 5. What was the concentration of hydrogen ions before adding the chemical, what was it afterwards, and did she add an acid or a base? Explain. ▪

Buffers minimize changes in pH

A **buffer** is any substance that tends to minimize the changes in pH that might otherwise occur when an acid or base is added to a solution. Buffers are essential to our ability to maintain homeostasis of pH in body fluids.

In biological solutions such as blood or urine, buffers are present as *pairs* of related molecules that have opposite effects. One of the pair is the acid form of the molecule (capable of donating an H^+ ion), and the other is the base form (capable of accepting an H^+ ion). When an acid is added and the number of H^+ ions increases, the base form of the buffer pair accepts some of the H^+ ions, minimizing the fall in pH that might otherwise occur. Conversely, when a base is added that might take up too many H^+ ions, the acid form of the buffer pair releases additional H^+ ions and thus minimizes the rise in pH. Buffer pairs are like absorbent sponges that can pick up excess water and then can be wrung out to release water when necessary.

One of the most important buffer pairs in body fluids such as blood is bicarbonate (HCO_3^-, the base form) and carbonic acid (H_2CO_3, the acid form). When blood becomes too acidic, bicarbonate accepts excess H^+ according to the following reaction:

$$HCO_3^- + H^+ \rightarrow H_2CO_3$$

When blood becomes too alkaline, carbonic acid donates H^+ by the reverse reaction:

$$HCO_3^- + H^+ \leftarrow H_2CO_3$$

In a biological solution such as blood, bicarbonate and carbonic acid take up and release H^+ all the time. Ultimately a chemical *equilibrium* is reached in which the rates of the two chemical reactions are the same, as represented by the following combined equation:

$$HCO_3^- + H^+ \leftrightarrow H_2CO_3$$

When excess acid is produced, the combined equation shifts to the right as the bicarbonate combines with H^+. The reverse is true for alkalinity.

There are many other buffers in the body as well. The more buffers that are present in a body fluid, the more stable the pH.

Recap Acids can donate hydrogen ions to a solution, whereas bases can accept hydrogen ions from a solution. The pH scale indicates the hydrogen ion concentration of a solution. The normal pH of blood is 7.4. Buffers help maintain a stable pH in body fluids. ▪

2.5 The organic molecules of living organisms

Organic molecules are molecules that contain carbon and other elements held together by covalent bonds. The name "organic" came about at a time when scientists believed that all organic molecules were created only by living organisms and all "inorganic" molecules came from nonliving matter. Today we know that organic molecules can be synthesized in the laboratory under the right conditions and that they probably existed on Earth before there was life.

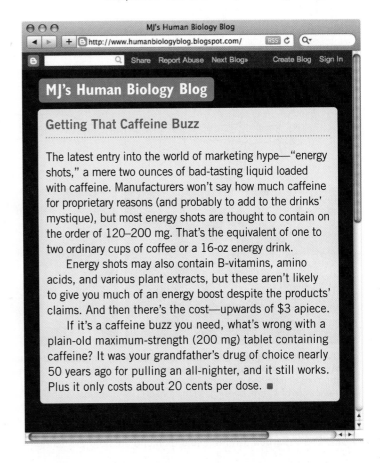

MJ's Human Biology Blog

http://www.humanbiologyblog.blogspot.com/

MJ's Human Biology Blog

Getting That Caffeine Buzz

The latest entry into the world of marketing hype—"energy shots," a mere two ounces of bad-tasting liquid loaded with caffeine. Manufacturers won't say how much caffeine for proprietary reasons (and probably to add to the drinks' mystique), but most energy shots are thought to contain on the order of 120–200 mg. That's the equivalent of one to two ordinary cups of coffee or a 16-oz energy drink.

Energy shots may also contain B-vitamins, amino acids, and various plant extracts, but these aren't likely to give you much of an energy boost despite the products' claims. And then there's the cost—upwards of $3 apiece.

If it's a caffeine buzz you need, what's wrong with a plain-old maximum-strength (200 mg) tablet containing caffeine? It was your grandfather's drug of choice nearly 50 years ago for pulling an all-nighter, and it still works. Plus it only costs about 20 cents per dose. ▪

a) Diamonds are formed only under conditions of extreme temperature and pressure. The structure of diamond resembles the steel framework of a large building; each atom is covalently bonded to four neighboring carbon atoms. This explains the hardness of diamonds.

b) Graphite is produced as a result of decay of older carbon-based substances. Its structure consists of layers of hexagonal rings of carbon atoms. Graphite is fairly soft (hence its use in pencils) because these layers of carbon atoms can slide past one another.

Figure 2.11 **Carbon.** Graphite and diamond are both elemental forms of carbon.

Carbon is the common building block of organic molecules

Carbon (Figure 2.11) is relatively rare in the natural world, representing less than 0.03% of Earth's crust. However, living organisms actively accumulate it. Carbon accounts for about 18% of body weight in humans.

Carbon is the common building block of all organic molecules because of the many ways it can form strong covalent bonds with other atoms. Carbon has six electrons, two in the first shell and four in the second. Because carbon is most stable when its second shell is filled with eight electrons, *its natural tendency is to form four covalent bonds with other molecules.* This makes carbon an ideal structural component, one that can branch in a multitude of directions.

Using the chemist's convention that a line between the chemical symbols of atoms represents a pair of shared electrons in a covalent bond, Figure 2.12 shows some of the many structural possibilities for carbon. Carbon can form covalent bonds with hydrogen, nitrogen, oxygen, or another carbon. It can form double covalent bonds with oxygen or another carbon. It can even form five- or six-membered carbon rings, with or without double bonds between carbons.

In addition to their complexity, there is almost no limit to the size of organic molecules derived from carbon. Some, called *macromolecules* (from the Greek *makros,* long), consist of thousands or even millions of smaller molecules.

Macromolecules are synthesized and broken down within the cell

Macromolecules are built (synthesized) within the cell itself. In a process called **dehydration synthesis** (also called the condensation reaction), smaller molecules called subunits are joined by covalent bonds, like pearls on a string. The name of the process accurately describes what is happening: Each time a subunit is added, the equivalent of a water molecule is removed ("dehydration") (Figure 2.13). The subunits needed to synthesize macromolecules come from the foods you eat and from the biochemical reactions in your body that break down other large molecules into smaller ones.

The synthesis of macromolecules from smaller molecules requires energy. That is one reason we need energy to survive and grow. It is no accident that children seem to eat enormous amounts of food. Growing children require energy to make the macromolecules necessary to create new cell membranes, muscle fibers, and other body tissues.

Some macromolecules are synthesized specifically for the purpose of storing energy within our cells. The ability to store energy internally allows organisms to survive even when food is not plentiful. Other macromolecules serve as structural components of cells or of extracellular (outside the cell) structures such as bone. Still others direct the many activities of the cell or serve as signaling molecules between cells.

$$O = C = O$$

a) In carbon dioxide, a carbon atom forms two covalent bonds with each oxygen atom.

b) Lipid molecules (a portion of one is shown here) contain long chains of carbon atoms covalently bound to hydrogen.

c) Carbon is the backbone of amino acids, the building blocks of protein. This amino acid is phenylalanine.

Figure 2.12 **Examples of the structural diversity of carbon.**

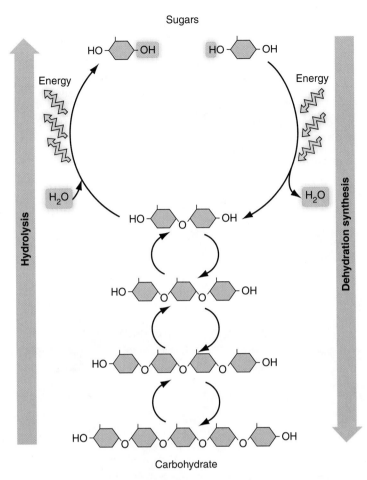

Figure 2.13 Dehydration synthesis and hydrolysis. The synthesis of larger molecules by dehydration synthesis requires energy, whereas the breakdown of molecules into smaller units by hydrolysis liberates stored energy. In this example the smallest units are simple sugars and the macromolecule is a carbohydrate.

Organic macromolecules are broken down by a process called **hydrolysis.** During hydrolysis the equivalent of a water molecule is added each time a covalent bond between single subunits in the chain is broken. Notice that hydrolysis is essentially the reverse of dehydration synthesis, and thus it should not surprise you that the breakdown of macromolecules releases energy. The energy was stored as potential energy in the covalent bonds between atoms. The body obtains much of its energy through hydrolysis of energy-storage molecules. Hydrolysis is also used to break down molecules of food during digestion, to recycle materials for reuse, and to get rid of substances that are no longer needed by the body.

Living organisms synthesize four classes of organic molecules, known as *carbohydrates, lipids, proteins,* and *nucleic acids.* The many different molecules within each class are constructed of the same handful of chemical elements. However, there is essentially no limit to the number of different molecules that could be created. No one knows for sure how many different organic molecules there are in humans. On a chemical level, the tremendous diversity among the many species of organisms on Earth is due to differences in their organic molecules, especially their proteins and nucleic acids.

Recap Carbon is a key element of organic molecules because of the multiple ways it can form strong covalent bonds with other molecules. Synthesizing organic molecules requires energy; breaking them down liberates energy. The four classes of organic molecules are carbohydrates, lipids, proteins, and nucleic acids. ■

2.6 Carbohydrates: Used for energy and structural support

A clue to the basic structure of carbohydrates is found in their name. Carbohydrates have a backbone of carbon atoms with hydrogen and oxygen attached in the same proportion as they appear in water (2-to-1); hence the carbon is "hydrated," or combined with water.

Most living organisms use carbohydrates for energy, and plants use at least one carbohydrate (cellulose) as structural support.

Monosaccharides are simple sugars

The simplest kind of carbohydrate is called a **monosaccharide** (meaning "one sugar"). Monosaccharides have relatively simple structures consisting of carbon, hydrogen, and oxygen in a 1-2-1 ratio. The most common monosaccharides contain five or six carbon atoms arranged in either a five-membered or six-membered ring (**Figure 2.14**).

a) The five-carbon monosaccharides ribose and deoxyribose.

b) Two 6-carbon monosaccharides (glucose and fructose) are joined together by dehydration synthesis, forming sucrose.

Figure 2.14 Monosaccharides. By convention, in a ringed structure the symbol C for carbon is often omitted because its presence is inferred by the union of two bond lines at an angle.

Ribose, deoxyribose, glucose, and *fructose* are four of the most important monosaccharides in humans. Ribose and deoxyribose (Figure 2.14a) are both five-carbon monosaccharides that are components of nucleotide molecules, discussed below. The only difference between the two is that deoxyribose has one less oxygen atom than ribose. Glucose, a six-carbon monosaccharide, is an important source of energy for cells. When more energy is available than can be used right away, glucose molecules can be linked together by dehydration synthesis to form larger carbohydrate molecules (Figure 2.14b).

Oligosaccharides: More than one monosaccharide linked together

Oligosaccharides are short strings of monosaccharides (*oligo* means "a few") linked together by dehydration synthesis. One common oligosaccharide is table sugar, or sucrose. Sucrose is also called a *disaccharide* because it consists of just two monosaccharides (glucose + fructose). Another is lactose (glucose + galactose), the most common disaccharide in human milk and an important source of energy for infants.

Some oligosaccharides are covalently bonded to certain cell-membrane proteins (called glycoproteins). Glycoproteins participate in linking adjacent cells together and in cell-cell recognition and communication.

✅ **Quick Check** Would you expect oligosaccharides to contain exactly the same ratio of carbon, hydrogen, and oxygen that monosaccharides have? If not, how would they differ? Explain. ▪

Polysaccharides store energy

Complex carbohydrates called **polysaccharides** (*poly* means "many") form when thousands of monosaccharides are joined together into straight or branched chains by dehydration synthesis. Polysaccharides are a convenient way for cells to stockpile extra energy by locking it in the bonds of the polysaccharide molecule.

The most important polysaccharides in living organisms consist of long chains of glucose monosaccharides. In animals the storage polysaccharide is **glycogen** (Figure 2.15), whereas in plants it is **starch.** The flour we obtain by grinding plant grains is high in starch, which we then utilize for our own energy needs by breaking it down to glucose. Any glucose not consumed for energy in the short term can be used to create glycogen or lipids and stored within our cells for later use.

Cellulose is a slightly different form of glucose polysaccharide. Plants use it for structural support rather than for energy storage. The nature of the chemical bonds in cellulose is such that most animals, including humans, cannot break cellulose down to glucose units (which is why we cannot digest wood). But there's plenty of energy locked in the chemical bonds of cellulose, as demonstrated by the heat generated by a wood fire.

Undigested cellulose in the food we eat contributes to the fiber or "roughage" in our diet. A certain amount of fiber is

a) Glycogen is formed by dehydration synthesis from glucose subunits.

b) A representation of the highly branched nature of glycogen.

c) A portion of an animal cell showing granules of stored glycogen (blue). The large pink structures are mitochondria.

Figure 2.15 **Glycogen is the storage carbohydrate in animals.**

✅ *Name two common polysaccharides made by plants. Are they also made of glucose? Can we digest both of them, and why or why not?*

thought to be beneficial because it increases the movement of wastes through the digestive tract. The more rapid excretion of wastes decreases the time of exposure to any carcinogens (cancer-causing agents) that may be in the waste material.

> **Recap** Carbohydrates contain carbon, hydrogen, and oxygen in a 1-2-1 ratio. Simple sugars such as glucose provide immediate energy for cells. Complex carbohydrates called polysaccharides store energy (in animals and plants) and provide structural support (in plants). ▪

2.7 Lipids: Insoluble in water

For biology, the most important physical characteristic of the class of organic molecules called **lipids** is that they are relatively insoluble , meaning they do not dissolve in water. The most important subclasses of lipids in your body are *triglycerides, phospholipids,* and *steroids.*

Triglycerides are energy-storage molecules

Triglycerides, also called neutral fats or just fats, are synthesized from a molecule of glycerol and three fatty acids (**Figure 2.16a**). **Fatty acids** are chains of hydrocarbons (usually about 16–18 carbons long) that end in a group of atoms known as a carboxyl group (COOH). Fats vary in the length of their fatty acid tails and the ratio of hydrogen atoms to carbon atoms in the tails.

Saturated fats have a full complement of two hydrogen atoms for each carbon in their tails (Figure 2.16b). In saturated fats, the tails are fairly straight, allowing them to pack closely together. As a result, saturated fats are generally solid at room temperature. Animal fats, such as butter and bacon grease,

are saturated fats. A diet rich in saturated fats is thought to contribute to the development of cardiovascular disease.

Unsaturated fats, also called *oils,* have fewer than two hydrogen atoms on one or more of the carbon atoms in the tails (Figure 2.16c). As a result, double bonds form between adjacent carbons, putting kinks in the tails and preventing the fats from associating closely together. Consequently, unsaturated fats (oils) are generally liquid at room temperature.

Triglycerides are stored in adipose (fat) tissue and are an important source of stored energy in our bodies. Most of the energy is located in the bonds between carbon and hydrogen in the fatty acid tails.

> ☑ **Quick Check** Cocoa butter is solid at room temperature; canola oil is liquid at room temperature. Which would you expect to contain more double bonds? Why? ▪

Phospholipids are the primary component of cell membranes

Phospholipids are a modified form of lipid. They are the primary structural component of cell membranes.

Like fats, phospholipids have a molecule of glycerol as the backbone, but they have only two fatty acid tails. Replacing the third fatty acid is a negatively charged phosphate group (PO_4^-) and another group that varies depending on the phospholipid but is generally positively charged (**Figure 2.17**). The presence of charged groups on one end gives the phospholipid a special property: one end of the molecule is polar and thus soluble (dissolves) in water, whereas the other end (represented by the two fatty acid tails) is neutral and therefore relatively insoluble in water.

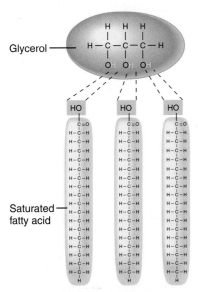

Glycerol

Saturated fatty acid

a) Triglycerides (neutral fats) are synthesized from glycerol and three fatty acids by dehydration synthesis.

b) Triglycerides with saturated fatty acids have straight tails, allowing them to pack closely together.

c) Triglycerides with unsaturated fatty acids have kinked tails, preventing them from packing closely together.

Figure 2.16 Triglycerides.

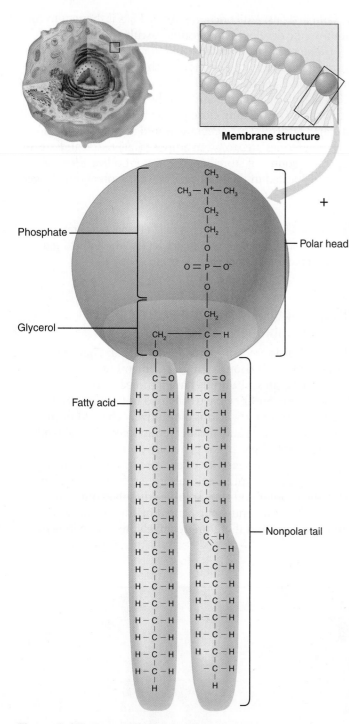

Figure 2.17 Phospholipids. Phospholipids are the primary constituent of animal cell membranes.

☑ *Why is the head of each phospholipid oriented toward the outer surface of each side of the membrane (toward water), instead of toward the interior? Put another way, what is stopping the phospholipids from flipping around?*

Steroids are composed of four rings

Steroids do not look at all like the lipids described previously but are classified as lipids because they are relatively insoluble in water. Steroids consist of a backbone of three 6-membered

a) Cholesterol:
A normal component of the cell membrane.

b) Estrogen (estradiol):
Female sex hormone synthesized from cholesterol.

c) Testosterone:
Male sex hormone synthesized from cholesterol.

Figure 2.18 Steroids. Steroids consist of a backbone of three 6-membered carbon rings and one 5-membered ring.

carbon rings and one 5-membered carbon ring to which any number of different groups may be attached.

One steroid you may be familiar with is **cholesterol** (Figure 2.18a). High levels of cholesterol in the blood are associated with cardiovascular disease. However, we all need a certain amount of cholesterol. It is a normal and essential structural component of animal cell membranes and is also the source of several important hormones, including the sex hormones estrogen and testosterone (Figure 2.18b and 2.18c). Our bodies manufacture cholesterol even though we generally get more than we need from our diet.

HBP **Web Animation** *Lipid Structure and Function* at www.humanbiology.com

↩ **Recap** Lipids (triglycerides, phospholipids, and steroids) are all relatively insoluble in water. Triglycerides are an important source of stored energy. Phospholipids, an important component of cell membranes, have a polar (water-soluble) head and two fatty acid (water-insoluble) tails. Steroids, such as cholesterol, have a four-ring structure. ■

2.8 Proteins: Complex structures constructed of amino acids

Proteins are macromolecules constructed from long strings of single units called **amino acids.** All human proteins are constructed from only 20 different amino acids (Figure 2.19). Each amino acid has an *amino group* (NH_3) on one end, a carboxyl group on the other, a C—H group in the middle, and an additional group (designated "R") that represents everything else. Some of the R groups are

Figure 2.19 The 20 amino acid building blocks of proteins. The portions of amino acids that make them different from each other, called R groups, are colored. The three-letter codes in parentheses designate the amino acids in written formulas.

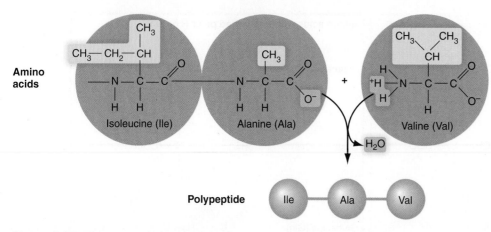

Figure 2.20 The synthesis of proteins. Proteins are created from amino acids by dehydration synthesis.

completely neutral, others are neutral but polar, and a few carry a net charge (either positive or negative). Differences in the charge and structure of the amino acids affect the shape and functions of the proteins constructed from them. Our bodies can synthesize (make) 11 of the amino acids if necessary. However, we generally get enough of most of them, including the 9 we cannot synthesize, in the food we eat.

Like complex carbohydrates and fats, proteins are formed by dehydration synthesis (**Figure 2.20**). A single string of 3 to 100 amino acids is called a **polypeptide.** A polypeptide is generally referred to as a **protein** when it is longer than 100 amino acids and has a complex structure and a function. Some proteins consist of several polypeptides linked together.

Protein function depends on structure

The function of every protein depends critically on its structure. We can define protein structure on at least three levels and sometimes four levels (**Figure 2.21**):

- *Primary structure.* The primary structure of a protein is represented by its amino acid sequence. In writing, each amino acid is indicated by a three-letter code (review Figure 2.19).
- *Secondary structure.* The secondary structure describes how the chain of amino acids is oriented in space. A common secondary structure of proteins is an *alpha (α) helix.* An α helix is a right-hand spiral that is stabilized by hydrogen bonds between amino acids at regular intervals. Another common secondary structure also stabilized by hydrogen bonds is a flat ribbon called a *beta (β) sheet.* A β sheet is formed when hydrogen bonds join two primary sequences of amino acids side by side. In addition to forming these two structures, proteins can coil into an almost infinite variety of shapes depending on which amino acids make up the sequence.
- *Tertiary structure.* Tertiary structure, the third level, refers to how the protein twists and folds to form a three-dimensional shape. The protein's three-dimensional

structure depends in part on its sequence of amino acids, because the locations of the polar and charged groups within the chain determine the locations of hydrogen bonds that hold the whole sequence together. In addition, occasionally a covalent bond called a disulfide (S—S) bond forms between the sulfur molecules of two cysteine amino acids (see Figure 2.19). Finally, proteins tend to fold in such a way that neutral amino acids are more likely to end up in the interior, whereas charged and polar amino acids are more likely to face the outside (aqueous environment). Proteins acquire their characteristic tertiary structure by a folding process that occurs either during synthesis or shortly thereafter.

- *Quaternary structure.* The quaternary (fourth) structure of some proteins refers to how many polypeptide chains make up the protein (if there is more than one) and how they associate with each other.

The human body has thousands of different proteins, each serving a different function. Some proteins are primarily for structural support. Others are involved in muscle contraction. Others form part of the cell membrane, where they help transmit information and materials into and out of cells. Still others, called *enzymes,* regulate the rates of biochemical reactions within cells (see next section).

Because the links that determine the secondary and tertiary structures of protein are relatively weak hydrogen bonds, they may be broken by nearby charged molecules. This means that *the shape of proteins can change* in the presence of charged or polar molecules. The ability to change shape is essential to the functions of certain proteins.

Protein structure can also be damaged, sometimes permanently, by high temperatures or changes in pH. **Denaturation** refers to permanent disruption of protein structure, leading to a loss of biological function. An egg becomes hard when it is exposed to high temperatures because the soluble proteins in the egg become denatured and clump together as a solid mass.

Most proteins are water soluble, meaning that they dissolve in water. There are exceptions, however. Many of the proteins that are part of our cell membranes either are insoluble in water or have water-insoluble regions. You will learn more about why this is important in Chapter 3.

 Web Animation *Protein Structure* at www.humanbiology.com

 Quick Check You've isolated an unknown macromolecule, and you are trying to identify it. So far, all you know is that it consists mostly of carbon and hydrogen, it doesn't contain any nitrogen at all, and it is insoluble in water. Is it most likely to be a protein, a carbohydrate, or a lipid? Why?

Primary structure
(sequence of amino acids)

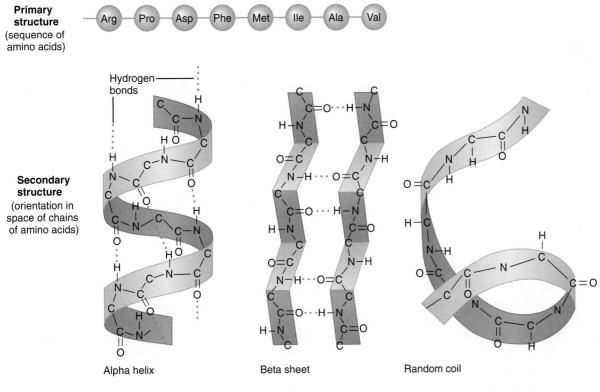

Secondary structure
(orientation in space of chains of amino acids)

Hydrogen bonds

Alpha helix   Beta sheet   Random coil

Tertiary structure
(three-dimensional shape)

Alpha helix

Beta sheet

Random coil

Quaternary structure
(number of polypeptide chains and their association)

Figure 2.21 **The structure of proteins.** In the diagrams of secondary structure the R groups have been omitted so that the basic backbone can be seen more easily.

Enzymes facilitate biochemical reactions

An **enzyme** is a protein that functions as a biological catalyst. A **catalyst** is a substance that speeds up the rate of a chemical reaction without itself being altered or consumed by the reaction. Enzymes help biochemical reactions to occur, but they do not change the final result of the reaction. That is, they can only speed reactions that would have happened anyway, although much more slowly. A chemical reaction that could take hours by itself might reach the same point in minutes or seconds in the presence of an enzyme.

Without help from thousands of enzymes, most biochemical reactions in our cells would occur too slowly to sustain life. Each enzyme facilitates a particular chemical reaction or group of reactions. Some enzymes break molecules apart; others join molecules together. In general, the enzyme takes one or more *reactants* (also called substrates) and turns them into one or more *products*. Enzymes serve as catalysts because, as proteins, they can change shape. The ability to change shape allows them to bind to other molecules and orient them so that they may interact. Figure 2.22 illustrates how a typical enzyme works.

Just how important are enzymes? As one example, the reason we can digest glycogen and starch is that we possess specific enzymes that break the chemical bonds between the glucose monosaccharides in these molecules (as you can demonstrate with the Try It Yourself box). In contrast, we cannot digest cellulose because we lack the right enzyme to break it apart. Termites can utilize cellulose only because their digestive systems harbor bacteria that have a cellulose-digesting enzyme.

The changeable shape of an enzyme shows why homeostasis within our cells is so important. Protein shape is in part determined by the chemical and physical environment inside a cell, including temperature, pH, and the concentrations of certain ions. Any deviation from homeostasis can affect the shapes and biological activities of dozens of different enzymes and thus alter the course of biochemical reactions within the cell.

🔄 **Recap** Proteins consist of strings of amino acids. The function of a protein relates to its shape, which is determined by its amino acid sequence and the twisting and folding of its chain of amino acids. Enzymes are proteins that facilitate biochemical reactions in the body. Without enzymes, many biochemical reactions would occur too slowly to sustain life. ∎

2.9 Nucleic acids store genetic information

Another important class of organic molecules is the *nucleic acids,* **DNA (deoxyribonucleic acid)** and **RNA (ribonucleic acid).** You have probably heard of such subjects as cloning, genetic engineering, and DNA "fingerprinting." These subjects relate to the nucleic acids, DNA and RNA.

DNA, the genetic material in living things, directs everything the cell does. It is both the organizational plan and the set of instructions for carrying the plan out. Because it directs and controls all of life's processes including growth, development, and reproduction, DNA is key to life itself. RNA, a closely related macromolecule, is responsible for carrying out the instructions of DNA, and in some cases, of regulating the activity of DNA itself. In some viruses RNA (rather than DNA) serves as the genetic material.

To fully appreciate the importance of DNA and RNA, consider that

- DNA contains the instructions for producing RNA.
- RNA contains the instructions for producing proteins.
- Proteins direct most of life's processes.

Both DNA and RNA are composed of smaller molecular subunits called **nucleotides.** Nucleotides consist of (1) a five-carbon sugar, (2) a single- or double-ringed structure containing nitrogen called a *base,* and (3) one or more phosphate groups. There are only eight different nucleotides, four in DNA and four in RNA.

Figure 2.23 shows the structures of the four nucleotides that make up DNA. Each nucleotide is composed of a five-carbon sugar molecule called *deoxyribose* (like the five-carbon sugar ribose but missing one oxygen atom), a phosphate group, and one of four different nitrogen-containing base molecules (*adenine, thymine, cytosine,* or *guanine*). In a single strand of DNA, these nucleotides are linked together by covalent bonds between the phosphate and sugar groups. The complete molecule of DNA is actually composed of two intertwined strands of nucleotides that are held together by weak hydrogen bonds (**Figure 2.24**). The sequence of one

Figure 2.22 Enzymes facilitate chemical reactions. This particular enzyme facilitates a dehydration synthesis reaction in which two reactants join to create one larger product plus a molecule of water. Note that the enzyme is not used up during the reaction.

Figure 2.23 The four nucleotides that compose DNA. The phosphate and sugar groups are identical in all four nucleotides.

smaller fragments of RNA. RNA is structurally like DNA, with a few exceptions (**Figure 2.25**):

- The sugar unit in all four of the nucleotides in RNA is *ribose* rather than deoxyribose (hence the name ribonucleic acid).
- One of the four nitrogen-containing base molecules is different (uracil is substituted for thymine).
- RNA is a single-stranded molecule, representing a complementary copy of a portion of only *one* strand of DNA.
- RNA is shorter, representing only the segment of DNA that codes for one or more proteins.

In Chapter 17 we discuss how RNA is used to make proteins, as well as RNA's possible role in regulating the activity of DNA.

HBP **Web Animation** *Nucleic Acids* at www.humanbiology.com

↶ **Recap** DNA and RNA are constructed of long strings of nucleotides. Double-stranded DNA represents the genetic code for life, and RNA, which is single-stranded, is responsible for carrying out those instructions. ∎

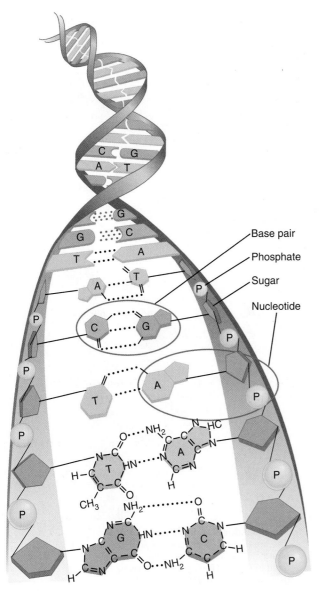

Figure 2.24 The double helical structure of DNA. The single strands of DNA are formed by dehydration synthesis. The two strands are held together by two hydrogen bonds between adenine and thymine and three hydrogen bonds between cytosine and guanine.

strand determines the sequence of the other (they are complementary strands), for adenine can form hydrogen bonds only with thymine (A with T) and cytosine can form bonds only with guanine (C with G).

The "code" for making a specific protein resides in the specific sequence of base pairs in one of the two strands of the DNA molecule. Notice that the entire genetic code is based entirely on the sequence of only four different molecular units (the four nucleotides). You will learn more about DNA, the genetic code, and inheritance in Chapters 17 and 19.

A single molecule of DNA carries the code for making a lot of different proteins. It is like an entire bookshelf of information, too big to be read all at once. To carry out their function, portions of the DNA molecule are transcribed into

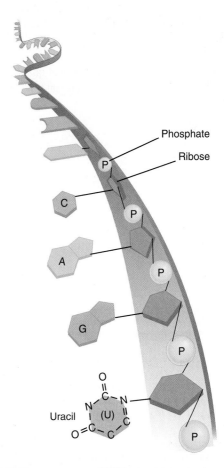

Figure 2.25 The structure of RNA. RNA is a single strand of nucleotides in which the base uracil substitutes for thymine. The sugar is ribose in RNA (as opposed to deoxyribose in DNA).

2.10 ATP carries energy

One additional related nucleotide with an important function is **ATP (adenosine triphosphate).** ATP is identical to the adenine-containing nucleotide in RNA except that it has two additional phosphate groups. ATP consists of an adenine base, the five-carbon sugar ribose (together they are called *adenosine*), and three phosphate groups (*triphosphate*) (Figure 2.26). ATP is a universal energy source for cells because the bonds between the phosphate groups contain a great deal of potential energy. Any time a cell needs energy for virtually any function, it can break the bond between the last two phosphate groups of ATP and release energy according to the following equation:

$$ATP \rightarrow ADP + P_i + energy$$

The breakdown of ATP produces **ADP (adenosine diphosphate)** plus an inorganic phosphate group (P_i), which is not attached to an organic molecule, plus energy that is now available to do work. The reaction is reversible, meaning that ATP is replenished by using another source of energy to reattach P_i to ADP. The energy to replenish ATP may come from stored energy in the food we eat, or from the breakdown of energy storage molecules such as glycogen or fat. You will learn more about ATP as an energy source when we discuss energy utilization by muscles (Chapter 6).

↺ **Recap** ATP is a nearly universal source of quick energy for cells. The energy is stored in the chemical bonds between phosphate groups. ■

a) **The structure of ATP.**

b) **The breakdown and synthesis of ATP.**
The breakdown (hydrolysis) of ATP yields energy for the cell. The reaction is reversible, meaning that ATP may be resynthesized using energy from other sources.

Figure 2.26 Adenosine triphosphate (ATP).

Chapter Summary

All matter consists of elements *p. 26*

- Atoms, the smallest functional unit of any element, contain a nucleus and a cloud of electrons.
- The protons and neutrons in an atom's nucleus account for most of its mass.
- Radioisotopes are unstable isotopes; an isotope has more or fewer neutrons than the usual number for that atom.

Atoms combine to form molecules *p. 28*

- Energy exists as either kinetic energy or potential energy.
- Three types of chemical bonds account for the structures of molecules: covalent, ionic, and hydrogen bonds. Covalent bonds are the strongest; hydrogen bonds are the weakest.
- Over 99% of your body weight consists of just six elements: oxygen, carbon, hydrogen, nitrogen, calcium, and phosphorus.

Life depends on water *p. 32*

- The polar nature of the water molecule accounts for its physical properties and for its unusually good qualities as a solvent for most other molecules and ions.
- Water is important in human temperature regulation.

The importance of hydrogen ions *p. 33*

- Molecules that can donate a hydrogen ion (H^+) are called acids. Molecules that can accept H^+ are called bases.
- The hydrogen ion concentration of a solution is expressed as pH.
- Buffers are pairs of molecules that tend to minimize changes in pH when an acid or base is added to a solution.

The organic molecules of living organisms *p. 35*

- The backbone of all organic molecules is carbon.
- Organic molecules are formed by a process called dehydration synthesis (requiring energy) and broken down by a process called hydrolysis (releasing energy).

Carbohydrates: Used for energy and structural support *p. 37*

- Monosaccharides, or simple sugars, are a source of quick energy for cells.
- Complex carbohydrates (polysaccharides) are formed by linking simple sugars (monosaccharides) together by dehydration synthesis.
- Carbohydrates are primarily energy-storage molecules. Plants use them for structural support as well.
- In animals the storage molecule is glycogen; in plants it is starch.

Lipids: Insoluble in water *p. 39*

- Lipids include fats and oils, phospholipids, and steroids. Lipids are insoluble in water.
- Fats store energy. Phospholipids and cholesterol are important structural components of the cell membrane. The sex hormones are steroids synthesized from cholesterol.

Proteins: Complex structures constructed of amino acids *p. 40*

- Proteins have unique three-dimensional structures that depend on their primary structure (their amino acid sequences). Living organisms construct a tremendous number of different proteins using just 20 different amino acids.
- The human body contains thousands of proteins, each with a different function.
- Enzymes are proteins that facilitate the rates of chemical reactions.

Nucleic acids store genetic information *p. 44*

- DNA is composed of two long strands of nucleotides intertwined into a double helix. DNA is constructed from just four different DNA nucleotides.
- RNA is a shorter single strand of RNA nucleotides, representing the code for one or more proteins.

ATP carries energy *p. 46*

- The nucleotide ATP is an energy source for cells. The energy is stored in the bonds between phosphate groups.

Terms You Should Know

acid, *34*
atom, *26*
ATP, *46*
base, *34*
catalyst, *43*
covalent bond, *29*
DNA, *44*
electron, *26*
enzyme, *43*

hydrogen bond, *30*
ionic bond, *30*
lipid, *39*
molecule, *28*
neutron, *26*
pH scale, *34*
protein, *42*
proton, *26*
RNA, *44*

Concept Review

Answers can be found at the Human Biology Place.
www.humanbiology.com

1. Describe the electrical charges and relative masses of protons, neutrons, and electrons.
2. Explain why two atoms of hydrogen tend to combine into a molecule of hydrogen gas (H_2).
3. Explain why polar and charged molecules tend to be soluble in water.
4. How is a covalent bond different from an ionic bond?
5. Compare and contrast potential energy and kinetic energy.
6. Distinguish between *saturated* and *unsaturated* fats.
7. Describe the role cholesterol plays in cells.
8. Explain why proteins come in an almost unlimited variety of shapes.
9. Discuss the importance of enzymes in living organisms.
10. Describe the role of ATP in energy transfer within a cell.

Test Yourself

Answers can be found in Appendix A.

1. If a molecule of starch is repeatedly hydrolyzed, which of the following would be the final product?
   a. glucose
   b. fructose
   c. ribose
   d. sucrose
   e. deoxyribose

2. Which of these molecules would be described as hydrophobic?
   a. glucose
   b. sodium chloride
   c. cholesterol
   d. DNA
   e. RNA

3. _____ bonds form between the oxygen and hydrogens within water molecules while _____ bonds form between different water molecules.
   a. hydrogen...... covalent
   b. covalent....... hydrogen
   c. ionic.......hydrogen
   d. hydrogen.......ionic
   e. covalent.......ionic

4. ^{13}C and ^{14}C are _____ of carbon.
   a. isotopes
   b. ions
   c. compounds
   d. molecules
   e. isomers

5. Which of the following substances has the lowest hydrogen ion concentration?
   a. water (pH 7)
   b. bleach (pH 13)
   c. baking soda (pH 9)
   d. vinegar (pH 3)
   e. black coffee (pH 5)

6. When sugar is dissolved in water, sugar is the _____ and water is the _____ .
   a. acid.........base
   b. base.........acid
   c. solvent......solute
   d. solute........solvent
   e. inorganic molecule.......organic molecule

7. A monosaccharide is to a polysaccharide as an amino acid is to a _____.
   a. nucleic acid
   b. carbohydrate
   c. protein
   d. nucleotide
   e. triglyceride

8. Which of these bonds is the easiest to disrupt, simply by raising the temperature?
   a. hydrogen bonds
   b. ionic bonds
   c. polar covalent bonds
   d. nonpolar covalent bonds
   e. peptide bonds

9. The primary structure of a protein is maintained by _____ bonds.
   a. hydrophilic
   b. hydrophobic
   c. hydrogen
   d. covalent
   e. ionic

10. DNA ultimately contains the instructions for the assembly of:
    a. proteins
    b. polysaccharides
    c. triglycerides
    d. nucleotides
    e. steroids

11. If one strand of DNA has the sequence A-A-C-T-G-T-G, what will be the nucleotide sequence of the complementary strand?
    a. A-A-C-T-G-T-G
    b. U-U-G-A-C-A-C
    c. T-T-G-U-C-U-C
    d. G-G-C-A-G-A-G
    e. T-T-G-A-C-A-C

12. Which of the following is true regarding the synthesis of a triglyceride?
    a. Three water molecules would be removed.
    b. The triglyceride would be hydrophilic.
    c. It would require three amino acids.
    d. It would require three monosaccharides.
    e. Hydrolysis reactions would be involved.

13. Which of the following is true regarding enzymes?
    a. The synthesis of an enzyme involves hydrolysis reactions.
    b. Enzymes provide energy for biochemical reactions.
    c. One enzyme can catalyze many different types of reactions.
    d. The instructions for the synthesis of an enzyme are found in DNA.
    e. Each enzyme molecule can be used only once.

14. Synthesis of proteins requires the input of energy which can be provided by:
    a. enzymes
    b. synthesis of ATP from ADP and P_i
    c. hydrolysis of ATP to form ADP and P_i
    d. amino acids
    e. isotopes

15. The element phosphorus (P) has an atomic number of 15 and a mass number of 31. Which of the following represents the numbers of subatomic particles in phosphorus?
    a. 15 protons, 15 electrons, 31 neutrons
    b. 15 protons, 16 electrons, 15 neutrons
    c. 31 protons, 31 electrons, 15 neutrons
    d. 15 protons, 15 electrons, 16 neutrons
    e. 16 protons, 16 electrons, 15 neutrons

Apply What You Know

Answers can be found at the Human Biology Place.
www.humanbiology.com

1. Athletes are sometimes advised to eat large amounts of complex carbohydrates (such as whole-wheat pasta) for a day or two before a competitive event. Explain the reasoning behind this.

2. Physicians become concerned about the potential for irreversible brain damage when body temperatures approach 105° F. Which of the four classes of macromolecules do you think is most likely affected by high temperatures? Explain.

3. Many people use cholesterol-lowering drugs to reduce their high cholesterol, because they know that a high cholesterol level is a risk factor for heart disease. Would it be advisable to take a little extra dose of these drugs to try to lower your cholesterol to below normal levels, just to be on the safe side?

4. In Miami when it's 90 degrees outside and very humid, the heat feels stifling. Yet most people report feeling fine in Arizona, where the humidity is generally low, even when it is 100 degrees. Why does the humidity have such an effect on our perception of comfort in terms of temperature?

5. Coca Cola is a very acidic drink; its pH is around 3. Blood has a pH of about 7. Yet when you drink a Coke the pH of the blood doesn't change measurably. Why is that?

6. Although normal physiological processes produce small amounts of free radicals, it is possible that your behavior, lifestyle, or environment might contribute to an increased production of free radicals. What behavioral or environmental risks do you think you have that might promote the formation of free radicals?

3

Structure and Function of Cells

Fluorescent light micrograph of fibroblast cells from connective tissue.

The Use of Human Stem Cells

What do boxing champion Muhammad Ali and actor Michael J. Fox have in common? They both suffer from Parkinson's disease, a debilitating neurological disorder. The key to curing Parkinson's disease and many other diseases and health problems including Alzheimer's disease, leukemia, diabetes, and spinal cord injuries may be **stem cells.**

A stem cell is a cell from which other types of more specialized cells originate (or stem). The ultimate stem cell is the fertilized egg, for *all* of the specialized cells of the body originate from it. The first eight cells of a human embryo are

also stem cells, since they have not yet begun to differentiate (become different from each other). But shortly after the eight-cell stage the cells begin to

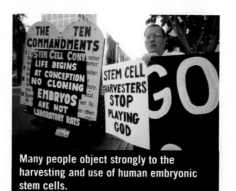

Many people object strongly to the harvesting and use of human embryonic stem cells.

specialize. Some become muscle; others become skin; still others become nerve cells in the brain.

Stem cells have several properties that make them desirable for research and for the treatment of certain diseases. They are easier to work with in the laboratory than adult cells because they don't adhere tightly to each other, and they grow better in culture (in controlled conditions, such as in a laboratory). They generally don't provoke a "tissue rejection" immune response in the patient because they are undifferentiated and thus not recognized as foreign cells. They are also easier to administer to a patient—usually they can be injected and allowed to migrate to their target site rather than having to be surgically transplanted. And most important, they still have the capacity to become the type of specialized cell the patient needs, under the direction of the patient's own cell division/differentiation control mechanisms.

Traditionally most human embryonic stem cells have come from very early-stage embryos. Currently, the only available embryos of this age are those created "in excess of clinical need" by *in vitro* fertilization at private fertility clinics. Only a few researchers have access to such embryos. However, cells used to treat specific diseases don't need to be completely undifferentiated. If nerve cells are needed, a good source of young cells is the very first embryonic neural tissue in the human fetus. Already, over 100 people with Parkinson's disease have received fetal nerve cell transplants worldwide, and some have shown measurable improvements in brain function. Nonliving human fetuses are widely available as a consequence of the more than one million legal abortions performed in the U.S. each year.

Controversy and Compromise

Not surprisingly, the use of human embryonic cells from fertilized eggs and undifferentiated fetal cells from legally aborted fetuses is highly controversial. On one hand, patient advocacy groups recognize the potential benefits of human embryonic and fetal cells and promote efforts to harvest and use them. On the other, some human rights groups object strongly to harvesting or using human embryonic stem cells or fetal cells under any circumstances, calling such research the destruction of precious human life. Both sides believe strongly in their position and both sides are active politically, and as a result politicians have been forced to take a stand.

According to guidelines developed under the Bush administration in 2001,

Muhammad Ali and Michael J. Fox, both sufferers of Parkinson's disease, support stem cell research.

federal funds could only be used to study stem cell lines derived from embryos before 2001. (A stem cell "line" is a group of identical cells grown from a single stem cell.) In effect, the 2001 federal guidelines prohibited the National Institutes of Health, which funds most biomedical research, from financing any stem cell research that might require the *future* death of a human embryo, but allows research on cells harvested from embryos in the past. However, the guidelines stopped short of prohibiting human stem cell research altogether; privately funded human embryonic stem cell research is still permitted under certain conditions.

The guidelines developed under President Bush were a political compromise at best. They allow selected federally funded research projects to go forward with the few stem cell lines that already existed at the time the law was passed (2001), while respecting concerns about the sanctity of human life. The guidelines were opposed by stem cell researchers, who contended that the limited number of cell lines available were not enough for the United States to stay in the forefront of this important research area. Some stem cell researchers moved to Britain, which had developed a facility for storing thousands of cell lines, or to other places in the world where stem cell research was not only allowed, but encouraged.

The election of President Obama changed the political environment yet again. Shortly after taking office in 2009,

President Obama signed an executive order lifting the restrictions on stem cell research laid down by President Bush. Federally funded researchers are now free to use the hundreds of stem cell lines in existence today, as well as new stem cell lines created by private funding in the future. But there is still a prohibition in place, called the Dickey-Wicker Amendment of 1996, which prohibits federal funding for research "in which human embryos are created, destroyed, discarded, or knowingly subjected to risk of injury or death." So while federally funded researchers will be able to use stem cell lines created by private funds (because the researchers themselves did not destroy any embryos), they will still be prohibited from creating their own new cell lines from human embryos. Obviously, the controversy over stem cells is not over.

What Is the Solution?

Both sides in this controversy actually do have something they can agree on; they both hope that someday we will not need to use stem cells from embryos or fetuses at all. The key may be the development of methods to create undifferentiated stem cells from fully differentiated adult cells—in essence, reversing the entire process of cellular differentiation and specialization. Several groups of scientists already claim to have done it successfully. But some scientists caution that new techniques for creating stem cell lines may not necessarily translate quickly into cures for specific diseases. It would be a shame if a well-meaning public, convinced that human embryonic stem cells are no longer necessary, came to accept laws that severely restrict human embryonic stem cell research in this country. Although the day may yet come when human embryonic stem cells truly aren't necessary, that day has not yet arrived. Just ask Michael J. Fox or Mohammed Ali.

Questions to consider

1 What is your opinion on this controversy? What basic beliefs do you hold that cause you to feel as you do?

2 Suppose that you and your spouse held frozen embryos at a private fertility clinic and that you knew you would never need them. Would you donate them for stem cell research? Why or why not?

The facts...

- Stem cells obtained from human embryos and fetuses have the potential to treat or cure diseases.
- Research using cells obtained from human embryos or fetuses is controversial.
- In the United States, stem cell research has been affected by changes in the political environment.
- In the distant future, human embryos and fetuses may no longer be needed as a source of stem cells.

Scientists first observed living cells under a microscope in 1674. Since then, countless observations and experiments have confirmed the **cell doctrine,** which consists of three basic principles:

1. All living things are composed of cells and cell products.
2. A single cell is the smallest unit that exhibits all the characteristics of life.
3. All cells come only from preexisting cells.

If we examine any part of the human body under a microscope, we find living cells and/or cell products. "Cell products" include materials composed of dead cells (such as the outer layer of your skin) and substances resulting from cellular activity (such as the hard crystalline elements of bone). There are no living units smaller than cells. All of our cells—all 100 trillion of them—are derived from earlier cells, going all the way back to our first cell, the fertilized egg. Even that original cell came from preexisting cells: the sperm and egg from our parents.

3.1 Cells are classified according to their internal organization

All cells are surrounded by an outer membrane called the **plasma membrane.** The plasma membrane encloses the material inside the cell, which is mostly water but also contains ions, enzymes, and other structures the cell requires to maintain life. All living cells are classified as either eukaryotes or prokaryotes, depending on their internal organization.

Eukaryotes have a nucleus, cytoplasm, and organelles

Human cells, like those of most species, are **eukaryotes** (*eu-* means "true" and *karyote* means "nucleus"). Nearly every eukaryotic cell has three basic structural components (**Figure 3.1a**):

1. A plasma membrane. The plasma membrane forms the outer covering of the cell.
2. A nucleus. Nucleus is a general term for "core." In the last chapter, we saw that chemists define a nucleus as the core of an atom. In biology, the nucleus is a membrane-bound compartment that houses the cell's genetic material and functions as its "information center." Most eukaryotic cells have one nucleus. There are a few exceptions, to be discussed in later chapters.
3. Cytoplasm ("cell material"). The cytoplasm includes everything inside the cell except the nucleus. It is composed of a soft, gel-like fluid called the *cytosol* ("cell solution"). The cytosol contains a variety of microscopic structures called **organelles** ("little organs") that carry out specialized functions, such as digesting nutrients or packaging cellular products.

Prokaryotes lack a nucleus and organelles

Prokaryotes (*pro-* means "before" and *karyote* means "nucleus") are the bacteria (kingdom Monera) (Figure 3.1b). Prokaryotes have a plasma membrane that is surrounded by a rigid cell wall. Their genetic material is concentrated in a particular region, but it is not specifically enclosed within a membrane-bound nucleus. Prokaryotes also lack most of the organelles found in eukaryotes. Nevertheless they are living organisms that fit the definition of a cell according to the cell doctrine.

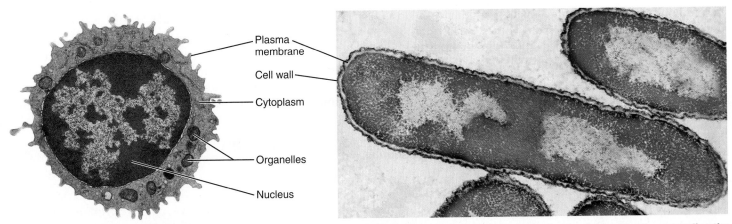

a) **A eukaryotic animal cell has a large nucleus and numerous small organelles.** The cytoplasm is enclosed by a flexible plasma membrane.

b) **Prokaryotic cells such as this bacterium have a rigid cell wall surrounding the plasma membrane.** The genetic material is not surrounded by a membrane, and there are no organelles in the cell. The elongated bacterium in the center of the photo is about to divide in two, as its genetic material is concentrated at both ends of the cell.

Figure 3.1 **Eukaryotes versus prokaryotes.**

In the rest of this chapter and throughout the book, we concentrate on the structure and function of eukaryotic cells. However, we discuss bacteria again in terms of how they can make us ill (Chapter 9) and in the context of evolution (Chapter 22).

3.2 Cell structure reflects cell function

Eukaryotic cells are remarkably alike in their structural features regardless of which organism they come from. This is because all cells carry out certain activities to maintain life, and there is a strong link between structure and function.

All cells must gather raw materials, excrete wastes, make macromolecules (the molecules of life), and grow and reproduce. These are not easy tasks. The specific activities carried out by a living cell (and the structures required to perform them) would rival those of any large city or even a country!

There is an outer structure that defines its border; an infrastructure for support; an information center; manufacturing facilities; refining, packaging, and shipping centers; transportation systems for supplying raw materials and energy; stockpiles of energy; and mechanisms for recycling or removing toxic waste. Cells even possess sophisticated defense mechanisms to combat invaders.

Most of the structural differences between cells reflect differences in function (Figure 3.2). Muscle cells contain numerous mitochondria that produce the energy for muscle contraction. Many nerve cells are long and thin; the longest nerve cells carry impulses all the way from your toes to your spinal cord. The cells that line the kidney tubules are cube shaped and tightly bound together, reflecting their role in the transport of water and other molecules. Essentially every cell has a specialized function of some sort or it would be of little use to the organism.

Cells that serve the same function are often remarkably similar between species. For example, a human nerve cell has more

a) **A portion of several muscle cells of the heart** ($\times$ 1,500).

b) **Nerve cells of the central nervous system** ($\times$ 830).

c) **Cells lining a tubule of a kidney** ($\times$ 250).

Figure 3.2 **Human cells vary in shape.**

in common (structurally and functionally) with a nerve cell in a cockroach than it does with a human liver cell. Furthermore, cells in a mouse are not that much different in size from those in an elephant; it's just that an elephant has more of them.

 Web Animation *Cell Structures* at www.humanbiology.com

Cells remain small to stay efficient

Despite their structural differences, all cells have at least one feature in common: they are small in one or more dimensions, requiring considerable magnification to be seen (Figure 3.3). Despite the incredible complexity of human cells, not one can be seen with the naked eye. Even

a) The light microscope (LM). Most of us have seen or used a light microscope. The light microscope uses visible light and a series of glass lenses to magnify a small sample as much as 1,000-fold. Focus can be difficult to maintain if objects are at different depths in a sample. Light does not transmit well through very dense or thick samples. Light microscopes have been in use for over 300 years.

b) The transmission electron microscope (TEM). A transmission electron microscope bombards the sample with a beam of electrons, some of which pass through the sample. Electrons behave like light waves, but since the wavelengths of electrons are shorter than visible light, the image has greater clarity at any magnification. A good electron microscope can magnify up to about 100,000 times, a hundred times greater than the light microscope. The images are two-dimensional (flat) because the sample must be very thin, but the magnification is sufficiently high that one can see the structural details of organelles within single cells.

c) The scanning electron microscope (SEM). The scanning electron microscope also focuses beams of electrons on the object. It produces what appears to be a three-dimensional view of the *surface* of an object. A narrow beam of electrons is scanned over the surface of an object that has been coated with a thin coat of metal so that the electron beam cannot penetrate. As the beam passes, the metal gives off secondary electrons, which can be recorded to produce a visual picture of the object's surface. Scanning electron microscopes have revealed stunning images of the relationships between cells and of the outer surfaces of cells. The scanning electron microscope can also magnify up to about 100,000 times.

Figure 3.3 Visualizing cells with microscopes. Photographs taken by the various methods of microscopy are called photomicrographs. All three of the photomicrographs shown here are of *Escherichia coli*, a normally harmless bacterium found in the digestive tract.

nerve cells that are over 3 feet long are so thin that we can't see them. Given the variety of life forms on Earth, why are there are so few giant cells? (An exception is the egg of some species.)

The answer shows that nature obeys certain simple and understandable principles:

- The total metabolic activities of a cell are proportional to its volume of cytoplasm, which is in effect its size. To support its activities, every cell needs raw materials in proportion to its size. Every cell also needs a way to get rid of its wastes.
- All raw materials, energy, and waste can enter or leave the cell only by crossing the plasma membrane.
- As objects get larger, their volume increases more than their surface area. For both spheres and cubes, for example, an eightfold increase in volume is accompanied by only a fourfold increase in surface area.

The larger a cell gets, then, the more likely that its growth and metabolism will be limited by its ability to supply itself across the plasma membrane (**Figure 3.4**). Put another way, the smaller a cell is, the more effectively it can obtain raw materials and get rid of wastes.

Some cells have numerous microscopic projections of the plasma membrane called *microvilli* (see Figure 3.4c) Microvilli are an effective way to increase surface area relative to volume. Plasma membranes with microvilli are especially common in cells that transport substances into and out of the body, such as the cells that line the digestive tract and the tubules of the kidneys.

↩ **Recap** Common features of nearly all eukaryotic cells are a plasma membrane, a nucleus, organelles, and the cytoplasm. Cells exchange materials with their environment across their plasma membrane. Cells are small, because this makes them more efficient at obtaining nutrients and expelling wastes. ▪

a) One large cell.

b) Eight small cells.

c) Cell with microvilli on one surface.

Figure 3.4 Cell size and plasma membrane shape affect surface area and volume. The volume is the same in these three groups of cells, but the ratio of volume to surface area is different in a), b), and c). Compared to the cell in a), the eight small cells in b) have twice the surface area. c) The surface area of any cell can be increased by the presence of microvilli.

3.3 A plasma membrane surrounds the cell

Consider a house. Its walls and roof are composed of special materials that prevent rain and wind from entering. They also form a barrier that allows the temperature inside the house to stay warmer or cooler than the temperature outside. At the same time, the house interacts with its environment. Windows allow light in; doors open and close to allow entry and exit. Water and power lines permit the regulated entry of water and energy, and sewer lines remove wastes. The house exchanges information with the outside world through mail slots, telephone lines, and computer cables.

The exterior structure of a living cell is its plasma membrane. Like the roof and walls of a house, the plasma membrane must permit the movement of some substances into and out of the cell yet restrict the movement of others. It must also allow the transfer of information across the membrane.

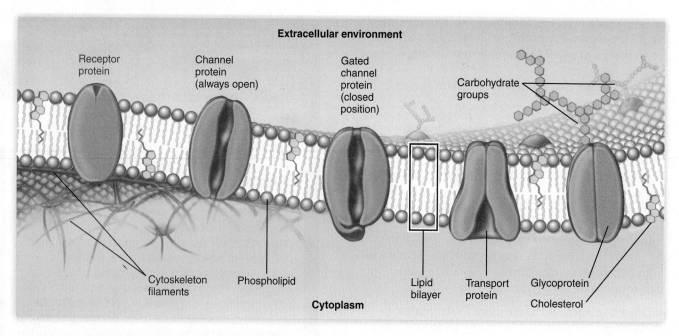

Figure 3.5 The plasma membrane. The plasma membrane is a phospholipid bilayer containing cholesterol and proteins. Cholesterol provides mechanical strength. The proteins transfer information, permit the passage of certain molecules, and provide structural support for the cytoskeleton.

The plasma membrane is a lipid bilayer

The plasma membrane is constructed of two layers of phospholipids, called a *lipid bilayer,* plus some cholesterol and various proteins (**Figure 3.5**). Each of the three components contributes to the membrane's overall structural and functional properties:

- *Phospholipids.* Recall that phospholipids are a particular type of lipid with a polar head and neutral nonpolar tails. In the plasma membrane the two layers of phospholipids are arranged so that the nonpolar tails meet in the center of the membrane. One layer of polar (water-soluble) heads faces the watery solution on the outside of the cell, and the other layer of polar heads faces the watery solution of the cell's cytoplasm.
- *Cholesterol.* Cholesterol increases the mechanical strength of the membrane by preventing it from becoming either too rigid or too flexible. It also prevents the phospholipids from moving around too much and helps to anchor the proteins within the membrane.
- *Proteins.* Various proteins are embedded in the phospholipid bilayer of the plasma membrane. Like the doors, windows, and wires of a house, they provide the means for transporting molecules and information across the plasma membrane. A few membrane proteins anchor the cell's internal scaffold-like support network. Some proteins span the entire membrane; others protrude from only one surface. Plasma membrane proteins generally have one region that is electrically neutral and another that is electrically charged (either + or −). The charged regions tend to extend out of the membrane and thus are in contact with water, whereas the neutral portions are often embedded within the phospholipid bilayer.

The phospholipid bilayer of the plasma membrane is only about 3.5 nanometers thick (a nanometer, abbreviated nm, is a billionth of a meter), too small to be seen in detail even with microscopes. To appreciate relative sizes, imagine as we did in Chapter 2 that a single sodium ion is the size of a penny. At this scale the phospholipid bilayer of a typical cell would be about 13 inches thick. It is no wonder that many substances are restricted from passing through the membrane unless there is some sort of channel or transport mechanism available.

Although we have likened the plasma membrane to the exterior of a house, there are two key differences. First, the plasma membrane of animal cells is not rigid. If you could touch a plasma membrane, it would probably feel like a wet sponge, giving way under your touch and springing back when you remove your hand. Most cells do maintain a certain shape, but it is mainly due to a supporting network of fibers inside the cell, the fluid within the cell, and the limitations imposed by contact with surrounding cells, not the stiffness of the plasma membrane itself.

Second, the phospholipids and proteins are not anchored to specific positions in the plasma membrane. Many proteins drift about in the lipid bilayer like icebergs floating on the surface of the sea. Imagine if you were to get up in the morning and find that the front door to your house had moved 3 feet to the left! The plasma membrane of animal cells is often described as a "fluid mosaic" to indicate that it is not a rigid structure and that the pattern of proteins within it constantly changes.

HBP **Web Animation** *Membrane Structure* at www.humanbiology.com

↩ Recap The plasma membrane is comprised of phospholipids, cholesterol, and proteins. The proteins transfer information and transport molecules across the membrane and provide structural support. ▪

3.4 Molecules cross the plasma membrane in several ways

The plasma membrane creates a barrier between the cell's external environment and the processes of life going on within. Life was impossible until these functions could be enclosed and concentrated in one place, keeping what was needed for growth and reproduction inside and limiting the entry of other materials. Molecules (and ions) cross the plasma membrane in three major ways: (1) passive transport (diffusion and osmosis), (2) active transport, and (3) endocytosis or exocytosis.

Passive transport: Principles of diffusion and osmosis

Passive transport is "passive" because it transports a molecule without requiring the cell to expend any energy. Passive transport relies on the mechanism of diffusion.

Diffusion Molecules in a gas or a liquid move about randomly, colliding with other molecules and changing direction. The movement of molecules from one region to another as the result of this random motion is known as **diffusion.**

If there are more molecules in one region than in another, then strictly by chance more molecules will tend to diffuse away from the area of high concentration and toward the region of low concentration. In other words, the *net* diffusion of molecules requires that there be a difference in concentration, called a *concentration gradient,* between two points. Once the concentration of molecules is the same throughout the solution, a state of equilibrium exists in which molecules diffuse randomly but equally in all directions.

Figure 3.6 illustrates diffusion by showing what happens when a concentrated solution of blue dye is placed in the

10 minutes   1 hour   24 hours

MJ's Human Biology Blog

Birth Dating Human Cells

How can scientists determine the age of human cells? How frequently are human cells replaced, if at all?

In 2005 scientists hit upon an ingenious method that takes advantage of a dark period in recent world history—the above-ground testing of nuclear weapons between the mid-1950s and 1963. Nuclear weapons testing resulted in a sharp spike in carbon-14 levels worldwide. The levels peaked in 1967 and have since declined as carbon-14 diffused and equilibrated with the atmosphere, the oceans, and the biosphere. Carbon is incorporated into the chemical components of all new cells, of course, including DNA. It turns out that the carbon-14 levels in nuclear DNA correspond very closely to the atmospheric levels at the time the DNA was synthesized. So by comparing the cells' nuclear DNA carbon-14 levels to a chart of atmospheric cabon-14 levels each year, one can determine the cells' birth date.

How does this help us determine cell turnover? Think about it: if all of the cells in a piece of tissue are the same age as the individual, then cells are not being replaced throughout life. But if the average cell age is much younger than the individual, then cell turnover must be relatively high. The scientists who developed the cell-dating technique report that neurons in the cerebral cortex (the most highly developed region of the brain) do not undergo significant replacement throughout life—you're born with all the cortical brainpower you're ever going to have. In contrast, cells lining the intestine are replaced frequently. ▪

Reference: "Retrospective Birth Dating of Cells in Humans." *Cell* 122: 133–143, 2005.

Figure 3.6 Diffusion. At time zero a concentrated solution of a blue dye was placed in the middle of a tube of water. To prevent the movement of water by bulk flow, a stabilizing agent had been added previously to the water. Over time the random motion of molecules in solution has caused some molecules of dye to move from the region of highest dye concentration to the region of low dye concentration. Although not visible, water has diffused in the opposite direction, from the region of highest water concentration (pure water) toward the region of lower water concentration (the region occupied by dye). Note that diffusion is effective only over short distances; it has taken 24 hours for the dye to move only 1 cm in any direction.

middle of a tube of water. Over time, the dissolved molecules diffuse away from their region of highest concentration and toward the regions of low concentration.

Water also diffuses from the region of its highest concentration toward the region of its lowest concentration. However, the concentration of water (the liquid, or solvent) in a solution is *opposite* to that of the molecules other than water (the solutes). The higher the concentration of solutes, the lower the concentration of water. Pure water is the solution with the highest possible concentration of water. Net diffusion of water is always toward the solution with the higher concentration of solutes and away from the solution with a higher concentration of water.

Osmosis Not all substances diffuse readily into and out of living cells. The plasma membrane is **selectively permeable,** meaning that it allows some substances to cross by diffusion but not others. It is highly permeable to water but not to all ions or molecules. The net diffusion of water across a selectively permeable membrane is called **osmosis.**

Figure 3.7 demonstrates the process of osmosis. In Figure 3.7a, a selectively permeable membrane—in this case permeable only to water—separates pure water from a solution of glucose in water. Although the glucose cannot diffuse, the water diffuses toward its region of lower concentration, from right to left. As osmosis occurs, the volume in the left chamber rises, creating a fluid pressure that begins to oppose the continued osmosis of water (Figure 3.7b). Eventually the movement of water from left to right (because of differences in fluid pressure) equals the movement from right to left (by osmosis), and there is no further net change in the volume of water on each side of the membrane.

The fluid pressure required to exactly oppose osmosis is called *osmotic pressure*. In Figure 3.7c osmotic pressure is represented by the extra weight of the higher column of water on the left than on the right.

 Web Animation *Diffusion and Osmosis* at www.humanbiolgoy.com

Passive transport moves with the concentration gradient

Most substances cross cell membranes by passive transport. Passive transport always proceeds "downhill" with respect to the concentration gradient, meaning that it relies on diffusion in some way. Three forms of passive transport across the cell membrane are (1) diffusion through the lipid bilayer, (2) diffusion through channels, and (3) facilitated transport (Figure 3.8).

Diffusion through the lipid bilayer The lipid bilayer structure of the plasma membrane allows the free passage of some molecules while restricting others. For instance, small uncharged nonpolar molecules can diffuse right through the lipid bilayer as if it did not exist. Such molecules simply dissolve in the lipid bilayer, passing through it like a ghost through a wall. Polar or electrically charged molecules, in contrast, cannot cross the lipid bilayer because they are not soluble in lipids.

Two important lipid-soluble molecules are oxygen (O_2), which diffuses into cells and is used up in the process of metabolism, and carbon dioxide (CO_2), a waste product of metabolism, which diffuses out of cells and is removed from the body by the lungs. Another substance that crosses the lipid bilayer by diffusion is urea, a neutral waste product removed from the body by the kidneys.

 Quick Check Do you think sodium (Na^+) or chloride (Cl^-) ions can diffuse directly through the lipid bilayer? How about water molecules? Explain. ▪

Answers to ☑ **questions can be found in Appendix A.**

Figure 3.7 Generation of osmotic pressure by osmosis. Starting in a), there is a net movement of water from right to left until the diffusion of water (osmosis) is opposed by movement of water due to osmotic pressure c).

 In the third figure, what would happen if you now add glucose to the right side—adding exactly the same total number of glucose molecules as are in the left side? Explain your answer.

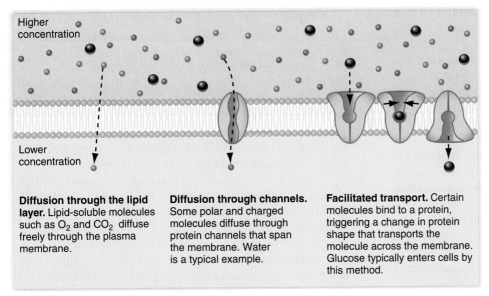

Higher concentration

Lower concentration

Diffusion through the lipid layer. Lipid-soluble molecules such as O_2 and CO_2 diffuse freely through the plasma membrane.

Diffusion through channels. Some polar and charged molecules diffuse through protein channels that span the membrane. Water is a typical example.

Facilitated transport. Certain molecules bind to a protein, triggering a change in protein shape that transports the molecule across the membrane. Glucose typically enters cells by this method.

Figure 3.8 The three forms of passive transport. All involve transport down a concentration gradient without the expenditure of additional energy.

Diffusion through channels Water and many ions diffuse through channels in the plasma membrane. The channels are constructed of proteins that span the entire lipid bilayer. The sizes and shapes of these protein channels, as well as the electrical charges on the various amino acid groups that line the channel, determine which molecules can pass through.

Some channels are open all the time (typical of water channels). The diffusion of any molecule through the membrane is largely determined by the number of channels through which the molecule can fit. Other channels are "gated," meaning that they can open and close under certain conditions. Gated channels are particularly important in regulating the transport of ions (sodium, potassium, and calcium) in cells that are electrically excitable, such as nerve cells (Chapter 11). Look at Figure 3.5, which represents a number of the proteins instrumental in transport.

Facilitated transport In **facilitated transport**, also called *facilitated diffusion*, the molecule does not pass through a channel at all. Instead it attaches to a membrane protein, triggering a change in the protein's shape or orientation that transfers the molecule to the other side of the membrane and releases it there. Once the molecule is released, the protein returns to its original form. A protein that carries a molecule across the plasma membrane in this manner, rather than opening a channel through it, is called a **transport protein** (carrier protein).

Facilitated transport is highly selective for particular substances. The direction of movement

is always from a region of higher concentration to one of lower concentration, and thus it does not require the cell to expend energy. The normal process of diffusion is simply being "facilitated" by the transport protein. Glucose and other simple sugars enter most cells by this method.

Active transport requires energy

All methods of passive transport allow substances to move only down their concentration gradients, in the direction they would normally diffuse if there were no barrier. However, **active transport** can move substances through the plasma membrane *against* their concentration gradient. Active transport allows a cell to accumulate essential molecules even when their concentration outside the cell is relatively low, and to get rid of molecules that it does not need. Active transport requires the expenditure of energy.

Like facilitated transport, active transport is accomplished by proteins that span the plasma membrane. The difference is that active transport proteins must have some source of energy to transport certain molecules. Some active transport proteins use the high-energy molecule ATP (adenosine triphosphate) for this purpose (**Figure 3.9a**). They break ATP down to ADP

a) In active transport using ATP, energy derived from the breakdown of ATP is used to change the shape of the carrier protein.

ADP + P_i

b) Some carrier proteins use energy derived from the downhill transport of one molecule to transport another molecule uphill. In this example, the energy to transport the square molecules comes from the facilitated transport of the spearhead molecules.

Figure 3.9 Active transport. A cell can employ active transport to move a molecule against a concentration gradient. Because this is an "uphill" effort, energy is required.

(adenosine diphosphate) and a phosphate group (P$_i$) and use the released energy to transport one or more molecules across the plasma membrane against their concentration gradient. Imagine that the active transport protein is a conveyor belt moving objects uphill, powered by a gasoline engine. In this analogy, ATP is the gasoline and ADP the exhaust (with one difference: the ADP "exhaust" in the cell can be recycled to ATP).

Proteins that actively transport molecules across the plasma membrane are sometimes called "pumps." Some pumps can transport several different molecules at once and even in both directions at the same time. One of the most important plasma membrane pumps is the **sodium-potassium pump**, which uses energy derived from breaking down ATP to transport sodium out of the cell and potassium into the cell.

Not all active transport pumps use ATP as the energy source. Some derive energy from the "downhill" facilitated transport of one molecule and use it to transport another molecule "uphill," against its concentration gradient (Figure 3.9b). This type of transport is analogous to an old-fashioned mill that grinds grain into flour by using energy derived from the downhill movement of water.

 Web Animation *Passive and Active Transport* at www.humanbiology.com

 Quick Check You've discovered a membrane protein that seems to be necessary for cells to transport fructose from higher to lower concentration. Predict whether this membrane protein requires ATP to function, and name the type of transport that it is most likely doing. Explain your reasoning. ▪

Endocytosis and exocytosis move materials in bulk

Most ions and small molecules move across the cell membrane by one or more of the passive and active transport mechanisms just described. However, some molecules are too big to be transported by these methods. To move large molecules or transport several kinds of molecules in bulk, some cells resort to endocytosis and/or exocytosis. These two processes are based on the same principle but have different directions of movement. **Endocytosis** moves materials into the cell, and **exocytosis** moves materials out of the cell.

Figure 3.10 shows the processes of endocytosis and exocytosis. In endocytosis, molecules dissolved in the extracellular fluid are surrounded by a pocket formed by an infolding of the plasma membrane. Eventually the pocket pinches off, forming a membrane-bound **vesicle** within the cell. To facilitate the selection of the right molecules for endocytosis, some vesicles have receptors on their surface that bind only to certain specific molecules. Insulin and certain enzymes enter cells by this method. Other vesicles are nonselective, engulfing whatever is in the extracellular fluid, such as nutrients and water. Vesicles of this type are often found in cells lining the digestive tract. Some white blood cells engulf and destroy whole bacteria by endocytosis (Chapter 9).

In exocytosis, a vesicle already present within the cell fuses with the plasma membrane and releases its contents into the fluid surrounding the cell. This is how certain cells release toxic waste products, get rid of indigestible material, or secrete their special products.

 Web Animation *Endocytosis and Exocytosis* at www.humanbiology.com

a) **Endocytosis.** In endocytosis, material is surrounded by the cell membrane and brought into the cell.

b) **Exocytosis.** In exocytosis, a membranous vesicle fuses with the plasma membrane, expelling its contents outside the cell.

Extracellular environment
Plasma membrane
Cytoplasm
Vesicle

c) **Photomicrograph showing various stages of endocytosis.**

Figure 3.10 Endocytosis and exocytosis.

Information can be transferred across the plasma membrane

Receptor proteins that span the plasma membrane can receive and transmit information across the membrane. The information received by receptor proteins generally causes something to happen within the cell even though no molecules cross the membrane.

Figure 3.11 illustrates how a receptor protein works. A molecule approaches the membrane and binds to a specific *receptor site* in a lock-and-key fashion. This binding triggers a series of biochemical events that ultimately cause changes within the cell. Receptor proteins are highly specific for a particular molecule or a group of similar molecules. For example, the receptor protein for the hormone insulin responds only to insulin and not to any other hormone. Furthermore, different cells have different sets of receptor proteins, which explains why some cells and tissues respond to a particular hormone and others do not.

We discuss receptor proteins and hormones in more detail in Chapter 13. For now, remember that certain molecules can influence what happens inside a cell merely by coming in contact with the cell's outer surface.

The sodium-potassium pump helps maintain cell volume

Probably the most critical task facing a cell is maintaining its volume. Why? Recall that the plasma membrane is soft and flexible. It cannot withstand much stretching or high fluid pressures. Furthermore, cells tend to accumulate certain materials depending on what is available in their *extracellular* environment (the area outside the cell, beyond the plasma membrane). Cells already contain a nucleus and organelles.

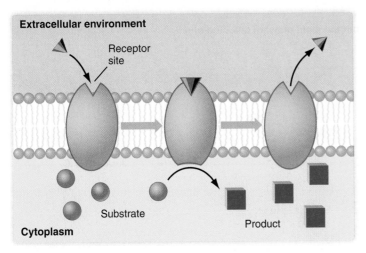

Figure 3.11 Receptor protein action. A specific molecule approaches a receptor site on a receptor protein and binds to it. The binding of the molecule to the receptor protein causes a series of chemical reactions within the cell. In this example, a particular cellular product (the squares) is produced from substrate molecules (the circles).

In addition, they produce or stockpile molecules, including amino acids, sugars, lipids, ions, and many others. These molecules are necessary for the cell to function normally, but they represent a lot of solute particles within the cell.

Because water can diffuse across the plasma membrane rather easily, you might expect that water would diffuse into the cell, toward the high cytoplasmic solute concentration. This inward diffusion would increase cell volume, eventually causing the cell to swell and even rupture.

The only way to avoid this is for the cell to keep the solute concentration in its cytoplasm identical to the solute concentration of the extracellular fluid. Then there is no net driving force for the diffusion of water. What the cell actually does is get rid of ions it *doesn't* need in large quantities (primarily sodium) in exchange for those it must stockpile. This is the primary function of a specialized protein embedded in the cell membrane called the sodium-potassium pump.

Figure 3.12a shows how the sodium-potassium pump works. The pump has three binding sites for sodium that are accessible from inside the cell. The binding of three cytoplasmic sodium ions triggers the breakdown of an ATP molecule to ADP and an inorganic phosphate. The energy released by ATP causes the pump to change shape, expelling the sodium ions and exposing two binding sites for potassium that are accessible only from outside the cell. The binding of potassium triggers another change of shape, and the potassium ions are transported into the cell.

One effect of this three sodium/two potassium exchange is to lower slightly the number of ions within the cell. More importantly, the plasma membrane is much more permeable to potassium than to sodium because it contains many potassium channels but very few sodium channels. Effectively, the cell keeps the number of sodium ions in the cytoplasm low by pumping them out again just as soon as a few leak in. The inward active transport of potassium does not increase the intracellular potassium concentration very much because they can leak back out so easily.

As Figure 3.12b shows, the sodium-potassium pump effectively controls cell volume. To reduce its volume, the cell increases the activity of the sodium-potassium pumps, getting rid of more sodium than usual. Water also exits to maintain osmotic equilibrium. To expand its volume, the cell lowers the activity of the pumps and retains water along with the extra sodium. Because potassium can diffuse so quickly no matter how much is pumped, the rate of potassium transport by the pumps is not relevant to the control of cell volume.

A single red blood cell may have over a hundred sodium-potassium pumps in its plasma membrane. In addition, the pump is essential to the ability of nerve cells to generate an electric current. We discuss the sodium-potassium pump in more detail in Chapter 11.

✓ **Quick Check** If a cell's sodium-potassium pumps are poisoned so that they stop working, will the cell tend to swell, shrink, or stay the same? Explain. ■

Extracellular fluid

1 Sodium ions bind to binding sites accessible only from the cytoplasm.

2 Binding of three cytoplasmic Na$^+$ to the sodium-potassium pump stimulates the breakdown of ATP.

Na$^+$

Na$^+$

Na$^+$

Na$^+$

Na$^+$

7 Most of the potassium diffuses out of the cell, but sodium diffuses in only very slowly.

K$^+$

K$^+$

K$^+$

Na$^+$

Na$^+$

ATP

ADP + P$_i$

3 Energy released by ATP causes the protein to change its shape, expelling the sodium ions.

Na$^+$

Na$^+$

Na$^+$

Na$^+$

K$^+$

Cytoplasm

K$^+$

K$^+$

6 Potassium is transported into the cell, and the sodium binding sites become exposed again.

K$^+$

K$^+$

4 The loss of sodium exposes two binding sites for potassium.

K$^+$

K$^+$

5 Potassium binding triggers another change of shape.

a) The cell membrane contains Na$^+$ - K$^+$ pumps, and also channels that permit the rapid outward diffusion of K$^+$ but only a slow inward diffusion of Na$^+$.

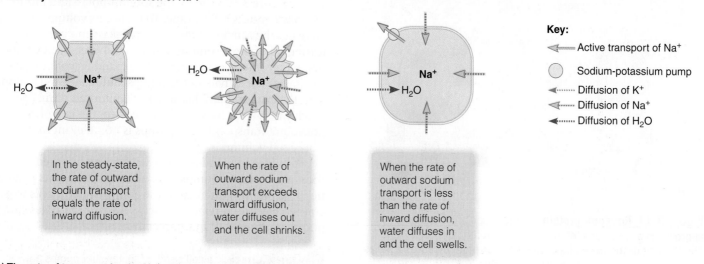

In the steady-state, the rate of outward sodium transport equals the rate of inward diffusion.

When the rate of outward sodium transport exceeds inward diffusion, water diffuses out and the cell shrinks.

When the rate of outward sodium transport is less than the rate of inward diffusion, water diffuses in and the cell swells.

Key:
- Active transport of Na$^+$
- Sodium-potassium pump
- Diffusion of K$^+$
- Diffusion of Na$^+$
- Diffusion of H$_2$O

b) The rate of transport by the Na$^+$ - K$^+$ pumps determines cell volume.

Figure 3.12 **Control of cell volume by the sodium-potassium pump (Na$^+$-K$^+$ pump).**

Isotonic extracellular fluid also maintains cell volume

Tonicity refers to the relative concentrations of solutes in two fluids (-*tonic* means "strength"). Because water can diffuse across the cell membrane so easily, the ability of a human cell to control its volume also depends on the tonicity of the extracellular fluid (**Figure 3.13**).

Extracellular fluid that is **isotonic** (Greek *isos*, equal) has the same solute concentration as the *intracellular* fluid. Cells maintain a normal volume in isotonic extracellular solutions because the concentration of water is the same inside and out. In humans, isotonic extracellular fluid is equivalent to about 9 grams of salt dissolved in a liter of solution. Regulatory mechanisms in the body ensure that the extracellular fluid solute concentration remains relatively constant at that level.

When cells are placed in a **hypertonic** solution, one with a concentration of solutes *higher* than the intracellular fluid, water diffuses out of the cells, and the cells shrink. Eventually this impairs normal function and the cells die.

Conversely, when cells are placed in a **hypotonic** solution with a *lower* concentration of solutes than intracellular fluid, water enters the cells and causes them to swell. Pure water is

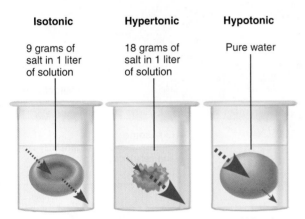

Isotonic	Hypertonic	Hypotonic
9 grams of salt in 1 liter of solution	18 grams of salt in 1 liter of solution	Pure water

a) Water movement into and out of human red blood cells placed in isotonic, hypertonic, and hypotonic solutions. The amount of water movement is indicated by the sizes of the arrows.

b) Scanning of electron micrographs of red blood cells placed in similar solutions.

Figure 3.13 The effect of extracellular fluid tonicity on cell volume.

 Suppose a woman runs a marathon on a hot summer day and becomes extremely dehydrated. Would you expect that her extracellular fluid would be isotonic, hypotonic, or hypertonic? Predict what might happen to her red blood cells.

the most hypotonic solution possible. Most human cells quickly swell, burst, and die when placed in pure water.

 Recap Molecules may move across the plasma membrane by diffusion, by passive or active transport, or by endocytosis/exocytosis. Sodium-potassium exchange pumps in the cell membrane are essential for the regulation of cell volume. In addition, homeostatic regulatory processes keep the tonicity of the extracellular fluid relatively constant. ▪

3.5 Internal structures carry out specific functions

So far we have discussed the plasma membrane that surrounds the cell. Now we move inside the cell, where we find a number of different membrane-bound and non-membrane-bound structures. The membrane-bound structures are called *organelles,* because they are like tiny organs in that they have a specific function to perform. **Figure 3.14** presents a cutaway view of an animal cell with its nucleus and organelles.

The nucleus controls the cell

The most conspicuous organelle of a living eukaryote is its *nucleus* (**Figure 3.15**). As the information center of a cell, the nucleus contains most of the cell's genetic material in the form of long molecules of DNA. (As you will learn in Chapter 17, mitochondria have their own DNA.) Ultimately, DNA controls nearly all the activities of a cell. Details of how DNA controls cellular function are discussed in Chapters 17 and 19.

The outer surface of the nucleus consists of a double-layered membrane, called the *nuclear membrane,* that keeps the DNA within the nucleus. The nuclear membrane is bridged by *nuclear pores* that are too small for DNA to pass through but that permit the passage of certain small proteins and RNA molecules.

Within the nucleus is a dense region called the **nucleolus,** where the components of ribosomes (RNA and ribosomal proteins) are synthesized. The components pass through the nuclear pores, to be assembled into ribosomes in the cytoplasm.

Ribosomes are responsible for protein synthesis

Ribosomes are small structures composed of RNA and certain proteins that are either floating freely in the cytosol or are attached to the endoplasmic reticulum, the cell organelle that synthesizes most biological molecules. Ribosomes are responsible for making specific proteins. They assemble amino acids into proteins by connecting the appropriate amino acids in the correct sequence according to an RNA template. We describe this process in more detail in Chapter 17.

Cytosol
Semifluid gel material inside the cell

Peroxisome
Destroys cellular toxic waste

Centrioles
Microtubular structures involved in cell division

Cytoskeleton
Structural framework of the cell

Smooth endoplasmic reticulum
Primary site of macromolecule synthesis other than proteins

Rough endoplasmic reticulum
Primary site of protein synthesis by ribosomes

Golgi apparatus
Refines, packages, and ships macromolecular products

Secretory vesicle
Membrane-bound shipping container

Ribosomes
Site of protein synthesis

Plasma membrane
Controls movement of materials into and out of cell

Nucleus
Information center for the cell. Contains DNA

Mitochondrion
Produces energy for the cell

Lysosome
Digests damaged organelles and cellular debris

Figure 3.14 A typical animal cell. Not shown in this cell are cilia, a flagellum, glycogen granules, or fat deposits, as only certain cells have these features.

Nucleolus

Nuclear membrane

Nuclear pores

Nuclear membrane

A transmission electron micrograph (× 6,000) of the nucleus of an animal cell

Figure 3.15 The nucleus. The nucleus contains the cell's genetic material. The nucleolus produces the protein and RNA components of ribosomes. These components exit the nucleus through nuclear pores.

Ribosomes that are attached to the endoplasmic reticulum release their proteins into the folds of the endoplasmic reticulum. Many of these proteins are packaged in membrane-bound vesicles, transported to the cell membrane, and secreted. Free-floating ribosomes generally produce proteins for immediate use by the cell, such as enzymes that serve as catalysts for chemical reactions within the cytoplasmic fluid.

The endoplasmic reticulum is the manufacturing center

The **endoplasmic reticulum (ER),** in conjunction with its attached ribosomes, synthesizes most of the chemical compounds made by the cell. If a cell were an industrial city, then the ER would be the city's steel mills, sawmills, and chemical plants. Like the output of a steel mill, the materials manufactured by the ER are often not in their final form. They are refined and packaged by the Golgi apparatus, discussed later.

Figure 3.16 shows the structure of the ER and its role in the manufacture of proteins and other materials. The ER is an extensively folded, membranous system surrounding a fluid-filled space. A portion of the ER connects to the nuclear membrane. Some regions of the ER's outer surface are dotted with ribosomes, giving those regions, called *rough ER*, a granular appearance. Regions without ribosomes are called *smooth ER*.

The rough ER is involved in the synthesis of proteins, as you may guess from the presence of ribosomes. Most of the proteins synthesized by the attached ribosomes are released into the fluid-filled space of the ER. Eventually they enter the smooth ER, where they are packaged for transfer to the Golgi apparatus.

The smooth ER synthesizes macromolecules other than protein. Most notable among these are the lipids, including some hormones. Numerous enzymes embedded in the inner surface of the ER membrane facilitate the chemical reactions necessary for macromolecule synthesis.

The smooth ER is also responsible for packaging the proteins and lipids for delivery to the Golgi apparatus. Newly synthesized proteins and lipids collect in the outermost layers of smooth ER. There, small portions of the fluid-filled space are surrounded by ER membrane and pinched off, forming vesicles that contain fluid, proteins, and lipids. The vesicles migrate to the Golgi apparatus, fuse with the Golgi apparatus membrane, and release their contents into the Golgi apparatus for further processing.

 Quick Check The cells of the pancreas make large quantities of proteins that are constantly secreted outside the cell. Would you expect a pancreatic cell to have a relatively large or small number of ribosomes, and would you expect it to have a lot of smooth or rough ER (compared to a cell that does not secrete proteins)? ■

Figure 3.16 The endoplasmic reticulum (ER). The rough ER is studded with ribosomes, where proteins are made. The smooth ER packages the proteins and other products of the ER and prepares them for shipment to the Golgi apparatus in vesicles.

The Golgi apparatus refines, packages, and ships

The **Golgi apparatus** is the cell's refining, packaging, and shipping center. To continue our analogy of the cell as an industrial city, here is where steel bars are shaped into nails and screws, raw lumber assembled into doors and window frames, and grain turned into bread.

Figure 3.17 diagrams the structure of the Golgi apparatus and the processes that occur there. In cross section the Golgi apparatus appears as a series of interconnected fluid-filled spaces surrounded by membrane, much like a stack of plates. Like the ER, it contains enzymes that further refine the products of the ER into final form.

The contents of the Golgi apparatus move outward by a slow but continuous process. At the outermost layer of the Golgi apparatus, the products are finally ready to be packaged into vesicles and shipped to their final destinations.

Vesicles: Membrane-bound storage and shipping containers

Vesicles are membrane-bound spheres that enclose something within the cell. Sometimes they contain it inside the cell, and sometimes they move it to another location. There are several types of vesicles, each with a different origin and purpose.

Vesicles that ship and store cellular products These vesicles enclose and transport the products of the ER and Golgi apparatus. Each vesicle contains only one product out of the thousands of substances made by the Golgi apparatus.

The contents of each vesicle depend on certain proteins in the vesicle membrane that act as "shipping labels." They determine which product is put into the vesicle and where the vesicle is sent. If a vesicle's products are not immediately needed it remains in the cell cytoplasm, like a box stored in a warehouse awaiting shipment.

Secretory vesicles Secretory vesicles contain products destined for export from the cell. They migrate to the plasma membrane and release their contents outside the cell by exocytosis. Because most secretory products are made in the Golgi apparatus, secretory vesicles generally derive from Golgi apparatus membrane.

Endocytotic vesicles These structures enclose bacteria and raw materials from the extracellular environment. They bring them into the cell by endocytosis.

Peroxisomes and lysosomes These vesicles contain enzymes so powerful that they must be kept within the vesicle to avoid damaging the rest of the cell. Both are produced by the Golgi apparatus. Figure 3.18 shows their functions.

The enzymes in **peroxisomes** destroy various toxic wastes produced in the cell, including hydrogen peroxide (H_2O_2). They also destroy compounds that have entered the cell from outside, such as alcohol. The detoxification process occurs entirely within the peroxisome.

Lysosomes (from the Greek *lysis*, dissolution, and *soma*, body) contain powerful digestive enzymes. Lysosomes fuse with endocytotic vesicles within the cell, digesting bacteria and other large objects. Lysosomes also perform certain housekeeping tasks, such as dissolving and removing damaged mitochondria and other cellular debris. When their

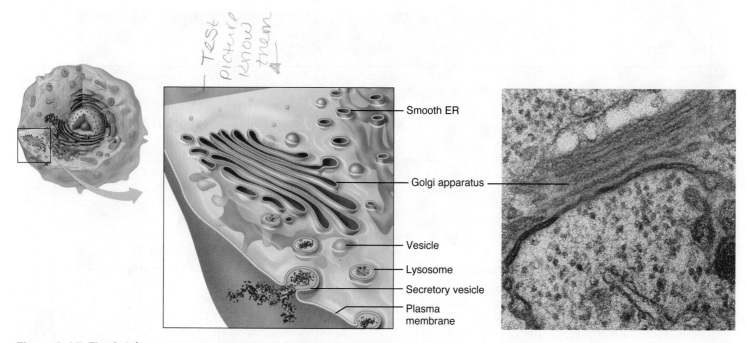

Smooth ER

Golgi apparatus

Vesicle

Lysosome

Secretory vesicle

Plasma membrane

Figure 3.17 The Golgi apparatus. The Golgi apparatus receives substances from the ER, refines them into final products, and packages them into vesicles for their final destinations.

Figure 3.19 shows a photograph of a single mitochondrion and diagrams its structure and function. A smooth outer membrane, similar to the plasma membrane, covers the entire surface. Within the outer membrane is an inner membrane with numerous folds that increase its surface area. The inner membrane and the fluid in its folds contain hundreds of protein enzymes, which serve as catalysts to break down chemical bonds in our food and release the energy. This process consumes oxygen and produces carbon dioxide.

The energy liberated within the mitochondria is used to create high-energy molecules such as ATP. ATP is then exported from the mitochondria to the cytosol, where it is available as a quick source of energy for the cell. Like electric power, ATP is useful for a variety of purposes. We have already seen one of its uses—providing the energy used to transport sodium and potassium across the plasma membrane.

✅ **Quick Check** While studying a human cell under a microscope, you spot a small, round organelle that seems to have a single membrane. Is it more likely to be a nucleus, ribosome, vesicle, or mitochondrion? What might it contain? ▪

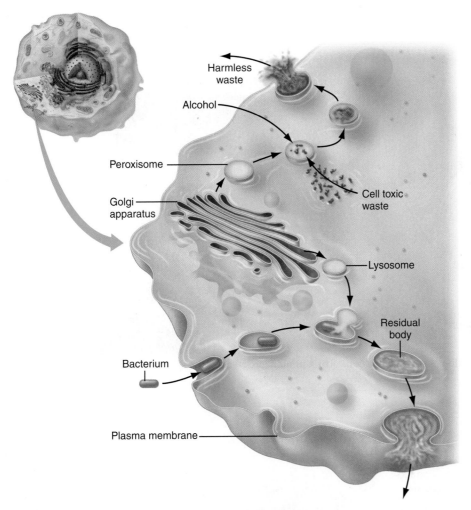

Figure 3.18 Lysosomes and peroxisomes. Lysosomes formed by the Golgi apparatus fuse with endocytotic vesicles containing a bacterium. Following digestion of the bacterium the residual waste is excreted by exocytosis. Peroxisomes take up toxic wastes (including alcohol) and degrade them to harmless waste, which is also excreted.

digestive task is complete they become "residual bodies," analogous to small bags of compacted waste. Residual bodies can be stored in the cell, but usually their contents are eliminated from the cell by exocytosis.

Mitochondria provide energy

Nearly all of a cell's functions require energy. Energy is available in the chemical bonds of the food we eat, but cells cannot use it directly. Most energy in ingested nutrients must be converted to a more usable form before it can power the chemical and physical activities of living cells.

Mitochondria are the organelles responsible for providing most of this usable energy; they are often called the cells' "power plants." Not surprisingly, their number within different cells varies widely according to the energy requirements of the cells. A cell with a high rate of energy consumption, such as a muscle cell, may contain over 1,000 mitochondria.

Fat and glycogen: Sources of energy

The mitochondria generally manufacture ATP as it is needed. To avoid the possibility of running out of fuel, some cells store energy in raw form. These energy stores are not enclosed in any membrane-bound container. They are more like large piles of coal on the ground, awaiting delivery to the power plants (mitochondria) for conversion to electricity (ATP).

Some cells store raw energy as lipids (fat). Our so-called fat cells are so specialized for this purpose that most of their volume consists of large droplets of stored lipids. Dieting and exercise tend to reduce the amount of stored fat—that is, they make the fat cells leaner. However, dieting and exercise do not reduce their number. The cells are available to store fat again, which is why it is so hard to keep lost weight off.

Other cells store energy as glycogen granules (review Figure 2.15). Muscle cells rely on glycogen granules rather than on fat deposits because the energy stored in the chemical bonds of glycogen can be used to produce ATP more quickly than the energy derived from fat.

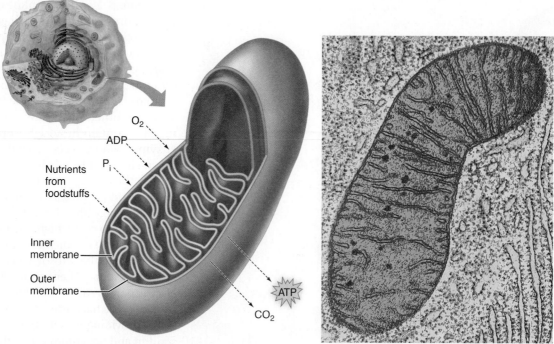

a) **The structure and overall function of a mitochondrion.**

b) **A photomicrograph of a mitochondrion.**

Figure 3.19 **Mitochondria.**

Recap The nucleus contains most of the cell's genetic material. Ribosomes are responsible for protein assembly. The endoplasmic reticulum manufactures most other cellular products in rough form. The Golgi apparatus refines cellular products and packages them into membrane-bound vesicles. Some vesicles store, ship, and secrete cellular products; others digest and remove toxic waste and cellular debris. Mitochondria manufacture ATP for the cell. ■

3.6 Cells have structures for support and movement

The soft plasma membrane is supported by an internal scaffolding that helps the cell maintain its shape. In addition, some cells have specialized structures to help them move around, and all cells contain structures that are involved in moving cellular components during cell division. Structural elements for support and movement include the cytoskeleton, cilia and flagella, and centrioles.

The cytoskeleton supports the cell

The **cytoskeleton** (Figure 3.20) consists of a loosely structured network of fibers called *microtubules* and *microfilaments*. As their names imply, microtubules are tiny hollow tubes, and microfilaments are thin solid fibers. Both are composed of protein. They attach to each other and to

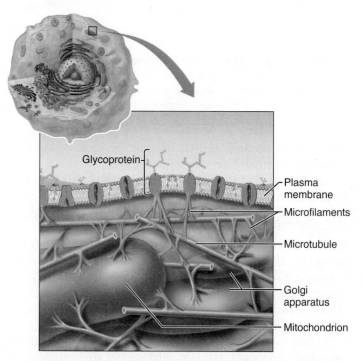

Figure 3.20 **The cytoskeleton.** The cytoskeleton consists of microtubules and microfilaments that attach to and support the cell's organelles and plasma membrane.

proteins in the plasma membrane, called *glycoproteins*, which typically have carbohydrate group components.

The cytoskeleton forms a framework for the soft plasma membrane, much as tent poles support a nylon tent. The cytoskeleton also supports and anchors the other structures within the cell.

Cilia and flagella are specialized for movement

A few cells have hairlike **cilia** (singular: cilium) or longer **flagella** (singular: flagellum) that extend from the surface. Cilia are generally only 2–10 microns long (1 micron equals one-millionth of a meter). In cells that have them, cilia are numerous. Cilia move materials along the surface of a cell with a brushing motion. They are common on the surfaces of cells that line the airways and in certain ducts within the body.

In humans, flagella (approximately 200 microns long) are found only on sperm cells (**Figure 3.21**). The whiplike movement of the flagellum moves the entire sperm cell from one place to another.

Cilia and flagella are similar in structure. They are composed primarily of protein microtubules held together by connecting elements and surrounded by a plasma membrane. Nine pairs of fused microtubules surround two single microtubules in the center. The entire structure bends when temporary linkages form between adjacent pairs of microtubules, causing the pairs to slide past each other. The formation and release of these temporary bonds requires energy in the form of ATP.

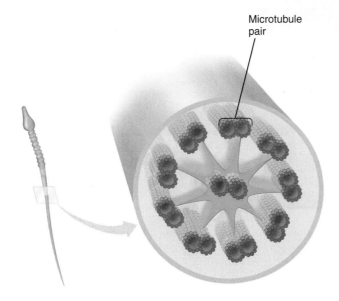

Microtubule pair

Figure 3.21 Flagella. A cross-sectional view of a flagellum showing that it is composed of nine pairs of microtubules surrounding a central pair.

Centrioles are involved in cell division

Centrioles are short, rodlike microtubular structures located near the nucleus. Centrioles are essential to the process of cell division because they participate in aligning and dividing the genetic material of the cell. We discuss them in Chapter 17 when we describe how a cell divides.

> **Recap** The cytoskeleton forms a supportive framework for the cell. Cilia and flagella are specialized for movement, and centrioles are essential to cell division. ∎

3.7 Cells use and transform matter and energy

Living cells can release the energy stored in the chemical bonds of molecules and use it to build, store, and break down still other molecules as required to maintain life. **Metabolism** is the sum of all of the chemical reactions in the organism.

In a single cell, thousands of different chemical reactions are possible at any one time. Some of these chemical reactions are organized as *metabolic pathways* in which one reaction follows after another in orderly and predictable patterns (**Figure 3.22**). Some metabolic pathways are linear, in which the **product** (or end material) from one chemical reaction becomes the **substrate** (starting material) for the next. Other metabolic pathways form a cycle in which substrate molecules enter and product molecules exit, but the basic chemical cycle repeats over and over again.

There are two basic types of metabolic pathways:

1. **Anabolism** (from the Greek *anabole*, a throwing up): molecules are assembled into larger molecules that contain more energy, a process that requires energy. The

a) **A linear pathway.**

b) **A cyclic pathway in which B through E repeat over and over again.**

Figure 3.22 Types of metabolic pathways.

assembly of a protein from many amino acids is an example of an anabolic pathway.

2. **Catabolism** (Greek *katabole*, a throwing down): larger molecules are broken down, a process that releases energy. The breakdown of glucose into water, carbon dioxide, and energy is an example of a catabolic pathway.

Two facts are important about metabolic pathways. First, nearly every chemical reaction requires a specific enzyme. The cell regulates and controls the rates of chemical reactions through the specificity and availability of key enzymes. Second, the metabolic activities of a living cell require a lot of energy. Energy is required for building the complex macromolecules found only in living organisms, such as proteins, DNA, cholesterol, and so on. Energy is also used to power cellular activities such as active transport and muscle contraction.

Cells get their energy by catabolism of molecules that serve as chemical stores of energy. The most immediately useful source of energy, a sort of "energy cash" if you will, is ATP. The energy in ATP is locked in the chemical bond between the second and third phosphate group. Every time the third phosphate group is removed from an ATP molecule, energy is released that the cell can use to do work. The reaction is reversible, meaning that the application of energy to ADP in the presence of a phosphate group can *phosphorylate* (add a phosphate group to) ADP again, re-creating ATP. The equation is written as:

$$ATP \leftrightarrow ADP + P_i + energy$$

In this equation, Pi is used as the abbreviation for the inorganic phosphate (PO_4^{-2}) to distinguish it from the chemical symbol for pure phosphorus (P).

Glucose provides the cell with energy

Cells can use a variety of fuels to make the ATP energy "cash" they need. The most readily available fuel generally is glucose, derived either from food recently eaten or from stored glycogen. However, if glucose is not available, cells may turn to stored fats and even proteins for fuel. Regardless of the fuel used, most of the ATP is produced by very similar metabolic pathways. Let's look at the use of glucose as a fuel source first, and then we'll describe briefly how other fuels are used.

Recall that glucose is a six-carbon sugar molecule with the chemical formula $C_6H_{12}O_6$. This seemingly simple little molecule packs a lot of potential energy in its chemical bonds. Just as a gallon of gas provides enough energy to power your car for 25–30 miles, a single glucose molecule provides the cell enough energy to produce approximately 36 molecules of ATP (Figure 3.23).

The production of ATP from glucose occurs in four stages, summarized in Figure 3.24:

Figure 3.23 **Glucose provides energy for the cell.** The complete catabolism of glucose uses oxygen, produces carbon dioxide and water, and generates 36 molecules of ATP that can be used to do cellular work.

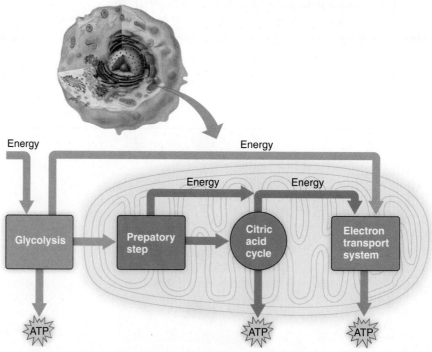

Figure 3.24 **Cellular respiration: An overview.** Glycolysis occurs in the cytoplasm. The products of glycolysis, two molecules of pyruvate, enter mitochondria. The preparatory step and the citric acid cycle result in the complete breakdown of the two pyruvate molecules but only limited ATP production. Most of the ATP is produced in the electron transport system, using energy harvested from the first three steps.

1. *Glycolysis:* The six-carbon glucose molecule is split into two 3-carbon pyruvate molecules. Energy is required to get the process started.
2. *The preparatory step:* In preparation for the citric acid cycle, pyruvate enters a mitochondrion. A series of chemical reactions yields a two-carbon molecule called acetyl CoA, plus some energy.
3. *The citric acid cycle:* An acetyl CoA molecule is broken down completely by mitochondrial enzymes, and its energy is released. Most of the energy is captured by certain high-energy electron transport molecules.
4. *The electron transport system:* Most of the energy derived from the original glucose molecule is used to phosphorylate ADP, producing high-energy ATP.

Glycolysis: Glucose is split into two pyruvate molecules

Figure 3.25 illustrates **glycolysis,** the first stage in the complete breakdown of glucose (*glyco-* means "sweet," referring to sugar, and *-lysis* means "to break"). Glycolyis occurs within the cell's cytoplasm, not within mitochondria. In the first five of the ten chemical reactions that constitute glycolysis, glucose is broken into two 3-carbon molecules called *glyceraldehyde-3-phosphate* (PGAL). Two of these steps require the input of energy (two molecules of ATP)—like putting a match to a bonfire.

Then, each of the two PGAL molecules is converted in a series of steps to *pyruvate.* This process requires several more enzymes. At several of these steps, enzymes carrying high-energy phosphate groups pass their phosphate groups directly to ADP, producing ATP. The process, called *substrate-level phosphorylation,* does not require oxygen. During glycolysis, four molecules of ATP are produced by substrate-level phosphorylation.

The rest of the potentially usable energy from glycolysis is released as high-energy hydrogen ions (H^+) and electrons (e^-). Most of these are picked up by a *coenzyme* called **NAD$^+$** (nicotinamide adenine dinucleotide). A **coenzyme** is a small molecule that assists an enzyme by transporting small molecular groups. In this case NAD$^+$ functions as an energy-carrying molecule. NAD$^+$ picks up one high-energy H^+ ion and two high-energy electrons to become the higher-energy molecule **NADH.**

So far, then, the net energy yield from glycolysis is only two molecules of ATP (four were formed, but two were used to get the process started) and two molecules of NADH. The energy carried by NADH is used to make ATP within the mitochondria, as we'll see later.

The preparatory step: Pyruvate is converted to acetyl CoA

At this point the pyruvate molecules enter the mitochondria, where all the rest of the ATP-generating reactions occur. In a series of steps preparatory to the citric acid cycle, a two-carbon *acetyl group* and a carbon dioxide (CO_2) waste molecule are formed from each pyruvate

Figure 3.25 Glycolysis. This initial breakdown of glucose to two molecules of glyceraldehyde-3-phosphate (PGAL) requires energy. Thereafter, energy is generated as the two PGAL molecules are further degraded to two molecules of pyruvate.

(**Figure 3.26**). Each acetyl group is then joined with another coenzyme (called *coenzyme A*) to form *acetyl CoA,* which delivers the acetyl group to the *citric acid cycle.* There is a net gain of an additional two NADH molecules from the conversion of two pyruvates to two acetyl CoA molecules in the preparatory step.

The citric acid cycle harvests energy

The **citric acid cycle,** also called the Krebs cycle for its discoverer, Hans Krebs, is a series of eight sequential steps in which each acetyl group is completely disassembled to CO_2 waste and

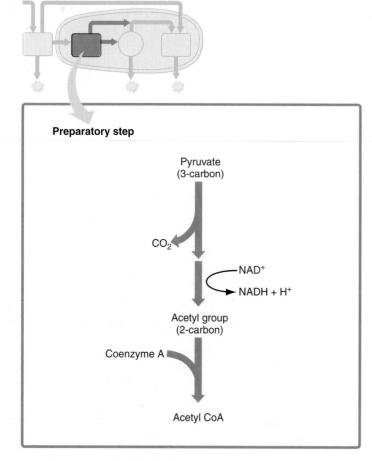

Figure 3.26 The preparatory step. Pyruvate is transported into a mitochondrion and catabolized to a two-carbon acetyl group. Energy is released, and the freed carbon is given off as carbon dioxide waste. The acetyl group combines with coenzyme A for delivery to the next step, the citric acid cycle.

various high-energy products (Figure 3.27). The citric acid cycle begins when acetyl CoA combines with the four-carbon fragment left over from the previous turn of the cycle (oxaloacetic acid) to produce *citric acid*, the six-carbon molecule for which the cycle is named. Citric acid is the starting substrate for seven reactions that end with oxaloacetic acid again. Each reaction is regulated by a different enzyme. The two carbons that are lost during the citric acid cycle are given off as CO_2 waste, and one ATP molecule is produced directly by substrate-level phosphorylation. The high-energy hydrogen ions and electrons removed at various points in the cycle are harvested in the form of three molecules of NADH and one molecule of another energy-carrying coenzyme called **FAD** (flavin adenine dinucleotide). FAD picks up two high-energy H^+ ions and two electrons to become **FADH$_2$.**

The electron transport system produces ATP So far, the glucose molecule has been completely dismantled and CO_2 has been generated as a waste product, but only four new ATP molecules have been generated. The rest of the energy is still in the electrons and hydrogen ions that are part of NADH and FADH$_2$.

At this point NADH and FADH$_2$ move to the inner membrane of the mitochondria and release their cargo to the **electron transport system** (Figure 3.28). Here, the energy-rich electrons are transferred sequentially from one protein carrier molecule to another. The sequential transfer is important because it allows the energy in the electrons to be released in manageable quantities. The sequence of events is as follows:

1. NADH and FADH$_2$ release the H^+ and high-energy electrons they acquired in the citric acid cycle to a carrier protein of the electron transport system.
2. The electrons pass from one protein carrier molecule to the next in the electron transport system.
3. Each time an electron is transferred, the carrier molecule acquires some of its energy and the electron loses energy. The carrier protein uses the energy to transport H^+ from the inner compartment of the mitochondria to the outer compartment.

The active transport of H^+ into the outer compartment of the mitochondria sets the stage for the actual production of ATP. Since the concentration of H^+ is now higher in the outer compartment than in the inner compartment, there is a concentration gradient that favors diffusion of H^+ back to the inner compartment. However, H^+ can diffuse only through special channels. These channels are actually an enzyme called **ATP synthase** that uses the energy derived from the diffusion of H^+ to catalyze the synthesis of ATP from ADP and P_i. Once it is formed, ATP leaves the inner compartment of the mitochondria via a special channel protein, to be used by the cell as an energy source.

As mentioned earlier, *phosphorylation* is the addition of a phosphate group. The process of producing ATP from ADP plus P_i, using the energy obtained as electrons are transferred from one molecule to another in the electron transport system, is called **oxidative phosphorylation.** The term *oxidative* indicates that the process uses oxygen and that electrons have been removed.

By the time they have reached the end of the electron transport system, most of the energy of the electrons and H^+ has been spent. At this point the spent hydrogen ions and electrons combine with oxygen to form water, a waste product. The low-energy NAD^+ and FAD molecules, now lacking hydrogen ions and electrons, are recycled and used again. The ability to recycle NAD^+, FAD, and ADP increases cellular efficiency because it means that these do not need to be synthesized anew each time.

Cellular metabolic processes that use oxygen and produce carbon dioxide in the process of making ATP are collectively termed **cellular respiration.** Cellular respiration takes place entirely within mitochondria. Although some ATP is made in the cytoplasm during glycolysis, glycolysis does not require oxygen. Indeed, glycolysis can proceed without oxygen because there are alternative metabolic

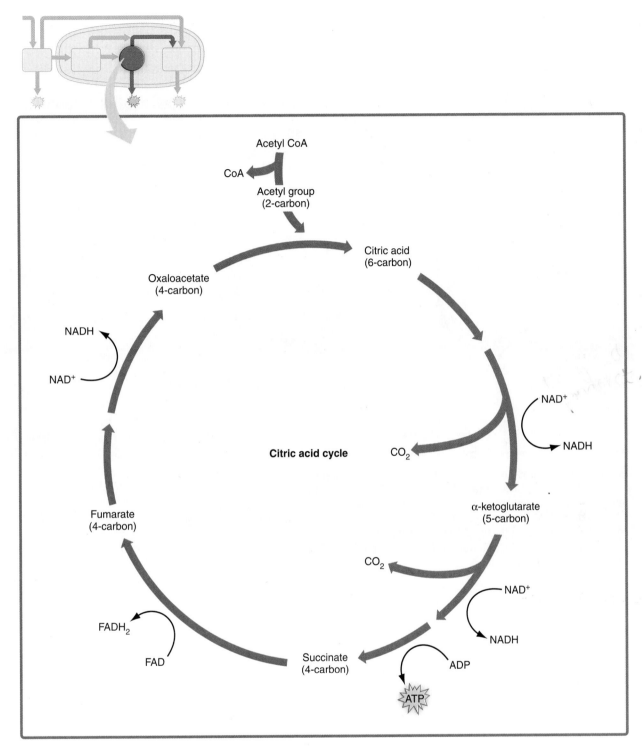

Figure 3.27 The citric acid cycle. A complete turn of the citric acid cycle produces two molecules of CO_2 waste, an ATP molecule, three molecules of NADH, and a molecule of $FADH_2$ for every 2-carbon acetyl group used as fuel. The cycle occurs twice for each molecule of glucose that undergoes glycolysis to produce two pyruvate molecules. The NADH and $FADH_2$ molecules carry high-energy hydrogen ions and electrons to the electron transport system, where their energy can be used to synthesize ATP.

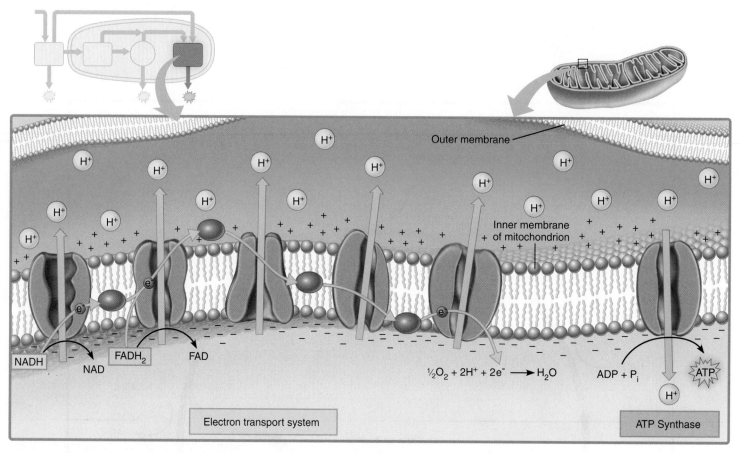

Figure 3.28 The electron transport system and oxidative phosphorylation. During electron transfer, the high-energy molecules of NADH (and FADH$_2$) give up electrons and hydrogen ions, releasing energy. The energy is used to transport hydrogen ions into the space between the two mitochondrial membranes. Diffusion of hydrogen ions back into the inner mitochondrial space provides the energy for the enzyme ATP synthase to synthesize ATP from ADP and inorganic phosphate (P$_i$). The process of using energy derived from the electron transport system to produce ATP by phosphorylation of ADP is called oxidative phosphorylation.

☑ *Explain in your own words what exactly is causing the ATP synthase to make ATP. Do NADH or FADH$_2$ interact directly with the ATP synthase?*

pathways that pyruvate can take besides entering the preparatory step and citric acid cycle. In contrast, the citric acid cycle *is* considered to be part of cellular respiration, because although it does not use oxygen directly, it would soon come to a complete halt in the absence of oxygen.

☑ **Quick Check** If ATP synthase completely stopped working, could glycolysis or the citric acid cycle still produce any ATP? Explain. ▪

Summary of energy production from glucose The most effective way to harvest the energy in any fuel, whether gasoline or glucose, is to release the energy *slowly*, under carefully controlled conditions. Holding a match to a gallon of gasoline results in a useless fire or even a

potentially dangerous explosion, but burning it drop by drop in the pistons of your car allows the energy to be converted to mechanical work. Similarly, in a living cell, the whole point of glycolysis and cellular respiration is to release the energy in the chemical bonds of glucose *slowly* so that the energy can be harvested effectively.

That is exactly what the cell does (**Figure 3.29**). The complete breakdown of glucose during glycolysis, the preparatory step, and the citric acid cycle are accomplished by a sequence of over 20 different chemical reactions, each controlled by an enzyme. The net yield of high-energy molecules (prior to the electron transport chain) is four molecules of ATP, ten molecules of NADH, and two molecules of FADH$_2$. Each NADH carries sufficient energy to produce about three ATP molecules in the electron transport system. FADH$_2$ carries less energy, enters the electron transport chain at a lower

a) **Most of the ATP generated during cellular respiration is synthesized in the electron transport system.**

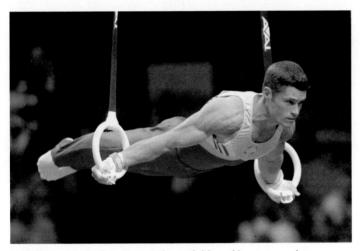

b) **Cellular respiration powers the activities of humans and many other organisms.**

Figure 3.29 **Cellular respiration: A recap.**

energy point, and so produces only about two ATP molecules per $FADH_2$ molecule. Add it all up and you should get about 38 ATP molecules per glucose molecule.

But there's a catch. Two of the NADH (the two produced during glycolysis) were produced in the cytoplasm. Yet the electron transport system utilizes NADH located in the inner compartment of the mitochondria. Since the inner mitochondrial membrane is impermeable to NADH, the cytosolic NADH molecules release their H^+ ions and electrons, which are then transported across by transport proteins. Once inside, they may be picked up again by other NAD^+

molecules. But every transfer and active transport results in a small energy loss—in this case it's the equivalent of about two ATP molecules. Therefore, the net maximum yield of energy from a single glucose molecule is closer to 36 ATP molecules. In fact it may be even lower than that because some of the energy may be used directly to do other work for the cell, but for the sake of simplicity we'll settle on a number of 36.

HBP **Web Animation** *Breaking Down Glucose for Energy* at www.humanbiology.com

Fats and proteins are additional energy sources

So far we have concentrated on the cellular catabolism of glucose. Normally the blood glucose concentration remains fairly constant even between meals because glycogen (the storage form of glucose in humans) is constantly being catabolized to replenish the glucose that is used by cells.

However, most of the body's energy reserves do not take the form of glycogen. In fact, the body stores only about 1% of its total energy reserves as glycogen; about 78% are stored as fats and 21% as protein. After glycogen, our bodies may utilize fats and proteins. Figure 3.30 illustrates the metabolic pathways for fat, glycogen, and protein.

Energy is constantly being transferred into and out of the body. Immediately after a meal, when plenty of glucose, lipids, and amino acids are readily available in the blood, we tend to use primarily glucose as an energy

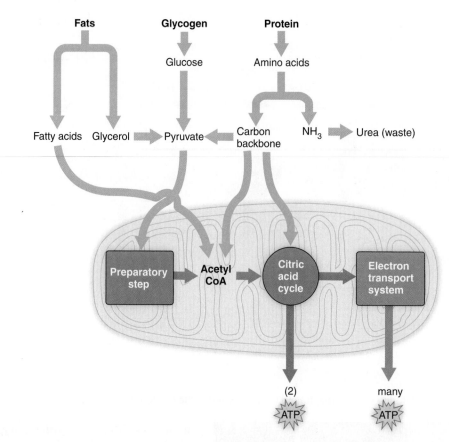

Figure 3.30 Metabolic pathways for fats, glycogen, and proteins as sources of cellular energy. All three sources produce pyruvate and acetyl groups. In addition, a few components of protein can enter the citric acid cycle directly.

source. When we eat more calories than we can use immediately, some of the excess energy goes to replenish the body's stores of glycogen and the rest is converted to fat and stored in fat tissue. After we have not eaten for hours, the body uses glycogen and eventually fats and even proteins for energy.

Pound for pound, fats carry more than twice the energy of carbohydrates such as glycogen. During fat catabolism, triglycerides (the storage form of fat) are first broken down to glycerol and fatty acids. The glycerol can be converted to glucose in the liver or it can be converted to pyruvic acid, which then enters the citric acid cycle. Enzymes break down the fatty acid tails to two-carbon acetyl groups, which also enter the citric acid cycle. Each molecule of triglyceride yields a great deal of ATP because the fatty acid tails are generally 16–18 carbons long and so generate many acetyl groups.

Proteins carry about the same amount of energy per pound as glycogen. Proteins are first broken down to amino acids, and the amine group (—NH$_2$) of each amino acid is removed. The carbon backbones then enter the citric acid cycle at various points, depending on the specific amino acid. The amine groups are converted to urea by the liver and then excreted in urine as waste.

Proteins serve primarily as enzymes and structural components of the body, not as a stored form of energy. Nevertheless, protein catabolism increases significantly during starvation. Prolonged protein catabolism causes muscle wasting, but at least it keeps the individual alive.

✓ **Quick Check** When people die from starvation, the actual cause of death is often a sudden heart attack. What causes the heart attack? ■

Anaerobic pathways make energy available without oxygen

As we have seen, cellular respiration requires oxygen to complete the chemical reactions of the citric acid cycle and the electron transport chain. However, a small amount of ATP *can* be made in humans by anaerobic metabolism (without oxygen) for at least brief periods of time. Glycolysis, for example, is an anaerobic metabolic pathway. In the absence of oxygen, glucose is broken down to pyruvate (glycolysis occurs), but then the pyruvate cannot proceed through the citric acid cycle and electron transport chain (**Figure 3.31**). Instead, the pyruvate is converted to *lactic acid*. The buildup of lactic acid is what causes the burning sensation and cramps associated with muscle fatigue when not enough oxygen is available to muscle tissue. When oxygen becomes available again, the lactic acid is metabolized by aerobic pathways.

Because glycolysis is the only step that can occur without oxygen, glucose (and glucose derived from glycogen) is the only fuel that can be used under anaerobic conditions. The amounts of ATP are very limited, however; only two molecules of ATP are produced per molecule of glucose instead of the usual 36.

Recap Metabolism refers to all of a cell's chemical processes. Metabolic pathways either create molecules and use energy (anabolism) or break them down and liberate energy (catabolism). The primary source of energy for a cell is ATP, produced within mitochondria by the complete breakdown of glucose to CO_2 and water. The process requires oxygen. A single molecule of glucose yields about 36 molecules of ATP. Fats and proteins can also be used to produce energy if necessary. ■

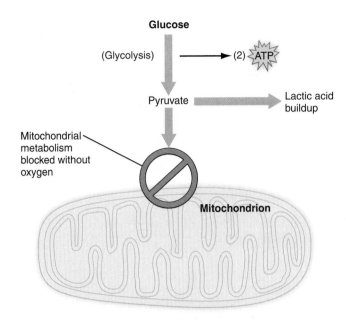

Figure 3.31 Anaerobic metabolism. In the absence of oxygen, glycolysis is the only ATP-producing step available. Glycolysis without oxygen results in lactic acid buildup.

Chapter Summary

Cells are classified according to their internal organization *p. 52*

- All cells have a plasma membrane that surrounds and encloses the cytoplasm.
- Eukaryotic cells have a nucleus.

Cell structure reflects cell function *p. 53*

- Limits to cell size are imposed by the mathematical relationship between cell volume and cell surface area.
- Various types of microscopes with magnifications up to 100,000-fold enable us to visualize cells and their structures.

A plasma membrane surrounds the cell *p. 55*

- The plasma membrane is a bilayer of phospholipids that also contains cholesterol and various proteins.

Molecules cross the plasma membrane in several ways *p. 57*

- Some molecules are transported across the plasma membrane passively (by diffusion), whereas others are transported by active processes requiring the expenditure of energy.
- Receptor proteins transfer information across the plasma membrane.
- The sodium-potassium pump is a plasma membrane protein with a critical role in the maintenance of cell volume.

Internal structures carry out specific functions *p. 63*

- The nucleus directs all of the cell's activities.
- Ribosomes, the endoplasmic reticulum, and the Golgi apparatus participate in the synthesis of life's molecules.

- Vesicles are membrane-bound spheres that transport, store, and ship cellular products and toxic or dangerous materials.
- Mitochondria make energy available for the cell in the form of the high-energy molecule ATP.

Cells have structures for support and movement *p. 68*

- A cytoskeleton of microtubules and microfilaments serves as structural support and anchors the various organelles.
- Cilia and flagella provide for movement in certain types of cells. Both cilia and flagella are made of pairs of protein microtubules.

Cells use and transform matter and energy *p. 69*

- The creation and destruction of molecules either requires energy or liberates energy.
- The most readily useful form of energy for cells is ATP.
- The production of ATP from glucose requires four consecutive stages: glycolysis, a preparatory step, the citric acid cycle, and the electron transport system.
- Cells can utilize glycogen, fats, or proteins for energy.
- Only a small amount of ATP can be made in the absence of oxygen.

Terms You Should Know

active transport, *59*
ATP synthase, *72*
citric acid cycle, *71*

cytoskeleton, *68*
diffusion, *57*
electron transport system, *72*

endocytosis/exocytosis, 60
endoplasmic reticulum (ER), 65
glycolysis, 71
Golgi apparatus, 66
metabolism, 69
mitochondria, 67
$NAD^+/NADH$, 71

osmosis, 58
passive transport, 57
plasma membrane, 52
ribosome, 63
sodium-potassium pump, 60
vesicles, 60

Concept Review

Answers can be found at the Human Biology Place.
www.humanbiology.com

1. Explain why being small is advantageous to a cell.
2. List the basic tenets of the *cell doctrine*.
3. Describe how phospholipids are oriented in the plasma membrane and why they orient naturally that way.
4. Define *passive transport* and name the three passive transport methods that are used to transport different molecules across the plasma membrane.
5. Compare and contrast endocytosis and exocytosis.
6. Describe the activity of the sodium-potassium pump and indicate its importance to the cell.
7. Explain what happens to cells placed in a high-salt or a low-salt environment, and why.
8. Define *vesicles* and name at least two different types of vesicles.
9. What are the four stages of ATP production from glucose, and which one yields the most ATP?
10. Describe what happens to a cell's ability to produce ATP when oxygen is not available.

Test Yourself

Answers can be found in Appendix A.

1. Which of the following adaptations would increase the surface area of a cell?
   a. increased number of mitochondria
   b. increased number of ribosomes
   c. presence of microvilli
   d. increased number of channel proteins
   e. changing a cell shape from cubical to spherical
2. Which of the following would be most likely to enter a cell by diffusion directly through the phospholipid bilayer?
   a. potassium ions
   b. steroid hormone
   c. hydrogen ions
   d. glucose
   e. sodium ions
3. Cells transport sodium ions out of the cell against the sodium concentration gradient. This is an example of:
   a. facilitated diffusion
   b. simple diffusion
   c. diffusion via channel proteins
   d. endocytosis
   e. active transport
4. Red blood cells placed into distilled water will:
   a. swell as water moves into the cells by osmosis
   b. shrink as sodium moves out of the cells by diffusion
   c. swell as water moves into the cells by active transport
   d. shrink as proteins move out of the cell by diffusion
   e. remain unchanged because of homeostatic mechanisms
5. Which organelles are most active during vigorous exercise?
   a. ribosomes
   b. endoplasmic reticulum
   c. mitochondria
   d. lysosomes
   e. cilia
6. Phagocytic white blood cells engulf and digest bacteria and cellular debris. Which organelles would be most involved in the digestion of the engulfed material?
   a. mitochondria
   b. lysosomes
   c. Golgi apparatus
   d. ribosomes
   e. endoplasmic reticulum
7. Some lymphocytes (white blood cells) synthesize and secrete defensive proteins known as antibodies. Which of the following represents the most likely path of these proteins from synthesis to secretion?
   a. endoplasmic reticulum—ribosomes—Golgi apparatus—vesicles—plasma membrane
   b. ribosomes—Golgi apparatus—vesicles—endoplasmic reticulum—plasma membrane
   c. ribosomes—Golgi apparatus—lysosome—endoplasmic reticulum—plasma membrane
   d. ribosomes—endoplasmic reticulum—Golgi apparatus—vesicles—plasma membrane
   e. ribosomes—lysosomes—endoplasmic reticulum—vesicles—Golgi apparatus
8. Which organelles would be active in liver cells that are detoxifying alcohol?
   a. Golgi apparatus
   b. lysosomes
   c. mitochondria
   d. endoplasmic reticulum
   e. peroxisomes
9. Cells lining the respiratory passages have numerous filamentous structures that sweep mucus and debris up and away from the lungs. These filamentous structures are:
   a. microtubules
   b. flagella
   c. cilia
   d. microfilaments
   e. microvilli
10. Which of the following is/are the most immediate source of energy for cellular work?
    a. glucose
    b. ATP
    c. glycogen

d. triglycerides

e. amino acid

11. All of the following cellular activities require energy except:

a. facilitated transport

b. active transport

c. movement

d. protein synthesis

e. cell division

12. Which of the following can occur within a cell in the absence of oxygen?

a. glycolysis

b. lactate production

c. citric acid cycle

d. electron transport system

e. both (a) and (b)

13. In which stage of cell respiration does oxygen play a role?

a. glycolysis

b. electron transport system

c. citric acid cycle

d. lactate production

e. transport of pyruvate into the mitochondria

14. Which of the following is/are recycled during cellular respiration and do not appear in the net equation?

a. NAD^+ and NADH

b. ATP, ADP, and P_i

c. glucose

d. carbon dioxide

e. oxygen

15. Most of the ATP produced during cell respiration is produced during:

a. glycolysis

b. citric acid cycle

c. electron transport system

d. lactate production

e. anaerobic pathways

Apply What You Know

Answers can be found at the Human Biology Place.
www.humanbiology.com

1. Imagine that you are shown two cells under the microscope. One is small, has lots of mitochondria, and contains numerous glycogen granules. The other is somewhat larger and has only a few mitochondria and no glycogen granules. Which cell do you think is more metabolically active? Explain your reasoning.

2. The sodium-potassium pump is a large protein molecule. Where do you think the sodium-potassium pumps are made in the cell, and how do you think they become inserted into the lipid bilayer of the plasma membrane?

3. Mitochondria resemble a bacterial cell in a number of ways. Some scientists hypothesize that mitochondria evolved from aerobic prokaryotes that were engulfed by anaerobic eukaryotes, and now both have evolved together in a mutually advantageous way. Can you think of an explanation for why it might have been advantageous for both cells to enter into such an arrangement?

4. You have decided that you need to lose a little weight. You have heard a lot about no-carbohydrate and low-carbohydrate diets, and you have decided to use one of these diet plans. Explain how a low-carbohydrate diet works. Can you think of any possible negative side effects of such a diet?

5. Recently a young man from Derby in the United Kingdom entered a contest and drank 26 pints of water in a very short time. He later died of complications due to hypotonic hydration, also known as water intoxication. How were his body's cells affected by the excess water, and how might that have contributed to his death?

6. You have been selected to serve on a jury for a trial involving a young man accused of public intoxication. His defense attorney argues that the alcohol found in his system was the result of natural fermentation; that he had just finished a grueling one-hour workout during which his body could not meet the oxygen demand, and that the excess lactic acid that was produced during the exercise was then converted to alcohol by the process of lactic acid fermentation. Should you believe the defense attorney? Explain why or why not.

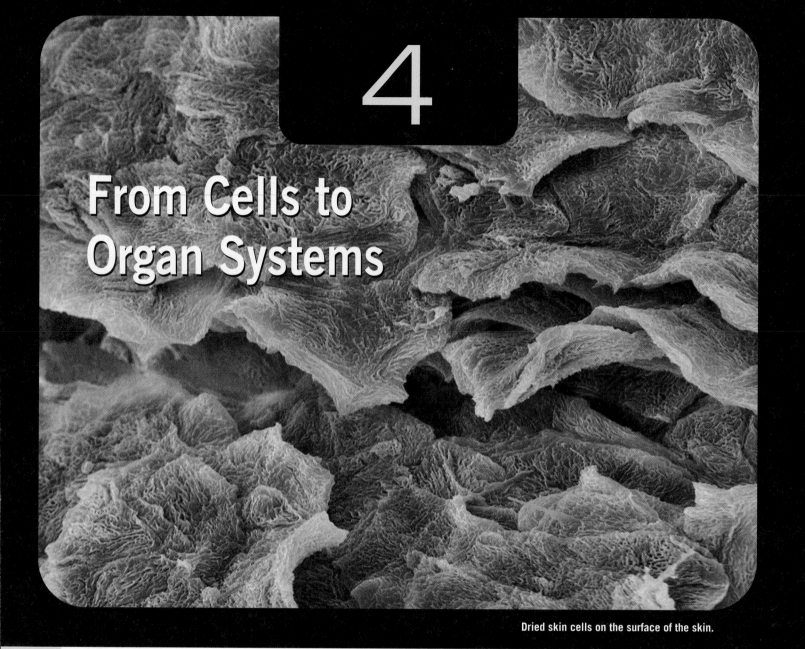

4

From Cells to Organ Systems

Dried skin cells on the surface of the skin.

Can Lipodissolve Melt Away Fat?

By now you are familiar with the chemistry of lipids (Chapter 3) and the structure and function of fat cells (this chapter). You understand that when your average daily caloric intake exceeds your average daily caloric expenditure, you gain weight. And yet all this knowledge hasn't helped, because you still have those annoying fat deposits in places where you don't want them! Perhaps you've tried dieting and exercise and still aren't happy with your body shape.

What if you could just melt away that unwanted fat? What if you could kill some of those pesky fat cells once

A human fat cell (× 3,000) filled with lipid droplets (yellow).

and for all? What if you could sculpt your body into perfect shape without ever having to diet or exercise again? It's a fantasy many of us have indulged in, and it's one of the reasons for the meteoric rise of a controversial cosmetic procedure called *Lipodissolve*.

What Is Lipodissolve?

Lipodissolve, also called injection lipolysis, is described by its promoters as a safe, effective, nonsurgical way to sculpt the body into perfect shape. Usually the technique is performed as a series of six injections two weeks apart directly into

Lipodissolve is injected directly into areas of excess fat.

subcutaneous fat deposits. The injections contain two active ingredients: phosphatidylcholine (a phospholipid) and deoxycholate (a bile acid). The combination of these two drugs, known as PCDC (for phosphotidylcholine deoxycholate), allegedly works by dissolving the bonds between the three fatty acids and the glycerol backbone that comprise a triglyceride molecule (see Chapter 3), the primary molecular form of stored fat in the body. PCDC also is thought to disrupt the cell membranes of fat cells in the vicinity of an injection, resulting in death of the fat cells themselves. Some Lipodissolve cocktails also include vitamins and plant extracts.

And at the moment, the safety and effectiveness of Lipodissolve are not backed by scientific data. The Food and Drug Administration (FDA) has called Lipodissolve just another example of "unapproved drugs for unapproved uses." Nevertheless Lipodissolve clinics can legally administer Lipodissolve because of a loophole in our drug regulatory laws. According to the law, a licensed doctor can legally and ethically prescribe compounded drugs (drugs made from more than one ingredient) specifically for individual patients, as long as all of the *ingredients* are approved by the FDA. The intent is to protect the sanctity of the doctor/patient relationship, allowing the doctor to choose what is best for that particular patient. Both of the compounds in a Lipodissolve injection have been approved by the FDA as ingredients in other drugs,

though neither was approved as a fat-dissolving drug.

Health clinics and spas have been quick to capitalize on the loophole, for at prices ranging from $400 to $1,500 per Lipodissolve treatment they stand to make a lot of money. Most patients who try the technique eventually sign up for treatment of several body parts. Women generally request treatment of their thighs and abdomen, whereas men are more likely to choose jaw lines and love handles. One of the distinct advantages of the technique is that because it does not involve surgery there is no recovery period. Not surprisingly, there are persistent rumors on various celebrity-watch Web sites that certain celebrities have tried the procedure or are thinking about it.

One of the first companies to offer the Lipodissolve technique in this country was Advanced LipoDissolve Center, later renamed fig. (short for figure). Fig.'s "dissolve to your beautiful shape" advertising campaign was so successful that the company grew to 18 body-shaping centers in eight states in just two years. The company went out of business in late 2007, but there are still plenty of other clinics and spas willing to perform the Lipodissolve procedure.

Is Lipodissolve Safe and Effective?

Critics of the Lipodissolve technique argue that the current law covering single doctor/patient relationships was never intended to allow mass marketing by chains of clinics or spas, or mass compounding of the PCDC drug combination. The American Society of Plastic Surgeons, The American Society of Aesthetic Plastic Surgery, and the American Society of Dermatologic Surgery have issued cautionary warnings against the technique, but that's understandable since all three societies represent surgeons who might be financially affected by Lipodissolve's increasing popularity. Lipodissolve proponents, on the other hand, counter by pointing out that there have been only anecdotal reports of minor unwanted side effects

associated with the technique, such as swelling, skin blistering, pain, and blackened skin in some patients. No deaths have been reported. However, because the technique has been performed in this country only since 2004, the long-term consequences of Lipodissolve therapy are still unknown. Some scientists worry that disruption of fat cell membranes at the site of injection might cause a sharp rise in cholesterol in the blood. After all, cholesterol is a normal constituent of cell membranes. In addition, little is known about how the Lipodissolve drugs are metabolized or removed from the body. Nor is it known whether they enter the circulatory system in sufficient quantities to affect other organs or tissues far from the injection site.

Does Lipodissolve work? The jury is still out on this one. Some patients see an improvement in body shape after Lipodissolve, others do not. One possible reason for differences in effectiveness between patients may be that there is no standard dose of Lipodissolve. The procedure is not regulated by the FDA, so physicians are free to try any combination of doses, injection intervals, and treatment sites they want.

Faced with the awkward and potentially dangerous situation of having patients flock to an unregulated procedure, the FDA decided to take action. Late in 2007 the FDA approved the first clinical trial of Lipodissolve—a double-blind, placebo-controlled prospective study that will follow patients for up to 46 weeks. The trial, to be conducted by the research arm of the American Society for Aesthetic Plastic Surgery, is expected to provide the first scientifically defensible data on the efficacy and safety of Lipodissolve. The results may not be known for several more years. In the meantime the FDA is keeping an eye on the situation. Consumers would be wise to remain skeptical about this procedure until more is known about it.

Questions to consider

1 Health clinics sometimes cite retrospective studies to support their claim that the Lipodissolve technique is safe. What is a "retrospective study"? How is it different from a controlled study?

2 How important is FDA approval and/or scientific evidence of safety to you? If someone were to pay for a Lipodissolve procedure for you, would you try it?

The facts...

- A nonsurgical cosmetic procedure for "melting away fat" is being heavily promoted by health clinics and spas.

- The procedure, called *Lipodissolve*, involves a series of injections of fat-dissolving drugs directly into local fat deposits.

- Lipodissolve is not approved as a fat-reducing therapy by the FDA. Its safety and efficacy are not yet backed by scientific evidence.

>> **In multicellular organisms, cells have special- ized functions.** These functions evolved (along with multicellularity) because they benefit the entire organism.

>> **Groups of cells with a common function are called tissues.** The four main tissue types are epithelial, connective, muscle, and nervous.

>> **Organs and organ systems each perform one or more essential complex functions for the organism.** Humans have 11 different organ systems; examples are the male and female reproductive systems (reproduction), skeletal system (structural support), and muscular system (movement).

>> **Multicellular organisms must maintain homeostasis** (constancy) of their internal environments. The maintenance of a constant internal environment compatible with life allows each living cell, regardless of its specialized function, to live far away from the external environment.

>> **Homeostasis is maintained by negative feedback control systems.** In negative feedback systems, any deviation from the desired condition is detected and corrected.

For nearly two-thirds of the history of life, or more than two billion years, all organisms consisted of just one cell. There are still plenty of single-celled organisms today; in fact they far outnumber multicellular organisms. Theirs is a simple, uncomplicated life. They get their raw materials and energy from the fluid in which they are bathed, they dump their wastes into that same fluid, and they reproduce by dividing in two.

There are, however, disadvantages to being a single cell. The single-celled organism is completely at the mercy of its immediate external environment for every require- ment of life. If the pond in which it lives dries up, it will die. If salt levels in the water rise, if the temperature gets too hot, or if its food runs out, it dies. There must be another way!

There is another way, and that is for cells to join together. In this chapter we look at how cells are organized into the tissues, organs, and organ systems that make up your body. We consider the structure and function of your skin as an example of an organ system. And we discuss how your cells, tissues, and organs work together to maintain the health and stability of your body.

4.1 Tissues are groups of cells with a common function

A *multicellular organism* consists of many cells that collectively share the functions of life. Advantages to multicellularity in- clude greater size (the better to eat, rather than be eaten) and the ability to seek out or maintain an environment conducive to life.

All cells in a multicellular organism have a specialized function that benefits the organism in some way. However, specialization is not enough. The specialized functions must be organized and integrated if they are to be useful. As an example, the activity of a single cell in your heart is insignificant because the cell is so small. The beating of your heart requires that hundreds of thousands of such cells be arranged end to end, so that their functions are coordinated to produce a single heartbeat.

Tissues are groups of specialized cells that are similar in structure and that perform common functions. There are four major types of tissues: epithelial, connective, muscle, and nervous.

4.2 Epithelial tissues cover body surfaces and cavities

Most **epithelial tissues** consist of sheets of cells that line or cover various surfaces and body cavities. Two epithelial tissues you know about are your skin and the lining of your mouth. Other epithelial tissues line the inner surfaces of your digestive tract, lungs, bladder, blood vessels, and the tubules of your kidneys.

Epithelial tissues are more than just linings. They protect underlying tissues. Often they are smooth to reduce friction; the smooth epithelial tissue lining your blood vessels helps blood flow more easily through your body, for instance. Some are highly specialized for transporting mate- rials. Epithelial tissues (and cells) absorb water and nutrients across your intestines into your blood. They also secrete waste products across the tubules of your kidneys so that you can eliminate them in urine.

A few epithelial tissues are *glandular epithelia* that form the body's glands. **Glands** are epithelial tissues that are specialized to synthesize and secrete a product. **Exocrine glands** (*exo-* means "outside" or "outward") se- crete their products into a hollow organ or duct. Examples of exocrine glands are the glands in your mouth that secrete saliva, sweat glands in your skin, and glands in your stomach that produce digestive acid. **Endocrine glands** (*endo-* means "within") secrete substances called *hormones* into the bloodstream. One endocrine gland is the thyroid gland, which secretes several hormones that help regulate your body's growth and metabolism. We describe various glands throughout the book where appropriate.

Epithelial tissues are classified according to cell shape

Biologists classify epithelial tissues into three types according to the shapes of the cells (**Figure 4.1**):

■ *Squamous epithelium* consists of one or more layers of flattened cells. (*Squama* means "platelike." Think of squamous epithelium as "squashed flat.") Squamous epithelium forms the outer surface of the skin and lines the inner surfaces of the blood vessels, lungs, mouth and throat, and vagina.

■ *Cuboidal epithelium* is composed of cube-shaped cells. Cuboidal epithelium forms the kidney tubules and also covers the surfaces of the ovaries.

■ *Columnar epithelium* is composed of tall, rectangular (column-shaped) cells. Columnar epithelium lines parts of the digestive tract, certain reproductive organs, and the larynx. Certain cells within columnar epithelium, called *goblet cells*, secrete mucus, a thick fluid that lubricates the tissues and traps bacteria, viruses, and irritating particles.

Epithelial tissues are classified not only by shape but also by the number of cell layers in the tissue. A *simple epithelium* is

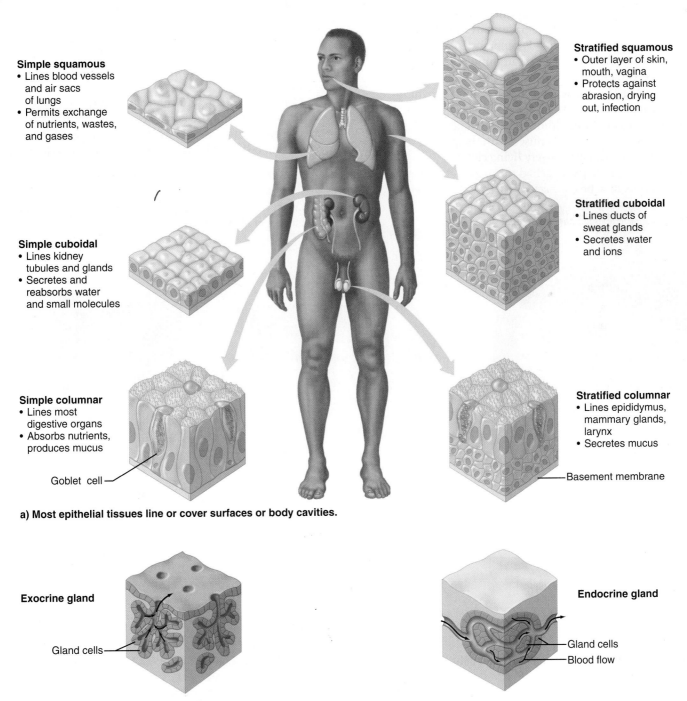

Simple squamous
• Lines blood vessels and air sacs of lungs
• Permits exchange of nutrients, wastes, and gases

Simple cuboidal
• Lines kidney tubules and glands
• Secretes and reabsorbs water and small molecules

Simple columnar
• Lines most digestive organs
• Absorbs nutrients, produces mucus

Goblet cell

Stratified squamous
• Outer layer of skin, mouth, vagina
• Protects against abrasion, drying out, infection

Stratified cuboidal
• Lines ducts of sweat glands
• Secretes water and ions

Stratified columnar
• Lines epididymus, mammary glands, larynx
• Secretes mucus

Basement membrane

a) Most epithelial tissues line or cover surfaces or body cavities.

Exocrine gland

Gland cells

Endocrine gland

Gland cells
Blood flow

b) Glandular epithelia secrete a product.

Figure 4.1 Types of epithelial tissues.

a single layer of cells, whereas a *stratified epithelium* consists of multiple layers (or strata). Simple epithelium is so thin that molecules can pass through it easily. Stratified epithelium is thicker and provides protection for underlying cells.

✅ **Quick Check** What sort of epithelium would you expect to find lining a part of the digestive tract that absorbs food molecules and also secretes mucus? Explain your answer. ▪

The basement membrane provides structural support

Directly beneath the cells of an epithelial tissue is a supporting noncellular layer called the **basement membrane** (see Figure 4.1a), and beneath that is generally a layer of connective tissue (described later). You can think of the basement membrane as the mortar that anchors the cells to the stronger connective tissue underneath. The basement membrane is composed primarily of protein secreted by the epithelial cells, and thus although noncellular, it is a cellular product. It should not be confused with the plasma membrane that is a part of every living cell.

In addition to being attached to a basement membrane, epithelial cells may be connected to each other by several

Answers to ✅ questions can be found in Appendix A.

types of **cell junctions** made up of various proteins. Three different types of junctions may hold the cells together, depending on the type of epithelial tissue (**Figure 4.2**):

- *Tight junctions* seal the plasma membranes of adjacent cells so tightly together that nothing can pass between the cells. Tight junctions are particularly important in epithelial layers that must control the movement of substances into or out of the body. Examples include the cells that line the digestive tract (which bring in nutrients) and the bladder (which stores urine), and the cells that form the tubules of the kidneys (which remove waste products from the body).
- *Adhesion junctions,* sometimes called "spot desmosomes," are looser in structure. The protein filaments of adhesion junctions allow for some movement between cells so that the tissues can stretch and bend. Adhesion junctions in the epithelium of your skin, for instance, allow you to move freely.
- *Gap junctions* represent connecting channels made of proteins that permit the movement of ions or water between two adjacent cells. They are commonly found in the epithelial cells in the liver, heart, and some muscle tissues.

↩ **Recap** Epithelial tissues line body surfaces and cavities, and form glands. They are classified according to cell shape (squamous, cuboidal, or columnar) and the number of cell layers (simple or stratified). ▪

a) **Tight junctions form leak-proof seals between cells.**

- Tight junction proteins
- Intercellular space

b) **Adhesion junctions anchor two cells together, yet allow flexibility of movement.**

- Protein filaments
- Intercellular space

c) **Gap junctions provide for the direct transfer of water and ions between adjacent cells.**

- Protein channel
- Intercellular space

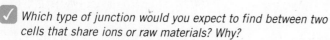

Figure 4.2 **Examples of junctions between cells.** Only one type of junction is generally present in any given tissue.

✅ *Which type of junction would you expect to find between two cells that share ions or raw materials? Why?*

MJ's Human Biology Blog

My Mother's Cells Within Me

Nearly all of us harbor a few cells that come from a close relative. The phenomenon, called microchimerism, occurs because the placenta is not a perfect barrier to formed cells—sometimes maternal cells make their way into the fetus, and vice versa. Apparently some of these cells live indefinitely in their new host, which is surprising since foreign cells are usually attacked and killed.

Scientists are still working out what the foreign cells may be doing in their host. In some cases they may differentiate into fully functional tissue cells in the host. For example, genetically female heart cells (presumably from the mother) have been found in males. In other cases these foreign cells may trigger immune disorders later in life, when the immune system of the host finally recognizes and attacks the foreign cells. Diseases that may have a microchimerism link include several autoimmune inflammatory diseases of connective tissue (scleroderma, lupus erythematosus, and rheumatoid arthritis), and perhaps even Type 1 diabetes, a disease characterized by poor regulation of blood sugar. ■

Reference: Nelson, J. Lee. Your Cells are My Cells. *Scientific American* Feb. 2008, pp. 72–79.

4.3 Connective tissue supports and connects body parts

Connective tissue supports the softer organs of the body against gravity and connects the parts of the body together. It also stores fat and produces the cells of blood.

Unlike epithelial tissue, most connective tissues have few living cells. Most of their structure consists of nonliving extracellular material, the *matrix,* that is synthesized by connective tissue cells and released into the space between them. The strength of connective tissue comes from the matrix, not from the living cells themselves. The few living cells rarely make contact with each other, and so direct cell-to-cell junctions are not present.

Connective tissues are so diverse that any classification system is really a matter of convenience (Table 4.1). Broadly, we can divide them into fibrous and specialized connective tissues.

Fibrous connective tissues provide strength and elasticity

Fibrous connective tissues connect various body parts, providing strength, support, and flexibility. Figure 4.3 shows the structural elements of fibrous connective tissue.

As indicated by their name, fibrous connective tissues consist of several types of fibers and cells embedded in a gel-like ground substance. **Collagen fibers,** made of protein, confer strength and are slightly flexible. Most fibrous connective tissues also contain thinner coiled **elastic fibers**, made primarily of the protein *elastin,* which can stretch without breaking. Some fibrous connective tissue also contains

Table 4.1 **Types of connective tissues**

Type	Structure	Attributes	Locations
Fibrous Connective Tissue			
Loose	Mostly collagen and elastin fibers in no particular pattern; more ground substance	Flexible but only moderately strong	Surrounds internal organs, muscles, blood vessels
Dense	Mostly collagen in a parallel arrangement of fibers; less ground substance	Strong	In tendons, ligaments, and the lower layers of skin
Elastic	High proportion of elastin fibers	Stretches and recoils easily	Surrounds hollow organs that change shape or size regularly
Reticular (lymphoid)	Mostly thin, interconnecting reticular fibers of collagen	Serves as a flexible internal framework	In soft organs such as liver, spleen, tonsils, and lymph glands
Special Connective Tissues			
Cartilage	Primarily collagen fibers in a ground substance containing a lot of water	Maintains shape and resists compression	Embryonic tissue that becomes bone. Also the nose, vertebral disks, and the lining of joint cavities
Bone	Primarily hard mineral deposits of calcium and phosphate	Very strong	Forms the skeleton
Blood	Blood cells, platelets, and blood fluid called *plasma*	Transports materials and assists in defense mechanisms	Within cardiovascular system
Adipose tissue	Primarily cells called adipocytes filled with fat deposits	Stores energy in the form of fat	Under the skin, around some internal organs

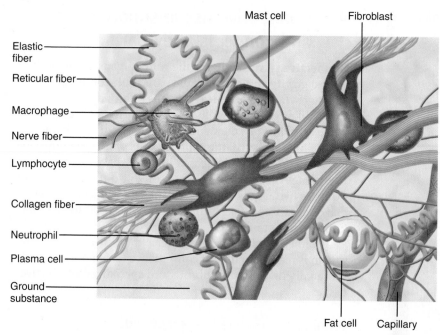

Mast cell — Fibroblast

Elastic fiber —
Reticular fiber —
Macrophage —
Nerve fiber —
Lymphocyte —
Collagen fiber —
Neutrophil —
Plasma cell —
Ground substance —

Fat cell — Capillary

Figure 4.3 Fibrous connective tissue. The main elements are three types of fibers (collagen, elastic, and reticular) and a variety of cells (fibroblasts, fat cells, mast cells, and several types of white blood cells) in a ground substance of polysaccharides, proteins, and water. Blood vessels and nerves pass through or are associated with connective tissue. Fibrous connective tissues vary in their relative proportions of cells and fibers, and also in fiber orientation.

thinner fibers of collagen, called **reticular fibers,** that interconnect with each other. The reticular fibers often serve as an internal structural framework for some of the "soft" organs such as the liver, spleen, and lymph nodes.

The various fibers are embedded in a *ground substance* consisting of water, polysaccharides, and proteins that ranges in consistency from gel-like to almost rubbery. The ground substance contains several types of cells, among them fat cells, mast cells, various white blood cells (macrophages, neutrophils, lymphocytes, and plasma cells), and most importantly,

fibroblasts. The fibroblasts are the cells responsible for producing and secreting the proteins that compose the collagen, elastic, and reticular fibers. The fat cells, of course, store fat, and both the mast cells and white blood cells are involved in the body's immune system (Chapter 9).

Fibrous connective tissues are subclassified according to the density and arrangement of their fiber types:

- *Loose connective tissue* (Figure 4.4a), also called *areolar* connective tissue, is the most common type. It surrounds many internal organs, muscles, and blood vessels. Loose connective tissue contains a few collagen fibers and elastic fibers in no particular pattern, giving it a great deal of flexibility but only a modest amount of strength.
- *Dense connective tissue* (Figure 4.4b), found in tendons, ligaments, and lower layers of skin, has more collagen fibers. The fibers are oriented primarily in one direction, especially in the tendons and ligaments in and around our joints. Dense connective tissue is the strongest connective tissue when pulled in the same direction as the orientation of the fibers, but it can tear if the stress comes from the side. There are very few blood vessels in dense connective tissue to supply the few living cells. This is why, if you strain a tendon or ligament, it can take a long time to heal.
- *Elastic connective tissue* surrounds organs that have to change shape or size regularly. Examples include the stomach, which must stretch to accommodate food; the bladder, which stretches to store urine; and the vocal cords, which vibrate to produce sounds. Elastic connective tissue contains a high proportion of elastic fibers, which stretch and recoil easily.

— Elastin fibers
— Fibroblast

— Collagen fibers

— Collagen fibers

— Nuclei of fibroblasts

a) Loose areolar connective tissue (× 160). In loose connective tissue the collagen and elastin fibers are arrayed in a random pattern.

b) Dense connective tissue (× 160). In dense connective tissue the fibers are primarily collagen fibers. In tendons and ligaments the fibers are oriented all in the same direction, with fibroblasts occupying narrow spaces between adjacent fibers.

Figure 4.4 Examples of fibrous connective tissues.

■ *Reticular connective tissue* (also called lymphoid tissue) serves as the internal framework of soft organs such as the liver and the tissues of the lymphatic system (spleen, tonsils, and lymph nodes). It consists of thin, branched reticular fibers (composed of collagen) that form an interconnected network.

✓ **Quick Check** People with a hereditary condition known as Ehlers-Danlos Syndrome (EDS) have hyperextensible joints that frequently dislocate, and extremely stretchy skin that tears easily. Develop a hypothesis for what type of tissue, and what protein in particular, might be the cause of EDS. ■

Specialized connective tissues serve special functions

The so-called *specialized connective tissues* are a diverse group that includes cartilage, bone, blood, and adipose tissue. Each is specialized to perform particular functions in the body.

Cartilage Cartilage is the transition tissue from which bone develops (Chapter 5). It also maintains the shape of certain body parts (such as the soft tip of your nose) and protects and cushions joints. Disks of cartilage separate and cushion the vertebrae in your backbone, for instance, and cartilage forms the tough, smooth surfaces that reduce friction in some body joints.

Like dense connective tissue, cartilage consists primarily of collagen fibers. The two tissues differ in that the ground substance of cartilage, which is produced by cells called *chondroblasts*, contains a great deal more water. This is why cartilage functions so well as a cushion. As cartilage develops, the cells become enclosed in small chambers called *lacunae* (**Figure 4.5a**). There are no blood vessels in cartilage, so the mature cells (called *chondrocytes*) obtain their nutrients only by diffusion through the ground substance from blood vessels located outside the cartilage. Consequently, cartilage is slow to heal when injured.

Bone Like cartilage, **bone** is a specialized connective tissue that contains only a few living cells. Most of the matrix of bone consists of hard mineral deposits of calcium and phosphate. However, unlike cartilage, bone contains numerous blood vessels, and for this reason it can heal within four to six weeks after being injured. We discuss bone in more detail in Chapter 5 when we discuss the skeletal system.

Blood Blood consists of cells suspended in a fluid matrix called *plasma*. It is considered a connective tissue because all blood cells derive from earlier cells (called *stem cells*) located within bone. Red blood cells transport oxygen and nutrients to body cells and carry away the waste products of the cells' metabolism. White blood cells function in the immune system that defends the body, and platelets participate in the mechanisms that cause blood to clot following an injury. You will learn more about the functions of blood in Chapter 7.

Adipose tissue Adipose tissue is highly specialized for fat storage (Figure 4.5b). It has few connective tissue fibers and almost no ground substance. Most of its volume is occupied by **adipocytes** (fat cells). Adipose tissue is located primarily under the skin, where it serves as a layer of insulation. It also forms a protective layer around internal organs such as the kidneys. The number of adipocytes you have is partly determined by your genetic inheritance. When you eat more food than your body can use, some of the excess energy is stored in your adipocytes as fat (the fat cells get "fatter"). When you lose weight the fat cells slim down, too. In other words, weight loss reduces the volume of each fat cell but it does not necessarily reduce the number of fat cells. *Lipodissolve* is a controversial technique that disrupts fat cells chemically (review the Current Issue at the beginning of this chapter).

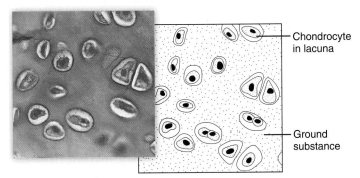

a) Cartilage from the trachea (× 300). Mature cartilage cells, called chondrocytes, become trapped in chambers called lacunae within the hard, rubbery ground substance. Ground substance is composed of collagen fibers, polysaccharides, proteins, and water.

b) Adipose tissue from the subcutaneous layer under the skin (× 140). Adipose tissue consists almost entirely of fat cells. The fat deposit within a fat cell can become so large that the nucleus is pushed to the side.

Figure 4.5 Examples of special connective tissues.

↩ **Recap** Fibrous connective tissues provide strength and elasticity and hold body parts together. Among the specialized connective tissues, cartilage and bone provide support, blood transports materials throughout the body, and adipose tissue stores energy in the form of fat. ▪

4.4 Muscle tissues contract to produce movement

Muscle tissue consists of cells that are specialized to shorten, or contract, resulting in movement of some kind. Muscle tissue is composed of tightly packed cells called muscle *fibers.* The fibers are generally long and thin and aligned parallel to each other (**Figure 4.6**). The cytoplasm of a muscle fiber contains proteins, which interact to make the cell contract.

There are three types of muscle tissue: *skeletal, cardiac,* and *smooth.* They vary somewhat in body location, structure, and function, but they all do essentially the same thing—when stimulated, they contract. We devote an entire chapter (Chapter 6) to muscles as an organ system. For now, we focus on differences between the three types of muscle tissue.

a) **Skeletal muscle (× 100).** Skeletal muscle cells are very long and have many nuclei.

— Nuclei

— Width of one muscle cell

b) **Cardiac muscle (× 225).** Cardiac muscle cells interconnect with each other.

— Intercalated disc

— Nucleus

c) **Sheet of smooth muscle (× 250).** Smooth muscle cells are thin and tapered.

— Smooth muscle cell

— Nucleus

Figure 4.6 Muscle tissue.

Skeletal muscles move body parts

Skeletal muscle tissue connects to tendons, which attach to bones. When skeletal muscles contract, they cause body parts to move. The individual fibers are thin cylinders too small to be seen with the naked eye, but they may be as long as the entire muscle (Figure 4.6a). Each muscle fiber has many nuclei, a phenomenon that comes about because many cells fuse end to end during development, producing one long fiber.

A skeletal muscle may contain thousands of individual fibers, all aligned parallel to each other. This parallel arrangement enables them to all pull together, shortening the muscle between its two points of attachment. Skeletal muscle is called *voluntary* muscle because we can exert conscious control over its activity.

Cardiac muscle cells activate each other

Cardiac muscle tissue (Greek *kardia*, the heart) is found only in the heart. The individual cells are much shorter than skeletal muscle fibers, and they have only one nucleus (Figure 4.6b). Like skeletal muscle, the cells are arranged parallel to each other. Cardiac muscle cells are short and blunt-ended, with gap junctions between the ends of adjoining cells. The gap junctions represent direct electrical connections between adjoining cells, so when one cell is activated it activates its neighbors down the line. Because of these gap junctions, the entire heart contracts in a coordinated fashion.

Cardiac muscle is considered involuntary because the heart can contract rhythmically entirely on its own, without any conscious thought on our part and without any stimulation by nerves.

✓ **Quick Check** Do you think skeletal muscle, like cardiac muscle, has gap junctions between adjacent cells? Why or why not? ◼

Smooth muscle surrounds hollow structures

Smooth muscle tissue surrounds hollow organs and tubes, including blood vessels, digestive tract, uterus, and bladder. These slim cells are much smaller than skeletal muscle cells and have only one nucleus, like cardiac muscle (Figure 4.6c). The cells are aligned roughly parallel to each other. In blood vessels they are generally aligned in a circular fashion around the vessel. When smooth muscle cells shorten, the diameter of the blood vessel is reduced.

Smooth muscle cells taper at both ends, and there are gap junctions between adjacent cells so that when one contracts, nearby cells also contract. Like cardiac muscle, smooth muscle is involuntary in that we cannot control its contractions consciously.

↩ **Recap** The common feature of all muscle tissues (skeletal, cardiac, and smooth) is that they contract, producing movement. ◼

4.5 Nervous tissue transmits impulses

Nervous tissue consists primarily of cells that are specialized for generating and transmitting electrical impulses throughout the body. It forms a rapid communication network for the body. Nervous tissue is located in the brain, the spinal cord, and the nerves that transmit information to and from various organs. Chapter 11 is devoted to the nervous system, so we describe nervous tissue only briefly here.

Nervous tissue cells that generate and transmit electrical impulses are called **neurons** (Figure 4.7). Neurons can be as long as the distance from your spinal cord to the tip of your toe. Neurons typically have three basic parts: (1) the *cell body* where the nucleus is located; (2) *dendrites,* numerous cytoplasmic extensions that extend from the cell body and receive signals from other neurons; and (3) a long extension called an *axon* that transmits electrical impulses over long distances.

Figure 4.7 **Nervous tissue: a neuron (×170).** The neuron is the functional unit of nervous tissue. The single neuron shown here is surrounded by numerous supporting cells called *glial cells*. The cell bodies of the glial cells do not stain well, but their nuclei are clearly visible.

Nervous tissue also includes another type of cell called a **glial cell** that does not transmit electrical impulses. Glial cells play a supporting role by surrounding and protecting neurons and supplying them with nutrients.

↩ **Recap** Nervous tissues serve as a communication network by generating and transmitting electrical impulses. ◼

4.6 Organs and organ systems perform complex functions

Many of the more complex functions of multicellular organisms (such as pumping blood or digesting food) cannot be carried out by one tissue type alone. **Organs** are structures composed of two or more tissue types joined together that perform a specific function or functions.

Your heart is an organ. Most of it consists of cardiac muscle, but there is also smooth muscle in the blood vessels that supply the cardiac muscle. The heart also contains nervous tissue that affects the rate at which the heart beats. It contains some connective tissue, primarily in the heart valves that open and close to control blood flow within the heart, and even a thin layer of epithelial tissue that lines the heart chambers. These tissues function together to pump blood, so together they constitute the organ known as the heart. Some organs have several functions. For example, the kidneys remove wastes and help control blood pressure.

The human body is organized by organ systems

Organ systems are groups of organs that together serve a broad function that is important to survival either of the individual organism (such as respiration, movement, or excretion of wastes) or of the species (reproduction). A good example is the organ system responsible for the digestion of food. Your

Integumentary System
- Protects us from injury, infection, and dehydration
- Participates in temperature control
- Receives sensory input from the external environment

Skeletal System
- Protects, supports, and anchors body parts
- Provides the structural framework for movement
- Produces blood cells
- Stores minerals

Muscular System
- Produces movement or resists movement
- Generates heat

Nervous System
- Detects both external and internal stimuli
- Controls and coordinates rapid responses to these stimuli
- Integrates the activities of other organ systems

Endocrine System
- Produces hormones that regulate many body functions
- Participates with the nervous system in integrative functions

Digestive System
- Provides the body with water and nutrients
- (The liver) synthesizes certain proteins and lipids for the body
- (The liver) inactivates many chemicals, including hormones, drugs, and poisons

The 11 organ systems of the human body.

Circulatory System
- Transports materials to and from all cells
- Participates in the maintenance of body temperature
- Participates in mechanisms of defense against disease and injury

Lymphatic System
- Returns excess tissue fluid to the circulatory system
- Participates in both general and specific (immune) defense responses

Respiratory System
- Exchanges gases (oxygen and carbon dioxide) between air and blood
- Participates in the production of sound (vocalization)

Urinary System
- Maintains the volume and composition of body fluids
- Excretes some waste products

Reproductive System
- Female: Produces eggs
- Female: Nurtures the fertilized egg, developing embryo, and fetus until birth
- Male: Produces sperm
- Male: Participates in the delivery of sperm to the female

digestive system includes your mouth, throat, esophagus, stomach, intestines, and even your liver, pancreas, and gallbladder. All of these organs must interact and be controlled and coordinated to accomplish their overall function.

The figures on pages 90 and 91 depict the 11 organ systems of the human body. Some organ systems perform several functions and so are discussed in several chapters in this book. For example, the lymphatic system has important functions related to defense against disease, the circulation of certain body fluids, and digestion. Other organ systems are covered in a single chapter.

Tissue membranes line body cavities

Some of the organs and organ systems are located in hollow cavities within the body (**Figure 4.8**). The large anterior cavity is divided into the thoracic cavity and abdominal cavity by the diaphragm between them. The thoracic cavity is in turn divided into two pleural cavities, each containing a lung, and the pericardial cavity, which encloses the heart. The lower part of the abdominal cavity is sometimes called the pelvic cavity. The smaller posterior cavity consists of the cranial cavity and the spinal cavity (vertebral canal). There

are many other smaller cavities as well, such as the synovial cavities in movable joints.

Tissue membranes consisting of a layer of epithelial tissue and a layer of connective tissue line each body cavity and form our skin. There are four major types of tissue membranes:

- *Serous membranes.* Line and lubricate body cavities to reduce friction between internal organs.
- *Mucous membranes.* Line the airways, digestive tract, and reproductive passages. Goblet cells within the epithelial layer secrete *mucus,* which lubricates the membrane's surface and entraps foreign particles.
- *Synovial membranes.* Line the very thin cavities between bones in movable joints. These membranes secrete a watery fluid that lubricates the joint.
- *Cutaneous membrane.* Our outer covering. You know it as skin, and it serves several functions discussed later in this chapter.

By now you may have noticed that "membrane" is a general term for a thin layer that covers or surrounds something. You have been introduced to three different

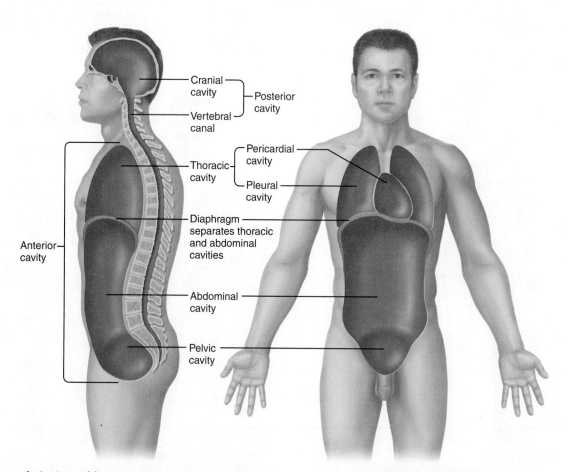

Figure 4.8 The main body cavities. The pelvic cavity and the abdominal cavity are continuous (not separated by a membrane).

membranes so far: the *plasma* membrane of phospholipids surrounding every cell, the *basement* membrane of extracellular material on which epithelial tissue rests, and *tissue* membranes consisting of several layers of tissue sandwiched together that cover or surround cavities, organs, and entire organ systems.

✅ **Quick Check** What kind of membrane would you expect to find lining a pleural cavity? Explain. ▪

Describing body position or direction

When describing parts of the body, biologists use precise terms to define position and direction. Generally speaking, an organ or even the entire body can be described by three planes known as the *midsagittal*, *frontal*, and *transverse* planes (**Figure 4.9**). These planes divide the body into left and right, front and back, and top and bottom, respectively. *Anterior* means "at or near the front" and *posterior* means "at or near the back." *Proximal* means "nearer (in closer proximity) to" any point of reference, usually the body trunk, and *distal* means "farther away." For example, your wrist is distal to your elbow. *Superior* means "situated above" or "directed upward," and *inferior* means "situated below" or "directed downward." There are dozens of such terms, each with a precise meaning, defined as they occur in this book.

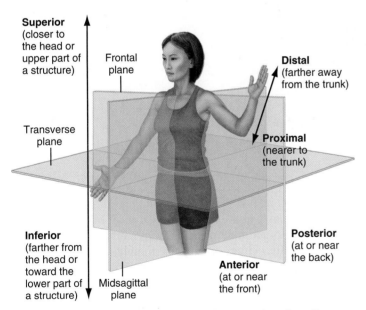

Figure 4.9 Planes of symmetry and terms used to describe position or direction in the human body. The frontal plane divides the body into anterior and posterior sections, the midsagittal plane divides it into left and right, and the transverse plane divides it into superior and inferior sections. Proximal and distal refer to points closer to or farther away from a point of reference, usually the trunk.

✅ *A friend who is an anatomist tells you that he has a blister on the distal part of his right leg, on the inferior surface. Where is the blister?*

↩ **Recap** An organ consists of several tissue types that join together to perform a specific function. An organ system is a group of organs that share a broad function important for survival. The body's hollow cavities are lined by tissue membranes that support, protect, and lubricate cavity surfaces. ▪

4.7 The skin as an organ system

The proper name for the skin and its accessory structures such as hair, nails, and glands is the **integumentary system** (from the Latin *integere*, meaning "to cover"). We describe the skin here as a representative organ system; other organ systems are covered later in the book.

Skin has many functions

The skin has several different functions related to its role as the outer covering of the body:

- Protection from dehydration (helps prevent our bodies from drying out)
- Protection from injury (such as abrasion)
- Defense against invasion by bacteria and viruses
- Regulation of body temperature
- Synthesis of an inactive form of vitamin D
- Sensation: provides information about the external world via receptors for touch, vibration, pain, and temperature

Skin consists of epidermis and dermis

Recall that skin is a tissue membrane, and that tissue membranes contain layers of epithelial and connective tissue. The outer layer of the skin's epithelial tissue is the **epidermis,** and the inner layer of connective tissue is the **dermis** (**Figure 4.10**).

The skin rests on a supportive layer called the *hypodermis* (*hypo-* means "under"), consisting of loose connective tissue containing fat cells. The hypodermis is flexible enough to allow the skin to move and bend. The fat cells in the hypodermis insulate against excessive heat loss and cushion against injury.

Epidermal cells are replaced constantly The epidermis consists of multiple layers of squamous epithelial cells. A key feature of the epidermis is that it is constantly being replaced as cells near the base of the epidermis divide repeatedly, pushing older cells toward the surface.

Two types of cell make up the epidermis: keratinocytes and melanocytes. The more numerous of the two cell types are **keratinocytes,** which produce a tough, waterproof protein called *keratin*. Actively dividing keratinocytes located near the base of the epidermis are sometimes called *basal cells*. As keratinocytes derived from the basal cells move toward the skin surface, they flatten and become squamous. Eventually they die and dry out, creating a nearly waterproof barrier that covers and protects the living cells below

Figure 4.10 The skin. The two layers of skin (epidermis and dermis) rest on a supportive layer (hypodermis). Although not part of the skin, the hypodermis serves important functions of cushioning and insulation.

(Figure 4.11). The rapid replacement of keratinocytes allows the skin to heal quickly after injury.

One reason the outer layers of epidermal cells die is that the epidermis lacks blood vessels, so as mature cells are pushed farther from the dermis they can no longer obtain nutrients. The dead cells of the outer layers are shed over time, accounting for the white flakes you sometimes find on your skin or on dark clothing, especially when your skin is dry.

Less numerous cells called **melanocytes** located near the base of the epidermis produce a dark-brown pigment called *melanin.* Melanin accumulates inside keratinocytes and protects us against the sun's ultraviolet radiation. Exposure to sunlight increases the activity of melanocytes, accounting for the ability of some people to develop a suntan. (For more on tanning, see Health & Wellness,

Suntans, Smoking, and Your Skin.) Because all humans have about the same number of melanocytes, racial differences in skin color reflect either differences in melanocyte activity or differences in the rate of breakdown of melanin once it is produced.

 Quick Check What would happen to your skin if your keratinocytes started dividing more rapidly than usual? Could the skin still perform its major functions? ▪

Fibers in dermis provide strength and elasticity The dermis is primarily dense connective tissue, consisting of collagen, elastic, and reticular fibers embedded in a ground substance of water, polysaccharides, and proteins. The fibers allow the skin to stretch when we move and give it strength to resist abrasion and tearing. Our skin becomes less flexible and

Dead cells of epidermis

Living cells of epidermis

Keratinocyte containing melanin

Melanocyte containing melanin granules

Dividing keratinocyte (basal cell)

Basement membrane

Blood vessel

Dermis with blood vessel

Figure 4.11 The epidermis. Living cells near the base of the epidermis divide, pushing more mature cells toward the surface. As cells migrate toward the surface they die and dry out, forming a tough, waterproof barrier. The cells of the epidermis are supplied only by blood vessels located in the dermis.

more wrinkled as we age because the number of fibers in the dermis decreases.

The surface of the dermis has many small projections called *papillae* that contain sensory nerve endings and small blood vessels. When the skin is rubbed excessively—such as when your shoes are too tight—the epidermis and dermis separate from each other and a fluid-filled blister develops between them.

The most abundant living cells in the dermis are the fibroblasts that produce the various fibers, but there are also mast cells, white blood cells, and occasional fat cells. Other structures in the dermis include:

- *Hair.* Each hair has a *shaft* above the skin's surface and a *root* below the surface. Hair is actually composed of several layers of cells enclosed in an outer layer of overlapping, dead, flattened keratinocytes. The root of a hair is surrounded by a sheath of several layers of cells called the *follicle.* The cells at the very base of the follicle are constantly dividing to form the hair root. As new hair cells are formed at the base, the hair root is pushed upward toward the skin's surface.

- *Smooth muscle.* Attached to the base of the hair follicle, it contracts when you are frightened or cold, causing your hair to become more erect.

- *Sebaceous glands.* Also known as oil glands, these secrete an oily fluid that moistens and softens hair and skin.

- *Sweat glands.* These produce sweat, a watery fluid containing dissolved ions, small amounts of metabolic wastes, and an antibiotic peptide called *dermicidin.* Sweat helps regulate body temperature and protects against bacteria.

- *Blood vessels.* These supply the cells of the dermis and epidermis with nutrients and remove their wastes. The blood vessels also help regulate body temperature. They

Suntans, Smoking, and Your Skin

Many of us think suntans are attractive. A suntan indicates we have leisure time to spend basking in the sun, and it gives the skin a healthy looking "glow." But are suntans really healthy?

The answer is clearly no, according to medical experts. Strong light rays can age your skin prematurely and increase your risk for skin cancer. The rays penetrate to the dermis and damage its collagen and elastin fibers. Elastin fibers clump together, leading at first to fine wrinkles and later to a wrinkled, leathery skin texture.

Prolonged exposure to light rays can also damage small blood vessels. Sometimes the vessels remain permanently dilated, leading to a condition called *telangiectasis,* or spider veins. Sunlight also damages the keratinocytes and melanocytes in the epidermis. The keratinocytes become rough and thickened and no longer fit together as a smooth interlocking layer. The melanocytes begin to produce melanin unevenly, leading to patches of darker pigmentation known as freckles, age spots, or liver spots.

What about tanning beds—are they okay? The tanning salon industry has been quite aggressive in trying to convince the public that tanning beds are not only safe, but that

a healthy tan is good for you. Experts disagree. According to skin cancer researchers, repeated exposure to light rays, whether from sunlight or from tanning beds, is a risk factor for melanoma. In addition, your skin is likely to age more quickly as you get older. Although a minimal amount of sunlight is necessary to activate vitamin D, the amount is far below what is necessary to cause a tan.

Perhaps you have heard that ultraviolet (UV) rays consist of "good" UVA rays that tan your skin and "bad" UVB rays that cause sunburn. Tanning lotions with high SPF (sun protection factor) numbers are designed to block the UVB rays, which at least prevents the acute damage and pain of a sunburn. However, UVA rays are also

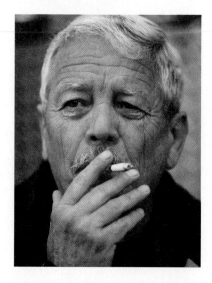

bad for you. They penetrate more deeply than the UVB rays and in fact cause most of the long-term changes that age skin prematurely.

What about smoking? Heavy smokers are nearly five times more likely to develop premature wrinkles. Smoking damages and thickens the elastin fibers in the dermis. It also dehydrates keratinocytes in the epidermis, causing the epidermis to develop a rough texture. Finally, smoking narrows blood vessels, reducing blood flow to the skin. As a result the skin of smokers heals more slowly from injury than the skin of nonsmokers. ▪

Repeated exposure to light rays is a risk factor for melanoma.

dilate to facilitate heat loss when we are too hot and constrict to prevent heat loss when we are too cool. The dermis also contains lymph vessels, which drain fluids and play a role in the immune system.

▪ *Sensory nerve endings.* These provide information about the outside environment. Separate receptors exist to detect heat, cold, light touch, deep pressure, and vibration. You will learn more about these nerve endings in Chapter 12.

As mentioned earlier, the skin synthesizes an inactive form of vitamin D. It is not known which cell type in the skin is responsible. But we do know that a cholesterol-like molecule in the skin becomes an inactive form of vitamin D when it is exposed to the ultraviolet rays of sunlight. The inactive form must then be modified in the liver and kidneys before it becomes active.

↺ **Recap** The skin is an organ because it consists of different tissues serving common functions. Functions of skin include protection, temperature regulation, vitamin D synthesis, and sensory reception. ▪

4.8 Multicellular organisms must maintain homeostasis

Although multicellularity offers many advantages to organisms, it presents certain disadvantages that must be overcome. For example, cells that are surrounded entirely by other cells can't obtain their nutrients directly from the organism's external environment and are constantly exposed to the waste products of neighboring cells.

The environment that surrounds the cells of a multicellular organism (their external environment) is the **internal environment** of the organism. The internal environment is a clear fluid called the **interstitial fluid** (the Latin noun *interstitium* means "the space between," in this case the space between cells). Every cell gets nutrients from the interstitial fluid around it and dumps wastes into it. In a multicellular organism, the interstitial fluid is the equivalent of the ocean, lake, or tiniest drop of fluid that surrounds and nourishes single-celled organisms.

Because every cell must receive all its requirements for life from the surrounding interstitial fluid, the composition of this fluid must be kept fairly constant to sustain life. In the long run, nutrients consumed by the cell must be replaced and wastes must be removed or the cell will die.

Relative constancy of the conditions within the internal environment is called **homeostasis** (*homeo-* means "unchanging" or "the same," and *-stasis* means "standing"). The maintenance of homeostasis is so important for life that multicellular organisms, including human beings, devote a significant portion of their total metabolic activities to it. Although small changes in the internal environment do occur from time to time, the activities of cells, tissues, organs, and organ systems are carefully integrated and regulated to keep these changes within acceptable limits.

Homeostasis is maintained by negative feedback

In living organisms, homeostasis is maintained by **negative feedback** control systems (**Figure 4.12**). Negative feedback control systems operate in such a way that deviations from the desired condition are automatically detected and counteracted. A negative feedback control system has the following components:

▪ A **controlled variable.** The focal point of any negative feedback control loop is the *controlled variable*. A controlled variable is any physical or chemical property that might vary from time to time and that must be controlled to maintain homeostasis. Examples of controlled variables are blood pressure, body temperature, and the concentration of glucose in blood.

▪ A **sensor** (or *receptor*). The sensor monitors the current value of the controlled variable and sends the information (via either nerves or hormones) to the control center.

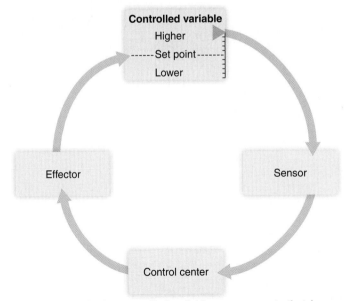

a) An *increase* in the controlled variable causes events that *lower* the controlled variable toward its set point again.

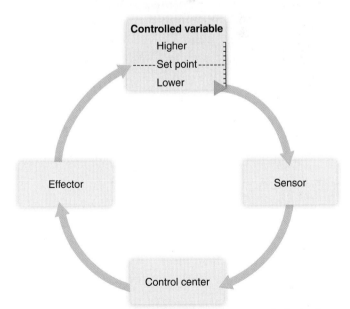

b) A *decrease* in the controlled variable causes events that *raise* the controlled variable toward its set point again.

Figure 4.12 Components of a negative feedback control system. The focal point of the control system is the controlled variable. A sensor monitors the controlled variable and sends signals to a control center, which compares the current value of the controlled variable with its set point. If the controlled variable and set point do not match, the control center sends signals to effectors that take action to reverse the difference between the controlled variable and its set point.

- A **control center.** The control center receives input from the sensor and compares it to the correct, internally set value of the controlled variable, sometimes called the **set point**. When the current value and the set point are not in agreement, the control center sends signals (again, via either nerves or hormones) to an effector.
- An **effector.** The effector takes the necessary action to correct the imbalance, in accordance with the signals it receives from the control center.

The cycle is called *negative* feedback because any change in the controlled variable triggers a series of events that ultimately opposes ("negates") the initial change, returning the variable to its set point. In other words, homeostasis is maintained.

✅ **Quick Check** Your refrigerator's ability to maintain a relatively steady cool temperature is another example of a feedback system. Is it a *negative* feedback system? Identify the major components of the system. ∎

Negative feedback helps maintain core body temperature

A prime example of negative feedback is the maintenance of homeostasis of your body temperature. In this case,

multiple organ systems participate in maintaining homeostasis (**Figure 4.13**).

The controlled variable is your core temperature, meaning the temperature near the center of your body. Temperature sensors in your skin and internal organs monitor core temperature. These sensors transmit signals via nerves to the control center, located in a region of your brain called the *hypothalamus*. The control center uses different combinations of effector mechanisms to raise or lower core temperature as needed.

When your core temperature falls *below* its set point, the hypothalamus:

- Sends more nerve impulses to blood vessels in the skin, causing the blood vessels to constrict. This restricts blood flow to your skin and reduces heat loss.
- Stimulates your skeletal muscles, causing brief bursts of muscular contraction known as shivering. Shivering generates heat.

When your core temperature rises *above* its set point, the hypothalamus:

- Sends fewer nerve impulses to blood vessels in the skin, causing the blood vessels to dilate. This increases blood flow to your skin and promotes heat loss.
- Activates your sweat glands. As perspiration evaporates from your skin, you lose heat.

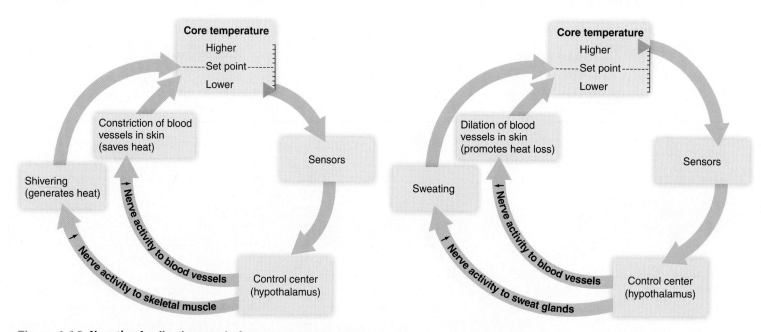

Figure 4.13 Negative feedback control of core temperature. Note that different combinations of effector mechanisms may be activated, depending on the direction of the initial change in core temperature.

✅ *In most people, hunger (and food intake) appears to be regulated in such a way that the amount of body fat stays surprisingly constant over time (despite people's efforts to lose weight). Draw a diagram showing how hunger and food intake might be regulated in a negative feedback system to keep body fat stores at a certain set point.*

Even when your core temperature is normal, your hypothalamus is transmitting some nerve impulses to the blood vessels in your skin. Small changes in temperature, then, can be handled effectively just by increasing or decreasing the normal number of signals. Only when the variations in temperature are large is sweating or shivering called into play.

From this example, we can make the following points about negative feedback control:

- Many sensors may be active at once. In this case, sensors throughout the body monitor body temperature.
- The control center integrates all of this incoming information and comes up with an appropriate response.
- There can be multiple effectors as well as multiple sensors, and they may belong to different organ systems.

Both skin and muscles function in returning body temperature to its set point.

 Quick Check People who have fevers will often get "chills"—feeling as if they are cold and shivering—even though body temperature is actually above normal. Later, when the fever "breaks," they will suddenly start sweating. Develop a hypothesis for why this happens. ▪

Positive feedback amplifies events

Positive feedback control systems are relatively uncommon in living organisms. In positive feedback, a change in the controlled variable sets in motion a series of events that *amplify* the original change, rather than returning it to normal. The process of childbirth once labor has started is governed by positive feedback mechanisms.

Obviously something must terminate positive feedback events. The contractions of childbirth end when the child is born. The important point is that positive feedback is not a mechanism for maintaining homeostasis.

 Web Animation *Homeostasis* at www.humanbiology.com

 Recap All multicellular organisms must maintain homeostasis of their internal environment. In a negative feedback control system, any change in a controlled variable sets in motion a series of events that reverse the change, maintaining homeostasis. ▪

Chapter Summary

Tissues are groups of cells with a common function *p. 82*

- The four main types of tissues are epithelial tissue, connective tissue, muscle tissue, and neural tissue.

Epithelial tissues cover body surfaces and cavities *p. 82*

- Epithelial tissues are sheets of cells that cover or line body surfaces and form the glands.
- Epithelial tissues are supported by a noncellular layer called the basement membrane.

Connective tissue supports and connects body parts *p. 85*

- Fibrous connective tissues contain several types of extracellular fibers and only a few living cells. They support and connect body parts.
- Cartilage, blood, bone, and adipose tissue are classified as special connective tissues.

Muscle tissues contract to produce movement *p. 88*

- Muscle tissue is composed of either skeletal, cardiac, or smooth muscle cells.
- Skeletal muscles are attached to bones by tendons.

Nervous tissue transmits impulses *p. 89*

- Neurons are specialized for conduction of electrical impulses.
- Glial cells surround and protect neurons and supply them with nutrients.

Organs and organ systems perform complex functions *p. 89*

- The human body is composed of 11 organ systems, each of which has at least one broad function.
- Membranes consisting of layers of epithelial and connective tissues line the body cavities.
- Positions of body parts are described on three planes: midsagittal, frontal, and transverse.

The skin as an organ system *p. 93*

- The skin functions as a protective barrier, participates in the maintenance of homeostasis, and provides us with sensory information about the external environment.
- Skin has two layers: an outer epithelial layer called the *epidermis* and an inner connective tissue layer called the *dermis*.
- Skin also contains nerves, blood vessels, glands, hair follicles, and smooth muscle.

Multicellular organisms must maintain homeostasis *p. 97*

- In a multicellular organism, the external environment of every cell is the internal environment of the organism.
- Relative constancy of the internal environment is called *homeostasis*.
- Homeostasis is maintained by negative feedback control systems.
- In a negative feedback control system, a change in the controlled variable sets in motion a sequence of events that tends to reverse (or negate) the initial change.
- In the regulation of body temperature, sensors located throughout the body send information about temperature to the control center, located in the hypothalamus of the brain.
- Possible responses to a change in body temperature include dilating or constricting the blood vessels to the skin, shivering (if temperature is too low), and sweating (if temperature is too high).

Terms You Should Know

basement membrane, *84*
cell junctions, *84*
connective tissue, *85*
controlled variable, *97*
dermis, *93*
endocrine gland, *82*
epidermis, *93*
epithelial tissue, *82*
exocrine gland, *82*

homeostasis, *97*
internal environment, *97*
muscle tissue, *88*
negative feedback, *97*
nervous tissue, *89*
neuron, *89*
organ system, *89*
set point, *98*

Concept Review

Answers can be found at the Human Biology Place.
www.humanbiology.com

1. Describe some advantages and disadvantages of multicellularity.
2. Name the four main types of tissues in the human body and list their main functions.
3. Describe the functions of the three types of cell junctions.
4. Distinguish between an organ and an organ system.
5. List the 11 organ systems of the body and give at least one function of each.
6. Define *interstitial fluid.*
7. Name the two cavities of the anterior body cavity that are separated from each other by the diaphragm.
8. Compare/contrast positive and negative feedback.
9. Discuss the purpose of *homeostasis* in the body.
10. Describe the function of a control center in a negative feedback control system.

Test Yourself

Answers can be found in Appendix A.

1. Collagen and elastin fibers are typically found in:
   a. connective tissue
   b. epithelial tissue
   c. intercellular junctions
   d. muscle tissue
   e. the epidermis

2. Cells in cardiac muscle are able to contract in a coordinated fashion because of communication made possible through:
   a. gap junctions
   b. spot desmosomes
   c. adhesion junctions
   d. tight junctions
   e. synapses

3. Which of the following membranes is not composed of cells?
   a. serous membrane
   b. synovial membrane
   c. cutaneous membrane
   d. basement membrane
   e. mucous membrane

4. Exocrine and endocrine glands are types of:
   a. loose connective tissue
   b. epithelial tissue
   c. squamous tissue
   d. fibrous connective tissue
   e. specialized connective tissue

5. The thoracic cavity is located _____ to the abdominal cavity.
   a. proximal
   b. superior
   c. inferior
   d. ventral
   e. distal

6. What property do all muscle tissues have in common?
   a. composed of collagen and elastin filaments
   b. under conscious control
   c. ability to form a communication network
   d. ability to contract
   e. cells are joined by tight junctions

7. Injury to bone heals more quickly than injury to cartilage because:
   a. bone marrow contains stem cells
   b. chondroblasts are only present during prenatal development
   c. the polysaccharides in cartilage ground substance can't be replaced
   d. there is a richer blood supply to bone
   e. bone has a higher mineral content

8. Which of the following tissues may be found in the skin?
   a. smooth muscle
   b. fibrous connective tissue
   c. nervous tissue
   d. epithelial tissue
   e. all of these tissues

9. Which of the following is responsible for the pigmentation of the skin?
   a. melanin
   b. keratin
   c. collagen
   d. sebum
   e. dermicidin

10. When a decrease in blood pressure is detected by the central nervous system, the central nervous system triggers several changes that will return the blood pressure to its set point. This is an example of:
    a. positive feedback
    b. thermoregulation

    c. negative feedback
    d. reverse feedback
    e. set point feedback

11. The presence of a full bladder triggers the bladder to contract. As a little urine is released, this causes more contractions which will completely empty the bladder. This is an example of:
    a. homeostatic regulation
    b. uroregulation
    c. negative feedback
    d. positive feedback
    e. reverse feedback

12. Which type of tissue stores triglycerides?
    a. muscle tissue
    b. loose connective tissue
    c. fibrous connective tissue
    d. columnar epithelial tissue
    e. adipose tissue

13. A substantial amount of nonliving extracellular material, also known as the matrix, characterizes all:
    a. muscle tissue
    b. epithelial tissue
    c. connective tissue
    d. nervous tissue
    e. membranes

14. All of the following organ systems may be involved in thermoregulation except:
    a. integumentary system
    b. muscular system
    c. circulatory system
    d. skeletal system
    e. nervous system

15. Reducing food intake may lead to weight loss by:
    a. decreasing the number of muscle cells
    b. decreasing the amount of fibrous connective tissue
    c. decreasing the volume of adipocytes
    d. decreasing the volume of muscle cells
    e. decreasing the number of adipocytes

Apply What You Know

Answers can be found at the Human Biology Place.
www.humanbiology.com

1. Your roommate says that the concept of homeostasis is being violated when the rate of respiration goes up during exercise, because the rate of respiration clearly is not being held constant. Explain to him where his thinking is faulty.

2. What do you think would be some of the problems associated with severe third-degree burns, in which both the epidermis and the dermis are severely damaged or destroyed?

3. Sherlock Holmes, the greatest fictional detective of all time, is talking to a woman in her late 40s, when he suddenly says, "I see, my dear madam, you must have enjoyed your cigarettes and your suntans." The woman is amazed, because she mentioned nothing about these two former favorite activities. What physical characteristics might Mr. Holmes have seen in this woman to indicate she was an avid sun-worshiper and cigarette smoker? And how do these characteristics develop?

4. Dieting is difficult. People who do manage to lose weight can gain it back if they're not careful. Are techniques such as liposuction or Lipodissolve a good way to keep the weight off permanently? Why or why not?

5. Fibrous connective tissue consists of ground substance and fibers that provide strength, support, and flexibility. Concrete is used to make tough, durable structures in construction projects. How is a concrete structure like or unlike fibrous connective tissue?

6. By definition, an organ is a structure composed of two or more tissue types that perform a specific function. Performance of that function often requires coordination among many cells. Why is it so important that cardiac muscle cells of the heart be synchronized (coordinated) so that they beat nearly all at once?

The Skeletal System

A colored X-ray of a foot on tiptoe.

A Black Market in Human Bones?

Alistair Cooke, famed host of the PBS series *Masterpiece Theatre,* died in 2004 at the age of 95. His body lay in a New York City funeral parlor for a few days awaiting cremation. But before Cooke's body was cremated, it was secretly carved up in a back room and his bones were removed. Authorities allege that his bones were then sold for a substantial profit, to be transplanted into patients in desperate need of tissue grafts.

Cooke's family, who had not given permission for his body parts to be donated, knew nothing of this until police contacted them after the funeral. Under-standably, they were appalled. But there

is an even more horrifying side to this story: Cooke died of lung cancer that had spread to his bones. Could his deadly cancer have been transmitted to the people who received his bone tissue? Its unlikely since bone products generally are sterilized (see below), but the answer may not be known for decades.

Recycling Body Parts: A Legitimate Industry

The processing of tissues taken from human corpses into products that can be transplanted into other people is a legitimate industry that serves urgent medical needs. The industry has evolved

Alistair Cooke, long-time host of *Masterpiece Theatre*. Cooke's body was subjected to secret, illegal bone harvesting after he died.

Donated bones are cleaned, sterilized, and shaped into bone products.

over the past several decades as harvesting and transplantation techniques have improved. Bones are used to repair fractures and replace cancerous bone. Bone pins and powdered bone are used in dental surgery; bone paste plugs holes. Tendons and ligaments are used to repair joints and tissues damaged by sports injuries, transplanted vertebrae relieve back pain, and veins and heart valves are used in heart surgeries. The bones, tendons, veins, and heart valves from just one corpse can be worth over $200,000 to surgeons, hospitals, and recipients.

Under federal law it is illegal to sell human body parts for a profit—they can only be donated, either by the patient while he or she is still alive or by the family after death. Several hundred licensed nonprofit tissue banks in the United States receive donated tissues and test them for infectious diseases such as HIV, syphilis, and the viruses that cause hepatitis (inflammation of the liver). To reduce the chances of tissue products transmitting disease, authorities impose strict guidelines that specify what types of tissues may be harvested, from whom they may be harvested, and how they must be processed. For instance, to prevent any risk of transmitting cancer, federal guidelines prohibit the use of bones from cancer patients for tissue implants.

After donation, bone tissue is shaped into usable forms, such as pins, plates, and powders. The final products are sterilized and shipped to hospitals and surgeons all over the country, where they are used in more than 600,000 surgical procedures every year. The patient pays all fees incurred in the handling, processing, testing, and shipping of the products, but the tissue banks themselves do not make a profit.

Illegal Body Parts Enter the Supply Chain

In the Alistair Cooke case, prosecutors alleged that Michael Mastromarino, an oral surgeon who had lost his license, arranged for a Brooklyn funeral parlor to deliver bodies to a secret operating room. There, Mastromarino and his accomplices removed body parts before the bodies were buried or cremated. Authorities say that the men paid the funeral parlor up to $1,000 per body and then sold the harvested tissues for up to $7,000 per body to a legitimate but unsuspecting tissue-processing company.

In some cases Mastromarino and his accomplices falsified records indicating the deceased's age and cause of death. Mr. Cooke died of cancer at the age of 95, but his records were falsified to indicate that he died at age 85 of a heart attack. They also allegedly looted body parts from a 43-year-old woman who had died of ovarian cancer; they then forged a signature on a consent form and listed the cause of death as a head injury. When investigators examined the corpse of one grandmother, they found that her leg bones had been removed and replaced with PVC pipes. Prosecutors eventually identified over a thousand corpses from which body parts were taken without permission between 2001 and 2005.

In 2008 Mr. Mastromarino plead guilty in a plea bargain that could reduce his jail time in exchange for providing information about others who were involved. He is expected to spend at least 18 years in prison. An accomplice and seven funeral home directors received lesser sentences.

Just as the Alistair Cooke case is not the first such incident, it is not likely to be the last. In 1999 the University of California at Irvine discovered that the director of its Willed Body Program was selling human spines to a Phoenix hospital for $5,000 apiece. And in 2008 the director of UCLA's Willed Body Program was sentenced to 4 years in prison for selling more than a million dollars worth of body parts. Regulators say that abuses such as these are most likely to occur when relatively poorly paid directors (including funeral home directors) have access to valuable body parts and when oversight is lax. UC Irvine and UCLA have both tightened their oversight procedures as a result of the scandals.

Only 22,000 cadavers are donated annually for body parts—not enough to supply the growing demand for human body parts and tissues. Done properly, the donation of a single cadaver to a nonprofit tissue-processing company can benefit several dozen patients. Patients should only have to pay the legitimate costs associated with the body parts processing industry—not the added fees paid to traffickers in illegal body parts. Safeguards need to be put in place to prevent abuses so that we can be assured of the legitimacy and the safety of the supply of human body parts.

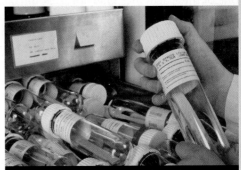
Human bones and bone products ready for shipment to hospitals.

Questions to consider

1 Do you approve of human bones being harvested from cadavers for processing into bone-based products for patients, provided the bones are legally obtained?

2 What steps do you think should be taken to curb abuses in the human body parts industry?

The facts...

- It is illegal to buy or sell human body parts for a profit. Patients or their families can donate body parts only to nonprofit tissue banks.

- Legitimately donated tissues are tested, sterilized, processed, and transplanted into patients who urgently need them.

- Only 22,000 cadavers are donated annually for body parts—not enough to meet the growing legitimate demand for human tissues and tissue products.

- The supply/demand imbalance may be contributing to a black market in body parts harvested illegally from cadavers.

The human body is capable of an awesome array of physical activities. With training, some individuals can run a mile in less than four minutes or lift more than their own weight. Exquisitely sensitive motor skills allow us to thread a needle, turn our head to focus on a single star, and throw a baseball into the strike zone. Considered individually, any one of these activities may not seem amazing, but for a single structure (the human body) to be capable of all of them is remarkable indeed. From an engineering standpoint it would be like designing a bulldozer that is strong enough to flatten a building, yet delicate enough to pick up a dime.

This chapter describes the skeletal system, the organ system for support, protection, and movement. We examine the structure and development of bones, and the way they remodel and repair themselves. We review how the bones fit together to make the skeleton. We take a look at how joints enable bones and muscles to work together. Finally, we consider what can go wrong with the skeletal system.

5.1 The skeletal system consists of connective tissue

The skeletal system comprises three types of connective tissue—bones, ligaments, and cartilage. *Bones* are the hard elements of the skeleton with which we are most familiar. *Ligaments* consist of dense fibrous connective tissue—they bind the bones to each other. *Cartilage* is a a specialized connective tissue consisting primarily of fibers of collagen and elastic in a gel-like fluid called *ground substance*. Cartilage has several functions, including reducing friction in joints.

Bones are the hard elements of the skeleton

Most of the mass of **bones** consists of nonliving extracellular crystals of calcium minerals that give bones their hard, rigid appearance and feel. But bone is actually a living tissue that contains several types of living cells involved in bone formation and remodeling, plus nerves and blood vessels. Indeed, bones bleed when cut during orthopedic surgery or when they break.

Bones perform five important functions. The first three—*support*, *protection*, and *movement*—are the same as the functions of the skeleton overall, which is, after all, primarily bone. The rigid support structure of bones is what allows us to sit and to stand upright. The bones of the skeleton also support, surround, and protect many of our soft internal organs, such as the lungs, liver, and spleen. The attachment of bones to muscles makes it possible for our bodies to move.

The fourth and fifth functions of bones—*blood cell formation* and *mineral storage*—are harder to remember, but they are just as important. Cells in certain bones are the only source of new red and white blood cells and platelets for blood. Without this production and supply function we would die within months. You will learn more about the formation of blood cells in Chapter 7. Bones also serve as an important long-term storage depot for two important minerals, calcium and phosphate. These two minerals can be drawn from bone when necessary, though excessive withdrawal may have consequences for bone composition and strengh.

Bone contains living cells

A typical long bone, so called because it is longer than it is wide, consists of a cylindrical shaft (called the *diaphysis*) with an enlarged knob called an *epiphysis* at each end (**Figure 5.1a**). Dense **compact bone** forms the shaft and covers each end. A central cavity in the shaft is filled with *yellow bone marrow*. Yellow bone marrow is primarily fat that can be utilized for energy.

The outer surface of the bone is covered by a tough layer of connective tissue, the *periosteum*, which contains specialized bone-forming cells. If an epiphysis of a long bone forms a movable joint with another bone, the joint surface is covered by a smooth layer of cartilage that reduces friction.

Inside each epiphysis is **spongy bone** (Figure 5.1b). Spongy bone is less dense than compact bone, allowing the bones to be light but strong. Spongy bone is a latticework of hard, relatively strong *trabeculae* (from Latin, meaning "little beams") composed of calcium minerals and living cells. In certain long bones, most notably the long bones of the upper arms and legs (humerus and femur, respectively), the spaces between the trabeculae are filled with *red bone marrow*. Special cells called *stem cells* in the red bone marrow are responsible for the production of red and white blood cells and platelets.

Taking a closer look (Figure 5.1c), we see that compact bone is made up largely of extracellular deposits of calcium

Central canal

Osteon

Osteocytes

c) A photograph of an osteon of compact bone showing osteocytes embeded within the solid structure.

Osteon

Osteocyte

Lacuna

Canalicula

d) A single osteocyte in a lacuna. Osteocytes remain in contact with each other by cytoplasmic extensions into the canaliculi between cells.

external structures

internal structures

(oxygen)

Epiphysis

Diaphysis

Epiphysis

Spongy bone (spaces contain red bone marrow)

Compact bone

Yellow bone marrow

Blood vessel

Periosteum

Central cavity (contains yellow bone marrow)

a) A partial cut through a long bone.

Spongy bone

Compact bone

Osteoblasts   Blood vessels and nerve in central canal

b) A closer view of a section of bone. Compact bone is a nearly solid structure with central canals for the blood vessels and nerves.

Figure 5.1 **Structure of bone.**

phosphate enclosing and surrounding living cells called **osteocytes** (from the Greek words for "bone" and "cells"). Osteocytes are arranged in rings in cylindrical structures called **osteons** (sometimes called *Haversian systems*). Osteocytes nearest the center of an osteon receive nutrients by diffusion from blood vessels that pass through a **central canal** (Haversian canal).

As bone develops and becomes hard, the osteocytes become trapped in hollow chambers called *lacunae* (Figure 5.1d). However, the osteocytes remain in direct contact with each other via thin canals called *canaliculi*. Within the canaliculi, extensions of the cell cytoplasm of adjacent osteocytes are joined together by gap junctions. (Gap junctions, as explained in Chapter 4, are channels that permit the movement of ions, water, and other molecules between two adjacent cells.) By exchanging nutrients across these gap

junctions, all the osteocytes can be supplied with nutrients even though most osteocytes are not located near a blood vessel. Waste products produced by the osteocytes are exchanged in the opposite direction and are removed from the bone by the blood vessels.

In spongy bone, osteocytes do not need to rely on central canals for nutrients and waste removal. The slender trabecular structure of spongy bone gives each osteocyte access to nearby blood vessels in red bone marrow.

✅ **Quick Check** If osteocytes did not have gap junctions in their cell membranes, would they be able to survive? Explain. ▪

Answers to ✅ questions can be found in Appendix A.

Ligaments hold bones together

Ligaments attach bone to bone. Ligaments consist of dense fibrous connective tissue, meaning that they are a regular array of closely packed collagen fibers all oriented in the same direction with a few fibroblasts in between (see Chapter 4). (Recall that fibroblasts are cells that produce and secrete the proteins that compose collagen, elastic, and reticular fibers.) Ligaments confer strength to certain joints while still permitting movement of the bones in relation to each other.

Cartilage lends support

Cartilage, as you already know, contains fibers of collagen and/or elastin in a ground substance of water and other materials. Cartilage is smoother and more flexible than bone. Cartilage is found where support under pressure is important and where some movement is necessary.

There are three types of cartilage in the human skeleton. *Fibrocartilage* consists primarily of collagen fibers arranged in thick bundles. It withstands both pressure and tension well. The intervertebral disks between the vertebrae, and also certain disk-like supportive structures in the knee joint called *menisci*, are made of fibrocartilage. *Hyaline cartilage* is a

smooth, almost glassy cartilage of thin collagen fibers. Hyaline cartilage forms the embryonic structures that later become the bones. It also covers the ends of mature bones in joints, creating a smooth, low-friction surface. *Elastic cartilage* is mostly elastin fibers, so it is highly flexible. It lends structure to the outer ear and to the epiglottis, a flap of tissue that covers the larynx during swallowing.

↩ Recap Bones contribute to support, movement, and protection. Bones also produce the blood cells and store minerals. Ligaments hold bones together, and cartilage provides support. ▪

5.2 Bone development begins in the embryo

In the earliest stages of fetal development, even before organs develop, the rudimentary models of future bones are created out of hyaline cartilage by cartilage-forming cells called **chondroblasts** (Figure 5.2a). After about two to three months of fetal development, the chondroblasts slowly die out and the cartilage models begin to dissolve

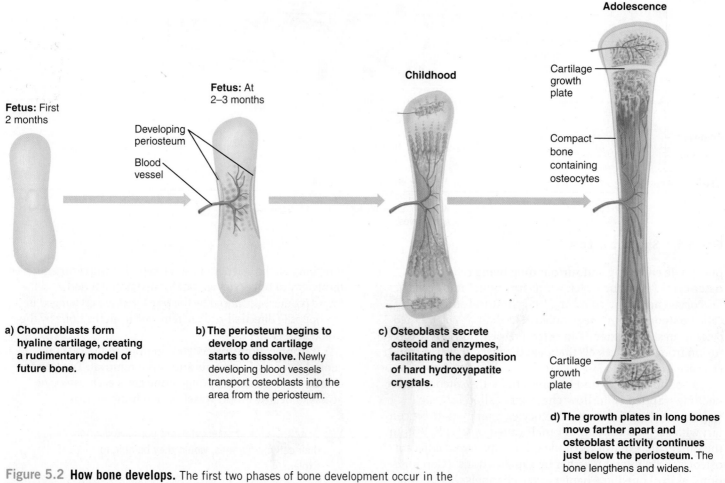

Adolescence

Fetus: First 2 months

Fetus: At 2–3 months

Developing periosteum

Blood vessel

Childhood

Cartilage growth plate

Compact bone containing osteocytes

Cartilage growth plate

a) **Chondroblasts form hyaline cartilage, creating a rudimentary model of future bone.**

b) **The periosteum begins to develop and cartilage starts to dissolve.** Newly developing blood vessels transport osteoblasts into the area from the periosteum.

c) **Osteoblasts secrete osteoid and enzymes, facilitating the deposition of hard hydroxyapatite crystals.**

d) **The growth plates in long bones move farther apart and osteoblast activity continues just below the periosteum.** The bone lengthens and widens.

Figure 5.2 How bone develops. The first two phases of bone development occur in the fetus. Bones continue to grow longer throughout childhood and adolescence because of growth at the growth plates.

and are replaced by bone. This process is called *ossification*. Although ossification is slightly different for flat bones and long bones, we will concentrate on the process for long bones.

After the chondroblasts die, the cartilage models they produced gradually break down inside the future shaft and epiphysis of the bone, making room for blood vessels to develop. The blood vessels carry bone-forming cells called **osteoblasts** (from the Greek words for "bone" and "to build") into the area from the developing periosteum (Figure 5.2b). The osteoblasts secrete a mixture of proteins (including collagen) called *osteoid*, which forms a matrix that provides internal structure and strength to bone. Osteoblasts also secrete enzymes that facilitate the crystallization of hard mineral salts of calcium phosphate, known as *hydroxyapatite*, around and between the osteoid matrix (Figure 5.2c). As more and more hydroxyapatite is deposited, the osteoblasts become embedded in the hardening bone tissue. In mature compact bone, approximately one-third of the structure is osteoid and two-thirds is crystals of hydroxyapatite.

Eventually the rate at which osteoblasts produce the osteoid matrix and stimulate the mineral deposits declines, and osteoblasts become mature osteocytes embedded in their individual lacunae. Mature osteocytes continue to maintain the bone matrix, however. Without them the matrix would slowly disintegrate.

Bones continue to lengthen throughout childhood and adolescence. This is because a narrow strip of cartilage called the **growth plate** (or *epiphyseal plate*) remains in each epiphysis (Figure 5.2d). Chondroblast activity (and hence the development of new cartilage as a model for the lengthening bone) is concentrated on the outside of the plate, whereas the conversion of the cartilage model to bone by osteoblasts is concentrated on the inside of the plate (Figure 5.3). In effect, the bone lengthens as the two growth plates migrate

farther and farther apart. Bones also grow in width as osteoblasts lay down more bone on the outer surface just below the periosteum.

 Web Animation *Bone Growth* at www.humanbiology.com

The bone development process is controlled by hormones, chemicals secreted by the endocrine glands. The most important hormone in preadolescents is growth hormone, which stimulates the bone-lengthening activity of the growth plate. During puberty the sex hormones (testosterone and estrogen) also stimulate the growth plate, at least initially. But at about age 18 in females and 21 in males these same sex hormones signal the growth plates to stop growing, and the cartilage is replaced by bone tissue. At this point the bones can no longer lengthen, though they can continue to grow in width.

Recap Bone-forming cells called osteoblasts produce a protein mixture (including collagen) that becomes bone's structural framework. They also secrete an enzyme that facilitates mineral deposition. ∎

Osteoblasts convert cartilage to bone at the inner surface

Chondroblasts deposit new cartilage at the outer surface

Joint cartilage

Growth plate

rotate smoothie

Figure 5.3 How long bones increase in length.

☑ *Mark the place(s) in this bone where osteoblasts are most active, and name the crystal that they are producing.*

5.3 Mature bone undergoes remodeling and repair

Even though bones stop growing longer, they do not remain the same throughout life. Bone is a dynamic tissue that undergoes constant replacement, remodeling, and repair. Remodeling may be so extensive that there is a noticeable change in bone shape over time, even in adults.

Bone remodeling and repair is in part due to a third type of bone cell called an **osteoclast** (from the Greek words for "bone" and "to break"). Osteoclasts cut through mature bone tissue, dissolving the hydroxyapatite and digesting the osteoid matrix in their path. The released calcium and phosphate ions enter the blood. The areas from which bone has been removed attract new osteoblasts, which lay down new osteoid matrixes and stimulate the deposition of new hydroxyapatite crystals.

Table 5.1 summarizes the four types of cells that contribute to bone development and maintenance.

Bones can change in shape, size, and strength

Over time, constant remodeling can actually change the shape of a bone. The key is that compression stress on a bone, such as the force of repeated jogging on the legs, causes tiny electrical currents within the bone. These electrical currents stimulate the bone-forming activity of osteoblasts. The compressive forces and the electric currents are greatest at the inside curvature of the long bone undergoing stress (Figure 5.4). Thus, in the normal course of bone turnover,

Compressive force

Bone removed here — New bone added here

a) The application of force to a slightly bent bone produces a greater compressive force on the inside curvature. Compressive force produces weak electrical currents which stimulate osteoblasts.

b) Over time, bone is deposited on the inside curvature and removed from the outside curvature.

c) The final result is a bone matched to the compressive force to which it is exposed.

Figure 5.4 **Bone remodeling.**

new bone is laid down in regions under high compressive stress and bone is resorbed in areas of low compressive stress. The final shape of a bone, then, tends to match the compressive forces to which it is exposed.

Weight-bearing exercise increases overall bone mass and strength. The effect is pronounced enough that the bones of trained athletes may be visibly thicker and heavier than those of nonathletes. You don't have to be a professional athlete to get this benefit, however. If you begin a regular program of any weight-bearing exercise, such as jogging or weight lifting, your bones will become denser and stronger as your osteoblasts produce more bone tissue.

The maintenance of homeostasis of bone structure depends on the precise balance of the activities of osteoclasts and osteoblasts. **Osteoporosis** is a common condition in which bones lose a great deal of mass (seemingly becoming

Table 5.1 **Cells involved in bone development and maintenance**

Type of cell	Function
Chondroblasts	Cartilage-forming cells that build a model of the future bone
Osteoblasts	Young bone-forming cells that cause the hard extracellular matrix of bone to develop
Osteocytes	Mature bone cells that maintain the structure of bone
Osteoclasts	Bone-dissolving cells

"porous") because of an imbalance over many years in the rates of activities of these two types of bone cells.

 Quick Check Professional bicyclists tend to lose bone density in their arms and back. Why does this happen given that they are otherwise in very good physical condition? ∎

Bone cells are regulated by hormones

Like bone growth, the rates of activities of osteoblasts and osteoclasts in adulthood are regulated by hormones that function to maintain calcium homeostasis. When blood levels of calcium fall below a given point, *parathyroid hormone (PTH)* stimulates the osteoclasts to secrete more bone-dissolving enzymes. The increased activity of osteoclasts causes more bone to be dissolved, releasing calcium and phosphate into the bloodstream. If calcium levels rise, then another hormone called *calcitonin* stimulates osteoblast activity, causing calcium and phosphate to be removed from blood and deposited in bone. Although the total bone mass of young adults doesn't change much, it's estimated that almost 10% of their bones may be remodeled and replaced each year. We discuss this and other types of hormonal regulation further in Chapter 13.

Quick Check Suppose a man is not getting sufficient calcium in his diet, such that his blood calcium level is chronically low. Would his PTH levels and calcitonin levels be low, normal, or high? Explain. ∎

Bones undergo repair

When you break (fracture) a bone, the blood vessels supplying the bone bleed into the area, producing a mass of clotted blood called a *hematoma*. Inflammation, swelling, and pain generally accompany the hematoma in the days immediately after a fracture. The repair process begins within days as fibroblasts migrate to the area. Some of the fibroblasts become chondroblasts, and together they produce a tough fibrocartilage bond called a *callus* between the two broken ends of the bone. A callus can be felt as a hard, raised ring at the point of the break. Then osteoclasts arrive and begin to remove dead fragments of the original bone and the blood cells of the hematoma. Finally, osteoblasts arrive to deposit osteoid matrix and encourage the crystallization of calcium phosphate minerals, converting the callus into bone. Eventually the temporary union becomes dense and hard again. Bones rarely break in the same place twice because the repaired union remains slightly thicker than the original bone.

The repair process can take weeks to months, depending on your age and the bone involved. In general, the repair process slows with age. Recently it has been discovered that the application of weak electrical currents to the area of a broken bone can increase the rate of healing. It is thought that electrical current works by attracting osteoclasts and osteoblasts to the area under repair.

 Web Animation *Bone Repair* at www.humanbiology.com

 Recap Healthy bone replacement and remodeling depend on the balance of activities of bone-resorbing osteoclasts and bone-forming osteoblasts. When a bone breaks, a fibrocartilage callus forms between the broken ends and is later replaced with bone. ∎

5.4 The skeleton protects, supports, and permits movement

Now that we have reviewed the dynamic nature of bone tissue, we turn to how all of those bones are classified and organized. Bones can be classified into four types based on shape: long, short, flat, and irregular. So far we have discussed *long bones,* which include the bones of the limbs and fingers. *Short bones* (the bones of the wrists) are approximately as wide as they are long. *Flat bones* (including the cranial bones, the sternum, and the ribs) are thin, flattened, and sometimes curved, with only a small amount of spongy bone sandwiched between two layers of compact bone. *Irregular bones* such as the coxal (hip) bones and the vertebrae include a variety of shapes that don't fit into the other categories. A few flat and irregular bones, including the sternum and the hip bones, contain red bone marrow that produces blood cells.

The 206 bones of the human body and the various connective tissues that hold them together make up the **skeleton** (Figure 5.5). The skeleton has three important functions. First, it serves as a structural framework for support of the soft organs. Second, it protects certain organs from physical injury. The brain, for example, is enclosed within the bones of the skull, and the heart and lungs are protected by a bony cage consisting of ribs, the sternum, and vertebrae. Third, because of the way that the bony elements of the skeleton are joined together at joints, the presence of the skeleton permits flexible movement of most parts of the body. This is particularly true of the hands, feet, legs, and arms.

The skeleton is organized into the *axial skeleton* and the *appendicular skeleton.*

The axial skeleton forms the midline of the body

The **axial skeleton** consists of the skull (including the maxilla and mandible), sternum, ribs, and vertebral column (including the sacrum) (see Figure 5.5).

The skull: Cranial and facial bones The human **skull** (cranium) comprises over two dozen bones that protect the brain and form the structure of the face. Figure 5.6 (on the next page) illustrates some of the more important bones of the skull.

The *cranial bones* are flat bones in the skull that enclose and protect the brain. Starting at the front of the skull, the *frontal bone* comprises the forehead and the upper ridges of the eye sockets. At the upper left and right sides of the skull are the two *parietal bones,* and forming the lower left and

Axial skeleton   **Appendicular skeleton**

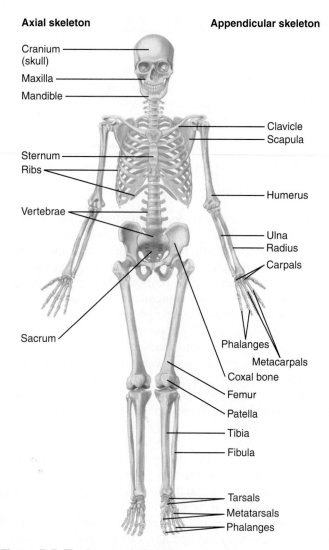

Cranium (skull)
Maxilla
Mandible
Clavicle
Scapula
Sternum
Ribs
Humerus
Vertebrae
Ulna
Radius
Carpals
Sacrum
Phalanges
Metacarpals
Coxal bone
Femur
Patella
Tibia
Fibula
Tarsals
Metatarsals
Phalanges

Figure 5.5 The human skeleton.

✓ *On this and subsequent figures, find the anatomical terms corresponding to the following common names: breastbone, collarbone, shoulder blade, hip bone, thighbone, shinbone.*

right sides are the two *temporal bones.* Each temporal bone is pierced by an opening into the ear canal that allows sounds to travel to the eardrum. Between the frontal bone and the temporal bones is the *sphenoid* bone, which forms the back of both eye sockets. The *ethmoid* bone contributes to the eye sockets and also helps support the nose.

The two small, narrow *nasal bones* underlie only the upper bridge of the nose; the rest of the fleshy protuberance called the nose is made up of cartilage and other connective tissue. Part of the space formed by the maxillary and nasal bones is the nasal cavity. The small *lacrimal bones,* at the inner eye sockets, are pierced by a tiny opening through which the tear ducts drain tears from the eye sockets into the nasal cavity.

The *mandible,* or lower jaw, contains the sockets that house the lower row of teeth. All the bones of the skull are joined tightly together except for the mandible, which attaches

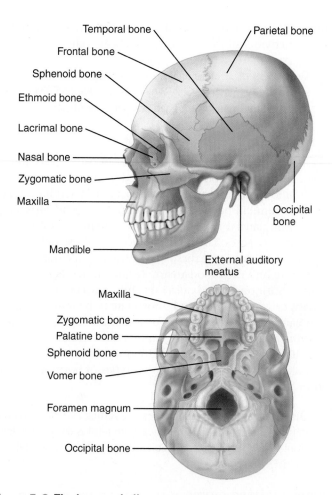

Temporal bone
Frontal bone
Sphenoid bone
Ethmoid bone
Lacrimal bone
Nasal bone
Zygomatic bone
Maxilla
Mandible
Parietal bone
Occipital bone
External auditory meatus

Maxilla
Zygomatic bone
Palatine bone
Sphenoid bone
Vomer bone
Foramen magnum
Occipital bone

Figure 5.6 The human skull. Except for the mandible, which has a hinged joint with the temporal bone, the bones of the skull are joined tightly together. Their function is protection, not movement.

to the temporal bone by a joint that, because it permits a substantial range of motion, allows us to speak and chew.

Curving underneath to form the back and base of the skull is the *occipital bone.* Near the base of the occipital bone is a large opening called the *foramen magnum* (Latin for "great opening"). This is where the vertebral column connects to the skull and the spinal cord enters the skull to communicate with the brain.

The *facial bones* compose the front of the skull. On either side of the nose are the two *maxilla* (maxillary) bones, which form part of the eye sockets and contain the sockets that anchor the upper row of teeth. The hard palate (the "roof" of the mouth) is formed by the maxilla bones and the two *palatine bones.* Behind the palatine bones is the *vomer bone,* which is part of the nasal septum that divides the nose into left and right halves. The two *zygomatic bones* form the cheekbones and the outer portion of the eye sockets.

Several of the cranial and facial bones contain air spaces called **sinuses,** which make the skull lighter and give the human voice its characteristic tone and resonance. Each sinus is lined with tissue that secretes mucus, a thick, sticky fluid that helps trap foreign particles in incoming air. The sinuses connect to the nasal cavity via small passageways through which the mucus normally drains. However, if you develop a cold or respiratory infection, the tissue lining your sinuses can become inflamed and block these passages. Sinus inflammation is called *sinusitis.* If fluid accumulates inside the sinuses, the resulting sensation of pressure may give you a "sinus headache."

The vertebral column: The body's main axis The **vertebral column** (the backbone or spine) is the main axis of the body (**Figure 5.7**). It supports the head, protects the spinal cord, and serves as the site of attachment for the four limbs and various muscles. It consists of a column of 33 irregular bones called *vertebrae* (singular: *vertebra*) that extends from the skull to the pelvis. When viewed from the side the vertebral column is somewhat curved, reflecting slight differences in structure and size of vertebrae in the various regions.

We classify the vertebral column into five anatomical regions:

- Cervical (neck)—7 vertebrae.
- Thoracic (the chest or thorax)—12 vertebrae.
- Lumbar (the lower portion or "small" of the back, which forms the lumbar curve of the spine)—5 vertebrae.
- Sacral (in the sacrum or upper pelvic region)—In the course of evolution, the 5 sacral vertebrae have become fused.
- Coccygeal (the coccyx or tailbone)—4 fused vertebrae. The coccyx is all that remains of the tails of our ancient ancestors. It is an example of a *vestigial* structure, meaning one that no longer has any function.

A closer look at vertebrae (**Figure 5.8a**) shows how they are stacked on each other and how they are joined. Vertebrae share two points of contact, called *articulations,* located

Figure 5.7 The vertebral column. Vertebrae are named and numbered according to their location. The vertebral column is moderately flexible because of the presence of joints and intervertebral disks.

a) **Healthy disks.**

b) **A herniated disk.**

Figure 5.8 Vertebrae.

behind their main body. There are also articulations with the ribs. The spinal cord passes through a hollow cavity between the articulations and the main body. Neighboring vertebrae are separated from each other by a flat, elastic, compressible **intervertebral disk** composed of a soft gelatinous center and a tough outer layer of fibrocartilage. Intervertebral disks serve as shock absorbers, protecting the delicate vertebrae from the impact of walking, jumping, and other movements. In conjunction with the vertebral joints, vertebral disks also permit a limited degree of movement. This lends the vertebral column greater flexibility, allowing us to bend forward, lean backward, and rotate the upper body.

An especially strong impact or sudden movement can compress an intervertebral disk, forcing the softer center to balloon outward, press against spinal nerves, and cause intense back pain. This condition is referred to as a "herniated" or "slipped disk" (Figure 5.8b), and it occurs most often in the lumbar vertebrae. Occasionally the disk may rupture, releasing its soft, pulpy contents. The pain that accompanies a herniated disk can be alleviated by surgery to remove the damaged disk, relieving the pressure against the nerve. However, surgical correction of a herniated disk reduces spinal flexibility somewhat because the two adjacent vertebrae must be fused together with bone grafts.

Generally the bony vertebral column does an effective job of shielding the softer spinal cord, which consists of nervous tissue that connects the brain to the rest of the body. However, injury to the vertebral column can damage the spinal cord or even sever it, resulting in partial or complete paralysis of the body below that point. Persons with suspected vertebral injuries should not be moved until a physician can assess the situation, because any twisting or bending could cause additional, perhaps permanent, damage to the spinal cord. You may have noticed that when athletes are injured on the field, they are instructed to lie absolutely still until a trainer and physician have examined them thoroughly.

The ribs and sternum: Protecting the chest cavity Humans have 12 pairs of **ribs** (Figure 5.9). One end of each rib branches from the thoracic region of the vertebral column. The other ends of the upper seven pairs attach via cartilage to the **sternum,** or breastbone, a flat blade-shaped bone composed of three separate bones that fuse during development. Rib pairs 8–10 are joined to the seventh rib by cartilage, and thus attach indirectly to the sternum. The bottom two pairs of ribs are called *floating ribs* because they do not attach to the sternum at all.

The ribs, sternum, and vertebral column form a protective *rib cage* that surrounds and shields the heart, lungs, and other organs of the chest (thoracic) cavity. The rib cage also helps us breathe, because muscles between the ribs lift them slightly during breathing, expanding the chest cavity and

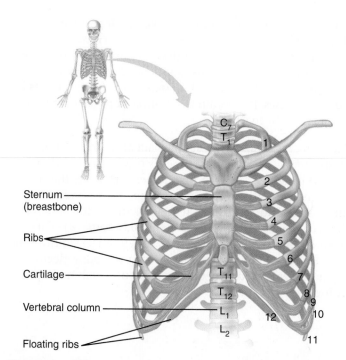

Figure 5.9 Ribs. The 12 pairs of ribs are numbered according to their attachment to the thoracic vertebrae. Only the first 7 pairs attach directly to the sternum.

 What function do the ribs and sternum have that other parts of the skeleton do not have? How might this explain the fact that the ribs and sternum are connected by flexible cartilage rather than by bone?

inflating the lungs. The base of the sternum is connected to the diaphragm, a muscle that is important to breathing (see Chapter 10).

 Quick Check Humans have more sacral vertebrae than most mammals do, and these sacral bones are fused into an unusually strong structure. Given what you know about the functions of the vertebral column, propose an explanation. ∎

The appendicular skeleton: Pectoral girdle, pelvic girdle, and limbs

Those parts of the body that attach to the axial skeleton are called appendages, from the Latin word meaning "to hang upon." The second division of the human skeleton, the **appendicular skeleton,** includes the arms, legs, and their attachments to the trunk, which are the pectoral and pelvic girdles.

The pectoral girdle lends flexibility to the upper limbs The **pectoral girdle,** a supportive frame for the upper limbs, consists of the right and left **clavicles** (collarbones) and right and left **scapulas** (shoulder blades) (Figure 5.10). The

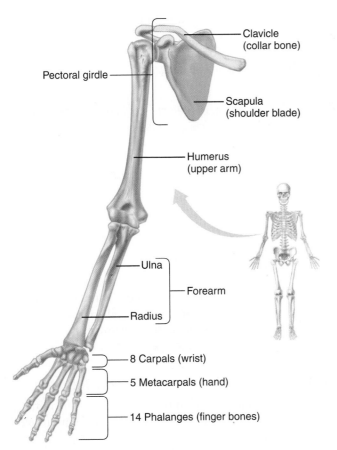

Clavicle
(collar bone)

Pectoral girdle

Scapula
(shoulder blade)

Humerus
(upper arm)

Ulna

Forearm

Radius

8 Carpals (wrist)

5 Metacarpals (hand)

14 Phalanges (finger bones)

Figure 5.10 Bones of the right side of the pectoral girdle and the right arm and hand.

clavicles extend across the top of the chest and attach to the scapulas, triangular bones in the upper back.

The arm and hand consist of 30 different bones. The upper end of the **humerus,** the long bone of the upper arm, fits into a socket in the scapula. The other end of the humerus meets with the **ulna** and **radius,** the two bones of the forearm, at the elbow. If you've ever hit your elbow and experienced a painful tingling, you know why this area is nicknamed the "funny bone"; you've just struck the ulnar nerve that travels along the elbow.

The lower ends of the ulna and radius meet the *carpal* bones, a group of eight small bones that makes up the wrist. The five *metacarpal* bones form the palm of the hand, and they join with the 14 *phalanges,* which form the fingers and thumb.

The pectoral girdle and arms are particularly well adapted to permit a wide range of motion. They connect to the rest of the body via muscles and tendons—a relatively loose method of attachment. This structure gives the upper body of humans a degree of dexterity unsurpassed among large animals. We can rotate our upper arms almost 360 degrees—a greater range of movement than with any other joint in the body. The upper arm can rotate in roughly a

circle, the arm can bend in one dimension and rotate, and the wrist and fingers can all bend and rotate to varying degrees. We also have "opposable thumbs," meaning we can place them opposite our other fingers. The opposable thumb has played an important role in our evolutionary history, as it makes it easier to grasp and manipulate tools and other objects.

We pay a price for this flexibility, because freedom of movement also means relative instability. If you fall on your arm, for example, you might dislocate your shoulder joint or crack a clavicle. In fact, the clavicle is one of the most frequently broken bones in the body.

Although our upper limbs are well adapted to a wide range of movements, too much of one kind of motion can be harmful. Repetitive motions—performing the same task over and over—can lead to health problems called *repetitive stress syndromes.* Depending on the part of the body that is overused, these injuries can take many forms. A well-known repetitive stress syndrome is *carpal tunnel syndrome,* a condition often due to repetitive typing at a computer keyboard. The carpal bones of the wrist are held together by a sheath of connective tissue. The blood vessels, nerves, and tendons to the hand and fingers pass through the sheath via the "carpal tunnel." Overuse of the fingers and hands produces swelling and inflammation of the tendons, which causes them to press against the nerve supplying the hand. The result may be pain, tingling, or numbness in the wrist and hand. Mild episodes of carpal tunnel syndrome respond to rest and pain relievers. Severe cases can be treated with surgery to relieve the pressure.

The pelvic girdle supports the body The **pelvic girdle** consists of the two **coxal bones** and the sacrum and coccyx of the vertebral column (**Figure 5.11** on the next page). The coxal bones attach to the sacral region of the vertebral column in back, then curve forward to meet in front at the *pubic symphysis,* where they are joined by cartilage. You can feel the upper curves of the coxal bones (the iliac region) as your hip bones. Together, these structures form the pelvis.

The primary function of the pelvic girdle is to support the weight of the upper body against the force of gravity. It also protects the organs inside the pelvic cavity and serves as a site of attachment for the legs. The structure of the pelvic girdle reflects a trade-off between dexterity and stability. Partly because the pelvic girdle and lower limbs are larger and more firmly connected to the rest of the body than the pectoral girdle and upper limbs, the lower limbs are less dexterous than the upper limbs.

In adult women the pelvic girdle is broader and shallower than it is in men, and the pelvic opening is wider. This allows for safe passage of a baby's head during labor and delivery. These characteristic differences appear during puberty when a woman's body begins to produce sex hormones. The sex hormones trigger a process of bone remodeling that shapes the female pelvic girdle to adapt for pregnancy and birth.

Coxal bones and
sacrum (pelvis)

Pubic
symphysis

Femur (upper leg)

Patella (knee cap)

Tibia

Lower leg

Fibula

7 Tarsals (ankle)

5 Metatarsals (foot)

14 Phalanges (toe bones)

Figure 5.11 Bones of the pelvic girdle and the left leg and foot.

The **femur** (thighbone) is the longest and strongest bone in the body. When you jog or jump, your femurs are exposed to forces of impact of several tons per square inch. The rounded upper end of each femur fits securely into a socket in a coxal bone, creating a stable joint that effectively supports the body while permitting movement. The lower end of the femur intersects at the knee joint with the larger of the two bones of the lower leg, the **tibia,** which in turn makes contact with the thinner **fibula.** The *patella,* or kneecap, is a triangle-shaped bone that protects and stabilizes the knee joint.

At the ankle, the tibia and fibula join with the seven *tarsal* bones that make up the ankle and heel. Five long bones, the *metatarsals,* form the foot. The 14 bones of the toes, like those of the fingers, are called *phalanges.*

🔄 **Recap** The skull and vertebral column protect the brain and spinal cord, the rib cage protects the organs of the chest cavity, and the pelvic girdle supports the body's weight and protects the pelvic organs. The upper limbs are capable of a wide range of motions (dexterous movement). The lower limbs are stronger but less dexterous than the upper limbs. ∎

5.5 Joints form connections between bones

We now turn to the structures and tissues that hold the skeleton together while still permitting us to move about freely: joints, ligaments, and tendons. **Joints,** also called articulations, are the points of contact between bones. Ligaments and tendons are connective tissues that stabilize many joints.

Joints vary from immovable to freely movable

Joints vary considerably from basically immovable to freely movable. Types of joints include fibrous, cartilaginous, and synovial joints.

Fibrous joints are immovable. At birth, the flat bones in a baby's skull are separated by relatively large spaces filled with fibrous connective tissue. These "soft spots," called *fontanels,* enable the baby's head to change shape slightly so that it can squeeze safely through the mother's pelvic opening during childbirth. The presence of joints also allows for brain growth and development after birth. During childhood these fibrous joints gradually harden. By the time we reach adulthood, the joints have become thin lines, or sutures, between skull bones. These immovable joints firmly connect the bones that protect and stabilize the skull and brain.

Cartilaginous joints, in which the bones are connected by hyaline cartilage, are slightly movable, allowing for some degree of flexibility. Examples include the cartilaginous joints that connect the vertebrae in the backbone, and those that attach the lower ribs to the sternum.

The most freely movable joints are **synovial joints,** in which the bones are separated by a thin fluid-filled cavity. The two bones of a synovial joint are fastened together and stabilized by ligaments. The interior of the cavity is lined with a *synovial membrane,* which secretes *synovial fluid* to lubricate and cushion the joint. To reduce friction even further, the articulating surfaces of the two bones are covered with a tough but smooth layer of hyaline cartilage. Together, the synovial membrane and the surrounding hyaline cartilage constitute the *joint capsule.*

Different types of synovial joints permit different kinds of movements. A *hinge joint,* such as the knee and elbow, gets its name because it allows movement in one plane like the hinges on a door. Figure 5.12 illustrates a human knee joint. The knee joint is strong enough to withstand hundreds of pounds of force, yet it is flexible enough to swing freely in one direction. To reduce friction, there are small disks of cartilage on either side of the knee called *menisci* (singular: *meniscus*). The knee joint also includes 13 small sacs of fluid, called *bursae* (singular: *bursa*), for additional cushioning. The entire joint is wrapped in strong ligaments that attach bone to bone and tendons that attach bone to muscle. Note the two *cruciate ligaments* (posterior and anterior) that join the tibia to the femur bone. The anterior cruciate ligament is sometimes injured when the knee is hit with great force from the side.

A second type of synovial joint, a *ball-and-socket joint,* permits an even wider range of movement. Examples include the joint between the femur and the coxal bone (see Figure 5.11), and between the humerus and the pectoral girdle (see Figure 5.10). In both cases, the rounded head of the bone fits into a socket, allowing movement in all planes.

Figure 5.13 (on the next page) illustrates the different types of movements made possible by hinge and ball-and-socket

joints. Note that you can rotate your arm and your leg because the shoulder and hip are ball-and-socket joints, but you cannot rotate the hinge joint in your knee.

Ligaments, tendons, and muscles strengthen and stabilize joints

Thanks to its design, a synovial joint can withstand tremendous pounding day after day, year after year without wearing out. But where does it get its strength? For that we turn to ligaments, tendons, and muscles. As we have seen, the bones of a synovial joint are held tightly together by ligaments. They are stabilized even more by **tendons,** another type of tough connective tissue, which join the bones to muscles. Ligaments and tendons contain collagen arranged in parallel fibers, making ligaments and tendons as strong and as flexible as a twisted nylon rope. In addition, muscle contraction strengthens and stabilizes certain joints at the very moment they need it the most.

To appreciate the role of muscle contraction in stabilizing a joint, try this simple experiment. Sit in a low chair, stretch your leg straight out in front of you with your heel

a) **A cutaway anterior view of the right knee with muscles, tendons, and the joint capsule removed and the bones pulled slightly apart so that the two menisci are visible.**

Femur — Hyaline cartilage — Posterior cruciate ligament — Anterior cruciate ligament — Meniscus — Ligament — Ligament — Meniscus — Fibula — Tibia — Ligament — Patella — Tendon

b) **A view of the knee with muscles, tendons, and ligaments in their normal position surrounding the intact joint capsule.** The combination of ligaments, tendons, and muscles holds the knee tightly together.

Thigh muscles — Tendon — Joint capsule — Ligaments — Patella — Ligaments

Figure 5.12 **The knee joint is a hinged synovial joint.**

 What is the difference between a ligament and a tendon? Hint: find all the ligaments and tendons in this figure and notice what they are attached to.

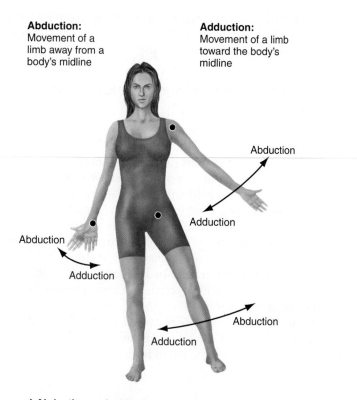

Abduction:
Movement of a
limb away from a
body's midline

Adduction:
Movement of a limb
toward the body's
midline

Abduction

Adduction

Abduction

Adduction

Abduction

Adduction

a) Abduction and adduction.

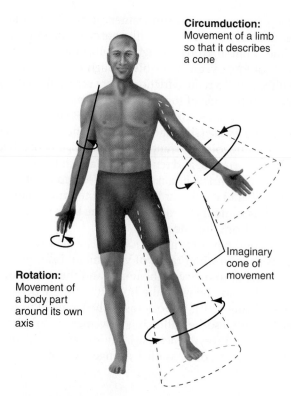

Circumduction:
Movement of a limb
so that it describes
a cone

Imaginary
cone of
movement

Rotation:
Movement of
a body part
around its own
axis

b) Rotation and circumduction.

Extension

Flexion   Hyperextension

Flexion:
Decreases the
angle of a joint

Flexion

Hyperextension:
Extension beyond
the anatomical
position

Extension

Flexion   Hyperextension

Extension

Extension:
Increases the
angle of a joint

c) Flexion, extension, and hyperextension.

Supination:
Rotation of the
forearm so palm
faces anteriorly

Pronation:
Rotation of the
forearm so palm
faces posteriorly

Humerus

Ulna   Ulna

Radius   Radius

d) Supination and pronation.

Figure 5.13 Types of movements made possible by synovial joints.

 Which of these types of movement can be produced by a hinge joint, a ball-and-socket joint, or a fibrous joint?

resting on the floor, and relax your muscles. Move your kneecap (patella) from side to side gently with your hand. Notice how easily you can shift it out of position. Now without changing position, tense the muscle of your thigh and again try to move your kneecap with your hand. See the difference? The patella is attached to the tibia by a ligament and to the muscles of the thigh by a tendon (Review Figure 5.12b). Contraction of the thigh muscle (as when you take a step while walking) puts tension on the tendon and the ligament. The increased tension holds the patella and the rest of the joint firmly in place. If you move your hand to just below the kneecap, you can feel the tightening of the patellar ligament as you alternately contract and relax your thigh muscle.

Recap Joints are the points of contact between bones. Fibrous joints are immovable in adults, cartilaginous joints permit some movement, and synovial joints are highly movable. Synovial joints are held together by ligaments and lubricated by synovial fluid. ▪

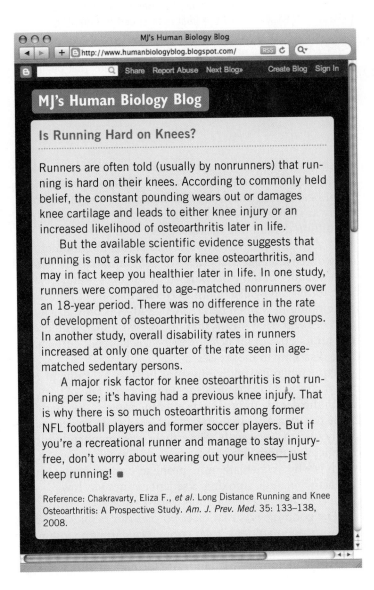

MJ's Human Biology Blog

Is Running Hard on Knees?

Runners are often told (usually by nonrunners) that running is hard on their knees. According to commonly held belief, the constant pounding wears out or damages knee cartilage and leads to either knee injury or an increased likelihood of osteoarthritis later in life.

But the available scientific evidence suggests that running is not a risk factor for knee osteoarthritis, and may in fact keep you healthier later in life. In one study, runners were compared to age-matched nonrunners over an 18-year period. There was no difference in the rate of development of osteoarthritis between the two groups. In another study, overall disability rates in runners increased at only one quarter of the rate seen in age-matched sedentary persons.

A major risk factor for knee osteoarthritis is not running per se; it's having had a previous knee injury. That is why there is so much osteoarthritis among former NFL football players and former soccer players. But if you're a recreational runner and manage to stay injury-free, don't worry about wearing out your knees—just keep running! ▪

Reference: Chakravarty, Eliza F., *et al.* Long Distance Running and Knee Osteoarthritis: A Prospective Study. *Am. J. Prev. Med.* 35: 133–138, 2008.

5.6 Diseases and disorders of the skeletal system

In this chapter we have already discussed several health conditions related to the skeletal system, including fractures and carpal tunnel syndrome. Now we look at several more.

Sprains mean damage to ligaments

A sprain is due to stretched or torn ligaments. Often it is accompanied by internal bleeding with subsequent bruising, swelling, and pain. The most common example is a sprained ankle. Sprains take a long time to heal because the ligaments have few cells and a poor blood supply. Minor sprains, in which the ligaments are only stretched, usually mend themselves with time. If a large ligament is torn completely, it generally does not heal by itself, and surgery may be necessary to remove it. Sometimes the joint can be stabilized with a piece of tendon or by repositioning other ligaments. Torn ligaments in the knee are particularly troublesome because they often leave the knee joint permanently unstable and prone to future injuries.

Quick Check Do you think a broken bone would take more or less time to heal than a sprained ligament? Why? ▪

Bursitis and tendinitis are caused by inflammation

Bursitis and tendinitis refer to inflammation of the bursae or tendons following injury. We discuss the inflammatory process in Chapter 9. Causes of bursitis and tendinitis may include tearing injuries to tendons, physical damage caused by blows to the joint, and even some bacterial infections. Like ligaments, tendons and the tissues lining the bursae are not well supplied with blood vessels, so they do not heal quickly. Treatment usually involves applying cold during the first 24 hours and heat after that, resting the injured area, and taking pain-relieving medications. "Tennis elbow" is a painful condition caused by either bursitis or tendinitis. Other common locations for pain include the knee, shoulder, and the Achilles tendon that pulls up the back of the heel.

Arthritis is inflammation of joints

By their nature, joints are exposed to high compressive forces and are prone to excessive wear caused by friction. "Arthritis" is a general term for joint inflammation. The most common type of arthritis is *osteoarthritis,* a degenerative ("wear-and-tear") condition that affects about 20 million Americans, most over age 45. In osteoarthritis the cartilage covering the ends of the bones wears out. With time the bone thickens and may form bony spurs, which further restrict joint movement. The result is increased friction between the bony surfaces, and the joint becomes inflamed and painful. Over-the-counter medications can reduce the inflammation and

Health & Wellness

Treating a Sprained Ankle

For a severe sprain, many physicians advise the frequent application of cold to the sprained area during the first 24 hours, followed by a switch to heat. Why the switch, and what is the logic behind the timing of cold versus heat?

The biggest immediate problem associated with a sprain is damage to small blood vessels and subsequent bleeding into the tissues. Most of the pain associated with a sprain is due to the bleeding and swelling, not damage to ligaments themselves. The immediate application of cold constricts blood vessels in the area and prevents most of the bleeding. The prescription is generally to cool the sprain for 30 minutes every hour or 45 minutes every hour and a half. In other words, keep the sprain cold for about half the time, for as long as you can stand it. The in-between periods ensure adequate blood flow for tissue metabolism. It's also a good idea to keep the ankle wrapped in

Treat sprains first with cold, then later with heat.

an elastic bandage and elevated between cooling treatments, to prevent swelling. If you're having trouble remembering all this, remember the acronym "RICE"— Rest, Ice, Compression, Elevation.

The key to a quick recovery from a sprain is rapid application of the RICE method. Athletes who try to "work through the pain" by continuing to compete while injured generally pay the price in a longer recovery time.

After 24 hours there shouldn't be any more bleeding from small vessels. The damage has been minimized, so now the goal is to speed the healing process. Heat dilates the blood vessels, improves the supply of nutrients to the area, and attracts blood cells that begin the process of tissue repair. ∎

---

pain, and surgical joint replacements for severe osteoarthritis are fairly routine today. Injections of hyaluronic acid, a component of hyaline cartilage, can also reduce arthritic knee pain. Many physicians advise people with osteoarthritis to exercise regularly, which helps preserve the joints' healthy range of motion. Several promising new treatments to reduce joint inflammation are still in the experimental stage.

Osteoarthritis should not be confused with *rheumatoid arthritis.* Rheumatoid arthritis also involves joint inflammation, but it is caused by the body's own immune system, which mistakenly attacks the joint tissues. We take a closer look at disorders of this type in Chapter 9.

✓ **Quick Check** A medical researcher is trying to develop a new drug that will help reverse arthritis. Which is likely to be most helpful: a drug that increases osteoclast activity, one that increases osteoblast activity, or one that increases chondroblast activity? Explain. ∎

Osteoporosis is caused by excessive bone loss

Osteoporosis is a condition caused by excessive bone loss over time (**Figure 5.14**), leading to brittle, easily broken bones. Symptoms include hunched posture (**Figure 5.15**),

difficulty walking, and an increased likelihood of bone fractures, especially of the spine and hip. Osteoporosis is a major health problem in the United States. Over 10 million Americans have the condition, and it accounts for more than 1.5 million debilitating fractures every year.

A very slow progressive bone loss occurs in both men and women after age 35 because of a slight imbalance between the rates of bone breakdown by osteoclasts and new bone formation by osteoblasts. Overall, the rate of bone loss in men (and in women before menopause) is only about 0.4% per year. That means that on average, a man will lose only about 20% of his bone mass by age 85—not enough to cause disability in most cases.

For women it's a different story, because a decline in estrogen after menopause leads to a more rapid rate of bone loss in the decade immediately after menopause—as high as 2–3% per year. After that, the rate of loss begins to decline slowly toward 0.4% again. Nevertheless, women tend to lose considerably more bone mass over a lifetime than men, which is why women are more prone to osteoporosis. Other risk factors include smoking, a sedentary lifestyle, low calcium intake, and being underweight.

The good news is that osteoporosis can be prevented. Two important strategies: get enough calcium and vitamin D, and maintain a consistent exercise program throughout your life. Calcium is crucial for the formation of new bone

a) A scanning electron micrograph (SEM) of normal bone.

b) SEM of a bone showing osteoporosis.

Figure 5.14 **Bone loss in osteoporosis.**

Figure 5.15 **Osteoporosis.** Osteoporosis can lead to repeated compression fractures of the spine and a permanent change in spine curvature.

tissue. Current recommendations call for a daily intake of about 1,000–1,500 mg per day for adults, but women who have gone through menopause may benefit from even higher intakes. Both men and women can benefit from weight-bearing exercise (such as walking) and strength training (such as lifting weights), because these activities increase bone mass. For women especially, estrogen replacement therapy after menopause can slow the rate of bone loss.

Several medications are available to treat osteoporosis. A class of drugs called biphosphonates (alendronate and risedronate) act by inhibiting the bone-resorbing function of osteoclasts. The FDA recently approved a new biphosphonate medication, Boniva Injection, that can be administered intravenously every three months. Teriparatide, a medication that is a fragment of the normal parathyroid hormone molecule, is the first osteoporosis medication that can actually stimulate the activity of the bone-forming osteoblasts.

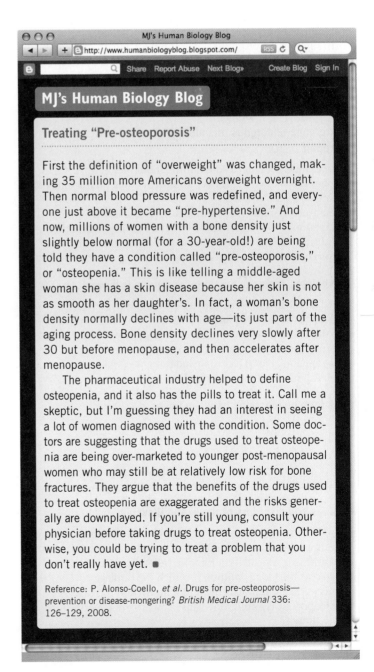

MJ's Human Biology Blog

http://www.humanbiologyblog.blogspot.com/

Share Report Abuse Next Blog» Create Blog Sign In

MJ's Human Biology Blog

Treating "Pre-osteoporosis"

First the definition of "overweight" was changed, making 35 million more Americans overweight overnight. Then normal blood pressure was redefined, and everyone just above it became "pre-hypertensive." And now, millions of women with a bone density just slightly below normal (for a 30-year-old!) are being told they have a condition called "pre-osteoporosis," or "osteopenia." This is like telling a middle-aged woman she has a skin disease because her skin is not as smooth as her daughter's. In fact, a woman's bone density normally declines with age—its just part of the aging process. Bone density declines very slowly after 30 but before menopause, and then accelerates after menopause.

The pharmaceutical industry helped to define osteopenia, and it also has the pills to treat it. Call me a skeptic, but I'm guessing they had an interest in seeing a lot of women diagnosed with the condition. Some doctors are suggesting that the drugs used to treat osteopenia are being over-marketed to younger post-menopausal women who may still be at relatively low risk for bone fractures. They argue that the benefits of the drugs used to treat osteopenia are exaggerated and the risks generally are downplayed. If you're still young, consult your physician before taking drugs to treat osteopenia. Otherwise, you could be trying to treat a problem that you don't really have yet. ■

Reference: P. Alonso-Coello, *et al*. Drugs for pre-osteoporosis—prevention or disease-mongering? *British Medical Journal* 336: 126–129, 2008.

Chapter Summary

The skeletal system consists of connective tissue *p. 104*

■ Connective tissues of the skeletal system are bones, ligaments, and cartilage.

■ Bone is a living tissue composed of cells and extracellular material.

■ Ligaments, composed of dense fibrous connective tissue, attach bones to each other.

■ Cartilage forms the intervertebral disks and lines the points of contact between bones.

Bone development begins in the embryo *p. 106*

■ After about two months of fetal development, rudimentary models of bones have been formed from cartilage.

■ Throughout the rest of fetal development and on into child-hood, bone-forming cells called *osteoblasts* replace the cartilage model with bone.

■ Growth in the length of long bones centers on growth plates in each epiphysis.

Mature bone undergoes remodeling and repair *p. 108*

■ Bone undergoes replacement throughout life.

■ Bones can change shape over time, depending on the forces to which they are exposed.

■ The process of bone repair includes: (1) the formation of a hematoma, (2) the formation of a fibrocartilage callus that binds the broken ends together, and (3) the eventual replacement of the callus with new bone.

The skeleton protects, supports, and permits movement *p. 109*

■ The axial skeleton is represented by the skull, the vertebral column, the sternum, and the ribs.

■ In the vertebral column, intervertebral disks of fibrocartilage absorb shock and permit limited movement.

■ The appendicular skeleton includes the pectoral girdle, the pelvic girdle, and the upper and lower limbs.

Joints form connections between bones *p. 114*

■ Three types of joints connect bones: fibrous, cartilaginous, and synovial.

■ Synovial joints are designed for movement without friction. They are lined with a synovial membrane and lubricated by synovial fluid.

Diseases and disorders of the skeletal system *p. 117*

■ Sprains are the result of stretched or torn ligaments. Bursitis and tendinitis are caused by injuries to the bursae and tendons.

■ Arthritis is a general term for joint inflammation.

■ Osteoarthritis is a condition in which the cartilage covering the ends of the bones wears out and joint friction increases.

■ Osteoporosis is a condition caused by progressive bone loss over time.

Terms You Should Know

appendicular skeleton, *112*
axial skeleton, *109*
bone, *104*
cartilage, *106*
central (Haversian) canal, *105*
chondroblast, *106*
compact bone, *104*
growth plate, *107*
intervertebral disk, *112*
joint, *114*
ligament, *106*
osteoblast, *107*
osteoclast, *108*
osteocyte, *105*
osteon, *105*
osteoporosis, *108*
spongy bone, *104*
tendon, *115*

Concept Review

Answers can be found at the Human Biology Place.
www.humanbiology.com

1. List the five functions of bone.
2. Describe the functions of red and yellow bone marrow.
3. Explain how the two growth plates in a long bone account for the ability of a long bone to lengthen.
4. Explain what might cause a long bone to slowly change shape over many years.
5. Describe the process of bone remodeling and how it can reshape bones to make them stronger.
6. Name the three anatomical regions of the vertebral column that are above the *sacral* and *coccygeal* regions.
7. Explain why it is important not to move someone who may have suffered an injury to the vertebral column until a medical assessment can be made.
8. Describe the features of synovial joints that reduce friction and prevent the joint from wearing out prematurely.
9. Distinguish between *flexion* and *extension*.
10. Define the differences between *osteoarthritis* and *rheumatoid arthritis*.

Test Yourself

Answers can be found in Appendix A.

1. Which of the following might result from a parathyroid tumor which causes oversecretion of parathyroid hormone?
   a. joint inflammation leading to osteoarthritis
   b. bone loss due to stimulation of osteoclasts
   c. bone growth due to stimulation of osteoblasts
   d. conversion of cartilage to bone

2. Steps in the repair of a bone fracture include (1) bone deposition by osteoblasts, (2) bone and debris removal by osteoclasts, (3) hematoma, and (4) formation of a fibrocartilage callus. In what order do these steps occur?
   a. 1–2–3–4
   b. 3–4–1–2
   c. 3–4–2–1
   d. 4–3–2–1

3. All of the following bones form part of the eye socket except:
   a. occipital bone
   b. lacrimal bone
   c. zygotmatic bone
   d. ethmoid bone

4. All of the following bones of the skull are stationary except:
   a. frontal bone
   b. mandible
   c. maxilla
   d. zygomatic bone

5. Which bones are found in both the hands and feet?
   a. carpals
   b. metacarpals
   c. tarsals
   d. phalanges

6. The movement of the thumb to trace a circle might best be described as:
   a. abduction
   b. rotation
   c. circumduction
   d. pronation

7. Synovial joints may include cartilage, ligaments, tendons, and synovial fluid. Which of these attach bones to other bones within the joint?
   a. synovial membrane
   b. ligaments
   c. tendons
   d. cartilage

8. Which of the following is an example of a cartilaginous joint?
   a. knee joint
   b. skull sutures
   c. pubic symphysis
   d. hip joint

9. All of the following are bones of the axial skeleton except:
   a. vertebrae
   b. ribs
   c. skull
   d. clavicle

10. Which of the following would be likely to prevent or slow the bone loss of osteoporosis?
    a. stimulate the activity of fibroblasts
    b. stimulate the activity of osteoblasts
    c. inhibit the activity of osteoclasts
    d. both (b) and (c)

11. Which of the following contains the richest population of the stem-cell precursors for red and white blood cells?
    a. red bone marrow
    b. yellow bone marrow
    c. osteoid
    d. hydroxyapatite

12. In the formation and development of bones within the fetus, which of these cell types functions earliest?
    a. osteocyte
    b. osteoblasts
    c. osteoclasts
    d. chondroblasts

13. Which of the following might be most helpful in determining whether an adolescent is no longer growing?
    a. measuring the length of the femur and humerus
    b. examining the growth plates near the ends of long bones
    c. examining bone density
    d. examining the fontanels in the skull

14. All of the following processes continue in the skeletal system throughout the life span except:
    a. bones continue to lengthen
    b. stem cells continue to form new blood cells
    c. bones continue to be remodeled
    d. bones continue to store minerals (calcium and phosphorus)

15. Which kind of joint is essentially immovable?
    a. hinge joint
    b. fibrous joint
    c. cartilaginous joint
    d. ball and socket joint

Apply What You Know

Answers can be found at the Human Biology Place.
www.humanbiology.com

1. Compare and contrast swimming and running as forms of exercise training in terms of how they might affect muscle mass, bone mass, and the possibility of injuries to joints.

2. The administration of growth hormone is sometimes used clinically to stimulate growth in unusually short children who are deficient in growth hormone. However, growth hormone is ineffective in unusually short but otherwise normal adults. What accounts for the difference?

3. Although sports are getting more and more competitive at younger and younger ages, in baseball it is not recommended that children learn to throw curveballs at too young an age. What is the problem with throwing a curveball?

4. You and a friend decide to volunteer to help build houses with Habitat for Humanity over your spring break. Although you have only rarely used a hammer, you take on the task helping to construct the frame of the house. After your break is over and you have returned to campus, you notice a sharp pain in your elbow every time you bend your arm. You seek medical advice, and the doctor tells you that you have tendonitis. What is tendonitis? What might have caused your condition?

5. You just graduated and got your first job as a forensic investigator. Your first case is a skeleton that was discovered in the desert. The pathologist examines the bones and tells you that the skeleton belonged to an adult man. How can the pathologist be certain by examining only the bones? What else might the pathologist be able to tell you by examining the bones of this skeleton?

6. Obesity is a common problem in this country, even among children. What changes would you expect to see in the skeletal system of a person who has been obese for a long time?

6

The Muscular System

Climber in Banff National Park, Canada.

Drug Abuse Among Athletes

Baseball slugger Barry Bonds is under indictment for lying under oath about using performance-enhancing drugs. Sprinter Marion Jones confessed to her drug use and offered a tearful apology, but was stripped of her five Olympic gold medals. High school athletes are routinely being tested for performance-enhancing drugs. What is going on here?

The short answer is that many performance-enhancing drugs do enhance athletic performance. They work in ways that are predictable and understood, based on human physiology. In an environment where just a hundredth of a second can make the difference between an Olympic gold medal and relative obscurity, the temptation to use these drugs is high.

Anabolic Steroids

Anabolic steroids and related compounds such as dehydroepiandrosterone (DHEA) and androstenedione ("Andro") are the most widely abused drugs in athletics today. Although anabolic steroids are banned by sports federations and school systems, many of them are available over the counter as the result of a 1994 federal law that was written to ensure access to herbal remedies (see the Current Issue in Chapter 2). In general, they are

Marion Jones after winning the gold in 2000, and in 2007 after admitting to using performance-enhancing drugs.

structurally and functionally related to the male sex steroid testosterone. And like testosterone, they make it easy for the user to increase his/her muscle mass. Muscle strength improves as well, leading to improved athletic performance in sports that require short bursts of energy.

How common is anabolic steroid use? A study funded by the National Institute on Drug Abuse found that 2.5% of 12th-graders, 1.5% of 10th-graders and 1.1% of 8th-graders had tried them. Information on steroid use by college and professional athletes is unreliable because the athletes are reluctant to talk about it. A few well-known athletes, including Arnold Schwarzenegger, sprinter Ben Johnson, and wrestler Hulk Hogan, have admitted to using them at one time or another.

A recent trend is the increased use of sophisticated "designer drugs" such as THG (tetrahydrogestrinone), designed specifically to avoid detection. THG is known as "the clear" to athletes because allegedly it couldn't be detected. And for years it wasn't detected—until an anonymous tipster sent a sample of the drug to a sport federation for testing. THG is so potent that it doesn't even have to be injected—just a couple of drops under the tongue are enough. In 2007 sprinter Marion Jones finally admitted that she started using THG in 1999 as she prepared for the 2000 Olympic games. She was stripped of her five medals from the 2000 Olympics and banned from participation in the 2008 Olympics in Beijing.

Aside from the obvious issue of fairness in athletic competition, anabolic steroids are banned by sports federations because of their side effects and possible health risks. Androgens have masculinizing effects in both sexes. Men may experience gynecomastia (enlargement of the breasts), shrinkage of the testicles, reduced sperm production, and impotence. In women, breast size and body fat decrease and the voice deepens. Women may lose scalp hair but gain body hair. Some of these changes

A young Barry Bonds playing for the Pirates, and years later, a bulkier Barry as the SF Giants' slugger.

are not reversible. Anabolic steroid abuse is also associated with irritability, hostility, and aggressive behavior ("roid rage"). Prolonged anabolic steroid abuse is associated with an increased risk of heart attack, stroke, and severe liver disease, including liver cancer. Although the number of cases of these diseases is fairly low (so far), the effects of steroid use/abuse may be underestimated because these diseases tend to come later in life. We just don't know what will happen to steroid abusers 30 years later.

Blood Enhancers

Marathoners and cyclists aren't interested in muscle mass; they're interested in maintaining a high level of sustained performance over long periods of time. For that, they need increased aerobic capacity. Their (banned) drug of choice is erythropoietin (EPO), a hormone produced by the kidneys that increases the production of red blood cells. EPO is available by prescription only for patients with *anemia* (too few red blood cells in the blood). But cyclists and marathon runners use it to improve their performance. It's all a matter of normal human physiology; EPO produces more red blood cells, which leads to a higher oxygen-carrying capacity, which in turn leads to a higher level of sustainable muscle activity and faster times.

But a health risk is associated with EPO abuse. Excessive production of red blood cells can raise the hematocrit (the percentage of the blood that is red cells) to dangerous levels. The blood becomes sludge-like, increasing the risk of high blood pressure, blood clots, and heart attacks. Statistically, one of the most common causes of death among professional cyclists is heart attack, although no deaths have ever officially been listed as having been caused by EPO.

It's hard to test for EPO abuse because EPO disappears from the blood within days, leaving behind an increased hematocrit and an improved endurance that lasts for a month or more. The cycling organizations are only able to curb EPO abuse by setting an upper limit for hematocrit of 50%; above that, EPO abuse is just assumed and the athlete is banned from competition. It is widely suspected that cyclists who choose to abuse EPO measure their hematocrit shortly before a race and then remove blood cells to just meet the 50% rule!

Next Up: Gene Doping

Within decades, it will probably be possible to use genetic engineering techniques to modify an athlete's genes for improved athletic performance. It's called *gene doping*. What if you could tinker with the genes that lead to the production of natural erythropoietin or testosterone, so that an athlete just naturally produces more of these hormones? What if you could alter muscle biochemistry so that muscles used energy more efficiently or more rapidly? What if you could insert genes that caused muscle cells to store up more ATP? These ideas are not so far-fetched. Nearly all experts on the subject are convinced that if gene doping hasn't been tried already, it soon will be. Gene doping will be extremely hard to detect or to prevent.

Have we lost our perspective for the role that sports should play in our lives?

Questions to consider

1 Do you think we should continue to try to prevent the use of drugs and genetic engineering in sports? Why or why not?

2 A friend who uses anabolic steroids says that there is no convincing scientific evidence that anabolic steroid use will lead to health problems such as heart disease or cancer later in life. Is he right? What would you say to him?

The facts...

- Performance-enhancing drugs such as anabolic steroids and erythropoietin (EPO) are used by some athletes because they improve certain types of athletic performance.

- Abuse of performance-enhancing drugs can lead to unwanted side effects, an increased risk of certain chronic diseases, and perhaps even premature death.

- Although most sports federations have banned the use of performance-enhancing drugs, enforcement has proven difficult.

- Soon it may be possible to use genetic engineering techniques to enhance athletic performance.

Muscle cells are found in every organ in the body and participate in every activity that requires movement. Together, they constitute nearly half of our body mass. The most obvious are the *skeletal muscles* that attach to the skeleton and give us strength and mobility. Skeletal muscles also sculpt the body and contribute to our sense of attractiveness and well-being. Some of the smallest skeletal muscles control the focus of our eyes; some of the largest are responsible for the shivering that helps keep us warm when it is cold. Nearly 40% of body weight in males and about 32% in females is skeletal muscle.

But there are two other types of muscle in the body besides skeletal muscle. Rhythmic contractions of the *cardiac muscle* of the heart pump blood throughout the body. Powerful intermittent contractions of *smooth muscle* in the walls of the uterus propel the child through the birth canal. Slower waves of smooth muscle contractions push food through the digestive tract and transport urine from the kidney to the bladder. Steady, sustained contractions of smooth muscle in the walls of blood vessels regulate blood flow to every living cell in the body.

How do muscles accomplish all this? In this chapter we examine the structure of muscles, learn how they function, and see how their activity is controlled. Finally, we'll see how muscles can become damaged and describe some diseases of muscle.

6.1 Muscles produce movement or generate tension

Some muscle movements are *voluntary,* meaning that we have conscious control over the movements they produce. An example is deliberately picking up an object. Other muscle movements, such as the pumping action of the heart or the maintenance of muscle tone in blood vessels, are *involuntary* in that they are generally beyond our conscious control. You cannot will your heart to stop beating.

We tend to think of muscles as producing movement, but another very important function of many muscles is to *resist* movement. The maintenance of posture while standing is a good example. If you faint, you collapse because you lose control over the muscles that support your upright posture. The maintenance of a constant blood vessel diameter even when blood pressure within the vessel changes is another example of how a muscle resists movement. These are cases where muscles generate a force that exactly opposes an equal but opposite force being applied to a body part.

Besides producing or resisting movement, muscles also generate heat. Under normal circumstances, contraction of our skeletal muscles accounts for over three-quarters of all the heat generated by the body. Heat generated by muscles is important in maintaining homeostasis of our body temperature, because normally our body temperature is higher than that of our surroundings. When we generate too much heat, temperature-control mechanisms are called into play that allow the body to get rid of it. However, if you spend time outdoors on a cold day you may notice that you start to shiver. Shivering occurs because your temperature-control mechanisms cause skeletal muscles to alternately contract and relax so as to generate more heat.

The fundamental activity of muscle is contraction

All three types of muscle cells have certain fundamental features in common. First, muscle cells are *excitable,* meaning that they contract in response to chemical and/or electrical signals from other organ systems. Second, all muscles have only one basic mechanism of action: they *contract* (shorten), and then they *relax,* returning to their original length. The movements of your limbs, the beating of your heart, and the regulation of the diameters of your blood vessels all depend on muscles.

The muscle type with which you are probably most familiar is skeletal muscle. In this chapter we concentrate on skeletal muscle because skeletal muscle, in conjunction with the skeleton, is responsible for voluntary movement. At the end of this chapter we briefly discuss cardiac and smooth muscle.

Skeletal muscles cause bones to move

Most **skeletal muscles** interact with the skeleton and cause bones to move (or prevent them from moving) relative to each other. All of the tasks accomplished by our skeletal muscles, whether shivering, threading a needle, lifting heavy weights, or even just standing completely still, are performed by skeletal muscles that are either contracting or relaxing. We have more than 600 skeletal muscles, often organized into pairs or groups. Hundreds of muscles, each controlled by nerves and acting either individually or in groups, produce all possible human motions. Muscle groups that work together to create the same movement are called *synergistic muscles.* Muscles that oppose each other are called *antagonistic muscles.* **Figure 6.1** summarizes some of the major muscles of the body and their actions.

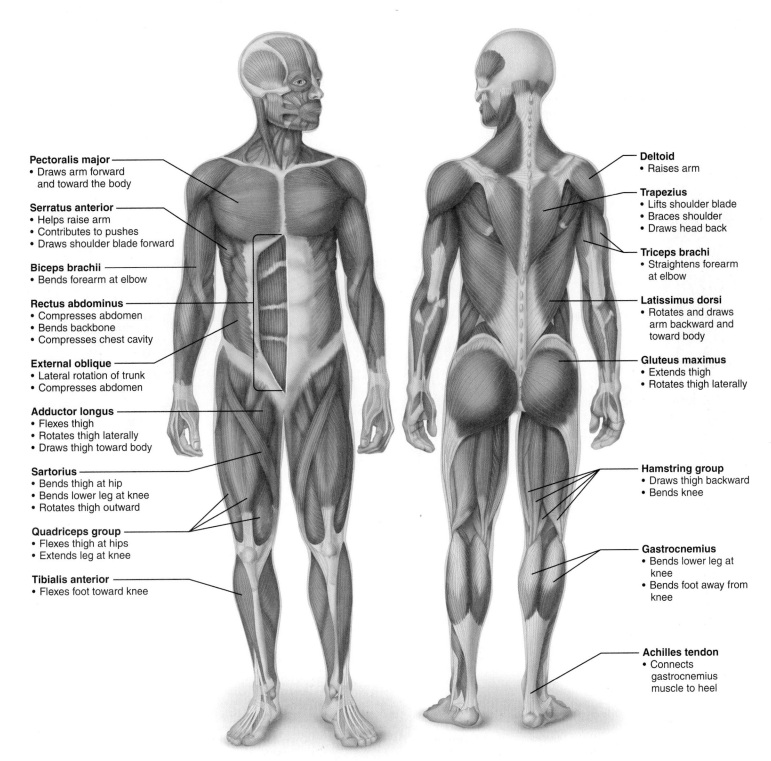

Pectoralis major
• Draws arm forward and toward the body

Serratus anterior
• Helps raise arm
• Contributes to pushes
• Draws shoulder blade forward

Biceps brachii
• Bends forearm at elbow

Rectus abdominus
• Compresses abdomen
• Bends backbone
• Compresses chest cavity

External oblique
• Lateral rotation of trunk
• Compresses abdomen

Adductor longus
• Flexes thigh
• Rotates thigh laterally
• Draws thigh toward body

Sartorius
• Bends thigh at hip
• Bends lower leg at knee
• Rotates thigh outward

Quadriceps group
• Flexes thigh at hips
• Extends leg at knee

Tibialis anterior
• Flexes foot toward knee

Deltoid
• Raises arm

Trapezius
• Lifts shoulder blade
• Braces shoulder
• Draws head back

Triceps brachi
• Straightens forearm at elbow

Latissimus dorsi
• Rotates and draws arm backward and toward body

Gluteus maximus
• Extends thigh
• Rotates thigh laterally

Hamstring group
• Draws thigh backward
• Bends knee

Gastrocnemius
• Bends lower leg at knee
• Bends foot away from knee

Achilles tendon
• Connects gastrocnemius muscle to heel

Figure 6.1 **Major skeletal muscle groups and their functions.**

Most skeletal muscles are attached to bones via tendons (Figure 6.2a), though a few are attached only to other muscles or to skin (such as the muscles that permit you to smile). Muscles join to the skeleton in such a way that each individual muscle produces a very specific movement of one bone relative to another. The skeleton is a complex set of levers that can be pulled in many directions by contracting or relaxing skeletal muscles. One end of a skeletal muscle, called its **origin,** joins to a bone that remains relatively stationary. The other end of the muscle, called its **insertion,** attaches to another bone across a joint. When the muscle contracts, the insertion is pulled toward the origin. The origin is generally closer to the midline of the body and the insertion is farther away.

Figure 6.2b shows how the two antagonistic muscles of the upper arm, the biceps and triceps, oppose each other to bend (flex) and straighten (extend) the forearm. When the triceps muscle relaxes and the biceps contracts, the combined action pulls on the forearm and flexes it. When the biceps relaxes and the triceps contracts, the combined action pulls the forearm down, extending it again.

A muscle is composed of many muscle cells

A single *muscle* (sometimes referred to as a "whole muscle") is a group of individual muscle cells, all with the same origin and insertion and all with the same function. A cross section of muscle (Figure 6.3) reveals that it is arranged in bundles called *fascicles,* each enclosed in a sheath of a type of fibrous connective tissue called *fascia.* Each fascicle contains anywhere from a few dozen to thousands of individual muscle cells, or *muscle fibers.* The outer surface of the whole muscle is covered by several more layers of fascia. At the ends of the muscle all of the

Scapula

Shoulder joint

Origins from scapula and humerus

Triceps muscle

Tendon

Insertion on ulna

Elbow joint

Ulna

Origins from scapula

Tendons

Humerus

Biceps muscle

Tendon

Insertion on radius

Radius

Triceps relaxes

Biceps contracts, pulling forearm up

Triceps contracts, pulling forearm down

Biceps relaxes

a) Origin and insertion. The point of attachment of a muscle to the stationary bone is its origin; the point of attachment to the movable bone is its insertion.

b) Movement. Antagonistic muscles produce opposite movements. The forearm bends when the biceps contracts and the triceps relaxes. The forearm straightens when the biceps relaxes and the triceps contracts.

Figure 6.2 **Movement of bones.**

Figure 6.3 **Muscle structure.** A muscle is arranged in bundles called fascicles, each composed of many muscle cells and each surrounded by a sheath of fascia. Surrounding the entire muscle are several more layers of fascia. The fascia join together to become the tendon, which attaches the muscle to bone.

a) **A single muscle cell contains many individual myofibrils and has more than one nucleus.**

b) **A photograph of portions of several skeletal muscle cells.**

Figure 6.4 **Muscle cells.**

fasciae (plural) come together, forming the tendons that attach the muscle to bone.

Individual muscle cells are tube shaped, larger, and usually longer than most other human cells. Some muscle cells are only a millimeter in length, whereas others may be as long as 30 centimeters—roughly the length of your thigh muscle. Taking a closer look at a single muscle cell (**Figure 6.4**), we see that each cell contains more than one nucleus. The nuclei are located just under the cell membrane because nearly the entire interior of the cell is packed with long cylindrical structures arranged in parallel, called **myofibrils.** The myofibrils are packed with contractile proteins called *actin* and *myosin,* discussed below. When myofibrils contract (shorten), the muscle cell also shortens.

✔️ **Quick Check** Beef, chicken, and fish are all "high protein" foods, and they are all primarily composed of muscle. Why are foods that are mostly muscle so high in protein content? ▪

The contractile unit is a sarcomere

Looking still closer at a single myofibril, we see a striated (or banded) appearance that repeats at regular intervals. Various

elements of the pattern stand out, but the one that is important for our discussion is a dark line called the *Z-line* (**Figure 6.5** on the next page). A segment of a myofibril from one Z-line to the next is called a **sarcomere.** A single myofibril within one muscle cell in your biceps muscle may contain over 100,000 sarcomeres arranged end to end. The microscopic shortening of these 100,000 sarcomeres all at once is what produces contraction (shortening) of the muscle cell and of the whole muscle. Understanding muscle shortening, then, is simply a matter of understanding how a single sarcomere works.

A sarcomere consists of two kinds of protein filaments. Thick filaments composed of a protein called **myosin** are interspersed at regular intervals within filaments of a different protein called **actin.** Notice that the actin filaments are structurally linked to the Z-line and that myosin filaments are located entirely within sarcomeres, stretching between two different actin filaments. As we will see, muscle contractions depend on the interaction between these actin and myosin filaments.

↩️ **Recap** Muscles either produce or resist movement. Their fundamental activity is contraction. A muscle is composed of many muscle cells arranged in parallel, each containing numerous myofibrils. The contractile unit in a myofibril is called a *sarcomere.* A sarcomere contains thick filaments of a protein called myosin and thin filaments of a protein called actin. ▪

Answers to ✔️ **questions can be found in Appendix A.**

a) **A closer view of a section of a myofibril showing that it is composed of sarcomeres joined end to end at the Z-line.**

b) **Sarcomeres contain thin filaments of actin that attach to the Z-lines and thicker filaments of myosin that span the gap between actin molecules.**

Thin filament (actin)

Thick filament (myosin)

d) **An electron micrograph cross section of a sarcomere in a region that contains both actin and myosin.**

Myosin

Actin

c) **A transmission electron micrograph (× 11,300) of a longitudial section of a sarcomere.** The rounded red objects are mitochondria.

Figure 6.5 Structure of a myofibril.

6.2 Individual muscle cells contract and relax

During a muscle contraction each sarcomere shortens just a little. Subtle though this action seems, it is also powerful. The contraction of an entire skeletal muscle depends on the simultaneous shortening of the tiny sarcomeres in its cells. There are four keys to understanding what makes a skeletal muscle cell contract and relax:

- A skeletal muscle cell must be activated by a nerve. It does not contract on its own.
- Nerve activation increases the concentration of calcium (Ca) in the vicinity of the contractile proteins.

- The presence of calcium permits contraction. The absence of calcium prevents contraction.
- When a muscle cell is no longer stimulated by a nerve, contraction ends.

Let's look at each point in more detail.

Nerves activate skeletal muscles

Skeletal muscle cells are stimulated to contract by certain nerve cells called **motor neurons.** The motor neurons secrete a chemical substance called *acetylcholine (ACh).* Acetylcholine is a **neurotransmitter,** a chemical released by nerve cells that has either an excitatory or inhibitory effect on another excitable cell (another nerve cell or a muscle cell). In the case of skeletal muscle, acetylcholine excites (activates) the cells.

The junction between a motor neuron and a skeletal muscle cell is called the **neuromuscular junction.** When an electrical impulse traveling in a motor neuron arrives at the neuromuscular junction, acetylcholine is released from the nerve terminal (Figure 6.6). The acetylcholine diffuses across the narrow space between the neuron and the muscle cell and binds to receptor sites on the muscle cell membrane. The binding of acetylcholine to the receptors causes the muscle cell membrane to generate an electrical impulse of its own that travels rapidly along the cell membrane in all directions. In addition, tubelike extensions of the cell membrane called *T tubules* (the T stands for transverse) transmit the electrical impulse deep into the interior of the cell. The function of the T tubules is to get the electrical impulse to all parts of the cell as quickly as possible.

Activation releases calcium

As shown in Figure 6.6, T tubules are in close contact with a series of membrane-bound chambers called the **sarcoplasmic reticulum** (*sarco-* is derived from a Greek word for "flesh" or "muscle"). The sarcoplasmic reticulum is similar to every other cell's smooth endoplasmic reticulum except that its shape is different, in part because it must fit into the small amount of space in the cell not occupied by myofibrils. The primary function of the sarcoplasmic reticulum is to store ionic calcium (Ca^{2+}).

Inside the muscle cell, an electrical impulse races down the T tubules to the sarcoplasmic reticulum. The arrival of an electrical impulse triggers the release of calcium ions from the sarcoplasmic reticulum. The calcium diffuses into the cell cytoplasm and then comes in contact with the myofibrils, where it sets in motion a chain of events that leads to contraction.

Calcium initiates the sliding filament mechanism

Muscles contract when sarcomeres shorten, and sarcomeres shorten when the thick and thin filaments slide past each other, a process known as the **sliding filament mechanism** of contraction. Taking a closer look at the arrangement of thick and thin filaments in a single sarcomere (Figure 6.7), we see

Figure 6.6 **How nerve activation leads to calcium release within a muscle cell.**

✓ *If a muscle cell's sarcoplasmic reticulum had little to no Ca²⁺, could the muscle cell still produce an electrical impulse, and could it still contract? Explain.*

a) Relaxed state. The myosin heads do not make contact with actin.

b) Contraction. The myosin heads form cross-bridges with actin and then bend, pulling the actin filaments toward the center of the sarcomere.

Figure 6.7 **Sliding filament mechanism of contraction.**

that every thin filament consists of two strands of actin molecules spiraling around each other. Thick filaments are composed of many individual molecules of myosin. Myosin molecules are shaped somewhat like a golf club, with a long shaft and a rounded head. Myosin shafts form the main part of the thick filaments. The heads stick out to the side, nearly touching the thin filaments of actin. When a muscle is relaxed, the myosin heads do not quite make contact with the thin filaments, however. Contraction occurs when the myosin heads make contact with the thin filaments, forming a *cross-bridge* between the two filaments. The formation of a cross-bridge causes the head to bend relative to the shaft, pulling the actin molecules toward the center of the sarcomere. The processes of cross-bridge formation and bending (the molecular events of contraction) require energy.

But what initiates the process of contraction? Put another way, what prevents contraction from occurring all the time? The answer is that contraction is inhibited unless calcium is present. An even closer look at a section of myosin and actin (**Figure 6.8**) shows why. Closely associated with the actin filaments are two other protein molecules called *troponin* and *tropomyosin* that together form the *troponin-tropomyosin protein complex*. In the absence of calcium, the troponin-tropomyosin protein complex interferes with the myosin binding sites on the actin molecule. Following an electrical impulse, calcium released from the sarcoplasmic reticulum binds to troponin, resulting in a shift in the position of the troponin-tropomyosin protein complex that exposes the myosin binding sites and

permits the formation of cross-bridges. At this point the myosin heads form cross-bridges with actin, undergo a bending process, and physically pull the actin filaments toward the center of the sarcomere from each end. With thousands of myosin cross-bridges doing this simultaneously, the result is a sliding movement of the thin filaments relative to the thick ones and a shortening of the sarcomere. As hundreds of thousands of sarcomeres shorten, individual muscle cells, and ultimately the whole muscle, shorten as well.

 Web Animation *Muscle Structure and Function* at www.humanbiology.com

 Quick Check Suppose a person had an unusual mutation in the troponin protein of his skeletal muscles, such that the troponin could not bind to calcium at all. Would this person's muscles be constantly contracted, constantly relaxed, or able to function normally? Explain. ▪

When nerve activation ends, contraction ends

Relaxation of a muscle cell occurs when nerve activity ends. In the absence of nerve activity, no more calcium is released from the sarcoplasmic reticulum. The calcium released as a result of prior electrical impulses is transported back into the sarcoplasmic reticulum by active transport, which requires energy in the form of ATP. As the calcium concentration in the myofibrils falls, the

a) **Resting sarcomere.** In the absence of calcium the muscle is relaxed because the myosin heads cannot form cross-bridges with actin.

b) **Cross-bridge attachment.** The binding of calcium to troponin causes a shift in the troponin-tropomyosin complex, allowing cross-bridges to form.

Figure 6.8 Role of calcium in contraction.

 Name two things that must happen for the myosin to stop binding to the actin (i.e., for the muscle to relax), and explain why each is necessary.

troponin-tropomyosin protein complex shifts back into its original position, preventing the binding of the myosin cross-bridges to actin. The sarcomere stretches passively to its original resting length, and the muscle cell relaxes.

Any factor that interferes with the process of nerve activation can disrupt muscle function. In the disorder *myasthenia gravis,* the body's immune system attacks and destroys acetylcholine receptors on the cell membrane of muscle cells. Affected muscles respond only weakly to nerve impulses or fail to respond at all. Most commonly impaired are the eye muscles, and many people with myasthenia gravis experience drooping eyelids and double vision. Muscles in the face and neck may also weaken, leading to problems with chewing, swallowing, and talking. Medications that facilitate the transmission of nerve impulses can help people with this condition.

✅ **Quick Check** Caffeine prolongs the lifespan of acetylcholine molecules in the motor junctions. Explain how this fact is related to caffeine's tendency to cause "jitters," such as hand tremors and other small involuntary contractions. ▪

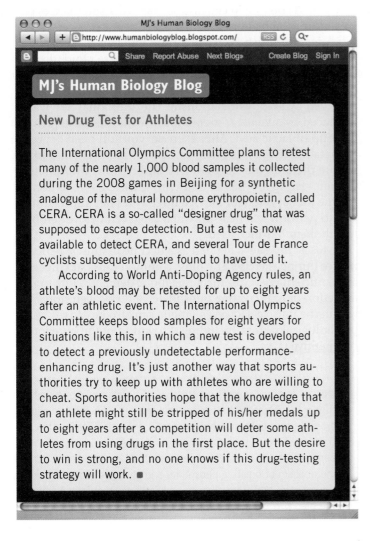

Muscles require energy to contract and to relax

Muscle contraction requires a great deal of energy. Like most cells, muscle cells use ATP as the energy source. In the presence of calcium, myosin acts as an enzyme, splitting ATP into ADP and inorganic phosphate and releasing energy to do work. The energy is used to "energize" the myosin head so that it can form a cross-bridge and undergo bending. Once the bending has occurred, another molecule of ATP binds to the myosin, which causes the myosin head to detach from actin. As long as calcium is present, the cycle of ATP breakdown, attachment, bending, and detachment is repeated over and over again in rapid succession. The result is a shortening of the sarcomere.

At the end of the contractile period (when nerve impulses end), energy from the breakdown of ATP is used to transport calcium back into the sarcoplasmic reticulum so that relaxation can occur. However, a second requirement for relaxation is that an intact molecule of ATP must bind to myosin before myosin can finally detach from actin. This last role of ATP is the explanation for *rigor mortis* (Latin, meaning "rigid death"), in which a body becomes stiff during the time period from about four hours to several days after death. Shortly after death, calcium begins to leak out of the sarcoplasmic reticulum, causing muscle contraction. The contractions use up the available ATP, but after death the ATP cannot be replenished. In the absence of ATP, the myosin heads cannot detach from actin, and so the muscles remain "locked" in the contracted state. Eventually the stiffness of rigor mortis decreases as the muscle cells degenerate.

Muscle cells store only enough ATP for about 10 seconds' worth of maximal activity. Once this is used up the cells must produce more ATP from other energy sources, including creatine phosphate, glycogen, glucose, and fatty acids.

An important pathway for producing ATP involves creatine phosphate (creatine-P), a high-energy molecule with an attached phosphate group. Creatine phosphate can transfer a phosphate group and energy to ADP and therefore create a new ATP molecule quickly. This reaction is reversible: if ATP is not needed to power muscle contractions, the excess ATP can be used to build a fresh supply of creatine phosphate, which is stored until needed. In recent years creatine phosphate loading has become common among bodybuilders and athletes, particularly those who need short-term bursts of power. Unfortunately, muscles cannot store much more creatine phosphate than they usually have, even without creatine phosphate loading. Creatine phosphate also seems to improve muscle performance in certain neuromuscular diseases, such as amyotrophic lateral sclerosis (ALS), also known as Lou Gehrig's Disease.

Table 6.1 **Energy sources for muscle**

Energy source	Quantity	Time of use	Comments
Stored ATP	Stored only in small quantities	About 10 seconds	ATP is only direct energy source. It must be replenished by the other energy sources.
Stored creatine phosphate	Three to five times amount of stored ATP	About 30 seconds	Converted quickly to ATP.
Stored glycogen	Variable; some muscles store large quantities	Primarily used during heavy exercise within the first 3–5 minutes	ATP yield depends on whether oxygen is available. One glucose molecule (derived from stored glycogen) yields only 2 ATP molecules in the absence of oxygen, but 36 ATP molecules in the presence of oxygen.
Aerobic metabolism	Not fueled by energy stored in muscle; oxygen and nutrients (glucose and fatty acids derived from fat) are constantly supplied by the blood	Always present; increases dramatically within several minutes of onset of exercise, when blood flow and respiration increase	High yield. Complete metabolism of one glucose molecule yields 36 ATP molecules.

The combination of previously available ATP plus stored creatine phosphate produces only enough energy for up to 30–40 seconds of heavy activity. Beyond that, muscles must rely on stored glycogen, a complex sugar (polysaccharide) composed of many smaller molecules of glucose. For the first three to five minutes of sustained activity, a muscle cell draws on its internal supply of stored glycogen. Glucose molecules are removed from the glycogen, and their energy is used to synthesize ATP. Part of the process of the breakdown of glucose can happen without oxygen (called anaerobic metabolism) fairly quickly, but anaerobic metabolism yields only two ATP molecules per glucose molecule. It also has the unfortunate side effect of producing lactic acid, which causes the burning sensation one feels right at the end of a heavy weight-lifting session, just as exhaustion sets in.

The most efficient long-term source of energy is the aerobic metabolism of glucose in the blood, fatty acids derived from stored fat in fat cells, and other high-energy molecules such as lactic acid. Aerobic metabolism, as you already know, takes place in mitochondria and requires oxygen. The next time you engage in strenuous exercise, notice that it may take a minute or two for your respiratory rate to increase dramatically. Your heart rate also increases because there is an increase in blood flow to the exercising muscle. The increased respiration and heart rate are signs that aerobic metabolism is now taking place. Until aerobic metabolism kicks in, however, your cells are relying on stored ATP, creatine phosphate, and anaerobic metabolism of glycogen.

Weight lifters can rely almost exclusively on stored energy because their muscles perform for relatively short periods of time. Long-distance runners start out by depending on stored energy, but within several minutes they are relying almost exclusively on aerobic metabolism. If they could not, they would collapse in exhaustion. Table 6.1 summarizes energy utilization by muscle.

After you finish exercising, note that you continue to breathe heavily for a period of time. These rapid, deep breaths help reverse your body's **oxygen debt,** incurred because your muscles used more ATP early on than was provided by aerobic metabolism. The additional ATP was produced by anaerobic metabolism, with the subsequent buildup of lactic acid. After exercise, you still need oxygen to metabolize the lactic acid by aerobic pathways and to restore the muscle's stores of ATP and creatine phosphate to their resting levels. The ability of muscle tissue to accumulate an oxygen debt and then repay it later allows muscles to perform at a near-maximal rate even before aerobic metabolism has increased.

Muscle **fatigue** is defined as a decline in muscle performance during exercise. The most common cause of fatigue is insufficient energy to meet metabolic demands, due to depletion of ATP, creatine phosphate, and glycogen stores within the muscle. However, fatigue can also be caused by psychological factors, including discomfort or the boredom of repetitive tasks.

↩ **Recap** Skeletal muscle contraction is initiated by nerves, which trigger the release of calcium within the cell. The calcium allows cross-bridges to form between myosin and actin, leading to contraction. Energy from the breakdown of ATP is required for contraction and for calcium transport. ATP is produced from metabolism of creatine phosphate and glycogen stores within the muscle, and from glucose and fatty acids obtained from the blood. ∎

Delayed Onset Muscle Soreness

Most of us are familiar with the feeling of stiffness and soreness that occurs a day or two after an unfamiliar form of exertion. The soreness is actually due to microscopic tears in myofibrils throughout the muscle. It is thought that the damage occurs because in the absence of regular use, some sarcomeres become like old, stiff rubber bands, unable to contract as well as they should. During exercise these old or injured sarcomeres stretch passively to the point that the thick and thin filaments no longer overlap, damaging the sarcomeres permanently. Exercises that cause muscles to be stretched (lengthened) while they are actively contracting (trying to shorten) are the most likely to cause damage and soreness. Examples

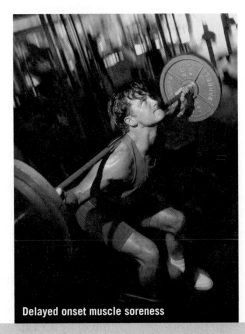

Delayed onset muscle soreness

include running downhill, lowering very heavy weights, and the downward motion of push-ups and squats.

The feeling of soreness after exercise begins a day or so after exercise and usually reaches a peak 1–3 days later. The soreness is caused by chemicals released during the repair process, which involves inflammation, swelling, and the release of chemical substances such as prostaglandins. With time, the damaged sarcomeres are removed completely and new sarcomeres take their place.

Once muscles become accustomed to a particular exercise, damage (and the accompanying soreness) no longer occurs. To minimize muscle injury and soreness, undertake any new exercise activity in moderation for the first few days. ∎

6.3 The activity of muscles can vary

The general functions of muscles are to move body parts or to maintain a certain body position. How well they carry out their functions depends on a number of factors, including whether bones actually move or not, the degree of nerve stimulation, the type of muscle fiber, and the degree to which exercise has improved muscle mass and aerobic capacity.

Isotonic versus isometric contractions: Movement versus static position

Most types of exercise include a combination of two different types of muscle contractions, called isotonic and isometric contractions.

Isotonic ("same" + "strength" or "tone") contractions occur whenever a muscle shortens while maintaining a constant force. An example of an isotonic contraction is the generation of enough muscle force to move an object or part of the skeleton. How heavy the object is doesn't matter, as long as parts of

the skeleton actually move. It could be just your empty hand, a pencil, a book, or a 100-pound barbell.

In *isometric* ("same" + "length") contractions, force is generated, muscle tension increases, and the muscle may even shorten a little as tendons are stretched slightly, but bones and objects do not move. As a result, isometric contractions do not cause body movement. Examples are tightening your abdominal muscles while sitting still, or straining to lift a weight too heavy to lift. Isometric contractions help to stabilize the skeleton. In fact, you contract your muscles isometrically whenever you stand, just to maintain an upright position. If you doubt it, think about how quickly you would fall down if you were to faint and lose control over your skeletal muscles. Isometric contractions are a useful way to strengthen muscles.

The degree of nerve activation influences force

A single muscle may consist of thousands of individual muscle cells. The individual cells in any muscle are organized into groups of cells that all work together. Each group of cells is controlled by a single nerve cell called a *motor neuron* (because it affects movement). The motor neuron and all of the muscle cells it controls are called a **motor unit** (Figure 6.9). A motor unit is the smallest functional unit of muscle contraction, because when the motor neuron is activated, all the muscle cells in that motor unit are activated together.

Our strength and ability to move effectively depend on how forcefully our muscles contract. The mechanical force that muscles generate when they contract is called **muscle tension.** How much tension is generated by a muscle depends on three factors:

- The number of muscle cells in each motor unit (motor unit size)
- The number of motor units active at any one time
- The frequency of stimulation of individual motor units

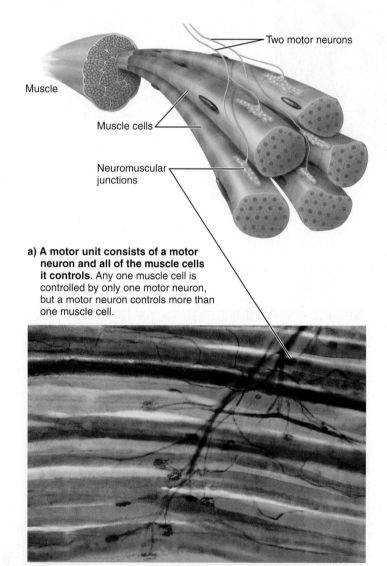

Two motor neurons

Muscle

Muscle cells

Neuromuscular junctions

a) A motor unit consists of a motor neuron and all of the muscle cells it controls. Any one muscle cell is controlled by only one motor neuron, but a motor neuron controls more than one muscle cell.

b) Photograph of the muscle cells in a motor unit, showing branches of the motor neuron and neuromuscular junctions.

Figure 6.9 Motor units.

Motor unit size can vary widely from one muscle to the next. The number of muscle cells per motor unit is a tradeoff between brute strength and fine control. Larger motor units generate more force but offer less control. In the thigh muscle, where strength is more important than fine control, a single motor unit may consist of as many as a thousand muscle cells. In muscles of the eye, where fine control is essential, a motor unit may consist of only 10 muscle cells.

According to the **all-or-none principle,** muscle cells are completely under the control of their motor neuron. Muscle cells never contract on their own. For an individual muscle cell, there is no such thing as a half-hearted contraction, and there is no such thing as disobeying an order. Muscle cells always respond with a complete cycle of contraction and relaxation (called a **twitch**) every time they are stimulated by an electrical impulse, called an *action potential*, from their motor neuron. You will learn more about action potentials in Chapter 11. For now you need only understand that they are the stimuli for muscle contraction.

Although individual motor units either are contracting or are relaxed, whole muscles generally maintain an intermediate level of force known as *muscle tone.* Muscle tone exists because, at any one time, some of the muscle's motor units are contracting while others are relaxed. The second factor that affects overall muscle force, then, is the number of motor units active at any one time. Increasing tone (or force) by activating more motor units is called **recruitment.** The maintenance of muscle tone depends on the nervous system.

The third factor that affects force generation by a muscle is the frequency of stimulation of individual motor units. To understand how frequency of stimulation influences force, we need to take a closer look at what happens when a muscle cell is stimulated by its motor neuron. Although we cannot easily study the contraction of single muscle cells in the laboratory, the study of whole muscles has revealed some important findings regarding the timing of the relationship between a stimulus and a twitch. A laboratory recording of muscle activity, called a *myogram,* reveals that the stimulus-twitch relationship has three stages (Figure 6.10):

1. *Latent period* (the time between stimulation and the start of contraction). This is the time it takes for the nerve impulse to travel to the sarcoplasmic reticulum, for calcium to be released, and for the myosin heads to bind to the actin filaments.
2. *Contraction* (the time during which the muscle actually shortens). Actin filaments are pulled toward the center of the sarcomere and myofibrils shorten.
3. *Relaxation* (muscle returns to its original length). Calcium is transported back into the sarcoplasmic reticulum, the troponin-tropomyosin protein complex shifts back into its original position, and the sarcomere stretches passively to its original length.

Figure 6.10 How frequency of stimulation affects muscle contractile force. A single stimulus produces, after a latent period, a contraction/relaxation cycle called a *twitch*. More than one stimulus in a short time may produce summation and ultimately a tetanic contraction (tetanus).

✓ *In what way is the tetanic contraction shown in this figure different from recruitment? In what way are they similar?*

A key point is that the contraction/relaxation cycle of the muscle twitch lasts longer than the stimulus that caused the contraction in the first place. If additional stimuli arrive at the muscle before the muscle has had a chance to transport calcium back into the sarcoplasmic reticulum and relax completely, the total force produced becomes greater than the force produced by one twitch alone. In effect, the force becomes greater because more calcium is present. Increasing muscle force by increasing the rate of stimulation of motor units is called **summation.**

There is a limit to summation, however. If stimulation becomes so frequent that the muscle cannot relax at all, it will remain in a state of maximum contraction called *tetanus*

or a *tetanic contraction.* On a myogram, tetanus appears as a straight horizontal line representing the fusion of the peaks and valleys of individual twitches. A tetanic contraction may lead eventually to muscle fatigue.

Table 6.2 summarizes the mechanism of muscle cell activation and contraction.

✓ **Quick Check** A friend of yours tells you that the "all-or-none" principle means that all the motor units in a muscle always contract simultaneously. Explain what is wrong with his reasoning. ▪

Slow-twitch versus fast-twitch fibers: Endurance versus strength

As we have seen, all muscle cells can obtain ATP through both aerobic and anaerobic pathways. Humans have two types of skeletal muscle fibers, called *slow-twitch* and *fast-twitch fibers*. The distinction is based on how quickly they can utilize ATP to produce a contraction, and whether they use primarily aerobic or anaerobic metabolic pathways. Most muscles contain a mixture of both slow-twitch and fast-twitch fibers. The ratio of fiber types in any one muscle depends primarily on the function of the muscle.

Slow-twitch fibers break down ATP slowly, and so they contract slowly. They tend to make ATP as they need it by aerobic metabolism. Slow-twitch fibers contain many mitochondria and are well supplied with blood vessels, so they draw more blood and oxygen than fast-twitch fibers. They store very little glycogen because they can obtain glucose and fatty acids quickly from the blood. They store

Table 6.2 Summary of activation and contraction of skeletal muscle

Action	Description	Additional facts
Motor neuron activation	A brief electrical impulse known as an action potential travels down the motor neuron from the central nervous system.	One motor neuron innervates more than one muscle cell.
Neurotransmitter release	At the neuromuscular junctions between a motor neuron and each of its muscle cells, a chemical neurotransmitter called *acetylcholine* is released.	The acetylcholine is removed quickly, so that stimulation of the muscle cell is short-lived.
Muscle cell activation	The release of acetylcholine causes an electrical impulse in the cell membrane of each muscle cell.	The electrical impulse in the muscle cell is similar to the electrical impulse in a nerve.
Calcium release	An impulse in the muscle cell membrane causes calcium to be released into the muscle cell cytoplasm from the sarcoplasmic reticulum.	The sarcoplasmic reticulum is a network of membrane-bound storage sacs in the muscle cell.
Muscle cell contraction	The presence of calcium allows the thick and thin filaments to attach to each other and to slide past each other. Energy in the form of ATP is required.	Muscle cell contraction lasts longer than neuron activation.
Muscle cell relaxation	Calcium is pumped back into the sarcoplasmic reticulum. The thick and thin filaments detach from each other, and the muscle relaxes.	Calcium transport back into the sarcoplasmic reticulum also requires ATP.

oxygen, however, in a molecule called *myoglobin*. The ability to maintain a temporary store of oxygen reduces the slow-twitch fiber's need for oxygen from the bloodstream. This is especially important during the early phases of an increase in activity, before blood flow to the muscle has increased. Myoglobin and the presence of numerous blood vessels make slow-twitch fibers reddish in color, so they are sometimes called "red" muscle.

Fast-twitch fibers can contract more quickly than slow-twitch fibers because they break down ATP more quickly. They have fewer mitochondria, fewer blood vessels, and little or no myoglobin compared to slow-twitch fibers, so they're called "white" muscle. Fast-twitch fibers store large amounts of glycogen and tend to rely heavily on creatine phosphate and anaerobic metabolism for quick bursts of high energy. Their contractions are rapid and powerful but cannot be sustained for long. Fast-twitch fibers depend on aerobic mechanisms for any activity that is sustained, but they have the capability of using anaerobic mechanisms for brief periods when bursts of power are needed. During periods of anaerobic activity they tend to accumulate lactic acid, which causes them to become fatigued quickly.

Which type of fiber is better? It depends on the activity. Because slow-twitch fibers offer more endurance, they are most useful for steady activities such as jogging, swimming, and biking. Slow-twitch fibers are also important for maintaining body posture. Many of the muscles of the leg and back, for example, contain a high percentage of slow-twitch fibers because they must contract for long periods to support us when we stand. Fast-twitch fibers are more often used to power brief, high-intensity activities such as sprinting for short distances, lifting weights, or swinging a tennis racquet. Muscles in our hands, for example, contain a high proportion of fast-twitch fibers, allowing the muscles to contract quickly and strongly when necessary.

The percentage of slow- and fast-twitch fibers varies not only from muscle to muscle but from person to person. The percentages are determined in part by inheritance, and they can influence athletic ability. For example, most world-class marathoners have a higher-than-average percentage of slow-twitch fibers in their legs.

☑ **Quick Check** Suppose a muscle biopsy done on an aspiring athlete shows that her leg muscles have an unusually dark red color, much darker than that of most other runners. Would you recommend that she train for sprints or for marathons, and why? ▪

Exercise training improves muscle mass, strength, and endurance

Although part of your athletic potential might be influenced by inheritance, a consistent, planned program of physical exercise (sometimes called *exercise training*) can improve your strength, endurance, and skill at any athletic endeavor (Figure 6.11). Whether primarily strength or endurance is improved by exercise training depends on the type and intensity of training. The

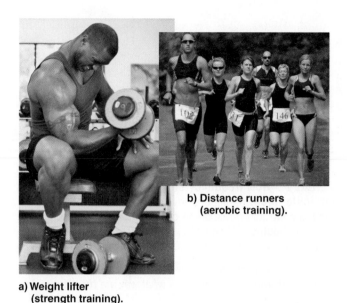

b) Distance runners (aerobic training).

a) Weight lifter (strength training).

Figure 6.11 **The effects of strength training versus aerobic training.**

MJ's Human Biology Blog

Beating the Testosterone Doping Test

According to a study by Swedish researchers, a common genetic deficiency might make it possible for some people to beat the standard testosterone doping test, even if they had been doping with the hormone. The researchers injected 55 male volunteers with testosterone and then tested them with the standard urine test for testosterone doping. More than 40% of all subjects with the genetic deficiency tested negative for testosterone doping over a 15-day testing period.

Two-thirds of the Asian population and about 10% of all Caucasians are deficient in the gene in question. The World Anti-Doping Agency is concerned, but it appears that there is little that they can do about it at this time. Genetic tests would reveal which athletes could beat the testosterone doping test, but genetic testing is not part of the standard anti-doping test for Olympic athletes. Individuals with the genetic deficiency may be able to use testosterone and get away with it, at least until the rules change. ▪

Reference: Schulz, Jenny Jakobsson *et al.* Doping Test Results Dependent on Genotype of Uridine Diphospho-Glucuronosyl Transferase 2B17, the Major Enzyme for Testosterone Glucuronidation. *J. Clin. Endo. Metab.* 93: 2500–2506, 2008.

two primary types of exercise training are strength (resistance) training and aerobic (endurance) training.

Strength training involves doing exercises that strengthen specific muscles, usually by providing some type of resistance that makes them work harder. Strength training is generally short, intense exercise such as weight lifting using free weights or weight machines. It builds more myofibrils, particularly in fast-twitch fibers, and causes the fast-twitch fibers to store more glycogen and creatine phosphate as quick energy sources. This increases the size of individual muscle cells and builds muscle mass and muscle strength, but it does not increase the number of muscle cells.

In general, the heavier the weight used, the more visible the increase in muscle size. However, this does not mean that strength training will necessarily build bulging biceps. The extent of muscle development depends on many factors, including the amount of resistance used, the duration and frequency of exercise, and your own genetic predisposition. However, even low to moderate weights can lead to noticeable improvements in muscle strength.

Aerobic training involves activities in which the body increases its oxygen intake to meet the increased demands for oxygen by muscles. Whereas resistance training strengthens muscles, aerobic training builds endurance. With aerobic training, the number of blood capillaries supplying muscle increases. In addition, the number of mitochondria in muscle cells and the amount of myoglobin available to store oxygen both increase. The muscle fibers themselves do not increase much in mass, nor do they increase in number. Aerobic exercise also improves the performance of the cardiovascular and respiratory systems. Less intense than strength training but carried out for prolonged periods, aerobic exercises include jogging, walking, biking, and swimming.

It's a good idea to combine any athletic activity with *stretching exercises.* Gentle stretching before exercise increases your heart rate gradually, pumping additional blood to your muscles and preparing you for more strenuous exertion. This lowers your risk of sprains and pulled muscles. After exercising, let your heart rate and breathing return gradually to normal as you walk slowly and do more stretching. Regular stretching improves joint mobility and range of motion. Whenever you stretch, do it gradually and hold each position for 30 seconds. You should feel a gentle pull in your muscles, but not pain. Try not to bounce, because abrupt stretches could cause your muscles to contract quickly in response, increasing the risk of injury.

Recap A motor unit consists of a motor neuron and all of the muscle cells it controls. Greater muscle force is produced by activation of more motor units and/or increased frequency of stimulation of motor units. Most muscles contain a combination of slow-twitch and fast-twitch fibers. Slow-twitch fibers rely on aerobic metabolism and are most useful for endurance. Fast-twitch fibers are most useful where strength is required. Exercise increases aerobic capacity, muscle mass, and muscle strength but does not increase the number of muscle cells. ▪

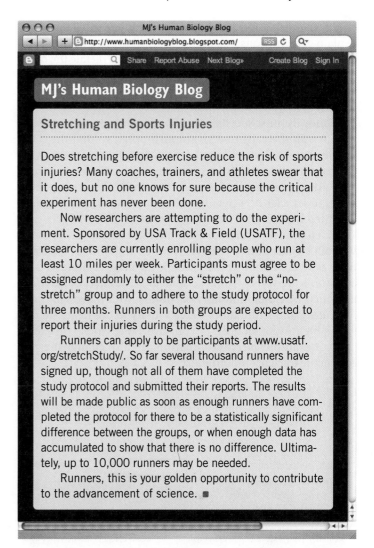

MJ's Human Biology Blog

Stretching and Sports Injuries

Does stretching before exercise reduce the risk of sports injuries? Many coaches, trainers, and athletes swear that it does, but no one knows for sure because the critical experiment has never been done.

Now researchers are attempting to do the experiment. Sponsored by USA Track & Field (USATF), the researchers are currently enrolling people who run at least 10 miles per week. Participants must agree to be assigned randomly to either the "stretch" or the "no-stretch" group and to adhere to the study protocol for three months. Runners in both groups are expected to report their injuries during the study period.

Runners can apply to be participants at www.usatf.org/stretchStudy/. So far several thousand runners have signed up, though not all of them have completed the study protocol and submitted their reports. The results will be made public as soon as enough runners have completed the protocol for there to be a statistically significant difference between the groups, or when enough data has accumulated to show that there is no difference. Ultimately, up to 10,000 runners may be needed.

Runners, this is your golden opportunity to contribute to the advancement of science. ▪

6.4 Cardiac and smooth muscles have special features

Most of the overall muscle mass of the body is skeletal muscle. Nevertheless, both cardiac and smooth muscle have unique features that suit them ideally for their roles in the body. All three types of muscle tissue are described briefly in Chapter 4. Here we look at some of the special attributes of cardiac and smooth muscle that set them apart from skeletal muscle.

How cardiac and smooth muscles are activated

Cardiac and smooth muscle are called *involuntary muscle* because we generally do not have voluntary control over them. Both cardiac and smooth muscles can contract entirely on their own, in the absence of stimulation by nerves.

Although all cardiac muscle cells are capable of beating spontaneously and establishing their own cycle of contraction and relaxation, those with the fastest rhythm are called *pacemaker cells* because the rest of the cells follow their faster pace. Cardiac muscle cells are joined at their blunt ends by structures called *intercalated discs* (Figure 6.12). The intercalated discs contain gap junctions that permit one cell to electrically stimulate the next one. In effect, the pacemaker cells dictate the rate of contraction of the whole heart, because their faster pace activates the slower cells before the slower cells would be activated by their own inherent rhythm.

Smooth muscle cells are also joined by gap junctions that permit the cells to activate each other, so that the whole tissue contracts together in a coordinated fashion. (In contrast to smooth and cardiac muscle, skeletal muscle cells are activated only by motor neurons. This is why the skeletal muscles of a person with a severed spinal cord are completely paralyzed below the point of injury.)

Even though cardiac and smooth muscle cells can contract without signals from the nerves, they do respond to nerve activity as well. In both types of muscle the nerves belong to the autonomic nervous system (Chapter 11). The effect of nerve activity may be either inhibitory or stimulatory. Changes in both inhibitory and stimulatory nerve activity to the heart are responsible for the increase in your heart rate when you exercise, for example. Nerve stimulation can also change the contractile force of smooth muscle.

✅ **Quick Check** If the gap junctions in heart muscle cells were eliminated, could the pacemaker cells still beat? Could they still set the pace of the entire heart? Explain. ∎

a) A view of several adjacent cardiac muscle cells showing their blunt shape and the intercalated discs that join them together.

— Adhesion junction

— Protein channel

— Gap junction

— Cell membranes of adjacent cells

b) A closer view showing that intercalated discs are bridged by gap junctions that permit direct electrical connections between cells.

Figure 6.12 Cardiac muscle cells.

Speed and sustainability of contraction

In terms of speed and sustainability of contraction, skeletal muscle is the fastest, cardiac muscle is of moderate speed, and smooth muscle is very slow.

Cardiac muscle cells go through rhythmic cycles of contraction and relaxation. The relaxation periods are necessary periods of rest so that the muscle doesn't fatigue.

Smooth muscle generally is partially contracted all the time. This makes it ideally suited for situations in which contractions need to be sustained. Nevertheless, it almost never fatigues because it contracts so slowly that its ATP usage is always less than its production capability. Smooth muscle is a key player in the homeostatic regulation of blood pressure because it can maintain the diameter of blood vessels indefinitely, adjusting them slightly as necessary.

Arrangement of myosin and actin filaments

Like skeletal muscle, cardiac muscle has a regular array of thick and thin filaments arranged in sarcomeres, so it too is called *striated muscle*. In contrast, the thick and thin filaments in smooth muscle are arranged in bundles that attach at various angles to the cell membrane. When the thick and thin filaments slide past each other, the points of attachment of the filaments are pulled toward each other and the cell gets shorter and fatter (Figure 6.13). Because its filaments are arranged in bundles rather than sarcomeres, smooth muscle lacks the striated appearance of skeletal and cardiac muscle. It is called "smooth" for this reason.

Table 6.3 summarizes defining characteristics of the three types of muscle.

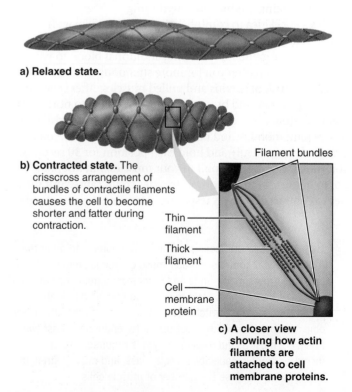

a) **Relaxed state.**

b) **Contracted state.** The crisscross arrangement of bundles of contractile filaments causes the cell to become shorter and fatter during contraction.

Filament bundles

Thin filament

Thick filament

Cell membrane protein

c) **A closer view showing how actin filaments are attached to cell membrane proteins.**

Figure 6.13 Smooth muscle.

Table 6.3 Defining characteristics of skeletal, cardiac, and smooth muscle

Defining characteristics	Skeletal muscle	Cardiac muscle	Smooth muscle
Location	Attached to bones (skeleton)	Found only in the heart	Found in the walls of blood vessels and in the walls of organs of the digestive, respiratory, urinary, and reproductive tracts
Function	Movement of the body. Prevention of movement of the body	Pumping of blood	Control of blood vessel diameter. Movement of contents in hollow organs
Anatomical description	Very large, cylindrical, multi-nucleated cells arranged in parallel bundles	Short cells with blunt, branched ends. Cells joined to others by intercalated discs and gap junctions	Small, spindle-shaped cells joined to each other by gap junctions
Initiation of contraction	Only by a nerve cell	Spontaneous (pacemaker cells), modifiable by nerves	Some contraction always maintained. Modifiable by nerves
Voluntary?	Yes	No	No
Gap junctions?	No	Yes	Yes
Speed and sustainability of contraction	Fast—50 milliseconds (0.05 second). Not sustainable	Moderate—150 milliseconds (0.15 second). Not sustainable	Slow—1–3 seconds. Sustainable indefinitely
Likelihood of fatigue	Varies widely depending on type of skeletal muscle and workload	Low. Relaxation between contractions reduces the likelihood	Generally does not fatigue
Striated?	Yes	Yes	No

Recap Unlike skeletal muscle, both cardiac and smooth muscle can contract in the absence of any nerve stimulation. Cardiac muscle contracts and then relaxes in a rhythmic cycle. Smooth muscle can sustain a contraction indefinitely without ever relaxing. ▪

6.5 Diseases and disorders of the muscular system

Throughout this chapter we have discussed a number of musculoskeletal health conditions. We'll finish by looking at several more.

Muscular dystrophy

Serious diseases of muscle are relatively uncommon, but foremost among them is **muscular dystrophy.** The term actually applies to several different hereditary diseases of muscle (*dystrophy* means "abnormal growth"). In *Duchenne muscular dystrophy,* a single defective gene results in the lack of a particular muscle cell protein. The normal gene, when present, directs the cell to produce a protein called *dystrophin* that is part of the muscle cell membrane. The function of dystrophin is to limit the inflow of calcium into muscle cells through calcium "leak" channels. People with muscular dystrophy lack dystrophin, and as a result too much calcium leaks into the muscle cell through the leak channels. The high intracellular calcium

concentration activates enzymes that damage muscle proteins and ultimately may kill the cell. The result is a loss of muscle fibers and muscle wasting. Eventually much of the muscle mass is replaced with fibrous connective tissue. Many people with muscular dystrophy die before age 30, usually because of failure of the heart muscle or the skeletal muscles used for breathing. At the moment there is no cure; however, it is an area of intense research interest, and progress is being made on several fronts.

Tetanus

Tetanus is caused by a bacterial infection. The disorder is called *tetanus* because this is the technical term for a maximal (tetanic) muscle contraction (see Figure 6.10). Generally the infection is acquired by a puncture wound to a muscle. The bacteria produce a toxin that overstimulates the nerves controlling muscle activity, resulting in tetanic contractions. The toxin affects a variety of skeletal muscles, but especially those of the jaws and neck. Jaw muscles may contract so forcefully that they seem locked shut (the origin of its common name, "lockjaw"). Untreated, tetanus may lead to death due to exhaustion or respiratory failure.

Quick Check Suppose a doctor is trying to treat a patient who has tetanus, and he has two drugs available: one that mimics the action of acetylcholine, and another that blocks the action of acetylcholine. Which would be better, and what is one dangerous possible side effect? ▪

Muscle cramps

Muscle cramps are painful, uncontrollable, reflex-mediated muscle contractions. They are thought to be caused by the dehydration and ion imbalances that sometimes occur with heavy exercise. The most likely culprit is a shift in potassium ions between the intracellular and extracellular fluid. Muscle cramps generally can be soothed by increasing the circulation to the affected muscle through gentle stretching and massage.

Pulled muscles

Pulled muscles, sometimes called torn muscles, result from stretching a muscle too far, causing some of the fibers to tear apart. Internal bleeding, swelling, and pain often accompany a pulled muscle.

Fasciitis

Fasciitis involves inflammation of the connective tissue sheath, or fascia, that surrounds a muscle (see Figure 6.3). It is usually caused by straining or tearing the fascia. Most often it affects the sole of the foot (plantar fasciitis), where it is a common cause of heel pain. Like tendons and ligaments, fascia mend slowly. Treatment includes resting the area and protecting it from pressure. Injections of corticosteroid drugs can relieve severe pain.

↩ **Recap** Muscular dystrophy is an inherited disease in which the absence of a single protein causes an abnormal leak of calcium into muscle cells. Ultimately the leak of calcium damages muscle cell proteins and kills muscle cells. Tetanus is caused by a bacterial infection that overstimulates nerves to muscles. ∎

Chapter Summary

Muscles produce movement or generate tension *p. 124*

- All muscles produce movement or maintain position by contracting (shortening in length).
- The ways in which skeletal muscles attach to the skeleton determine what particular motion they cause.
- Within a single myofibril in a single muscle cell (fiber), thousands of contractile units called *sarcomeres* are arranged end to end.

Individual muscle cells contract and relax *p. 128*

- Skeletal muscle cells contract only when activated by their motor nerve.
- Motor nerve activation causes calcium to be released from the sarcoplasmic reticulum of the muscle cell.
- In the presence of calcium, the thick (myosin) and thin (actin) filaments slide past each other, and the sarcomere shortens.
- ATP supplies the energy for the entire contraction/relaxation process.

The activity of muscles can vary *p. 133*

- An *isotonic contraction* occurs when a muscle shortens while maintaining a constant force; an *isometric contraction* occurs when tension is generated but the bones do not move.
- A motor unit is a single motor neuron and all of the skeletal muscle cells that it controls.
- The force generated by a muscle depends on the number of muscle cells in each motor unit, the number of motor units active at any one moment, and the frequency of stimulation of motor units.
- Muscle strength and endurance depend on the ratio of slow-twitch to fast-twitch fibers in the muscle and on the type and amount of exercise training.

Cardiac and smooth muscles have special features *p. 137*

- Cardiac and smooth muscles do not attach to bones.
- Both cardiac and smooth muscles can contract spontaneously, and both can be influenced by nerves of the autonomic nervous system.

- Cardiac muscle contracts rhythmically, with a period of relaxation between each contraction. Smooth muscle can maintain at least some contractile force indefinitely.

Diseases and disorders of the muscular system *p. 139*

- Muscular dystrophy is caused by inheritance of an abnormal gene.
- Pulled (torn) muscles occur when a muscle is stretched too far.

Terms You Should Know

actin, *127*	recruitment, *134*
all-or-none principle, *134*	sarcomere, *127*
fatigue, *132*	sarcoplasmic reticulum, *128*
motor neuron, *128*	sliding filament
motor unit, *134*	mechanism, *128*
myosin, *127*	summation, *135*
neurotransmitter, *128*	twitch, *134*
oxygen debt, *132*	

Concept Review

Answers can be found at the Human Biology Place.
www.humanbiology.com

1. Describe how muscle contraction can resist movement rather than cause movement.
2. Describe how a muscle's origin and insertion determine the specific body movement that will result from muscle contraction.
3. Describe the roles of calcium in muscle contraction.
4. Explain what causes *rigor mortis*.
5. Discuss some possible reasons for muscle fatigue.
6. Define *summation*, and explain why it occurs when a muscle is stimulated rapidly and repetitively.
7. Explain why a spinal cord injury in the neck completely paralyzes the skeletal muscles of the limbs, whereas the cardiac muscle of the heart still beats rhythmically.

8. Compare and contrast how a constant degree of moderate tension, or tone, is maintained by a skeletal muscle that maintains posture versus a smooth muscle that maintains blood vessel diameter.

9. Define a motor unit, and describe how the size and the number of motor units in a muscle affect muscle strength and fine motor control.

10. List the sources of energy that a muscle cell may use to make more ATP, both from within and from outside the cell.

Test Yourself

Answers can be found in Appendix A.

1. Muscles which oppose each other and produce opposite movements are described as:
   a. synergistic
   b. antagonistic
   c. cooperative
   d. oppositional

2. Which of the following choices arranges the structures (1) muscle fiber, (2) fascicle, (3) myofibril, and (4) muscle from the largest (most inclusive) to smallest?
   a. 1-2-3-4
   b. 2-3-1-4
   c. 4-2-1-3
   d. 4-2-3-1

3. All of the following are functions of the muscular system except:
   a. maintenance of body calcium stores
   b. resisting movement
   c. maintenance of body temperature
   d. movement

4. Which of the following happens during muscle contraction?
   a. actin filaments shorten
   b. myosin filaments shorten
   c. sarcomeres shorten
   d. both (a) and (b)

5. Botulism toxin inhibits the release of acetylcholine at the neuromuscular junctions. What effect does this have on the muscle activity?
   a. Muscles will contract continuously.
   b. Muscles will contract sporadically, without conscious control.
   c. Muscles will not contract because they will not receive nerve stimulation.
   d. There will be no effect on muscle activity.

6. The sliding filament mechanism describes the process during which:
   a. actin and myosin slide relative to each other
   b. sarcomeres slide relative to each other
   c. troponin and tropomyosin slide relative to each other
   d. muscle fibers slide past each other

7. What is the first and most direct energy source for muscle contraction?
   a. glucose
   b. ATP
   c. creatine phosphate
   d. glycogen

8. As you clasp your hands in front of you and push them toward each other, this is an example of:
   a. an isotonic contraction
   b. an isometric contraction
   c. a tetanic contraction
   d. aerobic training

9. All of the following may happen in response to exercise training except:
   a. increase in the number of myofibrils
   b. increase in the storage of glycogen and creatine phosphate
   c. increase in the number of muscle fibers
   d. increase in the number of mitochondria

10. Which of the following is/are characteristic of slow-twitch fibers?
    a. large amounts of glycogen storage
    b. myoglobin content enables oxygen storage
    c. numerous mitochondria
    d. both (b) and (c)

11. Which of the following is the site of calcium ion storage within muscles?
    a. T tubules
    b. sarcoplasmic reticulum
    c. actin filaments
    d. myosin filaments

12. What is the role of ATP in muscle function?
    a. ATP provides energy which enables myosin to form cross-bridges with actin.
    b. ATP enables myosin to detach from actin.
    c. ATP provides energy to transport calcium back into storage.
    d. all of the above

13. Which of the following would have motor units with the smallest number of muscle cells?
    a. thigh muscle
    b. muscles in fingers
    c. abdominal muscles
    d. muscles of the back

14. Which type(s) of muscle cells can contract the fastest?
    a. smooth muscle cells
    b. cardiac muscle cells
    c. skeletal muscle cells
    d. All muscle cells can exhibit the same speed of contraction.

15. Which type(s) of muscle cells can contract spontaneously?
    a. smooth muscle cells
    b. cardiac muscle cells
    c. skeletal muscle cells
    d. both (a) and (b)

Apply What You Know

Answers can be found at the Human Biology Place.
www.humanbiology.com

1. Why do you think it is generally accepted medical practice to get bedridden patients up and walking as soon as possible?

2. In what ways would you expect the training regimen for a sprinter to be different from that of a marathon runner, and why?

3. What would happen to a muscle if one of its tendons were torn? Would the muscle still be able to contract?

4. You and your friend are doing leg presses in the gym one day. As you extend your legs the stack of weights goes up, and when you bend your legs the stack goes down. Your friend says your muscles are actively pushing the weights up. Explain to him where he is wrong in his thinking.

5. You are outside on a cool fall day. You feel cool, but you think little of it until you notice yourself shivering. What is happening at the muscular level, and why is it at least partially effective in helping to maintain body temperature?

6. You have just joined an aerobics exercise class for the first time, and you have calculated your target heart rate. After class you notice that your heart rate remains high for a while and only slowly returns to normal. Explain why this occurs and what is happening at the physiological level. How will this response change over time if you maintain your exercise program consistently?

7. Some weight lifters like to consume various products containing creatine phosphate. Why would this be useful? Why would weight lifters benefit more than marathon runners from creatine phosphate?

7

Blood

Two red blood cells, a white blood cell, and *staphylococcus* bacteria.

Should You Bank Your Baby's Cord Blood?

When she was 15 and a sophomore in high school, Jaclyn Albanese was diagnosed with acute leukemia—a type of cancer of stem cells in bone marrow. The usual treatment is chemotherapy and radiation to kill the cancer cells (and normal stem cells), and then a bone marrow transplant to repopulate the bone marrow with stem cells. Traditionally these stem cells have come from bone marrow donated by a family member or an unrelated volunteer whose marrow is compatible. Compatibility is crucial because, as you will learn in this chapter, the body's immune cells recognize and attack foreign cells. Jaclyn had hoped to

get a bone marrow transplant from one of her relatives, but none of them was a close enough match. Fortunately for Jaclyn, she found compatible units of cord blood from an unrelated donor. They saved her life. But, if her parents had banked the cord blood from her delivery when she was born, she wouldn't have needed to search for donors at all.

What Is Cord Blood?

During pregnancy, the fetus is cushioned by a temporary organ called the placenta and connected to its mother by the umbilical cord. Blood vessels in the umbilical cord and placenta filter out toxic

Jaclyn Albanese

Blood in the umbilical cord contains stem cells similar to those found in bone marrow.

HLA-DR, each of which comes in two forms. The ideal match would be for the patient to have the same six forms as the donor (a 6/6 match). Good matches between unrelated donors are rare—siblings of the same parents match only 25% of the time, and only about 10% of all patients who need a bone marrow transplant are able to find a compatible match from among unrelated donors.

This is why cord blood has become such a precious commodity. The immune cells in cord blood are less mature than those in bone marrow, so cord blood transplants are less likely to cause transfusion reactions, and even when reactions do occur, they tend to be less severe. As a result, the match between donor and recipient does not need to be a perfect 6/6—matches of 5/6 or even 4/6 are sufficient. This opens a much wider field of possibilities. Jaclyn Albanese could not find a single compatible bone marrow donor, but she was able to locate two units of compatible cord blood. She had a cord blood transplant in 1999 just prior to her junior year in high school, and today she is a college graduate. Jaclyn is one of approximately 6,000 patients who have benefited from cord blood transplants to date.

Banking Privately or Publicly

Should you bank your baby's cord blood privately, or donate it to a public cord blood bank?

Private blood banks say that banking your baby's cord blood privately is like taking out a medical insurance policy. They argue that you might want to use your baby's cord blood stem cells to treat a future disease (such as leukemia) in your child or a close family member. By banking your baby's cord blood, you ensure that your child always has access to the "perfect match"—his or her own stem cells. They also point out that scientists are working on stem-cell therapies for a variety of other conditions such as diabetes and heart disease. The implication is that in 30 or 40 years, by the time your

newborn is at greater risk of developing these chronic ailments, he or she may be able to use these stem cells for treatments as yet undreamed of. So far they have convinced over 20,000 families to have their baby's cord blood collected, processed, tested, and stored, at an average cost of $1,700 plus an annual storage fee of about $125.

A cord blood collection unit.

Proponents of public blood banking argue that the likelihood of a baby born to a healthy family ever needing his or her stem cells range from one in 10,000 to one in 200,000. Is such an unlikely event worth hoarding your baby's blood? If you and your family are healthy, you might want to consider helping others by donating your child's cord blood to a public cord blood bank. To support public donations, in 2005 the federal government authorized $79 million in federal funds to collect and store cord blood from ethnically diverse donors. As a result, the National Marrow Donor Program now lists a national inventory of over 90,000 cord blood units. An inventory of 140,000 units would give all Americans an 80–90% chance of finding at least a 5/6 antigen match.

Jaclyn Albanese's parents did not have the choice of banking her cord blood privately when she was born—the technology didn't exist back then. Fortunately, she was able to rely on the generosity of others who donated their baby's cord blood to a public cord blood bank.

substances, deliver nutrients from the mother, and remove waste products from the fetus. After the baby is delivered, the mother's body expels the placenta and umbilical cord. A health professional cuts the cord, and the baby's circulatory system begins to function on its own.

Until recently, the placenta and cord were discarded after birth. However, these structures still contain about 50 ml of cord blood. In addition to containing the usual components of blood, cord blood is rich in stem cells from the fetus that are still relatively immature. These stem cells can be coaxed to divide repeatedly to produce immature blood cells, which in turn develop into platelets, red cells, and white cells.

Bone Marrow, Cord Blood, and Compatibility Issues

A good bone marrow transplant match between a donor and a patient involves three key antigens known as HLA-A, HLA-B, and

Questions to consider

1 Do you agree with the federal government's decision to allocate $79 million (about 40 cents per adult) for a public cord blood collection and storage network?

2 When (or if) you have a child, what will you do with its cord blood? Explain your decision.

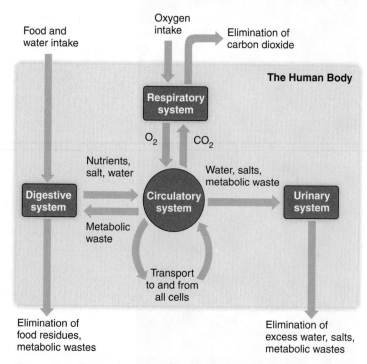

Figure 7.1 The transport role of the circulatory system. The cardiovascular system serves homeostasis by transporting nutrients and carrying off wastes from all parts of the body.

All cells in the body must obtain nutrients and get rid of wastes. How do they do these things? Diffusion to and from the fluid that surrounds them (interstitial fluid) is only part of the answer. If every cell drew its nutrients from interstitial fluid and there were no way to replenish the nutrients, the cells would soon starve. If every cell dumped its waste into the interstitial fluid and those wastes were not promptly removed, the cells would die in a sea of toxic waste. Cells need a system for keeping the oxygen content, the supply of nutrients, the concentrations of wastes, and the concentrations of every essential molecule and atom within acceptable limits. What they need is a system to maintain homeostasis of the interstitial fluid. In humans, that system is the circulatory system.

The **circulatory system** consists of the heart, the blood vessels, and the blood that circulates through them. As shown in Figure 7.1, the circulatory system plays a central role in supplying all cells with what they need and removing the substances they no longer need. The circulatory system ensures that blood flows throughout the entire body, bringing the necessary raw materials to the interstitial fluid surrounding every living cell and removing the waste. It picks up nutrients from the digestive system, exchanges gases with the respiratory system, and carries wastes and excess water and salts to the urinary system for removal from the body. It also carries some metabolic wastes to the liver for removal. Whenever a substance is transported over any distance within the body, the circulatory system is at work. Closely

coupled to the circulatory system—indeed, often considered part of it—is another system of fluid-filled vessels called the *lymphatic system*. We examine the lymphatic system further in Chapter 9 in connection with the immune system.

We discuss the heart and blood vessels as well as the transport functions of the lymphatic system in Chapter 8. In this chapter we concentrate on the composition and crucial functions of blood, the fluid that circulates within the heart and blood vessels.

7.1 The components and functions of blood

As noted in Chapter 4, **blood** is a specialized connective tissue. It consists of specialized cells and cell fragments suspended in a watery solution of molecules and ions. Blood carries out three crucial tasks for the body:

- *Transportation.* Blood transports all substances needed anywhere by the body, including oxygen from the lungs, nutrients from the digestive system, and hormones from the endocrine glands. Blood also transports the waste products of cellular metabolism away from body tissues to the organs that eliminate them from the body.
- *Regulation.* Blood helps to regulate body temperature, the volume of water in the body, and the pH of body fluids.
- *Defense.* Blood contains specialized defense cells that help protect against infections and illness, and it has the ability to prevent excessive blood loss through the clotting mechanism.

Together these functions are crucial for maintaining homeostasis. Blood is so effective at performing these functions that so far scientists' efforts to develop an artificial blood substitute have not been very successful. If someone needs blood, a transfusion of human blood is often the only solution.

Adult men average 5 to 6 liters of blood (about 1.5 gallons), and adult women average 4 to 5 liters. Differences between men and women reflect differences in body size. In general, blood represents approximately 8% of your body weight.

Blood is thicker and stickier than water. This is because some components of blood are denser (heavier) than water and because blood is roughly five times more viscous (viscosity is a measure of resistance to flow). The old saying that blood is thicker than water is true.

Despite its uniform color, blood carries a rich array of components. Table 7.1 summarizes the components of blood and their functions. They fall into two major categories: the liquid component (plasma) and the cellular component or formed elements (*red cells, white cells,* and *platelets*). If you spin a blood sample in a centrifuge (a high-speed rotation device that mimics and magnifies gravitational forces), formed elements sink to the bottom of a test tube because they are denser than plasma (Figure 7.2). Red blood cells (RBCs), representing the bulk of the formed elements, settle

Table 7.1 **Composition of blood**

Blood component	Examples and functions
Formed Elements (45%)	
Red blood cells	Transport oxygen to body tissues; transport carbon dioxide away from tissues.
White blood cells	Defend the body against invading organisms, abnormal cells.
Platelets	Take part in blood clotting as part of the body's defense mechanisms.
Plasma (55%)	
Water	The primary constituent of blood plasma.
Electrolytes (ions)	Sodium, potassium, chloride, bicarbonate, calcium, hydrogen, magnesium, others. Ions contribute to the control of cell function and volume, to the electrical charge across cells, and to the function of excitable cells (nerve and muscle). All ions must be kept at their normal concentrations for homeostasis to occur.
Proteins	Albumins maintain blood volume and transport electrolytes, hormones, and wastes. Globulins serve as antibodies and transport substances. Clotting proteins contribute to blood clotting.
Hormones	Insulin, growth hormones, testosterone, estrogen, others. Hormones are chemical messenger molecules that provide information needed to regulate specific body functions.
Gases	Oxygen is needed for metabolism; carbon dioxide is a waste product of metabolism. Both are dissolved in plasma as well as carried by RBCs.
Nutrients and wastes	Glucose, urea, many others. Nutrients, raw materials, and wastes (including heat) are transported by blood throughout the body.

a) **Whole blood.**

Whole blood

b) **Blood after being spun in centrifuge.**

Plasma (55%)

Platelets and WBC (1%)

RBC (44%)

c) **A table-top centrifuge.**

Figure 7.2 **Blood.** The formed elements sink to the bottom during centrifugation. The percentage of the blood that is red blood cells is called the hematocrit; in this sample the hematocrit is 44%.

to the bottom. White blood cells (WBCs) and platelets appear just above red blood cells in a thin, grayish white layer.

Plasma consists of water and dissolved solutes

The top layer of a centrifuged blood sample, representing about 55% of the total volume, consists of a pale yellow liquid called **plasma.** Plasma is the transport medium for blood cells and platelets. About 90% of plasma is water. The rest is dissolved proteins, hormones, more than 100 different small molecules (including amino acids, fats, small carbohydrates, vitamins, and various waste products of metabolism), and ions.

The largest group of solutes in plasma consists of **plasma proteins,** which serve a variety of functions. Important plasma proteins include albumins, globulins, and clotting proteins.

Nearly two-thirds of plasma proteins are **albumins,** which primarily serve to maintain the proper water balance between blood and the interstitial fluid. Manufactured in the liver, albumins also bind to certain molecules (such as bilirubin and fatty acids) and drugs (such as penicillin) and assist in their transport in blood.

Globulins (designated alpha, beta, and gamma) are a diverse group of proteins that transport various substances in the blood. Many beta globulins bind to lipid (fat) molecules, such as cholesterol. When a protein attaches to one of these molecules, it creates a complex called a *lipoprotein.* Two medically important lipoproteins are the low-density lipoproteins (LDLs) and high-density lipoproteins (HDLs), and medical exams often include taking a blood sample to measure LDL and HDL relative proportions. The LDLs are sometimes called "bad cholesterol," because high blood levels of these lipoproteins are associated with increased risk of cardiovascular health problems. High levels of HDLs often indicate a lower risk of cardiovascular disease. We discuss lipoproteins and the health implications of blood cholesterol levels in Chapter 8.

Gamma globulins function as part of the body's defense system, helping to protect against infections and illness. We take a closer look at them in Chapter 9.

Clotting proteins, a third group of plasma proteins, play an important role in the process of blood clotting. As we see later in this chapter, blood clotting minimizes blood loss and helps maintain homeostasis after injury.

In addition to plasma proteins, plasma transports a variety of other molecules, including ions (also called electrolytes), hormones, nutrients, waste products, and gases. Electrolytes such as sodium and potassium contribute to the control of cell function and cell volume. Hormones, which are chemical "messengers" from the endocrine system, transport information throughout the body. Nutrients such as carbohydrates, amino acids, vitamins, and other substances are absorbed from the digestive tract or produced by cells' metabolic reactions. Waste products in plasma include carbon dioxide, urea, and lactic acid. Gases dissolved in plasma include oxygen, which is necessary for metabolism, and carbon dioxide, a waste product of metabolism.

 Quick Check A medical researcher develops an artificial blood plasma that contains water, electrolytes, nutrients, various hormones, respiratory gases, and albumins. What is this artificial plasma missing, and what problems might result? ▪

Red blood cells transport oxygen and carbon dioxide

Just under half of the volume of whole blood consists of its formed elements. The most abundant are **red blood cells (RBCs),** also called **erythrocytes** ("red cells" in Greek). Red blood cells function primarily as carriers of oxygen and carbon dioxide. Each cubic millimeter of blood contains approximately 5 million red blood cells. They give blood its color and are the major reason why it is viscous.

Red blood cells offer a great example of how structure serves function. Red blood cells are small, flattened, doughnut-shaped disks whose centers are thinner than their edges (**Figure 7.3**). This is an unusual shape among human cells, but it has several advantages for RBCs. It makes them flexible, so they can bend and flex to squeeze through tiny blood vessels. It also means that no point within an RBC's cytoplasm is ever far from the cell surface, which facilitates the process of gas exchange.

Red blood cells are highly specialized to transport oxygen. Mature RBCs have no nucleus and essentially no organelles. They are essentially fluid-filled bags of plasma membrane, crammed with nearly 300 million molecules

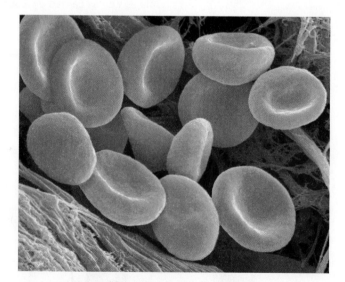

Figure 7.3 Red blood cells. Note that their flattened, biconcave shape gives them a sunken appearance.

Answers to **questions can be found in Appendix A.**

of an oxygen-binding protein called **hemoglobin.** Hemoglobin consists of four polypeptide chains, each containing a heme group (**Figure 7.4**). At the center of each heme group is an iron atom, which can readily form a bond with an oxygen molecule (O_2). In total, a single red blood cell can carry up to 1.2 billion molecules of oxygen. RBCs lack mitochondria and generate ATP by anaerobic pathways. So, they don't consume any of the oxygen they carry; they just transport it.

Several factors influence the binding of hemoglobin to oxygen. Hemoglobin binds oxygen most efficiently when the concentration of oxygen is relatively high and the pH is fairly neutral. These are precisely the conditions that prevail in the lungs. In the lungs, oxygen diffuses into blood plasma and then into red blood cells, where it attaches readily to the iron atoms in hemoglobin. The binding of O_2 by hemoglobin removes some of the O_2 from the plasma, making room for more O_2 to diffuse from the lungs into the plasma. Hemoglobin with four oxygen molecules attached, called *oxyhemoglobin,* has a characteristic bright red color.

The bond hemoglobin forms with oxygen must be temporary so that the oxygen can be released to the cells that need it. In body tissues that use oxygen in the course of their metabolic activities, the concentration of dissolved oxygen and the pH are both lower. Under these conditions, hemoglobin readily releases oxygen into body tissues, making it available to cells. Increased body heat also increases the rate at which hemoglobin releases oxygen. Hemoglobin that has given up its oxygen is called *deoxyhemoglobin.* Deoxyhemoglobin is characteristically dark purple, but because venous blood returning from the cells contains a mixture of oxyhemoglobin and deoxyhemoglobin, venous blood generally has a dark red or maroon color that is between red and purple.

Hemoglobin also transports some carbon dioxide (CO_2), a waste product of cellular metabolism. In tissues, where carbon dioxide levels are high, about 25% of the CO_2 binds to hemoglobin (at different sites than O_2). In the lungs, CO_2 detaches from hemoglobin and is eliminated through respiration. See Chapter 10 for more about gas transport, including how the rest of the CO_2 is transported.

✓ **Quick Check** Suppose a patient has an unusually low body temperature and his blood pH is unusually basic. How might this affect oxygen delivery to the body tissues? ▪

Hematocrit and hemoglobin reflect oxygen-carrying capacity

The percentage of blood that consists of red blood cells is called the **hematocrit** (review Figure 7.2). The hematocrit is a relative measure of the oxygen-carrying capacity of blood, and thus it is often of interest to the health care professional. The normal hematocrit range is 43–49% in men and 37–43% in women. A related number is the amount of hemoglobin in the blood, expressed in units of grams per 100 ml of blood (abbreviated Hb gm%). Normal values for hemoglobin are 14–18 gm% in men and 12–14 gm% in women.

An unusual hematocrit (or Hb gm%) may be cause for concern. A low hematocrit may signal *anemia* or other disorders of inadequate red blood cell production (see section 7.4). A high hematocrit can also be risky because excessive red blood cells thicken blood and increase the risk of blood clots. In rare cases a high hematocrit could signal *polycythemia,* a disorder of the bone marrow characterized by an overproduction of red blood cells. Polycythemia increases blood volume and blood viscosity, sometimes leading to headaches, blurred vision, and high blood pressure.

Some shifts in hematocrit (and hemoglobin) are normal and temporary. For example, if you visit the mountains on your next vacation and stay for at least several weeks, your hematocrit rises to compensate for lower levels of oxygen in the air you breathe. This is part of the normal homeostatic regulation of the oxygen-carrying capacity of the blood. After you return to your usual altitude, your hematocrit returns to normal.

✓ **Quick Check** Is the blood sample in Figure 7.2 more likely to be from a man or a woman? Why? ▪

All blood cells and platelets originate from stem cells

All blood cells and platelets originate from cells in the red marrow of certain bones. These cells, called **stem cells,** divide repeatedly throughout our lives, continually producing

Polypeptide chain

Heme group with iron atom

Figure 7.4 A hemoglobin molecule. Hemoglobin consists of four polypeptide chains folded together, each with a heme group containing a single iron atom. There are nearly 300 million of these molecules in every red blood cell.

immature blood cells. These immature cells develop into platelets, and the various types of mature red and white blood cells described in Figure 7.5.

RBCs have a short life span

Some stem cells develop into immature cells called *erythroblasts* ("red"+"immature"). Erythroblasts become filled with hemoglobin and develop into mature RBCs, or erythrocytes, in about a week. As they mature, these cells lose their nucleus and organelles, and so they cannot reproduce. Thus, all new RBCs must originate from dividing stem cells. Because they lack a nucleus and therefore cannot perform many standard cell activities (such as producing new proteins and phospholipids to renew their cell membranes), they wear out rather quickly. Red blood cells live for only about 120 days, but during that time they make nearly 3,000 round-trips a day, ferrying O_2 from the lungs to the tissues and CO_2 from the tissues back to the lungs. Because they live for such a short time, red blood cells must be produced throughout life—at the incredible rate of more than 2 million per second—just to keep the hematocrit constant.

Old and damaged RBCs are removed from the circulating blood and destroyed in the liver and spleen by large cells called **macrophages.** Macrophages are derived from *monocytes,* the largest of the white blood cells.

Macrophages surround, engulf, and digest the red blood cell. The process is called **phagocytosis.** The four peptide chains of the hemoglobin molecules are then dismantled into their constituent amino acids, and the amino acids are recycled to make new proteins. The iron atoms of the heme groups are returned to the red bone marrow, where they are used again in the production of new hemoglobin for new red blood cells. The heme groups (minus the iron) are converted by the liver to a yellowish pigment called *bilirubin.* If you've ever noticed how a bruise slowly changes color as it heals, from purple to blue to green to yellow, you have observed the chemical breakdown of the heme groups to bilirubin at the site of damage. Under normal circumstances,

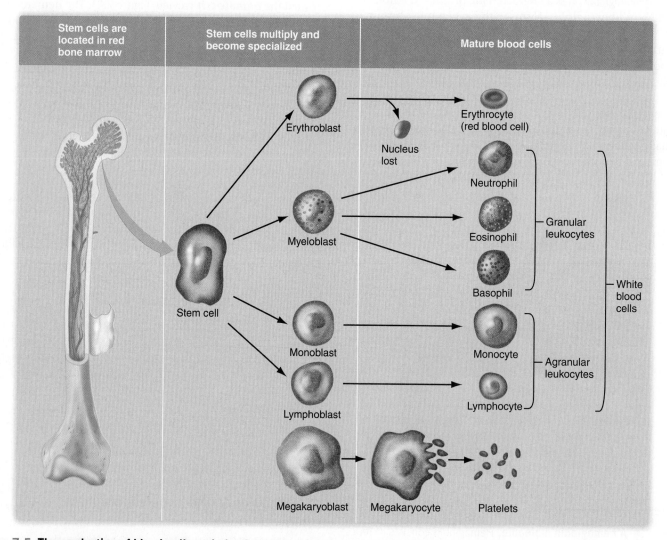

Figure 7.5 The production of blood cells and platelets. Blood cells have short life spans and must be continually replaced. Stem cells in the red marrow of bones continually divide and give rise to a variety of types of blood cells.

when hemoglobin is broken down in the liver, bilirubin mixes with bile secreted during digestion and passes into the intestines. This pigment contributes to the characteristic colors of urine and feces.

When the liver fails to secrete bilirubin into the bile properly or when the bile duct from the liver to the intestines is blocked, bilirubin may accumulate in blood plasma. High circulating levels of bilirubin make skin and mucous membranes look yellowish and can turn the whites of the eyes yellow. This condition is called *jaundice* (from *jaune*, French for "yellow"). Jaundice may also be caused by an increase in the rate of RBC breakdown.

RBC production is regulated by a hormone

Regulation of RBC production is a negative feedback control loop that maintains homeostasis (Figure 7.6). The number of RBCs is not regulated (there are no cells capable of

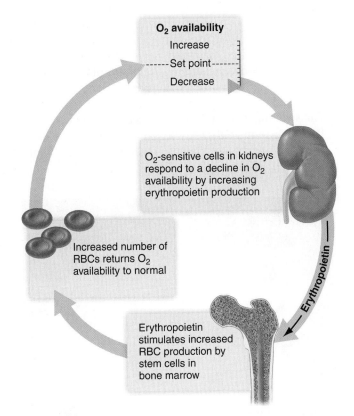

Figure 7.6 Negative feedback control of the availability of oxygen. Certain cells in the kidney are sensitive to the amount of oxygen available to them. When oxygen availability falls, these cells produce erythropoietin, a hormone that stimulates the bone marrow to produce red blood cells. The increase in red blood cells returns oxygen availability toward normal, which reduces the stimulus for further erythropoietin secretion. Ultimately, homeostasis of oxygen availability is achieved.

✓ *What might happen if bone marrow is diseased and is unable to make enough red blood cells (no matter how much erythropoietin is released)?*

counting the number of RBCs), only their effect—their ability to transport oxygen. Certain cells in the kidneys monitor the availability of oxygen. If oxygen availability falls for any reason, these cells cause the kidneys to secrete a hormone called **erythropoietin.** Erythropoietin is transported in the blood to the red bone marrow, where it stimulates stem cells to produce more red blood cells. When the oxygen-carrying capacity of blood returns to an appropriate level as monitored by kidney cells, the cells cut back on their production of erythropoietin, and RBC production returns to normal. Thus, the body maintains homeostasis of oxygen availability by adjusting the production rate of the RBCs that transport it.

Some people with kidney disease do not produce enough erythropoietin to regulate their RBC production properly. Fortunately, erythropoietin is now available commercially and can be administered to stimulate red cell production.

Some athletes have abused erythropoietin by injecting it to increase their RBC production and thus their blood oxygen-carrying capacity, a practice called **blood doping.** Three gold medalists at the 2002 Winter Olympics, Spain's Johann Mühlegg and Russia's Larissa Lazutina and Olga Danilova, were disqualified and stripped of their medals because of blood doping. All three skiers tested positive for darbepoetin, an erythropoietin-like drug that is 10 times more powerful than the natural hormone.

Blood doping can have serious health consequences. Excess red blood cells make blood more viscous, and so the heart must work harder to pump blood through the body. The dehydration that follows strenuous exercise can concentrate the blood even more, increasing the risk of blood clots, high blood pressure, heart attack, and stroke.

White blood cells defend the body

Approximately 1% of whole blood consists of **white blood cells (WBCs or leukocytes).** Larger than red blood cells, they are also more diverse in structure and function. They have a nucleus but no hemoglobin. Because they are translucent, they are difficult to identify under the microscope unless they have been stained. Each cubic millimeter of blood contains only about 7,000 of them, and there is only one WBC for every 700 RBCs. White blood cells play a number of crucial roles in defending against disease and injury.

Like red blood cells, white blood cells arise from stem cells in the red bone marrow. As shown in Figure 7.5, stem cells produce immature blood cells that develop into the various WBCs. There are two major categories of white blood cells: *granular leukocytes* (granulocytes) and *agranular leukocytes* (agranulocytes). Both types contain granules (actually vesicles) in their cytoplasm that are filled with proteins and enzymes to assist their defensive work. However, the granules of the agranular leukocytes are not visible when the cells are stained for viewing ("a−" means "without").

Most WBCs have a short life span. Many granular leukocytes die within a few hours to nine days, probably because of injuries sustained while fighting invading

MJ's Human Biology Blog

The Spleen Stores Monocytes

Shortly after severe tissue damage such as that caused by a heart attack or an infection, the number of a particular type of WBC called *monocytes* increases dramatically in the blood. These new monocytes appear too quickly to have been newly produced from stem cells in bone marrow. So where do they come from?

Apparently they come from the spleen. The spleen stores up to ten times as many monocytes as there are in the bloodstream at any one time. When a tissue is injured the spleen releases its stored monocytes, which then migrate to the site of injury and participate in the cleanup and repair process. It's a pretty efficient use of resources, when you think about it—a virtual army of monocytes is kept on standby, ready to be deployed when needed. ■

Reference: Swirski, Filip K., *et al*. Identification of a Splenic Reservoir: Monocytes and Their Deployment to Inflammatory Sites. *Science* 325: 612–616, July 31, 2009.

microorganisms. Monocytes may survive for several months; lymphocytes for several days to many years. Dead and injured WBCs are continually removed from the blood by the liver and spleen.

Circulating levels of white blood cells rise whenever the body is threatened by viruses, bacteria, or other challenges to health. When activated by tissue injury or microbes, each type of WBC seems able to produce chemicals that stimulate the production of new WBCs from the bone marrow and also the release of stored WBCs from the spleen (see the blog entry, this page).

Red cells remain entirely within the vascular system except in cases of tissue injury, but some white cells leave the vascular system and circulate in the tissue fluid between cells, or in the fluid in the lymphatic system. Because they can change their shape, they can squeeze between the cells that form the capillary walls. White blood cells, part of the body's defense system, are discussed in more detail in Chapter 9. Here we only describe the important characteristics of each type.

Granular leukocytes: Neutrophils, eosinophils, and basophils The granular leukocytes include neutrophils, eosinophils, and basophils. These names are based on their staining properties:

- **Neutrophils,** the most abundant type of granulocyte, account for about 60% of WBCs. (Their name—which means "neutral-loving"—reflects the fact that their granules do not significantly absorb either a red or blue stain.) The first white blood cells to combat infection, neutrophils surround and engulf foreign cells (Figure 7.7). They especially target bacteria and some fungi, and their numbers can rise dramatically during acute bacterial infections such as appendicitis or meningitis.

- **Eosinophils** make up a relatively small percentage (2–4%) of circulating white blood cells. (Their name comes from their tendency to stain readily with an acidic red stain called *eosin*.) Eosinophils have two important functions. The first is to defend the body against large parasites such as worms (hookworms, tapeworms, flukes, and pinworms, among others). These parasites are too big to be surrounded and engulfed through phagocytosis. Instead, clusters of eosinophils surround each parasite and bombard it with digestive enzymes. The second function of eosinophils involves releasing chemicals that moderate the severity of allergic reactions.

- **Basophils,** the rarest white blood cells, account for only 0.5% of leukocytes. (They are named for their tendency to stain readily with basic blue stains.) The granules in the cytoplasm of basophils contain histamine, a chemical that initiates the inflammatory response. When body tissues are injured, basophils secrete histamine, causing adjacent blood vessels to release blood plasma into the injured area. The plasma brings in nutrients, various cells, and chemicals to begin the process of tissue repair. The swelling, itching, and redness associated with inflammation may not feel pleasant, but they are part of the immune system's defenses against molecules that are perceived as threatening.

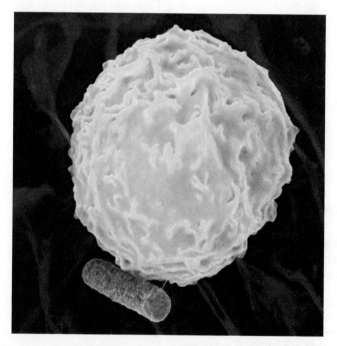

Figure 7.7 A neutrophil attacks a *Bacillus* bacterium. In the first stage of phagocytosis, the neutrophil has approached the bacterium using the protuberances on its surface. Next it will engulf and destroy the bacterium (Color SEM × 8,000).

Agranular leukocytes: Monocytes and lymphocytes The agranular leukocytes include monocytes and lymphocytes. The largest WBCs, **monocytes,** make up about 5% of circulating white blood cells. They can filter out of the bloodstream and take up residence in body tissues, where they differentiate into the macrophages that engulf invaders and dead cellular debris by phagocytosis. They also stimulate lymphocytes to defend the body. Monocytes seem especially active during chronic infections, such as tuberculosis, and against viruses and certain bacterial parasites.

Lymphocytes total about 30% of circulating white blood cells. They are found in the bloodstream, tonsils, spleen, lymph nodes, and thymus gland. They are classified into two types, *B lymphocytes* and *T lymphocytes* (or B cells and T cells). B lymphocytes give rise to *plasma cells* that produce *antibodies,* specialized proteins that defend against microorganisms and other foreign invaders. T lymphocytes target and destroy specific threats such as bacteria, viruses, and cancer cells. Both play a crucial role in the body's immune system, and we look at them in more detail in Chapter 9.

✅ **Quick Check** A friend sprains her ankle, and it soon becomes red, swollen, and sore. Which type of white blood cell is probably responsible for these symptoms, and what is its major function? ▪

Platelets are essential for blood clotting

Less than 1% of whole blood consists of **platelets**. Platelets are derived from megakaryocytes, which are large cells derived from stem cells in the bone marrow (review Figure 7.5). Megakaryocytes never circulate—they remain in the bone marrow. Platelets are just small pieces of megakaryocyte cytoplasm and cell membrane. Because platelets are not living cells, they last only about five to nine days in the circulation.

When a blood vessel is injured and leaks blood, platelets participate in the clotting process, thereby limiting the vascular and tissue damage. We examine the clotting process in the next section. Once the bleeding is stopped, platelets also participate in the repair process by releasing proteins that promote blood vessel growth and repair.

🌐 **Web Animation** *Blood* at www.humanbiology.com

↩ **Recap** Blood consists of a watery fluid containing cells, proteins, nutrients, cellular waste products, and ions. Red blood cells are specialized for transporting oxygen and carbon dioxide; white blood cells protect against disease. Blood cells arise from stem cells in bone marrow. Platelets, important in blood clotting, are small pieces of bone marrow cells called *megakaryocytes.* ▪

7.2 Hemostasis: Stopping blood loss

One of the most important properties of the circulatory system is its ability to limit blood loss following injury. **Hemostasis,** the natural process of stopping the flow or loss of blood,

proceeds in three stages: (1) vascular spasm, or intense contraction of blood vessels in the area, (2) formation of a platelet plug, and (3) blood clotting, also called *coagulation* (Figure 7.8). Once blood loss has stopped, tissue repair can begin.

Vascular spasms constrict blood vessels to reduce blood flow

When a blood vessel is damaged, smooth muscle in its wall undergoes spasms—intense contractions that constrict the vessels. If the vessels are medium-sized to large, the spasms

① **Vessel injury.** Damage to a blood vessel exposes the vessel muscle layers and the tissues to blood.

② **Vascular spasm.** The blood vessel contracts, reducing blood flow.

③ **Platelet plug formation.** Platelets adhere to each other and to the damaged vessel.

④ **Clot formation.** Soluble fibrinogen forms an insoluble mesh of fibrin, trapping RBCs and platelets.

Figure 7.8 **The stages of hemostasis.**

reduce immediate outflow of blood, minimizing the damage in preparation for later steps in hemostasis. If the vessels are small, the spasms press the inner walls together and may even stop the bleeding entirely. Vascular spasms generally last for about half an hour, long enough for the next two stages of hemostasis to occur.

Platelets stick together to seal a ruptured vessel

Normally, platelets circulate freely in blood. However, when the lining of a blood vessel breaks, exposing underlying proteins in the vessel wall, platelets swell, develop spiky extensions, and begin to clump together. They also become sticky and start adhering to the walls of the vessel and to

MJ's Human Biology Blog

Platelet-Rich Plasma Therapy Revisited

About a year ago in this blog I described PRP (platelet-rich plasma) therapy as an exciting and potentially effective new treatment for injuries to tendons and ligaments. Apparently athletes and other patients were asking for the treatment, even though insurance companies were reluctant to pay for it. I asked, "Does it work?" and mentioned that several clinical trials were currently under way to find out.

The results of several of the clinical trials are now in, and the results are not encouraging. In one study of patients with Achilles tendon injuries, PRP therapy was no more effective than an injection of saline (the control). In another study, of tennis elbow, PRP therapy appeared to be slightly more effective than injections of steroids (the standard treatment these days). However, some scientists have criticized the tennis elbow study for not having a control group. Steroid injections are known to reduce pain in the short term but to slow healing in the long term. So the jury is still out on whether PRP therapy for tennis elbow is actually better than no treatment at all.

No doubt, studies with other tendon/ligament injuries using different study protocols will be done in the future. In the meantime, enthusiasm for PRP therapy has cooled just a little. ■

References: De Vos, Robert J., *et al.* Plate-Rich Plasma Injection for Chronic Achilles Tendinopathy. *JAMA* 303: 144–149, 2010.
Peerbooms, Joost C., *et al.* Positive Effect of an Autologous Platelet Concentrate in Lateral Epicondylitis in a Double-Blind Randomized Controlled Trial. *Am. J. Sports Med.* 38, 255–262, 2010.

each other. More platelets congregate and undergo these same changes. The result is a platelet plug that seals the injured area. If the rupture is fairly small, a platelet plug may be able to close it within several seconds. This may be enough to stop the bleeding. If damage is more severe, blood clotting occurs.

A blood clot forms around the platelet plug

The third stage in hemostasis is the formation of a blood clot, during which the blood changes from a liquid to a gel. This involves a series of chemical reactions that ultimately produce a meshwork of protein fibers within the blood. At least 12 substances, known as clotting factors, participate in these reactions. We will focus on three clotting factors: prothrombin activator, thrombin, and fibrinogen.

Damage to blood vessels stimulates the vessels and nearby platelets to release *prothrombin activator*. This activates the conversion of *prothrombin*, a plasma protein, into an enzyme called **thrombin.** This reaction requires the presence of calcium ions (Ca^{2+}). Thrombin in turn facilitates the conversion of a soluble plasma protein, *fibrinogen*, into long insoluble threads of a protein called **fibrin.** The fibrin threads wind around the platelet plug at the wound site, forming an interlocking net of fibers that traps and holds platelets, blood cells, and various molecules against the opening (see Figure 7.8).

The mass of fibrin, platelets, and trapped red blood cells coalesces into an initial *clot* that reduces the flow of blood at the site of injury (**Figure 7.9**). This initial fibrin clot can form

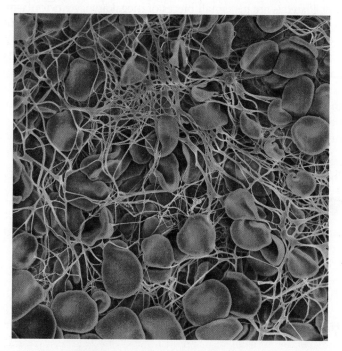

Figure 7.9 Magnified view of a developing clot, showing red blood cells trapped in a network of fibrin fibers.

Donating Blood

A rapid loss of 30% or more of blood volume strains the body's ability to maintain blood pressure and deliver oxygen to cells throughout the body. When this happens, survival may depend on receiving a gift of donated blood. In addition, blood is often needed for certain planned surgical procedures. Approximately 15 million units of blood are donated every year, and almost 5 million people receive donated blood. Most people who donate blood get nothing more (and nothing less) than the satisfaction of knowing they have helped someone in need.

To donate blood, you must be at least 17 years old (16 in some states) and weigh at least 110 pounds. You'll be given a physical examination and asked for your health history, including a confidential questionnaire about your sexual history and recent international travel. This is not done to embarrass you, but to ensure that it is safe for

you to give blood and that your blood will be safe for others.

The blood withdrawal procedure itself is relatively painless (a needle is inserted into an arm vein) and takes about 10–20 minutes. All needles used are brand-new and sterile—you cannot catch AIDS or any other bloodborne disease by donating blood. Afterward you'll be advised to drink and eat something, and avoid rigorous physical exercise for

Donating blood. The procedure is relatively painless, takes only a short time, and can save lives.

the rest of the day. This is not the best day to go mountain climbing, but just about anything less strenuous is OK.

Most donors are allowed to give only one unit of blood (1 pint, about 10% of your blood volume). This is not enough to affect you adversely. Usually the donated blood volume is replaced within several hours by any fluids that you drink. The liver replaces the lost plasma proteins within two days, and stem cells in bone marrow replace the lost RBCs in about a month.

What happens to the blood you give? Sometimes it is stored as whole blood, but more often it is separated into three components: packed cells, platelets, and plasma. Each component may be given to different recipients, meaning that your single "gift of life" can benefit several people.

To learn more about donating blood, contact the American Red Cross. ∎

in less than a minute. Shortly thereafter, platelets in the clot start to contract, tightening the clot and pulling the vessel walls together. Generally the entire process of blood clot formation and tightening takes less than an hour.

If any step in this process is blocked, even a minor cut or bruise can become life threatening. Consider *hemophilia*, an inherited condition caused by a deficiency of one or more clotting factors. People with the most common form of the condition, hemophilia A, lack a protein known as clotting factor VIII. When a vessel is breached, blood clots slowly or not at all. Even if the skin is not broken, severe bruising can spread into joints and muscles (Figure 7.10). Fifty years ago most people with hemophilia did not survive to adulthood, but today many bleeding episodes can be controlled by administering clotting factor VIII. Because clotting factor VIII was initially purified from donor blood, which at that time

Figure 7.10 Complications of hemophilia. People with hemophilia can suffer severe bruising (hemorrhaging) from only minor bumps or scrapes.

could not be tested for the HIV virus, in the past some hemophiliacs contracted AIDS. Today all blood is screened for HIV. In addition, the use of donor blood as a source of factor VIII has decreased because genetic engineering techniques have made it possible to produce factor VIII in the laboratory.

Certain medications can also interfere with hemostasis. If you cut yourself after taking aspirin, for example, you may notice that you bleed more than usual. This is because aspirin blocks platelet clumping and slows the formation of a platelet plug. If you plan to have surgery, your doctor will probably advise you to avoid taking aspirin for at least 7–10 days before the surgery.

Recap Damage to blood vessels causes the vessels to spasm (contract). Nearby platelets become sticky and adhere to each other, limiting blood loss. In addition, a series of chemical events causes the blood in the area to clot, or coagulate (form a gel). ▪

7.3 Human blood types

Blood transfusions—the administration of blood directly into the bloodstream of another person—may seem like a miracle of modern medicine, but the concept is not new. For over a century, physicians have tried to counteract severe blood loss by transfusing blood from one living person into another. Sometimes these attempts were successful. More often they were not, resulting in severe illness or even death for the recipient. Why did they save some lives but not others?

Today we know the success of blood transfusions depends largely on **blood type,** based primarily on the ABO blood group system. If you ever donate or receive blood, you will undergo testing to determine your blood type. If you receive blood from someone who does not belong to a compatible blood type, you could suffer a severe reaction.

To understand the concept of blood typing, we must first be familiar with antigens and antibodies. Our cells have certain surface proteins that the immune system can recognize and identify as "self"—in other words, belonging to us. These are like passwords that cause our immune system to ignore our own cells. Foreign cells carry different surface proteins, which the immune system recognizes as "nonself." An **antigen** (*anti* means "against," and the Greek word *gennan* means "to generate") is a nonself cell protein that stimulates the immune system of an organism to defend the organism. As part of this defense, the immune system produces an opposing protein called an **antibody** ("against" + "body").

Produced by lymphocytes, antibodies belong to the class of plasma proteins called gamma globulins, mentioned earlier. Antibodies mount a counterattack on antigens they recognize as nonself (**Figure 7.11**). There are many antibodies, each one specialized to attack one particular antigen. This response has been compared to a lock and key: only a specific antibody key can fit a specific antigen lock. Antibodies float freely in the blood and lymph until they encounter an invader with the matching antigen. They bind to the antigen molecule to form an antigen-antibody complex that marks the foreigner for destruction. The formation of an antigen-antibody complex often causes the foreign cells to clump together (Figure 7.11). Some antibodies also inactivate foreign cells by preventing them from entering human cells.

Antigens and antibodies are discussed in more detail in Chapter 9. For now, let's look at how their interactions relate to blood type and blood transfusions.

a) Antibody binds to antigen. Antibodies ignore the "self" surface proteins but bind to the antigen of the foreign cell.

b) An antigen-antibody complex forms. The formation of an antigen-antibody complex inactivates the foreign cells.

Figure 7.11 How antibodies recognize and inactivate foreign cells.

ABO blood typing is based on A and B antigens

Like other cells, red blood cells have surface proteins that allow the body to identify them as "self." The interactions between these antigens, and the development of antibodies against the antigens of foreign red blood cells, underlie the reactions that can occur after blood transfusions.

Red blood cells are classified according to the ABO blood group system, in which nearly all individuals be-long to one of four types: A, B, AB, or O. Type A blood has A antigens, type B blood has B antigens, type AB blood has *both* A and B antigens, and type O blood has neither (think of the O as a "zero"). In addition, all individuals have circulating antibodies (and the ability to make more antibodies) against any surface antigens different from their own; type A blood has type B antibodies, type B blood has type A antibodies, type O blood has both type A and B antibodies, and type AB blood has neither antibody. Figure 7.12 shows these various blood types and also indicates the relative incidences of each type in various populations. The antibodies appear early in life, regardless of whether a person has ever received a blood transfusion. These antibodies attack red blood cells with foreign antigens, damaging them and causing them to *agglutinate*, or clump together. If agglutination is extreme, the clumps may block blood vessels, causing organ damage or even death. In addition, hemoglobin released by damaged red blood cells can block the kidneys, leading to kidney failure. Any adverse effect of a blood transfusion is called a *transfusion reaction.*

If you have type A blood, you are restricted to receiving transfusions of either type A or type O blood because neither of them has a foreign (type B) antigen. A transfusion of type B or type AB blood would provoke your antibodies to mount an attack against the B antigen of the donated RBCs, causing them to agglutinate. Similarly, if you're type B, you cannot receive any blood with type A antigens (A or AB). People with type AB blood can generally receive transfusions not only from other AB individuals but from all three of the other blood types as well. People with type AB blood, however, can donate only to other type AB individuals. Type O persons can give blood to persons of A, B, or AB type, but they can receive blood only from type O. Notice that it is the antibodies of the recipient that generally cause the transfusion reaction. Though the donor blood may have antibodies against the recipient's RBCs, they rarely cause transfusion reactions because the volume of blood given is generally small compared to the volume of the recipient's blood.

✅ **Quick Check** Suppose a man has a rare mutation in his blood cell antigens, such that he has only a single unique blood antigen, "C." Nobody else in the world has the type C antigen and nobody has antigens that will react to it. Can he donate blood safely to anybody else? Explain. ∎

HBP **Web Animation** *Blood Types* at www.humanbiology.com

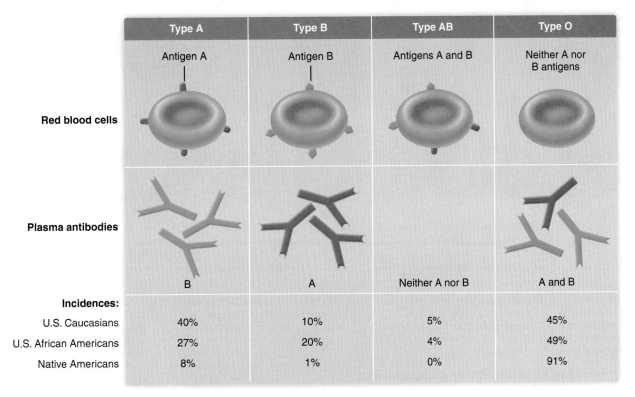

	Type A	Type B	Type AB	Type O
Red blood cells	Antigen A	Antigen B	Antigens A and B	Neither A nor B antigens
Plasma antibodies	B	A	Neither A nor B	A and B
Incidences:				
U.S. Caucasians	40%	10%	5%	45%
U.S. African Americans	27%	20%	4%	49%
Native Americans	8%	1%	0%	91%

Figure 7.12 **Characteristics of the four major blood types of the ABO typing system, showing their RBC surface antigens, antibodies, and relative incidences among various populations.**

Rh blood typing is based on Rh factor

Another red blood cell surface antigen, called **Rh factor** because it was first discovered in rhesus monkeys, is also important in blood transfusions. Approximately 85% of Americans are *Rh positive*, meaning they carry the Rh antigen on their red blood cells. About 15% are *Rh negative*—they do not have the Rh antigen, and consequently their immune systems respond to any foreign Rh antigen by making antibodies against it.

The Rh factor is a particular concern for Rh negative women who wish to have children. If an Rh negative woman becomes pregnant by an Rh positive man, the fetus may be Rh positive. If some of the fetus's Rh positive blood cells leak into the mother's blood, the mother starts producing anti-Rh antibodies. These maternal antibodies can cross the placenta and attack the fetus's red blood cells. The result may be *hemolytic disease of the newborn (HDN)*, a disorder characterized by a reduced number of red blood cells and toxic levels of hemoglobin breakdown products in the newborn. HDN can lead to mental retardation or even death.

The risk of HDN is much higher for the second and all subsequent Rh positive fetuses than for the first (Figure 7.13). This is because it takes days or even weeks for antibodies to be produced after the first exposure to an antigen. Although a few fetal cells may leak across the placenta during a normal pregnancy and so present a slight chance that the first fetus could be affected, the greatest chance of maternal exposure to fetal blood generally occurs right at childbirth, when the placenta detaches from the uterus. The antibodies that develop from maternal exposure during the woman's first delivery come too late to affect the first fetus. But the maternal immune system has learned its lesson, and it is ready and waiting to attack the blood of any subsequent Rh positive fetus (see section 9.7, Immune memory creates immunity).

To prevent this reaction, an Rh negative mother who *may* be carrying an Rh positive child is given an injection of anti-Rh antibodies (RhoGAM) at 28 weeks of gestation, just in case. Then, if the newborn *is* Rh positive, the mother is given a second injection no later than three days after childbirth. The injected antibodies quickly destroy any of the newborn's red blood cells that may have entered the woman's circulation during childbirth, before her immune system has time to react to them. The injected antibodies disappear in a short time.

In addition to its important medical applications, blood typing has many other uses. Because blood types are inherited, anthropologists can track early population migrations by tracing inheritance patterns. Blood typing is also used in criminal investigations to compare the blood of victims and perpetrators, and to eliminate or identify suspects on the basis of matching antigens. DNA tests can be done on blood samples to help determine paternity.

✅ **Quick Check** Will the immune system of an Rh *positive* woman attack blood cells from an Rh *negative* baby? Why or why not? ◾

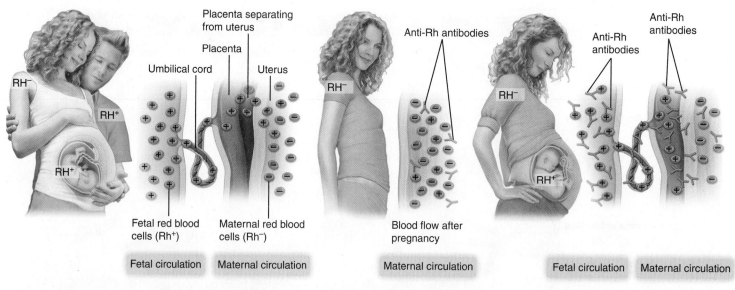

a) **When an Rh positive man fathers a child by an Rh negative woman, the fetus may inherit the Rh positive antigen.**

b) **During pregnancy or more commonly at childbirth, a small amount of fetal blood enters the mother's circulation.**

c) **Over the next several weeks the woman develops antibodies and an immune memory against the Rh antigen.**

d) **When the woman becomes pregnant with her second Rh positive child, her immune system quickly produces antibodies that attack the fetus' red blood cells.**

Labels within figure: RH⁻, RH⁺, RH⁺; Placenta separating from uterus; Placenta; Umbilical cord; Uterus; Fetal red blood cells (Rh⁺); Maternal red blood cells (Rh⁻); Fetal circulation; Maternal circulation; Anti-Rh antibodies; RH⁻; Blood flow after pregnancy; Maternal circulation; Anti-Rh antibodies; Anti-Rh antibodies; RH⁻; RH⁺; Fetal circulation; Maternal circulation.

Figure 7.13 How Rh factor incompatibility can affect a fetus.

 In which population—Caucasian, African American, or Native American—would it be least risky to do an emergency blood transfusion without blood typing either the donor or the recipient?

Blood typing and cross-matching ensure blood compatibility

Blood typing involves determining your ABO type and the presence or absence of the Rh factor. For example, if your blood type is "B-pos" (B+), you are type B and positive for the Rh factor. If you are "O-neg" (O−), you are type O and negative for the Rh factor.

ABO blood typing is done by adding plasma containing small amounts of anti-A and anti-B antibodies to drops of diluted blood. If the blood agglutinates, then it contains the antigens that match the antibodies (Figure 7.14).

AB+ individuals were once called *universal recipients* because they can generally receive blood from any other type. Type O− individuals were formerly called *universal donors* because their blood can usually be donated to any other type. However, because transfusion reactions can occur unexpectedly, the terms are now considered outdated. Why do transfusion reactions occur occasionally even when blood has been adequately typed for ABO blood type and the Rh factor? The reason is that there are over 100 other less common blood antigens in the human population, in addition to the very common A, B, and Rh antigens. Fortunately, most of them are fairly rare. To ensure that blood transfusions are absolutely safe, however, medical laboratories generally do

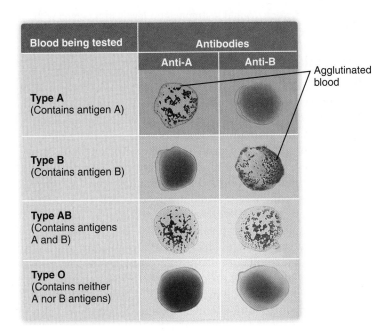

Blood being tested	Antibodies	
	Anti-A	Anti-B
Type A (Contains antigen A)		
Type B (Contains antigen B)		
Type AB (Contains antigens A and B)		
Type O (Contains neither A nor B antigens)		

Figure 7.14 **Blood typing for ABO blood types.** Anti-A and Anti-B antibodies are added to diluted samples of blood. If the red blood cells have the surface antigen that matches the antibody, the blood agglutinates. Type O blood does not agglutinate in response to either.

 Draw two more pictures illustrating what Rh-positive and Rh-negative blood will look like if they are each mixed with Rh antibodies.

blood typing and *cross-matching*. Cross-matching involves mixing small samples of donor blood with recipient plasma, and recipient blood with donor plasma, and examining both combinations for agglutination. If agglutination does not occur in either combination, the bloods are assumed to be a good match.

 Recap Blood types A, B, AB, and O are defined by the presence (or absence) of type A and/or type B surface antigens on red blood cells. In addition to blood type, all persons are classified according to the presence or absence of another red blood cell surface antigen called the Rh factor. Antibodies to the Rh factor can cause a serious immune reaction of a mother to her own fetus under certain circumstances. ▪

7.4 Blood disorders

Blood disorders include infections, several types of cancers, and disorders that affect the ability of the blood to deliver oxygen to the tissues or to clot properly when injury occurs. The effects of blood disorders are often widespread because blood passes through every organ in the body.

Blood poisoning: Infection of blood plasma

Normally blood is well defended by the immune system. Occasionally microorganisms invade the blood, overwhelm its defenses, and multiply rapidly in blood plasma. The organisms may be poisonous themselves, or they may secrete poisonous chemicals as by-products of their metabolism. The result is blood poisoning, also called *septicemia* or *toxemia*.

Blood poisoning may develop from infected wounds (especially deep puncture wounds), severe burns, urinary system infections, or major dental procedures. To help prevent it, wash wounds and burns thoroughly with soap and water. Consult your doctor immediately if you experience flushed skin, chills and fever, rapid heartbeat, or shallow breathing. Antibiotic drugs are usually effective against blood poisoning.

Mononucleosis: Contagious viral infection of lymphocytes

Mononucleosis is a contagious infection of lymphocytes in blood and lymph tissues caused by the Epstein-Barr virus, a relative of the virus that causes herpes. Most common during adolescence, "mono" is nicknamed the "kissing disease" because it's frequently spread through physical contact.

Symptoms of mononucleosis can mimic those of flu: fever, headache, sore throat, fatigue, and swollen tonsils and lymph nodes. A blood test reveals increased numbers of monocytes and lymphocytes. The disease is called mononucleosis because

MJ's Human Biology Blog

Cleansing Blood with Magnets

Researchers at Harvard University are working on an innovative method for treating blood infections—drawing the bacteria out of blood with magnets. They mix infected blood with tiny antibody-coated magnetic spheres only 1/8 the size of red blood cells. The antibodies on the spheres attach to bacteria in the blood, and then the spheres and the bacteria are drawn off together using a powerful magnetic field.

In initial experiments the method removed up to 80% of the bacteria in small samples (10–20 ml) of blood. However, several questions remain unanswered: 1) Can the method be tooled up to cleanse the larger volume of blood in human patients? 2) Will reducing the bacterial load in a patient's blood actually improve the patient's recovery? 3) What might happen to the patient if a few magnetic beads are not removed from the blood before it is returned to the patient?

It may be several years before we know if the method can be used safely and effectively to treat blood infections in human patients. ■

Reference: Yung, Chong W., *et al.* Micromagnetic-microfluidic blood cleansing device. *Lab on a Chip* 9: 1171–1177, 2009.

many of the lymphocytes enlarge and begin to resemble monocytes. There is no known cure for mononucleosis, but almost all patients recover on their own within four to six weeks. Extra rest and good nutrition help the body overcome the virus.

Anemia: Reduction in blood's oxygen-carrying capacity

Anemia is a general term for reduction in the oxygen-carrying capacity of blood. All causes of anemia produce similar symptoms: pale skin, headaches, fatigue, dizziness, difficulty breathing, and heart palpitations—the uncomfortable feeling that one's heart is beating too fast as it tries to compensate for the lack of oxygen delivery. Major types of anemia include:

- **Iron-deficiency anemia.** Recall that every hemoglobin molecule contains 4 molecules of iron. When the body is deficient in iron, hemoglobin cannot be synthesized properly. The result is fewer hemoglobin molecules per red blood cell, and thus a decreased ability to transport oxygen. Iron-deficiency anemia is the most common form of anemia worldwide. Usually it is due to too little iron in

the diet, but it can also be caused by an inability of the digestive tract to absorb iron properly. Generally it can be treated by taking pills that contain iron, or by eating foods rich in iron such as leafy green vegetables and meat.

- **Hemorrhagic anemia.** Anemia due to blood loss (hemorrhage) may be caused by injuries, bleeding ulcers, excessive menstrual flow, and even certain parasites. Treatment includes finding and treating the underlying cause of blood loss, if possible, and making sure one has enough iron in the diet to replenish the lost red blood cells.

- **Pernicious anemia.** Pernicious anemia is caused by a deficiency of vitamin B_{12} absorption by the digestive tract. Vitamin B_{12} is important for the production of normal red blood cells. Pernicious anemia can be treated by injections of B_{12}.

- **Hemolytic anemia.** Hemolytic anemia is the result of rupture (lysis) or early destruction of red blood cells. One cause is sickle-cell disease, an inherited disorder in which the red blood cells take on an abnormal sickle shape when the oxygen concentration is low. Because of their abnormal shape, sickled red blood cells become damaged as they travel through small blood vessels. Once damaged, they are destroyed by the body. Sickle-cell anemia is most prevalent in Africans who live near the equator and in African Americans, for reasons discussed in Chapter 19. Another common cause of hemolytic anemia is the parasite that causes malaria.

✅ **Quick Check** Would erythropoietin levels in a person with anemia be low, normal, or high? Why? ■

Leukemia: Uncontrolled production of white blood cells

Leukemia refers to any of several types of blood cancer. Their common characteristic is uncontrolled proliferation of abnormal or immature white blood cells in the bone marrow. Overproduction of abnormal WBCs crowds out the production of normal white blood cells, red cells, and platelets. Huge numbers of leukemia cells enter and circulate in the blood, interfering with normal organ function.

There are two major categories of leukemia: acute, which develops rapidly, and chronic. Both are thought to originate in the mutation of a white blood cell (a change in genetic structure) that results in uncontrolled cell division, producing billions of copies of the abnormal cell. Possible causes for the original mutation include viral infection or exposure to radiation or harmful chemicals. Genetic factors may also play a role.

Leukemia can produce a wide range of symptoms. Tissues may bruise easily because of insufficient production of platelets. Anemia may develop if the blood does not contain enough red blood cells. Bones may feel tender because the marrow is packed with immature white blood cells. Some people experience headaches or enlarged lymph nodes.

Treatment can cure leukemia in some cases and prolong life in others. Treatment generally involves radiation therapy and chemotherapy to destroy the rapidly proliferating cancer cells. This kills the normal stem cells as well, so transplants of bone marrow tissue are required to provide new stem cells. Cord blood transplants may be another option (see "Current Issue: Should You Bank Your Baby's Cord Blood?"). As with blood transfusions, all tissue must undergo testing to make sure the donor's antigens are compatible with those of the patient.

Multiple myeloma: Uncontrolled production of plasma cells

Like leukemia, *multiple myeloma* is a type of cancer. In this case, abnormal plasma cells in the bone marrow undergo uncontrolled division. Plasma cells are a type of lymphocyte responsible for making a specific antibody. The proliferating plasma cells manufacture too much of an abnormal, frequently incomplete antibody, impairing production of other antibodies and leaving the body vulnerable to infections. Bones become tender as healthy bone marrow is crowded out by malignant plasma cells. Levels of calcium in the blood soar as bone tissue is destroyed. Treatment includes anticancer drugs and radiation therapy.

Thrombocytopenia: Reduction in platelet number

Thrombocytopenia is a reduction in the number of platelets in the blood. Thrombocytopenia can occur for a number of reasons, such as viral infection, anemia, leukemia, other blood disorders, exposure to X-rays or radiation, and even as a reaction to certain drugs. Sometimes platelet levels decline for no apparent reason, in which case they often rise again after several weeks.

Symptoms include easy bruising or bleeding, nosebleeds, bleeding in the mouth, blood in urine, and heavy menstrual periods. Treatment of the underlying cause generally improves the condition. If it persists, surgical removal of the spleen often helps.

Recap Blood poisoning and mononucleosis are types of blood infection. Several factors, including iron deficiency or hemorrhage, can lead to a reduction in oxygen-carrying capacity of blood. Leukemia and multiple myeloma are blood cell cancers that arise when abnormal cells in the bone marrow divide uncontrollably. Thrombocytopenia, a disease of too few platelets, is characterized by easy bleeding or bruising. ∎

Chapter Summary

The components and functions of blood p. 144

- Blood consists of formed elements and plasma. Blood has transport, regulatory, and protective functions.
- Plasma contains numerous plasma proteins involved in transport, regulation of water balance, and protection. It also contains ions, hormones, nutrients, wastes, and gases.
- Erythrocytes (RBCs) are highly specialized for the transport of oxygen, but they also transport some carbon dioxide.
- Hemoglobin is the primary protein in red blood cells and gives blood its oxygen-carrying capacity.
- The formed elements of blood all originate from stem cells in red bone marrow.
- Leukocytes (WBCs) defend the body against disease and the effects of injury.
- RBCs and WBCs have short life spans and must continually be replaced. RBC production is stimulated when the body detects low oxygen levels in the blood.
- Platelets are cell products that participate in blood hemostasis.

Hemostasis: Stopping blood loss p. 151

- Hemostasis is a three-phase process that prevents blood loss through damaged vessels. The phases are (1) vascular spasm, (2) the formation of a platelet plug, and (3) blood clotting.

- During the formation of a blood clot, substances released by damaged blood vessels cause soluble proteins called *fibrinogen* to become insoluble protein threads called *fibrin*. The threads form an interlocking mesh of fibers, trapping blood cells and sealing ruptured vessels.

Human blood types p. 154

- Successful transfusion of blood from one person into another depends on compatibility of their blood types, which is determined by antibodies in plasma and surface antigens on red blood cells.
- Blood types are classified primarily on the basis of the ABO system and the presence or absence of the Rh factor.
- Rh factor in particular can affect certain pregnancies adversely.

Blood disorders p. 157

- Blood poisoning is a general term for infection of blood plasma by various microorganisms.
- Mononucleosis is a contagious viral disease of lymphocytes and lymphatic tissue.
- Anemia is a reduction in blood oxygen-carrying capacity for any number of reasons, including insufficient red blood cell or hemoglobin production, and excessive blood loss.
- Leukemia is a cancer characterized by uncontrolled production of abnormal leukocytes (white blood cells).

Terms You Should Know

anemia, *158*

blood type, *154*

erythrocyte (RBC), *146*

erythropoietin, *149*

fibrin, *152*

hematocrit, *147*

hemoglobin, *147*

hemostasis, *151*

leukocyte (WBC), *149*

phagocytosis, *148*

plasma, *146*

plasma proteins, *146*

platelet, *151*

Rh factor, *156*

stem cell, *147*

Concept Review

Answers can be found at the Human Biology Place.

www.humanbiology.com

1. Describe the functions of blood.
2. Describe the role of hemoglobin in the transport of oxygen and carbon dioxide.
3. Explain how the production of red blood cells is regulated to maintain homeostasis of the oxygen-carrying capacity of blood.
4. Define hematocrit, and explain why it is important.
5. Describe how damaged or dead RBCs and the hemoglobin they contain are removed from the blood.
6. Describe the difference between the actions of neutrophils and eosinophils.
7. Describe the mechanism of hemostasis.
8. List the four ABO blood types. For each one, list its red blood cell surface antigen(s) and plasma antibody (antibodies).
9. Describe Rh factor and its implications for pregnancy.
10. Compare and contrast the various causes of anemia.

Test Yourself

Answers can be found in Appendix A.

1. All of the following proteins are associated with blood. Which of these is found specifically inside red blood cells?
   a. prothrombin
   b. fibrinogen
   c. albumin
   d. hemoglobin

2. Which of the following blood components protects the individual from a variety of infectious agents such as bacteria and viruses?
   a. white blood cells
   b. platelets
   c. albumin
   d. red blood cells

3. Which of the following make(s) up the greatest volume of whole blood?
   a. platelets
   b. red blood cells
   c. plasma
   d. white blood cells

4. Which of the following influence(s) the bonding of oxygen to hemoglobin?
   a. pH
   b. oxygen concentration
   c. temperature
   d. all of the above

5. Jason has just spent four weeks in Rocky Mountain National Park, studying plants that grow above 10,000 feet elevation. Which of the following would be a likely change in his blood because of time spent at high elevation?
   a. increased number of red blood cells
   b. increased number of white blood cells
   c. increased number of platelets
   d. increased amount of globulins in the plasma

6. A person with Type A- (A-negative) blood will have:
   a. type A plasma antibodies
   b. type A antigens on the red blood cells
   c. Rh antigens on the red blood cells
   d. all of the above

7. A deficiency of platelets would result in:
   a. fatigue and dizziness
   b. bleeding and bruising
   c. increased susceptibility to infections
   d. all of the above

8. Which donor blood type would be most appropriate for transfusing an O− recipient?
   a. A−
   b. B−
   c. O−
   d. Any of these blood types could be successfully used for this recipient.

9. What do erythroblasts, myeloblasts, lymphoblasts, and megakaryoblasts have in common?
   a. They are immature cells that develop into white blood cells.
   b. They are immature cells that develop into red blood cells.
   c. They are found in the circulating blood.
   d. They are immature cells found in the bone marrow.

10. Jaundice is caused by the presence of _____ in the blood plasma, which is a breakdown product of _____.
    a. hemoglobin...red blood cells
    b. bilirubin...hemoglobin
    c. albumin... white blood cells
    d. prothrombin...platelets

11. Which white blood cells are present in the greatest number in the blood and are the body's first responders to infection?
    a. neutrophils
    b. lymphocytes
    c. platelets
    d. monocytes

12. The steps in the hemostasis process are (1) platelets become sticky and form a platelet plug, (2) prothrombin is converted to thrombin by prothrombin activator, (3) walls of a damaged blood vessel undergo spasms, (4) a clot forms from fibrin, platelets, and trapped red blood cells, and (5) fibrinogen is converted to fibrin. Which of the following choices represents the correct order of these steps?
    a. 1-3-2-4-5
    b. 3-1-2-5-4
    c. 3-2-1-4-5
    d. 3-5-4-1-2

13. Hemophilia results from:
    a. an insufficient number of red blood cells
    b. an insufficient number of platelets
    c. a lack of one or more plasma proteins involved in blood clotting
    d. an abnormal type of hemoglobin

14. Which of the following can lead to anemia?
    a. insufficient iron in the diet
    b. insufficient Vitamin B_{12} absorption from the digestive tract
    c. spending several weeks at a high altitude
    d. both (a) and (b)

15. Which property do red blood cells and platelets have in common?
    a. Both lack a nucleus.
    b. Both transport oxygen.
    c. They are found in approximately equal numbers in the circulating blood.
    d. Both are derived from erythroblasts.

Apply What You Know

Answers can be found at the Human Biology Place.
www.humanbiology.com

1. A 35-year-old white male is sent by his physician for a blood test. The lab results indicate his white blood cell count (number of WBCs per milliliter of blood) is 18,000. The typical WBC count for a man his age is 6,000–9,000, meaning his white blood cell count is considerably higher than normal. What may this mean?

2. One treatment for certain types of leukemia is to try to kill all of the stem cells in bone marrow through radiation and chemotherapy and then to give a bone marrow transplant from another person (a donor). Can just anyone be the donor? Who is most likely to be a good donor? Explain.

3. In the not too distant past, people with type O-negative blood were considered to be universal blood donors, and their blood was sought out during times of need. Explain what was meant by the term universal donor, why O-negative persons were considered to be universal donors, and why "universal blood donor" is now considered an outdated term.

4. The text states that when red blood cells reach actively metabolizing tissues, they release their cargo of oxygen because both the oxygen concentration and the pH are lower in metabolically active tissues than in the general circulation. The oxygen concentration is lower because actively metabolizing tissues are using oxygen at a rapid rate. But what causes the pH to fall? And how might a fall in pH cause the hemoglobin to release oxygen?

   Can you think of any other variables that might also lead to the release of oxygen by hemoglobin?

5. The term has just ended. Over the past three weeks you wrote three term papers, studied for finals, and went to work at your part-time job. You passed all your courses and never missed a day of work. Of course, you spent the last three weeks living on soda, frozen pizza, and Ramen noodles (who has time for real food with all this work?).

   Now that the term is over you notice that you are very tired, your skin is pale, you are experiencing headaches and dizziness, and even your breathing is difficult. You go to your doctor who diagnoses you with anemia. What type of anemia do you most likely have, how is it different from other forms of anemia, and what treatment will your doctor most likely suggest?

6. Coumadin is an anticoagulant drug that is sometimes given to patients who have just suffered a deep vein thrombosis, a pulmonary embolism, or a heart attack, or to patients with artificial heart valves. It helps reduce the chance of future clots and the further risk of embolism. The active ingredient in Coumadin is warfarin, a rat poison. How do you think the same compound can be used for these two very different purposes?

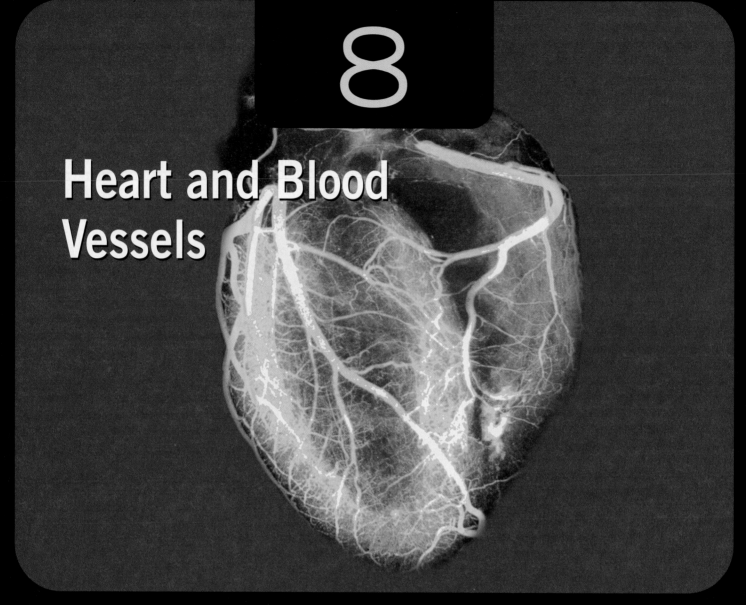

8

Heart and Blood Vessels

A colored arteriogram of a healthy heart.

Comparative Effectiveness Research

Mr. Reynolds has a heart problem. An angiogram shows that a short section of one of the main arteries supplying the left ventricle of his heart is narrowed, restricting blood flow to his heart muscle. His doctor tells him that he is at serious risk of a heart attack. The doctor explains that there are at least three techniques that could be used to restore blood flow to his heart: 1) balloon angioplasty, 2) placement of a coronary artery stent, or 3) a coronary artery bypass graft (CABG). Which would be best for Mr. Reynolds? They go over the options together, but to Mr. Reynolds it seems like comparing

apples to oranges, and he isn't sure he understands. He leaves the decision to his physician, whom he has known for 25 years. In the end the physician chooses the technique that has worked best for his previous patients.

The body of medical literature is now so vast and expanding so rapidly that even the best physicians can't know it all. This is where a relatively new field of medical science called "Comparative Effectiveness Research" (CER) comes in. CER focuses solely on analyzing the medical literature already available, in order to reach

scientifically sound judgments about the value (or lack of value) of specific medical tests, treatments, and disease prevention strategies. In essence, CER seeks to determine the best practices in medicine based on our current knowledge.

Changing How Medicine is Practiced

Consider how CER might benefit Mr. Reynolds' physician (and Mr. Reynolds, of course). By reviewing CER data, Mr. Reynolds' physician might learn that a stent is considered most effective for a middle-aged white male, but that there's an age-related tipping point; if the patient is over 55, balloon angioplasty is the better option. (Hmmm, how old *is* Mr. Reynolds this year?) CER might also be able to tell the physician whether the treatment of choice depends on the severity of the narrowing—if the degree of narrowing of a coronary artery is greater than 80%, for example, then the best option (again, for a middle-aged white male) would be a coronary artery bypass graft rather than balloon angioplasty. (What is the degree of narrowing in Mr. Reynolds, anyway?) Toss in other factors like gender, race, physical condition, body weight, smoker-versus-nonsmoker, and you can begin to see the full power of CER. In theory, CER could analyze multiple factors at once to arrive at the best treatment option for patients who are described by a particular combination of factors. Even the most experienced physicians don't carry *that* much information around in their heads!

Some politicians believe that little investment in CER now could pay for itself in reduced health care expenditures in the future. To jump-start a national CER program, Congress passed the "Comparative Effectiveness Research Act of 2009" and funded it with $1.1 billion as part of the economic stimulus package. To keep the program free of bias, the prestigious Institute of Medicine of the National Academies of Science was asked to come up with a list of 100 top priority topics for CER funding. Among the topics are comparisons of the most effective practices to treat or prevent a number of cardiovascular diseases and risk factors, including high blood pressure, coronary artery disease, heart failure, and abnormalities of heart electrical rhythm. This is not surprising, since cardiovascular diseases are the number one cause of death in the United States (cancer is second).

Doctoring, 1948.

Who Will Make Health Care Decisions?

CER could become a powerful tool for improving health care quality and lowering costs. Nevertheless, the CER Act of 2009 has stirred strong feelings among physicians, patients, politicians, and the health care industry because of the ways it could change how medicine is practiced. Physicians and patient advocacy groups worry that if "best practices" become defined by CER, doctors and patients could begin to lose the right to make decisions regarding treatment options. They fear that health care decisions may be dictated primarily by bureaucrats and insurance companies. In recognition of this concern, the legislation included language to the effect that the outcomes of CER research are not to be used to develop required or mandated guidelines when it comes to treatment decisions, payment, or insurance coverage, including Medicare/Medicaid coverage. In other words, physicians and patients can still use their best judgment in deciding upon the appropriate treatment option, regardless of what CER shows, at least for now.

But therein lies the big question: Will physicians and patients continue to be the decision makers in medical treatment decisions? Or is it inevitable that the old way

of practicing medicine is going to change? Do we really believe that health insurance companies, group health plans, and even Medicare/Medicaid will *not* find a way to use CER data to "influence" reimbursement policies, and hence treatment decisions? Would it be a good thing or a bad thing if they did?

Flash forward 25 years. It's you in the doctor's office now, and the doctor is telling you that your cardiac scan shows a 63% narrowing of a section of your left-anterior descending coronary artery. She swings around to her computer, taps a few keys, and reports that according to the latest data analysis from the Comparative Effectiveness Research Institute, the current most effective method for repair of your coronary artery is "Robotic Artificial Vessel Extension" (RAVE), and that your government-supported health insurance will pay for it. A few more taps and she informs you that there is an opening on the hospital's surgical schedule on Tuesday and that Dr. Sloan is available to do your surgery. She adds that 99.7% of Dr. Sloan's surgeries of this type have been successful, and that 94% of all patients with your condition are discharged from the hospital on the same day as the surgery. You go home to your wife with the good news, and tell her to go ahead and book that vacation to London next month.

Doctoring in the 21st century.

Questions to consider

1 Who do you want to help you decide which treatment options would be best for you? If it's not a specific person or position (doctor, patient representative, health insurance specialist), what information would you like to have available to you?

2 Do you think cost-effectiveness should be a part of any comparative effectiveness analysis of treatment or diagnostic options? Why or why not?

>> **The structure of blood vessels reflects their function.** Thick-walled arteries and arterioles transport blood to the tissues under high pressure; capillaries allow fluid exchange between blood and interstitial fluid; large thin-walled veins store most of the blood and return it to the heart.

>> **The heart is a pump comprised primarily of muscle.** The heart has no bones; its ability to pump blood depends on one-way valves and coordination of muscle contraction.

>> **Arterial blood pressure is held fairly constant by homeostatic control mechanisms.** Maintenance of a nearly constant arterial blood pressure allows local blood flow to each tissue to be regulated by local mechanisms.

>> **Cardiovascular disorders are the number one cause of death in the United States.** Cardiovascular disorders include heart attack, heart failure, stroke, and cardiac arrhythmias.

>> **Your risk of developing cardiovascular disease is affected by your lifestyle choices.** Risk factors include smoking, a lack of exercise, obesity, and chronic stress.

The heart is a pump, but it's a very special pump indeed. The heart is constructed entirely of living cells and cellular materials, yet it is capable of greater reliability than some of the best pumps ever built by humans. It can easily withstand 80–100 years of continuous service without ever stopping for repairs. Its output is also fully adjustable on demand, over a range of about 5–25 liters of blood per minute.

The heart and blood vessels are known collectively as the **cardiovascular system** (from the Greek *kardia*, heart, and the Latin *vasculum*, small vessel). The heart provides the power to move the blood, and the vascular system represents the network of branching conduit vessels through which the blood flows. The cardiovascular system is essential to life because it supplies every region of the body with just the right amount of blood. It is essential to maintaining homeostasis. We consider the blood vessels first.

8.1 Blood vessels transport blood

A branching network of blood vessels transports blood to all parts of the body. The network is so extensive that if our blood vessels were laid end to end, they would stretch 60,000 miles!

We classify the body's blood vessels into three major types: *arteries*, *capillaries*, and *veins*. Thick-walled arteries transport blood to body tissues under high pressure. Microscopic capillaries participate in exchanging solutes and water with the cells of the body. Thin-walled veins store blood and return it to the heart. **Figure 8.1** illustrates the structures of each type of blood vessel, described in more detail below.

Arteries transport blood away from the heart

As blood leaves the heart it is pumped into large, muscular, thick-walled **arteries.** Arteries transport blood away from the heart. The larger arteries have a thick layer of muscle because they must be able to withstand the high pressures generated by the heart. Arteries branch again and again, so the farther blood moves from the heart, the smaller in diameter the arteries become.

Large and medium-sized arteries are like thick garden hoses, stiff yet somewhat elastic (distensible). Arteries stretch a little in response to high pressure but are still strong enough to withstand high pressures year after year. The ability to stretch under pressure is important because a second function of arteries is to store the blood that is pumped into them with each beat of the heart and then provide it to the capillaries (at high pressure) even between heartbeats. The elastic recoil of arteries is the force that maintains the blood pressure between beats. Think of the arteries as analogous to a city's water system of branching, high-pressure pipes that provide nearly constant water pressure to every home.

The structure of the walls of large and medium-sized arteries is ideally suited to their functions. The vessel wall is a sandwich of three distinct layers surrounding the *lumen*, or hollow interior of the vessel:

1. The thin inner layer, the *endothelium*, is a layer of flattened, *squamous* epithelial cells. It is a continuation of the lining of the heart. The flattened cells fit closely together, creating a slick surface that keeps friction to a minimum and promotes smooth blood flow.
2. Just outside the endothelium is a layer composed primarily of smooth muscle with interwoven elastic connective tissue. In most arteries this is the thickest layer of the three. Tonic contraction of the smooth muscle of large and medium-sized arteries stiffens the arteries and helps them resist the high pressures within, but it does not constrict them enough to alter blood flow. The elastic tissue makes large and medium-sized arteries slightly distensible so they can stretch passively to accommodate the blood that enters with each heartbeat.
3. The outermost layer of large and medium-sized arteries consists of a tough supportive layer of connective tissue, primarily collagen. This sturdy casing anchors vessels to surrounding tissues and helps protect them from injury.

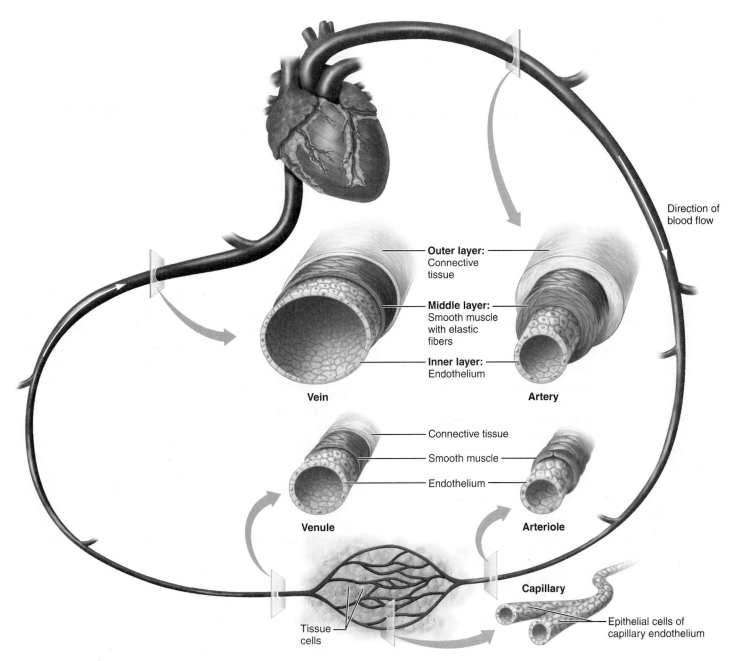

Figure 8.1 **The structures of blood vessels in the human body.**

The fact that arteries are constantly under high pressure places them at risk of injury. If the endothelium becomes damaged, blood may seep through the injured area and work its way between the two outer layers, splitting them apart. The result is an *aneurysm*, or ballooning of the artery wall. Some aneurysms cause the smooth muscle and endothelial layers to bulge inward as they develop, narrowing the lumen enough to reduce blood flow to an organ or region of the body. Others force the outer connective tissue layer to bulge outward. Sometimes aneurysms cause severe chest pain, but in other cases they are completely symptomless until they rupture or "blow out," causing massive internal bleeding and often death. If you've ever seen a water line burst, you know how quickly it can be

devastating. Aneurysms of the aorta (see section 8.2) kill an estimated 25,000 Americans every year. Actor John Ritter's sudden death in 2003 was caused by a ruptured aneurysm.

Aneurysms often take years to develop. During this time many can be detected and repaired surgically. Some physicians recommend that anyone with a family history of aneurysm should be examined, even if there are no symptoms. Doctors can sometimes detect inward-bulging aneurysms with a stethoscope (an instrument for listening to sounds inside the body) because flowing blood produces characteristic sounds as it passes through the narrowed area. A computerized tomography (CT) scan may also locate aneurysms before they rupture.

Arterioles and precapillary sphincters regulate blood flow

Eventually blood reaches the smallest arteries, called **arterioles** (literally, "little arteries"). The largest artery in the body, the *aorta*, is about 2.5 centimeters (roughly 1 inch) wide. In contrast, arterioles have a diameter of 0.3 millimeter or less, about the width of a piece of thread.

By the time blood flows through the arterioles, blood pressure has fallen considerably. Consequently, arterioles can be simpler in structure. Generally they lack the outermost layer of connective tissue, and their smooth muscle layer is not as thick. In addition to blood transport and storage, arterioles have a third function not shared by the larger arteries: they help regulate the amount of blood that flows to each capillary. They do this by contracting or relaxing the smooth muscle layer, altering the diameter of the arteriole lumen.

Right where an arteriole joins a capillary is a band of smooth muscle called the **precapillary sphincter** (Figure 8.2). The precapillary sphincters serve as gates that control blood flow into individual capillaries.

Contraction of vascular smooth muscle is called *vasoconstriction*. Vasoconstriction of arterioles and precapillary sphincters reduces their diameter and so reduces blood flow to the capillaries. Conversely, relaxation of vascular smooth muscle is called *vasodilation*. Vasodilation of arterioles and precapillary sphincters increases their diameter and so increases blood flow to the capillaries.

A wide variety of external and internal factors can produce vasoconstriction or vasodilation, including nerves,

hormones, and conditions in the local environment of the arterioles and precapillary sphincters. If you go outside on a cold day, you may notice that your fingers start to look pale. This is because vasoconstriction produced by nerves is narrowing your vessels to reduce heat loss from your body. On the other hand, hot weather will make your skin appear flushed as vasodilation occurs to speed up heat loss and cool you off. Emotions can also have an impact: vasodilation is partly responsible for the surge in blood flow that causes the penis or clitoris to become erect when we are sexually aroused. Later in this chapter we will talk more about how the cardiovascular system is regulated to maintain homeostasis.

Capillaries: Where blood exchanges substances with tissues

Arterioles connect to the smallest blood vessels, called **capillaries.** Capillaries are thin-walled vessels that average only about one-hundredth of a millimeter in diameter—not much wider than the red blood cells that travel through them. In fact, they are so narrow that RBCs often have to pass through them in single file or even bend to squeeze through.

Extensive networks of capillaries, called *capillary beds*, can be found in all areas of the body, which is why you are likely to bleed no matter where you cut yourself. The branching design of capillaries and their thin, porous walls allow blood to exchange oxygen, carbon dioxide, nutrients, and waste products with tissue cells. Capillary walls consist of a single layer of squamous epithelial cells. Microscopic pores pierce this layer, and the cells are separated by narrow slits. These openings are large enough to allow the exchange of fluid and other materials between blood and the interstitial fluid (the fluid that surrounds every living cell), yet small enough to retain red blood cells and most plasma proteins in the capillary (Figure 8.3). Some white blood cells can also squeeze between the cells in capillary walls and enter the tissue spaces.

In effect, capillaries function as biological strainers that permit selective exchange of substances with the interstitial fluid. In fact, capillaries are the *only* blood vessels that can exchange materials with the interstitial fluid.

Figure 8.4 illustrates the general pattern of how water and substances move across a capillary. At the beginning of a capillary, fluid is filtered out of the vessel into the interstitial fluid, accompanied by oxygen, nutrients, and raw materials needed by the cell. The filtered fluid is essentially like plasma except that it contains very little protein because most protein molecules are too large to be filtered. Filtration of fluid is driven by the blood pressure generated by the heart. Waste materials such as carbon dioxide and urea diffuse out of the cells and back into the blood.

Most of the filtered fluid is reabsorbed by diffusion back into the last half of the capillary before it joins a vein. The force for this reabsorption is the presence of protein in

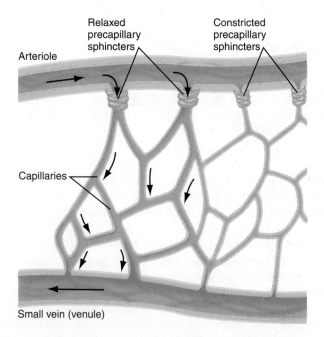

Relaxed precapillary sphincters

Constricted precapillary sphincters

Arteriole

Capillaries

Small vein (venule)

Figure 8.2 Precapillary sphincters control the flow of blood into individual capillaries. In this diagram the two precapillary sphincters on the right are vasoconstricted, reducing flow in that region. Arrows indicate direction of blood flow.

a) **A medium-magnification view showing a rich network of capillaries surrounding and interconnecting small arteries and veins.**

b) **A higher magnification showing a single branching capillary.** Notice the red blood cells traveling single file in the capillary.

c) **The structure of a capillary.**

Figure 8.3 **Capillaries.**

the blood but not in the interstitial fluid. In other words, water diffuses from an area of high water concentration (interstitial fluid) to an area of lower water concentration (blood plasma). However, the diffusional reabsorption of water does not quite match the pressure-induced filtration of water, so a small amount of filtered fluid remains in the interstitial space as excess interstitial fluid.

Figure 8.4 **The general pattern of movement between capillaries, the interstitial fluid, and cells.** For simplicity, only a single tissue cell is shown, but a single capillary may supply many nearby cells.

☑ *Why does most of the fluid move back into the capillary?*

☑ **Quick Check** Why doesn't exchange of gases and nutrients with the interstitial fluid occur in arteries and arterioles too, instead of just in capillaries? Put another way, what about the structure of an artery or an arteriole *prevents* such exchange from occurring? ▪

Lymphatic system helps maintain blood volume

Although the imbalance between the amount of plasma fluid filtered by the capillaries and the amount reabsorbed is not large, over the course of a day it would amount to about two or three liters. This excess plasma fluid must be returned to the cardiovascular system somehow, or soon all the plasma would end up in the interstitial fluid.

The excess plasma fluid is picked up by a system of blind-ended vessels called *lymphatic capillaries,* which branch throughout our body tissues and are part of the *lymphatic system.* The lymphatic system also picks up a few objects in the interstitial fluid that are too large to diffuse into capillaries. These include lipid droplets absorbed during digestion and invading organisms. Lymphatic capillaries transport the excess interstitial fluid and other objects to larger lymphatic vessels, which eventually return the fluid (called *lymph*) to veins near the heart. Along the way the lymphatic system intercepts the invading microorganisms.

We say more about the lymphatic system in Chapter 9 when we discuss the immune system. For now, just be aware

Answers to ☑ **questions can be found in Appendix A.**

that the lymphatic system, though technically not part of the cardiovascular system, plays a vital role in maintaining the proper volumes of blood and interstitial fluid.

☑ **Quick Check** There are certain parasitic worms that can enter the lymphatic system and completely block the lymphatic capillaries draining an arm or a leg. Predict what would happen to the arm or leg if this occurs. ▪

Veins return blood to the heart

From the capillaries, blood flows back to the heart through *venules* (small veins) and **veins** (see Figures 8.1 and 8.4). Like the walls of arteries, the walls of veins consist of three layers of tissue. However, the outer two layers of the walls of veins are much thinner than those of arteries. Veins also have a larger diameter lumen than arteries.

The anatomical differences between arteries and veins reflect their functional differences. As blood moves through the cardiovascular system, the blood pressure becomes lower and lower. The pressure in veins is only a small fraction of the pressure in arteries, so veins do not need nearly as much wall strength (provided by muscle and connective tissue) as arteries. The larger diameter and high distensibility of veins allows them to stretch (like thin balloons) to accommodate large volumes of blood at low pressures.

In addition to their transport function, then, veins serve as a blood volume reservoir for the entire cardiovascular system. Nearly two-thirds of all the blood in your body is in your veins. Thanks to their blood reservoir function, even if you become dehydrated or lose a little blood, your heart will still be able to pump enough blood to keep your blood pressure fairly constant.

The distensibility of veins can lead to problems in returning blood to the heart against the force of gravity. When you stand upright, blood tends to collect in the veins of your legs and feet. People who spend a lot of time on their feet may develop *varicose veins*, permanently swollen veins that look twisted and bumpy from pooled blood. Varicose veins can appear anywhere, but they are most common in the legs and feet. In severe cases the skin surrounding veins becomes dry and hard because the tissues are not receiving enough blood. Often varicose veins can be treated by injecting an irritating solution that shrivels the vessels and makes them less visible. This should not affect blood flow because surrounding undamaged veins take over and return blood to the heart.

Fortunately, three mechanisms assist the veins in returning blood to the heart: (1) contractions of skeletal muscles, (2) one-way valves inside the veins, and (3) movements associated with breathing. Let's look at each in turn.

Skeletal muscles squeeze veins On their path back to the heart, veins pass between many skeletal muscles. As we move and these muscles contract and relax, they press against veins and collapse them, pushing blood toward the heart. You may have noticed that you tire more easily when you stand still than when you walk around. This is because walking improves the return of blood to your heart and prevents fluid accumulation in your legs. It also increases blood flow and the supply of energy to your leg muscles.

One-way valves permit only one-way blood flow Most veins contain valves consisting of small folds of the inner layer that protrude into the lumen. The structure of these valves allows blood to flow in one direction only: toward the heart. They open passively to permit blood to move toward the heart and then close whenever blood begins to flow backward. Together, skeletal muscles and valves form what is called the "skeletal muscle pump" (Figure 8.5). Once blood has been pushed toward the heart by skeletal muscles or drained in that direction by gravity, it cannot drain back again because of these one-way valves. The opening and closing of venous valves is strictly dependent on differences in blood pressure on either side.

Valve
(open)

One-way
valves
(closed)

Calf muscles
relaxed

Calf muscles
contracted

Figure 8.5 The skeletal muscle pump. With the calf muscle relaxed, blood accumulates in the vein. Backflow is prevented by one-way valves. When the calf muscle contracts, skeletal muscles press on the vein, forcing blood toward the heart through the upper one-way valve. The lower one-way valve remains closed, preventing backflow.

Pressures associated with breathing push blood toward the heart The third mechanism that assists blood flow involves pressure changes in the thoracic (chest) and abdominal cavities during breathing. When we inhale, abdominal pressure increases and squeezes abdominal veins. At the same time, pressure within the thoracic cavity decreases, dilating thoracic veins. The result is to push blood from the abdomen into the chest and toward the heart. This effect is sometimes called the "respiratory pump."

Recap A branching system of thick-walled arteries distributes blood to every area of the body. Arterioles regulate blood flow to local regions, and precapillary sphincters regulate flow into individual capillaries. Capillaries consisting of a single layer of cells exchange materials with the interstitial fluid. The lymphatic system removes excess fluid. The thin-walled veins return blood to the heart and serve as a volume reservoir for blood. ▪

Figure 8.6 A human heart.

8.2 The heart pumps blood through the vessels

The heart, a muscular, cone-shaped organ slightly larger than your fist, is located in the thoracic cavity between the lungs and behind the sternum, or breastbone. As described in Chapter 6, the heart consists mostly of a special type of muscle called cardiac muscle. Unlike skeletal muscle, cardiac muscle does not connect to bone. Instead, it pumps ceaselessly in a squeezing motion to propel blood through the blood vessels.

Your heart pumps about 75 times every minute—and this does not include the times it speeds up to supply extra blood during exertion or stress. It never rests for more than two-thirds of a second. Over 70 years, this adds up to about 2.8 billion heartbeats, truly an impressive performance for any muscle. Under normal circumstances the heart's rate of pumping is controlled by the brain, but it can also beat on its own without any instructions from the brain at all.

The heart is mostly muscle

A human heart is shown in Figure 8.6. In its natural position within the chest cavity the heart is enclosed in a tough fibrous sac called the *pericardium* (not shown in **Figure 8.6**). The pericardium protects the heart, anchors it to surrounding structures, and prevents it from overfilling with blood. Separating the pericardium from the heart is a space called the *pericardial cavity*. The pericardial cavity contains a film of lubricating fluid that reduces friction and allows the heart and the pericardium to glide smoothly against each other when the heart contracts.

MJ's Human Biology Blog

Boosting Cardiac Repair Mechanisms

Conventional wisdom holds that a damaged heart cannot repair itself after a heart attack because the heart does not have the ability to produce new cardiac muscle cells throughout life. But now scientists have demonstrated that cardiac muscle cells *are* being replaced throughout life, though at a very slow rate—only about 1% of the heart muscle cells are replaced each year in young adults. The rate falls gradually to about half a percent per year by age 75. Over a lifetime, though, about 45% of the cardiac muscle cells present at birth will have been replaced.

One percent per year is not fast enough for the heart to repair itself under natural conditions after a heart attack. But the fact that it occurs at all is giving researchers new hope. If future research were to improve our understanding of how cardiac muscle cell replacement is regulated, perhaps new drugs or treatments could be developed that would jump-start the process after a heart attack.

It could be decades before any patients are actually helped by the new findings, but that's the way science goes . . . a little breakthrough here, a little breakthrough there, and pretty soon there's real progress! ▪

Reference: Evidence for Cardiomyocyte Renewal in Humans. *Science* 324: 98–102, April 3, 2009.

In cross section we see that the walls of the heart consist of three layers: the epicardium, myocardium, and endocardium (**Figure 8.7**). The outermost layer, the *epicardium*, is a thin layer of epithelial and connective tissue. The middle layer is the **myocardium.** This is a thick layer consisting mainly of cardiac muscle that forms the bulk of the heart. The myocardium is the layer that contracts every time the heart beats. As noted in Chapter 6, the structure of cardiac muscle cells allows electrical signals to flow directly from cell to cell. An electrical signal in one cardiac muscle cell can spread to adjacent cells, enabling large numbers of cells to contract as a coordinated unit. Every time the myocardium contracts, it squeezes the chambers inside the heart, pushing blood outward into the arteries. The innermost layer of the heart, the *endocardium*, is a thin endothelial layer resting on a layer of connective tissue. The endocardium is continuous with the endothelium that lines the blood vessels.

Occasionally one of the layers of the heart wall becomes inflamed. These conditions are named according to the location of the problem; *-itis* is a suffix that means "inflammation." Thus *pericarditis* refers to inflammation of the pericardium, *endocarditis* to an inflamed endocardium, and so on. A variety of factors can lead to inflammation in or around the heart wall, including infections, cancer, injuries, or complications from major surgery. Depending on the underlying cause, many cases respond well to antibiotic and anti-inflammatory drugs.

The heart has four chambers and four valves

Taking a closer look at the details of the structure of the heart, we see that it consists of four separate chambers (see Figure 8.7). The two chambers on the top are the **atria** (singular: *atrium*), and the two more-muscular bottom chambers are the **ventricles.** A muscular partition called the **septum** separates the right and left sides of the heart.

Blood returning to the heart from the body's tissues enters the heart at the *right atrium.* From the right atrium, the blood passes through a valve into the *right ventricle.* The right ventricle is more muscular than the right atrium because it pumps blood at considerable pressure through a second valve and into the artery leading to the lungs.

Blood returning from the lungs to the heart enters the *left atrium* and then passes through a third valve into the *left ventricle.* The very muscular left ventricle pumps blood through a fourth valve into the body's largest artery, the aorta. From the aorta, blood travels through the arteries and

Figure 8.7 A view of the heart showing major blood vessels, chambers, and valves.

arterioles to the systemic capillaries, venules, and veins and then back to the right atrium again.

The left ventricle is the most muscular of the heart's four chambers because it must do more work than any other chamber. The left ventricle must generate pressures higher than aortic blood pressure in order to pump blood into the aorta. (We'll see how high aortic pressure is in a minute.) The right ventricle has a thinner wall and does less work because the blood pressure in the arteries leading to the lungs is only about one-sixth that of the aorta.

Four heart valves enforce the heart's one-way flow pattern and prevent blood from flowing backward. The valves open and shut passively in response to changes in the pressure of blood on each side of the valve. The right and left **atrioventricular (AV) valves** located between the atria and their corresponding ventricle prevent blood from flowing back into the atria when the ventricles contract. The AV valves consist of thin connective tissue flaps (cusps) that project into the ventricles. The right AV valve is called the *tricuspid valve* because it has three flexible flaps. The left AV valve has two flaps, so it is referred to as the *bicuspid* or *mitral valve*. These valves are supported by strands of connective tissue called *chordae tendineae* that connect to muscular extensions of the ventricle walls called *papillary muscles*. Together, the chordae tendineae and papillary muscles prevent the valves from everting (opening backward) into the atria when the ventricles contract.

Two **semilunar valves** (the pulmonary and the aortic) prevent backflow into the ventricles from the main arteries leaving the heart when the heart relaxes. Each semilunar valve consists of three pocketlike flaps. The valves' name reflects the half-moon shape of these flaps (*semi* means "one-half"; *luna* comes from the Latin word for "moon").

Next, let's follow the overall pattern of blood flow through the body.

The pulmonary circuit provides for gas exchange

Have to know!

Figure 8.8 shows the general structure of the entire cardiovascular system. Note that the heart is pumping blood through the lungs (the **pulmonary circuit**) and through the rest of the body to all the cells (the **systemic circuit**) simultaneously. Each circuit has its own set of blood vessels. Let's follow the pulmonary circuit first:

1. When blood returns to the heart from the veins, it enters the right atrium. The blood that returns to the heart is deoxygenated—it has given up oxygen to tissue cells and taken up carbon dioxide.
2. From the right atrium, blood passes through the right atrioventricular valve into the right ventricle.
3. The right ventricle pumps blood through the pulmonary semilunar valve into the pulmonary trunk (the main pulmonary artery) leading to the lungs. The

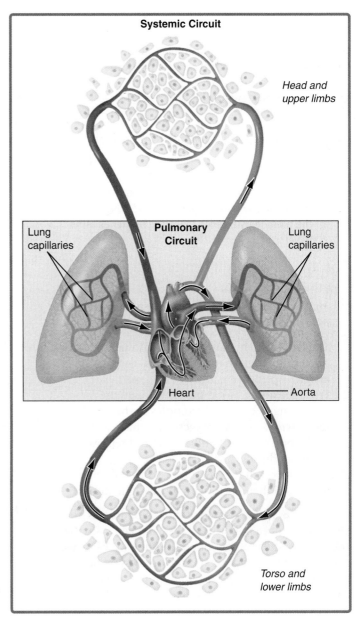

Figure 8.8 A schematic representation of the human cardiovascular system showing the separate pulmonary and systemic circuits. The systemic circuit includes the circulations of the head, torso, limbs, and internal organs, including the heart and some tissues in the lungs.

pulmonary trunk divides into the right and left pulmonary arteries, which supply the right and left lungs, respectively.
4. At the pulmonary capillaries, blood gives up carbon dioxide and receives a fresh supply of oxygen from the air we inhale. It is now oxygenated.
5. The freshly oxygenated blood flows into the pulmonary veins leading back to the heart. It enters the left atrium and flows through the left atrioventricular valve into the left ventricle.

The deoxygenated blood in the right side of the heart never mixes with oxygenated blood in the left. Deoxygenated blood leaving the right side of the heart must pass through the pulmonary circuit (where it picks up oxygen) before it reaches the left side of the heart.

 Web Animation *The Cardiovascular System* at www. humanbiology.com

✓ **Quick Check** Do the pulmonary arteries carry oxygenated blood or deoxygenated blood? What features of the pulmonary arteries make them arteries rather than veins? ▪

The systemic circuit serves the rest of the body

When blood enters the left ventricle, it begins the *systemic circuit*, which takes it to the rest of the body.

1. The left ventricle pumps blood through the aortic semilunar valve into the **aorta,** the largest artery.
2. From the aorta, blood travels through the branching arteries and arterioles to the capillaries, where it delivers oxygen and nutrients to all of the body's tissues and organs and removes waste products.
3. From the capillaries, blood flows to the venules, veins, and then back again to the right atrium.

Figure 8.9 shows some of the major arteries and veins of the human body. Arteries and veins serving the same vascular region often (but not always) have the same name and generally are located very near to each other. For example, a common iliac artery supplies blood to each leg, and a common iliac vein returns blood from the leg to the heart. However, carotid arteries supply the head, but jugular veins return the blood from the head.

As you might expect of such a hard-working muscle, the heart requires a great deal of oxygen and nutrients to fuel its own operations. Although the heart represents only about 1/200 of your body's weight, it requires roughly 1/20 of your total blood flow at rest. And although the heart is almost continuously filled with blood, the myocardium is too thick to be served by diffusion of oxygen and nutrients from the blood passing through. Thus the heart has its own set of blood vessels called the **coronary arteries** that supply the heart muscle (**Figure 8.10** on page 174). The coronary arteries branch from the aorta just above the aortic semilunar valve and encircle the heart's surface (the word coronary comes from the Latin *corona,* meaning "encircling

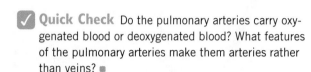

Figure 8.9 Some of the major arteries and veins in the human body. For simplicity, the lungs and most of the internal organs have been omitted.

like a crown"). From the surface they send branches inward to supply the myocardium. *Cardiac veins* collect the blood from the capillaries in the heart muscle and channel it back to the right atrium.

Cholesterol and Atherosclerosis

As noted in Chapter 2, cholesterol is a key component of all cell membranes and the precursor molecule for several hormones. All cells require a certain amount of it for normal functioning. However, too much cholesterol in the blood can lead to a condition called *atherosclerosis*—a thickening of an arterial vessel wall due to the buildup of fatty materials containing cholesterol. Left untreated, atherosclerosis contributes to heart attacks, strokes, aneurysms, and peripheral vascular disease.

Most of the cholesterol in the blood is bound to certain carrier proteins. Together, the cholesterol and the protein are called a *lipoprotein*. There are two types of lipoproteins, based on their densities. One of them, called *low-density lipoprotein (LDL)*, is considered "bad" in terms of atherosclerosis. When present in normal amounts, LDL transports cholesterol throughout the body and makes it available to cells. However, when there is too much LDL it begins to attach to the cells lining the arterial blood vessel wall and then makes its way into the cells. Once inside the cell, LDL triggers an inflammatory response that ultimately results in the buildup of fatty deposits called atherosclerotic *plaques* within the blood vessel wall. Eventually these plaques may rupture, causing blood clots to form that can occlude arteries and cause heart attacks and strokes.

High-density lipoproteins (HDLs), on the other hand, are considered "good" because they target cholesterol for removal. HDLs pick up free cholesterol and carry it to the liver, where it is detached from the protein, mixed with bile, and secreted into the small intestine. Some of the cholesterol in bile is excreted from the body with the feces, although some is reabsorbed, to be used again.

Risk factors for atherosclerosis include factors that raise blood cholesterol (obesity, sedentary lifestyle, and a high-fat diet), smoking, diabetes, hypertension, and a family history of atherosclerosis. Before age 45 men have a 10 times greater risk than women; however, women's risk rises after menopause.

According to the American Heart Association, a total cholesterol of under 200 mg/dl is considered desirable. Ideally, HDL should be greater than 60 mg/dl and LDL should be less than 100 mg/dl. A total cholesterol of greater than 240 mg/dl along with a high LDL and/or low HDL would be cause for concern.

If you're having trouble remembering which lipoprotein is bad for you and which is good, just remember that cholesterol is a lipid, and lipids are less dense than protein or water. So "low-density means more cholesterol," and therefore low-density lipoprotein (LDL) is the "bad" one.

Some degree of atherosclerosis is common with advancing age. However, lifestyle can make a big difference in how rapidly atherosclerosis develops and whether it becomes severe. At the end of this chapter we look at what you can do to lower your risk of atherosclerosis and other cardiovascular conditions. ▪

a) Cross-section of an artery narrowed by atherosclerotic plaque.

b) The same photo with the atherosclerotic plaque removed, showing how a normal artery would look.

Atherosclerosis.

The coronary arteries are relatively small in diameter. If they become partially or completely blocked, perhaps as a result of atherosclerosis, serious health problems can result. Later in this chapter and in the Health & Wellness feature we look at what happens when circulation to the heart is impaired.

✓ **Quick Check** Some babies are born with a heart defect in which the ventricles are connected to the wrong arteries—that is, the right ventricle sends blood to the aorta, and the left ventricle sends blood to the pulmonary trunk. What is the problem with this arrangement? ▪

Figure 8.10 Blood vessels of the heart.

The cardiac cycle: The heart contracts and relaxes

The pumping action of the heart is pulsatile rather than continuous, meaning that it delivers blood in separate and distinct pulses. A complete cardiac cycle involves contraction of the two atria, which forces blood into the ventricles, followed by contraction of the two ventricles, which pumps blood into the pulmonary artery and the aorta, followed by relaxation of the entire heart. The term **systole** refers to the period of contraction and **diastole** refers to the period of relaxation. The entire sequence of contraction and relaxation is called the **cardiac cycle.**

Every cardiac cycle consists of three steps, as shown in **Figure 8.11**. Starting with the heart as it first begins to contract:

1. *Atrial systole.* As contraction starts, the heart is already nearly filled with blood that entered the ventricles and atria passively during the previous diastole. Contraction of the heart begins with the atria. During atrial systole, both atria contract, raising blood pressure in the atria and giving the final "kick" that fills the two ventricles to capacity. Atrial systole also momentarily stops further inflow from the veins. Both atrioventricular valves are still open, and both semilunar valves are still closed.

2. *Ventricular systole.* The contraction that began in the atria spreads to the ventricles, and both ventricles contract simultaneously. The rapidly rising ventricular pressure produced by contraction of the ventricles causes the two AV valves to close, preventing blood from flowing backward into the atria and veins. At this time the atria relax and begin filling again. The pressure within the ventricles continues to rise until it is greater than the pressure in the arteries, at which point the pulmonary and aortic semilunar valves open and blood is ejected into the pulmonary trunk and the aorta. With each ventricular systole, about 60% of the blood in each ventricle is forcibly ejected.

3. *Diastole.* Both atria and both ventricles are relaxed throughout diastole. At this point pressure within the ventricles begins to fall. As soon as ventricular pressures fall below arterial pressures during early diastole, the pulmonary and aortic semilunar valves close, preventing backflow of arterial blood. Once ventricular pressure falls below blood pressure in the veins, the AV valves open and blood begins to flow passively into the heart.

A complete cardiac cycle occurs every 0.8 second or so. These cycles repeat, from birth to death, without ever stopping. Atrial systole lasts about 0.1 second; ventricular systole, about 0.3 second. During the remaining 0.4 second, the heart relaxes in diastole.

As each surge of blood enters the arteries during systole, the artery walls are stretched to accommodate the extra volume, and arterial pressure rises. Arteries recoil passively during diastole as blood continues to flow out of them

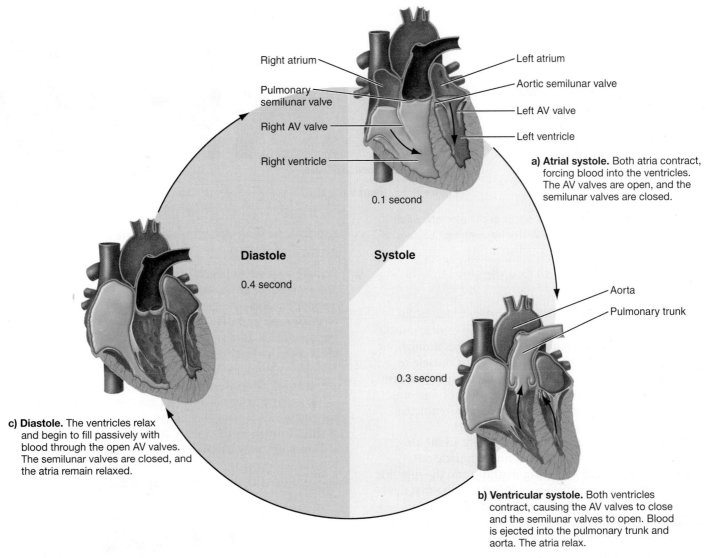

a) **Atrial systole.** Both atria contract, forcing blood into the ventricles. The AV valves are open, and the semilunar valves are closed.

0.1 second

Diastole

0.4 second

Systole

0.3 second

c) **Diastole.** The ventricles relax and begin to fill passively with blood through the open AV valves. The semilunar valves are closed, and the atria remain relaxed.

b) **Ventricular systole.** Both ventricles contract, causing the AV valves to close and the semilunar valves to open. Blood is ejected into the pulmonary trunk and aorta. The atria relax.

Right atrium
Pulmonary semilunar valve
Right AV valve
Right ventricle

Left atrium
Aortic semilunar valve
Left AV valve
Left ventricle

Aorta
Pulmonary trunk

Figure 8.11 The cardiac cycle.

✓ *What makes the AV and semilunar valves open and close?*

through the capillaries. You can feel this cycle of rapid expansion and recoil in the wall of an artery if it's located close to the skin's surface. This is called a *pulse.* A good place to detect a pulse is the radial artery (inside your wrist, just below the base of the thumb).

The heart is composed primarily of cardiac muscle, and like our skeletal muscles it benefits from regular exercise. During sustained (aerobic) exercise the heart beats more rapidly and more powerfully to sustain blood pressure in the face of increased blood flow to hard-working skeletal muscles. Over time this causes the heart to "bulk up" (hypertrophy) slightly. However, it is important not to overdo any exercise training regimen. If over exercised, the heart itself may become starved for oxygen, and heart muscle may be damaged. The usual guideline for safe but effective exercise training is to perform an activity that raises your heart rate to its "target heart rate" for at least 20 minutes, three times a week or more.

Heart sounds reflect closing heart valves

There is probably no more basic rhythm to which humans respond than the familiar "lub-DUB–lub-DUB" of the heart beating. We probably experience it, at least subconsciously, even before we are born. These *heart sounds* reflect events that occur during the cardiac cycle—specifically the closing of the heart valves. The "lub" signals the closure of the two AV valves during ventricular systole. The slightly louder "DUB" occurs when the aortic and pulmonary semilunar valves close during ventricular diastole. The sounds are due to vibrations in the heart chambers and blood vessels caused by the closing of the valves.

Blood flows silently as long as it flows smoothly. However, if blood encounters an obstruction, the disturbed flow can create unusual heart sounds called *murmurs.* Many murmurs result from incomplete closing of the heart valves due to unusually shaped valve flaps or stiffening of the flap tissue. If a valve does not close completely, some blood is forced through it during the cardiac cycle, creating a swishing noise that can be detected through a stethoscope. Murmurs are not necessarily a sign of disease, but physicians can diagnose a variety of heart conditions, including leaking or partially blocked valves, from their sound and timing. Even serious murmurs can often be treated with surgery to replace the defective valve with an artificial valve (Figure 8.12).

a) A pulmonary semilunar valve.

b) An artificial heart valve.

Figure 8.12 **Heart valves.**

✓ **Quick Check** Do any sounds occur at the moment when the atria contract? When the ventricles contract? Explain. ∎

Cardiac conduction system coordinates contraction

The coordinated sequence of the cardiac cycle is due to the *cardiac conduction system,* a group of specialized cardiac muscle cells that initiate and distribute electrical impulses throughout the heart. These impulses stimulate the heart muscle to contract in an orderly sequence that spreads from atria to ventricles. The cardiac conduction system consists of four structures: sinoatrial node, atrioventricular node, atrioventricular bundle and its two branches, and Purkinje fibers.

The stimulus that starts a heartbeat begins in the **sinoatrial (SA) node,** a small mass of cardiac muscle cells located near the junction of the right atrium and the superior vena cava. The SA node emits an electrical impulse that travels across both atria like ripples on a pond, stimulating waves of contraction (Figure 8.13). (As noted in Chapter 6 and Figure 6.12, cardiac muscle cells are connected by intercalated discs and gap junctions that let electrical signals flow directly from one cell to the next.) The SA node is properly called the *cardiac pacemaker* because it initiates the heartbeat. However, the cardiac pacemaker can be influenced by the brain to speed up or slow down, as we'll see in a minute.

The electrical impulse traveling across the atria eventually reaches another mass of muscle cells called the **atrioventricular (AV) node,** located between the atria and ventricles. The muscle fibers in this area are smaller in diameter, causing a slight delay of approximately 0.1 second, which temporarily slows the rate at which the impulse travels. This delay gives the atria time to contract and empty their blood into the ventricles before the ventricles contract.

From the AV node, the electrical signal sweeps to a group of conducting fibers in the septum between the two ventricles called the **atrioventricular (AV) bundle.** These fibers branch and extend into **Purkinje fibers,** smaller fibers that carry the impulse to all cells in the myocardium of the ventricles. Because the electrical impulse travels down the septum to the lower portion of the ventricles and then spreads rapidly upward through the Purkinje fibers, the lower part of the ventricles contract before the upper part. This lower-to-upper squeezing motion pushes blood into the pulmonary trunk and aorta.

Electrocardiogram records the heart's electrical activity

Because the body is largely water and water conducts electrical activity well, we can track the electrical activity of the heart as weak differences in voltage at the surface of the body. An **electrocardiogram (ECG** or **EKG)** is a record of the electrical impulses in the cardiac conduction system. An ECG involves placing electrodes on the skin at the chest, wrists, and ankles. The electrodes transmit the heart's electrical impulses, which are recorded as a continuous line on a screen or moving graph.

A healthy heart produces a characteristic pattern of voltage changes. A typical ECG tracks these changes as a series of three formations: P wave, QRS complex, and T wave. First is the small P wave, representing the electrical impulse

Sinoatrial (SA) node

Atrioventricular (AV) node

AV bundle

Bundle branches

Purkinje fibers

Figure 8.13 **The cardiac conduction system.** Electrical activity of the heart normally starts at the SA node, spreads across the atria to the AV node, and then progresses down the AV bundle and its branches to the purkinje fibers.

traveling across the atria (see Figure 8.13). Second is the QRS complex, representing the spread of the electrical impulse down the septum and around the two ventricles in the Purkinje fibers. It occurs just before the ventricles start to contract. Finally, the T wave is the result of the *end* of the electrical activity in the ventricles. At this time the ventricles relax.

If something goes wrong with the cardiac conduction system or if the heart muscle becomes damaged, abnormal heart electrical impulses and contractions may occur. An abnormality of the rhythm or rate of the heartbeat is called an *arrhythmia*. Arrhythmias take many forms. Occasional skipped heartbeats, for example, are fairly common and usually of no consequence. However, a type of rapid irregular ventricular contraction known as *ventricular fibrillation* (or "V-fib") is very quickly fatal unless treated immediately. Ventricular fibrillation is the leading cause of cardiac death in otherwise healthy people. In a hospital, ventricular fibrillation is treated by "cardioversion," in which a strong electrical current is applied to the chest to eliminate the abnormal fibrillating pattern and restore the normal rhythm.

Arrhythmias produce characteristic ECG tracings, and the ECG is a valuable tool for identifying the cause, type, and location of arrhythmias. **Figure 8.14** shows a patient undergoing an ECG, a normal ECG recording, and a recording from a patient in ventricular fibrillation. Arrhythmias less life-threatening than ventricular fibrillation can sometimes be treated with medications. In some cases an artificial pacemaker (a small generating unit that automatically stimulates the heart at set intervals) can be surgically implanted under the chest skin to normalize the heart rate.

Recap The heart wall consists of three layers; the epicardium, the myocardium, and the endocardium. The heart contains four chambers and four one-way valves. The right atrium and right ventricle pump blood to the lungs; the left atrium and left ventricle pump blood to the rest of the body. Each cardiac cycle is a repetitive sequence of contraction (systole) and relaxation (diastole). Contraction of the heart is coordinated by modified cardiac muscle cells that initiate and transmit electrical impulses through a specialized conduction system. An electrocardiogram (ECG) is a recording of the heart's electrical activity taken from the surface of the body. ∎

8.3 Blood exerts pressure against vessel walls

Blood pressure is the force that blood exerts on the wall of a blood vessel as a result of the pumping action of the heart.

Blood pressure is not the same in all blood vessels. **Figure 8.15** compares the pressures in the various segments of the vascular system. You can see from the highs and lows shown here that pressure is pulsatile in the arteries; that is, it varies with each beat of the heart. The highest pressure of the cycle, **systolic pressure,** is the pressure reached during ventricular systole when the ventricles contract to eject blood from the heart. The lowest pressure, **diastolic pressure,** occurs during ventricular diastole when the ventricles relax. Arteries store the energy generated by the heart during systole, and during diastole they use that stored energy to supply blood to the tissues.

a) **An ECG being recorded.**

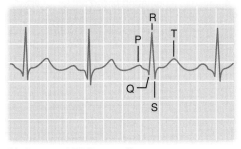

b) **A normal ECG recording.**

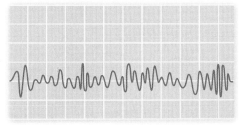

c) **Ventricular fibrillation.**

Figure 8.14 **The ECG is a tool for diagnosing heart arrhythmias.**

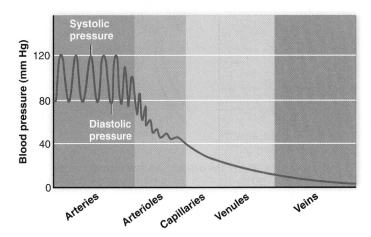

Figure 8.15 **Blood pressure in different segments of the vascular system.**

Maintenance of arterial blood pressure is crucial to drive the flow of blood throughout the body and all the way back to the heart. Recall that fluid always flows from a region of high pressure toward a region of lower pressure. By the time it reaches the capillaries, blood flow is steady rather than pulsatile, and pressure continues to fall as blood flows through venules and veins. The differences in the blood pressure of arteries, capillaries, and veins keep blood moving through the body.

Measuring Blood Pressure

When health professionals measure your blood pressure, they are assessing the pressure in your main arteries. From a clinical standpoint, blood pressure gives valuable clues about the relative volume of blood in the vessels, the condition or stiffness of the arteries, and the overall efficiency of the cardiovascular system. Trends in blood pressure over time are a useful indicator of cardiovascular changes.

Blood pressure is recorded as mm Hg (millimeters of mercury, because early equipment used a glass column filled with mercury to measure the pressure). In young, healthy individuals, systolic pressures of less than 120 mm Hg and diastolic pressures of less than 80 mm Hg are considered desirable. With advancing age there is a slight tendency for systolic blood pressure in particular to increase slightly, as a consequence of age-related stiffening of the arteries.

Blood pressure is measured with a *sphygmomanometer* (*sphygmo* comes from the Greek word for "pulse"; a manometer is a device for measuring fluid pressures). An inflatable cuff is placed over the brachial artery in your upper arm and connected to a pressure-measuring device (Figure 8.16). When the cuff is inflated to a pressure above systolic pressure, blood flow through the brachial artery stops because the high cuff pressure collapses the artery. The cuff is then deflated slowly while a health professional listens with a stethoscope for the sounds of blood flowing in your artery. As soon as pressure in the cuff falls below the peak of systolic pressure, some blood spurts briefly through the artery during the high point of the pressure pulse, making a characteristic light tapping sound that is audible through the stethoscope. The cuff pressure at which this happens is recorded as systolic pressure. As the cuff continues to deflate, eventually blood flow through the artery

b) **A schematic representation of the pulses of arterial blood pressure superimposed over the steadily declining cuff pressure.** Systolic pressure is recorded at cuff pressure 1 when sounds are first heard. Diastolic pressure is recorded at cuff pressure 2 when sounds cease.

a) **A clinician inflates the cuff with air and then allows the pressure in the cuff to fall gradually while using a stethoscope to listen for the sounds of blood movement through the artery.**

Figure 8.16 How blood pressure is measured.

✓ *Explain in your own words why the tapping noise is only heard when the cuff pressure is between systolic and diastolic pressure. In other words, what's making the noise?*

Table 8.1 Systolic and diastolic blood pressure

Blood pressure category	Systolic (mm Hg)		Diastolic (mm Hg)
Normal	Less than 120	and	Less than 80
Prehypertension	120–139	or	80–89
Hypertension, Stage 1	140–159	or	90–99
Hypertension, Stage 2	160 or higher	or	100 or higher

Source: National Institutes of Health, National Heart, Lung, and Blood Institute, March 2003.

becomes continuous and the tapping sound ceases. The point where the sound disappears is your diastolic pressure. This procedure yields two numbers, corresponding to your systolic and diastolic pressures. These represent the high and low points of blood pressure during the cardiac cycle.

Hypertension: High blood pressure can be dangerous

Blood pressure higher than normal is called **hypertension** (*hyper* comes from the Greek for "excess"). Table 8.1 presents the systolic and diastolic readings health professionals use to classify blood pressure as normal or hypertensive. Hypertension is a significant risk factor for cardiovascular disease, because the greater the pressure, the greater the strain on the cardiovascular system. Blood vessels react to the pounding by becoming hardened and scarred, which makes them less able to stretch during systole. Hypertension also places a greater strain on the heart, because the work it must do is directly proportional to the arterial pressure against which it must pump.

Hypertension is called "the silent killer" because usually it has no symptoms. The American Heart Association estimates that approximately 50 million Americans have hypertension and a third of them don't even realize it. If left untreated, hypertension increases the risk of serious health problems such as heart attack, heart failure, stroke, kidney damage, even damage to the tissues inside the eyes.

What causes hypertension? Many times it happens because blood vessels become narrowed from atherosclerosis. Certain other factors also increase the risk, as summarized in Table 8.2. The only sure way to diagnose it is to have your blood pressure measured.

Blood pressure varies from minute to minute even in healthy individuals. Simply getting up in the morning raises it, as do exercise, emotions, smoking cigarettes, eating, drinking, and many other factors. Even having your blood pressure measured can make you nervous enough for your blood pressure to rise—a situation that health professionals call "white coat hypertension." This is why physicians generally have you sit quietly while measuring your blood pressure.

If hypertension is suspected, your physician will probably measure your blood pressure on at least three different occasions before making a firm diagnosis. True hypertension is a *sustained* elevation in blood pressure above normal levels—a systolic pressure of 140 mm Hg or greater or a diastolic pressure

of 90 mm Hg or greater. Even blood pressure that is consistently just below the hypertensive level (prehypertension) may carry a slightly higher risk of health complications.

Generally, if systolic pressure is high, diastolic pressure will be, too. However, sometimes systolic pressure can register at above-normal levels while diastolic pressure remains normal, a condition called *isolated systolic hypertension*. Most common in older adults, it is diagnosed as a systolic pressure of 160 mm Hg or higher with a diastolic reading of less than 90 mm Hg. Like the more common form of hypertension, isolated systolic hypertension is associated with increased health problems.

At the end of this chapter we discuss what you can do to lower your risk of hypertension as well as other cardiovascular problems. If hypertension does develop, however, a number of medical treatments are available to lower blood pressure to a healthy level. It is important for people on antihypertensive drugs to take their medication consistently; even though hypertension has no symptoms, it increases the risk of related health problems.

Table 8.2 Risk factors for hypertension

Risk factor	Comments
Heredity	Family history of hypertension raises risk.
Age	Blood pressure tends to rise throughout life.
Race	African Americans have twice the incidence found in Caucasian Americans and Asian Americans.
Sex	Males are more likely than females to develop hypertension.
Obesity	The heart must pump harder to push blood through vessels.
High salt intake	In some individuals (but not others) a high salt intake raises blood pressure slightly.
Smoking	Smoking raises the blood concentration of epinephrine, a hormone that stimulates the heart.
Sedentary lifestyle	Not well understood. May be due to higher blood lipids or weight gain.
Persistent emotional stress	Emotional stress activates portions of the nervous system that elevate pressure.
Diabetes mellitus	Diabetics have a higher incidence of hypertension, for reasons not yet known.
Heavy alcohol consumption	Mechanisms unknown.
Oral contraceptives and certain medications	Mechanisms vary by medication.

Quick Check A nurse taking blood pressure in a patient hears a tapping sound begin when the cuff pressure is 141 mm Hg, and the sound ends when cuff pressure is 95 mm Hg. What are the systolic and diastolic pressures of this patient, and is this enough to tell you if this patient has normal blood pressure? ▪

Hypotension: When blood pressure is too low

It is also possible for blood pressure to fall below normal levels, a condition called *hypotension* (*hypo* comes from the Greek for "under"). Generally hypotension is a problem only if blood pressure falls enough to reduce blood flow to the brain, causing dizziness and fainting. Drops in blood pressure can follow abrupt changes in position, such as standing up suddenly. Other causes include severe burns or injuries involving heavy blood loss.

Recap Blood pressure is the force that blood exerts on the wall of a blood vessel. It is measured as two numbers corresponding to systolic and diastolic pressures. Hypertension (high blood pressure) is a serious risk factor for cardiovascular disease and other health problems. ▪

8.4 How the cardiovascular system is regulated

The overall function of the cardiovascular system is to provide every cell with precisely the right blood flow to meet its needs at all times. This might seem like a complicated process because different types of tissues have different needs, which change according to circumstances. In principle, however, the system is rather simple.

Consider for a minute how your community provides water to every house according to the needs of the occupants. Basically, municipal water systems provide constant water *pressure* to every house. With steady water pressure available, each household simply adjusts the *flow* of water into the house by turning faucets on and off.

The human cardiovascular system is based on similar principles. Consider the following key points:

- Homeostatic regulation of the cardiovascular system centers on *maintaining a constant arterial blood pressure.*
- A constant arterial blood pressure is *achieved* by regulating the heart rate and force of contraction (adjusting flow *into* the arteries) and by regulating the diameters of all the body's arterioles as a whole (adjusting overall flow *out* of the arteries).
- With arterial blood pressure held relatively constant, local blood flows are adjusted to meet local requirements.

As we will see, this overall regulation is achieved with nerves, hormones, and local factors coupled to metabolism.

Baroreceptors maintain arterial blood pressure

There is probably no more important regulated variable in the entire body than blood pressure. Without a relatively

constant blood pressure as the driving force for supplying blood to the capillaries, homeostasis simply would not be possible. Blood pressure does vary, but always within a range consistent with being able to provide blood to all cells.

To regulate blood pressure, the body must have some way of measuring it. In fact, several of the large arteries, including the aorta and the two carotid arteries, have certain regions called **baroreceptors** (the Greek *baro* denotes pressure). The baroreceptors regulate arterial blood pressure in the following manner:

1. When blood pressure rises, arterial blood vessels are stretched passively.
2. Stretch of baroreceptors in the carotid arteries and aorta causes them to send signals via nerves to an area of the brain called the *cardiovascular center.*
3. The cardiovascular center responds by sending signals via nerves to the heart and blood vessels.
4. The effect on the heart is to lower heart rate and the force of contraction. This reduces **cardiac output,** the amount of blood that the heart pumps into the aorta each minute.
5. The effect on arterioles is vasodilation, an increase in arteriole diameter. Vasodilation increases blood flow through all tissues.
6. The net effect of both reducing cardiac output and increasing flow through the tissues is to return arterial pressure to normal.

Exactly the opposite sequence of events occurs when arterial pressure falls *below* normal. When pressure falls and the arteries stretch less than normal, the baroreceptors send fewer nerve signals to the brain. The brain correctly interprets this as a fall in pressure and sends nerve signals that increase cardiac output and constrict arterioles, raising arterial blood pressure again.

All day long—every time you stand up, sit down, get excited, run for the bus—your blood pressure fluctuates up or down. But it is quickly brought back within normal range by a negative feedback loop initiated by baroreceptors.

Nerves and hormones adjust cardiac output

As mentioned before, cardiac output is the amount of blood pumped into the aorta in one minute. We calculate cardiac output by multiplying *heart rate* (number of heartbeats per minute) by *stroke volume* (volume of blood pumped out with each heartbeat). For a healthy adult at rest, the heart rate averages about 75 beats per minute, and the stroke volume averages about 70 milliliters per beat. Resting cardiac output, then, is about 5.25 liters per minute. Since our normal blood volume is only about 5 liters, in essence the entire blood supply passes through the heart every minute.

Regulation of cardiac output is centered in an area of the brain called the *medulla oblongata,* which is where the cardiovascular center is located. The medulla oblongata receives nerve signals from the baroreceptors, from receptors in muscles and joints, and from areas of the brain involved in emotions. In response to the inputs it receives, the cardiovascular center sends nerve signals to the heart in two sets of nerves. *Sympathetic nerves* stimulate the heart and cause it to beat faster, and *parasympathetic nerves* inhibit the heart and cause it to beat more slowly.

The hormones *epinephrine* (also called *adrenaline*) and *norepinephrine* stimulate the heart as well. Epinephrine and norepinephrine are secreted by the adrenal glands whenever the sympathetic nervous system is activated.

Sympathetic nerves also help to maintain blood pressure by controlling blood vessel diameters. Most blood vessels are under a constant state of partial constriction by the sympathetic nerves. An increase in sympathetic nerve activity further constricts blood vessels and raises blood pressure. Conversely, a decrease in sympathetic nerve activity dilates blood vessels and lowers pressure. Most blood vessels do not have parasympathetic nerves, so activation or deactivation of the parasympathetic nervous system has little effect on overall blood vessel diameters.

> ✓ **Quick Check** Suppose a girl suddenly develops hives—an allergic reaction that causes blood vessels to suddenly dilate all over the skin. How will her blood pressure likely change, and how will her body attempt to restore homeostasis? ▪

Local requirements dictate local blood flows

With arterial blood pressure held relatively constant, the flow of blood through each precapillary sphincter (and hence each capillary) can be adjusted according to need, just like turning a faucet on or off. How is the flow adjusted? When a particular tissue is metabolically active, such as when a muscle is contracting, it consumes more oxygen and nutrients. Increased metabolism also raises the production of carbon dioxide and other waste products. One or more of these changes associated with increased metabolism cause precapillary sphincters within the tissue to vasodilate, increasing flow. Scientists do not yet know the precise mechanisms of this vasodilation or the identity of all the chemical substances that influence the vasodilation, but we do know that it occurs. Figure 8.17 shows how an increase in cellular metabolism within a tissue increases the local blood flow to that tissue.

Look at the overall process and you can see how together these control mechanisms deliver blood efficiently to all tissues. First, nerves and hormones adjust cardiac output and the rate of blood flow through the vascular system as a whole, in an effort to maintain relatively constant arterial blood pressure. Second, any cell or part of the body that is not receiving enough blood can override the system locally and get exactly what it needs—no more, no less.

Some tissues and organs need a more consistent blood supply than others. That too is taken into account by our control system. Consider what would happen if you lost a substantial volume of blood due to injury, producing a precipitous fall in blood pressure. The negative feedback control of arterial blood pressure would stimulate your heart and

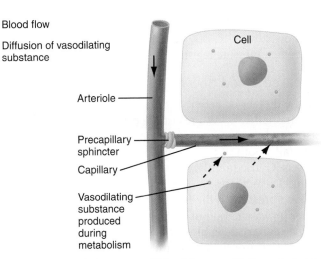

Blood flow

- - - ▶ Diffusion of vasodilating substance

Cell

Arteriole

Precapillary sphincter

Capillary

Vasodilating substance produced during metabolism

a) At rest, very little of the vasodilating substance would be produced, and flow would be minimal.

b) With increased metabolic activity, the presence of more of the substance in the interstitial space would cause the arteriole and precapillary sphincter to vasodilate, increasing flow.

Figure 8.17 How an increase in metabolism increases local blood flow.

constrict your arterioles, reducing flow to most organs in an effort to raise blood pressure. However, organs whose survival depends critically on a constant blood supply (your brain and heart, for example) can override the generalized vasoconstriction with their local control mechanisms. Organs whose metabolic activities are not required for their immediate survival (such as your kidneys and digestive tract) just remain vasoconstricted for a while. In effect, the limited available blood supply is shunted to essential organs.

You may wonder why, if these homeostatic controls are working properly, people ever develop hypertension. Every time blood pressure went up a little, wouldn't feedback mechanisms bring it back to normal? This is the great mystery of hypertension. Although scientists don't fully understand it yet, for some reason the body seems to adjust to high pressure once it has been sustained for a long time. Apparently the feedback system slowly resets itself to recognize the higher pressure as normal. The reason may be that hypertension tends to develop slowly over decades, so there is never a single defining moment when pressure registers as too high.

Exercise: Increased blood flow and cardiac output

During exercise the metabolic activity of the active skeletal muscles goes up dramatically. As a result the production of vasodilator waste products increases, and the local concentration of oxygen falls. Both of these contribute to dilation of the blood vessels. Consequently, blood flow through the active skeletal muscles increases.

To sustain blood pressure in the face of increased blood flow, the heart must increase its output. From what you know, you might guess that an increase in blood flow to muscle would cause a fall in blood pressure, which in turn would cause a baroreceptor-mediated reflex increase in cardiac output. However, blood pressure doesn't fall very much (if at all) during exercise—if anything, it rises a little. During exercise, the primary cause of increased cardiac output is sensory input from moving muscles and joints. This sensory input signals the cardiovascular center to stimulate the heart and increase cardiac output even before blood pressure can fall very much. In other words, during exercise the body anticipates the need for increased cardiac output and prevents blood pressure from falling in the first place.

In nonathletic people, cardiac output reaches a maximum of about 20–25 liters per minute during heavy exercise. However, trained athletes whose heart muscles have hypertrophied as a result of exercise training can reach cardiac outputs of up to 35 liters per minute, or almost seven times their resting cardiac output.

> **Recap** Homeostatic regulation of the cardiovascular system centers on maintaining a relatively constant arterial blood pressure. Arterial pressure is sensed by baroreceptors located in the carotid arteries and aorta. Two opposing sets of nerves (sympathetic and parasympathetic) and a hormone (epinephrine) adjust cardiac output and arteriole diameters to maintain arterial blood pressure fairly constant. Local factors regulate blood flow into individual capillaries by altering the diameters of precapillary sphincters. ▪

8.5 Cardiovascular disorders: A major health issue

Disorders affecting the cardiovascular system are a major health problem in Western countries. Cardiovascular disorders cause more than 700,000 deaths per year in the United States alone. They are the number one killer in the U.S., far ahead of the number two killer, cancer (approximately 550,000 deaths per year).

We have already discussed hypertension and aneurysms, and atherosclerosis. In the rest of this chapter we look at several other conditions, including arrhythmias, angina, heart attack, heart failure, embolism, and stroke. Finally, we examine

what you can do to reduce your own risk of cardiovascular health problems.

Angina: Chest pain warns of impaired blood flow

As a hard-working muscle, the heart requires a constant source of blood. Normally the coronary arteries and their branches provide all the blood the heart needs, even during sustained exercise. However, if these arteries become narrowed, blood flow to the heart may not be sufficient for the heart's demands. This may lead to *angina*, a sensation of pain and tightness in the chest. Often angina is accompanied by shortness of breath and a sensation of choking or suffocating (*angina* comes from the Latin word for "strangling"). Many angina episodes are triggered by physical exertion, emotional stress, cold weather, or eating heavy meals, because the heart requires more blood and oxygen at these times.

Angina is uncomfortable but usually temporary. Stopping to rest and taking several deep breaths can often relieve the discomfort. However, angina should never be ignored, because it is a sign of insufficient circulation to the heart. *Angiography* is a procedure that enables blood vessels to be visualized after they are filled with a contrast medium (a substance that is opaque to X-rays). Angiography allows health professionals to take X-ray pictures of blood vessels (called *angiograms*) and assess their condition (**Figure 8.18**).

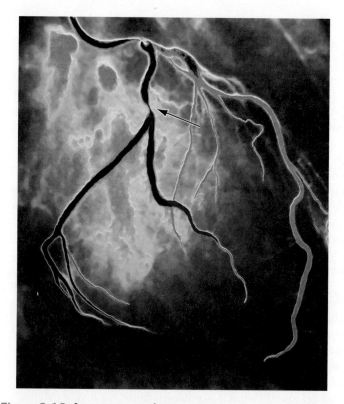

Figure 8.18 A coronary angiogram. A contrast medium (similar in function to a dye) is injected into the coronary arteries, causing them to become visible on an X-ray photograph. Note the area of narrowing in the coronary artery indicated by the arrow.

Certain medications can increase blood flow to the heart muscle. Another treatment for narrowed coronary arteries is *balloon angioplasty,* which involves threading into the blocked artery a slender flexible tube with a small balloon attached. When the balloon reaches the narrowest point of the vessel it is inflated briefly so that it presses against the fatty plaques that narrow the vessel lumen, flattening them and widening the lumen. Balloon angioplasty has a high success rate, although in some cases the vessel narrows again over time, requiring repeat treatments. A coronary artery bypass graft (see the next section) can yield longer-lasting benefits.

Heart attack: Permanent damage to heart tissue

If blood flow to an area of the heart is impaired long enough, the result is a **heart attack**—sudden death of an area of heart tissue due to oxygen starvation. (The clinical term is *myocardial infarction;* infarction refers to tissue death from inadequate blood supply.) Many people who suffer a heart attack have a previous history of angina. The classic symptoms of a major heart attack, especially in men but also in women, include intense chest pain, a sense of tightness or pressure on the chest that makes it hard to breathe, and pain that may radiate down the left arm. Women tend to experience nausea and jaw and back pain more frequently than men, but as these symptoms do not seem as immediately life-threatening, women's heart attacks are not diagnosed as soon as men's based on symptoms alone. The symptoms may come and go for several minutes at a time, leading many people to delay seeking medical attention.

A heart attack causes permanent damage to the heart. Because the body cannot replace cardiac muscle cells, damage to the heart impairs its ability to function. Most heart attack fatalities occur because of ventricular fibrillation, the serious heart arrhythmia brought about by damage to heart muscle.

Prompt treatment is crucial for recovery from a heart attack. If a heart attack is even suspected, the person should be rushed immediately to a hospital. The diagnosis of a heart attack is generally made on the basis of the ECG and the presence in the blood of certain enzymes that are released from dead and damaged heart cells. Health professionals can often control cardiac arrhythmias and other complications and administer clot-dissolving drugs to unblock vessels. The sooner treatment begins, the more successful it is likely to be.

Later (or even before a heart attack has occurred, if the narrowing of coronary vessels has been diagnosed by angiography), a *coronary artery bypass graft (CABG)* can be performed to improve coronary blood flow. In this procedure, a piece of blood vessel is removed from somewhere else in the body (often a leg vein is used) and grafted onto the blocked artery to bypass the damaged region (**Figure 8.19**). Over time the grafted vein thickens and takes on the characteristics of an artery.

Thanks to these and other treatments, the heart attack survival rate has risen dramatically. Eighty percent of heart attack survivors are back at work within three months.

Heart failure: The heart becomes less efficient

Generally our bodies maintain adequate arterial pressure because of the tight control mechanisms described earlier. However if the heart muscle becomes damaged for any reason, the heart may become weaker and less efficient at pumping blood, a condition called **heart failure**.

When the heart begins to pump less blood, blood backs up in the veins and pressure in the veins and capillaries rises. The high capillary blood pressure causes more fluid than usual to filter out of the capillaries and into the

Figure 8.19 Coronary artery bypass grafts. This is an example of a "triple CABG," in which three vein grafts have been placed to bypass three areas of atherosclerotic plaque. Arrows indicate additional blood flow to the area beyond the restricted artery.

interstitial space, causing fluid congestion. When that happens a person is said to have **congestive heart failure.** ("Congestive" refers to the buildup of interstitial fluid.)

People with congestive heart failure get out of breath when walking or climbing even a short flight of stairs. They may even have trouble breathing while lying down because the horizontal position results in even higher venous pressure and fluid accumulation in the lungs. Other symptoms include swollen ankles and legs, swollen neck veins, and weight gain from the extra fluid.

There are several reasons why a heart might begin to fail. Aging is one factor, but certainly not the only one because older people do not necessarily develop heart failure. Other possible causes include past heart attacks, leaking heart valves, heart valves that fail to open normally, uncontrolled hypertension, or serious arrhythmias. Lung conditions such as emphysema can raise blood pressure in the lungs and strain the heart.

Treatments for congestive heart failure focus on improving cardiac performance and efficiency while preventing the accumulation of interstitial fluid. Regular mild exercise promotes more efficient blood flow, and frequent resting with the feet elevated helps fluid drain from leg veins. Physicians may prescribe diuretics (drugs that help the body get rid of excess fluid), vasodilating drugs to expand blood vessels, or medications to help the heart muscle beat more forcefully.

Embolism: Blockage of a blood vessel

Embolism refers to the sudden blockage of a blood vessel by material floating in the bloodstream. Most often the obstacle (an *embolus*) is a blood clot that has broken away from a larger clot elsewhere in the body (often a vein) and lodged in an artery at a point where arterial vessels branch and get smaller in diameter. Other possible emboli include cholesterol deposits, tissue fragments, cancer cells, clumps of bacteria, or bubbles of air.

Embolism conditions are named according to the area of the body affected. A *pulmonary embolism* blocks an artery supplying blood to the lungs, causing sudden chest pain and shortness of breath. A *cerebral embolism* impairs circulation to the brain, possibly causing a stroke. A *cardiac embolism* can cause a heart attack.

Stroke: Damage to blood vessels in the brain

To function normally, the brain requires a steady blood supply—about 15% of the heart's output at rest. Any impairment of blood flow to the brain rapidly damages brain cells. A **stroke** (*cerebrovascular accident*) represents damage to part of the brain caused by an interruption to its blood supply. In effect, it is the brain equivalent of a heart attack. Strokes are the most common cause of brain injury

and a leading cause of death in Western nations. The two most common causes are an embolism blocking a vessel and rupture of a cerebral artery.

Symptoms of strokes appear suddenly and vary according to the area of the brain affected. They may include weakness or paralysis on one side of the body, fainting, inability to speak or slurred speech, difficulty in understanding speech, impaired vision, nausea, or a sudden loss of coordination.

Immediate medical care is crucial. If the stroke resulted from a cerebral embolism, clot-dissolving drugs or an embolectomy (surgery to remove the embolus) can save a life. Eliminating the clot as quickly as possible (within minutes or hours) can also limit the area of damage and reduce the severity of the permanent injury. If a ruptured artery is responsible, health providers may be able to surgically drain the excess blood.

Although some people recover well from a stroke or suffer only minor permanent effects, others do not recover much lost function at all despite intensive physical therapy over many months. The reason for generally poor recovery rates is that the body does not grow new nerve cells to replace damaged ones. Nevertheless, rehabilitation with skilled health professionals generally offers the best chance of at least a partial recovery. Recovery involves retraining nerve pathways that already exist so they can take over the functions of damaged nerve cells.

↺ **Recap** Cardiovascular disorders are the number one killer in the United States. Most disorders are caused either by conditions that result in failure of the heart as a pump or by conditions in which damage to blood vessels restricts flow or ruptures vessels. ▪

8.6 Reducing your risk of cardiovascular disease

Cardiovascular disorders are among the most preventable of chronic health conditions. Although some factors are beyond your control—such as sex, race, age, and genetic inheritance—your lifestyle choices can also affect your risk. Things you can do to reduce your risk include:

- Don't smoke, or if you do, quit. Smokers have more than twice the risk of heart attack that nonsmokers do, and smokers who suffer a heart attack are nearly four times as likely to die from it. Some researchers think second-hand smoke poses a risk as well.
- Watch your cholesterol levels. Cardiovascular risk rises with the blood cholesterol level. There is also evidence that high cholesterol increases risk even more when it is combined with other factors such as hypertension and tobacco smoke.

Chapter 8 Heart and Blood Vessels **185**

- Keep moving (Figure 8.20). Regular, moderate exercise lowers the risk of cardiovascular disease. This is not surprising because the heart is, after all, a muscle. Most physicians recommend exercising for at least 20 to 30 minutes, at least three times per week. Physical activity tends to lower blood pressure and cholesterol and makes it easier to maintain a healthy body weight. Always consult your physician before starting an exercise program.
- If your blood pressure is on the high side, seek treatment. As discussed earlier, untreated hypertension damages blood vessels and increases the workload on the heart.

In addition to these major risks, at least three other factors are associated with cardiovascular disease, although the precise link has not yet been determined. This is why doctors recommend the following:

- Maintain a healthy weight. It's not clear how obesity contributes to cardiovascular problems, but overweight people have a higher rate of heart disease and stroke even if they do not have other risk factors. One hypothesis is that increased weight strains the heart. Increased weight also has adverse effects on other risk factors such as blood cholesterol and hypertension.
- Keep diabetes under control. Diabetes mellitus is a disorder of blood sugar levels. Untreated diabetes damages blood vessels, but effective treatments reduce

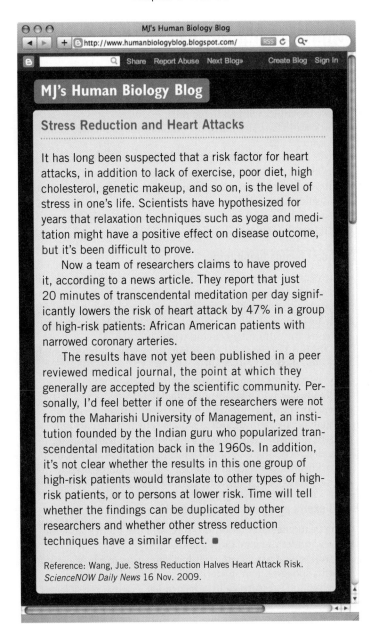

MJ's Human Biology Blog

Stress Reduction and Heart Attacks

It has long been suspected that a risk factor for heart attacks, in addition to lack of exercise, poor diet, high cholesterol, genetic makeup, and so on, is the level of stress in one's life. Scientists have hypothesized for years that relaxation techniques such as yoga and meditation might have a positive effect on disease outcome, but it's been difficult to prove.

Now a team of researchers claims to have proved it, according to a news article. They report that just 20 minutes of transcendental meditation per day significantly lowers the risk of heart attack by 47% in a group of high-risk patients: African American patients with narrowed coronary arteries.

The results have not yet been published in a peer reviewed medical journal, the point at which they generally are accepted by the scientific community. Personally, I'd feel better if one of the researchers were not from the Maharishi University of Management, an institution founded by the Indian guru who popularized transcendental meditation back in the 1960s. In addition, it's not clear whether the results in this one group of high-risk patients would translate to other types of high-risk patients, or to persons at lower risk. Time will tell whether the findings can be duplicated by other researchers and whether other stress reduction techniques have a similar effect.

Reference: Wang, Jue. Stress Reduction Halves Heart Attack Risk. *ScienceNOW Daily News* 16 Nov. 2009.

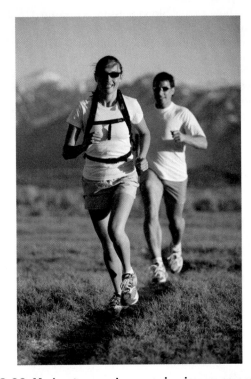

Figure 8.20 Moderate, regular exercise improves cardiovascular performance and lowers the risk of cardiovascular disease.

cardiovascular damage significantly. (For more on diabetes, see Chapter 13.)
- Avoid chronic stress. Again the mechanism is unclear, but there is an association between a person's perceived stress and behavior patterns and the development of cardiovascular disease. Stress may also affect other risk factors, for example how much a smoker smokes or whether a person starts smoking.

Recap You can reduce your risk of developing cardiovascular disease by not smoking, exercising regularly, watching your weight and cholesterol, and avoiding prolonged stress. If you have diabetes and/or hypertension, try to keep these conditions under control. ■

Chapter Summary

Blood vessels transport blood p. 164

■ The primary function of blood vessels is to bring blood into close proximity with all living cells.
■ Thick-walled arteries transport blood to the capillaries at high pressure.
■ Small arterioles and precapillary sphincters regulate the flow of blood into each capillary.
■ Thin-walled capillaries are the only vessels that exchange fluids and solutes with the interstitial fluid.
■ Distensible venules and veins store blood at low pressure and return it to the heart.

The heart pumps blood through the vessels p. 169

■ The heart is composed primarily of cardiac muscle. Structurally, it consists of four separate chambers and four one-way valves. Its primary function is to pump blood.
■ The heart pumps blood simultaneously through two separate circuits: the pulmonary circuit, where blood picks up oxygen and gets rid of carbon dioxide, and the systemic circuit, which supplies the rest of the body's cells.
■ The heart contracts and relaxes rhythmically. Contraction is called *systole*, and relaxation is called *diastole*.
■ The coordinated contraction of the heart is produced by a system of specialized cardiac muscle cells that initiate and distribute electrical impulses throughout the heart muscle.
■ An electrocardiogram, or ECG, records electrical activity of the heart from the surface of the body. An ECG can be used to diagnose certain cardiac arrhythmias and disorders.

Blood exerts pressure against vessel walls p. 177

■ The heart generates blood pressure, and the arteries store the blood under pressure during diastole.
■ Systolic and diastolic arterial blood pressures can be measured with a sphygmomanometer.
■ High blood pressure, called *hypertension*, is a major risk factor for cardiovascular disease.

How the cardiovascular system is regulated p. 180

■ The most important controlled variable in the cardiovascular system is arterial blood pressure.
■ Arterial blood pressure is monitored by stretch receptors located in certain large arteries.
■ Cardiac output and the diameters of the arterioles are regulated (controlled) to keep arterial blood pressure relatively constant.
■ With pressure held constant, local blood flows can be adjusted according to the metabolic needs of the tissues and cells in that area of the body.

Cardiovascular disorders: A major health issue p. 182

■ The heart muscle is always working. Impairment of blood flow to the heart can lead to a sense of pain and tightness in the chest (angina) and/or permanent damage to heart tissue (myocardial infarction, or heart attack).
■ Slowly developing, chronic failure of the heart as a pump can lead to excessive interstitial fluid, a condition known as *congestive heart failure*.
■ An embolism is the sudden blockage of a blood vessel by any object.

■ Strokes, also called *cerebrovascular accidents*, can be caused by either embolisms or rupture of blood vessels. The result is damage to a part of the brain when its blood supply is interrupted.

Reducing your risk of cardiovascular disease p. 184

■ Your chances of developing a cardiovascular disorder depend on certain risk factors. Some risk factors you cannot change, whereas others depend on the choices you make in life.

Terms You Should Know

artery, *164*
atrium, *170*
atrioventricular (AV) valves, *171*
capillary, *166*
diastole, *174*
myocardium, *170*
precapillary sphincter, *166*

pulmonary circuit, *171*
semilunar valves, *171*
sinoatrial (SA) node, *176*
systemic circuit, *171*
systole, *174*
vein, *168*
ventricle, *170*

Concept Review

Answers can be found at the Human Biology Place.
www.humanbiology.com

1. Why is it so important that the heart always get a consistent and adequate blood supply?
2. Compare and contrast the structures and functions of the three types of blood vessels.
3. Describe how arterial blood pressure is measured by the body.
4. Describe the function of the heart valves.
5. Compare the functions of the pulmonary circuit and the systemic circuit.
6. Describe the cardiac cycle of relaxation and contraction, and explain what causes each of the two heart sounds during the cycle.
7. Describe the function of the SA and AV nodes as they control contraction of the heart.
8. Of what functional value is the distensibility of veins? In other words, why not have them just as thick and stiff as arteries?
9. Name the one hormone that has a stimulatory effect on the heart (increases heart rate).
10. List ways to reduce the risk of cardiovascular disease.

Test Yourself

Answers can be found in Appendix A.

1. _____ carry blood away from the heart and _____ carry blood toward the heart.
   a. veins...arteries
   b. arteries...arterioles
   c. veins...capillaries
   d. arteries...veins
   e. arteries...capillaries

2. Which blood vessel is best suited for exchange of gases and nutrients with the surrounding tissue?
   a. artery   b. vein
   c. capillary   d. arteriole

3. Which of the choices represents the order of vessels through which blood passes after leaving the heart?
   a. artery – arteriole – capillary – venule – vein
   b. artery – capillary – arteriole – venule – vein
   c. vein – venule – capillary – arteriole – artery
   d. artery – vein – capillary – arteriole – venule

4. All of the following mechanisms assist in returning venous blood to the heart except:
   a. an increase in heart rate
   b. pressure changes in the abdominal and thoracic cavities due to breathing
   c. contraction of skeletal muscles in the legs
   d. one-way valves located inside veins

5. Which of the following represents the order of structures beginning inside the ventricle and traveling outward?
   a. pericardium – epicardium – myocardium – endocardium
   b. epicardium – myocardium – endocardium – pericardium
   c. endocardium – myocardium – epicardium – pericardium
   d. endocardium – pericardium – myocardium – epicardium

6. Which vein(s) carry oxygenated blood?
   a. superior vena cava and inferior vena cava
   b. right and left pulmonary veins
   c. aorta
   d. both (a) and (b)

7. Which of the following statements regarding the cardiac cycle is false?
   a. When the ventricles contract, the atrioventricular valves close.
   b. When the ventricles relax, the semilunar valves close.
   c. When the atria contract, the semilunar valves open.
   d. When the atria contract, the atrioventricular valves open.

8. A pacemaker is use to correct:
   a. coronary artery disease
   b. cardiac arrhythmias
   c. heart murmurs
   d. hypertension

9. As the blood travels through the circulatory system, the greatest drop in pressure occurs in:
   a. arteries
   b. arterioles
   c. capillaries
   d. venules

10. All of the following are part of the cardiac conduction system except:
    a. the chordae tendineae
    b. the Purkinje fibers
    c. the sinoatrial node
    d. the atrioventricular bundle

11. Which of the following is/are involved in regulating blood pressure?
    a. heart
    b. baroreceptors
    c. cardiovascular center in medulla oblongata
    d. all of the above

12. Which of the following would be an appropriate homeostatic response to a drop in blood pressure below what is normal?
    a. heart rate decreases
    b. vasoconstriction of arterioles
    c. force of cardiac contraction decreases
    d. both (a) and (b)

13. Which of the following might cause the heart rate to increase from 68 beats/min. to 120 beats/min?
    a. epinephrine
    b. stimulation via parasympathetic nerves
    c. stimulation via sympathetic nerves
    d. both (a) and (c)

14. Which of the following might be appropriately treated by administering clot-dissolving drugs?
    a. hypertension
    b. hypotension
    c. pulmonary embolism
    d. a hemorrhagic stroke

15. A heart attack occurs as a result of:
    a. prolonged hypotension
    b. narrowing or blockage of the coronary arteries
    c. improper closure of semilunar valves
    d. disruption of the cardiac conduction system

Apply What You Know

Answers can be found at the Human Biology Place.
www.humanbiology.com

1. During exercise, arterial blood pressure changes very little. However, cardiac output may double and blood flow to exercising muscle may go up 10-fold, while at the same time the blood flow to kidneys may decline by nearly 50%. Explain possible mechanisms that might account for these very different changes.

2. When a coronary artery bypass graft (CABG) is performed, the vessels used for the bypass grafts are usually veins taken from the patient's legs. Over time the grafted veins take on many of the characteristics of arteries; that is, they become thicker and stiffer. What might this suggest about the possible cause(s) of the structural differences between arteries and veins? Hypothesize what might happen if you took a section of artery and implanted it into a vein.

3. Soldiers have to stand in formation at full attention for long periods of time. Sometimes this can cause otherwise very fit and healthy young people to pass out during long inspections. What would cause this?

4. Workers who spend hours per day standing can develop circulation problems in their legs. A recommended solution is to wear graduated compression stockings. These stockings are tighter around the ankles and less tight higher up on the legs. Why does standing for long periods sometimes lead to circulatory problems, and how can wearing something tight on the legs help prevent this?

5. Inflammation of the pericardium can lead to a condition called *pericardial effusion,* in which fluids collect in the space surrounding the heart. What effect would this have on the functioning of the heart? How might the condition be treated?

6. Ventricular fibrillation is a potentially fatal condition where the cells of the heart are no longer coordinated in their contractions. Why is it important that contraction of the heart muscle occurs nearly all at once?

7. VSD or ventricular septal defect is a condition that accounts for half of all congenital cardiovascular anomalies. In this condition a hole exists in the septum, the muscular wall between the ventricles, which allows blood from the right and left side of the heart to mix. What sort of problems might this condition cause?

8. Last night you and your roommate were sound asleep when the phone rang. Your roommate, startled awake, jumped from her bed and rushed for the phone across the room. Before she reached the phone she suddenly felt dizzy and had to sit down to avoid fainting. What happened?

9

The Immune System and Mechanisms of Defense

Colorized TEM (× 25,000) of a mast cell containing granules of histamine (dark purple).

Current Issue

AIDS: A Crisis in Africa, a Challenge for the World

Africa is in the midst of an AIDS epidemic. What is AIDS, how bad is the epidemic, how will it affect us, and what (if anything) are we willing to do about it?

In Africa, AIDS Is Out of Control

AIDS stands for *acquired immune deficiency syndrome*. It is caused by a virus that attacks the immune system, called *human immunodeficiency virus* (HIV). Approximately two-thirds of the world's HIV-infected people live in sub-Saharan Africa. The United Nations estimates that 22.4 million people there are already infected with HIV. The region accounts for more than three-quarters of the world's AIDS deaths.

Patterns of AIDS infection and transmission in Africa differ from those of industrialized countries. In the United States, more men than women are infected with HIV, which is often transmitted via homosexual sex. In Africa, however, an estimated 60% of HIV-infected persons are women who have contracted the virus heterosexually. Studies in several African nations have found that females aged 15–19 are four to five times more likely to be infected than males their age. According to one report, in Africa older HIV-infected men coerce or pay impoverished girls to have sex in the mistaken belief that sex with a virgin will cure AIDS.

The problem of AIDS in sub-Saharan Africa is made worse by political, economic, and social instability. Of the area's 42 nations, almost a third are at war. In many of these countries the economies are weak, sanitation is poor, and even rudimentary medical services are often lacking. Malnutrition and starvation are widespread. Approximately 11 million children in sub-Saharan Africa have been orphaned by AIDS, war, or famine. Many of them are abandoned and impoverished, fending for themselves without adult supervision.

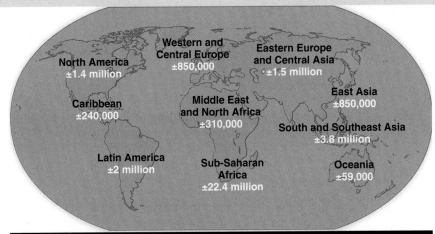

Western and Central Europe ±850,000

North America ±1.4 million

Eastern Europe and Central Asia ±1.5 million

East Asia ±850,000

Caribbean ±240,000

Middle East and North Africa ±310,000

South and Southeast Asia ±3.8 million

Latin America ±2 million

Sub-Saharan Africa ±22.4 million

Oceania ±59,000

Over two-thirds of all people living with HIV live in sub-Saharan Africa.

What can be done to help sub-Saharan Africa? Possibilities include:

- Providing accurate information on HIV transmission and helping people understand how their sexual practices may contribute to the spread of the disease.
- Seeking AIDS treatments and preventive measures that are inexpensive enough to be used effectively in poor countries.
- Providing economic assistance and encouraging political stability.
- Improving the delivery of health care.

These are all good ideas, of course. But any solution to the growing AIDS crisis in African countries must recognize that their problems are different from ours.

Roadblocks to Effective HIV Treatment

Strategies for treating HIV infection and preventing AIDS in sub-Saharan Africa face some significant obstacles not seen in North America. For example, improving the delivery of health care is difficult in rural areas without roads and bridges and in nations with too few medical personnel. Then there is the problem of identifying AIDS patients in the first place. As discussed in this chapter, people in North America who are suspected of having AIDS generally are tested for HIV. Most health professionals maintain careful records

documenting their patients' symptoms and treatments. In contrast, most African nations do not require that suspected AIDS patients be tested for HIV, and health professionals often can't monitor patients effectively or even pay for the tests to diagnose HIV infection. The diagnosis is made more difficult by the fact that other regional health problems, such as malnutrition, tuberculosis, and malaria, have symptoms that can resemble those of AIDS.

What about education? The type of public information campaigns used in North America may not be possible in countries with less developed media networks. Furthermore, social norms are different, and misinformation and denial are widespread. Sex education for both boys and girls and strategies to change men's sexual behavior are sorely needed. Providing male and female condoms at low cost might also help. In Thailand, a national program to promote condom use lowered the prevalence of HIV among 21-year-old military conscripts from 4% in the mid-1990s to less than 1% by 2002.

A final hurdle is the cost of treatment. Recent advances in AIDS treatment depend on new *antiretroviral* drugs (ARVs) that can cost up to $2,500 per month—more than many Africans earn in a year. High prices for new AIDS drugs reflect the enormous research and development costs that are

required by U.S. law to ensure drugs' safety and effectiveness. Foreign firms can manufacture and sell AIDS medicines at a fraction of the U.S. price because they do not have to recover the cost of drug development. In addition to selling the low-cost drugs abroad, foreign manufacturers sometimes smuggle them into North America. U.S. pharmaceutical firms claim that the knockoff ARVs endanger future research and the development of even better HIV treatments. They have struck back by taking legal action against distributors and purchasers of illegal AIDS medications.

This AIDS drug manufactured in Thailand costs less than $400/year.

The global economic downturn of 2009 only made things worse. AIDS treatment programs suffered financially along with the rest of the economy. In addition, increased poverty usually leads to poorer nutrition, even less access to health care, and risky social behaviors such as trading sex for money.

The result of all of these problems is that in sub-Saharan Africa, fewer than half of all HIV-infected people currently receive treatment. Furthermore, only 60% of all patients who begin HIV treatment are still receiving it two years later. This is unfortunate, for ARVs should be taken for the rest of the patient's life in order to fully protect against the development of AIDS.

The challenge of AIDS in sub-Saharan Africa is likely to remain with us for some time. We can deal with it now or we can deal with an even larger problem later.

Questions to consider

1 Should foreign pharmaceutical companies be allowed to manufacture ARVs at low cost and sell them to needy nations?

2 As a matter of public health, should governments ensure that AIDS medicines are available to those who need them?

The facts...

- **Two-thirds of the world's HIV-infected people live in sub-Saharan Africa.**
- **In sub-Saharan Africa, girls aged 15–19 are four to five times more likely to be HIV-infected than boys.**
- **Attempts to solve the AIDS crisis in Africa are hampered by poor infrastructure, inadequate health care, poor medical record keeping, and educational, political, and social issues.**
- **Many African nations cannot afford the high cost of current AIDS treatment drugs.**

>> **The health risk of a pathogen** (disease-causing organism) is determined by its **transmissibility** (how <u>easily</u> it can be passed from person to person), **mode of transmission** (<u>how</u> it is transmitted; through air, food, blood, etc.), and **virulence** (how <u>damaging</u> the disease is when one catches it).

>> **The immune system** has **nonspecific** (against many pathogens) and **specific** (against one pathogen) defense mechanisms.

>> **Nonspecific defense mechanisms** include immune system cells that engulf and digest foreign cells, chemicals that are toxic to foreign cells, proteins that interfere with viral reproduction, and the development of a fever.

>> **Specific defense mechanisms involve the production of antibodies and T cells** that recognize and inactivate one specific pathogen. Specific defense mechanisms have a memory component that is the basis of immunity.

>> **Inappropriate immune system activity** can lead to allergies and autoimmune diseases.

>> **AIDS (Acquired Immune Deficiency Syndrome) is caused by a virus that targets certain cells of the immune system,** thereby rendering the body incapable of fending off other infections.

- *Nonspecific defense mechanisms.* Nonspecific defenses help the body respond to generalized tissue damage and many of the more common or obvious pathogens, including most bacteria and some viruses.
- *Specific defense mechanisms.* These enable the body to recognize and kill specific bacteria and other foreign cells and to neutralize viruses. Our specific defense mechanisms employ sophisticated weaponry indeed. The specific defense mechanisms are also the basis of immunity from future disease.

All three mechanisms are operating night and day to protect us. Crucial to the nonspecific and specific defense mechanisms is the **immune system,** a complex group of cells, proteins, and structures of the lymphatic and circulatory systems. However, even the immune system is not perfect; it can only kill or neutralize pathogens or abnormal cells that it can recognize. This has implications for how the human body deals with certain pathogens.

What is the immune system on alert for? We start by looking at some of the kinds of pathogens that can invade our bodies.

9.1 Pathogens cause disease

Pathogens include bacteria, viruses, fungi, a few protozoa, and possibly prions. Larger parasites, including various worms, can also be pathogens, although they are relatively rare in industrialized countries. Because bacteria and viruses are by far the most numerous and problematic pathogens in Western nations, we focus on them as we describe how we protect and defend ourselves. We'll also discuss prions as a newly discovered class of pathogens.

Bacteria: Single-celled living organisms

As you learned in Chapter 3, **bacteria** (singular: *bacterium*) are single-celled organisms that do not have a nucleus or membrane-bound organelles. All the DNA in most bacteria is contained in just one chromosome, which usually forms a continuous loop that is anchored to the plasma membrane. Bacterial ribosomes are smaller than ours and float freely in the cytoplasm. The outer surface of bacteria is covered by a rigid cell wall that gives bacteria their distinctive shapes, including spheres, rods, and spirals (**Figure 9.1**).

Judging by their variety and numbers, bacteria are among the most successful organisms on Earth. Although they are smaller than the typical human cell, their small size is actually an advantage. Like all living organisms, bacteria need energy and raw materials to maintain life and to grow and divide. Their small size means that they have a high surface-to-volume ratio, a decided advantage when it comes to getting raw materials and getting rid of wastes by diffusion.

Like our own cells, bacteria use ATP as a direct energy source and amino acids for making proteins. They store energy as carbohydrates and fats. Where do they obtain those raw materials? Anywhere they can. Some bacteria break

The world is swarming with living organisms (bacteria) and even some nonliving entities (viruses and prions) too small to be seen with the naked eye. They're found on doorknobs, the money we handle, and our clothes. They're in the food we eat and the air we breathe. Most are harmless—indeed, some are highly beneficial. The ones that cause disease are called **pathogens** (from the Greek *pathos*, disease, plus *gennan*, to produce).

Pathogens come from outside the body. As a group they account for a large fraction of all human disease and suffering. But there are other challenges to our health that come from within. Mutations (alterations to a cell's DNA) may cause some of our own cells to become abnormal and may even lead to cancer. Fortunately, our bodies can recognize cells that become abnormal and dispatch most of them before they ever have the chance to develop further.

What ways do we have to protect ourselves? Our body's immune system and other general defense mechanisms include:

- *Barriers to entry or ways of expelling or neutralizing pathogens before they can do harm.* These include skin, stomach acid, tears, and such actions as vomiting and defecation.

a) SEM (× 2,000) of *Streptococcus*, a spherical bacterium that causes sore throats.

b) SEM (× 5,600) of *Escherichia coli*, a common intestinal bacteria that is usually harmless.

c) SEM (×12,000) of *Campylobacter jejuni*, a spiral-shaped bacterium that causes food poisoning.

Figure 9.1 Electron micrographs of the three common shapes of bacteria.

down raw sewage and cause the decomposition of dead animals and plants, thereby playing an essential role in the recycling of energy and raw materials. Others obtain nutrients from the soil and air.

Humans have learned to harness bacteria to produce commercial products, including antibiotic drugs, hormones, vaccines, and foods ranging from sauerkraut to soy sauce. Some bacteria even live within our digestive tract, drawing energy from the food we eat in exchange for manufacturing vitamins or controlling the populations of other, more harmful bacteria. Life as we know it would not be possible without these little organisms.

A few bacteria are pathogens, however. Pathogens rely on living human cells for their energy supply, and in the process they damage or kill the human cells. They cause pneumonia, tonsillitis, tuberculosis, botulism, toxic shock syndrome, syphilis, Lyme disease, and many other diseases. Although we

concentrate on pathogens in this chapter, do keep in mind that most bacteria are harmless, and many are even beneficial.

Bacterial infections are generally treated with **antibiotics**—chemotherapeutic agents that inhibit or abolish the growth of bacteria, fungi, and protozoa.

Viruses: Tiny infectious agents

Viruses are extremely small infectious agents, perhaps one-hundredth the size of a bacterium and one-thousandth the size of a typical eukaryotic cell (**Figure 9.2**). Structurally a virus is very simple, consisting solely of a small piece of genetic material (either RNA or DNA) surrounded by a protein coat. They have no organelles of their own, so they don't grow and they can't reproduce on their own.

Are viruses alive? Biologists are divided on this question. Most would say that they are not because they

a) Eukaryotic cell. Eukaryotic cells have a membrane-bound nucleus and well-defined membrane-bound organelles.

(labels: Mitochondrion, Nucleus, Golgi apparatus, Plasma membrane, Endoplasmic reticulum)

(Relative size of a bacterium)

b) Bacterial cell. Bacteria have a single strand of DNA and free-floating small ribosomes within their cytoplasm. Their plasma membrane is surrounded by a rigid cell wall.

(labels: Ribosomes, DNA, Cell wall, Plasma membrane)

(Relative size of a virus)

c) Viruses. Viruses consist of a protein coat surrounding either RNA or DNA.

(labels: RNA, Protein coat, DNA)

Figure 9.2 Size and structural differences between a eukaryotic cell, a bacterium (prokaryotic cell), and viruses.

cannot reproduce on their own. Viruses have no observable activity associated with life when they are not in contact with another living cell. However, when they enter a living cell, they take it over and use the cell's organelles to make more viruses.

Viruses have several ways of gaining entry into living cells. Most viruses that infect human cells are taken into the cell cytoplasm by endocytosis; once inside the cell the protein coats are dissolved and the viral genetic material is released for incorporation into the cell's genetic material. Other viruses merge their outer coat with the cell membrane and release their genetic contents into the cell's cytoplasm. Still other viruses attach to the outer surface of the cell membrane and inject just their genetic material into the cell, much as a needle and syringe inject drugs into the body. Regardless of the method of entry, the presence of the viral genetic material causes the cell to begin producing thousands of copies of the virus instead of carrying out its own metabolic activities. Sometimes the newly formed viruses are released by a type of budding from the cell membrane while the cell is still alive. In other cases the cell becomes so packed with viruses that it dies and bursts, releasing a huge number of viruses all at once.

Diseases caused by viruses range from serious—AIDS, hepatitis, encephalitis, rabies—to annoying—colds, warts, chicken pox. Viral infections can be minor for some people but serious for others. An otherwise healthy person may be ill for only a few days with a viral infection, whereas someone who is very young, very old, or in poor health may die.

Antibiotics generally don't work against viral infections. The best ways to cure a viral infection are either to prevent the viruses from entering living cells or to stop the host cell from producing more viruses once it's infected.

 Web Animation *Structure and Reproduction of Viruses* at www.humanbiology.com

Prions: Infectious proteins

In 1986, scientists identified a disease in British cattle that destroyed nerve cells in the animals' brain and spinal cord, causing the animals to stagger, jerk, tremble, and exhibit other bizarre behaviors. The press nicknamed it "mad cow disease." Then between 1994 and 1996, twelve Britons ages 19 to 39 developed signs of a new human disease called *variant Creutzfeldt-Jakob disease (vCJD)*. Alarmingly, researchers found that all of the vCJD patients had eaten beef from animals suspected of having mad cow disease. Later they confirmed that a *prion* was responsible for both the mad cow disease and the twelve cases of vCJD.

A **prion** is a misfolded form of a normal brain cell protein. But it is not just a misfolded protein—it is misfolded protein that can trigger the misfolding of nearby *normal* forms of the protein as well. Once prions enter a nerve cell, the misfolding process becomes self-propagating—one prion produces another, which produces another, and so on. Eventually so many prions accumulate within infected brain cells that the cells die and burst, releasing prions to infect other brain cells. The death of nerve cells accounts for

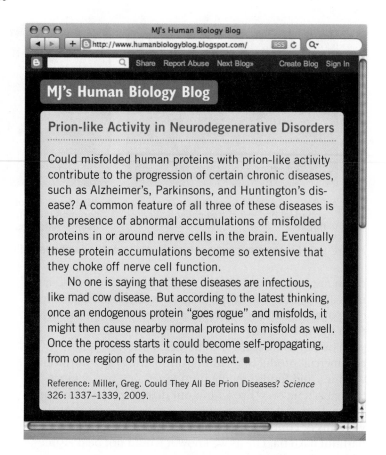

MJ's Human Biology Blog

Prion-like Activity in Neurodegenerative Disorders

Could misfolded human proteins with prion-like activity contribute to the progression of certain chronic diseases, such as Alzheimer's, Parkinsons, and Huntington's disease? A common feature of all three of these diseases is the presence of abnormal accumulations of misfolded proteins in or around nerve cells in the brain. Eventually these protein accumulations become so extensive that they choke off nerve cell function.

No one is saying that these diseases are infectious, like mad cow disease. But according to the latest thinking, once an endogenous protein "goes rogue" and misfolds, it might then cause nearby normal proteins to misfold as well. Once the process starts it could become self-propagating, from one region of the brain to the next. ■

Reference: Miller, Greg. Could They All Be Prion Diseases? *Science* 326: 1337–1339, 2009.

the debilitating neurological symptoms and progressive degeneration seen in both mad cow disease and human vCJD.

Prions are resistant to cooking, freezing, and even drying. There is no known cure for prion infection. Because infection occurs when humans (or cattle) eat prion-infected cattle tissues, the best way to prevent vCJD in humans is to limit the spread of mad cow disease in cattle. Global cooperation is making this possible. In 1994 the European Union banned the use of mammalian meat and bone meal products as cattle feed, and since that time the number of cases of mad cow disease has fallen dramatically.

☑ **Quick Check** Suppose you are studying a mysterious disease, and you discover that it is caused by a tiny pathogen that contains some nucleic acid and some protein, but does not have a plasma membrane. Is it most likely a bacterium, a virus, or a prion? Explain your answer. ■

Transmissibility, mode of transmission, and virulence determine health risk

Some pathogens are clearly more risky to human health than others. Factors that determine the danger of a particular pathogen include transmissibility (how easily it is passed from one person to another), mode of transmission (how it is transmitted), and virulence (how damaging the resulting disease is).

Answers to ☑ **questions can be found in Appendix A.**

For instance, the viruses that cause the common cold are easily transmitted from hands to mucous membranes, as well as in the fluid particles spread by a sneeze. However, these viruses tend not to be very virulent. The HIV virus that causes AIDS, by contrast, is only moderately transmissible. Its mode of transmission is limited to exchange of body fluids (blood, semen, breast milk, or vaginal secretions). However, the HIV virus is tremendously virulent, and this is what makes it so dangerous.

Imagine what would happen if a disease as transmissible as the flu and as virulent as AIDS were to arise in the human population. In fact, there have been such diseases, and they have caused deadly epidemics. Between 1348 and 1350 **bubonic plague,** a bacterial infection, killed an estimated 25–40% of the European population. A 1918 outbreak of influenza killed more than 20 million people worldwide.

Pathogens continue to challenge human defenses. A prime example is the Ebola virus that arose in Africa in 1976 and still presents a threat today. It is one of the most virulent pathogens known, killing more than 80% of an exposed population in less than two weeks.

Recap Like all cells, bacteria draw their energy and raw materials from their environment. Pathogenic bacteria get the materials they need from living cells, damaging or killing the cells in the process. A virus consists of a single strand of DNA or RNA surrounded by protein. Viruses use their DNA or RNA to force a living cell to make more copies of the virus. Prions are infectious proteins that cause normal proteins to misfold. ■

9.2 The lymphatic system defends the body

As noted in Chapter 8, the **lymphatic system** is closely associated with the cardiovascular system. The lymphatic system performs three important functions:

- It helps maintain the volume of blood in the cardiovascular system.
- It transports fats and fat-soluble vitamins absorbed from the digestive system to the cardiovascular system.
- It defends the body against infection.

In Chapter 8 we briefly described how the lymphatic system helps to maintain blood volume and interstitial fluid volume by returning excess fluid that has been filtered out of the capillaries back to the cardiovascular system. We discuss its role in transporting fats and vitamins when we describe the digestive system, in Chapter 14. In this chapter we turn to the third function: the role of the lymphatic system in protecting us from disease. Most of the cells of the immune system are housed in the lymphatic system, although they can also circulate in blood and enter the interstitial fluid. Here we describe the structural components of the system; in later sections we discuss how specific immune system cells carry out their function.

The basic components of the lymphatic system are a network of lymph vessels throughout the body, the lymph nodes, the spleen, the thymus gland, and the tonsils and adenoids (**Figure 9.3** on the next page).

Lymphatic vessels transport lymph

The lymphatic system begins as a network of small, blind-ended *lymphatic capillaries* in the vicinity of the cells and blood capillaries. Lymph capillaries have wide spaces between overlapping cells. Their structure allows them to take up substances (including bacteria) that are too large to enter a blood capillary.

The fluid in the lymphatic capillaries is *lymph*, a milky body fluid that contains white blood cells, proteins, fats, and the occasional bacterium and virus. Lymphatic capillaries merge to form the *lymphatic vessels*. Like veins, lymphatic vessels have walls consisting of three thin layers, and they contain one-way valves to prevent backflow of lymph. Also like veins, flow in lymphatic vessels is aided by skeletal muscle contractions and pressure changes in the chest during respiration. The lymphatic vessels merge to form larger and larger vessels, eventually creating two major lymphatic ducts: the *right lymphatic duct* and the *thoracic duct*. The two lymph ducts join the subclavian veins near the shoulders, thereby returning the lymph to the cardiovascular system.

MJ's Human Biology Blog

http://www.humanbiologyblog.blogspot.com/

Share   Report Abuse   Next Blog»   Create Blog   Sign In

MJ's Human Biology Blog

A Way to Cure HIV Infection?

Modern HIV treatment drugs suppress an active HIV infection well enough that an HIV-infected person can live a relatively normal life. But they don't cure an HIV infection because some viruses lie dormant inside living cells, out of reach of suppressive drug therapy. For that reason, high-cost suppressive therapy needs to be continued throughout the life of the patient, just so the drug is present whenever viruses do come out of hiding. What is needed is a way to get rid of the latent viruses lying dormant inside cells, once and for all.

A new approach is anti-latent therapy—therapy designed to prevent dormant viruses from staying dormant and hidden. The idea is to force any remaining dormant viruses to become active again so that they can be targeted and killed by suppressive therapy. A combination of anti-latent therapy and suppressive therapy just might wipe out an HIV infection completely. At least, that's the idea. ■

Reference: Richman, Douglas D. *et al.* The Challenge of Finding a Cure for HIV Infection. *Science* 323: 1304–1307, 2009.

Thymus gland

Heart

Lymphocytes mature in thymus.

Nasal passages
Adenoids
Tongue
Tonsils
Trachea

Tonsils protect the throat.

Lymph flow
Lymph node
Lymph vessels

Macrophages cleanse lymph; lymphocytes activate defense mechanisms.

Red pulp
White pulp
Spleen

Blood flow

Spleen removes damaged blood cells and microorganisms from blood.

Blood capillary
Lymphatic capillary
Cells

Lymphatic vessels transport fluid, bacteria, and viruses.

Figure 9.3 The lymphatic system. The lymphatic system consists of a network of lymphatic vessels throughout the body, lymph nodes, the thymus gland, tonsils and adenoids, and the spleen.

Lymph nodes cleanse the lymph

Located at intervals along the lymphatic vessels are small organs called **lymph nodes**. Lymph nodes remove microorganisms, cellular debris, and abnormal cells from the lymph before returning it to the cardiovascular system. There are hundreds of lymph nodes, clustered in the areas of the digestive tract, neck, armpits, and groin (**Figure 9.4**). They vary in diameter from about 1 millimeter to 2.5 centimeters. Each node is enclosed in a dense capsule of connective tissue pierced by lymphatic vessels. Inside each node are connective tissue and two types of white blood cells, *macrophages* and *lymphocytes*, which identify microorganisms and remove them. (Macrophages and lymphocytes are discussed in greater detail in later sections.)

The lymphatic vessels carry lymph into and out of each node (see Figure 9.3). Valves within these vessels ensure that lymph flows only in one direction. As the fluid flows through a node, the macrophages destroy foreign cells by phagocytosis (Chapter 7), and the lymphocytes activate other defense mechanisms. The cleansed lymph fluid flows out of the node and continues on its path to the veins.

The spleen cleanses blood

The largest lymphatic organ, the **spleen,** is a soft, fist-sized mass located in the upper-left abdominal cavity. The spleen is covered with a dense capsule of connective tissue interspersed with smooth muscle cells. Inside the organ are two types of tissue, called *red pulp* and *white pulp*.

The spleen has two main functions: it controls the quality of circulating red blood cells by removing the old and damaged ones, and it helps fight infection. The red pulp contains macrophages that scavenge and break down microorganisms as well as old and damaged red blood cells and platelets. The cleansed blood is then stored in the red pulp. Your body can call on this reserve for extra blood in case of blood loss or a fall in blood pressure, or whenever you need extra oxygen-carrying capacity. The white pulp contains primarily lymphocytes searching for foreign pathogens; it does not store blood. Notice that the main distinction between the spleen and lymph nodes is *which* fluid they cleanse—the spleen cleanses the blood, and the lymph nodes cleanse lymph. Together, they keep the circulating body fluids relatively free of damaged cells and microorganisms.

A number of diseases, such as infectious mononucleosis and leukemia, cause the spleen to enlarge. The swollen spleen can sometimes be felt as a lump in the upper-left abdomen. A strong blow to the abdomen can rupture the spleen, causing severe internal bleeding. In this case surgical removal of the spleen may be necessary to forestall a fatal hemorrhage. We can live without a spleen because its functions are shared by the lymph glands, liver, and red bone marrow. However, people who have had their spleen removed surgically are often a little more vulnerable to infections.

Thymus gland hormones cause T lymphocytes to mature

The **thymus gland** is located in the lower neck, behind the sternum and just above the heart. Encased in connective tissue, the gland contains lymphocytes and epithelial cells. The thymus gland secretes two hormones, thymosin and thymopoietin, that cause certain lymphocytes called *T lymphocytes* (T cells) to mature and take an active role in specific defenses.

The size and activity level of the thymus gland varies with age. It is largest and most active during childhood. During adolescence it stops growing and then slowly starts to shrink. By that time our defense mechanisms are typically well established. In old age the thymus gland may disappear entirely, to be replaced by fibrous and fatty tissue.

Tonsils protect the throat

The *tonsils* are masses of lymphatic tissue near the entrance to the throat. Lymphocytes in the tonsils gather and filter out many of the microorganisms that enter the throat in food or air.

We actually have several tonsils, and some are not readily visible. The familiar tonsils at the back of the throat are the largest and most often infected. When they become infected, the resulting inflammation is called *tonsillitis*. If the infection becomes serious, the tissues can be surgically removed.

Lymphatic tissue called the *adenoids* lies at the back of the nasal passages. The adenoids tend to enlarge during early childhood, but in most people they start to shrink after age 5 and usually disappear by puberty. In some cases they continue to enlarge and obstruct airflow from nose to throat. This can cause mouth breathing, a nasal voice, and snoring.

Figure 9.4 Lymph node. Color scanning electron micrograph (× 1,000) of a section through a lymph node, showing macrophages (pink) and lymphocytes (yellow) lying in wait to attack and destroy foreign and damaged cells. A few red blood cells (red) are also present.

Like the tonsils, the adenoids can be surgically removed if they grow large enough to cause problems.

Recap The lymphatic system helps protect us from disease. Macrophages and lymphocytes within the lymph nodes identify microorganisms and remove them. The spleen removes damaged red blood cells and foreign cells from blood. The thymus gland secretes hormones that help T lymphocytes mature. Cells in the tonsils gather and remove microorganisms that enter the throat. ■

9.3 Keeping pathogens out: The first line of defense

Our bodies have several ways to prevent entry of pathogens. We also have ways to neutralize them, inhibit their growth, or expel them when they do get in. Here are some of the body's defense strategies.

Skin: An effective deterrent

The most important barrier against entry of any pathogen into our bodies is the skin. Skin has four key attributes that make it such an effective barrier: (1) its structure, (2) the fact that it is constantly being replaced, (3) its acidic pH, and (4) the production of an antibiotic by sweat glands.

As described in Chapter 4, the outermost layers of the skin's epidermis consist of dead, dried-out epithelial cells. These cells contain a fibrous protein called *keratin*, which is also a primary component of fingernails and hair. Once the cells have died and the water has evaporated, the keratin forms a dry, tough, somewhat elastic barrier to the entry of microorganisms. A second key feature of skin is that it is continually being renewed throughout life. Dead cells shed from the surface are replaced by new cells at the base of the epidermis. Any pathogens deposited on the surface are shed along with the dead cells. Third, healthy skin has a pH of about 5 to 6, primarily because of the sweat produced by sweat glands. This relatively low (acidic) pH makes skin a hostile environment for many microorganisms. Finally, sweat glands produce a natural antibiotic peptide called *dermicidin* that can kill a range of harmful bacteria.

For further proof of skin's effectiveness as a barrier to infection, look at what happens when you get a tiny cut or scratch in your skin. If the damage reaches the moist layers of living cells underneath the skin, you may see signs of infection in the area within a few days. One of the most critical problems in treating patients with extensive burns is the infections that often result from the loss of the barrier function of skin.

Impeding pathogen entry in areas not covered by skin

Most successful pathogens enter the body at places where we do not have skin. They enter through the mucous membranes that line the digestive, urinary, respiratory, and reproductive tracts, where they can take advantage of moist surfaces in direct contact with living cells. They enter around the eyes or in the

ears. However, even these areas have ways to impede pathogen entry. Defenses include tears, saliva, earwax, digestive and vaginal acids, and mucus, which impede entry; the ability to remove pathogens by vomiting, urination, and defecation; and even competition created by nonpathogenic bacteria that normally live in (and on) the body.

Tears, saliva, and earwax Although we may not think of tears as a defense mechanism, they perform a valuable service by lubricating the eyes and washing away particles. Tears and saliva both contain **lysozyme,** an enzyme that kills many bacteria. In addition, saliva lubricates the delicate tissues inside the mouth so that they do not dry out and crack. It also rinses microorganisms safely from the mouth into the stomach, where most of them are killed by stomach acid. Earwax traps small particles and microorganisms.

Mucus Mucus is a thick, gel-like material secreted by cells at various surfaces of the body, including the lining of the digestive tract and the branching airways of the respiratory system. Microorganisms that come into contact with the sticky mucus become mired and cannot gain access to the cells beneath. In addition, the cells of the airways have tiny hairlike projections, called *cilia*, that beat constantly in a wavelike motion to sweep mucus upward into the throat. There we get rid of the mucus by coughing or swallowing it. Sometimes we remove mucus and microorganisms by sneezing, which is also one of the primary ways we pass microorganisms to other people (**Figure 9.5**).

Digestive and vaginal acids Undiluted digestive acid is strong enough to kill nearly all pathogens that enter the digestive tract on an empty stomach. Only one strain of bacteria, *Helicobacter pylori,* has actually evolved to thrive in the highly acidic environment of the stomach. *H. pylori* is now known to contribute to many cases of stomach ulcers (see Chapter 14).

Vaginal secretions are slightly acidic, too, though not nearly as acidic as stomach secretions.

Figure 9.5 Sneezing removes mucus and microorganisms from the body. Sneezing also promotes the transmission of certain respiratory infections from person to person.

Vomiting, urination, and defecation Vomiting, though unpleasant, is certainly an effective way of ridding the body of toxic or infected stomach contents.

Generally speaking, the urinary system does not have a resident population of bacteria. Urine is usually slightly acidic, and in addition the constant flushing action of urination tends to keep bacterial populations low. Urine pH can vary from fairly acidic to slightly basic, depending on diet. Some physicians advise patients with bladder or urethral infections to drink cranberry juice, which is acidic. The increased acidity of the urine inhibits bacterial growth, and the increased urine volume flushes the bacteria out.

The movement of feces and the act of defecation also help remove microorganisms from the digestive tract. When we become ill, the muscles in the intestinal wall may start to contract more vigorously, and the intestine may secrete additional fluid into the feces. The result is *diarrhea*—increased fluidity, frequency, or volume of bowel movements. Unpleasant though diarrhea may be, mild cases serve a useful function by speeding the removal of pathogens.

Resident bacteria Certain strains of beneficial bacteria normally live in the mucous membranes lining the vagina and the digestive tract. They help control population levels of more harmful organisms by competing successfully against them for food. They may also make the body less vulnerable to pathogens. For example, *Lactobacillus* bacteria in the vagina produce a substance that lowers vaginal pH to levels that many fungi and bacteria cannot tolerate.

One might ask how any beneficial bacteria ever get to the small and large intestine if they have to pass through the stomach first. The answer is that following a meal the stomach contents are not so acidic because food both dilutes and buffers the stomach acid, so some bacteria pass through the stomach with the food we eat.

🔄 **Recap** Various mechanisms create an inhospitable environment for pathogenic microorganisms. Skin is a dry outer barrier. Tears, saliva, earwax, and mucus trap pathogens or wash them away. Acidic conditions kill them or inhibit their growth; urination, defecation, and vomiting forcibly expel them; and resident bacteria compete with pathogens for food. ■

9.4 Nonspecific defenses: The second line of defense

If pathogens manage to breach our physical and chemical barriers and start to kill or damage cells, we have a problem of a different sort. Now the body must actively seek out the pathogens and get rid of them. It must also clean up the injured area and repair the damage.

Our second line of defense includes a varied group of defense mechanisms. We refer to them as *nonspecific* because they do not target specific pathogens. Instead, they appear in response to all types of health challenges without discriminating between them. Table 9.1 (next page) summarizes nonspecific defenses, which include phagocytes, natural killer cells, the inflammatory response, the complement system, interferons, and fever.

Phagocytes engulf foreign cells

As noted in Chapter 7, *phagocytes* are white blood cells that destroy foreign cells through the process of **phagocytosis.** As illustrated in Figure 9.6, a phagocyte first captures a bacterium

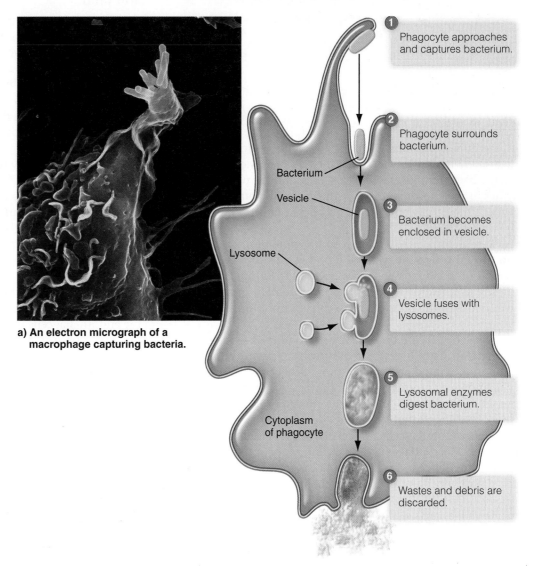

a) **An electron micrograph of a macrophage capturing bacteria.**

① Phagocyte approaches and captures bacterium.

② Phagocyte surrounds bacterium.

Bacterium

Vesicle

③ Bacterium becomes enclosed in vesicle.

Lysosome

④ Vesicle fuses with lysosomes.

⑤ Lysosomal enzymes digest bacterium.

Cytoplasm of phagocyte

⑥ Wastes and debris are discarded.

b) **Steps in the process.**

Figure 9.6 **Phagocytosis.**

Table 9.1 The second line of defense: Nonspecific defense

Defense	Functions
Phagocytes	Neutrophils and macrophages engulf and digest foreign cells.
	Eosinophils bombard large parasites with digestive enzymes and phagocytize foreign proteins.
Natural killer cells	Release chemicals that disintegrate cell membranes of tumor cells and virus-infected cells.
Inflammatory response	Four components include redness, warmth, swelling, and pain.
	Attracts phagocytes and promotes tissue healing.
Complement system	A group of proteins that assists other defense mechanisms. Enhances inflammation and phagocytosis, kills pathogens.
Interferons	Stimulate the production of proteins that interfere with viral reproduction.
Fever	Modest fever makes internal environment less hospitable to pathogens, fosters ability to fight infections.

with its cytoplasmic extensions. Then it draws the bacterium in, eventually engulfing it (endocytosis) and enclosing it in a membrane-bound vesicle. Inside the cell, the vesicle containing the bacterium fuses with lysosomes, and the powerful enzymes in the lysosomes dissolve the bacterial membranes. Once digestion is complete the phagocyte jettisons the bacterial wastes by exocytosis.

Recall from Chapter 7 that some white blood cells can filter through the walls of blood vessels into tissue spaces, attracted by substances released by injured cells at the site of infection. Other phagocytes remain in connective tissues of the lymph nodes, spleen, liver, lungs, and brain.

Neutrophils are the first white blood cells to respond to infection. They digest and destroy bacteria and some fungi in the blood and tissue fluids. Other white blood cells known as monocytes leave the vascular system, enter the tissue fluids, and develop into **macrophages** (from the Greek for "large eater") that can engulf and digest large numbers of foreign cells, especially viruses and bacterial parasites. Macrophages serve a cleanup function by scavenging old blood cells, dead tissue fragments, and cellular debris. They also release chemicals that stimulate the production of more white blood cells. Technically, macrophages are no longer white blood cells because they are no longer in blood.

When invaders are too big to be engulfed and digested by phagocytosis, other white blood cells called **eosinophils** take action. Eosinophils cluster around large parasites such as flukes and pinworms and bombard them with digestive enzymes. Eosinophils also engulf and digest certain foreign proteins.

When the body is actively fighting an infection, the mortality rate of white blood cells rises dramatically. Tissue fluid, dead phagocytes and microorganisms, and cellular debris accumulate at the infection site, producing a characteristic discharge called *pus.* If pus becomes trapped and cannot drain, the body may wall it off with connective tissue. The result is an *abscess.* Common places for abscesses to form include the breast (mastitis), the gums (dental abscesses), and more rarely the liver or brain. Many abscesses subside after being drained, while others require antibiotic drugs or surgical removal.

Inflammation: Redness, warmth, swelling, and pain

Any type of tissue injury—whether infection, burns, irritating chemicals, or physical trauma—triggers a series of related events collectively called the *inflammatory response,* or **inflammation.** Inflammation has four outward signs: redness, warmth, swelling, and pain. Although these may not sound like positive developments, the events that cause these signs prevent the damage from spreading, dispose of cellular debris and pathogens, and set the stage for tissue-repair mechanisms.

The inflammatory response starts whenever tissues are injured (**Figure 9.7**). The release of chemicals from damaged cells sounds the alarm for the process to begin. These chemicals stimulate **mast cells,** which are connective tissue cells specialized to release **histamine.** Histamine promotes vasodilation of neighboring small blood vessels. White blood cells called **basophils** also secrete histamine.

Recall that most white blood cells are too large to cross capillary walls. As histamine dilates blood vessels, however, the endothelial cells in vessel walls pull slightly apart, and the vessels become more permeable. This allows additional phagocytes to squeeze through capillary walls into the interstitial fluid. There they attack foreign organisms and damaged cells. After destroying pathogens, some phagocytes travel to the lymphatic system, where their presence activates lymphocytes to initiate specific defense mechanisms (discussed later).

Vasodilation brings more blood into the injured area, making it red and warm. The rising temperature increases phagocyte activity. The increased leakiness of capillary walls allows more fluid to seep into tissue spaces, causing swelling. The extra fluid dilutes pathogens and toxins and brings in clotting proteins that form a fibrin mesh to wall off the damaged

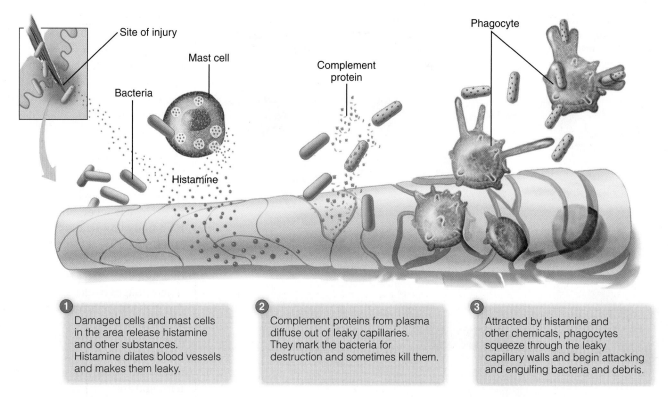

Figure 9.7 **The inflammatory response.**

1. Damaged cells and mast cells in the area release histamine and other substances. Histamine dilates blood vessels and makes them leaky.

2. Complement proteins from plasma diffuse out of leaky capillaries. They mark the bacteria for destruction and sometimes kill them.

3. Attracted by histamine and other chemicals, phagocytes squeeze through the leaky capillary walls and begin attacking and engulfing bacteria and debris.

area from healthy tissue. As a bonus, the fluid carries in extra oxygen and nutrients to promote tissue healing and carries away dead cells, microorganisms, and other debris from the area.

Swollen tissues press against nearby nerve endings. This swelling, plus the sensitizing effects of inflammatory chemicals, creates the sensation of pain that accompanies inflammation. However, even pain can be positive. The discomfort hinders active movement and forces the injured person to rest, facilitating the healing process.

✅ **Quick Check** Antihistamines are drugs that block the effect of histamine. Why do antihistamines help alleviate the "stuffy nose" and nasal congestion of a cold? ▪

HBP **Web Animation** *The Inflammatory Response* at www.humanbiology.com

Natural killer cells target tumors and virus-infected cells

Natural killer (NK) cells are a group of white blood cells (lymphocytes) that destroy tumor cells and cells infected by viruses. NK cells are able to recognize certain changes that take place in the plasma membranes of tumor cells and virus-infected cells. The name "natural killer" reflects the fact that NK cells are nonspecific killers, unlike other killer cells discussed later in this chapter that target only specific enemies.

NK cells are not phagocytes. Instead, they release chemicals that break down their targets' cell membranes. Soon after an NK attack, the target cell's membrane develops holes, and its nucleus rapidly disintegrates. NK cells also secrete substances that enhance the inflammatory response.

The complement system assists other defense mechanisms

The **complement system,** or *complement,* comprises at least 20 plasma proteins that circulate in the blood and complement, or assist, other defense mechanisms. Normally these proteins circulate in an inactive state. When activated by the presence of an infection, however, they become a potent defense force. Once one protein is activated it activates another, leading to a cascade of reactions. Each protein in the complement system can activate many others, creating a powerful "domino effect."

Some activated complement proteins join to form large protein complexes that create holes in bacterial cell

1 Activated complement proteins form complexes of proteins that create holes in the bacterial cell wall.

Complement proteins

Cell wall of bacteria

2 Water and salts diffuse into the bacterium through the holes.

Water and salts

Bacterium

3 The bacterium swells and eventually bursts.

Photomicrograph of an intact bacterium

A bacterium after lysis by activated complement proteins

Figure 9.8 How activated complement proteins kill bacteria.

walls. Fluids and salts leak in through these holes, until eventually the bacterium swells and bursts (**Figure 9.8**). Other activated complement proteins bind to bacterial cell membranes, marking them for destruction by phagocytes. Still others stimulate mast cells to release histamine or serve as chemical attractants to draw additional phagocytes to the infection.

Interferons interfere with viral reproduction

One of the most interesting defense mechanisms is an early warning system between virus-infected and still-healthy cells. As mentioned earlier, viruses cannot reproduce on their own. Instead, they invade body cells and use the cells' machinery to make more viruses.

Cells that become infected by viruses secrete a group of proteins called **interferons**. Interferons diffuse to nearby healthy cells, bind to their cell membranes, and stimulate the healthy cells to produce proteins that interfere with the synthesis of viral proteins, making it harder for the viruses to infect the protected cells.

Interferons are now being produced in pharmaceutical laboratories. At least one interferon protein (*alpha interferon*) has shown promise against certain viral diseases, including genital warts, hepatitis B, and one form of leukemia.

Fever raises body temperature

A final weapon in our second line of defense is **fever**, an abnormally high body temperature. Your body's "thermostat" is set to approximately 98.6°F (37°C), with a normal range of about 97–99°F (36–37.2°C). When macrophages detect and attack bacteria, viruses, or other foreign substances, they release certain chemicals into the bloodstream. These chemicals, called *pyrogens,* cause the brain to reset your thermostat to a higher temperature.

There is a tendency to treat all fevers as if they were a problem. But a modest fever may be beneficial because it makes our internal environment less hospitable to pathogens and enhances the body's ability to fight infections. Fever increases the metabolic rate of body cells, speeding up both defense mechanisms and tissue-repair processes. When the infection is gone, the process reverses. Macrophages stop releasing pyrogens, the thermostat setting returns to normal, and your fever "breaks."

However, high fevers can be dangerous. As described in Chapter 2, the chemical bonds that give a protein its shape are relatively weak. Consequently, the shape (and function) of some proteins can be affected by high temperatures. It's a good idea to monitor the course of any fever, particularly in children and older adults. Health professionals recommend seeking medical advice for any fever that lasts longer than two days or rises above 100°F.

Recap Nonspecific defense mechanisms involve a general attack against all foreign and damaged cells. Neutrophils and macrophages engulf and digest bacteria and damaged cells, and eosinophils bombard larger organisms (too large to be engulfed) with digestive enzymes. The inflammatory response attracts phagocytes and promotes tissue healing. Interferons interfere with viral reproduction, and a modest fever enhances our ability to fight infections. ■

9.5 Specific defense mechanisms: The third line of defense

Even if foreign cells manage to bypass physical and chemical barriers and overcome nonspecific defenses, they must still cope with the body's third line of defense, the most sophisticated weapon of all. The *immune system* comprises cells, proteins, and the lymphatic system, all working together to detect and kill particular pathogens and abnormal body cells. The activities of the immune system are collectively called the **immune response.** Because the immune system targets specific enemies, we refer to these operations as *specific defense mechanisms.*

The immune response has three important characteristics:

- It recognizes and targets specific pathogens or foreign substances.
- It has a "memory," the capability to store information from past exposures so that it can respond more quickly to later invasions by the same pathogen.
- It protects the entire body; the resulting immunity is not limited to the site of infection.

The key to specific defenses is the body's ability to distinguish between its own cells and those of foreign invaders. Among its own cells, it must also be able to distinguish between those that are healthy, those that are abnormal (such as cancer cells), and those that are dead or dying.

The immune system targets antigens

An **antigen** is any substance that mobilizes the immune system and provokes an immune response. Generally antigens are large protein or polysaccharide molecules. In much the same way that a key fits a lock, each antigen has a unique shape, and every bacterium or virus has a different one. The immune system responds to each uniquely shaped antigen by producing specific *antibodies* to attack and inactivate the antigen (and the cell carrying it).

All antigens are located only on the outer surface of a cell or virus. Hence the immune system cannot detect viruses (or viral DNA/RNA) once they are safely inside a living human cell.

Human cells also have surface proteins that can act as antigens under the right circumstances. Your cells have a unique set of proteins on their surfaces that your immune system uses to recognize that the cells belong to you. These *self markers* are known as **major histocompatibility complex (MHC) proteins.** Your MHC proteins are unique to you by virtue of your unique set of genes. Normally they signal your immune system to bypass your own cells. They are a sort of password, the equivalent of a cellular fingerprint. Your immune system "reads" the password and leaves your cells alone.

However, the same MHC proteins that define your cells as belonging to you would be read as *nonself markers* in another person. In other words, your MHC proteins would be antigens in another person. Abnormal and cancerous cells in your own body also have MHC proteins that are not recognized as "self." The immune system targets all antigens,

including those on pathogens and foreign and damaged human cells, for destruction.

 Quick Check Why does pregnancy pose a problem for the mother's immune system? ▪

Lymphocytes are central to specific defenses

Lymphocytes play crucial roles in our specific defense mechanisms. As described in Chapter 7 (see Figure 7.5), lymphocytes are white blood cells originating from stem cells in bone marrow. They are fairly small white blood cells with a single nucleus that fills nearly the entire cell. They total about 30% of circulating white blood cells. Lymphocytes are found in the bloodstream, tonsils, spleen, lymph nodes, and thymus gland.

There are two types: B lymphocytes and T lymphocytes, also called **B cells** and **T cells.** (Their names are based on where they mature: B cells mature in bone marrow; T cells in the thymus gland.) Although both types of lymphocytes can recognize and target antigen-bearing cells, they go about this task in different ways.

B cells are responsible for **antibody-mediated immunity.** B cells produce **antibodies**—proteins that bind with and neutralize specific antigens. They release antibodies into the lymph, bloodstream, and tissue fluid, where they circulate throughout the body. Antibody-mediated immunity works best against viruses, bacteria, and foreign molecules that are soluble in blood and lymph.

T cells are responsible for **cell-mediated immunity,** which depends on the actions of several types of T cells. Unlike B cells, T cells do not produce antibodies. Instead, some T cells directly attack foreign cells that carry antigens. Other T cells release proteins that help coordinate other aspects of the immune response, including the actions of T cells, B cells, and macrophages. Cell-mediated immunity protects us against parasites, bacteria, viruses, fungi, cancerous cells, and cells perceived as foreign (including, unfortunately, transplanted tissue—see section 9.8, Tissue rejection: A medical challenge). T cells can identify and kill infected human cells even before the cells have a chance to release new bacteria or viruses into the blood.

 Quick Check Suppose a baby were born without a functional thymus gland. Could this child still produce antibodies? Explain your answer. ▪

B cells: Antibody-mediated immunity

In adults, B cells mature in the bone marrow. As they mature, they develop unique surface receptors (with the same structure as an antibody) that allow them to recognize specific antigens. Then they travel in the bloodstream to the lymph nodes, spleen, and tonsils, where they remain inactive until they encounter a foreign cell with that particular antigen. When a B cell with just the right surface receptor encounters the appropriate antigen, its surface receptors bind to the antigen (**Figure 9.9**). This activates the B cell to grow and then multiply rapidly, producing more B cells exactly like the original and bearing the same surface receptors. The resulting identical cells, all descended from the same cell, are called *clones.*

Although the B cells themselves tend to remain in the lymphatic system, most of the cells of the clone are called **plasma cells** because they begin to secrete their antibodies into the lymph fluid and ultimately into the blood plasma.

A typical plasma cell can make antibody molecules at a staggering rate—about 2,000 molecules per second. A plasma cell maintains this frantic pace for a few days and then dies, but its antibodies continue to circulate in blood and lymph.

When the antibodies encounter matching antigens, they bind to them and create an *antigen-antibody complex.* Antibodies specialize in recognizing certain proteins; thus one particular antibody can bind to one particular antigen (**Figure 9.10**). The formation of an antigen-antibody complex marks the antigen (and the foreign cell that carries it) for destruction either by activated complement proteins or by phagocytes. Some antibodies also inactivate pathogens by causing the cells to agglutinate (clump together), preventing them from entering human cells and causing disease.

Some of the clone cells become **memory cells,** long-lived cells that remain inactive until that same antigen reappears in the body at some future date. Memory cells store information about the pathogen; if there is a second exposure, the immune response is even faster than the first time. Upon exposure, these memory cells quickly become plasma cells

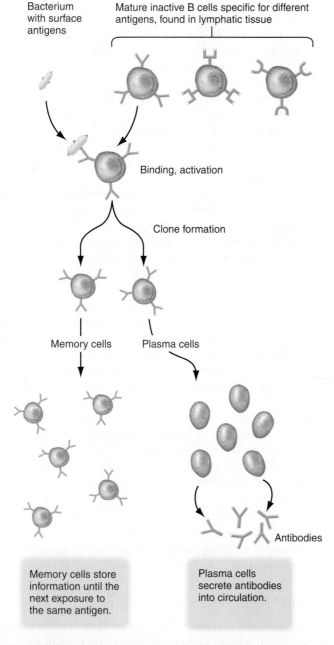

Figure 9.9 The production of antibodies by B cells. The surface antigen of a pathogen binds to the matching receptor on a mature, inactive B cell in lymphatic tissue. The B cell becomes activated and grows larger. The enlarged cell begins to divide rapidly, forming a clone. Some of the clone cells become memory cells; others become plasma cells. Memory cells lie in wait for the next exposure to the antigen. Plasma cells secrete antibodies into the lymph fluid.

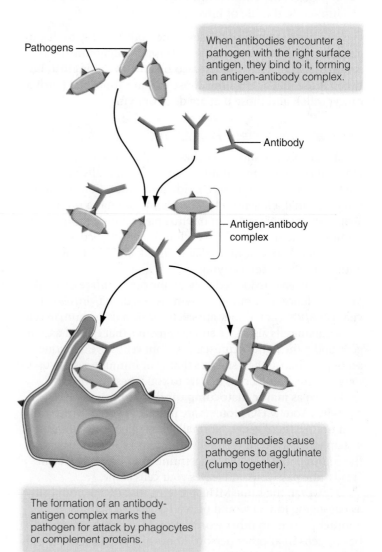

Figure 9.10 How antibodies inactivate pathogens.

and start to secrete antibodies. Memory cells are the basis for long-term immunity.

The five classes of antibodies

Antibodies belong to the class of blood plasma proteins called *gamma globulins.* Because they play such a crucial role in immunity, the term **immunoglobulin (Ig)** is often used. There are five classes of immunoglobulins, each designated by a different letter: IgG, IgM, IgA, IgD, and IgE. Each type has a different size, location in the body, and function:

- *IgG (75% of immunoglobulins).* This is the most common class. Found in blood, lymph, intestines, and tissue fluid, the long-lived IgG antibodies activate the complement system and neutralize many toxins. They are the only antibodies that cross the placenta during pregnancy and pass on the mother's acquired immunities to the fetus.
- *IgM (5–10%).* IgM antibodies are the first to be released during immune responses. Found in blood and lymph, they activate the complement system and cause foreign cells to agglutinate. ABO blood cell antibodies belong to this class.
- *IgA (15%).* IgA antibodies enter areas of the body covered by mucous membranes, such as the digestive, reproductive, and respiratory tracts. There they neutralize infectious pathogens. They are also present in mother's milk and are transmitted to the infant during breast-feeding.
- *IgD (less than 1%).* IgD antibodies are in blood, lymph, and B cells. Their function is not clear, but they may play a role in activating B cells.
- *IgE (approximately 0.1%).* The rarest of the immunoglobulins, IgE antibodies are in B cells, mast cells, and basophils. They activate the inflammatory response by triggering the release of histamine. They are also the troublemakers behind allergic responses (covered in section 9.9).

Antibodies' structure enables them to bind to specific antigens

An antigen provides all the information the immune system needs to know about a foreign substance. Essentially, antigens are the locks on an enemy's doors. The immune system can identify the lock, produce the antibody key, and then send in immune system cells to open the door and neutralize the invader. How does this happen?

All antibodies share the same basic structure, represented by an IgG antibody (**Figure 9.11**). Each IgG antibody (or surface receptor, if it is attached to a B cell) consists of four linked polypeptide chains arranged in a Y shape. The two larger chains are called "heavy" chains, and the two smaller ones are called "light" chains. Each of the four chains has a constant region that forms the trunk and two branches and a variable region that represents the antigen-binding site. Because it has a unique amino acid

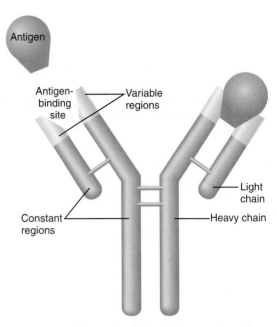

Figure 9.11 Structure of an antibody (or B cell surface receptor). An antibody consists of four peptide chains linked together to form a Y shape. Only the lighter-colored ends of each chain vary. These variations determine the specificity of each antibody for only one antigen.

sequence, each variable region has a unique shape that fits only one specific antigen.

The constant regions are similar for all antibodies in one class, although they differ from those of other classes. The IgG antibody depicted in Figure 9.11 is a single Y-shaped molecule with two binding sites. Larger antibodies in the IgM class consist of five Y-shaped molecules linked together, with 10 binding sites.

☑ **Quick Check** Explain, in terms of details of antibody structure, why a memory cell that protects you against hepatitis cannot protect you against the common cold. ■

T cells: Cell-mediated immunity

There are two basic functional differences between B cells and T cells. First, B cells produce circulating antibodies; T cells either release chemicals that stimulate other cells of the immune system, or they directly attack the foreign cell and kill it. Second, T cell receptors cannot recognize whole antigens—they only react to small fragments of antigens. Thus it is necessary for a certain amount of antigen processing to occur before T cells can do their job. The antigen must be presented to them in a form that they can recognize.

Certain cells—including macrophages and activated B cells—fulfill this role by acting as **antigen-presenting cells (APCs)** that engulf foreign particles, partially digest them, and display fragments of the antigens on their surfaces

(Figure 9.12). After an APC engulfs a pathogen, it partially degrades the antigen inside a vesicle. The vesicle containing antigen fragments joins with another vesicle containing MHC molecules, the molecules that become the cell's self markers. The MHC binds to the antigen pieces and moves to the cell surface, to be displayed as an antigen-MHC complex. Essentially, the cell "presents" a fragment of the antigen for T cells to recognize, along with its own cell-surface self marker.

T cells develop from stem cells in bone marrow but migrate to the thymus gland, where they become mature but remain inactive. During maturation they also develop one of two sets of surface proteins, CD4 or CD8. These proteins determine what type of T cell they will become. CD4 T cells will become helper and memory cells, and CD8 T cells will become cytotoxic and suppressor cells.

Helper T cells stimulate other immune cells When a T cell with CD4 receptors encounters an antigen-presenting cell displaying a fragment of an antigen, it differentiates into a **helper T cell** (Figure 9.13). The new helper T cell undergoes mitosis (see Chapter 17), quickly producing a clone of identical helper T cells. Because all the cells in the clone carry the same receptors, they all recognize the same antigen.

Most of the cells in the helper T cell clone begin secreting a class of signaling molecules called **cytokines.** Among them are proteins that stimulate the actions of T cells and macrophages, and substances that promote development of other immune cells. Cytokines released by helper T cells stimulate other immune cells such as phagocytes, natural

Figure 9.13 **The activation and clonal expansion of helper T cells.** Activation occurs when a mature but inactive helper T cell comes in contact with an antigen-presenting cell displaying the appropriate antigen fragment. After clonal expansion, the clones produce cytokines. A few activated clones become memory T cells.

 Why does the clonal expansion occur? That is, why can't the activated T cells simply start producing cytokines without undergoing clonal expansion?

Figure 9.12 **How a macrophage acts as an antigen-presenting cell (APC).**

Major histocompatibility complex protein (MHC)

Antigen

Pathogen

① The macrophage engulfs a pathogen.

Lysosome

Vesicle with MHC molecules

② Lysosomes partially digest the pathogen.

③ A vesicle containing MHC molecules binds to the digestive vesicle.

④ The MHC molecules and a fragment of the antigen form an antigen-MHC complex.

⑤ The antigen-MHC complex is displayed on the surface of the cell when the vesicle fuses with the cell membrane and releases its digestive products.

Antigen-MHC complex

 What other immune system cells will probably be stimulated by the antigen fragments presented by this macrophage? What will these cells do in response?

killer cells, and T cells with CD8 receptors. They also attract other types of white blood cells to the area, enhancing non-specific defenses. They even activate B cells, creating an important link between antibody-mediated and cell-mediated immunity.

The role of helper T cells and their cytokines in directing the activities of other immune cells is crucial to an effective immune response—without them the immune response would be severely impaired or nonexistent. The reason AIDS is so devastating is that HIV destroys helper T cells, and thus weakens the body's ability to mount a cell-mediated immune response to a wide variety of other diseases. (See section 9.10, Immune deficiency: The special case of AIDS.)

Cytotoxic T cells kill abnormal and foreign cells When a mature T cell with CD8 receptors meets an antigen-producing cell that displays an antigen fragment, the T cell begins to produce a clone of **cytotoxic T cells** (also called *killer T cells*). These are the only T cells that directly attack and destroy other cells (Figure 9.14).

Once activated, cytotoxic T cells roam throughout the body. They circulate through blood, lymph, and lymphatic tissues in search of cells that display the antigens they recognize. Or they may migrate to a tumor or site of infection, where they release chemicals that are toxic to abnormal cells.

Figure 9.15 (next page) illustrates cytotoxic T cells in action. When a cytotoxic T cell locates and binds to a target cell, secretory vesicles release a protein called *perforin* into the space between the two cells. The perforin molecules assemble themselves into a pore in the target cell, allowing water and salts to enter. That alone should eventually kill the cell in much the same way that activated complement protein does (review Figure 9.8). But just to make sure, the cytotoxic T cell also releases granzyme, a toxic enzyme that is small enough to pass through the pore. The cytotoxic T cell then detaches from the target cell and goes off in search of other prey.

Several promising medical treatments involve harnessing the defensive capabilities of cytokines, specifically interferons. Genetically engineered gamma interferon is used to treat the chronic viral disorder hepatitis C. Another type of interferon has been moderately successful for treating multiple sclerosis, and a third is being used to treat certain types of cancer.

Memory T cells reactivate during later exposures Some activated T cells become memory cells, retaining receptors for the antigen that originally stimulated their production. If that antigen is presented to them again, the memory cells are reactivated. Some form new helper T cells that multiply quickly to marshal an immune response. Others form a new army of cytotoxic T cells to attack and destroy. Like memory B cells, memory T cells are an important factor that distinguishes specific defenses from nonspecific defense mechanisms.

Table 9.2 (next page) summarizes the various cells and proteins involved in specific defense mechanisms.

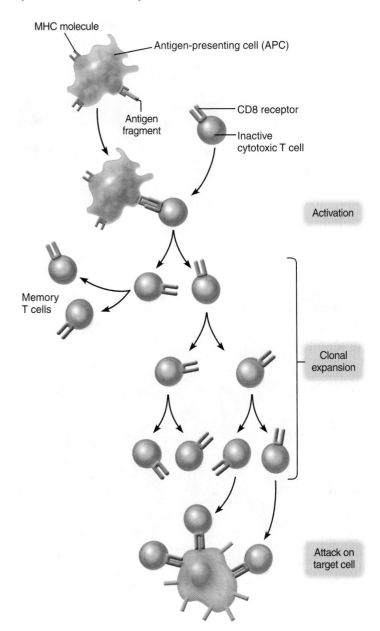

Figure 9.14 The activation and clonal expansion of cytotoxic T cells. Following activation and clonal expansion, the clones directly attack and kill cells carrying the antigen fragment they recognize.

↩ **Recap** An antigen is any substance that provokes an immune response. When activated by first exposure to a specific antigen, lymphocytes called B cells quickly produce antibodies against the antigen. They also produce a few long-lived memory cells that remain inactive until the next exposure to the same antigen. Other lymphocytes called T cells mature in the thymus gland. Helper T cells stimulate other immune cells, cytotoxic T cells attack abnormal and foreign cells, and memory T cells store information until the next exposure to the same antigen. ■

Cytotoxic T cell

Target cell

a) Cytotoxic T cells (blue) attaching to a target cell (pink).

Cytotoxic T cell

Vesicle

Perforin

Granzyme

Cytotoxic T cell membrane

Intercellular space

1 Intact target cell membrane

2 Perforin pore partially assembled

3 Completed pore; granzyme passing through

Target cell

b) How cytotoxic T cells kill a target cell.

Figure 9.15 **Cell-mediated immunity in action.**

 Web Animation *Antibody- and Cell-Mediated Immunity* at www.humanbiology.com

9.6 Immune memory creates immunity

When you are first exposed to an antigen, your immune system protects you with the wealth of defense mechanisms described so far. Your first exposure to a particular antigen generates a *primary immune response*. As we have seen, this involves recognition of the antigen, and production and proliferation of B and T cells.

Typically the primary immune response has a lag time of three to six days after the antigen first appears. During this period B cells specific to that antigen multiply and develop

Table 9.2 Cells and proteins involved in specific defenses

Cell/Protein	Function
B cells	Mature in bone marrow. Responsible for antibody-mediated immunity.
Plasma cells	Produce and secrete specific antibodies.
Memory B cells	Store information. Upon subsequent exposure, become plasma cells and secrete antibodies.
Immunoglobulins	Five classes of antibodies. Every antibody has a unique shape that fits one specific antigen.
T cells	Mature in thymus. Responsible for cell-mediated immunity.
Helper T cells	Produce cytokines. Enhance immune responses by stimulating other immune cells.
Cytotoxic T cells	Attack and destroy abnormal cells.
Memory T cells	Store information. Upon subsequent exposure, become helper and cytotoxic T cells.
Cytokines	A class of signaling molecules that stimulate various immune system activities.

into plasma cells. Antibody concentrations rise, typically reaching their peak about 10–12 days after first exposure. Then they start to level off (**Figure 9.16**).

However, as you have learned, B and T cells create a population of memory cells. The presence of these memory cells is the basis for **immunity** from disease. (The Latin word *immunis* means "safe" or "free from.") Subsequent exposure to the pathogen elicits a *secondary immune response* that is faster, longer lasting, and more effective than the first. Within hours after second exposure to an antigen, memory cells bind to the pathogen. New armies of T and plasma cells form, and within a few days antibody concentrations rise rapidly to much higher levels than in the primary response. Notice that antibody levels remain much higher in the body after second exposure.

Memory cells are long-lived, and many retain their ability to generate a secondary immune response over a lifetime. The secondary immune response can be so effective that you don't even realize you've been exposed to the pathogen a second time. At worst, you may experience only a fleeting sensation of feeling unwell. Other memory cells, such as the ones for the bacterial infection that causes tetanus, need to be reactivated every 10 years or so.

Given this immunity, though, why is it possible to get a cold or the flu over and over, sometimes several times a year? One reason is that there are more than 100 different viruses that can cause colds and flu. Even if your latest respiratory ailment feels like the previous one, it may actually be due to an entirely different pathogen. Furthermore, the viruses that cause

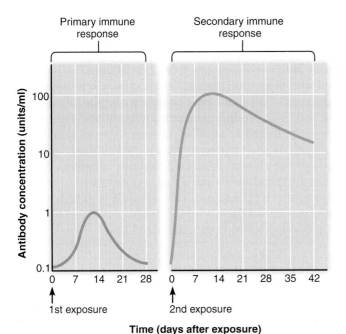

Figure 9.16 The basis of immunity. The antibody response to a first exposure to antigen (the primary immune response) declines in about a month. A second exposure to the same antigen results in a response that is more rapid in onset, larger, and longer lasting. On this graph the abscissa scale is broken because it may be months or even years before second exposure. Note that the concentration of antibody during a secondary response may be 100 times higher than the primary response.

☑ *Suppose a person lacks the capability to produce any memory cells, but otherwise has a normal immune system. Sketch new lines on each of the graphs showing what this person's antibody response will lack during a first exposure and a second exposure.*

colds and flu evolve so rapidly that they are essentially different each year. Their antigens change enough that each one requires a different antibody, and each exposure triggers a primary response. Rapid evolution is their survival mechanism. Our survival mechanism is a good immune system.

↺ **Recap** First exposure to a specific antigen generates a primary immune response. Subsequent exposure to the same antigen elicits a secondary immune response that is faster, longer lasting, and more effective than the primary immune response. ■

9.7 Medical assistance in the war against pathogens

Our natural defenses against pathogens are remarkable. Nevertheless, we humans have taken matters into our own hands by developing the science of medicine. We have been able to conceive of and produce other sophisticated weaponry to help us combat pathogens. Important milestones in human health include the development of active and passive methods of **immunization,** which help the body resist specific pathogens; the production of monoclonal antibodies; and the discovery of antibiotics.

Active immunization: An effective weapon against pathogens

It is said that an ounce of prevention is worth a pound of cure, and this is certainly true when dealing with pathogens. The best weapon against them is to give the body a dose of the antigen in advance so that the immune system will mount a primary immune response against it. Then, if exposed to that microorganism in the environment, the body is already primed with the appropriate antibodies and memory cells. The immune system can react swiftly with a secondary response, effectively shielding you from the danger of the disease and discomfort of its symptoms.

The process of activating the body's immune system in advance is called **active immunization**. This involves administering an antigen-containing preparation called a **vaccine.** Most vaccines are produced from dead or weakened pathogens. An example is the oral polio vaccine (the Sabin vaccine), made from weakened poliovirus. Other vaccines are made from organisms that have been genetically altered to produce a particular antigen.

Of course, vaccines created from dead or weakened pathogens have their limitations. First, there are issues of safety, time, and expense. Living but weakened pathogens generally make better vaccines because they elicit a greater immune response. However, a vaccine that contains weakened pathogens has a slight potential to cause disease itself. This has happened, although rarely, with the polio vaccine. It takes a great deal of time, money, and research to verify the safety and effectiveness of a vaccine. Second, a vaccine confers immunity against only one pathogen, so a different vaccine is needed for every virus. This is why doctors may recommend getting a flu vaccine each time a new flu strain appears (nearly every year). Third, vaccines are not particularly effective after a pathogen has struck; that is, they do not cure an already existing disease.

Nonetheless, vaccines are an effective supplement to our natural defense mechanisms. Active immunization generally produces long-lived immunity that can protect us for many years. The widespread practice of vaccination has greatly reduced many diseases such as polio, measles, and whooping cough.

In the United States, immunization of adults has lagged behind that of children. It's estimated that more than 50,000 Americans die each year from infections, including pneumonia, hepatitis, and influenza, that could have been prevented with timely vaccines.

In many countries, vaccines are too costly and difficult to administer: generally they must be injected by a health care worker with some basic level of training. To get around this problem, some researchers are developing potatoes or bananas that are genetically modified to produce vaccines against diseases such as hepatitis B, measles, and the diarrhea-causing Norwalk virus. An oral vaccine against Norwalk virus has already undergone limited testing in humans.

Passive immunization can help against existing or anticipated infections

To fight an existing or even anticipated infection, a person can be given antibodies prepared in advance from a human or animal donor with immunity to that illness. Usually this takes the form of a *gamma globulin* shot (serum containing primarily IgG antibodies). The procedure is called **passive immunization.** In essence, the patient is given the antibodies that his/her own immune system might produce if there were enough time.

Passive immunization has the advantage of being somewhat effective against an existing infection. It can be administered to prevent illness in someone who has been unexpectedly exposed to a pathogen, and it confers at least some short-term immunity. However, protection is not as long-lasting as active immunization following vaccine administration because the administered antibodies disappear from the circulation quickly. Passive immunization also can't confer long-term immunity against a second exposure, because the person's own B cells aren't activated and so memory cells for the pathogen do not develop.

Passive immunization has been used effectively against certain common viral infections including those that cause hepatitis B and measles, bacterial infections such as tetanus, and Rh incompatibility. Passive immunization of the fetus and newborn also occurs naturally across the placenta and through breast-feeding.

✅ **Quick Check** In New York City in the late 1800s, children ill with diphtheria were often given an extract from the blood of horses that had previously had diphtheria. Was this an example of active or passive immunization, and did it provide permanent or temporary protection? ▪

Monoclonal antibodies: Laboratory-created for commercial use

An antibody preparation used to confer passive immunity in a patient is actually a mixture of many different antibody molecules, because a single pathogen can have many different antigens on its surface. **Monoclonal antibodies,** on the other hand, are antibodies produced in the laboratory from cloned descendants of a single hybrid B cell. As such, monoclonal antibodies are relatively pure preparations of antibodies specific for a single antigen. Monoclonal antibodies are proving useful in research, testing, and cancer treatments because they are pure and they can be produced cheaply in large quantities.

Figure 9.17 summarizes a technique for preparing monoclonal antibodies. Typically, B cells are removed from a mouse's spleen after the mouse has been immunized with a specific antigen to stimulate B cell production. The B cells are fused with myeloma (cancer) cells to create *hybridoma* (hybrid cancer) cells that have desirable traits of both parent cells: they each produce a specific antibody, and they proliferate with cancer-like rapidity. As these hybridoma cells grow in culture, those that produce the desired antibody are separated out and cloned, producing millions of copies. The antibodies they produce are harvested and processed to create preparations of pure monoclonal antibodies. (The term *monoclonal* means these antibody molecules derive from a group of cells cloned from a single cell.)

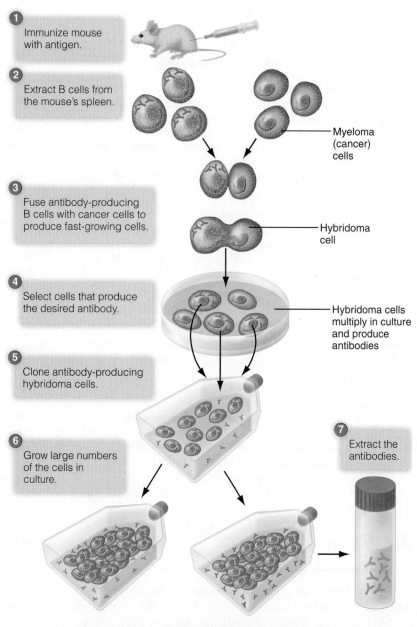

1 Immunize mouse with antigen.

2 Extract B cells from the mouse's spleen.

Myeloma (cancer) cells

3 Fuse antibody-producing B cells with cancer cells to produce fast-growing cells.

Hybridoma cell

4 Select cells that produce the desired antibody.

Hybridoma cells multiply in culture and produce antibodies

5 Clone antibody-producing hybridoma cells.

6 Grow large numbers of the cells in culture.

7 Extract the antibodies.

Figure 9.17 How monoclonal antibodies are produced.

Monoclonal antibodies have a number of commercial applications, including home pregnancy tests, screening for prostate cancer, and diagnostic testing for hepatitis, influenza, and HIV/AIDS. Monoclonal antibody tests tend to be more specific and more rapid than conventional diagnostic tests. In the future it may be possible to use monoclonal antibodies to deliver anticancer drugs directly to cancer cells. The first step would be to bond an anticancer drug to monoclonal antibodies prepared against the cancer. Upon injection into the patient, the antibodies would deliver the drug directly to the cancer cells, sparing nearby healthy tissue.

Antibiotics combat bacteria

Literally, *antibiotic* means "against life." Antibiotics kill bacteria or inhibit their growth. The first antibiotics were derived from extracts of molds and fungi, but today most antibiotics are synthesized by pharmaceutical companies.

There are hundreds of antibiotics in use today, and they work in dozens of ways. In general, they take advantage of the following differences between bacteria and human cells: (1) bacteria have a thick cell wall, human cells do not; (2) bacterial DNA is not safely enclosed in a nucleus, human DNA is; (3) bacterial ribosomes are smaller than human ribosomes; and (4) bacterial rate of protein synthesis is very rapid as they grow and divide. So penicillin focuses on the difference in cell walls and blocks the synthesis of bacterial cell walls, and streptomycin inhibits bacterial protein synthesis by altering the shape of the smaller bacterial ribosomes.

Some antibiotics combat only certain types of bacteria. Others, called *broad-spectrum antibiotics*, are effective against several groups of bacteria. By definition, however, antibiotics are ineffective against viruses. Recall that viruses do not reproduce on their own; they replicate only when they are inside living cells. Using antibiotics to fight viral infections such as colds or the flu is wasted effort and contributes to the growing health problem of bacterial resistance to antibiotics.

↺ Recap Active immunization with a vaccine produces a primary immune response and readies the immune system for a secondary immune response. Administration of prepared antibodies (passive immunization) can be effective against existing infections but does not confer long-term immunity. Most antibiotics kill bacteria by interfering with bacterial protein synthesis or bacterial cell wall synthesis. ■

9.8 Tissue rejection: A medical challenge

Because the goal of the immune system is to protect the body from invasion by nonself cells, it is not surprising that it attacks foreign human tissues with vigor. This phenomenon is called *tissue rejection*. As described in Chapter 7, a type of tissue rejection called a *transfusion reaction* can be fatal. But red blood cells carry only a dozen or so self markers; most body cells have far

more. Therefore, it is even more difficult to match donor and recipient tissues than it is to match blood types.

Surgical techniques for performing many organ transplants are really not that difficult. Historically, the major stumbling block to widespread transplantation of most organs has been the effectiveness of the recipient's immune response. In the normal immune response, cytotoxic T cells swiftly attack and destroy the foreign cells.

Before a transplant is even attempted, then, the donor's and recipient's ABO and other blood group antigens must first be determined. Next, donor and recipient tissues are tested to compare MHC antigens, because cytotoxic T cells target foreign MHC proteins. The closer the relationship between donor and recipient the better, because their MHC antigens are likely to be similar. Although successful transplants can be done between unrelated people, at least a 75% match between tissues is essential. After surgery the patient must take *immunosuppressive drugs* that block the immune response, such as corticosteroid drugs to suppress inflammation or cytotoxic medications that kill rapidly dividing cells (to block activated lymphocytes).

Immunosuppressive therapy can dramatically prolong the lives of transplant patients, but it brings other risks. An impaired immune system cannot protect the body as effectively against pathogens and abnormal cells, so these patients are vulnerable to infections and certain cancers. The key to a successful transplant is to suppress the immune system enough to prevent rejection while preserving as much immune function as possible. Antibiotic drugs can help control infections as they arise.

In recent years, three factors have made organ transplants a viable option for many people: (1) improvements in immunosuppressive drugs, (2) better techniques for cross-matching (or "typing") tissue, and (3) national sharing of information and donor organs through organ-bank systems. The organ-bank system allows patients to receive the best matches possible regardless of where they live.

↺ Recap The major obstacle to organ transplantation is the recipient's immune response, as cytotoxic T cells usually attack all foreign cells. ■

9.9 Inappropriate immune system activity causes problems

Conditions characterized by inappropriate immune system activity include allergies and autoimmune disorders such as lupus erythematosus and rheumatoid arthritis.

Allergies: A hypersensitive immune system

Many of us—perhaps 10% of North Americans—suffer from allergies. Some allergies are relatively minor; others are quite severe. Examples include hay fever, poison ivy rashes, and severe reactions to specific foods or drugs. Some allergic reactions even require hospitalization.

An **allergy** is an inappropriate response of the immune system to an **allergen** (any substance that causes an allergic reaction). The key word is "inappropriate": the allergen is not a dangerous pathogen, but the body reacts as if it were.

Recall that there are five classes of immunoglobulins. As shown in **Figure 9.18**, the culprits involved in allergic reactions are those in the IgE group. At some point, exposure to an allergen triggers a primary immune response, causing B cells to produce specific IgE antibodies. The IgE antibodies bind to mast cells (found primarily in connective tissue) and to circulating basophils.

When the same allergen enters the body a second time, it binds to the IgE antibodies on mast cells and basophils, causing them to release histamine. The result is an allergic reaction, a typical inflammatory response that includes warmth,

redness, swelling, and pain in the area of contact with the allergen. Histamine also increases secretion of mucus in the region. Every time the body is exposed to this allergen, the body reacts as if it has an injury or infection, even though it doesn't.

Some allergens affect only the areas exposed. Other allergens, including food allergens and bee sting venom, are absorbed or injected into the bloodstream. These substances are carried quickly to mast cells throughout the body, including connective tissue in the respiratory, digestive, and circulatory systems. Such allergens often elicit a *systemic response*, meaning they affect several organ systems. Systemic responses include constriction of smooth muscle in the lungs and digestive system and dilation of blood vessels.

Symptoms of a severe systemic allergic reaction can include difficulty breathing (caused by constricted airways), severe stomach cramps (muscle contractions), swelling throughout the body (increased capillary permeability), and circulatory collapse with a life-threatening fall in blood pressure (dilated arterioles). This is known as *anaphylactic shock*. Anyone who appears to be suffering from anaphylactic shock should be rushed to a hospital, because the reaction can be fatal. The symptoms often begin suddenly, and doctors advise people with a history of strong allergic reactions to carry an emergency kit with them at all times. The kit contains a self-injected hypodermic syringe of epinephrine, a hormone that dilates the airway and constricts peripheral blood vessels, preventing shock.

Most allergies, however, are more annoying than dangerous. Antihistamines—drugs that block the effects of histamine—are often effective treatments for mild to moderate reactions. Allergy shots can help by causing the body to produce large numbers of IgG antibodies, which combine with the allergen and block its attachment to IgE.

✓ **Quick Check** Why do allergic reactions often get worse with each successive exposure to the antigen? ∎

Autoimmune disorders: Defective recognition of "self"

On rare occasions, the immune system's remarkable ability to distinguish self from nonself fails. When that happens, the immune system may produce antibodies and cytotoxic T cells that target its own cells. Conditions in which this happens are called **autoimmune disorders.**

B cell

Allergen

1 Exposure to an allergen causes B cells to produce specific IgE antibodies.

IgE antibodies

Binding sites for IgE

Mast cell or basophil

Vesicles containing histamine

2 The IgE antibodies bind to mast cells and basophils, sensitizing them to future exposures to the same allergen.

Allergens specific for IgE

3 The next exposure to the allergen causes mast cells and basophils to release histamine.

4 Histamine causes a localized or systemic inflammatory response.

Histamine

Figure 9.18 How an allergic reaction develops.

Approximately 5% of adults in North America, two-thirds of them women, have some type of autoimmune disorder. We don't yet know all the details of how these diseases arise. In some cases, certain antigens simply are never exposed to the immune system as it undergoes fetal development. These antigens were never programmed into the system as self, so when tissue damage exposes them, the mature immune system responds as if they are foreign. In other cases, antibodies produced against a foreign antigen may cross-react with the person's own tissues.

At the moment there are no cures for autoimmune disorders. Treatments include therapies that depress the body's defense mechanisms and relieve the symptoms. Autoimmune conditions include a wide range of diseases, including multiple sclerosis, a progressive disorder of the central nervous system (see Chapter 11), and Type 1 diabetes mellitus, which targets cells in the pancreas (see Chapter 13). We will look at two more conditions: lupus erythematosus and rheumatoid arthritis.

Lupus erythematosus: Inflamed connective tissue Lupus erythematosus (or *lupus*) is an autoimmune disorder in which the body attacks its own connective tissue. One type of lupus, called discoid lupus erythematosus, primarily affects areas of the skin exposed to sunlight. More serious is systemic lupus erythematosus, which may affect various tissues and organs including the heart, blood vessels, lungs, kidneys, joints, and brain.

Lupus often starts as a red skin rash on the face or head. Other symptoms include fever, fatigue, joint pain, and weight loss. Spreading inflammation can lead to osteoarthritis (see Chapter 5), pericarditis (see Chapter 8), or pleurisy (inflammation of the lining of the lungs).

Lupus affects nine times as many women as men. Typically it occurs during childbearing age and is more common in certain racial groups such as African Americans, West Indians, and Chinese. Medications can reduce the inflammation and alleviate the symptoms.

Rheumatoid arthritis: Inflamed synovial membranes Rheumatoid arthritis is a type of arthritis involving inflammation of the synovial membrane that lines certain joints (see Chapter 5). In rheumatoid arthritis, B cells produce antibodies against a protein in the cartilage of synovial membranes. The resulting immune response releases inflammatory chemicals that cause further tissue damage. At first, fingers, wrists, toes, or other joints become painful and stiff. Over time the inflammation destroys joint cartilage and the neighboring bone. Eventually, bony tissue begins to break down and fuse, resulting in deformities (**Figure 9.19**) and reduced range of motion. The disease is intermittent, but with each recurrence the damage is progressively worse.

Pain-relieving medications can help many people with rheumatoid arthritis, as can regular mild exercise and physical therapy to improve range of motion. Powerful drugs that neutralize chemicals in the inflammatory response can prevent

Figure 9.19 **Rheumatoid arthritis.**

joints from becoming deformed. Surgery to replace damaged joints with artificial joints can restore the ability to move, and prevent painful disabilities.

🔁 **Recap** An allergy is an inappropriate inflammatory response. An autoimmune disorder occurs when the immune system fails to distinguish self from nonself cells and begins to attack the body's own cells. Examples of autoimmune disorders are lupus erythematosus (inflammation of connective tissue) and rheumatoid arthritis (inflammation of synovial membranes). ▪

9.10 Immune deficiency: The special case of AIDS

Immune deficiency is a general term for an immune system that is not functioning properly. One immune deficiency disease is *severe combined immunodeficiency disease (SCID).*

For the rare person who inherits SCID, even a minor infection can become life threatening. People with SCID have too few functional lymphocytes to defend the body against infections.

By far the most common and best-known severe immune deficiency condition is **AIDS (acquired immune deficiency syndrome).** A *syndrome* is a medical term for a group of symptoms that occur together, and *acquired* means that one catches it—in this case by becoming infected with the virus called **HIV (human immunodeficiency virus).**

HIV targets helper T cells of the immune system

Figure 9.20 shows the structure of HIV. The virus consists of nothing more than single-stranded RNA and enzymes, wrapped in two protein coats and a phospholipid membrane with protein spikes. It has no nucleus and no organelles. Like other viruses, HIV infects by entering a cell and using the cell's machinery to reproduce. HIV targets helper T cells, gaining entry by attaching to CD4 receptors.

HIV belongs to a particular class of viruses, called *retroviruses,* that have a unique way of replicating (**Figure 9.21**). Retroviruses first attach to the CD4 receptor of a helper T cell. The attachment fuses the retrovirus's envelope with the cell's membrane, releasing the viral RNA and enzymes into the cell. Under the influence of the viral enzymes and using the viral RNA as a template (a pattern), the host cell is forced to make a single strand of DNA complementary to the viral RNA, and from it a second strand of DNA complementary to the first. The new double-stranded DNA fragment is then inserted into the cell's DNA. The cell, not recognizing the DNA as foreign, uses it to produce more viral RNA and proteins, which are then assembled into thousands of new viruses within the cell. The sheer magnitude of viral replication so saps the T cell's energy that eventually it dies and ruptures, releasing the viral copies. The new viruses move on to infect other helper T cells.

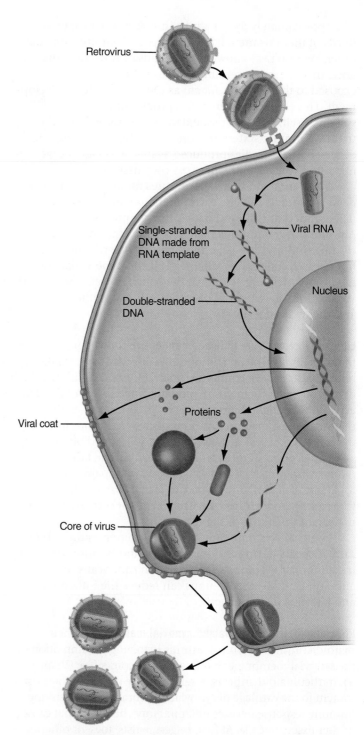

Figure 9.21 How HIV replicates. After binding to a helper T cell's CD4 receptor, the virus injects its RNA and enzymes into the cell and uses them to force the cell to make single- and then double-stranded DNA. The DNA is inserted into the cell's own DNA within the nucleus, where it directs the cell to make more protein coat and viral RNA and protein core. The coats and cores join, exiting the cell as new viruses.

HIV is transmitted in body fluids

HIV is a fragile virus that cannot survive dry conditions for even a short time. This means that it cannot be transmitted through the air, by casual contact, or by doorknobs or toilet

Figure 9.20 The structure of HIV.

seats. However, HIV can be transmitted in at least four body fluids—blood, semen, breast milk, and vaginal secretions. It is not known to be transmitted by contact with urine, feces, saliva, perspiration, tears, or nasal secretions unless those fluids contain blood.

Most commonly, HIV is transmitted by sexual contact or contaminated hypodermic needles. In addition, infected mothers can pass it to their children during pregnancy, labor and delivery, or when breast-feeding. It has been estimated, for example, that a healthy newborn who is breastfed by an infected mother has about a 20% chance of contracting HIV from its mother's milk. A significant number of cases of HIV infection resulted from blood transfusions until the mid-1980s, when it became routine to test blood for HIV antibodies.

Because transmission of HIV generally requires direct contact with body fluids, HIV is considered relatively hard to transmit from person to person (at least when compared to such common diseases as the flu). However, unlike the flu, HIV is extremely virulent unless it is aggressively treated with an expensive regimen of drugs—over 90% of all U.S. citizens diagnosed with AIDS by 1988 are now dead.

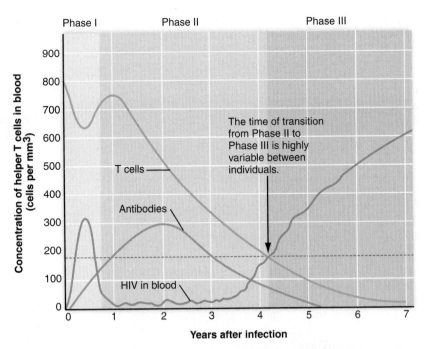

Figure 9.22 Time course of the progression toward AIDS after HIV infection. Phase I is characterized by flulike symptoms followed by apparent recovery. There is a brief spike in HIV in the blood. Within six months HIV-specific antibodies appear. In Phase II there is a slow decline in the number of T cells. Most of the HIV is harbored inside cells. Phase III (AIDS) is characterized by a T cell count of less than 200 per cubic millimeter of blood, a decline in HIV-specific antibodies, a rise in HIV in blood, opportunistic infections, and certain cancers.

AIDS develops slowly

The symptoms of HIV infection and the subsequent development of full-blown AIDS progress slowly. The course of the disease falls into three phases (Figure 9.22).

Phase I Phase I lasts anywhere from a few weeks to a few years after initial exposure to the virus. There is a brief spike in HIV in the blood, followed by typical flulike symptoms of swollen lymph nodes, chills and fever, fatigue, and body aches. The immune system's T cell population may decline briefly then rebound as the body begins to produce more cells and antibodies against the virus. The presence of antibodies against the virus is the basis for a diagnosis of HIV infection, and a person having these antibodies is said to be "HIV-positive." However, he or she will not yet have the disease syndrome called *AIDS*.

Unless they suspect they have been exposed to HIV, most people would not associate Phase I symptoms with HIV infection and so would not think to be tested. The antibodies do not destroy the virus entirely because many of the virus particles remain inside cells, where antibodies and immune cells cannot reach them.

Phase II In Phase II the virus begins to do its damage, wiping out more and more of the helper T cells. The loss of T cells makes the person more vulnerable to *opportunistic infections*—infections that take advantage of the weakened immune system to establish themselves in the body. During Phase II people *may* have persistent or recurrent flulike

symptoms, but they may have no symptoms at all if they do not have an opportunistic infection. If they have not been tested for HIV antibodies they may not even know they are infected.

Two-thirds to three-quarters of all people who test positive for antibodies to HIV do not exhibit symptoms associated with AIDS. During Phases I and II, many people pass the virus on to others without realizing it. Those they infect may transmit the virus to still others. Phase II can occur within 6 months, but on average it takes about 10 years to progress to Phase III. Left untreated, 95% of people in Phase II progress to Phase III.

Phase III Once the number of helper T cells of an HIV-positive person falls below 200 per cubic millimeter of blood *and* the person has an opportunistic infection or type of cancer associated with HIV infection, the person is said to have AIDS. Infections and cancers associated with AIDS include pneumonia, meningitis, tuberculosis, encephalitis, Kaposi's sarcoma, and non-Hodgkin's lymphoma, among others.

Notice that AIDS may not appear until years after initial HIV infection. Untreated AIDS is nearly always fatal.

 Web Animation *Effects of HIV on the Immune System* at www.humanbiology.com

The AIDS epidemic: A global health issue

AIDS was first described in 1981 when the Centers for Disease Control in Atlanta began to notice a disturbing similarity between cases involving a strange collection of symptoms (this illustrates the advantage of having a central clearinghouse for information). It is now believed that HIV first infected humans in the 1960s in Africa after "jumping species" from other primates to humans.

The worldwide damage done by HIV so far is truly astonishing. By the end of 1990 there were almost 8 million HIV-infected people worldwide. Today, more than 33 million people are living with HIV infection or AIDS, representing nearly 1% of the adult population worldwide. New infections are occurring at a rate of nearly 7,000 people per day. Fully two-thirds of all HIV-infected people reside in sub-Saharan Africa, where the disease is thought to have originated. The areas of the world where the AIDS epidemic is increasing most rapidly include sub-Saharan Africa, South and Southeast Asia, and Eastern Europe.

So far, more than 25 million people have died of AIDS. In recent years the death toll from AIDS has been hovering around 2 million people per year.

The picture is slightly better in the United States, thanks to a well-developed AIDS reporting system and the availability of good health care. The number of deaths from AIDS in the United States has stabilized at just over 20,000 per year after peaking in 1995 at nearly 60,000 per year (**Figure 9.23a**). Part

of the decline in death rates since 1995 is due to new drugs that suppress the active component of an HIV infection and keep infected people alive much longer. However, to be completely effective the drugs need to be taken for the rest of the patient's lifetime.

Although the death rate from AIDS has stabilized in the United States, the number of persons living with HIV/AIDS continues to climb (Figure 9.23b). There are about three times as many new diagnoses of HIV/AIDS each year as there are deaths from AIDS. The treatment of persons living with HIV/AIDS is likely to place a heavy financial burden on our health care system in the future, for these patients will need HIV suppressive drugs that currently cost $10,000 to $30,000 per year.

Risky behaviors increase your chances of getting AIDS

Because your behavior will ultimately determine your risk for AIDS, it's worth knowing what constitutes "risky behavior" in terms of HIV infection. Table 9.3 shows how HIV is contracted in the United States among adults and adolescents (people above 12 years of age).

In the United States, more than three-fourths of all new cases of AIDS in 2007 were men. For men, the most risky behaviors are engaging in sex with other males or sharing needles during intravenous drug use. Some men mistakenly believe that they cannot get AIDS if their only

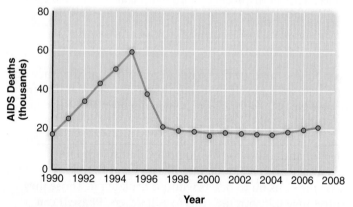

a) Estimated number of deaths due to HIV/AIDS, 1990–2007.

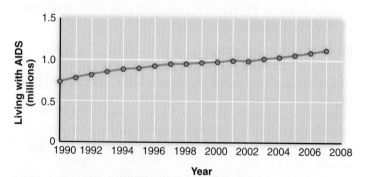

b) Estimated number of people living with HIV/AIDS, 1990–2007.

Figure 9.23 HIV/AIDS in the United States.

Source: UNAIDS, Epidemiological Fact Sheet on HIV and AIDS, United States of America, 2008 update.

Table 9.3 U.S. AIDS cases in adults and adolescents in 2007, by sex and exposure category

Exposure category	Percent of total
Males	
Sex with other men	61
Injected drug use	8
Sex with other men and injected drug use	5
High-risk heterosexual contact:	7
(Sex with injection drug user)	(1)
(Sex with HIV-infected person)	(6)
Risk factor not reported or identified	19
Male total: 48,347 cases	**100**
Females	
Injected drug use	14
High-risk heterosexual contact:	46
(Sex with injection drug user)	(6)
(Sex with bisexual male)	(2)
(Sex with HIV-infected person)	(37)
Risk factor not reported or identified	39
Female total: 14,226 cases	**100**

Source: Centers for Disease Control and Prevention, HIV/AIDS Surveillance Report, 2007.

risky behavior is heterosexual sex. They are wrong; 7% of all new infections in males in 2007 resulted from heterosexual sex.

The primary risk for women is having sex with HIV-infected men. The other primary risk category for women is sharing needles during injected drug use.

Sex can be safer

Given the risks of contracting HIV/AIDS, it would be wise to make sex as safe as possible. Below are several *safer sex guidelines* and some of the evidence for them. The suggestions do not guarantee complete safety, which is why they are called recommendations for "safer" rather than "safe" sex.

- *Abstain from sex.* Eliminating sexual contact eliminates risk, and some people choose abstinence for ethical as well as practical reasons. However, abstinence may not be considered desirable or practical. In that case, your safest course is to follow as many of the remaining guidelines as possible.
- *Reduce the number of sexual partners.* Evidence shows that this is particularly effective among homosexual men. The evidence is not yet conclusive regarding the relationship between HIV infection and number of partners in heterosexual transmission, perhaps because the median number of partners is generally lower. Nevertheless, it makes sense that fewer partners translates to lower risk.
- *Choose a sexual partner with low-risk behavior.* This is especially important. A partner who has a history of injecting drugs or a man who has sex with other men is a high-risk partner. One study reported that choosing a partner who is not in any high-risk group lowers the risk of AIDS almost 5,000-fold.
- *Avoid certain high-risk sexual practices.* Any sexual behavior that increases the risk of direct blood contact, such as anal-genital sex, should be considered risky. Oral-genital sex is less risky but not risk-free.
- *Use latex or polyurethane condoms or other barriers.* In laboratory tests, latex or polyurethane condoms prevent the passage of HIV. How effective they are in actual use is not certain, but because they are about 90% effective as a birth control method, a reasonable hypothesis (not yet tested) is that they reduce the risk of HIV by about the same amount. Condoms are effective only if used consistently and correctly. Natural skin condoms are not as safe as latex condoms. The use of dental dams is recommended for oral-vaginal sex.
- *Use nonoxynol-9, a spermicidal agent.* In laboratory tests, nonoxynol-9 has been shown to inhibit the growth of HIV. Be aware, however, that the evidence is conflicting regarding whether nonoxynol-9 offers any protection during sexual contact.
- *Get tested (and have your partner tested).* Notice, however, that the HIV tests currently available are designed to

detect the human *antibodies* to HIV, not the virus itself. It can take up to six months after exposure to produce enough antibodies for the antibody test to become positive. For this reason, retesting at least six months after the last possible exposure is a good idea. Evidence shows that close to 90% of HIV-infected persons will have a positive HIV antibody test result within six months. A negative test at six months indicates substantially lower risk, but it does not provide complete assurance. A few individuals may not test positive for three or more years.

New treatments offer hope

As yet there is no cure for AIDS, but researchers are investigating more than a hundred drugs to treat the condition. Some drugs, such as AZT, inhibit the enzymes the virus needs to replicate inside the cell. Others, such as ritonavir and saquinavir, are protease inhibitors (protease is an enzyme required to assemble viral proteins).

In 2007 the FDA approved maraviroc (brand name Celsentri), the first new oral class of HIV treatment in more than a decade. Developed by Pfizer, maraviroc works by inhibiting the entry of the virus into healthy T cells. If the virus can't enter the T cell, it can't take over its metabolic machinery and force it to make more viruses. Maraviroc is expected to be a blockbuster drug for Pfizer, with annual sales of $500 million by 2011.

In addition to treating full-blown AIDS, doctors are starting to treat people who are infected with HIV but don't yet show the symptoms of AIDS. This approach may prevent the widespread destruction of the immune system that precedes the onset of symptoms. Many health professionals advise people to be tested for HIV so that asymptomatic cases can be detected and treated early.

Some researchers believe that safe and effective vaccines offer the only real hope of conquering HIV. Several dozen potential vaccines are already being tested on human volunteers around the world, but so far none has proved effective enough to warrant widespread use.

The production of vaccines is complicated by the fact that HIV mutates rather quickly. There are already several strains of the virus, each of which would need a separate vaccine. In addition, HIV is so dangerous that it is considered too risky to produce vaccines from whole but weakened viruses, the way many other vaccines are produced. Most HIV vaccines are produced from other viruses engineered to contain pieces of the HIV virus, or from pieces of the HIV viral genetic material.

↻ Recap AIDS (acquired immune deficiency syndrome) is a devastating disorder of the immune system caused by a virus (HIV) that attacks helper T cells. HIV is transmitted in body fluids, typically through sexual contact, blood transfusions, contaminated needles, or breast-feeding. ∎

Chapter Summary

Pathogens cause disease p. 190

- Pathogens include bacteria, viruses, fungi, protozoa, worms, and possibly prions.
- Bacteria have very little internal structure and are covered by a rigid outer cell wall.
- Viruses cannot reproduce on their own. Viral reproduction requires a living host cell.
- Prions are misfolded proteins that replicate by causing a normal protein to misfold.
- The danger from a particular pathogen depends on *how* it is transmitted, how *easily* it is transmitted, and how *damaging* the resulting disease is.

The lymphatic system defends the body p. 193

- The lymphatic system consists of vessels, lymph nodes, the spleen, the thymus gland, and the tonsils.
- The lymphatic system helps protect us against disease.
- Phagocytic cells in the spleen, lymph nodes, and tonsils engulf and kill microorganisms.
- The thymus gland secretes hormones that help T lymphocytes mature.

Keeping pathogens out: The first line of defense p. 196

- Skin is an effective barrier to the entry of microorganisms.
- Tears, saliva, mucus, and earwax trap organisms and/or wash them away.
- Digestive acid in the stomach kills many microorganisms.
- Vomiting, defecation, and urination physically remove microorganisms after entry.

Nonspecific defenses: The second line of defense p. 197

- Phagocytes surround and engulf microorganisms and damaged cells.
- Inflammation has four outward signs: redness, warmth, swelling, and pain.
- Natural killer cells kill their targets by releasing damaging chemicals.
- Circulating proteins of the complement system either kill microorganisms directly or mark them for destruction.
- Interferons are proteins that interfere with viral reproduction.
- Fever raises body temperature, creating a hostile environment for some microorganisms.

Specific defense mechanisms: The third line of defense p. 201

- Cells of the immune system can distinguish foreign or damaged cells from our own healthy cells.
- All cells have cell-surface markers called *MHC proteins* that identify the cells as "self."
- An antigen is a substance that stimulates the immune system and provokes an immune response.
- B cells produce antibodies against foreign antigens.
- T cells of several types release chemicals that enhance the immune response and kill foreign cells directly.

Immune memory creates immunity p. 206

- Information about an antigen is stored in memory cells after first exposure.
- The second exposure to the antigen produces a much greater immune response than the first.
- The rapidity of the second response is the basis of immunity from disease.

Medical assistance in the war against pathogens p. 207

- Vaccines immunize the body in advance against a particular disease.
- Injected antibodies provide temporary immunity and are of some benefit against an existing infection.
- Monoclonal antibodies are used primarily in medical tests.
- Antibiotics are effective against bacteria, but not against viruses.

Tissue rejection: A medical challenge p. 209

- The phenomenon of tissue rejection is a normal consequence of the body's ability to recognize self from nonself.
- Immunosuppressive drugs, the ability to test for various antigens, and organ donor matching programs have increased the success rate of organ transplantation in humans.

Inappropriate immune system activity causes problems p. 209

- Allergies occur when the immune system responds excessively to foreign particles that are not otherwise harmful.
- Autoimmune disorders develop when a person's immune system attacks the person's own cells as if they were foreign.

Immune deficiency: The special case of AIDS p. 211

- AIDS is caused by a virus (HIV).
- The disease can take years to develop after initial HIV infection, but it is nearly always fatal.
- Worldwide, the number of cases of HIV infection and of AIDS is still rising rapidly.
- The chances of contracting AIDS can be reduced (but never eliminated completely) by practicing "safer" sex.

Terms You Should Know

active immunization, *207*
AIDS (acquired immune deficiency syndrome), *212*
antibiotics, *191*
antibody, *201*
antibody-mediated immunity, *201*
antigen, *201*
bacteria, *190*
B cell, *201*
cell-mediated immunity, *201*
cytokines, *204*
HIV (human immunodeficiency virus), *212*
immune response, *201*
inflammation, *198*
interferon, *200*
macrophage, *198*
monoclonal antibodies, *208*
passive immunization, *208*
pathogen, *190*
phagocytosis, *197*
prion, *192*
T cell, *201*
vaccine, *207*
virus, *191*

Concept Review

Answers can be found at the Human Biology Place.
www.humanbiology.com

1. Define the term *pathogen* and give some examples.
2. Describe what is unusual about viruses that causes some people to question whether they are living things.
3. Explain how prions affect human health.
4. Describe the functions of organs of the lymphatic system.
5. List the four components of an inflammatory response.
6. Describe in general terms the distinction between *nonspecific* and *specific* defense mechanisms.
7. Describe the concept of an antigen and how it relates to self and nonself markers.

8. Describe how cells that belong to a particular individual are identified so that the individual's immune system doesn't attack them.
9. Explain how cytotoxic T cells kill target cells.
10. Describe how a vaccine produces immunity from a specific disease.

Test Yourself

Answers can be found in Appendix A.

1. In which of the following ways are bacterial cells similar to human cells?
   a. Bacterial cells have cell walls.
   b. Bacterial cells have a single, circular chromosome.
   c. Bacterial cells use ATP to fuel cellular activities.
   d. Bacterial cells lack mitochondria.

2. Which of the following statements about viruses is true?
   a. Viruses require a host cell in which to reproduce.
   b. Viruses are very small bacteria.
   c. Viral infections can generally be controlled with antibiotics.
   d. Viruses are composed of protein only.

3. Which of the following pathogenic agents causes a self-propagating misfolding of proteins in nerve cells?
   a. bacteria   b. prions
   c. viruses   d. helminths (worms)

4. Consider the following group of diseases: hepatitis, chicken pox, warts, and measles. What do these diseases have in common?
   a. They are all caused by bacteria.
   b. They are readily treated with antibiotics.
   c. They are all caused by viruses.
   d. They are very common in patients infected with HIV.

5. Which of the following is true regarding prion diseases?
   a. They are caused by a deadly type of virus.
   b. They can readily be treated with antibiotics.
   c. They can be prevented by vaccinations.
   d. They cause accumulation of misfolded proteins in brain cells.

6. Which of the following is a benefit of resident bacteria?
   a. Resident bacteria cause the stomach to be acidic.
   b. Resident bacteria produce antiviral compounds that prevent viral infections.
   c. Resident bacteria can out-compete harmful bacteria and lower the incidence of infection.
   d. Resident bacteria digest cellulose within the human digestive tract.

7. DiGeorge syndrome is a congenital disease that results in a poorly developed, non-functioning thymus gland. Which of the following would be a likely problem experienced by a baby with DiGeorge syndrome?
   a. lack of B cells   b. lack of antibodies
   c. lack of T cells   d. lack of macrophages

8. The following are steps in phagocytosis: (1) Bacterium is digested by lysosomal enzymes, (2) phagocyte approaches bacterium, (3) phagocytic vesicle fuses with lysosome, and (4) phagocyte engulfs bacterium, forming a phagocytic vesicle. In which order do these steps occur?
   a. 4-2-3-1   b. 2-4-3-1
   c. 2-3-4-1   d. 4-1-3-2

9. In which of the following choices is the cell correctly matched with its function?
   a. eosinophil: produces antibodies
   b. B lymphocyte: directly attacks foreign cells
   c. basophil: secretes histamine
   d. T lymphocyte: phagocytizes bacteria

10. Each of the following processes helps combat infection except:
    a. inflammation   b. fever
    c. autoimmunity   d. antibody production

11. The primary immune response is:
    a. faster than the secondary immune response
    b. longer lasting than the secondary immune response
    c. less effective than the secondary immune response
    d. due to the presence of memory cells

12. Compared to active immunization, passive immunization:
    a. provides immediate protection
    b. is longer lasting
    c. creates a large number of memory cells
    d. may occasionally cause the disease it is intended to prevent

13. Which of the following increases the likelihood of successful organ transplant?
    a. matching the ABO blood group antigens
    b. matching the MHC tissue antigens
    c. administration of immunosuppressive drugs
    d. all of these choices

14. Which of the following does not belong with the others?
    a. lupus erythematosus   b. rheumatoid arthritis
    c. anaphylactic shock   d. Type I diabetes mellitus

15. Which of the following statements about HIV is true?
    a. Latex condoms are 100% effective in blocking transmission of HIV.
    b. Most individuals who progress to Phase II of HIV infection remain in Phase II and never progress to Phase III (AIDS).
    c. HIV specifically impairs the cell-mediated immune response.
    d. Anti-HIV medications such as AZT and maraviroc can cure HIV infection.

Apply What You Know

Answers can be found at the Human Biology Place.
www.humanbiology.com

1. When you get a minor infection in a small cut in the skin, sometimes soaking it in hot water speeds the healing process. How might the heat help?
2. Explain why antibiotics don't work against viruses.
3. Everyone knows that bacteria can cause disease. Suppose that we could actually remove all bacteria from our bodies all at once. Would that be a good thing or a bad thing?
4. In 1918 a pandemic of a deadly flu strain killed upward of 30 million people. If that same flu strain were to come around again, would we be better off or worse off than those who were alive in 1918?
5. Researchers have been working on an effective vaccine for gonorrheal infections for some years. One promising vaccine is delivered via a spray into the nasal cavity. This is clearly not the site of gonococcal infection. Why would one administer a vaccine meant to protect the reproductive tract into the nasal cavity?
6. Why is it that most people get the chicken pox only once, but they can get a cold or the flu over and over again throughout a lifetime?
7. The immune system is supposed to defend us from harmful microorganisms. Why doesn't it always work? In other words, why do some people still get sick and die?

10

The Respiratory System: Exchange of Gases

False color SEM (×150) of a section of lung tissue.

Limiting Exposure to Secondhand Smoke

It's tough being a smoker these days. First, there's a $1.01 federal excise tax and an average of $1.20 in state taxes on every pack of cigarettes. Then there's that Surgeon General's warning on the pack itself. And to top it off, there are fewer and fewer places where smoking is even permitted. In 2003 only 5 states (California, Connecticut, Delaware, Maine, and New York) had smoke-free workplaces. By 2010 at least 20 states had banned smoking in workplaces, restaurants, and bars, even if the restaurant or bar could provide separately ventilated rooms. A few laws even limit smoking within a certain distance of a building. Where *can* a smoker smoke these days?

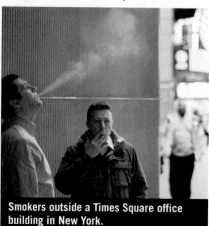

Smokers outside a Times Square office building in New York.

Is Secondhand Smoke a Health Problem?

We all know that smoking is bad for your health. Cigarette smoke contains a whole host of known or suspected carcinogens. Numerous scientific studies have shown convincingly that smoking is the single leading cause of lung cancer in the world. Hardly anyone would dispute these facts today. But is the secondhand smoke that comes from a burning cigarette or is exhaled by a smoker also bad for your health? The evidence is rather limited. According to the National Institutes of Health report on carcinogens, secondhand

Cigarette smoke is classied as a carcinogen.

smoke (also called environmental tobacco smoke) is listed as a "known human carcinogen" because there are now conclusive published studies indicating that women who live with smoking husbands or who work with smokers are at increased risk for lung cancer. There is also substantial evidence that children who live with smokers are at increased risk for decreased lung function, increased lung infections, and asthma.

The dilemma for lawmakers is that no one has yet accurately quantified or proven the risks of secondhand smoke at the kinds of occasional exposure levels that might be typical. Just because long-term exposure to a high concentration of secondhand smoke is bad for you does not prove that an occasional whiff of secondhand smoke is equally bad. There could be a big difference between working or living in an enclosed space with a smoker (there are 28,800 seconds in an 8-hour workday) and catching a whiff of smoke for a few seconds in an outdoor park. Estimates of the health risks (including deaths) from low exposures to secondhand smoke generally are derived by extrapolation—admittedly, not a very convincing scientific technique.

Making Public Policy Without Scientific Evidence

Smokers argue that laws seeking to prohibit even a whiff of smoke ever being detected by nonsmokers represent a disproportionate response to a relatively minor health problem, without regard for the scientific evidence. Are they correct? Consider the following:

- Twenty states ban smoking entirely in workplaces, restaurants, and bars. Not one of them even allows smokers to smoke in separately ventilated rooms, away from nonsmokers.
- Some antismoking laws prohibit smoking closer than 50 feet to a window or door of a building's exterior.
- Thirty-six percent of all respondents to an online poll believe that smoking should be banned in outdoor parks and playgrounds.

It has been argued that the real intent of these laws is to reduce smoking by smokers. This is called "soft paternalism"—a political philosophy that government should "influence" you to do what is right, because you know what is right but just don't have the willpower to make the right choice yourself. Warning labels on cigarette packs and high cigarette taxes are examples of soft paternalism. The justification is that there is good scientific evidence that smoking causes lung cancer. We might choose to ban cigarettes outright, but for other reasons we don't.

Lawmakers who seek to limit exposure to *secondhand* smoke, however, face a problem of justification. Although many of us might not like to catch even a whiff of smoke outdoors, at the moment there is no scientific evidence that brief exposure to outdoor secondhand smoke does any harm.

A purely political decision to limit exposure to secondhand smoke does not need to be based on scientific evidence, of course. One could argue that some choices are unacceptable to a society even *without* scientific proof. Prostitution, child pornography, and discrimination based on race might fall into this category, but does smoking?

Most people would probably agree that smoking should be banned from places where it might cause harm to nonsmokers. But how do policymakers decide just where that is? So far, science has not provided any satisfactory answers. Perhaps the bigger question is, what are our motives in supporting and passing laws designed to limit exposure to secondhand smoke? The real agenda may not be about protecting nonsmokers from secondhand smoke at all; it may be about trying to influence smokers to quit. If that's the agenda, let's be honest about it.

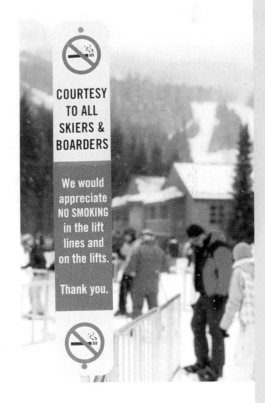

COURTESY TO ALL SKIERS & BOARDERS

We would appreciate NO SMOKING in the lift lines and on the lifts.

Thank you.

Questions to consider

1 Should smokers have the right to smoke? Why or why not?

2 Nonsmokers claim they have a right to smoke-free air. Do you agree?

3 Where would you allow smoking, and where would you ban it? On what do you base your opinions?

The facts...

- **Under conditions of heavy or prolonged exposure, secondhand smoke can cause cancer in women and is associated with increased respiratory health risks in children.**
- **It is not known whether light or occasional exposure to secondhand smoke also poses a health risk.**
- **Smoking is increasingly being banned in workplaces, in restaurants, and even in public places outdoors.**
- **Smokers argue that antismoking laws take away their right to smoke.**

>> **The function of the respiratory system is to facilitate the diffusion of gases (O_2 and CO_2) between the atmosphere and blood.**

>> **The actual structure across which gases are exchanged is only two living cell layers thick.** It is essential that this delicate structure be kept moist, clean, and free of micro-organisms.

>> **Breathing requires physical effort by skeletal muscles.** The muscles of breathing include the dome-shaped *diaphragm* and the *intercostal muscles* between the ribs.

>> **Breathing is controlled by powerful feedback control mechanisms** that seek to maintain homeostasis of blood O_2 and CO_2. That's why your breathing increases automatically during exercise, and why you cannot hold your breath indefinitely even if you try.

>> **Red blood cells are essential for transporting O_2 from the lungs to all living cells.** Blood plasma alone could not carry enough O_2 for our survival.

Your dinner companion is turning blue. One minute ago he was laughing and drinking and spearing his steak with enthusiasm, and now he can't breathe or talk and is frantically pulling at his collar.

Choking is an emergency, but one that is seldom handled by emergency medical personnel. The victim's fate is almost always decided before professional help can arrive. Unless someone in the immediate vicinity can quickly intervene to relieve the choking, there's a good chance the person won't survive.

We can live for days without nutrients or water, but will die within minutes if denied oxygen. Of all the exchanges of materials between an animal and its environment, the exchange of respiratory gases is almost always the most urgent.

What is the physical basis of this urgency? All cells use energy and certain raw materials in order to survive, grow, and reproduce. To do this, cells rely on aerobic metabolism, meaning that they use up oxygen (O) and create carbon dioxide (CO_2) as a waste product. As discussed in Chapter 3, oxygen is the final electron acceptor in cellular respiration, essential for the production of ATP for most cells. Without oxygen, essential cellular processes cannot continue. Removal of carbon dioxide from the tissues is a pressing concern as

well, because its accumulation in the body has many harmful effects.

The primary function of the **respiratory system** is to exchange these gases (oxygen and carbon dioxide) with the air. We extract oxygen from the air we breathe, and when we exhale we get rid of carbon dioxide, the waste product of our metabolism. The oxygen in the air we breathe comes from plants. Plants absorb carbon dioxide through small pores in their leaves and use it in their own energy-producing process called *photosynthesis*. In the process, they produce oxygen for their (and coincidentally our) use. (We consider this and other examples of ecosystems and the delicate balance of life on Earth in Chapters 23 and 24.)

10.1 Respiration takes place throughout the body

The term *respiration* encompasses four processes:

- *Breathing* (also called **ventilation**). The movement of air into and out of the lungs.
- *External respiration.* The exchange of gases between inhaled air and blood.
- *Internal respiration.* The exchange of gases between the blood and tissue fluids.
- *Cellular respiration.* The process of using oxygen to produce ATP within cells. Cellular respiration generates carbon dioxide as a waste product.

Breathing is facilitated by the respiratory system and its associated bones, muscles, and nerves. External respiration takes place within the lungs, and internal respiration and cellular respiration take place in the tissues throughout the body.

The respiratory system in humans and most animals has another function in addition to gas exchange, and that is the production of sound (vocalization). The production of sound is an important mechanism that infants use to signal their adult caregivers. Sound production has, of course, also proved valuable in information exchange and cultural development, and thus contributes to the long-term survival of our species.

10.2 The respiratory system consists of upper and lower respiratory tracts

The respiratory system (**Figure 10.1**) consists of (1) a system of passageways for getting air to and from the lungs and (2) the lungs themselves, where gas exchange actually occurs. Also important to respiration are the bones, muscles, and components of the nervous system that cause air to move into and out of the lungs.

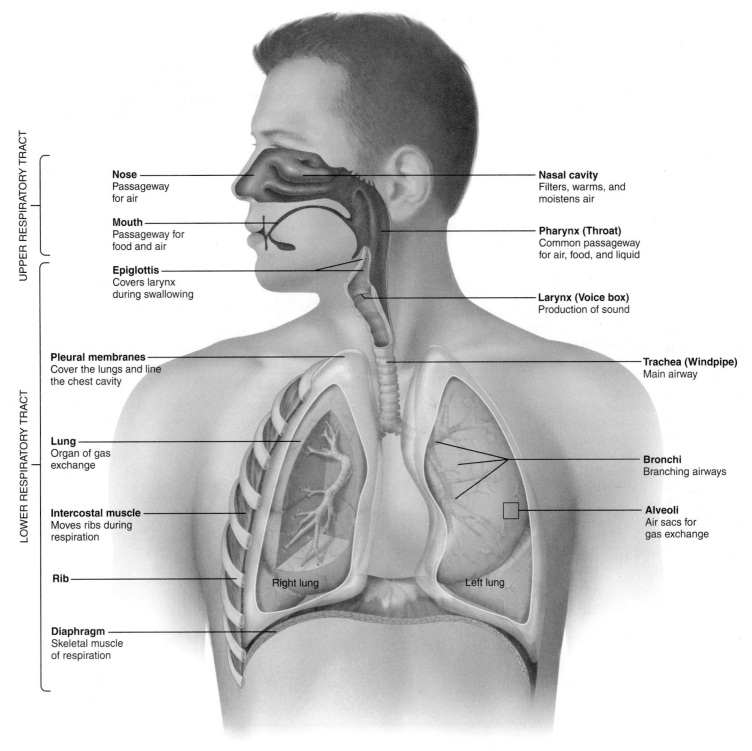

Figure 10.1 The human respiratory system. The functions of each of the anatomical structures are included.

For the sake of convenience, the respiratory system can be divided into the upper and lower respiratory tracts. The *upper respiratory tract* comprises the nose (including the nasal cavity) and pharynx—structures above the "Adam's apple" in your neck. The *lower respiratory tract* starts with the larynx and includes the trachea, the two bronchi that branch from the trachea, and the lungs themselves.

The upper respiratory tract filters, warms, and humidifies air

When you inhale, air enters through your nose or your mouth (**Figure 10.2**). Your *nose* is to be appreciated, as it does more than serve as a passageway for respiration. The nose also

- Contains receptors for the sense of smell.
- Filters inhaled air and screens out some foreign particles.
- Moistens and warms incoming air.
- Provides a resonating chamber that helps give your voice its characteristic tone.

The visible portion of the nose is known as the **external nose**. The internal portion of the nose is called the **nasal cavity.** The external nose consists of cartilage in the front and two nasal bones behind the cartilage. The nose varies in size and shape from person to person, primarily as a result of individual differences in the cartilage tissue.

The external nose and nasal cavity are divided into two chambers by the nasal septum. Air enters through the nostrils, the two openings at the base of the external nose, where it is partially filtered by nose hairs, then flows into the nasal cavity. This cavity is lined with moist epithelial tissue that is well supplied with blood vessels. The blood vessels help to warm incoming air and the epithelial tissue secretes mucus, which humidifies the air. The epithelium is also covered with tiny hairlike projections called *cilia*.

The mucus in the nasal cavity traps dust, pathogens, and other particles in the air before they get any farther into the respiratory tract. The cilia beat in a coordinated motion, creating a gentle current that moves the particle-loaded mucus toward the back of the nasal cavity and pharynx. There we cough it out, or swallow it to be digested by powerful digestive acids in the stomach. Ordinarily we are unaware of our nasal cilia as they carry on this important task. However, exposure to cold temperatures can slow down their activity, allowing

mucus to pool in the nasal cavity and drip from the nostrils. This is why your nose "runs" in cold weather.

As described in Chapter 5, air spaces called *sinuses* inside the skull are also lined with tissue that secretes mucus and helps trap foreign particles. The sinuses drain into the nasal cavity via small passageways. Two tear ducts, carrying fluid away from the eyes, drain into the nasal cavity as well. This is why excess production of tears, perhaps due to strong emotions or irritating particles in your eyes, also makes your nose "runny."

Incoming air next enters the **pharynx** (throat), which connects the mouth and nasal cavity to the larynx (voice box). The upper pharynx extends from the nasal cavity to the roof of the mouth. Into it open the two *auditory tubes* (eustachian tubes) that drain the middle ear cavities and equalize air pressure between the middle ear and outside air. The lower pharynx is a common passageway for both food and air. Food passes through on its way to the esophagus, and air flows through to the lower respiratory tract.

✓ **Quick Check** Why can upper respiratory tract infections sometimes cause ear infections? ■

The lower respiratory tract exchanges gases

The lower respiratory tract includes the larynx, the trachea, the bronchi, and the lungs with their bronchioles and alveoli (**Figure 10.3**).

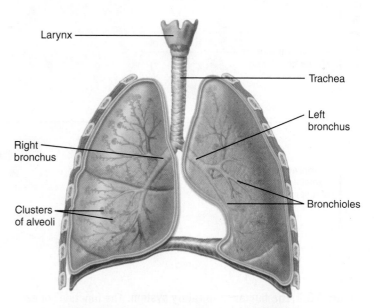

Figure 10.3 Components of the lower respiratory tract.
Individual alveoli are too small to be seen clearly in this figure.

✓ *List, in order, all of the anatomical structures that a molecule of oxygen passes through as it moves from the nose to a pulmonary capillary. In how many of these structures does gas exchange occur?*

Know

Sinuses
Nasal cavity
External nose
Opening of the auditory tube
Nostril
Pharynx
Tongue
Epiglottis
Glottis
Larynx
Trachea
Esophagus

Figure 10.2 Components of the upper respiratory tract.

Answers to ✓ questions can be found in Appendix A.

The larynx produces sound The **larynx,** or voice box, extends for about 5 cm (2 inches) below the pharynx. The larynx serves to

- Maintain an open airway.
- Route food and air into the appropriate channels.
- Assist in the production of sound.

The larynx contains two important structures: the epiglottis and the vocal cords (**Figure 10.4**). The **epiglottis** is a flexible flap of cartilage located at the opening to the larynx. When air is flowing into the larynx, the epiglottis remains open. But when we swallow food or liquids, the epiglottis tips to block the opening temporarily. This "switching mechanism" routes food and beverages into the esophagus and digestive system, rather than into the trachea. This is why it is impossible to talk while you are swallowing.

The **vocal cords** consist of two folds of connective tissue that extend across the airway. They surround the opening to the airway, called the **glottis.** The vocal cords are supported by ligaments and enclosed within a cartilaginous structure nicknamed the "Adam's apple."

We produce most sounds by vibration of the vocal cords, although we can also make a few sounds by moving our tongue and teeth. The tone of the sounds produced by the vocal cords depends on how tightly the vocal cords are stretched, which is controlled by skeletal muscle. When we are not talking, the vocal cords are relaxed and open (Figure 10.4a). When we start to talk they stretch tightly across the tracheal opening, and the flow of air past them causes them to vibrate (Figure 10.4b).

Like any string instrument, cords that are relatively short yield higher-pitched tones than longer cords do. Also, the tighter the vocal cords stretch, the higher the tone they produce. Men tend to have deeper voices than women and a more prominent Adam's apple, due to testosterone that causes the larynx to enlarge at puberty. We can exert some control over the volume and pitch of the voice by adjusting the tension on our vocal cords. (You may have noticed that when you're nervous, your voice becomes higher pitched.) The resulting vibrations travel through the air as sound waves.

Most of us can be recognized by the distinctive quality of our voices. In addition to the shape and size of vocal cords, individual differences in voice are determined by many components of the respiratory tract and mouth, including the pharynx, nose and nasal cavity, tongue, and teeth. Muscles in the pharynx, tongue, soft palate, and lips cooperate to create recognizable sounds. The pharynx, nose, and nasal sinuses serve as resonating chambers to amplify and enhance vocal tone.

MJ's Human Biology Blog

Snus—Smokeless Tobacco Made Easy

Now available over the internet and in stores: Snus (rhymes with loose), a nicotine-containing smokeless tobacco product that originated in Sweden. Snus comes in colorful tins and is packaged in small tea bag–like packets. Its real advantage over traditional smokeless chewing tobacco is that no spitting is necessary—the small amount of juice produced can just be swallowed. It's likely to become popular among teens who want their tobacco use to go undetected, patrons of bars and restaurants where smoking is not permitted, and smokers who want to quit.

The tobacco companies are looking to attract a whole new generation of tobacco users and to shore up profits in the face of declining cigarette sales. R.J. Reynolds launched a nationwide marketing campaign for "Camel Snus" in 2009. As usual, the company denies that it aims its marketing campaign toward underage users.

So far there is no evidence that Snus use may be a risk factor for cancers of the mouth and throat. Nevertheless, it is a tobacco product. It's worth remembering that there were no clear health risks associated with cigarettes, either, when they first became popular in the 1940s. ■

a) Position of the vocal cords during quiet breathing.

b) Position of the vocal cords during sound production.

Figure 10.4 Structures associated with the production of sound.

Speech has played an important role in our evolutionary history and the development of human culture. Although sound production does not play a role in the homeostasis of respiratory gases, animals have evolved to take advantage of the energy available in the moving air to accomplish this important function.

The trachea transports air As air continues down the respiratory tract, it passes to the **trachea,** the "windpipe" that extends from the larynx to the left and right bronchi. The trachea consists of a series of C-shaped, incomplete rings of cartilage held together by connective tissue and muscle. As shown in Figure 10.5, each cartilage ring extends only three-quarters of the circumference

Epithelial tissue

Connective tissue

Cartilage rings

Smooth muscle

a) Relaxed state. The maximum diameter facilitates air movement in and out.

b) During the cough reflex, the smooth muscle contracts briefly, reducing the diameter of the trachea. Combined with contraction of the abdominal muscles, this increases the velocity of air movement, forcibly expelling irritants or mucus from the trachea.

Figure 10.5 The trachea. The trachea consists of smooth muscle and layers of epithelial and connective tissue held open by tough, flexible C-shaped bands of cartilage.

of the trachea. The rings of cartilage keep the trachea open at all times, but because they are not complete circles they permit the trachea to change diameter slightly when we cough or breathe heavily.

Like the nasal cavity, the trachea is lined with cilia-covered epithelial tissue that secretes mucus. The mucus traps foreign particles and the cilia move them upward, away from the lungs.

If a foreign object lodges in the trachea, respiration is interrupted and choking occurs. If the airway is completely blocked, death can occur within minutes. Choking often happens when a person carries on an animated conversation while eating. Beyond good manners, the risk of choking provides a good reason not to eat and talk at the same time.

Choking typically stimulates receptors in the throat that trigger the *cough reflex*. This is a sudden expulsion of air from the lungs in an attempt to dislodge foreign material (Figure 10.5b). If the object blocks the airway completely before the person has finished inhaling, there may not be much air in the lungs. This will make the obstacle more difficult to remove . If the object blocks air flow only partially, it may be possible to dislodge it by inhaling slowly, then coughing.

Bronchi branch into the lungs The trachea branches into two airways called the right and left **bronchi** (singular: *bronchus*) as it enters the lung cavity (refer to Figure 10.3). Like the branches of a tree, the two bronchi divide into a network of smaller and smaller bronchi. The walls of bronchi contain fibrous connective tissue and smooth muscle reinforced with cartilage. As the branching airways get smaller and smaller, the amount of cartilage declines. By definition, the smaller airways that lack cartilage are called **bronchioles.** The smallest bronchioles are 1 mm or less in diameter and consist primarily of a thin layer of smooth muscle surrounded by a small amount of elastic connective tissue.

The bronchi and bronchioles have several other functions in addition to air transport. They also clean the air, warm it to body temperature, and saturate it with water vapor before it reaches the delicate gas-exchange surfaces of the lungs. The air is warmed and humidified by contact with the moist surfaces of the cells lining the bronchi and bronchioles. With the exception of the very smallest bronchioles, the bronchi and bronchioles are lined with ciliated epithelial cells and occasional mucus-secreting cells. The thin, watery mucus produced by the mucus-secreting cells traps dust, bacteria, and other small particles. The ciliated cells then sweep the accumulated mucus and trapped material upward toward the pharynx so that it can be swallowed.

Tobacco smoke contains chemicals and particles that irritate the respiratory tract. Mucus production increases in response, but the smoke impairs the activity of the cilia. Continued smoking destroys the cilia, allowing mucus and debris from the smoke to accumulate in the airway. "Smoker's cough" refers to the violent coughing necessary to dislodge the mucus from the airway. Mucus pooling leads to frequent infections because pathogens and irritants remain in the respiratory tract. It also increases the risk of bronchitis, emphysema, and lung cancer (**Figure 10.6**) (see section 10.6, Disorders of the respiratory system).

From the nose and mouth to the tiniest bronchioles in the lungs, none of the airways we have described so far participate in gas exchange. Essentially, they are all tubes for getting air to the lungs, where gas exchange actually occurs.

The lungs are organs of gas exchange The lungs are organs consisting of supportive tissue enclosing the bronchi, bronchioles, blood vessels, and the areas where gas exchange occurs. They occupy most of the thoracic cavity. There are two lungs, one on the right side and one on the left, separated from each other by the heart (**Figure 10.7**). The shape of the lungs follows the contours of the rib cage and the thoracic cavity. The base of each lung is broad and shaped to fit against the convex surface of the diaphragm.

Each lung is enclosed in two layers of thin epithelial membranes called the **pleural membranes.** One of these layers represents the outer lung surface and the other lines the thoracic cavity. The pleural membranes are separated by a small space, called the *pleural cavity*, that contains a very small amount of watery fluid. The fluid reduces friction between the pleural membranes as the lungs and chest wall move during breathing. Inflammation of the pleural membranes, a condition called pleurisy, can reduce the secretion of pleural fluid, increase friction, and cause pain during breathing. Pleurisy can be a symptom of pneumonia (see section 10.6).

a) Healthy airway.

b) Smoker's airway.

Figure 10.6 **Effects of smoking on the cilia of the airways.**

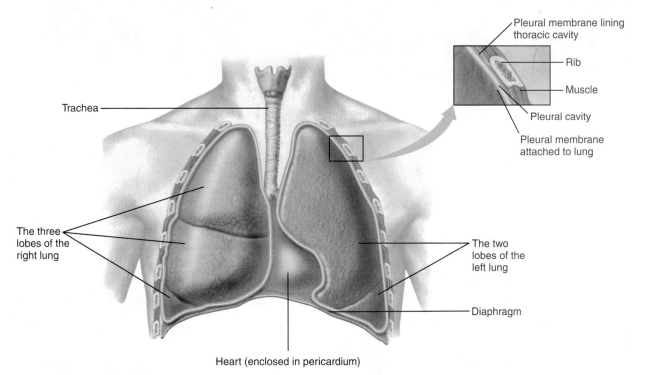

Figure 10.7 **The lungs, pleural membranes, and pleural cavity.** The pleural cavity has been expanded so that it can be seen in the drawing. In reality, it is no more than a very thin, watery space that reduces friction between the two pleural membranes.

Lungs consist of several lobes, three in the right lung and two in the left. Each lobe contains a branching tree of bronchioles and blood vessels. The lobes can function fairly independently of each other, so it is possible to surgically remove a lobe or two without totally eliminating lung function.

Gas exchange occurs in alveoli If you could touch a living lung, you would find that it is very soft and frothy. In fact, most of it is air. The lungs are basically a system of branching airways that end in 300 million tiny air-filled sacs called **alveoli** (singular: *alveolus*). It is here that gas exchange takes place (**Figure 10.8**). Alveoli are arranged in clusters at the end of every terminal bronchiole, like grapes clustered on a stem. A single alveolus is a thin bubble of living squamous epithelial cells only one cell layer thick. Their combined surface area is nearly 800 square feet, approximately 40 times the area of our skin. This tremendous surface area

and the thinness of the squamous type of epithelium facilitate gas exchange with nearby capillaries.

Within each alveolus, certain epithelial cells secrete a lipoprotein called *surfactant* that coats the interior of the alveoli and reduces surface tension. Surface tension is due to the attraction of water molecules toward each other (review Chapter 2). Without surfactant, the force of surface tension could collapse the alveoli. This can occur in infants who are born prematurely, because the surfactant-secreting cells in their lungs are underdeveloped. Called *infant respiratory distress syndrome*, the condition is treated with surfactant replacement therapy.

Pulmonary capillaries bring blood and air into close contact
As described in Chapter 8, the right ventricle of the heart pumps deoxygenated blood into the pulmonary trunk, which splits into the left and right pulmonary arteries. The pulmonary arteries divide into smaller and smaller

a) Bronchioles end in clusters of alveoli, each surrounded by capillaries. CO_2 and O_2 are exchanged across the capillary and alveolar walls by diffusion.

b) **Photo of the surface of alveoli covered with capillaries.**

Figure 10.8 **Gas exchange between the blood and alveoli.**

arteries and arterioles, eventually terminating in a capillary bed called the *pulmonary capillaries*. In the pulmonary capillaries, blood comes into very close proximity to the air in the alveoli. Only two living cells (the squamous epithelial cell of the alveolus and the cell of the capillary wall) separate blood from air at this point. A series of venules and veins collects the oxygenated blood from the pulmonary capillaries and returns the blood to the left side of the heart. From there it can be transported to all parts of the body.

The close contact between air and blood and the large air surface area of the lungs suggests that the lungs might be useful as an alternative method for delivering medications to the bloodstream. Currently pharmaceutical researchers are experimenting with fine mists or powders that can be inhaled deep into the lungs. The research has begun to pay off. The first inhalable form of insulin (a dry powder) was approved by the Food and Drug Administration in 2006.

↺ Recap The respiratory system is specialized for the exchange of oxygen and carbon dioxide with the air. Sound is produced by vibration of the vocal cords of the larynx as air passes through the glottis. The trachea, or "windpipe," branches into the right and left bronchi. The bronchi and bronchioles filter, warm, and humidify the incoming air. The lungs are organs containing a branching system of bronchi and bronchioles, blood vessels, and 300 million alveoli. Gas exchange occurs between the alveoli and pulmonary capillaries. ▪

10.3 The process of breathing involves a pressure gradient

Breathing involves getting air into and out of the lungs in a cyclic manner, and that requires muscular effort. However, the lungs themselves don't have any skeletal muscle tissue. The lungs expand passively because the surrounding bones and muscles expand the size of the chest cavity.

The bones and muscles of respiration include the ribs, the intercostal muscles between the ribs, and the main muscle of respiration, called the **diaphragm**, a broad sheet of muscle that separates the thoracic cavity from the abdominal cavity (review Figure 10.1). The intercostal muscles and the diaphragm are skeletal muscles.

Inspiration brings in air, expiration expels it

To understand why air moves into and out of the lungs in a cyclic manner, we need to understand the following general principles of gas pressure and of how gases move:

- Gas pressure is caused by colliding molecules of gas.
- When the volume of a closed space *increases*, the molecules of gas in that space are farther away from each other, and the pressure inside the space *decreases*. Conversely, when the volume in a closed space *decreases*, the gas pressure *increases*.

- Gases flow from areas of *higher* pressure to areas of *lower* pressure.

As we have seen, the lungs are air-filled structures consisting almost entirely of bronchioles, alveoli, and blood vessels. Lacking skeletal muscle, they cannot expand (increase in volume) or contract (decrease in volume) on their own. The lungs expand and contract only because they are compliant (stretchable) and because they are surrounded by the pleural cavity, which is airtight and sealed. If the volume of the pleural cavity expands, the lungs will expand with it.

Inspiration (inhalation) pulls air into the respiratory system as lung volume expands, and *expiration* (exhalation) pushes air out as lung volume declines again. Let's look at a cycle of inspiration and expiration, starting from the relaxed state at the end of a previous expiration (**Figure 10.9** on the next page):

1. *Relaxed state.* At rest, both the diaphragm and the intercostal muscles are relaxed. The relaxed diaphragm appears dome shaped.

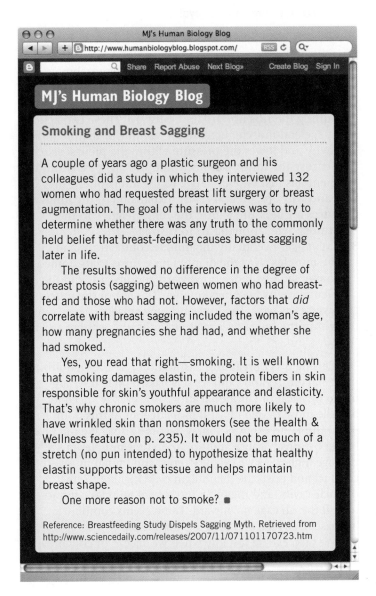

MJ's Human Biology Blog

Smoking and Breast Sagging

A couple of years ago a plastic surgeon and his colleagues did a study in which they interviewed 132 women who had requested breast lift surgery or breast augmentation. The goal of the interviews was to try to determine whether there was any truth to the commonly held belief that breast-feeding causes breast sagging later in life.

The results showed no difference in the degree of breast ptosis (sagging) between women who had breast-fed and those who had not. However, factors that *did* correlate with breast sagging included the woman's age, how many pregnancies she had had, and whether she had smoked.

Yes, you read that right—smoking. It is well known that smoking damages elastin, the protein fibers in skin responsible for skin's youthful appearance and elasticity. That's why chronic smokers are much more likely to have wrinkled skin than nonsmokers (see the Health & Wellness feature on p. 235). It would not be much of a stretch (no pun intended) to hypothesize that healthy elastin supports breast tissue and helps maintain breast shape.

One more reason not to smoke? ▪

Reference: Breastfeeding Study Dispels Sagging Myth. Retrieved from http://www.sciencedaily.com/releases/2007/11/071101170723.htm

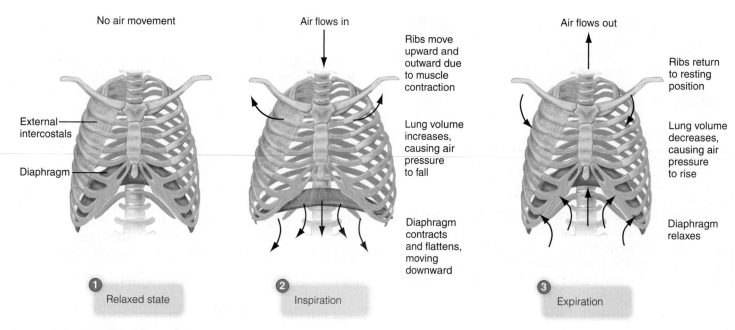

No air movement

Air flows in

Air flows out

Ribs move upward and outward due to muscle contraction

Ribs return to resting position

External intercostals

Lung volume increases, causing air pressure to fall

Lung volume decreases, causing air pressure to rise

Diaphragm

Diaphragm contracts and flattens, moving downward

Diaphragm relaxes

1 Relaxed state

2 Inspiration

3 Expiration

Figure 10.9 **The respiratory cycle.**

2. *Inspiration.* As inspiration begins, the diaphragm contracts, flattening it and pulling its center downward. At the same time, the intercostal muscles contract, pulling the ribs upward and outward. These two actions of skeletal muscle increase the volume of the pleural cavity and lower the pressure within the pleural space. Because the lungs are elastic and the pressure around them has just fallen relative to the atmosphere, they expand with the pleural cavity. Expansion of the lungs reduces air pressure within the lungs relative to the atmosphere, allowing air to rush in.

3. *Expiration.* Eventually the muscle contractions end. As the muscles relax the diaphragm returns to its domed shape, the ribs move downward and inward, and the pleural cavity becomes smaller. The rest of the process reverses as well. The lungs become smaller, so pressure within the lungs rises relative to the atmosphere and air flows out.

During quiet breathing, inspiration is active (requiring muscular effort) and expiration is passive. When we are under physical or emotional stress, however, we need to breathe more frequently and more deeply. At this point both inspiration and expiration may become active. We can take bigger breaths because additional rib cage muscles raise the rib cage higher. As we exhale deeply, abdominal muscles contract and push the diaphragm even higher into the thoracic cavity, and the inner intercostal muscles contract to pull the rib cage downward. These events combine to increase the speed and force of respiration.

We also exhale forcibly when we sneeze or cough. During sneezing and coughing the abdominal muscles contract suddenly, raising abdominal pressure. The rapid increase in abdominal pressure pushes the relaxed diaphragm upward against the pleural cavity, forcing air out of the lungs.

Lung volumes and vital capacity measure lung function

At rest, you take about 12 breaths every minute. Each breath represents a **tidal volume** of air of approximately 500 milliliters (ml), or about a pint (**Figure 10.10a**). On average, only about 350 ml of each breath actually reach the alveoli and become involved in gas exchange. The other 150 ml remain in the airways, and because it does not participate in gas exchange, this air is referred to as *dead space volume*.

The maximal volume that you can exhale after a maximal inhalation is called your **vital capacity.** Your vital capacity is about 4,800 ml, almost 10 times your normal tidal volume at rest. The amount of additional air that can be inhaled beyond the tidal volume (about 3,100 ml) is called the *inspiratory reserve volume*, and the amount of the air that we can forcibly exhale beyond the tidal volume (about 1,200 ml) is the *expiratory reserve volume*. No matter how forcibly you exhale, some air always remains in your lungs. This is your *residual volume*, approximately 1,200 ml.

A device called a *spirometer* can measure lung capacity (Figure 10.10b). The measurements are made by having a person breathe normally into the device (to measure tidal volume) and then take a maximum breath and exhale it forcibly and as completely as possible (to measure vital capacity). Lung volumes and rates of change of volume are useful in diagnosing various lung diseases. For example, emphysema is a condition in which the smaller airways lose

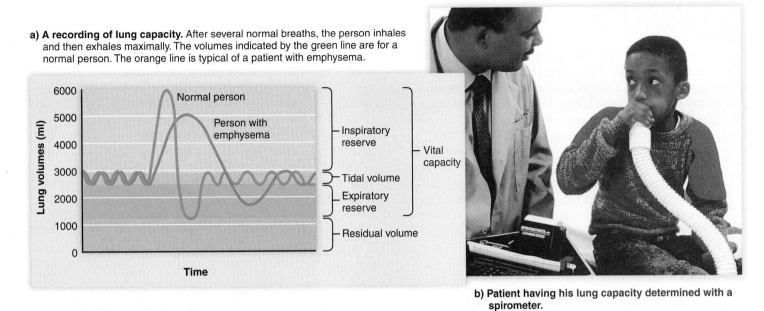

a) **A recording of lung capacity.** After several normal breaths, the person inhales and then exhales maximally. The volumes indicated by the green line are for a normal person. The orange line is typical of a patient with emphysema.

b) **Patient having his lung capacity determined with a spirometer.**

Figure 10.10 **Measurement of lung capacity.**

✓ *When a person who is initially at rest begins exercising, do you think tidal volume will increase, decrease, or stay the same? How about vital capacity? Explain.*

elasticity, causing them to collapse during expiration and impairing the ability to exhale naturally. A spirometer recording of someone with emphysema might show a prolonged period of expiration after a maximum inspiration because of resistance to air outflow.

↩ **Recap** Inspiration is an active process (requiring energy) that occurs when the diaphragm and intercostal muscles contract. Normally expiration is passive, but it can become active when we forcibly exhale, cough, or sneeze. Although we normally take breaths of about 500 ml, the maximum breath we can inhale and then forcibly exhale is about 4,800 ml. Some air, called the residual volume, remains in the lungs even at the end of expiration. ▪

10.4 Gas exchange and transport occur passively

So far, we have focused on the first process in respiration: breathing. Once air enters the alveoli, gas exchange and transport occur. In this section we review some basic principles governing the diffusion of gases to set the stage for our discussion of external and internal respiration (the second and third processes of respiration). We also describe how the gases are transported by blood. The fourth process of respiration, cellular respiration, is the use of oxygen by cells in the production of energy. Cellular respiration is described in Chapter 3.

Gases diffuse according to their partial pressures

Earth is surrounded by an atmosphere of gases. Like liquids, gases have mass and are attracted to the earth by gravity. Though it doesn't really feel like we are pressed down by a heavy weight of gases, in fact the atmosphere (air) exerts a total atmospheric pressure at sea level of about 760 mm Hg (millimeters of mercury). A normal atmospheric pressure of 760 mm Hg means that the pressure of the atmosphere will cause a column of mercury in a vacuum to rise 760 mm, or about 2.5 feet. The pressure seems like zero to us because the pressure inside our lungs is the same as atmospheric pressure, at least when we are resting between breaths.

The primary gases of Earth's atmosphere are nitrogen (78%) and oxygen (21%), with a trace amount of carbon dioxide (about 0.04%) and less than 1% of all other gases combined. In a mixture of gases, each gas exerts a **partial pressure** that is proportional to its percentage of the total gas composition. Partial pressure is represented by P and, like atmospheric pressure, it is measured in mm Hg. The pressure of the atmosphere is thus the sum of the partial pressures of each of the gases found in the atmosphere at sea level. Because we know the percentages of each gas in our atmosphere, we can find the partial pressure of each. For example, the partial pressure of oxygen (P_{O_2}) in air is about 160 mm Hg (760 mm Hg × 0.21).

Because partial pressures in a mixture of gases are directly proportional to concentrations, a gas will always diffuse down its *partial pressure gradient*, from a region of higher partial pressure to a region of lower partial pressure. As we shall see, the exchanges of O_2 and CO_2, both between the alveoli and the blood and between the blood and the tissues, are purely passive. No consumption of ATP is involved; changes in partial pressures are entirely responsible for the exchange and transport of these gases.

✔ **Quick Check** At the top of Mount Everest, the atmospheric pressure is usually around 260 mm Hg, and the percentages of nitrogen, oxygen, and other gases are the same as they are at sea level. What is the partial pressure of oxygen at the summit of Everest? ▪

External respiration: The exchange of gases between air and blood

A comparison of the partial pressures of O_2 and CO_2 in inhaled air, in the alveoli, and in the blood in the lungs illustrates how external respiration takes place. As stated above, the partial pressure of the oxygen (P_{O_2}) in the air we breathe is about 160 mm Hg (**Figure 10.11a**). As there is very little CO_2 in the air, the partial pressure of inspired CO_2 is negligible. The partial pressures of alveolar air are not, however, the same as those of inspired air. This is because only about 1/8 of the air is actually exchanged with each breath, so most of the air in the lungs is actually "old" air that has already undergone some gas exchange. Consequently, the partial pressures of O_2 and CO_2 in the alveoli average about 104 and 40 mm Hg, respectively.

When venous (deoxygenated) blood with a P_{O_2} of only 40 mm Hg and a P_{CO_2} of 46 mm Hg arrives at the pulmonary capillaries from the pulmonary arteries, O_2 diffuses from the alveoli into the capillaries, and CO_2 diffuses in the opposite direction (Figure 10.11b). As a result, the P_{O_2} of oxygenated (arterial) blood leaving the lungs rises to 100 mm Hg and the P_{CO_2} falls to 40 mm Hg (Figure 10.11c). The oxygenated blood is carried in the pulmonary veins to the heart and then throughout the body in the arterial blood vessels. The CO_2 that diffuses into the alveoli is exhaled, along with some water vapor.

Notice that the definition of venous blood is that it is deoxygenated, not that it happens to be in a vein. The pulmonary arteries transport venous blood to the lungs, and the pulmonary veins transport arterial blood to the heart.

✔ **Quick Check** Suppose an airplane flying at high altitude loses cabin pressure, such that the partial pressure of oxygen in the air in the passengers' alveoli drops to 35 mm Hg (while the partial pressure of oxygen of their venous blood remains normal). Which way will oxygen diffuse—from the venous blood to air, or from air to the venous blood—and why? ▪

Internal respiration: The exchange of gases with tissue fluids

The body's cells get their supply of O_2 for cellular respiration from the interstitial fluid that surrounds them (Figure 10.11d).

Because the cells are constantly drawing oxygen from the interstitial fluid (again, by diffusion), the interstitial fluid P_{O_2} is usually quite a bit lower than that of arterial blood (less than 40 mm Hg). As blood enters the capillaries, then, O_2 diffuses from the capillaries into the interstitial fluid, replenishing the O_2 that has been used by the cells. CO_2 diffuses in the opposite direction, from the cell into the interstitial fluid and then into the capillary blood.

Both external and internal respiration occur entirely by diffusion. The partial pressure gradients that permit diffusion are maintained by breathing, blood transport, and cellular respiration. The net effect of all these processes is that homeostasis of the concentrations of O_2 and CO_2 in the vicinity of the cells is generally well maintained.

Hemoglobin transports most oxygen molecules

Our discussion of external and internal respiration brings us to an important aspect of the overall subject of gas exchange, and that is how the gases are transported between the lungs and tissues in the blood. As noted in Chapter 7, an important function of blood is to carry oxygen from the lungs to body tissues; now we examine how that happens.

Oxygen is transported in blood in two ways: either it is bound to hemoglobin (Hb) in red blood cells, or it is dissolved in blood plasma (**Figure 10.12** on page 232). The presence of hemoglobin is absolutely essential for the adequate transport of O_2 because O_2 is not very soluble in water. Only about 2% of all O_2 is dissolved in the watery component of blood known as blood plasma. Most of it—98%—is taken out of the watery component by virtue of its binding to hemoglobin molecules. Without hemoglobin, the tissues would not be able to receive enough oxygen to sustain life.

As described in Chapter 7, *hemoglobin* is a large protein molecule consisting of four polypeptide chains, each of which is associated with an iron-containing heme group that can bind oxygen. Because there are four heme groups, each hemoglobin molecule can bind four oxygen molecules at a time, forming **oxyhemoglobin** (HbO_2). We can represent this reaction as:

$$Hb \quad + \quad O_2 \quad \rightarrow \quad HbO_2$$
hemoglobin   oxygen   oxyhemoglobin

This reaction is reversible and highly dependent on the partial pressures of O_2 in plasma. When the P_{O_2} rises (in the lungs), oxygen attaches to hemoglobin and is transported in arterial blood. When the P_{O_2} falls (at the tissues), oxygen detaches from hemoglobin.

Several other factors affect O_2 attachment to hemoglobin as well. Hemoglobin binds O_2 most efficiently in conditions of fairly neutral pH and relatively cool temperatures—similar to conditions existing in the lungs. Body regions having warmer temperatures and lowered pH—such as in body tissues—reduce hemoglobin's affinity for binding O_2. Consequently, O_2 and hemoglobin tend to combine in the lungs, facilitating the transport of oxygen to the tissues, and to detach in body tissues, making O_2 available to cells.

Figure 10.11 Partial pressures. All partial pressures are expressed in units of mm Hg. Differences in partial pressures account for the diffusion of O_2 and CO_2 between the lungs and blood, and between blood and the body's tissues.

✓ *Does PO_2 increase, decrease, or stay the same as a molecule passes from (1) the alveolar air to (2) a pulmonary capillary as it leaves the lungs to (3) the interstitial fluid? Explain why this trend is important.*

a) Approximately 98% of all the O_2 transported to the tissues by arterial blood is bound to hemoglobin (Hb) within red blood cells.

b) Most of the CO_2 transported to the lungs by venous blood is converted to bicarbonate (HCO_3^-) within red blood cells and then carried in plasma as dissolved HCO_3^-. The rest is either dissolved in plasma as CO_2 or transported within the red blood cells bound to hemoglobin.

Figure 10.12 **How oxygen and carbon dioxide are transported in blood.**

Hemoglobin's affinity for oxygen is also greatly reduced by carbon monoxide, a colorless, odorless, highly poisonous gas (see the Health & Wellness box on page 235).

✓ **Quick Check** People who have insufficient iron in their diets will often become chronically oxygen-deprived, even to the point of passing out, despite having normal levels of O_2 dissolved in the blood plasma. Explain how they can be oxygen-deprived if they have a normal O_2 concentration in the blood plasma. ▪

Most CO_2 is transported in plasma as bicarbonate

Cellular metabolism in body tissues continuously produces carbon dioxide as a waste product. As we learned in Chapter 7, one function of blood is to transport CO_2 away from tissues and back to the lungs, where it can be removed from the body.

Because the partial pressure of CO_2 is higher in the tissues than it is in blood, CO_2 readily diffuses from tissues into the bloodstream. Once in the blood, CO_2 is transported in three ways: dissolved in blood plasma, bound to hemoglobin, or in the form of bicarbonate (see Figure 10.12).

Only about 10% of the CO_2 remains dissolved in blood plasma. Another 20% binds with hemoglobin to

form **carbaminohemoglobin** ($HbCO_2$). This reaction is represented as:

$$Hb \quad + \quad CO_2 \quad \rightarrow \quad HbCO_2$$
hemoglobin   carbon dioxide   carbaminohemoglobin

Hemoglobin can transport O_2 and CO_2 molecules simultaneously because the two gases attach to different sites on the hemoglobin molecule. O_2 combines with heme, CO_2 with globin.

About 70% of all the CO_2 produced by the tissues is converted to bicarbonate (HCO_3^-) prior to transport. When bicarbonate is produced, CO_2 combines with water (H_2O) to become carbonic acid (H_2CO_3). This first reaction is catalyzed by an enzyme called *carbonic anhydrase*. The carbonic acid immediately breaks apart into bicarbonate and hydrogen (H) ions, as follows:

$$CO_2 \quad + \quad H_2O \quad \rightarrow \quad H_2CO_3 \quad \rightarrow \quad HCO_3^- \quad + \quad H^+$$
carbon dioxide   water   carbonic acid   bicarbonate   hydrogen

The formation of bicarbonate from CO_2 occurs primarily inside red blood cells because this is where the carbonic anhydrase enzyme is located. However, most of the bicarbonate quickly diffuses out of red blood cells and is transported back to the lungs dissolved in plasma.

Some of the hydrogen ions formed along with bicarbonate stay inside the red blood cells and bind to hemoglobin. Their attachment to hemoglobin weakens the attachment between hemoglobin and oxygen molecules and causes hemoglobin to release more O_2. This is the chemistry behind the previously mentioned effect of pH on oxygen binding. The overall effect is that the presence of CO_2 (an indication that cellular metabolism has taken place) actually enhances the delivery of O_2 to the very sites where it is most likely to be needed.

At the lungs, dissolved CO_2 diffuses out of the blood and into the alveolar air. The loss of CO_2 from the blood plasma causes the P_{CO_2} to fall, which in turn causes the chemical reaction that formed bicarbonate in the first place to reverse, as follows:

$$\underset{\text{bicarbonate}}{HCO_3^-} + \underset{\text{hydrogen}}{H^+} \rightarrow \underset{\substack{\text{carbonic} \\ \text{acid}}}{H_2CO_3} \rightarrow \underset{\substack{\text{carbon} \\ \text{dioxide}}}{CO_2} + \underset{\text{water}}{H_2O}$$

Thus, as CO_2 is removed by breathing, the bicarbonate and hydrogen ions formed in the peripheral tissues to transport CO_2 are removed as well.

 Recap In a mixture of gases such as air, each gas exerts a partial pressure. Gases diffuse according to differences in their partial pressures. Diffusion accounts for both external and internal respiration. Nearly all of the oxygen in blood is bound to hemoglobin in red blood cells. Most carbon dioxide produced in the tissues is transported in blood plasma as bicarbonate. ∎

HBP **Web Animation** *The Human Respiratory System* at www.humanbiology.com

10.5 The nervous system regulates breathing

As mentioned earlier, breathing depends on the contractions of skeletal muscles. Skeletal muscles are activated only by motor neurons, as described in Chapter 5. It follows, then, that respiration is controlled by the nervous system. The nervous system regulates the rate and depth of breathing in order to maintain homeostasis of the concentrations of certain key variables, most notably the concentrations of CO_2, H^+, and O_2. In addition, we can exert a certain amount of conscious control over breathing if we wish.

A respiratory center establishes rhythm of breathing

The basic cyclic pattern of inspiration and expiration and the rate at which we breathe are established in an area near the base of the brain called the *medulla oblongata* (**Figure 10.13** on the next page). Within this area, called the **respiratory** **center,** groups of nerve cells automatically generate a cyclic pattern of electrical impulses every 4–5 seconds. The impulses travel along nerves to the diaphragm and the intercostal muscles and stimulate those muscles to contract. As these respiratory muscles contract, the rib cage expands, the diaphragm is pulled downward, and we inhale. As inhalation proceeds, the respiratory center receives sensory input from stretch receptors in the lungs. These receptors monitor the degree of inflation of the lungs and serve to limit inhalation and initiate exhalation. When nerve impulses from the respiratory center to the muscles end, the respiratory muscles relax, the rib cage returns to its original size, the diaphragm moves upward again, and we exhale.

Any disorder that interferes with the transmission of these nerve impulses can affect breathing. Consider amyotrophic lateral sclerosis (ALS, also known as Lou Gehrig's disease, for the famous baseball player who succumbed to it). In ALS the nerves to skeletal muscle become damaged and no longer conduct impulses properly. Over time the skeletal muscles, including the diaphragm and intercostal muscles, weaken and waste away from lack of use. Most ALS patients die within five years of initial diagnosis. Although ALS is not a respiratory condition per se, the immediate cause of death is usually respiratory failure.

Chemical receptors monitor CO_2, H^+, and O_2 levels

The body modifies, or regulates, the rate and depth of the breathing pattern to maintain homeostasis. Under normal circumstances the regulation of breathing centers on maintaining homeostasis of CO_2, H^+, and O_2, with the main emphasis on CO_2.

The sensory mechanisms for detecting changes in CO_2 levels are actually indirect. Rather than detecting changes in CO_2 levels, certain cells in the medulla oblongata can detect changes in the H^+ concentration of the cerebrospinal fluid (the interstitial fluid around the cells in the brain). This is pertinent to control of CO_2 because, as you may recall, a rise in CO_2 concentration is accompanied by a rise in the hydrogen ion concentration according to the following reaction:

$$CO_2 + H_2O \rightarrow H_2CO_3 \rightarrow HCO_3^- + H^+$$

Thus, when the P_{CO_2} of arterial blood rises, the concentration of H^+ in cerebrospinal fluid also rises. Receptor cells detecting an elevated H^+ concentration transmit signals to the respiratory center, causing it to increase the rate of the cyclic pattern of impulses and the number of impulses per cycle. As a result, we breathe more frequently and more deeply, exhaling more CO_2 and lowering blood levels of the gas back to normal. Our normal arterial P_{CO_2} is maintained at about 40 mm Hg because the normal regulation of respiration keeps it there. Any rise above 40 mm Hg stimulates breathing and any fall below 40 mm Hg inhibits it.

Certain other receptor cells respond to blood P_{O_2} rather than P_{CO_2}. These receptors are located in small structures associated with the carotid arteries (the large arteries to the head)

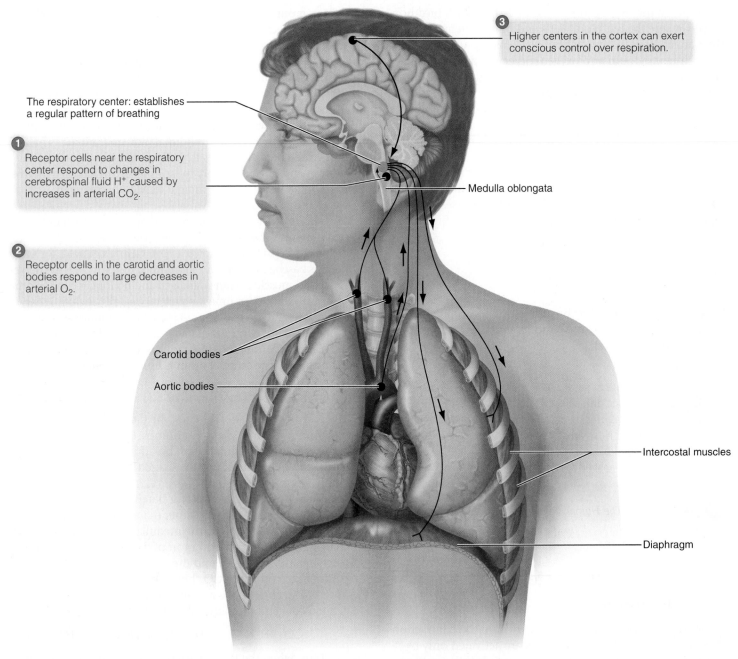

The respiratory center: establishes a regular pattern of breathing

3 Higher centers in the cortex can exert conscious control over respiration.

1 Receptor cells near the respiratory center respond to changes in cerebrospinal fluid H^+ caused by increases in arterial CO_2.

Medulla oblongata

2 Receptor cells in the carotid and aortic bodies respond to large decreases in arterial O_2.

Carotid bodies

Aortic bodies

Intercostal muscles

Diaphragm

Figure 10.13 Regulation of breathing. The basic pattern of breathing is established by a respiratory center within the medulla oblongata. Normally, the rate and depth of respiration are set primarily to regulate blood CO_2 levels, but respiration can also be stimulated by a substantial drop in blood O_2 levels. In addition, respiration can be controlled (at least for short periods of time) by conscious control.

and the aorta, called the *carotid* and *aortic bodies*. Normally, P_{O_2} in arterial blood is about 100 mm Hg. If P_{O_2} drops below about 80 mm Hg, these receptors signal the respiratory center to increase the rate and depth of respiration in order to raise blood O_2 levels back toward normal. The carotid and aortic bodies also can be stimulated by an increase in the blood concentrations of H^+ and CO_2 if they are great enough.

It is important to note that the receptors for O_2 in the carotid and aortic bodies become activated only when arterial P_{O_2} falls by at least 20%. In contrast, even a 2–3% change in P_{CO_2} stimulates respiration. Under normal circumstances, then, respiration is controlled entirely by the receptors that regulate CO_2 by responding to cerebrospinal fluid H^+. It just so happens that the regulation of CO_2 also keeps O_2

Health & Wellness

Carbon Monoxide: An Invisible, Odorless Killer

On January 2, 2006, an explosion roared through the Sago Mine in West Virginia, trapping 13 coal miners 260 feet underground. The men retreated deeper into the mine and, following standard safety guidelines, hung up a fabric barrier that was supposed to block poisonous gases. Rescue workers labored for a day and a half to reach them. Tragically, all but one of them were found dead.

What killed the miners? Not the explosion itself, but a by-product of the combustion, carbon monoxide (CO). It is a colorless, odorless, highly poisonous gas produced by burning carbon-containing substances such as coal, natural gas, gasoline, wood, and oil.

Carbon monoxide has a high affinity for hemoglobin and competes with oxygen for hemoglobin binding sites. CO binds to hemoglobin almost 200 times more tightly than oxygen and, once bound, remains attached for many hours. Even trace amounts significantly reduce the oxygen-transporting capacity of blood. High concentrations reduce the oxygen-carrying capacity of hemoglobin

so much that body tissues, including those of the brain, literally starve.

Early symptoms of CO poisoning include flushed skin (especially the lips), dizziness, headache, nausea, and feeling faint. Continued exposure leads to unconsciousness, brain damage, and death. The leading cause of accidental poisoning in the United States, CO kills about 500 people and contributes to 40,000 emergency room visits every year. About 2,000 people per year in the United States commit suicide by inhaling it. Common sources of CO include

Randall McCloy, Jr., sole survivor of the Sago Mine explosion.

industrial pollution, automobile exhaust, furnaces and space heaters, and cigarette smoke.

If you are exposed to CO it's vital that you get away from the gas and into fresh air as soon as possible. People with severe cases of CO poisoning should be taken to a hospital, where pressurized oxygen or transfusions of red blood cells can be administered to increase blood oxygen levels. Depending on the degree of exposure, survivors can show permanent heart or brain impairment. In one study, almost 40% of CO survivors with heart damage died within seven years, compared to 15% of those without heart damage.

Fortunately, taking a few precautions can prevent CO poisoning. Keep your gas or oil furnace and your car's exhaust system in good repair, and consider installing a CO detector in your home. Never run a car's engine in an enclosed space. Don't smoke—cigarette smoke contains measurable amounts of CO that are delivered directly to your lungs and blood. As much as possible, avoid spending a lot of time in smoke-filled rooms and high-traffic urban areas. ∎

---

within the normal range. In other words, the rate and depth of normal breathing is set by the need to get rid of CO_2, not to obtain O_2.

There are circumstances, however, when the O_2 receptors do come into play. These include certain disease states, drug overdoses, and breathing at high altitudes where the partial pressure of O_2 is much lower than normal.

✓ **Quick Check** Some medical conditions can cause an abnormally low concentration of H^+ ions in the blood. Explain what effect this will have on respiratory rate, and why. ∎

We can exert some conscious control

Finally, there is another way to regulate breathing, and that is by conscious control. Conscious control resides in higher brain centers, most notably the cortex. The ability to modify our breathing allows us to speak and sing. You can choose to hold your breath for a minute or even to hyperventilate (breathe rapidly) for a short while.

But if you don't think automatic regulatory mechanisms are powerful, just try to hold your breath indefinitely. You will find your conscious control overpowered by the automatic regulatory mechanisms described above.

↩ **Recap** The respiratory center in the brain establishes a regular pattern of cyclic breathing. The rate and depth of breathing are then adjusted by regulatory mechanisms that monitor arterial concentrations of CO_2, H^+, and O_2. Conscious control can modify regulatory control but cannot override it completely. ■

10.6 Disorders of the respiratory system

Many factors can lead to disorders of respiration. Among them are conditions that reduce air flow or gas exchange, infections by microorganisms, cancer, diseases of other organs such as congestive heart failure, and genetic diseases.

Reduced air flow or gas exchange impedes respiratory function

Respiration depends on the flow of air between the atmosphere and the alveoli and on the diffusional exchange of gases across the alveolar and capillary walls. Any factor that impairs these activities impedes respiratory function.

Asthma: Spasmodic contraction of bronchi **Asthma** is characterized by spasmodic contraction of bronchial muscle, bronchial swelling, and increased production of mucus. An asthma attack causes partial closure of the bronchi, making breathing difficult. It is a recurrent, chronic lung disorder that affects 17 million people in North America. Its incidence is on the rise both in the United States and around the world.

Symptoms of an asthma attack include coughing while exercising, shortness of breath, wheezing, and a sense of tightness in the chest. People with asthma often wheeze when they exhale. The symptoms can be triggered by any number of causes, including viruses, air particles, allergies (such as allergies to pollen, house dust, and animal fur), exercise (especially in cold temperatures), tobacco smoke, and air pollution. The symptoms may come and go.

Most asthma attacks are caused by a hyperactive immune system. When a person with asthma breathes in allergens such as pollen or tobacco smoke, the body reacts with excessive production of immunoglobulin E. The IgE stimulates mast cells in the lungs to release chemical weapons such as histamine, leading to excessive inflammation and constriction of bronchiolar smooth muscle.

Drugs are available to dilate the bronchi (bronchodilators), reduce the inflammation (corticosteroids), and restore normal breathing. Treatments also focus on preventing attacks by isolating the cause and avoiding it when possible.

Emphysema: Alveoli become permanently impaired **Emphysema** is a chronic disorder involving damage to the alveoli. It begins with destruction of connective tissue in the smaller airways. As a result, the airways become less elastic, do not stay open properly, and tend to collapse during expiration. The high pressures in the lungs caused by the inability to exhale naturally through the collapsed airways eventually damage the fragile alveoli. The result is a permanent reduction in the surface area available for diffusion, and eventually breathlessness and reduced capacity to exchange gases across the lung.

At least one form of emphysema is inherited, but most cases are associated with smoking or long-term exposure to air pollutants. The difference between asthma and emphysema is that asthma is an episodic, recurrent condition of increased airway resistance that largely goes away between episodes, whereas emphysema involves permanent damage to airways that eventually destroys alveoli.

Bronchitis: Inflammation of the bronchi **Bronchitis** refers to inflammation of the bronchi, resulting in a persistent cough that produces large quantities of phlegm. Bronchitis may be acute (comes on suddenly and clears up within a couple of weeks) or chronic (persists over a long period and recurs over several years). Both forms are more common in smokers and in people who live in highly polluted areas.

Symptoms include wheezing, breathlessness, and a persistent cough that yields yellowish or greenish phlegm. Sometimes there is fever and a feeling of discomfort behind the sternum. Acute bronchitis can be treated by humidifying the lungs (using a home humidifier or inhaling steam), drinking plenty of fluids, and taking antibiotics if the infection is caused by bacteria. People with chronic bronchitis may need further testing to rule out other health conditions. Bronchodilators can widen the bronchi; oxygen may be prescribed to raise blood oxygen levels.

Cystic fibrosis: An inherited condition **Cystic fibrosis** is an inherited condition in which a single defective gene causes the mucus-producing cells in the lungs to produce a thick, sticky mucus. The disease affects other organ systems as well. In the lungs, the abnormally thick mucus impedes air flow and also provides a site for the growth of bacteria. People with cystic fibrosis tend to get frequent infections of the airways. Treatment of the disease includes consistent physical therapy to try to dislodge the mucus and keep the airways open. Several promising new drugs are now on the market for this disease. For more on cystic fibrosis, see the Health Watch box in Chapter 19.

✓ **Quick Check** For the past three years, a friend of yours has had symptoms of breathlessness, a tendency to wheeze, and a chronic cough that produces large amounts of yellowish phlegm. He is a smoker. Name the most likely cause of his symptoms, and explain your reasoning. ■

Microorganisms can cause respiratory disorders

The lungs are particularly prone to infections. One of the reasons is that they are moist, warm, and covered in a thin layer of fluid, exactly the conditions favored by microorganisms. Besides the common cold and the flu (both caused by viruses), more serious diseases caused by microorganisms include pneumonia, tuberculosis, and botulism.

MJ's Human Biology Blog

Living with Cystic Fibrosis

Sixty years ago, before doctors knew very much about cystic fibrosis, most children with the disease died before school age. Today people with the disorder are living well into their 30s. Today we know that cystic fibrosis is a genetic disorder, and we know what causes the symptoms.

What we don't have is a cure. Prolonging the life of cystic fibrosis patients is largely based on improvement of care, including medicines that alleviate some manifestations of the disease and physical therapy to keep the lungs clear. What works for one patient doesn't always work as well for the next.

An effective tool in the improvement of care for cystic fibrosis patients has been the establishment of a national registry of patients. The registry, which is managed by the Cystic Fibrosis Foundation, currently collects and stores patient records from more than 100 cystic fibrosis treatment centers across the country. The pooled data has proven very useful in learning what works best for which kinds of patients, and why. It's like having the collective experience of thousands of doctors right at your fingertips.

The concept of using a national registry of patients to improve patient care has proven so successful that other patient groups are copying it. ■

Colds and the flu: Common viral respiratory tract infections

Nearly everyone has had a cold or the flu at some time in their life. Both are caused by viruses. Colds (sometimes called an *upper respiratory infection*, or *URI*) are generally caused by viruses of the rhinovirus or coronavirus families; both are highly contagious but not very virulent. The primary symptoms are coughing, runny nose, nasal congestion, and sneezing. The flu is caused by viruses of the influenza family, and although the symptoms are generally more severe, the flu is also not very virulent. Symptoms of the flu include sore throat, fever, and a cough, sometimes accompanied by aches and chills, muscle pains, and headache. The flu and a cold are easily mistaken for each other, except that colds are generally not accompanied by a fever. In any case it hardly matters because there is no medical treatment for either of them. Rest and plenty of fluids are the best prescription. Only rarely, colds or the flu lead to inflammation of the lungs and pneumonia, accompanied by a bacterial infection that requires antibiotics (see below).

Colds and the flu can be caught over and over again throughout a person's lifetime. The reason is that these viral infections evolve rapidly, so that each year they are just a little bit different from the previous year and so are not recognized by the immune system.

Pneumonia: Infection inflames the lungs In **pneumonia** the lungs become inflamed from an infection, usually caused by viruses or bacteria. The alveoli secrete excess fluid, impairing the exchange of oxygen and carbon dioxide. Symptoms typically include fever, chills, shortness of breath, and a cough that produces yellowish-green phlegm and sometimes blood. Some people experience chest pain when breathing due to inflammation of the membranes that line the chest cavity and cover the lungs.

In North America, pneumonia ranks among the top 10 causes of death, primarily because it is a frequent complication of many serious illnesses. Treatment depends on the microorganism involved; if it is bacterial, antibiotics may be effective. In severe cases, oxygen therapy and artificial ventilation may be necessary. However, most people who develop pneumonia recover completely within a few weeks.

Tuberculosis: Bacterial infection scars the lungs
Tuberculosis (TB) is an infectious disease caused by the bacterium *Mycobacterium tuberculosis*. People pass the infection in airborne droplets by coughing or sneezing. The bacteria enter the lungs and multiply to form an infected "focus." In most cases, the immune system fights off the infection, although it may leave a scar on the lungs. In perhaps 5% of cases, however, the infection spreads via lymphatic vessels to the lymph nodes and may enter the bloodstream. Sometimes the bacteria become dormant for many years, then reactivate later to cause more lung damage.

Major symptoms include coughing (sometimes bringing up blood), chest pain, shortness of breath, fever, night sweats, loss of appetite, and weight loss. A chest X-ray usually reveals lung damage, such as cavities in the lungs or old infections that have healed, leaving scarred lung tissue. A skin test called the tuberculin test can indicate whether someone has been exposed to the infection.

A century ago, tuberculosis was a major cause of death worldwide. With the development of antibiotics the incidence of tuberculosis declined precipitously, and most patients in industrialized countries now recover fully. The disease remains a major health problem in undeveloped nations, however, and recently the incidence of tuberculosis has increased in industrialized countries as well. Many authorities attribute this increased prevalence to immigration of people from developing nations. Worldwide, the increased prevalence of AIDS may also be a factor. Furthermore, some strains of tuberculosis are becoming resistant to antibiotics.

Botulism: Poisoning by bacterial toxin **Botulism** is a form of poisoning caused by a bacterium, *Clostridium botulinum*, occasionally found in improperly cooked or preserved foods.

The bacterium produces a powerful toxin that blocks the transmission of nerve signals to skeletal muscles, including the diaphragm and intercostal muscles.

Symptoms of botulism poisoning usually appear 8 to 36 hours after eating the contaminated food. They can include difficulty swallowing and speaking, double vision, nausea, and vomiting. If not treated, botulism can be fatal because it paralyzes the respiratory muscles.

✓ **Quick Check** A patient comes into a hospital with a bloody cough, shortness of breath, chest pain, and a fever. Name two possible causes for these symptoms, and describe at least two tests that might be helpful in arriving at the correct diagnosis. ▪

Lung cancer is caused by proliferation of abnormal cells

Cancer is the uncontrolled growth of abnormal cells. In the lung, cancerous cells crowd out normal cells and eventually impede normal function. Lung cancer can impair not only the movement of air in the airways but also the exchange of gases in the alveoli and the flow of blood in pulmonary blood vessels (**Figure 10.14**). It accounts for one-third of all cancer deaths in the United States.

Lung cancer takes years to develop and is strongly associated with smoking. More than 90% of lung cancer patients are current or former smokers, and some of those who don't smoke were exposed to secondhand smoke from other people. Two other important causes are radon gas (formed in rock as a breakdown product of uranium, it seeps into uncirculated air in a building's foundation or basement) and chemicals in the workplace, the best known of which is asbestos.

Symptoms of lung cancer include chronic cough, wheezing, chest pain, and coughing up blood. Most people don't go to the doctor for every trivial cough, but any cough that is persistent or accompanied by chest pain—especially coughs that bring up blood—should be checked out immediately. If cancer is caught early it may be possible to remove the cancerous lobe or region of the lung, or to eradicate the tumor with radiation therapy and chemotherapy.

Lung cancer is highly preventable. It is a good idea to not smoke, to know the conditions of your work environment, and, depending on where you live, to have your home inspected for radon gas. For more information on all types of cancer, see Chapter 18.

Pneumothorax and atalectasis: A failure of gas exchange

A pneumothorax is collapse of one or more lobes of the lungs. The most common cause is a penetrating wound of the chest that allows air into the pleural cavity around the lungs, but it can also occur spontaneously as the result of disease or injury to a lung. Pneumothorax can be a life-threatening event, for the inability to inflate the lung results in reduced exchange of oxygen and carbon dioxide. Treatment requires repairing the damage to the chest wall or lung and removing the air from the pleural cavity.

Atalectasis refers to a lack of gas exchange within the lung as a result of alveolar collapse or a buildup of fluid within alveoli. In either case there is no exchange of gases between the atmosphere and the blood in regions affected. Atalectasis is sometimes a complication of surgery, but it can also occur when the amount of surfactant is deficient. Treatment involves finding and reversing the underlying cause. Post-surgical patients in general are encouraged to take deep breaths, cough, and get up and start walking as soon as possible to avoid atalectasis. Sometimes positive pressure ventilation helps to force alveoli open again.

Congestive heart failure impairs lung function

In Chapter 8 we discussed *congestive heart failure* as a cardiovascular condition in which the heart gradually becomes less efficient. Even though it starts as a heart disorder, it can end by impairing lung function as well.

Recall that in congestive heart failure the heart begins to fail as a pump. When the left side of the heart fails, blood backs up in the pulmonary blood vessels behind that side of the heart. The result is a rise in blood pressure in the pulmonary vessels. When pulmonary capillary pressure increases, the balance of physical pressure and osmotic forces across the capillary wall favors fluid loss from the capillary. As a result, fluid builds up in the interstitial spaces between capillaries and alveoli and sometimes within alveoli themselves. This increases the diffusional distance and reduces diffusion of gases. Treatments focus on reducing this fluid buildup by helping the body get rid of fluid and improving the heart's pumping action.

↺ **Recap** The lungs are prone to damage by environmental pollutants, tobacco smoke, and infections by microorganisms. Cases of both asthma and tuberculosis are on the rise. ▪

a) Normal lung.

b) Cancerous lung.

Figure 10.14 **Lung cancer.**

Chapter Summary

Respiration takes place throughout the body *p. 220*

- Respiration encompasses four processes: breathing, external respiration, internal respiration, and cellular respiration.
- External respiration occurs in the lungs; internal respiration and cellular respiration take place in the tissues.

The respiratory system consists of upper and lower respiratory tracts *p. 220*

- The respiratory system includes the air passageways to the lungs and the lungs themselves.
- Bones and skeletal muscles support the respiratory system and participate in breathing.
- The upper respiratory tract consists of the nose and pharynx. The upper respiratory tract filters, warms, and humidifies the air we breathe.
- The lower respiratory tract consists of the larynx, the trachea, the main bronchi, and the lungs.
- Within the lungs, the bronchi branch many times, becoming smaller airways called bronchioles that end in air-filled sacs called *alveoli*.
- The tremendous surface area of the alveoli, coupled with the thinness of the respiratory membrane, facilitates gas exchange with the pulmonary capillaries.

The process of breathing involves a pressure gradient *p. 227*

- Inspiration occurs as the lungs expand due to the action of the diaphragm and the intercostal muscles; expiration occurs when these muscles relax.
- When the lungs expand, the pressure within them falls relative to atmospheric pressure and air rushes in; during expiration, the lungs become smaller and increasing pressure within them forces air out.
- During normal breathing, inspiration is active (requiring energy) and expiration is passive.
- Normally we breathe at about 12 breaths per minute with a tidal volume of 500 ml per breath.
- Vital capacity is the maximum amount of air a person can exhale after a maximal inhalation.

Gas exchange and transport occur passively *p. 229*

- The diffusion of a gas is dependent on a partial pressure gradient, which is equivalent to a concentration gradient.
- External and internal respiration are both processes that occur entirely by diffusion.
- Nearly all (98%) of the oxygen transported by blood is bound to hemoglobin in red blood cells.
- Although some carbon dioxide is transported as dissolved CO_2 or is bound to hemoglobin, most CO_2 (70%) is converted to bicarbonate and then transported in plasma.

The nervous system regulates breathing *p. 233*

- A respiratory center in the medulla oblongata of the brain establishes a regular cycle of inhalation and exhalation.
- Under normal conditions, the rate and depth of breathing is adjusted primarily to maintain homeostasis of arterial blood P_{CO_2}.

- Regulation of respiration by O_2 comes into play only when the P_{O_2} concentration falls by more than 20%, such as in disease states or at high altitude.
- We can exert some conscious control over breathing.

Disorders of the respiratory system *p. 236*

- Asthma is episodic, spasmodic contractions of the bronchi that impede air flow.
- Emphysema is a chronic disorder characterized by high resistance to air flow and destruction of alveoli.
- The lungs are prone to infections because their surface is kept moist and warm in order to facilitate gas exchange. Colds and the flu can occur nearly every year. Tuberculosis is an infectious disease caused by a bacterium. In pneumonia, infected alveoli secrete excess fluid, impairing gas exchange.
- Pneumothorax and atalectasis are conditions characterized by a failure of gas exchange, either due to collapse of a lung (pneumothorax) or alveolar collapse or filling of alveoli with fluid (atalectasis).
- Lung disease can be a secondary condition resulting from impairment of another organ, as in congestive heart failure.

Terms You Should Know

alveoli, *226*
asthma, *236*
bronchi, *224*
bronchioles, *224*
bronchitis, *236*
carbaminohemoglobin, *232*
diaphragm, *227*
emphysema, *236*
epiglottis, *223*
glottis, *223*
larynx, *223*

oxyhemoglobin, *230*
partial pressure, *229*
pharynx, *222*
pleural membranes, *225*
respiratory center, *233*
tidal volume, *228*
trachea, *224*
tuberculosis, *237*
ventilation, *220*
vital capacity, *228*
vocal cords, *223*

Concept Review

Answers can be found at the Human Biology Place.
www.humanbiology.com

1. List at least three functions of your nose.
2. Explain why men's voices are lower than women's.
3. Explain why smokers sometimes have a chronic cough.
4. Distinguish between *bronchi* and *bronchioles*.
5. Explain what happens to gas pressure in a closed container when the size of the container becomes larger and smaller.
6. Describe the sequence of pressure changes that occurs during inspiration and expiration.
7. Define *partial pressure* and explain its importance to gas diffusion.
8. Explain why the partial pressure of oxygen in the alveoli is always lower than the partial pressure of oxygen in the air we breathe.
9. Explain (or use a diagram to show) how bicarbonate is formed from dissolved CO_2 in the region of metabolizing tissues.
10. 10. Describe how heart failure can lead to a decrease in the ability of O_2 and CO_2 to diffuse between the blood and the alveoli.

Test Yourself

Answers can be found in Appendix A.

1. Which of the following lists the order of structures through which air will pass during inspiration?
   a. nose, pharynx, larynx, trachea, bronchi, bronchioles, alveoli
   b. nose, pharynx, trachea, larynx, bronchioles, bronchi, alveoli
   c. alveoli, bronchioles, bronchi, trachea, larynx, pharynx, nose
   d. nose, pharynx, larynx, bronchi, bronchioles, trachea, alveoli

2. During acute epiglottitis, the epiglottis swells to several times its normal size. This would interfere directly with:
   a. internal respiration
   b. external respiration
   c. breathing
   d. gas exchange in the alveoli

3. Much of the respiratory tract is lined with ciliated epithelium. The exception is:
   a. the alveoli
   b. the bronchi
   c. the nasal cavity
   d. the trachea

4. The bronchioles are:
   a. kept open by cartilage rings
   b. larger than the bronchi
   c. located between the bronchi and the alveoli
   d. an important site of gas exchange in the lungs

5. Why do smokers experience a higher incidence of respiratory infection than nonsmokers?
   a. Smoking introduces disease-causing bacteria.
   b. The nicotine and tar in smoke promote bacterial growth.
   c. Smoking inhibits mucus production.
   d. Smoking diminishes the number and action of cilia.

6. Unlike capillaries in the systemic circuit, in the pulmonary circuit, _____ blood enters the capillaries from the arterioles and _____ blood leaves the capillaries for the venules.
   a. oxygenated…oxygenated
   b. oxygenated…deoxygenated
   c. deoxygenated…xygenated
   d. deoxygenated…deoxygenated

7. Botulism toxin interferes with the transmission of nerve impulses (action potentials) to skeletal muscles. This interferes directly with:
   a. external respiration
   b. gas exchanges
   c. cellular respiration
   d. breathing

8. During a breathing cycle, as the diaphragm and intercostals muscles _____, the volume of the pleural cavity _____ and air moves _____ the lungs.
   a. contract…decreases…into
   b. contract…increases…into
   c. relax…increases…out of
   d. relax…decreases…into

9. Which of the following has the highest P_{O_2}?
   a. alveolar air
   b. pulmonary artery
   c. interstitial fluid
   d. pulmonary vein

10. All of the following conditions favor the attachment of O_2 to hemoglobin except:
    a. high P_{O_2}
    b. cooler temperature
    c. neutral pH
    d. presence of carbon monoxide

11. In the blood, O_2 is transported primarily _____ while CO_2 is transported primarily _____.
    a. as oxyhemoglobin…as carbaminohemoglobin
    b. dissolved in plasma…as carbaminohemoglobin
    c. as oxyhemoglobin…as bicarbonate in plasma
    d. as oxyhemoglobin…dissolved in plasma

12. Which of the following is characterized by loss of elasticity in the bronchioles and permanently damaged alveoli?
    a. bronchitis
    b. emphysema
    c. asthma
    d. pneumonia

13. Which of the following is an inherited condition characterized by the production of abnormally thick mucus that can interfere with air flow within the lungs?
    a. asthma
    b. emphysema
    c. cystic fibrosis
    d. congestive heart failure

14. Which of the following will happen when someone holds their breath?
    a. The P_{CO_2} will increase.
    b. The P_{O_2} will increase.
    c. The H^+ concentration will decrease.
    d. All of the above.

15. All of the following respiratory disorders are due to infectious microorganisms except:
    a. tuberculosis
    b. emphysema
    c. influenza
    d. the common cold

Apply What You Know

Answers can be found at the Human Biology Place.
www.humanbiology.com

1. Why would someone administering a Breathalyzer test for alcohol ask the person being tested to give one deep breath rather than many shallow ones?

2. At a murder trial a pathologist is asked whether the dead baby he examined had been alive at the time of birth, or was stillborn. How can the pathologist tell?

3. A lifetime of heavy smoking often leads to destruction of the cilia of the cells that line the bronchi and bronchioles. How does this contribute to an increased chance of lung infections?

4. Laryngitis is an inflammation of the larynx. Why does someone with laryngitis have a hard time speaking?

5. If you take a sharp blow to the stomach you can be said to have the "wind knocked out of you." Why would an unexpected blow to the stomach leave someone unable to breathe for a moment?

6. When a hole is created in the chest wall, the lung collapses, even if the lung is not damaged. A collapsed lung is called a pneumothorax. Why does the lung collapse if it is not damaged? Will the patient be able to breathe?

7. A person who gets overly excited or has an anxiety attack may begin to hyperventilate. This can lead to a feeling of dizziness, light-headedness, and muscle weakness. One remedy is to have the person breathe into a paper bag. Why would this help?

11

The Nervous System: Integration and Control

Colored SEM (× 2,700) of two nerve cells from the cerebellum of the human brain.

Medically Induced Coma

Ariel Sharon, former prime minister of Israel, suffered a massive stroke and deteriorated rapidly. Randal McCloy, the sole survivor of a West Virginia mining accident (see the Health Watch box in Chapter 10), was close to death from carbon monoxide poisoning when he was rescued. Minutes after Mexican boxer Ruben Contreras stopped fighting in the sixth round of a scheduled eight-round bout, he had a seizure and collapsed. One month after being bitten by a bat, Wisconsin teenager Jeanna Giese almost died from rabies, a viral infection.

What do these people have in common? Although their circumstances were different, each suffered some type of injury to the brain. And in each case they were placed in a medically induced coma in a last-ditch effort to save their lives.

A Treatment of Last Resort for Brain Injuries

In a typical brain injury, fluid leaks out of capillaries in the injured area, causing swelling. Because the brain is enclosed in a rigid skull, the swelling leads to a rise in fluid pressure within

the skull that constricts the capillaries, cutting off blood flow and reducing oxygen delivery. Deprived of oxygen, healthy brain cells that could have survived the injury begin to die within minutes. In other words, in many cases it's the swelling that results from a brain injury that causes most of the damage, not the initial injury itself.

Treatment for brain injury generally centers on limiting the degree of swelling and subsequent oxygen deprivation. The patient is rushed to a hospital, where a breathing tube is inserted into the airway

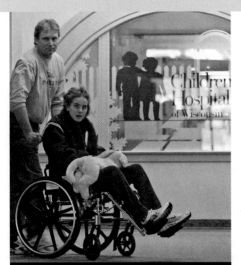

Jeanna Giese was placed in a coma after contracting rabies. She leaves the hospital as the first unvaccinated victim in the world to survive the disease.

so that a respirator can take over the breathing function. Physicians may inject a drug called *mannitol* to reduce brain swelling, or cool the patient's body to 90°F to lower the metabolic rate. Sometimes part of the skull is removed surgically to give brain tissue room to swell.

Medically induced coma is a treatment of last resort. Physicians administer a large dose of a barbiturate or sedative drug, which induces the coma. An electroencephalogram (EEG) recording of brain electrical activity shows flat waves with no brain activity at all, and the person literally appears to be dead. EEG monitoring continues, and drug levels are adjusted accordingly to ensure that the brain stays "asleep."

Why is a medically induced coma thought to be a good idea in some extreme cases? Over half of the brain's metabolic activity (and hence requirement for oxygen) is related to the electrical activity of brain cells; the remainder is for basic cell functions required by all cells. In theory, drastically reducing brain neural activity by

inducing a coma should reduce the brain's consumption of oxygen and glucose by half. A lower metabolic rate should lower blood flow and blood pressure within the capillaries of the brain and that, in turn, should reduce brain swelling. Medically induced coma is thought to buy time for the brain to heal.

But Does It Work?

Medicating an already critically ill individual into a coma is likely to be risky. An immobilized patient may develop atelectasis (collapsed alveoli or fluid-filled alveoli in the lungs) and pneumonia, for it's not possible to ask a comatose patient to get up and walk or even to take a deep breath or cough to avoid developing these conditions. A complete lack of activity may lead to blood clots that can be fatal if they migrate to the lungs. Some patients develop persistent muscle weakness or paralysis that can linger for weeks even after they come out of the coma. Furthermore, it is impossible for physicians to assess the extent of brain damage until the person is aroused.

Although there are several good scientific studies showing that medically induced coma does reduce swelling and fluid pressure in the brain in patients after head trauma, it is not clear that medically induced coma actually improves patient outcomes or quality of life. One retrospective study reported the outcomes of 38 patients who had been placed in medically induced comas after head injuries. Over half of the patients died during intensive care for their head injuries and four others were left severely disabled or remained in a chronic vegetative state. Only six patients had returned to work. Although these data might suggest that medically induced coma is of little or no benefit, the results are not conclusive because they do not separate

the negative effects of the head injuries from the effects (positive or negative) of the medically induced coma. A carefully controlled prospective study of a group of brain injury patients, half of whom were placed in a medically induced coma and half of whom were not, would have to be undertaken to answer this question. So far, the study has not been done.

And yet, there are success stories occasionally in the headlines. The physicians who placed Jeanna Giese in a coma for eight days are glad they took the risk. Today Ms. Giese is the only unvaccinated person ever to have survived rabies once the symptoms of the disease had developed. Randal McCloy survived his mining accident, came out of a medically induced coma, and has returned home. And Ruben Contreras was well enough to attend a boxing match less than four months after his seizure and subsequent coma. These are anecdotes, of course, but they're intriguing enough to cause physicians and medical researchers to wonder—does medically induced coma really work?

Boxer Ruben Contreras throws a punch in a match against Brian Viloria. Contreras suffered a seizure after the match and was placed in a coma during his treatment.

Questions to consider

1 If someone you love were to suffer a severe head injury, would you want that person placed in a medically induced coma? Would you want to be placed in one yourself? Why or why not?

2 What would you ask the physician in order to help you make an informed decision?

The facts...

- Medically induced coma is used to reduce brain metabolism and brain swelling in certain patients in an effort to limit or prevent brain damage.

- Physicians induce the coma by administering drugs that essentially put the brain to sleep.

- Medically induced coma remains a risky, last-ditch choice: while it can buy time for the brain to recover, it may lead to unrelated medical complications.

>> **The nervous system is the body's main control system.** The nervous system receives input from a variety of sources. It controls our body's physical movements, maintains homeostasis of many internal variables, and even initiates our higher thought processes and emotions.

>> **The unit of function in the nervous system is a neuron** (nerve cell). Neurons generate and transmit electrical impulses called *action potentials*.

>> **Neurons communicate with each other chemically.** Information is passed from one neuron to another via the release of chemical substances called *neurotransmitters*.

>> **The brain has the capacity for long-term memory.** Long-term memory involves permanent chemical and physical changes within the brain.

>> **Psychoactive drugs** (drugs that affect states of consciousness, emotions, or behavior) all act by altering communication between neurons in the brain. They do so by influencing the concentrations or actions of neurotransmitters.

Y**ou're on your way to class.** It is cold but you are not wearing your coat, and you are hungry. Your mind is elsewhere as you think about your psychology exam coming up. As you cross the street you see a doughnut shop and you smell doughnuts and coffee in the air. Suddenly you hear sounds—the screech of a skidding truck behind you and the blaring of the truck's horn. Quickly you whirl around, glance at the approaching truck, and leap to safety on the sidewalk. The doughnuts and the psychology exam are forgotten momentarily.

Your nervous system is constantly receiving input from all kinds of sources. It's also sorting through a huge amount of stored information in your brain's large memory banks, looking for the probable meaning of all sensory inputs, which in this scenario includes the meaning of screeching truck tires. The nervous system adds up and makes sense of this seemingly unrelated information quickly and enables you to react. How does this happen?

Here are four characteristics of the nervous system:

- The nervous system receives information from many different senses simultaneously.
- The nervous system integrates information. Integration is the process of taking different pieces of information from many different sources and assembling the pieces

into a whole that makes sense. This process entails sifting through mountains of data and coming up with a plan or course of action.

- The nervous system is very fast. It can receive information, integrate it, and produce a response within tenths of a second. In this chapter we learn about the special properties of the nervous system that give it such speed.
- The nervous system can initiate specific responses, including muscle contraction, glandular secretion, and even conscious thought and emotions.

Most of the functions of the nervous system are completely automatic and can be carried on simultaneously without requiring our attention or conscious decision making. However, the nervous system can also bring selected information to the level of conscious awareness. Working with the endocrine system (Chapter 13), the nervous system maintains homeostasis. It also allows us to feel emotions, to be aware of ourselves, and to exert conscious control over the incredible diversity of our physical movements.

11.1 The nervous system has two principal parts

The nervous system includes the **central nervous system (CNS)** and the **peripheral nervous system (PNS)** (Figure 11.1). The CNS consists of the brain and the spinal cord. It receives, processes, stores, and transfers information. The PNS includes the components of the nervous system that lie outside the CNS. The PNS has two functional subdivisions: the sensory division of the PNS carries information to the brain and spinal cord, and the motor division of the PNS carries information from the CNS to other parts of the body.

The motor division of the peripheral nervous system is further subdivided along functional lines. The *somatic division* of the PNS controls skeletal muscles, and the *autonomic division* of the PNS controls smooth muscles, cardiac muscles, and glands. In turn, the autonomic division has two subdivisions called the *sympathetic* and *parasympathetic* divisions. In general, the actions of the sympathetic and parasympathetic divisions oppose each other. They work antagonistically to accomplish the automatic, subconscious maintenance of homeostasis.

We examine each of these components in more detail later in this chapter. But first, we discuss the structure and function of neurons and of the neuroglial cells that accompany and support them.

Recap The nervous system has two major subdivisions: the central nervous system (CNS), consisting of the brain and spinal cord, and the peripheral nervous system (PNS), which includes all parts of the nervous system that lie outside the CNS. The motor division of the PNS has a somatic division, which controls skeletal muscles, and an autonomic division, which controls smooth muscles, cardiac muscles, and glands. ■

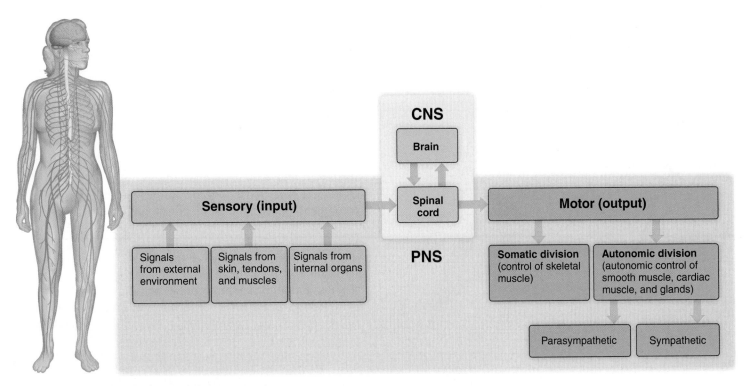

Figure 11.1 Components of the nervous system. The CNS receives input from the sensory component of the PNS, integrates and organizes the information, and then sends output to the periphery via the motor components of the PNS.

11.2 Neurons are the communication cells of the nervous system

Neurons are cells specialized for communication. They generate and conduct electrical impulses, also called *action potentials,* from one part of the body to another. The longest neurons extend all the way from your toes to your spinal cord. These are single cells, remember.

There are three types of neurons in the nervous system, as shown in Figure 11.2 (next page).

- **Sensory neurons** of the PNS are specialized to respond to a certain type of stimulus, such as pressure or light. They transmit information about this stimulus to the CNS in the form of electrical impulses. In other words, sensory neurons provide input to the CNS.
- **Interneurons** within the CNS transmit impulses between components of the CNS. Interneurons receive input from sensory neurons, integrate this information, and influence the functioning of other neurons.
- **Motor neurons** of the PNS transmit impulses away from the CNS. They carry the nervous system's output, still in the form of electrical impulses, to all of the tissues and organs of the body.

All neurons consist of a cell body, one or more dendrites, and an axon. The main body of a neuron is called the **cell body.** The nucleus, with its content of DNA, is located in the cell body, as are the mitochondria and other cell organelles.

Slender extensions of the cell body, called **dendrites,** receive information from receptors or incoming impulses from other neurons (see Figure 11.2). Interneurons and motor neurons have numerous dendrites that are fairly short and extend in many directions from the cell body. Sensory neurons are an exception, for their dendrites connect directly to an axon.

An **axon** is a long, slender tube of cell membrane containing a small amount of cytoplasm. Axons are specialized to conduct electrical impulses. Axons of sensory neurons originate from a dendrite, whereas the axons of interneurons and motor neurons originate from the point of union with the cell body, called the *axon hillock* (see Figure 11.2). At its other end, the axon branches into slender extensions called *axon terminals.* Each axon terminal ends in a small, rounded tip called an *axon bulb.*

Typically, an interneuron or motor neuron receives incoming information from other neurons at its dendrites or cell body. If the incoming information is of the right kind and is strong enough, the neuron responds by generating an electrical impulse of its own at its axon hillock. In contrast, in a sensory neuron the impulse is initiated where the dendrite joins the axon. The impulse is then transmitted from one end of the axon to the other, bypassing the cell body entirely. We talk more about different types of receptors in Chapter 12 (Sensory Mechanisms).

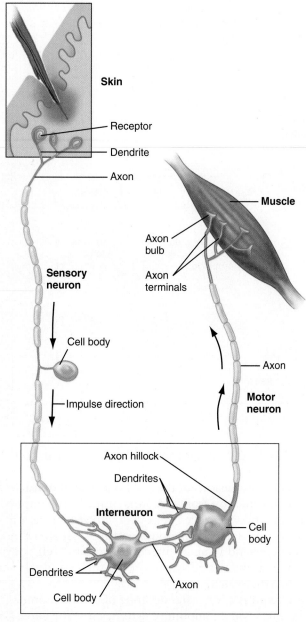

Figure 11.2 Types of neurons in the nervous system. The flow of information begins at a receptor near a dendritic ending of a sensory neuron and ends at the axon bulbs of a motor neuron. In this example the receptors are in skin. Interneurons are located entirely within the central nervous system and typically have short axons.

🔄 **Recap** Neurons generate and transmit electrical impulses from one part of the body to another. Sensory neurons transmit impulses to the CNS. Interneurons transmit impulses between components of the CNS. Motor neurons transmit impulses away from the CNS to muscles and glands. ▪

11.3 Neurons initiate action potentials

The function of a neuron is to transmit information from one part of the body to another in the form of electrical impulses. Next, we look at how a neuron initiates and conducts an impulse.

Sodium-potassium pump maintains resting potential

As described in Chapter 3, the sodium-potassium pump is essential to the control of cell volume by virtue of its ability to remove osmotic particles (sodium ions) from the cell. Now we will learn that the same sodium-potassium pump is also essential for the development and maintenance of an electrical charge across the cell membrane. How does it help to create and maintain an electrical charge?

Recall that the sodium-potassium pump transports three sodium ions (Na^+) out of the cell for every two potassium ions (K^+) transported in. Each time the sodium-potassium pump goes through a cycle, the net effect is the removal of one osmotic particle (a Na^+ ion) and also one positive charge (because the Na^+ ion is positively charged). In other words, the sodium-potassium pump removes both osmotic particles and positive charges from the cell at the same time. As a result of the activity of the sodium-potassium pumps and also the presence of negatively charged protein molecules trapped within the cell, the cell cytoplasm has a slight excess of negative charge compared to the interstitial fluid. The difference in charge results in a small but measurable difference in voltage across the cell membrane, called the *membrane potential.*

In a neuron capable of action potentials but not generating one at the moment, the normal membrane potential at rest is its **resting (membrane) potential.** The resting potential of a neuron is about −70 millivolts (mV), meaning that the inside of the neuron is negatively charged compared to the outside. **Figure 11.3** illustrates the role of the sodium-potassium pump and the processes of diffusion of Na^+ and K^+ (through channels) in maintaining the resting potential.

The concentration of sodium is much higher in the interstitial fluid than it is in the cytoplasm, whereas the inverse is true for potassium. Because sodium is always leaking into the cell and potassium is leaking out by passive diffusion, you might think that the concentration differences between the inside and outside would disappear over time. This doesn't happen because the active transport of sodium and potassium by the pump in the opposite directions exactly balances the rates of leakage. At rest, the ionic concentration differences are maintained and the membrane potential stays at −70 mV, inside negative. As we'll see in a moment, the large concentration differences for sodium and potassium across the axon

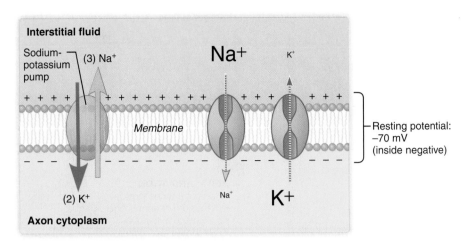

Figure 11.3 Maintenance of the resting membrane potential. The sodium-potassium pump actively transports three Na⁺ ions out of the cell in exchange for two K⁺ ions. Most of the K⁺ leaks out again. However, the inward leak of sodium is so slow that the pump effectively is able to exclude most Na⁺ from the cell. The exclusion of positive ions (Na⁺) from the cell creates a voltage difference (an electrical potential) across the cell membrane of about −70 mV (inside negative). The relative concentrations of Na⁺ and K⁺ in the interstitial fluid and the cytoplasm are indicated by the sizes of the letters associated with the diffusion channels.

membrane enable the neuron to generate impulses simply by opening and closing ion channels.

☑ **Quick Check** Suppose that a neuron has unusual Na/K pumps that move *four* Na⁺ out for every two K⁺ in. Predict whether the resting potential of this neuron will be more negative, less negative, or the same as a normal cell, and explain your reasoning. ■

Graded potentials alter the resting potential

The resting potential of a neuron undergoes a change when an impulse arrives from another neuron. Every incoming impulse produces, in a series of steps discussed in a moment, a small transient change in the resting potential in a local area of the membrane. Depending on the type of signal and its strength, the change might be to *depolarize* the membrane (move the voltage closer to zero), or *hyperpolarize* it (make it even more negative). These transient local changes in resting potential are called **graded potentials** because they can vary in size. They occur only at a single region on the membrane, fading away at increasing distances from that region like the ripples made by a raindrop on water. A neuron may receive hundreds of incoming signals from other neurons, so hundreds of these graded potentials may occur at the same time.

Answers to ☑ questions can be found in Appendix A.

A key feature of graded potentials is that they can add up in space and time, meaning that many incoming signals from other neurons produce a bigger change in membrane potential than does one impulse alone. This is known as **summation.**

An action potential is a sudden reversal of membrane voltage

If the sum of all graded potentials is sufficiently strong to reach a certain triggering membrane voltage called the **threshold,** an action potential results. An **action potential** is a sudden, temporary reversal of the voltage difference across the cell membrane. Once an action potential is initiated, it sweeps rapidly down the axon at a constant amplitude and rate of speed until it reaches the axon terminals.

Figure 11.4 depicts the resting membrane potential, graded potentials and how they summate, and an action potential.

Functionally, an action potential (impulse) is the only form in which information is transmitted long distance by the nervous system. An action potential is like a single monotonic "beep" in a touch-tone phone or a single on-off pulse of current. Its meaning is provided only by where it came from and where it goes—that is, by how the nervous system is wired.

Figure 11.4 The resting membrane potential, graded potentials, and an action potential in a neuron. Stimulation by other neurons is indicated by the arrows. Note that graded potentials can either depolarize or hyperpolarize the membrane and that they can summate. If the sum of all graded potentials is sufficient to depolarize the membrane to threshold, an action potential occurs.

☑ *Extend the graph to draw what you think will happen if this neuron now receives an even stronger depolarizing stimulus (say, a series of four depolarizing graded potentials) that result in another action potential. Will this new action potential be higher and/or longer lasting than the first one?*

Action potentials occur because the axon membrane contains voltage-sensitive ion channels that open and close sequentially once threshold has been reached. An action potential occurs as a sequence of three events: (1) depolarization, (2) repolarization, and (3) reestablishment of the resting potential (**Figure 11.5**).

1. *Depolarization: sodium moves into the axon.* When threshold is exceeded, voltage-sensitive Na$^+$ channels in the axon's membrane open briefly and Na$^+$ ions diffuse rapidly into the cytoplasm of the axon. This influx of positive ions causes *depolarization*, meaning that the membrane potential shifts from negative (-70 mV) to positive (about $+30$ mV).

2. *Repolarization: potassium moves out of the axon.* After a short delay the Na$^+$ channels close automatically. But the reversal of the membrane polarity triggers the opening of K$^+$ channels. This allows more K$^+$ ions than usual to diffuse rapidly out of the cell. The loss of positive ions from the cell leads to *repolarization*, meaning that the interior of the axon becomes negative again.

3. *Reestablishment of the resting potential.* Because the K$^+$ channels are slow to close, there is a brief overshoot of membrane voltage during which the interior of the axon is slightly hyperpolarized. Shortly after the K$^+$ channels close, the resting potential is reestablished. At this point the axon is prepared to receive another action potential. The entire sequence of three steps takes about 3 milliseconds.

While an action potential is under way, an axon can't generate

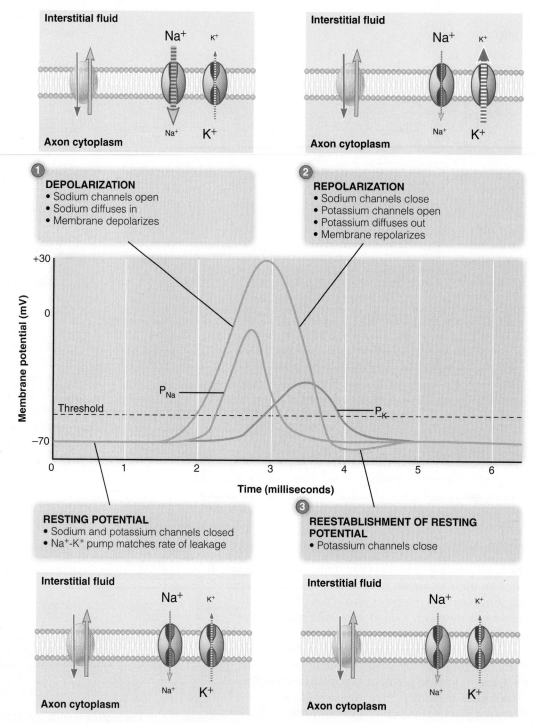

Figure 11.5 Propagation of an action potential. This is a graph of changes in potential over time recorded from a single point on the axon. At rest, the sodium-potassium pump transports sodium and potassium, and the sodium and potassium channels are essentially closed. **(1)** *Depolarization*: once threshold is exceeded, voltage-sensitive sodium channels open briefly and the membrane permeability to sodium (P$_{Na}$) increases. Sodium diffuses rapidly into the axon, depolarizing the membrane. **(2)** *Repolarization*: sodium channels close. Potassium channels open and the membrane permeability to potassium (P$_K$) increases. Potassium diffuses outward, repolarizing the membrane. **(3)** *Reestablishment of resting potential*: after a brief hyperpolarization caused by a delay in the full closure of potassium channels, the membrane potential returns to its normal resting value.

✓ *Explain in your own words why the blue line goes down exactly when the green line goes up.*

another action potential. This period is called the *absolute refractory period*. The presence of an absolute refractory period ensures that action potentials always travel in one direction only, because an area of the axon that has just experienced an action potential cannot produce another one right away.

The absolute refractory period is followed by a brief *relative refractory period* during which it is harder than usual to generate the next action potential. In practical terms, this means that by encroaching into the relative refractory period, very strong stimuli can generate more frequent action potentials than relatively weak stimuli can. The absolute and relative refractory periods are shown in **Figure 11.6**.

Whether or not a neuron is generating action potentials, the sodium-potassium pumps continue to maintain the normal concentrations of Na^+ and K^+ and the normal resting potential. The actual number of Na^+ and K^+ ions that diffuse across the membrane during a single action potential is so small compared to the activity of the sodium-potassium pumps that an axon can transmit many action potentials in a short period of time without adversely affecting the subsequent resting membrane potential.

 Web Animation *The Nerve Impulse* at www.humanbiology.com

✓ **Quick Check** Suppose a neuron is placed in a solution that has no Na^+ ions, but is a normal extracellular solution in other respects (normal concentrations of osmotic particles, normal charge). If the neuron is brought to threshold, will it be able to generate an action potential? Explain your answer. ▪

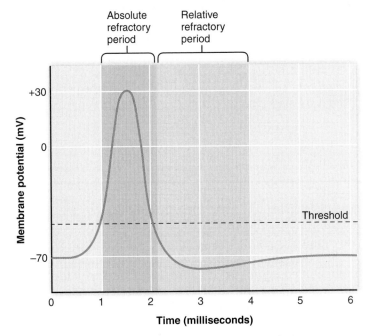

Figure 11.6 Absolute and relative refractory periods.

Action potentials are all-or-none and self-propagating

As we have seen, an action potential does not occur unless a certain threshold membrane voltage is reached. Once it is triggered, however, it always looks exactly the same in form and voltage. Even if the graded stimulus that causes it is significantly higher than the threshold level, the form and amplitude of the action potential are always the same. Consequently an action potential is an "all-or-none" phenomenon: either it occurs or it does not. It's like firing a gun. A certain amount of pressure—the threshold level—is required to make the gun fire. Pressing the trigger too lightly will not make it fire, and pressing it harder than necessary will not cause the bullet to leave the gun any faster.

An action potential is also a self-propagating event, meaning that it continues to propagate itself in the next region of the axon. This is because the local spread of electrical current forward from the region of the axon that is currently undergoing an action potential brings the next region of the axon to threshold. Thus the action potential is propagated continuously down the axon like a moving wave, at a constant rate of speed and amplitude.

If all action potentials are alike, how does a neuron transmit information about stimulus intensity? Simple, really: it's encoded by the number of action potentials generated and transmitted per time unit, ranging from near zero to perhaps several hundred per second. Stronger stimuli generate *more* action potentials per time unit, not faster or bigger ones.

Although the speed of travel of an action potential is always constant for a given neuron, the speed in different neurons can vary from about 5 mph up to 250 mph. Action potential speed is greater in larger-diameter axons and in axons covered by an insulating sheath, as we'll see in a minute.

↺ **Recap** A neuron's resting potential of about −70 millivolts is maintained by the constant action of the sodium-potassium pump. Impulses arriving from other neurons can cause small, local changes in the neuron's membrane potential called *graded potentials*. The sum of all graded potentials may initiate a self-propagating, all-or-none action potential in a neuron. An action potential involves three events: depolarization, repolarization, and reestablishment of the resting potential. ▪

11.4 Neuroglial cells support and protect neurons

Only about 20% of cells in the human nervous system are neurons. The rest are **neuroglial cells,** which provide physical support and protection to neurons and help maintain healthy concentrations of important chemicals in the fluid surrounding them. (Neuroglia derives from the Greek words for "neuron" and "glue.") Neuroglial cells do not generate or transmit impulses.

In the peripheral nervous system, many neuron axons are enclosed and protected by specialized neuroglial cells called *Schwann cells.* Schwann cells produce a fatty insulating material called *myelin.* During development, individual Schwann cells wrap themselves around a short segment of an axon many times as a sort of insulating blanket, creating a shiny white protective layer around the axon called a **myelin sheath.** Between adjacent Schwann cells are short uninsulated gaps called *nodes of Ranvier,* where the surface of the axon is still exposed (**Figure 11.7**). Neurons that have

In saltatory conduction, the nerve impulse jumps from node to node.

a) A myelinated motor neuron of the peripheral nervous system.

Node of Ranvier

Schwann cell

Myelin sheath

Neuron axon

b) TEM cross section of part of an axon (yellow) and its surrounding myelin sheath (beige).

Figure 11.7 A myelinated motor neuron of the peripheral nervous system. The axon of the motor neuron is wrapped by individual Schwann cells, which produce an insulating substance called *myelin.* Between adjacent Schwann cells are uninsulated sections called *nodes of Ranvier.* In myelinated neurons, electrical impulses appear to jump rapidly from node to node as a result of local current flows both within and outside the neuron (shown by thin, black arrows).

axons wrapped in a sheath of myelin are called **myelinated neurons.**

The myelin sheath around the axon serves three important functions:

- *It saves the neuron energy.* The insulating layer of myelin prevents some of the slow inward leak of sodium and outward leak of potassium that would otherwise occur. These leaks normally have to be replaced by active transport processes requiring energy (see section 11.3).
- *It speeds up the transmission of impulses.* The myelin sheath prevents nearly all leakage of charged ions across the axon membrane except where the axon is bare (at the nodes of Ranvier). When an action potential occurs at a node of Ranvier, then, the local depolarizing current spreads much farther within the axon—all the way to the next node of Ranvier. The rate of travel of this local current within the axon is even faster than the rate of travel of a continuously propagated action potential along an unmyelinated axon. At the next node of Ranvier, the local depolarizing current brings the axon membrane to threshold, triggering a new action potential. In effect, a succession of action potentials appears to jump from node to node at a very fast rate. This "leaping" pattern of conduction along myelinated neurons is called **saltatory conduction** (from *saltare,* Latin for "dance"). Continuously propagated action potentials in unmyelinated neurons travel at a speed of only about 5 mph (2.3 meters per second). In contrast, saltatory conduction in myelinated neurons can reach speeds of up to 250 mph (about 110 meters per second).
- *It helps damaged or severed axons of the peripheral nervous system regenerate.* If a neuron axon is severed, the portion of the axon distal to the cell body may degenerate. However, the cut end of the axon still attached to the cell body can regrow through the channel formed by the sheath, eventually reconnecting with the cell that it serves. Depending on the length of the axon, the regeneration process can take anywhere from a few weeks to more than a year. On average, an axon in the peripheral nervous system grows by nearly 1/16 inch (approximately 1.5 millimeters) per day.

In the central nervous system, protective sheaths of myelin are produced by another type of neuroglial cell called an *oligodendrocyte.* Unlike the sheath formed by Schwann cells, however, the sheath formed by oligodendrocytes degenerates once the axon it protects is destroyed. Consequently, neurons of the central nervous system do not regenerate after injury. This is why spinal cord injuries and CNS disorders such as multiple sclerosis result in a permanent change or loss of function (see section 11.13, Disorders of the nervous system).

In the autoimmune disorder *multiple sclerosis* (MS), the sheaths of myelinated neurons in the brain and spinal cord become progressively damaged until they form hardened (sclerotic) scar tissue. These sclerotic areas can no longer insulate the neurons effectively, disrupting and slowing the transmission of impulses. People with MS can experience a variety of symptoms depending on which areas of the CNS are damaged. Common symptoms include muscle weakness, visual impairment, and urinary incontinence. The disease usually appears in people between 20 and 40 years of age. Although the disease is generally progressive, people with MS may experience periods of remission.

A similar but rarer disease is *amyotrophic lateral sclerosis* (ALS). In ALS the sclerotic areas typically begin in regions of the spinal cord involved in the motor control of skeletal muscle. The primary symptom of ALS is progressive weakening and wasting of skeletal muscle tissue, most notably the diaphragm and intercostal muscles. As mentioned in Chapter 10, people with ALS usually die from respiratory failure.

↩ **Recap** Neuroglial cells support and protect neurons. Neuroglial cells called *Schwann cells* (in the PNS) and *oligodendrocytes* (in the CNS) form myelin sheaths that protect axons and speed transmission of impulses. ■

HBP **Web Animation** *Myelinated Neurons and Saltatory Conduction* at www.humanbiology.com

11.5 Information is transferred from a neuron to its target

Once an action potential reaches the axon terminals of a neuron, the information inherent in it must be converted to another form for transmittal to its target (muscle cell, gland cell, or another neuron). In essence, the action potential causes the release of a chemical that crosses a specialized junction between the two cells called a **synapse.** This chemical substance is called a **neurotransmitter** because it transmits a signal from a neuron to its target. The entire process of transmission from a neuron to its target is called *synaptic transmission.* We describe in Chapter 6 how synaptic transmission occurs between a neuron and skeletal muscle. Now we'll see how it occurs between neurons.

Neurotransmitter is released

Figure 11.8 illustrates the structure of a typical synapse and the events that occur during synaptic transmission. At a synapse, the *presynaptic membrane* is the cell membrane of the neuron that is sending the information. The *postsynaptic membrane* is the membrane of the cell that is about to receive the information. The small fluid-filled gap that separates the presynaptic and postsynaptic membranes is the *synaptic cleft.*

Each axon terminal of the presynaptic neuron ends in an axon bulb that contains neurotransmitter stored in membrane-bound vesicles. The events that occur during synaptic transmission follow a particular pattern.

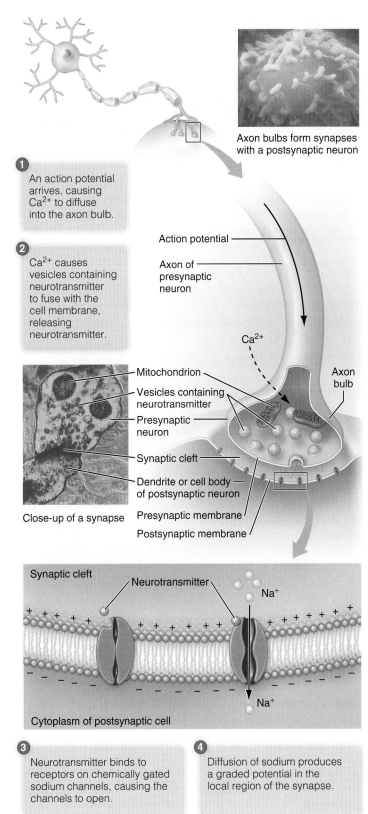

Axon bulbs form synapses with a postsynaptic neuron

1 An action potential arrives, causing Ca^{2+} to diffuse into the axon bulb.

2 Ca^{2+} causes vesicles containing neurotransmitter to fuse with the cell membrane, releasing neurotransmitter.

Action potential

Axon of presynaptic neuron

Ca^{2+}

Axon bulb

Mitochondrion

Vesicles containing neurotransmitter

Presynaptic neuron

Synaptic cleft

Dendrite or cell body of postsynaptic neuron

Presynaptic membrane

Postsynaptic membrane

Close-up of a synapse

Synaptic cleft

Neurotransmitter

Na^+

Na^+

Cytoplasm of postsynaptic cell

3 Neurotransmitter binds to receptors on chemically gated sodium channels, causing the channels to open.

4 Diffusion of sodium produces a graded potential in the local region of the synapse.

Figure 11.8 **Summary of synaptic transmission.**

1. An action potential arrives at the axon bulb, causing calcium (Ca^{2+}) channels in the presynaptic membrane to open. Ca^{2+} diffuses into the axon bulb.
2. The presence of Ca^{2+} causes the vesicles to fuse with the presynaptic membrane and release their contents of neurotransmitter directly into the synaptic cleft. Because the synaptic cleft is so narrow (about 0.02 micrometer, or one-millionth of an inch), the neurotransmitter molecules reach the postsynaptic membrane by diffusion.
3. Molecules of neurotransmitter bind to receptors on the postsynaptic membrane. This causes certain chemically gated channels, such as those for sodium, to open.
4. Sodium ions diffuse inward, producing a graded depolarization of the postsynaptic membrane in the area of the synapse.

Note that graded potentials are caused by the opening of *chemically* sensitive ion channels (the chemical being a neurotransmitter), rather than voltage-sensitive ones.

✓ **Quick Check** If there were no available Ca^{2+} in the synaptic cleft, could the presynaptic neuron still produce action potentials and release neurotransmitter? Explain your answer. ▪

Neurotransmitters exert excitatory or inhibitory effects

The response of the postsynaptic cell to neurotransmitter depends on several factors, including the type of neurotransmitter, its concentration in the synaptic cleft, and the types of receptors and chemically sensitive ion channels in the postsynaptic membrane.

To date, scientists have identified more than 50 chemicals that can function as neurotransmitters. All are stored in vesicles within the axon bulb and released into the synaptic cleft in response to an action potential. All produce a graded potential in the postsynaptic cell, though the effects may be quite different because they can affect different ion channels.

Neurotransmitters are classified as excitatory, inhibitory, or both. Excitatory neurotransmitters depolarize the postsynaptic cell, causing threshold to be approached or exceeded. Therefore, excitatory neurotransmitters encourage the generation of new impulses in the postsynaptic neuron. Inhibitory neurotransmitters cause the postsynaptic cell to hyperpolarize, meaning that the cell's interior becomes even more negative than before. Hyperpolarization makes it harder for threshold to be reached, so inhibitory neurotransmitters tend to prevent the generation of action potentials in the postsynaptic neuron. Some neurotransmitters can be excitatory or inhibitory, depending on the type of receptor to which they bind on the postsynaptic membrane.

Table 11.1 summarizes the actions of several common neurotransmitters.

The effects of neurotransmitters are relatively short-lived because the neurotransmitter remains in the synaptic cleft for only a short time. Prompt removal of neurotransmitter causes neural signals to be terminated almost as rapidly as they are initiated. Only then can the next message be received and recognized. The neurotransmitter may be removed from the synaptic cleft in any or all of three ways: (1) it may be taken back up again by the presynaptic neuron and repackaged into membrane-bound vesicles, to be used again; (2) it may be destroyed by enzymes in the synaptic cleft; or (3) it may diffuse away from the synaptic cleft into the general circulation, where it will ultimately be destroyed.

 Web Animation *The Synapse* at www.humanbiology.com

Table 11.1 **Actions of common neurotransmitters**

Neurotransmitter	Sites where released	Principal actions
Acetylcholine	Neuromuscular junctions, autonomic nervous systems, brain	Excitatory on skeletal muscles; excitatory or inhibitory at other sites, depending on receptors
Norepinephrine	Areas of brain and spinal cord, autonomic nervous system	Excitatory or inhibitory, depending on receptors; plays a role in emotions
Serotonin	Areas of brain, spinal cord	Usually inhibitory; involved in moods, sleep cycle, appetite
Dopamine	Areas of brain, parts of peripheral nervous system	Excitatory or inhibitory, depending on receptors; plays a role in emotions
Glutamate	Areas of brain, spinal cord	Usually excitatory; major excitatory neurotransmitter in brain
Endorphins	Many areas in brain, spinal cord	Natural opiates that inhibit pain; usually inhibitory
Gamma-aminobutyric acid	Areas of brain, spinal cord	Usually inhibitory; principal inhibitory neurotransmitter in brain
Somatostatin	Areas of brain, pancreas	Usually inhibitory; inhibits pancreatic release of growth hormone

Postsynaptic neurons integrate and process information

The conversion of the signal from electrical (action potential) to chemical (neurotransmitter) allows the postsynaptic cell to do a lot of integration and information processing. This is because a single all-or-none action potential in the presynaptic neuron does not always cause a corresponding all-or-none action potential in the postsynaptic neuron, as it does in a muscle cell. A single action potential produces only a small graded potential in the postsynaptic neuron, and a lot of these graded potentials generally are required for threshold to be reached.

One way for threshold to be reached in a postsynaptic neuron is for the presynaptic neuron (or a few neurons) to increase their frequency of stimulation, sending lots of action potentials in a short time (hundreds per minute). In addition to stimulus intensity, other factors affecting whether threshold is reached in the postsynaptic neuron include (1) how many other neurons form synapses with it, and (2) whether those synaptic connections are stimulatory or inhibitory. Typically, one neuron receives input from many neurons, a phenomenon known as *convergence*. In turn, its action potential goes to many other neurons, a phenomenon known as *divergence* (Figure 11.9).

Figure 11.9 Neural information processing by convergence and divergence. Convergence occurs when several presynaptic cells form synapses with a single postsynaptic cell. Divergence occurs when one nerve cell provides information to a variety of postsynaptic cells.

Though it would be incorrect to say that neurons "interpret" information—that would imply that a neuron has a brain—the effect of convergence is that neurons may integrate and process thousands of simultaneous incoming stimulatory and inhibitory signals before generating and transmitting their own action potentials. They also reroute the information to many different destinations. Individual neurons do not see, smell, or hear, yet their combined actions allow us to experience those complex sensations.

Consider again the example given at the beginning of this chapter, in which you were able to ignore certain incoming signals—the smell of doughnuts, the sound of your shoes on pavement, the sensation of being cold—and focus on others, such as the noises made by the truck and its horn. You tapped your memory banks for the probable meaning of these sounds (danger approaching!), and then coordinated your muscles to turn around, see the truck, and jump out of the way—all in a fraction of a second. This is a complex set of events, and it is all done with neurons, action potentials, and neurotransmitters.

Muscle cells are also target cells of presynaptic neurons. As described in Chapter 6, however, skeletal muscle cells do not process and integrate information. There are two reasons for this. First, a muscle cell receives input from only one neuron. There are no converging neural inputs to process. Second, the neuromuscular junction is quite large, with many points of contact between neuron and muscle cell. Consequently, threshold is reached in a skeletal muscle cell every time the motor neuron sends even a single action potential. Functionally, this means that the nervous system has absolute and final control over the activity of skeletal muscle cells.

↩ Recap A neuron transmits information to another cell across a synapse. At a synapse, neurotransmitter released from a presynaptic neuron diffuses across the synaptic cleft and binds to receptors on the postsynaptic membrane, opening ion channels and causing a graded potential in the postsynaptic cell. Postsynaptic neurons may integrate incoming signals from many different presynaptic neurons. ▪

11.6 The PNS relays information between tissues and the CNS

Now that we have seen how neurons generate and transmit action potentials, we look at how this information travels throughout the nervous system. As stated earlier, the nervous system has two major subdivisions: the central nervous system (CNS), which includes the spinal cord and brain, and the peripheral nervous system (PNS), which consists of nerves that transmit information between body tissues and the CNS.

Nerves carry signals to and from the CNS

A single nerve axon is like a single fiber-optic fiber in that both transmit specific kinds of information from one location to another. By the same token, a nerve is like a large fiber-optic cable strung between two cities. A **nerve** consists of the axons of many neurons, all wrapped together in a protective sheath of connective tissue and all coming from and going to the same place. The function of each nerve depends on where it originates and the organs or tissues to which it travels.

The peripheral nervous system consists of cranial nerves and spinal nerves. The 12 pairs of **cranial nerves** connect directly with the brain. Cranial nerves carry action potentials between the brain and the muscles, glands, and receptors of the head, neck, and thoracic and abdominal cavities.

The 31 pairs of **spinal nerves** connect with the spinal cord. Each spinal nerve attaches to the spinal cord via two short branches of the spinal cord called the *dorsal root* and the *ventral root*. The dorsal root contains the sensory neurons that transmit incoming action potentials from body tissues to the spinal cord. The ventral root contains motor neurons that carry action potentials away from the spinal cord to the rest of the body. Thus each spinal nerve carries both sensory and motor (incoming and outgoing) information.

 Web Animation *Cranial and Spinal Nerves* at www.humanbiology.com

Sensory neurons provide information to the CNS

The nervous system must have good information if it is to take appropriate action. That information arrives at the CNS as action potentials traveling in sensory neurons located throughout the body. Because information provided by sensory neurons to the CNS may affect the motor output of both the somatic and autonomic divisions of the PNS, the sensory (input) component of the PNS cannot be classified as belonging specifically to either the somatic or autonomic motor (output) divisions.

The somatic division controls skeletal muscles

Recall that the motor (or output) neurons of the peripheral nervous system are organized into the somatic division and the autonomic division. The **somatic division** controls voluntary and involuntary skeletal muscle movement. Somatic motor neurons transmit information from the spinal cord to skeletal muscles.

As described in Chapter 6, a single somatic motor neuron controls several skeletal muscle cells (together

they form a motor unit), and every time the neuron sends an action potential down its axon, all the muscle cells of that motor unit experience an action potential and contract. Motor neurons are activated either by conscious control from the brain or by an involuntary response called a *reflex*.

Spinal reflexes are involuntary responses that are mediated primarily by the spinal cord and spinal nerves, with little or no involvement of the brain. **Figure 11.10** illustrates two such spinal reflexes. Stepping on a sharp object, for example, activates pain receptors, producing action potentials in sensory neurons traveling to your spinal cord. The sensory neurons stimulate interneurons within the spinal cord, which in turn transmit signals to motor neurons in your leg, causing you to flex the appropriate muscles to lift your foot. This is called the *flexor (withdrawal) reflex*. Meanwhile, other interneurons that cross over to the opposite side of the spinal cord are also stimulated by the sensory neuron. These interneurons stimulate motor neurons to extend, rather than flex, certain skeletal muscles in your other leg. After all, it wouldn't do to flex both legs and fall on the sharp object! This second reflex is called a *crossed extensor reflex*.

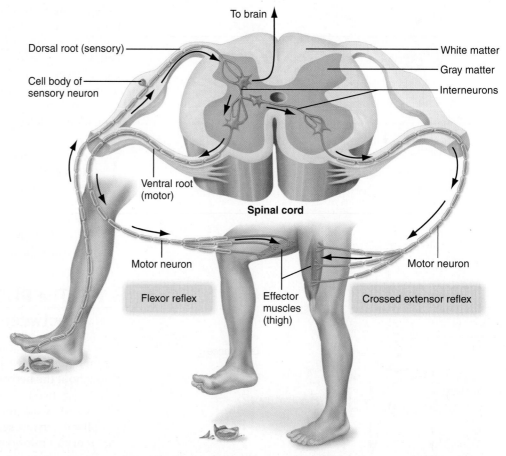

Figure 11.10 Spinal reflexes. In this example of withdrawal and crossed extension reflexes, a painful stimulus activates sensory nerves in a foot. Sensory nerves stimulate interneurons within the spinal cord that activate motor neurons to muscles on both sides of the body. Note that the motor effects are different on the two sides of the body: withdrawal (flexion) on the stimulus side and extension on the opposite side. Other interneurons relay the sensory information to the brain.

Notice that these events did not require conscious thought. A simple spinal reflex is processed primarily at the level of the spinal cord, allowing a rather complex series of actions to occur quickly. Some sensory information does ascend to the brain as well, allowing the brain to integrate the sensory information from your foot with all the other information it is receiving and determine your next course of action (such as sitting down and examining your foot). But most of the spinal reflex itself occurs without any involvement of the brain.

Another type of spinal reflex is a *stretch reflex*. In a stretch reflex, stretch receptors in a skeletal muscle stimulate sensory nerves, which carry the signals to the spinal cord. In the spinal cord, motor neurons to the stretched muscle are activated. At the same time, motor neurons to any antagonistic muscles are inhibited so that contraction will be effective. Stretch reflexes play an important role in maintaining upright posture and movement, as they allow us to stand and move without having to concentrate on our actions.

Stretch reflexes associated with posture and movement are often modified slightly by signals from higher brain centers. Physicians can test such reflexes to determine whether the spinal cord and areas of the brain involved in stretch reflexes are intact and functioning properly. An example is the well-known knee-jerk reflex. When the patellar tendon is tapped lightly below the kneecap, the resulting slight stretch of the thigh muscles initiates a spinal reflex that ultimately results in contraction of the thigh muscles and an upward movement of the foot and lower leg (**Figure 11.11**). If the

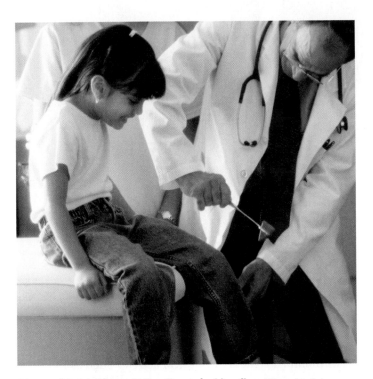

Figure 11.11 The patellar (knee-jerk) reflex. The physician tests the integrity of the patellar reflex to verify that certain sensory and motor nerves, the spinal cord, areas of the brain, and skeletal muscle are all functioning normally.

neural pathways are not intact or if the brain is not modifying the reflex properly, the reflex may be either less vigorous or more vigorous than normal.

 Quick Check Compare and contrast a flexor reflex and a stretch reflex. What are the major similarities and differences? ▪

The autonomic division controls automatic body functions

Motor neurons carrying information to smooth or cardiac muscle and to all other tissues and organs except skeletal muscle belong to the **autonomic division** (*autonomic* derives from the Greek words for "self" and "law"). The autonomic division of the peripheral nervous system carries signals from the CNS to the periphery that control "automatic" functions of the body's internal organs. Many of these activities occur without our ever being aware of them.

Unlike the somatic division in which one neuron extends from the CNS to the target cell, the autonomic division requires two neurons to transmit information from the CNS to the target cell. The cell bodies of the first neurons (called *preganglionic neurons*) lie within the central nervous system. The axons of many of these neurons go to structures called *ganglia* (singular: *ganglion*) that lie outside the CNS. Ganglia are clusters of neuron cell bodies of the second neurons (called *postganglionic neurons*). Axons of the postganglionic neurons extend to the far reaches of the body to communicate with internal glands and organs.

The sympathetic and parasympathetic divisions oppose each other

There are two types of motor neurons in the autonomic division: sympathetic and parasympathetic. Both function automatically, usually without our conscious awareness, to regulate body functions. They innervate smooth muscle, cardiac muscle, glands, the epithelial cells of the kidneys, and virtually every internal organ except skeletal muscle, bone, and the nervous system itself.

In most organs, the actions of the sympathetic and parasympathetic divisions have very opposite effects, as shown in **Figure 11.12** (next page). Which division predominates at any one time determines what happens in your body. And with two opposing systems active at all times, control can be very precise. It is as if you were driving a car with one foot on the gas pedal and one on the brake—you could speed up by hitting the gas pedal, releasing the brake pedal a little, or both.

The sympathetic division arouses the body Preganglionic motor neurons of the **sympathetic division** originate in the thoracic and lumbar regions of the spinal cord. Many of their axons extend to a chain of sympathetic ganglia that lie alongside the

SYMPATHETIC DIVISION

PARASYMPATHETIC DIVISION

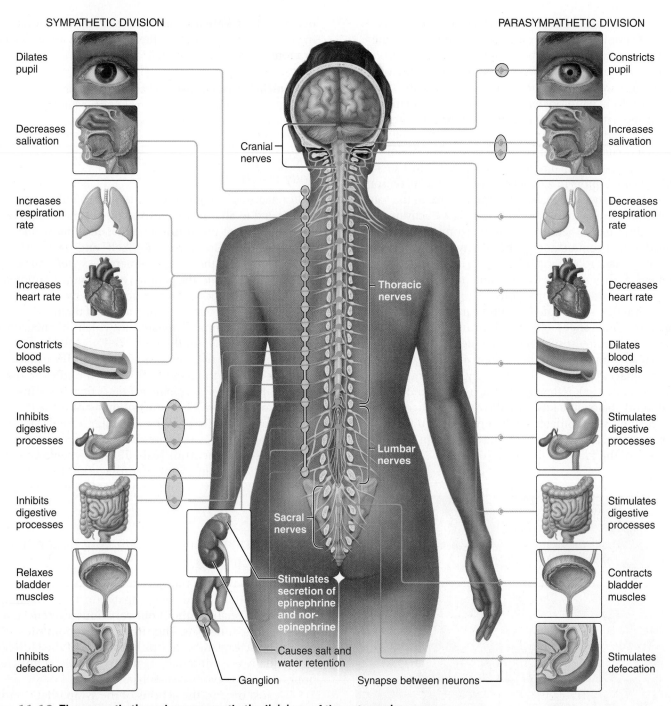

Dilates pupil

Decreases salivation

Increases respiration rate

Increases heart rate

Constricts blood vessels

Inhibits digestive processes

Inhibits digestive processes

Relaxes bladder muscles

Inhibits defecation

Cranial nerves

Thoracic nerves

Lumbar nerves

Sacral nerves

Stimulates secretion of epinephrine and nor-epinephrine

Causes salt and water retention

Ganglion

Synapse between neurons

Constricts pupil

Increases salivation

Decreases respiration rate

Decreases heart rate

Dilates blood vessels

Stimulates digestive processes

Stimulates digestive processes

Contracts bladder muscles

Stimulates defecation

Figure 11.12 The sympathetic and parasympathetic divisions of the autonomic nervous system. The parasympathetic division predominates in the body's relaxed state. The sympathetic division is associated with fight-or-flight responses. The nerves of the parasympathetic division arise from cranial and sacral segments, while the nerves of the sympathetic division arise in the thoracic and lumbar segments of the spinal cord. Ganglia (groups of nerve cell bodies) are shown in yellow.

spinal cord, where they form synapses with the postganglionic cells. Because these ganglia are connected to each other, the sympathetic division tends to produce a unified response in all organs at once. Axons of the postganglionic neurons branch out to innervate all the organs. The neurotransmitter released by sympathetic postganglionic neurons in the target organs is *norepinephrine.*

The sympathetic division transmits signals that prepare the body for emergencies and situations requiring high levels of mental alertness or physical activity, such as fighting or running away from danger ("fight or flight"), and even play and sexual activity. It increases heart rate and respiration, raises blood pressure, and dilates the pupils—effects that help you detect and respond quickly to changes in your environment. The sympathetic division also reduces blood flow to organs that do not help you cope with an immediate crisis, such as the intestines and kidneys, and inhibits less important body functions such as digestion and production of saliva (this is why your mouth feels dry when you are anxious).

The effects of the sympathetic division are sometimes called the *fight-or-flight response*, reflecting its association with emergency situations.

The parasympathetic division predominates during relaxation
Preganglionic neurons of the **parasympathetic division** originate either in the brain, becoming part of the outflow of certain cranial nerves, or from the sacral region of the spinal cord. The ganglia where the preganglionic neurons synapse with the postganglionic neurons is generally some distance from the CNS and may even be in the target organ itself. The neurotransmitter released by parasympathetic postganglionic neurons at their target organs is acetylcholine, the same neurotransmitter used at the neuromuscular junction.

In contrast to the sympathetic division, the parasympathetic division predominates in situations in which you are relaxed. It transmits signals that lower heart rate and respiration, increase digestion, and permit defecation and urination, among other actions. The parasympathetic division exerts calming, restorative effects ("rest and repose") that counteract the fight-or-flight stimulation of the sympathetic division. Parasympathetic nerves are also responsible for the vasodilation that causes erection in males and swelling of the labia and erection of the clitoris in females.

> ✓ **Quick Check** Suppose a person took a drug that blocked the release of norepinephrine from all peripheral nerves. Which division of the peripheral nervous system would be affected, and what would be some of the likely symptoms? ▪

The sympathetic and parasympathetic divisions work antagonistically to maintain homeostasis As we have seen, the actions of the sympathetic and parasympathetic divisions oppose each other. They work antagonistically to accomplish the automatic, subconscious maintenance of homeostasis. Both systems are intimately involved in feedback loops that control homeostasis throughout the body.

An example is the control of blood pressure. As described in Chapter 8, certain sensory nerves are activated by baroreceptors. When blood pressure rises, the baroreceptors are activated, and the sensory nerves send more frequent action potentials to the brain. The brain integrates this information with the other information it is receiving. Then it decreases the number of action potentials traveling in sympathetic nerves and increases the number of action potentials traveling in parasympathetic nerves serving the heart and blood vessels. As a result, heart rate slows, blood vessels dilate, and blood pressure falls until it is back to normal.

The sympathetic and parasympathetic divisions can also cause a widespread response, such as arousing the entire body. They can even carry out many localized homeostatic control mechanisms simultaneously and independently. Table 11.2 summarizes the functions of the somatic and autonomic divisions of the PNS.

> ↺ **Recap** Twelve pairs of cranial nerves transmit action potentials between the brain and the head, neck, and thoracic and abdominal cavities. Thirty-one pairs of spinal nerves transmit action potentials to and from the spinal cord to all regions of the body. Sensory neurons provide information (input) to the CNS. The somatic output division of the PNS controls voluntary (conscious) and involuntary (reflex) skeletal muscle movement. The autonomic division of the PNS includes sympathetic and parasympathetic components that regulate automatic body functions to maintain homeostasis. ▪

🌐 **Web Animation** *The Autonomic Nervous System* at www.humanbiology.com

11.7 The brain and spinal cord constitute the CNS

A neuron does not think or love, yet humans do. Scientists are still trying to unravel the biology of emotion and thought, but we do know that they originate in the central nervous system, specifically in the brain.

The central nervous system, consisting of brain and spinal cord, is "central" both in location and action. It is where integration, or processing, of information occurs. It is

Table 11.2 The somatic and autonomic divisions of the PNS

	Somatic division	Autonomic division	
		Sympathetic	*Parasympathetic*
Functions	Serves skeletal muscles	Fight or flight; arouses body to deal with situations involving physical activity, mental alertness	Relaxes body; promotes digestion and other basic functions
Neurotransmitter	Acetylcholine	Norepinephrine	Acetylcholine
Number of neurons required to reach the target cell	One	Two (preganglionic and postganglionic)	Two (preganglionic and postganglionic)

where information is received and where complex outputs are initiated. It's where billions of action potentials traveling in millions of neurons all come together as a conscious thought. That is not a very satisfactory explanation for "thinking," but it may be the best we can do.

Bone, meninges, and the blood-brain barrier protect the CNS

The central nervous system is vital to proper functioning of the entire body, so it is no surprise that it is well protected against physical injury and disease.

First, the CNS is protected by bone. The brain is encased in the skull, and the spinal cord is enclosed in a hollow channel within the vertebrae. This shell of bone around the central nervous system helps shield it from physical trauma.

Second, the CNS is enclosed by three membranes of connective tissue, called **meninges,** named (from outermost

to innermost layers) the *dura mater,* the *arachnoid,* and the *pia mater.* These three meninges protect the neurons of the CNS and the blood vessels that service them.

Third, the CNS is bathed in its own special liquid, called **cerebrospinal fluid,** which fills the space between the arachnoid and the pia mater. In addition to serving as a liquid shock absorber around the brain and spinal cord, cerebrospinal fluid tends to isolate the central nervous system from infections. Inflammation of the meninges due to a bacterial or viral infection is called *meningitis.* Meningitis can be serious because it may spread to the nervous tissue of the CNS.

Cerebrospinal fluid is somewhat similar to the interstitial fluid that bathes all cells, but it does not exchange substances as freely with blood. It is secreted from specialized capillaries into four hollow cavities (called *ventricles*) in the brain. From the ventricles, it flows through a system of fluid-filled cavities and canals, bathing and surrounding the neurons in the brain and spinal cord (Figure 11.13).

a) Anterior view.   **b) Sagittal view.**

Figure 11.13 **The ventricles of the brain and circulation of cerebrospinal fluid.** Cerebrospinal fluid is secreted into the four brain ventricles (left, right, third, and fourth). From there it flows through a series of cavities, bathing the brain and spinal cord.

Most capillaries in the body are fairly leaky because there are narrow slits between adjacent capillary cells. However, the cells of the capillaries that produce cerebrospinal fluid are fused tightly together. Consequently, substances must pass through rather than between the capillary cells in order to get from blood to cerebrospinal fluid. Lipid-soluble substances such as oxygen and carbon dioxide can diffuse easily through the lipid bilayer of the cell membrane. Certain important molecules such as glucose are actively transported. However, most charged molecules and large objects, such as proteins, viruses, and bacteria, generally are prevented from entering the cerebrospinal fluid.

This functional barrier between blood and brain is called the **blood-brain barrier.** Thanks to this isolation, bacterial and viral infections of the brain and spinal cord are rare, but when these illnesses occur they are generally serious and difficult to treat. Many of our best antibiotics are not lipid soluble, which means they cannot cross the blood-brain barrier and so are ineffective against CNS infections. Lipid-soluble substances that cross the blood-brain barrier rather easily include alcohol, caffeine, nicotine, cocaine, and general anesthetics.

 Quick Check Antibodies are large, charged protein molecules that are produced by the immune system to help fight infections. Do you think antibodies can cross the blood-brain barrier? Why or why not? ■

The spinal cord relays information

The **spinal cord** serves as the superhighway for action potentials traveling between the brain and the rest of the body. As we have seen, the spinal cord can also process and act on certain information on its own via spinal reflexes, without necessarily involving the brain. However, its power to act independently is limited, and such actions never reach the level of conscious awareness.

Approximately the diameter of your thumb, the spinal cord extends from the base of the skull to the area of about the second lumbar vertebra, or about 17 inches (**Figure 11.14**). It is protected by the vertebral column. The outer portions of the spinal cord consist primarily of bundles of axons, which in the CNS are called *nerve tracts* rather than nerves. Because these axons are generally myelinated, giving them a whitish appearance, the areas of the cord occupied by these ascending

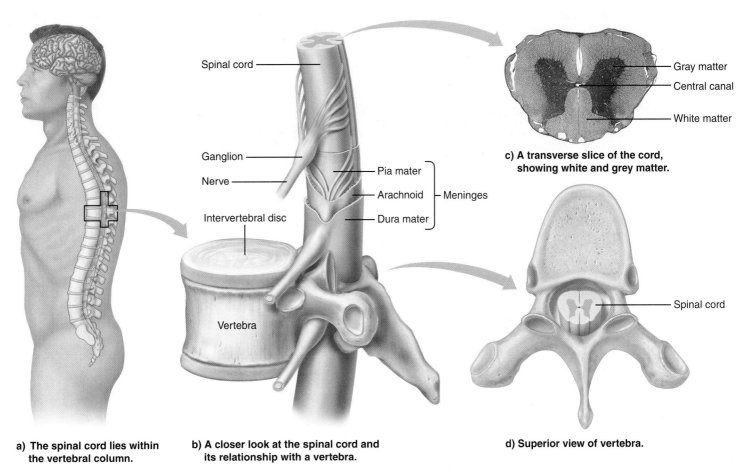

a) **The spinal cord lies within the vertebral column.**

b) **A closer look at the spinal cord and its relationship with a vertebra.**

c) **A transverse slice of the cord, showing white and grey matter.**

d) **Superior view of vertebra.**

Figure 11.14 **The human spinal cord.**

(sensory) and descending (motor) nerve tracts are called *white matter*. Neurons of the PNS enter and leave the spinal cord at regular intervals via the dorsal (sensory) and ventral (motor) horns that fuse to form spinal nerves.

Near the center of the spinal cord is a region occupied primarily by the cell bodies, dendrites, and axons of neurons of the CNS, and also neuroglial cells. These structures are not myelinated, so the area they occupy is referred to as *gray matter*. Within the gray matter, sensory and motor neurons synapse with neurons of the CNS that transmit signals up the spinal cord to the brain.

Recap The CNS consists of the brain and spinal cord. It is well protected by bone, the meninges, and cerebrospinal fluid. The capillaries that produce cerebrospinal fluid have special properties that result in a functional blood-brain barrier. The spinal cord transmits information between the brain and the rest of the body. It contains white matter (myelinated axons) and gray matter (unmyelinated neural structures). ■

11.8 The brain processes and acts on information

The **brain** is the command center of the body. The brain receives information in the form of action potentials from various nerves and the spinal cord, integrates it, and generates the appropriate response.

Within the brain, certain areas specialize in integrating different types of information or performing certain motor tasks. Action potentials derived from various receptors, such as light and sound, arrive at the brain in different sensory neurons that specialize in processing those signals. For example, one area of the brain responds to visual stimuli and another area responds to stimuli that help us maintain our balance.

Three major anatomical and functional divisions of the brain have been identified:

- The *hindbrain* coordinates basic, automatic, and vital tasks.
- The *midbrain* helps coordinate muscle groups and responses to sights and sounds.
- The *forebrain* receives and integrates sensory input from the external environment and determines most of our more complex behavior.

Figure 11.15 illustrates these major divisions and their components.

The hindbrain: Movement and automatic functions

The **hindbrain** is connected to the spinal cord. From an evolutionary standpoint the hindbrain is the oldest, most primitive brain division, and the one most similar among animals. Important components of the hindbrain include the medulla oblongata, the cerebellum, and the pons.

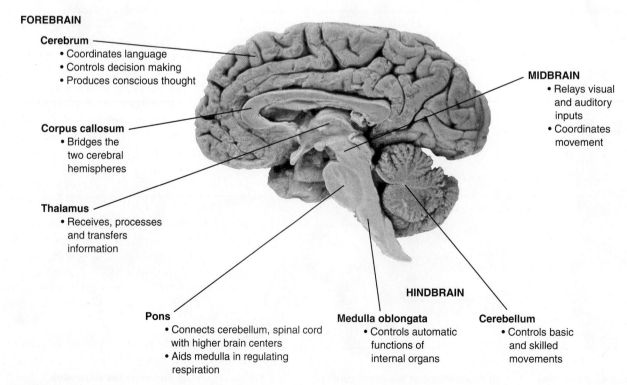

FOREBRAIN

Cerebrum
- Coordinates language
- Controls decision making
- Produces conscious thought

Corpus callosum
- Bridges the two cerebral hemispheres

Thalamus
- Receives, processes and transfers information

MIDBRAIN
- Relays visual and auditory inputs
- Coordinates movement

Pons
- Connects cerebellum, spinal cord with higher brain centers
- Aids medulla in regulating respiration

HINDBRAIN

Medulla oblongata
- Controls automatic functions of internal organs

Cerebellum
- Controls basic and skilled movements

Figure 11.15 The brain. A sagittal section (front to back on a vertical plane) of a human brain near the midline.

The medulla oblongata controls automatic functions The **medulla oblongata** connects to the spinal cord. It contains areas that control vital automatic functions of internal organs. Within the medulla oblongata, the *cardiovascular center* regulates heart rate and blood pressure, and the *respiratory center* integrates information about blood levels of oxygen and carbon dioxide, and adjusts respiration accordingly. Other areas coordinate reflexes such as coughing, vomiting, sneezing, and swallowing.

All information passing between the higher areas of the brain and the spinal cord must pass through the medulla oblongata. An interesting feature of the way the body is "wired" is that the motor nerves from one side of the forebrain cross over to the other side of the body in the medulla oblongata. As a result, the left side of the brain controls the right side of the body, and vice versa.

The cerebellum coordinates basic movements Located just behind the medulla oblongata, the **cerebellum** coordinates basic body movements that are below the level of conscious control. For example, the cerebellum ensures that antagonistic muscles do not contract at the same time. The cerebellum also stores and replicates (without our conscious thought) whole sequences of skilled movements, such as tying a shoe, shifting a car, or hitting a home run (hitters sometimes say they just "pull the trigger" and the rest happens automatically). To coordinate these skilled activities, the cerebellum receives sensory input from many sources, including joint and muscle receptors, balance and position receptors in the ear, and visual receptors.

Excessive consumption of alcohol acts on the cerebellum and disrupts its normal functions. This is why people who are drunk can't coordinate their movements or walk a straight line.

The pons aids information flow Just above and partially surrounding the medulla oblongata, the *pons* (from the Latin for "bridge") connects higher brain centers and the spinal cord. The pons contains groups of axons that extend from the cerebellum to the rest of the CNS, and is especially important in coordinating the flow of information between the cerebellum and higher brain centers. In addition, the pons aids the medulla oblongata in regulating respiration.

 Quick Check If a man's medulla oblongata were temporarily rendered nonfunctional for just one hour, do you think he could survive? How about if the cerebellum were rendered nonfunctional for an hour? Explain. ∎

The midbrain: Vision and hearing

Most functions of the **midbrain** relate to vision and hearing. Visual and auditory sensory inputs pass through the midbrain before being relayed to higher brain centers. The midbrain coordinates movements of the head related to vision and hearing, such as turning toward a sudden sound

or flash of light, and it controls movement of the eyes and size of the pupils. The midbrain also monitors the unconscious movement of our skeletal muscles so that their actions are smooth and coordinated.

A group of neurons collectively called the *reticular formation* extends through the medulla oblongata, the pons, and the midbrain. The reticular formation works with the cerebellum to coordinate the skeletal muscle activity that maintains posture, balance, and muscle tone. Neurons in the reticular formation are also responsible for maintaining our level of wakefulness, as explained later in this chapter.

The forebrain: Emotions and conscious thought

The **forebrain** determines our most complex behavior, including emotions and conscious thought. Important areas in the forebrain are the hypothalamus and thalamus, limbic system, and cerebrum. It also includes two glands, the pineal and the pituitary, which are described in Chapter 13.

The hypothalamus and thalamus maintain homeostasis and process information The **hypothalamus** is a small region at the base of the forebrain that coordinates some automatic functions of the pituitary gland, such as water and solute balance, temperature control, carbohydrate metabolism, and production of breast milk. It plays an important role in regulating homeostasis because it monitors sensory signals relating to sight, smell, taste, noise, and body temperature. A "hunger center" located in the hypothalamus alerts us when we need to eat, and a "thirst center" signals us to drink something when body fluids become too concentrated.

The larger **thalamus**, located just above the hypothalamus, is primarily a receiving, processing, and transfer center. It integrates and relays some outgoing motor activity as well. The thalamus accepts sensory signals from other parts of the body and brain and channels them to the cerebrum to be interpreted. As sensory information—for example, pain, pressure, and temperature changes—arrives at the thalamus, we are conscious of the sensation but cannot identify where it originates. After the information reaches the cerebrum, we become aware of which part of the body is experiencing the sensation.

The limbic system is involved in emotions and memory The *limbic system* refers to a group of neuronal pathways that connect parts of the thalamus, hypothalamus, and inner portions of the cerebrum. The limbic system is associated with emotions and memory, and we will look at it in more detail later in this chapter when we discuss memory and behavior.

The cerebrum deals with higher functions In humans, the large **cerebrum** is the most highly developed of all brain regions. It deals with many of the functions that we associate with being human: language, decision making, and conscious thought.

The cerebrum (**Figure 11.16**) consists of left and right cerebral hemispheres joined in the middle by a network of nerve tracts called the *corpus callosum*. The corpus callosum enables the two hemispheres to share sensory and motor

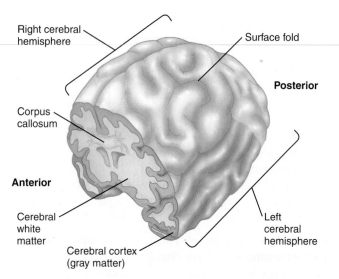

Right cerebral hemisphere

Surface fold

Posterior

Corpus callosum

Anterior

Cerebral white matter

Cerebral cortex (gray matter)

Left cerebral hemisphere

a) The cerebral cortex (gray matter; outer layer, shown in four colors) consists of interneurons that integrate and process information. White matter (inner core) consists of ascending and descending nerve tracts. The two separate hemispheres are joined by the corpus callosum.

Parietal lobe
• Interprets sensory information from skin

Occipital lobe
• Processes visual information

Frontal lobe
• Initiates motor activity
• Responsible for speech
• Conscious thought

Temporal lobe
• Interprets auditory information
• Comprehends language
• Perceptual judgment

b) The functions of the four lobes of the cerebral cortex are location-specific.

Figure 11.16 **The cerebrum.**

 Explain how a blow to the back of the head can sometimes cause blindness, even if the eyes are undamaged.

information. The outer layer of the cerebrum is called the **cerebral cortex** (Figure 11.16a). The cerebral cortex is primarily gray matter, consisting of CNS neurons with unmyelinated axons and their associated neuroglial cells. It has numerous folds on its surface, like a walnut. The inner portion of the cerebrum is mostly white matter containing myelinated nerve axons that connect lower brain areas to the cerebral cortex.

The structure of the cerebrum—an inner section of ascending and descending axons and an outer layer of cells—makes it ideally suited to direct incoming information to the proper region of the brain, integrate the information and process it at the highest level, and then route outgoing motor activity to the appropriate areas of the body.

A closer look at the cerebral cortex Although the cerebral cortex is only about 6 millimeters (about 1/4 inch) thick, it contains roughly 50 to 80 billion neuron cell bodies. It is the center for integrating and interpreting sensory information, but it is much more than that. The cerebral cortex is responsible for memory storage, abstract thought, conscious awareness, and conscious control of skeletal muscle.

All of these functions are location-specific. Functionally, the cortex is divided into four lobes (Figure 11.16b). At the back of the brain, the *occipital lobe* processes visual information. The *temporal lobe* interprets auditory information, comprehends spoken and written language, and is responsible for perceptual judgment. The *parietal lobe* receives and interprets sensory information from the skin, such as temperature, touch, pain, and pressure. Finally, toward the front of the cortex is the *frontal lobe*, perhaps the most highly developed of all brain regions. The frontal lobe initiates motor activity and is also responsible for speech and conscious thought. All four lobes are involved in memory storage.

Notice that the region for speech (frontal lobe) is anatomically separate from the region for understanding spoken language (temporal lobe). The reasons for this separation are not known.

Two areas of the brain that have been studied extensively are the region in the parietal lobe that receives sensory input from the skin (called the *primary somatosensory area*) and the corresponding region of the frontal lobe that initiates motor activity (called the *primary motor area*). Researchers have found very specific regions within each of these areas corresponding to specific parts of the body (**Figure 11.17**). Body parts that are extremely sensitive, such as the lips and genitals, involve more neurons and consequently larger portions of the primary somatosensory area, whereas less sensitive body parts, such as a thigh, involve fewer neurons and smaller portions. Similarly, body parts that can perform intricate and precise movements, such as the fingers, involve more neurons in the primary motor area than do less dexterous regions, such as an elbow.

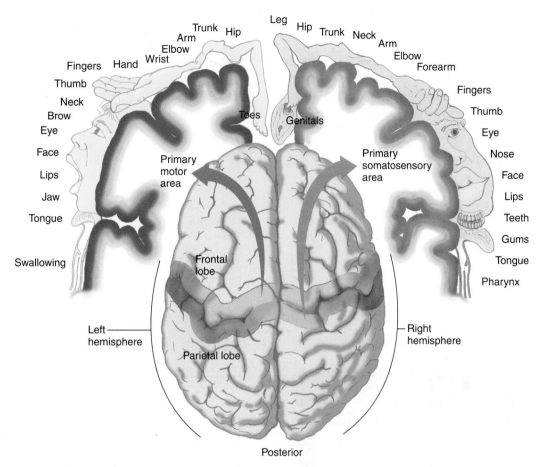

Figure 11.17 Primary somatosensory and motor areas of the cerebral cortex. Note the close similarity of the maps of the two areas. The larger the anatomical representation on the diagram, the more precisely controlled or more sensitive the area. The lips, face, and tongue are particularly well represented.

The location-specific nature of information processing explains why damage to different brain regions can cause such varied, seemingly contradictory, effects. This is often apparent in stroke victims. One stroke patient might be unable to speak yet still be fully capable of understanding speech. Another could have trouble controlling basic skeletal muscle movements but retain the capacity for abstract thought. Physicians use their understanding of these known location-specific functions to diagnose which areas of the brain have been affected.

Recap The brain has three divisions: hindbrain, midbrain, and forebrain. The evolutionarily oldest, more primitive hindbrain controls most automatic activities. The midbrain functions in vision, hearing, and movement. Structures in the forebrain help maintain homeostasis, process information, control emotions and memory, and are responsible for language, decision making, and conscious thought. ∎

11.9 Brain activity continues during sleep

Why do we sleep? This may seem like a simple question, but scientists are still looking for the answer. The obvious hypothesis is that we sleep to get rest. However, brain activity continues even during deep sleep.

Levels of sleep and wakefulness are controlled by the *reticular activating system* (RAS), a group of neurons in the reticular formation. Some neurons of the RAS transmit a steady stream of action potentials to the cerebrum, keeping us awake and alert. A "sleep center" in the RAS releases the neurotransmitter serotonin, which induces sleep by inhibiting the neurons that arouse the brain. Norepinephrine, secreted from another area in the brain, inhibits the effects of serotonin and makes us feel more alert.

We can study what happens to the brain during sleep with *electroencephalograms* (EEGs), recordings of the brain's electrical activity as measured at the body's surface. As **Figure 11.18** shows, EEG readings for different states of sleep show characteristic patterns.

- *Stage 1*. Stage 1 is a transitional phase between wakefulness and sleep. The pupils gradually constrict, breathing slows, and heart rate slows. On an EEG, Stage 1 appears as random small waves, only slightly different from wakefulness.
- *Stage 2*. During this phase, skeletal muscles relax, and eye and body movements usually cease. On an EEG, Stage 2 produces sharp waves known as "sleep spindles."
- *Stage 3*. Sleep deepens as heart rate and respiration slow even more. On an EEG, this stage and Stage 4 produce large, slow oscillations, which is why they are sometimes called "slow wave sleep."
- *Stage 4*. Sleepers are difficult to awaken; heart rate and respiration are at their slowest, and body temperature falls.

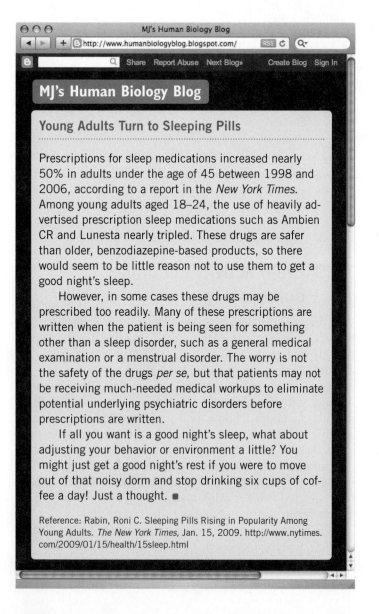

Young Adults Turn to Sleeping Pills

Prescriptions for sleep medications increased nearly 50% in adults under the age of 45 between 1998 and 2006, according to a report in the *New York Times*. Among young adults aged 18–24, the use of heavily advertised prescription sleep medications such as Ambien CR and Lunesta nearly tripled. These drugs are safer than older, benzodiazepine-based products, so there would seem to be little reason not to use them to get a good night's sleep.

However, in some cases these drugs may be prescribed too readily. Many of these prescriptions are written when the patient is being seen for something other than a sleep disorder, such as a general medical examination or a menstrual disorder. The worry is not the safety of the drugs *per se*, but that patients may not be receiving much-needed medical workups to eliminate potential underlying psychiatric disorders before prescriptions are written.

If all you want is a good night's sleep, what about adjusting your behavior or environment a little? You might just get a good night's rest if you were to move out of that noisy dorm and stop drinking six cups of coffee a day! Just a thought. ■

Reference: Rabin, Roni C. Sleeping Pills Rising in Popularity Among Young Adults. *The New York Times*, Jan. 15, 2009. http://www.nytimes.com/2009/01/15/health/15sleep.html

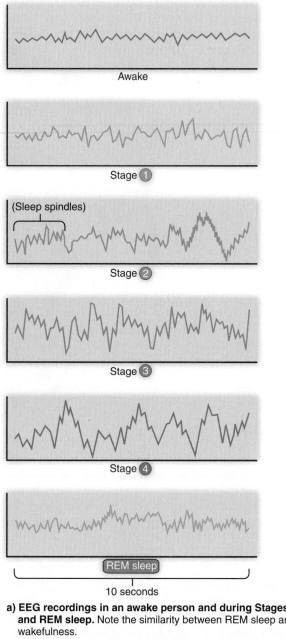

a) EEG recordings in an awake person and during Stages 1-4 and REM sleep. Note the similarity between REM sleep and wakefulness.

b) Patterns of sleep during a typical night.

Figure 11.18 Brain activity patterns during wakefulness and sleep.

- *REM (rapid eye movement) sleep.* This is when we dream. Heart rate, respiration, and blood flow to the brain increase. Most of the body's muscles go limp, preventing us from responding to our dreams. Paradoxically, the eyes move rapidly under closed eyelids. EEG patterns show a level of brain activity comparable to that of wakefulness.

Although a typical night's sleep involves cycling through these phases, the length of time spent in each stage can vary. REM sleep returns about every 90 minutes throughout the night and lasts a little longer each time. Toward morning a single REM period may last up to 20 minutes.

Although it is not clear exactly why we sleep, it is clear that we need a certain amount of sleep. Persons deprived of REM sleep become moody, irritable, and depressed and may even suffer hallucinations if deprived long enough.

🔄 **Recap** The brain remains active even during sleep. Stages of sleep have been correlated with brain wave patterns. The reticular activating system controls levels of sleep and wakefulness. ▪

11.10 The limbic system is the site of emotions and basic behaviors

Emotions and basic patterns of behavior originate in the **limbic system** (Figure 11.19). Limbic means "border." The name was originally used to describe those structures that bordered the basal regions of the cerebrum. As we learned about the function of these structures, however, the name came to

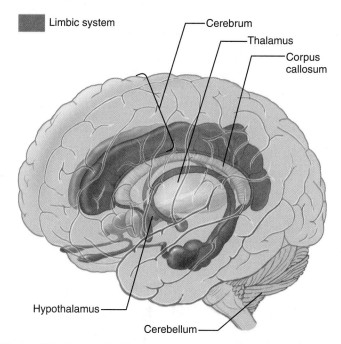

◾ Limbic system

— Cerebrum

— Thalamus

— Corpus callosum

Hypothalamus —

Cerebellum —

Figure 11.19 The limbic system. The limbic system is a collection of neuronal areas near the base of the cerebrum that are involved in emotions, basic behaviors, and short-term memory.

describe all of the neuronal structures that, together, control emotional behavior and motivational drives. When different areas of the limbic system are stimulated, we experience strong emotions such as fear, anger, sorrow, or love. These reactions, both pleasant and unpleasant, influence our behavior.

You may have noticed that odors have the power to evoke emotions and vivid recollections. This is because pathways for the sense of smell also pass through the limbic system, creating a strong association between olfactory information and memory.

Recall that the hypothalamus monitors and regulates homeostasis of many basic body functions. The hypothalamus also serves as a gateway to and from the limbic system and thus exerts some control over basic self-gratifying behaviors, such as satisfying hunger, thirst, and sexual desire. The thalamus receives and organizes information regarding these behaviors and transmits it to the cerebrum. The cerebrum integrates this information with the other information it is receiving, and then adjusts behavior accordingly.

Thus the intensity with which we feel emotions and the extent to which we act on them depends on modification by the cerebrum. To appreciate the importance of cerebral modification, just imagine what would happen if our most basic urges were no longer influenced by social rules or codes of conduct!

🔄 **Recap** A collection of structures bordering the cerebrum called the limbic system controls our emotions and behavior. The activities of the limbic system are monitored by the hypothalamus and modified by the cerebrum. ▪

11.11 Memory involves storing and retrieving information

Memory involves storing information and retrieving it later as needed. Memory has two stages: short-term ("working") memory involves retrieval of information that was stored within the past few hours, and long-term memory is the ability to retrieve information days or years later.

There is a clear difference between how the brain manages the two stages of memory. Short-term memory occurs in the limbic system. When you receive new sensory information (such as reading an unfamiliar telephone number), the stimulus triggers a quick burst of action potentials in the limbic system. You can remember the number for a few seconds or minutes, because neurons that have just fired are easier to fire again in the short term. The pathway becomes entrained, if only for a brief time. If the number is not important, you quickly forget it because the enhancement of the limbic pathway is not long lasting. That particular piece of information does not move into long-term memory.

However, if the number is important to you, as evidenced by the fact that you say it out loud several times, write it down and then read it to yourself, or use it repeatedly over several weeks, the information may be transmitted to your cerebral cortex for storage in long-term memory centers. During this process, neurons undergo a permanent chemical or physical change. Evidence from experimental animals suggests that long-term memory storage creates additional synapses between connecting neurons, enabling those circuits to be activated more easily in the future.

⮌ Recap Short-term memory involves quick bursts of action potentials in the limbic system. Long-term memory resides in the cerebral cortex and involves permanent changes in neurons and synapses. ■

11.12 Psychoactive drugs affect higher brain functions

A drug is any substance introduced into the body for the purpose of causing a physiological change. Psychoactive drugs are those that affect states of consciousness, emotions, or behavior—in other words, the higher brain functions. Some psychoactive drugs are legal, such as alcohol and nicotine. Others are not legal but have found their way into the fabric of our society.

Most drugs, including psychoactive substances, were originally discovered or designed to treat disease. The nonmedical use of drugs is sometimes euphemistically called "recreational" use, but the term "illicit" is certainly a more accurate reflection of the truth.

All psychoactive drugs are able to cross the blood-brain barrier (not all drugs can do so). They work by influencing the concentrations or actions of brain neurotransmitters. Recall that normally the effects of neurotransmitters are short-lived because the neurotransmitter remains in the synaptic cleft for only a brief time. Psychoactive drugs either bind to neurotransmitters or affect their release, action, or reuptake (reabsorption) in some way. This changes normal patterns of electrical activity in the brain.

Depending on which neurotransmitter they influence and where they act, psychoactive drugs can exert a variety of effects. For example, dopamine is one of the most important neurotransmitters in areas of the brain associated with pleasure. Cocaine blocks the reuptake of dopamine back into the presynaptic bulb, leading to too much dopamine in the synaptic cleft and therefore excessive stimulation of postsynaptic neurons. Cocaine initially allows the user to experience pleasurable effects longer than normal.

Psychoactive drugs can lead to *psychological dependence*, meaning that users begin to crave the feelings associated with a drug and alter their behavior to obtain it. Eventually tolerance develops—they need to use more of it, or use it more frequently, to achieve the same effect. Tolerance occurs in part because the liver produces enzymes that detoxify many drugs, and repeated drug use increases circulating levels of these detoxifying enzymes. Soon the original dose is no longer enough to create the desired effect, or

sometimes any effect at all. A second reason for tolerance, in the case of cocaine, is that if too much dopamine remains in the synaptic cleft for prolonged periods, the body starts to produce less of it. So with repeated cocaine use, a person needs more and more drug to maintain even a normal amount of dopamine neurotransmitter in the synaptic cleft. Eventually, cocaine addicts become incapable of experiencing pleasure without the drug and cannot even maintain a normal emotional state. At this point, severe anxiety and depression set in.

Tolerance leads to *addiction*, the need to continue obtaining and using a substance despite one's better judgment and good intentions. Essentially, choice is lost. Addiction has serious negative consequences for both emotional and physical health. Eventually, most addicts find it impossible to concentrate and hold a steady job. At the same time, they must spend large amounts of money to support their habit. Personal relationships suffer, and many addicts lose their families, their employment, and their physical health. In addition, they face the probability of *withdrawal*, the symptoms that occur if the drug is removed suddenly.

Cocaine and nicotine are both highly addictive. Alcohol is addictive in some people but not in others, for reasons we do not yet fully understand. One of the most addictive street drugs is crystal methamphetamine (also known as "crank," "ice," "tina," or "glass").

The federal government keeps track of illicit drug use through an annual survey. **Figure 11.20** shows reported illicit drug use during the month prior to the survey by persons aged 12 and older. Illicit drug use peaks in 18- to 20-year-olds and then declines with age. The most commonly used illicit drugs in 18- to 25-year-olds are marijuana, prescription

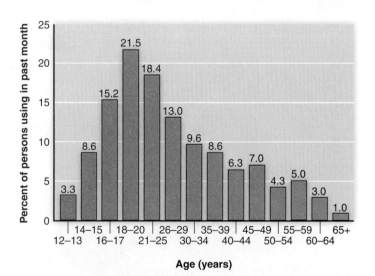

Figure 11.20 Past month illicit drug use among persons aged 12 and older, 2008.

Source: Substance Abuse and Mental Health Services Administration, 2009. Overview of Findings from the 2008 National Survey on Drug Use and Health.

pain relievers, cocaine, and hallucinogens (primarily ecstasy, or MDMA), in that order.

Over half of all 18- to 25-year-olds report having used marijuana at least once in their lives. However, only about 15% report having used cocaine or ecstasy, and fewer than 4% have used methamphetamine in the past year.

 Recap Psychoactive drugs work by altering either the concentration or action of brain neurotransmitters. Some lead to psychological dependence, tolerance, and addiction. ∎

HBP **Web Animation** *Psychoactive Drugs and the Brain* at www.humanbiology.com

11.13 Disorders of the nervous system

Because the nervous system is so complex and controls so many body functions, disorders of the nervous system can be particularly debilitating. We have already discussed stroke in the context of cardiovascular disease. Here we discuss several classes of disorders that affect the nervous system, including trauma, infections, disorders of neural or synaptic transmission, and brain tumors.

Trauma

Trauma (physical injury) to either the brain or spinal cord can be particularly dangerous because the brain controls so many body functions and the spinal cord is the information pathway to virtually all the organs.

Concussion disrupts electrical activity A violent blow to the head or neck can cause *concussion*, a brief period of unconsciousness in which electrical activity of brain neurons is briefly disrupted. After the person "comes to," he or she may experience blurred vision, confusion, and nausea and vomiting. Although most concussions are not associated with permanent injury, one of the biggest dangers is a *subdural hematoma*, or hemorrhage (bleeding) into the space between the meninges.

Because the brain is encased in a rigid skull, any bleeding increases pressure within the skull. Eventually the trapped blood presses against healthy brain tissue and disrupts function. Symptoms can include drowsiness, headache, and growing weakness of one side of the body. Depending on how rapidly bleeding occurs, symptoms may take weeks or even months to appear. Sometimes surgeons drill through the skull to drain the hematoma, relieve the pressure, and repair damaged blood vessels.

Spinal cord injuries impair sensation and function Violent trauma to vertebrae may press against the spinal cord, tear it, or even sever it completely. Damage to the spinal cord usually impairs sensation and body function below the injured area, so the extent of injury depends on the site of damage. Injuries below the neck can cause paraplegia (weakness or paralysis of the legs and sometimes part of the trunk). Injuries to the neck may lead to quadriplegia (weakness or paralysis of the legs, arms, and trunk). The higher the spinal cord injury, the more likely it is to be fatal. Spinal cord injuries may also cause problems with bladder or bowel control because the autonomic division no longer functions properly.

Researchers are investigating many exciting treatments for spinal cord injuries. For more information, see the Health & Wellness, *Repairing Spinal Cord Injuries* on page 268.

Infections

The brain and spinal cord are spared most of the infections that affect the rest of the body because of the blood-brain barrier. Nevertheless, certain viruses and bacteria do occasionally infect the brain or spinal cord.

Encephalitis: Inflammation of the brain Encephalitis refers to inflammation of the brain, often caused by a viral infection. Frequent sources of encephalitis include the herpes simplex virus (which causes cold sores) and a virus transmitted through mosquito bites. HIV (human immunodeficiency virus) infection is responsible for a growing number of cases.

Symptoms of encephalitis can vary, but often include headache, fever, and fatigue, progressing to hallucinations, confusion, and disturbances in speech, memory, and behavior. Some patients also experience epileptic seizures. Treatment varies depending on the cause, but antiviral drugs may help.

Meningitis: Inflammation of the meninges Meningitis involves inflammation of the meninges, usually due to viral or bacterial infection. Symptoms include headache, fever, nausea and vomiting, sensitivity to light, and a stiff neck. Mild cases may resemble the flu. Viral meningitis tends to be relatively mild and usually improves within a few weeks. Bacterial meningitis can be life threatening and requires antibiotics.

Rabies: Infectious viral disease Rabies is an infectious viral brain disease of mammals, especially dogs, wolves, raccoons, bats, skunks, foxes, and cats. It is transmitted to humans by direct contact, usually through a bite or a lick over broken skin. The virus makes its way through the sensory neurons in the region of the bite to the spinal cord and then to the brain, where it multiplies and kills cells.

Signs of rabies include swollen lymph glands, painful swallowing, vomiting, choking, spasms of throat and chest muscles, fever, and mental derangement. The symptoms generally appear one to six months after initial infection, followed by coma and death within two to twenty days.

If an animal has bitten you, wash the wound thoroughly and consult a physician. If possible, have the animal tested for rabies. If there is any possibility it could be rabid, your doctor will advise you to be immunized, because untreated rabies may be fatal. The rabies vaccine is most effective if administered within a few days after the bite, but it may still work when given weeks or months later.

Health & Wellness

Repairing Spinal Cord Injuries

On September 7, 2007, Buffalo Bills football player Kevin Everett suffered a fracture and dislocation of the cervical spine while making a tackle during a game against the Denver Broncos. As he was taken off the field, Kevin moved only his eyes. Two days after the accident Kevin still could barely feel or move his arms and legs. Doctors gave him a "statistically very small" chance of ever walking again.

Injuries like Kevin Everett's generally crush the spinal cord axons rather than actually severing the cord. Although the nerves are initially intact, over the next few weeks many of the damaged neurons die or lose the ability to function.

Until recently it was believed that damaged neurons could not regenerate. However, researchers are investigating a number of exciting possibilities for helping spinal cord patients recover lost function. Current areas of research or treatment include

- *Limiting cell injury and death.* At least three mechanisms of cell injury and death are involved. Researchers are exploring ways to block these injury and death pathways to preserve as many neurons and oligodendrocytes as possible.
- *Controlling inflammation.* Ironically, the inflammatory process, which normally promotes healing, makes matters worse by causing secondary damage to cells that were not injured initially. Anti-inflammatory drugs such as methylprednisolone can help, especially if given within eight hours of the injury.
- *Blocking certain molecules that prevent regeneration.* The CNS of adult mammals contains inhibitory molecules that block neurons from continually growing and dividing inappropriately in healthy adults. After an injury, however, they become an obstacle to tissue repair. Scientists are working on ways to neutralize these molecules.
- *Stimulating axons to grow.* Some researchers are working with growth factors and guidance molecules to stimulate damaged axons to "sprout" new extensions across the injured area.
- *Replacing lost cells with new tissue.* Stem cells, which have the potential to differentiate into many types of cells, might be stimulated to develop into neurons and neuroglial cells for transplant. See the Current Issue in Chapter 3 for a detailed discussion of stem cell research.
- *Bridging the injury site or removing barriers to regrowth.* Some researchers are using artificial materials to support and guide neurons as they regenerate across an injured area. Other groups are experimenting with grafts of neural tissue, or even proteins that can assemble into nanofibers smaller than a human hair. Still others are experimenting with enzymes that dissolve scar tissue and allow neurons to grow through the affected area.
- *Stimulating muscles electronically.* Computer-controlled functional electrical stimulation (FES) systems help patients move by activating their muscles with electrodes.

Some of this research is beginning to pay off. Kevin Everett attended the Super Bowl on January 31, 2008, and was able to walk under his own power during an appearance on the *Oprah* show that same day. Today, 90% of people with spinal cord injuries survive and live a near-normal life span. Unfortunately, fewer than half of spinal cord injury victims are able to return to the workforce. Perhaps with further advances that number will rise. ∎

Kevin Everett suffered a fracture of the spine on September 7, 2007. He was walking again just five months later.

Brain tumors: Abnormal growths

A brain tumor is an abnormal growth in or on the brain. Brain tumors are not necessarily cancerous, but even noncancerous growths can cause problems if they press against neighboring tissue and increase pressure within the skull. Over time, the rising pressure disrupts normal brain function. Depending on the location and size of the tumor, symptoms include headache, vomiting, visual impairment, and confusion. Some people also experience muscle weakness, difficulty speaking, or epileptic seizures.

Some brain tumors originate from the brain itself, but most are seeded by cancer that spreads from other

MJ's Human Biology Blog

Your Cell Phone's Radiation Emission

Does the radiation emitted by cell phones cause brain cancers? Although most of the scientific evidence suggests that it doesn't, the media always seems to find a physician or a patient or even a scientist who is willing to talk about it. The debate has been going on since cell phones first came out, and it's not likely to go away any time soon.

Are you concerned? How much radiation does your cell phone emit, anyway—a lot, or a little? The Federal Trade Commission sets an upper limit of 1.6 W/kg (watts per kilogram) for certification, but commercial phones on the market vary all the way from 1.6 down to 0.14—more than a 10-fold difference.

For a complete listing of the radiation output of all current phone models, go to http://reviews.cnet.com, go to "Cell phone reviews," then click on "Cell phone radiation charts." ∎

parts of the body. For more information on cancer, see Chapter 18.

Some brain tumors can be surgically removed. If tumors are inaccessible or too large to be removed, radiation and chemotherapy treatments may shrink them or limit their growth.

Disorders of neural and synaptic transmission

The nervous system depends on action potentials traveling in neurons and the chemical transfer of signals from one neuron to another across a synapse. It should come as no surprise, then, that problems involving neural or synaptic transmission can cause disease. The symptoms of these disorders depend on which nerves and/or neurotransmitters are affected.

Epilepsy: Recurring episodes of abnormal electrical activity

Epilepsy is a disorder characterized by transient, recurring bouts of abnormal brain cell activity. Fatigue, stress, or patterns of flashing lights may trigger epileptic attacks, known as seizures. Seizures vary widely, because virtually any activity controlled by the brain may be affected. Some epileptic seizures are almost not noticeable: perhaps a short episode of staring or nonpurposeful movements, a shaking limb, or repeated hand rubbing. Some seizures result in changes in the way things smell, taste, or sound to the patient or cause temporary speech disturbances. These seizures end quickly without causing any loss of consciousness. Other seizures affect larger areas of the brain and thus have more noticeable

symptoms, such as sudden collapse, convulsions (stiffening and shaking of the extremities or the entire body), or temporary loss of consciousness, from which the person recovers fully after a short time. Epilepsy usually appears during childhood or adolescence. Some cases result from infections (such as meningitis or encephalitis), head injuries, brain tumors, or stroke, but often the cause is not known.

Alzheimer's disease: Shortage of acetylcholine In contrast to Parkinson's disease, which affects primarily physical activity, Alzheimer's disease is a disorder of mental impairment, especially impairment of memory. One in every ten people over age 65 has the disorder.

Alzheimer's disease begins with memory lapses and progresses to severe memory loss, particularly of recent events; long-term memory is affected more slowly. Patients can also become disoriented, demented, and develop personality changes. Eventually they can no longer function independently.

The disease primarily affects neurons in the limbic system and frontal lobe, which use acetylcholine as their neurotransmitter. As the disease progresses, the supply of acetylcholine becomes insufficient for normal neurotransmission. The study of affected brains has revealed definite anatomical changes, including abnormal, tangled neurons and unusually large deposits of an abnormal protein called *beta amyloid*. Along with these anatomical changes, metabolic activity in the frontal and temporal lobes declines, as do levels of the enzyme that produces acetylcholine (**Figure 11.21**). Drugs that enhance the brain's production of acetylcholine can slow the progression of the disease, but as yet there is no cure.

The cause of Alzheimer's is still unknown, but possibilities include infections and impaired circulation to the brain.

a) A PET scan of a healthy brain.

b) A PET scan of a patient with Alzheimer's disease, showing decreased activity and an irregular pattern.

Figure 11.21 Changes in brain metabolic activity in Alzheimer's disease. A positron emission tomography (PET) scan reveals differences in metabolic activity of tissue. Reds and yellows represent high metabolic activity, blues and greens represent lower activity.

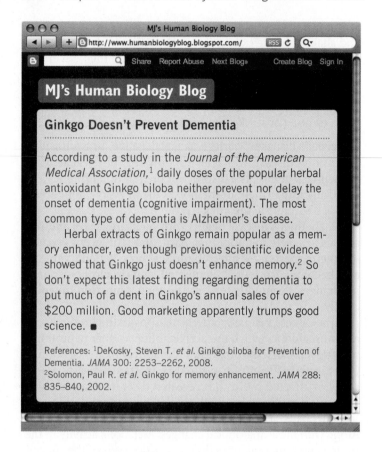

Ginkgo Doesn't Prevent Dementia

According to a study in the *Journal of the American Medical Association*,[1] daily doses of the popular herbal antioxidant Ginkgo biloba neither prevent nor delay the onset of dementia (cognitive impairment). The most common type of dementia is Alzheimer's disease.

Herbal extracts of Ginkgo remain popular as a memory enhancer, even though previous scientific evidence showed that Ginkgo just doesn't enhance memory.[2] So don't expect this latest finding regarding dementia to put much of a dent in Ginkgo's annual sales of over $200 million. Good marketing apparently trumps good science. ∎

References: [1]DeKosky, Steven T. *et al.* Ginkgo biloba for Prevention of Dementia. *JAMA* 300: 2253–2262, 2008. [2]Solomon, Paul R. *et al.* Ginkgo for memory enhancement. *JAMA* 288: 835–840, 2002.

There appears to be a genetic component, as a family history of the disease increases the risk. Recent studies suggest that an abnormal gene on chromosome 21 for the beta amyloid precursor protein may be involved.

Parkinson's disease: Loss of dopamine-releasing neurons

Parkinson's disease is a progressive, degenerative disorder that strikes nearly 50,000 people a year in North America, most over age 55. Symptoms include stiff muscles and muscle tremors in the hands and feet. Eventually people with Parkinson's lose mobility; they may also become mentally impaired and depressed.

We now know that Parkinson's disease is caused by degeneration of dopamine-releasing neurons in the area of the midbrain that coordinates muscle movement. The shortage of dopamine impairs the ability to perform smooth, coordinated motions. The symptoms can be improved by taking L-dopa, a drug that the body converts into dopamine, but this does not slow the loss of neurons. As the disease progresses, even L-dopa becomes ineffective.

However, there is an experimental and highly controversial treatment for patients with Parkinson's disease that might replace lost cells: receiving transplants of fetal neurons directly into the brain. Experiments with animals have shown that transplanted fetal neurons will survive and grow in the recipient animal. In humans, the fetal cells would come from aborted fetuses, which is why the technique generates such controversy. For an in-depth discussion of stem cell research and related social and ethical issues, see the Current Issue in Chapter 3.

Recap Disorders of the nervous system are often serious because of the central role of the nervous system in regulating other body systems. Common disorders include physical trauma and infections, as well as acquired or genetic diseases that disrupt neural transmission in some way, such as epilepsy, Parkinson's disease, and Alzheimer's disease. ∎

Chapter Summary

The nervous system has two principal parts *p. 244*

- The central nervous system receives, processes, and stores information, and the peripheral nervous system provides sensory input and motor output.
- The PNS has somatic and autonomic divisions.
- The autonomic division of the PNS has sympathetic and parasympathetic divisions.

Neurons are the communication cells of the nervous system *p. 245*

- Neurons generally consist of a cell body, several short extensions called dendrites, and a long, thin extension called an axon.

Neurons initiate action potentials *p. 246*

- Incoming signals from other neurons produce short-lived graded potentials in a neuron that alter its resting potential slightly.
- If the sum of all graded potentials exceeds threshold, voltage-sensitive sodium channels open briefly and an action potential is initiated.
- The three components of an action potential are (1) depolarization, in which sodium channels open briefly and then close, (2) repolarization, in which potassium channels open briefly and then close, and (3) reestablishment of the resting potential by the continuous action of the sodium-potassium pump.

Neuroglial cells support and protect neurons *p. 249*

- Schwann cells produce myelin, an insulating material that increases conduction velocities in neurons.
- Multiple sclerosis is a disease in which the myelin sheath of neurons becomes damaged.

Information is transferred from a neuron to its target *p. 251*

- An action potential in a presynaptic neuron causes ion channels for Ca^{2+} to open in the membrane of the axon bulb.
- Ca^{2+} diffuses into the axon bulb and causes a chemical neurotransmitter to be released into the synaptic cleft.
- The neurotransmitter binds to receptors on the postsynaptic cell membrane, causing a graded potential in the postsynaptic cell.
- Information processing depends on differences in stimulus intensity and on converging and diverging neural connections.

The PNS relays information between tissues and the CNS *p. 253*

- The peripheral nervous system consists of 12 pairs of cranial nerves and 31 pairs of spinal nerves. Each nerve may contain a mixture of sensory, somatic motor, and autonomic motor nerve axons.
- The somatic motor division of the peripheral nervous system controls skeletal muscle.
- The autonomic motor division of the peripheral nervous system controls automatic functions and helps to maintain homeostasis.

The brain and spinal cord constitute the CNS *p. 257*

- The brain and spinal cord are protected by bone, by three layers of connective tissue called meninges, and by cerebrospinal fluid.
- The spinal cord consists of ascending (going to the brain) and descending (coming from the brain) nerve tracts, plus neurons and neuroglial cells of the CNS.

The brain processes and acts on information *p. 260*

- The brain receives sensory information, integrates and stores it, and generates the appropriate response.
- The medulla oblongata contains areas that control blood pressure and respiration.
- The cerebellum coordinates basic and certain skilled body movements.
- The hypothalamus coordinates certain automatic functions and participates in the maintenance of homeostasis.
- The cerebrum is involved in language, decision making, and conscious thought. It is also the site of long-term memory storage.

Brain activity continues during sleep *p. 263*

- The brain exhibits characteristic cyclic patterns of activity during sleep.
- The functional importance of sleep is unknown.

The limbic system is the site of emotions and basic behaviors *p. 265*

- The limbic system is the center of emotions such as fear, anger, sorrow, and love and basic behaviors such as seeking food or sexual gratification.

Memory involves storing and retrieving information *p. 265*

- Short-term memory resides in the limbic system. Long-term memory involves permanent changes to neurons in the cerebral cortex.

Psychoactive drugs affect higher brain functions *p. 266*

- Psychoactive drugs act either by altering the concentrations of brain neurotransmitters or by altering the responses of postsynaptic neurons to them.

- Addiction to psychoactive drugs is characterized by the compulsion to use the drug, often at increasing doses, despite good intentions not to use and the likelihood of negative consequences.

Disorders of the nervous system *p. 267*

- Head trauma can be dangerous because any bruising or bleeding within the confined space of the skull presses against the brain.
- Spinal cord injury can interrupt neural signals traveling up and down the spinal cord, resulting in a loss of sensation and paralysis of muscle.
- Rabies is an infectious viral disease that is transmitted by the bite of an infected animal. The virus travels to the spinal cord and brain via the axon of a sensory nerve.
- Alzheimer's disease is characterized by a decline in the amount of the neurotransmitter acetylcholine in the limbic system and frontal lobe. Although the cause of Alzheimer's is unknown, the disease is associated with deposits of an abnormal protein in the brain.

Terms You Should Know

action potential, *247*	motor neuron, *245*
autonomic division, *255*	myelin sheath, *250*
axon, *245*	nerve, *254*
blood-brain barrier, *259*	neuroglial cell, *249*
central nervous system (CNS), *244*	neuron, *245*
cerebellum, *261*	neurotransmitter, *251*
cerebral cortex, *262*	parasympathetic division, *257*
cerebrospinal fluid, *258*	peripheral nervous system (PNS), *244*
cerebrum, *261*	resting potential, *246*
dendrite, *245*	sensory neuron, *245*
forebrain, *261*	somatic division, *254*
graded potential, *247*	synapse, *251*
hindbrain, *260*	sympathetic division, *255*
interneuron, *245*	threshold, *247*
midbrain, *261*	

Concept Review

Answers can be found at the Human Biology Place.
www.humanbiology.com

1. Distinguish between the central nervous system and the peripheral nervous system.
2. Explain what is meant by summation of graded potentials and explain how summation may lead to an action potential.
3. Describe, in terms of the opening and closing of ion channels and the subsequent diffusion of ions, how the depolarization and repolarization phases of an action potential take place.
4. Describe the functions of neuroglial cells.
5. Describe the role of neurotransmitters in the function of neurons.
6. List the three major divisions of the brain and identify their general functional roles.

7. Explain what is meant by the statement that the functions of the cerebral cortex are "location-specific."
8. Define REM sleep and indicate the physiological changes that might be observed during REM sleep.
9. Describe why a subdural hematoma is so much more dangerous than a hematoma in a muscle (known commonly as a bruise).
10. Explain why an injury to the spinal cord may impair sensations and motor movement below, but not above, the point of injury.

Test Yourself

Answers can be found in Appendix A.

1. Which of the following statements correctly characterizes an action potential?
   a. Action potentials are initiated by depolarization of the membrane to threshold.
   b. Action potentials reverse the membrane potential so that the interior is negatively charged and the exterior is positively charged.
   c. A stronger action potential will travel faster than a weaker action potential.
   d. Action potentials are more likely to result when the membrane is hyperpolarized.

2. Which of the following influence(s) the speed of an action potential?
   a. the presence of a myelin sheath
   b. the extent of depolarization that initiates the action potential
   c. the diameter of the axon
   d. both the presence of a myelin sheath and the diameter of the axon

3. Which of the following cell types makes up the majority of cells in the nervous system?
   a. sensory neurons
   b. motor neurons
   c. neuroglial cells
   d. interneurons

4. What do multiple sclerosis and ALS (amyotropic lateral sclerosis) have in common?
   a. They both are more common in young men than in young women.
   b. They both impair the initiation of action potentials.
   c. They both damage the myelin sheath, slowing action potential transmission.
   d. They both impair nerve function in the peripheral nervous system.

5. All of the following are directly involved in synaptic transmission except
   a. voltage-sensitive ion channels on the post-synaptic membrane
   b. chemical-sensitive ion channels on the post-synaptic membrane
   c. Ca^{2+}
   d. neurotransmitters

6. An _____ neurotransmitter causes _____ of the post-synaptic membrane.
   a. inhibitory…depolarization
   b. excitatory…depolarization
   c. inhibitory…hyperpolarization
   d. both (b) and (c) are correct

7. Information coming into the central nervous system arrives via _____ while information going from the central nervous system to the muscles, glands, and organs travels via _____.
   a. motor neurons…sensory neurons
   b. neuroglial cells…motor neurons
   c. sensory neurons…motor neurons
   d. sensory neurons…interneurons

8. The brain and spinal cord make up the:
   a. sensory nervous system
   b. central nervous system
   c. sympathetic nervous system
   d. parasympathetic nervous system

9. A spinal reflex requires the participation of each of the following except:
   a. cerebral cortex
   b. sensory neuron
   c. interneuron
   d. motor neuron

10. Which of the following would cause an acceleration of heart rate, an increase in blood pressure, and a slowing of digestive processes?
    a. motor division of the somatic nervous system
    b. sensory branch of the central nervous system
    c. parasympathetic division of the autonomic nervous system
    d. sympathetic division of the autonomic nervous system

11. Which part of the brain is responsible for regulating the heart and respiratory rates and blood pressure?
    a. cerebral cortex
    b. medulla oblongata
    c. cerebellum
    d. thalamus

12. The coordination necessary for a musician to play the piano depends on control exerted by the:
    a. thalamus
    b. cerebral cortex
    c. cerebellum
    d. medulla oblongata

13. All of the following are associated with storage of long-term memory except:
    a. limbic system
    b. cerebral cortex
    c. permanent changes in neurons
    d. formation of new synapses

14. Which of the following is characteristic of psychoactive drugs?
    a. They do not readily cross the blood-brain barrier.
    b. They alter levels of neurotransmitters.
    c. They can alter the amplitude of action potentials.
    d. They alter the speed with which action potentials travel.

15. Which of the following statements regarding infections of the nervous system is accurate?
    a. Most cases of encephalitis are caused by bacteria.
    b. Rabies can be treated with antibiotics.
    c. Bacterial meningitis is generally more severe than viral meningitis.
    d. The blood-brain barrier is not effective at preventing microorganisms from entering the central nervous system.

Apply What You Know

Answers can be found at the Human Biology Place.
www.humanbiology.com

1. Explain in terms of brain anatomy and function how it would be possible for people to remember whole events in the distant past yet not be able to recall what they had for breakfast.

2. What do you suppose would happen to a person's behavior and emotional expression if the neural connections between the limbic system and the cerebral cortex were severed?

3. A friend is washing dishes one evening and gets a deep cut from a sharp knife hidden in the soap suds. Several weeks after the visit to the ER, the stitches are removed. Your friend mentions that one part of his finger is numb; he cannot feel anything when he touches one spot. Will this numbness ever go away?

4. You're feeding a semi-wild cat outside the campus cafeteria when it accidentally nips you, drawing blood. Your friends advise you to report to the campus medical center immediately for a rabies vaccination. Should you go for the vaccination?

5. Overnight, your roommate develops a high fever. She says she has a severe headache and complains of neck stiffness. An hour later she seems confused. You take her to the hospital and the physician recommends a spinal tap—a test in which a sample of cerebrospinal fluid is removed and analyzed. What might the physician be looking for?

6. Some states still administer the death penalty by lethal injection. The injection may contain an anesthetic (for pain) combined with a high dose of potassium to stop the heart. Why would a large dose of potassium stop the heart?

7. At a party someone offers you an ecstasy pill. Aside from the legal issues involved, what are the short-term and long-term effects of trying it just one time?

8. Certain bacteria, such as those that cause botulism or tetanus, produce toxins that interfere with neuromuscular transmission. From what you know of neuromuscular transmission, how might such toxins work?

12

Sensory Mechanisms

Colorized SEM of the surface of the tongue, showing papillae (pointing toward the back of the tongue).

DWD: Driving While Distracted

Linda Doyle died on September 3, 2008, when her Toyota Rav4 was struck broadside by a Ford pickup truck in an intersection in Oklahoma City. The 20-year-old driver of the truck, a college student named Christopher Hill, admitted to police that he ran a red light at 45 miles per hour because he had been distracted by a cell phone call. Asked what color the light was when he went through it, he said, "I never saw it."

Your brain receives sensory inputs from many different sources at once. It copes with all this sensory input by focusing on input considered most important at the moment, sometimes at the expense of other input. Anything that distracts you while driving is likely to lower your ability to drive safely. Distractions may be visual (taking your eyes off the road), cognitive (thinking about something else), or manual (taking your hands off the wheel). Typical distractions include texting or talking on the phone, putting on makeup, eating, reading, or using a navigation system. Some distractions, such as texting, are especially risky because they involve all three types of distractions at once (visual, cognitive, *and* manual).

According to the National Highway Traffic Safety Administration (NHTSA), more than half a million injuries and nearly 6,000 deaths a year are due to car crashes involving distracted drivers. The highest proportion of distracted drivers involved in fatal crashes are under 20 years old.

Driving while on the phone is the most common distraction. The NHTSA recommends that drivers not use their

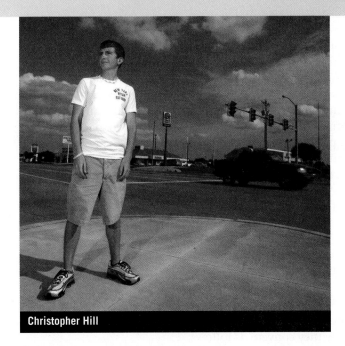

Christopher Hill

cell phones while driving except in an emergency, but it doesn't appear that anyone is listening. According to the NHTSA, at any given moment during daylight hours nearly 800,000 drivers are on their cell phones while driving

Reacting to the statistics and to high-profile deaths like Linda Doyle's, some states are beginning to take action. Six states (California, Oregon, Washington, New Jersey, New York, and Connecticut) now prohibit all drivers from using handheld cell phones while driving. Nineteen states prohibit text messaging. Seventeen states and the District of Columbia have placed special restrictions on cell phone use by drivers with learner's permits or drivers under a certain age. Other states are likely to follow suit.

The use of handheld cell phones is not illegal yet in Oklahoma but Linda Doyle's daughter, Jennifer Smith, continues to press lawmakers for change. She's also suing Samsung, maker of the phone used by the driver, and the Sprint Nextel service that provided the phone coverage in the area. The suit alleges that the companies failed to warn consumers about the dangers of driving while using their product. Sprint Nextel says that it includes adequate safety messages in its packaging, user manuals, advertising, and even on its Web site.

According to legal experts, Ms. Smith's lawsuit faces an uphill battle because crashes are caused by drivers, not cell phones, and because deep down inside, most drivers understand that talking on the phone while driving is a distraction but they continue to do it anyway. Lawyers for Samsung and Sprint Nextel are likely to argue that the companies had no responsibility to Linda Doyle because she did not use their phone or phone service; their only responsibility is to the young man who was driving the other vehicle, and he's not suing them.

Nevertheless, momentum is building to do something about the number of accidents being attributed to distracted drivers. "Every single time someone takes their eyes or their focus off the road—even for just a few seconds—they put their lives and the lives of others in danger," said Transportation Secretary Ray LaHood at a national summit on distracted driving. "Distracted driving is unsafe, irresponsible and in a split second, its consequences can be devastating."

Christopher Hill, the driver of the pickup truck that killed Linda Doyle, and Jennifer Smith, her daughter, would undoubtedly agree. Christopher Hill plead guilty to negligent homicide (a misdemeanor) and was sentenced to five years of probation plus 240 hours of community service. He deeply regrets the accident and took responsibility for his actions, saying it was his choice to talk on the phone while driving. Linda Doyle's family says they have forgiven him. These days he talks to schools and community groups about the dangers of driving while distracted. Jennifer Smith, for her part, is now a spokesperson for FocusDriven, the first national nonprofit organization for increasing awareness about the dangers of driving while distracted.

Both Christopher Hill and Jennifer Smith say they no longer use their cell phones while driving.

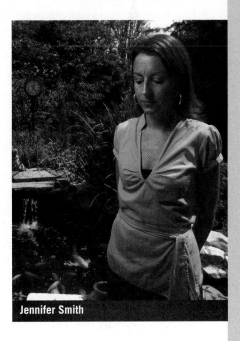

Jennifer Smith

Questions to consider

1 Do you think that texting while driving should be illegal nationwide? What about talking on a handheld phone?

2 When you cross a state line, how do you know whether texting or using a phone is illegal in the state you are entering? Should you be told?

3 In several states (Utah and New Hampshire), using a handheld phone is a violation only if the driver commits another moving offense while on the phone. Is that a reasonable solution?

The facts...

- More than 6,000 deaths and 500,000 injuries per year are attributed to driving while distracted.
- At any given moment during daylight hours, more than 800,000 drivers are talking on their cell phones.
- The highest proportion of distracted drivers is under 20 years old.
- Laws against driving while distracted (texting, talking on a handheld phone) vary from state to state, from complete bans to no ban at all. There is no national standard.

>> **Sensory mechanisms provide information** about the world around us and also about conditions within our own bodies.

>> **Various types of receptors convert different types of sensory stimuli (physical touch, odors, light, etc.) into action potentials.** The brain interprets incoming action potentials correctly by where they go in the brain.

>> **Some receptors adapt quickly,** which is why you stop noticing some stimuli after a while or they may seem less intense.

>> **Receptors are located throughout the body.** They provide us with information about body position, touch, temperature, vibration, pressure, and especially pain.

>> **The five special senses (taste, smell, hearing, balance, and vision) originate from special areas of the body.** Four of them (taste, smell, hearing, and vision) provide us with detailed information about the external world.

Look around you at the complex and colorful world you inhabit. Then, after reading this paragraph, close your eyes and concentrate for a moment on what you can hear. Identify the sources of any sounds you hear and try to judge their direction and distance from you. Describe any odors that you smell, and make a guess at the temperature in the room. Now concentrate on your own body. Move your arms and legs about, and then identify what type of clothing you are wearing. Finally, visualize in your mind the precise position of every part of your body, right down to your fingers and toes.

Your body's sensory mechanisms are constantly providing the brain with detailed information about the world around you, and even about the body itself. In this chapter we learn how different kinds of sensory information are received by the body, how that information is converted to nerve impulses, and how the nerve impulses are transmitted to the brain in such a way that the brain can make sense of it all.

12.1 Receptors receive and convert stimuli

Sensory input that causes some change within or outside the body is called a **stimulus** (plural: *stimuli*). The stimulus is often a form of physical energy such as heat, pressure, or sound waves, but it can also be a chemical. A **receptor**—a

structure specialized to receive certain stimuli—accepts the stimulus and converts its energy into another form.

Some receptors are evolved from the dendritic structures of a sensory neuron. They convert the stimulus into a graded potential that, if it is powerful enough, initiates an impulse within the sensory neuron. Other receptors are parts of cells that produce graded potentials and release a neurotransmitter, stimulating a nearby sensory neuron. In the end, the effect is the same—generation of an impulse in a sensory neuron.

When the central nervous system receives these impulses, we often experience a **sensation,** meaning that we become consciously aware of the stimulus. A sensation is different from **perception,** which means understanding what the sensation means. As an example, hearing the sound of thunder is a sensation; the belief that a storm is approaching is a perception.

Receptors are classified according to stimulus

Receptors are classified according to the type of stimulus energy they convert, as follows:

- **Mechanoreceptors** respond to forms of mechanical energy, such as waves of sound, changes in fluid pressure, physical touch or pressure, stretching, or forces generated by gravity and acceleration.
- **Thermoreceptors** respond to heat or cold.
- **Pain receptors** respond to tissue damage or excessive pressure or temperature.

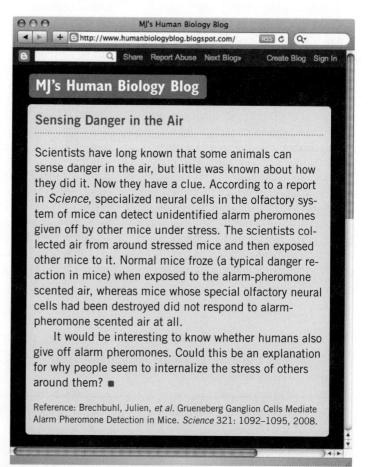

MJ's Human Biology Blog

http://www.humanbiologyblog.blogspot.com/

Share   Report Abuse   Next Blog»   Create Blog   Sign In

MJ's Human Biology Blog

Sensing Danger in the Air

Scientists have long known that some animals can sense danger in the air, but little was known about how they did it. Now they have a clue. According to a report in *Science*, specialized neural cells in the olfactory system of mice can detect unidentified alarm pheromones given off by other mice under stress. The scientists collected air from around stressed mice and then exposed other mice to it. Normal mice froze (a typical danger reaction in mice) when exposed to the alarm-pheromone scented air, whereas mice whose special olfactory neural cells had been destroyed did not respond to alarm-pheromone scented air at all.

It would be interesting to know whether humans also give off alarm pheromones. Could this be an explanation for why people seem to internalize the stress of others around them? ∎

Reference: Brechbuhl, Julien, *et al.* Grueneberg Ganglion Cells Mediate Alarm Pheromone Detection in Mice. *Science* 321: 1092–1095, 2008.

Table 12.1 **Types of receptors**

Type of receptor	Examples	Sensation
Mechanoreceptors		
Touch, pressure	Unencapsulated dendritic endings, Meissner's corpuscles, Pacinian corpuscles, Ruffini endings, Merkel disks	Touch and pressure on skin or body hair
Vibration	Pacinian corpuscles	Vibration on skin
Stretch	Muscle spindles Tendon receptors Joint receptors	Muscle length Tendon tension Joint position
Hearing	Hair cells in cochlea of the inner ear	Vibration of fluid originating from sound waves
Balance	Hair cells in semicircular canals of the inner ear. Hair cells in the utricle and saccule of the inner ear	Rotational movement of the head. Static position and linear acceleration and deceleration of the head
Thermoreceptors		
Temperature	Receptors in unencapsulated dendritic endings in skin	Heat or lack of heat (cold)
Pain receptors		
Pain	Receptors in unencapsulated dendritic endings in skin and internal organs	Pain caused by excessive pressure, heat, light, or chemical injury
Chemoreceptors		
Taste	Taste receptors in taste buds	Salty, sweet, sour, and bitter tastes
Smell	Olfactory receptors of nasal passages	Over a thousand different chemical smells
Photoreceptors		
Vision	Rods of the retina Cones of the retina	Light (highly sensitive black-and-white vision) Light (color vision)

- **Chemoreceptors** respond to the presence of chemicals in the nearby area.
- **Photoreceptors** respond to light.

Many of these receptors contribute to sensations, although some (such as joint receptors) contribute no more than a general sense of where our limbs are located. Table 12.1 lists examples of each type of receptor and the sensations they give us.

A few receptors are "silent" in the sense that we are not consciously aware of their actions. These receptors function in negative feedback loops that maintain homeostasis inside the body. They include stretch receptors for monitoring and regulating blood pressure and fluid volumes, and chemoreceptors for regulating the chemical composition of our internal environment. They aren't discussed in detail here, as we discuss most of them in the context of the homeostatic control of various organ systems.

The CNS interprets nerve impulses based on origin and frequency

How can cells in the central nervous system (CNS) interpret some incoming impulses as images and others as sounds? How do they distinguish a loud sound from a soft one and bright lights from dim?

As described in Chapter 11, nerve impulses are transmitted from receptors to specific brain areas. For example, impulses generated by visual stimuli travel in sensory neurons whose axons go directly to brain regions associated with vision. All incoming impulses traveling in these neurons are interpreted as light, regardless of how they were initiated. This is why, when you are hit in the eye, you may "see stars." The blow to your eye triggers impulses in visual sensory neurons. When these signals arrive in the visual area of the brain, they are interpreted as light.

Stronger stimuli activate more receptors and trigger a greater frequency of impulses in sensory neurons. In effect, the central nervous system gets all the information it needs by monitoring where impulses originate and how frequently they arrive.

✓ **Quick Check** Suppose the axon of a sound-detecting neuron in a person's ear were somehow redirected so that it connected to the visual area of the brain. How will this person likely perceive the action potentials produced by this neuron? Explain. ▪

Answers to ✓ questions can be found in Appendix A.

Some receptors adapt to continuing stimuli

The central nervous system can ignore one sensation to concentrate on others. For example, you ignore the feel of your clothing when you are more interested in other things. Some sensory inputs are ignored after a while because of **receptor adaptation,** in which the sensory neuron stops sending impulses even though the original stimulus is still present. To demonstrate, try touching a few hairs on your forearm. Notice that the initial feeling of light pressure goes away within a second or so, even though you maintain the same amount of pressure.

Receptors in the skin for light touch and pressure adapt rather quickly. This confers a survival advantage, because they can keep the central nervous system informed of *changes* in these stimuli, without constantly bombarding it with relatively unimportant stimuli. Olfactory (smell) receptors also adapt rapidly. Olfactory adaptation can be hazardous to people if they are continually exposed to low levels of hazardous chemicals that they can no longer smell.

Other receptors—pain receptors, joint and muscle receptors that monitor the position of our limbs, and essentially all of the silent receptors involved in homeostatic feedback control loops—adapt slowly or not at all. Lack of adaptation in these receptors is also important to survival. Persistent sensations such as pain alert us to possible tissue damage from illness or injury and prompt us to take appropriate action. Persistent activity of silent receptors is essential to our ability to maintain homeostasis.

Somatic sensations and special senses provide sensory information

The sensations (or senses) provided by receptors are categorized as either somatic or special. The **somatic sensations** originate from receptors present at more than one location in the body (*soma* is the Greek word for "body"). The somatic sensations include temperature, touch, vibration, pressure, pain, and awareness of body movements and position. The five **special senses** (taste, smell, hearing, balance, and vision) originate from receptors that are restricted to particular areas of the body, such as the ears and eyes. The special senses deliver highly specialized information about the external world.

We look at the somatic sensations first and then discuss the five special senses in detail.

↺ **Recap** Receptors receive a physical or chemical stimulus, causing nerve impulses to be generated in sensory neurons. Receptors are classified according to the type of stimulus they convert. The five classes of receptors are mechanoreceptors, thermoreceptors, pain receptors, chemoreceptors, and photoreceptors. Stronger stimuli activate more receptors and trigger a greater frequency of action potentials in sensory neurons. Some receptors adapt rapidly; others adapt slowly or not at all. ∎

12.2 Somatic sensations arise from receptors throughout the body

The somatic sensations of temperature, touch, pressure, vibration, pain, and awareness of body movements and position are essential to help us coordinate muscle movements, avoid danger, and maintain body temperature. Some somatic sensations contribute to pleasurable feelings as well, such as our response to the gentle touch of a loved one.

Receptors that detect the somatic sensations are located throughout the body in skin, joints, skeletal muscles, tendons, and internal organs. Sensory neurons linked to these receptors send their impulses to the brain, specifically to the primary somatosensory area of the parietal lobe of the cerebral cortex (review Figure 11.17). As you already know, body parts with the greatest sensory sensitivity, such as the mouth and fingers, involve more neurons in the somatosensory area, whereas less sensitive body parts involve fewer neurons.

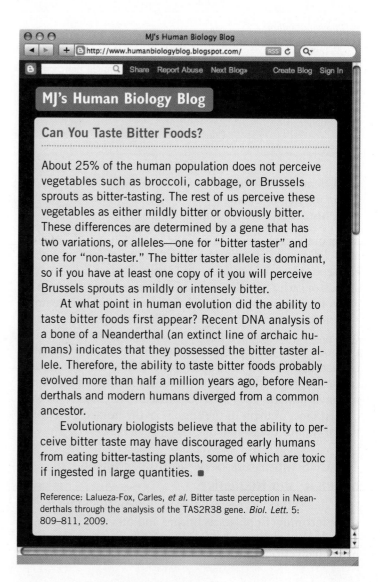

MJ's Human Biology Blog

Can You Taste Bitter Foods?

About 25% of the human population does not perceive vegetables such as broccoli, cabbage, or Brussels sprouts as bitter-tasting. The rest of us perceive these vegetables as either mildly bitter or obviously bitter. These differences are determined by a gene that has two variations, or alleles—one for "bitter taster" and one for "non-taster." The bitter taster allele is dominant, so if you have at least one copy of it you will perceive Brussels sprouts as mildly or intensely bitter.

At what point in human evolution did the ability to taste bitter foods first appear? Recent DNA analysis of a bone of a Neanderthal (an extinct line of archaic humans) indicates that they possessed the bitter taster allele. Therefore, the ability to taste bitter foods probably evolved more than half a million years ago, before Neanderthals and modern humans diverged from a common ancestor.

Evolutionary biologists believe that the ability to perceive bitter taste may have discouraged early humans from eating bitter-tasting plants, some of which are toxic if ingested in large quantities. ∎

Reference: Lalueza-Fox, Carles, *et al*. Bitter taste perception in Neanderthals through the analysis of the TAS2R38 gene. *Biol. Lett.* 5: 809–811, 2009.

The somatosensory area processes the information and sends it to the nearby primary motor area in the frontal lobe. If necessary, impulses are then generated in motor (output) neurons of the peripheral nervous system to cause body movement.

Mechanoreceptors detect touch, pressure, and vibration

Receptors for sensing touch, pressure, and vibration are called *mechanoreceptors*. Though they may take several different forms, all mechanoreceptors are the modified dendritic endings of sensory neurons. Any force that deforms the plasma membrane of the dendritic ending produces a typical graded potential. If the graded potential (or the sum of multiple graded potentials) is large enough to exceed threshold, the sensory neuron initiates an impulse (Chapter 11).

The various receptors for touch, pressure, and vibration differ in their locations, the intensity of the stimulus that must be applied in order to generate an impulse (pressure is a stronger mechanical force than light touch), and the degree to which they adapt. Vibration-sensitive receptors adapt so quickly that it takes a rapidly changing physical deformation (many on-and-off cycles per second) to keep them stimulated. Vibration-sensitive receptors are particularly useful for providing information about potentially harmful flying insects because those receptors respond to the kinds of vibrations generated by insects' wings.

Figure 12.1 depicts the skin with several types of receptors for detecting somatic sensations. They are

- **Unencapsulated dendritic endings.** Naked dendritic endings of sensory neurons around hairs and near the skin surface signal pain, light pressure, and changes in temperature.
- **Merkel disks.** Merkel disks are modified unencapsulated dendritic endings that detect light touch and pressure.
- **Meissner's corpuscles.** Encapsulated touch receptors located close to the skin surface detect the beginning and the end of light pressure and touch.
- **Ruffini endings.** These encapsulated receptors respond continually to ongoing pressure.

- **Pacinian corpuscles.** Several kinds of dendritic endings are encapsulated (enclosed) in epithelial or connective tissue. The best known, called *Pacinian corpuscles,* are encapsulated receptors located in the dermis that respond to either deep pressure or high-frequency vibration. The encapsulation structure of the Pacinian corpuscle permits the mechanoreceptor to adapt rapidly, even to a continuously applied stimulus (Figure 12.2 on next page).

Mechanoreceptors indicate limb position, muscle length, and tension

With your eyes closed, try to describe the positions of your arms and legs. You can probably describe their position rather well. You may even be able to tell which muscle groups are contracting. You do this with a variety of mechanoreceptors in joints (for joint position), in skeletal muscles (for muscle length), and in tendons (for tension). Perhaps the best known of these mechanoreceptors are the specialized structures for monitoring muscle length, called **muscle spindles.** For most joints, muscle length determines joint position because of the way the muscle is attached to the bones.

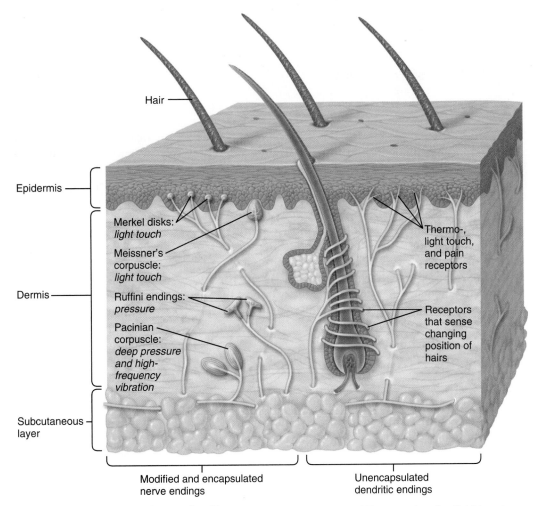

Figure 12.1 Sensory receptors in skin. Thermoreceptors, some of the receptors for light touch, and the receptors for pain are the free endings of sensory neurons. Other receptors for touch, pressure, and vibration are modified or encapsulated endings of sensory neurons.

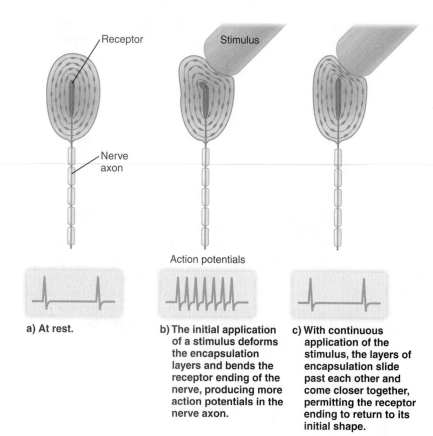

a) At rest.

b) **The initial application of a stimulus deforms the encapsulation layers and bends the receptor ending of the nerve, producing more action potentials in the nerve axon.**

c) **With continuous application of the stimulus, the layers of encapsulation slide past each other and come closer together, permitting the receptor ending to return to its initial shape.**

Figure 12.2 The Pacinian corpuscle.

A muscle spindle is a small bundle of modified skeletal muscle cells located within a skeletal muscle (**Figure 12.3**). Muscle spindles are innervated by sensory nerves whose dendritic endings are mechanoreceptors that respond to stretch (lengthening). One type of dendritic ending wraps around the middle of a muscle spindle cell. Another has several branches that attach nearer the end of the muscle spindle cell. When the whole muscle is stretched, the receptors attached to the muscle spindle cells are stretched as well. Mechanical distortion of the mechanoreceptors produces local graded potentials in the dendritic endings and (if threshold is passed) an action potential.

This is exactly what happens in the classic patellar stretch reflex, described in Chapter 11. When a physician taps your patellar tendon (at the front of your knee) with a hammer, the patellar tendon and quadriceps (thigh) muscle are stretched, stretching the muscle spindle as well. Stretch of the muscle causes sensory neurons innervating the muscle spindle to transmit action potentials to the spinal cord. Within the cord, motor nerves to the quadriceps are activated, causing the quadriceps to contract. All this occurs in a fraction of a second.

Tendons, which connect muscle to bone, also have mechanoreceptors. Tendon receptors respond to tension,

but they cannot distinguish between tension produced by passive stretch and tension produced by active muscle contraction. Tendon receptors may play a protective role in preventing injury by extremely high tensions. For example, if you were to fall out of a tree but land on your feet, the impact when you hit the ground would produce high levels of tension, activating your tendon stretch receptors. The sensory information they provide inhibits muscle contraction, so you would crumple to the ground rather than stay upright. This reflex-induced collapse protects your muscles from tearing and your bones from breaking.

✅ **Quick Check** A friend is trying to touch his toes (with his legs straight) by bouncing his upper body up and down. This sort of bouncing typically stimulates muscle spindles, but not tendon receptors. Explain whether this is a good method of stretching the hamstring muscles in the back of the legs. ■

Thermoreceptors detect temperature

Near the skin's surface, thermoreceptors for heat and cold provide useful information about the external environment. These receptors adapt quickly, allowing us to monitor changes in temperature accurately and yet adjust sensory input so it becomes more bearable. For example, when you step into a hot shower the water may seem uncomfortably warm at first, but you soon become used to it.

Other thermoreceptors located in abdominal and thoracic organs throughout the body monitor internal temperature (also called core temperature). In keeping with their role in maintaining homeostasis, these thermoreceptors do not adapt rapidly.

Pain receptors signal discomfort

As much as we dislike pain, the ability to perceive pain is essential for survival. Pain warns us to avoid certain stimuli and informs us of injuries. Pain receptors are unencapsulated endings that respond to injury from excessive physical pressure, heat, light, or chemicals. Pain is described as either fast or slow, depending on its characteristics.

Fast pain, also called *sharp* or *acute pain*, occurs as soon as a tenth of a second after the stimulus. Receptors for fast pain generally respond to physical pressure or heat and usually are located near the surface of the body. They inform us of stimuli to be avoided—for example, stepping on a nail or touching a hot burner on the stove. The reflex withdrawal response to fast pain is rapid and strong.

In contrast, *slow pain* generally arises from muscles or internal organs. Slow pain, which may not appear until

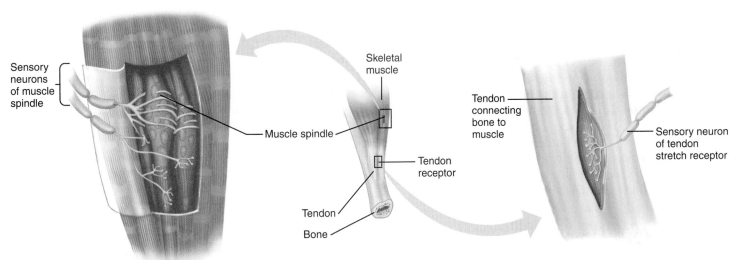

a) Muscle spindle. A muscle spindle responds to muscle length. Passive stretch of a muscle stretches the muscle spindle, stimulating mechanoreceptors in the nerve endings of the sensory neurons. Conversely, muscle contraction shortens the muscle spindle, reducing muscle spindle mechanoreceptor stimulation.

b) Tendon receptor. A tendon receptor responds to tension in tendons. When a muscle contracts and also when it is stretched passively, tension on the tendon increases, activating tendon receptors.

Figure 12.3 Muscle spindles and tendon receptors.

seconds or even minutes after injury, is due to activation of chemically sensitive pain receptors by chemicals released from damaged tissue. Slow pain from internal organs is often perceived as originating from an area of the body completely removed from the actual source. This phenomenon, called **referred pain,** happens because action potentials from internal pain receptors are transmitted to the brain by the same spinal neurons that transmit action potentials from pain receptors in the skin and skeletal muscles. The brain has no way of knowing the exact source of the pain, so it assigns the pain to another location.

Referred pain is so common that physicians use it to diagnose certain disorders of internal organs. For example, a heart attack usually manifests itself as pain in the left shoulder and down the left arm. **Figure 12.4** shows where pain from internal organs may be felt. The close association between areas of referred pain from the heart and esophagus is why it can be difficult, by sensations alone, to distinguish a heart attack from the less serious problem of esophageal spasm.

Pain receptors generally do not adapt. From a survival standpoint this is beneficial, because the continued presence of pain reminds us to avoid doing whatever causes it. However, it also means that people with chronic diseases or disabilities can end up experiencing constant discomfort.

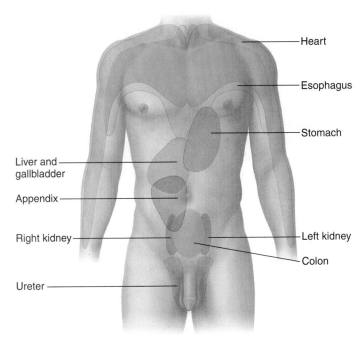

Figure 12.4 Locations of referred pain from various internal organs. Note that the areas of overlap for esophageal and heart pain are so extensive that it is often difficult to tell them apart by location.

🔁 **Recap** Somatic sensations include temperature, touch, pressure, vibration, pain, and awareness of body movements and position. Mechanoreceptors in skeletal muscle, tendons, and joints provide information about the position of our limbs and the amount of tension on muscles. Thermoreceptors near the skin surface monitor the temperature of the external environment, and thermoreceptors in internal organs maintain homeostasis of body temperature. Fast pain arises from receptors near the body surface, whereas slow pain originates in muscles or internal organs. Pain is important to survival because it warns us of injury. ∎

12.3 Taste and smell depend on chemoreceptors

We turn now to the five special senses: taste, smell, hearing, balance, and vision. As mentioned earlier, these senses originate from receptors located in special areas of the body.

The first senses we consider, taste and smell, are both due to the action of chemoreceptors. The rich fragrance of a rose, the less pleasing odor of a skunk, the tart taste of an orange—all depend on our ability to convert a chemical stimulus into the only language the central nervous system can understand: action potentials.

Taste: Chemoreceptors bind with dissolved substances

The tongue has a rough texture because its surface is composed of numerous small projections, called *papillae*, surrounded by deep folds (**Figure 12.5**). About 10,000 **taste buds** are located at the surface of many of these folds. A taste bud is a cluster of about 25 taste cells and 25 supporting cells that separate the taste cells from each other.

The exposed tip of a taste cell has hairs that project into the mouth. The hairs contain chemoreceptors that are specific for certain chemicals (called *tastants*). When we eat or drink, tastants dissolve in our saliva and bind to the chemoreceptor. The binding of tastant to chemoreceptor causes the taste cell to release a neurotransmitter, which stimulates a nearby sensory neuron to produce action potentials. Note that the taste cell itself is not a sensory neuron. It takes two cells (the taste cell and a sensory neuron) to convert the chemical stimulus into a nerve impulse, but the ultimate effect is the same as if it were one cell.

The four commonly accepted primary taste qualities are sweet, sour, salty, and bitter. In recent years a fifth primary taste quality for savory foods, called *umami*, has been added to the list. Taste receptors for umami allow us to appreciate foods like parmesan cheese, mushrooms, beef, soy sauce, and Worcestershire sauce. Although an individual taste receptor and its associated sensory neuron seem to be "tuned" to respond most strongly to only one of the taste qualities, they do respond weakly to the other taste qualities as well.

Most of the taste receptors are located around the edges and at the front and back of the tongue. Although all receptor types can be found in each of these regions, there is some evidence that their distribution is not entirely even. The tongue is slightly more sensitive to sweet tastes on the tip, sour on the sides, salty on the tip and sides, and bitter toward the back (the distribution of umami receptors is not yet well documented). Therefore, to gain the maximum taste sensation, we move food around in our mouth before swallowing. The combined sensitivities of our taste receptors allow us to distinguish hundreds of different flavors, far beyond the basic five qualities. On the other hand, if we don't want to taste something, we pass it over the center of the tongue and swallow quickly.

It is probably no accident that bitter taste receptors are heavily concentrated at the very back of the tongue. Bitterness is often a sign of an inedible or even poisonous substance, and the location of these receptors makes it difficult to bypass them.

☑ **Quick Check** If you blot your tongue completely dry, you will find that you temporarily lose the ability to taste dry foods such as sugar or salt. Why? ▪

Smell: Chemoreceptors bind with odorants

The sense of smell also relies on a chemoreceptor mechanism. There are two fundamental differences between taste and

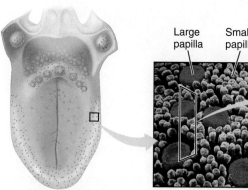

a) The tongue. Most of the receptors for taste are located at the edges and at the front and back of the tongue.

b) A view (approximately × 75) of the surface of the tongue showing that it is covered with large and small papillae.

c) A sagittal section (approximately × 200) through several large papillae showing the location of the taste buds along their sides, just beneath the tongue's surface.

d) A taste bud is composed of a group of taste cells and supporting cells. Each taste cell connects to a sensory neuron.

Figure 12.5 The structure of the receptors for taste.

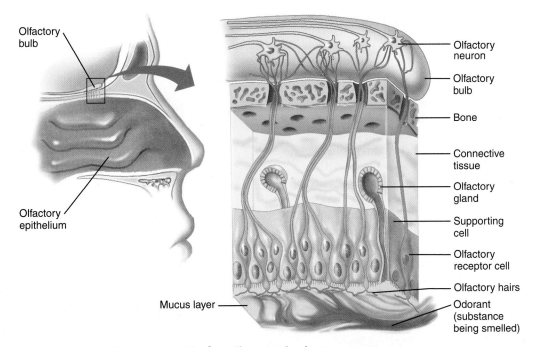

Figure 12.6 Olfactory receptors and the mucus-producing olfactory glands. Odorants become trapped in mucus and then bind to the olfactory hairs of receptor cells.

smell: (1) The receptors for smell are located on true sensory neurons, and (2) there are receptors for over 1,000 different "odorant" chemicals, as opposed to just 4 tastant classes. With these 1,000 different receptor types and our sense of taste we can distinguish as many as 10,000 taste and smell sensations.

Odors are detected by **olfactory receptor cells** located in the upper part of the nasal passages. The receptor cells are sensory neurons, each with a modified dendritic ending that branches to become several olfactory hairs (**Figure 12.6**). The olfactory hairs extend into a layer of mucus covering the surface of the nasal passages. The mucus, produced by nearby olfactory glands, keeps these hairs from drying out.

Gaseous and airborne odorants enter the nasal passages, become dissolved in the mucus, bind to chemoreceptors located on the olfactory hairs, and cause the olfactory receptor cell to generate impulses. Olfactory receptor cells synapse with olfactory neurons in a nearby area of the brain called the *olfactory bulb,* where information is partially integrated and then passed to higher brain centers.

Our sense of smell complements our sense of taste because when food is chewed it releases chemicals that come into contact with olfactory receptors. The combined sensations create an effect that is interpreted at higher brain levels (**Figure 12.7**). When we have a cold, most foods taste less appetizing. In part this is because increased production of mucus prevents odorant molecules from reaching olfactory receptors, impairing our sense of smell.

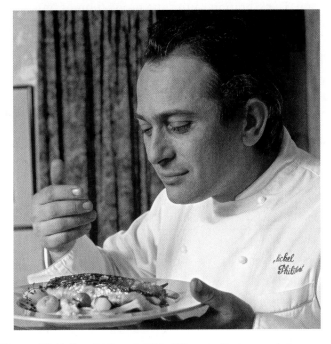

Figure 12.7 Smell. Over 1,000 different olfactory receptors contribute to our sense of smell and taste.

⤺ **Recap** Taste buds contain chemoreceptor cells that bind dissolved chemicals. The five currently accepted basic taste qualities are sweet, salty, sour, bitter, and umami. Our sense of smell augments and complements our sense of taste, allowing us to experience over 10,000 combinations of taste and smell. ▪

12.4 Hearing: Mechanoreceptors detect sound waves

Sounds are waves of compressed air. Sound intensity is a physical property of sound that refers to the energy of the sound, whereas loudness refers to our subjective interpretation of sound. Both intensity and perceived loudness are related to the amplitude of the sound waves, and both are measured in units called decibels (dB). The threshold of human hearing (barely audible) is set at 0 dB. A 10-dB increase represents a 10-fold increase in sound intensity (sound energy), but only a doubling of perceived loudness. In other words, a sound of 30 dB has 100 times more energy than a 10-dB sound, but humans would perceive it as only four times as loud. Hearing loss can result from exposure to sounds louder than 85 dB. **Table 12.2** lists the decibels associated with some common sounds.

The tone or pitch of a sound is determined by its frequency, the number of wave cycles that pass a given point per second (cycles/sec). Frequency is expressed in *hertz* (Hz). Higher tones are those of a higher frequency (**Figure 12.8**). The frequency range for normal human hearing is from 20 to 20,000 Hz.

When waves of sound strike the ear they cause vibration-sensitive mechanoreceptors deep within the ear to vibrate. The vibration causes the production of impulses that travel to the brain for interpretation.

The outer ear channels sound waves

The human ear consists of three functionally different regions: the outer, middle, and inner ear (**Figure 12.9**).

The **outer ear** consists of the pinna, or visible portion of the ear, and the auditory canal. Sound waves arrive at the pinna and are directed into the auditory canal, which

Table 12.2 Common noises and your hearing

Sound	Decibels	How it affects your hearing
Rustling leaves	20	Does not cause hearing loss
Normal conversation	50–60	Does not cause hearing loss
Vacuum cleaner	70	Does not cause hearing loss
Power lawn mower	85–100	Cumulative exposure causes hearing loss; newer mowers may be quieter
Live rock concert	110–130	Definite risk of permanent hearing loss
Jackhammer	120–150	Definite risk of permanent hearing loss
Shotgun	140	Repeated exposure likely to cause permanent hearing loss
Jet engine	150	Exposure likely to cause permanent hearing loss

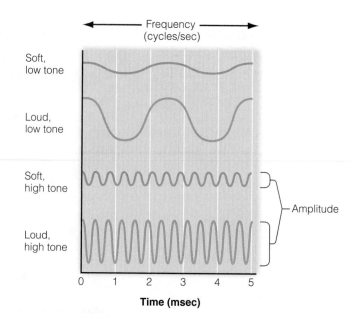

Figure 12.8 Properties of sound waves. Intensity (loudness) is determined by sound wave amplitude. Tone is determined by sound wave frequency.

☑ *Draw one more graph, to the same scale, illustrating a sound wave of extremely high pitch and intermediate loudness as compared to the sounds illustrated in this figure.*

channels them to the **tympanic membrane (eardrum)**. The tympanic membrane serves as a partition between the outer and middle ears.

The middle ear amplifies sound

The **middle ear** consists of an air-filled chamber within the temporal bone of the skull, bridged by three small bones called the malleus ("hammer"), incus ("anvil"), and stapes ("stirrup"). When sound waves strike the tympanic membrane, it moves back and forth slightly (vibrates). Vibration of the tympanic membrane makes the malleus, incus, and stapes vibrate. The stapes touches a smaller membrane, called the **oval window,** causing it to vibrate, too.

All the vibrating force of the much larger tympanic membrane is now concentrated on the smaller oval window. In effect, the middle ear amplifies the sound waves severalfold. Amplification is essential because the oval window must vibrate with sufficient force to produce pressure waves in the watery fluid of the inner ear.

The air-filled middle ear is kept at atmospheric pressure by the **auditory tube** (eustachian tube), a narrow tube that runs from the middle ear chamber to the throat. Although the tube is normally held closed by the muscles along its sides, it opens briefly when you swallow or yawn to equalize air pressure in the middle ear with that of the surrounding atmosphere. When you change altitude quickly, your ears "pop" as the auditory tube opens briefly and air pressures are suddenly equalized. When you have a cold or allergy, inflammation and swelling may prevent the auditory tube from

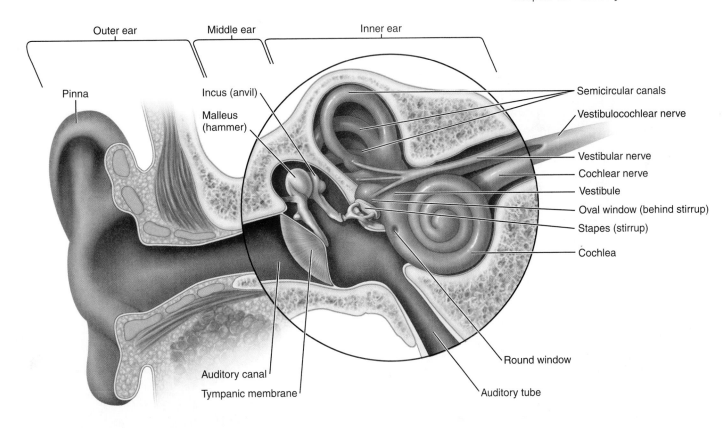

Figure 12.9 Structure of the human ear. The outer ear receives and channels sound, the middle ear amplifies sound, and the cochlea of the inner ear converts the sound to action potentials. The vestibule and the three semicircular canals are responsible for balance.

equalizing pressures normally, which is why airplane trips can be painful when you have a cold or allergies. Children tend to get a lot of earaches (middle ear infections) via the auditory tube until the skull elongates, changing the angle of the tube to the ear.

✅ **Quick Check** Suppose human ears had no outer ear and no middle ear, and just had the inner ear—that is, with the oval window in direct contact with the external air. In what way would our sense of hearing probably be *most* different? ▪

The inner ear sorts and converts sounds

The **inner ear** sorts sounds by tone and converts them into impulses. It consists of the **cochlea,** where sound is converted, and the *vestibular apparatus* consisting of the vestibule and the three semicircular canals, which does not contribute to hearing at all. We will discuss the vestibular apparatus later in connection with the sense of balance.

The cochlea is a coiled structure shaped a bit like a snail. **Figure 12.10a** (next page) shows what the cochlea would

look like if it were uncoiled. It is a tapered tube containing two interconnected outer canals called the *vestibular canal* and the *tympanic canal,* surrounding a third, closed fluid-filled space called the *cochlear duct.* The base of the cochlear duct is formed by the **basilar membrane.** The basilar membrane supports a population of about 15,000 **hair cells,** the mechanoreceptor cells of the ear. Hair cells have hairlike projections that are embedded in an overhanging structure called the **tectorial membrane** (the Latin word for "roof" is *tectum*), which is composed of a firm, gelatinous, noncellular material. Together, the hair cells and the tectorial membrane are called the *organ of Corti,* the organ that converts pressure waves to impulses.

When sound waves strike the oval window, it generates pressure waves in the watery fluid in the vestibular canal. These waves travel around the cochlea to the tympanic canal. Eventually the waves strike another membrane called the *round window,* which bulges inward and outward in synchrony with the oval window, dissipating some of the pressure.

Some of the pressure waves take a shortcut from the vestibular canal to the tympanic canal via the cochlear duct. These pressure waves cause the basilar membrane to vibrate like the strings of a musical instrument. The slight

a) **The cochlea as it might appear if it were uncoiled.** Higher-frequency sounds are converted to impulses near the oval window, whereas lower-frequency sounds are converted near the tapered tip.

b) **A cross section through the cochlea.** Pressure waves passing from the vestibular canal to the tympanic canal through the cochlear duct cause the basilar membrane to vibrate.

c) **A section through part of the cochlea, showing hair cells with their hairs embedded in the tectorial membrane.** Vibration of the basilar membrane bends the hairs, ultimately generating impulses in sensory neurons.

d) **SEM (× 4,200) of healthy hair cells.**

Figure 12.10 **Structure and function of the cochlea.**

 Suppose the hair cells at the very beginning of the cochlea began malfunctioning, such that they began sending constant action potentials (even if no vibrations were occurring anywhere in the cochlea). What sort of perception would probably result, and why?

movement of the basilar membrane causes the hair cells above the membrane to vibrate as well. Because the hairs of the hair cells are embedded in the less movable tectorial membrane, vibration of the basilar membrane causes the hairs to bend. This physical movement of the hairs causes the hair cells to release more or less neurotransmitter, depending on which way the hairs are bent. The release of neurotransmitter in turn generates impulses in nearby sensory neurons (Figure 12.11). The impulses travel in neurons of the auditory nerve, which joins the vestibular

nerve to become the vestibulocochlear nerve leading to the brain.

Vibrations of fluid in the vestibular and tympanic canals are of the same frequency as the original sound waves, which may be of many different tones at once. But because the fibers of the basilar membrane are different lengths at different regions of the cochlea, sound waves are converted to vibrations of the basilar membrane in a location-specific manner according to tone. High-pitched tones vibrate the basilar membrane closer to the base of the cochlea (near the

a) Hair cell at rest.

b) Hairs moving in one direction.

c) Hairs moving in other direction.

Figure 12.11 **How hair cells function as mechanoreceptors.** The hairs are actually modified cilia or microvilli.

12.5 The inner ear plays an essential role in balance

Maintaining your balance against gravity requires integration of multiple sensory inputs. Those sensory inputs include signals from the joint receptors, muscle spindles, and tendon receptors described previously, as well as from special structures associated with the inner ear. Even visual input is involved. The inner ear structures, described next, provide information about rotational movement, position, and linear acceleration of the head.

Looking again at the inner ear (Figure 12.12a), we see that next to

oval and round windows), whereas low-pitched tones vibrate the basilar membrane closer to the cochlear tip. The identical impulses generated in sensory neurons at these different locations are interpreted by the brain as sounds of specific tones on the basis of where they originated on the basilar membrane. This is yet another example of the location-specific nature of information processing by the brain.

the cochlea is the **vestibular apparatus,** a system of fluid-filled canals and chambers. The vestibular apparatus consists of three *semicircular canals* for sensing rotational movement of the head and body, and an area called the *vestibule,* which senses static (nonmoving) position and linear acceleration and deceleration.

The semicircular canals and the vestibule have mechanoreceptor-type hair cells embedded in a gel-like material. When the head moves or changes position, the gel also moves, although a bit more slowly because of inertia. The movement of the gel bends the hairs, which causes the hair cells to release a neurotransmitter that ultimately generates impulses in sensory neurons of the vestibular nerve.

🔄 **Recap** The outer ear receives and directs sound to the middle ear. The three bones of the middle ear amplify the sound and pass it to the cochlea. Vibrations of fluid in the cochlea cause the hairs of receptor hair cells to bend, generating action potentials in nearby neurons. ■

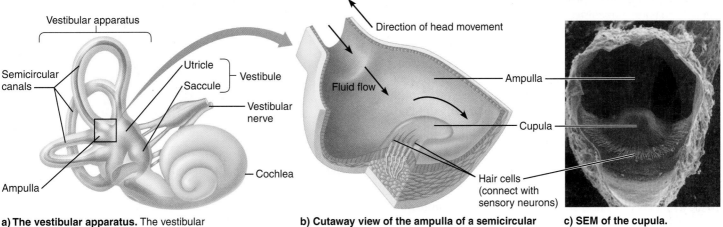

a) The vestibular apparatus. The vestibular apparatus (the organ of balance) is part of the inner ear, which also includes the cochlea.

b) Cutaway view of the ampulla of a semicircular canal. Rotational movement of the head causes fluid movement in the opposite direction in the semicircular canal. The fluid pushes against the cupula, bending it and activating the mechanoreceptor hairs of the hair cells.

c) SEM of the cupula.

Figure 12.12 **The vestibular apparatus.**

Sensing rotational movement

The three semicircular canals are fluid-filled tubes of bone. Near the end of each canal is a bulging region called the *ampulla*, containing a soft, gel-like dome called the *cupula* (Figure 12.12b and c). Hairs of mechanoreceptor sensory neurons are embedded in the cupula.

When you turn your head, the fluid in your semicircular canal moves more slowly than your head because of inertia. The shifting position of the fluid bends the soft cupula and the hairs within it, producing impulses. The structure of the hairs enables them to detect the direction of the movement, in addition to movement in general. Because there are three semicircular canals and each is oriented on a different plane, together they can detect any and every possible rotational movement of the head.

✅ **Quick Check** If you spin in place many times and then suddenly stop spinning, you will experience a sensation as if you are spinning in the *other* direction (even though you are now standing still). Propose an explanation for this phenomenon. ▪

Sensing head position and acceleration

To sense the position of the head relative to the Earth, we need a mechanism that is sensitive to gravity. The two fluid-filled chambers of the vestibule, called the **utricle** and the **saccule,** each contain mechanoreceptor hair cells, gel, and hard crystals of bonelike material called **otoliths** (literally, "ear stones") (**Figure 12.13**).

The otoliths are embedded in the gel near the gel's surface, unattached to the bone wall of the vestibule. The otoliths are indeed like little rocks, and they are considerably heavier than the gel itself. When the head tips forward, for instance, the otoliths slide toward the center of the gravitational pull, causing the gel to slip and bending the hairs of hair cells (Figure 12.13a, b). Again, this produces impulses in nearby sensory neurons that are transmitted to the brain for interpretation.

The utricle and saccule respond to linear acceleration and deceleration as well as static position. When you are in a car that is accelerating rapidly, you feel like you are being pushed backward in your seat. This is the force of acceleration acting against your own inertia. Likewise, your otoliths are being pushed backward, causing the hair cells to bend (Figure 12.13c). The otoliths respond only to acceleration and deceleration (a change in linear velocity), and not to a constant rate of speed in one direction.

The unpleasant sensation we call "motion sickness" represents a physiological response to abnormal body motion and visual experience. Motion sickness may be due to conflicting sensory inputs from the vestibular apparatus, the eyes, and various receptors in muscles, tendons, and joints throughout the body. The CNS is unable to integrate these conflicting signals into an understandable whole, producing the characteristic symptoms of nausea, cold sweats, hyperventilation, and headache.

Table 12.3 summarizes the structures and functions of the ear.

a) A cutaway view of the vestibule with the head at rest.

b) Head tilted. A change of head position causes a shift in position of crystals of bonelike material called *otoliths*, which ultimately stimulates hair cells.

c) Linear acceleration. Linear acceleration also causes otoliths to shift position.

d) SEM of otoliths.

Figure 12.13 **How the utricle and saccule sense static head position and linear acceleration.**

↩ **Recap** Rotational movement of the head is sensed as the result of movement of fluid in the three semicircular canals of the inner ear. Head position and linear acceleration are sensed as the result of movement of otoliths (ear stones) in the utricle and saccule of the inner ear. ▪

 Web Animation *The Human Ear* at www.humanbiology.com

Table 12.3 **Structures and functions of the ear**

Region of ear	Functions	Structures
Outer ear	Receives and channels sound	Pinna, auditory canal
Middle ear	Amplifies sound	Eardrum, maleus, incus, stapes, oval window, auditory tube
Inner ear	Sorts sounds by tone and converts them into impulses	Cochlea, vestibular canal, tympanic canal, cochlear duct, basilar membrane, hair cells, tectorial membrane, organ of Corti, round window
	Senses rotational movements and converts them into impulses	Semicircular canals, ampulla, cupula, hair cells
	Senses static (nonmoving) position and linear acceleration and deceleration; converts them into impulses	Vestibule, utricle, saccule, otoliths, hair cells

12.6 Vision: Detecting and interpreting visual stimuli

Light is a form of electromagnetic radiation that travels in waves at a speed of 186,000 miles per second. Our eyes allow us to receive and process light. They enable us to detect light from objects both nearby and distant and from sources either dim or bright, from the candle in your hand to a star light-years away. But first we must collect and focus light onto specialized cells in our eyes called *photoreceptors*.

Structure of the eye

The structure and functional components of the eye are illustrated in **Figure 12.14** and summarized in

Figure 12.14 Structure of the eye. Light enters through the transparent cornea and passes through aqueous humor, the pupil, the lens, and vitreous humor before striking the retina at the back of the eye. The macula is the central area of the retina with the highest visual acuity. The nerves and blood vessels exit the eye through the optic disk.

Table 12.4. A tough outer coat known as the sclera, or "white of the eye," covers the outer surface except in the very front, where it is continuous with the clear **cornea.**

Light passes through the cornea and a space filled with fluid called *aqueous humor* that nourishes and cushions the cornea and lens. Light then either strikes the **iris,** a colored, disk-shaped muscle that determines how much light enters the eye, or passes through the **pupil,** the adjustable opening in the center of the iris. Light that passes through the pupil strikes the **lens,** a transparent, flexible structure attached by connective tissue fibers to a ring of circularly arranged smooth muscle called the *ciliary muscle.*

After passing through the main chamber of the eye filled with *vitreous humor,* light encounters the layers at the back and sides of the eye. This is the **retina,** comprising primarily photoreceptor cells, neurons, and a few blood vessels. Between the retina and the sclera at the back of the eye lies the choroid, consisting of pigmented cells and blood vessels. The pigmented cells absorb light not sensed by photoreceptors so that the image does not become distorted by reflected light, and the blood vessels nourish the retina. At the back of the eyeball is the **optic nerve,** which carries information to the thalamus, to be forwarded to the visual cortex for interpretation. Finally, skeletal muscles surround the eye and control its movements, so we can choose exactly where to look.

Several sites on the retina deserve special mention. The **macula** is the central region of the retina, where photoreceptor density is the highest. When we want high visual acuity we look directly at an object, focusing the image on the macula. At the very center of the macula is a small pit called the *fovea centralis* that is lined with highly packed photoreceptors.

The **optic disk** is the area where the axons of the optic nerve and associated blood vessels exit the eye, so there are no photoreceptors there at all. The optic disk leaves us with a "blind spot" in each eye. To find the blind spots, try the exercise in Figure 12.15.

Regulating the amount of light and focusing the image

The iris adjusts the amount of light entering the eye with two sets of smooth muscle. When bright light strikes the eye, contraction of muscles arranged circularly around the pupil causes the pupil to contract. Otherwise, the intensity of daylight would overwhelm our photoreceptors and temporarily blind us. In dim light, contraction of smooth muscles arranged radially around the pupil causes the pupil to dilate.

Nerves control each set of muscles. When examining an unconscious or injured person, a physician may shine a light in each eye. The pupils should contract in response to light and dilate when the light is removed. Pupils that are "fixed and dilated" (unresponsive to light) are a bad sign, possibly indicating more widespread failure of the nervous system.

Light entering the eye is focused by the cornea and the lens. The cornea, which is curved, is actually responsible for bending most of the incoming light. However, the curvature of the cornea is not adjustable. Our ability to regulate the degree to which incoming light is bent, and therefore our ability to change focus between near and far objects, is accomplished solely by adjusting the curvature of the lens. This is done by the ciliary muscle.

Table 12.4 Parts of the eye and their functions

Eye structure	Functions
Sclera	Covers and protects eyeball
Cornea	Bends incoming light (focuses light)
Aqueous humor	Nourishes and cushions cornea and lens
Iris	Adjusts amount of incoming light
Lens	Regulates focus
Ciliary muscle	Adjusts curvature of the lens
Vitreous humor	Transmits light to retina
Retina	Absorbs light and converts it into impulses
Rods	Photoreceptors responsible for black-and-white vision in dim light
Cones	Photoreceptors responsible for color vision and high visual acuity
Macula	Central region of the retina with the highest density of photoreceptors
Optic disk	"Blind spot" where optic nerve exits eye
Optic nerve	Transmits impulses to the brain
Choroid	Nourishes retina and absorbs light not absorbed by retinal photoreceptors

Figure 12.15 Find your blind spot. To find the blind spot in your right eye, position the figure about four inches directly in front of you. Cover your left eye. While focusing your right eye on the square, slowly move your head away from the figure. Eventually the circle will disappear, because at a certain distance from your right eye, its image will strike the blind spot. You can do the same demonstration with your left eye by covering your right eye and focusing on the circle; the square will disappear.

When the ciliary muscle contracts, the inner radius of the muscle shrinks, reducing the tension on the fibers attached to the lens (Figure 12.16a). This allows the lens to bulge, and we focus on a near object. When the ciliary muscle relaxes, the ring of muscle increases the tension on the lens, stretching and flattening it and bringing more distant objects into focus (Figure 12.16b). **Accommodation** refers to the adjustment of lens curvature so we can focus on either near or far objects.

The light rays from each point of an object are bent and focused so that the image created on the retina is inverted (upside down). However, the brain interprets the image as right side up.

As we get older, the lens tends to stiffen and cannot resume a bulging shape even when the ciliary muscles are maximally contracted. The result is *presbyopia*, the inability to focus on nearby objects, which generally appears after age 40. People with presbyopia must hold reading material farther away to see the print clearly. Reading glasses can easily correct presbyopia.

✅ **Quick Check** Suppose a person's lenses were removed entirely. Would the person become entirely blind, able to see some light (but completely unfocused), able to see blurry images, or able to see sharp images? Explain. ▪

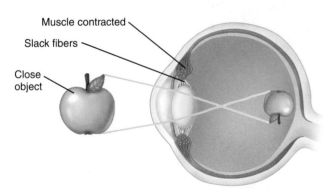

a) **For near objects, contraction of ciliary muscles reduces the tension on the fibers and allows the lens to bulge.**

b) **For focus on distant objects, relaxation of ciliary muscles increases tension on the fibers, which pull on the lens and flatten it.**

Figure 12.16 Accommodation. Notice that the images created on the retina are upside down.

Eyeball shape affects focus

Differences in the shape of the eyeball can affect the ability to focus properly (Figure 12.17).

Myopia is a common inherited condition in which the eyeball is slightly longer than normal (more rarely, myopia occurs when the ciliary muscle contracts too strongly). Even when the lens is flattened maximally, distant objects focus in front of the retina. People with myopia can see nearby objects, but distant objects are out of focus, which is why it is called *nearsightedness*. Concave lenses, which bend incoming light so it focuses on the retina, can correct nearsightedness.

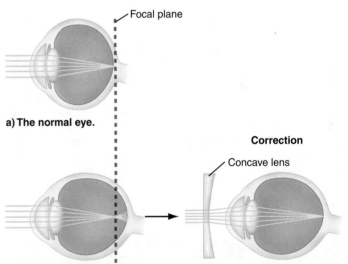

a) **The normal eye.**

Correction

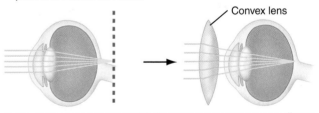

b) **Nearsightedness (myopia).** Nearsighted persons can see near objects clearly, but distant objects are out of focus because the focal point is in front of the retina.

c) **Farsightedness (hyperopia).** Farsighted persons can see distant objects clearly, but near objects are out of focus.

d) **Astigmatism.** Astigmatism is due to an abnormal curvature of either the cornea or the lens.

Figure 12.17 Examples of abnormal vision. All three types of abnormal vision are correctable.

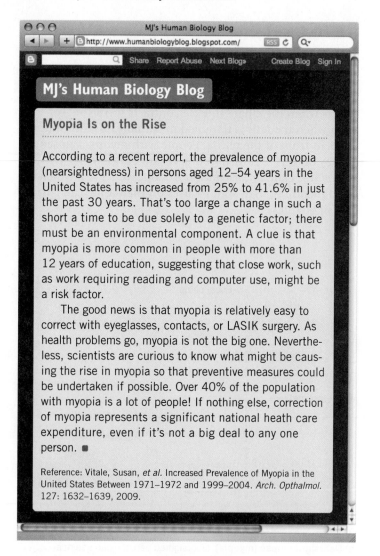

MJ's Human Biology Blog

Myopia Is on the Rise

According to a recent report, the prevalence of myopia (nearsightedness) in persons aged 12–54 years in the United States has increased from 25% to 41.6% in just the past 30 years. That's too large a change in such a short a time to be due solely to a genetic factor; there must be an environmental component. A clue is that myopia is more common in people with more than 12 years of education, suggesting that close work, such as work requiring reading and computer use, might be a risk factor.

The good news is that myopia is relatively easy to correct with eyeglasses, contacts, or LASIK surgery. As health problems go, myopia is not the big one. Nevertheless, scientists are curious to know what might be causing the rise in myopia so that preventive measures could be undertaken if possible. Over 40% of the population with myopia is a lot of people! If nothing else, correction of myopia represents a significant national heath care expenditure, even if it's not a big deal to any one person. ∎

Reference: Vitale, Susan, *et al.* Increased Prevalence of Myopia in the United States Between 1971–1972 and 1999–2004. *Arch. Opthalmol.* 127: 1632–1639, 2009.

Less common is the opposite condition, *hyperopia* (farsightedness), which occurs when the eyeball is too short and nearby objects focus behind the retina. Farsighted people can see distant objects clearly, but nearby objects are out of focus. Convex lenses correct farsightedness.

Astigmatism refers to blurred vision caused by irregularities of the shape of the cornea or lens. The result is that light is scattered and may not focus evenly on the retina. Astigmatism can be corrected with specially ground lenses that exactly compensate for the irregularities of the cornea and lens.

Light is converted into action potentials

Like other sensory organs, the eyes convert a stimulus (light) into impulses. In the eye this process occurs in the retina. The unique structure of the retina allows us to see in color, adapt to varying light intensities, and perceive images of the world around us.

Figure 12.18 shows a close-up of the retina, which consists of four layers:

1. The outermost layer consists of pigmented cells that, along with the choroid, absorb light not captured by the photoreceptor cells.
2. Next is a layer of photoreceptor cells called **rods** and **cones** because of their shapes.
3. The rods and cones synapse with the third layer, a layer of neurons called *bipolar cells*. Bipolar cells partially process and integrate information and then pass it on to the fourth layer.
4. The innermost layer consists of ganglion cells. Ganglion cells are also neurons. Their long axons become the optic nerve going to the brain.

Notice that light must pass through the third and fourth layers to reach the photoreceptor cells, but this does not significantly hinder the ability of the photoreceptor cells to receive and convert light.

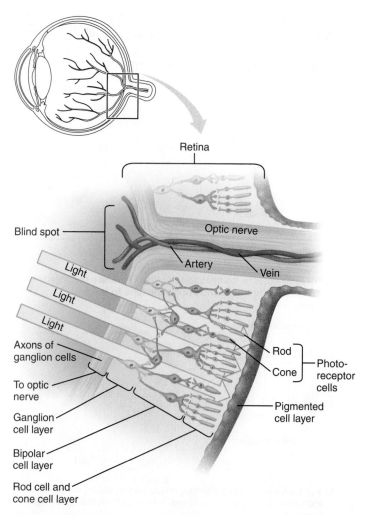

Figure 12.18 Structure of the retina. Light must pass between the axons of the ganglion cells and through the ganglion and bipolar cell layers to reach the photoreceptor cells, the rods and cones.

LASIK to Correct Vision Problems

Nearsightedness, farsightedness, and astigmatism are all visual problems that can be corrected with eyeglasses or contact lenses of one sort or another. But if you want a more permanent solution, you might consider LASIK surgery.

LASIK stands for laser in-situ keratomileusis, a generally safe and effective surgical technique in which lasers are used to permanently change the shape of the cornea. The technique can be done under local anesthesia in the ophthalmologist's office. The surgeon creates a thin flap of corneal tissue, folds it back, and then removes a small amount of corneal tissue from the cornea to reshape the cornea and correct the visual problem. The flap of tissue is returned to its original position and the eye is allowed to heal.

Most patients who undergo LASIK surgery will have vision close to 20/20 (normal) once the eye has healed completely. Normal vision is not assured, but most patients come pretty close to it. Some people experience halos of light around objects or are sensitive to glare after surgery, but these side effects generally are temporary. Every patient is different, of course, but complications from the surgery are rare. Answers to questions about *your* chances of attaining normal vision with LASIK should be addressed to your surgeon.

Not everyone is a good candidate for LASIK. People whose eyesight is still changing are generally advised to wait until their prescription for eyeglasses or contacts has been stable for at least two

years. In addition, LASIK does not correct presbyopia, the condition for which one uses reading glasses. Nevertheless, many people who wear glasses or contacts all the time for nearsightedness, farsightedness, and astigmatism experience the freedom of being able to get rid of their corrective lenses entirely. ∎

Rods and cones respond to light

Figure 12.19 takes a closer look at the photoreceptor cells, rods, and cones. One end of the cell consists of a series of flattened disks arranged to form either a rod- or cone-shaped structure, as the names imply. The flattened disks contain numerous molecules of a particular light-sensitive protein called a *photopigment.*

A photopigment protein undergoes a change of shape when it is exposed to energy in the form of light. The change in photopigment molecule shape causes the photoreceptor (rod or cone) to close some of its sodium channels and reduce the amount of neurotransmitter it normally releases. Because the neurotransmitter released by rods or cones normally inhibits the bipolar cells, light ultimately increases the activity of the bipolar cells, which in turn activate the ganglion cells (see Figure 12.18).

There are approximately 120 million rods and 6 million cones, but only 1 million ganglion cells with axons going to the brain. Clearly, a great deal of convergence and summation of signals occurs at each level of neuron transfer.

Rods provide vision in dim light

Rods all have the same photopigment, called **rhodopsin.** Rhodopsin is much more sensitive to light than the

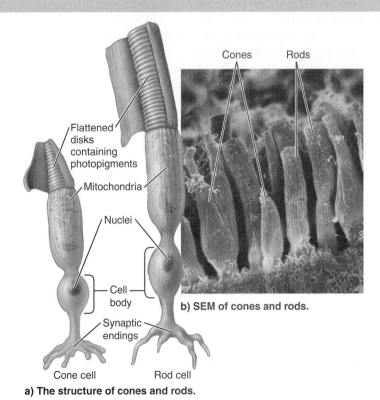

Flattened disks containing photopigments

Mitochondria

Nuclei

Cell body

Synaptic endings

Cone cell   Rod cell

a) The structure of cones and rods.

Cones   Rods

b) SEM of cones and rods.

Figure 12.19 The photoreceptor cells, cones and rods.
The photopigments in the flattened disks are sensitive to light.

photopigments in cones, and therefore in dim light our vision is dependent primarily on rods. However, rods do not give us color vision, which is why objects appear less colorful in dim light.

As we have seen, there are about 20 times more rods than cones. If we imagine that all 120 million rods converge on only half of the ganglion cells (one-half million), there would be 240 rods converging on a single ganglion cell. This convergence increases our ability to see in very dim light, but at the expense of acuity (accuracy and detail). As a result, our vision in very dim light is not very colorful and not very detailed, but at least we can see.

Rods and cones are not distributed evenly on the retina. Regions of the retina farthest away from the fovea have the highest ratio of rods to cones. If you want to see a dim star at night, don't look directly at it—look off to the side just a little.

Cones provide color vision and accurate images

Not all animals perceive color, but humans do. We are able to see colors because we have three different kinds of cones: red, green, and blue. Each contains a photopigment that absorbs the energy of red, green, or blue light particularly well.

Our ability to distinguish a variety of colors is due to the way the brain interprets the ratios of impulses coming from the ganglion cells connected to the three types of cones. When all three types are activated by all different wavelengths, we perceive white light. The perception of black is no light at all.

Cones also are responsible for visual acuity. However, cones require stronger light to be activated because the cone photopigments are much less sensitive to light than the rhodopsin in rods. This is why your ability to distinguish between colors declines in dim light and it becomes difficult to tell whether a dark-colored car is green or red. In dim light you are seeing primarily with your rods, which can detect only black and white.

HBP **Web Animation** *The Human Eye* at www.humanbiology.com

✓ **Quick Check** Do you think that animals active only at night, such as most owls, can see in color? Explain your answer. ■

Visual receptors adapt

From your own experience you know that vision adapts to changing light conditions over several seconds or minutes. Adaptation seems to take longer when going from bright to dim light than it does in the opposite direction. Adaptation depends on rapid adjustment of the pupil by the iris and on adaptation by the rods.

The absorption of light by rhodopsin "uses up" the photopigment temporarily. Light energy actually breaks the

rhodopsin into two molecules. These molecules can be resynthesized into rhodopsin again, but only over a period of minutes.

When you have been out in bright light, most of the rhodopsin in the light-sensitive rods has been broken down. Then, when you enter a dim room and the cones no longer are functioning, there is not enough intact rhodopsin for good vision in dim light. With time (5–15 minutes) the rhodopsin is resynthesized and available again.

Conversely, when you go out into sunlight from a dimly lit room, the light seems very bright because you have the maximal amount of photopigment available in both rods and cones. Most of the rhodopsin in the rods is quickly used up, however, and you revert to using primarily cones in bright light.

↩ **Recap** Light passes through the cornea, the lens, and the aqueous fluid of the eye before striking the photoreceptors in the retina. The iris regulates the amount of light entering the eye, and the cornea and the lens adjust the focus. Light energy alters the structure of photopigments—rhodopsin in rods, and red, blue, or green photopigments in cones. Rods provide vision in dim light. Cones are responsible for color vision and detailed images. ■

12.7 Disorders of sensory mechanisms

Disorders of the ears

Deafness: loss of hearing Deafness can have many causes. Deafness caused by damage to hair cells (**Figure 12.20**) is called **nerve deafness** because sounds cannot be converted into impulses in sensory nerves. Nerve deafness is generally the result of frequent exposure to loud sounds. Damage to the tympanic membrane or the bones of the middle ear, on the other hand, is called **conduction deafness.** In conduction

a) Normal healthy hair cells of the cochlea of a guinea pig.

b) Hair cells after 24 hours of exposure to loud sounds.

Figure 12.20 Damage to hair cells in the ears.

deafness sound waves simply are not transferred to the inner ear at all. Conduction deafness is frequently due to arthritis of the middle ear bones.

Some types of hearing loss may be partially corrected with a hearing aid, which amplifies incoming sounds, but even amplification won't be very effective if hair cells are missing. Some people with nerve deafness benefit from a cochlear implant, a tiny implanted microprocessor that converts sound waves into electrical signals.

Otitis media: Inflammation of the middle ear A common cause of earache is *otitis media,* or inflammation of the middle ear. Usually this results from an upper respiratory tract infection that extends up the auditory tube. The tube may become blocked and trap fluid in the middle ear. Otitis media usually responds to antibiotics.

Ménière's syndrome: Inner ear condition impairs hearing and balance Ménière's syndrome is a chronic condition of the inner ear. The cause is unknown, but it may be due to excess fluid in the cochlea and semicircular canals. Symptoms include repeated episodes of dizziness and nausea accompanied by progressive hearing loss. Balance is usually affected to the point that patients find it difficult to stay upright. Mild cases can be treated with motion sickness medications. Severe cases may benefit from surgery to drain excess fluid from the inner ear.

Disorders of the eyes

Retinal detachment: Retina separates from choroid If the thin retina is torn, the vitreous humor may leak through the tear and peel the retina away from the choroid, like peeling wallpaper from a wall (Figure 12.21a). The most common cause of retinal detachment in healthy persons is a blow to the head. The detached region of the retina loses most of its blood supply and its ability to focus an image properly. Symptoms include flashes of light, blurred vision, or loss of peripheral vision. Retinal detachment is an emergency, but prompt surgery can usually repair the damage.

Cataracts: The lens becomes opaque Cataracts are a decrease in the normal transparency of a lens (Figure 12.21b). Normally, the lens contains a particular kind of protein folded in such a way that it is transparent. If the delivery of nutrients to the lens is not sufficient, these proteins may denature, or clump, making the lens opaque. Some cataracts are congenital, but most are age-related or secondary to diabetes or other diseases.

About 5.5 million people in the United States (1 in 50) have a cataract impairing their vision. The treatment, which is usually successful, involves removing the lens surgically and replacing it with an artificial lens.

Glaucoma: Pressure inside the eye rises As mentioned previously, aqueous humor fills the chamber between the cornea and the lens. Capillaries located near the ciliary muscles of the eye produce aqueous humor, which brings nutrients and oxygen to the lens and cornea and carries away

a) Retinal detachment. In this ophthalmoscope view of a patient's eye the retina has peeled away from the choroid (upper third of the eye). At lower center is the yellow optic disk, through which the blood vessels travel.

b) Cataracts. The cataract in this person's right eye was caused by a chronic inflammatory disease process.

Figure 12.21 Common eye disorders.

wastes. Normally, aqueous humor is drained from the eye and returned to the venous blood by a vessel called the *canal of Schlemm* (review Figure 12.14). Glaucoma is a condition in which this drainage vessel becomes blocked. The excess fluid increases pressure inside the eye and compresses blood vessels supplying the retina. Cells of the retina or optic nerve may eventually die, impairing vision and ultimately leading to blindness. The impairment of vision often starts as a loss of peripheral vision (Figure 12.22a and b on the next page).

Glaucoma generally develops slowly and painlessly over several years, without any outward symptoms. About 2 million people in the United States (1 in 136) are visually impaired by glaucoma. Most patients do not realize they have it until they have already lost some vision. Less commonly, glaucoma develops acutely (within hours) with symptoms of blurred vision and red, swollen, sometimes painful eyes. Acute glaucoma is a medical emergency that requires immediate attention by an eye doctor.

a) Normal vision.

b) Glaucoma.

c) Age-related macular degeneration.

Figure 12.22 Visual impairments caused by glaucoma and age-related macular degeneration.

Both chronic (slow-onset) and acute glaucoma can be detected with a simple test that measures the pressure in the eye. If it is detected early enough, glaucoma can be controlled with drugs or surgery before any permanent damage is done. However, any vision that has already been lost cannot be restored. This is why a test of pressure in the eye is usually included in any comprehensive eye examination.

Age-related macular degeneration Age-related macular degeneration (AMD) is a disease of visual impairment caused by detachment of the retina and degeneration of photoreceptor cells in the macular region of the retina. The most common cause of AMD is accumulation of cellular debris between the choroid and the retina. Less commonly, it may be caused by abnormal growth of blood vessels in the region. In both cases the result is a loss of visual acuity in the center of the visual field, making it difficult to recognize faces or to read (Figure 12.22c). AMD affects about 1.75 million people in the United States (1 in 155). Because it is the leading cause of loss of vision in persons over 60, its prevalence is expected to rise dramatically as the U.S. population continues to age, unless a cure is found. There is no effective cure yet for most cases of AMD, although various vascular growth factors and vitamin treatments show promise at slowing or delaying the progression of the disease.

Color blindness: Inability to distinguish the full range of colors Color blindness is a general term for the inability to distinguish the full range of colors. Most forms of color blindness are caused by deficient numbers of particular types of cones. Less common are conditions in which one of the three cones is completely missing.

People with red-green color blindness, the most common form, are deficient in either red cones or green cones. They have trouble distinguishing between red and green, or they perceive them as the same color. The color they see is red or green, depending on which type of cone is missing.

The inability to perceive any color at all is quite rare in humans because it occurs only when two of the three cones are completely missing.

Color blindness is often inherited. Red-green color blindness is an X-linked recessive trait that is more common

in men than in women (see Chapter 19). It affects approximately 8% of Caucasian males but only 1% of Caucasian females, and is less common in other races. Traffic lights always have the red ("stop") light at the top and the green ("go") light at the bottom so that color-blind people can tell them apart by position. Color blindness is easily tested with a series of colored test plates (Figure 12.23).

Recap Earaches are most often caused by infections of the middle ear. A detached retina is a retina that has separated from the underlying choroid. Glaucoma is a condition in which fluid pressure inside the eye rises, constricting the blood vessels that normally supply the living cells. Age-related macular degeneration is the most common cause of loss of vision in older persons. ■

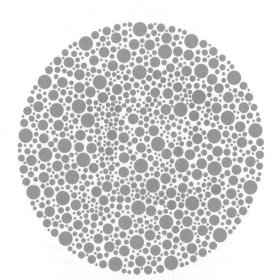

Figure 12.23 Test plate for diagnosing color blindness. A person with red-green color blindness would not be able to see the number 12 in this test plate. The original plate should be used for accurate testing of color blindness.

☑ *Suppose a woman had no blue cones at all in her retinas, but had normal numbers of the other types of rods and cones. Could she see the "12" in this picture? Explain.*

Chapter Summary

Receptors receive and convert stimuli p. 276

- Receptors convert the energy of a stimulus into impulses in sensory neurons.
- The brain interprets incoming impulses on the basis of where the sensory neurons originate.
- Some types of receptors adapt to continuous stimulation, whereas others (most notably pain receptors) do not.
- The somatic senses (temperature, touch, vibration, pressure, pain, and awareness of body position and movements) arise from receptors throughout the body.
- The five special senses (taste, smell, hearing, balance, and vision) arise from only certain specialized areas of the body.

Somatic sensations arise from receptors throughout the body p. 278

- The skin contains receptors for heat, cold, vibration, pressure, light touch, and pain. Receptors for temperature and for pain are also located in some internal organs.
- Receptors in joints sense joint position.
- Muscle spindles sense muscle length, and tendon receptors sense tendon (and muscle) tension.

Taste and smell depend on chemoreceptors p. 282

- Chemoreceptors for taste are located in structures called *taste buds*, most of which are on the tongue.
- The receptors for smell (called *olfactory receptors*) are located in the nasal passages. Humans have over 1,000 different olfactory receptor types.

Hearing: Mechanoreceptors detect sound waves p. 284

- The outer ear channels sound waves to the tympanic membrane, the middle ear amplifies the sound waves, and the cochlea of the inner ear converts sound waves to impulses.
- Sound waves of different tones are converted to impulses at different locations in the cochlea.

The inner ear plays an essential role in balance p. 287

- The vestibular apparatus of the inner ear consists of three semicircular canals and the vestibule.
- The semicircular canals are structures that sense rotational movement of the head on three planes.
- Structures of the vestibule sense static (nonmoving) head position and linear acceleration.

Vision: Detecting and interpreting visual stimuli p. 289

- The eye is the organ of vision, with structures for moving the eye (muscles), regulating the amount of light entering the eye (iris), focusing the visual image (lens), and converting light to impulses (retina).
- The photoreceptor cells in the retina are called *rods* and *cones*.
- Rods are 20 times more numerous than cones. They give us vision in dim light, but not in color and not with the highest detail.

- The three types of cones have different photopigments, so each type responds best to a different wavelength (color) of light. Cones are most effective in bright light, and they give us color vision and the sharpest image.
- Photoreceptors adapt to conditions of varying light intensities (brightness). It takes about 15 minutes to fully adapt to very dim light.

Disorders of sensory mechanisms p. 294

- Inflammation of the middle ear is a common cause of earache.
- Common eye disorders include retinal detachment, cataracts, glaucoma, age-related macular degeneration, and color blindness.

Terms You Should Know

accommodation, *291*
auditory tube, *284*
basilar membrane, *285*
chemoreceptors, *277*
cochlea, *285*
cones, *292*
cornea, *290*
hair cell, *285*
iris, *290*
lens, *290*
macula, *290*
mechanoreceptors, *276*
muscle spindle, *279*
olfactory receptor cell, *283*

optic disk, *290*
photoreceptor, *277*
receptor adaptation, *278*
retina, *290*
rhodopsin, *293*
rods, *292*
somatic sensations, *278*
special senses, *278*
taste bud, *282*
thermoreceptors, *276*
tympanic membrane (eardrum), *284*
vestibular apparatus, *287*

Concept Review

Answers can be found at the Human Biology Place.
www.humanbiology.com

1. List the five classifications of receptors in terms of types of stimulus they convert.
2. Describe how receptor adaptation works.
3. Describe what information the somatic sensations provide.
4. Name the five special senses.
5. Compare and contrast fast pain, slow pain, and referred pain.
6. Describe the overall function of the three bones (the malleus, the incus, and the stapes) of the middle ear.
7. The eye contains both skeletal muscle and smooth muscle. Describe where each is located and indicate their functions.
8. Compare and contrast rods and cones.
9. Explain why it may take up to 15 minutes to become fully adapted to very dim light conditions.
10. Explain what causes glaucoma.

Test Yourself

Answers can be found in Appendix A.

1. What do the receptors in the skin for light touch and the receptors in the nose for smell have in common?
   a. They both respond to pressure.
   b. They both respond to chemicals.
   c. They send action potentials to the same area of the brain.
   d. They both adapt very quickly.

2. Which of the following are NOT found among the somatosensory receptors?
   a. mechanoreceptors
   b. thermoreceptors
   c. chemoreceptors
   d. pain receptors

3. A more powerful stimulus will:
   a. result in action potentials that travel more quickly
   b. result in activation of more receptors
   c. result in more frequent action potentials
   d. both (b) and (c)

4. Receptor adaptation results when:
   a. sensory neurons stop sending action potentials
   b. a stimulus is discontinued
   c. action potentials from different neurons cancel one another out
   d. the action potentials do not reach the brain

5. In what order do the structures of the ear come into play when hearing?
   a. malleus – tympanic membrane – cochlea – stapes – cochlear nerve
   b. auditory tube – malleus – cochlea – incus – cochlear nerve
   c. auditory canal – tympanic membrane – malleus – cochlea – cochlear nerve
   d. auditory canal – cochlea – tympanic membrane – cochlear nerve – malleus

6. Different pitched sounds will result in stimulation of receptors located in different regions of:
   a. hair cells
   b. round and oval windows
   c. the vestibular apparatus
   d. the cochlea

7. Which of the following correctly traces the path of light through the eye?
   a. cornea – aqueous humor – pupil – lens – vitreous humor – retina
   b. aqueous humor – pupil – cornea – lens – vitreous humor – retina
   c. pupil – cornea – vitreous humor – lens – aqueous humor – retina
   d. cornea – vitreous humor – lens – pupil – aqueous humor – retina

8. When we adjust our head and eye position to carefully look at something we are directing the image to focus on:
   a. the optic disk
   b. the macula
   c. the periphery of the retina
   d. the optic nerve

9. Which of the following will regulate the amount of light entering the eye?
   a. the cornea
   b. the lens
   c. the iris
   d. the retina

10. Which of the following is/are involved in focusing light?
    a. lens
    b. pupil
    c. cornea
    d. both (a) and (c)

11. Which of the following correctly describes the distribution of rods and cones in the retina?
    a. Rods and cones are evenly distributed throughout the retina.
    b. There are more cones on the periphery of the retina.
    c. Cones are more highly concentrated in the fovea centralis.
    d. Rods are more highly concentrated in the optic disk.

12. Which of the following statements explains why vision is less sharp in dim light than in bright light?
    a. Cones are more responsive in dim light.
    b. There are more rods than cones in the eye.
    c. Many rods converge on a single ganglion.
    d. The dilated pupil lets in more light than the lens can focus.

13. Conduction deafness would most likely be caused by damage to:
    a. the cochlear nerve
    b. the malleus, incus, and/or stapes
    c. the hair cells in the cochlea
    d. the vestibular apparatus

14. Which of the following would most readily be treated with antibiotics?
    a. cataracts
    b. otitis media
    c. glaucoma
    d. Ménière's syndrome

15. Which of the following can be corrected with corrective lenses?
    a. myopia
    b. color blindness
    c. hyperopia
    d. both (a) and (c)

Apply What You Know

Answers can be found at the Human Biology Place.
www.humanbiology.com

1. With the eyes closed, would an astronaut in outer space be able to detect lateral movement of the head (would the semicircular canals be functioning normally)? Would the astronaut be able to detect static head position (would the utricle and saccule be functioning normally)? Explain.

2. Why do you suppose that you are not normally aware of the blind spot in each eye?

3. Our sense of smell is closely linked to our feelings and emotions. Bad odors usually evoke negative feelings, whereas pleasant odors give us a general sense of well-being or may be associated with good memories (perfume makers know this). The house of a pet owner can have an odor that a guest finds objectionable but the owner may not even be aware of. With a long enough visit even the guest may not notice the pet odors. Why is this?

4. Walk into any drugstore and you'll find dozens of over-the-counter products for reducing pain. Why is the experience of pain so common and so problematic?

5. An amateur boxer reported on an online forum that he saw lights flash before his eyes after being hit on the top of the head. This was unusual since he's normally hit in the chin or ear and knows what it's like to brownout during a knockout blow. Once the lights faded he was able to continue the fight. What happened when he was hit on the top of the head?

6. Try taking a really, really deep breath. If done correctly there should be a moment of sharp pain in your chest. What's causing this?

7. What do you think might cause motion sickness, the combination of nausea and dizziness that some people get after a roller-coaster ride or when riding in a boat or a car? And why don't gymnasts get motion sick when they're performing?

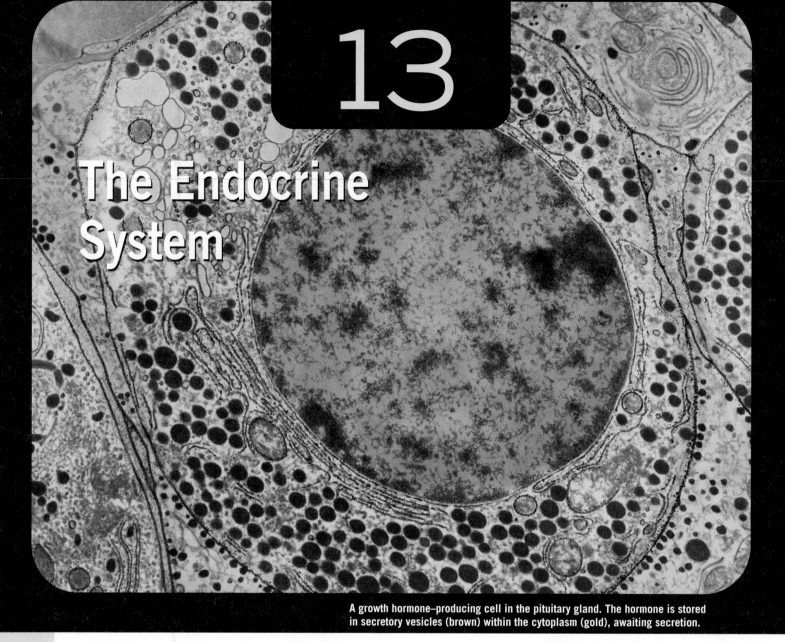

13

The Endocrine System

A growth hormone–producing cell in the pituitary gland. The hormone is stored in secretory vesicles (brown) within the cytoplasm (gold), awaiting secretion.

Dealing with Type 2 Diabetes

According to the World Health Organization (WHO), 30 million people worldwide had diabetes in 1985. By 2000 the number was up to 171 million, and by 2030 it is estimated to be a whopping 366 million. In the United States one-third of all children born today are projected to develop diabetes during their lifetime, according to the U.S. Centers for Disease Control and Prevention (CDC). Nearly all of the increase is expected to be in *Type 2 diabetes*, characterized by *resistance to insulin* and consequently an inability to control blood glucose levels. (In contrast, in *Type 1 diabetes* the pancreas does not *produce* enough insulin—see Section 13.10, this chapter.)

Why Is Type 2 Diabetes on the Rise?

Basically, the human body is not well suited for a sedentary lifestyle combined with a steady diet of high-calorie, high-sugar foods. Humans evolved from prehuman ancestors over several hundred thousand years. Over most of that time, humans existed only as small bands that traveled widely in search of food and shelter. Abundant food was a rare event, and finding food and obtaining shelter was often hard physical work. Under such

The easy availability of high-sugar, high-fat foods places children and adults alike at increased risk for diabetes.

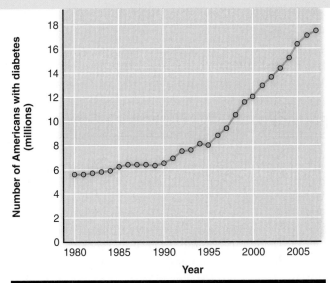

The number of Americans with diabetes has tripled since 1980.

conditions it was entirely appropriate, when high-calorie foods rich in carbohydrates were available, for a burst of insulin from the pancreas to help the body's cells take up the glucose quickly so that it could be used or converted to storage products for future use.

In just the last few centuries, food has become readily available to most people in the world. With high-calorie foods always in abundance and the body's cells stuffed to capacity with glucose and fat, our cells slowly develop a *resistance to insulin* (the hallmark of Type 2 diabetes) to try to slow down the influx of glucose into cells. This raises blood glucose levels even further. For a time, the pancreas secretes even more insulin in an effort to overcome the developing insulin resistance, but eventually the pancreas may begin to fail and insulin may decline. So Type 2 diabetics may or may not have enough insulin, but what they do have is resistance to insulin and blood sugar concentrations out of control.

Prevention, Early Detection, and Management Would Reduce Costs

The major risk factors for Type 2 diabetes are obesity, unhealthy diets, and insufficient exercise. Modest changes in

lifestyle combined with early detection and careful management of diabetes could reduce the social, economic, and health impacts of the disease dramatically.

The true cost of untreated diabetes includes the many debilitating problems it may cause, including cardiovascular disorders, kidney failure, blindness, and neural diseases. For example, in the United States approximately 40,000 diabetics develop kidney disease each year. Many of them progress to kidney failure, at which point they require treatment at a cost of over $50,000 per year (assuming the treatment, called *dialysis,* is available to them). Their lives and the lives of their families begin to revolve around managing their disease as they spend several half-days a week in a hospital or dialysis unit. Eventually they may need a kidney transplant, which can cost $100,000 or more. The total cost of treating the long-term complications of diabetes in the United States is estimated to be $174 billion/year—the equivalent of $1,300 for every U.S. taxpayer.

Because early detection and management are so effective in limiting the ultimate cost of the disease, a fasting blood glucose concentration and a glucose tolerance test (GTT) are now standard components of most medical workups. However, early detection may not be possible in areas of the world where basic health care is not readily available. Over a third of all diabetics do not even know they have the disease.

Even when the diagnosis is made early, motivating patients to rigorously monitor and control their blood sugar can be a problem. It's hard to convince a newly diagnosed diabetic of the need to make major lifestyle changes when they seem so

healthy (at least at first). Nobody enjoys pricking a finger several times a day and drawing blood to measure glucose levels. And who among us is willing to give up sweets, lose 5–10 pounds (and keep it off), and start exercising nearly every day?

Shortcomings in Our Health Care System

A concerted effort by health professionals and governments aimed at promoting healthy lifestyles, combined with early detection and effective management of those who do develop high blood sugar levels, may be the best ways to combat this disease in the long run. Unfortunately, the U.S. health care reimbursement system does not take this into account. A short example will show you why.

Hospitals, doctors, and health care centers make their money through payments from government sources (such as Medicare and Medicaid) and private insurers. At current reimbursement rates, they generally lose money on preventive care and disease management, but make money every time a diabetic is hospitalized for a major complication such as a kidney transplant or an amputation. One recent study estimated that New York health care centers lost $40–$55 every time a diabetic met with a diabetes educator to learn how to monitor blood sugar properly or administer insulin or other treatment, and another $40–$55 when the patient met with a nutritionist to discuss controlling blood sugar through weight loss and a proper diet. In contrast, hospitals earned $1,500 to over $10,000 for every amputation, depending on the type of amputation and its complications. Is it any wonder that hospitals are more likely to hire a surgeon and promote their center as a surgical center of excellence, rather than add nutritionists or health educators to their staff?

Something needs to be done. The question is: what?

Questions to consider

1 Would you be in favor of government-supported screening of all people for high blood sugar? Should it be mandatory?

2 What should be the government's role in paying for education, counseling, and follow-up of patients with diabetes? What about private insurance companies?

The facts...

- Diabetes is on the rise around the globe, from 30 million people in 1985 to over 171 million today.

- Many cases of diabetes could be prevented by lifestyle changes that include a healthy diet, moderate weight loss, and exercise.

- Early detection and disease management would reduce the economic, social, and medical impact of diabetes dramatically.

- Our health care system focuses on the long-term complications of diabetes because this is more profitable. Hospitals actually lose money on preventive care.

>> The endocrine system produces chemical messenger molecules called *hormones* and secretes them into the blood.

>> Hormones only act on certain cells, called the hormone's *target cells*, because only a hormone's target cells have specific receptors for the hormone.

>> Steroid and nonsteroid hormones have different mechanisms of action. Steroid hormones enter the target cell before binding to a receptor, whereas nonsteroid hormones bind to receptors at the outer surface of the cell membrane.

>> Some hormones participate in negative feedback loops that maintain homeostasis. Other hormones regulate major life-changing processes such as growth and sexual maturation.

>> The pituitary gland is sometimes called *the master gland*. It secretes eight different hormones that in turn regulate the production of many other hormones.

At about ages 11–13, something strange and wonderful happens to the human body and mind. The body grows rapidly, and secondary sex characteristics develop. Many boys get interested in girls, and many girls get interested in boys with an intensity that sometimes seems to border on obsession, when only months before they couldn't care less.

Welcome to the endocrine system, an internal system of powerful chemicals that controls a host of body functions. In addition to sexual maturation and sexual desire, the endocrine system regulates salt and water balance, blood pressure, stress responses, digestion, cellular metabolism, the production of red blood cells, and overall organ growth and development.

13.1 The endocrine system produces hormones

The **endocrine system** is a collection of specialized cells, tissues, and glands that produce and secrete circulating chemical messenger molecules called **hormones**. Hormones are secreted by **endocrine glands**—ductless organs that secrete their products into interstitial fluid, lymph, and blood (*endocrine* means "secreted internally"). In contrast, **exocrine glands** secrete products such as mucus, sweat, tears, and digestive fluids into ducts that empty into the appropriate sites.

Approximately 50 known hormones circulate in the human bloodstream, and new ones are still being discovered.

Hormones are bloodborne units of information, just as nerve impulses are units of information carried in nerves. Some hormones participate in feedback control loops regulating various body functions. We need these hormones to maintain homeostasis. Other hormones produce specific effects, such as contraction of the uterus at birth, growth during childhood, and the development of sexual characteristics at puberty. We need specific hormones to carry out each of these very specific functions.

Figure 13.1 shows the various tissues and organs that make up the endocrine system and the most important hormones they secrete. The many functions of these hormones are discussed in this chapter and in other chapters where appropriate.

The endocrine system has certain characteristics that set it apart from the nervous system as a communications system.

1. *Hormones of the endocrine system reach nearly every living cell.* This attribute gives the endocrine system a distinct advantage over the nervous system. Delivery of hormones to nearly all cells occurs because hormones circulate in the blood, and almost every cell in the body is near a blood vessel. (An exception is the central nervous system, which is effectively isolated from exposure to most systemic hormones by the blood-brain barrier.) With the endocrine system there is no need for the direct connections (the hard wiring, so to speak) required by the nervous system.

2. *Each hormone acts only on certain cells.* As all hormones circulate together in the same well-mixed blood, how can one hormone specifically regulate red blood cell production whereas another regulates blood calcium concentration? The answer is that each hormone acts only on a certain group of cells, called its **target cells,** because only the hormone's target cells have the appropriate receptor to fit it. When a hormone binds to a receptor on its target cell, a change occurs within the cell. The cell may grow, divide, or change its metabolism in some way. All other cells of the body fail to respond to the hormone because they lack the appropriate receptor. As an analogy, consider that your car is a cell, the car's ignition switch is the receptor, and the car key is the hormone. Your car key fits only your car (the "target" car) and no other. Fitting your key into the ignition and turning it causes the car to start. Gas is consumed, engine parts rotate, and heat is generated.

3. *Endocrine control tends to be slower than nervous system control.* There is a price for using the cardiovascular system as the message delivery system, and that is a reduction in speed—a lengthening of the response time. Nerve impulses can travel from head to toe in a fraction of a second. It takes longer to secrete a hormone into the bloodstream, have the hormone circulate to its

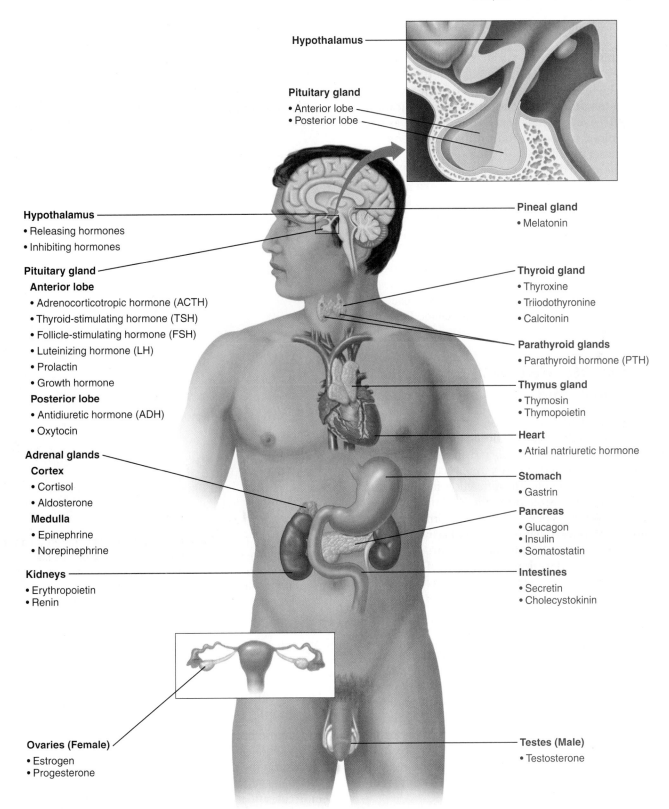

Hypothalamus

Pituitary gland
- Anterior lobe
- Posterior lobe

Hypothalamus
- Releasing hormones
- Inhibiting hormones

Pituitary gland
  Anterior lobe
  - Adrenocorticotropic hormone (ACTH)
  - Thyroid-stimulating hormone (TSH)
  - Follicle-stimulating hormone (FSH)
  - Luteinizing hormone (LH)
  - Prolactin
  - Growth hormone
  Posterior lobe
  - Antidiuretic hormone (ADH)
  - Oxytocin

Adrenal glands
  Cortex
  - Cortisol
  - Aldosterone
  Medulla
  - Epinephrine
  - Norepinephrine

Kidneys
- Erythropoietin
- Renin

Ovaries (Female)
- Estrogen
- Progesterone

Pineal gland
- Melatonin

Thyroid gland
- Thyroxine
- Triiodothyronine
- Calcitonin

Parathyroid glands
- Parathyroid hormone (PTH)

Thymus gland
- Thymosin
- Thymopoietin

Heart
- Atrial natriuretic hormone

Stomach
- Gastrin

Pancreas
- Glucagon
- Insulin
- Somatostatin

Intestines
- Secretin
- Cholecystokinin

Testes (Male)
- Testosterone

Figure 13.1 Components of the human endocrine system. Note that some organs with nonendocrine primary functions (the heart, stomach, intestines, and kidneys) produce and secrete hormones as secondary functions.

target, and exert its effect. At the very fastest, this takes 20 seconds or so.

Given this difference in speed, it is not surprising that reflexes that prompt us to avoid a hot flame are controlled by the nervous system and not the endocrine system. On the other hand, endocrine communication is highly effective for longer-term controls, such as regulation of blood pressure, production of red blood cells, and onset of puberty.

4. *The endocrine and nervous systems can (and often do) interact with each other.* The timing of growth and sexual maturation, for example, involves a complex sequence of changes in both neural and endocrine signals, and the release of some hormones depends on input from sensory neurons.

↩ **Recap** Hormones secreted by glands of the endocrine system act only on target cells with appropriate receptors. Hormones reach their targets via the circulatory system, making endocrine system control slower than nervous system control. The two systems frequently interact. ■

13.2 Hormones are classified as steroid or nonsteroid

Hormones are classified into two basic categories based on their structure and mechanism of action. **Steroid hormones** are structurally related to cholesterol, and all are lipid soluble. **Nonsteroid hormones,** because they are structurally related to proteins, are lipid insoluble. The differences in lipid solubility of steroid and nonsteroid hormones explain the differences in their mechanisms of action. Steroid hormones enter the cell, bind to an intracellular receptor, and activate genes that produce new proteins. Nonsteroid hormones bind to receptors on the cell's surface, initiating a series of events that ultimately alters cellular activity in some way, even though the hormones never entered the cell at all.

Steroid hormones enter target cells

Recall that the cell membrane is primarily composed of a bilayer of phospholipids. Because steroid hormones are related to cholesterol and therefore are lipid soluble, they can easily diffuse right across both the cell membrane and the nuclear membrane. Once inside the cell, steroid hormones bind to specific hormone receptors, forming a hormone-receptor complex either within the nucleus (**Figure 13.2**) or within the cytoplasm (not shown). If the hormone-receptor complex was formed in the cytoplasm, it too can diffuse into the nucleus.

Once inside the nucleus, the hormone-receptor complex attaches to DNA, activating specific genes. Gene activation causes the formation of messenger RNA, which then leaves the nucleus and directs the synthesis of certain proteins. The proteins then carry out the cellular response to the hormone, whatever it might be.

Steroid hormones tend to be slower acting than non-steroid hormones. Starting from the time the steroid hormone

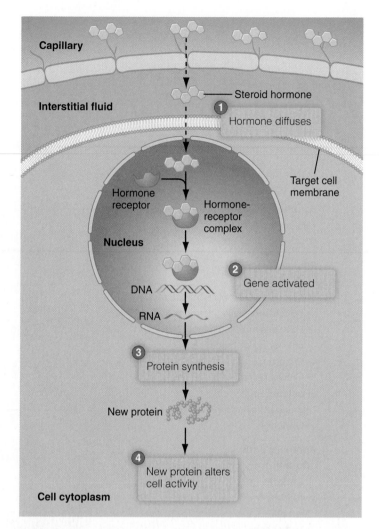

Figure 13.2 Mechanism of steroid hormone action on a target cell. Lipid-soluble steroid hormones diffuse across the cell and nuclear membranes into the nucleus, where they bind to hormone receptors that activate genes. Gene activation results in the production of a specific protein.

first enters the cell, it can take minutes or even hours to produce a new protein.

Nonsteroid hormones bind to receptors on target cell membranes

Nonsteroid hormones cannot enter the target cell because they are not lipid soluble. Instead they bind to receptors located on the outer surface of the cell membrane. The receptors are generally associated with, or are part of, protein molecules floating in the phospholipid bilayer of the cell membrane (Chapter 3). The binding of hormone to receptor causes a change in the shape of the membrane protein, which in turn initiates a change within the cell. It's like turning the lights on in a room by flipping a switch located in the hall outside, without ever having to enter the room.

Some nonsteroid hormones cause ion channels in the cell membrane to open or close, similar to the action of a neurotransmitter. More commonly, hormone-receptor binding converts an inactive molecule within the cell into an active molecule. The activated molecule is called a **second messenger,** because it carries the message provided by the hormone (the "first messenger") without the hormone ever entering the cell. A common second messenger is cyclic AMP, produced from ATP when the hormone binds to the receptor. Cyclic AMP then activates an enzyme, which activates another, which in turn activates another, and so on (**Figure 13.3**). At each successive activation the effect is amplified, so that even a small amount of hormone can produce a significant cellular change.

In fact, amplification is the norm for both steroid and nonsteroid hormones. If a hormone were a lighted match, the effect of the hormone would be the equivalent of a bonfire.

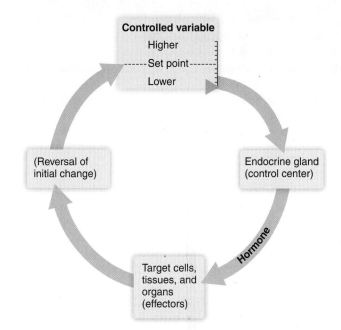

Figure 13.3 Mechanism of nonsteroid hormone action on a target cell. Nonsteroid hormones bind to receptors in the cell membrane, leading to the conversion of ATP to cyclic AMP (the second messenger) within the cell. Cyclic AMP initiates a cascade of enzyme activations, amplifying the original hormonal signal and generating a cellular response.

✓ *If 1 hormone molecule results in production of 10 cyclic AMPs, each cyclic AMP then activates 10 molecules of Enzyme 1, each enzyme activates 10 copies of the next enzyme, and the final enzyme makes 100 copies of the "final product." How many molecules of the final product will be produced by arrival of a single hormone molecule at the cell surface? How does this explain why hormones can circulate in very low concentrations in the blood and still have an effect?*

Hormones start something small (a change in protein synthesis or activation of an enzyme), but the final effect can be as dramatic as the development of secondary sexual characteristics.

Most nonsteroid hormones tend to be faster-acting than steroid hormones because they activate enzymes that already exist within the cell in inactive form. An entire cascade of existing enzymes can be activated within seconds or minutes.

✓ **Quick Check** Would you expect a steroid hormone to result in production of a second messenger? Why or why not? ■

HBP **Web Animation** *How Hormones Influence Target Cells* at www.humanbiology.com

Hormones participate in negative feedback loops

As messenger molecules, some hormones participate in internal homeostatic control mechanisms and control vital physiological processes. It is essential to regulate carefully the rate at which each hormone is secreted, so that its concentration in blood is just right to carry out its function.

As described in Chapter 4, homeostasis is generally maintained by a negative feedback control loop. In a negative feedback loop involving a hormone, the endocrine gland is the control center, the hormone represents the pathway between the control center and the effectors, and the effectors are the hormone's target cells, tissues, or organs (**Figure 13.4**). As the

Figure 13.4 A negative feedback loop involving a hormone. In response to a change in the controlled variable, the endocrine gland releases a hormone that acts on target cells to return the controlled variable to its normal state.

Answers to ✓ **questions can be found in Appendix A.**

figure shows, a negative feedback loop involving an endocrine gland and a hormone is a stable, self-adjusting mechanism for maintaining homeostasis of the controlled variable, because any change in the controlled variable sets in motion a response that reverses that change. (Figure 7.6 shows an example of a hormonal negative feedback loop.)

Not all negative feedback loops involving hormones are quite as simple as the one depicted in Figure 13.4. In some cases the real control center is the brain, which activates an endocrine gland via nerves. Nevertheless the effect is the same—reversal of the initial change in the controlled variable.

A few hormones are secreted in response to specific environmental cues or for specific purposes, such as to avoid danger or to initiate puberty. This type of secretion is not part of a negative feedback loop.

We turn now to some key endocrine glands and their hormones. Some of the hormones mentioned here are covered in more depth in other chapters, when the function of specific organ systems is discussed.

 Web Animation *Hormonal Feedback Loops* at www.humanbiology.com

MJ's Human Biology Blog

Athlete Caught Doping with GH

British rugby player Terry Newton has the distinction of being the first athlete ever caught doping with growth hormone to boost muscle mass. He's been banned from rugby competition for two years.

Until recently, athletes had assumed that human growth hormone doping was undetectable, because the synthetic form of growth hormone looks exactly like natural growth hormone. But then scientists discovered that the body actually produces several different forms of the hormone, and that by examining the ratio of these different forms in blood they could tell if the pure synthetic form had been administered. Anti-doping agencies also began requiring random blood tests rather than just tests after athletic events, because the hormone disappears from the blood just days after administration. It was a random test that caught Mr. Newton.

The U.S. National Football League and Major League Baseball have shown interest in the test but so far it has not been instituted, in part because it requires a blood sample. ■

Reference: Travis, John. Growth Hormone Test Finally Nabs First Doper. *Science* 327: 1185, 2010.

Recap Many hormones participate in negative feedback loops that maintain homeostasis. Steroid hormones enter the target cell, activate specific genes, and cause the production of new proteins. Nonsteroid hormones bind to a cell membrane receptor that either opens or closes ion channels or activates a second messenger within the cell. ■

13.3 The hypothalamus and the pituitary gland

As described in Chapter 11, the **hypothalamus** is a small region in the brain that serves as a homeostatic control center. It is an important link between the nervous system and the endocrine system. It receives neural input about certain internal environmental conditions such as water and solute balance, temperature, and carbohydrate metabolism. The hypothalamus produces two hormones of its own, and it monitors and controls the hormone secretions of the pituitary gland, located directly beneath it and connected to it by a thin stalk of tissue (review Figure 13.1).

The **pituitary gland** is sometimes called the "master gland" because it secretes eight different hormones that in turn regulate many of the other endocrine glands (Table 13.1). It consists of two lobes: a posterior lobe that appears to be a bulging extension of the connecting stalk, and a larger, more distinct anterior lobe (Figure 13.5).

The posterior pituitary stores ADH and oxytocin

The interaction between the hypothalamus and posterior lobe of the pituitary gland is an example of the close working relationship between the nervous and endocrine systems. Cells in the hypothalamus called **neuroendocrine cells** essentially function as both nerve cells and endocrine cells, because they can generate nerve impulses and secrete hormones into blood vessels.

Figure 13.5 illustrates the posterior pituitary lobe and its connections to the hypothalamus. Neuroendocrine cells with cell bodies in the hypothalamus have axons that extend down the stalk connecting to the posterior pituitary. The neuron cell bodies make either antidiuretic hormone (ADH) or oxytocin in the hypothalamus and then transport the hormones down the axon for storage in axon endings in the posterior pituitary.

When the hypothalamus is stimulated to release the hormones, the neuroendocrine cells send impulses down the axon, causing hormone to be secreted into nearby capillaries. The hormones circulate to their target cells.

ADH regulates water balance Antidiuretic hormone reduces the amount of water lost in urine, helping to regulate the body's overall water balance. The primary target cells for ADH are in the kidneys, where it causes changes in cell permeability

Table 13.1 **Hormones of the pituitary gland**

Hormone	Abbreviation	Main targets	Primary actions
Posterior Pituitary			
Antidiuretic hormone	ADH	Kidneys	Reduces amount of water lost in urine
Oxytocin	—	Uterus	Induces uterine contractions
		Mammary glands	Induces ejection of milk from mammary glands
Anterior Pituitary			
Adrenocorticotropic hormone	ACTH	Adrenal cortex	Stimulates adrenal cortex to release glucocorticoids
Thyroid-stimulating hormone	TSH	Thyroid gland	Stimulates synthesis and secretion of thyroid hormones
Follicle-stimulating hormone	FSH	Ovaries	In females, stimulates egg maturation and the secretion of estrogen
		Testes	In males, stimulates the formation of sperm
Luteinizing hormone	LH	Ovaries	In females, stimulates ovulation (egg release) and the secretion of progesterone
		Testes	In males, stimulates testosterone secretion
Prolactin	PRL	Mammary glands	Stimulates the development of mammary gland cells and production of milk
Growth hormone	GH	Most cells	Stimulates growth in young individuals; plays multiple roles in cell division, protein synthesis, and metabolism in adults

Figure 13.5 Posterior pituitary lobe and hypothalamus.
Individual neuroendocrine cells of the hypothalamus and the posterior pituitary secrete either ADH or oxytocin. The hormones enter the blood supply of the posterior pituitary and make their way to the general circulation. ADH acts on the kidneys to regulate water balance in the body. Oxytocin primarily acts on the mammary glands and uterus muscles of lactating and pregnant women, respectively.

to water as part of the mechanism for conserving water (Chapter 15).

When the hypothalamus detects low concentrations of water (equivalent to a high concentration of solute particles) in the blood, neuroendocrine cells release ADH from the posterior pituitary. ADH circulates in blood to the kidneys, where it stimulates the reabsorption of water. When water concentrations rise to normal levels, the hypothalamus is no longer stimulated to release ADH, and ADH secretion declines.

☑ **Quick Check** What would happen to your ADH levels if you drank a large amount of water, and what effect would this have on the kidneys? ■

Oxytocin causes uterine contractions and milk ejection The primary target cells for oxytocin are in the uterus and the mammary glands (breasts) of pregnant and lactating females, respectively. Oxytocin is secreted at high concentrations during childbirth and also when an infant nurses at its mother's breast

(Figure 13.6). It stimulates contractions of the uterus during childbirth and ejection of milk into the ducts of the mammary glands during suckling.

The stimulation of oxytocin secretion by nursing is an example of a *neuroendocrine reflex,* in which a nervous system stimulus (activation of sensory mechanoreceptors in the mother's breast) is responsible for secretion of a hormone. When a baby suckles, sensory receptors in the nipple are stimulated. Impulses are transmitted to neuroendocrine cells in the hypothalamus, which release oxytocin from the posterior pituitary. The hormone circulates to breast tissue, binds with receptors, and causes contractions of smooth muscle that eject milk.

Oxytocin is also present in males. In both sexes it is thought to play a role in sexual arousal and the feeling of sexual satisfaction after orgasm.

The anterior pituitary produces six key hormones

The anterior lobe of the pituitary gland produces the following six key anterior pituitary hormones:

- Adrenocorticotropic hormone (ACTH), also called corticotropin
- Thyroid-stimulating hormone (TSH), also called thyrotropin
- Follicle-stimulating hormone (FSH)
- Luteinizing hormone (LH)
- Prolactin (PRL)
- Growth hormone (GH)

Each anterior pituitary hormone is produced and secreted by a separate cell type, and each hormone is regulated by separate mechanisms. The first four hormones (ACTH, TSH, FSH, and LH) act by stimulating the release of yet other hormones from other endocrine glands.

The secretion of anterior pituitary hormones depends partly on the hypothalamus, but the mechanism differs from that of the posterior pituitary. The connection between the hypothalamus and anterior pituitary is endocrine, not neural; it relies on specific releasing and inhibiting hormones secreted by neuroendocrine cells in the hypothalamus (Figure 13.7). These releasing and inhibiting hormones, in turn, alter the secretion of the six anterior pituitary hormones. Releasing and inhibiting hormones are produced in miniscule quantities. However, they are secreted directly into a special blood supply that runs directly between the hypothalamus and anterior pituitary (called the pituitary portal system), so their concentrations are sufficient to regulate anterior pituitary hormone secretions.

ACTH stimulates the adrenal cortex The names of hormones generally offer a clue to their function or site of action. The target cells for ACTH are located in the outer layer, or cortex, of the adrenal gland (in *adrenocorticotropic* the suffix *-tropic* means "acting upon").

3 **Neuroendocrine cells** release oxytocin when stimulated.

Spinal cord

2 **Sensory and spinal nerves** carry impulses to the neuroendocrine cells of hypothalamus.

4 **Oxytocin** is transported by blood to mammary glands.

Milk ejected

1 **Nursing stimulates** nerve receptors in nipple.

Neuroendocrine cell

Hypothalamus

Anterior pituitary

Posterior pituitary

Oxytocin

Figure 13.6 The control of oxytocin secretion by nursing. The act of suckling triggers a series of events that ultimately eject milk from the nipple.

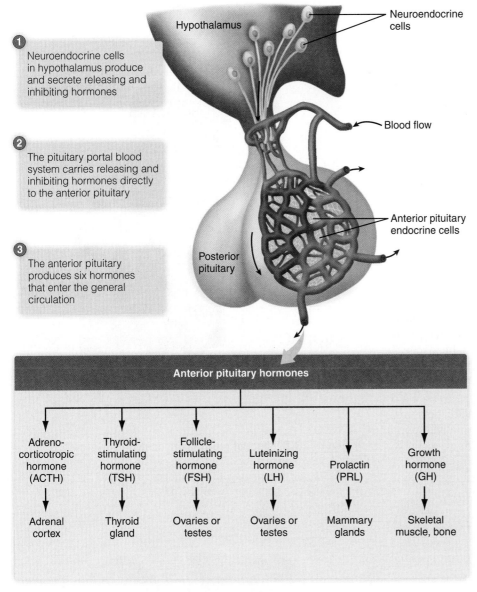

① Neuroendocrine cells in hypothalamus produce and secrete releasing and inhibiting hormones

② The pituitary portal blood system carries releasing and inhibiting hormones directly to the anterior pituitary

③ The anterior pituitary produces six hormones that enter the general circulation

Hypothalamus

Neuroendocrine cells

Blood flow

Anterior pituitary endocrine cells

Posterior pituitary

Anterior pituitary hormones

Adreno-corticotropic hormone (ACTH)	Thyroid-stimulating hormone (TSH)	Follicle-stimulating hormone (FSH)	Luteinizing hormone (LH)	Prolactin (PRL)	Growth hormone (GH)
Adrenal cortex	Thyroid gland	Ovaries or testes	Ovaries or testes	Mammary glands	Skeletal muscle, bone

Figure 13.7 The relationship between the hypothalamus and the anterior pituitary gland. Neuroendocrine cells in the hypothalamus secrete releasing hormones and inhibiting hormones into a system of capillaries that goes directly to the anterior pituitary. Specific hypothalamic hormones inhibit or stimulate each anterior pituitary hormone.

ACTH stimulates the adrenal cortex to release another group of hormones called *glucocorticoids*, which are steroid hormones involved in stress-related conditions and the control of glucose metabolism (see section 13.5). ACTH secretion is regulated by a negative feedback loop involving a releasing hormone from the hypothalamus and the concentration of glucocorticoids in blood.

TSH acts on the thyroid gland
TSH stimulates the thyroid gland to synthesize and release thyroid hormones. There is a TSH-releasing hormone from the hypothalamus, mentioned earlier.

FSH and LH stimulate the reproductive organs
FSH and LH are called *gonadotropins* because they stimulate the growth, development, and function of the reproductive organs (gonads) in both males and females. In females, FSH induces egg

development, LH promotes ovulation (egg release), and both hormones stimulate the secretion of the ovarian hormone estrogen. In addition, LH stimulates the secretion of the ovarian hormone progesterone after ovulation. In males, FSH induces sperm development and LH stimulates the production of the hormone testosterone by the testes.

FSH and LH are essentially absent until about age 10–13. A major increase in the production of these two hormones is what initiates sexual maturation and the development of secondary sex characteristics (puberty).

Prolactin (PRL): Mammary glands and milk production
Prolactin's primary function is to stimulate the development of mammary gland cells and the production of milk. Prolactin is present in males but its function is unknown. In females, the prolactin concentration in blood rises toward the end of pregnancy because estrogen stimulates the hypothalamus to step up production of *prolactin-releasing hormone,* one of those releasing hormones mentioned earlier. When estrogen levels decline after birth, prolactin-releasing hormone and prolactin levels are kept high by a sensory neural reflex associated with nursing.

When a woman is not in late pregnancy or breast-feeding, prolactin secretion is suppressed by *prolactin-inhibiting hormone,* another of those releasing/inhibiting hormones from the hypothalamus. It is now thought that prolactin-inhibiting hormone is actually identical to the neurotransmitter, dopamine.

 Quick Check Suppose a pregnant woman has a mutant version of oxytocin that cannot bind to oxytocin receptors, but she produces normal prolactin. Will she still be able to give birth, produce milk, and nurse her baby? ∎

Growth hormone (GH): Widespread effects on growth
The effects of growth hormone are so widespread that it is difficult to define a specific target cell or action. The effects of GH are most obvious in bone and muscle. In general, GH influences cells in ways that promote growth. Specific actions include stimulating protein synthesis and cell division, and using body fat as a source of energy.

GH is present throughout life, but most of its growth-promoting effects occur during childhood and adolescence. At these times it causes a dramatic increase in the mass and length of muscles and bones. Abnormal GH concentrations during development can lead to permanent changes in the final height of an adult. In adults it continues to help regulate metabolism by promoting the utilization of fats and amino acids as cellular fuel and by helping to maintain blood glucose levels.

 Web Animation *The Hypothalamus and Pituitary* at www. humanbiology.com

Pituitary disorders: Hypersecretion or hyposecretion

As you might guess, maintaining just the right quantities of hormones in the blood is a delicate balance that can sometimes go awry. Many diseases and disorders are chronic conditions of too much hormone (hypersecretion) or too little hormone (hyposecretion). Some are inherited. To illustrate the types of things that can go wrong, we focus on hyposecretion of one posterior pituitary hormone (ADH) and both hyposecretion and hypersecretion of an anterior pituitary hormone (GH).

ADH affects maintenance of water balance by the kidneys. Hyposecretion of ADH causes the kidneys to lose too much water, leading to a disorder called *diabetes insipidus*. Symptoms include excessive urination, dehydration, thirst, headache, and dry mouth. Some cases of diabetes insipidus are caused by head trauma or brain surgery that disrupts the normal production of ADH.

Even when ADH hormone levels are normal, diabetes insipidus can still occur if the kidney cells that normally respond to ADH are lacking the specific ADH receptors. As a general rule, then, an endocrine control loop can be disrupted either by lack of a hormone or by lack of target cell receptors for that hormone. Either way, the system does not function effectively. Endocrine disorders due to defects in target cell receptors are quite common and are often inherited.

Growth hormone has much longer-acting effects than ADH. Recall that in children, growth hormone stimulates the bone-lengthening activity of the growth plate, so that bones grow longer along with the child's overall growth. It is not surprising, therefore, that abnormalities of growth hormone secretion are most damaging when they occur before puberty. Hyposecretion of growth hormone during childhood can result in *pituitary dwarfism*. Conversely, hypersecretion of growth hormone during childhood and adolescence produces *gigantism* (**Figure 13.8**).

Dwarfism can be prevented if children are diagnosed early and treated with growth hormone throughout childhood. However, once adulthood is reached, dwarfism cannot be overcome by administering growth hormone. The sex steroid hormones cause the cartilaginous growth plates at the ends of long bones to be replaced by bone during puberty, and after that time, bones cannot grow longer.

Excessive production of growth hormone in adults over long periods produces a condition called *acromegaly*. The features of acromegaly are quite different from gigantism, because bones can no longer lengthen in adults. Instead, under the influence of growth hormone there is gradual thickening of the bones in the face, hands, and feet. The most notable visible effect is a change in facial features (**Figure 13.9**). Excessive

Figure 13.8 Effect of growth hormone on body growth. The tall person exhibits the traits of gigantism caused by excessive secretion of growth hormone during childhood. The short person has pituitary dwarfism, the result of too little growth hormone during childhood.

Figure 13.9 Effect of excessive growth hormone as an adult. This woman developed acromegaly as a result of too much growth hormone as an adult. She was symptom-free at 16.

production of growth hormone in adults usually results from a tumor of either the pituitary or the hypothalamus. Acromegaly is no longer common because growth hormone–secreting tumors are now usually diagnosed and treated early enough to prevent the development of the condition. Excessive growth hormone secretion is diagnosed by measuring growth hormone levels.

> ↻ **Recap** ADH acts on the kidneys to regulate water balance. Oxytocin stimulates uterine contractions and milk ejection in pregnant and lactating females, respectively. In both sexes, oxytocin contributes to feelings of sexual satisfaction. The six hormones produced by the anterior pituitary are ACTH, TSH, FSH, LH, prolactin (PRL), and GH. The release of each hormone is controlled in part by a releasing or inhibiting hormone from the hypothalamus. ■

13.4 The pancreas secretes glucagon, insulin, and somatostatin

The **pancreas** is both an endocrine gland (secreting hormones into the blood) and an exocrine gland (secreting enzymes, fluids, and ions into the digestive tract to aid in digestion). In this chapter we deal with its endocrine function.

The endocrine cells of the pancreas are located in small clusters scattered throughout the pancreas called the *islets of Langerhans* (an islet is a little island), or just the pancreatic islets. The pancreatic islets contain three types of cells and produce three hormones, all of them having to do with the regulation of blood sugar (glucose). One hormone raises blood sugar, the second lowers it, and the third inhibits the secretion of the other two hormones.

1. Alpha cells secrete **glucagon,** which raises blood sugar. As blood glucose levels decline between meals, glucagon is secreted into the bloodstream. In the liver, glucagon causes the breakdown of glycogen (a stored form of sugar) to glucose, raising blood glucose levels. Glucagon is particularly important between meals for increasing the supply of glucose to cells that need it.

2. Beta cells secrete **insulin,** which lowers blood sugar. After a meal, blood glucose levels rise as sugars are absorbed from the digestive tract. The high glucose concentration stimulates the beta cells to secrete insulin into the blood, where it does the opposite of glucagon. Insulin promotes the uptake of glucose by cells of the liver, muscle, and fat tissue. It also promotes the conversion of glucose into glycogen in the liver, both glycogen and proteins in muscle, and fats in adipose tissue.

3. Delta cells secrete somatostatin. The physiological significance of pancreatic somatostatin is not well understood, though it appears to inhibit the secretion of both glucagon and insulin. Somatostatin is also secreted by the hypothalamus, where it is the inhibitory hormone for growth hormone secretion, and the digestive tract, where it helps regulate several digestive secretions. The example of somatostatin reinforces the principle that the function of a hormone in any particular tissue or organ depends on the types of target cells present in those locations.

Figure 13.10 illustrates how the pancreas might respond to a meal. In actual fact, insulin and glucagon are both secreted

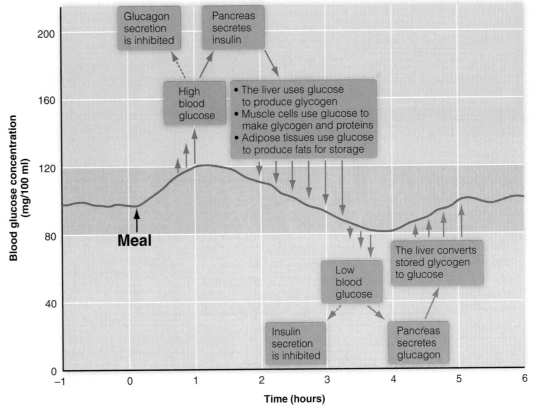

Figure 13.10 How the pancreas responds to a meal. The normal range for blood glucose concentration is 80–120 mg/100 ml. As blood glucose concentration rises, the secretion of insulin increases and the secretion of glucagon decreases, promoting the uptake and storage of glucose. If glucose concentration falls to near its normal lower limit, the secretion of glucagon increases and the secretion of insulin is inhibited. In the steady state, both hormones are secreted in modest amounts.

Glucose Monitoring Devices Are Inaccurate

Most diabetic patients measure their blood glucose level on a regular basis, as part of a daily regimen to try to maintain their blood glucose within normal limits. Many diabetics rely on home glucose monitors that are relatively cheap and easy to use. Trouble is, they're not very accurate. For example, one study reported that five different popular home glucose monitors gave readings that differed by as much as 30%.

Consumers probably are not aware that the devices can be this inaccurate. The Food and Drug Administration (FDA) is understandably concerned, and it's pressuring the international organization that sets standards for the devices to tighten up the standards. If they don't, the FDA may do it for them.

But this raises an interesting dilemma. Re-engineering and manufacturing the devices would make them more accurate, but it could also increase their price. That's the manufacturers' arguments against new standards for accuracy. And if an increase in price caused some people to stop using them altogether, have we really gained anything? There's always a trade-off; always a choice to be made. ■

Reference: Kinberly, M.M. et al. Variability among five over-the-counter blood glucose monitors. *Clin. Chim. Acta.* 364: 292–297, 2006.

in modest amounts all of the time. Adjusting the blood sugar is just a matter of altering the relative secretion rates of these two hormones—more insulin and less glucagon when blood sugar is high, and the inverse when blood sugar is low.

Recap Hormones secreted by the pancreas include glucagon, insulin, and somatostatin. Glucagon raises blood glucose levels by increasing production of glucose from glycogen. Insulin lowers glucose levels by facilitating glucose uptake and storage. Somatostatin inhibits the secretion of both glucagon and insulin. ■

13.5 The adrenal glands comprise the cortex and medulla

The **adrenal glands** are two small endocrine organs located just above the kidneys. Each gland has an outer layer called the **adrenal cortex,** and an inner core called the adrenal medulla. The cortex and medulla function as two separate endocrine glands. (**Cortex** and **medulla** are general terms for outer and inner regions; you'll hear the terms again in the context of the kidneys, and you're already aware of the cerebral cortex of the brain.)

The adrenal cortex: Glucocorticoids and mineralocorticoids

The adrenal cortex produces small amounts of the sex hormones estrogen and testosterone (most of the sex hormones are produced by the ovaries and testes, discussed later) and two classes of steroid hormones called glucocorticoids and mineralocorticoids. Their names give a clue to their functions: glucocorticoids help regulate blood glucose levels, and mineralocorticoids help regulate the minerals sodium and potassium.

Glucocorticoids: Cortisol The adrenal cortex produces a group of glucocorticoids with nearly identical structures. **Cortisol** accounts for approximately 95% of these glucocorticoids. During prolonged fasting, cortisol assists in maintaining blood glucose levels by promoting the utilization of fats and by increasing the breakdown of protein to amino acids in muscle. The free amino acids are then converted to new glucose by the liver. In addition to its role in controlling blood glucose, cortisol is also important in suppressing inflammatory responses that occur following infection or injury.

Cortisol secretion is controlled by a typical feedback loop. A releasing factor from the hypothalamus stimulates the secretion of ACTH from the anterior pituitary gland, which in turn stimulates the adrenal cortex to secrete cortisol. When cortisol reaches its upper limit of normal blood level, it inhibits further secretion of the releasing hormone from the hypothalamus and of ACTH from the anterior pituitary.

A fall in blood glucose concentration can override the normal feedback control mechanism and increase cortisol secretion significantly. So, too, can physical injury and emotional stress. All of these stimuli act at the level of the hypothalamus, which secretes more releasing hormone, ultimately raising both ACTH and cortisol. The elevated cortisol concentration helps to restore the blood glucose concentration to normal and makes more amino acids available to repair damaged tissues. It also reduces inflammation by decreasing fluid loss from capillaries.

Cortisol and cortisol-like drugs can be administered topically to treat inflammation or injury. When used for prolonged periods, however, these drugs suppress the immune system.

Quick Check Develop a hypothesis to explain why cortisol increases in response to stressful experiences such as physical injury and emotional stress. That is, how would the effects of cortisol help your body adjust to a stressful situation? ■

Mineralocorticoids: Aldosterone The other hormones produced by the adrenal cortex are the mineralocorticoids, the most abundant of which is **aldosterone.** Aldosterone is the hormone primarily responsible for regulating the amounts of sodium and potassium in the body. With ADH, it helps maintain body water balance.

Adrenal secretion of aldosterone increases whenever the total amount of sodium and water in the body is too low or when too much potassium is present (see Chapter 15 for details). In turn, aldosterone acts on cells in the kidneys to promote sodium reabsorption and potassium excretion, precisely what you would expect it to do if it were part of a homeostatic control loop for these ions.

The adrenal medulla: Epinephrine and norepinephrine

The **adrenal medulla** produces the nonsteroid hormones **epinephrine** (adrenaline) and **norepinephrine** (noradrenaline). These hormones play roles in metabolism and controlling blood pressure and heart activity.

Epinephrine and norepinephrine are mentioned in Chapter 11 (the nervous system) because they are also neurotransmitters. Function, not structure, determines whether they act as neurotransmitters or hormones. When norepinephrine and epinephrine are released into the blood and act on distant target cells, they function as hormones.

Like the hypothalamus and the posterior pituitary, the adrenal medulla is really a neuroendocrine organ. Epinephrine and norepinephrine are synthesized and stored in specialized cells called chromaffin cells in the adrenal medulla that are innervated by sympathetic nerves. When sympathetic nerves are activated, the chromaffin cells secrete epinephrine and norepinephrine (**Figure 13.11**).

Once released, the two hormones participate in a wide range of actions that are essentially like the fight-or-flight responses of the sympathetic division. They raise blood glucose levels, increase heart rate and force of contraction, increase respiration rate, and constrict or dilate blood vessels in many organs, all actions that prepare the body for emergency activity.

You may have noticed that approximately 30 seconds to a minute after you are frightened by a dangerous situation such as avoiding an auto accident, your heart starts to pound, you breathe faster, and you feel "pumped up," alert, and a little shaky. These changes may occur even though the danger has already passed. You are experiencing an "adrenal rush" due to intense and rapid stimulation of your adrenal glands by the sympathetic nervous system. The adrenal hormonal response occurs about half a minute after the event, because it takes that long for the two hormones to be released, circulate to all tissues, bind

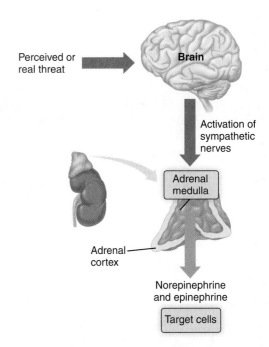

Figure 13.11 Secretion of norepinephrine and epinephrine by the adrenal medulla. The secretion of these two hormones is closely linked with the activity of the sympathetic division of the autonomic nervous system (ANS). Neural pathways are noted in dark brown; hormonal pathways, in light brown.

to receptors on their target cells, and initiate the appropriate intracellular cascade of events. The response is also short-lived, lasting only a couple of minutes at most. Norepinephrine and epinephrine from the adrenal glands enhance the action of the sympathetic nervous system because the hormones reach more cells and are released in large amounts.

↻ Recap The main secretory products of the adrenal cortex are cortisol and aldosterone. Cortisol raises blood glucose levels and suppresses inflammatory responses, and aldosterone regulates water balance by promoting sodium reabsorption. The hormones secreted by the adrenal medulla (epinephrine and norepinephrine) contribute to the fight-or-flight response initiated by the sympathetic nervous system. ∎

13.6 Thyroid and parathyroid glands

The thyroid and parathyroid glands are anatomically linked. The **thyroid gland** is situated just below the larynx at the front of the trachea, and the two lobes of the thyroid gland

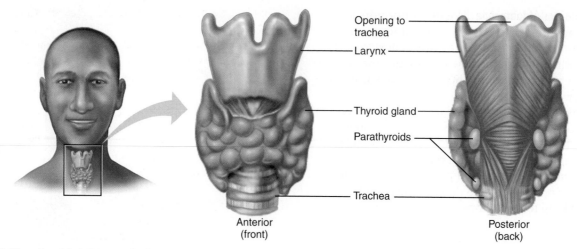

Figure 13.12 **The thyroid and parathyroid glands.**

wrap part of the way around the trachea (Figure 13.12). The four small **parathyroid glands** are embedded in the back of the thyroid.

The thyroid and parathyroid glands are also functionally linked; both help to regulate calcium balance. In addition, the thyroid has a separate and important role in controlling metabolism.

The two main hormones produced by the thyroid gland are thyroxine and calcitonin. The parathyroid glands produce parathyroid hormone.

The thyroid gland: Thyroxine speeds cellular metabolism

The thyroid gland produces two very similar hormones called **thyroxine** (T_4) and triiodothyronine (T_3). Thyroxine and triiodothyronine are identical except that thyroxine contains four molecules of iodine (hence the abbreviation T_4) whereas triiodothyronine contains only three. Because they are so similar we consider them together as thyroxine.

Structurally, thyroxine is not a steroid hormone. Nevertheless, it acts like a steroid because it is lipid soluble. It crosses the cell and nuclear membranes, binds to nuclear receptors, and activates genes as its mode of action. Those genes carry the code for the various enzymes that regulate our metabolic rate.

Thyroxine increases the production and use of ATP from glucose in nearly all body cells. When thyroxine concentration increases, the basal metabolic rate (BMR) also increases (see section 14.11). Conversely,

decreasing the blood concentration of thyroxine reduces energy utilization and BMR.

The basal rate of thyroxine secretion is regulated by a typical negative feedback loop (Figure 13.13). Any fall in thyroxine concentration would increase the secretion of the hypothalamic releasing hormone for TSH, which would increase pituitary TSH concentration, ultimately increasing

Figure 13.13 **Negative feedback control of thyroxine secretion.** A fall in plasma thyroxine concentration causes the hypothalamus and anterior pituitary to increase their secretion of the releasing hormone and TSH. The increase in TSH stimulates thyroid gland production and secretion of thyroxine, returning the plasma concentration to normal. Conversely, a rise in thyroxine causes a fall in the releasing hormone and TSH. Cold temperatures alter the feedback loop (raise plasma thyroxine concentration) by increasing hypothalamic secretion of the releasing hormone.

thyroxine secretion and driving thyroxine concentration back toward normal. Superimposed on this normal feedback loop, conditions that increase the body's energy requirements (such as exposure to cold temperatures) can stimulate the hypothalamus and raise the steady-state concentrations of *all* hormones in the loop.

✅ **Quick Check** Do you think thyroxine levels would rise or fall in people who are starving (or who are dieting)? Why? ■

Iodine deficiency can cause goiter Production of active thyroid hormones requires iodine. In fact, the main reason we need iodine in our diet is to ensure adequate production of thyroxine. Iodine deficiency can result in a thyroid-deficiency disease. When thyroxine is absent or abnormally low, the normal feedback inhibitory controls on the hypothalamus and pituitary are missing, so the hypothalamus and pituitary continue to secrete large quantities of the releasing hormone and TSH unchecked.

The high TSH levels stimulate the thyroid gland to grow to enormous size in an effort to get the thyroid to make more hormone, which it cannot do because it lacks iodine. An enlarged thyroid, as a result of iodine deficiency or other factors, is called a goiter (**Figure 13.14**). Goiter is less common today in industrialized nations because iodine is added to table salt. However, it is still a public health problem in certain areas of the world, most notably in Africa, China, and mountainous regions where rain has leached iodine from the soil.

Figure 13.14 A goiter caused by dietary iodine deficiency.

Calcitonin promotes bone growth Calcitonin, the other main hormone of the thyroid gland, is produced by a separate group of thyroid cells. Calcitonin decreases the rate of bone resorption by inhibiting the activity of osteoclasts. It also stimulates the uptake of calcium by bone. These two actions tip the normal balance of bone deposition/resorption toward deposition, which over time increases bone mass. Calcitonin is part of a negative feedback control mechanism for calcium, in which high blood calcium levels stimulate calcitonin release and low blood calcium levels inhibit its release.

Calcitonin is particularly important for the growth and development of bones in children. Once adulthood is reached, bone responsiveness to calcitonin declines. In adults, calcitonin is not essential for the normal regulation of blood calcium concentration. However, calcitonin may play a role in protecting the skeleton when calcium demand is high, such as during pregnancy.

Parathyroid hormone (PTH) controls blood calcium levels

The parathyroid glands produce only one hormone, **parathyroid hormone (PTH),** which

1. Removes calcium and phosphate from bone,
2. Increases absorption of calcium by the digestive tract,
3. Causes the kidneys to retain calcium and excrete phosphate.

Together these actions tend to raise the blood calcium concentration (and coincidentally lower blood phosphate concentration). As you might expect for a negative feedback control system, the secretion of PTH is stimulated by low blood calcium levels and inhibited by high blood calcium levels. **Figure 13.15** (next page) shows the overall regulation of blood calcium concentration.

The effect of PTH on calcium absorption by the digestive tract is actually indirect, through vitamin D. Vitamin D is required for calcium absorption by the intestines. However, vitamin D circulates in an inactive form until it is activated by a biochemical alteration that occurs in the kidneys. PTH is required for the renal activation step. An increase in PTH concentration increases the renal activation of vitamin D, which in turn enhances calcium absorption from the digestive tract.

As mentioned previously, calcitonin plays little or no role in the normal maintenance of blood calcium concentration once adulthood is reached. In adults, the primary regulator of blood calcium concentration is PTH.

↩ **Recap** Production of the thyroid gland hormones (T_4 and T_3) requires iodine. T_3 and T_4 speed cellular metabolism and have widespread effects on growth in children and basal metabolic rate (BMR) in adults. Calcitonin (from the thyroid) lowers blood calcium levels and PTH (from the parathyroids) raises blood calcium. ■

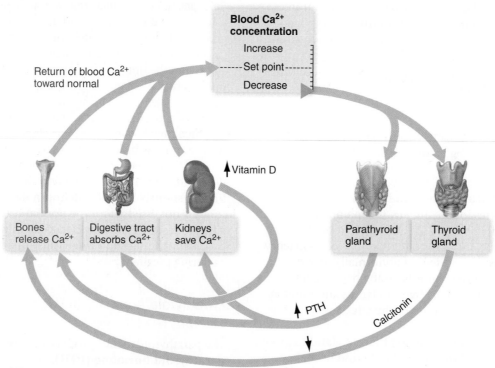

a) **Responses to a fall in blood calcium.**

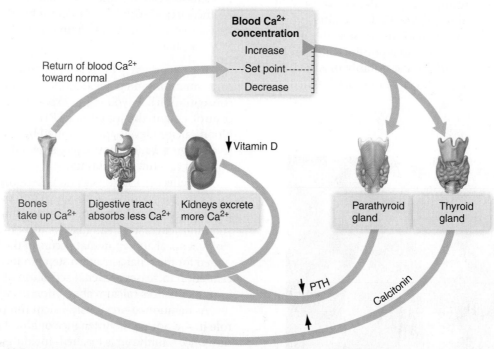

b) **Responses to a rise in blood calcium.** The importance of calcitonin is greatest in children, declining once adulthood is reached.

Figure 13.15 The homeostatic regulation of blood calcium concentration.

✓ *Draw a diagram similar to part b illustrating how blood glucose is brought back to normal after it rises following a meal. (Refer to Figure 13.10 for a reminder of the principles involved.)*

13.7 Testes and ovaries produce sex hormones

The human gonads are the testes of males and the ovaries of females. These organs are responsible for the production of sperm and eggs, respectively. Both organs are also endocrine glands in that they produce the steroid sex hormones.

Testes produce testosterone

The testes, located in the scrotum, produce androgens, the male sex hormones. The primary androgen produced by males is testosterone.

Before birth, testosterone production by the fetal testes is responsible for the development of the external male genitalia. Testosterone production declines essentially to zero between birth and puberty. During puberty the anterior pituitary gland begins to release luteinizing hormone (LH), which stimulates the testes to resume testosterone production.

In males, testosterone regulates the development and normal functioning of sperm, the male reproductive organs, and male sex drive. It is also responsible for the spurt of bone and muscle growth at puberty and the development of male secondary sex characteristics such as body hair and a deepening voice. LH, FSH, and testosterone continue to maintain the male reproductive system after puberty.

The adrenal glands in both sexes produce a small amount of a testosterone-like androgen called dihydroepiandrosterone (DHEA). DHEA has no demonstrable effect in males because they have an abundance of the more powerful testosterone. In females, DHEA from the adrenal glands is responsible for many of the same actions as testosterone in males, including enhancement of female pubertal growth, the development of axillary (armpit) and pubic hair, and the development and maintenance of the female sex drive.

Ovaries produce estrogen and progesterone

The ovaries, located in the abdomen, produce the female sex hormones known collectively as estrogens (17b-estradiol, estrone, and estriol), and also progesterone. The estrogens are often commonly referred to just as estrogen.

During puberty, the anterior pituitary starts to release luteinizing hormone (LH) and follicle-stimulating hormone (FSH). These hormones stimulate the ovaries to begin secreting estrogen and progesterone.

Estrogen initiates the development of female secondary sex characteristics such as breast development, widening pelvis, and distribution of body fat. Both estrogen and progesterone regulate the menstrual cycle, the monthly changes in the tissue of the uterus that prepare the female body for pregnancy. Release of LH, FSH, estrogen, and progesterone continues throughout a woman's reproductive years.

We discuss male and female hormones and reproductive systems in greater detail in Chapter 16.

13.8 Other glands and organs also secrete hormones

Two lesser-known glands (the thymus and pineal glands) and several organs (the heart, the digestive system, and the kidneys) also produce hormones. The hormones of the heart, digestive system, and kidneys are described briefly here and then discussed in more detail in later chapters.

Thymus gland hormones aid the immune system

The **thymus gland** is located between the lungs, behind the breastbone and near the heart. Sometimes it is considered part of the lymphatic system. For years the function of the thymus gland went unrecognized because its removal in an adult has little effect.

We now know that the thymus gland secretes peptide hormones called *thymosin* and *thymopoietin*, which help lymphocytes develop into mature T cells. The thymus is relatively important in children, but it shrinks and reduces its secretion in adults, especially after age 30.

The pineal gland secretes melatonin

The **pineal gland** is a pea-sized gland located deep within the brain, in the roof of the third ventricle. Its name derives from the fact that it is shaped like a small pine cone (*pineas* in Latin). More than 200 million years ago the pineal gland of our ancestors was a photosensitive area, or "third eye," located near the skin's surface. Although it is now shielded from light by the skull, it still retains its photosensitivity because it receives input indirectly from the eyes via the optic nerve and nerve pathways in the brain.

The pineal secretes the hormone **melatonin** (not to be confused with the skin pigment, melanin) in a cyclic manner, coupled to the daily cycle of light and dark. Melatonin is sometimes called the "hormone of darkness" because its rate of secretion rises nearly 10-fold at night and then falls again during daylight. Its secretion appears to be regulated by the absence or presence of visual cues. During the day, nerve impulses from the retina inhibit its release.

We are just beginning to understand what melatonin does. Researchers believe that it may be important in synchronizing the body's rhythms to the daily light/dark cycle (the *circadian cycle*). Some scientists suggest that declining sensitivity to melatonin may contribute to the onset of puberty; evidence in favor of this hypothesis is that destruction of the pineal gland by disease is associated with early puberty. Other proposed roles include inducing sleep, slowing the aging process, and enhancing immunity.

Melatonin is readily available over the counter as a pill. Some people take it as a natural sleep aid, although its effectiveness is controversial.

Table 13.2 (next page) lists all of the primary endocrine organs other than the hypothalamus and the pituitary, and summarizes their targets and hormonal actions.

Table 13.2 Hormones of endocrine glands other than the hypothalamus and pituitary

Gland	Hormone(s)	Main targets	Primary actions
Pancreas	Glucagon	Liver	Raises blood sugar levels
	Insulin	Liver, muscle, adipose tissue	Lowers blood sugar levels
	Somatostatin	Pancreas, hypothalamus, digestive tract	Inhibits release of glucagon, insulin, growth hormone, and digestive secretions
Adrenal cortex	Glucocorticoids (including cortisol)	Most cells	Raise blood glucose levels and promote breakdown of fats and proteins; suppress inflammation
	Mineralocorticoids (including aldosterone)	Kidneys	Regulate body water balance by promoting sodium reabsorption, potassium excretion
Adrenal medulla	Epinephrine and norepinephrine	Liver, muscle, adipose tissue	Raise blood sugar; increase heart rate and force of contraction; dilate or constrict blood vessels, increase respiration
Thyroid	Thyroxine (T_4) and triiodothyronine (T_3)	Most cells	Regulate the rate of cellular metabolism; affect growth and development
	Calcitonin	Bone	Lowers blood calcium levels
Parathyroid glands	Parathyroid hormone (PTH)	Bone, digestive tract, kidneys	Raises blood calcium levels
Testes	Testosterone	Most cells	Development of sperm, male secondary sex characteristics, and reproductive structures
Ovaries	Estrogen, progesterone	Most cells	Development of female secondary sex characteristics and reproductive structure; regulation of menstrual cycle
Thymus	Thymosin, thymopoietin	Lymphocytes	Help lymphocytes mature, especially in children
Pineal	Melatonin	Many cells	Synchronizes body rhythms; may be involved in onset of puberty

☑ **Quick Check** During winter, when days are short, many office workers never see the sun at all, because they spend their days indoors working under relatively dim office lights. How might this affect their melatonin levels and the 24-hour pattern of melatonin secretion? Explain your answer. ∎

Endocrine functions of the heart, the digestive system, and the kidneys

The heart, the digestive system, and the kidneys all have functions unrelated to the endocrine system. Nevertheless, all three of these organs secrete at least one hormone. The hormones they secrete and the functions of the hormones are listed below and summarized in Table 13.3.

- *Atrial natriuretic hormone (ANH)* is a peptide (nonsteroid) hormone secreted by the atria of the heart that

helps regulate blood pressure. When blood pressure rises, ANH increases the rate at which sodium and water are excreted in urine. This decreases blood volume and lowers blood pressure. We discuss how ANH acts on the kidneys in Chapter 15.

- *Gastrin, secretin,* and *cholecystokinin* are all hormones secreted by the digestive system. They have effects on the stomach, pancreas, and gallbladder (see Chapter 14).

- *Erythropoietin* and *renin* are secreted by the kidneys. Erythropoietin is a hormone that stimulates the production of red blood cells in bone marrow (see Chapter 7). Renin is actually an enzyme, but it is generally included in any discussion of hormones because it is a critical component of the *renin-angiotensin system*, which functions as a hormone. The other critical component of the renin-angiotensin system comes from the liver. The renin-angiotensin system is discussed in Chapter 15.

Table 13.3 Hormones of the digestive tract, kidneys, and heart

System or organ	Hormone(s)	Main targets	Primary actions
Digestive tract	Gastrin, secretin, cholecystokinin	Stomach, pancreas, gallbladder	Stimulate activities of the stomach, pancreas, and gallbladder
Kidneys	Erythropoietin	Bone marrow	Stimulates red blood cell production
Kidneys and liver	Renin-angiotensin system	Adrenal cortex, blood vessels	Stimulates aldosterone secretion, constricts blood vessels
Heart	Atrial natriuretic hormone (ANH)	Kidney, blood vessels	Increases sodium excretion, dilates blood vessels

13.9 Other chemical messengers

Some chemical messengers can function in ways similar to hormones but are not considered hormones because they are not secreted into the blood. Their actions are primarily local (near their site of release). You already know about neurotransmitters, one group of chemical messengers that act on nearby cells.

Such nonhormonal chemical messengers include histamine, prostaglandins, nitric oxide, and various growth factors and growth-limiting factors, plus dozens of other molecules about which we still know very little (Table 13.4). Most of these substances are short-lived because they are either quickly destroyed or taken back up by the cells that produced them.

Histamine is important in inflammation

Histamine is discussed in Chapter 9. Mast cells release histamine into the local interstitial fluid in response to tissue injury or the presence of an allergen. In turn, the histamine increases local secretion of mucus by mucus-secreting cells, dilates blood vessels, and increases the leakiness of capillaries, all part of a constellation of symptoms called inflammation.

Histamine is largely responsible for the runny nose of a cold and the nasal congestion of allergies. This is why over-the-counter drugs for colds and allergies generally contain an antihistamine compound.

Prostaglandins: Local control of blood flow

Prostaglandins are a group of chemicals derived from a fatty acid precursor. Their name comes from the fact that they were first discovered in semen and were thought to originate from the prostate gland, but they have since been found in nearly every tissue in the body.

Prostaglandins have a variety of functions, including local control of blood flow. Some prostaglandins constrict blood vessels; others dilate blood vessels. They also contribute to the inflammatory response and are involved in blood clotting at the site of an injury.

Nitric oxide has multiple functions

Nitric oxide (NO) is a gas that has only recently received attention as a chemical messenger. NO was originally called "endothelial-derived relaxing factor" because an extract of the endothelial cells lining blood vessels caused blood vessels to relax, or dilate. We now know that NO is produced in many other tissues as well.

Nitric oxide functions include regulating local blood flow in many tissues of the body, controlling penile erection, and regulating smooth muscle contraction in the digestive tract. Macrophages use it as a "chemical warfare" agent against bacteria, and it also interferes with clotting mechanisms. It even functions as a neurotransmitter in the brain.

Growth factors regulate tissue growth

Chemical messenger molecules called growth factors modulate development of specific tissues at the local level. The list includes nerve growth factor, epidermal growth factor, fibroblast growth factor, platelet-derived growth factor, tumor angiogenesis growth factors, insulin-like growth factor, and many more. Growth factors influence when a cell will divide and even in what direction it will grow. For example, nerve growth factors help to determine the direction of neuron axon growth so that nerves develop properly.

Recap Various chemical messengers (histamine, prostaglandins, nitric oxide, and growth factors) act primarily locally. Among their actions are regulating local blood flows, participating in local defense responses, and regulating growth of specific types of tissue. ∎

13.10 Disorders of the endocrine system

Because the endocrine system is one of the two primary systems for controlling body functions, any disruption to it can have dramatic and widespread effects on the body. We have already discussed several endocrine disorders earlier in this chapter. Here we look at several more.

Diabetes mellitus: Inadequate control of blood sugar

Diabetes mellitus means "a high flow of urine with a sweet taste." Called diabetes for short, it is a disease of sugar regulation. The common feature of the several types of diabetes is an inability to get glucose into cells where it can be used. The result is an abnormally high glucose concentration in the blood—so high that it overwhelms the kidney's ability to reabsorb all that is filtered. Glucose and excessive water appear in the urine (hence the name of the disease), causing dehydration and thirst. In addition to these symptoms, many diabetics experience fatigue, frequent infections, blurred vision, cuts that are slow to heal, and tingling in the feet or hands.

Table 13.4 **Other chemical messengers**

Messenger(s)	Primary actions
Histamine	Important in inflammatory response; increases secretion of mucus, dilates blood vessels, increases leakiness of capillaries
Prostaglandins	Control local blood flow; constrict blood vessels, dilate blood vessels, contribute to inflammatory response and blood clotting
Nitric oxide (NO)	Regulates local blood flow, controls penile erection, regulates smooth muscle contraction in digestive tract, acts as a "chemical warfare" agent against bacteria, interferes with clotting mechanisms, acts as a neurotransmitter in brain
Growth factors	Modulate growth of tissues at local level

Without the ability to metabolize glucose properly, the body turns to the metabolism of fat and proteins. The abnormal metabolism of glucose, fats, and proteins causes most of the other medical problems associated with diabetes, including cardiovascular and neural diseases, renal failure, blindness, and damage to small blood vessels that may lead to the need for amputations. There are two types of diabetes.

- *Type 1 diabetes* (5–10% of all cases) is caused by the failure of the pancreas to produce enough insulin. It is called insulin-dependent diabetes because the person depends on daily injections of insulin to allow glucose to enter the cells. Typically it is not governed by lifestyle; rather, it is an autoimmune disorder in which the person's own immune system attacks the insulin-producing cells in the pancreas. It is thought to have a strong genetic component (close relatives of a person with Type 1 diabetes are at increased risk), but it may also require an environmental trigger, such as a viral infection. Typically it appears in childhood or adolescence.

- *Type 2 diabetes (non-insulin-dependent diabetes)* represents 90–95% of all cases. Its hallmark is *insulin resistance*—cells fail to respond adequately to insulin even when it is present. Type 2 diabetes usually occurs in adults beyond the age of 40; however, in recent years it has been occurring more frequently and at younger ages (see the Current Issue at the chapter opening). Type 2 diabetes may have a genetic component, but lifestyle factors are thought to be the major determinants.

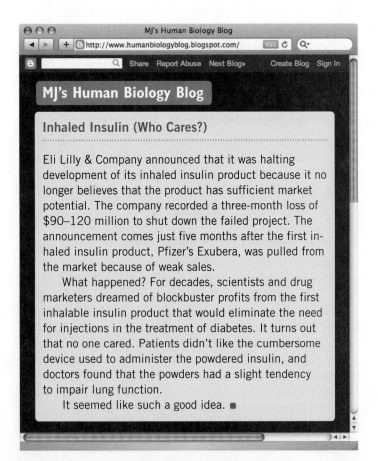

MJ's Human Biology Blog

Inhaled Insulin (Who Cares?)

Eli Lilly & Company announced that it was halting development of its inhaled insulin product because it no longer believes that the product has sufficient market potential. The company recorded a three-month loss of $90–120 million to shut down the failed project. The announcement comes just five months after the first inhaled insulin product, Pfizer's Exubera, was pulled from the market because of weak sales.

What happened? For decades, scientists and drug marketers dreamed of blockbuster profits from the first inhalable insulin product that would eliminate the need for injections in the treatment of diabetes. It turns out that no one cared. Patients didn't like the cumbersome device used to administer the powdered insulin, and doctors found that the powders had a slight tendency to impair lung function.

It seemed like such a good idea. ∎

Some Type 2 diabetics and "prediabetics" whose blood sugars are just approaching the diabetic range can control the disease entirely through lifestyle changes—maintaining a healthy weight, eating a nutritious diet, and exercising regularly. Others take insulin, especially to supplement their lifestyle changes when necessary. Other treatments include oral drugs that stimulate the pancreas to secrete more insulin, drugs that improve the action of the body's natural insulin, and drugs that increase the uptake of glucose by liver and muscle cells. More permanent solutions, such as implanting an artificial pancreas or transplanting human pancreatic cells, are under investigation.

 Quick Check In general, do you think it would be easier to treat disorders that result from failure to *produce* a hormone (such as Type I diabetes) or disorders that result from failure to *respond* to a hormone (such as Type II diabetes)? Why? ∎

Hypothyroidism: Underactive thyroid gland

Hypothyroidism refers to underactivity of the thyroid gland (hyposecretion of thyroid hormones). Mild cases of hypothyroidism may not have any signs, but more severe deficiencies can cause a variety of symptoms.

In children, insufficient thyroxine production for any reason can slow body growth, alter brain development, and delay the onset of puberty. Left untreated, this deficiency can result in cretinism, a condition of mental retardation and stunted growth.

In adults, insufficient thyroxine can lead to myxedema, a condition characterized by edema (swelling) under the skin, lethargy, weight gain, low BMR, and low body temperature. Hyposecretion of the hormone can be treated with thyroxine pills.

Hyperthyroidism: Overactive thyroid gland

Hyperthyroidism involves an overactive thyroid gland and hypersecretion of thyroid hormones. Too much thyroxine increases BMR and causes hyperactivity, nervousness, agitation, and weight loss.

The most common form of hyperthyroidism is Graves' disease, an autoimmune disorder in which a person's antibodies stimulate the thyroid to produce too much thyroxine. Graves' disease is often accompanied by protruding eyes caused by fluid accumulation behind the eyes.

Addison's disease: Too little cortisol and aldosterone

Addison's disease is caused by failure of the adrenal cortex to secrete sufficient cortisol and aldosterone. Lack of cortisol lowers blood glucose levels, and lack of aldosterone lowers blood sodium. Addison's disease tends to develop slowly, with chronic symptoms of fatigue, weakness, abdominal pain, weight loss, and a characteristic "bronzed" skin color. It can be successfully treated with medications that replace the missing hormones.

Cushing's syndrome: Too much cortisol

The symptoms of Cushing's syndrome are due to the exaggerated effects of too much cortisol, including (1) excessive production of glucose from glycogen and protein, and (2) retention of too much salt and water. Blood glucose concentration rises, and muscle mass decreases because of the utilization of protein to make glucose. Some of the extra glucose is converted to body fat, but only in certain areas of the body, including the face, abdomen, and the back of the neck (Figure 13.16). Symptoms of Cushing's syndrome include muscle weakness and fatigue, edema (swelling due to too much fluid), and high blood pressure. The disease can be caused by tumors of the adrenal gland or the ACTH-secreting cells of the pituitary. It can also be due to excessive use of cortisol and cortisol-like drugs (cortisone, prednisone, dexamethasone, and others) to control chronic inflammatory conditions such as allergies and arthritis.

a) Before the disease.

b) During the disease a characteristic hump of fat develops at the back of the neck.

Figure 13.16 Cushing's syndrome.

Chapter Summary

The endocrine system produces hormones *p. 302*

- The endocrine system is a collection of glands that secrete hormones into the blood.
- Hormones are chemical messengers secreted by the endocrine system that circulate in the blood and act on specific target cells.

Hormones are classified as steroid or nonsteroid *p. 304*

- Steroid hormones are lipid soluble; they enter the target cell in order to function.
- Nonsteroid hormones generally are not lipid soluble. They bind to receptors on the outer surface of the cell membrane of the target cell.

The hypothalamus and the pituitary gland *p. 306*

- The hypothalamus of the brain and the pituitary gland of the endocrine system link the nervous and endocrine systems.
- The posterior lobe secretes two hormones (ADH and oxytocin) that are produced by the hypothalamus.
- The anterior lobe of the pituitary produces and secretes six hormones (ACTH, TSH, FSH, LH, PRL, and GH). The first four stimulate the release of other hormones from other endocrine glands.

- Abnormalities of ADH secretion lead to disorders of water balance.
- Hypersecretion of growth hormone in a child leads to gigantism; hyposecretion causes dwarfism. In an adult, hypersecretion of growth hormone causes acromegaly.

The pancreas secretes glucagon, insulin, and somatostatin *p. 311*

- Within the pancreas, clusters of cells called the pancreatic islets produce and secrete three hormones (glucagon, insulin, and somatostatin).
- Glucagon raises blood glucose levels, insulin lowers blood glucose levels, and somatostatin appears to inhibit the secretion of glucagon and insulin.

The adrenal glands comprise the cortex and medulla *p. 312*

- The outer zone of the adrenal gland, called the cortex, secretes steroid hormones that affect glucose metabolism (glucocorticoids) and salt balance (mineralocorticoids).
- The inner zone of the adrenal gland, called the medulla, is actually a neuroendocrine organ. The adrenal medulla secretes epinephrine and norepinephrine into the blood.

Thyroid and parathyroid glands *p. 313*

- The thyroid gland produces thyroxine (T_4), triiodothyronine (T_3), and calcitonin.
- The four small parathyroid glands produce PTH, which, along with calcitonin, is responsible for the regulation of blood calcium concentration.

Testes and ovaries produce sex hormones *p. 317*

- The primary male sex hormone is testosterone, produced by the testes.
- The ovaries produce the two female sex hormones, estrogen and progesterone.

Other glands and organs also secrete hormones *p. 317*

- The thymus gland produces two hormones (thymosin and thymopoietin) that help lymphocytes mature into T cells.
- Melatonin from the pineal gland synchronizes the body's daily rhythms to the daily light/dark cycle.
- The heart secretes atrial natriuretic hormone, the kidneys secrete renin and erythropoietin, and the digestive tract secretes gastrin, secretin, and cholecystokinin.

Other chemical messengers *p. 319*

- Some substances not defined as hormones (because they are not specifically secreted into the bloodstream) nevertheless serve as chemical messengers.
- Most of these chemical messengers tend to act locally, near their site of secretion. They include histamine, prostaglandins, nitric oxide, and growth factors.

Disorders of the endocrine system *p. 319*

- Because the endocrine system controls so many body functions, disruptions can have widespread effects.
- Hypothyroidism can lead to cretinism (stunted growth and mental retardation) in children and myxedema in adults. Graves' disease is the most common form of hyperthyroidism.
- Insufficient cortisol and aldosterone can cause Addison's disease, and hypersecretion of cortisol results in Cushing's syndrome.

Terms You Should Know

adrenal cortex, *312*
adrenal medulla, *313*
aldosterone, *313*
calcitonin, *315*
cortisol, *312*
endocrine gland, *302*
epinephrine, *313*
glucagon, *311*
hormone, *302*
hypothalamus, *306*
insulin, *311*
neuroendocrine cells, *306*
nonsteroid hormone, *304*

norepinephrine, *313*
parathyroid gland, *314*
parathyroid hormone, *315*
pineal gland, *317*
pituitary gland, *306*
second messenger, *305*
steroid hormone, *304*
target cell, *302*
thymus gland, *317*
thyroid gland, *313*
thyroxine, *314*

Concept Review

Answers can be found at the Human Biology Place.
www.humanbiology.com

1. List the functions of the six hormones secreted by the anterior lobe of the pituitary gland.

2. Explain the characteristics of the endocrine system that make its reactions different from those of the nervous system.
3. Compare and contrast the mechanisms by which steroid and nonsteroid hormones act.
4. Define when you would call norepinephrine a neurotransmitter and when you would call it a hormone.
5. Explain why a 30-year-old short person cannot be made taller by taking growth hormone.
6. Describe the function of the two hormones from the thymus gland. Why is this function likely to be more important in children than in adults?
7. Explain how the pineal gland can be sensitive to light, even though it is located deep within the brain.
8. Describe why substances such as histamine and nitric oxide, although they are chemical messengers, are not considered true hormones.
9. Name the hormone that is released from the posterior pituitary of lactating females when the infant nurses.
10. Describe the role of the pineal gland in the circadian cycle.

Test Yourself

Answers can be found in Appendix A.

1. _____ hormones enter target cells and bind to intracellular receptors while _____ hormones bind to cell membrane receptors and never enter the target cells.
   a. Nonsteroid…steroid
   b. Protein…peptide
   c. Steroid…nonsteroid
   d. Steroid…carbohydrate

2. Which of the following statements comparing steroid and non-steroid hormones is true?
   a. Steroid hormones typically act through second messengers such as cAMP.
   b. Nonsteroid hormones usually are faster acting than steroid hormones.
   c. Steroid hormones are secreted by the anterior pituitary gland, and nonsteroid hormones are secreted by the posterior pituitary gland.
   d. Steroid hormones are produced only after puberty, whereas nonsteroid hormones are produced throughout the lifespan.

3. Secretion of hormones by the anterior pituitary gland is closely regulated by the:
   a. hypothalamus
   b. posterior pituitary gland
   c. cerebellum
   d. adrenal cortex

4. Which two hormones most directly regulate lactation?
   a. estrogen and progesterone
   b. estrogen and prolactin
   c. progesterone and oxytocin
   d. oxytocin and prolactin

5. All of the following represent hormone pairs with antagonistic activities except:
   a. estrogen…progesterone
   b. insulin…glucagon
   c. parathyroid hormone…calcitonin
   d. aldosterone…ANH

6. Which of the following hormones/chemical messengers is produced exclusively in males?
   a. prostaglandins
   b. melatonin
   c. testosterone
   d. none of these choices

7. Hormones secreted by the anterior pituitary gland control the activity of all of the following glands except:
   a. adrenal medulla
   b. thyroid gland
   c. testes
   d. ovaries

8. Which of the following hormones is/are involved in normal functioning of the male reproductive system?
   a. testosterone
   b. LH
   c. FSH
   d. all of these choices

9. Which of these endocrine disorders is associated with the inability of the pancreas to produce insulin?
   a. diabetes insipidus
   b. Addison's disease
   c. type 1 diabetes mellitus
   d. type 2 diabetes mellitus

10. Which of these statements comparing the nervous system and the endocrine system is true?
    a. The endocrine system exclusively controls development, whereas the nervous system controls homeostasis.
    b. The nervous system can access and communicate with all cells in the body, whereas the access of the endocrine system to different cells and tissues is more limited.
    c. The nervous system enables more rapid communication and signaling than does the endocrine system.
    d. The nervous system is fully functioning at birth, whereas the endocrine system doesn't begin to function until puberty.

11. Which of the following statements about the adrenal cortex or its hormones is true?
    a. The adrenal cortex secretes a mixture of steroid and nonsteroid hormones.
    b. The adrenal cortex secretes hormones that may decrease the inflammatory response and suppress immune responses.
    c. The secretion of hormones by the adrenal cortex is controlled by the sympathetic division of the nervous system.
    d. Hypersecretion of cortisol and aldosterone is seen in Addison's disease.

12. Which of the following statements about glucagon is true?
    a. Glucagon causes glycogen to be broken down to form glucose.
    b. Glucagon secretion is highest following a heavy meal.
    c. Glucagon promotes the uptake of glucose by liver and muscle cells.
    d. Glucagon works with cortisol to lower blood glucose levels.

13. Which of the following would be least likely to be affected by a tumor of the pituitary gland?
    a. cortisol secretion
    b. thyroid hormone secretion
    c. lactation
    d. secretion of epinephrine by the adrenal medulla

14. Which of the following is incorrectly paired?
    a. insulin…diabetes insipidus
    b. growth hormone…acromegaly
    c. thyroid…Grave's disease
    d. cortisol…Cushing's syndrome

15. Which of the following would be triggered by a drop in blood Ca^{2+} levels?
    a. increase in PTH secretion
    b. decrease in calcitonin secretion
    c. increase in ADH secretion
    d. both (a) and (b)

Apply What You Know

Answers can be found at the Human Biology Place.
www.humanbiology.com

1. If the biological effectiveness of a hormone were greatly reduced because its target cells lacked the appropriate receptors, would you expect the hormone concentration to be high, normal, or low? Explain. (*Hint:* It may help to draw the components of a negative feedback loop.)

2. Explain why an injection of epinephrine to combat an acute immune reaction (such as an allergic response to a bee sting) has an almost immediate effect, whereas an injection of a steroid hormone can take hours to have an effect.

3. From what you read in section 13.5, draw the negative feedback control loop for cortisol secretion. Then indicate on your diagram how stress or injury would affect the feedback loop. Check your diagram with a classmate or your teacher—if you can draw it and explain it, you understand it!

4. When a pregnant woman goes into labor, a series of strong muscle contractions force the baby from the uterus through the birth canal. Labor can be painful. Normally, such pain would initiate a negative feedback loop that would inhibit muscle contractions to prevent such pain, but that doesn't happen during labor. Why not?

5. Diabetics have to be particularly careful about sugar intake. Some are able to control the condition through careful diet. What sorts of foods should diabetics avoid, what foods should they try to get more of, and what can they do if they make a mistake?

6. A student finishes up the fall semester with several term papers and a number of big exams. After the semester she goes home for the holidays and comes down with a bad cold. The following year the same thing happens; she's sick for the holidays. During her senior year the fall semester isn't too bad, but applications to graduate school are due and she spends most of her semester break filling out forms and nervously worrying about what she will do next year. She's again sick for the beginning of the next term. Is this student just unlucky, or is something else at work?

14

The Digestive System and Nutrition

Colored SEM (× 20) of the lining of the stomach. The large folds flatten out when the stomach is full. The small indents are gastric pits.

Is "Overweight" Overstated?

If you have a body mass index (BMI) of greater than 25 but under 30, then you're officially considered overweight by the federal government. The government believes that your health may be at risk, and it's out to save you. You're likely to feel like you're being targeted by a concerted campaign to get you to eat better, exercise more, and above all, slim down! It's all very well meaning, of course, but there's a problem: Some of the statements made about the dangers of being overweight are based on poor science and weak statistics.

Several recent books have challenged the conventional wisdom that

too much body fat is bad for your health. But they also have very little data to back up their claims—they're just telling us what we want to hear. It's time for a little scientific skepticism.

The hard fact is that nearly every scientific study conducted on the relationship between body fat and health risk has found a positive association between BMIs above the normal range and the risk of certain diseases. Those diseases include hypertension, type 2 diabetes, coronary heart disease, stroke, gallbladder disease, osteoarthritis, and several types of cancers. Being slightly overweight is *sometimes*

a risk factor, but being obese generally *always* is.

The question becomes, then, "How much body fat is too much for *me?*" That's where the truth is unknown and misinformation begins. The answer depends on the specific disease you're concerned about and your lifestyle and genetic risk factors, not just on your BMI. To understand this, we need to understand the BMI scale; when it's useful and when it's not.

BMI and the Definition of "Overweight"

You and I tend to judge our fatness by getting on a bathroom scale. BMI is a bit

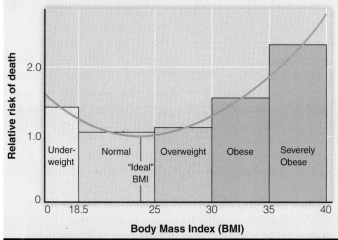

A representation of the typical J-shaped relationship between BMI and relative risk of death in a large population. The lowest risk of death is at a BMI of 24.

more precise because it takes height into account (see Figure 14.18, this chapter). It is the simplest comparative measure of fatness. Its value is that it is easy to obtain, even as a self-report over the phone or on a questionnaire.

Before 1980 the BMI scale primarily was used by life insurance companies to predict mortality risk (risk of death). By then it was known that there's a normal, predictable U- or J-shaped relationship between risk of death and BMI (see graph), with people at the low and high ends of the scale at greater relative risk of dying in a given year.

It wasn't long before medical researchers began to use the BMI scale to look for the health risks associated with high BMIs within large populations. Literally hundreds of studies, some performed well and some not, came to the same conclusion—that especially at very high BMIs there is a positive relationship between BMI and certain chronic diseases.

But if that was true, shouldn't we be doing something about it? If only everyone were at that ideal BMI of 24, imagine all the lives that would be saved! It wasn't long before the government declared that women with a BMI over 27 and men with a BMI over 28 were officially "overweight," and overweight was lumped with obesity as a health risk. A concerted educational campaign was begun to get us to eat better, exercise more often, and above all, lose that weight. The overzealous authors of one study even declared that "adults should try to maintain a BMI between 18.5 and 21.9 to minimize their risk of disease."

But the weight of the average U.S. adult kept rising, and government concern rose with it. In 1998 the federal government lowered its definition of overweight to include everyone over a BMI of 25. Overnight, another 29 million Americans were labeled overweight. And that's why by 2000, over 64% of the U.S. population was overweight or obese.

Overweight versus Obese— Different Levels of Risk

The first problem with any population-based definition of overweight is that it is not a good predictor of an individual's level of fitness or health. People who have a high muscle mass or who are just genetically built with larger body frames are likely to be labeled overweight or even obese by the BMI scale. The BMI may be a useful research tool for epidemiologists and public policymakers as they study whole populations, but it should not be used to make a recommendation about your personal lifestyle.

The second problem with the current definition of overweight is that it may have been oversold as a risk factor. Look again at the graph depicting mortality versus BMI. As a group, the overweight category has essentially no greater risk of dying than the normal group. To be fair, a population with a BMI in the overweight range *is* at increased risk for some of those specific diseases listed above, such as diabetes, hypertension, and heart disease. But as you might expect, the risks are not as great

Regardless of her BMI, tennis champion Serena Williams would best be described as "fit."

as the risks associated with obesity or severe obesity. It's a matter of degree.

Is the government right to worry that obesity continues to rise in this country and to try to do something about it through public education? The answer is a resounding "yes." If every U.S. adult could somehow magically lose 10 pounds, would the incidence of hypertension, diabetes, and mortality go down? The answer again is "yes," but probably because it would reduce the number of obese or severely obese individuals, not because it would improve the health of people just barely overweight. If you've been labeled as overweight but you exercise regularly and eat properly, then lose 10 pounds if you want to, but don't do it because you think you'll live longer.

The facts...

- According to government statistics, over 64% of the U.S. adult population is either overweight or obese.

- Statements about our national weight problem tend to lump overweight and obesity together as one problem.

- Obesity increases the risk of mortality and of certain chronic diseases. The health risk of being just slightly overweight is less severe.

- BMI is not particularly useful in assessing one's individual health.

Questions to consider

1 Do you agree with the government's definition of overweight as BMIs above 25? Why or why not?

2 What do you think is government's proper role (if any) when it comes to noncommunicable health risks such as obesity?

>> **The digestive system and its accessory organs digest and absorb nearly everything we eat and drink,** regardless of how much we eat or drink. The leftover non-absorbed waste products and bacteria are stored until they are eliminated as feces.

>> **The stomach stores ingested food and water until it can be delivered to the small intestine.** It also secretes a strong acid—hydrochloric acid (HCl)—that breaks down proteins and also kills most bacteria.

>> **Nutrients and water are absorbed in the small intestine** and to a lesser extent, the large intestine. Enzymes from the pancreas and the small intestine break down carbohydrates and fats so that they can be absorbed. The inner surface of the small intestine has many *villi* and *microvilli,* which markedly increase the surface area for absorption.

>> **Our nutritional requirements for good health include certain vitamins and minerals.** Some of these are not stored within the body for long periods, so we need a consistent supply.

>> **To maintain a constant body weight, energy intake must equal energy expenditure.** Any consistent imbalance in energy intake verus energy expenditure over time leads to weight gain or loss. Exercise can dramatically change how many calories we expend. To lose weight we must either eat less, exercise more, or do a little of both.

>> **A healthy diet contains a variety of grains, fruits, vegetables, and low-fat milk products.** Saturated fats, foods containing cholesterol, and refined sugars should be consumed in moderation.

S top for a moment and make a list of what you ate for lunch—a hamburger, fries, and a soft drink? Pizza and bottled water? A salad, apple, and milk?

All cells need **nutrients**—substances in food that are required for growth, reproduction, and the maintenance of health. Quite simply, the cells of your body must live on whatever you choose to eat. Right now the nutrients in your last meal (as well as nutrients stored in your body from previous meals) are being used to fuel cellular activities, build

cell components, and serve other vital functions. Your cells draw these nutrients from your blood, and your blood obtains them from your digestive system or from nutrient storage pools.

In many ways the digestive system is a highly efficient disassembly line. It takes in food and processes it, breaking it into small pieces and digesting the fragments with enzymes and strong chemicals. Then it passes the mixture down the line, absorbing its nutrients along the way, and stores the leftover waste until it can be eliminated from the body.

Does what you eat matter? Yes! Poor nutrition is associated with diseases ranging from cancer to cavities. Good nutrition, on the other hand, improves overall health and lowers the risk of health problems. In this chapter we follow the digestive process and the fate of your food. Then we discuss how you can apply this information to promote a healthy diet and body weight.

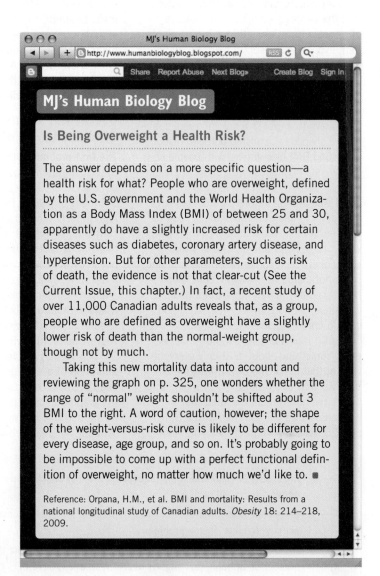

MJ's Human Biology Blog

Is Being Overweight a Health Risk?

The answer depends on a more specific question—a health risk for what? People who are overweight, defined by the U.S. government and the World Health Organization as a Body Mass Index (BMI) of between 25 and 30, apparently do have a slightly increased risk for certain diseases such as diabetes, coronary artery disease, and hypertension. But for other parameters, such as risk of death, the evidence is not that clear-cut (See the Current Issue, this chapter.) In fact, a recent study of over 11,000 Canadian adults reveals that, as a group, people who are defined as overweight have a slightly lower risk of death than the normal-weight group, though not by much.

Taking this new mortality data into account and reviewing the graph on p. 325, one wonders whether the range of "normal" weight shouldn't be shifted about 3 BMI to the right. A word of caution, however; the shape of the weight-versus-risk curve is likely to be different for every disease, age group, and so on. It's probably going to be impossible to come up with a perfect functional definition of overweight, no matter how much we'd like to. ∎

Reference: Orpana, H.M., et al. BMI and mortality: Results from a national longitudinal study of Canadian adults. *Obesity* 18: 214–218, 2009.

14.1 The digestive system brings nutrients into the body

The **digestive system** consists of all the organs that share the common function of getting nutrients into the body. It includes a series of hollow organs extending from the mouth to the anus: the mouth, pharynx, esophagus, stomach, small intestine, large intestine, rectum, and anus. These organs form a hollow tube called the **gastrointestinal (GI) tract** ("gastric" is the adjective for stomach). The space within this hollow tube—the area through which food and liquids travel—is called the *lumen.*

The digestive system also includes four accessory organs—the salivary glands, liver, gallbladder, and pancreas. Figure 14.1 summarizes the functions of the organs and accessory organs of the digestive system.

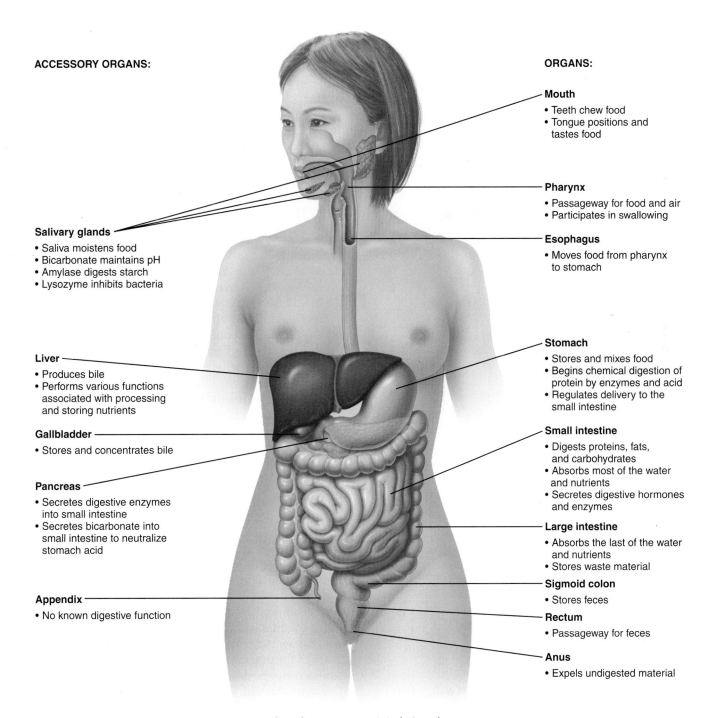

ACCESSORY ORGANS:

Salivary glands
• Saliva moistens food
• Bicarbonate maintains pH
• Amylase digests starch
• Lysozyme inhibits bacteria

Liver
• Produces bile
• Performs various functions associated with processing and storing nutrients

Gallbladder
• Stores and concentrates bile

Pancreas
• Secretes digestive enzymes into small intestine
• Secretes bicarbonate into small intestine to neutralize stomach acid

Appendix
• No known digestive function

ORGANS:

Mouth
• Teeth chew food
• Tongue positions and tastes food

Pharynx
• Passageway for food and air
• Participates in swallowing

Esophagus
• Moves food from pharynx to stomach

Stomach
• Stores and mixes food
• Begins chemical digestion of protein by enzymes and acid
• Regulates delivery to the small intestine

Small intestine
• Digests proteins, fats, and carbohydrates
• Absorbs most of the water and nutrients
• Secretes digestive hormones and enzymes

Large intestine
• Absorbs the last of the water and nutrients
• Stores waste material

Sigmoid colon
• Stores feces

Rectum
• Passageway for feces

Anus
• Expels undigested material

Figure 14.1 **Organs and accessory organs of the digestive system and their functions.**

The walls of the GI tract are composed of four layers

From the esophagus to the anus, the walls of the GI tract share common structural features (Figure 14.2). The walls of the GI tract consist of four layers of tissue.

1. **Mucosa.** The innermost tissue layer (the mucous membrane in contact with the lumen) is the mucosa. All nutrients must cross the mucosa to enter the blood.
2. **Submucosa.** Next to the mucosa is a layer of connective tissue containing blood vessels, lymph vessels, and nerves, called the submucosa. Components of food that are absorbed across the mucosa enter the blood and lymph vessels of the submucosa.
3. **Muscularis.** The third layer of GI tract tissue, called the muscularis, is responsible for motility or movement. The muscularis consists of two or three sublayers of smooth muscle. In general, the fibers of the inner sublayer are

oriented in a circular fashion around the lumen, whereas those in the outer sublayer are arranged lengthwise, parallel to the long axis of the digestive tube. The exception is the stomach, which has a diagonal (oblique) sublayer of muscle inside the other two.

4. **Serosa.** The outermost layer of the GI tract wall, or serosa, is a thin connective tissue sheath that surrounds and protects the other three layers and attaches the digestive system to the walls of the body cavities.

Some of the organs of the GI tract are separated from each other by thick rings of circular smooth muscle called *sphincters.* When these sphincters contract they can close off the passageway between organs.

Five basic processes accomplish digestive system function

In practical terms, the digestive system is a long disassembly line that starts with huge chunks of raw material (we call it food) and takes them apart so that the nutrients in the food can be absorbed into the body. The digestive system accomplishes this task with five basic processes.

1. *Mechanical processing and movement.* Chewing breaks food into smaller pieces, and two types of movement (motility) mix the contents of the lumen and propel it forward.
2. *Secretion.* Fluid, digestive enzymes, acid, alkali, bile, and mucus are all secreted into the GI tract at various places. In addition, several hormones that regulate digestion are secreted into the bloodstream.
3. *Digestion.* The contents of the lumen are broken down both mechanically and chemically into smaller and smaller particles, culminating in nutrient molecules.
4. *Absorption.* Nutrient molecules pass across the mucosal layer of the GI tract and into the blood or lymph.
5. *Elimination.* Undigested material is eliminated from the body via the anus.

Two types of motility aid digestive processes

The smooth muscle of the GI tract produces two kinds of motility, called peristalsis and segmentation. The functions of peristalsis and segmentation are quite different.

Peristalsis propels food forward (Figure 14.3a). Peristalsis begins when a lump of food (called a bolus) stretches a portion of the GI tract, causing the smooth muscle in front of the bolus to relax and the muscle behind it to contract. The contractions push the food forward, stretching the next part of the tube and causing muscle relaxation in front and contraction behind. The peristaltic wave of contraction ripples through the organs of the GI tract, mixing the contents of the stomach and pushing the contents of the esophagus and intestines forward. Peristalsis occurs in all parts of the GI

Vein —— Lymph vessel
Artery —
Nerve —

Serosa
• Connective tissue outer covering
• Protects and anchors the digestive tract

Longitudinal layer

Circular layer

Lumen

Mucosa
• Mucous membrane layer
• Lines the digestive tract

Muscularis
• Two layers of smooth muscle
• Responsible for motility of the digestive tract

Submucosa
• Connective tissue layer
• Contains blood vessels, lymph vessels, and nerves

Figure 14.2 The four tissue layers of the GI tract wall.

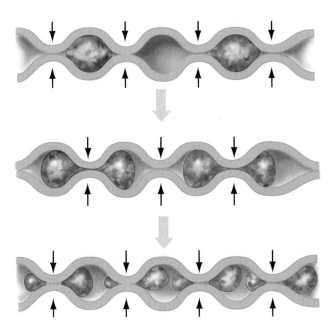

a) Peristalsis.

b) Segmentation.

Figure 14.3 **Motility of the gastrointestinal tract.** a) Peristalsis in the esophagus. The three sequential diagrams show how peristalsis moves the contents of the lumen onward. b) Segmentation in the small intestine. Segmentation mixes the contents of the lumen.

tract, but is especially prevalent in the esophagus, where it transports food rapidly to the stomach.

Segmentation mixes food (Figure 14.3b). In segmentation, short sections of smooth muscle contract and relax in seemingly random fashion. The result is a back-and-forth mixing of the contents of the lumen. Food particles are pressed against the mucosa, enabling the body to absorb their nutrients. Segmentation occurs primarily in the small intestine as food is digested and absorbed.

Recap The digestive system consists of organs and accessory organs that share the function of bringing nutrients into the body. The wall of the GI tract consists of four tissue layers: the mucosa, the submucosa, the muscularis, and the serosa. The five basic processes of digestion are (1) mechanical processing and movement, (2) secretion, (3) digestion, (4) absorption, and (5) elimination. ▪

14.2 The mouth processes food for swallowing

The mouth, or oral cavity, is the entrance to the GI tract. Digestion begins in the mouth with the process of chewing, which breaks food into smaller and smaller chunks. Essentially, the mouth functions as an effective food processor.

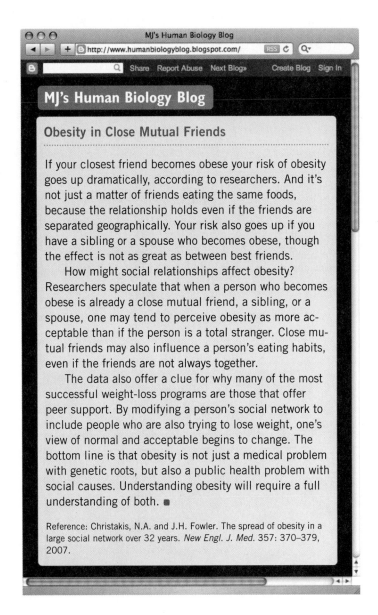

MJ's Human Biology Blog

Obesity in Close Mutual Friends

If your closest friend becomes obese your risk of obesity goes up dramatically, according to researchers. And it's not just a matter of friends eating the same foods, because the relationship holds even if the friends are separated geographically. Your risk also goes up if you have a sibling or a spouse who becomes obese, though the effect is not as great as between best friends.

How might social relationships affect obesity? Researchers speculate that when a person who becomes obese is already a close mutual friend, a sibling, or a spouse, one may tend to perceive obesity as more acceptable than if the person is a total stranger. Close mutual friends may also influence a person's eating habits, even if the friends are not always together.

The data also offer a clue for why many of the most successful weight-loss programs are those that offer peer support. By modifying a person's social network to include people who are also trying to lose weight, one's view of normal and acceptable begins to change. The bottom line is that obesity is not just a medical problem with genetic roots, but also a public health problem with social causes. Understanding obesity will require a full understanding of both. ▪

Reference: Christakis, N.A. and J.H. Fowler. The spread of obesity in a large social network over 32 years. *New Engl. J. Med.* 357: 370–379, 2007.

Teeth bite and chew food

The teeth chew food into pieces small enough to swallow. There are four types of teeth, each specialized for a different purpose (**Figure 14.4a**). The sharp-edged incisors cut food and the pointed canines tear it. The flat surfaces of the pre-molars and molars are well adapted to grinding and crushing food.

Children have only 20 teeth. They develop by about age 2 and are gradually replaced by permanent teeth. Permanent teeth usually develop by late adolescence, except for the wisdom teeth (third molars), which generally appear by age 25. Most adults have 32 permanent teeth.

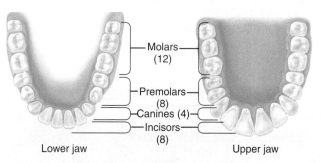

a) Locations and types of adult human teeth.

Molars
(12)

Premolars
(8)

Canines (4)

Incisors
(8)

Lower jaw   Upper jaw

Enamel

Dentin

Gingiva (gum)

Pulp cavity

Root canal

Periodontal membrane

Bone

Crown

Root

b) Anatomy of a tooth.

c) Bacteria on a tooth's surface (approx. × 8,000).

Figure 14.4 Teeth.

Each tooth (Figure 14.4b) consists of a visible region called the crown and a region below the gum line called the root. The crown is covered by a layer of enamel, an extremely hard nonliving compound of calcium and phosphate. Beneath the enamel is a bonelike living layer called dentin. The soft innermost pulp cavity contains the blood vessels that supply the dentin, as well as the nerves that cause so much pain when a tooth is infected or injured. The entire tooth sits in a socket in the jawbone that is lined with periodontal membrane.

Our mouths contain large numbers of bacteria that flourish on the food that remains between the teeth (Figure 14.4c). During their metabolism these bacteria release acids that can dissolve enamel, creating cavities or *dental caries*. If not treated, cavities deepen, eroding the dentin and pulp cavity and causing a toothache. Tooth decay may inflame the soft gum tissue (gingiva) around the tooth, causing *gingivitis*. Decay that inflames the periodontal membrane leads to *periodontitis*. However, good dental hygiene—including regular exams and teeth cleaning—can prevent most dental problems from becoming serious.

The tongue positions and tastes food

Chewing would be inefficient without the muscular tongue, which positions food over the teeth and mashes it against the roof of the mouth. The tongue consists of skeletal muscle enclosed in mucous membrane, so we have voluntary control over its movements. The tongue contributes to the sense of taste (Chapter 12) and also is important for speech.

Saliva begins the process of digestion

Three pairs of **salivary glands** produce a watery fluid called *saliva*. The parotid gland lies near the back of the jaw, and the smaller sublingual and submandibular glands are located just below the lower jaw and below the tongue, respectively (**Figure 14.5**). All three glands connect to the oral cavity via ducts.

Saliva moistens food, making it easier to chew and swallow. Saliva contains four main ingredients, each with important functions. One is *mucin*, a mucus-like protein that holds food particles together so they can be swallowed more easily. An enzyme called *salivary amylase* begins the process of digesting carbohydrates. *Bicarbonate* (HCO_3^-) in saliva maintains the pH of the mouth between 6.5 and 7.5, the range over which salivary amylase is most effective. Salivary bicarbonate may also help protect your teeth against those acid-producing bacteria. Saliva also contains small amounts of an enzyme called *lysozyme*, which inhibits bacterial growth.

Recap The four kinds of teeth (molars, premolars, canines, and incisors) mechanically digest chunks of food. Salivary glands secrete saliva, which moistens food, begins the chemical digestion of carbohydrates, maintains the pH of the mouth, and protects the teeth against bacteria. ∎

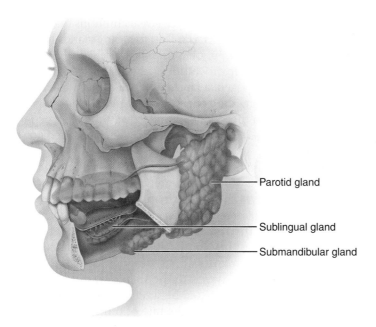

Figure 14.5 **The salivary glands.** The three pairs of glands deliver saliva to the mouth via salivary ducts.

14.3 The pharynx and esophagus deliver food to the stomach

After we have chewed our food and mixed it with saliva, the tongue pushes it into the **pharynx,** or throat, for swallowing. The act of swallowing involves a sequence of events that is coordinated with a temporary halt in breathing.

First, voluntary movements of the tongue and jaws push a bolus of food into the pharynx (**Figure 14.6a**). (As described in Chapter 10, the pharynx is a common passageway for food and air.) The presence of the food stimulates receptors in the pharynx and initiates the second, involuntary phase, the "swallowing reflex." The soft palate rises to close off the passageway into the nasal cavity and the larynx rises slightly. The epiglottis bends to close the airway to the trachea temporarily, so we do not inhale the

food. Meanwhile, the tongue pushes the food back even farther, sliding it past the epiglottis and into the esophagus (Figure 14.6b). Once started, the swallowing reflex is involuntary and cannot be stopped. The initiation of swallowing is the last voluntary act in the process until defecation.

Just beyond the pharynx is the **esophagus,** a muscular tube consisting of both skeletal and smooth muscle that connects the pharynx to the stomach. The lining of the esophagus produces lubricating mucus that helps food slide easily. Occasionally thick foods such as mashed potatoes or peanut butter stick briefly to the mucosa of the esophagus, causing a painfully strong reflex contraction behind the food. If the contraction does not dislodge the food, drinking liquid may help.

Gravity assists peristalsis in propelling food. However, peristaltic contractions enable the esophagus to transport food even against gravity, such as when we are lying down.

The lower esophageal sphincter, located at the base of the esophagus, opens briefly as food arrives and closes after it passes into the stomach. The sphincter prevents reflux of the stomach's contents back into the esophagus. Occasionally this sphincter malfunctions, resulting in the backflow of acidic stomach fluid into the esophagus. This condition, known as *acid reflux*, is responsible for the burning sensation known as "heartburn." Acid reflux becomes more common with weight gain, pregnancy, and age. Occasionally it indicates a *hiatal hernia*, a condition in which part of the stomach protrudes upward into the chest through an opening (hiatus) in the diaphragm muscle. Mild or temporary acid reflux is usually not serious and often improves with weight loss. However, prolonged acid reflux may cause esophageal ulcers because stomach acid can erode the mucosa of the esophagus.

Recap Swallowing begins with voluntary movements of the tongue; the presence of food initiates an involuntary swallowing reflex. Peristalsis and gravity transfer food through the esophagus to the stomach. ∎

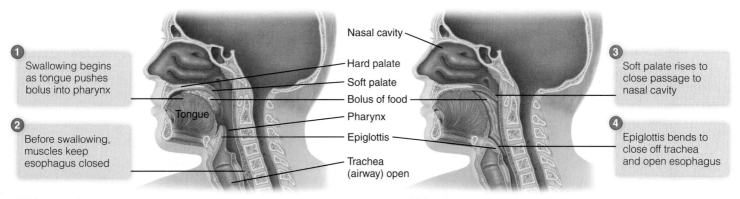

a) Voluntary phase.

b) Involuntary phase.

Figure 14.6 **Swallowing.**

14.4 The stomach stores food, digests protein, and regulates delivery

The **stomach** is a muscular, expandable sac that performs the following three important functions:

- *Food storage.* Humans tend to eat meals several times a day. The stomach stores the food until it can be digested and absorbed. The stomach shrinks when empty, then expands to 1–3 liters of capacity when we eat.
- *Digestion.* The stomach is more than a storage bag, however. It also digests proteins, using strong acid and protein-digesting enzymes. The strong acid also kills most bacteria. Muscle contractions mix these secretions with food, assist in mechanically breaking apart food particles, and push the mixture into the small intestine.
- *Regulation of delivery.* The stomach regulates the rate at which food is delivered to the small intestine.

Gastric juice breaks down proteins

The walls of the stomach consist of the usual four layers: mucosa, submucosa, muscularis, and serosa (**Figure 14.7a**).

A closer look at the mucosal layer reveals millions of small openings called gastric pits that lead to gastric glands below the surface (Figure 14.7b, c). Some of the cells lining the glands secrete either hydrochloric acid (HCl) or mucus, but most secrete pepsinogen, a large precursor molecule that becomes a protein-digesting enzyme called **pepsin** once it is exposed to acid in the stomach. Collectively, the HCl, pepsinogen, and fluid that are secreted into the glands are known as gastric juice.

The stomach typically produces 1–2 liters of gastric juice per day, most of it immediately after meals. The acid in gastric juice gives the stomach an acidic pH of approximately 2. The pepsin and acid in gastric juice dissolve the connective tissue in food and digest proteins and peptides into amino acids so they can be absorbed in the small intestine. The watery mixture of partially digested food and gastric juice that is delivered to the small intestine is called **chyme.** The pyloric sphincter, between the stomach and the small intestine, regulates the rate of transport of chyme into the small intestine.

If gastric juice is powerful enough to digest proteins, why doesn't it digest the stomach too? The reason is that some of

Oblique layer
Circular layer
Longitudinal layer

Mucosa
Submucosa
Muscularis
Serosa

Mucus
Gastric pits
Mucus-secreting cells
Acid-secreting cell
Pepsinogen-secreting cell

a) Stomach wall.

b) Gastric glands.

c) Gastric pits.

Figure 14.7 The stomach.

the cells lining the stomach and the gastric glands continuously produce a protective barrier of mucus. Normally the stomach contents are in contact with mucus, not living cells. If the mucous layer becomes damaged, however, this leaves the underlying tissue vulnerable. An open (sometimes bleeding) sore called a **peptic ulcer** (from the Greek word *peptein*, digest) may form. Peptic ulcers occasionally occur in the esophagus and upper part of the small intestine as well.

In addition to pepsinogen, HCl, and mucus, some mucosal cells secrete *intrinsic factor*, a protein that binds to vitamin B_{12} so that it can be absorbed in the small intestine. Finally, certain cells in the gastric glands secrete a hormone called *gastrin* into the bloodstream. We will talk about gastrin when we discuss the regulation of the digestive process.

> ✅ **Quick Check** Suppose the acid-secreting cells were not working properly and did not secrete any HCl. Would the stomach still produce pepsin? Which of the stomach's major functions would be affected? ◼

Stomach contractions mix food and push it forward

While your stomach is empty, muscle contractions keep it small. When you eat a meal, contractions cease and the stomach relaxes and stretches to accommodate the food. Stretching signals peristalsis to increase.

Each wave of peristalsis starts at the lower esophageal sphincter and moves toward the pyloric sphincter, becoming stronger as it proceeds. The peristaltic wave pushes the chyme forward and then, when there is nowhere else for it to go, backward again in a squeezing, mixing motion (**Figure 14.8**). Each contraction propels about a tablespoon of chyme into the small intestine before the pyloric sphincter closes

Answers to ✅ **questions can be found in Appendix A.**

momentarily. A peristaltic contraction occurs every 15–25 seconds. Have you ever noticed your stomach gurgling at this rate after a meal? You're hearing peristalsis in action.

It takes two to six hours for the stomach to empty completely after a meal. Most forceful when the stomach is full, peristalsis declines as the stomach empties. Chyme with a high acid or fat content stimulates the release of hormones that slow stomach peristalsis, giving the small intestine more time to absorb the nutrients.

The stomach does not absorb nutrients because it lacks the required cellular transporting mechanisms and because its inner lining is coated with mucus. Exceptions to this rule are alcohol and aspirin, both of which are small lipid-soluble substances that can cross the mucus barrier and be absorbed into the bloodstream directly from the stomach. Nevertheless, this is not the primary absorptive pathway for alcohol; most alcohol is absorbed in the intestine along with the nutrients.

> ↩ **Recap** The stomach stores food, digests it, and regulates its delivery to the small intestine. Gastric juice dissolves connective tissue, large proteins, and peptides in food. The presence of food stretches the stomach and increases peristalsis. Peristaltic contractions mix the chyme and push it gradually into the small intestine. ◼

14.5 The small intestine digests food and absorbs nutrients and water

The process of digestion continues in the small intestine, so-named because it is smaller in diameter than the final segment of the digestive tract, the large intestine. The small intestine has two major functions.

1. *Digestion.* The stomach partially digests proteins to smaller peptides, under the influence of strong acids and pepsin.

Lower esophageal sphincter

Stomach

Pyloric sphincter

Key: ⟶ = Location and direction of peristalsis

⟶ = Movement of stomach contents

Figure 14.8 Peristalsis of the stomach. A peristaltic wave of contraction occurs in the stomach approximately every 15–25 seconds. Peristalsis mixes the contents of the stomach and forces a small amount of chyme into the small intestine with each contraction.

Protein digestion continues in the small intestine, but here we also digest carbohydrates and lipids. Digestion of proteins, carbohydrates, and lipids in the small intestine involves neutralizing the highly acidic gastric juice and adding additional digestive enzymes from the intestine and pancreas.

2. *Absorption.* Eventually the proteins, carbohydrates, and lipids in food are broken down to single amino acids, monosaccharides, fatty acids, and glycerol, which are small enough to be transported across mucosal cells into the blood. Nearly 90% of the absorbable nutrients and water is absorbed in the small intestine.

The small intestine consists of three different regions. The first region, the **duodenum,** is only about 10 inches long, but it is here that most of the digestion takes place. The products of digestion are absorbed primarily in the other two segments, the *jejunum* and the *ileum,* which together are about 10 feet long.

The structure of the small intestine wall makes it well suited for absorption (**Figure 14.9**). The mucosa contains large folds covered with microscopic projections called **villi** (singular: *villus*). Each epithelial cell of the villi has dozens of even smaller, cytoplasmic projections called *microvilli*. The microvilli give the mucosal surface a velvety appearance, which is why they are sometimes called the "brush border." Combined, the folds, villi, and microvilli enlarge the surface area of the small intestine by more than 500 times, increasing its ability to absorb nutrients. At the center of each villus are capillaries and a small lymph vessel called a lacteal, which transport nutrients to larger blood vessels and lymph vessels.

> **Recap** The small intestine has two major functions: 1) digesting proteins, carbohydrates, and lipids, and 2) absorbing approximately 90% of the nutrients and water we consume. Projections called villi in the mucosa increase the small intestine's surface area for absorption. ∎

14.6 Accessory organs aid digestion and absorption

As mentioned earlier, the digestive system has four accessory organs: salivary glands, pancreas, gallbladder, and liver. We have already discussed the salivary glands. Before discussing absorption in more detail, let's look at the roles of the other three accessory organs.

The pancreas secretes enzymes and NaHCO₃

The **pancreas,** an elongated organ that lies just behind the stomach, has both endocrine and exocrine functions (**Figure 14.10**). In Chapter 13 we saw that as an endocrine gland the pancreas secretes hormones that regulate blood glucose levels. In this chapter we focus on its exocrine role. The pancreas produces and secretes the following:

■ *Digestive enzymes.* Pancreatic enzymes include *proteases* (enzymes that digest proteins), such as trypsin, chymotrypsin,

Folds

a) **The wall of the small intestine contains numerous folds that increase its surface area.**

Villi
Mucosa
Submucosa
Muscularis
Serosa

b) **Each fold is covered with smaller folds called villi.**

Epithelial cell
Lacteal
Mucus-producing cell
Arteriole
Venule
Lymph vessel

Microvilli

d) **A single cell with its outer membrane of microvilli.**

c) **A close-up view of a villus showing the single layer of mucosal cells covering the surface and the centrally located lymph vessel (lacteal) and blood vessels.**

e) **Human intestinal villi (× 270).**

Figure 14.9 The small intestine.

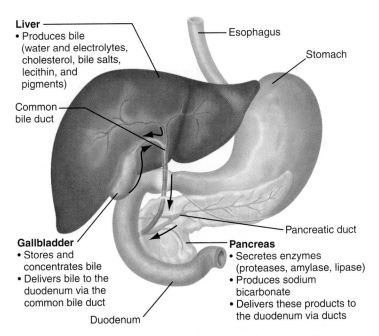

Liver
• Produces bile (water and electrolytes, cholesterol, bile salts, lecithin, and pigments)

Common bile duct

Esophagus

Stomach

Pancreatic duct

Gallbladder
• Stores and concentrates bile
• Delivers bile to the duodenum via the common bile duct

Pancreas
• Secretes enzymes (proteases, amylase, lipase)
• Produces sodium bicarbonate
• Delivers these products to the duodenum via ducts

Duodenum

Figure 14.10 Locations and digestive functions of the liver, gallbladder, and pancreas. These three accessory organs are connected to the small intestine via ducts. The common bile duct joins one of the two pancreatic ducts just before it enters the duodenum. Arrows indicate direction of movement of secretory products.

and carboxypeptidase; *pancreatic amylase,* which continues the digestion of carbohydrates only partially accomplished by salivary amylase; and *lipase,* a lipid-digesting enzyme.

■ *Sodium bicarbonate* (NaHCO$_3$). Unlike pepsin, which is effective in acid conditions, pancreatic enzymes work best at a more neutral pH. NaHCO$_3$ functions to neutralize stomach acid.

Two pancreatic ducts deliver these secretions to the duodenum, where they facilitate the process of digestion. Table 14.1 lists major digestive enzymes in the gastrointestinal tract.

MJ's Human Biology Blog

http://www.humanbiologyblog.blogspot.com/

Share   Report Abuse   Next Blog»   Create Blog   Sign In

California Bans *Trans* Fats

The state of California has banned the use of *trans* fats in most food products. The ban goes into effect for all restaurant products in 2010 and for all retail baked goods in 2011. Packaged foods will not be affected, however.

Trans fats are created by bubbling hydrogen through liquid oil at high temperature. The resultant partially hydrogenated oil is a solid at room temperature. *Trans* fats prolong the shelf life and (some say) improve the flavor of foods. They were popular as a deep-frying oil until it became apparent that they raise the levels of low-density lipoproteins (the bad cholesterol), thereby potentially contributing to heart disease. Some restaurant chains, most notably McDonalds, have already discontinued the use of *trans* fats in their deep fryers, and other chains are following suit.

The California ban raises an interesting question: Whose responsibility is it to legislate our health? The California Restaurant Association argued (unsuccessfully) that it should be the federal government, not the states—otherwise, restaurants with outlets in many states could face a wide array of different rules.

But if the government won't act, should the states be allowed to? Until California's law is challenged in federal courts, the answer is "Yes." What do YOU think? ■

Table 14.1 Major enzymes of digestion

Enzyme	Source	Where active	Substance digested	Breakdown products
Carbohydrate Digestion				
Salivary amylase	Salivary glands	Mouth	Polysaccharides	Disaccharides
Pancreatic amylase	Pancreas	Small intestine	Polysaccharides	Disaccharides
Intestinal enzymes	Small intestine	Small intestine	Disaccharides	Monosaccharides (e.g., glucose)
Protein Digestion				
Pepsin	Stomach	Stomach	Proteins	Peptides
Trypsin	Pancreas	Small intestine	Proteins	Peptides
Chymotrypsin	Pancreas	Small intestine	Proteins	Peptides
Carboxypeptidase	Pancreas	Small intestine	Peptides	Amino acids
Intestinal enzymes	Small intestine	Small intestine	Peptides	Amino acids
Lipid Digestion				
Lipase	Pancreas	Small intestine	Triglycerides	Free fatty acids, monoglyceride

The liver produces bile and performs many other functions

The liver is a large organ located in the upper right abdominal cavity. The liver performs many significant functions, some of which are associated with digestion.

The liver's primary digestive function is to facilitate the digestion and absorption of lipids by producing **bile.** Bile is a watery mixture containing electrolytes, cholesterol, bile salts derived from cholesterol, a phospholipid called lecithin, and pigments (primarily bilirubin) derived from the breakdown of hemoglobin. The bile salts emulsify lipids in the small intestine; that is, they break them into smaller and smaller droplets. Eventually the droplets are small enough to be digested by lipases (lipid-digesting enzymes) from the pancreas.

An important feature of the vascular anatomy of the GI tract is the **hepatic portal system** (**Figure 14.11**). In general terms, a *portal system* carries blood from one capillary bed to another: we have seen a portal system before, in the vascular connection between the hypothalamus and the anterior pituitary gland. In the digestive system the hepatic portal system carries nutrient-rich blood directly from the digestive organs to the liver (*hepatos* is the Greek word for "liver") via the hepatic portal vein. Therefore, the liver is ideally located to begin processing and storing nutrients for the body just as soon as digestion and absorption have begun. After passing through the liver the blood is returned to the general circulation.

The liver serves a number of other functions that maintain homeostasis.

- Storing fat-soluble vitamins (A, D, E, and K) and iron
- Storing glucose as glycogen after a meal, and converting glycogen to glucose between meals
- Manufacturing plasma proteins, such as albumin and fibrinogen, from amino acids
- Synthesizing and storing some lipids
- Inactivating many chemicals, including alcohol, hormones, drugs, and poisons
- Converting ammonia (NH_3), a toxic waste product of metabolism, into less toxic urea
- Destroying worn-out red blood cells

Because of its central role in so many functions, liver injury can be particularly dangerous. Overexposure to toxic chemicals, medications, or alcohol can damage the liver because it takes up these substances to "detoxify" them, killing some liver cells in the process. Long-term exposure, such as prolonged alcohol abuse, can destroy enough cells to permanently impair liver function, a condition known as *cirrhosis.*

✅ **Quick Check** Many people eat the livers of domestic animals such as cows and chickens. Considering the contents of the liver (see above), do you think eating liver is generally a nutritious food choice? Can you think of any situations in which it might be dangerous? ■

The gallbladder stores bile until needed

The bile produced by the liver flows through ducts to the **gallbladder.** The gallbladder concentrates bile by removing most of the water and stores it until after a meal, when it is secreted into the small intestine via the bile duct, which joins the pancreatic duct.

↩ **Recap** The pancreas secretes digestive enzymes and sodium bicarbonate. The sodium bicarbonate neutralizes stomach acid, making the digestive enzymes more effective. The liver produces bile, which is stored in the gallbladder until after a meal. The liver also produces plasma proteins; inactivates toxic chemicals; destroys old red blood cells; stores vitamins, iron, and certain products of metabolism; and performs other functions important for homeostasis. ■

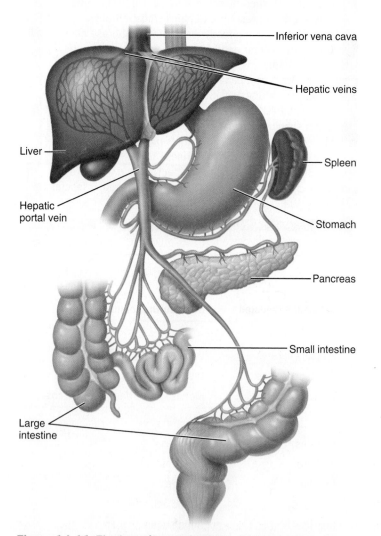

Liver

Hepatic portal vein

Large intestine

Inferior vena cava

Hepatic veins

Spleen

Stomach

Pancreas

Small intestine

Figure 14.11 The hepatic portal system. The hepatic portal system carries nutrient-rich venous blood from the stomach, intestines, pancreas, and spleen directly to the liver before the blood is returned to the general circulation. Only the first part of the large intestine is shown.

14.7 The large intestine absorbs nutrients and eliminates wastes

By the time the contents of the digestive tract reach the large intestine, most of the nutrients and water have been absorbed. The **large intestine** absorbs most of the remaining nutrients and water and stores the now nearly solid waste material until it can be eliminated.

Although the large intestine is larger in diameter than the small intestine, it is only about half as long. It begins at a pouch called the cecum, which receives the chyme from the small intestine (Figure 14.12). A small fingerlike pouch, the **appendix,** extends from the cecum. The appendix has no known digestive function, but we become acutely aware of its presence if it becomes inflamed or infected, a condition called appendicitis.

Most of the large intestine consists of four regions collectively called the **colon.** The ascending colon rises along the right side of the body (the left side of Figure 14.12, viewed from the front), the transverse colon crosses over to the left side, and the descending colon passes down the left side to the short final segment, the sigmoid colon. Feces are stored in the sigmoid colon until defecation, when they pass through the rectum to the anus.

In addition to indigestible material, the feces contain about 5% bacteria by weight. Many strains of bacteria thrive on the leftover material in the colon that we cannot digest. Some of these bacteria release by-products that are useful to us, such as vitamin K, which is important for blood clotting. They also produce less helpful substances such as intestinal gas, a by-product of their metabolism as they break down food.

Defecation is controlled by a neural reflex. Normally the anus is kept closed by contraction of a ring of smooth muscle called the *internal anal sphincter*. But when feces enter the rectum and the rectum is stretched, a neural reflex causes the internal anal sphincter to relax and the rectum to contract, expelling the feces.

We can prevent defecation by voluntarily contracting the *external anal sphincter*, a ring of skeletal muscle under our conscious control. "Potty training" of children involves learning to control the external anal sphincter. Until they learn how to control it consciously, children defecate whenever the defecation reflex occurs.

⤺ Recap The large intestine absorbs the last of the water, ions, and some nutrients and stores feces until defecation occurs. ∎

(HBP) Web Animation *The Digestive System* at www.humanbiology.com

14.8 How nutrients are absorbed

Once food has been digested, how does your body absorb the nutrients? The mechanism depends on the type of nutrient.

Proteins and carbohydrates are absorbed by active transport

In the small intestine, enzymes from the pancreas and enzymes secreted by the mucosal layer of the stomach and from the small intestine itself break down proteins into amino acids, completing the digestion of proteins that began in the stomach. The amino acids are then actively transported into the mucosal cells. Eventually they move by facilitated diffusion out of the mucosal cells and make their way into the capillaries (Figure 14.13).

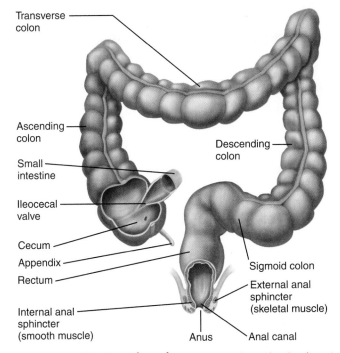

Figure 14.12 The large intestine. The large intestine begins at the cecum and ends at the anus.

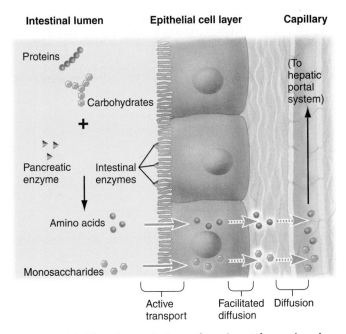

Figure 14.13 Digestion and absorption of proteins and carbohydrates in the small intestine. Digestion depends on enzymes from the pancreas and enzymes attached to the surface of the epithelial cells of the intestine. The amino acid and carbohydrate products of digestion cross the epithelial cell layer of the mucosa and enter a capillary for transport to the liver.

Carbohydrate digestion begins in the mouth, where salivary amylase breaks down polysaccharides into disaccharides (review Table 14.1). It is completed in the small intestine with the addition of pancreatic amylase and enzymes from the small intestine. Together these enzymes break down the remaining carbohydrates into monosaccharides (simple sugars such as glucose). Monosaccharides follow transport pathways that are similar to those for amino acids. However, they use different active transport proteins.

Lipids are broken down, then reassembled

Recall that bile salts emulsify lipids into small fat droplets, which are then digested by pancreatic and intestinal lipases. The products of lipid digestion are fatty acids and monoglycerides. Because they are nonpolar, the fatty acids and monoglycerides quickly dissolve in *micelles*—small droplets composed of bile salts and lecithin that have a polar outer surface and a nonpolar inner core. The function of micelles is to transport fatty acids and monoglycerides to the outer surface of the mucosal cells so that they can be absorbed into the cells (**Figure 14.14**).

Once inside the cells, the fatty acids and monoglycerides recombine into triglycerides. Clusters of triglycerides are then coated with proteins to form water-soluble droplets called *chylomicrons*. Chylomicrons are released from the cell by exocytosis. However, they are too large to enter the capillaries directly. Instead, they enter the more permeable lacteals and travel in the lymph vessels until the lymph is returned to the venous blood vessels near the heart.

Water is absorbed by osmosis

As nutrients are absorbed in the small intestine (or when you drink pure water), the concentration of the water in the intestinal lumen becomes higher than in the intestinal cells or in the blood. A higher concentration of water in the lumen represents a strong driving force for the diffusion of water through the epithelial layer of cells of the small intestine and into the blood. The capacity for water absorption by the small intestine is nearly limitless, which explains why you do not experience diarrhea every time you drink a lot of fluid.

Water absorption continues in the large intestine, but here the capacity is limited. Conditions that cause the small intestine to deliver too much food residue to the large intestine or that speed up the rate of movement through the large intestine can lead to diarrhea (watery feces). A common cause of diarrhea is a bacterial infection in the small intestine.

The opposite problem is constipation, in which the food residue, now called feces, remains in the large intestine and colon so long that too much water is absorbed. The feces become dry and hard, making defecation difficult. Constipation can result from stress, lack of exercise, or insufficient fiber (indigestible material) in the diet.

Vitamins and minerals follow a variety of paths

How vitamins are absorbed depends on whether they are fat soluble or water soluble. Fat-soluble vitamins dissolve in the micelles and are absorbed by diffusion across the lipid membrane of the mucosal cell layer, just like the components of lipids. Water-soluble vitamins are absorbed either by active transport or diffusion through channels or pores. Minerals (ions) such as sodium, potassium, calcium, phosphate, sulfate, and magnesium are electrically charged and hence not lipid soluble. These are either actively transported or are absorbed by diffusion via specific transport proteins, pores, or channels.

In addition to the nutrients we ingest in our food, the body digests and reabsorbs the components of the digestive secretions themselves. Nearly 9 liters of gastric juice, pancreatic juice, digestive enzymes, and bile are produced every day. The water and minerals in the digestive secretions are

Intestinal lumen   **Epithelial cell layer**   **Lacteal**

Figure 14.14 Digestion and absorption of fats in the small intestine. Bile salts emulsify large fat droplets into smaller droplets so that lipases can digest the fats into fatty acids and monoglycerides. The fatty acids and monoglycerides dissolve in micelles composed of bile salts and lecithin. Micelles near the cell's surface release their fatty acids and monoglycerides, which then diffuse across the cell membrane. In the cell the fatty acids and monoglycerides are resynthesized to triglycerides and coated with protein. The protein-coated triglycerides are called chylomicrons. Chylomicrons exit the cell by exocytosis and enter a lacteal for transport to the circulation.

reabsorbed by the normal mechanisms for these nutrients. Enzymes are digested to their component amino acids, and the amino acids are then reabsorbed. Bile salts are reabsorbed, returned to the liver, and used again.

> ↺ **Recap** Proteins and carbohydrates are transported into the cells lining the small intestine by active transport processes, then diffuse into capillaries. The products of lipid digestion are transported to the mucosa in micelles, diffuse into the cell, and recombine into lipids within the cell. Then they are coated with protein to become chylomicrons that enter the lymph. The intestines also absorb water, vitamins, minerals, and digestive secretions. ■

14.9 Endocrine and nervous systems regulate digestion

The digestive system is most active when food or chyme is present and fairly inactive when it is not. Regulation involves altering the motility and secretions of various organs so that each operates as efficiently as possible.

Most regulatory mechanisms operate to maintain a constant internal environment. Regulation of the digestive system, on the other hand, promotes rapid, efficient digestion and absorption of whatever is delivered to the system, regardless of homeostasis. The digestive process actually alters the internal environment temporarily because all those absorbed nutrients enter the blood in a relatively short time—a few hours.

Regulation depends on volume and content of food

The endocrine system and nervous system regulate digestion according to both the volume and content of food. Because most digestion and absorption occurs in the stomach and small intestine, most regulatory processes involve those organs.

When the stomach stretches to accommodate food, neural reflexes increase stomach peristalsis and secretion of gastric juice. Stretching and the presence of protein stimulate the stomach to release the hormone **gastrin,** which triggers the release of more gastric juice.

When chyme arrives at the small intestine, stretching of the duodenum increases segmentation to mix the chyme. The duodenum also secretes two hormones into the bloodstream:

secretin and cholecystokinin. Acid in chyme triggers release of **secretin,** which stimulates the pancreas to secrete water and bicarbonate to neutralize acid. Fat and protein stimulate release of **cholecystokinin (CCK),** which signals the pancreas to secrete more digestive enzymes. CCK and stretching of the duodenum also stimulate the gallbladder to contract and release bile (Table 14.2).

> ✓ **Quick Check** Do carbohydrates result in the release of any hormones from the duodenum? Where in the digestive system are carbohydrates first broken down? ■

Secretin, CCK, and stretching of the small intestine inhibit stomach motility and stomach secretions. In other words, if chyme flows too quickly from the stomach, the small intestine will slow stomach activity accordingly.

Gastrin and a neural reflex involving stretching of the stomach increase motility of the large intestine after eating. This is why people often feel the urge to defecate after their first meal of the day.

Nutrients are used or stored until needed

Once all those nutrients have been absorbed, the body must use them. Will they be consumed for immediate energy, stored until later, or combined with more nutrients to create other molecules? Regulation of organic metabolism involves interactions between virtually all organs in the body. Key organs are the pancreas with its two main endocrine secretions (insulin and glucagon) and the liver with its multiple roles in overall metabolism.

Figure 14.15 (next page) summarizes how the body utilizes organic molecules. Three concepts are evident from this figure.

1. Depending on which molecules are in short supply and which are in excess at the moment, there can be a great deal of interconversion of one to another. We first introduced this concept in connection with cellular metabolism in Chapter 3. Lipids, carbohydrates, and proteins can all be converted to storage forms of lipid or carbohydrate, then recycled according to the body's needs.
2. When we consume more energy-containing nutrients than we use, our bodies store the excess for future use. Over time, these mounting reserves can increase body weight.

Table 14.2 Nervous and endocrine systems' regulation of the digestive processes

Organ	Nervous system	Endocrine system
Stomach	Stretching increases peristalsis and secretion of gastric juice.	Stretching and protein trigger release of gastrin, which stimulates gastric glands.
Small intestine	Stretching increases segmentation. Stretching inhibits stomach motility and stomach secretions.	Acid stimulates release of secretin, which stimulates pancreas to secrete bicarbonate, neutralizing acid. Fat and protein stimulate release of cholecystokinin (CCK), which stimulates the pancreas to secrete enzymes and the gallbladder to release bile. CCK and secretin inhibit stomach motility and stomach secretions.
Large intestine	Stomach stretching increases motility.	Gastrin increases motility.

Figure 14.15 The pathways of organic metabolism.

3. When we consume fewer energy-containing nutrients than we use, the body draws on these storage forms of energy to make up the difference. If we do this on a regular basis, we lose weight.

In the rest of this chapter we discuss nutrition, energy balance, and healthy body weight.

Recap The endocrine and nervous systems regulate digestion based on content and volume of food. Regulatory mechanisms include neural reflexes involving organ stretching and release of the hormones gastrin, secretin, and cholecystokinin. Amino acids and the products of lipid and carbohydrate digestion can all be converted to storage forms of lipid or carbohydrate, to be retrieved and used when needed. ∎

14.10 Nutrition: You are what you eat

Nutrition is the interaction between an organism and its food. Because nearly all nutrients enter the body through the digestive system, it is fair to say that "you are what you eat." What nutrients do we actually need, and why?

MyPyramid plan offers a personalized approach

Figure 14.16 presents one answer to these questions. Developed by the Center for Nutrition Policy and Promotion at the U.S. Department of Agriculture (USDA), *MyPyramid* is a comprehensive, personalized approach that includes physical activity as well as healthy nutrition. MyPyramid divides foods into six groups and gives you recommendations on what to eat from each group.

MyPyramid provides many options to help you prepare a personalized nutrition plan. The system includes food plans appropriate to different Calorie levels of food intake, based on the latest *Dietary Guidelines for Americans* issued by the USDA and the Department of Health and Human Services. If you enter your age, gender, and activity level on the MyPyramid Web site, the system will match you with the best plan for your needs. It will suggest how many Calories you need per day and recommend specific amounts of each food group as well as discretionary Calories such as alcohol and added sugar. You can print worksheets to help you set nutrition goals and monitor your progress. The feature, MyPyramid Tracker, lets you compare your current eating and physical activity patterns to recommendations in the *Dietary Guidelines* and track them for up to a year to follow your progress.

Most nutritionists would agree with the general components of a healthy diet listed in MyPyramid. However, some of the specific recommendations of MyPyramid have been challenged by critics, who charge that the pyramid was built under intense pressure from lobbying groups such as the National Dairy Council and the National Cattleman's Beef Association. For example, MyPyramid recommends that to prevent osteoporosis, adults should drink three glasses of low-fat milk per day. That's over 300 calories just in milk, when not everyone needs to change their diet to prevent osteoporosis. MyPyramid also lumps meat (beef, poultry, and fish) and beans together in the protein

MyPyramid

STEPS TO A HEALTHIER YOU

MyPyramid.gov

USDA

GRAINS	VEGETABLES	FRUITS	OILS	MILK	MEAT & BEANS
Make half your grains whole	Vary your veggies	Focus on fruits	Know your fats	Get your calcium-rich foods	Go lean on protein
Eat at least 3 oz. of whole-grain breads, cereal, crackers, rice, or pasta every day.

Look for "whole" before the grain name on the list of ingredients. | Eat more dark-green veggies.

Eat more orange veggies.

Eat more dry beans and peas. | Eat a variety of fruits.

Choose fresh, frozen, canned, or dried fruit.

Go easy on fruit juices. | Make the most of your fat sources from fish, nuts, and vegetable oils.

Limit your solid fats like butter, stick margarine, shortening, and lard. | Go low-fat or fat-free.

If you don't or can't consume milk, choose lactose-free products or other calcium-rich sources. | Choose low-fat or lean meats and poultry.

Bake it, broil it, or grill it.

Vary your choices—with more fish, beans, peas, nuts, and seeds. |

Figure 14.16 MyPyramid. MyPyramid is a visual representation of the recommended dietary guidelines of the Center for Nutrition Policy and Promotion at the U.S. Department of Agriculture (USDA). The pyramid indicates the recommended consumption of foods from six different groups, taking into account both nutritional needs and concerns about the effects of diet on nutrition-related disease. The steps indicate the importance of regular physical activity. MyPyramid is designed to be highly interactive.

category, ignoring the mounting evidence that replacing red meat with a combination of fish, poultry, and beans is a healthier choice. Nevertheless, MyPyramid is still a good place to start if you want basic information about a healthy diet overall.

General recommendations for a healthy diet include the following:

- Eat a variety of foods.
- Maintain a healthy weight.
- Eat plenty of fruits, vegetables, and whole-grain products.
- Choose a diet low in cholesterol and saturated fat.
- Use sugar only in moderation.
- Consume salt and sodium in moderation (less than 2,300 milligrams, or about 1 teaspoon of salt per day).
- If you drink alcoholic beverages, do so in moderation. How much is "moderate"? The equivalent of up to one drink per day for women and up to two drinks per day for men.

We consider the three basic nutrient groups first—carbohydrates, lipids, and proteins—before discussing vitamins, minerals, and the importance of fiber.

Carbohydrates: A major energy source

Carbohydrates are one of the body's main sources of energy, and many nutritionists recommend that approximately 45–65% of our Calorie intake come from carbohydrates. (A Calorie is a measure of energy, discussed in more detail in section 14.11.)

Carbohydrates may be either simple or complex (Table 14.3). Simple carbohydrates (sugars) are found in many natural foods such as fruit and honey. Refined sugars, such as granulated sugar and corn syrup, have had most other nutrients removed, so they are less nutritious than their natural counterparts. Complex carbohydrates, such as starch or glycogen, consist of many sugar units linked together. Whole foods containing complex carbohydrates are better for us than refined sugars because they release sugars more slowly and also contribute fiber, vitamins, and minerals. In the body, stored starch and glycogen are broken down to glucose, one of the premier sources of energy.

It is difficult to tell exactly how much refined sugar we eat because so much is disguised on food labels as "corn sweeteners," "dextrose," or "fructose" (levulose). The USDA estimates the average North American consumes more than 142 pounds of sweeteners and refined sugar annually—almost 3 pounds each week.

Table 14.3 Good sources of natural carbohydrates

Simple carbohydrates	Complex carbohydrates
Fruit	**Starchy vegetables**
Citrus	Potatoes
Berries	Corn
Melons	Peas
	Beans
Unsweetened fruit juice	**Grains**
Honey	Whole-grain breads and cereals
	Whole-grain pasta
	Rice
	Oatmeal
	Wheat bran

Lipids: Essential cell components and energy sources

Lipids (including fats) are essential components of every living cell. Phospholipids and cholesterol make up most of the cell membrane. Cholesterol also forms the backbone of steroid hormones and is used to synthesize bile. Fat stores energy, cushions organs, insulates the body under the skin, and stores several vitamins. Most of the fats in food are triglycerides, which consist of three fatty acids attached to a glycerol molecule. Recall from Chapter 2 that fats are classified according to the ratio of hydrogen to carbon atoms in their fatty acids.

- *Saturated fats.* Saturated fats have two hydrogen atoms for every carbon atom in their fatty acid tails. They tend to be solids at room temperature. Saturated fats are found primarily in meat and dairy products and in a few plant sources such as coconut and palm kernel oil. They tend to raise blood levels of LDL cholesterol, the "bad" cholesterol that is associated with atherosclerosis and heart disease (Chapter 8).
- *Unsaturated fats.* Unsaturated fats are missing one or more pairs of hydrogen atoms in their fatty acid tails. Every missing pair of hydrogen atoms leads to a double bond between adjacent carbons and a kink, or bend, in the tail, making it more difficult for adjacent molecules to form bonds with each other. Consequently, unsaturated fats are liquids (oils) at room temperature. In general, unsaturated fats are considered healthier than saturated fats because they tend to lower LDL cholesterol levels. Olive, canola, safflower, and corn oils are all unsaturated fats derived from plants. Certain cold-water fish (salmon, trout, and sardines) are rich in omega-3 fatty acids, polyunsaturated fatty acids that have been linked to a reduced risk of heart disease.

Vegetable oils can be turned back into solids at room temperature by partial hydrogenation of the fatty acid tails. However, the process of partial hydrogenation reconfigures the positions of some of the remaining unpaired hydrogen atoms, producing *trans* fats. *Trans* fats were once very popular with fast-food restaurants as a deep-frying oil because they increase the shelf life and flavor stability of foods. *Trans* fats are used in commercial baked goods (cookies, crackers, and snacks) and are also found in vegetable shortening and margarine. However, like saturated fats, *trans* fats tend to raise LDL cholesterol and thus may increase the risk of cardiovascular disease. Food manufacturers now list the percentage of *trans* fats in the Nutrition Facts panel on product labels, and many fast-food restaurants no longer use *trans* fats in their deep-fryers.

The liver can synthesize cholesterol and most of the lipids the body needs, but it cannot produce every last one of them. The few fatty acids that our bodies cannot make, called *essential fatty acids,* must be consumed as food. Two examples are linoleic and linolenic acids, present in corn and olives, both of which contribute to proper cell membrane structure. A teaspoon of corn or olive oil, or its equivalent in food, is enough to satisfy our daily requirements.

Although lipids are essential to human health, most of us consume far more than we need. Nutritionists recommend that lipids account for no more than 20–35% of Calories consumed per day. In addition to watching the amount of fat we eat, though, it's important to pay attention to the type. Diets high in saturated fat, cholesterol, and *trans* fats place us at risk for cardiovascular disease and certain cancers. Less than 10% of Calories consumed daily should come from saturated fats, with less than 300 milligrams per day of cholesterol and as little *trans* fat as possible.

✓ **Quick Check** Why are *cold-water* fish particularly likely to have stores of *unsaturated* fats in their bodies? (Hint: Almost all of these fish are "cold-blooded," meaning their body temperature is the same as the temperature of the water in which they live.) ■

Complete proteins contain every amino acid

Like lipids, **proteins** are vital components of every cell. They make up the enzymes that direct metabolism, they serve as receptor and transport molecules, and they build our muscle fibers. A few are hormones.

Despite their variety, all proteins are composed of just 20 different amino acids. The body can make 12 of these amino acids. The 8 that the body cannot produce (isoleucine, leucine, lysine, methionine, phenylalanine, threonine, tryptophan, and valine) are called **essential amino acids** because they must be ingested in food. Two more that the body can make (histidine and arginine) are sometimes considered essential in children because their rapidly growing bodies cannot synthesize these amino acids fast enough.

A **complete protein** contains all 20 of the amino acids in proportions that meet our nutritional needs. Most animal proteins are complete, but nearly all plant proteins (with the exception of soybeans) lack one or more of the essential amino acids. Vegetarians must be careful to choose the right combinations of plant-based foods to obtain all amino acids. A combination of foods from any two of the three rightmost columns of Figure 14.17 will provide the necessary balance of all essential amino acids. Foods such as trail mix (roasted soybeans and nuts), hummus (chick peas and sesame seeds), a bean burrito in a corn tortilla, and red beans and rice do just that.

Approximately 15% of our Calories should come from protein. In many parts of the world protein deficiency is a serious health problem. Every enzyme has a unique amino acid sequence, so if even one amino acid is missing in the diet, the body may be unable to produce crucial enzymes at the proper time in development. Protein deficiencies during pregnancy and childhood can retard growth and alter physical and mental performance. Unfortunately, foods with complete proteins such as meat and milk are often expensive in poor societies, and many people remain uninformed about ways to achieve a balanced diet of plant proteins.

✓ **Quick Check** Do you think eggs and milk contain complete proteins? Why or why not? (Hint: What are eggs and milk used for in nature?) ■

Complete proteins	Incomplete proteins: low in lysine	Incomplete proteins: low in methionine	Incomplete proteins: low in tryptophan
Legumes: Soybean Tofu Soy milk	**Legumes:** Peanuts	**Legumes:** Beans (dried) Black-eyed peas Chickpeas Lentils Lima beans Mung beans Peanuts	**Legumes:** Beans (dried) Garbanzos Lima beans Mung beans Peanuts
Cereal grains: Wheat germ	**Cereal grains:** Barley Corn Oats Rice Rye Wheat		**Fresh vegetables:** Green peas Mushrooms Swiss chard
Dairy products: Milk Cheeses Yogurt Eggs	**Nuts, seeds:** Almonds Cashews Coconut English walnuts Hazelnuts Pecans Sunflower seeds Sesame seeds	**Fresh vegetables:** Asparagus Broccoli Green peas Mushrooms Potatoes	**Nuts, seeds:** Almonds English walnuts
Animal products: Meats Fish		**Nuts, seeds:** Hazelnuts	

Figure 14.17 Getting the essential amino acids. Most plants are low in one or more of the essential amino acids. A vegetarian should choose a combination of foods from any two of the three rightmost columns to ensure an adequate intake of all essential amino acids.

Vitamins are essential for normal function

In addition to carbohydrates, lipids, and proteins, humans need **vitamins,** a group of at least 13 chemicals that are essential for normal functioning. The body can produce only a few vitamins; our skin synthesizes vitamin D when exposed to sunlight, and bacteria living in the colon manufacture vitamins K, B_6, and biotin. We must obtain all others from our food and absorb them in the digestive tract. Table 14.4 (next page) lists the 13 vitamins along with their sources, functions, and signs of deficiency or excess.

Vitamins fall into two groups: fat soluble and water soluble. The distinction reflects how a vitamin is absorbed and stored and how steady a supply of it you need. Fat-soluble vitamins are absorbed more readily if there is fat in the diet. They tend to be absorbed along with the components of fat, and they are stored in fat tissue and released as needed. Body stores of some fat-soluble vitamins may last for years, so it is rarely necessary to take supplements.

Water-soluble vitamins are absorbed more readily than fat-soluble vitamins, but they are stored only briefly and rapidly excreted in urine. Thus we need to consume foods containing water-soluble vitamins on a regular basis.

✓ **Quick Check** Vegans—people who do not eat meat, dairy products, or eggs—are particularly vulnerable to a deficiency of a particular vitamin. Looking at Table 14.4, can you tell which one? Explain. ◾

Minerals: Elements essential for body processes

Minerals are the atoms of certain chemical elements that are also essential for body processes. They are the ions in blood plasma and cell cytoplasm (sodium, potassium, chloride, and many others). They represent most of the chemical structure of bone (calcium, phosphorus, and oxygen). They also contribute to the activity of nerves and muscles (sodium, potassium, and calcium), among many other functions. Twenty-one minerals are considered essential for animals. Nine of the 21 (arsenic, chromium, cobalt, manganese, molybdenum, nickel, selenium, silicon, and vanadium) are called trace minerals because they represent less than 0.01% of your body weight. Table 14.5 (on page 345) lists the 12 most important minerals.

How much do we need of each vitamin and mineral? The National Research Council publishes the current best estimate, the **Recommended Dietary Allowance (RDA).** Most healthy people can achieve the RDA without taking supplements if they eat a balanced diet of whole foods. That's a big "if" for some of us, considering our diets. Scientific studies suggest that supplements may also benefit certain groups, such as newborns, the elderly, or people taking medications that interfere with nutrient absorption. Calcium is advised for postmenopausal women to prevent the bone loss of osteoporosis.

Never take massive doses of any vitamin or mineral unless prescribed by your doctor. As the "excess" columns show, more is not always better.

Fiber benefits the colon

Fiber—found in many vegetables, fruit, and grains—is indigestible material. Even though our bodies cannot digest it, we need a certain amount in our diet. Fiber is beneficial because it makes feces bulky and helps them pass more efficiently through the colon.

A low-fiber diet can lead to chronic constipation, *hemorrhoids* (swollen veins in the lining of the anus, often caused by straining during defecation), and a disorder called *diverticulosis* (see section 14.12). A low-fiber diet has also been associated with a higher risk of colon cancer, perhaps because cancer-causing substances remain in the colon longer (see Chapter 18). Doctors recommend eating 20–35 grams of fiber every day—considerably more than the average 12–17 grams that most North Americans consume.

Table 14.4 Vitamins

Vitamin	Common sources	Primary functions	Signs of severe, prolonged deficiency	Signs of extreme excess
Fat-Soluble Vitamins				
A	Fortified dairy products, egg yolk, liver, yellow fruits, yellow and dark green leafy vegetables	Synthesis of visual pigments and the development of bones and teeth	Night blindness, dry scaly skin, increased susceptibility to infections	Damage to bone and liver, blurred vision
D	Fish oil, fortified milk, egg yolk; also produced by skin	Promotes absorption of calcium, required for healthy bones and teeth	Bone deformities in children, bone weakening in adults	Calcium deposits in tissues, leading to cardiovascular and kidney damage
E	Vegetable oils, nuts, whole grains	Thought to be an antioxidant that prevents cell membrane damage	Damage to red blood cells and nerves	Generally nontoxic
K	Green vegetables, also produced by bacteria in colon	Formation of certain proteins involved in blood clotting	Defective blood clotting, abnormal bleeding	Liver damage and anemia
Water-Soluble Vitamins				
B$_1$ (thiamine)	Whole grains, legumes, eggs, lean meats	Coenzyme in carbohydrate metabolism	Damage to heart and nerves, pain	None reported
B$_2$ (riboflavin)	Dairy products, eggs, meat, whole grains, green leafy vegetables	Coenzyme in carbohydrate metabolism	Skin lesions	Generally nontoxic
Niacin	Nuts, meats, grains, green leafy vegetables	Coenzyme in carbohydrate metabolism	Causes a disease called pellagra (damage to skin, digestive tract, and nervous system)	Skin flushing, potential liver damage
B$_6$	High-protein foods in general; also made by colonic bacteria	Coenzyme in amino acid metabolism	Muscle, skin, and nerve damage, anemia	Poor coordination, numbness in feet
Folic acid (B$_9$)	Green vegetables, nuts, grains, legumes, orange juice	Coenzyme in nucleic acid and amino acid metabolism	Anemia, digestive disturbances	Masks vitamin B$_{12}$ deficiency
B$_{12}$	Animal products	Coenzyme in nucleic acid metabolism	Anemia, nervous system damage	Thought to be nontoxic
Pantothenic acid	Widely distributed in foods, especially animal products	Coenzyme in glucose metabolism, fatty acid and steroid synthesis	Fatigue, tingling in hands and feet, intestinal disturbances	Generally nontoxic except occasional diarrhea
Biotin	Widely distributed in foods; also made by colonic bacteria	Coenzyme in amino acid metabolism, fat and glycogen formation	Scaly skin (dermatitis)	Thought to be nontoxic
C (ascorbic acid)	Fruits and vegetables, especially citrus fruits, cantaloupe, broccoli, and cabbage	Antioxidant; needed for collagen formation; important for bone and teeth formation; improves iron absorption	A disease called scurvy, poor wound healing, impaired immune responses	Digestive upsets

Recap A healthy diet includes plenty of fruits, vegetables, and whole grains and adequate fiber. Saturated fat, sugar, salt, and alcohol should be ingested only in modest amounts. Some essential fatty acids and eight essential amino acids must be ingested to meet the body's nutritional requirements because the body cannot synthesize them. Complete proteins contain all 20 amino acids. Most animal proteins are complete, but most plant proteins are incomplete. All minerals and nearly all vitamins must be obtained from food. ■

Table 14.5 **Major minerals in the human body**

Mineral	Common sources	Primary functions	Signs of severe, prolonged deficiency	Signs of extreme excess
Calcium (Ca)	Dairy products, dark green vegetables, legumes	Bone and teeth formation, nerve and muscle action, blood clotting	Decreased bone mass, stunted growth	Kidney damage, impaired absorption of other minerals
Chloride (Cl)	Table salt	Role in acid-base balance, acid formation in stomach, body water balance	Poor appetite and growth, muscle cramps	Contributes to high blood pressure in susceptible people
Copper (Cu)	Seafood, nuts, legumes	Synthesis of hemoglobin, melanin	Anemia, bone and blood vessel changes	Nausea, liver damage
Fluoride (F)	Fluoridated water, tea, seafood	Maintenance of teeth and perhaps bone	Tooth decay	Digestive upset, mottling of teeth
Iodine (I)	Iodized salt, marine fish and shellfish	Required for thyroid hormone	Enlarged thyroid (goiter), metabolic disorders	Goiter
Iron (Fe)	Green leafy vegetables, whole grains, legumes, meats, eggs	Required for hemoglobin, myoglobin, and certain enzymes	Iron-deficiency anemia, impaired immune function	*Acute:* shock and death. *Chronic:* liver and heart failure
Magnesium (Mg)	Whole grains, green leafy vegetables	Coenzyme in several enzymes	Nerve and muscle disturbances	Impaired nerve function
Phosphorus (P)	Meat, poultry, whole grains	Component of bones, teeth, phospholipids, ATP, nucleic acids	Loss of minerals from bone, muscle weakness	Depressed absorption of some minerals, abnormal bone deposition
Potassium (K)	Widespread in the diet, especially meats, grains	Muscle and nerve function; a major component of intracellular fluid	Muscle weakness	*Acute:* cardiac arrhythmias and death. *Chronic:* muscle weakness
Sodium (Na)	Table salt; widespread in the diet	Muscle and nerve function; major component of body water	Muscle cramps	High blood pressure in susceptible people
Sulfur (S)	Meat, dairy products	Component of many proteins	None reported	None reported
Zinc (Zn)	Whole grains, meats, seafood	Component of many enzymes	Impaired growth, reproductive failure, impaired immune function	*Acute:* nausea, vomiting, diarrhea. *Chronic:* anemia and impaired immune function

14.11 Weight control: Energy consumed versus energy spent

The body requires energy to fuel metabolic processes and other activities. When we digest nutrients, we can put the energy they contain to use.

Energy is measured in units called calories. Technically, a calorie is the amount of energy needed to raise the temperature of 1 gram of water by 1° Celsius. Because this is not much energy in biological terms, scientists use a kilocalorie (1,000 calories) to measure the nutrient content of food and the energy used to perform biological activities. A kilocalorie is commonly referred to as a **Calorie** with a capital "C." When people "count calories" they really mean the big Calories, and that is the unit we use in this chapter.

BMR: Determining how many Calories we need

If we want to maintain a stable body weight, the number of Calories we consume must equal the number we use. Your daily caloric energy needs are determined by your **basal metabolic rate (BMR),** the energy your body needs to perform essential activities such as breathing and maintaining organ function. BMR can be influenced by the following factors:

- *Gender and body composition.* BMR is higher in males. Muscle tissue consumes more energy than fat tissue; because men generally have more muscle than women, they have a higher BMR. Also, highly muscular men and women have a faster BMR than their less muscular counterparts.
- *Age.* BMR declines over time.
- *Health.* Some health conditions (such as fever, infections, and hyperthyroidism) increase BMR. Other conditions (such as hypothyroidism) lower it.

■ *Stress.* Norepinephrine and epinephrine raise BMR.

■ *Food intake.* Eating increases the metabolic rate, whereas fasting and extreme dieting decrease it. This is why crash diets often fail to lower body weight permanently. Although weight may fall temporarily, a slower BMR makes it difficult to keep the pounds from returning.

■ *Genetics.* Rates of metabolism vary between individuals, independent of the factors listed above. Genetics plays a strong role in determining your BMR, although precisely how has not been fully determined.

How can we get a rough estimate of our own BMR? The formula is roughly one Calorie per kg body wt/hr for males, 0.9 Calorie per kg body wt/hr for females. So if you're male, divide your weight in pounds by 2.2 (lb/kg) to get your weight in kg; then multiply by 24 (hrs/day) to get Calories per day. Females would divide their weight in pounds by 2.2, then multiply by 0.9 before multiplying by 24 hrs/day. Try it yourself and you'll see that a 200-lb man has an estimated BMR of 2,182 Calories per day. A 130-lb woman would have an estimated BMR of 1,276 Calories/day.

By definition, your BMR only indicates your *basal* metabolic energy needs. Most people's energy expenditure is 50—100% higher than their BMR unless they spend all day in bed. The additional energy is expended to do work (physical activity).

Energy balance and body weight

Healthy weight control involves balancing energy intake against energy expenditure. Notice that the three classes of nutrients have quite different Caloric contents. Fat contains about 9 Calories per gram, but carbohydrates and proteins contain only 4. Thus, although fat is a good energy source, a high-fat diet can easily contain more Calories than we need. This is one reason why controlling fat intake is so important.

When we eat more Calories than we use, the excess energy is stored in specialized cells as fat. The number of fat cells a person has is largely determined by the time he or she is an adult. This may be why some people find losing weight so difficult and others do not. Research suggests that overweight people have two to three times more fat cells than normal-weight individuals, so when they diet and shrink their fat reserves in each cell, their bodies respond as if they are starving. Dieting is difficult for chronically overweight persons because they are constantly fighting the body's own weight-control system, which responds as if their excess weight were normal.

Physical activity: An efficient way to use Calories

Although the BMR stays fairly constant, the amount of exercise we get can dramatically change the total number of Calories we burn each day. Table 14.6 compares the number of Calories spent per hour by a 100-, 150-, or 200-pound person performing various types of exercise. Note that heavier people do more work per hour for the same level of activity. To lose one pound of fat, we must use up about 3,500 Calories.

The best approach to weight loss is a gradual one. Nutritionists recommend reducing caloric intake by a small amount each day while gradually increasing physical activity.

Table 14.6 Approximate number of Calories burned per hour by various activities

Activity	100-lb Person	150-lb Person	200-lb Person
Bicycling, 6 mph	160	240	312
Bicycling, 12 mph	270	410	534
Jogging, 5.5 mph	440	660	962
Jogging, 10 mph	850	1,280	1,664
Jumping rope	500	750	1,000
Swimming, 25 yds/min	185	275	358
Swimming, 50 yds/min	325	500	650
Walking, 2 mph	160	240	312
Walking, 4.5 mph	295	440	572
Tennis (singles)	265	400	535

Source: American Heart Association

Of course, exercise has many benefits in addition to weight loss. It improves cardiovascular health, strengthens bones, tones muscles, and promotes a general sense of well-being. Your best strategy is a healthful diet combined with moderate exercise.

✅ **Quick Check** Suppose two women start a weight-loss program. They are the same weight, they decide to eat the same diet, and they both start an exercise program to do exactly 200 Calories' worth of physical activity per day. However, they choose different forms of exercise: One woman lifts weights to build stronger muscles, while the other person walks, which in her case does not result in muscle growth. Which person is likely to lose more body fat? Why? ■

Healthy weight improves overall health

Why do we worry about caloric intake and body weight, anyway? A primary reason is that there is a link between obesity and health status. According to the federal Economic Research Service, nutrition-related health conditions cost the United States $80 billion annually in medical expenses and lost productivity. Numerous studies reveal a direct correlation between obesity and the incidence of heart disease, diabetes (see the Current Issue at the beginning of the chapter), cancer, arthritis, and other health problems. The connection is strong enough that government groups and insurance companies regularly publish charts that use a person's height and weight to define **body mass index (BMI).** Read your own BMI from the chart in Figure 14.18.

According to the government, a BMI of between 18.5 and 25 is considered healthy, 25 to 30 is considered overweight, and 30 or higher represents obesity. These numbers should be considered a general guide only because they are based on population statistics and do not take into account such factors as bone structure, fitness, or gender. BMI may overestimate the amount of body fat in individuals who are very muscular or

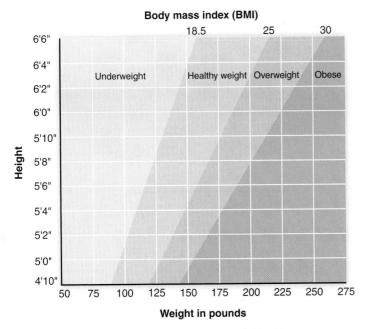

Body mass index (BMI)

Figure 14.18 **Find your BMI.** Find your height in the left column of the chart and read right to find your weight, shown at the bottom. The government's weight categories (Underweight, Healthy weight, Overweight, or Obese) are indicated by colors. Their corresponding BMI ranges are shown at the top of the chart. The weight categories are government definitions based on large population groups. They are not a good measure of an individual's health or fitness.

have a heavy bone structure. It may also underestimate the amount of body fat in people who have lost muscle mass. Thus, BMI charts do not provide definitive information about the appropriate weight status for any one individual. Within each range, the higher weights generally apply to men and the lower weights to women.

🔄 **Recap** Weight control involves balancing energy consumed in food against energy spent. Calculating your BMR helps you estimate how many Calories you need each day. Increasing physical activity is an efficient way to increase Calorie expenditure. The best strategy for losing weight combines a healthy diet with moderate regular exercise. ∎

14.12 Disorders of the digestive system

Although common, many digestive problems are not necessarily life threatening. One of the most common conditions worldwide is *food poisoning,* caused by food and beverages contaminated with bacteria or their toxic products. Diarrhea and vomiting often accompany food poisoning. *Food allergies* can also cause diarrhea, vomiting, and generalized allergic responses throughout the body. Common food allergens include shellfish, wheat, peanuts, and eggs.

Disorders of the GI tract

Lactose intolerance: Difficulty digesting milk Human infants are born with the enzyme lactase in their small intestines for digesting lactose, the primary sugar in milk. However, many adults gradually lose the enzyme, and with it their ability to digest lactose. The result is lactose intolerance. Symptoms of lactose intolerance include diarrhea, gas, bloating, and abdominal cramps after ingesting milk products. Diarrhea occurs because the undigested lactose causes fluid to be retained in the digestive tract. The gas, bloating, and abdominal cramps are due to bacterial fermentation of the lactose, which produces gases.

Lactose-intolerant people can eat cheese or yogurt because the lactose in these milk products has already been digested. Lactose-free milk is also available.

Peptic ulcers: sores in the stomach Peptic ulcers are painful erosions of the mucosal lining of the stomach or duodenum. Most peptic ulcers are associated with infection by one of the few bacteria that can live in the acidic environment of the stomach, called *Helicobacter pylori.* The bacterial infection leads to chronic inflammation, an increase in gastric acid secretion, and damage to the mucosal lining. Peptic ulcers can also be caused by excessive use of aspirin or non-steroidal anti-inflammatory drugs (NSAIDs). NSAIDs block the production of the mucus that protects the gastric mucosa from gastric acid.

After treatment to eliminate any *H. pylori* infection, most peptic ulcers heal on their own. Antacids may also be prescribed as needed. The use of aspirin and NSAIDs should be discontinued if possible.

☑ **Quick Check** Why do ulcers typically occur in the stomach or duodenum, but not in the rest of the small intestine or the large intestine? ∎

Celiac disease (gluten intolerance) When people with celiac disease (also known as sprue) eat gluten, a protein found primarily in wheat, rye, and barley, their immune systems respond by damaging or destroying the villi that line the small intestine. The result is malabsorption of nutrients of all kinds, not just gluten. Celiac disease is an inherited disorder with a prevalence of about 1 in 130 persons. Symptoms vary widely, from acute abdominal pain and vomiting to chronic fatigue, depression, and eventually malnutrition, depending on a person's age, how much gluten they eat, and how sensitive they are to it.

Celiac disease often goes undiagnosed at first because its symptoms are so varied and so similar to other diseases and conditions. A definitive diagnosis is made by a blood test for a certain autoantibody in the patient's blood. The only effective treatment is a gluten-free diet.

Diverticulosis: Weakness in the wall of the large intestine
Diverticulosis gets its name from its characteristic diverticula (small sacs produced when the mucosal lining of the large intestine protrudes through the other layers of the intestinal wall). Most people with diverticulosis do not experience discomfort, but sometimes the diverticula become infected

or inflamed, in which case it is called *diverticulitis*. Antibiotics usually resolve the problem.

Inadequate dietary fiber is thought to contribute to the development of diverticulosis. A low-fiber diet produces smaller feces, narrowing the colon and making its contractions more powerful. This increases the pressure on colon walls, forcing weak areas outward and forming diverticula. The disease is more prevalent with advancing age. Although it is rare in someone under 40, it may be present in up to two-thirds of all 85-year-olds.

Colon polyps: Noncancerous growths A polyp is a noncancerous growth that projects from a mucous membrane. Polyps can develop in many areas of the body, including the colon. The majority of colon polyps do not become cancerous, but because most colon cancers start as polyps, doctors recommend removing them. Polyps can be detected and removed in a *colonoscopy*, an outpatient procedure in which a flexible fiber-optic scope is inserted into the colon through the anus.

Disorders of the accessory organs

Hepatitis: Inflammation of the liver **Hepatitis** refers to inflammation of the liver, generally caused by viruses or toxic substances. Researchers have identified at least five viruses that cause hepatitis. The most common are hepatitis A, B, and C.

Hepatitis A is transmitted by contaminated food or water and causes a brief illness from which most people recover completely. A vaccine is available for hepatitis A. Although the vaccine is not considered necessary for most residents of industrialized nations, it may be a good idea for travelers to underdeveloped countries.

Hepatitis B travels in blood or body fluids, so it is usually passed via contaminated needles, blood transfusions, or sexual contact with infected individuals. Approximately 300,000 new cases are diagnosed each year. If not treated, hepatitis B can lead to liver failure. Symptoms include jaundice (the skin takes on a yellowish color due to accumulated bile pigments in the blood), nausea, fatigue, abdominal pain, and arthritis. A vaccine is available for hepatitis B. Federal law requires all health care workers to be vaccinated.

Hepatitis C is also transmitted in infected blood, usually through contaminated needles or blood transfusions. By 1992 U.S. blood banks began routine testing for hepatitis C, so the risk of contracting it from transfusions has fallen. However, researchers estimate that as many as 4 million Americans may already be infected, and many of them have no symptoms. Hepatitis C may remain dormant for years but still damage the liver. Severe cases can lead to chronic hepatitis, cirrhosis, or liver cancer. Doctors recommend testing for anyone who has had kidney dialysis (Chapter 15) or an organ transplant before 1992. Also at risk are those who have injected drugs or had sexual contact with a hepatitis C carrier.

Gallstones can obstruct bile flow The gallbladder normally concentrates bile about 10-fold by removing about 90% of the water. Excessive cholesterol in the bile may precipitate out of solution with calcium and bile salts, forming hard

Figure 14.19 Gallstones.

crystals called gallstones (**Figure 14.19**). Only about 20% of gallstones ever cause problems, but if the crystals grow large enough they can obstruct bile flow and cause intense pain, especially after a meal. Treatments include drugs to dissolve the crystals, ultrasound vibrations or laser treatments to break the stones apart, or surgery to remove the gallbladder.

Malnutrition: Too many or too few nutrients

Malnutrition refers to conditions in which human development and function are compromised by an unbalanced or insufficient diet. Malnutrition can be caused either by overnutrition or undernutrition. Overnutrition can lead to obesity. Worldwide, however, undernourishment is far more common. The United Nations estimates that 800 million people, or about 13% of the world's population, are undernourished.

Deficiencies of one or more nutrients can produce specific effects. For example, years of vitamin A deficiency lead to eye damage and night blindness. In children, undernutrition stunts growth and increases susceptibility to infection. Severe undernutrition, or starvation, is still the leading cause of malnutrition worldwide. Nearly 20 million people, most of them children, die every year of starvation or related diseases.

Obesity: A worldwide epidemic?

Estimates of the prevalence of obesity vary depending on how it's defined and how the data are collected. Nevertheless, the World Health Organization calls obesity a global epidemic. In the United States, obesity has increased from 12.6% of the population in 1990 to 34% in 2006. **Figure 14.20** compares the figures state by state. Obesity is rising even in some less industrialized countries: Five percent of women in Papua New Guinea are obese today, though obesity was virtually unseen in 1980.

Body weight is determined by both internal factors (such as genetic makeup) and environmental factors (such as activity level and availability of food). There's a lot of interest in genetic factors these days, such as "fat genes" that might make it harder to maintain a healthy weight. But genetics cannot account for an entire population's weight gain, because genes don't evolve that quickly. It took us millions of

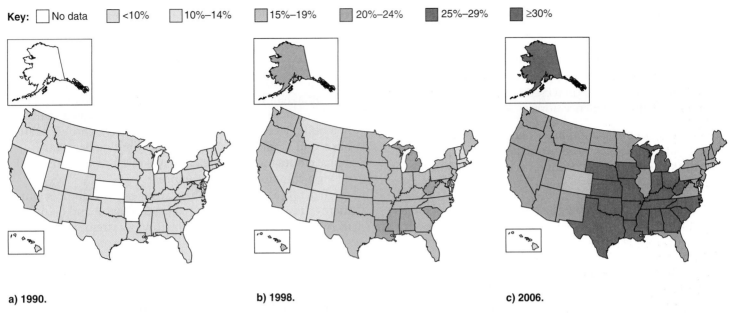

Key: ☐ No data ☐ <10% ☐ 10%–14% ☐ 15%–19% ☐ 20%–24% ☐ 25%–29% ☐ ≥30%

a) 1990.　　　b) 1998.　　　c) 2006.

Figure 14.20 Obesity trends among U.S. adults, 1990, 1998, and 2006. These maps show the percentage of the population classified as obese in each state. Obesity is defined as a body mass index greater than 30.

Source: Behavioral Risk Factor Surveillance System, Centers for Disease Control.

☑ *Which area(s) of the United States tend(s) to show the most obesity? Can you think of any local dietary or activity patterns that might explain this trend?*

years to become what we are, and our collective gene pool cannot change in just 20 years.

We must look to the environment to explain this global rise in obesity. Modernization has produced an environment that favors an imbalance between caloric intake and expenditure. Computers, television, cars, desk jobs—all combine to produce a more sedentary lifestyle. At the same time, food has become relatively cheap and available, which has apparently encouraged us to eat and drink more. In 1970 Americans consumed an average of 2,234 Calories per day; by 2003 this had risen to 2,757 Calories daily. Additional fats and oils account for 42% of the increase, or an extra 216 Calories per person each day.

↩ **Recap** Disorders of the GI tract and accessory organs include lactose intolerance, diverticulosis, colon polyps, gallstones, and hepatitis. Malnutrition can be caused by over- or undernutrition. Whereas starvation is the leading cause of malnutrition in underdeveloped countries, obesity is increasing in industrialized nations. ▪

14.13 Eating disorders: Anorexia nervosa and bulimia

Several conditions involving digestion and nutrition are really not digestive disorders at all but eating disorders involving the nervous system. Although eating disorders occur around the world, they are most common in women living in industrialized Western countries. Two examples are anorexia nervosa and bulimia.

Anorexia nervosa is a condition in which a person diets excessively or stops eating altogether, even to the point of starvation and death (**Figure 14.21**). Symptoms include the following:

▪ Refusal to maintain healthy body weight; people with anorexia often weigh less than 85% of their ideal weight.

Figure 14.21 Anorexia nervosa. Anorexia nervosa is a form of starvation based on an intense fear of gaining weight.

- Intense fear of gaining weight, even though underweight.
- Distorted perception or preoccupation with body weight or shape.
- In premenopausal women, the absence of at least three consecutive menstrual cycles. Severe undernutrition interferes with the hormonal cycles of menstruation. Male anorexics also experience hormonal abnormalities.

Bulimia is a binge-and-purge condition in which someone eats and deliberately vomits or takes other steps to minimize the Calories ingested. Symptoms of bulimia include the following:

- Recurrent episodes of binge eating. An episode of binge eating involves both (1) eating large amounts of food and (2) feeling a lack of control over eating.
- Taking recurrent inappropriate steps to prevent weight gain, such as self-induced vomiting; misusing laxatives, diuretics, enemas, or other medications; fasting; or exercising excessively.
- Binge eating and compensatory behaviors that occur, on average, at least twice a week for three months.

- Preoccupation with body shape and weight. However, unlike anorexics, some bulimics maintain a normal weight.

Both anorexia and bulimia play havoc with the body and mind. Anorexics become malnourished and suffer insomnia, hair loss, fatigue, and moodiness. Over time, they lose bone mass and develop osteoporosis. Repeated trauma to a bulimic's digestive system leads to ulcers, chronic heartburn, and rectal bleeding. Recurrent vomiting damages gums, erodes tooth enamel, and makes salivary glands swell, giving a chipmunklike appearance.

Doctors are not sure what causes eating disorders, but there appear to be deep-rooted psychological and cultural factors. Young women with a history of extreme dieting are 18 times more likely to develop eating disorders than those who do not have a history of dieting. Many people with eating disorders also suffer from depression and anxiety. Effective treatment of eating disorders generally requires a team of professionals who can address the patient's medical, psychiatric, dental, psychological, and nutritional needs.

Chapter Summary

The digestive system brings nutrients into the body *p. 327*

- The digestive system consists of the gastrointestinal (GI) tract and four accessory organs: the salivary glands, liver, gallbladder, and pancreas.
- The five basic processes of the digestive system are motility, secretion, digestion, absorption, and excretion.
- Two types of motility in the GI tract are peristalsis, which propels food forward, and segmentation, which mixes the contents.

The mouth processes food for swallowing *p. 329*

- The 32 teeth of adults include incisors, canines, premolars, and molars. Teeth cut, tear, grind, and crush food.
- Saliva moistens food, begins the digestion of starch, and helps to protect against bacteria.

The pharynx and esophagus deliver food to the stomach *p. 331*

- Swallowing is a reflex that is initiated by voluntary movements of the tongue. Once started, swallowing is involuntary.
- The sole function of the esophagus is to get food from the mouth to the stomach.
- The lower esophageal sphincter prevents reflux of stomach contents.

The stomach stores food, digests protein, and regulates delivery *p. 332*

- The stomach stores ingested food until it can be delivered to the small intestine.
- Glands in the mucosa of the stomach secrete gastric juice into the lumen, beginning the process of protein digestion.
- Peristalsis of the stomach mixes the food and pushes it toward the small intestine.

The small intestine digests food and absorbs nutrients and water *p. 333*

- Digestion occurs primarily in the first part of the small intestine, called the duodenum.
- The jejunum and ileum of the small intestine absorb most of the products of digestion.
- The small intestine has a very large surface area because of its many folds, villi, and microvilli.

Accessory organs aid digestion and absorption *p. 334*

- The pancreas secretes fluid containing bicarbonate and digestive enzymes into the small intestine.
- The liver produces bile and participates in homeostasis in a variety of ways.
- All of the venous blood from the GI tract is routed directly to the liver.
- The gallbladder stores bile from the liver and concentrates it by removal of most of the water.

The large intestine absorbs nutrients and eliminates wastes *p. 337*

- The large intestine absorbs most remaining nutrients and water, and vitamin K produced by bacteria. It also stores wastes.
- Defecation is generally controlled by a neural reflex, but it can be overridden by conscious control.

How nutrients are absorbed *p. 337*

- Amino acids and simple sugars are actively transported into the mucosal cells that line the small and large intestines.
- The products of fat digestion enter the mucosal cells by diffusion, then reform into triglycerides. Triglycerides are coated with protein and move into lymph vessels for transport to the blood.
- Water is absorbed by osmosis.
- Vitamins and minerals follow a variety of specific pathways.

Endocrine and nervous systems regulate digestion *p. 339*

- The volume and content of food play a large part in regulating digestive processes.
- Stretching of the stomach increases stomach peristalsis and the secretion of gastric juice.
- Stretching of the small intestine inhibits gastric motility, increases intestinal segmentation, and causes the secretion of two digestive enzymes, secretin and cholecystokinin.
- Acid in the small intestine triggers the secretion of pancreatic juice containing bicarbonate.

Nutrition: You are what you eat *p. 340*

- Good nutrition requires a variety of foods weighted toward fruits, vegetables, and whole-grain products.
- The human body needs certain nutritional components that it cannot make, including a few fatty acids, eight amino acids, 13 vitamins, and all essential minerals.

Weight control: Energy consumed versus energy spent *p. 345*

- The basal metabolic rate (BMR) represents the daily energy needs of the body for all essential activities except physical activity.
- Fat contains over twice as many Calories per gram as carbohydrate or protein.
- A person must use 3,500 Calories more than he or she ingests to lose a pound of body fat.

Disorders of the digestive system *p. 347*

- Lactose intolerance is caused by the lack of the enzyme (lactase) that normally digests lactose.
- Hepatitis, or inflammation of the liver, can be caused by several different viruses.
- Starvation is the most common form of malnutrition in the world. However, the incidence of obesity is rising in the United States and other industrialized nations.

Eating disorders: Anorexia nervosa and bulimia *p. 349*

- Two serious eating disorders are anorexia nervosa, in which the person diets excessively or stops eating entirely, and bulimia, a binge-and-purge condition in which the person eats and deliberately vomits or takes other steps to minimize caloric intake.
- The most effective treatments are multidisciplinary, addressing medical, psychiatric, dental, psychological, and nutritional needs.

Terms You Should Know

anorexia nervosa, *349*
basal metabolic rate (BMR), *345*
bile, *336*
bulimia, *350*
Calorie, *345*
cholecystokinin (CCK), *339*
chyme, *332*
colon, *337*
complete protein, *342*
esophagus, *331*
essential amino acids, *342*
gallbladder, *336*
gastrin, *339*
gastrointestinal (GI) tract, *327*
hepatic portal system, *336*
hepatitis, *348*
large intestine, *337*
mucosa, *328*
muscularis, *328*
nutrient, *326*
pancreas, *334*
pepsin, *332*
peptic ulcer, *333*
peristalsis, *328*
Recommended Dietary Allowance (RDA), *343*
salivary gland, *330*
secretin, *339*
segmentation, *329*
serosa, *328*
submucosa, *328*
villus, *334*
vitamin, *343*

Concept Review

Answers can be found at the Human Biology Place.
www.humanbiology.com

1. Describe the organs of the GI tract including the four accessory organs.
2. Name the five basic processes associated with carrying out the function of the GI tract.
3. Describe the role saliva plays in initiating the digestive process.
4. Compare and contrast *peristalsis* and *segmentation*.
5. Indicate how many teeth adult humans have.
6. Describe the events that occur in the stomach that contribute to the digestive process.
7. Indicate where in the GI tract most of the absorption of nutrients occurs.
8. Rank the major nutrient groups (carbohydrates, lipids, or proteins) according to Calories per gram.

9. Indicate what fraction of our daily caloric expenditure is accounted for by our basal metabolic rate (BMR).

10. Explain why it is harder to get an adequate supply of all amino acids from a vegetarian diet than from meat.

Test Yourself

Answers can be found in Appendix A.

1. Which of the GI tract tissue layers is most responsible for peristalsis and segmentation?
   a. serosa
   b. muscularis
   c. submucosa
   d. mucosa

2. All of the following are secreted into the lumen of the digestive tract during the process of digestion except:
   a. acid
   b. digestive enzymes
   c. hormones
   d. bile

3. All of the following are found in saliva except:
   a. mucin
   b. salivary amylase
   c. pepsin
   d. bicarbonate

4. Which of the following might be true of an individual unable to synthesize and secrete intrinsic factor?
   a. They would be unable to absorb adequate amounts of vitamin B_{12}.
   b. They would be unable to digest lactose found in dairy products.
   c. They would be unable to absorb amino acids in the small intestine.
   d. They would be unable to digest and absorb triglycerides.

5. Which of the following nutrients are digested within the small intestine?
   a. proteins
   b. lipids
   c. carbohydrates
   d. all of these choices

6. The region of the digestive tract most responsible for absorption of nutrients is:
   a. the stomach
   b. the small intestine
   c. the large intestine
   d. the liver

7. Which of the following lists structures in order from smallest to largest?
   a. small intestine, folds, microvilli, villi
   b. villi, microvilli, folds, small intestine
   c. microvilli, folds, villi, small intestine
   d. microvilli, villi, folds, small intestine

8. What do the enzymes pepsin, chymotrypsin, trypsin, and carboxypeptidase have in common?
   a. They are secreted and active within the stomach.
   b. They are pancreatic enzymes.
   c. They are secreted and active within the small intestine.
   d. They are enzymes that participate in protein digestion.

9. Once nutrients are absorbed into the circulatory system, the next major organ they pass through is the:
   a. heart
   b. liver
   c. kidneys
   d. brain

10. Which of the following terms does not belong with the others?
    a. monoglycerides
    b. micelle
    c. peptide
    d. bile

11. Each of the following is absorbed into capillaries except:
    a. triglycerides
    b. amino acids
    c. monosaccharides
    d. water

12. Which of the following foods would be the best source of complex carbohydrates?
    a. strawberries
    b. honey
    c. orange
    d. oatmeal

13. Which of the following statements about *trans* fats is correct?
    a. *Trans* fats are found in animal products such as meat and dairy.
    b. *Trans* fats will lower LDL cholesterol levels.
    c. *Trans* fats are produced by the hydrogenation of unsaturated fats in vegetable oils.
    d. Salmon, trout, and sardines are rich in healthy *trans* fats.

14. Each of the following is a source of complete protein except:
    a. milk
    b. soy products
    c. peanuts
    d. eggs

15. Which of the following is associated with celiac disease?
    a. inability to digest and absorb proteins in dairy products
    b. damage to the villi resulting from the consumption of wheat products
    c. erosion of the wall of the stomach and duodenum due to hyperacidity
    d. chronic constipation due to low intestinal motility

Apply What You Know

Answers can be found at the Human Biology Place.
www.humanbiology.com

1. A meal containing a lot of fat takes longer to be emptied out of the stomach than a low-fat meal. Why?

2. Procter & Gamble produces a calorie-free fat substitute called Olean (olestra) that is used to deep-fry foods such as potato chips. Olestra is a fat-soluble molecule that is neither digested nor absorbed. There is a concern that eating too much olestra might lead to certain vitamin deficiencies. What types of vitamins do you think would be affected by the use of olestra, and how might they be affected?

3. Other side effects of the consumption of too much olestra include gas, diarrhea, and a sudden need to defecate. What could be a cause of these?

4. Many prescription medications are supposed to be taken with a meal. Others are only to be taken on an empty stomach. Why would there be such differences?

5. Sometimes, the cure for gallstones is to surgically remove the gallbladder. What effect would removal of the gallbladder have on digestion?

6. People with lactose intolerance cannot digest the common dairy sugar lactose. Besides being careful not to eat dairy foods, is there anything else a lactose-intolerant person can do?

7. Cholera is a dangerous infectious disease that is common in places that do not have a steady supply of clean drinking water. Severe untreated cholera can be fatal within a matter of days. Cholera is caused by a bacterial toxin that causes severe diarrhea. What do you think might cause death from cholera, if the only symptom is diarrhea?

The Urinary System

SEM of kidney glomeruli (red) and the blood vessels that supply them.
The glomerular capsule and the nephron tubules have been removed.

How Should We Allocate Scarce Kidneys?

Over 60% of all organs transplanted every year are kidneys. For years now, a serious imbalance has been developing between the numbers of kidneys available for transplant and the number of patients needing them (see graph). Where do kidneys for transplant come from, and who decides who gets a kidney (or any other transplantable organ, for that matter) when one becomes available?

About 40% of all donated kidneys come from living donors. Patient and donor must have the same blood type and be tested for six key tissue antigens that determine the closeness of the

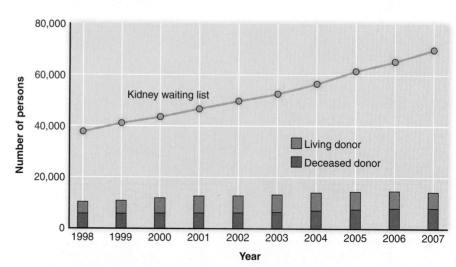

immunological match. Living donors are usually close relatives, but even among close relatives the chances of a perfect match are not all that good. Matches between unrelated donors are rare. *The responsibility for finding a living donor rests with the patient.* There is no national registry of living unrelated potential donors.

If a patient cannot find his/her own living donor or if he/she needs an organ that cannot be taken from a living donor, such as a heart or a pancreas, the only source is a deceased person (a cadaver). Organs may be harvested from a cadaver only when permission has been granted by the deceased or by the deceased's relatives. Persons who wish to donate their organs after death can make their wishes known by signing and carrying a uniform donor card available through most state motor vehicle departments or on the Web. Fewer than half of all adults have signed an organ donor card.

How Are Cadaver Kidneys Currently Allocated?

Cadaver kidneys are allocated to patients according to federal rules established by the Department of Health and Human Services (DHHS). Under the rules, patients who need a cadaveric kidney place themselves on the transplant list of one of 11 regional Organ Procurement Organizations (OPO), linked together by the United Network for Organ Sharing (UNOS), a national private nonprofit corporation. UNOS maintains the national waiting lists of potential transplant recipients. When a cadaver kidney becomes available in a particular region it is tested for blood type and the six antigens. Perfect matches—the same blood type and the same six antigens—are so rare that a patient anywhere in the country has first priority for a perfectly matched kidney.

For all partial matches, the kidney is first made available to all patients in the region of the OPO that procured the organ. Only if none of the patients in that region is considered suitable is the kidney made

Madolena (in yellow) chose to give one of her kidneys to a total stranger, Tracy (center). Here they meet for the first time.

available (by UNOS) to patients in other OPOs. A patient's income is not supposed to be a consideration, but people who are willing to pay for the transplant themselves can put themselves on the transplant lists of several OPOs simultaneously (insurance generally pays for a transplant only in the patient's primary OPO).

Critics charge that the current system is unfair because organs do not always go to the sickest patients, who need them the most. The current system may also discriminate against minorities and the poor because they live in regions that have lower rates of organ donation.

Developments to Watch

Currently it is against the law to sell a kidney—they can only be donated. Studies show that only about 50% of families of a deceased potential donor give their consent even when asked. To increase this number, the American Medical Association suggests that families who donate cadaveric kidneys from a loved one should be "compensated" in some way, but so far it hasn't happened. The DHHS is considering proposals to compensate living kidney donors as well.

At least 21 countries have adopted "presumed consent" laws. Under presumed consent, organs are presumed to be available for donation unless the donor (or family) explicitly states otherwise. Presumed

consent laws have dramatically increased donations in countries that have adopted them. In the United States, however, presumed consent laws have not gained widespread acceptance, because they can be viewed as a subtle form of government coercion or religious discrimination.

Some patients are finding their own unrelated living donors by turning to the Internet. The first commercial Web site to match living organ donor volunteers with potential patients was MatchingDonors.com. Patients who need an organ make an appeal on the site, and potential donors read through them and choose a patient to help, provided the immunological match is good. Donors receive no financial rewards for their donation, but perhaps money can't buy the ultimate gift. As one donor put it, "I wanted to do something nice for somebody." Another site, called LifeSharers, is for people who agree to make their organs available first to others who have also agreed to donate their organs when they die, in exchange for the chance of receiving an organ from the network should they themselves need one.

Organ gifting via the Internet is not such a good idea, according to some ethicists. They point out that when matches are made on sites such as MatchingDonors.com, organs are more likely to go to patients with the best story or the best photo, not the sickest patients.

At least five states (Kentucky, Louisiana, Oklahoma, South Carolina, and Wisconsin) have passed laws mandating that organs donated in their states must be offered first to in-state patients. Whether these laws will stand up under challenge is an open question. The battle for scarce organs continues.

Questions to consider

1 How do you think we should allocate cadaveric kidneys and other organs? What improvements, if any, would you suggest to the present system?

2 Would you be willing to donate your kidneys or other organs after death? Have you completed a donor card?

3 Should the government set up a national registry to match living donors with unrelated patients? Why or why not?

The facts...

- There are not enough donated cadaveric kidneys to meet the growing need for kidney transplants.
- The responsibility for finding a living donor rests with the patient.
- Under current federal allocation rules, where you live may affect your chances of receiving a cadaveric kidney.
- Some patients are using Web sites to locate unrelated living donors.

» **The primary organs of the urinary system are the kidneys.** The kidneys play a major role in the *maintenance of homeostasis*. They regulate the ionic composition and total solute concentration of the blood and contribute to the control of fluid volume and blood pressure.

» **The kidneys produce urine.** Urine is just a waste product. It consists of excess water and ions, metabolic wastes, and sometimes drugs, toxic chemicals, and vitamins.

» **The kidneys are endocrine organs.** They secrete a hormone called *erythropoietin* that regulates the production of new red blood cells. They also secrete an enzyme called *renin* that is part of the renin-angiotensin hormonal system.

» **The unit of function in a kidney is called a *nephron*.** Nephrons have the capability of producing either a dilute or a concentrated urine as needed to maintain proper fluid volume and composition.

» **When kidneys fail, the only option may be a kidney transplant** from another living person or from a cadaver.

Drop by drop—slowly but steadily, the kidneys go about the business of making urine. Minutes or even hours pass as the urinary bladder slowly fills. Finally, although you hadn't been thinking about it before, the thought comes to your mind that you need to find a bathroom, and before long you can think of little else.

We are sometimes aware of our heartbeat, we can deliberately control our muscle movements, and we are often preoccupied by our stomach, especially when it's empty. But do we ever give our kidneys a second thought? They are forgotten organs, tucked away at the back of the abdominal cavity. With urine as their primary product, the kidneys may not seem to serve a very glamorous function, but they play a vital role in maintaining the constancy of the internal environment. Urine itself serves no purpose; it is simply the end product of the regulation of the internal environment, the waste that is discarded.

As a recurrent theme in physiology, we have discussed homeostasis in terms of regulating body temperature, blood pressure, blood gases, and many other controlled variables in connection with the functions of other organ systems. This chapter describes the central role of the urinary system in maintaining the constancy of the composition and volume of the body fluids.

15.1 The urinary system contributes to homeostasis

Excretion refers to processes that remove wastes and excess materials from the body. Let's briefly review the excretory organs and systems involved in managing metabolic wastes and maintaining homeostasis of water and solutes (Figure 15.1). The digestive system provides the body with nutrients and water, and excretes food residues. The lungs take in oxygen and excrete carbon dioxide gas. The skin gets rid of heat and is also a site of water and salt loss, especially when we need to get rid of heat quickly. The liver destroys or inactivates numerous substances, some of which are excreted as bile in the feces. But the key organs in maintaining homeostasis of body fluids are the kidneys.

The **urinary system** consists of the kidneys, the ureters, the bladder, and the urethra. The two **kidneys** produce urine. The other components of the urinary system just transport and store urine until it is eliminated from the body.

Urine is primarily waste water and waste solutes. Among the solutes are various ions, drugs, vitamins, toxic chemicals,

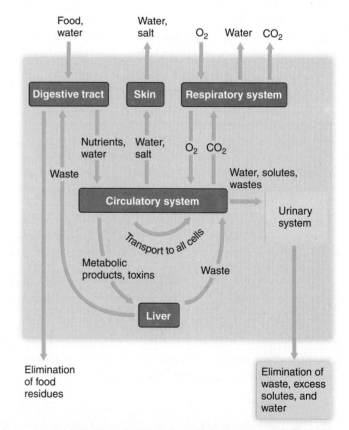

Figure 15.1 Organ systems involved in removing wastes and maintaining homeostasis of water and solutes. With the large tan box representing the body, this diagram maps the inflow and outflow of key compounds we consume. The kidneys of the urinary system are the organs primarily responsible for the maintenance of homeostasis of water and solutes and for the excretion of most waste products.

and virtually every small waste molecule produced anywhere in the body. But the kidneys do much more than just get rid of liquid garbage. They also carefully regulate how much water and salt are excreted in order to maintain homeostasis of fluid volume and blood pressure. Finally, they retain within the body all three classes of nutrients (carbohydrates, lipids, and proteins), leaving their management to other organs.

The kidneys regulate water levels

Water is the most abundant molecule in your body, representing about 60% of body weight. Normally you exchange about 2½ liters of water per day with the external environment (Table 15.1). We consume most of our water in food and beverages, but we also produce about 300 milliliters as part of cellular metabolism. Meanwhile, we lose water steadily through evaporation from lungs and skin (not counting sweat), and through defecation.

Homeostasis is maintained only when water intake equals water output, as it does in Table 15.1. But you know from experience that water intake can vary tremendously, and so can loss through sweat. When you take in a lot of water or when you become severely dehydrated on a hot day, it's up to the kidneys to excrete the excess water or to conserve as much water as possible. The kidneys have a tremendous capacity to adjust water excretion as necessary, from a minimum of about 1/2 liter per day to nearly 1 liter per hour.

The kidneys regulate nitrogenous wastes and other solutes

The primary solutes excreted by the kidneys are nitrogenous wastes, excess ions, and trace amounts of other substances, as described below.

The normal metabolism of proteins leaves us with excess nitrogenous wastes, which must be excreted by the kidneys. The metabolism of protein initially liberates ammonia (NH_3). Ammonia is toxic to cells, but the liver quickly detoxifies it by combining two ammonia molecules with a molecule of carbon dioxide to produce **urea** ($H_2N-CO-NH_2$) plus water. Most of the urea is excreted by the urinary system.

Dozens of different ions are ingested in excess with food or liberated from nutrients during metabolism. The most abundant ions in the body are sodium (Na^+) and chloride

(Cl^-), both of which are important in determining the volume of the extracellular fluids, including blood. The volume of blood, in turn, affects blood pressure. Other important ions include potassium (K^+), which maintains electrical charges across membranes; calcium (Ca^{2+}), important in nerve and muscle activity; and hydrogen (H^+), which maintains acid-base balance. The kidneys regulate the urinary excretion of these ions to maintain homeostasis of each one.

Trace amounts of many other substances are excreted in direct proportion to their daily rate of production. Among them are creatinine, a waste product produced during the metabolism of creatine phosphate (an energy source in muscle), and various waste products that give urine its characteristic yellow color.

Recap The urinary system maintains a constant internal environment by regulating water balance and body levels of nitrogenous wastes, ions, and other substances. It filters metabolic wastes from the blood and excretes them in urine. The major nitrogenous waste product is urea. ∎

MJ's Human Biology Blog

Using Urine to Diagnose Disease

Your body gets rid of over a thousand different small metabolic waste molecules, called metabolites, by excreting them in the urine. In certain diseases one or more of these metabolites may be elevated in urine. Scientists now believe that through "metabolomics"—the study of metabolites—it may be possible to diagnose certain diseases even before symptoms appear in the patient.

The concept is simple; once you know the normal range of excretion of metabolites in urine, a change in the excretion of one or more metabolites might signal the presence of a specific disease. For example, scientists at the University of Michigan measured 1,126 metabolites in urine and found that one of them in particular, called sarcosine, showed promise as a marker for metastatic prostate cancer.

Unlike blood, urine is easily collected in large quantities by non-invasive techniques. (Put more bluntly, you just pee into a cup!) As the field of metabolomics comes of age, perhaps some day it will be possible to screen for dozens of diseases at once. ∎

Reference: Sreekumar, Arun et al. Metabolic Profiles Delineate Potential Role for Sarcosine in Prostate Cancer Progression. *Nature* 457: 910–914, 2009.

Table 15.1 Sources of water gain and loss per day

Water gain (ml/day)		Water loss (ml/day)	
Drinking fluids	1,000	Urine	1,500
Water in food	1,200	Evaporative loss (lungs)	500
Metabolic production	300	Evaporative loss (skin)	350
		Feces	150
Total	2,500	Total	2,500

15.2 Organs of the urinary system

Table 15.2 lists the four organs of the urinary system and their principal functions. In the following discussion, function will be covered after anatomy.

Kidneys: The principal urinary organs

The kidneys are located on either side of the vertebral column, near the posterior body wall (**Figure 15.2a**). Each kidney is a dark reddish-brown organ about the size of your fist and shaped like a kidney bean. A renal artery and a renal vein connect each kidney to the aorta and inferior vena cava, respectively (*renal* comes from the Latin word *ren,* "kidney").

Table 15.2 **Functions of organs in urinary system**

Organ	Functions
Kidneys	Excrete metabolic wastes, especially urea
	Maintain water and salt homeostasis
	Help regulate acid-base balance
	Help regulate blood pressure (produce renin, an enzyme)
	Control red blood cell production (produce erythropoietin, a hormone)
	Activate vitamin D
Ureters	Transport urine to bladder
Urinary bladder	Stores urine until excretion
Urethra	Transports urine to outside the body

Seen in a longitudinal section (Figure 15.2b), each kidney consists of inner pyramid-shaped zones of dense tissue (called renal pyramids) that constitute the **medulla,** and an outer zone called the **cortex.** At the center of the kidney is a hollow space, the **renal pelvis,** where urine collects after it is formed.

A closer look at a section of the renal cortex and medulla reveals that it contains long, thin, tubular structures called nephrons (Figure 15.2c). Nephrons share a common final section called the collecting duct, through which urine is delivered to the renal pelvis.

Besides being the primary organs of the urinary system, the kidneys have many other functions related to maintaining homeostasis, which are discussed later in this chapter.

Ureters transport urine to the bladder

The renal pelvis of each kidney is continuous with a **ureter,** a muscular tube that transports urine to the bladder. Peristaltic waves of smooth muscle contraction, occurring every 10–15 seconds, move urine along the 10-inch length of the ureters to the bladder.

Urinary bladder stores urine

The **urinary bladder** stores urine. The bladder consists of three layers of smooth muscle lined on the inside by epithelial cells. Typically the bladder can hold about 600–1,000 ml of urine, though volumes that large may feel uncomfortable.

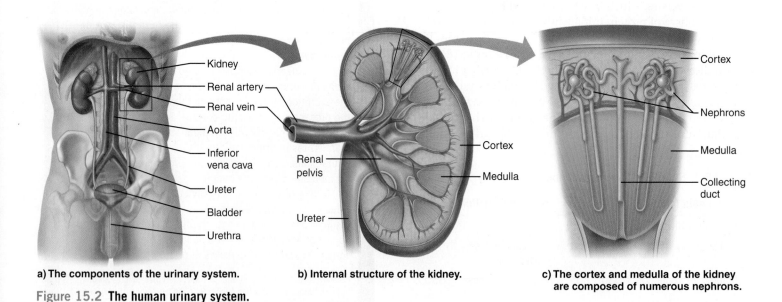

a) The components of the urinary system.

b) Internal structure of the kidney.

c) The cortex and medulla of the kidney are composed of numerous nephrons.

Figure 15.2 **The human urinary system.**

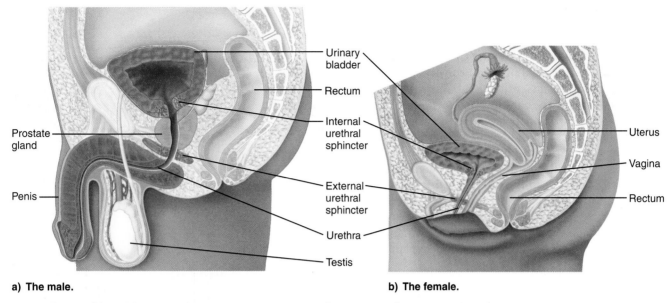

a) The male.

b) The female.

Figure 15.3 The positions of the bladder, the urethra, and associated organs in males and in females.

Women generally have a smaller bladder capacity than men because their bladders are slightly compressed by the nearby uterus (**Figure 15.3**).

✓ **Quick Check** A woman has a bladder infection, and goes to her doctor for treatment. The doctor asks if the symptoms include any back pain. Why did the doctor ask about back pain? ■

Urethra carries urine from the body

During urination, urine passes through the **urethra,** a single muscular tube that extends from the bladder to the body's external opening. Until then the bladder is prevented from emptying by the internal urethral sphincter, where the bladder joins the urethra, and the external urethral sphincter farther down the urethra. The urethra is about 8 inches long in men and about 1.5 inches long in women.

↩ **Recap** Organs of the urinary system include the kidneys, ureters, bladder, and urethra. The kidneys are the principal urinary organs, although they have several homeostatic functions as well. The ureters transport urine to the bladder, where it is stored until carried by the urethra to the body's external opening. ■

Answers to ✓ **questions can be found in Appendix A.**

MJ's Human Biology Blog

http://www.humanbiologyblog.blogspot.com/

Share Report Abuse Next Blog»   Create Blog Sign In

MJ's Human Biology Blog

Buying/Selling Kidneys

A Brooklyn businessman named Levy-Izhak Rosenbaum was arrested in New Jersey in 2009 for allegedly trying to broker the purchase of a kidney for $160,000. According to the criminal complaint filed against him, Mr. Rosenbaum told an undercover agent that he could arrange for a live kidney donor from Israel, and that they would then fabricate a fictitious "relationship" between the donor and the recipient so that the hospital in the United States would not become suspicious.

The Rosenbaum case is just one example of the shadowy black market in human organs worldwide. The World Health Organization estimates that about 10% of the more than 60,000 kidneys transplanted each year worldwide come from living donors who have sold their kidneys strictly for money. The temptation is hard to resist, especially for donors from poor countries where the choice may come down to selling a kidney or selling a child. The practice is not even illegal in some countries (Pakistan is an example), and as a result those countries are rapidly developing thriving "transplant tourism" enterprises.

What, if anything, could be done about the shortage of organs for transplantation? ■

15.3 Nephrons produce urine

Each kidney contains approximately a million small functional units called **nephrons** (from the Greek word for kidney, *nephros*). An individual nephron consists of a thin, hollow tube of epithelial cells, called a tubule, plus the blood vessels that supply the tubule. The function of the nephron is to produce urine. However, nephrons don't just pick the waste molecules out of blood and excrete them. Instead, they remove about 180 liters of fluid from the blood every day (about 2½ times your body weight) and then return almost all of it to the blood, leaving just a small amount of fluid behind in the tubule to be excreted as urine. If you took a similar approach to cleaning your room, you would take everything out of the room and then put it all back except for the dust and waste paper you wanted to discard.

The tubule filters fluid and reabsorbs substances

Figure 15.4 shows the tubular portion of a nephron. The nephron begins with a cup of tissue that looks like a deflated ball with one side pushed in, called the **glomerular capsule** (sometimes called Bowman's capsule). The glomerular capsule surrounds and encloses a network of capillaries called the **glomerulus,** which is part of the blood supply of the nephron.

The process of urine formation begins when plasma fluid is filtered out of the capillaries of the glomerulus and into the space between the two layers of the glomerular capsule. From the glomerular capsule the tubule continues as a long, thin tube with four distinct regions: proximal tubule, loop of Henle, distal tubule, and collecting duct. The **proximal tubule** (*proximal* means nearest to) starts at the glomerular capsule and ends at the renal medulla. The hairpin-shaped **loop of Henle** (loop of the nephron) extends into the medulla as the *descending limb* and then loops back up to the vicinity of the glomerular capsule as the *ascending limb*. After it passes the glomerular capsule, the tubule is called the **distal tubule** (*distal* means more distant from). Finally, the distal tubules of up to a thousand nephrons join to become a **collecting duct.** The collecting duct extends from the cortex to the renal pelvis, where the urine is finally deposited.

As we will see, the tubule and collecting duct are not just passive conduits. As urine flows through the tubule and collecting duct, their cells modify the urine content substantially by reabsorbing water and ions. Reabsorption prevents the loss of valuable substances from the body and adjusts the composition of blood and body fluids.

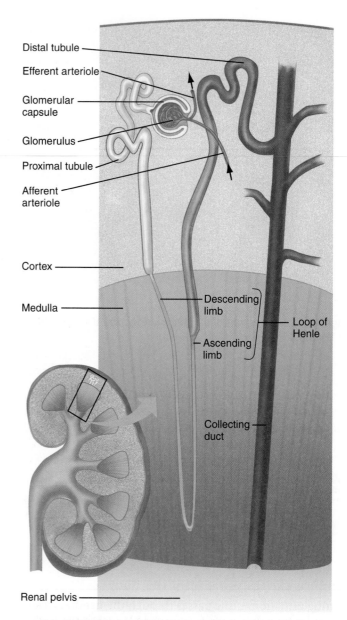

Figure 15.4 Tubular regions of a nephron. A portion of the glomerular capsule has been cut away to expose the blood vessels (the glomerulus) enclosed by the capsule. The proximal tubule begins at the glomerular capsule and ends where the tubule narrows and enters the medulla. The loop of Henle consists of descending and ascending limbs; it ends where the tubule passes the glomerular capsule. The distal tubule connects to a collecting duct, which is shared by many nephrons.

Special blood vessels supply the tubule

Figure 15.5 provides a closer look at the blood vessels serving the nephrons. The renal artery supplying a kidney branches many times to serve those million nephrons.

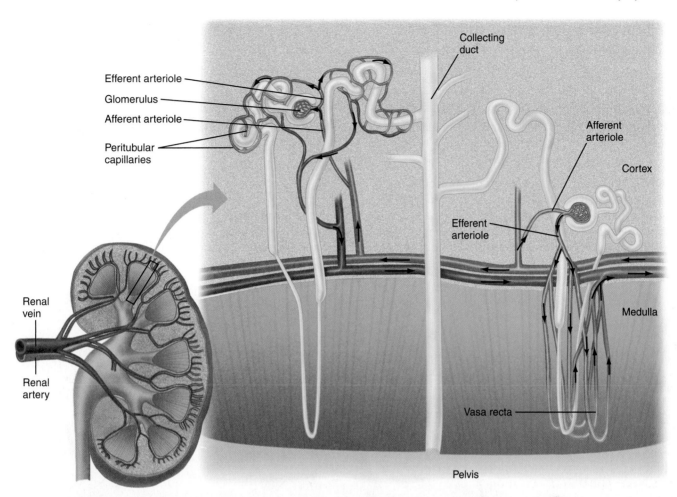

Efferent arteriole

Glomerulus

Afferent arteriole

Peritubular capillaries

Collecting duct

Afferent arteriole

Cortex

Efferent arteriole

Renal vein

Renal artery

Medulla

Vasa recta

Pelvis

a) Main blood vessels in the kidney. The renal artery and renal vein branch many times to deliver blood to each glomerulus.

b) Post-glomerular blood vessels. The efferent arteriole of most nephrons, such as the one shown here on the left, divides to become the peritubular capillaries, which supply proximal and distal tubules in the cortex. In a few nephrons, such as the one on the right, the efferent arteriole descends into the medulla to become the vasa recta, which supply loops of Henle.

Figure 15.5 Relationships between the tubular and vascular components of nephrons.

Ultimately, every nephron is supplied by a single arteriole, called the **afferent arteriole** (*afferent* means "directed toward"). The afferent arteriole enters a glomerular capsule and then divides many times to become the network of capillaries that constitutes the glomerulus. Here plasma fluid and solutes are filtered from the blood into the capsular space.

The glomerular capillaries rejoin to become the **efferent** ("directed away") **arteriole,** which carries filtered blood from the glomerulus. The efferent arteriole divides again into another capillary network that surrounds the proximal and distal tubules in the cortex, called the **peritubular capillaries** (*peri-* means "around"). The peritubular capillaries remove water, ions, and nutrients, which are reabsorbed by the proximal and distal tubules. The efferent arterioles of a few nephrons descend into the medulla and

divide into long, thin capillaries called the **vasa recta** (Latin for "straight vessels") that supply the loop of Henle and collecting duct. Eventually the filtered blood flows into progressively larger veins that become the single renal vein leading to the inferior vena cava.

Recap A nephron is the functional unit of a kidney. A nephron tubule consists of a glomerular capsule, where fluid is filtered, and four regions in which the filtrate is modified before it becomes urine: proximal tubule, loop of Henle, distal tubule, and collecting duct. Blood flows to the glomerulus via the renal artery and afferent arterioles. Peritubular capillaries carry the blood to the proximal and distal tubules, and vasa recta supply the loops of Henle and collecting ducts. ∎

15.4 Formation of urine: Filtration, reabsorption, and secretion

As we have noted, the urinary system regulates the excretion of water and ions to achieve homeostasis of fluid volume and composition. It excretes certain wastes while retaining precious nutrients. How do the kidneys select what to retain and what (and how much) to excrete in urine?

The formation of urine involves three processes:

1. *Glomerular filtration.* The movement of a protein-free solution of fluid and solutes from the glomerulus into the space within the glomerular capsule.
2. *Tubular reabsorption.* The return of most of the fluid and solutes back into the peritubular capillaries or vasa recta.

3. *Tubular secretion.* The addition of certain solutes from the peritubular capillaries or vasa recta into the tubule.

The fluid and solutes that remain in the tubule constitute the urine, which is eventually excreted. **Figure 15.6** summarizes the relationship between these three processes. As the figure indicates, the amount of any substance excreted in urine is equal to the amount filtered, minus the amount reabsorbed back into the blood, plus the amount secreted into the tubule.

Glomerular filtration filters fluid from capillaries

Urine formation starts with **glomerular filtration,** the process of filtering a large quantity of protein-free plasma fluid from the glomerular capillaries into the glomerular

Figure 15.6 The three processes that contribute to the formation of urine.

 Name two substances that are reabsorbed and one that is not. Explain why it makes sense to absorb some molecules and not others.

space. The filtration barrier consists of two cell types: modified tubular epithelial cells called **podocytes** that cover and surround the outside surface of the capillaries, and the capillary cells themselves (Figure 15.7a). Fluid first passes through pores in the capillary cells (Figure 15.7b), and then through tiny slits between cytoplasmic extensions of the podocytes in order to enter the glomerular space.

The podocytes and the glomerular capillary cells are highly specialized for filtration of a large volume of fluid. Together, they are nearly 100 times more permeable to water and small solutes than the capillaries in most other vascular beds. However, they are also highly selective; they are *less* permeable than other capillaries to large proteins and whole cells. As a result, the filtered fluid, called the glomerular filtrate, contains water and all of the small solutes in the same concentrations found in blood plasma, but it contains no large proteins or blood cells.

Together, the two kidneys produce about half a cup of glomerular filtrate per minute, a whopping 180 liters per day. Contrast that amount with the daily urine excretion of about 1.5 liters, and you can see how efficient the kidneys are at reabsorbing substances that are filtered.

Glomerular filtration is driven by high blood pressure in the glomerular capillaries. As described in Chapter 8, the filtration of fluid out of capillaries is driven by the capillary blood pressure. In most capillaries the inward and outward movements of fluid are nearly equal because the blood pressure is opposed by similar osmotic pressure generated by the proteins in blood. The difference is that the blood pressure in the glomerular capillaries is about twice that in any other capillary. This high pressure forces large amounts of glomerular filtrate into the glomerular space. The kidneys themselves do not have to expend energy to produce the filtrate.

The rate of filtration is regulated in two ways:

- Under resting conditions, pressure-sensitive cells in the arterioles and flow-sensitive cells in the tubule walls can release chemicals to adjust the diameter of the afferent arterioles. These feedback mechanisms maintain a relatively constant rate of glomerular filtration, allowing the kidneys to carry out their regulatory functions.
- During times of stress (such as after an injury or while running a marathon), blood flow to the kidneys falls substantially as blood is redistributed to more critical organs. The sympathetic division of the autonomic nervous system constricts afferent and efferent arterioles, reducing blood flow and the rates of glomerular filtration and urine formation. The kidneys are unharmed because they do not

a) **The outer surface of several glomerular capillaries.**

Glomerular capsule

Glomerular space

Glomerulus

Blood flow

Movement of glomerular filtrate

Afferent arteriole

Efferent arteriole

Podocyte

Filtrate

Proximal tubule

Capillary wall

b) **A highly magnified view of the inner surface of a single glomerular capillary, revealing its porous sievelike structure.**

Figure 15.7 **Glomerular filtration.**

need a high blood flow to survive, and the body can cut back on urine production temporarily without ill effects.

If the delicate sievelike structure of the glomerular capillaries is disrupted, proteins may cross the filtration barrier into the tubule fluid. This is *proteinuria*, the appearance of protein in urine. Persistent proteinuria is a sign of glomerular damage, perhaps by toxins or sustained high blood pressure. However, temporary proteinuria can occur after strenuous exercise, even in healthy people. The reason for this is unknown, but it may be that reduced renal blood flow during exercise allows blood to remain in contact with the glomerular barrier longer, letting more plasma proteins leak through. Exercise-induced proteinuria is not dangerous and goes away within a day or so.

Tubular reabsorption returns filtered water and solutes to blood

Tubular reabsorption, the second step in urine formation, returns filtered water and solutes from the tubule into the blood of the peritubular capillaries or vasa recta. As filtrate flows through the tubule, major nutrients are almost completely reabsorbed—all of the filtered glucose, amino acids, and bicarbonate, and more than 99% of the water and sodium. About 50% of the urea is also reabsorbed. The final product—urine—contains just enough water and sodium to balance the daily net gain from all other sources, the equivalent of all the urea produced in one day's metabolism, and trace amounts of other ions and wastes. Some waste products, such as creatinine, are not reabsorbed at all once they are filtered (Table 15.3).

Most tubular reabsorption occurs in the proximal tubule, leaving the fine-tuning and regulation of reabsorption to the more distal regions. A good example is water. The proximal tubule always reabsorbs about 65–70% of the water, and the

loop of Henle reabsorbs another 25%, both completely unregulated. The distal tubule and collecting duct together reabsorb less than 10%, yet it is here that water excretion is effectively regulated.

To be reabsorbed, substances must cross the layer of epithelial cells of the tubules to reach blood capillaries. Like the epithelial cells of the digestive tract, the cell membrane on the lumen side of proximal tubular cells has a "brush border" of microvilli to increase the surface area for reabsorption (Figure 15.8). The process of reabsorption begins in the proximal tubule with the active transport of sodium from inside the cell toward the peritubular capillaries that surround the tubule. This is an active process that requires energy in the form of ATP. Other steps in the reabsorptive process, shown in Figure 15.9, follow from this one event:

1. The active transport of sodium out of the proximal tubular cell decreases the intracellular concentration of sodium, setting up a driving force for sodium to enter the cell by facilitated transport.
2. Because sodium is a positively charged ion, its movement drives the diffusion of chloride across the tubular cell (negative charges must follow positive ones to maintain electrical neutrality).
3. The transport of solutes (Na^+ and Cl^-) across the tubular cell creates a concentration gradient for the diffusion of water, so water is also reabsorbed. Water reabsorption occurs through special water channels in membrane proteins called **aquaporins.**
4. The facilitated transport of sodium inward across the luminal cell membrane provides the energy for the secondary active transport of glucose and amino acids into the cell. Once inside the cell, the glucose and amino acids diffuse into the interstitial fluid and eventually into the peritubular capillaries.

Table 15.3 Amounts of various substances filtered and excreted or reabsorbed

Substance	Amount filtered per day	Amount excreted in urine per day	Proportion reabsorbed
Water	180 L	1–2 L	99%
Sodium (Na^+)	620 g	4 g	99%
Chloride (Cl^-)	720 g	6 g	99%
Potassium (K^+)	30 g	2 g	93%
Bicarbonate (HCO_3^-)	275 g	0	100%
Glucose	180 g	0	100%
Urea*	52 g*	26 g	50%
Creatinine*	1.6 g*	1.6 g	0

*The plasma concentrations of these substances and the amounts filtered per day vary with diet, age, and level of physical activity.

A section of a proximal tubule, showing the microvilli that form the brush border of the inner (luminal) surface.

Figure 15.8 **The proximal tubule.**

Figure 15.9 Basic mechanisms of reabsorption in the proximal tubule. The key to the entire process is the active transport of sodium (Na^+) across the tubular cell membrane on the capillary side of the cell. This step requires energy in the form of ATP. The active transport of sodium keeps the intracellular sodium concentration low, which permits sodium to enter into the cell from the luminal side by facilitated transport. Negatively charged chloride ions follow in order to maintain electrical neutrality, and water follows because the reabsorption of sodium and chloride produces an osmotic driving force for the diffusion of water. Finally, the facilitated transport of sodium across the luminal membrane is used as an energy source for the reabsorption of glucose and amino acids.

Note that the reabsorption of water, salt, glucose, and amino acids in the proximal tubule all depend on just one metabolic energy-using process: the primary active transport of sodium out of the tubular epithelial cell. Reabsorption in the more distal regions of the tubule, although less in amount, also relies on the active transport of sodium as the energy-utilizing step.

Tubular secretion removes other substances from blood

A few substances undergo **tubular secretion**—that is, they move from the capillaries (either peritubular capillaries or vasa recta) into the tubule to be excreted. Tubular secretion may occur by either active transport or passive diffusion, depending on the substance being secreted.

Tubular secretion is critical for removing or regulating levels of certain chemicals, including toxic and foreign substances and drugs. The list of substances secreted into the tubule includes penicillin, cocaine, marijuana, many food preservatives, and some pesticides.

Some substances that occur naturally in the body are secreted just to maintain normal homeostatic levels. The proximal tubule secretes hydrogen (H^+) and ammonium (NH_4^-) to regulate the body's acid-base balance, and the distal tubule secretes potassium (K^+) to maintain healthy levels of that mineral.

Recap In the glomerulus, glomerular filtration separates some of the plasma fluid and small solutes from larger proteins and blood cells. Tubular reabsorption returns nearly all the filtered water and sodium and all the major nutrients back to the peritubular capillaries or vasa recta. Tubular secretion removes toxic, foreign, and excess substances from the peritubular capillaries or vasa recta. Tubular secretion is essential to the regulation of acid-base balance, potassium balance, and the excretion of certain wastes. ▪

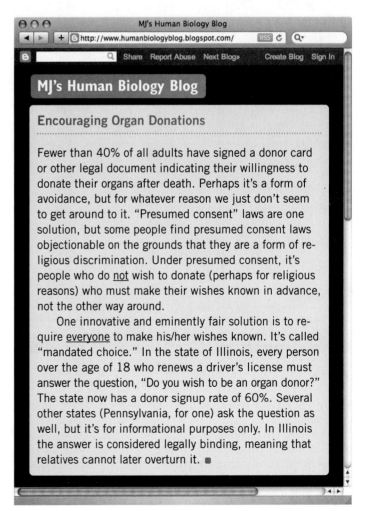

MJ's Human Biology Blog

Encouraging Organ Donations

Fewer than 40% of all adults have signed a donor card or other legal document indicating their willingness to donate their organs after death. Perhaps it's a form of avoidance, but for whatever reason we just don't seem to get around to it. "Presumed consent" laws are one solution, but some people find presumed consent laws objectionable on the grounds that they are a form of religious discrimination. Under presumed consent, it's people who do <u>not</u> wish to donate (perhaps for religious reasons) who must make their wishes known in advance, not the other way around.

One innovative and eminently fair solution is to require <u>everyone</u> to make his/her wishes known. It's called "mandated choice." In the state of Illinois, every person over the age of 18 who renews a driver's license must answer the question, "Do you wish to be an organ donor?" The state now has a donor signup rate of 60%. Several other states (Pennsylvania, for one) ask the question as well, but it's for informational purposes only. In Illinois the answer is considered legally binding, meaning that relatives cannot later overturn it. ▪

We urinate less when we are dehydrated because the kidneys are reabsorbing more of the glomerular filtrate water than usual. The process of reabsorbing water is regulated by **antidiuretic hormone (ADH)** from the posterior pituitary gland (see Chapter 13). ADH increases the permeability of the collecting duct to water. The formation of concentrated urine follows a process similar to that just described for the formation of dilute urine until the urine reaches the collecting duct. In the presence of ADH, more water is reabsorbed because it diffuses out of the collecting duct toward the higher solute concentration in the medulla. Consequently only a small volume of highly concentrated urine is excreted (**Figure 15.11**).

HBP **Web Animation** *The Urinary System* at www.humanbiology.com

The production of concentrated urine depends on a **countercurrent exchange mechanism.** Because of the

hairpin arrangement of the loops of Henle, fluid flows in opposite directions in the two sides of the loop (*countercurrent* flow). The countercurrent direction of flow and the close anatomical association between the two sides of the loop allow events happening on one side of the loop to influence conditions on the other, and even for water and solutes to be *exchanged* between the two sides if permitted by permeability characteristics of the tubule. Countercurrent exchange is critical to the development of the high concentration gradient of the inner medulla because it permits a small difference in composition between the tubular fluid and the medullary interstitial fluid to be multiplied severalfold along the length of the loop (from the top of the medulla to the bottom). Countercurrent flow and exchange in the vasa recta is also important, because it removes solutes and water from the medulla without dissipating the medullary solute concentration from top to bottom (**Figure 15.12**).

Recap Production of dilute urine requires the reabsorption of salt without the concurrent reabsorption of water in the ascending limb of the loop of Henle, the distal tubule, and the collecting duct. The formation of concentrated urine requires antidiuretic hormone (ADH). In the presence of ADH, most of the water is reabsorbed from the collecting duct, leaving a small volume of concentrated urine to be excreted. ■

15.6 Urination depends on a reflex

Urination depends on a neural reflex called the *micturition reflex.* This involves the two urethral sphincters and the bladder. The internal urethral sphincter, which consists of smooth muscle, is an extension of the smooth muscle of the bladder. It remains contracted unless the bladder is emptying. The external urethral sphincter is skeletal muscle and therefore under voluntary control.

Normally the external urethral sphincter is kept closed by tonic activity of somatic motor neurons controlled by the brain. As the bladder fills with a cup (roughly 250 ml) or more of urine, it starts to stretch. Stretching stimulates sensory nerves, which send signals to the spinal cord. Nerves in the spinal cord initiate an involuntary (parasympathetic autonomic) reflex that contracts the smooth muscle of the bladder and relaxes the internal urethral sphincter. Stretch receptor input also goes to the brain, which decreases the activity of somatic motor neurons to the external urethral sphincter, allowing it to relax so that urine can flow through.

Humans don't usually have the opportunity to urinate whenever the urge strikes. Fortunately the brain can voluntarily override the micturition reflex by increasing the activity of the somatic nerves that control the external urethral sphincter. This delays urination, for extended periods if necessary. When we finally decide to urinate, the somatic nerve activity is shut off, the external urethral sphincter relaxes, and urination occurs. Voluntary prevention of urination becomes increasingly difficult as the bladder approaches maximum capacity.

Figure 15.11 The formation of a low volume of concentrated urine. In the presence of the hormone ADH the collecting duct becomes permeable to water. Consequently, most of the water in the collecting duct is reabsorbed by passive diffusion as the fluid passes through the medulla with its high solute concentration.

 Would you expect a kidney that is making concentrated urine to expend more energy (more ATPs) than when it is making dilute urine? Why or why not?

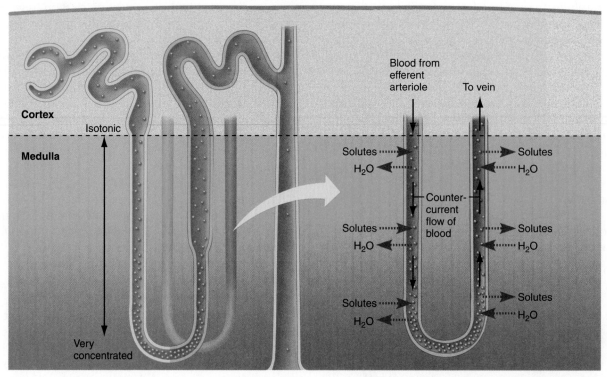

Figure 15.12 **Countercurrent exchange in the vasa recta.** The vasa recta are closely associated with the tubular segments in the medulla. The vasa recta remove the water and solutes that accumulate in the medulla as a result of reabsorption from the loop of Henle and the collecting duct. Like most capillaries, the vasa recta are highly permeable to water and solutes and rapidly equilibrate with their surrounding fluid. The countercurrent flow of blood in the two parallel vessels of the vasa recta permits the rapid exchange of water and solutes by diffusion, at the same time preserving the medullary concentration gradient from top to bottom.

✓ *Animals that live in deserts tend to have very long loops of Henle, much longer than those of humans. Why?*

↺ **Recap** Urination depends on the neural micturition reflex; bladder stretching initiates involuntary relaxation of the internal urethral sphincter. The brain can override the reflex by voluntary contraction of the external urethral sphincter. ▪

15.7 The kidneys maintain homeostasis in many ways

The kidneys are much more than simply the organs that make urine. As mentioned earlier, they play other important roles in the body as well. The kidneys:

- contribute to the maintenance of water balance
- contribute to the maintenance of salt balance
- secrete an enzyme involved in the control of blood volume and blood pressure
- maintain acid-base balance and blood pH

- control production of red blood cells
- activate an inactive form of vitamin D

The kidneys are responsible for maintaining homeostasis (by excretion) of virtually every type of ion in the body, considering that ions cannot be created or destroyed by metabolic processes. Because the kidneys secrete hormones, they are also endocrine organs. And by chemically altering the vitamin D molecule so that it can function as a vitamin, they play a role in metabolism as well. The common theme in these activities is the maintenance of homeostasis.

ADH regulates water balance

As mentioned earlier, blood volume is determined in part by the amount of water in the body. Changes in blood volume are likely to cause parallel changes in blood pressure, so it is important that blood volume be maintained within a fairly narrow range. But it is also important that blood solute concentration be kept constant as well. If blood solute concentration can be kept constant under a wide variety of circumstances, then the control of blood volume can be determined by the regulation of salt balance, as we'll see below.

The kidneys, the hypothalamus of the brain, and the posterior pituitary gland of the endocrine system share the function of maintaining water balance. This balance is achieved by a negative feedback loop involving the hormone ADH (Figure 15.13).

When the concentration of blood solutes rises (too little water compared to the amount of solutes in the blood), the ADH-producing neurons in the hypothalamus cause ADH to be secreted from the posterior pituitary into the blood. ADH circulates to the kidney, where it increases the permeability of the collecting duct to water. This allows more water to be reabsorbed, reducing the amount of water excreted in the urine.

In addition to its effect on ADH, a rise in the blood solute concentration triggers a sensation of thirst, encouraging us to drink more water. The increased water intake combined with the reduced loss of water in the urine lowers the blood solute concentration toward normal again.

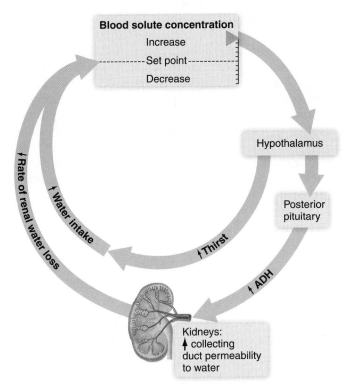

Figure 15.13 Negative feedback loop for the control of blood solute concentration. A rise in blood solute concentration triggers thirst and increases ADH secretion. ADH causes the kidneys to form a more concentrated urine and decrease their excretion of water. An increase in water intake coupled with a reduction in water loss from the kidneys dilutes the blood and returns blood solute concentration toward normal. Although only the responses to an increased solute concentration are shown here, the control loop responds equally well (with opposite responses) to a fall in blood solute concentration.

Normally some ADH is present in the blood. This allows the control mechanism to respond to excess water, too. When there is too much water and blood solute levels fall, the hypothalamus signals the posterior pituitary to reduce its secretion of ADH. The fall in ADH concentration makes the renal collecting duct less permeable to water, so the excess water is excreted. We also feel less thirsty when we already have more than enough water.

The term *diuresis* refers to a high urine flow rate. A *diuretic* is any substance that increases the formation and excretion of urine. Diuretic drugs, such as furosemide (Lasix), are prescribed to reduce blood volume and blood pressure in certain patients, such as people with congestive heart failure or hypertension.

Caffeine is a mild diuretic because it inhibits sodium reabsorption, and as sodium is excreted it takes water with it. Alcohol is also a diuretic, although by a different mechanism: it inhibits ADH release. The result is that the renal tubules become less permeable to water, and more water is excreted. If you drink excessive amounts of alcohol you may feel dehydrated and thirsty the next day.

Aldosterone regulates salt balance

The control of blood volume depends critically on maintaining the body's salt balance. Sodium (salt) excretion by the kidneys is regulated by **aldosterone,** a steroid hormone from the adrenal gland. Aldosterone causes more sodium to be reabsorbed from the distal tubule and collecting duct, back into the blood. High concentrations of aldosterone cause nearly all of the filtered sodium to be reabsorbed, so that less than 50 milligrams (0.05 gram) per day is excreted in the urine. Low levels of the hormone allow as much as 20–25 grams of sodium to be excreted each day. When we consider that the average North American consumes about 10 grams of sodium daily, we see that aldosterone provides more than enough control over sodium excretion.

Given what aldosterone does, we might think that its concentration would be controlled by blood sodium levels. This is partly true, for blood sodium concentration is a weak stimulus for aldosterone secretion. However, the real controller of aldosterone secretion is the *renin-angiotensin system.*

✅ **Quick Check** You've just eaten a large pepperoni-and-cheese pizza that has an enormous amount of sodium. (You hardly drank anything, so your blood pressure does not change detectably.) Will this cause your ADH and aldosterone levels to increase or decrease? What will your kidneys do in response? ■

The renin-angiotensin system controls blood volume and blood pressure

The **renin-angiotensin system** does not fit the classical definition of a hormone, for there is no one gland that releases a hormone into the blood. Rather, the production of a biologically active peptide (angiotensin II) that functions like a hormone requires three different organs (kidneys, liver, and lungs). The renin-angiotensin system is the primary hormonal controller of blood pressure and blood volume, both directly (via vasoconstriction) and indirectly (via the secretion of aldosterone).

The controlling component of the renin-angiotensin system is an enzyme called **renin**. Renin is synthesized and stored in specialized cells of the afferent arteriole in a region near the glomerulus called the *juxtaglomerular apparatus* (**Figure 15.14.**), the region where the afferent and efferent arterioles are in close contact with the distal tubule. When blood pressure falls in the afferent arteriole or when the sympathetic nerves to the kidney are stimulated (as they would be when systemic blood volume or blood pressure falls), renin is secreted into the blood of the afferent arteriole.

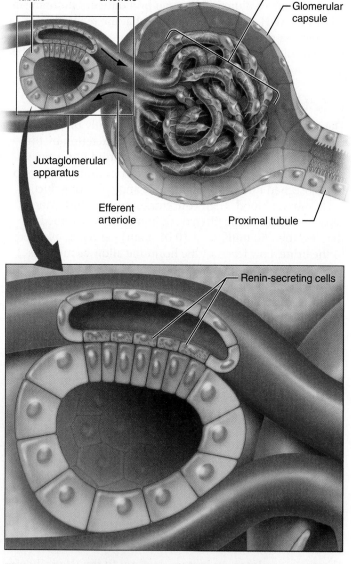

Figure 15.14 The juxtaglomerular apparatus. The juxtaglomerular apparatus is a region of contact between the afferent arteriole, the efferent arteriole, and the distal tubule. The afferent arteriolar cells of the juxtaglomerular apparatus synthesize and secrete renin.

How the secretion of renin ultimately leads to a return of blood volume and blood pressure to normal is illustrated in **Figure 15.15**. In the bloodstream, renin attaches to and cleaves a large inactive protein molecule produced by the liver called *angiotensinogen*. The product of angiotensinogen cleavage is a 10-amino-acid fragment called *angiotensin I*. Angiotensin I is biologically inactive, but when it passes through the lungs another enzyme found primarily in lung tissue called *angiotensin-converting enzyme* (ACE) chops off another two amino acids, producing a peptide of just eight amino acids called *angiotensin II*.

Angiotensin II is a biologically active peptide with several important effects. First, angiotensin II causes constriction of small blood vessels (primarily arterioles), returning blood pressure toward normal. Angiotensin II–induced vasoconstriction can occur within minutes of the secretion of renin.

Second, angiotensin II is a powerful stimulus for the release of aldosterone from the adrenal glands. From this point on, you already know the actions of the other hormones involved; aldosterone causes the kidneys to save more salt, which raises blood solute concentration and triggers ADH-induced water retention as well. Overall, the increased retention of water and salt by the kidneys coupled with water and salt in the diet result in a slow but sure return of blood volume back to normal within a matter of hours.

On a historical note, knowledge of how the renin-angiotensin system regulates blood pressure and blood volume led to the development of one of the most important classes of drugs for the treatment of high blood pressure, known as *ACE inhibitors*. As the name implies, ACE inhibitors inhibit the action of angiotensin converting enzyme (ACE) in the lungs. In the presence of ACE inhibitors, the normal production of angiotensin II is blocked. Without angiotensin II aldosterone concentration falls, sodium and water excretion by the kidneys increases, and blood volume is reduced slightly. In addition, blood vessels dilate (because angiotensin II–induced vasoconstriction declines), and so blood pressure falls. The effects are just what are needed to control hypertension.

Atrial natriuretic hormone protects against blood volume excess

There is another controller of renal sodium excretion unrelated to aldosterone; a peptide hormone called *atrial natriuretic hormone (ANH)*. (A "natriuretic" is a substance that increases the excretion of sodium.) When the atria of the heart are stretched by an excessively high blood volume, they secrete ANH into the bloodstream. ANH inhibits sodium reabsorption in the distal tubules and collecting ducts of the kidneys, which leads to increased sodium excretion. Once again, ADH causes water to follow salt in order to keep the blood solute concentration constant. In short, the effect of ANH is opposite to that of aldosterone; ANH protects the body against salt and water excess, aldosterone protects the body against salt and water deficit.

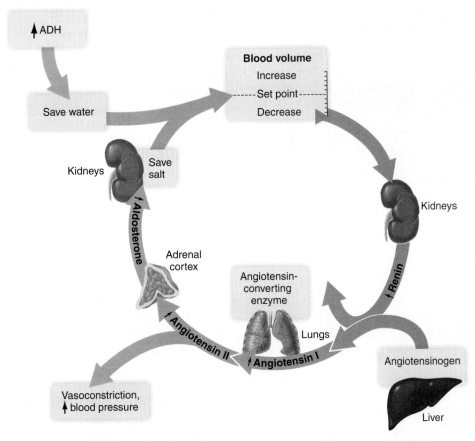

Figure 15.15 Regulation of blood volume by renin, aldosterone, and ADH. A decline in blood volume causes the kidneys to secrete renin. Renin activates a cascade of events that ultimately causes the kidneys to save salt and water, returning blood volume to normal.

✔️ *What makes the ADH increase? Does it increase before or after the other events shown in the diagram?*

✔️ **Quick Check** A man loses a large amount of blood in an accident. How will this affect secretion of renin, aldosterone, and ANH, and what will happen in response? ◾

Kidneys help maintain acid-base balance and blood pH

Recall that *acids* are molecules that can donate hydrogen ions (H^+) and *bases* are molecules that can accept a hydrogen ion. Many of the body's metabolic reactions generate H^+. If not eliminated, these acids would accumulate—a dangerous situation, because blood pH must remain in a relatively narrow range of 7.35 to 7.45. Any change in blood pH by more than a few tenths of a unit beyond this range could be fatal.

The kidneys participate in the maintenance of acid-base balance, a role they share with various buffers in the body and with the lungs. The kidneys do this in two ways (**Figure 15.16**):

■ *Reabsorption of filtered bicarbonate.* Kidney tubular cells secrete H^+ into the tubular lumen. Most of the secreted H^+ combines with bicarbonate (HCO_3^-) filtered by the

glomerulus, creating carbon dioxide (CO_2) and water. The CO_2 and water are then reabsorbed back into the cell, where they dissociate into H^+ and HCO_3^- again. The H^+ is secreted once again, but the HCO_3^- diffuses into the blood capillary. In effect, the secretion of H^+ is a mechanism for reabsorbing, or recovering, nearly all of the HCO_3^- that is filtered. When nearly all of the filtered HCO_3^- is recovered as CO_2, any additional H^+ that is secreted is buffered by negatively charged ions in the urine (phosphate and sulfate) and eventually excreted in the urine.

■ *Renal excretion of acid as ammonium.* Kidney tubule cells produce ammonium (NH_4^+) and HCO_3^- as part of their normal metabolism of the amino acid glutamine within the cell. The NH_4^+ is transported into the tubule lumen and then excreted in the urine. The new HCO_3^- diffuses into the blood to replace a HCO_3^- lost from the lung during respiratory control of CO_2 concentration.

Ordinarily our diets provide us with a small amount of excess H^+ that must be excreted if we are to stay in balance. Most of our daily production of H^+ results from the creation of CO_2 during metabolism, which is handled by the lungs. The renal excretion of H^+ with sulfate, phosphate, and ammonia exactly offsets the excess acid intake per day.

✔️ **Quick Check** If you climb a very high mountain, the low air pressure of high altitude will probably cause you to hyperventilate slightly. This causes a condition called respiratory alkalosis, in which blood pH becomes too high. Should the kidney excrete more or fewer H^+ ions in this situation? Explain. ◾

Erythropoietin stimulates production of red blood cells

As mentioned in Chapter 7, the kidneys regulate the production of red blood cells in the bone marrow. They do this by secreting a hormone, **erythropoietin.**

When certain cells scattered throughout the kidneys sense a fall in the amount of oxygen available to them, they release erythropoietin, which then stimulates the bone marrow to produce more red blood cells. The presence of more red blood cells increases the oxygen-carrying capacity of the blood. Once oxygen is again being delivered normally, the cells stop activating erythropoietin, and RBC production

a) Reabsorption of filtered bicarbonate.
Hydrogen ions within the cell are secreted into the lumen, where they combine with filtered bicarbonate to form CO_2 and water. The CO_2 and water diffuse into the cell and dissociate back into a hydrogen ion and bicarbonate. The hydrogen ion is secreted again, but the bicarbonate diffuses from the cell back into the blood.

b) Excretion of ammonium. Kidney cells produce ammonium and bicarbonate during cellular metabolism of glutamine. The ammonium is secreted into the tubule and is excreted in the urine, but the new bicarbonate diffuses into the blood.

Figure 15.16 Renal maintenance of acid-base balance.

slows. The activation of erythropoietin by certain kidney cells constitutes a negative feedback loop for controlling oxygen availability to all cells, because it keeps the oxygen-carrying capacity of the blood within normal limits.

Kidneys activate vitamin D

Vitamin D is important for absorbing calcium and phosphate from the digestive tract and developing healthy bones and teeth. We absorb some vitamin D from food, but the body also manufactures it in a three-step process involving the skin, liver, and kidneys.

Vitamin D synthesis begins when ultraviolet rays in sunlight strike a steroid molecule in skin that is similar to cholesterol, producing an inactive form of vitamin D. The inactive vitamin D is transported to the liver, where it is chemically altered, then carried to the kidneys, where it is converted to its active form under the influence of parathyroid hormone (PTH) from the parathyroid gland. During kidney failure this final step may not occur. This is why people with long-standing kidney failure often suffer from vitamin D deficiency.

🔄 **Recap** ADH controls blood total solute concentration, aldosterone controls blood sodium concentration, and the combination of renin-angiotensin, aldosterone, and ANF controls blood volume. The kidneys help maintain the body's acid-base balance and blood pH by reabsorbing all filtered bicarbonate and by excreting H^+. Decreased oxygen delivery to the kidneys triggers the release of erythropoietin, which stimulates the production of red blood cells by bone marrow. The kidneys activate an inactive form of vitamin D ∎

15.8 Disorders of the urinary system

The good news about urinary system disorders is that the kidneys have an enormous reserve of function. You can survive quite nicely on one kidney, and even on one kidney that is functioning only half as well as it should. Nevertheless, sometimes urinary system problems do occur. They include kidney stones, urinary tract infections, and renal failure.

Kidney stones can block urine flow

Sometimes minerals in urine crystallize in the renal pelvis and form solid masses called *kidney stones*. Most stones are less than 5 millimeters (1/5 inch) in diameter and are excreted in urine with no trouble. Others may grow larger and obstruct a ureter, blocking urine flow and causing great pain. Kidney stones can be removed surgically or crushed with ultrasonic shock waves, after which the pulverized fragments can be excreted with less pain.

Urinary tract infections are often caused by bacteria

A *urinary tract infection (UTI)* refers to the presence of microbes in urine or an infection in any part of the urinary system. Symptoms include swelling and redness around the urethral opening, a burning sensation or pain while urinating, difficulty urinating, bed-wetting, low back pain, and sometimes visible blood and pus in urine.

Most UTIs are caused by bacteria that make their way up the urethra from the body surface. Although UTIs can occur in men, they are more common in women because women's urethras are shorter, allowing organisms to reach the bladder more easily. From the urethra and bladder, microbes can travel up the ureters to the kidneys. This is why it is important to treat all urinary tract infections immediately. Most cases can be cured with antibiotics.

Acute and chronic renal failure impair kidney function

Kidney impairments that are short term and possibly correctable are called *acute renal failure*. Conditions that might impair kidney function temporarily include sustained decreases in blood pressure to below the pressure required for filtration, large kidney stones in the renal pelvis, infections, transfusion reactions, burns, severe injuries, and toxic drugs or chemicals. A surprising number of common medications are toxic to the kidneys if taken too long or at high doses. The kidneys are particularly vulnerable because of their reabsorptive efficiency, which tends to concentrate harmful substances in the urine. The toxicity problem may be prevented by drinking lots of water and excreting a more dilute urine.

Chronic renal failure, also known as end-stage renal disease (ESRD), is defined as long-term irreversible damage leading to at least a 60% reduction in functioning nephrons and failure of the kidneys to function properly. People with chronic renal failure may have less than 10% of the normal filtering capacity of the kidneys, and some essentially have no renal function at all.

Any of the factors that cause acute renal failure can, if not corrected, progress to chronic failure. In addition, approximately 40% of people with Type 1 (insulin-dependent) diabetes eventually develop renal failure, especially if the diabetes has not been well controlled. Approximately 13 million Americans have some form of chronic renal failure.

Dialysis cleanses the blood artificially

The body cannot replace irreparably damaged nephrons. This leaves two options for people with severe renal failure: dialysis or a kidney transplant.

Dialysis attempts to duplicate the functions of healthy kidneys. A dialysis technique that can be done at home without a kidney machine is called *continuous ambulatory peritoneal dialysis (CAPD)*. In this procedure, dialysis fluid (a fluid similar to extracellular fluid) is placed directly into the peritoneal cavity through an access port permanently implanted in the abdominal wall. The fluid is left in the peritoneal cavity for several hours, during which time it exchanges wastes and ions with the capillaries. Then the fluid is drained and discarded. If done on a regular basis, CAPD can be modestly effective. It is convenient and allows freedom of movement. However, there is a risk of infection because of the access port through the abdominal wall.

In another type of dialysis, called *hemodialysis,* the patient's blood is circulated through an artificial kidney machine consisting of semipermeable membranes in contact with a large volume of clean fluid (**Figure 15.17**). Metabolic wastes and excess ions diffuse from the blood into the dialysis fluid, which is discarded. The procedure is costly (over $50,000 per year) and time consuming. Typically the patient must go to a medical site called a dialysis unit for treatment several times each week, with each treatment taking a whole morning or afternoon.

Kidney transplants are a permanent solution to renal failure

Many people currently on renal dialysis will remain on dialysis for the rest of their lives. Although dialysis is a life-saving technique, it is not the perfect solution. It is difficult, if not impossible, to achieve complete homeostasis of all ions and wastes by artificial means. Furthermore, dialysis does not replace the renal hormones.

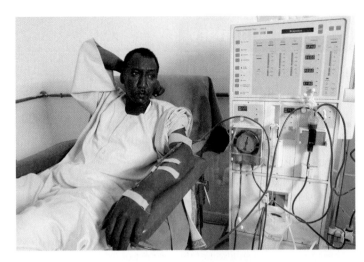

Figure 15.17 A patient undergoing hemodialysis.

The best hope for chronic renal failure patients is to receive a donated kidney, either from a living person or from a cadaver. When kidney transplants began in the 1960s most donated kidneys came from living close relatives, because the biggest challenge was to find a good immunological match so that the recipient's body would not reject the foreign kidney. With the advent of better drugs to suppress the immune system, improved tissue-matching techniques, national data banks of patients, and rapid jet transport, it is now feasible for the donor kidney to come from a cadaver. The main bottleneck now is that not enough people have offered to donate their organs after death. Thousands of people currently on waiting lists will die before they can receive a kidney.

If you want to help, show your willingness to donate your organs after death by completing an organ donor card. In many states, driver's licenses have a donor card on the back. You can also download a card from the Internet.

Relatives of patients who desperately need a kidney understandably ask, "If I turn out to be a good immunological match, is it safe for me to donate one of my kidneys?" The answer is a cautious yes. Aside from the usual risks of major surgery, past donors who have only one remaining kidney apparently do not suffer any significant deficit in overall kidney performance, even over decades. Nevertheless, the decision to donate a precious organ is not to be taken lightly. Any person considering donating one of their own kidneys to a relative should discuss it frankly with their family and their doctor.

Recap Acute or chronic renal failure can result from prolonged changes in blood pressure, disease, large kidney stones, transfusion reactions, burns, injuries, toxic substances, and other conditions such as diabetes. When kidneys fail, dialysis (either CAPD or hemodialysis) may be needed to keep the patient alive. Kidney transplants are the best hope for people in renal failure. Currently there is a severe shortage of kidneys available for transplant. ▪

Chapter Summary

The urinary system contributes to homeostasis p. 356

- The kidneys, the lungs, the liver, and the skin all participate in the maintenance of homeostasis.
- The kidneys are the primary regulators of water balance and most excess solutes, especially inorganic ions and urea.

Organs of the urinary system p. 358

- The urinary system consists of those organs that produce, transport, store, and excrete urine. The urinary system includes the kidneys, the ureters, the bladder, and the urethra.
- Functions of the kidneys include regulation of the volume and composition of body fluids, excretion of wastes, regulation of blood pressure, regulation of the production of red blood cells, and the activation of vitamin D.

Nephrons produce urine p. 360

- The functional unit of the kidneys is the nephron. Each nephron consists of a tubular component and the blood vessels that supply it.
- The tubular components of a nephron are the glomerular capsule, proximal tubule, loop of Henle, distal tubule, and collecting duct. The collecting duct is shared by many nephrons.
- A tuft of capillaries called the glomerulus is enclosed within each glomerular capsule. Peritubular capillaries supply proximal and distal tubules, and vasa recta supply loops of Henle and collecting ducts.

Formation of urine: Filtration, reabsorption, and secretion p. 362

- The formation of urine involves three processes: glomerular filtration, tubular reabsorption, and tubular secretion.
- Approximately 180 liters per day of protein-free plasma fluid is filtered across the glomerulus and into the glomerular space. Filtration is driven by high blood pressure in the glomerular capillaries.
- Ninety-nine percent of all filtered water and salt and all of the filtered bicarbonate, glucose, and amino acids are reabsorbed in tubular reabsorption. The active transport of sodium provides the driving force for the reabsorption of nearly all other substances.
- Tubular secretion is a minor process relative to reabsorption but is critical for the regulation of acid-base balance and for the removal of certain toxic wastes.

The kidneys can produce dilute or concentrated urine p. 366

- The ability of the kidneys to form either dilute or concentrated urine depends on the high solute concentration in the renal medulla and the ability to alter the permeability of the collecting duct to water.
- Dilute urine is formed in the absence of the hormone ADH. In the absence of ADH, reabsorption of salt without reabsorption of water continues in the collecting duct.
- Concentrated urine is formed when ADH increases the collecting duct's permeability to water, allowing water to diffuse toward the high solute concentration in the medulla.
- The ability of the kidneys to produce concentrated urine is dependent on a countercurrent mechanism that exists in the hairpin arrangements of the loops of Henle and the vasa recta.

Urination depends on a reflex p. 367

- Urination is caused by the micturition reflex, a neural reflex initiated when the bladder is stretched.
- Urination can be prevented by higher (voluntary) neural signals from the brain.

The kidneys maintain homeostasis in many ways p. 368

- Water balance is maintained by a negative feedback loop involving ADH. The main stimulus for the secretion of ADH is an increase in the solute concentration of the blood.
- Blood volume is regulated by maintenance of the body's salt balance, which is controlled primarily by a negative feedback loop involving the renin-angiotensin system and aldosterone.
- A hormone from the heart called atrial natriuretic peptide increases sodium excretion by the kidneys.
- The kidneys secrete (and excrete) H^+ and NH_4^+ in amounts equal to the net gain of acid per day (other than as CO_2). They also secrete H^+ as part of the mechanism for the reabsorption of all filtered HCO_3^-.
- The kidneys synthesize and secrete erythropoietin, the hormone responsible for the regulation of red blood cell production.
- The kidneys are required for the activation of vitamin D.

Disorders of the urinary system p. 372

- The kidneys are vulnerable to damage by toxic substances, infections, sustained decreases in blood pressure, or blockage of urinary outflow.
- Nephrons that are damaged beyond repair are not replaced.
- Hemodialysis is an artificial procedure for cleansing the blood of wastes and excess solutes.
- Renal transplantation is technically easy and usually quite successful, but there is a shortage of donor kidneys.

Terms You Should Know

afferent arteriole, 361
aldosterone, 369
antidiuretic hormone (ADH), 367
collecting duct, 360
countercurrent exchange mechanism, 367
dialysis, 373
distal tubule, 360
efferent arteriole, 361
glomerular capsule, 360
glomerular filtration, 362

glomerulus, 360
loop of Henle, 360
nephron, 360
peritubular capillary, 361
proximal tubule, 360
renin, 370
tubular reabsorption, 364
tubular secretion, 365
urea, 357
ureter, 358
urethra, 359
vasa recta, 361

Concept Review

Answers can be found at the Human Biology Place.
www.humanbiology.com

1. Describe in general terms the kidneys' role in maintaining homeostasis.
2. Describe the function of each of the organs of the urinary system.
3. Give the names and functions of each of the tubular sections of the nephron.
4. Indicate where vasa recta are located, and list the two tubular regions of the nephron they serve.
5. Describe the driving force responsible for filtration of fluid from the glomerular capillaries into the glomerular space.
6. Name the one region of the tubule where water permeability is *regulated*.
7. Indicate the primary stimulus for the secretion of the hormone ADH.
8. Describe the function of the hormone aldosterone.

9. Describe how chronic renal failure differs from acute renal failure.
10. Discuss the treatment options available to someone in chronic renal failure (end-stage renal disease).

Test Yourself

Answers can be found in Appendix A.

1. Which of the following organ systems is/are involved in excretion?
   a. urinary system
   b. integumentary system
   c. respiratory system
   d. all of these choices
2. Three of the choices below are organ systems directly involved in excretion of wastes. Which system is not directly involved in excretion, but interconnects the other three?
   a. urinary system
   b. respiratory system
   c. circulatory system
   d. digestive system
3. The _____ detoxifies ammonia, producing _____ which is excreted by the _____.
   a. urinary system...urine...kidneys
   b. liver...urea...kidneys
   c. urinary system...urea...liver
   d. liver...carbon dioxide...respiratory system
4. The nitrogenous wastes excreted by the kidneys are produced through the metabolism of:
   a. carbohydrates
   b. proteins
   c. triglycerides
   d. cholesterol
5. Which of the following indicates the correct order of the structures through which urine would pass during formation and excretion?
   a. nephron, renal pelvis, ureter, bladder, urethra
   b. renal pelvis, ureter, nephron, urethra, bladder
   c. nephron, renal pelvis, urethra, bladder, ureter
   d. renal pelvis, nephron, ureter, bladder, uretrha
6. Which structure is correctly paired with its function?
   a. ureter: formation of urine
   b. nephron: storage of urine
   c. urethra: transport of urine
   d. bladder: concentration of urine
7. Protein is generally not found in the urine because:
   a. It does not pass through the glomerular filter into the space within the glomerular capsule.
   b. It is reabsorbed in the ascending limb of the Loop of Henle.
   c. It is reabsorbed within the collecting duct.
   d. It is metabolized by the cells that line the nephron.
8. Blood enters the glomerulus through the _____ and exits the glomerulus through the _____.
   a. efferent arteriole...afferent arteriole
   b. efferent arteriole...distal tubule
   c. proximal tubule ... efferent arteriole
   d. afferent arteriole...efferent arteriole
9. Which of the following would not be found in the glomerular filtrate of a healthy individual?
   a. water
   b. hydrogen ions
   c. glucose
   d. hemoglobin
10. Which of the following is most responsible for the production of the glomerular filtrate?
    a. diffusion
    b. active transport
    c. glomerular capillary blood pressure
    d. facilitated diffusion
11. In tubular absorption, which of the following processes essentially drives all of the others?
    a. active transport of Na^+ from the proximal tubular cell toward a capillary
    b. facilitated transport of Na^+ from the tubular lumen into the proximal tubular cell
    c. diffusion of water from the proximal tubular cell toward a capillary
    d. diffusion of Cl^- from the tubular lumen into the proximal tubular cell
12. All of the following would happen if an individual was perspiring heavily and becoming dehydrated except:
    a. ADH hormone would be secreted.
    b. The kidneys would reabsorb more water.
    c. The collecting ducts would become less permeable to water.
    d. Less urine would be produced.
13. How does ADH decrease the loss of water through the kidneys?
    a. ADH increases water reaborption in the collecting ducts.
    b. ADH decreases the amount of blood flowing through the kidneys.
    c. ADH increases water reabsorption in the proximal tubule.
    d. ADH decreases water excretion in the Loop of Henle.
14. ACE inhibitors decrease blood pressure by:
    a. inhibiting the activity of ADH
    b. decreasing the amount of angiotensin II
    c. altering the permeability of the renal tubules to water
    d. inhibiting the activity of atrial natriuretic hormone
15. All of the following statements about urinary tract infections are true except:
    a. Most urinary tract infections are caused by bacteria.
    b. The shorter urethra predisposes women to more UTIs than men.
    c. Urinary tract infections are generally treated with antibiotics.
    d. Urinary tract infections usually start in the kidney and spread to the bladder.

Apply What You Know

Answers can be found at the Human Biology Place.
www.humanbiology.com

1. Do your kidneys have to work harder (expend more metabolic energy) to get rid of the extra salt and water after you binge on salty snacks and liquid beverages? Explain your reasoning.
2. Explain the mechanism for why you feel thirsty after heavy exercise accompanied by sweating.
3. What does the production of a low volume of dark yellow urine indicate?
4. The kidneys filter a large volume of fluid, putting it on a path to ultimate disposal, and then reabsorb nearly all of it back into the body. Only a few waste products are actively secreted into the tubule. Why are the kidneys set up in this way? Wouldn't it be easier and more energetically efficient to selectively secrete only the waste molecules and the exact right amount of excess water and ions directly into an excretory tube or the bladder?
5. Why aren't children already "potty trained"—able to hold urination and defecation as well as adults can—when they are born?
6. A possible complication of a strep throat infection is post-streptococcal glomerulonephritis. In this disease, antibodies produced against the streptococcus invader also cause inflammation of glomeruli in the kidneys and damage to the glomerular filtration barrier. What might be the symptoms of damage to the glomerular filtration barrier, and what effect would you expect this to have on kidney function?

16

Reproductive Systems

Human sperm on the inner surface of the uterus (the endometrium). Some of the cells of the endometrium are ciliated.

Would You Like a Boy or a Girl?

Your parents probably did not know whether you were a boy or a girl until the very moment you were born. Ultrasound, typically done between 18 to 26 weeks of pregnancy if it was done at all, could detect gender only about 80% of the time.

Today, expectant parents may be able to discover the gender of a fetus as early as five weeks. They may even be able to select gender (and several other genetic traits) ahead of time.

Early Gender Detection—Good Idea or Not?

A biotech company called Acu-Gen Bio-lab, Inc. now offers a gender testing service and kit called Baby Gender Mentor™. According to the company, Baby Gender Mentor can determine the gender of an unborn child with 99.9% accuracy as early as five weeks after fertilization. For $275, the service analyzes a small blood sample from the mother and provides confidential results via e-mail, generally within 48 hours. Baby Gender Mentor is based on the principle that a small amount of fetal DNA is generally present in the mother's blood plasma. The presence of Y-chromosomal material in the mother's blood plasma indicates a boy; the presence of only X-chromosomal material indicates a girl.

(We discuss the X and Y chromosomes in Chapter 19.)

A word of caution is due here. Baby Gender Mentor is not regulated by the FDA, and the company won't release the "scientific information" on which their 99.9% accuracy claim is based. Blog sites are reporting that the test is not nearly as accurate as claimed by the company.

Some ethicists question whether knowing a fetus's gender as early as five weeks serves any purpose, other than providing couples with an opportunity to have an elective abortion if the fetus is not of the desired gender. They have a point. In India, where there is a strong preference for male children, prenatal testing and selective abortions are thought to be responsible for a deficit of over 500,00 girls every year.

Gender Selection Is Now Possible

If you would prefer to have a child of a specific gender without resorting to an elective abortion there's a relatively easy way to at least increase the odds. A procedure called *sperm sorting* is based on the fact that the X (female) chromosome is larger and carries more DNA than the Y (male) chromosome. First, the sperm in a semen sample from the would-be father are labeled with a fluorescent dye that attaches to DNA. Because sperm with the female (X) chromosome have more DNA, they fluoresce more brightly than sperm with the male (Y) chromosome. The sperm are then sorted on the basis of light intensity by a machine called a *flow cytometer*. After sorting, either the "female-enriched" or the "male-enriched" sperm sample is inserted into the mother using artificial insemination. For couples who want a girl, sperm sorting produces girls 91% of the time; for boys the odds are a bit lower (only 76%) but that's still a lot better than random chance

Could a couple choose their next child's gender with absolute certainty? Yes, they could, but the techniques involved are expensive and would require the services of a fertility clinic. According to the standard techniques at most fertility clinics, eggs and sperm would be collected from the prospective couple and mixed together. Once the early-stage embryos had begun to develop, each one would be tested for the presence or absence of the Y-chromosome by a technique called *preimplantation genetic diagnosis* (PGD). The presence of a Y-chromosome would indicate a male embryo, whereas the absence of a Y-chromosome would indicate a female embryo. A couple wanting only a boy, then, could choose to have only a boy-embryo implanted into the prospective mother.

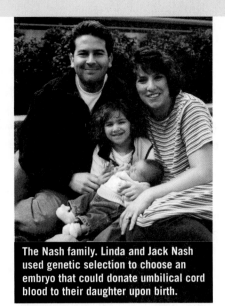

The Nash family. Linda and Jack Nash used genetic selection to choose an embryo that could donate umbilical cord blood to their daughter upon birth.

The American Society of Reproductive Medicine does not endorse PGD for gender selection. Nevertheless, a number of fertility clinics will perform PGD for this purpose. To avoid controversy over the ethics of gender selection, the clinics may describe gender selection as a way to select against certain diseases (females don't get prostate cancer for example, and males are at very low risk for breast cancer).

Beyond Gender Selection

In addition to gender selection, PDG can be used to test for genetic disorders. It can also be used to select embryos for certain specific genotypes, including being a compatible tissue match for a sibling.

Molly Nash was born with Fanconi's anemia, a rare, incurable disease in which the bone marrow does not produce blood cells normally. Typically people with this condition die in their early 20s. Fanconi's anemia is a genetic disorder. Although neither of Molly's parents have the disease, they both carry the gene that causes it. In a desperate attempt to save their daughter, Molly's parents turned to a fertility clinic that performed PGD. Eggs were harvested

from the mother's ovaries, fertilized, and then tested for the abnormal Fanconi gene, among other genetic disorders. Embryos that lacked the fatal gene were screened further to determine whether they would be a safe tissue donor match for Molly. Only embryos that lacked the fatal gene and that were a good tissue donor match for Molly were implanted into the mother's uterus. After four attempts Molly's mother gave birth to a healthy boy, whom the couple named Adam. Shortly after Adam's birth, blood-forming stem cells harvested from Adam's umbilical cord blood were infused into Molly (see the Current Issue about umbilical cord blood in Chapter 7). Apparently, enough of these cells became blood-forming stem cells in Molly that she was cured. Today both children are healthy, and Molly has started school.

When the Nash story first hit the press it caused a firestorm of controversy. The media labeled Adam the "savior sibling" and predicted dire effects on his future mental health from knowing that he had been brought into being to save his sister. Molly's parents insisted that they wanted more children anyway, so this procedure enabled them to have a healthy child and also save their daughter's life. However, many people questioned whether it is ethical to have a child for the purpose of being a tissue donor. And of course there is always the issue of what happens to all the embryos not selected by Molly's parents because they were not a good match for Molly.

There's no doubt that as reproductive technologies improve, our ability to screen for genetic abnormalities will also improve. But beyond just avoiding genetic diseases, what about using PGD to influence physical characteristics such as hair, skin, and eye color, intelligence, even musical ability? If we could pick and choose from a menu of specific traits for our offspring, would we want to?

Questions to consider

1 If you and your partner decided to have a child together, would you want to know its gender before it is born?

2 Would you try to select the gender of your next child by one of the methods described here? Explain why or why not.

3 Would you be willing to select a second child specifically to help an earlier child, as the Nashes did?

The facts...

- It is now possible to determine the gender of a fetus just five weeks after fertilization.
- Sperm sorting can improve the odds of having a child of the desired gender. *In vitro* fertilization and pre-implantation testing can guarantee it.
- As our ability to specify gender and other genetic traits develops, so does the possibility of being able to "customize" children genetically.

Key concepts

>> **The male reproductive system** can deliver over 100 million sperm at once. **The female reproductive system** usually releases and nurtures just one egg each month.

>> **In females, a cyclic pattern of changes controlled by hormones prepares one egg for release each month.** Another parallel cycle prepares the uterus to receive and nurture the egg if fertilization occurs.

>> **In both males and females, sexual excitement leads to arousal and orgasm.** In males orgasm leads to ejaculation, the expulsion of sperm.

>> **Birth control (contraceptive) methods and devices include** 1) altering female hormones to disrupt maturation or fertilization of the egg, 2) condoms or diaphragms that prevent sperm and egg from making contact, 3) spermicides that kill sperm, 4) intrauterine devices that prevent implantation of the egg, and 5) surgical sterilization.

>> **Sexually transmitted diseases (STDs) range from mildly irritating to deadly.** Some are difficult or even impossible to cure. For a sexually active person the best defenses against the damaging effects of STDs are know your partner, use physical barriers (condoms) during risky behavior, and get tested if you think you are at risk.

MJ's Human Biology Blog

The Dark Side of Gender Preference

In China, 120 boys are born for every 100 girls. Experts think that the recent trend (since 1980) is the result of China's "one-child" population-control policy, combined with a traditional society in which the eldest boy inherits the family land and is responsible for taking care of aging parents. So if a couple is planning on having just one child, usually they will decide it should be a boy. The boy/girl sex ratio is highest in rural areas; in cities, the earning power of women tends to be accompanied by a more normal sex ratio.

What can be done about it? One approach is educational programs to encourage couples to value daughters equally with sons; another is to allow families whose first child is a girl to have a second child; still another is to encourage young couples to live with her parents rather than his. The government is even offering cash payments for girls in some regions. ∎

Reference: Hvistendahl, Mara. Making Every Baby Girl Count. *Science* 323: 1464–1466, 2009.

Most of the organ systems we have considered so far in this book are not exactly topics of daily interest to most people. How many songs have you heard praising the urinary system or digestive processes? How many movies show close-ups of people just breathing? Certainly some of the subjects in this chapter, such as differences in male and female anatomy and in patterns of sexual arousal, get more than their share of attention. The beauty and mystery of sex are a source of endless fascination to us.

On a practical level, sexual attractiveness and sexual arousal are normal events in the human reproductive process—a process that requires the **reproductive systems.** The male and female reproductive systems consist of the tissues and organs that participate in creating a new human being. At first glance, it may seem that the male and female reproductive systems are completely dissimilar. After all, the external and internal structures look different from each other. Even

secondary sexual characteristics, such as body form, muscle mass, hair patterns, and distribution of body fat, aren't the same. From an evolutionary standpoint, there must have been a tremendous advantage to having the male and female reproductive systems evolve in this manner.

However, there are many similarities between male and female reproductive systems. Both systems consist of primary reproductive organs, accessory glands, and ducts. Both men and women experience similar phases of sexual desire. Each system has highly specific functions that allow it to complement the other perfectly. And ultimately, both systems share the same goal: to join a sperm and an egg in an environment (the woman's body) that enables the fertilized egg to become an individual.

16.1 The male reproductive system delivers sperm

Figure 16.1 illustrates the organs of the male reproductive system. The male reproductive system evolved to deliver the male reproductive cells, called **sperm,** to the female.

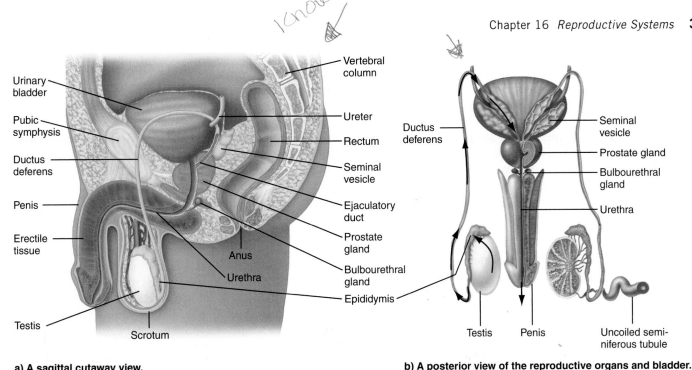

a) A sagittal cutaway view.

Urinary bladder

Pubic symphysis

Ductus deferens

Penis

Erectile tissue

Testis

Scrotum

Vertebral column

Ureter

Rectum

Seminal vesicle

Ejaculatory duct

Prostate gland

Anus

Urethra

Bulbourethral gland

Epididymis

b) A posterior view of the reproductive organs and bladder.

Ductus deferens

Seminal vesicle

Prostate gland

Bulbourethral gland

Urethra

Testis

Penis

Uncoiled semi-niferous tubule

Figure 16.1 **The male reproductive system.**

As we will see, the chance of any one sperm fertilizing a female reproductive cell (an **egg**) is extremely low. The reproductive strategy of the male, then, is to deliver huge numbers of sperm at a time to improve the odds that one will reach the egg.

Testes produce sperm

The sites of sperm production in the male are the paired **testes** (singular: *testis*). Shortly before birth the testes descend into the **scrotum,** an outpouching of skin and smooth muscle. The scrotum regulates the temperature of the developing sperm within the testes, for sperm develop best at temperatures a few degrees cooler than body temperature. When the temperature within the scrotum falls below about 95°F the smooth muscle of the scrotum contracts, bringing the testes closer to the warmth of the body. When the temperature warms to above 95°F the smooth muscle relaxes and the scrotum and testes hang away from the body.

Each testis is only about 2 inches long, but it contains over 100 yards of tightly packed **seminiferous tubules,** where the sperm are produced. The seminiferous tubules join to become the **epididymis,** a single coiled duct just outside the testis. The epididymis joins the long **ductus (vas) deferens,** which eventually joins the duct from the seminal vesicle to become the **ejaculatory duct.** The newly formed sperm are not fully mature (they cannot yet swim) when they leave the seminiferous tubules. Their ability to swim develops in both the epididymis and the ductus deferens. The epididymis and the ductus deferens also store the sperm until ejaculation.

When the male reaches sexual climax and ejaculates, rhythmic contractions of smooth muscle propel the sperm through the short ejaculatory duct and finally through the urethra, which passes through the penis. (As described in Chapter 15, the urethra is a pathway for urine as well.)

The **penis** is the male organ of sexual intercourse. Its function is to deliver sperm internally to the female, safely away from the harsh external environment. The penis contains erectile tissues that permit **erection**—an increase in length, diameter, and stiffness of the penis that facilitates its entry into the vagina.

Accessory glands help sperm survive

Sperm must endure the rapid journey from epididymis to vagina and then continue to live within the woman for many hours. To improve their chances of survival, the male delivers sperm in a thick, whitish mixture of fluids called **semen.** In addition to sperm, semen contains secretions from the seminal vesicles, the prostate gland, and the bulbourethral gland.

The **seminal vesicles** produce seminal fluid, a watery mixture containing fructose and prostaglandins that represents about 60% of the volume of semen. Fructose, a carbohydrate, provides the sperm with a source of energy. The prostaglandins are thought to induce muscle contractions in the female reproductive system that help sperm travel more effectively. The **prostate gland** contributes an alkaline fluid. The vagina is generally fairly acid (pH 3.5–4.0), which helps prevent infections but is too acidic for sperm. The prostate secretions temporarily raise the vaginal pH to about 6, optimal for sperm. Finally, the **bulbourethral glands** secrete mucus into the urethra during sexual arousal. The mucus washes away traces of acidic urine in the urethra before the sperm arrive and also provides lubrication for sexual intercourse. The sperm are not mixed with seminal and prostate fluids until the moment of ejaculation.

Table 16.1 summarizes the organs and glands of the male reproductive system.

Sperm production requires several cell divisions

Let's take a closer look at the process of sperm production. As mentioned, it takes place in the seminiferous tubules within each testis (**Figure 16.2**). Toward the outer surface of each tubule are undifferentiated cells (cells that have not yet become any kind of specialized cell) called *spermatogonia* (singular: *spermatogonium*). Spermatogonia are *diploid* cells, meaning they have 46 chromosomes, the normal human number.

The formation of the gametes (sperm and eggs) involves a series of cell divisions called *mitosis* and *meiosis*. In Chapter 17, we discuss meiosis and mitosis in more detail. The important point to remember here is that during the two successive cell divisions of meiosis (meiosis I and II), the diploid number of chromosomes in the developing gametes is halved. **Gametes** are called *haploid* cells (from the Greek *haploos*, "single") because they contain only 23 chromosomes, rather than the usual 46. This ensures that when a sperm and egg unite, the embryo will have the proper (diploid) number of chromosomes again.

Spermatogonia divide several times during the course of sperm development (Figure 16.2b). The first division, by mitosis, ensures a constant supply of *primary spermatocytes*, each with the diploid number of chromosomes. Primary spermatocytes then undergo two successive cell divisions during meiosis to become *secondary spermatocytes* and finally *spermatids*. Spermatids, which are haploid cells, mature slowly to become the male gametes, or sperm. The entire process of sperm formation and maturation takes about 9–10 weeks. During this time the cells are surrounded and nourished by the large **Sertoli cells** that make up most of the bulk of the seminiferous tubules (Figure 16.2c).

A mature sperm has a head, midpiece, and tail (Figure 16.2d). The head contains the DNA, organized into chromosomes. The head is covered by an *acrosome*, a cap containing enzymes that help the sperm penetrate the egg. The sperm's whiplike tail moves it forward, powered by energy produced by mitochondria in the midpiece.

Tens of millions of sperm are produced every day after puberty in young men. A typical ejaculation may contain 100–300 million sperm, but only one sperm will fertilize an egg.

Table 16.1 Summary of the male reproductive organs and glands

Organ	Function
Testis (2)	Produces sperm, testosterone, and inhibin
Scrotum	Keeps the testes at the proper temperature
Epididymis (2)	Site of sperm maturation and storage
Ductus deferens (2)	Duct for sperm maturation, storage, and transport
Ejaculatory duct (2)	Duct for transporting sperm and glandular secretions
Penis	Erectile organ of sexual intercourse
Accessory Glands	
Seminal vesicle (2)	Secretes fructose and most of the seminal fluid
Prostate gland	Secretes watery alkaline fluid to raise vaginal pH
Bulbourethral gland (2)	Secretes lubricating mucus

MJ's Human Biology Blog

http://www.humanbiologyblog.blogspot.com/

Share Report Abuse Next Blog» Create Blog Sign In

MJ's Human Biology Blog

Infertility Patients Favor Stem Cell Research

Most couples that have had to resort to *in vitro fertilization* (IVF) techniques in order to have a child are in favor of stem cell research. When asked in a national survey what they might choose to do with their frozen embryos left over after they have successfully had a child, 60% reported that they were "somewhat likely" or "very likely" to donate them for stem cell research. In contrast, less than a quarter of the respondents planned to discard their frozen embryos, or expressed a desire to donate them to another couple.

Infertility patients are especially aware of the advances in science that have made it possible for them to have children. Perhaps they are just more grateful than most, but apparently most of them have resolved any internal moral dilemma over what to do with their leftover embryos. It is interesting, however, that most of them would rather donate their embryos to research than to know that their biological child was being raised by another couple. ∎

Reference: Lyerly, A. D. and R. R. Faden. Willingness to Donate Frozen Embryos for Stem Cell Research. *Science* 317: 46–47, 2007.

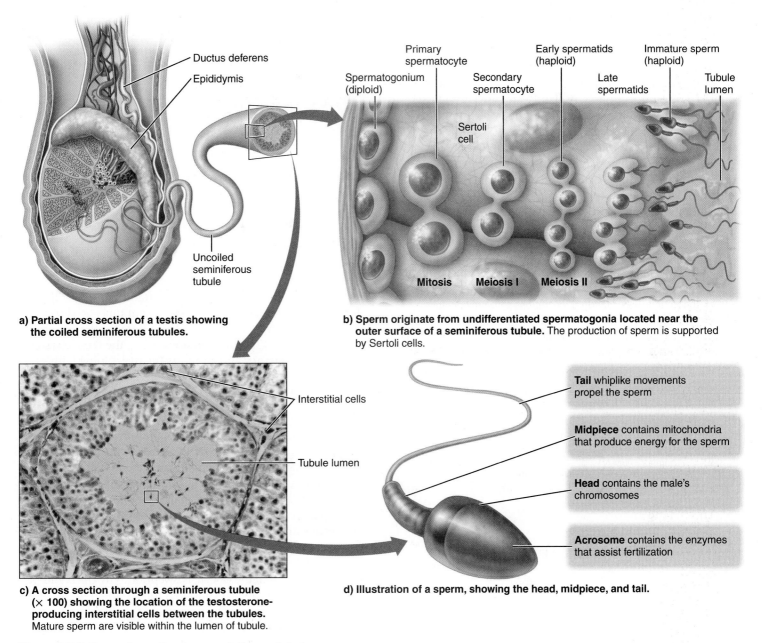

a) **Partial cross section of a testis showing the coiled seminiferous tubules.**

b) **Sperm originate from undifferentiated spermatogonia located near the outer surface of a seminiferous tubule.** The production of sperm is supported by Sertoli cells.

c) **A cross section through a seminiferous tubule (× 100) showing the location of the testosterone-producing interstitial cells between the tubules.** Mature sperm are visible within the lumen of tubule.

d) **Illustration of a sperm, showing the head, midpiece, and tail.**

Tail whiplike movements propel the sperm

Midpiece contains mitochondria that produce energy for the sperm

Head contains the male's chromosomes

Acrosome contains the enzymes that assist fertilization

Figure 16.2 Sperm formation in a seminiferous tubule.

☑ *Which gives rise to more sperm, a spermatogonium or a primary spermatocyte? Explain.*

Testosterone affects male reproductive capacity

Male reproductive capacity depends on **testosterone,** a steroid hormone produced by **interstitial cells** located between the seminiferous tubules within the testes (see Figure 16.2c). Testosterone controls the growth and function of the male reproductive tissues, stimulates aggressive and sexual behavior, and causes the development of secondary sexual characteristics at puberty, such as facial hair and deepening voice. Within the testes, testosterone stimulates the

undifferentiated spermatogonia to begin dividing and the Sertoli cells to support the developing sperm. Thus, it determines the rate of sperm formation.

The production and secretion of testosterone depends on three hormones: **gonadotropin-releasing hormone (GnRH)** from the hypothalamus, and luteinizing hormone (LH) and follicle-stimulating hormone (FSH) from the anterior pituitary gland. (As described in Chapter 13, LH and FSH are called gonadotropins because they stimulate the reproductive organs—the gonads.) Historically, the names for LH and FSH came from their functions in the female, but they are active hormones in the male as well.

Answers to ☑ **questions can be found in Appendix A.**

Negative feedback loops involving the hypothalamus, the pituitary, and the testes maintain a fairly constant blood concentration of testosterone, and thus a consistent rate of sperm production (**Figure 16.3**). When blood concentrations of testosterone fall, the hypothalamus secretes GnRH, which stimulates the anterior pituitary gland to release LH and FSH. LH stimulates the secretion of testosterone by the interstitial cells. The role of FSH is less clear, but it seems to enhance sperm formation indirectly by stimulating the Sertoli cells.

Excessively high concentrations of testosterone inhibit the secretion of GnRH and also LH and FSH. In addition, when Sertoli cells are highly active they secrete a hormone called **inhibin** that directly inhibits the secretion of FSH.

↩ **Recap** The male reproductive system comprises the testes, the penis, and associated ducts and glands. Semen consists of sperm and three glandular secretions that provide energy and the proper pH environment for the sperm and also lubrication for sexual intercourse. Millions of sperm form every day throughout a man's life; a typical ejaculate contains up to 300 million. Testosterone stimulates the growth and function of

the male reproductive system, and encourages aggressive and sexual behavior. Blood levels of testosterone are regulated by a negative feedback loop involving GnRH from the hypothalamus and LH and FSH from the anterior pituitary. ▪

🌐 **Web Animation** *The Male Reproductive System* at www.humanbiology.com

16.2 The female reproductive system produces eggs and supports pregnancy

The female reproductive system evolved to perform more functions than that of the male. Eggs are precious, so the system releases only one or two every month. In addition, because it can never be known for certain if an egg will be fertilized, the female reproductive system must go through cyclic changes every month just to prepare for the possibility of fertilization. If fertilization does occur, the female system adjusts to pregnancy and supports the developing fetus.

Ovaries release oocytes and secrete hormones

Figure 16.4a illustrates the organs of the female reproductive system. The primary female reproductive organs are the two **ovaries,** which release the female gametes, immature eggs called *oocytes,* at regular intervals during the reproductive years. The ovaries also secrete the female sex steroid hormones **estrogen** and **progesterone.** (As mentioned in Chapter 13, the term estrogen is commonly used to refer to three hormones known collectively as *estrogens*—17β-estradiol, estrone, and estriol.)

Once released, the oocyte is swept into the open end of an **oviduct** (also called a Fallopian tube), which leads from the ovary to the uterus. The open end of the oviduct has fingerlike projections called fimbriae that help move the oocyte into the oviduct. If fertilization by a sperm occurs, it usually takes place in the upper third of the oviduct. The unfertilized oocyte or fertilized egg moves down the oviduct over the course of about 3–4 days to the uterus.

✓ **Quick Check** Embryos sometimes implant somewhere other than the uterus, most often in the oviduct before it reaches the uterus, and sometimes in the abdominal cavity. How could an embryo end up in the abdominal cavity? ▪

The uterus nurtures the developing embryo

The **uterus** is a hollow, pear-shaped organ where the embryo grows and develops. (Human development is described in Chapter 21.) The walls of the uterus comprise two layers of tissue (Figure 16.4b).

Figure 16.3 Feedback control of blood testosterone concentration and sperm production. Negative feedback control loops effectively maintain testosterone concentration (and hence sperm production) relatively constant over time. Negative feedback control occurs at both the hypothalamus and the anterior pituitary.

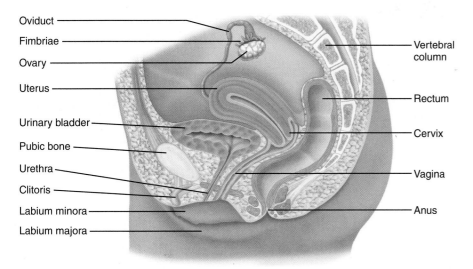

Oviduct
Fimbriae
Ovary
Uterus
Urinary bladder
Pubic bone
Urethra
Clitoris
Labium minora
Labium majora

Vertebral column
Rectum
Cervix
Vagina
Anus

a) Sagittal section showing the components of the system in relation to other structures.

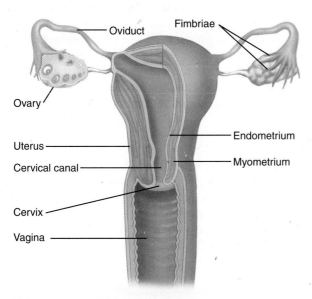

Oviduct
Fimbriae
Ovary
Uterus
Cervical canal
Cervix
Vagina
Endometrium
Myometrium

b) Anterior view of internal organs.

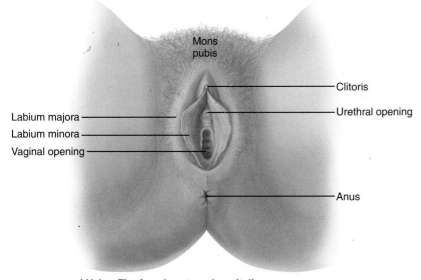

Mons pubis
Clitoris
Urethral opening
Labium majora
Labium minora
Vaginal opening
Anus

c) Vulva: The female external genitalia.

Figure 16.4 The female reproductive system. The labia have been pulled to the sides to expose the vaginal and urethral openings.

The inner layer of the uterus, the **endometrium,** is a lining of epithelial tissue, glands, connective tissue, and blood vessels. A fertilized egg attaches to the endometrium in a process called *implantation*. After implantation the endometrium helps to form the *placenta*. The placenta provides nourishment to the developing fetus and also provides for waste removal and gas exchange.

 Web Animation *Preparation of the Endometrium for Implantation* at www.humanbiology.com

The outer layer of the uterus, or **myometrium,** consists of thick layers of smooth muscle. The myometrium stretches during pregnancy to accommodate the growing fetus. It also provides the muscular force to expel the mature fetus during labor.

The narrow lower part of the uterus is the **cervix.** An opening in the cervix permits sperm to enter the uterus and allows the fetus to exit during birth.

The vagina: Organ of sexual intercourse and birth canal

The cervix joins the **vagina,** a hollow muscular organ of sexual intercourse and also the birth canal. Glands within the vagina produce lubricating secretions during sexual arousal. A thin ring of tissue called the *hymen* may partially cover the opening to the vagina. The hymen is ruptured by the first sexual intercourse, or sometimes by the insertion of tampons or vigorous physical activity.

The vagina is continuous at the body surface with the female external genitalia, collectively called the **vulva** (Figure 16.4c). An outer, larger pair of fat-padded skin folds called the **labia majora** (in Latin, "major lips") surround and enclose the **labia minora** (singular: *labium*), a highly vascular but smaller pair of folds. The **clitoris,** a small organ partly enclosed by the labia minora, is important in the female sexual response. The clitoris and the penis originate from the same tissue during early fetal development, and like the penis, the clitoris contains erectile tissue that is highly sensitive to stimulation. Between the clitoris and the vaginal opening lies

Table 16.2 Summary of the components of the female reproductive system

Component	Function
Ovary (2)	Site of storage and development of oocytes
Oviduct (2)	Duct for transporting oocyte from ovary to uterus; also site of fertilization if it occurs
Uterus	Hollow chamber in which embryo develops
Cervix	Lower part of the uterus that opens into the vagina
Vagina	Organ of sexual intercourse; produces lubricating fluids; also the birth canal
Clitoris	Organ of sexual arousal

the urethral opening. Table 16.2 summarizes the main components of the female reproductive system.

Mammary glands nourish the infant

Human breasts contain **mammary glands,** modified sweat glands that technically are part of the skin, or integumentary system. Though breasts are not required to produce a viable fetus, they undergo developmental changes during puberty in women and they support the infant's development, so we consider them here.

A *nipple* at the center of each breast contains smooth muscles that can contract, causing the nipple to become erect in response to sexual stimulation, cold temperatures, or a nursing infant. Surrounding the nipple is the pigmented *areola*. Internally, the breasts contain the mammary glands, which are specialized to produce milk (**Figure 16.5**). The mammary glands consist of hundreds of small milk-producing lobules. Contractile cells around each lobule allow the milk to be released, and ducts deliver it to the nipple. Most of the breast consists of adipose tissue, so breast size does not indicate the potential for milk production. A network of fibrous connective tissue supports the glands and adipose tissue.

At puberty, female mammary glands enlarge under the control of the hormone estrogen. Both estrogen and progesterone prepare the glands for **lactation,** or the production of milk, late in pregnancy. At birth, prolactin stimulates lactation and oxytocin stimulates the contractions that release

milk during nursing. In males the mammary glands are *vestigial,* meaning that they have no function.

🔄 **Recap** The ovaries secrete estrogen and progesterone, store immature oocytes, and (usually) release one oocyte at a time at intervals of about 28 days. The oocyte travels through the oviduct to the uterus, where implantation occurs if the egg has been fertilized. The vagina is the female organ of sexual intercourse and the birth canal; around its opening are the structures of the vulva. Mammary glands are accessory organs that produce and store milk. ∎

16.3 Menstrual cycle consists of ovarian and uterine cycles

Every month, the ovaries and uterus go through a pattern of changes called the **menstrual cycle.** The menstrual cycle lasts about 28 days and is controlled by hormones. Menstrual cycles begin at puberty and generally continue throughout the reproductive years, except during pregnancy.

A complete menstrual cycle consists of two linked cycles, the *ovarian cycle* and the *uterine cycle.* Together, they periodically release oocytes and prepare the uterus to receive a fertilized egg.

The ovarian cycle: Oocytes mature and are released

The **ovarian cycle** is a regular pattern of growth, maturation, and release of oocytes from the ovary. At birth, a female has approximately one million *primary oocytes* already formed and stored in each ovary, and no more are ever produced. Each primary oocyte has already developed partway through meiosis at birth, but at birth all further development halts

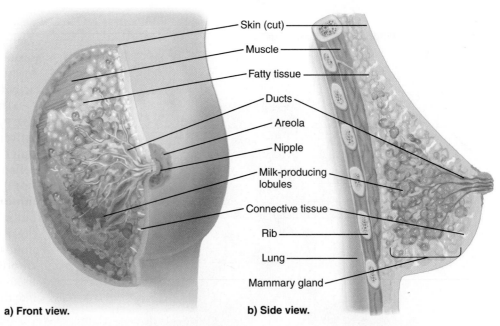

Skin (cut)
Muscle
Fatty tissue
Ducts
Areola
Nipple
Milk-producing lobules
Connective tissue
Rib
Lung
Mammary gland

a) Front view.

b) Side view.

Figure 16.5 The female breast. Anatomical relationships between the milk-producing mammary glands, adipose tissue, and muscle.

until after puberty. By puberty 85% of them have been resorbed, leaving only about 300,000 in both ovaries.

Each month perhaps a dozen primary oocytes start the development process, but typically only one completes it. Only about 400 to 500 oocytes are released during a woman's lifetime, a number that contrasts sharply with the 100–300 million sperm released in a single ejaculation.

Figure 16.6 illustrates the events of the ovarian cycle:

1. A primary oocyte is surrounded by a layer of *granulosa cells* that nourish it. Together, a primary oocyte and its granulosa cells constitute an immature **follicle.** At the beginning of the ovarian cycle, an increase in GnRH from the hypothalamus slowly increases the secretion of FSH and LH by the anterior pituitary. These two hormones stimulate the follicle to grow, enlarge, and change form. FSH and LH also stimulate the follicle (and surrounding ovarian cells) to secrete estrogen.

2. The granulosa cells begin to divide and to secrete a sugary material (glycoproteins) that becomes the *zona pellucida,* a noncellular coating around the oocyte.

3. A fluid-filled space called the *antrum* develops within the follicle, and some of the granulosa cells begin to secrete estrogen and progesterone.

4. The primary oocyte completes stage I of meiosis to yield a *secondary oocyte* and a nonfunctional *polar body.* The mature follicle, sometimes called a Graafian follicle, consists of a secondary oocyte, a polar body, and numerous granulosa cells, all surrounded by ovarian connective tissue cells.

5. Several days before the midpoint of the ovarian cycle, rising levels of estrogen produced by the follicle and surrounding ovarian cells stimulate the pituitary to release a surge of LH. The LH surge triggers **ovulation** at the midpoint of the cycle—the follicle wall balloons outward and ruptures, releasing the secondary oocyte with its polar body, zona pellucida, and surrounding granulosa cells into the extracellular fluid.

At this point the secondary oocyte enters the oviduct and is swept toward the uterus (and toward any sperm that may be swimming toward it) by peristaltic contractions of the oviduct. If a sperm makes contact with the

Figure 16.6 The ovarian cycle. Approximately a dozen primary follicles start this process each month, but generally only one completes it. For any particular primary follicle the events take place in one location, but for clarity the events are shown as if they migrate around the ovary in a clockwise fashion.

secondary oocyte and penetrates its outer membrane, the secondary oocyte quickly completes meiosis to produce a mature egg, or *ovum*, which unites with the sperm. Chapter 21 discusses the developmental events that follow the union of sperm and egg.

6. But the ovarian cycle is not over yet. A **corpus luteum** forms from what is left of the ruptured follicle (the name "luteinizing hormone" comes from the fact that LH causes the corpus luteum to form). The corpus luteum has an important function: It secretes large amounts of progesterone and estrogen. These hormones prepare the endometrium of the uterus for the possible arrival of a fertilized egg that would need the proper environment for further development. The high levels of progesterone and estrogen also signal the hypothalamus to inhibit FSH secretion, temporarily preventing other follicles from developing.

7. If fertilization does not occur, the corpus luteum degenerates in about 12 days. At this point estrogen and progesterone levels fall rapidly, and the ovarian cycle starts again.

If pregnancy does occur, a layer of tissue derived from the embryo called the *chorion* begins to secrete a hormone called *human chorionic gonadotropin (hCG)*. Home-use pregnancy test

MJ's Human Biology Blog

Is She a Woman?

After 18-year-old South African Caster Semenya turned in the fastest time of the year in a women's 800-meter event in 2009, track and field's governing body ordered an investigation into her gender. They're demanding "proof" that she's actually a female, so they've assembled and instructed a committee to determine Semenya's gender. Let's see, now, a geneticist, an endocrinologist, a gynecologist, and a psychologist are getting together to determine the sex of…

It sounds like the beginning of a bad joke, but it's not. Legitimate physiological and anatomical "gender confusion" can be caused by sex hormone imbalances, tissue unresponsiveness to sex hormones, genes that don't turn on properly during fetal development, or even sex-determining genes located on the wrong chromosome (in rare cases, an XY male can develop the characteristics of a female).

Sports federations should develop rules that require gender determination (for the purposes of sports competition) before an athletic event, not afterward. One can only wonder about the emotional damage about to be inflicted unnecessarily on young Caster Semenya. ■

kits work by detecting hCG in urine. The presence of hCG signals the corpus luteum to produce progesterone and estrogen for another 9–10 weeks. By 10 weeks the developing placenta secretes enough progesterone and estrogen to support the pregnancy, and the corpus luteum degenerates. The high levels of estrogen and progesterone that are maintained throughout pregnancy ensure that ovulation does not occur during pregnancy. Pregnancy and fetal development are discussed further in Chapter 21.

✓ **Quick Check** A woman has a pituitary tumor removed, and afterward she is unable to produce LH. Will she still have primary oocytes and can she still ovulate? Explain. ■

The uterine cycle prepares the uterus for pregnancy

The **uterine cycle** is a series of structural and functional changes that occur in the endometrium of the uterus as it prepares each month for the possibility that a fertilized egg may arrive. The uterine cycle is linked to the ovarian cycle. Because the beginning of the cycle is defined as the first day of menstruation, we begin there. Note that although a complete cycle generally lasts about 28 days, many women have cycles that are shorter or longer. **Figure 16.7** summarizes the ovarian, uterine, and hormonal events of a single menstrual cycle of 28 days:

- *Days 1–5: Menstrual phase.* If no fertilized egg from the previous cycle has arrived, body levels of estrogen and progesterone decline precipitously. Without these hormones the uterus cannot maintain its endometrial lining, which has thickened to prepare for a potential pregnancy. **Menstruation** is the process by which the endometrial lining disintegrates and its small blood vessels rupture. The shed tissue and blood pass through the vagina and are excreted. The period of visible menstrual flow, which lasts about 3–5 days, is called the *menstrual period.*
- *Days 6–14: Proliferative phase.* Triggered primarily by rising levels of estrogen from the next developing follicle, the endometrial lining begins to proliferate. It becomes thicker, more vascular, and more glandular. In addition, the lining of the cervix produces a thin, watery mucus that facilitates the passage of sperm, if present, into the uterus from the vagina.
- *Ovulation.* Ovulation occurs midway through the cycle, around the 14th day.
- *Days 15–28: Secretory phase.* Production of estrogen and especially progesterone by the corpus luteum causes the endometrium to continue proliferating, until it becomes almost three times thicker than it was after menstruation. Uterine glands mature and begin producing glycogen as a potential energy source for the embryo. The cervical mucus becomes thick and viscous, forming a "cervical plug" that prevents sperm and bacteria from entering the uterus. These structural changes in the endometrium prepare the uterus for a developing embryo.

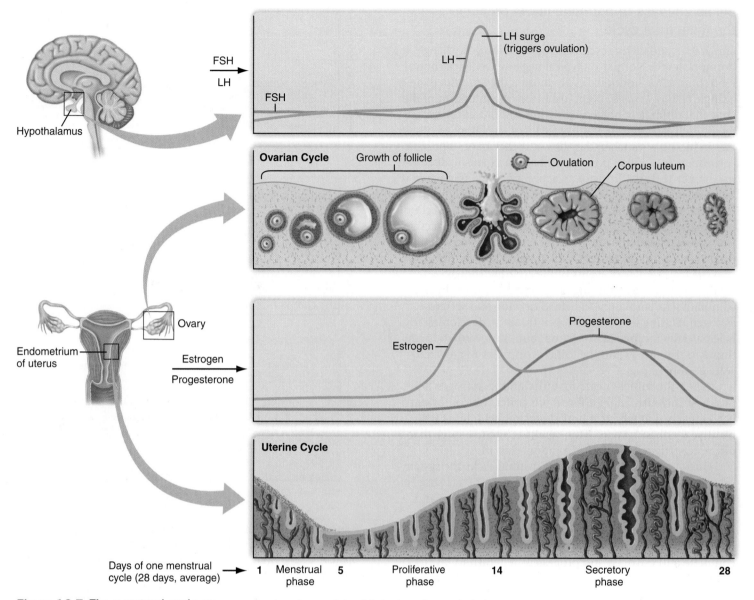

Figure 16.7 The menstrual cycle. The menstrual cycle consists of linked ovarian and uterine cycles. The ovarian cycle is a series of changes associated with oocyte maturation, which depends on the blood levels of FSH and LH. The uterine cycle involves changes in the endometrial lining of the uterus, which depend on the blood levels of estrogen and progesterone.

✔ *What makes progesterone (purple line) rise during the second half of the menstrual cycle, and what makes it fall at the end of the cycle? (Hint: Where is this progesterone coming from?)*

If fertilization does not occur, degeneration of the corpus luteum at about 12 days after ovulation results in declining estrogen and progesterone levels, and another menstrual cycle begins.

The menstrual cycle is not a pleasant experience for some women, to put it mildly. It involves rapid hormonal and physical changes that can sometimes cause pain or mood changes. *Premenstrual syndrome (PMS)* is a group of symptoms often associated with the premenstrual period, including food cravings, mood swings, anxiety, back and joint pain, water retention, and headaches. It generally develops in the

second half of the cycle, after ovulation, and lasts until menstruation begins. Exercise and medications that reduce pain and water retention can help.

Painful menstruation is known as *dysmenorrhea*. The endometrium of the uterus produces prostaglandins, which can trigger contractions and cramping of the uterine smooth muscle. Aspirin and ibuprofen may offer some relief from dysmenorrhea because these drugs inhibit prostaglandin formation. Although ovulation itself is usually not uncomfortable, some women experience a brief sharp pain when the follicle ruptures.

Cyclic changes in hormone levels produce the menstrual cycle

The ovarian and uterine cycles are regulated by cyclic changes in hormone levels. Normally a typical negative feedback control loop controls the hormonal cycles most of the time. But there is a single *positive feedback* event just before the midpoint of the cycle that ultimately triggers ovulation and causes the cycle to repeat itself each month. Positive feedback regulatory events are rare in biology, but when they do occur they have important consequences. Another positive feedback mechanism, discussed in Chapter 11, is the firing of an action potential in a nerve—once the membrane potential is depolarized to threshold, the membrane potential rapidly depolarizes further and an action potential is generated. The cyclic nature of the menstrual cycle is somewhat analogous to an action potential except that it occurs over a month rather than a fraction of a second.

The key regulatory events of the cycle are shown in **Figure 16.8**. At the beginning of the cycle, LH rises slowly, stimulating an immature follicle to develop and secrete estrogen. During this time estrogen from the developing follicle partially inhibits LH secretion, but LH nevertheless continues to rise slowly and the follicle continues to grow and secrete more estrogen.

When the follicle nears maturity and its estrogen secretion is sufficiently high, a threshold concentration of estrogen is reached at which it now *stimulates* LH release (the positive feedback event). This results in the LH surge that ultimately triggers ovulation at about day 14.

During the latter half of the menstrual cycle, the system returns to typical negative feedback control. The LH surge is short-lived because high but steady levels of estrogen and progesterone from the corpus luteum inhibit LH and FSH secretion. Without LH and FSH, the next follicle does not develop. This prevents a second ovulation until the cycle is complete and estrogen and progesterone return to low levels again.

From a functional standpoint, hormonal regulation of the menstrual cycle makes perfect sense. Because only a mature follicle produces enough estrogen to trigger the LH surge, ovulation is triggered at the precise moment that the follicle is ready. And once ovulation occurs, the start of another cycle is delayed until the body determines whether or not pregnancy has taken place. If there is no pregnancy, then estrogen and progesterone return to low levels within about two weeks and the cycle begins again with a slowly rising LH concentration.

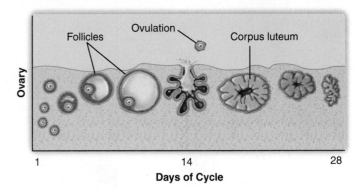

Figure 16.8 Regulation of the menstrual cycle. During the first half of the cycle LH and FSH stimulate follicular growth and development. The growing follicle secretes estrogen, which partially inhibits LH secretion. Shortly before the midpoint of the cycle, estrogen from the mature follicle triggers the LH surge, which in turn triggers ovulation. In the second half of the cycle, estrogen and progesterone from the corpus luteum again inhibit LH and FSH secretion, preventing maturation of another follicle until it is determined (by regression of the corpus luteum and a decline in estrogen and progesterone concentrations) that pregnancy has not occurred. Then the cycle can begin again.

✔️ *Are there any similarities between hormonal control of the menstrual cycle in women, and hormonal control of reproduction in men? (Hint: Review Figure 16.3.)*

↩️ **Recap** During the ovarian cycle, a primary oocyte within a developing follicle divides once to form a secondary oocyte. The follicle ruptures, releases the oocyte, and forms the corpus luteum that secretes progesterone and estrogen. Rising levels of estrogen cause the endometrium to proliferate. If pregnancy does not occur, hormone levels fall and the endometrial layer disintegrates and is shed, a process known as menstruation. Ovulation is triggered by a surge of LH, which in turn is caused by the positive feedback effect of a high concentration of estrogen from the maturing follicle. During the second half of the menstrual cycle, sustained high levels of estrogen and progesterone from the corpus luteum inhibit further ovulation. ▪

Health & Wellness

Erectile Dysfunction and Viagra Abuse

Healthy, young adult men generally do not have any difficulty achieving and sustaining an erection sufficient for sexual intercourse, except on rare occasions. But as they get older many men find that it takes longer to become erect, more stimulation is needed, and the penis becomes less rigid than when they were younger. The occasions when an erection cannot be achieved or sustained may become more common. This is fairly normal. Some decline in sexual function is expected with age.

Erectile dysfunction, on the other hand, is a medical condition. Formerly called impotence, *erectile dysfunction* is defined as the *repeated* inability to develop and maintain an erection sufficient for satisfactory sexual intercourse or activity. By that measure, erectile dysfunction is rare in 20-year-olds, reaches a prevalence of about 5% by age 40, and rises to 15–25% after age 65.

Until 1998, many of the standard treatments for erectile dysfunction were almost worse than the condition. They included vascular surgery, penile implants, vacuum devices to pull blood into the penis, and injections of a drug into the base of the penis with a needle and syringe. Only men with true erectile dysfunction were tempted to try such treatments.

But then in 1998 Viagra (sildenafil) came on the market, followed by Levitra and Cialis. Although all three drugs were developed to treat a medical condition common only in older men, suddenly men of all ages were being diagnosed with erectile dysfunction. Physicians and health care officials are concerned that an increasing number of men are trying these drugs in the belief that they have "recreational" value.

All three drugs relax vascular smooth muscle in the penis, promoting blood inflow to the penis and increasing the chances that sexual stimulation will produce an erection. Note that these drugs do not automatically *cause* an erection—they *allow* one to occur following normal sexual stimulation. They are not aphrodisiacs.

Like most medications these drugs can have side effects, which may include a sudden loss of vision or a drop in blood

pressure. Side effects are more likely if the drugs are taken with cardiovascular medications or street drugs such as ecstasy, methamphetamines, and ketamine.

The bottom line: Viagra, Levitra, and Cialis are not recreational drugs. If you don't need them, don't take them. A drug that increases blood flow to the penis is not a good way to improve your sex life. ∎

16.4 Human sexual response, intercourse, and fertilization

The human sexual response is a series of events that coordinate sexual function to accomplish intercourse and fertilization. The response can have up to four phases in both men and women:

1. *Excitement.* Increased sexual awareness and arousal
2. *Plateau.* Intense and continuing arousal
3. *Orgasm.* The peak of sexual sensations
4. *Resolution.* The abatement of arousal

The male sexual response

In men, excitement is accompanied by pleasurable sensations from the penis and by an erection. Erection occurs when neural activity relaxes (dilates) arterioles leading into vascular compartments within the penis. The compartments fill with blood, and as they fill they press on the veins draining the penis and reduce the rate of venous outflow. As more blood enters the penis and less blood leaves it, the penis swells, lengthens, and stiffens. Heart rate and breathing increase, as does the tone of skeletal muscle. Sights, sounds, memories, and other psychological stimuli; kissing or other physical contact; and especially physical contact with the penis all contribute to excitement. Arousal continues during the plateau phase, which can last for a few seconds or many minutes.

At some point sexual arousal becomes so great that **orgasm** occurs. An orgasm is a brief, intensely pleasurable reflex event that accomplishes **ejaculation,** or the expulsion of semen. During ejaculation, sympathetic nerve activity contracts smooth muscle in the seminal vesicles, epididymis, bulbourethral glands, and prostate to move glandular secretions into the urethra. The internal urethral sphincter closes tightly to prevent urine from leaking into the urethra, and the skeletal muscles at the base of the penis contract rhythmically, forcing semen out of the urethra in several spurts. In a typical ejaculation, the male expels about 3–4 ml (a teaspoon or more) of semen.

During the resolution phase, the erection subsides and breathing and heart rate return to normal. Generally a refractory period follows, during which another erection and orgasm cannot be achieved. The refractory period may last from several minutes up to several hours.

The female sexual response

Women experience the same four phases of the sexual response. As in men, the excitement phase is triggered by sights, sounds, psychological stimuli, and physical stimulation, especially of the clitoris and breasts. Stimulation initiates neural reflexes that dilate blood vessels in the labia, clitoris, and nipples. During arousal the vagina and the area around the labia secrete lubricating fluids. Without these lubricating fluids intercourse can be difficult and sometimes painful for both partners.

The female orgasm, like that of the male, consists of rhythmic and involuntary muscular contractions. Contractions of the uterus and vagina are accompanied by intense feelings of pleasure and sensations of warmth and relaxation. The rate of orgasm is more variable in females than in males. Some women do not experience orgasms frequently, whereas others experience multiple orgasms within one period of sexual contact. Female orgasm is not required for fertilization or pregnancy, nor is arousal, for that matter.

Fertilization: One sperm penetrates the egg

During ejaculation, the male may deposit several hundred million sperm inside the vagina, near the cervix. The sperm have a long way to go, however, before one of them encounters an egg. Random swimming takes some sperm through the cervical opening, through the uterus, and up the two oviducts. The journey can take hours or even a day or more. Fertilization occurs when a sperm encounters an oocyte in an oviduct and penetrates the zona pellucida (Figure 16.9). Some sperm may be viable for up to five days in the female reproductive tract, although most do not last more than two days.

Recap Women and men experience the same four phases of sexual responsiveness. Sexual arousal in the male results in penile erection that leads to orgasm and ejaculation. Females experience sexual arousal and pleasurable orgasms marked by rhythmic muscular contractions. During ejaculation, the male deposits several hundred million sperm in the vagina. Fertilization of the egg by a single sperm occurs within five days, if it occurs at all. ∎

 Web Animation *The Female Reproductive System* at www.humanbiology.com

Figure 16.9 **Colored SEM (× 900) of sperm in contact with an egg.** Only one sperm will fertilize the egg.

16.5 Birth control methods: Controlling fertility

There are many reasons why couples or individuals might choose to control fertility, and many ways to do it. Birth control (contraceptive) methods vary widely in effectiveness (Table 16.3). Some methods reduce the chance of pregnancy from just a single intercourse; others virtually eliminate the chances of pregnancy for a lifetime. An understanding of the advantages and the limitations of each method will help you make an informed choice, should you choose to control your own fertility for any period of time.

Table 16.3 **Approximate failure rates of various contraceptive methods**

Contraceptive method	Annual number of pregnancies expected per 100 women
Spermicide alone	30
Female condom	20
Diaphragm with spermicide	15
"Morning after" pills	15
Male condom	11–16
Hormonal contraceptives	
Oral contraceptives	5
Skin patch	5
Vaginal contraceptive ring	5
Injected hormonal contraceptives	1
Intrauterine devices	1
Sterilization surgery—men	1
Sterilization surgery—women	1

Source: Adapted from *Birth Control Guide*, FDA Office of Women's Health, 2007.

Abstinence: Not having intercourse

The only completely effective method of birth control is to not have sexual intercourse at all. Abstinence, as it is called, works for many people for short periods of time, and throughout life for others. Abstinence is natural in that no intervention or artificial method is employed. When abstinence is not the option of choice, there are still many ways to reduce the probability of pregnancy.

Surgical sterilization: Vasectomy and tubal ligation

Both men and women may opt for surgical sterilization (Figure 16.10). Male surgical sterilization is called *vasectomy*. It can generally be performed in a doctor's office under local anesthesia. The doctor makes small incisions in the scrotum, locates each ductus deferens, and ties them in two places. The segments between the two ties are then removed, preventing sperm from reaching the urethra.

After a vasectomy the testes continue to produce testosterone and sperm, and sexual interest and all secondary sexual characteristics remain unchanged. With no access to the urethra, however, the sperm eventually are resorbed, just as they are if ejaculation never takes place. Generally it is impossible to detect a change in semen volume per ejaculation. If done properly, surgical male sterilization is close to foolproof. However, to be on the safe side, the surgeon will probably recommend doing a sperm count several months after the operation.

Female sterilization involves a *tubal ligation*. The doctor makes a small incision in the woman's abdominal wall, locates each oviduct, and ties (ligates) each at two sites. The sections between the ties are cut away, leaving the oocyte no way to reach the uterus or sperm to reach the oocyte. In a newer version of a tubal ligation, called a *hysteroscopy*, a flexible tube is inserted through the vagina and uterus and into the oviduct. The tube emits electrical current that seals shut the oviduct. Both tubal ligation and hysteroscopy are highly ef-

fective but require surgery that is slightly more difficult than in the male. In addition, tubal ligation may leave two small but possibly visible scars.

A drawback to surgical sterilization in both men and women is that although it *may* be possible to reverse, sterilization generally is permanent. Sterilization should be carefully considered before it is chosen as a method of birth control.

✅ **Quick Check** Would a man be sterile the day after a vasectomy, and would a woman be sterile the day after a tubal ligation? Explain. ∎

Hormonal methods: Pills, injections, patches, and rings

Manipulation of hormone levels provides reasonably effective and safe methods of birth control. The most common method is oral contraceptives, or *birth control pills*. Although the exact mix of hormones in the pills varies, they all administer synthetic progesterone and estrogen in amounts that inhibit the release of FSH and LH. That way, follicles do not mature and ovulation does not take place. It is important to take a pill at the same time every day; if pills are not taken regularly, ovulation (and pregnancy) may occur.

Most birth control pills are taken for three weeks and then replaced with placebo pills for a week so that normal menstrual periods occur each month. However, many physicians have questioned the need for regular monthly periods. In 2003 the FDA approved the first *continuous birth control pill (Seasonale)* that reduces the number of menstrual periods to only four per year. In 2007 the FDA approved a pill that eliminates all periods for a full year (*Lybrel*). Continuous birth control pills are particularly attractive to women who routinely suffer from PMS.

Roughly a third of all women in the United States use birth control pills for at least some period of their lives. Oral contraceptives have several side effects, some beneficial and others harmful. Some women report that oral contraceptives reduce cramps and menstrual flow, and they may offer some protection against cancers of the ovaries and uterus. On the negative side, oral contraceptives can cause acne, headaches, fluid retention, high blood pressure, or blood clots. Women who smoke have a greater risk of blood clots and vascular problems from birth control pills.

To avoid the drawbacks of taking—or forgetting to take—a daily pill, some women opt for hormone injections, skin patches, vaginal rings, or implants. *Depo-Provera* is a progesterone injection that lasts three months; *Lunelle* is a monthly injection of progesterone and estrogen. *Ortho Evra* is a patch that slowly releases hormones through the skin into the bloodstream. The user applies a fresh patch every week for three weeks straight, followed by one week

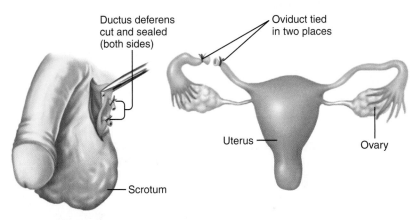

a) Vasectomy. Small incisions are made in the scrotum, and each ductus deferens is tied in two places and cut.

b) Tubal ligation. An incision is made in a woman's abdominal wall, the two oviducts are located, and each is tied in two places and cut.

Figure 16.10 Surgical sterilization.

with no patch, during which she menstruates. *NuvaRing,* a flexible ring about 2 inches in diameter, is inserted into the vagina, where it slowly releases hormones. The ring remains in place for three weeks, then is removed for one week to allow menstruation. Injections, patches, and rings can all have side effects similar to those of hormone pills, including weight gain, headaches, and irregular menstrual periods. The latest in contraception is *Implanon,* a matchstick-size rod containing synthetic progesterone that is implanted under the skin. Implanon provides effective contraception for up to three years.

Hormonal methods of contraception have the advantage of not interfering with sexual activity or foreplay in any way. However, none of them protect against sexually transmitted disease.

A variety of contraceptive methods is shown in **Figure 16.11**.

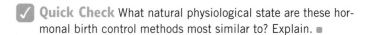 **Web Animation** *Ovulation and Hormonal Birth Control Methods* at www.humanbiology.com

 Quick Check What natural physiological state are these hormonal birth control methods most similar to? Explain. ▪

IUDs are inserted into the uterus

Intrauterine devices, or *IUDs,* are small pieces of plastic or metal that are inserted into the uterus by a health care provider. They should be removed only by a health care provider.

IUDs create a mild chronic inflammation that prevents either fertilization or implantation of the fertilized egg in the uterine wall. On the positive side, they are relatively effective and, like hormonal methods, do not interfere with sexual activity. However, risks include uterine cramping and bleeding, infection, and possible damage to the uterus. IUDs offer no protection from sexually transmitted diseases.

An IUD called Mirena contains small amounts of a progesterone-like steroid. The steroid decreases the possibility of conception and also reduces menstrual flow, a significant benefit to women who normally experience heavy bleeding during their periods. Mirena is effective for up to five years.

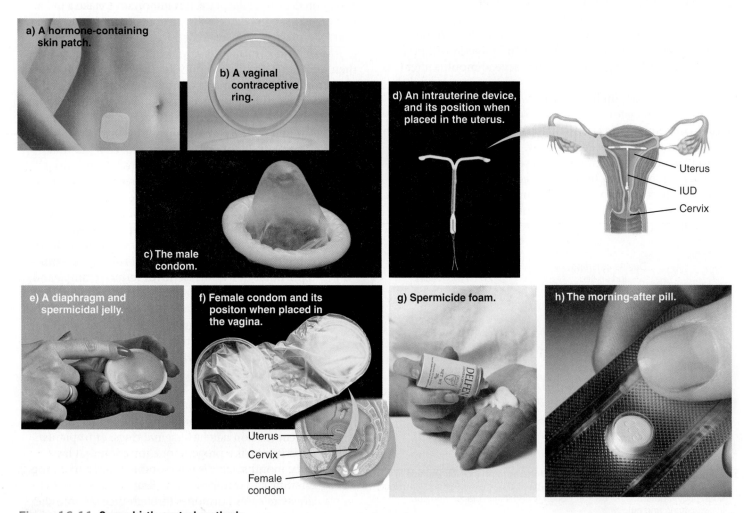

Figure 16.11 Some birth control methods.

MJ's Human Biology Blog

Birth Control Method Failures

On page 390 there's a table of data provided by the U.S. Food and Drug Administration on the failure rates of various contraceptive methods. I've always felt that the reported failure rate for condoms of 11–16% seemed awfully high. Condoms are supposed to work, right?

Now there's an explanation. Recent evidence shows that for both condoms and withdrawal, there's a big difference between "perfect use" and "typical use." If a couple actually uses a condom every time (no exceptions, folks), the annual failure rate is only about 2%. For withdrawal every time, the failure rate is about 4%. The problem is that in the heat of passion some couples "forget," or convince themselves that not using a birth control method just this once won't be a big deal. Under these more typical use conditions, the failure rates of condoms and of withdrawal are indeed closer to 20% per year.

If you're going to rely on these contraceptive methods, don't cheat! ∎

Reference: R.K. Jones. Better Than Nothing or Savvy Risk-Reduction Practice? The Importance of Withdrawal. *Contraception* 79: 407–410, June 2009.

Diaphragms and cervical caps block the cervix

Diaphragms and *cervical caps* are latex devices that a woman inserts into her vagina. They cover the cervical opening and prevent sperm from entering the uterus.

Both types of devices must be fitted initially to the user's cervix. A disadvantage is that they must be inserted shortly before intercourse and then removed sometime later, making sexual activity less spontaneous. However, they are fairly effective, especially if inserted correctly and used with chemical spermicides. Diaphragms and cervical caps do not protect against sexually transmitted disease.

Chemical spermicides kill sperm

Chemical *spermicides*, which destroy sperm, are available in a wide variety of forms—foams, creams, jellies, and douches. Their effectiveness varies. All reduce the risk of pregnancy, but none is 100% effective. Most are inserted into the vagina shortly before intercourse, except douches, which are used after intercourse (and are less effective). Some kill organisms that cause sexually transmitted diseases; others do not.

Condoms trap ejaculated sperm

Despite their decidedly low-tech approach (a condom is just a balloon, really), *condoms* remain one of the most popular forms of birth control. They are fairly effective if used properly. Combining them with a chemical spermicide makes them even more effective. They also offer some protection against sexually transmitted diseases.

The male condom is a soft sheath, made of latex or animal membrane, that encloses the penis and traps ejaculated sperm. There is also a female condom that is essentially a polyurethane liner that fits inside the vagina. Condoms made of latex (but not animal membrane) have the tremendous advantage of offering some protection against disease, including AIDS.

Withdrawal and periodic abstinence

Two other methods, withdrawal and periodic abstinence, are not very effective in preventing pregnancy. In fact, for a sexually active couple they are so ineffective in the long run that some people believe they should not be classified as birth control at all. Couples who are serious about preventing pregnancy should consider more effective techniques.

In principle, the *withdrawal* method means that the man withdraws his penis from the vagina at just the right moment before ejaculation, so that ejaculation occurs outside the vagina. In practice, it takes skill and commitment on the part of the male and a lot of trust on the part of the female. In addition, some sperm can be flushed into the ejaculatory duct even before ejaculation, so there is a possibility of fertilization even if the man does not ejaculate.

People who practice *periodic abstinence* (sometimes called the *rhythm method*) rely on the fact that fertilization is possible for only a limited time in each menstrual cycle, defined by the life span of sperm and the time during which an oocyte can be fertilized after ovulation. As its name suggests, periodic abstinence requires the woman to avoid intercourse for about an eight-day span every month in the middle of her cycle, from approximately five days before ovulation to three days after ovulation. It is possible to determine the day of ovulation by carefully monitoring body temperature, because the temperature rises slightly—several tenths of a degree—at ovulation and stays elevated for about three days. However, it is difficult to determine which five days *precede* ovulation, because obviously it has not yet occurred.

Pills that can be used after intercourse

When intercourse has occurred in the absence of any birth control method, couples have several options that may either prevent pregnancy or terminate it early on. A product called *Preven* is a series of four pills that, if started within 72 hours *after* intercourse, may prevent a pregnancy entirely. Available only by prescription, Preven contains high doses of progesterone and estrogen. It is thought to work by preventing ovulation if it has not already occurred. It may also prevent fertilization even after ovulation, but this is not well documented. Another after-intercourse pill, called *Plan B*, contains progesterone only. Both Plan B and Preven are commonly called "emergency contraception" or "morning after" pills. Plan B is now available without a prescription to persons aged 17 and older.

More controversial is *Mifeprex* (mifepristone, formerly called RU-486), a drug that *blocks* the action of progesterone. By blocking the action of progesterone, Mifeprex can prevent the endometrial lining from proliferating. If taken in the first few days after intercourse, it can prevent pregnancy by preventing implantation of the pre-embryo in the uterine lining. However, Mifeprex is more commonly prescribed *after* a pregnancy has been identified, because its main advantage is that it is effective up to seven weeks after pregnancy. By causing the endometrial lining to regress to the point that it can no longer nourish the fetus, Mifeprex essentially causes a very early-stage chemical abortion. For this reason, Mifeprex has generally been labeled an "abortion pill" rather than an emergency contraception. Mefeprex first became available in France in 1988, but concern over its mode of action caused so much controversy that it was not approved for sale in the United States until 2000.

Elective abortion

An *elective abortion* is an elective termination of a pregnancy. In common usage the term "elective" is generally understood without being stated. Aside from the very early-stage use of Mifeprex, an abortion can be performed in several ways, including vacuum suctioning of the uterus, surgical scraping of the uterine lining, or infusion of a strong saline solution that causes the fetus to be rejected. Sometimes abortions are performed in cases of incest or rape, or when continuing the pregnancy would endanger the mother's health. Some women choose to have an abortion if medical tests reveal a nonviable fetus.

Abortions are highly controversial. Some people support a woman's right to choose her own reproductive destiny, whereas others support an embryo's right to life. The beliefs on both sides are often based on strongly held moral and religious convictions.

The future in birth control

Several new forms of birth control are still in the research and development stage. Possibilities include a male birth control pill that reduces sperm production, and several vaccines for women. For instance, a vaccine is being developed to immunize women against human chorionic gonadotropin, the hormone that maintains the corpus luteum so implantation can occur. Also under development is a vaccine against sperm.

As it stands today, women have most of the choices. It's worth knowing what the options are so that one's choices can be well-informed.

Recap Surgical sterilization should be considered a permanent method of birth control. Hormonal methods—pills, injections, patches, and rings—are also relatively effective but can have side effects. Physical barriers (diaphragms, cervical caps, and condoms) and chemical spermicides are moderately effective; a few afford some protection against disease. IUDs are fairly effective against pregnancy but do not protect against disease. Withdrawal and periodic abstinence are not effective forms of birth control in the long term. Abortion is an elective but controversial procedure that terminates a pregnancy. ■

16.6 Infertility: Inability to conceive

If both partners are healthy, pregnancy is usually not difficult to achieve: The annual pregnancy rate without any birth control averages 85–90%. Timing can improve the odds. For example, if a couple wants to increase their chance of having a child, they might have intercourse near the middle of the menstrual cycle, around the time of ovulation.

But some couples find it difficult to become pregnant. Next we describe conditions that can impair fertility and solutions that are available.

Infertility can have many causes

A couple is considered infertile if they fail to achieve pregnancy after a year of trying. By this definition, about 15% of all couples in the Unite States are infertile. Infertility is about equally attributable to men and women.

Male infertility is the insufficiency or lack of normal, healthy sperm. The chance of any one sperm achieving reproductive success is so low that men whose sperm count is below 60 million per ejaculation are, for all practical purposes, infertile. Reasons for a low sperm count include low testosterone levels, immune disorders that attack the sperm, radiation, certain drugs such as anabolic steroids, and diseases such as mumps and gonorrhea.

Sperm function best at temperatures slightly cooler than body temperature. One simple measure that may raise the viable sperm count is to switch from tight-fitting underwear, which hold the testes close to the body, to looser-fitting boxer shorts.

Causes of female infertility are more variable. One of the most common reasons is *pelvic inflammatory disease (PID)*, a general term for any extensive bacterial infection of the internal female reproductive organs. PID that involves the oviducts can result in scar tissue that seals the oviducts shut, preventing passage of either eggs or sperm (**Figure 16.12**). Abnormal production of LH or FSH may limit the development of oocytes into follicles. Irregular menstrual cycles can make it difficult to time the period of ovulation. Some women produce strongly acid vaginal secretions or thick cervical mucus that damage sperm or make it difficult for sperm to move toward the egg. Uterine tumors may interfere with implantation.

Some women (1–3%) develop *endometriosis,* a condition in which endometrial tissue migrates up the oviduct during a menstrual period and implants on other organs such as the ovaries, the walls of the bladder, or the colon. There it grows, stimulated each month by the normal hormonal cycle. Endometriosis can cause pain and infertility, but can sometimes be corrected by surgery, drugs, or hormone therapy.

Female reproductive capacity is more affected by age than male reproductive capacity. By the mid-40s, women begin to run out of oocytes. In addition, the ovaries become less responsive to LH and FSH, leading eventually to menopause and the end of menstrual cycles. Oocytes ovulated later in life are also more likely to have been damaged by years of exposure to radiation, chemicals, and disease.

A common cause of failure to achieve reproductive success even when the couple is fertile is spontaneous abortion, or *miscarriage,* defined as the loss of a fetus before it can survive outside the uterus. Perhaps as many as one-third of all pregnancies end in miscarriages, some of them so quickly that a woman has no idea she ever conceived.

 Quick Check A woman who has had several previous miscarriages will sometimes be prescribed progesterone pills if she becomes pregnant again. Why? ▪

Enhancing fertility

Several options are available for infertile couples who wish to have a child. None guarantee success and most are expensive. It is not unusual for couples to spend tens of thousands of dollars, even $200,000 or more, to conceive a child by some of these methods. Some couples choose to hire a surrogate mother.

Artificial insemination By far the easiest, most common, and least expensive technique is *artificial insemination.* Sperm are placed with a syringe into the vagina or uterus, as close to the time of ovulation as possible. Artificial insemination is the method of choice when the man's sperm count is low, because his sperm can be collected over a period of time and all delivered at once. It is also a good option for men who produce no sperm, in which case the woman may receive sperm from an anonymous donor who has sold his semen to a "sperm bank," and for single women who make the personal choice to become a parent. However it is not an option when the problem is a woman's failure to conceive. Over 20,000 babies are born in the United States each year as a result of artificial insemination.

Artificial reproductive technologies Artificial reproductive technologies (ART) refers to all techniques in which both sperm and eggs are handled outside the body. ART techniques are the method of choice for couples who are unable to conceive because the woman's oviducts are damaged in such a way that sperm cannot reach the egg. The most usual cause is scarring of the oviduct due to a prior pelvic inflammatory disease.

In all of the various ART techniques, a physician first harvests immature eggs from the woman using a needle guided by ultrasound. Generally the woman is stimulated with fertility-enhancing drugs in advance to ensure that a sufficient number of eggs are ready for harvesting. Sperm is collected from the man. From this point on the methods vary slightly.

The simplest and least expensive ART technique is *in vitro fertilization* (IVF). In IVF the eggs are allowed to mature in laboratory glassware under sterile conditions and then sperm are added. The method is called *in vitro* fertilization because the eggs are fertilized outside the body, that is, under glass (*vitreus* is Latin for "glass"). Two to four days later, when it is clear that fertilization has

Figure 16.12 Pelvic inflammatory disease can make women sterile. Bacteria that migrate from the vagina into the oviducts can inflame the oviducts, leading to scarring and either partial or complete closure of the oviduct.

Oviduct

Blockage caused by scar tissue

Ovary

Uterus

Vagina

occurred because cells are dividing (**Figure 16.13**), the embryo is inserted into the woman's uterus via the vagina. IVF with vaginal insertion results in a live birth only about 20% of the time, so some couples try it several times before achieving pregnancy.

Two variations on the IVF method that improve success rates are *gamete intrafallopian transfer (GIFT)* and *zygote intrafallopian transfer (ZIFT)*. In GIFT, unfertilized eggs and sperm are placed directly into an oviduct (Fallopian tube) through a small incision in the woman's abdomen, without waiting for an embryo to develop. In ZIFT the egg is first fertilized outside the body and then placed into the oviduct (below any point of scarring, of course).

There are now over 430 ART clinics in the United States that can help infertile couples achieve their dream of having a child. The overall success rate at achieving a live birth from an ART cycle (a cycle of fertility drugs, egg collection, and insertion) is now about 30%. Because the chances of pregnancy in any one cycle are low, some women elect to have several embryos transferred at once. Consequently, multiple births are common among women undergoing ART techniques; about a quarter of all live births are twins. Over 50,000 babies are born each year in the United States as a result of ART techniques.

If the woman cannot produce healthy eggs because of an ovarian problem, ART techniques can be performend with eggs donated by another woman. It's also possible to freeze unused embryos from a cycle so they can be used later if the first GIFT or ZIFT attempt isn't successful.

Fertility-enhancing drugs Medications are available to boost the production of developing eggs. Sometimes these drugs are tried before IVF, or given to women who are preparing for IVF so that several eggs may be harvested at once. Occasionally multiple pregnancies can occur as a result of the medications themselves. When you hear news reports of multiple births of four or more babies, the mother was almost certainly taking fertility-enhancing drugs.

Surrogate motherhood Some couples choose to pay another woman, called a *surrogate mother,* to become pregnant and bear a baby for them. The prospective parents may contribute sperm, eggs, or both.

Surrogate motherhood raises a number of difficult legal issues that the couple and surrogate mother should explore before entering into an agreement. The surrogate must confirm legally that she will give up the baby for adoption to the parents. However, there have not yet been enough court cases to define legal precedents should conflicts arise between the surrogate mother and the prospective parents at a later date.

Recap Male infertility is an insufficiency or lack of sperm. Causes of female infertility are variable and include failure to ovulate, damage to oviducts, pelvic inflammatory disease, secretions that impair sperm function, uterine tumors, endometriosis, age-related changes, and miscarriages. The choice of options to improve fertility depends on the cause of the infertility. Options include artificial insemination, *in vitro* fertilization, gamete intrafallopian transfer, zygote intrafallopian transfer, fertility-enhancing drugs, and surrogate motherhood. ▪

16.7 Sexually transmitted diseases

Disorders of the reproductive systems fall into four classes: (1) infertility, (2) complications of pregnancy, (3) cancers and tumors of the reproductive organs, and (4) sexually transmitted diseases. We have already considered infertility and approaches to overcoming it. We discuss potential complications of pregnancy in Chapter 21, and we cover cancer in detail in Chapter 18. In the rest of this chapter, we focus on sexually transmitted diseases.

Sexually transmitted diseases (STDs) are transmitted by sexual contact, whether genital, oral-genital, or anal-genital. STDs are on the rise worldwide, apparently because young people are becoming sexually active earlier, and birth control reduces the fear of pregnancy. In the United States alone, over 12 million people contract an STD each year. A quarter of them are teenagers.

STDs are not just a social embarrassment. Some cause serious organ damage, and a few are deadly. Many don't actually affect the reproductive organs, even though they are transmitted sexually. Some STDs are difficult if not impossible to cure, so an ounce of prevention is definitely worth a pound of (attempted) cure.

Now that we have studied the human reproductive systems, you can see why STDs are so common. Most disease organisms cannot live long outside a warm, moist environment. That's why their most common sites of entry are those that are warm and moist, such as the digestive tract, respiratory tract, and, of course, the reproductive organs. Some organisms,

Figure 16.13 A human embryo at the eight-cell stage, approximately 2-1/2 days after *in vitro* fertilization.

such as the virus that causes AIDS, travel in body fluids. Others, like the pubic lice known as "crabs," simply take the opportunity afforded by intimate contact to hop from one human to another.

STD-causing organisms include viruses, bacteria, fungi, protozoa, and even arthropods. Table 16.4 summarizes common STDs, including causes, symptoms, potential complications, and treatments.

Table 16.4 Sexually transmitted diseases

Disease	Organism	Symptoms	Complications	Comments
Gonorrhea	*Neisseria gonorrhoeae* bacterium	Men: Pus from penis, painful urination Women: Painful urination, vaginal discharge	Men: Inflamed testes and sterility Women: Pelvic inflammatory disease, scarring, sterility	May be asymptomatic, or symptoms may disappear even though disease remains; can be treated with antibiotics
Syphilis	*Treponema pallidum* bacterium	Primary: Chancre at infection site	Secondary: Non-itchy rash, hair loss, gray areas of infection on mucous membranes Tertiary: Widespread damage to cardiovascular and nervous systems, blindness, skin ulcers	Symptoms may disappear spontaneously between phases; can be treated with antibiotics
Chlamydia	*Chlamydia trachomatis* bacterium	Men: Discharge from penis, burning during urination Women: Vaginal discharge, burning, and itching	Pelvic inflammatory disease, urethral infections, sterility, complications of pregnancy	Can be treated with antibiotics
AIDS (also see Chapter 9)	Human immunodeficiency virus (HIV)	Fatigue, fever, chills, night sweats, swollen lymph nodes or spleen, diarrhea, loss of appetite, weight loss	Infections, pneumonia, meningitis, tuberculosis, encephalitis, cancer	Asymptomatic phase can last for years, even though disease is still contagious; no cure
Hepatitis B (also see Chapter 14)	Hepatitis B virus	Nausea, fatigue, jaundice, abdominal pain, arthritis	Cirrhosis, liver failure	Preventable with a vaccine
Genital herpes	Herpes simplex virus types 1 and 2	Genital blisters, painful urination, fever, swollen lymph nodes in groin	Symptoms disappear during remission, then reappear; contagious during active phases	Drugs available to control active phase
Genital warts	Human papillomavirus (HPV)	Warts on penis, labia, anus, or in vagina or cervix	Risk factor for cancers of cervix or penis	Warts can be removed by surgery, freezing, or drugs. Preventable with a vaccine
Yeast infections	*Candida albicans*	Men: Discharge from penis, painful urination Women: Vaginal pain, inflammation, cheesy discharge	Can develop in women without sexual contact, but is transmitted sexually to men	Can be treated with antifungal drugs
Trichomoniasis	*Trichomonas vaginalis* protozoan	Men: Discharge from inflamed penis Women: Frothy, foul-smelling vaginal discharge	Recurrent urethral infections	Can be treated with drugs
Pubic lice	*Phthirus pubis*	Intense itching, skin irritation in pubic area; lice are small but visible	Can also be transmitted through infected sheets, clothing	Can be treated with drugs

Bacterial STDs: Gonorrhea, syphilis, and chlamydia

Gonorrhea Caused by the bacterium *Neisseria gonorrhoeae*, gonorrhea is easily transmitted—roughly 50% of women and 20% of men contract it after a single exposure. It can be transmitted to the mouth and throat by oral-genital contact, or to the mouth or eyes through hand contact. An infected mother can pass gonorrhea to her child during birth, causing a potentially blinding eye infection.

In men, early symptoms include discharge of pus from the penis and painful urination, but the symptoms may disappear even though the disease remains. Women may experience a burning sensation during urination and vaginal discharge that may not appear particularly abnormal. However, nearly 20% of infected men and up to 80% of infected women show no symptoms, at least for periods of time, and so do not seek medical advice. Left untreated, gonorrhea can lead to inflammation, scarring, and sterility.

Gonorrhea is far less common than it was just 20 years ago (**Figure 16.14**). Although gonorrhea generally responds to antibiotics, in recent years antibiotic-resistant strains have emerged.

Syphilis The bacterium *Treponema pallidum* causes syphilis, one of the most dangerous STDs if left untreated. Fortunately it is no longer very common. Syphilis develops in three phases, often separated by periods in which the disease

seems to disappear. One to eight weeks after infection, the disease enters its *primary phase*. A hard, dry, bacteria-filled sore, called a *chancre*, appears at the infection site. The chancre is not painful and disappears spontaneously after several weeks; infected women may not even notice it.

During the *secondary phase*, which can last several years, the bacteria invade lymph nodes, blood vessels, mucous membranes, bones, and the nervous system. Symptoms include a widespread but non-itchy rash, hair loss, and gray patches representing areas of infection on mucous membranes. Again the symptoms may disappear, and sometimes the symptoms end there. However, some people progress to the *tertiary phase* characterized by widespread damage to the cardiovascular and nervous systems, blindness, skin ulcers, and eventual death. A child of a syphilis-infected woman can be born blind, malformed, or dead because the bacteria cross the placenta into the developing fetus. Syphilis can be treated with penicillin.

Chlamydia Chlamydia is a bacterial infection caused by *Chlamydia trachomatis*. Although it is easily cured with antibiotics, it often goes undiagnosed and untreated because the symptoms are so mild. Men may experience a discharge from the penis and a burning sensation during urination, and women may have a vaginal discharge and a burning and itching sensation.

Women are diagnosed with chlamydia three times as often as men. Women between 15 and 24 years old are especially at risk, with reported rates of infection nearly 10 times that of the general population (**Figure 16.15**).

The eventual health consequences for women can be severe despite the apparent mildness of the symptoms. If left untreated, chlamydia can cause pelvic inflammatory disease, permanent scarring of the fallopian tubes, complications of

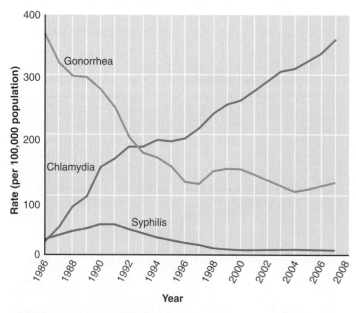

Figure 16.14 Relative rates of infection for the three most common bacterial STDs. Data are expressed in terms of numbers of infection per 100,000 population per year in the United States, since 1986. Syphilis rates refer to primary and secondary syphilis.

✓ *Why do you think chlamydia is increasing, instead of declining like the other two STDs?*

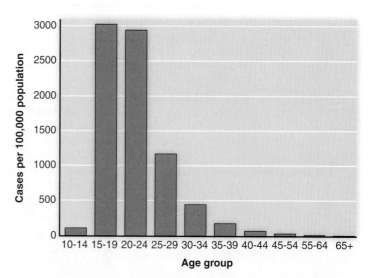

Figure 16.15 Rates of chlamydia infection in women, by age. Young women ages 15–24 are particularly likely to contract chlamydia.

pregnancy, and sterility. It can also infect a newborn's eyes and lungs at birth, leading to pneumonia. Health officials recommend that all sexually active women under the age of 26 be tested for chlamydia every year.

Viral STDs: HIV, hepatitis B, genital herpes, and HPV

The government does not keep accurate statistics on most of the viral STDs. Some of them are very dangerous; others are just very common.

HIV Human immunodeficiency virus (HIV), the virus that causes *acquired immune deficiency syndrome (AIDS)*, is one of the most dangerous STDs of our time. One of the most dangerous features of AIDS is that it can be transmitted from person to person long before symptoms of the disease appear.

HIV slowly destroys the immune system over a period of years. Early symptoms include fatigue, fever and chills, night sweats, swollen lymph nodes or spleen, diarrhea, loss of appetite, and weight loss. As the immune system grows weaker, patients experience numerous infections and complications that lead eventually to death. There are drugs that can keep it in remission, but it cannot be cured. AIDS and its global impact were discussed in detail in Chapter 9.

Hepatitis B Several viruses can cause *hepatitis* (inflammation of the liver). The hepatitis B virus is transmitted in blood or body fluids during unprotected sex. It is more contagious than HIV but not as deadly. A vaccine for hepatitis B became widely available in 1991, and since that time the incidence of hepatitis B has declined. The government requires all health care workers to be vaccinated against hepatitis B because of their risk of being stuck with a needle on the job.

Genital herpes Genital herpes is caused by *herpes simplex* viruses. Generally it is caused by herpes simplex virus type 2 (HSV-2) as a result of sexual contact or skin-to-skin

contact. Less commonly, the type 1 virus (HSV-1) that causes cold sores or blisters around the mouth may be spread to the genital area during oral sex.

Although the government does not keep accurate statistics on genital herpes, it is thought that over 8% of the population is infected. Most infected people have no or minimal symptoms. The primary symptoms are blisters around the genitals or rectum which eventually break, leaving painful sores that are slow to heal (**Figure 16.16**). The blisters may recur from time to time, but usually their number and frequency decrease over a period of years. The virus can stay in the body indefinitely.

There is no cure for genital herpes, but drug treatments are available that can suppress the active (and contagious) phase. Like so many STDs, genital herpes can infect infants' eyes during birth and cause blindness if not treated.

Human papillomavirus (HPV) Infection with 1 of about 30 HPV viruses can cause **warts** in the genital area, including the penis, the vulva (the area outside the vagina), around the anus, and in the vagina or cervix. Genital warts can be removed by several methods including laser surgery, freezing with liquid nitrogen, or certain drugs, but they have a tendency to reappear.

HPV is so common that most women have been infected with HPV by the time they are 50. The infection generally goes away on its own in less than two years. Most people who are infected will have no symptoms at all (fewer than 5% will get warts), so they don't worry about it much. However, it is now known that certain strains of HPV increase the risk of cancers of the cervix or the penis, especially when the infections are prolonged.

In 2006 the FDA approved Gardasil, a vaccine against the strains of HPV that cause cervical cancer. To be effective, however, Gardasil must be given *before* infection. Doctors and the CDC (Centers for Disease Control) strongly recommend that all girls be vaccinated with Gardasil at about age 11 or 12, before their first sexual contact.

a) Genital herpes on the external genitalia of the female. Genital herpes is generally caused by HSV-2.

b) Genital herpes on the penis of the male.

c) Oral herpes. Cold sores on the lips (also called fever blisters) generally are caused by HSV-1.

Figure 16.16 Herpes simplex virus.

The recommendation has sparked controversy in some circles, but the fact remains that if most girls were vaccinated, cervical cancer would be nearly eliminated over the next 50 years.

Other STDs: Yeast infections, trichomoniasis, and pubic lice

Yeast infections *Candida albicans* is a yeast (a type of fungus) that is always present in the mouth, colon, and vagina. Normally it is kept in check by competition from other organisms. Sometimes, however, the *Candida* population may grow out of control in the vagina, leading to pain, inflammation, and a thick, cheesy vaginal discharge. A woman with an active yeast infection can pass it to her partner. Infected men sometimes develop a discharge from the penis accompanied by painful urination. It is easily treated by antifungal drugs.

Trichomoniasis Another condition that can cause *vaginitis* (inflammation of the vagina) is *trichomoniasis*, caused by the protozoan *Trichomonas vaginalis*. In women it causes a frothy, foul-smelling vaginal discharge; in men it inflames the penis and causes a discharge. Trichomoniasis can be transmitted even when there are no symptoms, and reinfection is common if partners are not treated simultaneously. It is treatable with drugs.

Pubic lice Pubic lice are tiny arthropods related to spiders and crabs (**Figure 16.17**). Commonly called "crabs," pubic lice live in hair but especially prefer pubic hair, jumping from one host to the next during sexual contact. They can also be transmitted by infected bed sheets or clothes. The adult lice lay eggs near the base of the hair, and the eggs hatch in a few days. Pubic lice nourish themselves by

Figure 16.17 The pubic louse, *Phthirus pubis*.
(Approximately × 30)

sucking blood from their host, causing intense itching and skin irritation; the adults are very small but visible with the naked eye. They can be killed with anti-lice treatments, but all clothes and bedding should be thoroughly washed in hot water to prevent reinfestation.

 Quick Check Overall, which sex do you think is more likely to get STDs—women or men? Why? ■

Protecting yourself against STDs

We've just reviewed a group of diseases with symptoms ranging from painful urination to brain damage and outcomes ranging from full recovery to death. The good news is that you can probably avoid contracting an STD, but you need to know what to do and, more importantly, be willing to do it.

In Chapter 9 we present a number of safer-sex recommendations that can lower your risk of STDs. Many of them are commonsense precautions:

- Select your partner wisely. A monogamous relationship with someone you know and trust is your best defense.
- Communicate with your partner. If you have questions or concerns, share them, and be willing to do the same yourself.
- Use suitable barriers, depending on the sexual activity in which you engage. Be aware, though, that barrier methods of birth control are not a cure-all for risky behavior. For example, condoms are only marginally effective in preventing the transfer of HIV, HPV, and genital herpes, and they will not protect at all against pubic lice.

Finally, know your own health. If you or your partner think you are at risk for or may have a certain disease, get tested. It's confidential, it's not difficult, and it can save you a lot of embarrassment. It may even save your life.

Recap Major bacterial STDs include gonorrhea, syphilis, and chlamydia. The most dangerous viral STD is HIV. Hepatitis B can be prevented by a vaccine. Genital herpes is irritating but not particularly deadly. HPV can cause warts and is a risk factor for cervical cancer—it, too, can be prevented by a vaccine. Yeast, normally present in the vagina, can multiply and cause a yeast infection. Pubic lice are tiny arthropods that are transmitted during intimate contact or by contact with clothes or bedding. You can reduce your risk of contracting an STD with a little effort. Choose your partner wisely, use a barrier method of birth control, and, if you suspect you have a disease, get tested promptly. ■

Chapter Summary

The male reproductive system delivers sperm *p. 378*

- Sperm are produced in the male reproductive organs, called the testes, and stored in the epididymis and ductus deferens.
- Semen contains sperm and the secretions of three glands: the seminal vesicles, the prostate gland, and the bulbourethral glands.
- Tens of millions of sperm are formed every day throughout the male's adult life. The production of sperm is under the control of three hormones: testosterone, LH, and FSH.

The female reproductive system produces eggs and supports pregnancy *p. 382*

- The female reproductive organs (the ovaries) produce mature oocytes and release them one at a time on a cyclic basis.
- Fertilization of the oocyte (if it occurs) takes place in the oviduct.
- The fertilized egg makes its way to the uterus, where it implants and begins to develop into a fetus.
- The vagina contains glands that produce lubricating fluid during sexual arousal.
- The hormone estrogen causes the mammary glands to enlarge at puberty.

Menstrual cycle consists of ovarian and uterine cycles *p. 384*

- The cyclic changes in the female reproductive system are called the menstrual cycle. The menstrual cycle consists of an ovarian cycle that produces mature oocytes, and a uterine cycle in which the uterus prepares for pregnancy.
- The menstrual cycle is controlled by the hormones estrogen, progesterone, LH, and FSH. A surge in LH secretion triggers ovulation.

Human sexual response, intercourse, and fertilization *p. 389*

- Both males and females are aroused by certain stimuli and respond in ways that facilitate intercourse and ejaculation. In males, the penis swells and hardens. In females, glandular secretions provide lubrication.
- Both males and females experience a pleasurable reflex event called an orgasm. Orgasm is accompanied by ejaculation, or the expulsion of semen, in the male.
- Sperm deposited in the vagina during intercourse make their way through the cervix and uterus and migrate up the oviducts to the egg. Only one fertilizes the egg.

Birth control methods: Controlling fertility *p. 390*

- The most effective methods of preventing pregnancy are abstinence and female or male surgical sterilization. Sterilization should be considered permanent, though it may be reversible.
- Other effective methods of birth control include intrauterine devices (IUDs) and manipulation of hormone levels with pills, patches, injections, or vaginal rings.
- Moderately effective methods of birth control include condoms, cervical caps and diaphragms, and various spermicides. The "natural" methods (withdrawal and periodic abstinence) are the least effective.
- Morning-after pills are now available to be used as emergency contraceptives up to 72 hours after intercourse.

Infertility: Inability to conceive *p. 394*

- Infertility is defined as the failure to achieve a pregnancy after a year of trying. Nearly 15% of all couples are infertile.
- For male infertility, the primary option is artificial insemination. When the male has a low sperm count but still some viable sperm, the sperm can be concentrated before insemination. In many cases, donor sperm are used.
- Female infertility may be overcome by several methods. In *in vitro* fertilization, egg and sperm are combined under laboratory conditions. The fertilized egg is allowed to develop for several days and then inserted into the uterus. In GIFT, the egg and sperm are inserted directly into an oviduct immediately after collection. In ZIFT, the egg and sperm are combined first, and then the fertilized egg is inserted into the oviduct.

Sexually transmitted diseases *p. 396*

- The common feature of sexually transmitted diseases (STDs) is that they are transmitted during sexual contact. Their disease effects are not necessarily on the reproductive system.
- Bacterial STDs include gonorrhea, syphilis, and chlamydia. Syphilis is the most dangerous; chlamydia, the most common. All are treatable with antibiotics.
- Viral STDs include HIV (the virus that causes AIDS), hepatitis B, genital herpes, and genital warts (HPV). HIV is particularly deadly, and there is as yet no cure. There are now vaccines for hepatitis B and HPV.
- Avoiding the diseases caused by STDs is a matter of reducing your risk of exposure and paying attention to (and taking responsibility for) your own health.

Terms You Should Know

cervix, *383*
clitoris, *383*
corpus luteum, *386*
ductus (vas) deferens, *379*
egg, *379*
endometrium, *383*
epididymis, *379*
erection, *379*
estrogen, *382*
follicle, *385*
gametes, *380*
menstrual cycle, *384*
menstruation, *386*
orgasm, *389*

ovarian cycle, *384*
ovaries, *382*
oviduct, *382*
ovulation, *385*
penis, *379*
progesterone, *382*
semen, *379*
sperm, *378*
testes, *379*
testosterone, *381*
uterine cycle, *386*
uterus, *382*
vagina, *383*

Concept Review

1. Name the three accessory glands of the male reproductive system.
2. Describe the development of a sperm cell.
3. Explain where fertilization takes place in the female reproductive system.
4. Describe the phases of the uterine cycle.
5. Describe what is happening in the uterus that leads to menstruation, and explain why menstruation does not occur during pregnancy.
6. Compare and contrast the male and female sexual responses.
7. Describe how in vitro fertilization is done.
8. Name three viral STDs.
9. Describe birth control options, and list those that afford at least some protection against STDs.
10. Describe the various STDs of worldwide concern.

Test Yourself

1. Which of the following might interfere with a sperm's ability to swim?
   a. an acrosome with defective enzymes
   b. a defective flagella
   c. a defective midpiece lacking an appropriate number of mitochondria
   d. both choice (b) and (c)

2. Anabolic steroids are similar in structure and function to testosterone. Which of the following would be a likely outcome in a male taking anabolic steroids?
   a. Testosterone production by the interstitial cells would increase.
   b. Release of GnRH by the hypothalamus would increase.
   c. Testosterone production by the interstitial cells would decrease.
   d. Testosterone production would be unaffected.

3. Which of the following structures is correctly matched with its function?
   a. myometrium: contracts during labor and child birth
   b. cervix: transports oocyte from ovary to uterus
   c. oviduct: transports oocyte from uterus to vagina
   d. endometrium: birth canal

4. Which of the following statements about oocytes and ovulation is correct?
   a. The production of primary oocytes continues in women from the onset of puberty until the onset of menopause.
   b. Meiotic divisions are complete at the time of ovulation.
   c. At ovulation, the structure released from the follicle includes a secondary oocyte and polar body.
   d. All of these statements are correct.

5. Which of the following would provide the most accurate prediction of ovulation?
   a. a surge of progesterone
   b. a sudden drop in FSH
   c. a sudden drop in estrogen
   d. a surge of LH

6. What is the role of the corpus luteum during early pregnancy?
   a. It produces the hormones that will support the early pregnancy.
   b. The fertilized egg will implant in the corpus luteum.
   c. The corpus luteum nourishes the rapidly growing embryo.
   d. The corpus luteum protects the fertilized egg during its transport to the uterus.

7. Which of the following lists the female reproductive structures that sperm will pass through on thir way to fertilize an oocyte, in the correct order?
   a. cervix…vagina…oviduct…ovary
   b. cervix…oviduct…vagina…uterus
   c. uterus…cervix…vagina…ovary
   d. vagina…cervix…uterus…oviduct

8. The prescription drugs used currently to treat erectile dysfunction primarily affect:
   a. testosterone levels
   b. blood flow to the penis
   c. contraction of skeletal muscles
   d. sympathetic nervous system

9. Which two means of birth control are most similar in the way they prevent pregnancy?
   a. tubal ligation and IUDs
   b. oral contraceptives and spermicides
   c. tubal ligation and condoms
   d. oral contraceptives and hormone patch

10. The most widely used hormonal methods of birth control:
    a. prevent fertilization by killing sperm
    b. prevent ovulation by inhibiting release of FSH and LH
    c. prevent implantation into the endometrium
    d. block transport of fertilized egg to uterus

11. If a woman is infertile as a result of oviducts blocked by the scarring of pelvic inflammatory disease, which of the following methods would enable her to become pregnant?
    a. artificial insemination
    b. hormonal treatments
    c. *in vitro* fertilization
    d. all of the above

12. Which of the following STDs is most likely to infect a fetus and cause birth defects?
    a. syphilis
    b. gonorrhea
    c. chlamydia
    d. trichomoniasis

13. Which of the following STDs can be prevented by vaccination?
    a. genital herpes
    b. syphilis
    c. gonorrhea
    d. hepatitis B

14. Which of the following STDs is associated with the development of cervical cancer?
    a. genital herpes
    b. human papillomavirus infection
    c. hepatitis B infection
    d. trichomoniasis
15. Pelvic inflammatory disease may be associated with:
    a. chlamydia
    b. syphilis
    c. gonorrhea
    d. both (a) and (c)

Apply What You Know

Answers can be found at the Human Biology Place.
www.humanbiology.com

1. Male sexual responsiveness and secondary sex characteristics remain unchanged after a vasectomy. Explain why this is so.

2. What keeps the ovarian and uterine cycles always exactly in phase with each other, even though the length of the menstrual cycle can vary slightly?
3. Why is it important for a new mother to breastfeed her baby immediately after birth if she wishes to breastfeed the baby once she gets home?
4. What provides the force that pushes the baby out of the uterus during childbirth?
5. Both women and men can have an STD and not have any symptoms. What might be the long-term consequences of an untreated STD?
6. If a woman does not have an intact hymen does that mean she's not a virgin?

17

Cell Reproduction and Differentiation

Immunofluorescent light micrograph of a mammalian cell during anaphase of cell division. Paired sister chromatids (blue) and the microtubules of the mitotic spindle (green) are being pulled apart by the centrioles.

Should We Clone Humans?

In 1997 researchers announced they had cloned an adult sheep. The clone, named Dolly, was genetically identical to the adult from which she was derived. The feat shocked the world, because until that time it was presumed that a fully differentiated cell from an adult could never be made to go back and start over again. Since then, cloning techniques have produced a variety of pet and farm animals, but so far no humans.

A clone is a copy. The popular media generally think of a clone only as a copy of an adult organism, but scientists use the term to describe the copying of cells, genes, and molecules as well. They've been doing it successfully for years. Using genetic engineering techniques, scientists now routinely insert human genes into bacteria, yeast, and even large animals to produce cloned human hormones in large quantities. Human erythropoietin (to treat anemia), insulin (for diabetes), and growth hormone

British embryologist Professor Ian Wilmut with "Dolly," the first cloned sheep.

(for endocrine disorders) are produced this way. Human cells can also be cloned to produce specialized tissues. Scientists who grow layers of skin cells to treat badly burned patients are practicing a form of cloning.

Should we allow the cloning of adult humans? If you and your partner were infertile, would you want to clone (copy) just one of you? Could we stop human cloning now even if we wanted to? The answer to the first question is a societal decision, up to all of us. The answer to the second question would be a personal choice, but whether you will have that choice depends on how society answers the first question. And the answer to the third question is a political and law enforcement issue.

Adult animals generally are cloned by a technique known as *somatic cell nuclear transfer* (see Section 17.6). Basically, the nucleus of a cell from an adult animal is fused with a fertilized egg from which the nucleus has been removed, producing an egg with the genetic makeup of the adult. When implanted into a surrogate mother, the egg will develop into a genetic twin of the adult, born of a different generation. The adult and its clone will *look* like identical twins separated by a big age difference, but keep in mind that individuals are shaped by their experiences and their environments as well as by their genes.

Except for the novelty factor of actually accomplishing human cloning for the first time, human cloning is likely to be of little practical value in the near future. It's easy enough to produce babies the natural way, or by *in vitro* fertilization (IVF) techniques. And there are some good reasons *not* to allow human cloning, at least at the present time.

Biological Issues

With sexual reproduction, sperm or eggs with damaged chromosomes generally don't survive. This eliminates potentially harmful mutations from the human gene pool before fertilization takes place. With cloning by

Snuppy, the first successfully cloned Afghan hound, sits with his genetic father.

somatic cell nuclear transfer, however, any defects in the cell used for cloning would be passed on to the clone. Indeed, some cloned animals show genetic and developmental abnormalities and have poor survival rates. Dolly, the famous sheep who started it all, had to be euthanized at age 6 after developing early arthritis and a lung tumor. There is a very real possibility that the first human clones would be born with birth defects. Who would be responsible, and how would we react if a substantial number of human clones were created that required expensive medical care? Is there any reason to take the risk?

Scientists are unsure whether such problems are due to accumulated mutations, unstable donor cells, or errors in the cloning procedure itself. Better information will probably come from further experiments with animals.

Political and Legal Issues

When Dolly was born in 1997, there were no laws specifically addressing human cloning because they weren't considered necessary. That changed quickly under President Bush, who placed a ban on the use of federal funds for the creation of human embryos for research purposes.

The government did, however, allow the use of federal funds for research using human stem cell lines that were already in existence at the time of the ban.

Under the Obama administration, the rules governing use of human cell lines derived from embryos have been relaxed somewhat (see the Current Issue feature in Chapter 3). As a result, research in the area of *therapeutic cloning* (cloning for the benefit of human patients) is picking up again (see Section 17.7). Nevertheless the official view of the U.S. mission to the United Nations is that all human *reproductive cloning* (cloning an adult human for reproductive purposes) should be banned.

The political and ethical debate in the United States has had a dampening effect on cloning research in the country. Nevertheless, research continues in other nations where it is legal and regulated, such as South Korea, China, Japan, and the United Kingdom. Researchers in South Korea made global headlines awhile back when it was discovered that they had successfully produced human blastocysts (early-stage human embryos) in about 10% of their cloning attempts. Researchers in the United Kingdom have achieved a comparable track record. Although these blastocysts were produced solely for the purposes of therapeutic cloning—their cells were to be harvested only to produce cell lines for therapeutic uses in patients—these advances increase the likelihood of human cloning as well. No cloned humans have been born yet. A few scientists have claimed publicly that they are planning to clone a human, but they have not followed through. These are clear warning signs that regulation and oversight are sorely needed.

Successful human cloning is not likely to be far off. The time to deal with the social, moral, ethical, and legal issues is now, not later.

Questions to consider

1 Do you think the potential benefits of therapeutic cloning justify research with human stem cells derived from human embryos? Defend your position.

2 Should research leading to cloning adult humans be banned? Why or why not? If so, how would you do it?

>> **All cells in every living organism were created by division of previously existing cells.** The exception is a fertilized egg, produced from the union of two cells (sperm and egg).

>> **Before a living cell divides in two, its genetic material must be copied completely and accurately.**

>> **In most multicellular organisms, cells *differentiate* during development,** taking on specialized forms and functions.

>> **Cell division and differentiation are highly regulated.** Some cells rarely (if ever) divide in an adult organism; others continue to divide throughout life.

Figure 17.1 The cell cycle. The cell grows throughout all three phases of interphase, though most of the growth is in G_1. During the short mitotic phase, the nucleus divides first (mitosis) and then the cell divides (cytokinesis).

Cells reproduce by dividing in two. Indeed, cell division (cell reproduction) is one of the defining features of life. For single-celled organisms, cell division is the equivalent of organism reproduction. For multicellular organisms, cell division and growth allow the organism to grow in size.

You started life as just a single cell created by the union of your father's sperm and your mother's egg. By the time you were born you were already composed of over 10 trillion cells, each created by cell division. Cell division continues throughout childhood and adolescence as the body grows and matures. Even after adulthood, some cells continue to divide in order to replace cells that die or become damaged. For instance, your 25–30 trillion red blood cells are completely replaced every 120 days, which works out to an astonishing 175 million cell divisions every minute.

The mechanisms of cell division are the same in all eukaryotes (organisms with cell nuclei) regardless of organism size. In this chapter we see how the cell's genetic material is copied to prepare for cell division, how the cell divides in two, and how cell growth and division are regulated. Finally, we discuss how cells differentiate to take on special forms and functions.

17.1 The cell cycle creates new cells

The creation of new cells from existing cells involves a repetitive sequence of events known as the **cell cycle** (Figure 17.1). The cell cycle consists of two distinct periods, called *interphase* and the *mitotic phase.*

Interphase is a long growth period during which the cell grows and DNA is duplicated in preparation for the next division. Interphase is divided into three subphases as follows:

- **G_1 phase.** The cell cycle begins with G_1 ("first gap"), the period between the last cell division and DNA synthesis. At the start of G_1 the cell is at its smallest size. G_1 is a period of very active cell growth.
- **S phase.** The S stands for "synthesis" of DNA. During the S phase the cell's chromosomes are duplicated. Growth continues throughout the S phase, although at a slower pace.
- **G_2 phase.** During G_2 ("second gap"), the cell continues to grow slowly as it prepares for cell division.

The **mitotic phase** is a much shorter period during which the nucleus and then the cytoplasm divide. The mitotic phase consists of (1) *mitosis,* during which the duplicated DNA is divided into two sets and the nucleus divides, and (2) *cytokinesis,* during which the cytoplasm divides and two new cells, called "daughter" cells, are formed. (The two new cells are called daughter cells regardless of the person's sex.) This cycle repeats itself over and over.

In mammals, a complete cell cycle takes about 18–24 hours if the cells are actively dividing. (Cell cycles as short as 2 hours have been recorded in some nonmammalian species.) DNA replication during the S phase in mammalian cells takes about 7–8 hours. Mitosis and cytokinesis generally take only about 30–45 minutes.

Most of the cells in your body eventually enter a nongrowing, nondividing state called **G₀**. In tissues that do not need a steady supply of new cells, G_0 is a healthy condition that stops cells from proliferating unnecessarily. Cells in some tissues can be called back from G_0 to begin dividing again, but other tissues cannot restart division. Neurons and osteocytes, for example, generally remain in G_0 forever after adolescence.

In Chapter 18 we discuss cancer, in which the process of cell division spirals out of control.

> **Recap** Cell reproduction is required for growth and to replace cells throughout life. The cell cycle consists of a long growth phase (interphase), during which the cell's DNA is replicated, and a shorter phase (mitotic phase), during which the nucleus and then the cell cytoplasm divide. ◾

17.2 Replication, transcription, and translation: An overview

As described in Chapter 2, **DNA (deoxyribonucleic acid)** is a double-stranded string of nucleotides intertwined into a helical shape. Three billion base pairs of human DNA are packed onto 46 separate structures called **chromosomes**, which organize and arrange the DNA within the nucleus. Chromosomes also contain proteins called *histones* that confer a certain structure to the chromosome molecule. The number of chromosomes varies among species of organisms, but it is always the same for all the individuals of a species. Humans have 46 chromosomes in their cell nuclei.

Throughout most of the cell cycle the chromosomes are not visible because they are so long and thin. During interphase they appear only as a diffuse, grainy substance known collectively as *chromatin* material. However, just before the nucleus divides, the duplicated chromosomes condense briefly into a shorter, compact shape that can be seen under the microscope (**Figure 17.2**). Chromosomes are moved around a lot during nuclear division, and their compactness at this stage helps prevent the delicate chromatin strands from breaking. Each visible duplicated chromosome consists of two identical *sister chromatids* held together by a *centromere*.

Mitochondria also contain a very small amount of DNA. The single human mitochondrial DNA molecule is not organized into a chromosome and contains only 37 genes, about 5% of the genes needed for mitochondrial function. Scientists believe that mitochondrial DNA is a holdover from a time in evolutionary history when mitochondria were independent organisms. At some

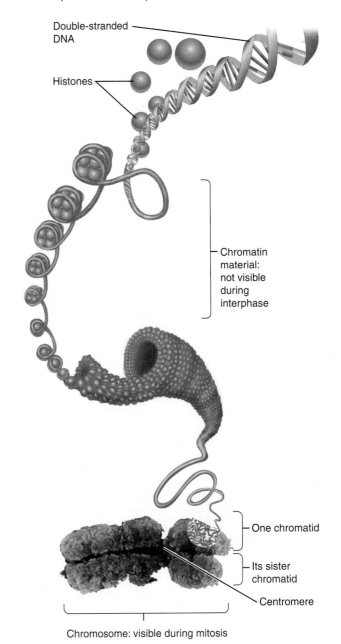

Double-stranded DNA

Histones

Chromatin material: not visible during interphase

One chromatid

Its sister chromatid

Centromere

Chromosome: visible during mitosis

Figure 17.2 The structure of a chromosome. DNA combines with histone proteins to become long, thin strands of chromatin material. During most of the cell cycle the chromatin material of a chromosome is not visible within the cell. During mitosis the chromatin material condenses and the duplicated chromosome becomes visible as two identical sister chromatids held together by a centromere. The two sister chromatids separate during nuclear division.

point, mitochondria were incorporated into eukaryotic cells and became dependent on them.

Because DNA represents essentially all instructions for life, every cell in the organism must contain exactly the same set of DNA. It follows that the DNA must be duplicated every time a cell divides. DNA **replication** is the process of copying the DNA prior to cell division.

A **gene** is a short segment of DNA that contains the code, or recipe, for one or more proteins. A single gene is the smallest functional unit of DNA. There are approximately 20,000 genes on the 46 chromosomes, each one located at a precise position on a particular chromosome. The chromosomes are too large to pass through the nuclear pores, so they remain within the nucleus. For a gene to be useful to the cell, its code must first be converted to a form that can leave the nucleus and enter the cytoplasm. This is the function of single-stranded **ribonucleic acid (RNA),** more specifically a certain form of RNA called **messenger RNA (mRNA),** described in more detail later.

Transcription is the process by which the DNA code of a single gene is converted into a complementary single strand of mRNA. Messenger RNA is small enough to pass through the nuclear pores and enter the cytoplasm. In the cytoplasm, mRNA is used as a template (or pattern) to create a precise sequence of amino acids that constitutes one particular protein. The process of converting the mRNA template code into one or more proteins is called **translation.**

Figure 17.3 summarizes replication, transcription, and translation. We now look more closely at these three processes.

Figure 17.3 Replication, transcription, and translation. DNA is replicated before every cell division. Within the nucleus the genetic code for a protein is first transcribed into mRNA. mRNA transfers the code to the cytoplasm, where it is translated into the correct sequence of amino acids for one or more proteins.

Answers to ✓ **questions can be found in Appendix A.**

✓ **Quick Check** Explain in your own words the difference between genes, chromosomes, chromatin, and chromatids. ∎

Replication: Copying DNA before cell division

Before a cell can divide, its DNA must be duplicated so that the two new cells resulting from the division each have an identical set of genes. This process, called *replication*, begins with the uncoiling of the DNA helices. As described in Chapter 2, the nucleotides in each strand of the double-helical DNA molecule consist of a sugar group, a phosphate group, and one of four bases (adenine, thymine, guanine, or cytosine). The double helix is formed as the bases pair up, thymine with adenine and guanine with cytosine. This precise pairing facilitates the replication of DNA, because upon the uncoiling and "unzipping" of the two strands, each single strand can serve as a template for the creation of a new complementary strand. The precise nature of the base pairing of the nucleotides (T–A and G–C) ensures that the new complementary strand is exactly like the original complementary strand (**Figure 17.4a**). The new nucleotides are positioned and linked together by enzymes called *DNA polymerases*.

The unzipping of the DNA strands does not proceed linearly from one end of the DNA molecule to the other. Instead, the DNA strands become detached from one another at multiple locations along the length of the molecule. Were this not so, replication would take too long, considering the huge number of nucleotides involved. Because there are 3 billion nucleotide pairs in the human set of DNA, 6 billion nucleotides have to be added to the unzipped single strands in order to replicate the DNA entirely. Even if the cell replicated all 46 DNA molecules simultaneously, it would have to add an average of 5,000 nucleotides per second to each developing new strand of DNA to get the job done during the 7–8 hours of the S phase. That is an incredibly fast rate of unzipping and complementary base pairing—too fast when accuracy is essential.

What actually happens during replication is much more efficient. Certain enzymes bind at various points along the DNA molecule and gently unwind and "unzip" a portion of the DNA, creating a "replication bubble" in the two strands. New complementary strands begin to form at these bubbles. Replication proceeds outward in opposite directions in each strand until the expanding replication bubbles join (Figure 17.4b). Because replication occurs at thousands of sites along the molecule at once, it can conclude much more quickly.

Once replication is complete the two identical sister chromatids remain attached at a single point called the *centromere* (refer back to Figure 17.2). The centromere holds the sister chromatids together until they are physically pulled apart during mitosis. **Figure 17.5** illustrates how the

Keys:

= Cytosine
= Adenine
= Guanine
= Thymine

Replication bubble

Parent DNA molecule

2 complete daughter DNA molecules

Parent strands

New complementary strands

b) The unwinding and the formation of new strands occur simultaneously at many sites on the DNA molecule. The sites of replication expand outward until they join. For simplicity the two strands are shown as parallel in (b), but in actuality they form a helical shape as shown in (a).

Parent strand

New (daughter) strands forming

Parent strand

a) The double-stranded DNA unwinds, and each single strand serves as a template for a new complementary strand.

Figure 17.4 **How DNA replicates.**

structure of a chromosome changes during the various stages of the cell cycle.

✓ **Quick Check** Roughly how many copies of the DNA polymerase enzyme are present in a cell during replication—one, two, a few dozen, or several thousand? Explain your answer. ▪

Mutations are alterations in DNA

Alterations in DNA are called **mutations.** Mutations may result from mistakes made during DNA replication. In addition, chemicals or physical forces can damage a segment of DNA before it is replicated. Unless these errors are detected and corrected before the next replication, they may be passed on to future cells or could even prevent the DNA from being copied at all. In other words, the DNA needs to be in as good a shape as possible before it is replicated.

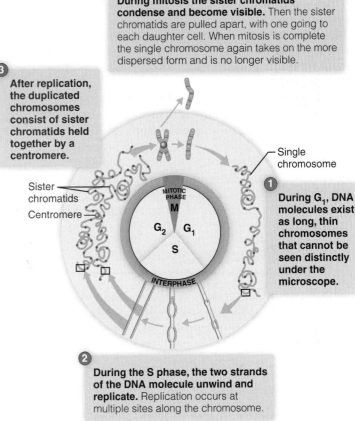

④ **During mitosis the sister chromatids condense and become visible.** Then the sister chromatids are pulled apart, with one going to each daughter cell. When mitosis is complete the single chromosome again takes on the more dispersed form and is no longer visible.

③ **After replication, the duplicated chromosomes consist of sister chromatids held together by a centromere.**

Sister chromatids

Centromere

Single chromosome

① **During G₁, DNA molecules exist as long, thin chromosomes that cannot be seen distinctly under the microscope.**

MITOTIC PHASE
M
G₂ G₁
S
INTERPHASE

② **During the S phase, the two strands of the DNA molecule unwind and replicate.** Replication occurs at multiple sites along the chromosome.

Figure 17.5 **The structure of DNA throughout the cell cycle.**

When the DNA of a somatic cell (a cell of the body other than a gamete) is severely damaged, it may not be possible to repair all the errors. In this case, substantial changes in cell function and even life-threatening cancers may result. For example, skin cancer can be caused by damage to DNA from excessive exposure to sunlight. Mutations of somatic cells, fortunately, are not passed to the offspring.

On the other hand, mutations that occur in the gametes (sperm or egg) may be passed to future generations. These are the changes that lead to evolutionary change, including changes that cause species to diverge. Generally speaking, these heritable mutations occur very slowly. The DNA of humans and chimpanzees, for example, differs by only about 1% even though our evolutionary paths diverged over 5 million years ago. Two humans differ by perhaps one base pair in a thousand, but that is still 3 million base pair differences, more than enough to account for all of our individual variations.

Mechanisms of DNA repair

DNA repair mechanisms play a crucial role in the survival of an organism and its species. These mechanisms are quite efficient unless they are overwhelmed by massive damage.

DNA repair involves recognizing the errors, cutting out and replacing the damaged section or incorrect nucleotide base, and reconnecting the DNA backbone. The process utilizes numerous *DNA repair enzymes.* One enzyme might recognize certain types of damage and cut out the impaired nucleotide, another might insert the correct nucleotide, and yet another might reconnect the adjoining ends of the DNA. The repair process is most active in the hour between DNA replication and the beginning of mitosis. Repair prior to cell division ensures that the best possible copy of DNA is passed on to each daughter cell.

DNA repair mechanisms are directed by certain genes that code for the repair enzymes. Thus the genetic code plays a vital role in repairing itself. If the damaged genes are the ones that control the repair process, DNA errors may accumulate more quickly than normal. Researchers now know that mutations of certain genes that direct DNA repair are associated with an increased risk of several cancers, including forms of colon and breast cancer.

Transcription: Converting a gene's code into mRNA

A gene is "expressed" when it is called into action and the protein for which it codes is produced. An essential step in this process is transcription, which involves converting the DNA code of a single gene into a complementary single strand of mRNA.

Transcription is similar to the process of DNA replication, with the following exceptions:

■ Only the segment of DNA representing a single gene unwinds, as opposed to the entire molecule.

■ RNA is single-stranded, so that only one of the two strands of DNA actually carries the genetic code specifying the synthesis of RNA.

■ One of the four complementary base pairs of RNA is different from those of DNA: uracil replaces thymine.

■ The sugar group of RNA is ribose rather than deoxyribose.

Once it is fully transcribed the RNA molecule is released from the DNA strand and the two DNA strands entwine around each other again. Figure 17.6a illustrates the process of transcription.

How is the specific gene to be transcribed identified on the chromosome? A unique base sequence called the *promoter* marks the beginning of every gene. An enzyme called *RNA polymerase* attaches to the promoter, starts the DNA unwinding process, and assists in attaching the appropriate RNA nucleotides to the growing chain. The process ends at another base sequence that identifies the end of the gene.

The RNA molecule first transcribed from DNA, called the *primary transcript,* is not yet functional because most of the DNA base sequences of a gene do not code for

DNA Mutations Between Generations

How often do errors in DNA replication occur between generations—i.e., how many times are new mutations found in a child that were not present in a parent? These new mutations created between generations would represent a "human mutation rate." Undoubtedly they would contribute to human evolutionary change.

The human mutation rate per generation has been difficult to measure, but several different laboratories have now come up with numbers that are remarkably similar. According to the most recent estimate, the answer is about 60 DNA nucleotide base errors per generation. That sounds like a lot, but given that there are 3 billion nucleotide base pairs (6 billion bases) in the DNA in a human cell, that's just one error in every 100 million bases. DNA replication and repair mechanisms are remarkably accurate, it seems.

And that's why human evolution has occurred so slowly, over several million years. ■

Reference: Roach, J. C. et al. Analysis of Genetic Inheritance in a Family Quartet by Whole-Genome Sequencing. *Science Express,* March 11, 2010, pp. 1–3.

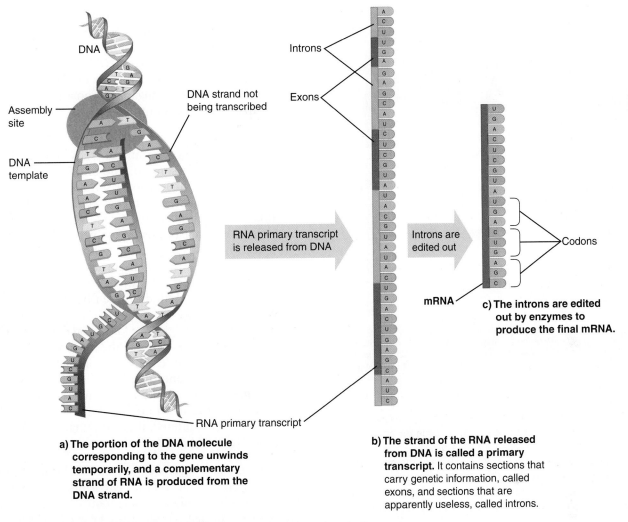

a) **The portion of the DNA molecule corresponding to the gene unwinds temporarily, and a complementary strand of RNA is produced from the DNA strand.**

b) **The strand of the RNA released from DNA is called a primary transcript.** It contains sections that carry genetic information, called exons, and sections that are apparently useless, called introns.

c) **The introns are edited out by enzymes to produce the final mRNA.**

Figure 17.6 Transcription of a gene into mRNA. Transcription takes place in the nucleus.

 Mark the place where you would expect to find the promoter. (Hint: Notice that transcription is already well under way.)

anything at all. The primary transcript is "edited" by enzymes that snip out the sections that do not carry any useful genetic information, called **introns,** leaving only the sequences that do carry genetic information, called **exons** (Figure 17.6b–c). The importance of the introns is unclear, though there is now some evidence that they may allow the exon fragments to be joined in different ways so that a single gene could code for different mRNA molecules in different cells. The RNA molecule produced from the primary transcript is called *messenger RNA* (mRNA) because it contains a message, in the form of a template, that can be "translated" into a specific sequence of amino acids that constitutes a particular protein.

The message is encrypted as a *triplet code,* so-called because three successive bases of mRNA, called a **codon,** each code for one of the 20 amino acids. Because there are four possible nucleotide bases and the code is a triplet code, there are precisely 64 different possible codons (4 × 4 × 4). We

know exactly which amino acids are specified by the 64 possible codons of mRNA (**Figure 17.7** on the next page). Note the following two important points:

1. Several different codons can specify the same amino acid because there are more possible codons than there are amino acids. For example, both the UUU and UUC codons code for the same amino acid, phenylalanine (Phe).
2. The codon AUG specifies a "start" code and three others specify "stop" codes. These are needed to specify where to begin and end the protein.

HBP **Web Animation** *Transcription* at www.humanbiology.com

 Quick Check Some hormones act by binding to the promoters of particular genes, in such a way that RNA polymerase can attach much more easily to that gene. What effect would this have on whether or not the gene is "expressed"? ■

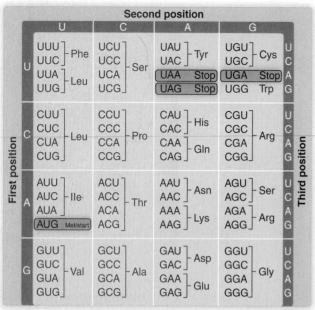

Second position

Figure 17.7 **The genetic code of mRNA.** Three successive bases represent a codon, and all but three codons encode for an amino acid. The letters correspond to the four RNA bases: uracil (U), cytosine (C), adenine (A), and guanine (G). To read the chart, start with the letter on the far-left side, then move across to one of the letters at the top, then down the right side to the appropriate letter. AUG is the "start" codon and the only codon for the amino acid methionine; there are three "stop" codons.

 A strand of DNA has the following sequence: TAC GGG CAT ATT. If it is transcribed, what will be the nucleotide sequence of the resulting mRNA, and what will be the amino acid sequence of the final peptide?

Translation: Making a protein from RNA

Translation is the process of using mRNA as a template to convert, or "translate," the mRNA code into a precise sequence of amino acids that composes a specific protein.

The genes of DNA actually produce at least three different kinds of RNA, each with a specific role in the production of proteins. Only mRNA carries the recipe for a new protein. The other two kinds of RNA are called *transfer RNA* and *ribosomal RNA*. Each has an important role in protein synthesis.

Transfer RNA (tRNA) molecules are small molecules that carry the code for just one amino acid. They also carry an *anticodon*, a base triplet that is the complementary sequence to a codon of mRNA. The function of tRNA is to capture single amino acids and then bring them to the appropriate spot on the mRNA chain.

Once they have been captured and brought to the mRNA molecule, the amino acids must be connected together into one long chain in order to become a protein. This is the

function of a ribosome. A **ribosome** consists of two subunits composed of **ribosomal RNA (rRNA)** and proteins. A ribosome has binding sites for both mRNA and tRNA. It also contains the enzymes that connect the amino acids together. The function of a ribosome, then, is to hold the mRNA and tRNA in place while joining the amino acids.

Translation occurs in three steps, as illustrated in Figure 17.8:

1. *Initiation.* A particular initiator tRNA binds to the smaller of two ribosome subunits and to the mRNA molecule. The tRNA and the ribosome subunit move along the mRNA until they encounter a "start" (AUG) codon. At this point they are joined by a larger ribosomal subunit to form the intact ribosome, which holds the mRNA in place while the tRNAs bring amino acids to it.
2. *Elongation.* The chain of amino acids lengthens one amino acid at a time. A tRNA molecule carrying the next appropriate amino acid binds to the ribosome and to mRNA. As the mRNA passes between the two ribosomal subunits, the ribosome catalyzes the formation of the bond between the newest amino acid and the previous amino acid in the growing chain. The tRNA molecule is then released to find another amino acid.
3. *Termination.* There is no tRNA anticodon corresponding to a "stop" codon on mRNA. When a "stop" mRNA codon is encountered, the ribosomal subunits and the newly formed peptide chain detach from the mRNA.

To use a cooking analogy, the mRNA brings the recipe, the tRNAs find and deliver the ingredients, and the ribosome is the cook who creates the final product. What happens next to the newly formed peptide chain depends on its function. Some enter the endoplasmic reticulum for further processing, packaging, and shipping, whereas others remain in the cytoplasm. Many become parts of new structural units, organelles, cell membrane components, and enzymes that the cell will need as it grows and then divides into two new cells.

HBP **Web Animation** *Translation* at www.humanbiology.com

HBP **Web Animation** *The Genetic Code* at www.humanbiology.com

Recap Before a cell divides, its DNA is replicated. During *replication* the two strands of DNA unwind and separate from each other. Enzymes called DNA polymerases add new nucleotides to each of the original strands, producing two complete molecules of DNA from one. Mutations may result from mistakes in DNA replication or from physical or chemical damage. Repair mechanisms remove and replace damaged DNA, if possible, before replication. During *transcription*, short portions of DNA representing single genes are converted into a readable and transportable mRNA code. *Translation* is a three-step process (initiation, elongation, and termination) by which proteins are assembled from amino acid building blocks according to an mRNA code. Assembly takes place on a ribosome. ∎

Figure 17.8 The three steps of translation. Translation takes place in the cytoplasm.

✓ *Suppose the gene that produced this mRNA happens to mutate so that one of the codons in the middle of the mRNA becomes a stop codon. What will happen the next time the gene is transcribed and translated? Will any protein be produced from the mutant mRNA? Explain.*

17.3 Cell reproduction: One cell becomes two

Let's review what we have learned so far about the preparations for cell division. DNA is replicated during the S phase of the cell cycle so there are two complete copies. The genes of DNA are used to produce specific proteins. These proteins in turn direct the cell's activities and form many of its structural units as the cell grows.

We turn now to the mitotic phase of the cell cycle, when first the nucleus and then the cell cytoplasm divide. The division of the nucleus is called **mitosis.** It is followed by **cytokinesis,** the division of the cytoplasm.

Mitosis: Daughter cells are identical to the parent cell

Mitosis is the process of nuclear division in which the sister chromatids of each duplicated chromosome are separated from each other. Following mitosis, each daughter cell has a complete set of DNA and is identical to the parent cell. Mitosis ensures that all cells of a complex organism have the same set of DNA.

Mitosis (**Figure 17.9** on the next page) includes a sequence of phases that compose most of the mitotic phase of the cell cycle. The phases are defined by the structural changes taking place in the cell as the DNA divides and two new nuclei form. Although each stage has a name and certain defining characteristics, mitosis is really a seamless series of events.

MITOTIC PHASE

Figure 17.9 Mitosis. Mitosis accounts for most of the mitotic phase of the cell cycle, overlapping slightly with cytokinesis, the division of the cell into two cells. The duplicated chromosomes become visible in prophase, line up in metaphase, are pulled apart in anaphase, and uncoil and become surrounded by nuclear membranes in telophase. All 46 duplicated chromosomes undergo this process in humans, although only 4 are depicted in the drawings.

Prophase In late G_2 of the interphase, the strands of replicated DNA and associated proteins begin to condense and coil tightly. Eventually they become visible as a duplicated chromosome consisting of sister chromatids held together by the centromere. **Prophase** begins when the duplicated chromosomes first become visible.

Recall that the cell has a cytoskeleton that helps hold it together. During prophase the tubular elements of the cytoskeleton come apart and are reassembled between the pairs of centrioles. This newly formed parallel arrangement of microtubules, called the *mitotic spindle*, will become the structure that causes the two identical sets of DNA to divide. The

pairs of centrioles drift apart and the mitotic spindle lengthens between them. The nuclear membrane dissolves, and the centrioles move to opposite sides of the cell. This allows the mitotic spindle to cross through the middle of the cell. The centromere in each duplicated chromosome develops into two separate structures (one for each of the duplicate DNA molecules) that attach to microtubules originating from opposite sides of the cell. Forces within the mitotic spindle draw the duplicated chromosomes toward the center of the cell.

Metaphase During **metaphase,** the duplicated chromosomes align on one plane at the center of the cell. Cells often seem to rest at this stage, and metaphase can occupy as much as 20 minutes of the entire one-hour process of mitosis. In fact, the duplicate DNA molecules are being pulled in opposite directions by equal forces (like a tug-of-war), but no separation occurs because they are still held together by the centromeres.

Anaphase During **anaphase,** the duplicate DNA molecules separate and move toward opposite sides of the cell. This is the shortest phase of mitosis, lasting only a minute or so. The centromere abruptly comes apart, and the now separated DNA molecules are pulled in opposite directions by the micro-tubules to which they are attached. The process of separation requires energy in the form of ATP and utilizes certain proteins that act as "motors" to tow the microtubules in opposite directions. Cytokinesis usually begins at this time.

Telophase **Telophase** begins when the two sets of chromosomes arrive at opposite poles of the cell. At this point the mitotic spindle comes apart, and new nuclear membranes form around the chromosomes. The chromosomes uncoil and revert to the extended form in which they are no longer visible under a microscope.

 Quick Check Suppose a chromosome in one of your skin cells has acquired a mutation in its centromere, such that its chromatids cannot separate from each other. During which stage of mitosis will this cause a problem, and what do you think will happen? ▪

Cytokinesis divides one cell into two identical cells

Cytokinesis is the process by which a cell divides to produce two daughter cells. Certain physical properties of the cell facilitate this process. A contractile ring of protein filaments of the same type found in muscle forms just inside the cell membrane. The contractile ring tightens, forming a cleavage furrow and then pinching the cell in two (**Figure 17.10**). The ring is

Cleavage furrow

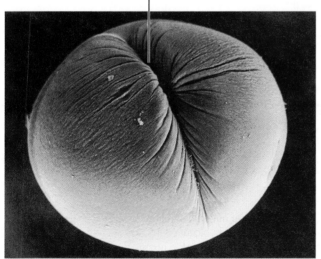

Figure 17.10 Cytokinesis. A cleavage furrow forms as the ring of contractile filaments in the cell tightens.

perpendicular to the long axis of the mitotic spindle, which ensures that one nucleus is enclosed in each of the daughter cells.

Cytokinesis is an example of the resourcefulness and efficiency of life's processes. The contractile ring is assembled just before it is needed from the remnants of the cytoskeleton that dissolved during mitosis. After the cell divides, the ring itself is disassembled, presumably to form new cytoskeletons for the daughter cells. In effect, cells use a "just-in-time" materials delivery system coupled to an efficient recycling system. Living cells were doing it long before the business community invented a name for it.

 Web Animation *Mitosis* at www.humanbiology.com

✓ **Quick Check** Each of your skeletal muscle cells has many nuclei within just one very long cell. Propose a simple modification of the cell cycle that could result in such a large, multinucleated cell. ▪

Mitosis produces diploid cells, and meiosis produces haploid cells

In humans, all cells in the body except those developing into sperm or egg cells have 46 chromosomes. Human cells with 46 chromosomes are called **diploid** cells (in Greek, *diploos* means "twofold") because the 46 chromosomes actually represent 23 *pairs* of chromosomes. Diploid cells reproduce by first replicating the 46 chromosomes and then undergoing mitosis (one nuclear division), so that the two daughter cells end up with 46 chromosomes that are identical to those of the parent cell.

In diploid cells, one of each pair of chromosomes came from each parent. There are 22 pairs of **autosomes** (chromosomes other than the sex chromosomes) plus the sex chromosomes X and Y. The 22 pairs of autosomes plus the two sex chromosomes XX (in the female) are called homologous chromosomes because they look identical under the microscope and they have copies of the same genes in the same location (*homologous* means "same form and function"). Only the X and Y sex chromosomes in the male look different from each other as a pair.

Sperm and eggs are **haploid** cells, meaning that they have only one set of 23 chromosomes. Haploid sperm and eggs are created by **meiosis,** a sequence of two successive nuclear divisions in which the human genes are mixed, reshuffled, and reduced by half. Once fertilization occurs, the fertilized egg and all subsequent cells have the diploid number of chromosomes again.

 Web Animation *The Human Life Cycle* at www.humanbiology.com

Meiosis: Preparing for sexual reproduction

The two successive nuclear and cell divisions of meiosis are called *meiosis I* and *meiosis II*. Both meiosis I and II can be subdivided into four stages: prophase, metaphase, anaphase, and telophase. **Figure 17.11** (next page) illustrates the events that occur in each stage.

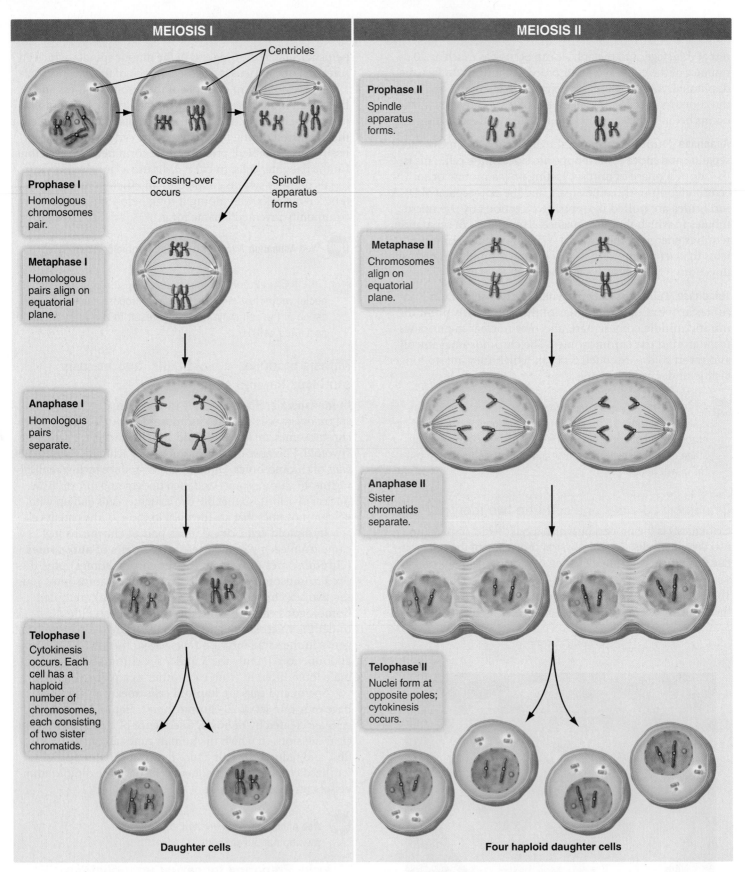

Figure 17.11 Meiosis. Meiosis is a process of two nuclear divisions in which the chromosomal material is reshuffled and reduced by half during the formation of sperm and eggs. The blue and purple colors indicate homologous chromosomes that were inherited from each of the person's parents. During prophase of meiosis I, pairs of duplicated homologous chromosomes pair up and exchange sections of DNA by a process known as crossing-over. During the rest of meiosis I, the homologous pairs are separated from each other. In meiosis II the duplicates of each chromosome are separated from each other, just as they are in mitosis. The result is four daughter cells, none of which are alike.

Meiosis I Before meiosis begins, the precursor cell undergoes a typical S phase of interphase in which the DNA is replicated and the 46 chromosomes are duplicated. But then something very different happens, compared to mitosis:

1. During prophase of meiosis I the duplicated homologous chromosomes pair up and swap sections of DNA (genes) by a process called *crossing-over*. Consequently, the homologous chromosomes now contain a recombination of the genes of both of the person's parents.
2. During the rest of meiosis I (metaphase through telophase), the homologous pairs of chromosomes are separated from each other, rather than the duplicates of each pair.

Meiosis II Meiosis II proceeds like mitosis except that the chromosomes are not duplicated again. During meiosis II, the 23 duplicated chromosomes line up and the sister chromatids are separated from each other. However, because of crossing-over during meiosis I, none of the four haploid daughter cells are exactly alike.

 Web Animation *Meiosis* at www.humanbiology.com

 Web Animation *Comparing Mitosis and Meiosis* at www.humanbiology.com

 Quick Check A normal diploid cell from a chimpanzee has 48 chromosomes. How many chromosomes will be in a chimpanzee cell just before meiosis begins, after meiosis I, and after meiosis II? During which of these stages will the chromosomes have two chromatids? ▪

Sex differences in meiosis: Four sperm versus one egg

There is a fundamental difference in how meiosis proceeds in males and females, and it has to do with the different functions of sperm and eggs. In males there is only a slim chance that any one sperm will ever reach an egg. It is important, therefore, that lots of sperm be produced. Meiosis in males produces four equal-sized but genetically different sperm from every precursor cell (**Figure 17.12**).

In females, each egg is precious. If fertilized, an egg will need a lot of energy to grow and develop. A large egg with plenty of cytoplasm and lots of organelles has a better chance of surviving the early stages of development. In females, therefore, as much of the cytoplasm as possible is reserved for only one daughter cell at each cell division. The smaller cell produced at each division is called a *polar body*. The polar body produced during meiosis I may or may not divide again during meiosis II, but in any event the two (or three) polar bodies eventually degenerate.

In females, meiosis II is not completed until a sperm penetrates the egg. The secondary oocyte released from

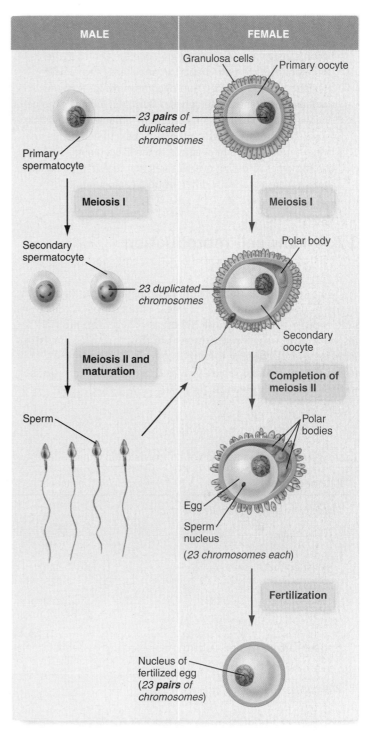

Figure 17.12 Sex differences in meiosis. In males, meiosis produces four different small sperm. In females, nearly all of the cytoplasm is reserved for the future egg at each stage of meiosis. Meiosis I in females thus results in a secondary oocyte and a polar body. Meiosis II is not completed in females until fertilization. The polar bodies and granulosa cells eventually disintegrate.

the ovary stops development at metaphase of meiosis II. Once penetration occurs the secondary oocyte quickly completes meiosis II and the sperm and egg nuclei then unite.

⮌ **Recap** *Mitosis* is a four-phase process in which the cell nucleus divides into two nuclei. The duplicated chromosomes become visible in prophase, align along the center of the cell in metaphase, are pulled apart in anaphase, and move to opposite sides of the cell and become surrounded by new nuclear membranes in telophase. During *cytokinesis* the cell divides into two cells, each containing a nucleus. *Meiosis* is a two-step process in which the nucleus and cell divide twice, producing sperm or eggs with the haploid number of chromosomes. Crossing-over ensures that the sperm (and also the eggs) are genetically different from each other. ∎

17.4 How cell reproduction is regulated

Although division is the norm for most cell types, not all cells in the body divide at the same rate. Some groups of cells divide rapidly throughout life, whereas others stop after adolescence. Still others reproduce at highly variable speeds, increasing or decreasing their rates of cell division in response to regulatory signals. Table 17.1 lists examples of variations in the rates of cell division.

What causes such different and sometimes changing rates of cell divisions? Although we do not yet have all the answers, we do know that cells have an internal control mechanism that undergoes cyclic changes, and that the control mechanism can be stopped at certain checkpoints by signals from inside or outside the cell.

Consider a washing machine or a dishwasher. Once a wash cycle is started, a whole sequence of timed events occurs until the entire cycle is finished. The same thing happens in a living cell. Under normal circumstances, progression through the G_1, S, and G_2 phases is controlled by cyclic fluctuations in the concentrations of certain proteins called *cyclins*. Cyclins in turn activate certain regulatory proteins that initiate specific events within the cell, such as DNA replication or formation of the mitotic spindle.

Most washing machine control devices can stop the cycle at certain checkpoints, based on input received from internal sensors. If the washing machine does not fill with water, the cycle may stop right before agitation would normally begin. If a load of clothes is out of balance, the spin cycle may stop prematurely. Cells also have internal surveillance and control mechanisms to ensure that the cell is ready for the next phase before it proceeds

Table 17.1 Variation in rates of cell division by human cells

Rate of division	Comments
Divide constantly and rapidly throughout life	
Skin cells	Skin is continually being formed from deep layers. The outermost layer of dead cells is constantly being sloughed off.
Most epithelial cells	Epithelial cells lining the inner surfaces of body organs such as the digestive tract and the lungs are exposed to frequent damage and must be replaced.
Bone marrow cells	Stem cells in the bone marrow produce red blood cells (RBCs) and white blood cells (WBCs) throughout life. WBC production can be increased as part of the immune response.
Spermatogonia (after puberty)	Spermatogonia divide to produce sperm throughout life in the adult male. The rate declines with age.
Will divide under certain circumstances	
Liver cells	Liver cells don't normally divide in adulthood but will do so if part of the liver is removed.
Epithelial cells surrounding the egg	Called granulosa cells, they are quiescent for most of adult life. They begin to divide as the follicle matures.
Normally do not divide in adulthood	
Nerve cells	Although there are exceptions, most human nerve cells apparently do not divide in adulthood.
Osteocytes	Mature bone cells called osteocytes become trapped within the hard crystalline matrix of bone.
Muscle cells	The commonly held view is that most muscle cells do not divide in adulthood, or divide very slowly.

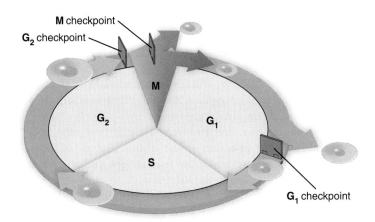

Figure 17.13 Cell cycle checkpoints. The cell cycle stops at several well-defined checkpoints unless the cell is ready for the next phase.

(Figure 17.13). Has all of the DNA been replicated, and has it been checked for errors? Has the mitotic spindle been formed properly? Is the cell large enough to be divided in two? Unless the cell surveillance system gives the go-ahead, the cell may stop at one of several checkpoints. The most important checkpoint seems to be near the end of G_1, but there are also checkpoints near the end of G_2 and at metaphase of mitosis.

Finally, the cell cycle can be influenced by conditions outside the cell. The cell cycle may stop at a checkpoint if certain nutrients are not available, if a particular hormone is not present, or if certain growth factors are not supplied by other cells. For example, a connective tissue cell called a fibroblast will not divide unless *platelet-derived growth factor* (derived, obviously, from platelets) is present. Even the presence of other cells affects the cell cycle. When cells come into contact with each other, for instance, they release a substance that inhibits cell division. Contact inhibition is an effective negative feedback mechanism to regulate tissue growth and organ size.

> **Recap** An internal cyclic control mechanism regulates the cell cycle. The cycle can be stopped at certain checkpoints by internal surveillance systems and is influenced by conditions outside the cell. ∎

17.5 Environmental factors influence cell differentiation

All of the cells in your adult body originated from a single cell, the fertilized egg, and all of those cells (except those destined to become your sperm or eggs) have exactly the same set of DNA. Yet your body is composed of lots of different kinds of specialized cells. If all your cells have the same exact set of genes, why don't all your cells stay identical, like identical twins? What causes the cells to change form and function?

Differentiation is the process by which a cell becomes different from its parent or sister cell. At various developmental stages of life, cells differentiate because they begin to express different genes. The examples below show that differentiation is probably due to environmental influences.

Differentiation during early development

Let's examine the earliest stage of human development, starting from a fertilized egg. The fertilized egg is a relatively large cell. Soon after fertilization, it begins to divide, becoming two cells, then four, then eight. During these early divisions, up to about 16 or 32 cells, the cells do not grow. Consequently each cell gets progressively smaller as a ball of identical cells develops (Figure 17.14).

In a ball of eight or fewer cells each cell is exposed to the same environment—the surrounding fluid plus contact with cells just like themselves. Once the ball becomes more than about eight cells the situation changes, for now only the cells at the outer surface of the ball are exposed to the surrounding fluid. The rest are exposed only to the extracellular fluid between the cells. At this point the environment that

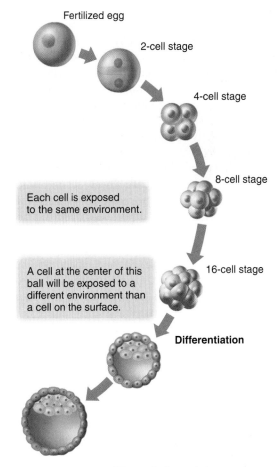

Figure 17.14 How early differentiation might occur. Beyond about the eight-cell stage, the cells begin to be exposed to different environments. Differences in environments may trigger different gene expressions, leading to differentiation.

surrounds a cell at the center of the ball becomes different from the environment of a cell at the surface. Cells at the center may be exposed to a lower O_2 content, higher levels of CO_2, or a different pH, or they may be influenced by products secreted by the other cells. Any one of these or other environmental factors could activate different genes, leading to different developmental pathways for the cells at the center versus the cells at the surface.

Notice that no "thought" or feedback control mechanisms are needed to account for early differentiation; it can be accounted for solely by differences in the physical and chemical environment of individual cells. Cell differentiation begins long before homeostatic feedback mechanisms ever develop in the organism.

Scientists have known for some time that it is possible to **clone** (make copies of) an organism by "embryo splitting"; that is, by splitting the eight-cell ball that develops from a fertilized egg into eight separate cells. At that stage, all eight cells are equally capable of becoming complete organisms, presumably because they have not yet differentiated. After about 16 cells, however, cloning by embryo splitting no longer works.

Differentiation later in development

Differentiation at later stages is similar in principle to that at earlier stages. Each cell is shaped by two factors: the developmental history of the cells that came before it and its local environment. Genes are turned on at certain stages of development and then turned off, some never to be used again.

For example, early in development some cells begin to express the genes that cause them to become epithelial cells. Others become connective tissue, nerves, or muscle cells. Those that are destined to become muscle tissue differentiate even further, becoming either skeletal, smooth, or cardiac muscle.

In the end, even cells that are closely related in function and structure can differ in hundreds of ways. Each difference in gene expression, added to those that occurred before, eventually results in a very specialized cell. For example, your smooth and skeletal muscle cells differ in their metabolic pathways (how they use energy), speed and strength of contraction, shape and size, and internal arrangement of muscle fibers. They got that way by an ever-increasing cascade of selective gene expressions.

The mechanisms of differentiation explain why the developing fetus is more vulnerable to genetic damage than an adult. Alteration of a gene that plays a critical role in early development can be so disruptive to embryonic differentiation that the embryo may not survive. Later in development the embryo may be able to survive the damage but be left with marked physical deformities. In an adult, alteration of a single gene may be similarly dangerous, or it could have little or no visible effect. It all depends on what the proteins coded by the gene do.

These are some substances that can harm a fetus:

- Cigarette smoke. Smoking can retard fetal growth and may contribute to spontaneous abortions.
- Alcohol. Babies born to mothers who abuse alcohol may have fetal alcohol syndrome. These babies may have heart defects, limb abnormalities, and delays in motor and language development.
- Prescription and over-the-counter medications. If a woman becomes pregnant, she should review with her physician all the drugs she takes.
- Illegal drugs. Cocaine and heroin use during pregnancy can kill a fetus or cause abnormalities. The use of marijuana is associated with poor fetal weight gain and behavioral abnormalities in the newborn.
- Chemicals in air, water, and soil. Examples include lead, mercury, pesticides, and industrial solvents such as polychlorinated biphenyls (PCBs) and toluene.
- Radiation, for instance, from radon gas or pelvic irradiation during cancer treatment. Video display terminals and natural radiation from cosmic rays are not thought to pose any risk.

Intrauterine infections can also be dangerous. The most damaging are HIV infection, syphilis, and rubella (German measles).

↩ **Recap** Differentiation is the process whereby cells become different from each other. During early development, environmental influences trigger differentiation. Some genes are expressed only at certain stages of development, so genetic mutations during early development can be particularly damaging. In adults, hundreds of different genes may be expressed by various types of specialized cells. ■

17.6 Cloning an organism requires an undifferentiated cell

Making copies of (cloning) entire organisms, called **reproductive cloning**, has proven to be much more difficult than cloning specific molecules by genetic engineering techniques. Reproductive cloning requires a completely undifferentiated cell—a cell as close to a fertilized egg as possible—as the starting point. The two techniques for accomplishing reproductive cloning are called *embryo splitting* and *somatic cell nuclear transfer*.

Embryo splitting: producing identical offspring

In *embryo splitting*, an egg is fertilized *in vitro* and allowed to divide to the eight-cell stage. At this stage the cells are all still identical because they have not yet begun to differentiate. The cells are then separated, and each is implanted into a different surrogate mother (**Figure 17.15**). Each of the cells will develop normally, producing an organism that is an exact genetic copy of all the others. This type of reproductive cloning is currently quite feasible; indeed, it is being used effectively to produce valuable farm animals. The embryos

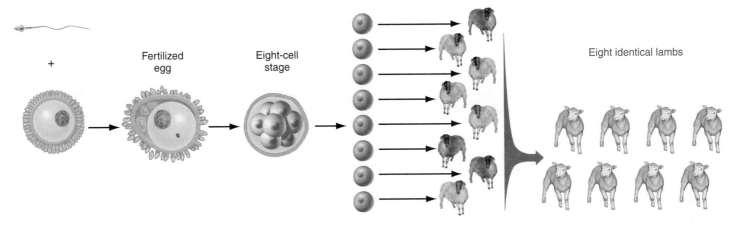

Figure 17.15 Cloning by embryo splitting. Clones produced by embryo splitting are genetically identical to each other but are not exact copies of either parent.

can even be shipped to other countries and implanted in less valuable surrogate mothers, for herd improvement.

Although the offspring produced by embryo splitting are clones of each other, they are not clones of either parent. Genetically, they are like eight identical children, different from each parent because they received some of their genes from each one.

So far, no one has attempted to produce identical human octuplets by embryo splitting. The techniques required are not that difficult—it could be done if someone really wanted to.

✔ **Quick Check** Suppose you're working at a zoo that has the last living pair of Arabian oryx—an endangered antelope. The female can only produce one calf per year, and you're worried that she can't produce enough calves for the species to survive. Outline a strategy using embryo splitting to try to save this species. What is one obvious problem with this approach? ∎

Somatic Cell Nuclear Transfer: Cloning an Adult

A *somatic cell* is any cell in the body except a germ (sex) cell. Somatic cells have the full set of instructions for creating a complete, unique individual; germ cells contain only half a set of DNA.

Cloning an organism by *somatic cell nuclear transfer* yields a true clone of an adult organism. This was the technique used to produce Dolly the sheep back in 1997 (see the *Current Issue* at the beginning of the chapter). The feat shocked the world, because at that time no one thought that a fully differentiated cell could be used to "start over." Basically, the technique involves combining a somatic cell from an adult with an *enucleated* fertilized egg (an egg with its nucleus removed). The nucleus of the adult cell contains the set of instructions for making a copy (clone) of the adult, and the enucleated egg represents the proper environment for carrying out those instructions.

Figure 17.16 shows how it is done in practice. An egg is removed from a donor animal and the nucleus is carefully

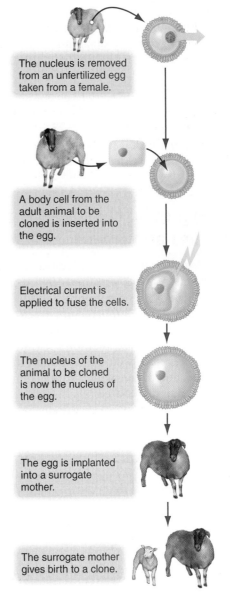

The nucleus is removed from an unfertilized egg taken from a female.

A body cell from the adult animal to be cloned is inserted into the egg.

Electrical current is applied to fuse the cells.

The nucleus of the animal to be cloned is now the nucleus of the egg.

The egg is implanted into a surrogate mother.

The surrogate mother gives birth to a clone.

Figure 17.16 Cloning by somatic cell nuclear transfer. The clone is genetically identical to the animal that provided the body cell.

removed. A somatic cell is removed from the animal to be cloned and inserted into the enucleated egg. Finally, electric current is used to fuse the two cells into one. The new cell represents a fertilized egg, but its nucleus is from the adult animal being cloned. The egg is then inserted into a surrogate mother and allowed to develop normally. Because the DNA is from the adult organism being cloned, the organism is a true clone of the adult organism from which the somatic cell was taken.

The popular press tends to view cloning by somatic cell nuclear transfer as science fiction, but it is not. It is feasible. It has been done successfully in animals, though not with the same success as embryo splitting. It could be done with humans, though so far it hasn't. It probably will be done in your lifetime unless severe legal sanctions are put in place to prevent it.

Dolly died at a relatively young age for a sheep. Scientists worry that somatic cells taken from an adult may already have suffered so much mutational damage that clones might routinely live shorter lives than the original adult. More experience over time with cloned farm animals may answer that question.

> **Recap** A clone is a copy. Cloning by *embryo splitting* produces genetically identical offspring that are different from either parent. Cloning by *somatic cell nuclear transfer* produces a genetic twin of the adult that was cloned. So far, humans have not been cloned by either method. ∎

17.7 Therapeutic cloning: Creating tissues and organs

Therapeutic cloning is the cloning of human cells specifically for treating patients. As a first step, therapeutic cloning research is aimed at taking undifferentiated or partially differentiated cells of a certain type from human embryos (for example, early-stage neural cells) and cloning them for use in patients with specific diseases, such as Parkinson's disease. The ultimate goal of therapeutic cloning would be to be able to take a single cell from a patient and coax it along a path of differentiation and cell division until it produced just the right cells, a certain tissue, or even a whole new organ specifically for that patient. The new cells, tissue, or organ would be ideal because there would be no danger of tissue rejection.

As a practical treatment for human disease, therapeutic cloning is still a long way off. Scientists still need to work out the mechanisms that control cell differentiation pathways, and how to control them. Nevertheless, some progress is being made. It is likely that the first success story will be the ability to grow layers of human skin cells for use in burn patients.

Unlike the cloning of adult humans, the goals of therapeutic cloning generally do not meet with much societal resistance. The problem is the source of undifferentiated cells for use in research. Currently the best cells are *embryonic stem cells* (ES cells) harvested from early-stage human embryos. Unfortunately, the embryos are destroyed in the process. To eliminate the need for human embryos, scientists are working on techniques for producing *induced pluripotent stem cells* (iPS cells) from fully differentiated cells. The idea is to put fully differentiated cells back into a state from which they can once again differentiate into one of many cell types, given the right conditions.

Therapeutic cloning has the potential to transform medical treatment in the future. But the more we learn about cell division and differentiation and the closer we get to therapeutic cloning, the closer we also get to being able to clone adults. We need to think about how to use and control this new technology.

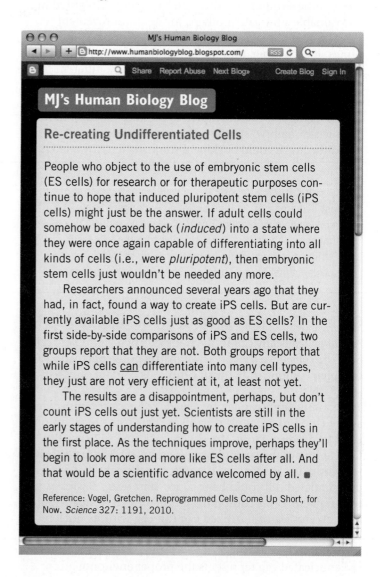

MJ's Human Biology Blog

Re-creating Undifferentiated Cells

People who object to the use of embryonic stem cells (ES cells) for research or for therapeutic purposes continue to hope that induced pluripotent stem cells (iPS cells) might just be the answer. If adult cells could somehow be coaxed back (*induced*) into a state where they were once again capable of differentiating into all kinds of cells (i.e., were *pluripotent*), then embryonic stem cells just wouldn't be needed any more.

Researchers announced several years ago that they had, in fact, found a way to create iPS cells. But are currently available iPS cells just as good as ES cells? In the first side-by-side comparisons of iPS and ES cells, two groups report that they are not. Both groups report that while iPS cells <u>can</u> differentiate into many cell types, they just are not very efficient at it, at least not yet.

The results are a disappointment, perhaps, but don't count iPS cells out just yet. Scientists are still in the early stages of understanding how to create iPS cells in the first place. As the techniques improve, perhaps they'll begin to look more and more like ES cells after all. And that would be a scientific advance welcomed by all. ∎

Reference: Vogel, Gretchen. Reprogrammed Cells Come Up Short, for Now. *Science* 327: 1191, 2010.

Chapter Summary

The cell cycle creates new cells *p. 406*

■ Cells reproduce by a repetitive cycle called the cell cycle in which one cell grows and then divides in two.

■ The cell cycle has two primary phases: interphase and the mitotic phase. During interphase the cell grows and the DNA is replicated. During the mitotic phase, first the nucleus and then the cell divide in two.

Replication, transcription, and translation: An overview *p. 407*

■ Human DNA consists of 46 separate molecules, packed with histones into structures called chromosomes.

■ A gene is the smallest functional unit of a chromosome. A gene contains the code for making one or more specific proteins.

■ DNA replication is a process in which the two strands of DNA separate and a new complementary copy is made of each strand.

■ DNA is repaired when it is damaged and is checked for errors after it is replicated.

■ For a gene to be expressed, the strand of DNA with that gene must be transcribed to create a complementary strand of mRNA. The mRNA leaves the nucleus, attaches to a ribosome, and serves as the template for protein synthesis.

■ Translation is the process of making a protein using the mRNA as a template. Three successive bases of mRNA, called a codon, code for a particular amino acid.

■ The amino acid building blocks for the protein are captured in the cytoplasm by tRNA, brought to the mRNA, and attached to each other by ribosomal enzymes.

Cell reproduction: One cell becomes two *p. 413*

■ Mitosis is a sequence of events in which the replicated chromosomes are separated to form two new genetically identical nuclei.

■ Cytokinesis is the process whereby the cell divides into two new cells, each with one of the new nuclei produced by mitosis and roughly half of the cell's organelles and mass.

■ Meiosis is a sequence of two cell divisions that produces haploid cells. Meiosis occurs only in cells destined to become sperm or egg. Crossing-over during meiosis mixes the genes of homologous chromosomes, and subsequent cell divisions reduce the number of chromosomes by half.

How cell reproduction is regulated *p. 418*

■ Cell reproduction is regulated in part by selective gene expression. Selective gene expression is controlled by regulatory genes.

■ The cell cycle may be influenced by the physical and chemical environments both inside and outside the cell.

Environmental factors influence cell differentiation *p. 419*

■ Differentiation is the process by which cells become different from each other, acquiring specialized forms and functions.

■ Because all cells have the same set of genes, differentiation in the early embryo must be triggered by environmental influences.

■ Cell differentiation later in development can be influenced by environmental cues, but it also depends on the developmental history of the cells that preceded it.

Cloning an organism requires an undifferentiated cell *p. 420*

■ *Embryo splitting* can produce up to eight identical offspring.

■ *Somatic cell nuclear transfer* produces a clone of an adult animal.

Therapeutic cloning: Creating tissues and organs *p. 422*

■ *Therapeutic cloning* is the cloning of human cells for the purpose of treating human disease.

■ The ultimate goal of therapeutic cloning is to be able to create cells, tissues, or even whole new organs for human patients.

Terms You Should Know

anaphase, *415*
cell cycle, *406*
chromosome, *407*
clone, *420*
codon, *411*
cytokinesis, *413*
differentiation, *419*
diploid, *415*
gene, *408*
haploid, *415*
homologous chromosomes, *415*
interphase, *406*
introns, *411*
metaphase, *414*

messenger RNA, *408*
mitosis, *413*
mitotic phase, *406*
mutation, *409*
prophase, *414*
replication, *407*
reproductive cloning, *420*
ribosomal RNA (rRNA), *412*
ribosome, *412*
telophase, *415*
therapeutic cloning, *422*
transcription, *408*
transfer RNA (tRNA), *412*
translation, *408*

Concept Review

Answers can be found at the Human Biology Place.
www.humanbiology.com

1. Describe how DNA is replicated before cell division.
2. Compare and contrast the processes of transcription and translation.
3. Explain what mutations are and the role of DNA repair mechanisms.
4. Name the four phases of mitosis and describe briefly what is happening to the chromosomes during each phase.
5. Explain why only one large egg is formed during meiosis in the female, whereas four sperm are formed during meiosis in the male.
6. Describe what is meant by *selective gene expression* and why it is important to how a cell functions.
7. Explain how factors present in the environment can influence cell differentiation.
8. Describe how ribosomes contribute to the formation of a protein.
9. Compare and contrast the roles of introns and exons.
10. Describe how the cell cycle is regulated.

Test Yourself

Answers can be found in Appendix A.

1. What would be the outcome if a cell completed mitosis but did not undergo cytokinesis?
   a. The cell would have two nuclei.
   b. The cell would have one nucleus, but twice as many chromosomes.
   c. The cell would be cancerous.
   d. The cell would die.

2. Which of the following cell types are most likely to remain in G_0?
   a. neurons
   b. skin cells
   c. bone marrow cells
   d. cells lining the GI tract

3. If the human genome was considered to be a large cookbook, which of the following would represent the individual recipes?
   a. chromosomes
   b. genes
   c. nucleotides
   d. nitrogenous bases

4. During which stages of the cell cycle do chromosomes consist of two sister chromatids?
   a. G_1
   b. G_2
   c. prophase of mitosis
   d. both (b) and (c)

5. Which of the following are listed in order from largest, most inclusive, to smallest?
   a. genome – chromosome – nucleotide – gene
   b. gene – nucleotide – chromosome – genome
   c. chromosome – genome – nucleotide – gene
   d. genome – chromosome – gene – nucleotide

6. What might be a likely outcome if a mutation occurred in a promoter region of a gene, such that it was no longer recognized by RNA polymerase?
   a. The DNA would not be replicated.
   b. The protein encoded by that gene would not be synthesized.
   c. The cell would not be able to produce any proteins.
   d. The gene would be transcribed normally.

7. Which is likely to be the shortest chain of nucleotides?
   a. the DNA of a gene
   b. the primary transcript of the gene
   c. the mRNA
   d. DNA, the primary transcript of the gene, and mRNA all contain the same number of nucleotides.

8. How many different amino acids could be encoded if the genetic code was a doublet code (2 bases) instead of a triplet code?
   a. 2
   b. 4
   c. 8
   d. 16

9. What is the most likely target of cell surveillance as a cell approaches the M (metaphase) checkpoint?
   a. Has all of the DNA been replicated?
   b. Has the DNA been checked for errors following replication?
   c. Is each chromosome properly joined to the mitotic spindle?
   d. Is the cell large enough?

10. Why do cells within an organism differentiate, such that one cell may eventually become a liver cell and another will become an epithelial cell?
    a. Cells differentiate because of differences in gene expression.
    b. Cells differentiate because different genes within them encode for different proteins.
    c. Cells differentiate because of crossing-over, a random process.
    d. Cells differentiate because of differences in genetic code.

11. Which method of cloning is most similar to the way identical (monozygotic) twins are formed?
    a. therapeutic cloning
    b. embryo splitting
    c. somatic cell nuclear transfer
    d. production on induced pluripotent stem cells

12. At which point in the cell cycle are cells the smallest?
    a. the beginning of G_1
    b. S
    c. G_2
    d. the beginning of mitosis

13. By the end of meiosis II, a cell that had entered meiosis I with 32 chromosomes would have produced _____ daughter cells which are genetically _____ and which each has _____ chromosomes.
    a. 2…identical…32
    b. 2…identical…16
    c. 4…non-identical…16
    d. 4…identical…32

14. Transcription occurs in/at the _____ and produces _____ from a _____ template.
    a. nucleus…RNA…DNA
    b. nucleus…DNA…RNA
    c. ribosomes…RNA…protein
    d. ribosomes…protein…RNA

15. How does the production of sperm differ from the production of eggs?
    a. Sperm production involves meiosis whereas egg production involves only mitosis.
    b. Meiosis during sperm production produces four sperm cells, whereas meiosis during egg production produces only one functional egg cell.
    c. Sperm are haploid whereas eggs are diploid.
    d. Sperm production begins during fetal development whereas the meiotic process that begins egg production doesn't begin until puberty.

Apply What You Know

Answers can be found at the Human Biology Place.
www.humanbiology.com

1. Recall that the base pairing in DNA is always A–T and C–G. Suppose that during a DNA replication a C is accidentally paired with a T. Explain why it is important that the error be detected and corrected before the cell divides again, by illustrating what would happen at the next cell division if the error were not corrected.

2. The RNA code is a "triplet code," meaning that three bases code for a single amino acid. Would a "doublet code" be sufficient? Explain.

3. Are gene mutations that occur during fetal development always fatal? Why or why not?

4. Why do so many commercials for various pharmaceutical drugs contain warnings for women who are pregnant or are trying to become pregnant?

5. Mitochondria contain their own DNA that is inherited independently of nuclear DNA. From which parent did you inherit your mitochondrial DNA?

6. Bacteria can reproduce by simple cell division. Sexual reproduction is far more complicated and difficult; success is not guaranteed. What is the main advantage of sexual reproduction?

18

Cancer: Uncontrolled Cell Division and Differentiation

Colorized chest X-ray showing a cancerous lung tumor (red/yellow).

Current Issue

Voluntary Breast and Ovary Removal

René Syler, former anchorwoman of CBS's The Early Show, faced a difficult decision. Both her mother and her father had suffered from breast cancer (breast cancer is rare in males). She had had numerous mammograms and several biopsies of suspected cancer tissue. Although she did not yet have cancer, it felt as if she were just waiting for it to happen. Her breasts were shrunken and scarred from the biopsies, and she was scared and just tired of it all. So René made a decision to take control of her own health. She said, "Instead of asking, 'Is this the year I get breast cancer?' I had to say,

'I will not get breast cancer this year. Or ever.'"[1] On January 2, 2007 she had both breasts removed.

Could You Do It?

Would you choose to have your breasts and ovaries removed surgically (called prophylactic mastectomy and oophorectomy) in order to avoid breast and ovarian cancer? Women can now make that decision while they are still in good health, thanks to our increased

---

[1] René Syler. Defending My Life. O, The Oprah Magazine, April 15, 2007.

understanding of the causes of cancer and our ability to test for some of the known risk factors. It is now known that mutations in at least two genes, called BRCA1 (short for breast cancer 1) and BRCA2, are strongly associated with breast and ovarian cancer. In their normal forms these two genes function to suppress cell growth; only the mutated forms increase the risk of cancer.

Mutated forms of BRCA genes are relatively rare—only 0.5% of all women have a mutated form—but for those who do, the risk of cancer goes up dramatically. For all women the lifetime risk of developing breast cancer is 12%, according to the

René Syler, former anchorwoman of CBS's. The Early Show, had both breasts removed and reconstructed in 2007 as a preemptive strike against breast cancer.

National Cancer Institute. For women with a mutated BRCA gene the risk of breast cancer is about 60%, or five times as high. For ovarian cancer the lifetime risk is 1.4% for all women versus 15-40% for women with a mutated BRCA gene.

Testing for mutated forms of BRCA1 and BRCA2 genes is relatively easy, requiring only a blood sample. That means that women can find out whether they are at high risk for breast and ovarian cancers long before the cancers might actually develop.

Should you be tested for the BRCA genes just as a precaution? That depends on whether you would really want to know the answer, and what (if anything) you would choose to do if you did know you had the BRCA genes. If you are convinced that you could never have your breasts and ovaries removed while you are still healthy, then perhaps you would be better off not knowing. Your decision might also be influenced by whether or not you have other risk factors. Although fewer than 0.5% of all women have a mutated BRCA gene, your chances of having a mutated gene go up five- to tenfold if you have a family history of women who developed breast or ovarian cancer. In fact, both of the BRCA genes can be inherited from either of your parents (in men, the BRCA genes are associated with slightly increased risks of breast and prostate cancer).

If you inherited a mutated BRCA gene, couldn't you just wait to see if cancer ever developed in a breast or ovary, and then (and only then) have it removed? After all, if you're still of reproductive age you may want to have children. The problem is that ovarian cancer is particularly deadly. There is no easy screening test for ovarian cancer and there are generally no symptoms in the early stages, so it may not be diagnosed until it is already in an advanced stage. The five-year survival rate after a diagnosis of ovarian cancer is still under 40%.

After oophorectomy, the risk of developing ovarian cancer essentially disappears. The risk of developing breast cancer after mastectomy drops by 90% if you have a mutated BRCA gene, but it doesn't go completely to zero. There is still a very small risk of developing breast cancer even after surgery because breast tissue is located not just in the breast but also in surrounding tissues, making it nearly impossible to remove it all. There are other treatment options, too, but they're not quite as effective as surgery. Some physicians recommend a drug called tamoxifen for women at high risk who choose not to have surgery—it reduces the cancer risk by about 50%.

What Is Peace of Mind Worth?

And what about just peace of mind? Surgeons report that 10% of women in their 40s who undergo surgery for breast cancer are now asking to have their healthy breast removed as well. For these cancer survivors, not having to worry about mammograms, biopsies, or breast surgery ever again may be the motivating factor. Some surgeons worry that these cancer survivors may not fully understand that if they are not carrying a mutated BRCA gene, removal of the healthy breast apparently has no effect on long-term survival. But who's to say what their motivation is, or to question their choices?

We've been talking about probabilities of cancer and even death in an impersonal way, as if this were a game of numbers. But this is not a game. Each of us may have to make difficult choices for ourselves or for a loved one some day. We'll be forced to balance personal feelings against impersonal odds. René Syler did not have mutated forms of either the BRCA1 or BRCA2 gene, but her family history and her own personal history of pre-cancerous changes in her breast tissue put her at high risk. So René Syler made a choice. She says she's happy with what she calls her "teenaged breasts" and with the knowledge that she'll never need another mammogram or biopsy. It may not be a choice you could make, but she still thinks it was the right choice for her.

Questions to consider

1 Blood tests for BRCA1 and BRCA2 are readily available. Are you planning to be tested? Why or why not?

2 If you knew that you had a mutated BRCA gene, would it change what you do or how you live?

3 What would you want your mother/wife/girlfriend/daughter to do if it she had a mutated BRCA gene?

The facts...

- Heritable mutations of two genes known as **BRCA1** and **BRCA2** are strongly associated with breast and ovarian cancer.
- Ovarian cancer often goes undetected until it is too late. The five-year survival rate is still under 40%.
- Some women are opting for prophylactic breast and ovary removal while they are still healthy.

>> **The essence of all cancers is a loss of control of cell division and differentiation.** As cancer cells divide out of control and even invade other tissues, they may prevent the proper functioning of organs and organ systems.

>> **Cell growth, division, and differentiation are controlled by specific regulatory genes.** When these regulatory genes mutate or are damaged, cancer may result.

>> **Newer imaging and diagnostic testing techniques and sophisticated cancer treatments are slowly leading to gains against some cancers.** But no single cure for cancer is likely in the foreseeable future.

>> **Many cancers could be prevented by changes in lifestyle.** Tobacco smoke accounts for over 30% of all cancer deaths.

The word *cancer* may conjure up images of pain, suffering, and lingering death. We may associate the word with visits to the hospital with our parents when we were children; we may recall years of long-term care of a grandparent at home; we may remember a friend or loved one no longer with us. A diagnosis of cancer sounds like a death sentence.

And yet, people do recover from cancer. There are known risk factors for most types of cancers and things we can do to reduce our risk. There are treatments for many cancers, some of which work rather well. The death rate from some cancers is even declining as we learn more about the disease.

We know of at least 100 different types of cancers, but all are the same in one respect—they all are diseases of cell division and differentiation. Once that is understood, cancer makes some sense. In this chapter we examine the biological basis for how cancer develops and spreads. Then we discuss how cancer is diagnosed and look at new advances in treatment and prevention.

To understand how cancer develops we must first understand two key characteristics of normal cells.

■ Normal cells have regulatory mechanisms that keep their rates of cell division appropriately in check. They may divide more frequently at some times than at others, but their rates of cell division are carefully controlled by regulatory mechanisms that may involve an internal clock and various hormones. In addition, their rates of division are generally inhibited by signals from nearby cells.

■ Normal cells generally remain in one location throughout their lifetime because they adhere to their nearby neighbors. (Blood cells are an exception, because they must travel throughout the body to carry out their normal function.)

18.1 Tumors can be benign or cancerous

Sometimes even normal cells increase their rate of cell division as part of their normal physiological function. Examples are the growth of the lining of the uterus during the menstrual cycle and the growth of breast tissue during pregnancy. In general, any substantial increase in the rate of cell division is called **hyperplasia** ("increased" plus Greek *plasis,* "formation").

In the progression of normal cells toward cancer, however, hyperplasia is not necessarily a normal event. Instead, it may signal a loss of control over cell division. Eventually, these rapidly dividing cells will develop into a discrete mass called a **tumor,** or *neoplasm* (meaning literally "new growth").

Not all tumors are cancers. Tumors that remain in one place as a single well-defined mass of cells are *benign tumors* (from the Latin *benignus,* "well disposed") (Figure 18.1). The cells in a benign tumor still have most of the structural features of the cells from which they originated, and in particular they stay together and may be surrounded by a layer of connective tissue. Benign tumors tend to threaten health only if they become so large that they crowd normal cells, which can happen sometimes in the brain. In most cases surgery can easily remove benign tumors because they are so well contained. Many moles on the skin are a form of benign tumor so inconsequential that usually we don't bother to remove them. Occasionally, a previously benign tumor may change into something more damaging. For example, certain moles have the potential to develop into melanoma, a form of skin cancer.

Figure 18.1 Development of a benign tumor. A benign tumor begins when a cell (in this case an epithelial cell) becomes genetically altered and begins to divide more frequently than normal. The mass enlarges but stays well contained, generally by a capsule of normal cells.

✓ *Should benign tumors always be removed? Explain why or why not.*

Answers to ✓ questions can be found in Appendix A.

18.2 Cancerous cells lose control of their functions and structures

If and when cells progress toward cancer, they undergo a series of structural changes in addition to dividing more frequently than normal. The nucleus often becomes larger, there is less cytoplasm, and the cells lose their specialized functions and structures. An abnormal structural change is called **dysplasia** (from *dys*, abnormal, plus *plas*, growth). Figure 18.2 shows how dysplasia may lead to cancer.

Doctors have learned to recognize characteristic patterns of changes that cells typically go through as they get more and more out of control. Dysplasia is often a sign that tumor cells are "precancerous," meaning they are altering in ways that may herald the possibility of cancer. The tumor itself becomes more and more disorganized as cells pile on top of each other in seemingly random fashion.

An example of a precancerous skin lesion is *actinic keratosis* (Figure 18.3). Actinic keratoses generally appear as dry, scaly, or rough skin lesions on areas of the skin exposed to the sun. Often they are felt more easily than seen, although sometimes they appear as reddish patches. They may grow slowly and even disappear for a time, only to return later. Only 10–20% of all actinic keratoses ever become cancerous. Fortunately they are easily removed by chemical peeling, freezing, or scraping.

Figure 18.3 Actinic keratosis. Actinic keratosis is a pre-cancerous skin lesion caused by repeated exposure to sunlight. Actinic keratosis lesions usually do not progress to cancer.

A tumor is defined as a **cancer** when at least some of its cells completely lose all semblance of organization, structure, and regulatory control. As long as the entire tumor remains in one place it is known as an ***in situ* cancer.** *In situ* cancers can often be successfully removed surgically if they are detected early enough.

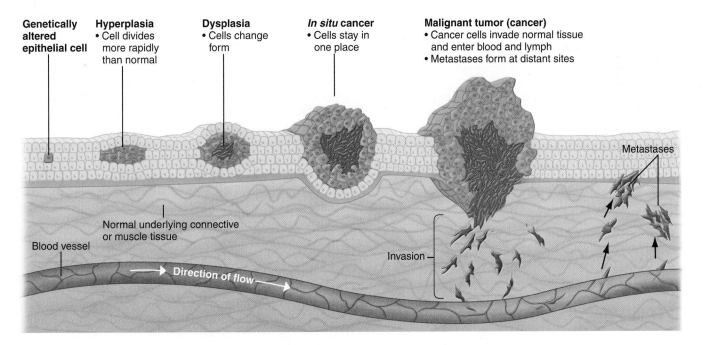

Figure 18.2 The development of a malignant tumor (cancer). A genetically altered cell begins to divide more frequently than normal, resulting in an increased number of cells (hyperplasia). Additional genetic alterations cause some cells to change form (dysplasia) and to lose all semblance of organization or control over cell division and function. As long as the mass stays in one place it is called an *in situ* cancer. However, additional changes in the cells may cause them to break away from the tumor and invade normal tissues or enter the blood or lymph. These invading cells may set up new colonies of cancer (called metastases) at distant sites. A cancer that invades and metastasizes is called a malignant tumor.

Table 18.1 A summary of the characteristics of benign tumors and cancers

	Benign tumor	*In situ* tumor (cancer)	Malignant tumor (cancer)
Frequency of cell division	More rapid than normal	Rapid and completely out of control	Rapid and completely out of control
Cell structure and function	Slightly abnormal	Increasingly abnormal	Very abnormal
Tumor organization	A single mass generally surrounded by a capsule	Still a single mass, but increasingly disorganized and not always surrounded by a capsule	Some cells break away to invade underlying normal tissues or to enter the blood or lymph. New tumor colonies (metastases) become established at distant sites.

If cancer cells go through additional changes that cause them not to adhere to each other in the normal fashion, the result may be **metastasis**—the spread of cancer to another organ or region of the body. Metastasis occurs when cancer cells break away from the main tumor, travel via the blood or lymph, and develop into new colonies of cancer cells at distant sites. Cancers that *metastasize* and cancers that invade normal tissues until they compromise organ function are called **malignant tumors** (from the Latin *malignus*, "ill-disposed"). Table 18.1 compares benign, *in situ*, and malignant tumors.

A malignant, metastasizing tumor is like a forest fire that is out of control and throwing off sparks. The sparks start new fires in sometimes distant locations, which also spread uncontrollably. Metastasizing cancer continues to spread until it completely overruns the functions of tissues, organs, and organ systems. Ultimately it may kill the entire organism, just as a forest fire may not die out until it has consumed the whole forest.

Cancer is extremely difficult to treat, for obvious reasons. One in three people in the United States will experience cancer in their lifetime, and one in four will die of it. Cancer ranks second only to heart disease as a cause of death. The good news is that most incidences of cancer could be prevented, and later in this chapter we discuss how you can lower your risk.

Recap Some tumors are benign, but when tumor cells change form dramatically and divide uncontrollably, the tumor is called cancer. Cancer becomes malignant when cells invade and metastasize, starting new tumors at distant sites. ∎

18.3 How cancer develops

For cancer to develop, at least two things must happen simultaneously:

- The cell must grow and divide uncontrollably, ignoring inhibitory signals from nearby cells to stop dividing.
- The cell must undergo physical changes that allow it to break away from surrounding cells. These events begin when some of the cell's genes become abnormal and no longer function properly.

Mutant forms of proto-oncogenes, tumor suppressor genes, and mutator genes contribute to cancer

Several types of genes control the various activities of a cell. *Structural genes* code for the proteins necessary for cell growth, division, differentiation, and even cell adhesion (the adherence of cells in the same tissue to each other). *Regulatory genes*, on the other hand, code for proteins that either activate or repress the expression of these structural genes.

Some regulatory genes code for proteins secreted by cells in order to influence nearby cells. These proteins are called *growth factors* or *growth inhibitors*, depending on which it is that they do. Growth inhibitors are especially important in regulating cell division because they influence cells in a particular tissue to stop dividing when enough cells are present.

Under normal circumstances, two classes of regulatory genes control the cell's activities: proto-oncogenes and tumor suppressor genes. **Proto-oncogenes** are normal regulatory genes that *promote* cell growth, differentiation, division, or adhesion (*proto*, "first"; *onco-* means "mass, bulk," that is, a tumor). Mutated or damaged proto-oncogenes that contribute to cancer are called **oncogenes.** Some oncogenes drive the internal rate of cell growth and division faster than normal. Others produce damaged protein receptors that fail to heed inhibitory growth signals from other cells. However, it is important to recognize that one oncogene alone is not sufficient to cause cancer. Because so many cellular processes must be disrupted at once and each process may be controlled by multiple genes, cancer develops only when multiple oncogenes are present.

Tumor suppressor genes are regulatory genes that under normal conditions apply the brakes to unchecked cell growth, division, differentiation, or adhesion. They tend to *inhibit* these cell activities when cells are stressed or when conditions are not right for cell division. When they become damaged, tumor suppressor genes contribute to cancer because the cell activities they normally regulate are then likely to continue unchecked.

One tumor suppressor gene, called p53, has received a lot of attention lately. The p53 gene codes for a protein found in most tissues that prevents damaged or stressed cells from dividing. In fact its primary role may be to inhibit division by cells that already show cancerous features. If the p53 gene itself becomes damaged, a variety of cancers may develop more easily. To date, mutated p53 genes have been identified in tumors of the cervix, colon, lung, skin, bladder, and breast.

A third class of gene may also contribute to cancer. In their unmutated form, **mutator genes** are involved in DNA repair during DNA replication. When these genes mutate, the cell becomes increasingly prone to errors in DNA replication. As a result, the cell may accumulate mutations in other genes at a faster rate than normal.

 Web Animation *Cancer* at www.humanbiology.com

✅ **Quick Check** A gene for a growth factor (a protein involved in normal cell growth) mutates so that it now produces growth factor constantly, triggering abnormal cell growth. What is the best term for the mutated gene: proto-oncogene, oncogene, tumor suppressor gene, or mutator gene? Why? ■

Human Gene Patents Invalidated

Basing his decision on legal arguments that human genes are products of nature and hence cannot be patented, in March of 2010 a U.S. District Court judge invalidated several patents held by a company called Myriad Genetics on the human breast cancer genes, BRCA1 and BRCA2. Myriad Genetics sells a kit used to test for the two genes for about $3,000. The company tried to argue that their method of isolating the genes changed the genes, and thus made them patentable. But the judge ruled that such an argument was just a trick to circumvent the prohibition on the direct patenting of DNA.

About 20% of human genes have already been patented. A number of biotech companies were planning to make a hefty profit by developing and selling patent-protected genetic tests or by selling the patent rights to others. Whether all of those patents could ultimately be invalidated is unclear. With potentially billions of dollars at stake, the decision is likely to be appealed.

Medical and research organizations and patients are pleased by the decision. If it holds up on appeal it should provide wider access to human genes currently under patent protection, and ultimately make genetic tests such as those for BRCA1 and BRCA2 less expensive. ■

Reference: Marshall, Eliot. U.S. Judge Rejects Breast Cancer Gene Patents. *Science* 328: 25, 2010.

A variety of factors can lead to cancer

A car analogy demonstrates how a cell progresses toward cancer. At first your new car runs very well. As time goes on a few parts wear out, but you manage to keep it in good repair with regular service checkups. But as the car gets older, the need for repairs gets more frequent, and you aren't always able to keep up with them. Now the suspension needs replacing, the tires are worn out, the accelerator pedal sticks occasionally, and the brakes are thin. No single defect is enough to cause an accident, but together they spell trouble. There may come a moment when you may not be able to control the car, and you'll have an accident.

The same sort of progression occurs within cells. In fact, the single most important factor in the development of cancer may be age. Skin cancer is uncommon before adulthood, even though most severe sunburns occur during childhood. Recall from Chapter 4 that the outer layers of your skin are continually being replaced by basal cells located at the base of the epidermis. The basal cells of a child's skin haven't been through very many cell divisions yet—they are still in good repair. But decades later, after thousands more cell divisions and a few more sunburns, those same basal cells are about worn out. Cellular repair mechanisms fail more frequently; mechanisms controlling cell division become less effective. First there's dysplasia, then cancer. The cells are out of control.

In addition, your parents may carry a few of the cancer-promoting mutations that have accumulated in the precursor cells to human eggs and sperm during the course of human evolution. These rare heritable mutations may not be enough to guarantee cancer, but they do increase your *heritable susceptibility*. If six mutations are required to produce cancer, for example, you could inherit five mutations and still be cancer free. This is why, even in families with a history of cancer, some siblings develop the disease and others do not.

The multigene basis of cancer also explains why it is possible to develop cancer even if there is no family history. Remember that we all inherit a slightly different combination of our parents' genes, and if you happen to get mutated genes from both parents, you could be at greater risk than either of them. (We discuss genetic inheritance in more detail in Chapter 19.)

The process of transforming a normal cell into a cancerous one is called *carcinogenesis*. A **carcinogen** is any substance or physical factor that causes cancer. The word *carcinogen* comes from the recognition by early physicians that certain skin cancers had a crablike appearance (Greek *karkinos,* "crab"; *genein,* "to produce"). Some of the factors that contribute to the probability that a normal cell will progress toward cancer are described below.

Viruses and bacteria Some viruses and bacteria may contribute to certain cancers. As described in Chapter 9, viruses reproduce by inserting their DNA into the host's DNA. If the viral sequence of DNA impairs the function of a normal gene, perhaps by preventing a regulatory gene from being turned off after it is turned on, this will increase the cancer risk.

Viruses thought to contribute to cancer include:

- Human papillomavirus (HPV; cancers of the cervix and penis)
- Hepatitis B and hepatitis C viruses (liver cancer)
- HIV (Kaposi's sarcoma, non-Hodgkin's lymphoma)
- Epstein-Barr virus (Hodgkin's disease and non-Hodgkin's lymphoma)
- Human T-cell leukemia/lymphoma virus (HTLV-1; T-cell non-Hodgkin's lymphoma)

The *Helicobacter pylori* bacterium, which causes ulcers, may contribute to stomach cancer as well. Although this list may sound frightening, viruses and bacteria probably account for fewer than 15% of all cancers.

Chemicals in the environment Some chemical carcinogens damage DNA directly. Others do not themselves damage DNA, but their presence increases the potency of still other carcinogens. The list of chemical carcinogens in the environment includes many chemicals used in (or produced by) industrial processes, such as coal tar, soot, asbestos, benzene, vinyl chloride, and some pesticides and dyes (Figure 18.4). Table 18.2 lists examples of viral, chemical, and radiation carcinogens, ranked by likelihood of exposure.

Tobacco About 2% of all cancers are thought to be caused by industrial pollutants. In contrast, tobacco is by far the single most lethal carcinogen in the United States, accounting for over 30% of all cancer deaths. Smoking dramatically increases your risk of cancer of the lungs, mouth, pharynx, pancreas, and

Figure 18.4 Environmental carcinogens. This agricultural worker is wearing protective clothing and an air-filtering mask while spraying.

bladder. Smokers are not the only ones at risk; passive smoking (inhaling secondhand smoke) can cause lung cancer too.

Smokeless tobacco (chewing tobacco or snuff) is also a carcinogen. It is a major risk factor for cancers of the mouth, pharynx, and esophagus.

Radiation Sources of radiation include the sun and radioactive radon gas, which is released by rocks, soil, and groundwater at intensities that vary in different geographic areas. Cellular telephones, household appliances, and electric power lines also emit trace amounts of radiation.

Table 18.2 Examples of carcinogens

Carcinogen	Source of exposure or types of persons exposed	Type of cancer	Exposure of general population
Tobacco (smoking and smokeless)	Smokers, chewers, people exposed to secondhand smoke	Lung, mouth, pharynx, bladder, cervix, colon, pancreas	Common
Diesel exhaust	Buses and trucks, miners, railroad yard workers	Lung	Common
Benzene	Paints, dyes, furniture finishes	Bone marrow (leukemias)	Common
Pesticides	Agricultural workers	Lung	Common
Ultraviolet light	Sunlight	Skin cancers	Common
Ionizing radiation	Radioactive materials, medical and dental procedures	Bone marrow (leukemias)	Common
Human papillomaviruses	Sexually transmitted	Cervix, penis	Common
Hepatitis B virus	Sexually transmitted	Liver	Less common
HIV	Sexually transmitted	Kaposi's sarcoma	Less common
Hydrocarbons in soot, tar smoke	Firefighters, chimney cleaners	Skin, lung	Uncommon
Asbestos	Shipyard, demolition, and insulation workers, brake linings	Lung, epithelial linings of body cavities	Uncommon
Radon	Mine workers, basements of houses	Lung	Uncommon

From time to time, the carcinogenic potential of radiation receives a great deal of media attention. In truth, all sources of radiation combined cause only about 2% of all cancer deaths, and most of these are due to skin cancer caused by the sun's ultraviolet rays.

The high-frequency ultraviolet B rays in sunlight cause over 80% of all skin cancers, including a particularly dangerous cancer of the melanin-producing cells of the skin called **melanoma.** Frequent sunburns, especially during childhood, increase the risk of melanoma. Fair-complexioned people who sunburn quickly are more vulnerable than those who tan easily (Figure 18.5). However, cumulative exposure to ultraviolet radiation, even without burning, raises the risk of skin cancer. Ultraviolet radiation from sunlamps and tanning booths is just as dangerous as radiation from natural sunlight.

Diet How much does diet contribute to cancer? Although difficult to define precisely, its role appears substantial. Dietary factors may be involved in up to 30% of cancers, rivaling smoking. In particular, red meat and saturated animal fat raise the risk of cancers of the colon, rectum, and prostate gland. Obesity in general is a risk factor for several types of cancer, and a high salt intake may contribute to stomach cancer. Excessive consumption of alcohol is associated with breast, rectal, colon, and liver cancer.

Certain fungi and plants produce carcinogens. One such carcinogen is *aflatoxin,* a by-product of a fungus that infects ground nuts. Aflatoxin may be present in raw peanut butter. Commercial peanut butters are processed to eliminate the fungus.

In recent years researchers have become interested in what might be *missing* from the diet that could contribute to cancer. Recent evidence suggests that fruits and vegetables in particular may be mildly protective. A compound in broccoli sprouts has received a lot of attention lately as a cancer risk–reducing agent. Dietary analysis is complicated because there are literally thousands of compounds in the foods we eat.

Figure 18.5 Risky behavior. Persons who were often sunburned as children have an increased risk for skin cancer later in life. Persons with lighter complexions are at greater risk.

Internal factors It is difficult to estimate how significant internal factors might be to the development of cancer. In theory, errors introduced during DNA replication and damage caused by certain chemicals within the body could contribute to the disease.

In particular, scientists are studying the effects of *free radicals,* highly reactive fragments of molecules that are produced by the body's biochemical processes. Free radicals are a normal by-product of cellular metabolism, and ordinarily they are detoxified by peroxisomes (Chapter 3). However, if detoxification processes become less efficient for any reason, free radicals may accumulate in the body and damage other molecules, including DNA. Some researchers believe that free radicals play a role in the aging process as well.

Certain vitamins known as *antioxidants* appear to neutralize free radicals, although the mechanism by which they do this is still controversial. In particular, vitamins A, C, and E seem to exert antioxidant benefits. Whether antioxidants affect the development of cancer or the aging process, however, is still an open question.

 Quick Check People who live in basements have a higher risk of lung cancer than people who live aboveground. Which of the above influences on cancer might explain this? ▪

The immune system plays an important role in cancer prevention

Under normal circumstances, the immune system plays a critical role in defending against damaged or defective cells, including cancer cells. Recall that all cells display proteins that identify them as "self" so that the immune system leaves them alone. As cells change in ways that make them cancerous they may stop displaying "self" proteins, or they may display completely different proteins that mark them as abnormal. This makes them subject to attack and destruction by the immune system, at least as long as it is functioning normally.

Suppression of the immune system by drugs, viruses such as HIV, or even mental states of anxiety, stress, or depression may allow certain cancers to develop more easily. Some cancers apparently can suppress the immune system, and others may have mechanisms for disguising themselves from attack by the immune system. Figure 18.6 (next page) presents an overview of how cancer develops.

Recap Proto-oncogenes and tumor suppressor genes normally control the rate of cell division. Mutator genes are involved in DNA repair. Mutations of any of these genes may contribute to cancer. Aside from heritable susceptibility, factors that may contribute to the development of cancer include viruses and bacteria, environmental chemicals, tobacco, radiation, dietary factors, and alcohol. Free radicals created during cellular metabolism may also contribute to cancer. The immune system normally protects us from cancer cells by killing them before they spread. Immune system suppression allows cancers to develop more easily. ▪

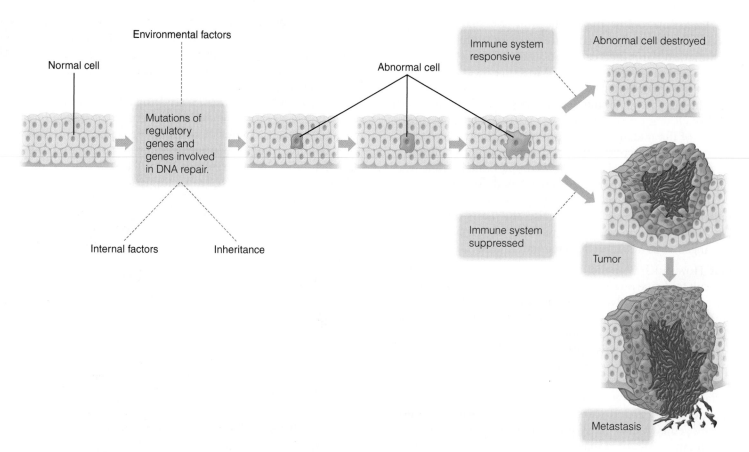

Figure 18.6 Overview of the development of cancer. The progression to cancer requires multiple genetic changes in regulatory genes, some of which may be inherited and some of which may be caused by environmental or internal factors. A cancer cell reproduces rapidly (develops into a cancer) and may invade and colonize distant tissues (metastasize). Suppression of the immune system appears to be involved in many cancers.

18.4 Advances in diagnosis enable early detection

In 1971 President Nixon signed into law the National Cancer Act, declaring "war" on cancer. At that time many researchers hoped we could overcome cancer with early diagnosis and better treatments.

The sobering fact is that the overall death rate from all cancers has decreased only 6% in the past 55 years. The main reason has been a dramatic increase in lung cancer in older Americans (Figure 18.7). These were the teens and young adults of the 1940s and 1950s, when smoking was first popularized and even encouraged. We are now paying the price with much higher rates of cancer in these older Americans, compared to these same age groups in 1950. As these older generations are replaced with generations who grew up with anti-smoking campaigns, lung cancer rates may begin to decline.

Although we have made significant progress against certain types of cancer, much more remains to be done. Early detection (diagnosis) of cancer is an important key to success. Early detection means that therapy can begin sooner, hopefully before the disease has had a chance to metastasize. Prompt treatment can even cure some cancers.

Tumor imaging: X-rays, PET, and MRI

The classic approach to diagnosing a tumor is to view its image, either on an X-ray (Figure 18.8) or with a newer method. Computer-enhanced imaging is now able to detect changes that visual imaging may miss. Advanced imaging techniques such as *positron-emission tomography (PET)* and *magnetic resonance imaging (MRI)* show physiological differences between tissues in places that cannot be diagnosed effectively with X-rays.

PET scans employ radioactive substances to create three-dimensional images showing the metabolic activity of body structures. In addition to diagnosing cancer, PET scans are used to study blood flow, heart function, and brain activity.

MRI scans use short bursts of a powerful magnetic field to produce cross-sectional images of body structures. For example, MRI detects differences in water and chemical composition between tissues on the basis of changes in magnetic fields. MRI can detect tumors because tumors often have a slightly different consistency and composition than normal tissue. It can detect tumors even when they are hidden behind bone, such as tumors of the brain (Figure 18.9). Unlike X-rays and PET scans, MRI does not expose the patient to radiation.

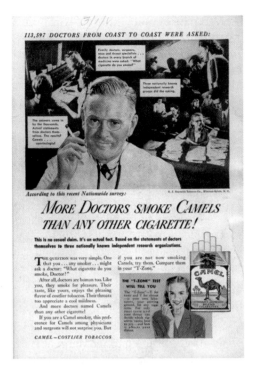

a) Ads promoting smoking were common in the 1940s and 1950s.

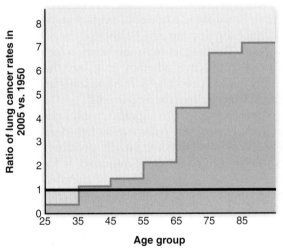

b) Lung cancer in 2005 compared to 1950, by age group.
A ratio of 1 (bold line) denotes the same rate of cancer in 2005 as in 1950. Ratios greater than one indicate higher rates of cancer in 2005 than in 1950 for those age groups; a ratio less than one indicates a lower rate of cancer in 2005 than in 1950 in that age group.

Figure 18.7 Smoking and lung cancer.

✓ *According to this graph, was a 70-year-old living in 2005 more likely or less likely to have lung cancer than a 70-year-old living in 1950? How about a 30-year-old in 2005 versus in 1950?*

Genetic testing can identify mutated genes

As we learn more about which human characteristics are transmitted by specific genes, genetic testing for diseases, including cancers, becomes a real possibility. Already hundreds of new genes and their mutated counterparts have been identified, and tests are being devised to detect them.

Figure 18.8 An X-ray of the breast, called a mammogram. A tumor is clearly visible in this X-ray.

Figure 18.9 An MRI of the brain. The magenta and red area to the right is a brain tumor.

Such tests are highly controversial, however, and society will have to choose how to use them. One of the concerns about genetic testing is that it may be possible to test for mutated genes that have no cure. Some people believe that such knowledge may not be in the patient's best interest. In addition, it is difficult to convince people that in many cases a mutated gene is nothing more than a risk factor. It does not guarantee development of a disease.

Enzyme tests may detect cancer markers

An enzyme called *telomerase* is rarely found in normal cells but is present in nearly all cancer cells. Some researchers propose that a sensitive test for telomerase in body fluids could detect early stages of cancer. An enzyme test could also be used to screen large populations or monitor the progression of the disease in known cancer patients. Researchers are exploring other potential cancer markers as well.

↩ **Recap** Early diagnosis and prompt treatment of cancer are vital. Diagnostic techniques include X-rays, positron-emission tomography (PET), magnetic resonance imaging (MRI), genetic testing, and enzyme markers. ▪

18.5 Cancer treatments

Many cancers are treatable, especially if discovered early. Although some forms of the disease have higher survival rates than others, current treatments cure approximately 50% of all cancer cases.

Conventional cancer treatments: Surgery, radiation, and chemotherapy

Classic treatments for cancer include surgery, radiation, chemotherapy, or a combination of two or more of these. Better imaging techniques have benefited the surgical treatment of cancer, and techniques for focusing external sources of radiation directly onto the cancer have improved as well. These focused sources of intense radiation are designed to damage and kill cancer cells in a specific location. It is sometimes possible to implant radioactive materials in the region of a tumor.

Healthy cells recover from radiation more readily than cancer cells, but radiation therapy generally does injure or kill some normal tissue, too. A drawback of both surgery and radiation is that they may miss small groups of metastasized cells, allowing the cancer to reappear later.

✓ **Quick Check** Children with cancer that are treated with radiation treatments have an increased risk of developing other cancers later in life. Why? ▪

Chemotherapy is the administration of cytotoxic (cell-damaging) chemicals to destroy cancer cells. Chemotherapy addresses some of the limitations of surgery and radiation. Some chemotherapeutic drugs act everywhere in the body, and for that reason they must inflict only acceptable amounts of damage to normal cells. Some chemotherapy chemicals stop cells from dividing by interfering with their ability to replicate DNA. Chemicals that inhibit cell division act primarily on cancer cells because cancer cells divide very rapidly. But such chemicals also damage or kill many normal cells that divide rapidly, such as bone marrow cells and cells lining the digestive tract. Damage to cells in the digestive tract is the reason so many chemotherapy patients experience nausea.

In addition to nausea and vomiting, common side effects of chemotherapy include hair loss, anemia (a reduction in the number of red blood cells), and an inability to fight infections (due to fewer white blood cells). A major problem is that many tumors become resistant to chemotherapeutic drugs, just as bacteria become resistant to antibiotics. Treatment then becomes a battle of trying to destroy abnormal cells with stronger and stronger chemicals without killing the patient in the process. Sometimes combination therapy, in which surgery is followed by radiation and combinations of chemotherapeutic drugs, works best.

✓ **Quick Check** Why does chemotherapy cause anemia, increased risk of infections, and hair loss? Hint: What kind of tissues undergo a rate of cell division that would cause these symptoms, and what do these tissues have in common? ▪

Magnetism and photodynamic therapy target malignant cells

To avoid the side effects of standard radiation and chemotherapy, researchers are developing techniques to target malignant cells more precisely while sparing healthy tissue. Promising developments include the following:

▪ *Magnetism:* Positioning a powerful magnet at the tumor site and injecting tiny metallic beads into the patient's bloodstream. The magnet pulls the beads, which have been coated with a chemotherapy drug, into the tumor, where they kill the cancer cells. Magnetism is undergoing clinical trials in liver cancer patients.

▪ *Photodynamic therapy:* Targeting cancer with light-sensitive drugs and lasers. The patient takes light-sensitive drugs that are drawn into the rapidly dividing cancer cells. Laser light of a particular frequency is focused on the tumor, where it triggers a series of chemical events that kill malignant cells. Photodynamic therapy has been approved for several years to treat tumors in the esophagus and lung, and new refinements in lasers make it promising for other cancers too (**Figure 18.10**).

Immunotherapy promotes immune response

Immunotherapy attempts to boost the general responsiveness of the immune system so that it can fight cancer more effectively. Recent efforts have focused on finding specific antigen molecules that are present on cancer cells but not on normal cells, and using them to produce antibodies that target cancer cells for destruction by the immune system. This approach is showing particular promise. Taking it one step further, researchers are working to attach radioactive molecules or chemotherapeutic drugs to the antibodies so that treatments can be delivered only to cancer cells and spare normal cells. Researchers are working to create vaccines that prevent specific cancers.

Researchers are also testing other vaccines created from a patient's own cancer cells. When injected into the patient, the

Figure 18.10 Photodynamic therapy. Before surgery, this patient was injected with a light-sensitive drug that is taken up by cancer cells. During surgery to remove the cancer, laser light (seen here traveling in a fiber-optic cable) is aimed directly at cells in the region of the excised cancer. Activation of the drug by laser light leads to a toxic reaction that kills any remaining cancer cells, but not any normal cells.

modified cancer cells appear to stimulate the person's immune defenses to recognize and combat the abnormal cells.

"Starving" cancer by inhibiting angiogenesis

Tumors grow and divide rapidly, so they require a great deal of energy. We now know that something within tumors promotes *angiogenesis* (the growth of new blood vessels) to serve their energy needs. Without angiogenesis, the tumor would reach about the size of a pea and stop growing. At least a dozen proteins stimulate angiogenesis, some of which are released by tumor cells, and several other proteins are known to inhibit angiogenesis. Interest is very high in *anti-angiogenic* (against blood vessel growth) drugs, because they may be able to starve tumors by limiting their blood supply. Anti-angiogenic drugs now on the market include Avastin, a blockbuster drug now approved for the treatment of a variety of cancers, and Revlimid, recently approved for the treatment of multiple myeloma, a cancer of antibody-producing white blood cells in bone.

Interestingly, researchers have found that anti-angiogenic drugs may not work just by starving tumors, at least at first. As they begin to inhibit the growth of the abnormal tangle of blood vessels normally found in a tumor, they actually normalize the vasculature and improve blood flow for a time. And that time (perhaps a month) may be a window of opportunity for delivering anti-tumor chemicals to the tumor via the bloodstream.

 Quick Check Patients on Avastin are advised to stop taking the drug temporarily if they are planning to have surgery. Why? ■

Molecular treatments target defective genes

Sophisticated molecular strategies target the defective genes that lead to cancer. Some researchers are attempting to inactivate specific genes, or the proteins they encode, to slow the runaway cell division. Future techniques may include *gene therapy*, in which defective genes are repaired or replaced with their normal counterparts. A key target for gene therapy is the p53 tumor suppressor gene that, when defective, contributes to many cancers.

↩ **Recap** The mainstays of cancer treatment are surgery, radiation, and chemotherapy. Recent advances include magnetism, photodynamic therapy, immunotherapy, drugs to inhibit angiogenesis, and molecular treatments that focus on specific defective genes. ■

18.6 The 10 most common cancers

Table 18.3 ranks the 10 most common cancers by incidence. Each has its own risk factors, warning signs, methods of detection, treatments, and (most important of all) survival rate.

In both men and women, the most frequent cancers are those of the skin, lung, and colon and rectum. In addition, prostate cancer is common in men and breast cancer is common in women.

Table 18.3 The top 10 cancers ranked by estimated incidence (numbers of new cases per year)

Cancer	New cases per year	Deaths per year
Skin (melanoma and nonmelanoma)	>1,000,000*	11,590
Lung	219,000	159,400
Breast	213,000 (women) 1,700 (men)	40,200 (women) 440 (men)
Prostate	192,300	27,400
Colon and rectum	147,000	50,000
Lymphoma	74,000	20,800
Urinary bladder	71,000	14,300
Kidney	58,000	13,000
Uterus	53,000	12,000
Leukemia	45,000	21,900

Source: American Cancer Society, Cancer Facts & Figures 2009, American Cancer Society, Inc.; 2009.

*Estimate—nonmelanoma cases are often not reported.

The most deadly cancers—in terms of the number of deaths—are lung, colon and rectum, and breast cancer. Lung cancer kills more people than the next top four cancers combined.

Skin cancer: Look for changes in your skin

Skin cancers are generally classified as either melanoma or nonmelanomas (all other skin cancers). Although most of the nonmelanomas are not particularly deadly if detected early, they should still be removed as soon as they are detected. In contrast, melanoma can be very serious:

- The most common of the nonmelanomas is *basal cell* carcinoma (cancer). Recall that basal cells located at the base of the epithelium divide throughout life to produce the squamous epithelial cells that form the outer layer of the skin. Basal cell cancer occurs when the basal cells begin to divide out of control. Basal cell cancer usually appears as a pink or flesh-colored bump with a smooth texture. Sometimes the bump bleeds or crusts. Basal cell cancer rarely metastasizes but should still be removed. If not treated it can eventually spread to underlying tissues.
- *Squamous cell* cancer arises from the epithelial cells produced by the basal cells. It consists of pink scaly patches or nodules that may ulcerate and crust. Squamous cell cancer can metastasize, although slowly.
- *Melanoma* is the least common of the three main skin cancers, but it is also the deadliest because it is likely to metastasize fairly quickly. It arises from abnormal melanocytes, the cells that produce the dark-brown pigment called melanin. Look for dark spots or patches on your skin and evaluate them according to the "ABCD" rule (**Figure 18.11**): A = *Asymmetry*—the two halves of the spot or patch don't match; B = *Border*—the border is irregular in shape; C = *Color*—color varies or is intensely black; and D = *Diameter*—diameter is greater than 6 millimeters, or about the size of a pea. Itchiness, scaliness, oozing, or bleeding may also be significant signs.

Early detection of skin cancers is important. Both basal and squamous cell cancers have a better than 95% cure rate if removed promptly because they metastasize slowly, if at all. Early detection is especially important for melanomas; the five-year survival rate (percentage of persons diagnosed who are still alive after five years) is 98% if the melanoma is detected while it is still localized, but only 16% if it has already metastasized.

✓ **Quick Check** An interesting fact about melanocytes is that, in embryonic life, they originate near the spinal cord and then migrate to reach their final locations in the skin. No other skin cells migrate in this way. How might this be related to the fact that melanomas are the most dangerous form of skin cancer? ∎

a) **A: Asymmetry.**

b) **B: Border.**

c) **C: Color.**

d) **D: Diameter.**

Figure 18.11 Melanoma. Melanomas generally are diagnosed by the ABCD rule.

Lung cancer: Smoking is leading risk factor

Most cases of lung cancer are preventable, because by far the greatest risk factor is cigarette smoking. All other risk factors, which include exposure to smoke, radiation, and industrial chemicals such as asbestos, are relatively minor by comparison.

Unfortunately there is no simple screening test for lung cancer, and so it is generally not detected early. Most of the symptoms of lung cancer (persistent cough, recurrent pneumonia or bronchitis, voice change) are relatively non-specific.

By the time blood appears in the sputum and the patient experiences chest pain, the cancer may already be at an advanced stage. Early detection improves the odds, but the one-year survival rate for all stages of diagnosis is still only 41%, and the overall five-year survival rate is only 15%.

Treatment options depend on when the cancer is detected. When the cancer is detected early and is still localized, surgery is the treatment of choice. However, in most cases the disease has already spread by the time it is detected. For the more advanced stages, radiation therapy and chemotherapy, sometimes combined with surgery, are the only options.

Current research efforts are directed at finding earlier, more effective detection methods. A large clinical trial is currently under way to determine whether screening high-risk individuals with low-dose computed tomography (CT) scans can reduce lung cancer deaths. Such scans are not cheap, however. Other research is aimed at finding molecular markers for lung cancer in sputum so that inexpensive screening tests could detect the disease at an earlier stage.

Breast cancer: Early detection pays off

Breast cancer is almost exclusively a cancer of women (but note that it can occur in men; approximately 2,000 new cases of male breast cancer are diagnosed each year). Breast cancer is usually first diagnosed by a mammogram (an X-ray of breast tissue) or when the patient or health care provider detects an abnormal lump during a physical exam (Table 18.4).

Table 18.4 Recommendations for cancer screening

Cancer	Gender	Test	Age and frequency	Description
All	Men and women	Cancer-related checkup	20–40: Every 3 years 40+: Every year	General exam by health professional that includes health counseling and screening as appropriate for malignant and nonmalignant diseases
Colon and Rectum*	Men and women	Fecal occult blood test or immunochemical test	50+: Every year	Checks for hidden blood in feces
		Flexible sigmoidoscopy	50+: Every 5 years	Examining interior of rectum and sigmoid colon with flexible fiber-optic scope
		Colonoscopy	50+: Every 10 years	Examining interior of colon with flexible fiber-optic scope
		CT colonography	50+: Every 5 years	Virtual colonoscopy
Breast	Women	Self-examination	Optional	
		Clinical examination	20–40: Every 3 years 40+: Every year	Physical exam by health professional
		Mammogram	40+: Every year	X-ray of breast tissues
Cervix and Endometrium	Women	Pelvic exam	18+ and all sexually active women: Every year	Physical exam by health professional
		Pap test	18+ and all sexually active women: Every year 30+: every two years	Smear of cells taken from cervix by health professional
		Endometrial tissue sample	At menopause. Women who have any risk factors should be screened every year starting at age 35.	Sample of uterine tissue taken by health professional
Prostate	Men	Digital rectal exam	50+: Every year. Men who have any risk factors should be screened every year starting at age 45 or earlier.	Physical exam by health professional
		PSA (prostate-specific antigen)	50+: optional	Blood test that detects elevated levels of PSA

*For colon and rectum, any one of the tests listed is sufficient.

Note: People who have one or more risk factors for any cancers should begin screening at earlier ages and have tests done more often. Consult your physician for advice.
Data based on American Cancer Society Prevention and Detection Guidelines.

Breast Self-Examination and Routine Mammograms

For decades now, many women have been taught how to do regular breast self-examinations (BSEs) to detect suspicious lumps that might be breast cancer. They've also been told that they should have an annual mammogram (breast X-ray) after age 40. These recommendations are being called into question by new findings in comparative effectiveness research (see Current Issue, Chapter 8).

For example, the U.S. Preventive Services Task Force (USPSTF) now recommends against teaching BSE as a way to detect breast cancer. The USPSTF systematically reviews the benefits and harms of preventive care services and then tries to come up with a net value assessment. In the case of BSE, the USPSTF reviewed the latest published data and concluded that (a) regular BSE does not lower the mortality rate from breast cancer, and (b) women who perform BSE tend to have more imaging procedures and biopsies than women who don't. These procedures are expensive and are themselves associated with minor health risks, such as increased exposure to radiation and infections. Overall, the risk/ benefit ratio for BSE is on the side of net risk.

The USPSTF also recommends that women who are not at risk for breast cancer should delay having routine yearly screening mammograms until age 50, as opposed to the previous recommendation of age 40. Breast tissue is more sensitive to radiation damage before menopause than after menopause, and the chances of detecting cancer by a mammogram are much lower in younger women because breast cancer is just less common in younger women. According to the USPSTF, for routine screening mammograms the shift from net risk to net benefit is closer to age 50 than age 40.

The new USPSTF recommendations came out in late 2009, updating their 2002 recommendations. So far, the American Cancer Society (ACS) has not changed their 2003 recommendation that women over 40 have annual mammograms. However, BSE is now considered optional by the ACS, meaning that it's acceptable not to do them. ■

Reference: U.S. Preventive Task Force. Screening for Breast Cancer: U.S. Preventive Services Task Force Recommendation Statement. *Ann. Int. Med.* 151: 716–726, 2009.

Recent research suggests that there are at least two inherited susceptibility genes for breast cancer, called BRCA1 and BRCA2. Fortunately, mutated forms of these genes are relatively rare. A major risk factor is age; only one in 200 women will get breast cancer before age 40, but one out of 26 will get breast cancer in her 60s and one in eight will get it in her lifetime. Other risk factors include early onset of menstruation or late onset of menopause, obesity after menopause, use of oral contraceptives, and hormone replacement therapy after menopause.

Environmental and dietary factors may also play a role. Early detection dramatically increases the chance of survival: the 5-year survival rate for breast tumors that have not yet metastasized is 98%, but it falls to 27% for women who have metastases in distant tissues at diagnosis. Overall, the 10-year survival rate for all diagnoses is now 81%.

✓ **Quick Check** Use the data in Table 18.3 to calculate breast cancer deaths per year, in men and then in women, as a *percentage* of newly diagnosed cases. In which gender is a newly diagnosed patient more likely to die? Can you explain these results? ■

Prostate cancer: Most common after age 50

Symptoms of prostate cancer include difficulty urinating or inability to urinate, blood in the urine, and pain, usually in the pelvic area.

The biggest risk for prostate cancer is advancing age. Physicians recommend yearly testing for men aged 50 and older. Prostate cancer is generally initially detected by a digital rectal exam or by a *prostate-specific antigen* (PSA) test, which detects elevated blood levels of a protein produced by the prostate gland when prostate cancer is present. The initial diagnosis is generally confirmed by a biopsy and examination of the prostate cells.

Treatment options include surgery, radiation therapy, and hormones. Aggressive treatment in older men is controversial because prostate cancer generally grows so slowly that it is rarely the cause of death. The five-year survival rate is now almost 99%, up from 69% just 25 years ago. The dramatic improvement is due to earlier diagnoses and improved treatments.

Cancers of colon and rectum: Tests can detect them early

An obvious worrisome sign of colon or rectum cancer is blood in the stool or rectal bleeding, though bleeding can also result from noncancerous causes. Risk factors include sedentary lifestyle, obesity, smoking, a family history of colorectal cancer, and a low-fiber, high-fat diet with not enough fruits and vegetables.

Most colorectal tumors start as **polyps,** small benign growths that develop from the colon lining (**Figure 18.12**).

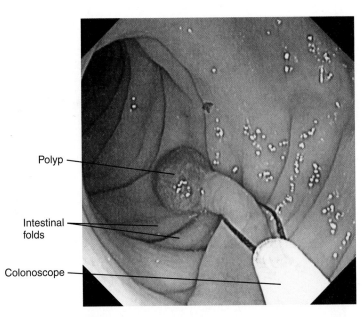

Figure 18.12 Polyps. A polyp in the colon being removed with a wire snare.

The majority of colon polyps never turn malignant, and even when they do, the process usually takes years. This underscores the importance of regular exams to detect polyps and remove them before they can become dangerous. For everyone over age 50, the American Cancer Society recommends a *colonoscopy* (examining the interior of the complete colon with a flexible fiber-optic scope) every 10 years. People who have any risk factors should begin screening earlier and undergo testing more often. Polyps can generally be removed rather easily with a small wire snare or by electrocautery (electric heat).

The five-year survival rate for all cases of colorectal cancer has improved to 64%, largely because of early detection and treatment. For people with distant metastases at the time of diagnosis, the five-year survival rate drops to just 11%. Here is another case where early detection is extremely beneficial. However, fewer than 40% of colorectal cancers are detected at an early stage because rates of screening are still low.

Lymphoma: Cancers of lymphoid tissues

Lymphoma is the general term for cancers of the lymphoid tissues, which include Hodgkin's disease and non-Hodgkin's lymphoma. Symptoms include enlarged lymph nodes, intermittent fever, itching, weight loss, and night sweats.

Risk factors are not entirely clear but seem to relate to altered immune function. People at risk include organ transplant recipients, those with HIV or human T-cell leukemia/lymphoma virus-1 (HTLV-1) infection, and perhaps workers with high occupational exposure to herbicides. Treatments generally are radiation or chemotherapy. Treatments involving specific antibodies to lymphoma cells can be effective. Bone marrow transplant may be necessary after high-dose radiation therapy.

Urinary bladder cancer: Surgery is often successful if done early

Blood in the urine is an important sign, although there may be other reasons for it. Smoking doubles the risk of bladder cancer. Living in an urban area, exposure to arsenic in the water supply, or occupational exposure to rubber, leather, or dye are also risks. Bladder cancer is usually diagnosed by microscopic examination of cells in the urine, and by direct visualization of the bladder wall with a cytoscope (a thin tube with a lens) inserted into the bladder via the ureter. Surgery combined with chemotherapy is generally successful if the cancer is detected while it is in the layer of cells in which it began (five-year survival rate—97%) or still localized to the bladder (73%). However, the five-year survival rate declines to just 6% if the cancer has already metastasized to distant sites by the time it is detected.

✓ **Quick Check** Propose a hypothesis that could explain why smoking increases the risk of bladder cancer, despite the fact that the bladder is never exposed to smoke. ∎

MJ's Human Biology Blog

The PSA Test for Prostate Cancer

When the prostate-specific antigen (PSA) test was first introduced in 1987, scientists thought it might lead to as much as a 50% reduction in prostate cancer deaths. But now it seems that it may not save lives after all. In a study of 77,000 U.S. men, the 10-year death rate from prostate cancer was the same in a group who had an annual PSA test for six years plus a digital rectal examination for four years, compared to a control group who were never tested over the same time period. The study may need to be continued for another decade or so to determine whether the PSA test has any usefulness over the longer term, however.

How can a test that accurately detects prostate tumors not save lives? Apparently the answer is that most prostate cancers are so slow-growing that older men are likely to die of something else first, even if they do have a diagnosis of prostate cancer. For older men, skipping the PSA test altogether may someday be a safe option.

By the way, a digital rectal examination is not some sort of digital readout. It's a physician's gloved digit, or finger. ∎

Reference: Andriole, G. R., et al. Mortality Results from a Randomized Prostate-Cancer Screening Trial. *New Engl. J. Med.* 360: 1310–1319, 2009.

Kidney cancer: Detected during examination for a renal-related problem

Risk factors for kidney (renal) cancer include heritable gene mutations, smoking, and exposure to certain toxic agents such as asbestos and cadmium. Indeed, cancer-causing agents and toxic chemicals are more likely to affect the kidneys than many other organs because as the kidneys concentrate the urine, they also concentrate the toxic agents in it. Other risk factors that you can do nothing about are age (renal cancer occurs primarily after middle age) and gender (males are twice as likely as females to get it).

There are no blood or urine tests to screen for renal cancer. It is usually detected during examination for a suspected renal-related problem such as low back pain or blood in the urine, and then confirmed by one of several imaging techniques, such as a CT scan or ultrasound. The main treatment for renal cancer is surgical removal of the affected kidney, especially when the other kidney is still normal. If surgery is not an option, chemotherapy or radiation therapy may be used. However, chemotherapy is not particularly effective against kidney cancer.

Cancer of the uterus: Unusual uterine bleeding is major symptom

Uterine cancer includes cancers of the endometrium and cervix. The primary symptoms of endometrial cancer are abnormal bleeding or spotting. Pain is a late manifestation of the disease. High lifetime exposure to estrogen is the major risk factor for endometrial cancer. Reasons for high exposure include early onset of menstrual periods, late menopause, obesity, not having children, and/or estrogen replacement therapy after menopause (progesterone added to estrogen is thought to offset the risk related to estrogen alone). Pregnancy and the use of oral contraceptives seem to offer some protection from endometrial cancer.

Cervical cancer, in contrast, is strongly correlated to sexually transmitted infection with the human papillomavirus (HPV). Women who have had many sexual partners (or whose partners have had many sexual partners) are at higher risk for cervical cancer. Cigarette smoking also increases the odds. Cervical cancer is usually detected by a Pap test (examination of a smear of cervical cells; the name comes from Papanicolaou, the doctor who devised the test). The Pap test is not very effective at detecting cancer of the endometrium, which is generally detected by a pelvic exam.

Treatments for both endometrial and cervical cancer include surgery, radiation, hormones, or chemotherapy. The one-year survival rate is 92% for endometrial cancer and 88% for cervical cancer.

Cervical cancer may be one of the first cancers to be largely preventable by a vaccine. In clinical trials in more than 17,000 girls and women, the vaccine, called Gardasil, was nearly 100% effective in blocking the strains of the

MJ's Human Biology Blog

A DNA Test for Cervical Cancer

It was bound to happen eventually. The annual Pap smear, the gold standard for over 40 years for detecting cervical cancer, may soon be replaced by a modern, much more specific DNA test for the human papillomavirus that is responsible for most cervical cancers. The cost of the test is currently around $20–30, but the test is so accurate and specific that women may only need the test once every 5–10 years.

A comparative study of the effectiveness of the new DNA test (HPV test) versus the Pap smear (cytologic test) in preventing cervical cancer deaths was conducted in India with funds provided by the Bill and Belinda Gates Foundation. After just eight years, the study showed that the DNA test reduced cervical cancer deaths by nearly 50% compared to the Pap test. Significantly, not one woman whose DNA test for HPV was negative died of cervical cancer during the study period. ■

Reference: Sankaranarayanan, R. et al. HPV Screening for Cervical Cancer in Rural India. *New Engl. J. Med.* 360:1385–1394, 2009.

human papillomavirus that cause most cases of cervical cancer and genital warts.

Leukemia: Chemotherapy is often effective

Leukemias are cancers of immature white blood cells within bone marrow. In its more advanced stages the bone marrow becomes completely filled with cancerous cells. Eventually the production of normal blood cells declines, leading to anemia and a reduced resistance to infections. The causes of leukemia are largely unknown, but may be linked to Down syndrome, excessive exposure to ionizing radiation, and benzene. Certain leukemias are caused by the HTLV-1 virus.

Although leukemia is commonly thought of as a childhood disease, more adults than children develop leukemias. The symptoms of leukemia are fairly general (fatigue, weight loss, increased incidence of infections), and as a result leukemias are not always diagnosed early. A firm diagnosis is made with a blood test and a bone biopsy.

The standard treatment for leukemias is chemotherapy. Sometimes chemotherapy is followed by a bone marrow transplant to replace the normal blood-forming stem cells eliminated by chemotherapy. The survival rates for certain types of leukemias have improved over the years, especially among children.

18.7 Most cancers can be prevented

Some cancers are linked to the genes we inherit from our parents, such as the BRCA genes that increase the risk of breast and ovarian cancer. For heritable cancers, early detection is not only vital but increasingly possible. Early detection of risk factors will allow treatment options to begin earlier. However, true prevention of heritable cancers will have to wait for advances in gene therapy and vaccines.

Nevertheless, most incidences of cancer are thought to be preventable. At least 60% of all cancer cases other than nonmelanoma skin cancers are thought to be caused by just two factors—smoking and poor diet. In addition, most nonmelanoma skin cancers could be prevented by reducing our exposure to the sun.

If we really want to make a dent in the cancer death rate, we need to work on public education: encouraging people to exercise, control their weight, eat a healthy diet, limit alcohol, avoid too much sun, and above all, not smoke. Researchers are studying what constitutes a healthy diet and why certain dietary components are harmful or beneficial.

What can you do to reduce your own risk of cancer?

■ Know your family history. Because cancer generally involves multiple genetic defects, you may have inherited one or more risk factors. Of course, this does not guarantee you will develop cancer. However, until we know of and can test for all such defects, you can at least assess your potential by examining your family's history of cancer and discussing it with your physician.

■ Know your own body. Suspicious lumps in a breast or a testicle should be reported to your physician. Be on the lookout for changes in the skin that might represent a melanoma (refer to Figure 18.11).

■ Get regular medical screenings for cancer. As we've seen, early detection is best, and some slow-growing cancers can be cured if caught early. Your physician may suggest other tests based on your family history, gender, and age, in addition to those described in Table 18.4.

■ Avoid direct sunlight between 10 a.m. and 4 p.m., wear a broad-brimmed hat, and apply sunscreen with an SPF (sun protection factor) of 15 or greater. Also avoid sunlamps and tanning salons.

■ Watch your diet and your weight. There are no absolutes here, but the general recommendation is a diet high in fruits, vegetables, legumes (peas and beans), and whole grains, and low in saturated fats, red meat, and salt. Unsaturated fats, such as fish and olive oil, appear to be beneficial (Figure 18.13).

■ Don't smoke. Here is a personal choice you can make that will dramatically lower your risk. Smoking is considered *the* single largest preventable risk factor for cancer. It accounts for fully 85% of all lung cancers and also contributes to cancers of the bladder, mouth, colon, cervix, pancreas, and pharynx. Ultimately, we may save more lives by persuading people not to smoke than by any amount of early detection.

■ If you consume alcohol, drink in moderation. One to two alcoholic beverages per day *may* reduce the chance of cardiovascular disease, but the trade-off is that alcohol consumption appears to raise the risk of certain cancers. It is particularly risky when combined with smoking.

■ Stay informed. There's a wealth of information on cancer in general, on specific types of cancer, and on the effects of nutrition and diet. Good resources include your physician, the American Cancer Society, the National Cancer Institute, or any good library.

Figure 18.13 **A healthy meal.**

Chapter Summary

Tumors can be benign or cancerous *p. 428*

- Normal cells have two key characteristics: (1) their rates of division are kept under control, and (2) most remain in one location in the body.
- A mass of cells that is dividing more rapidly than normal is called a tumor. Some tumors are benign.

Cancerous cells lose control of their functions and structures *p. 429*

- Cancer develops when cells divide uncontrollably, undergo physical changes, and no longer adhere to each other.
- Eventually some cancer cells invade nearby tissues. Others metastasize, colonizing distant sites.
- Cancer is the second leading cause of death in the United States.

How cancer develops *p. 430*

- Mutations of proto-oncogenes, tumor suppressor genes, and/or mutator genes may contribute to cancer.
- Inheritance of one or more mutated genes from your parents may increase your risk of developing certain cancers.
- Known carcinogens include some viruses and bacteria, environmental chemicals, tobacco, radiation, and dietary factors including alcohol consumption. Internal factors, such as faulty DNA replication and internally produced chemicals, can also contribute to cancer.
- Cancer is a multifactorial disease. Perhaps six or more mutations may need to be present before a cell becomes cancerous.

Advances in diagnosis enable early detection *p. 434*

- Most cancers are diagnosed when a tumor is detected by imaging techniques such as X-rays, PET, or MRI.
- Enzyme tests may also identify cancer.
- Genetic testing can determine the presence of a specific known mutated gene, but the presence of the mutated gene does not necessarily mean that cancer will develop.

Cancer treatments *p. 436*

- Many cancers are treatable; early diagnosis is important.
- Conventional treatments for cancer include surgical removal of the tumor, radiation, and chemotherapy.
- Newer treatments include magnetism and photodynamic therapy to target malignant cells precisely, immunotherapy to activate the patient's immune system, drugs to inhibit angiogenesis and cut off the tumor's blood supply, and molecular treatments that target specific genes.

The 10 most common cancers *p. 437*

- In both sexes, the most frequently occurring cancer is cancer of the skin.
- In addition, prostate cancer is common in men; breast cancer is common in women.
- Lung cancer causes more deaths than any other type of cancer.

Most cancers can be prevented *p. 443*

- Some cancers will occur despite our best efforts because we cannot control inherited risk factors.

- However, most cancers can be prevented. The single most effective way to reduce cancer deaths is to reduce the rate of smoking.
- A healthy diet will also decrease your cancer risk. Eat lots of fruits, vegetables, legumes, and grains; reduce your intake of saturated fat, red meat, and salt.
- Other preventive strategies include knowing your family's health history, getting regular medical screenings, maintaining a healthy weight, avoiding direct sunlight and sunlamps, drinking alcohol in moderation if at all, and staying informed.

Terms You Should Know

cancer, *429*
carcinogen, *431*
chemotherapy, *436*
dysplasia, *429*
hyperplasia, *428*
immunotherapy, *436*
in situ cancer, *429*
leukemia, *442*
lymphoma, *441*

malignant tumor, *430*
melanoma, *433*
metastasis, *430*
mutator gene, *431*
oncogene, *430*
proto-oncogene, *430*
tumor, *428*
tumor suppressor
  gene, *430*

Concept Review

Answers can be found at the Human Biology Place.
www.humanbiology.com

1. Compare and contrast a benign tumor and a malignant tumor.
2. Describe the key changes in cancer cells that allow them to metastasize.
3. Describe the impact a carcinogen has on a cell.
4. Explain why we have not yet made much progress against lung cancer, compared to other types of cancer.
5. Explain what causes most skin cancers.
6. List two dietary factors associated with an increased risk of cancers of the colon and rectum.
7. Describe the role the immune system plays in preventing different types of cancer.
8. Describe how tumors are diagnosed.
9. Describe how antibodies against cancer cells could improve the delivery of chemotherapeutic drugs.
10. List the things one can do to help prevent cancer.

Test Yourself

Answers can be found in Appendix A.

1. Which of the following represents the correct order of steps leading to the development of malignant cancer?
   a. dysplasia…hyperplasia…*in situ* cancer…metastasis and/or invasion
   b. hyperplasia…dysplasia…metastasis and/or invasion…*in situ* cancer
   c. dysplasia…*in situ* cancer…hyperplasia…metastasis and/or invasion
   d. hyperplasia…dysplasia…*in situ* cancer…metastasis and/or invasion

2. Metastasis is enabled by the:
   a. nervous system
   b. lymphatic system
   c. circulatory system
   d. both (b) and (c)

3. A mutation in which of the following types of genes would likely allow cell division to proceed even if conditions are not suitable or if the cells are already showing cancerous tendencies?
   a. mutator genes
   b. oncogenes
   c. tumor suppressor genes
   d. proto-oncogenes

4. Which of the following statements regarding skin cancer is true?
   a. Chemical carcinogens contribute more to the development of skin cancer than does UV radiation.
   b. Sun exposure during childhood does not impact the development of skin cancer later in life.
   c. Tanning booths enable safer tanning because the UV exposure from lamps is not likely to contribute to the development of skin cancer.
   d. Skin cancer can develop from cumulative sun exposure, even in the absence of sunburns.

5. Which of the following foods is associated with an increased risk of cancer development?
   a. unprocessed peanut butter
   b. red and yellow peppers
   c. broccoli, kale, cauliflower
   d. none of these choices

6. Which of the following imaging techniques expose(s) the body to radiation?
   a. X-rays
   b. MRI
   c. PET scans
   d. both (a) and (c)

7. Which of the following cancer treatments would be most effective at targeting both the original cancer as well as metastases?
   a. surgery
   b. focused sources of radiation
   c. chemotherapy
   d. photodynamic therapy

8. Hair loss, anemia, nausea, and depletion of white blood cells are side effects most often associated with:
   a. immunotherapy
   b. chemotherapy
   c. radiation therapy
   d. angiogenesis inhibition

9. Which of the following cancer treatments prevents growth of blood vessels that support growing tumors?
   a. immunotherapy
   b. anti-angiogenic drugs
   c. gene therapy
   d. radiation therapy

10. The ABCD rule refers to the evaluation of:
    a. melanoma
    b. squamous cell carcinoma
    c. basal cell carcinoma
    d. all types of skin cancer

11. The most common cause(s) of cancer deaths in the United States is/are:
    a. viruses
    b. radiation
    c. smoking
    d. poor diet and lack of exercise

12. Which of the following statements about breast cancer is true?
    a. Most women have a mutated form of either the BRCA1 or the BRCA2 gene.
    b. Post-menopausal hormone replacement therapy can reduce the risk of development of breast cancer.
    c. The risk of breast cancer declines with advancing age.
    d. Early onset of menstruation and/or late onset of menopause is a risk factor for the development of breast cancer.

13. Smoking increases the risk of developing:
    a. lung cancer
    b. bladder cancer
    c. cervical cancer
    d. all of these cancers

14. Which of the following would be useful in detecting a tumor within the urinary bladder?
    a. X-ray
    b. cytoscope
    c. colonoscope
    d. a blood test for a urinary bladder–specific antigen

15. Which cancer appears to be prevented by a vaccine?
    a. endometrial cancer
    b. cervical cancer
    c. ovarian cancer
    d. all of these cancers

Apply What You Know

Answers can be found at the Human Biology Place.
www.humanbiology.com

1. Why do you suppose that the death rate from lung cancer has increased much more in women than in men over the past 35 years? What information would you like to have that might support or disprove your hypothesis?

2. If you can't do anything about the genes you inherit, does it do any good to know your family history for particular cancers? Explain your reasoning.

3. The overall death rate from cancer dropped only 6% between 1950 and 2005. Why do you think we seem to have made so little progress against cancer?

4. From time to time we hear stories in the press that cell phones may cause certain types of brain cancer. Then someone else says the evidence is not yet clear. From what you know of the scientific method, why do you think this issue has been so hard to resolve?

5. Common therapies for cancer include chemotherapy and radiation treatments. How effective are these therapies, and why do they both have such serious side effects?

6. The first cancer that can be nearly completely prevented by a vaccine is cervical cancer. How does the vaccine work? If a vaccine can be developed against cervical cancer, why hasn't someone developed a vaccine against the number one cancer killer—lung cancer?

19

Genetics and Inheritance

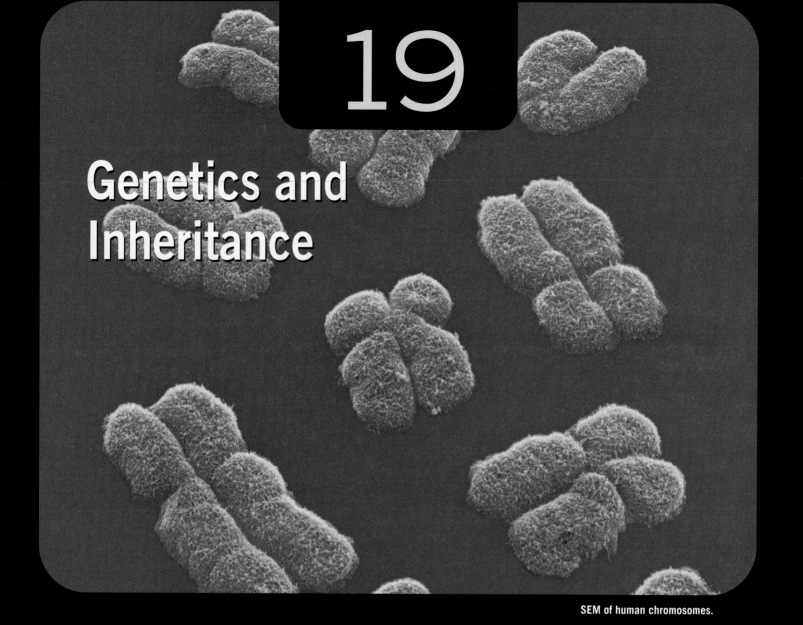

SEM of human chromosomes.

The Promises and Perils of Genetic Testing

Jeremy is having trouble in school, so his mother takes him to the doctor. After numerous tests and trips to several different doctors, genetic testing finally reveals that Jeremy has fragile X syndrome, an inherited form of mental retardation. His mother's insurance company promptly cancels Jeremy's health coverage on the basis that Jeremy has a "preexisting condition" that was not revealed to the company when his mother took out the insurance policy five years ago. After trying and failing to find an insurance company that will insure both of them, his mother is forced to quit her job so that she can become eligible for Medicaid.

Genetic testing (also called DNA-based testing) refers to the analysis of an individual's DNA to determine the risk for a particular health condition or to confirm the diagnosis of genetic disease. Genetic testing is increasingly available to anyone who wants it. For example, DNADirect.com sells a variety of genetic tests over the internet. The company even offers consultations with genetic counselors (at $250/hr).

Genetic Testing Offers Many Benefits

Genetic testing can provide undeniable benefits. Genetic testing can confirm the diagnosis of a genetic disorder even before symptoms appear, enabling patients to start treatment early. A negative gene test can rule out the likelihood of ever having a particular inherited disorder, saving a person years of stress and worry.

Genetic testing can also help couples make informed reproductive choices. Couples can be tested to determine whether they are carriers of recessive disorders, enabling them to make an informed decision about whether or not to have a child at all. For couples who are already expecting a child, prenatal genetic testing can be used to test fetal cells for approximately 40 genetic

Fetal cells harvested during amniocentesis can now be tested for over 40 inherited genetic disorders.

defects. Early genetic testing of the fetus gives parents time to prepare emotionally or perhaps to terminate the pregnancy, depending on their beliefs and the severity of the disorder.

As mentioned earlier, some gene mutations do not necessarily cause a disease but are associated with a higher risk of it. Suppose you learn that you carry a gene that predisposes you to colon cancer. The American Cancer Society recommends that most people have their first colonoscopy at age 50. Given your genes, you might decide to have your first colonoscopy at age 35 instead. If you're at known high risk, earlier testing could save your life.

Social and Ethical Issues

Privacy Who should have access to your genetic data—you? your physician? the company that performed the test? the insurance company that paid for it? Parts of your existing medical records are probably already available to your insurance company, as well as the physicians and staff of the various medical facilities you have visited over the years. Should your genetic data become part of this record?

Eligibility for health insurance As Jeremy's mother found out the hard way, many health insurance policies exclude preexisting conditions. If genetic testing revealed that you carried a gene that might increase your risk of a particular disease (such as cancer) later in life, it's conceivable that an insurer might try to exclude coverage for that disease from your policy. This could discourage you from having a genetic test.

But before we blame the insurers, consider that no private insurance company can lose money consistently and stay in business for long. The costs of health care are soaring. To an insurance company, information that helps them understand your liklihood of developing a disease, even if it's in the distant future, is like money in the bank.

Employment discrimination Some years ago, managers at the Burlington Northern Santa Fe Railroad became concerned about employees' high rate of insurance claims for carpal tunnel syndrome, and so they instructed company-paid physicians to secretly test employees for a rare genetic condition that can cause the syndrome. Employees were not informed that their DNA was being tested, and managers threatened to fire one staff member who refused the test.

In another case a social worker mentioned at work that her mother died of Huntington disease. One week later she was fired, despite a recent outstanding performance review.

These true stories demonstrate the potential for employment discrimination, even though it's now against the law.

Ethnic or racial discrimination Sickle-cell anemia is most common in people of African heritage. Tay-Sachs disease occurs most frequently in Ashkenazi Jews. People of Northern European descent are most likely to develop hemochromatosis, a genetic disorder that causes the body to absorb excess iron. Does this mean we should test all African Americans for sickle-cell anemia, all Jews for Tay-Sachs, or all Northern Europeans for hemochromatosis? Should we require members of high-risk groups to be tested if they are getting married or considering having children? Where do we want to draw the line between individual privacy and public health?

Psychological and emotional impact Discovering that you have a genetic predisposition for colon cancer might motivate you to take preventive steps. On the other hand, some health professionals are cautious about recommending widespread genetic testing of healthy people as a future health-predicting tool. They worry that the discovery that a person is likely to develop a genetic disease in the future might do more harm than good, especially if there is nothing that can be done about it.

Genetic testing and the law In 2008 the U.S. Congress passed the Genetic Information Nondiscrimination Act (GINA). GINA prevents insurers from setting higher insurance premiums or denying coverage to healthy people based on the results of genetic tests. GINA also makes it illegal for employers to use genetic information in making decisions about hiring or firing a person. However, GINA does not address other privacy issues, such as a genetic testing company selling genetic information to other companies or academic researchers.

The science of genetic testing is progressing far more rapidly than our solutions for the social and ethical challenges that it raises. Expect continued debate on these issues in the future.

Questions to consider

1 If you might be a carrier of a gene for an unpreventable condition, would you want to know? Why or why not?

2 Should genetic counseling be required before a person is allowed to purchase a genetic test? What about counseling afterward, if anything negative is found?

The facts...

- Genetic testing is a complex process for diagnosing genetic disorders by the direct examination of a person's DNA.

- Genetic testing is used to diagnose or rule out genetic disorders, help couples make informed reproductive choices, and prevent future health problems. Genetic testing raises difficult questions of privacy, eligibility for insurance, discrimination, and psychological or emotional harm.

>> **A *gene* is a segment of DNA that carries the code for making one or more proteins.** Humans have about 22,000 different genes on 23 pairs of chromosomes.

>> **Each person has two identical or nearly identical copies, called *alleles*, of most genes.** One allele was inherited from each parent. Differences between alleles are the result of rare mutations over millions of years of evolution.

>> **Your complete set of alleles, called your genotype, determines your particular physical and functional traits,** and even your susceptibility to certain diseases.

>> **When two alleles in an individual are different, one is usually *dominant* and the other is *recessive*.** A trait expressed by a recessive allele will only be present when there is no dominant allele, that is, when both alleles are recessive.

>> **The X and Y chromosomes, also called the sex chromosomes, carry the genes that determine the gender of the individual.** Females·have two X chromosomes and males have an X and a Y chromosome.

19.1 Your genotype is the genetic basis of your phenotype

As described in Chapter 17, chromosomes are partially composed of DNA, and genes are DNA sequences that contain instructions for building proteins. Humans have 23 pairs of chromosomes −22 pairs of **autosomes,** plus a pair of sex chromosomes. The autosomes and the pair of X-chromosomes in a female are sometimes called *homologous chromosomes* because they look alike and they each have a copy of the same gene at the same location, or *locus*, on the chromosome (Figure 19.1). Humans inherit one of each pair of autosomes and one sex chromosome from each parent, giving us two copies of each gene.

Although autosomes may look alike, closer inspection reveals that they are not completely identical. Small differences in DNA sequence may exist between any pair of autosomes. These sequence differences may occur within genes, and when they do, they produce alternative versions of genes called **alleles.** Because they are different, alleles may code for different proteins with slightly different structures. As discussed in Chapter 2, alterations in protein structure can affect how the protein functions. For example, the ability to roll your tongue is influenced by a single gene, but there are two alleles for that gene: One allele enables you to curl your tongue, the other allele does not.

If an individual possesses two identical alleles of a particular gene, the person is said to be **homozygous** for that gene (*homo-* means "same," and *zygous* means "yoked," or joined together). A person who has two different alleles of a gene is **heterozygous.** In many cases there are more than two alleles of a particular gene in the human population, although a given individual can inherit only two of them. For example, there are three alleles for blood type: the A, B, and O alleles.

s we have seen, the information necessary for the development of a new human being is contained in the DNA within the nucleus of the fertilized egg. This information is found in the form of **genes,** DNA sequences that code for one or more specific proteins. An organism's **genome** is the sum total of all of its DNA, even though not all of that DNA codes for proteins. **Genetics** is the study of genes and their transmission from one generation to the next.

The word **inheritance** often conjures up notions of money or possessions. The broader definition of inheritance, however, is something received from an ancestor or another person. In terms of genetics, each of us inherits one complete set of genes from our mother and one set from our father. Although everyone inherits two complete sets of genes, the sets vary slightly from individual to individual. Those variations help to account for the differences between us. In this chapter, we explore how genes are passed from generation to generation, and how genes affect our traits, or distinguishing features, such as our appearance, health, and possibly even our thoughts and actions.

Pair of autosomes. Each automsome carries the same genes at the locus

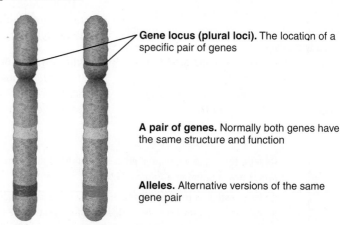

Gene locus (plural loci). The location of a specific pair of genes

A pair of genes. Normally both genes have the same structure and function

Alleles. Alternative versions of the same gene pair

Figure 19.1 **Autosomes.**

Where did these different alleles come from? Most likely, different alleles resulted from millions of years of **mutations,** or random changes to the DNA sequence, of cells destined to become sperm or eggs. Because those mutations were not corrected before cell division and meiosis, and because they did not cause the sperm or egg to be nonviable, they were passed on to the next generation as slight variations of the original gene, and they remain in the human population to this day. Collectively, all the various genes and their alleles in the human population are known as the *human gene pool.* There are enough different alleles among our approximately 22,000 genes to account for the uniqueness of each individual.

Your complete set of alleles is called your **genotype.** Your genotype has a profound influence on your **phenotype,** or the observable physical and functional traits that characterize you. Among your phenotypic traits are your hair, eye, and skin color; your height and body shape; and your ability (or inability) to curl your tongue. Many phenotypic traits, such as blood type and susceptibility to disease, are less observable. Your phenotype is determined not only by the alleles you inherit from your parents (your genotype), but also by environmental factors and lifestyle choices such as how much you are exposed to the sun, whether you exercise, and how much and what you eat.

⟲ Recap We inherit two alleles of every gene found on autosomes; these alleles may be identical or slightly different. The sum of our alleles is our genotype, and the physical and functional expression of those alleles is our phenotype. ▪

19.2 Genetic inheritance follows certain patterns

Punnett square analysis predicts patterns of inheritance

One easy way to designate two alleles of the same gene is to assign them uppercase (such as A) and lowercase (a) letters. Recall that people with the same two alleles of a gene (either aa or AA) are homozygous for that gene and people with different alleles (Aa) are heterozygous.

A **Punnett square** provides a simple way to represent patterns of inheritance of alleles and to predict the probability that a particular genotype will be inherited (**Figure 19.2**).

To create a Punnett square, place the possible alleles of the male gametes (produced during meiosis, Chapter 17) on one axis and the possible alleles of the female gametes on the other. Bear in mind that although the parents are diploid—they contain two copies of each chromosome—they will each donate *only one* of each chromosome to their offspring. By combining the letters for the sperm and egg in the appropriate squares, which represent the possible combinations that might occur during fertilization, we can see all of the possible genotypes of the offspring and the ratios of each possible genotype.

Mendel established the basic principles of genetics

In the 1850s, Gregor Mendel, a university-educated Austrian monk who specialized in natural history, described many of

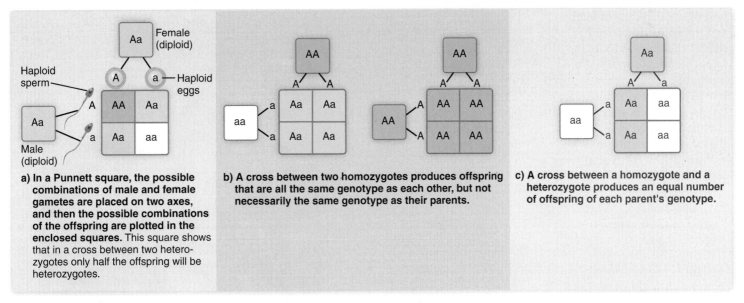

a) **In a Punnett square, the possible combinations of male and female gametes are placed on two axes, and then the possible combinations of the offspring are plotted in the enclosed squares.** This square shows that in a cross between two heterozygotes only half the offspring will be heterozygotes.

b) **A cross between two homozygotes produces offspring that are all the same genotype as each other, but not necessarily the same genotype as their parents.**

c) **A cross between a homozygote and a heterozygote produces an equal number of offspring of each parent's genotype.**

Figure 19.2 Punnett squares. Punnett squares are used to predict the outcomes of a particular combination of parental alleles.

✓ *Suppose the parents shown in the first Punnett square have a baby. How likely is it that this baby will have the genotype AA? How about Aa? More generally, how likely is each of the four "cells" in a Punnett square?*

the fundamental principles of inheritance through a number of experiments involving controlled breeding of garden peas. He chose pea plants as a "model" organism because they were cheap and easy to grow, and matings between plants could be set up easily. Many biologists use model organisms to investigate the fundamental principles of biology because they can do experiments such as setting up matings that would be unethical to do with human subjects. Mendel followed the inheritance of traits such as flower color, pea color, pea texture, and plant height from one generation to the next. He collected an extensive set of data over a seven-year period, and proposed that there were discrete "factors" of heredity that united during fertilization and then separated again in the formation of sperm and eggs. He concluded that pea plants inherit two units of each factor, one unit from each parent. Today, we call these factors genes and we know that they are found on chromosomes. Even though Mendel knew nothing about chromosomes, DNA, or meiosis, his pioneering studies with pea plants provided great insight into the principles of genetics in many complex organisms, and his rules of inheritance hold for humans as well as pea plants.

To better understand the laws that govern the passage of traits from generation to generation, let's consider one of Mendel's experiments in detail. In this experiment, Mendel performed a one-trait cross, where he followed the inheritance of a single trait: pea color. Pea plants can produce either green or yellow peas; when Mendel mated, or crossed, a plant with green peas to a plant with yellow peas, he found that all of the offspring produced yellow peas. These peas were identical to those produced by the parent with yellow peas; it seemed that the green pea trait had disappeared completely in these offspring. But was the green trait really gone? To check, Mendel crossed two of these yellow pea–producing offspring together. To his surprise, Mendel found that only about 75% of this second group of offspring produced yellow peas, while the remaining 25% of these offspring produced green peas. Thus, the green pea trait hadn't disappeared; it was as if the green trait had simply "skipped" a generation. Mendel deduced that the information needed to produce green peas was still present in the first group of yellow pea–producing offspring but was somehow hidden. Based on this and other experiments, Mendel reasoned that each pea plant must have two alleles of the gene for pea color, and that each parent contributes only one allele of the gene to the offspring (Figure 19.3).

This is now known as Mendel's first rule of inheritance, the **law of segregation:** When gametes are formed in the parents, the alleles separate from each other so that each gamete gets only one allele of each gene. Importantly, which allele a parent contributes to a gamete is determined in a completely random manner; a heterozygous parent harboring both the green and the yellow alleles has a 50% chance of donating the green allele and a 50% chance of donating the yellow allele to each gamete. Today we understand that gametes are haploid, with only half the number of chromosomes and genes of the parent (review the process of meiosis, Chapter 17).

Mendel's law of segregation doesn't apply only to pea plants; it applies to many complex diploid organisms, including humans. For example, let's consider hairlines (Figure 19.4). Human hairline pattern is controlled by a single gene that has two alleles. One allele causes a person to have a widow's

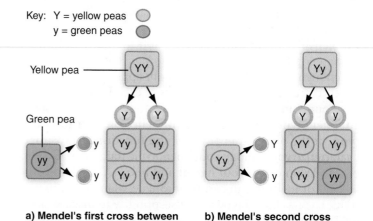

a) Mendel's first cross between homozygous yellow-pea plants (YY) and homozygous green-pea plants (yy) yielded all yellow-pea plants.

b) Mendel's second cross between two of the offspring of his first cross yielded 75% yellow-pea and 25% green-pea plants.

Figure 19.3 **Mendel's one-trait crosses in pea plants.**

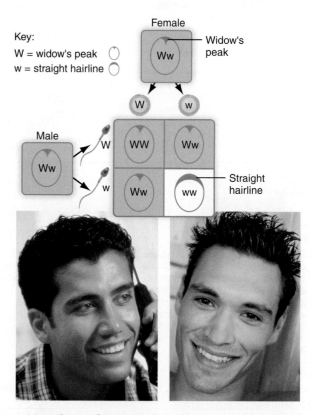

Figure 19.4 **One-trait crosses in humans.** The allele for the widow's peak (W) is dominant. A mating between two parents heterozygous (Ww) for the widow's peak trait can produce homozygous dominant (WW), heterozygous (Ww), and homozygous recessive (ww) children. Only the homozygous recessive genotype will have the straight hairline phenotype.

peak, while the other allele causes people to have straight hairlines. If two people heterozygous for the hairline pattern gene have a child, each parent has a 50% chance of donating the widow's peak allele to the child and a 50% chance of donating the straight hairline allele to the child. Mendel's law of segregation, or the fact that the parents randomly donate one allele of each gene to their sperm or egg cells, is a major reason that so much phenotypic variation exists among individuals in the human population.

Dominant alleles are expressed over recessive alleles

In the experiment described above, Mendel observed that the green pea trait "skipped" a generation. Why did this happen? Mendel reasoned that the yellow allele behaves in a **dominant** manner with respect to the green allele, meaning that when both alleles are present in the same plant, the plant will only show the yellow phenotype. The green allele is thus **recessive** to the yellow allele, and is only visible when no yellow alleles are present. Figure 19.3 shows the inheritance pattern of pea color that Mendel observed. It is customary to denote the dominant allele by an uppercase letter (in this case, Y) and the recessive allele by a lowercase letter (y).

Many human genes have dominant and recessive alleles as well. The hairline pattern gene has a dominant allele (W) that produces the widow's peak phenotype, and a recessive allele (w) that causes the straight hairline phenotype. If a person homozygous for the widow's peak allele were to have children with a person homozygous for the straight hairline allele, all of their children would be heterozygous for the gene, and would show the dominant widow's peak phenotype. But, if one of these heterozygous children were to mate with another person heterozygous for the hairline pattern gene, as shown in Figure 19.4, the recessive straight hairline phenotype could reappear in their offspring. It would appear that the straight hairline trait had skipped a generation, just like the green pea trait in Mendel's experiment.

Most recessive alleles—such as the allele for a straight hairline and also the allele for attached earlobes (Figure 19.5)—do not confer a distinct advantage or disadvantage to the homozygous recessive person. They just originated as mutations at some point in our evolutionary history, and they remain in the human gene pool because they do no harm.

However, some recessive alleles result in the absence of a functionally important protein. An example is cystic fibrosis (see Health & Wellness feature), which afflicts only homozygous recessive individuals. These alleles no longer code for a specific essential protein because their nucleotide sequence has been altered slightly from the dominant form. In principle, the

Attached earlobes (Johnny Depp, left) and unattached earlobes (George Clooney, right).

Figure 19.5 Harmless dominant/recessive traits and alleles. The allele for unattached earlobes is dominant over that for attached earlobes. Johnny Depp (on the left) is homozygous for the recessive allele, while George Clooney (on the right) is either homozygous for the dominant allele or is heterozygous.

frequency of truly harmful recessive alleles in a population is kept in check by the occasional premature death of homozygous recessives, but the allele can persist in the population in heterozygotes.

Finally, the term dominant refers only to how an allele behaves in combination with a recessive allele in a heterozygote. Dominance has nothing to do with the frequency with which an allele is found in a population; in fact, some dominant alleles are quite rare. Consider the number of human fingers and toes. Most people are homozygous recessive (pp) for the gene controlling the number of fingers and toes, and they have four fingers and a thumb on each hand and five toes on each foot. But some people possess a dominant allele (Pp or PP), and the presence of this allele causes them to have extra fingers and toes on each hand and foot (**Figure 19.6a**), a condition known as polydactyly. Even

a) A human infant with polydactyly.   **b) A polydactyl cat.**

Figure 19.6 Dominant alleles are not always common in a population. Polydactyly, or the presence of extra digits, is an uncommon phenotype in both humans and cats caused by a dominant allele.

Cystic Fibrosis

Cystic fibrosis (CF), the most common fatal genetic disease in North America today, is caused by a recessive allele of a gene on chromosome 7. Because it is recessive, parents may have no idea that they are both carriers for cystic fibrosis until they have a child with the disease. Approximately 1 in 23 Caucasians carries the recessive allele and about 1 Caucasian child in 2,000 has the disease. Among African Americans, only 1 child in 17,000 is born with the disease.

The recessive allele produces a defective version of a protein that normally aids the transport of chloride across the cell membrane of epithelial cells. Impaired chloride transport affects the secretion of salt and water across cells that line the ducts in several organs. As a result, organs that normally produce a thin, watery mucus secrete a thick, sticky mucus instead.

In the lungs, the thick mucus clogs the bronchioles and leads to chronic inflammation and infections that can be fatal. In the pancreas, the mucus blocks ducts and prevents pancreatic enzymes from reaching the intestines. The sweat glands are also affected, but ordinarily this does not cause major problems.

Typical symptoms of cystic fibrosis include persistent coughing, wheezing, or pneumonia, excessive appetite but poor weight gain, and foul-smelling, bulky feces. The disease is usually diagnosed by age 3 by testing for high salt levels in perspiration. A DNA-based blood test can screen for variations on the gene responsible.

Cystic fibrosis. This child with cystic fibrosis is receiving vigorous percussion therapy ("clap therapy") to loosen mucus in the lungs.

Management of cystic fibrosis focuses on reducing the damage caused by the symptoms, often a constant battle throughout life. Regular, vigorous chest physical therapy helps to dislodge mucus in the lungs, and antibiotics are often needed to treat lung infections. Researchers are working on gene therapy techniques that insert copies of the normal gene into viruses, such as those that cause colds, and then deliver the modified viruses to the airway. (For more on gene therapy, see Chapter 20.) Improved treatments and better understanding of the disease have increased the average life expectancy for patients with cystic fibrosis from just 5 years before we knew much about the disease, to over 37 years today. In addition, genetic testing can now identify couples who are at risk of having a child with the disease. Do not be surprised if cystic fibrosis is cured in our lifetime, or at least treated so effectively that patients live long and healthy lives. ◼

---

though this allele behaves in a dominant fashion with respect to the allele for the normal number of fingers and toes, it is extremely rare in the population. Interestingly, there is a dominant allele in cats that causes affected individuals to have extra toes, as well (Figure 19.6b).

✅ **Quick Check** In humans and other mammals, red hair is recessive to dark hair. Suppose a husband and wife are both heterozygous for the red hair color gene. What are the husband's and wife's own hair colors? If they have a baby, what is the probability that their child will have red hair? ◼

Answers to ✅ **questions can be found in Appendix A.**

Two-trait crosses: Independent assortment of genes for different traits

So far we have followed the inheritance patterns of single genes throughout multiple generations. But what happens when we follow *two* genes? Are they inherited as a set, or can they be inherited independently of each other? To answer this question, let's consider the human hairline pattern gene and a second gene that controls whether a person's earlobes hang freely or are attached (Figure 19.7). For the earlobe gene, the allele for free earlobes (E) is dominant over the allele for attached earlobes (e). If a person with a widow's peak and free earlobes who is homozygous dominant for both genes (EEWW) mates with a person with a straight hairline and attached earlobes who is homozygous recessive for both genes (eeww), all of their children will be heterozygous

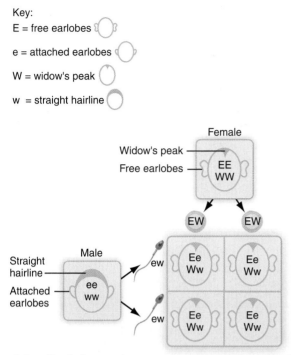

Key:

E = free earlobes

e = attached earlobes

W = widow's peak

w = straight hairline

Female

Widow's peak

Free earlobes

EE
WW

EW   EW

Male

Straight
hairline

Attached
earlobes

ee
ww

ew

ew

| Ee WW | Ee WW |
| Ee WW | Ee WW |

a) A mating between a homozygous person with a widow's peak and free earlobes (EEWW) and a homozygous person with a straight hairline and attached earlobes (eeww). All of the offspring will have the dominant widow's peak and free earlobes phenotypes.

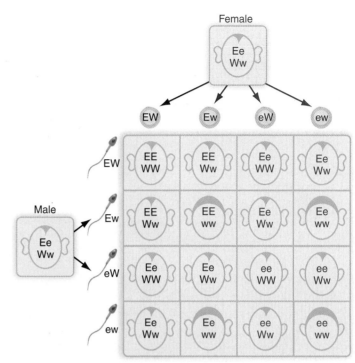

Female

Ee
Ww

EW   Ew   eW   ew

Male

Ee
Ww

EW

Ew

eW

ew

EE WW	EE Ww	Ee WW	Ee Ww
EE Ww	EE ww	Ee Ww	Ee ww
Ee WW	Ee Ww	ee WW	ee Ww
Ee Ww	Ee ww	ee Ww	ee ww

b) A mating between two heterozygous people with widow's peaks and free earlobes (EeWw). Because the alleles for the two traits assort independently, some of the offspring show one dominant and one recessive trait.

Figure 19.7 A two-trait cross showing independent assortment of alleles for different traits.

What is the ratio of the four possible phenotypes in the second mating (part b)? If you observed this same ratio in the offspring of an actual mating, what could you conclude about the parents?

for both genes (EeWw), and will have a widow's peak and free earlobes (Figure 19.7a).

But what happens if two people heterozygous for both genes mate? Would their children have either both dominant or both recessive phenotypes, or are other combinations possible? As shown in Figure 19.7b, a mating between two heterozygotes for both genes can indeed produce children with both dominant traits and children with both recessive traits, but other combinations are also possible. Some offspring may have the dominant widow's peak and the recessive attached earlobe phenotypes, and others may have the recessive straight hairline and the dominant free earlobe phenotypes.

When predicting the results of a two-trait cross, remember that each parent donates only one allele of each gene, and that the genes for different traits assort independently of each other during the formation of sperm and egg. That is, a parent can donate a dominant hairline pattern allele and a dominant earlobe allele, a recessive hair pattern allele and a recessive earlobe allele, or a mixture of the two. Each of these outcomes is equally probable because the hairline pattern and earlobe alleles are donated to egg and sperm independently of each other. The outcome of a two-trait cross can be predicted either by using either a

large Punnett square as shown in Figure 19.7b, or simply by multiplying the ratios from small Punnett squares as shown in **Figure 19.8** (next page). The latter works because the two events are independent of each other.

HBP **Web Animation** *One- and Two-Trait Crosses* at www.humanbiology.com

Mendel saw similar results in pea plants when performing two-trait crosses, and he formulated his **law of independent assortment,** which states that the alleles of different genes are distributed to egg and sperm cells independently of each other during meiosis. But, on this point Mendel was only partly right. Recall that he knew nothing about chromosomes or meiosis. We now know that the 22,000 or so genes in the human genome do not simply float around independently of each other in the cell's nucleus. Instead, genes are structurally part of the 23 pairs of chromosomes found in the cell's nucleus. Only genes that are located on different chromosomes always assort truly independently. Alleles located on the same chromosome, called linked alleles, may or may not be inherited together, as discussed later in this chapter.

Figure 19.8 Another method of calculating phenotypic ratios from two-trait crosses. Instead of drawing the large Punnett square in Figure 19.7b, each trait can be considered individually. The phenotypic ratios obtained can then be multiplied together to get the same results shown in Figure 19.7b.

Recap A Punnett square is a useful method for predicting the ratios of possible offspring genotypes from a particular mating. Mendel's *law of segregation* holds that when sperm and eggs form, the two alleles of each gene separate from each other so that each sperm or egg receives only one allele of each gene. Where two alleles of each gene exist in the human gene pool, one is often dominant over the other and controls the phenotype of a heterozygote. According to Mendel's *law of independent assortment*, during the formation of sperm and egg, alleles for different traits are distributed independently of each other provided they are located on different chromosomes. ∎

19.3 Other dominance patterns

Incomplete dominance: Heterozygotes have an intermediate phenotype

Some alleles do not follow the dominant/recessive pattern described above. In **incomplete dominance,** the heterozygous genotype results in a phenotype that is intermediate between the two homozygous phenotypes. An example of incomplete dominance in Caucasian humans is the trait of having straight, wavy, or curly hair (**Figure 19.9**). Another example is *familial hypercholesterolemia*. People homozygous for familial hypercholesterolemia may have blood cholesterol concentrations more than six times the normal level. The homozygous condition is quite rare (around one in a million people) because most homozygous individuals have heart attacks in childhood or adolescence and die by age 20. In contrast, the heterozygous condition is fairly common (about one in 500 people). People heterozygous for the disease may have blood cholesterol concentrations that are two to three times the normal level. With diagnosis and treatment, most can live fairly normal, healthy lives.

Codominance: Both gene products are equally expressed

In **codominance** the products of both alleles are expressed equally. One important example of codominance is the gene for blood type. There are three alleles for this gene: A, B, and O. The O allele is recessive; when combined in a heterozygous person with either the A allele or the B allele, the person will have A or B blood type. But the A and B alleles are codominant. If an individual is heterozygous with an A allele and a B allele, that person will have type

AB blood with *both* A and B antigens on his or her red blood cells, rather than a mixed or intermediate antigen (see Figure 19.10). Blood phenotype has important implications for which types of blood can be given to which patients, as seen in Chapter 7.

hh curly hair

HH straight hair

Hh wavy hair

Figure 19.9 Incomplete dominance. Wavy hair (Hh) is an intermediate phenotype between straight (HH) and curly (hh) hair because neither allele is dominant.

Another example of codominance is **sickle-cell anemia.** Sickle-cell anemia is a disease caused by one of two alleles involved in the production of hemoglobin for red blood cells. One of the two alleles codes for a hemoglobin molecule (designated Hb^S) that differs from the normal hemoglobin molecule (designated Hb^A) by just one amino acid. People homozygous for *sickle-cell anemia* (Hb^SHb^S) have only Hb^S hemoglobin in their red blood cells, whereas heterozygous individuals, who are said to have *sickle-cell trait* (Hb^AHb^S), have equal amounts of each type of hemoglobin. As shown in Figure 19.11a, a mating between two individuals with sickle-cell trait can result in children who are phenotypically normal (Hb^AHb^A), children with sickle-cell trait (Hb^AHb^S), and children with sickle-cell anemia (Hb^SHb^S).

In people with sickle-cell anemia, breathing air with even slightly lower oxygen concentrations than normal can cause the Hb^S hemoglobin to crystallize within the red blood cells. When this happens their red blood cells take on a crescent, or sickle, shape (Figure 19.11b). Sickled cells rupture easily because they are more fragile than normal red blood cells. They are also less flexible, so they tend to clog small blood vessels and block blood flow, and they carry less oxygen. The resultant oxygen deprivation of tissues and organs produces symptoms of fatigue, chest pain, bone and joint pain, swollen hands and feet, jaundice, and visual problems due to retinal damage. If sickle-cell anemia is left untreated, long-term complications can include stroke, blindness, pneumonia, and other health problems. Many people with sickle-cell anemia do not live into their 30s. In contrast, people with sickle-cell trait rarely suffer from any symptoms. The half of their hemoglobin that is normal is generally sufficient for them to lead normal lives. However, they may have mild episodes of red blood cell sickling when oxygen concentrations are lower than normal, such as at high altitudes or in airplanes in flight.

	Type A	Type B	Type AB	Type O
Red blood cells	Antigen A	Antigen B	Antigens A and B	Neither A nor B antigens
Possible genotypes	AA AO	BB BO	AB	OO

Figure 19.10 Blood type: An example of codominance. Characteristics of the four major blood types are shown, including their surface antigens and the genotypes that cause each blood type phenotype.

✓ *A woman with blood type AB and a man with blood type O have a baby. What is the probability that the baby has blood type O, like his father? Explain your answer.*

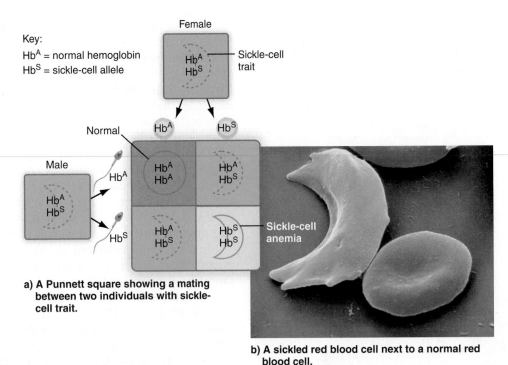

a) A Punnett square showing a mating between two individuals with sickle-cell trait.

b) A sickled red blood cell next to a normal red blood cell.

Figure 19.11 Sickle-cell anemia: an example of codominance. If two people who have sickle-cell trait have children, each child will have a 25% chance of being phenotypically normal, a 50% chance of having sickle-cell trait, and a 25% chance of having sickle-cell anemia.

Sickle-cell anemia affects primarily Africans (and their descendants on other continents) and Caucasians of Mediterranean descent. In East Africa, sickle-cell anemia is fairly common and nearly 40% of the black population has sickle-cell trait. In contrast, only about 1 in 500 African Americans is born with sickle-cell anemia and only 8% have sickle-cell trait. Why the big difference?

It turns out that the sickle-cell allele protects against malaria, a deadly parasitic disease common in tropical climates. The malaria parasite spends part of its life cycle in humans (in the liver and in red blood cells) and the other part in mosquitoes. When the malaria parasite enters the red blood cell of a person with sickle-cell trait or sickle-cell disease, the cell sickles and eventually ruptures, killing the parasite before it can reproduce.

Malaria kills one million people every year worldwide. Where malaria exists, having sickle-cell trait (but not sickle-cell anemia) is a distinct advantage. Were it not for malaria, we could surmise that sickle-cell anemia and sickle-cell trait would become rare in Africa, just as they are becoming rarer in the United States.

Because our phenotype includes all of our unique characteristics from our physical appearance to our biochemical makeup, the definition of an inheritance pattern as either incompletely dominant or codominant depends on the context. At the biochemical level, sickle-cell trait is an example of codominance because both the normal hemoglobin (Hb^A) and the sickle cell hemoglobin (Hb^S)

are produced in equal amounts. However, at the level of the whole person, sickle-cell trait is an example of incomplete dominance, because the person with sickle-cell trait exhibits a disease pattern that is intermediate between the two homozygous conditions. Understanding the concept is more important than an argument over whether sickle-cell trait is one or the other.

HBP **Web Animation** *Codominance and Incomplete Dominance* at www.humanbiology.com

 Recap People heterozygous with incompletely dominant alleles have a phenotype that is intermediate between the phenotype of either homozygous genotype. People heterozygous with codominant alleles express the products of both alleles equally. Examples of codominance are AB blood type and sickle-cell anemia. ▪

19.4 Other factors influence inheritance patterns and phenotype

Polygenic inheritance: Phenotype is influenced by many genes

Many traits result from not just one pair of genes, but many genes acting simultaneously. Inheritance of phenotypic traits that depend on many genes is called **polygenic inheritance**. Eye color, for example, is controlled by at least three

Figure 19.12 Polygenic inheritance. Eye color is determined by at least three genes, not just one. Different combinations of the alleles of the genes account for a range of eye colors.

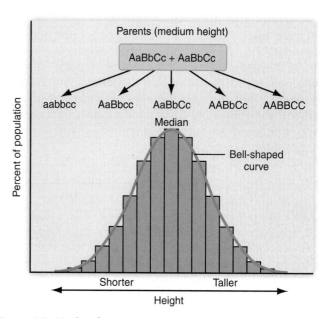

Figure 19.13 Continuous range of variation in traits governed by polygenic inheritance. Polygenic traits, such as height, appear in the population as a continuous range of phenotypes, with more people in the middle than at either extreme. In this hypothetical example, three genes influence height. Heterozygotes for all three genes have an intermediate (average) height, and can produce a wide range of heights among their offspring, depending on how many tall alleles and short alleles each child inherits.

genes, yielding a range of different eye color phenotypes from nearly black to light blue (**Figure 19.12**). Both skin and eye color are determined by the amount of a light-absorbing pigment called melanin in the skin and iris. The striking differences in eye color occur because in people with relatively low amounts of melanin we see different wavelengths of light (green, gray, or blue) reflected from the iris.

Physical characteristics such as height, body size, and shape are also examples of polygenic inheritance. Traits governed by polygenic inheritance generally are distributed in the population as a continuous range of values, with more people at the middle (or median) and fewer people at the extremes. As an example of a trait influenced by multiple genes, let's consider height. We will simplify matters and assume that height is influenced by only three genes, each of which has two alleles. Each gene has a tall allele (A, B, or C) that behaves in an incompletely dominant manner with respect to the short alleles (a, b, or c). In this example, an individual homozygous for tall alleles for all three genes (AABBCC) would be very tall, whereas an individual homozygous for short alleles for all three genes (aabbcc) would be very short. Heterozygotes for all three genes (AaBbCc) would have an intermediate phenotype and would be of medium height. Now, consider a cross between individuals of medium height. Most of their offspring would be similar in height to their parents, inheriting a mix of tall and short alleles, but by random chance, some offspring might inherit mostly tall or mostly short alleles, and a few might be very tall or very petite (**Figure 19.13**). This phenotypic distribution occurs throughout the human population, and a plot of the numbers of individuals at each value for the trait typically forms a normal (bell-shaped) curve.

We are just beginning to understand that many health conditions may be caused or at least influenced by polygenic inheritance. Examples include cancer, high blood pressure, heart disease, and stroke.

Both genotype and the environment affect phenotype

Phenotype is determined in part by our genotype and in part by environmental influences. A prime example of an environmental influence is the effect of diet on height and body size. Analysis of human populations reveals a clear trend toward increased height and weight in certain populations, particularly in developed countries. The changes in height and weight have sometimes occurred within one generation—too short a time to be due to changes in the gene pool. The trend is primarily due to improvements in diet and nutrition, especially early in childhood.

Except for a few rare genetic diseases such as Huntington disease (more on this later in the chapter), our genotype is not the sole determinant of whether we will actually develop a heritable disease (that is, have the disease phenotype). Usually we inherit only a slightly increased *risk* of developing the disease, with those risks modifiable by environmental factors and our own actions. If you know that certain heritable conditions or diseases run in your family, it would be prudent to reduce any environmental risk factors associated with that disease. For example, if your family has a history of skin cancer, limit your exposure to the sun. If your family has a history of heart disease due to high cholesterol, have your cholesterol measured regularly and change your diet or begin other corrective measures if your cholesterol levels are above normal.

This is really the old nature-versus-nurture question: Are we the way we are because of our genes or because of our environment? The answer is both. Your genes carry the instructions for all your proteins, but the environment can influence how genes are expressed and how they contribute to your phenotype. There is no one answer to the nature-versus-nurture dichotomy. But because you cannot change your genes, your best defense against any possible future disease is to take proper care of yourself.

 Quick Check Suppose you're studying male house finches, and you've gotten curious about why some male house finches are dark red whereas others are a washed-out pale pink. Briefly outline one experiment that could help determine whether this variation in feather color is due to genetic factors, environmental factors, or both. ■

Linked alleles may or may not be inherited together

As previously mentioned, alleles of different genes are distributed to egg and sperm cells independently of each other during meiosis. But, because humans have only 23 pairs of

chromosomes, it doesn't always work this way. Many alleles for different traits are often inherited together because they are physically joined on the same chromosome. These alleles are called **linked alleles,** and the closer they are located to each other on the chromosome the more likely they are to be inherited together. However, they are not *always* inherited together because of crossing-over, which partially reshuffles alleles across each pair of autosomes during meiosis (review meiosis in Chapter 17). How often two linked alleles are inherited together is directly proportional to how close together they are and inversely proportional to how often crossing-over occurs (the more crossing-over occurs, the less likely the linked alleles will be inherited together).

To get a feel for how often linked alleles might be inherited together, consider that the average pair of chromosomes may have 2,000 or more genes, and that crossing-over occurs approximately 30 times per pair of chromosomes. Alleles located close to each other, then, may be inherited together quite often. Scientists have learned that they can map the positions of genes by studying how often particular linked alleles are inherited together, a technique called *linkage mapping.*

Adding it all up, there are three major sources of genetic variability as a result of sexual reproduction:

■ Independent assortment of alleles located on different chromosomes

■ Partial shuffling of linked alleles as a result of crossing-over between autosomes

■ Random fertilization of an egg by a sperm

 Recap In polygenic inheritance, many different genes influence a single phenotype, such as height or eye color. In addition, some phenotypes can be influenced by environmental factors such as diet. Alleles located close together on the same chromosome are inherited together, except when crossing-over occurs. ■

19.5 Sex-linked inheritance: X and Y chromosomes carry different genes

Chromosomes are identifiable in cells only just before cell division. At this time each chromosome can be identified by its characteristic size, centromere location, and distinct banding pattern. A composite display of all the chromosomes of an organism is called a **karyotype.**

The karyotype shown in **Figure 19.14** shows the 23 pairs of human chromosomes. Of these, 22 are matched pairs (autosomes). The last pair of chromosomes, however, do not look alike and do not match up. These are the sex chromosomes, X and Y. The sex chromosomes are not a homologous pair. They look different from each other, and they function differently because they carry different genes. They represent a pair only in that you get one

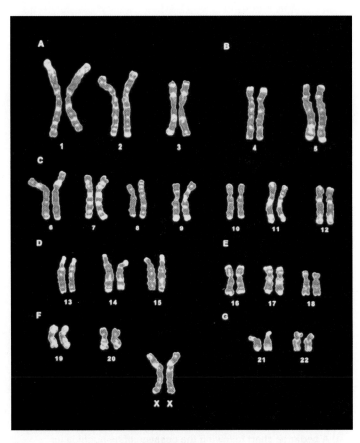

Figure 19.14 The Human karyotype. Chromosomes are identified and paired according to their size, centromere location, and characteristic banding patterns. A numbered arrangement of all 23 pairs is called the human karyotype. This karyotype is from a female, as indicated by the two X chromosomes. A male would have an X chromosome and a smaller Y chromosome.

from each parent as a consequence of meiosis. As seen in Figure 19.15, the sperm from the father determines the gender of the offspring. Females have two X chromosomes and will donate one of them to the offspring, whereas males have an X and a Y chromosome, only one of which is donated to the offspring. If the male donates an X chromosome, the fertilized egg will have two X chromosomes and will develop into a female. If the male donates a Y chromosome, the fertilized egg has an X and a Y chromosome and will develop into a male.

Sex-linked inheritance depends on genes located on sex chromosomes

Gender determination by inheritance of sex chromosomes is not without cost, especially to males. The advantage of having pairs of autosomes is that you have two copies of each gene—a backup file, if you will. If something goes wrong with one copy, the other can still carry out its function.

In sexually reproducing animals in which the Y chromosome determines maleness, females have a homologous pair of X chromosomes but males do not. This means that males have a greater susceptibility for diseases associated with

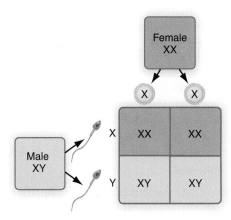

Figure 19.15 Sex determination in humans. The sex of the embryo is determined by the male's sperm, about 50% of which carry an X chromosome and 50% of which carry a Y chromosome. Maleness is determined by the inheritance of a Y chromosome.

recessive alleles on the sex chromosomes. In fact, males are more likely to display any recessive trait found on the X chromosome, not just those that cause disease.

Sex-linked inheritance refers to inheritance patterns that depend on genes located on the sex chromosomes. Sex-linked inheritance is *X-linked* if the gene is located only on the X chromosome and *Y-linked* if the gene is located only on the Y chromosome. There are plenty of examples of X-linked inheritance because many of the genes on the X chromosome are not related to sex determination at all. However, there are relatively few well-documented examples of Y-linked inheritance. The reason is that the Y chromosome is relatively small and many of its genes relate to "maleness." Genes on the Y chromosome apparently influence differentiation of the male sex organs, production of sperm, and the development of secondary sex characteristics, but not much else.

In the female, X-linked inheritance behaves just like inheritance of autosomes with paired genes. Not so in the male, because he inherits only one X chromosome. In the male, X-linked genotype and phenotype are both determined solely by the single X chromosome he inherited from his mother.

The best known example of an X-linked disease is hemophilia, also known as bleeder's disease. Hemophiliacs lack a blood-clotting factor that is encoded by an X-linked gene with two alleles. The recessive allele (X^h) is a mutant that cannot produce the clotting factor, whereas the dominant allele (X^H) produces the clotting factor. Individuals have the disease only if they do *not* have at least one normal dominant allele (that is, if they are an X^hX^h female or an X^hY male).

The inheritance of hemophilia and other X-linked recessive diseases follows this pattern:

- Many more males have the disease than females. Females can be protected by inheriting at least one normal allele (X^H) out of two, whereas males inherit only one allele because they inherit only one X chromosome.

■ The disease is passed to sons solely through their mothers even though the mothers may only be *carriers*. Carriers are heterozygous ($X^H X^h$); therefore, they can pass on the recessive hemophilia allele even though they do not suffer from the disease themselves. Statistically, half of the sons of carrier mothers will have the disease and half of the daughters will be carriers.

■ A father cannot pass the disease to a son, but his daughters will all be carriers (unless the mother is also a carrier or has the disease, in which case the daughters may also have the disease).

The inheritance pattern of a genetic disorder can be followed through many generations by using a pedigree chart to map out a family's lineage. **Figure 19.16** shows a pedigree following the hemophilia trait. Muscular dystrophy and red-green color blindness are also X-linked recessive conditions that follow this inheritance pattern. Nearly all X-linked diseases are caused by a recessive allele rather than a dominant one.

 Web Animation *Sex-Linked Traits* at www.humanbiology.com

 Quick Check Can a father ever pass an X-linked trait to his son(s)? Why or why not? ■

Sex-influenced traits are affected by actions of sex genes

A few genes that you might suspect are located on the sex chromosomes are actually located on one of the 22 autosomal pairs of chromosomes. A good example is a gene that influences baldness. The normal allele (b^+) and the allele for baldness (b) can be present in both men and women. However, the baldness allele is recessive in women. Women must be (bb) and probably have other baldness genes or factors as well before their hair loss is significant. Men become very bald if they are (bb) and develop significant pattern baldness even if they are heterozygous (bb^+).

Why the difference between men and women? Testosterone strongly stimulates the expression of the baldness allele, in effect converting it from a recessive to a dominant allele in males. This is an example of a *sex-influenced* phenotype, one not inherited with the sex chromosomes *per se* but influenced by the actions of the genes on the sex chromosomes. In fact, a number of genes in addition to this particular baldness gene influence the presence and pattern of hair on the head.

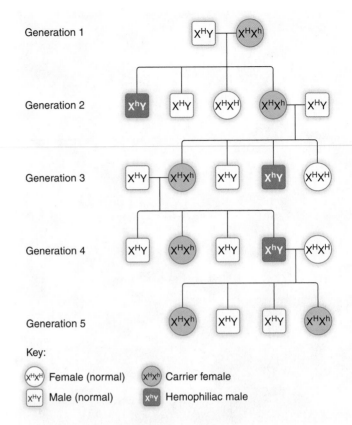

a) **A pedigree chart following the passage of hemophilia for five generations.** Female carriers pass the hemophilia allele to half their daughters and the disease to half their sons. Males with the disease pass the hemophilia allele to all their daughters (if they survive long enough to have children), but never to their sons.

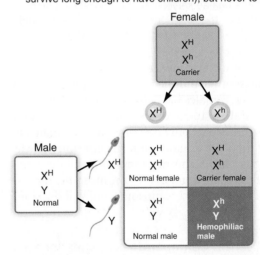

b) **A Punnett square showing the possible outcomes of the mating in Generation 1.**

Figure 19.16 A typical inheritance pattern for hemophelia, an X-linked recessive disease.

 Suppose that one of the women in Generation 5 marries and has a daughter (not a son) who has full-fledged hemophilia. How could this happen?

🔄 **Recap** In humans, the sex chromosome carried by the male sperm determines the gender of the fertilized embryo. Males have only one X chromosome, and are more susceptible to sex-linked recessive diseases than females are. Sex-linked inheritance depends on genes located on the sex chromosomes; sex-influenced traits depend on genes located on autosomes that are influenced by the actions of sex-linked genes. ■

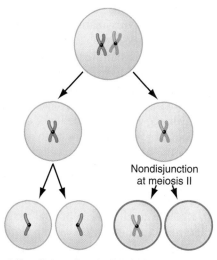

a) Normal meiosis. Duplicated homologous chromosomes separate during meiosis I, sister chromatids separate during meiosis II.

b) Nondisjunction during meiosis I. The duplicated homologous chromosomes fail to separate from each other.

c) Nondisjunction during meiosis II. Sister chromatids fail to separate from each other.

Figure 19.17 Nondisjunction during meiosis. Abnormal gametes have red outlines. This figure shows just one of the 23 pairs of human chromosomes.

19.6 Chromosomes may be altered in number or structure

During mitosis of diploid cells and meiosis of the precursor cells of sperm and eggs, duplicated chromosomes must separate appropriately into two different cells. Just as duplication of DNA is not always perfect, neither is separation of the chromosomes.

Failure of homologous chromosomes or sister chromatids to separate properly is called **nondisjunction.** Sometimes both sister chromatids go to one of the two cells formed during mitosis. Occasionally homologous chromosomes or sister chromatids fail to separate during meiosis (Figure 19.17), resulting in an alteration in chromosome number of sperm or egg cells. Very rarely, a piece of a chromosome may break off and be lost, or it may reattach to another chromosome where it doesn't belong.

For the most part we never see these chromosomal alterations. If mitosis goes awry, the two daughter cells probably just die and their place is taken by other dividing cells. Alterations that arise during meiosis are more serious because they have at least the potential to alter the development of an entire organism. However, it appears that we never see most of these either because so many of the genes on the chromosomes are essential for embryonic development. Any sperm, egg, or developing embryo with errors as great as an extra or missing chromosome is unlikely to survive. Most die early in embryonic development before we are even aware of their presence.

Down syndrome: Three copies of chromosome 21

A few alterations of autosomal chromosome number do result in live births. The most common alteration of autosomal chromosome number is *Down syndrome.* There are actually three types of Down syndrome, but the overwhelming majority (95% of all cases) is caused by having three copies of chromosome 21 (trisomy 21) (Figure 19.18). Other disorders resulting from altered autosomal chromosomal number are Edwards syndrome (trisomy 18) and Patau syndrome (trisomy 13).

People with Down syndrome are typically short with round faces and other distinctive physical traits; most are friendly, cheerful, and affectionate. They generally are slow

Figure 19.18 A person with Down syndrome. Down syndrome is caused by inheriting an extra copy of chromosome 21. Note that alterations in chromosome number can happen for any of the chromosomes, but most alterations are not seen in the population because they are lethal during fetal development.

to develop mentally and are prone to respiratory complications or heart defects that can lead to death at a young age. However, with improved care many now live well into their adult years.

Down syndrome affects approximately one in every 1,000 live births in the United States. The risk of giving birth to a Down syndrome baby rises markedly with the mother's increasing age. The chance is around one in 1,300 for mothers under 30, one in 100 at age 40, and one in 25 for mothers over age 45. Recent evidence suggests that paternal age may also be a factor when both parents are 35 or older.

Some pregnant women elect to undergo fetal testing to detect Down syndrome and other disorders. In amniocentesis, a sample of amniotic fluid containing fetal cells is withdrawn and tested. Another option is chorionic villi sampling, in which a small sample of fetal tissue is suctioned from the placenta for examination. Fetal cells are examined for the presence of third copies of chromosome 21 (Down syndrome), third copies of chromosomes 18 and 13, and for alterations of sex chromosome number. Chromosomal damage can also sometimes be detected.

 Quick Check Is Down syndrome caused by a gene mutation? Explain. ▪

Alterations of the number of sex chromosomes

Nondisjunction of the sex chromosomes can produce a variety of combinations of sex chromosome number, several of which are fairly common in the human population. In general, an individual with at least one Y chromosome will be essentially a male phenotype. An individual lacking a Y chromosome will be a female phenotype, regardless of the number of X chromosomes. The four most common alterations of sex chromosome number are the following:

XYY—Jacob syndrome XYY individuals are males who tend to be tall but otherwise fairly normal, except that some show impaired mental function. At one time it was thought that XYY males had a tendency toward violent criminal behavior, but careful study has shown this is not true. Whether their rate of minor, nonviolent criminality tends to be higher than normal is still a disputed issue.

XXY—Klinefelter syndrome
People with Klinefelter syndrome are also tall male phenotype. They are sterile, may show mild mental impairment, and may develop enlarged breasts because of the extra X chromosome (**Figure 19.19a**).

XXX—Trisomy-X syndrome
Trisomy-X syndrome individuals are female phenotype. Typically they are nearly normal except for a tendency toward mild mental retardation.

a) **Klinefelter syndrome (XXY).**   b) **Turner syndrome (XO).**

Figure 19.19 Alterations of sex chromosome number.

XO—Turner syndrome Individuals with only one X chromosome (XO) are phenotypically female. However, they tend to be short with slightly altered body form and small breasts (Figure 19.19b). Most are not mentally retarded and can lead normal lives, though they are sterile and may have shortened life expectancies. Turner syndrome is relatively rare compared to the other three because the XO embryo is more likely to be spontaneously aborted.

Table 19.1 summarizes the most common alterations of chromosomal number and gives their frequencies of occurrence.

 Quick Check Given the information above, what exactly causes an XY individual to be male? That is, is it the presence of the Y that causes maleness, or the number of X chromosomes? ▪

Table 19.1 **Some common alterations of chromosome number**

Chromosome genotype	Sexual phenotype	Common name	Approximate frequency among live births
Trisomy 21	Male or female	Down syndrome	1/1,000
Trisomy 18	Male or female	Edwards syndrome	1/6,000
Trisomy 13	Male or female	Patau syndrome	1/5,000
XYY	Male	Double-Y syndrome	1/2,000
XXY	Male	Klinefelter syndrome	1/2,000
XXX	Female	Trisomy-X syndrome	1/2,000
XO	Female	Turner syndrome	1/10,000

Deletions and translocations alter chromosome structure

A **deletion** occurs when a piece of chromosome breaks off and is lost. In general we cannot afford to lose genes, and therefore nearly all deletions are lethal to the sperm, egg, or embryo. There are a few rare conditions in which a chromosome deletion results in a live birth. An example is *cri-du-chat* (French, meaning "cat-cry") syndrome, caused by a deletion from chromosome 5. Babies born with this syndrome are often mentally and physically retarded, with a characteristic kittenlike cry due to a small larynx.

Translocation occurs when a piece of a chromosome breaks off but reattaches at another site, either on the same chromosome or another chromosome. Conceptually you might think that translocations are not a problem because all of the genes are still present, just in different locations. However, translocations result in subtle changes in gene expression and therefore their ability to function. For instance, translocation appears to increase the risk of certain cancers, including one form of leukemia.

> **Recap** Nondisjunction leads to altered numbers of chromosomes in daughter cells. Down syndrome is usually caused by inheriting three copies of chromosome 21. When nondisjunction occurs in the sex chromosomes, usually people with at least one Y chromosome are male phenotype and people without a Y chromosome are female phenotype. A deletion occurs when part of a chromosome is lost; deletions rarely result in a live birth. ▪

19.7 Many inherited genetic disorders involve recessive alleles

People express inherited genetic disorders caused by recessive alleles only if they inherit two of the defective alleles. People who inherit only one defective allele can pass the gene to their children, but they do not get the disease themselves because their good allele is dominant. For this reason, recessive disease-causing alleles are likely to remain in a population more readily than dominant ones.

In some cases more than one gene pair may contribute to a disease. Hypertension and heart disease are examples of diseases that apparently have a heritable component but aren't associated with any single gene pair.

One of the better known recessive genetic disorders is cystic fibrosis, discussed in the Health & Wellness feature on page 452. In this section we discuss two other genetic disorders caused by recessive alleles (phenylketonuria and Tay-Sachs disease), as well as one (Huntington disease) that is caused by a dominant allele.

Phenylketonuria is caused by a missing enzyme

Phenylketonuria (PKU) is a human inherited disease in which homozygous recessive individuals are unable to make an enzyme that is needed for the normal metabolism of phenylalanine, an amino acid. If phenylalanine accumulates, some of it is used in other metabolic pathways, including one in which excess phenylalanine is converted to phenylpyruvic acid instead. Phenylpyruvic acid is toxic at high concentrations and can lead to mental retardation, slow growth rate, and early death.

Phenylketonuria occurs in about 1 in 12,000 births in Caucasians, so it is considerably less prevalent than cystic fibrosis. The disease is caused by mutation of the gene on chromosome 1 that is responsible for producing the enzyme phenylalanine hydroxylase.

Fortunately, people homozygous for the recessive allele are not affected before birth because the mother's enzymes metabolize phenylalanine for the fetus. After birth, the blood level of phenylalanine is very much dependent on the amount of phenylalanine in the diet. Given the seriousness of mental retardation, all states now require that newborns be tested for PKU by a simple blood test called the Guthrie test, which detects excess phenylalanine in the blood.

Treatment of PKU requires limiting the dietary intake of phenylalanine. This is not easy because most proteins contain phenylalanine, but it can be done for about $5,000 per year by substituting a carefully controlled mixture of amino acids for most protein. In addition, individuals with phenylketonuria cannot use the artificial sweetener NutraSweet because it contains aspartame, a chemical consisting of just two amino acids, phenylalanine and aspartic acid, linked together. When aspartame is broken down, people with PKU may experience a dangerous increase in phenylalanine.

> ✓ **Quick Check** Suppose a mother and father have a child with PKU, though neither parent exhibits the disorder. What is the probability that their next child will also have PKU? Explain your answer. ▪

Tay-Sachs disease leads to brain dysfunction

Tay-Sachs disease is another enzyme deficiency disease, this one caused by a recessive gene located on chromosome 15. Individuals homozygous for the recessive allele are unable to make an enzyme that is responsible for the metabolism of a particular type of lipid, called a sphingolipid, found primarily in the lysosomes of brain cells. Without this enzyme, the sphingolipid accumulates in brain cells, causing cerebral degeneration. Infants seem unaffected at birth, but by 4–8 months of age motor function and brain function begin to decline. The child gradually develops seizures, becomes blind and paralyzed, and usually dies by age 3 or 4. There is no known cure or treatment.

Tay-Sachs disease is rare in the general population. However, it is fairly common among Ashkenazi Jews of Central European descent, with about 1 in 3,500 children having the disease. Carriers of the disease remain unaffected because they have sufficient enzyme activity for normal function. Tests are available to determine whether either member of a couple is a carrier of the disease.

Huntington disease is caused by a dominant-lethal allele

Huntington disease is marked by progressive nerve degeneration leading to physical and mental disability and death. Afflicted individuals begin to show symptoms in their 30s, and most die in their 40s or 50s.

Huntington disease follows the simple rules of inheritance of a single pair of alleles, one or both of which may be the dominant (HD) or recessive (hd) form. However, in this case it is the *dominant* allele (HD) that causes Huntington disease. Anyone with even one HD allele will ultimately develop the disorder. The disease is fatal, and there is no known cure at the moment. For this reason HD is called a **dominant-lethal** allele.

Dominant-lethal alleles are by their nature uncommon because they tend to eliminate themselves from the population, especially if they cause disease before the affected individual's reproductive years. Huntington disease is unusual in this regard. The disease remains in the human population because in the past, individuals who carried the HD allele had no way of knowing they would develop Huntington disease, and so they would pass the allele on to half their children before they developed any symptoms. Folksinger/songwriter Woody Guthrie and his mother both died of Huntington disease, and Woody could have passed it to his son, Arlo (Figure 19.20). Today there is a test for the HD allele. Here is an example where genetic testing and

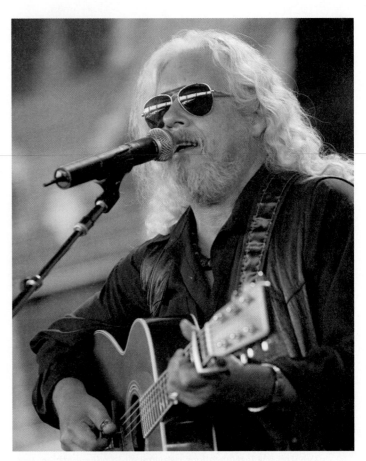

Figure 19.20 Genetic testing revealed that folksinger Arlo Guthrie would not suffer from Huntington disease as his father Woody Guthrie did.

counseling could have a dramatic effect on the prevalence of a disease in the future. (See the Current Issue box in this chapter for a discussion of the biological, social, and ethical issues raised by genetic testing.)

Genetic testing for Huntington disease and other inherited disorders is one of the great benefits of the *Human Genome Project.* Engaging scientists from around the world, this massive endeavor involved locating each of the approximately 22,000 genes in the human genome and identifying their functions. Researchers have already successfully identified the genes responsible for several disorders, including forms of breast cancer, skin cancer, and osteoporosis. This information holds promise for developing ways to prevent new cases of genetic diseases and treating existing cases with new technologies. One possibility might be to correct genetic defects with *gene therapy.* We discuss gene therapy in more detail in Chapter 20.

Recap Recessive alleles such as those for phenylketonuria and Tay-Sachs disease cause disease only if the individual is homozygous recessive. Dominant disease-causing alleles are unusual because they tend to be eliminated from the population. Huntington disease is a rare example of a dominant-lethal disease. ■

○○○   MJ's Human Biology Blog
◄ ►   +   http://www.humanbiologyblog.blogspot.com/   RSS ♻ Q▾

Ⓑ   Q   Share   Report Abuse   Next Blog»   Create Blog   Sign In

MJ's Human Biology Blog

Genetic Screening Tests

What a difference a decade makes. Back in 2000, most U.S. states screened newborn infants for just four genetic disorders. Today, all 50 states screen for at least 21 of the 29 serious genetic or functional disorders that the American College of Medical Genetics recommends for testing. The list includes cystic fibrosis, sickle cell anemia, hypothyroidism, phenylketonuria, and a host of other conditions so rare you've probably never heard of them.

Lists of the screening programs and tests provided by all 50 states can be found on the *National Newborn Screening and Genetics Resource Center* Web site, genes-r-us.uthscsa.edu. For information about the recommended tests, go to the *March of Dimes* Web site, www.marchofdimes.com. ■

19.8 Genes code for proteins, not for specific behaviors

Genes represent the set of instructions for a human being. The actions of genes can be very specific, such as determining hair patterns or earlobe shape. In terms of behavior, the logical conclusion might be that specific genes determine specific behaviors. Indeed, we read in the popular press of a heritable "schizophrenia gene." We hear of studies that show certain groups of criminals may share a common genetic defect; that some people may have a gene that predisposes them to depression; that there may even be a "happiness gene." Is any of this true?

The bottom line is that genes code for specific proteins. Specific genes code for proteins that give you curly hair or freckles. Other genes (and their protein products) may strongly *influence* whole patterns of behavior or moods, such as a tendency toward anxiety or aggression. However, genes do not necessarily cause *specific* behaviors. They do not lead to specific thoughts, and there is no one gene that causes you to turn to the left and another one that causes you to turn to the right. Nor are there any genes that cause people to commit specific crimes, as far as we know.

The key to understanding how genes influence behaviors lies in understanding the many roles of proteins. Proteins may act as hormones, neurotransmitters, enzymes, or intracellular messengers. They make up many of the structural components of cells, and they control cell functions. Some proteins influence a broad range of other genes and proteins; others have very specific effects within only certain target cells. Together, groups of genes and their protein products apparently do influence broad patterns of behavior, such as feeding, mating, or learning. They may also influence moods. Nevertheless, there is no evidence that specific genes or their protein products cause depression, lead to complete happiness, or produce specific physical behaviors.

Recap Genes influence patterns of behavior. They do not cause specific behaviors.

Chapter Summary

Your genotype is the genetic basis of your phenotype *p. 448*

- The DNA we inherit from both parents represents the complete set of instructions for creating a human being.
- The two copies of every gene (one from each parent) can differ very slightly because of mutations. Different versions of a gene are called alleles.
- Our genotype is the complete set of alleles that we inherit; our phenotype refers to the physical and functional expression of those alleles.

Genetic inheritance follows certain patterns *p. 449*

- A Punnett square can be used to show patterns of inheritance for one or two pairs of alleles.
- A common pattern of phenotypic inheritance is one in which one allele exerts complete dominance over the other.
- Genes for different traits assort independently of each other.

Other dominance patterns *p. 454*

- In incomplete dominance, heterozygotes exhibit an intermediate phenotype.
- In codominance, heterozygotes express both phenotypes equally.

Other factors influence inheritance patterns and phenotype *p. 456*

- In polygenic inheritance, multiple genes influence a single trait.
- Our phenotype is determined both by our genotype and by our environment.
- Alleles located on the same chromosome that may be inherited together are called linked alleles.

Sex-linked inheritance: X and Y chromosomes carry different genes *p. 458*

- Maleness is determined early during development by genes on the Y (male) sex chromosome. In the absence of this chromosome, the individual becomes a female.
- Sex-linked inheritance refers to inheritance of genes located on the sex chromosomes. Most of these genes have nothing to do with sex determination.
- Diseases associated with sex-linked inheritance include hemophilia, color blindness, and muscular dystrophy.

Chromosomes may be altered in number or structure *p. 461*

- Nondisjunction refers to the failure of homologous chromosomes or sister chromatids to separate properly during mitosis or meiosis; this causes altered chromosome number in the daughter cells.
- Down syndrome is a condition usually caused by the inheritance of three copies of chromosome 21. The primary risk factor is the age of the parents, especially the mother.
- Deletion of even part of a chromosome is usually fatal to a sperm, egg, or embryo. Translocation of a chromosome segment may alter a gene's ability to function properly.

Many inherited genetic disorders involve recessive alleles *p. 463*

- Most autosomal heritable genetic diseases are caused by recessive alleles.
- Phenylketonuria, Tay-Sachs disease, and cystic fibrosis are diseases caused by recessive alleles.
- Huntington disease is caused by a dominant-lethal allele. It has persisted in the human population because people who inherit it do not show symptoms until after reaching reproductive age.

Genes code for proteins, not for specific behaviors *p. 465*

■ Genes influence only general patterns of behavior or mood. They do not determine specific behaviors.

Terms You Should Know

allele, *448*
codominance, *454*
deletion, *463*
dominant, *451*
dominant-lethal, *464*
genotype, *449*
heterozygous, *448*
homozygous, *448*
incomplete dominance, *454*
inheritance, *448*
karyotype, *458*
law of independent
    assortment, *453*

law of segregation, *450*
mutation, *449*
nondisjunction, *461*
phenotype, *449*
polygenic inheritance, *456*
Punnett square, *449*
recessive, *451*
sex-linked inheritance, *459*
sickle-cell anemia, *455*
translocation, *463*

Concept Review

Answers can be found at the Human Biology Place.
www.humanbiology.com

1. Where do we get the two copies we have of every gene?
2. What is the difference between a gene and an allele?
3. Distinguish between genotype and phenotype.
4. Describe the contributions of Mendel to the field of genetics.
5. Explain how alterations of chromosome number and structure can occur.
6. Distinguish between the concepts of dominance, incomplete dominance, and codominance.
7. Describe how genetic variability is accomplished via sexual reproduction.
8. Describe what is meant by sex-linked inheritance.
9. Explain why lethal diseases caused by dominant alleles are so rare.
10. Explain how genes affect mood or behavior.

Test Yourself

Answers can be found in Appendix A.

1. All of the following statements about homologous chromosomes are true except:
   a. One of each pair of homologous chromosomes comes from each parent.
   b. The alleles on homologous chromosomes are identical to each other.
   c. They have genes for the same traits at the same loci.
   d. Homologous chromosomes physically pair up and may exchange parts during meiosis.

2. How many different kinds of gametes (with respect to a particular gene locus) are formed by an individual who is homozygous at that locus?
   a. 1
   b. 2
   c. 4
   d. It can't be determined from this information.

3. Which of the following statements correctly expresses the relationship between dominant and recessive alleles?
   a. A dominant allele is always the preferred or most beneficial allele.
   b. A dominant allele is only expressed when it is homozygous.
   c. A dominant allele masks the expression of a recessive allele in a heterozygote.
   d. A dominant allele is always the most numerous allele in the population.

4. Which of the following would provide an exception to Mendel's law of independent assortment?
   a. adjacent genes on the same chromosome
   b. dominant genes
   c. recessive genes
   d. genes located on different chromosomes

5. Two parents are heterozygous for familial hypercholesterolemia, having blood cholesterol levels two to three times the normal level. What is the probability that their child will be unaffected by this disorder?
   a. 0%
   b. 25%
   c. 50%
   d. 100%

6. What is the likelihood that a parent with Type AB blood would have a child with Type O blood?
   a. 50%
   b. 25%
   c. 0%
   d. This can't be determined without knowing the blood type of the other parent.

7. All of the following human traits are determined by the interaction of multiple genes and alleles except:
   a. eye color
   b. hair color
   c. height
   d. sickle cell anemia

8. Which of the following results in the separation (or unlinking) of linked genes?
   a. crossing-over during meiosis I
   b. independent assortment during mitosis
   c. segregation during meiosis
   d. separation of sister chromatids during mitosis

9. What is the likelihood that a man with hemophilia mating with a woman who is homozygous for the normal form of the blood clotting factor gene will have a son with hemophilia?
   a. 100%
   b. 50%
   c. 25%
   d. 0%

10. Which of the following could be detected by preparing a karyotype?
    a. sickle cell anemia
    b. cystic fibrosis
    c. Klinefelter syndrome
    d. hemophilia

11. What tool is used to determine the probabilities of different genetic outcomes from various crosses?
    a. preparation of a karyotype
    b. chorionic villus testing
    c. biochemical testing of the amniotic fluid
    d. Punnett square analysis

12. Which of the following statements is/are true?
    a. Genotype determines phenotype.
    b. Phenotype determines genotype.
    c. Environment can affect phenotype.
    d. Both (a) and (c) are true.

13. What is the basis for the tremendous genetic diversity resulting from sexual reproduction?
    a. independent assortment during gamete production
    b. crossing-over and shuffling of linked genes during meiosis
    c. random fertilization
    d. all of these choices

14. Examination of a karyotype reveals 23 pairs of chromosomes, with each homologous pair illustrating similar size and centromere location. Which of the following can be concluded from this karyotype?
    a. The individual has Down syndrome.
    b. The individual is a female.
    c. The individual does not have cystic fibrosis.
    d. The individual does not have sickle cell anemia.

15. Which of the following events or processes can result in Patau syndrome, Turner Syndrome, or Klinefelter syndrome?
    a. nondisjunction
    b. deletion of part of a chromosome
    c. crossing-over
    d. independent assortment

Apply What You Know

1. What fraction of the offspring of two wavy-haired Caucasians would have wavy hair? Explain using a Punnett square. *Hint*: The allele for curly hair exhibits incomplete dominance over the allele for straight hair.

2. Why is it that the range of resting blood pressures of humans is best represented by a bell-shaped curve covering a continuous range rather than by two populations, one with high blood pressure and one with normal blood pressure?

3. How is it that each parent is able to contribute only a single allele of each gene to his/her offspring, rather than both alleles?

4. Geneticists often study patterns of gene transfer in a variety of model organisms, including plants, fruit flies, and even worms. If they're really interested in patterns of inheritance in humans, why don't they use humans or at least larger animals more similar to humans, such as pigs or even primates?

5. Can a father who is blood type A and a mother who is blood type B have a child who is blood type O?

6. Hemophilia is a sex-linked trait that mothers pass on to male offspring. Is it possible for females to have hemophilia too? How would such a thing happen?

7. Nondisjunction during meiosis can lead to the formation of gametes (sperm or eggs) with extra copies of one or more chromosomes. Normally, fertilizations involving extra copies of chromosomes simply do not result in a live birth because too many developmental events are altered in some way. The most common extra-chromosomal condition that *does* lead to live births is Down syndrome (trisomy 21). Why do you suppose that trisomy 21 is more common than other trisomies?

20

DNA Technology and Genetic Engineering

Soybeans. Over 90% of the U. S. soybean crop is genetically modified.

Current Issue

Genetically Engineered Plants

By now you have almost certainly eaten food products that came from *genetically engineered* (GE) plants. GE plants are plants that have been genetically engineered to contain specific genes for a specific purpose. They have been cultivated commercially since 1994. But should these plants have been developed in the first place? Now that we have them, what should we do about it?

Nearly all GE plants are *transgenic*, meaning that they contain foreign genes taken from some other species of organism (a few GE plants are *cisgenic*—containing genes or gene deletions from the same species). Nearly all of the current commercially important GE plants contain genes that either allow the plant to produce its own insecticide (insect-killer) or that make the plant resistant to one of the most common herbicides (weed-killers).

GE Plants Offer Many Benefits

Now that GE crops have been around a while, it is clear that they offer many benefits. GE crops have reduced the cost of crop production for farmers because they can spray less often and use fewer toxic chemicals. This in turn is a net benefit to the environment, both in terms of the reduced use of toxic chemicals and in the reduced need for fuel for farm equipment. An even greater benefit to the environment may be in the reduced need to till fields to control weeds. Less tilling to control weeds means less soil erosion, and therefore less soil and fertilizer runoff into streams and rivers. Farmers and food processors in the United States seem to have accepted the benefits of GE crops, as judged by the trends in crop and processed foods production. It is estimated that 70–75% of all processed foods in U.S. supermarkets contain GE ingredients. In 2009, 85% of the U.S. corn crop and a whopping 91% of the soybean crop came from GE plants.

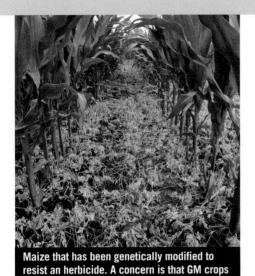

Maize that has been genetically modified to resist an herbicide. A concern is that GM crops could cross-pollinate with non-GM crops.

In theory, consumers should benefit from GE plants as well, as the reduced cost of crop production is passed on up the food production chain. GE plants with direct benefits to consumers are less common yet, though many are in the development stage. Scientists are working on plants that will produce edible vaccines, valuable human proteins, and even drugs.

GE Plants Are Not Without Risks

Many people worry that GE plants are not as beneficial as they might seem, and they are concerned that GE plants may not have been tested rigorously for safety before they were widely adapted. Although resistance to GE plants seems to be dying down in the United States, farmers and consumers in Europe still have their doubts. The concerns are legitimate, even if the potential risks have not yet been proven.

Are GE foods safe to eat?

In the United States, the safety of GE foods is monitored by the FDA. So far there is no scientific evidence that any currently approved GE food product is dangerous, despite the fact that millions of us are eating them every day. Nevertheless, we would be wise to remain vigilant. A recent report by the U.S. General Accounting Office concluded that the FDA's current regulation of GE foods is adequate but could be improved. Suggested improvements include making the FDA evaluation process more straightforward and less burdensome, and randomly checking the food companies' data.

Do GE plants endanger the environment?

Some environmental concerns about GE plants are that they may endanger non-harmful insects, that they will cross-pollinate with conventional non-GE food plants, and that cross-pollination of herbicide-resistant GE plants with weeds will create "superweeds" resistant to herbicides.

Some non-harmful insects might be killed as a consequence of ingesting too much pollen from insect-resistant GE crops. Such a consequence, however, should be weighed against the benefits of not having to spray so often with insecticides, which is also likely to kill non-harmful insects.

More problematic is the possibility of cross-pollination. Studies show that transgenes have already spread from GE corn in the United States to conventional maize in Mexico, and that other food crops in the United States have been contaminated by their GE counterparts as well. Whether this might create a problem in the future is unknown. However, research is already beginning to address the issue. "Terminator technologies" are under development that could render the seeds of GE plants sterile in the second generation, or that could prevent GE plants from expressing the novel gene unless the plant is treated with a special "gene unlocking" chemical. However, social scientists worry that terminator technologies will prevent poor farmers from benefiting from GE crops because they will not be able to save seed for planting the next year—they'll have to buy GE seed (or the unlocking chemical) each year from the agribusiness that holds the patent.

As for the development of superweeds, scientists agree that it could happen, just as many bacteria are becoming resistant to antibiotics. One way to slow the evolution of resistant weeds would be to spray with different kinds of herbicides occasionally.

Public Resistance Inhibits GE Crop Development

Despite the general findings of a lack of significant harm by GE crops so far, public resistance remains high in some countries. In Europe, protesters still conduct raids on GE crops, destroying them before harvest. The European Union requires food producers to monitor GE crops at all stages of production, and any product that contains more than 0.9% GE ingredients must be labeled accordingly. In Australia and New Zealand, all GE foods intended for sale must undergo a premarket safety assessment by a government agency. In some cases these regulations effectively amount to trade barriers for GE food products, most of which come from the United States or Canada. The resistance has had a chilling effect on GE foods research, development, and commercialization.

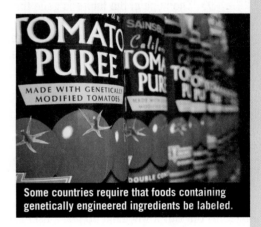

Some countries require that foods containing genetically engineered ingredients be labeled.

Angry rhetoric will not solve this dilemma. GE crops appear to be here to stay. We need to learn how to harness this new technology for the public good. We need simplified but effective regulations that protect consumers and the environment from potentially harmful practices, while at the same time permitting practices that may be beneficial.

Questions to consider

1 Do you approve of the widespread planting of GE food crops? Do you feel comfortable eating them?

2 GE foods are created for a variety of reasons, from resistance to herbicides to the delivery of vaccines and vitamins. Should we permit the development of some GE foods and not others, and if so, which ones? Defend your position.

The facts...

- Genetically engineered (GE) plants contain genes from other plants, bacteria, or even animals.
- Over 85% of all of the corn and soybeans in the United States come from GE plants.
- The use of GE crops has improved crop yields, lowered crop production costs, and reduced some aspects of environmental damage.
- Whether GE crops pose any long-term health or environmental risks is not yet fully known.
- GE food testing and labeling requirements in some countries threaten the export of GE food crops from the United States and Canada.

>> **DNA can now be accurately sequenced in the laboratory.** The entire genomes of many species, including humans, have already been completely sequenced.

>> **Individuals can be positively identified from just a small sample of their DNA.** The technique, called DNA fingerprinting, is now widely used in criminal investigations.

>> **DNA from one species can be inserted into the genome of another species**, producing *transgenic* organisms. Human genes have been inserted into bacteria and large animals (goats, sheep, cows) to produce valuable human proteins.

>> **Transgenic plants can be produced with a wide variety of features**, including increased resistance to drought and disease, improved vitamin content, and even the ability to produce edible vaccines against certain human diseases.

>> **The hope of the future is gene therapy**—the ability to insert human genes into human cells to treat or cure human diseases. Gene therapy techniques have not yet been perfected.

Knowledge for the sake of knowledge alone has never been enough for humans—we have always felt the need to put our knowledge to good use. The technical application of biological knowledge for human purposes is called **biotechnology.**

The recent explosion of knowledge regarding DNA and the human genome has spawned a new field of biotechnology called **recombinant DNA technology.** As the phrase implies, we now know how to take DNA apart, analyze its structure, and recombine it in new ways, producing molecules of DNA that did not exist previously in nature. We also know how to cut and splice genes into bacteria, plants, and animals, causing those organisms to make proteins they never made before. We are even beginning to understand how to insert, alter, or replace genes in humans to correct certain genetic diseases and treat cancer. Manipulation of the genetic makeup of cells or whole organisms, including humans, is known as **genetic engineering.** Genetic engineering is still in its infancy, but it holds great promise for the future.

As with all new technologies, biotechnology comes with new risks and new responsibilities. If we tinker with the structure of DNA, could we accidentally produce harmful organisms? How will we even know what the potential risks are until something happens? What kinds of oversight and regulation are needed for these new technologies? It would be tempting to prohibit or at least limit some of these new techniques because of their possible risks, but their potential benefits are just too great to ignore.

In this chapter we discuss technical and practical aspects of the new technologies so that we may make better judgments regarding their application. First, we look at how researchers determine the sequence of a piece of DNA. Then we look at how DNA can be combined in new ways and how to make more of it. We discuss how genes are located and identified and how they are being used to alter living organisms. Finally, we discuss the promises and pitfalls of genetic engineering.

20.1 DNA sequencing reveals structure of DNA

Biologists have known for some time how to purify the DNA from a cell. But how do they determine the precise sequence of the base pairs that make up individual strands of DNA?

First, they place millions of identical copies of a single-stranded piece of DNA to be sequenced into a test tube. Then they add short single-stranded pieces of DNA called **primers** that bind to one end of the DNA to be sequenced. The primer serves as the beginning point for synthesis of a new strand of DNA complementary to the strand to be sequenced. Next, they add a mixture of the four nucleotides (T, A, G, and C) that normally constitute DNA, plus a mixture of four nucleotides (designated T*, A*, G*, and C*) that have been modified so that as soon as one of them is added to the growing DNA chain, further chain synthesis stops. Each of the four types of modified nucleotides has also been labeled with a different fluorescent label. Finally, they add an enzyme called **DNA polymerase** that facilitates the addition of nucleotides one by one to the growing strands. Synthesis of DNA begins.

During synthesis, either a normal or a modified nucleotide appropriate for a given point in the chain is added. But because synthesis always stops as soon as a modified (labeled) nucleotide is added, the final result is a mixture of pieces of DNA of varying lengths, each ending with a single labeled nucleotide (**Figure 20.1**, step 1).

Next, the pieces are placed on a column (or a flat slab) of gel and subjected to a process called **gel electrophoresis.** Gel electrophoresis creates an electrical field that causes the DNA pieces to migrate through the gel. Smaller pieces move more quickly through the gel than larger pieces. A laser connected to a computer detects the four different fluorescent labels as they come off the gel (Figure 20.1, step 2). The result is a graphic display of the nucleotides in the pieces, arranged in order by the size of the pieces (Figure 20.1, step 3). This represents the sequence of a single strand of

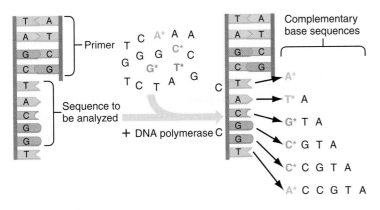

1 In the first step, short DNA seqences complementary to the sequence to be analyzed are produced, each ending in a labeled nucleotide.

A photo of a DNA sequencing gel

2 In the second step, the short DNA sequnces produced by the first step are separated by gel electrophoresis and then analyzed by a laser.

3 The result is a printout of the sequence of the complementary strand of DNA.

Figure 20.1 DNA sequencing.

✓ *A technician is preparing a sample for step 1, and he accidentally forgets to add the non-fluorescent nucleotides (the ordinary, uncolored As, Ts, Gs, and Cs shown in the diagram for step 1). Will any DNA be produced? What will the printout in step 3 show? (Hint: Think about what will happen when the DNA polymerase adds the first nucleotide.)*

DNA complementary to the original fragment of DNA. The sequence of the original unknown fragment of DNA can then be easily calculated on the basis of the known complementary base pairing in DNA.

🔁 **Recap** DNA sequencing can determine the nucleotide sequence of short pieces of DNA. Gel electrophoresis separates DNA segments by length. ◾

20.2 DNA can be cloned in the laboratory

For billions of years nature has been recombining DNA every time an organism reproduces. For thousands of years humans have been domesticating and selectively breeding plants and animals to induce nature to recombine DNA in ways useful to us. In just the last few decades we have learned enough about DNA to be able to cut, splice, and create huge quantities of it at will. In effect, we now have the capacity to develop

organisms never before created in nature and perhaps even the ability to modify or fix defective human genes.

Recombinant DNA technology: Isolating and cloning genes

Cutting, splicing, and copying (cloning) DNA and the genes it contains is *recombinant DNA technology*. The goal of recombinant DNA technology is to transfer pieces of DNA (and the genes the DNA contains) from one organism into another. Most commonly it is used to insert specific genes (including human genes) into bacteria so that the bacteria can be induced to produce useful protein products. Recombinant DNA technology requires specialized tools and components, including restriction enzymes, DNA ligases, plasmids, and bacteria.

Restriction enzymes are naturally occurring enzymes in some bacteria that break the bonds between specific neighboring base pairs in a DNA strand. In nature they protect bacteria from viral invasion by cutting up—restricting—the DNA of the invading virus, but they will cut DNA from any source. They are like specialized scissors that cut a line of text only after a specific word. There are many different restriction enzymes in nature, but the most useful for recombinant DNA technology are those that make their cut in *palindromic* double-stranded

DNA nucleotide sequences, such as CTTAAG in one strand paired with GAATTC in the other (**Figure 20.2**). (A **palindrome** is a sequence of letters or words that reads the same backward as forward, such as the word, "racecar".) A cut in a palindromic DNA sequence leaves two short single-stranded ends that are complementary to each other and also to any other DNA that is cut with the same enzyme.

DNA ligases are enzymes that bind fragments of DNA back together after the restriction enzymes have cut them.

Plasmids are small, circular, self-replicating DNA molecules found in bacteria. They are not part of the normal bacterial chromosome, but they are important to the bacterium because they contain certain genes needed for bacterial replication. Plasmids are useful because they can be made to incorporate a foreign piece of DNA. Once the plasmid is reinserted into a bacterium, it is copied every time the bacterium reproduces. By reproducing the bacteria in the laboratory, then, scientists can manufacture unlimited copies of a recombinant DNA plasmid.

Figure 20.3 illustrates the technique for producing recombinant DNA using human DNA:

1. **Isolate DNA plasmids and the human DNA of interest.** The DNA of interest is first isolated and purified from a sample of tissue. DNA plasmids are also prepared from bacteria.
2. **Cut both DNAs with the same restriction enzyme.** The restriction enzyme cuts the DNA of interest and the plasmid DNA only at specific palindromic nucleotide sequences, leaving single-stranded ends that will match up with each other.
3. **Mix the human DNA fragments with the cut plasmids.** The DNA fragments begin to join together with the

plasmids by complementary base pairing of the single-stranded cut ends of the fragments.
4. **Add DNA ligase to complete the connections.** DNA ligase joins the plasmid and human DNA strands together. The circular bacterial plasmid now contains human DNA.
5. **Introduce the new plasmid into bacteria.** Relatively harmless bacteria such as *E. coli* generally are used for this purpose.
6. **Select the bacteria containing the human gene of interest and allow them to reproduce.** Many different plasmids will have been formed by the technique described above, each containing a different fragment of human DNA. Therefore the bacteria that are carrying the gene of interest must first be identified and isolated before cloning the bacteria (and their human genes) in large numbers. One way to do this is by adding labeled antibodies against the protein expressed by the gene to small colonies of bacteria. If the gene (and its protein product) is present in the bacterial colony, the antibody will bind to the protein and the bacterial colony of interest can be identified.

This, in essence, is the technique for cloning genes. Taking it one step further, if you can get the bacteria to express (activate) the gene located in the DNA fragment, the bacteria may churn out virtually limitless quantities of a particular protein.

Figure 20.2 Cutting DNA with restriction enzymes. Restriction enzymes are used to cut DNA only at particular palindromic sequences, leaving single-stranded cut ends that are complementary to each other.

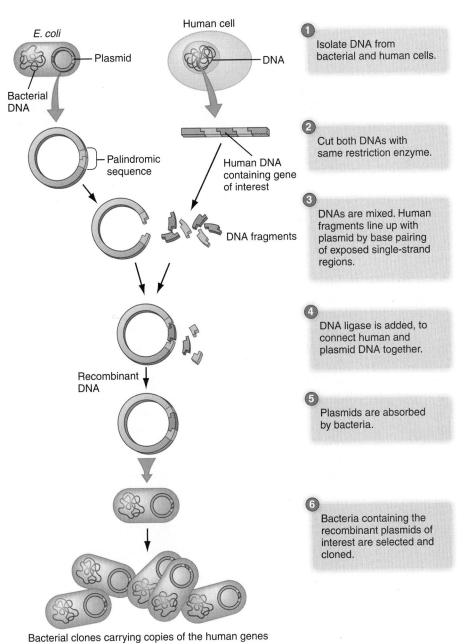

1. Isolate DNA from bacterial and human cells.

2. Cut both DNAs with same restriction enzyme.

3. DNAs are mixed. Human fragments line up with plasmid by base pairing of exposed single-strand regions.

4. DNA ligase is added, to connect human and plasmid DNA together.

5. Plasmids are absorbed by bacteria.

6. Bacteria containing the recombinant plasmids of interest are selected and cloned.

Figure 20.3 Recombinant DNA technique for producing clones of a gene or the protein product of a gene.

✓ *What would happen if the human DNA sample did not happen to contain any palindromic sequences? At the end of the procedure, would the bacteria have any plasmids, and would the plasmids have any human DNA?*

✓ **Quick Check** Suppose you're trying to clone a particular human gene. You follow the procedure outlined above, but to your disappointment it appears that none of the resulting bacteria have the complete human gene. Several of them have small pieces of the gene, however. What happened? What could you do next? ▪

 Web Animation *Recombinant DNA* at www.humanbiology.com

Cloning DNA fragments: The polymerase chain reaction

What if you only had a very small fragment of DNA and you wanted to determine the specific species or even the specific individual from which it came? Identifying the source of a sample of DNA requires a larger sample than is usually available from just a few cells or a small drop of blood. Fortunately, a technique called the **polymerase chain reaction (PCR)** can be used to make millions of copies of a small fragment of DNA very quickly. However, PCR is not a useful technique for copying (cloning) whole genes and the proteins they produce, because the copies of small segments of DNA produced by PCR lack the regulatory genes and proteins required to activate genes.

Conceptually, making multiple copies of DNA by PCR is fairly straightforward (**Figure 20.4** on next page). First the two strands of a short segment of DNA are unwound by gentle heating. Then they are mixed with (1) primers that are complementary to one end of each strand, (2) nucleotides, necessary to create the new complementary strands of DNA, and (3) DNA polymerase, the enzyme that catalyzes the attachment of nucleotides to the growing complementary DNA strand. When the mixture is cooled slightly, the primers bind to the ends of the two single strands. The primers represent the starting point for replication of each strand. The nucleotides then attach to the growing complementary chain in sequence, assisted by DNA polymerase. Once both strands of DNA are completely replicated, the heating and cooling sequence is repeated again. Each heating and cooling cycle doubles the amount of the desired DNA sample. Repeat the reaction just 20 times and you have over a million identical copies of the original fragment of DNA.

HBP **Web Animation** *The Polymerase Chain Reaction (PCR)* at www.humanbiology.com

✓ **Quick Check** To do PCR on a certain piece of DNA, there are two important things that you must already know (in advance) about the DNA's sequence. Can you figure out what two parts of the sequence you must know in advance, and why? ▪

Sample of double-stranded DNA

DNA is unwound by gentle heating; the single strands will serve as templates for new strands.

The single strands of DNA are mixed with primers able to bind to one end. Nucleotides and DNA polymerase are added.

As the mixture cools, the primers pair with the ends of the template strands.

Nucleotides are attached in sequence, and the DNA strands replicate. The number of double-stranded DNA molecules doubles.

The mixture is reheated, unwinding the double-stranded DNA molecules again.

The mixture is cooled again. Primers again pair with the ends of each strand, and nucleotides again are attached in sequence.

Replication again doubles the number of double-stranded DNA molecules.

Figure 20.4 The polymerase chain reaction (PCR) for amplifying a small sample of a segment of DNA.

Identifying the source of DNA: DNA fingerprinting

DNA fingerprinting is a technique for identifying the source of a fragment of DNA after it has been sufficiently copied (cloned) by PCR. The most common use of DNA fingerprinting is in the positive identification of suspects in a criminal investigation. The technique takes advantage of two facts: (1) between the useful genes in DNA are long repeating sequences of nucleic acids that code for nothing at all, and (2) the lengths of these "junk" sequences are highly variable between individuals.

In practice, after a fragment of DNA from a crime scene has been sufficiently copied by PCR, restriction enzymes are used to cut the DNA into pieces. Because the lengths of the repeating junk sequences between the cuts vary between individuals, the sizes of the pieces are different for every individual. These pieces are amplified by PCR and then labeled so that they can be visualized. Next, they are placed on a thin layer of gel and subjected to gel electrophoresis, which separates them according to size. A printout of the pattern of the DNA fragments on the gel is called an *electropherogram*, or a **DNA fingerprint** (Figure 20.5).

In addition to establishing the identity of individuals who were at a crime scene, DNA fingerprinting can be used to identify unknown deceased individuals, determine paternity (who's the father?), and trace ancestral relationships. Law enforcement agencies have used DNA fingerprinting to trace the

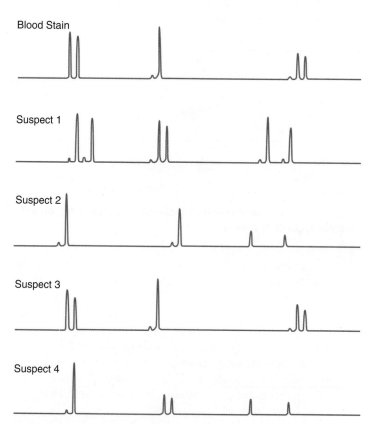

Blood Stain

Suspect 1

Suspect 2

Suspect 3

Suspect 4

Figure 20.5 DNA fingerprints. This group of DNA fingerprints includes one from a blood sample found at a crime scene and four from different people. Can you identify the match?

illegal trade in endangered live animals, animal meat, and ivory. In human evolutionary studies it has been used to establish evolutionary relationships between fossils, and in biology it has been used to study the mating relationships between animals in a population.

> ↩ **Recap** Recombinant DNA technology is used to cut and splice DNA in order to clone entire functional genes in bacteria. Multiple copies of DNA fragments can be produced by the polymerase chain reaction (PCR) technique. DNA fingerprinting is a technique for identifying fragments of DNA as having come from a specific individual. ∎

20.3 Genetic engineering creates transgenic organisms

The ability to produce **transgenic** organisms—organisms that have been genetically engineered so that they carry one or more foreign genes from a different species—has spawned a new frontier for life scientists. Next, we describe some of the exciting applications of genetic engineering techniques in the twenty-first century.

Transgenic bacteria have many uses

In the past few decades we have created new strains of transgenic bacteria specifically for our own purposes. Bacteria are rapidly becoming the workhorses of the genetic engineering industry because they readily take up plasmids containing foreign genes, and because their reproductive cycles are so short. Once a gene that codes for a desirable protein is introduced into bacteria by recombinant DNA techniques, it is only a matter of selecting those bacteria that express the gene (and thus make the protein) and then reproducing them endlessly to obtain nearly a limitless supply of the protein.

One of the very first applications of transgenic bacteria involved the manufacture of essential human proteins, most notably hormones. Recall that hormones are signaling molecules produced in very small amounts only by specific endocrine organs. The lack of a single hormone can be devastating, even life threatening. Take insulin, for example. People with insulin-dependent diabetes mellitus need to inject this hormone daily just to survive. Prior to genetic engineering, insulin had to be purified by extraction from the pancreas glands of slaughtered pigs and cattle, making it expensive and difficult to obtain. Today, insulin is produced by transgenic bacteria grown in huge vats (**Figure 20.6**). Transgenic bacteria also produce human growth hormone and erythropoietin. Some nonhormone human proteins are also being manufactured, including tissue plasminogen activator (tPA) for preventing or reversing blood clots, and human blood clotting factor VIII for treating hemophilia.

Transgenic bacteria can also be used to produce vaccines. Recall that vaccines prepare the immune system for a possible future attack by a disease-causing organism (bacterium or virus). Most vaccines are made from weakened or killed versions of the same organism that causes the disease, because

Figure 20.6 Transgenic bacteria can produce a human hormone. These vats are producing human insulin using *E. coli* bacteria.

in nature only these organisms carry the antigens that enable the immune system to recognize them. Unfortunately, a vaccine made from a weakened disease-causing organism carries a slight risk of transmitting the very disease it was designed to protect against. For example, a few people have developed polio from the polio vaccine. A second drawback is that vaccines are sometimes in short supply.

Vaccines are produced in genetically engineered bacteria by first inserting into a harmless bacterium a gene that encodes for a surface antigen protein of the disease-causing organism. The goal is to get the harmless bacteria to produce the surface antigen of the disease-causing organism, for use as a vaccine. Because the vaccine is comprised solely of purified antigen—not the whole disease-causing organism—there is no chance of causing the disease.

Although producing a vaccine by genetic engineering sounds easy in principle, in practice it is quite time-consuming and expensive. Producing a genetically engineered vaccine requires successfully finding and transferring just the right gene into harmless bacteria, and then getting the bacteria to produce the desired protein in large amounts. To complicate matters, some disease-causing organisms evolve so rapidly that a vaccine produced one year may not work the next. Nevertheless, a genetically engineered vaccine against hepatitis B is now on the market and several other vaccines are under investigation, including one against malaria. However, work on several potential vaccines against HIV had to be halted recently because the vaccines just didn't prevent HIV infection. Successful production of vaccines by genetic engineering will continue to be a challenge.

Current practical uses of transgenic bacteria include producing enzymes in industrial quantities (for example, amylase to break down starch to glucose), manufacturing citric acid and ethanol, and producing drugs for human use. Transgenic bacteria are also used to clean up toxic wastes and oil pollutants, remove sulfur from coal, and even monitor hazardous waste sites (**Figure 20.7** on the next page).

Figure 20.7 "Critters on a chip." The *Pseudomonas fluorescens* bacteria in this photo have been genetically engineered to emit light when they eat and digest naphthalene, a potential environmental pollutant. Some of the bacteria have been grown on small silicon chips (black rectangles) that are designed to convert the light to electrical current and send it to an electronic receiver. Bacteria and microelectronic devices such as this could be used to monitor hazardous waste sites.

 Quick Check When animal cells (including human cells) make proteins, they often modify the proteins in several ways. Bacteria cannot perform these modifications. How might this affect the use of transgenic bacteria to create human proteins? ▪

Transgenic plants: More vitamins and better pest resistance

It is somewhat more difficult to create transgenic plants than transgenic bacteria. Typically, the DNA of interest is inserted into a plasmid in a particular bacterium that infects plants. When the bacterium is incubated with embryonic plant cells, a few of the plant cells may take up the DNA-laced plasmid, and a few of those may incorporate the recombinant DNA into their own genetic material. Other techniques include shocking plant cells with high voltage in the presence of recombinant DNA, and shooting the recombinant DNA into plant cells at high velocity. Once the individual plant cells with the desired DNA are identified, they are used to regenerate entire plants with the desired traits.

Genetic engineering has produced tomato plants that resist freezing, tomatoes that last longer on the shelf, crops that are resistant to insect pests, and a commonly used herbicide. Genetic engineering has also yielded plants that produce larger leaves for better photosynthesis and faster growth (**Figure 20.8**) and "golden rice" containing high levels of beta-carotene (**Figure 20.9**). Beta-carotene is converted to vitamin A in the human body, and vitamin A is in short supply in the diets of many people, especially in Latin America and Asia. Researchers are also developing plants that mature earlier, develop more extensive root systems for better drought resistance, have short, stiff stems to better resist being blown down by the wind, and even are capable of nitrogen fixation, the process whereby nitrogen from the air is converted to a form that can be used by

Figure 20.8 Genetically engineered plants. The leaf on the right is from a poplar tree that was genetically engineered to grow larger. The leaf on the left is from a normal poplar tree.

✓ *Do you think genetic engineering is fundamentally different from processes that occur in nature? Put another way, is it possible that a plant with the leaf on the right could evolve naturally? Explain your answer.*

plants. Success in this latter area would reduce the reliance on nitrogen fertilizers.

Transgenic plants are also being used to produce edible vaccines against infectious diseases. Researchers have already produced a hepatitis B vaccine in raw potatoes, and they are investigating bananas as a source of vaccines against hepatitis B, cholera, and viruses that cause diarrhea. Bananas containing edible vaccines would be of great value because they are palatable to children, do not require cooking, and are easy to grow in tropical and subtropical climates.

Transgenic plants can even be made to incorporate human genes and produce human proteins. For example, efforts are currently under way to produce commercially viable quantities of

Figure 20.9 Transgenic plants can improve human nutrition. The golden rice on the right is high in beta-carotene, a nutrient in short supply in many areas of the world.

human serum albumin in potatoes and tobacco plants. Serum albumin is an essential component of fluid replacement therapy.

Some people are concerned about the safety of transgenic plants, particularly when the development of new plants seems to be happening so quickly compared to their regulation and safety testing. Protests against so-called Franken foods are becoming more commonplace, especially in Europe. Look for genetic engineering of plants to come under increased political scrutiny and public debate in the future.

Aside from the issue of safety, there is concern that the widespread use of genetically engineered food crops could lead to occasional but widespread crop failures. The problem is that genetically engineered plant lines are sometimes produced from just one or a very few original embryonic plants that took up the gene of interest and then were divided over and over again in the laboratory. Such plants are clones—all of the individual plants are genetically identical. Therefore if a new pest or a brief change in environmental or climatic conditions were to damage the plants, the entire crop could fail. In a population of crop plants produced by sexual reproduction, on the other hand, genetic variation leads to variable responses to pests and environmental conditions because some plants are better able to withstand the challenge than others.

✔️ **Quick Check** Explain in your own words why it is more difficult to create a transgenic plant than to create a transgenic bacterium. ▪

Transgenic animals: A bigger challenge

Producing transgenic animals is even harder than producing transgenic plants. First, animal cells do not take up plasmids as do bacteria and plant cells, so introducing foreign DNA into an animal cell is a greater challenge. Second, the techniques for cloning animals from a single cell or group of cells are much more difficult than for cloning plants.

Producing a transgenic animal typically begins with inserting the DNA of interest into fertilized eggs. One way is to microinject DNA into each egg individually. Another is to agitate fertilized eggs in the presence of DNA and tiny silicon needles. The needles create holes in the eggs, allowing the DNA to enter. The eggs are then reimplanted into a female for gestation. The number of transgenic animals that can be produced at any one time is limited because usually only a few mature eggs are available at a time and fewer than 10% of the eggs incorporate the recombinant DNA into their own.

Despite these drawbacks, the production of transgenic animals is moving forward. To date, the gene for bovine growth hormone (bGH) has been inserted into cows, pigs, and sheep in order to create faster-growing and larger animals for food production. Transgenic animals are also used to study specific human diseases. For example, transgenic mice have been created that over-express the beta-amyloid protein thought to be responsible for Alzheimer's disease in humans.

Pharmaceutical companies are inserting genes that code for useful human proteins into goats, sheep, and cows. Large animals are preferred because it may be possible to obtain sufficient quantities of the protein from the female's milk (**Figure 20.10**), rather than having to repeatedly bleed or

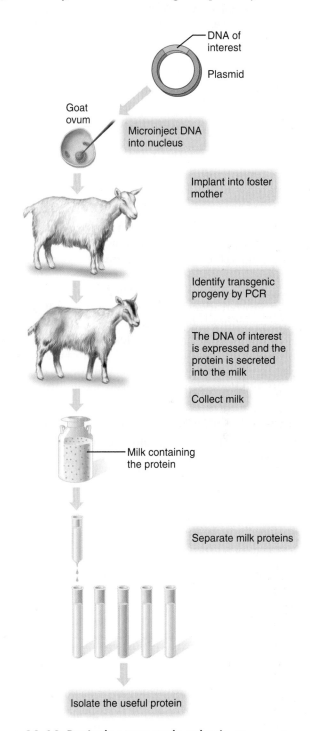

Figure 20.10 Producing transgenic animals. One way to produce a transgenic animal is to microinject the DNA of interest into an animal's egg and then implant the egg into another female for development. Genes that code for useful human proteins are inserted into the eggs of large mammals such as goats, and then the human protein of interest is collected from the animal's milk once the transgenic animal matures.

even sacrifice the animal to obtain the protein from blood. The process of producing proteins in farm animals for medical uses is called "gene pharming," aptly enough (**Figure 20.11**).

Recap Bacteria are useful for genetic engineering because they readily take up plasmids containing foreign DNA and their reproductive cycles are short. Transgenic bacteria can produce human proteins, human hormones, and even vaccines. Transgenic plants and animals can provide new agricultural crops and useful human proteins. ▪

20.4 Gene therapy: The hope of the future?

Nothing quite fires the imagination like the notion of **gene therapy**—the insertion of human genes into human cells to treat or correct disease. The potential for successful gene therapy is often mentioned as one of the possible benefits of the Human Genome Project. We now know the specific gene mutation and its chromosomal location for many genetic diseases, including cystic fibrosis, severe combined immunodeficiency disease, hemophilia, familial hypercholesterolemia, and others. The great hope of the future is that with gene therapy, these mutations could be fixed.

But a word of caution is in order: Despite considerable effort, successful gene therapy is still more of a dream than a reality. In this section we consider some of the obstacles to success and describe what has been accomplished so far.

Gene therapy must overcome many obstacles

We have already discussed how difficult it is to produce an animal by genetic engineering. However, it is much easier to create an entire transgenic animal than to correct a genetic disease in a human adult, or even a child.

Figure 20.11 Gene pharming. These goats have been genetically engineered to produce milk containing antithrombin III, a human protein that can be used to prevent blood clots during surgery.

First, in an individual with trillions of already differentiated and matured cells, it has proven exceedingly difficult to get the recombinant DNA containing the gene of interest into enough of the right cells. Recall that transgenic animals are created by inserting the gene into a *single* cell (the fertilized egg) that has been isolated and is therefore available to the researcher for microinjection—and even then the procedure works less than 10% of the time. How would we ever insert the gene that codes for a single missing protein, for example, into every cell in the body, or at least into the cells that normally produce that protein? How would we deliver the gene that codes for a specific missing hormone into the millions of cells in a particular endocrine gland that normally produces that hormone? Ideally, what we need are delivery systems that deliver recombinant DNA efficiently to all body cells or to specific tissues or cell types. No such delivery systems exist yet.

Second, even if we could correct a specific genetic disorder in an individual, it does not follow that the disease would also be corrected in his or her offspring. If we managed to cure cystic fibrosis in an individual by inserting the appropriate normal gene into enough lung cells to make a difference, that person's children could still inherit the disease—unless we were also able to insert the normal gene into the cells that become sperm or eggs, called germ cells. In fact, if we *do* become able to treat genetic diseases in individuals who might otherwise die before reproductive age, we may well increase the prevalence of genetic diseases in the population.

We might be forced to use genetic engineering to correct more and more genetic disorders in the future.

☑ **Quick Check** Suppose a woman has a disease that is due to a single defective gene and that is certain to be passed on to her children. Yet she and her husband want to have children. In theory, might it be possible to do gene therapy on a human embryo from this woman and her husband, so that every cell in the resulting baby would be completely "cured"? If so, what techniques might work, and what might be some of the possible problems? ▪

Vectors transfer genes into human cells

Although in principle it would be ideal if we could deliver a gene so that it replaces a damaged or missing gene in all of a person's cells, in practice this is not necessary. All we really need to do is get the gene into enough living cells to produce enough of the missing protein to prevent the disease. There are two strategies for doing this, both of which take advantage of transporters, or *vectors*, capable of delivering genes into human cells. The best vectors are a class of viruses called retroviruses. Retroviruses splice their own RNA-based genetic code permanently into the DNA of the cells they infect. This makes them ideal gene-transfer vectors as long as they are rendered harmless first.

In one gene-transfer method, human cells from a particular target tissue are removed from the body and exposed to retroviruses containing the human gene of interest. Then the cells are returned to the patient from whom they came (**Figure 20.12**). The hope is that the virus-infected cells will incorporate themselves back into the tissue from which they

MJ's Human Biology Blog

http://www.humanbiologyblog.blogspot.com/

Share Report Abuse Next Blog» Create Blog Sign In

MJ's Human Biology Blog

That's One Small Step for Gene Therapy...

French researchers report that they have successfully used gene therapy to treat beta-thalassemia in a 19-year-old male patient. Beta-thalassemia is a genetic blood disorder in which a defect in the gene coding for the beta-globin chain of hemoglobin results in persistent and life-threatening anemia and dangerously high blood iron levels. Two years after the treatment, according to the researchers, the young man no longer needs regular monthly blood transfusions and appears to be in good health.

The French team has the approval of French authorities to treat more patients with the same inherited disorder. The hope is that someday they'll be able to successfully treat one of the most common of all genetic blood disorders—sickle cell anemia. ▪

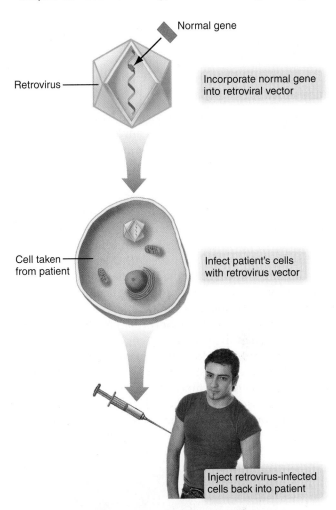

Figure 20.12 Gene therapy using a retroviral vector. Cells are removed from the human patient, exposed to retroviruses that contain the code for a human gene of interest, and then returned to the patient. The goal is for the "infected" cells to produce enough of the missing protein to correct symptoms of the disease.

originated, except that now they will express the previously missing protein. In another technique, the retroviruses with their human gene payload are injected directly into the patient, or even directly into the desired tissue if possible. Again, the hope is that some of the cells will take up the retroviruses and incorporate the human gene into the cell's own DNA.

Retroviruses have their drawbacks as gene delivery vectors, however. One problem is that retroviruses generally insert foreign DNA into a cell's DNA only when the cell is dividing. Another is that they insert their genetic material randomly in the genome, so they might disrupt the function of other human genes. To get around these drawbacks, some researchers are experimenting with nonviral methods of gene transfer. These methods include embedding plasmids containing the genes in small fatty spheres called liposomes, coating them with amino acid polymers, or even injecting plain ("naked") DNA. Such techniques are still highly experimental.

✅ **Quick Check** Gene therapy that uses retroviruses tends to be more successful if patients are given drugs that suppress their immune systems. Why? ▪

Success with SCID gives hope

Given the technical problems associated with gene therapy, it's a wonder we have accomplished anything at all. Although failures have outnumbered successes so far, every small advance opens new avenues for research and increases the chances of success in the future.

The first patient treated for a disease by gene therapy was Ashanthi DeSilva (**Figure 20.13**). In 1990 Ashanthi, then 4 years old, suffered from severe combined immunodeficiency disease (SCID), an inherited disorder in which the lack of an enzyme called *adenosine deaminase (ADA)* leads to a marked deficiency of B and T cells of the immune system and an increased susceptibility to infections. In the landmark treatment of Ashanthi, T cells isolated from her blood were grown in the laboratory with a viral vector containing the gene for the enzyme. When the genetically engineered T cells were reintroduced into her blood they expressed the gene, ADA was produced, and her condition improved. However, the improvement was only temporary because mature T cells have a finite life span. Ashanthi and most other SCID patients treated by gene therapy have had to supplement their

Figure 20.13 The first beneficiary of gene therapy. Ashanthi DeSilva was the first patient to successfully receive gene therapy for severe combined immunodeficiency disease. The restoration of her immune system function was only partial, but the experience gained from her treatment paved the way for more recent advances in gene therapy for others. This photo was taken on the tenth anniversary of the event.

treatment with regular doses of ADA (now produced by genetic engineering techniques).

Other researchers have tried isolating and treating blood-forming stem cells from the umbilical cord of infants with SCID and then reintroducing them into the infants, in the hope of getting the stem cells themselves to produce corrected T cells throughout the patients' lives. The first results were only mildly encouraging: a few of the patients' T cells did produce the enzyme, but not enough to prevent the disease without additional treatment with regular doses of a synthetic form of the missing enzyme.

As of this writing, several SCID patients have finally been treated successfully with gene therapy alone. The time span from first attempt to first real success was approximately 10 years.

✅ **Quick Check** Several severe combined immunodeficiency disease (SCID) patients in gene therapy trials have developed leukemia, a type of cancer. Why would gene therapy increase the risk of cancer? ▪

Research targets cystic fibrosis and cancer

Recall that cystic fibrosis is a genetic disease caused by a mutant allele of a single gene. Because the disease primarily affects the respiratory system, researchers are experimenting with delivering the normal gene in a viral vector via a nasal spray. The results so far have been only slightly encouraging. Although about 5% of the cells of the respiratory system take up the normal gene, this is not enough to prevent the disease. Nor is it clear that the gene will be expressed by those cells. Nevertheless, the results are promising enough to warrant the search for more effective delivery vectors.

Some types of cancer may soon be treatable by gene therapy. One promising approach is to add genes for interleukins (which activate the immune system) to cancer cells in the laboratory. The cancer cells are then returned to the patient. As the cancer cells divide, the interleukins they produce stimulate the patient's own immune system to recognize and attack the cancer cells. Another method is to incorporate the gene for a protein of a foreign cell into cancer cells so that they are specifically targeted for destruction by the immune system.

Another promising approach is based on an understanding of why cells die when they become damaged. When normal cells become damaged they undergo a process called apoptosis (programmed cell death). Apoptosis is nature's way of getting rid of damaged cells quickly. Scientists have discovered that certain proteins, including one called mda-7, are responsible for initiating apoptosis. However, cancer cells seem to avoid apoptosis. In one experiment, researchers inserted the mda-7 protein into a viral vector and injected it directly into malignant tumors. The tumors began to secrete mda-7, killing many cancerous cells and stimulating the patients' immune system.

Gene therapy of somatic cells holds the promise of treating specific diseases in specific individuals. As with any standard medical treatment, the effect would last only for the lifetime of the individual patient. However, gene therapy targeting the germ cells that lead to sperm and eggs would be another matter entirely, for it would alter the human genome for all future generations. We should start thinking about the legal, ethical, and moral implications of germ cell gene therapy now, before it becomes a real issue.

Recap Gene therapy is still more of a hope than a reality. Even if we are able to correct genetic disease in an individual, that person's children could still inherit the disorder. Gene therapy depends on vectors to deliver corrected DNA into a patient's cells. Although there have been a few successes so far, gene therapy is still in its infancy. ◼

Chapter Summary

DNA sequencing reveals structure of DNA *p. 470*

- DNA sequencing can now be done in the laboratory.
- The technique involves synthesizing a complementary new strand of DNA to a single strand that is to be sequenced, and then determining its sequence.
- An enzyme called DNA polymerase is used to add new nucleotides to the growing strand.
- A technique called *gel electrophoresis* is used to sort DNA strands by size.

DNA can be cloned in the laboratory *p. 471*

- Recombinant DNA technology refers to techniques for splicing segments of DNA into the DNA of other organisms.
- The technique involves cutting and splicing the DNA of interest into a plasmid, and then delivering the plasmid into the organism of interest.
- The *polymerase chain reaction* makes multiple copies of short segments of DNA.
- *DNA fingerprinting* can identify fragments of DNA as having come from a specific individual.

Genetic engineering creates transgenic organisms *p. 475*

- Bacteria are commonly used to make human proteins and hormones because they readily take up recombinant DNA plasmids and can be grown easily in the laboratory.
- Creating transgenic plant crops has proven easy because many plants can be regenerated from individual transgenic plant cells.
- Many transgenic plant crops are already in use in the agricultural industry. Their long-term safety is largely untested.
- Transgenic animals are more difficult to produce, and they can be produced only one at a time.

Gene therapy: The hope of the future? *p. 478*

- Human gene therapy has proven difficult because it is hard to deliver the appropriate recombinant DNA to the right human cells.

- Gene therapy has shown some success in treating SCID.
- Some cancers may be treatable by gene therapy techniques in the near future.

Terms You Should Know

biotechnology, *470*
DNA fingerprinting, *474*
DNA ligase, *472*
DNA polymerase, *470*
gel electrophoresis, *470*
gene therapy, *478*
genetic engineering, *470*
palindrome, *472*
plasmid, *472*
polymerase chain reaction (PCR), *473*
primer, *470*
recombinant DNA technology, *470*
restriction enzyme, *471*
transgenic, *475*

Concept Review

Answers can be found at the Human Biology Place.
www.humanbiology.com

1. Explain how DNA is sequenced.
2. Describe the enzymes used in recombinant DNA techniques.
3. Describe the process of producing recombinant DNA.
4. Discuss the polymerase chain reaction technique and explain its advantages.
5. Describe how one could insert a human gene into bacteria.
6. Explain why bacteria are so frequently used in the genetic engineering industry, and discuss some of the products that can result.
7. Give two reasons why it is harder to produce large numbers of transgenic animals than it is to produce transgenic plant crops.
8. Discuss the obstacles to successful gene therapy.
9. Discuss how gene therapy is presently used to treat genetic diseases.
10. List the diseases that gene therapy may be used to cure.

Test Yourself

Answers can be found in Appendix A.

1. Which of the following facilitates joining a piece of human DNA with bacterial plasmid DNA?
   a. DNA polymerase
   b. DNA ligase
   c. RNA polymerase
   d. restriction enzymes

2. Which of the following acts like a molecular copying machine, rapidly making large numbers of specific sequences of DNA?
   a. DNA fingerprinting
   b. DNA ligase
   c. DNA sequencing
   d. polymerase chain reaction

3. Which of the following is used to separate fragments of DNA by size?
   a. gel electrophoresis
   b. polymerase chain reaction
   c. DNA insertion into a plasmid
   d. DNA splicing

4. Genetically engineered human insulin, human growth hormone, and human clotting factor VIII are made by:
   a. gel electrophoresis
   b. polymerase chain reaction
   c. transgenic bacteria
   d. DNA fingerprinting

5. Which of the following vaccines is now produced by recombinant DNA technology?
   a. MMR vaccine
   b. hepatitis B vaccine
   c. polio vaccine
   d. tetanus vaccine

6. All of the following statements about transgenic plants are true except:
   a. Genes have been successfully introduced into plants to improve the nutritional value.
   b. Transgenic plants have been developed that are resistant to pests such as larval insects.
   c. Human genes cannot be incorporated into transgenic plants.
   d. Transgenic plants are being studied as a way to produce edible vaccines.

7. "Gene pharming" refers to:
   a. using transgenic farm animals to produce human proteins for medical uses
   b. producing transgenic plants with improved nutritional value
   c. producing drugs using transgenic plants
   d. using transgenic plants to produce vaccines

8. Which of the following poses the greatest challenge to gene therapy?
   a. identifying the chromosomal location of the defective gene
   b. obtaining a normal allele of the defective gene
   c. producing multiple copies of the normal allele to replace the defective allele
   d. delivering the normal allele into the cells of the recipient

9. What would be the advantage of introducing genes into germ cells rather than somatic cells?
   a. Genes can be more readily inserted into germ cells.
   b. Introduced genes are more often lost from somatic cells.
   c. Genes introduced into germ cells are more readily expressed.
   d. Genes introduced into germ cells would be passed on to gametes and future generations.

10. The role of restriction enzymes in bacterial cells is:
    a. to assist in DNA replication
    b. to process primary transcripts into mRNA
    c. to destroy invading viruses by cutting up their DNA
    d. to join pieces of DNA

11. Which of the following is needed for sequencing DNA?
    a. DNA polymerase
    b. gel electrophoresis
    c. primers
    d. all of these choices

12. A tiny amount of blood was recovered from a crime scene. What technique will enable comparison of this DNA to that of a potential suspect?
    a. DNA fingerprinting
    b. DNA cloning
    c. DNA sequencing
    d. gene therapy

13. Which technique will allow amplification of the DNA from the crime scene, in order to provide a large enough specimen?
    a. gel electrophoresis
    b. polymerase chain reaction
    c. transgenic introduction
    d. DNA sequencing

14. A gene for a bacterial protein with insecticide properties has been introduced into a plant. This plant is now referred to as:
    a. an insectivorous plant
    b. a transgenic plant
    c. a genetically modified plant
    d. both (b) and (c)

15. Bacteria have been genetically modified to do all of the following except:
    a. produce human hormones
    b. digest toxic wastes
    c. treat sickle cell anemia
    d. produce vaccines

Apply What You Know

Answers can be found at the Human Biology Place.
www.humanbiology.com

1. What do you think are the chances that someone will try to use genetic engineering techniques to produce a biological weapon of mass destruction? Should this affect our research into genetic engineering, and if so, how? Explain your reasoning.

2. Imagine a future in which the respiratory effects of cystic fibrosis, a genetic disorder, could be cured by a gene therapy technique in which the normal gene is delivered to the patient via a nasal spray. You are the patient's genetic counselor.

Explain to the patient why the cure will not alter his or her future child's chances of having the disease.

3. Scientists in a Hollywood film used prehistoric dinosaur DNA to make real dinosaurs. Although recreating a complete dinosaur is not yet within the realm of possibility, with present techniques it would at least be feasible to analyze a dinosaur's DNA (if some were available). What techniques would be used for such an analysis?

4. Pharmaceutical companies produce human insulin using genetic engineering techniques. The drug Humulin was one of the first recombinant DNA drugs ever approved by the FDA. How would such a drug be produced?

5. Why are vaccines produced by genetic engineering techniques (such as the Recombivax vaccine against hepatitis B) safer than vaccines produced from killed or weakened (attenuated) viruses?

6. There is interest in producing vaccines in foods. What would be the advantages of having a vaccine embedded in food?

7. Transgenic plants have the potential to improve food production around the world. They also have the possibility of being more nutritious than their naturally occurring cousins. Still, there are drawbacks. How could global climate change affect transgenic crops?

21

Development and Aging

Staying fit may improve quality of life as we age.

Current Issue

Who Should Make Life and Death Decisions for You?

If you suffered severe brain damage and could survive only on artificial life support, would you want to continue living? Who do you think will take care of you and make decisions related to your life—and death? If you think you're too young to worry about such issues, consider the story of Terri Schiavo.

In February 1990, 26-year-old Terri suffered cardiac arrest, possibly caused by potassium imbalance due to bulimia. She was rushed to a hospital. However, lack of oxygen had damaged her brain permanently, and she lapsed into a coma. A feeding tube was inserted through the skin and into her stomach to provide fluids and nutrition. Over the next four years Terri was transferred to various medical centers, where she received ongoing testing, speech and occupational therapy, and other attempts at rehabilitation.

A Legal Battle and National Debate Over the Right to Die

In June 1990 Michael Schiavo, Terri's husband, was appointed her legal guardian.

Family photo of Terri Schiavo before she had a heart attack.

At the time Terri's parents, Robert and Mary Schindler, did not object. But eight years later, a decision made by Michael Schiavo regarding Terri's life became the focus of a bitter court battle that escalated into a national debate about the right to live versus the right to die.

In 1998, when Terri had been in a coma for eight years, Michael petitioned the court to authorize the removal of Terri's feeding tube and allow her to die. Immediately thereafter, Robert and Mary Schindler counterpetitioned to stop Michael from ending their daughter's life.

If Terri had made a *living will* (a legal document that expresses one's preferences regarding life support and resuscitation), the decision to terminate her life support might have been relatively straightforward. In this case, however, the decision became an immensely difficult undertaking. What would Terri have wanted? When a person's wishes are not clear, who should make the decision? Do we have an ethical obligation to keep someone alive as long as possible? And when, exactly, is someone no longer "alive"?

In Terri's court case, Michael and two relatives testified that prior to her heart attack, Terri had commented about a seriously ill person, "I wouldn't want to live like that. I would want to just die." In response, Terri's parents told the court that Terri was a devout Catholic who would want to live. They insisted their daughter had the right to live; her husband insisted she had the right to die. A medical report submitted to the court indicated that Terri was in a "persistent vegetative state" with no chance of improvement.

In 2000 a circuit court judge ruled that the feeding tube should be removed, on the grounds that Terri would have chosen to do so. Her parents initiated a series of legal appeals, countered by appeals from Michael Schiavo. In 2001 the feeding tube was removed, but it was reinserted two days later when the Schindlers filed a lawsuit.

A supporter lights a prayer candle for Terri.

By the time Terri's feeding tube was removed again in 2003, the case had caught the public's (and politicians') attention. Eventually the argument went all the way to the U.S. Supreme Court, which declined to hear the case.

Terri Schiavo's feeding tube was removed for the last time on March 18, 2005. On March 31, at age 41, she died. An autopsy showed that her brain had shrunk to 50% of its normal weight and confirmed the medical diagnosis of persistent vegetative state. She could not have recovered.

Protecting Yourself from a Similar Fate

Given the choice, many of us would choose not to live 15 years in a persistent vegetative state. How can we protect ourselves from such a fate? As mentioned previously, one way is to sign a living will, also known as an "advance health care directive." A living will describes what medical procedures you would want performed on you, and under what circumstances. It is intended to be used only in the event that you become unable to make decisions for yourself. However, a comprehensive living will is difficult to write because it is hard to envision virtually every possibility that might befall you. Furthermore, in a situation in which your wishes come in conflict with good medical practice, doctors can refuse to honor it.

One way around the uncertainty of a living will is to combine it with a *durable power of attorney for health care*, also called a "health care proxy appointment." This is a legal document in which you give a named person the right to make all medical decisions on your behalf should you be unable to make them for yourself. Combined with a living will, a durable power of attorney offers reasonable assurance that your wishes will be carried out. But you've got to trust that person, and you need to be sure that they understand your wishes well.

Currently more advanced types of directives are being studied as well. One type, called a *Lifecare Advance Directive*, focuses more on general health outcomes than on lists of specific medical conditions and treatments. A Lifecare Advance Directive is less likely to come in conflict with good medical practice, but on the other hand it may be too vague to describe a specific situation. Some legal experts are recommending that you use all three types (living will, durable power of attorney, and lifecare advance directive) if you want virtual certainty that your wishes will be carried out. Skeptics say that all of these advance directives are more about protecting the medical and legal professions than about doing the right thing for the patient.

And what if you continue to do nothing and then become unable to express your wishes, as happened to Terri Schiavo? In order of priority, decision-making authority generally rests first with your spouse, then with your parents. Heaven help you if they disagree, or if you and your spouse have split up but have not yet gotten a divorce! The more you make your wishes clear, the better off you'll be.

Questions to consider

1 If you were in Terri Schiavo's situation, would you want to be allowed to die?

2 Do you have a living will or a durable power of attorney? If not, who do you think will make decisions for you if you are unable to speak for yourself?

Life is a journey that begins at fertilization and ends around the time of our last breath. It's an interesting and complicated process in which one developmental stage follows the next in a carefully choreographed dance.

The changes that occur during development are nothing short of amazing: in just nine months a single undifferentiated cell develops into a human baby with complex organ systems comprising more than 10 trillion cells. In less than 20 years, this helpless baby becomes a fully functioning adult capable of producing still more human babies. Then, starting at about age 40, there is a slow decline of function for reasons we don't yet fully understand, ending inevitably in death.

The amazing journey of life is what this chapter is all about. We begin at the beginning when sperm meets egg, and end at the ending when life finally ceases.

21.1 Fertilization begins when sperm and egg unite

The possibility that fertilization will occur begins with intercourse. Fertilization is a process that begins when a sperm and egg unite and ends when the **zygote** (the new diploid cell) is formed.

The journeys of egg and sperm

It's remarkable that sperm and egg ever find each other at all, given the journey that each must take. An egg is released from one of the two ovaries at ovulation, and the sperm are deposited in the vagina near the cervix during intercourse. The egg moves slowly and passively down the oviduct, propelled by cilia that line the tube and sweep it gently downward.

Sperm are not much more than a head containing the all-important DNA, a midpiece containing mitochondria for energy production during the sperm's long journey, and a flagellum for movement (Figure 21.1a). The several hundred million sperm encounter many hazards on their journey toward the egg. First they must pass through the mucus that blocks the cervical opening and cross the vast (to them) expanse of the uterus. Then they must locate and enter the correct oviduct (half don't). Sperm swim randomly at a rate of 1/4 inch per minute. Along the way they must tolerate the strongly acidic pH of the vagina, and avoid bacteria and the occasional white blood cells roaming the uterine lining.

A sperm's journey can take hours, and only a few thousand or a few hundred make it successfully to the upper oviduct; the rest are lost along the way and die. Fertilization typically takes place in the upper third of the oviduct approximately 6–24 hours after intercourse, provided that an egg is present (Figure 21.1b). If no egg is present, the sperm may live for a couple of days and then die. If no sperm arrive, the egg dies in about 24 hours.

One sperm fertilizes the egg

The odds of success for a single sperm are still not very good, for only one of the several hundred million sperm will fertilize the egg. This is important, for otherwise the zygote would end up with an abnormal number of chromosomes. The process of fertilization has evolved to ensure that only one sperm can succeed. Recall that before fertilization, the egg is really a secondary oocyte that has started but not finished the second stage of meiosis (Figure 21.1c). The second stage of meiosis is not completed until a sperm makes contact with the egg and fertilization begins. At this point the secondary oocyte is surrounded by a protective covering called the zona pellucida and by a layer of granulosa cells derived from the follicle called the corona radiata. The egg is relatively large (nearly 2,000 times the mass of a sperm) because it contains a great deal of cytoplasm. The cytoplasm of the egg must support nearly two weeks' worth of cell divisions until the "pre-embryo," as it is called, makes contact with the uterine lining and begins to receive nutrients from the mother.

When a sperm encounters the egg, the tip of the sperm head, called the acrosome, releases powerful enzymes (Figure 21.2). These enzymes digest a path for the sperm between the granulosa cells of the corona radiata and through the zona pellucida to the oocyte plasma membrane. Several sperm may be making this journey at the same time. When the first sperm makes contact with the oocyte plasma

a) The male gamete, or sperm. The size of a sperm relative to the secondary oocyte has been greatly exaggerated.

b) Fertilization. Fertilization generally takes place in the upper third of the oviduct.

c) The female gamete. The female gamete is a secondary oocyte that is in an arrested state of stage II of meiosis.

Figure 21.1 **Female and male gametes and fertilization.**

Figure 21.2 **Fertilization.** Fertilization begins when a sperm makes contact with the secondary oocyte and releases enzymes that digest a path through the zona pellucida. Once a sperm makes contact with the oocyte plasma membrane and enters the oocyte, granules are released from the oocyte that make the zona pellucida impenetrable to other sperm.

a) **A sperm nucleus has just entered the secondary oocyte, triggering completion of meiosis II.**

b) **Completion of meiosis yields an ovum (egg) with a haploid nucleus and another polar body.** If it has not already regressed, the original polar body may divide as well.

c) **The nuclei of the ovum and sperm fuse, and the polar bodies die.** A single diploid cell called the zygote is formed, and fertilization is complete.

Figure 21.3 **Completion of fertilization.**

membrane, special protein "keys" of the sperm recognize receptor protein "locks" in the oocyte plasma membrane, ensuring that only human sperm can penetrate a human egg. The combination of lock and key causes the plasma membranes of egg and sperm to fuse so the nucleus of the sperm can enter the egg.

Fertilization begins when the sperm's nucleus enters the egg. The entry of one sperm triggers the release of enzymes from granules located just inside the egg. These enzymes produce changes in the zona pellucida that make it impenetrable to all other sperm. One sperm nucleus, and only one, will enter the egg.

The process of fertilization is not yet complete, however, because the "egg" is still a secondary oocyte. Entry of the sperm nucleus triggers the completion of meiosis II and formation of the haploid **ovum,** or mature egg, and another polar body. Fertilization is considered complete when the haploid nuclei of sperm and mature ovum join, forming a single diploid cell (the zygote) with 46 chromosomes (**Figure 21.3**). Some people refer to the end of fertilization as conception and to the product of conception as the conceptus. The follicular cells that surrounded the secondary oocyte are shed and the polar bodies die, leaving just the fertilized cell, or zygote.

✓ **Quick Check** An egg can occasionally break out of its zona pellucida before the sperm reach it. If no zona pellucida is present when the sperm reach the egg, what might go wrong during fertilization? ■

Answers to ✓ **questions can be found in Appendix A.**

Twins may be fraternal or identical

Twins occur once in about every 90 births. Twins are the most common form of multiple births. Occasionally we hear of sextuplets or septuplets (six or seven newborns at once), but these are rare and almost always result from the use of fertility drugs. Even natural triplets are rare (1 in 8,000 births).

Two different processes lead to twins (**Figure 21.4**). **Fraternal twins** arise from the ovulation of more than one oocyte in a particular cycle. The oocytes are as different genetically as any two oocytes ovulated at different cycles, and each is fertilized by a different sperm. Fraternal twins are as different as any children by the same parents. Indeed, because they are derived from different sperm they can be of different genders.

Identical twins, sometimes called maternal twins, are always genetically identical. Identical twins arise from a single zygote. Recall that up until about the 8-cell stage, all cells of

a) **Fraternal twins arise when two eggs are ovulated and fertilized in the same monthly cycle.**

b) **Identiclal twins arise when a zygote divides in two during development.**

Figure 21.4 **Twins.**

Prenatal Diagnostic Techniques

Fifty years ago, when a woman became pregnant there was little that the parents or physician could do except hope that the baby would be born healthy. Late in pregnancy the physician could hear the fetal heartbeat, but that was about it. Given that nearly 7% of all newborns had a birth defect of some kind, pregnancy was a stressful period of wait-and-see. Today, fetal cells can be collected for genetic testing fairly early in fetal development, even early enough to permit a couple to terminate a pregnancy if that is their wish after reviewing the test results.

In **amniocentesis,** a needle is inserted through the abdominal wall and a sample of amniotic fluid containing fetal cells is collected. The harvested fetal cells are then grown in the laboratory until there are enough of them to examine the chromosomes and perform certain biochemical tests. Amniocentesis is relatively safe and may be recommended if there is a family history of a genetic disease or the risk of Down syndrome. Nearly 40 different fetal defects can be diagnosed using amniocentesis. A drawback is that amniocentesis cannot be performed until about the fifteenth week of gestation, and test results may not be available until the middle of the second trimester. This is close to the time that termination of pregnancy will no longer be an option.

In **chorionic villi sampling,** a thin, flexible tube attached to a syringe is inserted through the vagina and into the uterus, and a small sample of chorionic villi tissue is collected. Chorionic villi tissue originates from the fetus (review Figure 21.8), so most of the cells are fetal cells. The great advantage of chorionic villi sampling is that it can be done as early as five weeks of pregnancy. Because there is a slight but real risk of injury to the embryo, the technique is recommended only when a family history indicates a risk of genetic defects.

Percutaneous umbilical blood sampling (PUBS) is a technique for collecting fetal blood directly from the umbilical cord. Under ultrasound guidance, a needle is inserted through the abdominal wall and into a vein in the umbilical cord. Cord blood obtained by PUBS can be tested for sickle-cell anemia, hemophilia, anemia, and Rh problems, as well as genetic abnormalities. The risk of miscarriage is slightly higher than with amniocentesis or chorionic villi sampling, so again, this technique may be advisable only when there is a known risk of certain diseases. ∎

---

the developing pre-embryo are identical. If the ball of cells breaks into two groups before differentiation has begun, two complete and similar individuals may be formed. Identical twins are always of the same gender and are usually closely alike in phenotypic appearance.

↩ **Recap** Sperm deposited in the vagina must swim through the uterus and up the correct oviduct to meet the egg. Fertilization begins when one sperm's nucleus enters the oocyte and ends when the haploid nuclei of sperm and egg fuse, creating a new diploid cell. Fraternal twins result from the fertilization of two separate eggs. Identical twins occur when a single fertilized egg divides in two before differentiation has begun. ∎

21.2 Development: Cleavage, morphogenesis, differentiation, and growth

During development, rapid and dramatic changes in size and form take place. This involves four processes.

- **Cleavage. Cleavage** is a series of cell divisions without cell growth or differentiation during the first four days following fertilization. Cleavage produces a ball of identical cells that is about the same size as the original zygote. Growth does not occur because the ball of cells is traveling down the oviduct at this time. With no attachment to the mother, the only energy available is that stored in the cytoplasm of the cells plus a little glycogen found within the oviduct itself.

- **Morphogenesis.** Throughout development the organism undergoes dramatic changes in shape and form. Starting as a ball of identical cells at day four, the organism becomes (in succession) several layers of different types of cells, a pre-embryo with a tail and head, an embryo with recognizable human features, and finally a fetus with a nearly complete human form. This process of physical change is called **morphogenesis** (*morpho* meaning "form; shape" and *genesis* meaning "origin; production").

- **Differentiation.** Individual cells, too, are beginning to take on specialized forms and functions, a process we know as differentiation (review Chapter 17). Differentiation of cells, and the development of organs and organ systems as a result of cell differentiation, is the primary cause of morphogenesis.

- **Growth.** Starting about the time that the developing organism becomes embedded in the endometrial lining of the uterus and begins to receive nutrients from the mother, the organism begins to grow in size. The growth of a human infant from fertilization to birth is truly spectacular—from a single cell too small to be seen to over 10 trillion cells with a combined weight of 6 or 7 pounds. Every time the cells divide, the two daughter cells double in mass to prepare for the next cell division. The only time cell division is not accompanied by cell growth is the initial period of cleavage.

Pregnancy is considered to comprise three periods called trimesters. Each trimester is approximately three months long, and characteristic events in embryonic and fetal development take place during each.

↺ **Recap** The four processes associated with development are (1) cleavage, a series of cell divisions producing a ball of identical cells; (2) morphogenesis, a sequence of physical changes; (3) differentiation, as cells assume specialized forms and functions; and (4) growth in size. Pregnancy is divided into three trimesters. ∎

21.3 Pre-embryonic development: The first two weeks

The period of prenatal ("before birth") human development can be divided into three stages called pre-embryonic, embryonic, and fetal development. Characteristic processes and changes take place in each stage.

During pre-embryonic development (the first two weeks), the conceptus makes its way down the oviduct, embeds itself in the endometrial lining of the uterus, begins to receive nutrients from the mother, and starts to grow. Cells differentiate into several tissue layers, and the first stages of morphogenesis begin.

Throughout the pre-embryonic period, the conceptus is known informally as a **pre-embryo** because many (if not most) of the cells are destined to become part of the placenta, not the embryo. While the conceptus is still in the oviduct, a series of successive cleavages yields a ball of about 32 identical cells called the **morula** (**Figure 21.5**). The term morula means "little mulberry," aptly describing its spherical, clustered appearance.

On about the fourth day the morula is swept into the uterus, where it undergoes the first stages of differentiation and morphogenesis. Over the next several days it becomes a **blastocyst,** a hollow ball comprising (1) an outer sphere of cells called a trophoblast, (2) a hollow central cavity, and (3) a group of cells called the inner cell mass. Only the inner cell mass is destined to become the embryo.

About day six or seven, the trophoblast cells make contact with the endometrial lining and secrete enzymes that digest endometrial cells. The creation of a path for the blastocyst causes it to burrow inward (**Figure 21.6**). The process by which the blastocyst becomes buried within the endometrium is called **implantation.**

Figure 21.5 Pre-embryonic development leading up to implantation. After a series of cleavages the morula enters the uterus and develops into a blastocyst comprising an outer layer of cells called the trophoblast, a hollow cavity, and an inner mass of cells. The blastocyst attaches to the uterine wall.

✓ *How is cleavage different from the normal cycles of cell division seen in adult tissues? (Hint: Compare the final size of the blastocyst with the original zygote.)*

During the second week the inner cell mass begins to separate from the surface of the blastocyst, creating a second hollow cavity. The second cavity will become the amniotic cavity filled with amniotic fluid. At this point the cell mass, now called the **embryonic disk,** differentiates into two cell types called ectoderm and endoderm. The appearance of an amniotic cavity and of ectoderm and endoderm in the embryonic disk marks the end of the pre-embryonic period.

Pre-embryonic development is a hazardous time in which a lot can go wrong. Usually a woman is not even aware that she is pregnant at this time, and so she may continue risky behaviors such as smoking and drinking alcohol.

On rare occasions the blastocyst implants outside the uterine cavity, resulting in an **ectopic pregnancy.** Most ectopic pregnancies occur in the oviduct, but on rare occasions they may occur in the abdominal cavity, in an ovary, or near the cervix. Ectopic pregnancies usually are not successful because sites other than the uterus either are not large enough to accommodate a full-term baby or result in poor placental development. Occasionally an ectopic pregnancy in an oviduct is successful, but only if the infant is delivered surgically as soon as it is able to survive with intensive care in a hospital.

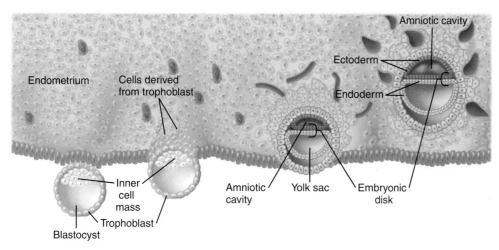

Figure 21.6 Implantation and the end of pre-embryonic development. Implantation takes place during days 7–14, at the same time that the inner cell mass of the blastocyst is developing into the embryonic disk. The inner cell mass separates from the outer ball to become the embryonic disk, comprising two cell types. The embryonic disk is destined to become the embryo.

✓ *What does the trophoblast secrete that enables implantation to occur?*

↩ **Recap** During pre-embryonic development, successive cleavages yield a morula. Early stages of differentiation and morphogenesis cause the morula to become a blastocyst, which implants in the lining of the uterus. The embryonic disk is destined to become the embryo. ▪

21.4 Embryonic development: Weeks three to eight

From about the beginning of week three until the end of week eight the developing human is called an **embryo.** During this time growth, differentiation, and morphogenesis are especially rapid. All the organs and organ systems are established, though most of them are not fully functional. By the end of the embryonic period the embryo has taken on distinctly human features but is still only an inch long.

Tissues and organs derive from three germ layers The embryonic period begins when a third cell layer, called the mesoderm, appears between the other two layers in the embryonic disk (Figure 21.7). These three primary tissues, called the germ layers, represent the precursor (or germ) cells for the four basic tissue types introduced in Chapter 4 (epithelial, muscle, connective, and nervous) and all organs and organ systems in the body. Differentiation and morphogenesis follow a predictable pattern, so we know which tissues and organs are derived from each germ layer. The three germ layers are

- **Ectoderm.** The ectoderm is the outermost layer, the one exposed to the amniotic cavity. Tissues derived from the ectoderm become the epidermis of the skin, the nervous system, hair, nails, enamel of teeth, parts of the eye, and several other organs and tissues (Table 21.1).

- **Mesoderm.** The middle layer, or mesoderm, becomes muscle, connective tissue and bone, kidneys and ureters, bone marrow, testes or ovaries, the lining of the blood vessels, and other organs and tissues.
- **Endoderm.** The innermost layer is the endoderm. It gives rise to the liver and pancreas, the alveoli of the lungs, the linings of the urinary bladder, urethra and vagina, and several glands.

✓ **Quick Check** A certain type of birth defect involves an absence of important nerves in various organs. In many of these patients, the skin and hair is unusually pigmented as well. Which germ layer is the likely cause of this birth defect? Explain your reasoning. ▪

Extra-embryonic membranes

Early in embryonic development, four different extra-embryonic membranes form that extend out from or surround the embryo: the amnion, allantois, yolk sac, and chorion. Most of the components of these membranes are either resorbed during later development or discarded at birth.

The innermost layer is the **amnion,** also known as the "bag of waters" (Figure 21.8a). The amnion lines the amniotic cavity, which is filled with **amniotic fluid.** The amniotic fluid is derived from the mother's interstitial fluid and is in continuous exchange with it. Later in development when the kidneys of the fetus form, the fetus will urinate into the amniotic fluid and the urine will be removed via exchange with the mother's

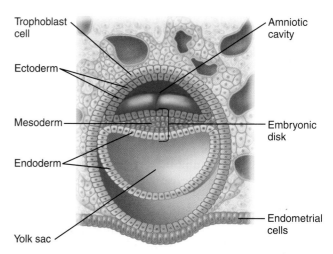

Figure 21.7 The three primary germ layers. The appearance of a third type of cell layer (mesoderm) between the ectoderm and endoderm in the embryonic disk marks the start of the embryonic period.

Table 21.1 **Tissues, organs, and organ systems derived from the three primary germ layers**

Ectoderm	Mesoderm	Endoderm
Epidermis of skin, including hair, nails, and glands	Dermis of skin	Epithelial tissues lining the digestive tract (except mouth and anus), vagina, bladder, and urethra
Mammary glands	All connective tissue, cartilage, and bone	
The nervous system, including the brain, spinal cord, and all nerves	All muscle tissue (heart, skeletal, smooth)	Alveoli of lungs
	Bone marrow (blood cells)	Liver and pancreas
Cornea, retina, and lens of eye	Kidneys and ureters	Thyroid, parathyroid, and thymus glands
Enamel of teeth	Testes, ovaries, and reproductive ducts	Anterior pituitary gland
Posterior pituitary gland	Lining of blood vessels	Tonsils
Adrenal medulla	Lymphatic vessels	Portions of the inner ear
Epithelial lining of nose, mouth, and anus	Adrenal cortex	

a) **The fetus is bathed in amniotic fluid within the amnion.** Its only connection to the mother is the umbilical cord.

b) **A closer view of portions of the placenta and umbilical cord, showing how nutrients and gases are exchanged between maternal and fetal blood without mixing.**

Figure 21.8 **The placenta and umbilical cord.**

blood. The amnion and amniotic fluid absorb physical shocks, insulate the fetus, and keep it from drying out.

The allantois is a temporary membrane that helps form the blood vessels of the umbilical cord. It degenerates during the second month of development.

The yolk sac forms a small sac that hangs from the embryo's ventral surface. In many species the yolk sac serves a nutritive function, but in humans that function has been taken over by the placenta, discussed shortly. Nevertheless, the yolk sac is important in humans because it becomes part of the fetal digestive tract. It also produces fetal blood

cells until that job is taken over by other tissues, and it is the source of the germ cells that migrate into the gonads (testes and ovaries) and give rise to the gametes (sperm and eggs).

The outermost layer, the **chorion,** is derived primarily from the trophoblast of the early blastocyst. The chorion forms structures that will be part of the exchange mechanism in the placenta. It is also the source of **human chorionic gonadotropin** (hCG), a hormone that supports pregnancy for the first three months until the placenta begins producing enough progesterone and estrogen. The presence (or absence) of hCG in the urine is the basis of most home pregnancy tests.

The placenta and umbilical cord

As the embryo grows and develops, its relationship with the mother becomes more complex. The embryo cannot supply itself with nutrients or get rid of its own wastes, so exchange vessels form between the embryo and the mother for the exchange of nutrients and wastes without direct mixing of their blood. This is the function of the placenta and the umbilical cord. The **placenta** (Figure 21.8b) is the entire structure that forms from embryonic tissue (chorion and chorionic villi) and maternal tissue (endometrium). The **umbilical cord** is the two-way lifeline that connects the placenta to the embryo's circulation.

The placenta develops because the cells of the chorion, just like the cells of the trophoblast from which they derive, secrete enzymes that eat away at the endometrial tissues and capillaries in the vicinity of the embryo. These enzymes rupture the capillaries, causing the formation of small blood-filled cavities. The developing chorion extends fingerlike processes called chorionic villi into these pools of maternal blood. Each chorionic villus contains small capillaries that are connected to umbilical arteries and veins that are part of the circulation of the fetus. In other words, the chorion damages the endometrium to cause local bleeding, and then it taps that bleeding as a source of nutrients and oxygen for the embryo and as a place to get rid of embryonic wastes, including carbon dioxide. In effect, an infant behaves like a parasite within the mother.

The placenta is an effective filter. It permits the exchange of nutrients, gases, and antibodies between the maternal and embryonic circulations but not the exchange of large proteins or blood cells. The placenta allows the embryo to take advantage of the mother's organ systems until its own organs develop and become functional.

However, the placenta may permit certain toxic substances and agents of disease to cross over to the fetus as well. Examples include alcohol, cocaine, the HIV virus, and a wide variety of prescription and nonprescription drugs. Many of these substances can do a great deal of harm to the embryo when it is in the earliest stages of differentiation and morphogenesis, even though some of the prescription drugs may be therapeutic for the mother (review Chapter 17). Upon learning she is pregnant, a woman should review all her prescription and nonprescription drugs with her physician.

The placenta is also an endocrine organ. Initially the placenta secretes hCG so that the corpus luteum will continue to secrete the progesterone and estrogen necessary to maintain the pregnancy (Chapter 16). Later the placenta secretes its own progesterone and estrogen, and the corpus luteum regresses. Progesterone and estrogen promote the growth of the myometrium in preparation for intense contractions at birth, maintain the endometrial lining so

menstruation does not occur, inhibit uterine contractions during pregnancy, and help form a thick mucous plug over the cervix to inhibit uterine infections.

The two umbilical arteries and single umbilical vein are considered part of the embryonic circulation, meaning the umbilical vein carries blood *toward* the embryonic heart and the umbilical arteries carry blood *away* from the embryonic heart and back to the placenta. Because the exchange of nutrients by the placenta requires a functional embryonic heart, the placenta and umbilical vessels do not become fully functional until the embryonic heart develops at about five weeks. Until then the developing chorion supplies the nutrient needs of the embryo by diffusion.

 Quick Check Name the four extra-embryonic membranes and state the major function of each. Which one plays the largest role in the formation of the placenta? ▪

The embryo develops rapidly

At two weeks the embryo is fully embedded in the endometrium, which provides it with nutrients via the developing chorion. At about this time the embryo begins to take shape as cells of the three primary tissue layers migrate to other locations and begin to form the rudimentary organs and organ systems. First a small groove called the primitive streak appears in the flat, round embryonic disk, and the embryonic disk begins to elongate along one axis (**Figure 21.9**).

At three weeks certain precursors to embryonic structures emerge. A neural groove of ectoderm forms that will later become the brain and spinal cord. Meanwhile, the mesoderm begins to separate into several segments called somites that will become most of the bone, muscle, and skin. Prominent bumps called the pharyngeal arches appear at one end; they are destined to become part of the face, neck, and mouth. By the end of week four the heart is beginning to develop, the head begins to take shape, the position of the eyes becomes apparent, the neural groove has closed into a neural tube, and four limb buds and a tail appear (**Figure 21.10**).

The patterns of cell differentiation and organism morphogenesis are so consistent among vertebrates that even an expert would have a hard time telling whether a four-week-old embryo of unknown origin was destined to become a fish, a mammal, or a bird. Because early differentiation and morphogenesis follow such a common path, it is likely that these patterns developed early in the history of the vertebrates.

Weeks five through eight mark the transition from a general vertebrate form to one that is recognizably human. The head grows in relation to the rest of the body, the eyes and ears are visible, and the four limbs are formed with distinct fingers and toes. A cartilaginous skeleton forms. The heart and circulatory system complete their development

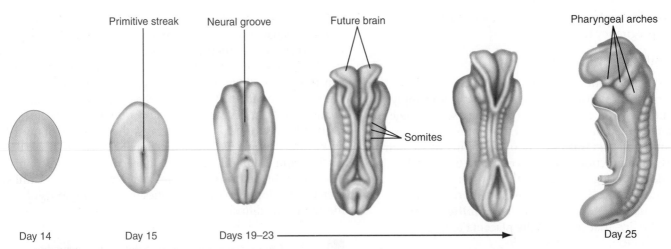

Figure 21.9 **Embryonic development during the third and early fourth week.** Day 14 represents the flat embryonic disk as it would look from above (from the amniotic cavity). On day 15 the embryo begins to elongate and the primitive streak of ectoderm appears. Days 19–23 mark the appearance of a neural groove of ectoderm. Somites develop in the mesoderm that will become bone, muscle, and skin. By day 25, pharyngeal arches that will contribute to structures of the head become visible. The embryo on day 25 is viewed from the side.

and the umbilical cord and placenta become functional, so that now blood circulates throughout the fetus. However, the blood cells for the fetal blood are produced by the yolk sac because the blood-forming tissues of the fetus are not yet mature. Nutrients and wastes are now exchanged more efficiently and in greater quantities. The tail regresses. At the end of the embryonic period, the embryo is an inch long and weighs just 1 gram (Figure 21.11).

Although it is difficult to estimate how many embryos fail to survive to the end of the embryonic period, some estimates place the number as high as 20%. Spontaneous termination of pregnancy followed by expulsion of the embryo is called a **miscarriage,** or spontaneous abortion. Sometimes a woman never even realizes she was pregnant. Miscarriages are probably nature's way of weeding out embryos with genetic disorders that might prevent them from developing normally.

Figure 21.10 **Week four of development.** The future head is prominent, the position of the eyes becomes apparent, the heart and neural tube are forming, and limb buds appear.

gases between embryo and mother, and secretes hormones. The umbilical cord joins the embryo to the placenta. By the fifth week the embryo is becoming distinctly human in form, and by eight weeks it is an inch long. ∎

 Web Animation *Embryonic Development* at www.humanbiology.com

Recap By the beginning of embryonic development the embryo comprises three primary germ layers, called ectoderm, mesoderm, and endoderm, that ultimately give rise to fetal tissues and organs. Four extra-embryonic membranes (amnion, allantois, yolk sac, and chorion) serve varying supportive functions. The placenta exchanges nutrients and

21.5 Gender development begins at six weeks

Male and female external and internal genitalia do not begin to develop until about six weeks. Until that time the embryo remains "sexually indifferent," meaning that it still has the capacity to develop into either a male or a female. However, the final outcome (male or female) is not in

Figure 21.11 **Week eight of development.** The embryo is now distinctly human in appearance. The face, limbs, hands, and feet are well formed. The circulatory system is functional, and blood now circulates throughout the embryo and to and from the placenta. Actual length is 1 inch.

doubt; it was determined at the moment of conception by the single sperm that fertilized the egg.

The development of either male or female external genitalia is shown in Figure 21.12. By six weeks an undifferentiated urogenital groove has developed, topped by a bud and surrounded by labioscrotal swellings. After six weeks, the presence of a Y chromosome causes the embryo to start developing male sexual characteristics. The process begins when a gene on the Y chromosome called SRY (for Sex-determining Region Y) is switched on. SRY encodes for a protein called *testis-determining factor* that directs the initial development of the testes internally. Shortly thereafter the testes begin to secrete testosterone, which in turn stimulates the development of the male internal ducts and external genitalia. (The embryonic testes also secrete a second hormone called anti-Mullerian hormone that

suppresses the development of the female internal and external genitalia.) In males the undifferentiated urogenital groove closes up to become the urethra within the penis. The penis elongates, and the previously undifferentiated bud becomes the head of the penis. The urogenital swellings develop into the scrotum into which the testes later descend.

In the absence of a Y chromosome, female internal and external genitalia develop and the embryo becomes a female. The urogenital groove expands to become the vagina, and the undifferentiated bud becomes the clitoris. Meanwhile the ovaries, fallopian tubes, and uterus all develop internally. The development of female internal and external genitalia is the "default" condition; it occurs in the absence of any hormones, whenever a Y chromosome is *not* present. In other words, gender is determined by the presence or absence of a Y chromosome, not by whether the embryo has one or two X chromosomes.

Recap Gender development begins at about 6 weeks. The presence of a Y chromosome signals the embryo to develop into a male; the absence of a Y chromosome causes the embryo to develop into a female. ▪

Figure 21.12 **Development of male or female external genitalia.**

✓ *Occasionally an embryo has two X chromosomes and a Y chromosome (three sex chromosomes in total). What will happen to this embryo during sexual differentiation? Explain your answer.*

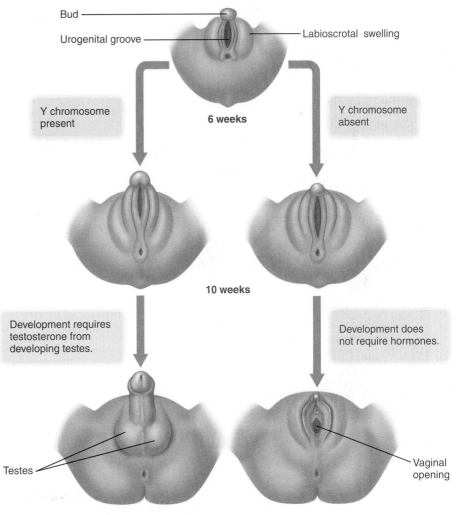

Bud
Urogenital groove
Labioscrotal swelling

Y chromosome present

6 weeks

Y chromosome absent

10 weeks

Development requires testosterone from developing testes.

Development does not require hormones.

Testes

Vaginal opening

Birth approaching

21.6 Fetal development: Nine weeks to birth

After the eighth week the developing human is called a fetus. This is the time when fetal organ systems begin to function and the bones start to calcify. The mother, too, undergoes obvious changes in physical shape. Next, we examine each of these stages in detail.

Months three and four

During the third month the kidneys develop sufficiently for the fetus to begin eliminating some wastes as urine into the amniotic fluid. The limbs are well developed, the cartilaginous skeleton begins to be replaced with bone, and the teeth form. The spleen participates briefly in the production of red blood cells, a job that will soon be taken over by the liver and bone marrow. The liver begins to function, and the genitalia are well enough developed that sex can be determined. The end of the third month marks the end of the first trimester.

During the fourth month the liver and bone marrow begin producing blood cells. The face takes on nearly its final form as the eyes and ears become fixed in their permanent locations (**Figure 21.13**). In the female fetus, follicles are forming in the ovaries.

Overall growth is rapid during this time. By the end of the fourth month the fetus is already 6 inches long and weighs about 170 grams (6 ounces).

Months five and six

By the fifth month the nervous system and skeletal muscles are sufficiently mature for the fetus to begin moving. These movements are called quickening, and the mother may feel them for the first time. The skin, which is well formed although red and wrinkled, is covered by soft, downy hair.

Figure 21.13 The fetus and placenta at four months of development. Note the amnion and the clear amniotic fluid surrounding the fetus. Actual length is 6 inches.

The heartbeat is now loud enough to be heard with a stethoscope. Skeletal hardening continues.

The sixth month marks the first point at which the fetus could, with the best neonatal care available, survive outside the uterus. At this stage the fetus seems to respond to external sounds. Most importantly, the lungs begin to produce surfactant, the substance that reduces surface tension in the lungs and permits the alveoli to fill with air at birth.

The end of the sixth month marks the end of the second trimester. At this point the fetus weighs nearly 700 grams, or approximately 1½ pounds.

Months seven through nine

The seventh through ninth months (the third trimester) are a period of continued rapid growth and maturation in preparation for birth. The eyes open and close spontaneously and can be conditioned to respond to environmental sounds. Activity increases, as if the fetus is seeking a more comfortable position in the increasingly restrictive space of the uterus. Usually the fetus moves to a position in which the head is positioned downward, near the cervix. The skin begins to lose some of its reddish color and its coat of downy hair. In the male the testes descend into the scrotal sac. Although neither the lungs nor the digestive systems have had a chance to function yet, both are now ready.

By nine months the fetus is about 20 inches long and weighs approximately 6–7½ pounds. Birth usually occurs at about 38 weeks of development.

Recap The period of fetal development extends from nine weeks to birth at 38 weeks. Growth is rapid, with the mature fetus weighing approximately 6–7½ pounds at birth. The fetus begins to move at about five months, and life outside the womb is at least possible by about six months when the lungs begin to produce surfactant. ▪

21.7 Birth and the early postnatal period

Other than perhaps conception and death, there is no period in our lives marked by greater developmental change than the hours during and immediately after birth. Within minutes, the newborn makes the shift from relying solely on the maternal circulation for nutrient and gas exchange to depending on its own circulation and lungs. The digestive tract, too, starts to function with the first swallow of milk. Let's take a quick look at the changes that occur during birth and the postnatal ("after birth") period.

Labor ends in delivery

Birth involves a sequence of events that we call labor (it's hard work!), which ends in the delivery of the newborn into the world.

Labor begins as a result of a series of events that is triggered by maturation of the fetal pituitary gland, which serves as the timing device to indicate that the fetus is now ready for birth.

The mature fetal pituitary gland begins to release ACTH, which stimulates the fetal adrenal gland to secrete steroid hormones that cause the placenta to increase its production of estrogen and decrease the production of progesterone. Estrogen, in turn, increases the number of oxytocin receptors and stimulates production of prostaglandins. Together, the increased estrogen, prostaglandins, and oxytocin receptors (along with the mother's oxytocin) stimulate the uterus to contract rhythmically.

A positive feedback cycle begins: rhythmic contractions of the uterus cause the release of still more oxytocin from the maternal pituitary, which increases contractions still further, and so on. As labor progresses the periods of contraction get closer together, last longer, and become stronger, creating enough force to push the fetus toward the cervix and eventually through the vagina.

The period of labor and delivery lasts about 24 hours for the first birth and slightly less time for subsequent births. Labor and delivery are divided into three phases: dilation, expulsion, and afterbirth (**Figure 21.14**).

■ **Stage 1—dilation.** The first phase of labor can last 6–12 hours. During this time the rate, duration, and strength of contractions increase over time, pushing the head of the fetus against the cervix. The cervix itself is drawn back toward the uterus, widening the cervical opening and expelling the mucus plug. Continued pressure of the fetus against the cervix widens the cervical opening even further until eventually it is large enough to accommodate the baby's head, or about 10 centimeters. At about this time the pressure of the baby's head ruptures the amnion, releasing the amniotic fluid. This "breaking of the water" is a normal sign that delivery is proceeding. The physician keeps a close watch on the stage of dilation by measuring the degree of cervical dilation at regular intervals.

■ **Stage 2—expulsion.** The period of expulsion extends from full cervical dilation through actual delivery. During this time uterine contractions strengthen and the woman experiences an intense urge to assist the expulsion with voluntary contractions as well. To make the birth easier,

a) **9-month-old fetus.** As birth approaches, the fetus usually is positioned with the head down and toward the cervix.

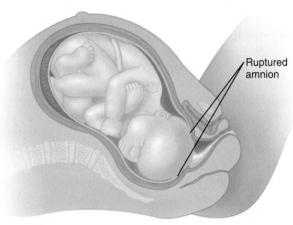

b) **Stage 1: dilation.** The cervical opening widens. The amnion may break at this stage.

c) **Stage 2: expulsion.** The fetus passes headfirst through the cervical canal and the vagina.

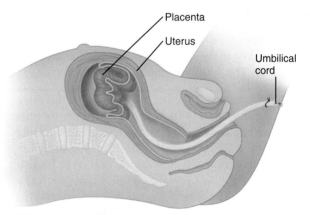

d) **Stage 3: afterbirth.** The placenta detaches from the uterus and is expelled along with the remainder of the umbilical cord.

Figure 21.14 The stages of birth.

[handwritten note: Know the heart →]

so
ep   e
co
va   of
the   as
the   cus
fro
The
bab   nd
the   its
own
■ Stag
birth
the p
cord _____, conectively called the afterbirth. The
afterbirth stage usually lasts less than half an hour after
the birth of the infant.

Cesarean delivery: Surgical delivery of a baby

Some deliveries are not allowed to proceed as described above.
Surgical delivery of a baby is called a **cesarean delivery,** or C-
section. In a C-section, an incision is made through the abdominal wall and uterus so the baby can be removed quickly.

A C-section is often performed when the position or size
of the fetus could make a vaginal delivery dangerous. Possible reasons include a fetus that is too large for the birth
canal, improper position of the fetus with the legs near the
cervix, maternal exhaustion from a long delivery, or signs of
fetal distress such as an elevated fetal heart rate.

In the past, if a C-section was performed it was common
practice to do all subsequent deliveries by C-section as well.
Today most women who have a C-section can still opt for a
vaginal delivery with the next child if they wish.

The transition from fetus to newborn

Somewhere between the point when its head emerges and the
umbilical cord is clamped, the fetus must become capable of
sustaining life on its own. The first necessity is for the newborn
(also called the neonate for the first 28 days) to start breathing,
preferably even before its umbilical connection to the mother is
severed. Second, over the first few days of life, its cardiovascular
system undergoes a series of remarkable anatomical changes
that reflect the loss of the placental connection for gas and nutrient exchange. The newborn is now entirely on its own.

Taking the first breath When the infant emerges, the three
hundred million alveoli in its lungs have never been inflated.
After all, the infant essentially has been living underwater
(more correctly, under amniotic fluid). The first inflation is
critical and not all that easy. If you've ever tried to blow up a
really stiff balloon, you know that it takes a lot of effort to
start the inflation. The reason for this phenomenon is that
the surface tension generated by any small sphere is greater
than the surface tension generated by a larger sphere. The
first inflation of the alveoli (representing small spheres) is
facilitated by substances called pulmonary surfactants that
reduce alveolar surface tension.

But what causes the infant to take that first breath? During
labor, the placental connection to the mother begins to separate,
reducing gas exchange with the fetus. Clamping the umbilical
cord stops gas exchange entirely. This situation is equivalent to
someone covering your mouth and nose completely. Within
seconds the carbon dioxide concentration in the fetus rises to
the point that the respiratory centers in its brain stimulate
respiration. With the enormous effort of one who is being
smothered, the infant takes its first breath and begins to cry.

Changes in the cardiovascular system The fetal cardiovascular
system is different from the cardiovascular system described in
Chapter 8. While the fetus is still in the womb the fetal lungs
are not yet useful for gas exchange. The fetus receives its
nutrients and exchanges gases via blood that travels to and
from the fetus in the umbilical cord. Immediately after birth
the umbilical circulation ceases, and pulmonary circulation
and gas exchange becomes crucial to survival.

Let us follow the circulation of blood between mother
and fetus, and also within the fetus, starting at the point that
oxygen-laden, nutrient-rich blood from the placenta enters
the fetus via the umbilical vein (Figure 21.15a). Some of the
blood is carried to the fetal liver either directly or via a
branch of the umbilical vein that joins the hepatic portal
vein. But the fetal liver is not yet fully developed and cannot
handle the entire umbilical blood flow. Most of the blood
bypasses the liver and is carried directly to the inferior vena
cava by the ductus venosus. In the inferior vena cava, the nutrient-rich blood mixes with the venous blood of the fetus.

Most of the blood that enters the fetal heart must bypass
the fetal lungs because they too are not fully developed. Some
of the blood passes from the right atrium to the left atrium
through the foramen ovale, and some is shunted from the pulmonary artery directly to the aorta via the ductus arteriosus.

Even though the fetal lungs are not yet functional and
the digestive tract is not receiving nutrients, the aortic blood
still has sufficient oxygen and nutrients (brought to the fetus
via the umbilical vein) to supply all the fetal tissues. Some of
the fetal arterial blood returns to the placenta via the umbilical arteries. In the placenta the blood again picks up
nutrients and oxygen and gets rid of carbon dioxide
generated by fetal metabolism.

The situation changes dramatically at birth, for the umbilical circulation is immediately cut off. Over time the umbilical vein, umbilical arteries, and ductus venosus regress to
become vestigial connective tissue. The foramen ovale and
ductus arteriosus close off in the first few days or weeks after
birth as well, so that all the cardiac output passes through
the lungs for efficient gas exchange (Figure 21.15b). Nutrients are now absorbed by the newborn's digestive tract and
pass through the liver via the hepatic portal vein. The only
reminder of the intimate connection between fetus and
mother is the umbilicus (navel).

To summarize the unique features of the fetal circulation:

■ Blood enters the fetus via the umbilical vein. Some
  of the blood flows through the liver, but most of it

■ High oxygenation   ■ Moderate oxygenation   ■ Low oxygenation   ■ Very low oxygenation

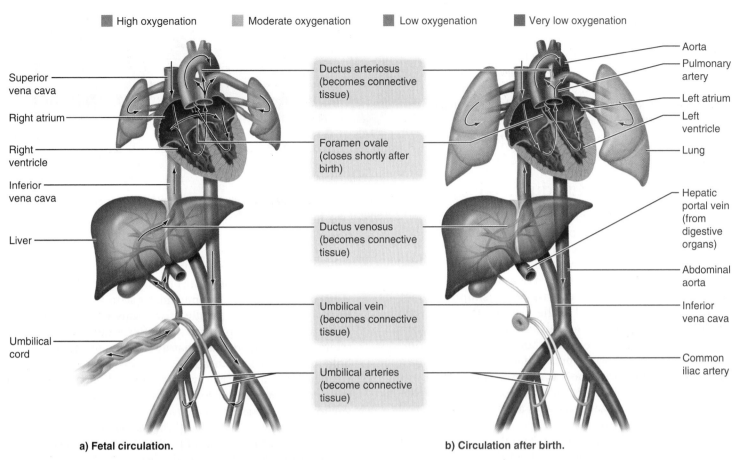

a) Fetal circulation.

b) Circulation after birth.

Figure 21.15 The cardiovascular systems of the fetus and of the newborn. In the fetus the immature lungs are not used for gas exchange, so most of the blood returning to the heart bypasses the lungs by flowing through either the foramen ovale or the ductus arteriosus. Gas exchange occurs via the placenta, so special fetal blood vessels connect the fetus to the placenta. These unique pathways for blood circulation in the fetus close shortly after birth.

✓ *What would happen if either the ductus arteriosus or the foramen ovale did not close after birth?*

bypasses the liver and joins the inferior vena cava of the fetus via the ductus venosus (venous duct).

■ Blood leaves the fetus via two umbilical arteries that originate from arteries in the lower extremities.

■ In the fetus there is a hole between the two atrial chambers of the heart called the foramen ovale (oval opening). The hole permits some blood to pass from right atrium to left atrium, bypassing the pulmonary circulation. The hole closes after birth.

■ In the fetus there is a shunt, or shortcut, from the pulmonary artery directly to the aorta called the ductus arteriosus (arterial duct). Consequently, most of the blood pumped by the right ventricle into the pulmonary artery bypasses the lungs. The ductus arteriosis closes after birth.

✓ **Quick Check** Infants born prematurely often have not yet produced surfactants. What problems might this cause, and how could it be treated? ■

Lactation produces milk to nourish the newborn

The intimate relationship between mother and child does not end at birth, for the human infant must rely entirely on its mother (or someone else) for all its nutritional needs. During pregnancy, high concentrations of estrogen and progesterone cause the woman's breasts to enlarge in preparation for lactation (milk production). However, the breasts do not secrete milk before childbirth because estrogen and progesterone prevent the action of prolactin, the hormone that actually stimulates milk production.

During the first day or so after birth, the breasts produce a watery milk called **colostrum** that is rich in antibodies but low in fat and lactose. The antibodies help to protect the newborn until its own immune system matures. Later, the high fat and lactose content of milk help the infant to grow rapidly.

A fourth hormone, oxytocin, is responsible for the contractions that deliver colostrum or milk only when needed. Oxytocin is released during childbirth and also as a result of a neural reflex every time the infant nurses. The expulsion of milk from the breast during suckling is called milk ejection.

⮌ **Recap** The period of labor and delivery is composed of three stages; dilation, expulsion, and afterbirth. The dilation phase may last 6–12 hours. At birth a sharp increase in carbon dioxide causes the newborn to take the first breath. Shortly after birth, anatomical changes in the newborn route all blood through the lungs. In the mother, prolactin stimulates lactation, and oxytocin stimulates the release of milk during suckling. ▪

21.8 From birth to adulthood

The long process of human development continues after birth. Because development is a continuum, experts do not always agree on how to categorize its stages. Some categories may seem somewhat arbitrary, such as the distinction (if there is one) between adulthood and old age. At the risk of oversimplifying, we define the neonatal period as the first month, infancy as months 2–15, childhood as the period from infancy to adolescence, adolescence as the transitional period from childhood to adulthood (approximately 12 to 18 or 20 years of age), and adulthood as 20 years and beyond.

Table 21.2 summarizes the stages of human development. Bear in mind that these ages are approximate, intended as a guide to orient you with regard to your progression through life. We concentrate primarily on biophysical development, leaving human psychosocial and cognitive (learning) development to the fields of psychology and education. Figure 21.16 depicts the changes in body form that accompany human development to adulthood.

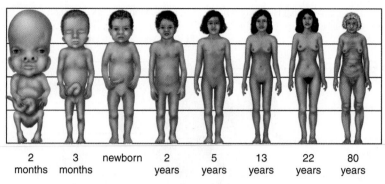

2 months	3 months	newborn	2 years	5 years	13 years	22 years	80 years

Figure 21.16 Changes in body form and relative proportion throughout prenatal and postnatal growth and development.

The neonatal period: A helpless time

The human neonate (newborn) is fairly helpless. In particular, the nervous system and the muscular system are relatively immature at birth. Consequently, movements are uncoordinated, weak, and governed primarily by reflexes rather than conscious control. The neonate hardly has enough strength to hold its head up, so its head must be constantly supported whenever it is carried. The eyes are open but often they seem to wander about, unable to focus. Neonates are aware of their environment and can respond to it, but they are unable to retain any long-term memories. (What is *your* first memory?) They hear a narrower range of pitches than adults and are more sensitive to higher-pitched tones.

What neonates do best is suckle, urinate, and defecate. The digestive system can absorb the constituents of milk but is not yet ready for solid foods. The stools are soft, and defecation and urination occur as reflexes.

Infancy: Rapid development and maturation of organ systems

Infancy is a time of rapid change. In just 15 months the infant more than triples in weight to about 22–24 pounds. The bones harden gradually, and there is a disproportionately large increase in muscle mass and strength. The brain also grows rapidly, with nearly half of all brain growth after birth occurring in infancy. Most of the growth occurs in the cerebral cortex, the area associated with sensory perception, motor function, speech, and learning. Most of the myelination of nerves occurs in infancy as well. The first teeth appear at about six months, and all 20 baby teeth are generally present by about one year. Most infants begin eating solid foods by about six months to a year.

By about 14 months the combined maturation of the musculoskeletal and nervous systems allows the human infant to begin to walk on its own. A rapidly growing body needs lots of rest, and infants sleep a lot. Infants can easily sleep 10 hours a night and still take several naps during the day.

The immune system consistently lags behind most other systems in its development. Most vaccinations are ineffective in infants because their immune systems cannot produce the

Table 21.2 A summary of the stages of human development

Stage	Age
Prenatal (after fertilization)	
Pre-embryo	From fertilization to 2 weeks
Zygote	Days 1–3
Morula	Days 4–7
Blastocyst	Days 8–14
Embryo	Weeks 3–8
Fetus	Weeks 9–40 (birth)
Postnatal (after birth)	
Neonate	The first month
Infant	Months 2–15
Child	16 months to 12 years
Adolescent	12–20 years
Adult	Beyond 20 years
Old age	Aging accelerates after 40 years

appropriate antibodies. The schedule of vaccinations in infants and children is designed so that each vaccine is given only after the immune system becomes able to respond to that vaccine.

Childhood: Continued development and growth

The long period of childhood, from about 15 months to 12 years of age, involves continued growth of all systems. The brain grows to 95% of its final size during childhood. The immune system continues its slow process of maturation throughout childhood and even into adolescence. Body weight increases to an average of about 100 pounds in both boys and girls as the result of periods of slow growth interspersed with growth spurts. Both muscle strength and fine motor coordination improve.

In addition, body form alters as the long bones of the arms and legs lengthen. A lumbar curve develops in the small of the back, and abdominal musculature strengthens, so that the childhood "pot belly" disappears.

By the end of childhood the body form is distinctly adult like, though still sexually immature. Most organ systems are fully functional or nearly so (although not necessarily their full adult size). The lone exceptions are the male and female reproductive systems, which have not yet matured.

Adolescence: The transition to adulthood

One of the most challenging times of life is the transition from childhood to adulthood. The final growth spurt, one of the largest, occurs in adolescence. The growth spurt begins at different times in different individuals, so there is a wide range of normal body weights and degrees of sexual maturity in the 12- to 14-year-old age group. Rapid growth can lead to awkwardness and a temporary loss of coordination.

Both the skeletal and muscular systems experience their greatest rates of growth during adolescence. Other organs increase in size as well. Notably, the lungs more than double in mass, and the kidneys and stomach increase by more than 50%. The brain, however, gains only about 5%.

Adolescence is marked by maturation of the reproductive systems and the human sexual response, an event known as **puberty.** Puberty is defined as the onset of menstrual periods (menarche) in the female at some time between 9½ and 15½ years and by reflex discharge of semen during sleep (nocturnal emissions) in the male sometime between 11½ and 17½ years.

The concentrations of all the reproductive hormones are low in children, but they begin to rise during the period of sexual maturation (known as the pubertal period). The pubertal period is initiated by maturation of certain neurons of the hypothalamus, which begin to secrete a releasing hormone called GnRH (Gonadotropin Releasing Hormone). The initiation of GnRH secretion by the hypothalamus stimulates the secretion of luteinizing hormone (LH) and follicle-stimulating hormone (FSH) from the anterior pituitary, which in turn stimulate the production of sex hormones and the maturation of sex organs and secondary sexual characteristics.

The brain is one of the last organs to reach full maturity. The areas of the cerebral cortex that regulate impulse and emotion don't fully mature until about 18 years of age. This may be why teenagers, who may seem to be responsible in most of their decisions, still sometimes make bad judgments when they are under intense peer pressure.

Recap Maturation of organ systems occurs at different rates. At birth, the neonate is physically helpless and cannot form memories. Infants begin to eat solid foods and to walk. Brain growth is nearly complete by the end of childhood. Muscle strength and coordination continue to improve throughout childhood, and the body changes to a more adult form. Adolescence is accompanied by a growth spurt and by maturation of the reproductive systems. ∎

21.9 Aging takes place over time

Physiologically, at least, humans reach their peak in early adulthood at about 20 years of age. Most body systems are fairly stable from 20 to about 40, and then the process that we call aging seems to accelerate. No matter whether we exercise regularly, eat right, and in general take care of ourselves, there still seems to be at least some "natural" aging process, just like the carefully timed sequence of events that occurs in human development. To be sure, exercise and a proper diet do help, but in the end we all must face the fact that we will not be the same at age 80 as we were at 20. The big question, of course, is why we age and whether the aging process can be halted or at least slowed.

Aging is the process of change associated with the passage of time (**Figure 21.17**). Although the word does not necessarily imply deterioration (after all, some wines get better as they age), human aging invariably leads in that direction. Aging as we use the term in biology could be used synonymously with senescence, the progressive deterioration of multiple organs and organ systems over time. Human senescence (or aging) has no apparent or obvious cause—that is, it cannot be blamed on a particular disease or event.

a) Rolling Stones' Mick Jagger at age 23, in 1967.

b) Jagger at age 60 on December 12, 2003, the night he became "Sir Mick."

Figure 21.17 Aging.

A common statistic is that the average life span in the United States has increased from 47 years in 1900 to the upper 70s today. Does that mean we are aging more slowly now? Not at all. **Longevity**—how long a person lives—does not depend solely on the aging process. Most experts believe that our increase in longevity is largely due to a reduction in deaths by accident or disease, not to a slowdown in the aging process itself.

In fact, it appears that nothing we have done (so far!) has decelerated the aging process. If we could solve every known disease and prevent every accidental death, would we still come face to face with a natural endpoint of life? Nobody knows for sure, but so far the answer seems to be a qualified yes. No one has beaten the aging process yet, and there are very few individuals with documented ages of over 100 years.

What causes aging?

There are three hypotheses regarding the causes of aging. The first is that there is some sort of internal genetically determined program that dictates the timing of cell death. The second is that accumulated cell damage or errors eventually limit cells' ability to repair themselves. A third hypothesis is that the decline in function of a critical body system may lead to the parallel senescence of other systems. There is evidence to support each hypothesis.

An internal cellular program counts cell divisions It has long been known that even when cells are grown under laboratory conditions that are ideal for cell growth and division, most cells will not continue to divide indefinitely. Most cells go through approximately 50–90 successive divisions and then just stop. Cancer cells are the exception, for they will continue to divide indefinitely.

Why do normal cells stop dividing? Apparently every strand of DNA ends with a long tail of disposable bits of DNA called **telomeres** that do not code for any genes. The telomeres are like a roll of tickets, and the idea is that every time the DNA is replicated prior to cell division a ticket (telomere) is removed (Figure 21.18). When the cell runs out of telomeres, cell division begins to chew away at vital genetic information. Eventually enough vital genetic information is lost that the cell stops dividing and dies.

However, cancer cells manufacture an enzyme called telomerase that builds new telomeres to replace the ones lost during cell division. There has been a lot of excitement about telomerase lately as a result of these findings. Some researchers have even suggested that telomerase may be the magic bullet that could reverse or prevent the aging process, or that an antitelomerase substance might offer the cure for cancer.

These ideas remain far-fetched at the moment; understanding why a single cell stops dividing does not tell us how to arrest the aging process at the level of the whole body. After all, aging is not just a decline in cell division but in cell function as well.

Cells become damaged beyond repair The second hypothesis holds that over time, accumulated damage to DNA and errors in DNA replication become so great that they can no longer be countered by the cell's normal repair mechanisms. The damage is thought to be caused by certain toxic by-products of metabolism, such as oxygen free radicals. Eventually, like a ticking time bomb, the damage becomes too great to allow survival.

According to proponents of this hypothesis, how long cells live is a function of the balance between how often damage occurs and how much of it can be repaired. Cells that undergo frequent injury and are slow to repair the damage have shorter lives than cells that are rarely damaged and repair themselves efficiently.

This hypothesis is supported by the finding that severe caloric restriction (reduced food intake) in certain animals prolongs life. One possibility is that by lowering body temperature just a little, caloric restriction may slow metabolism and reduce the rate at which damaging metabolic by-products form. Some people believe that taking antioxidant vitamins may slow the aging process and prolong life, but there is not yet enough long-term evidence to know for sure if this is true.

Aging is a whole-body process Why does aging seem to happen to virtually all body systems at once, instead of just one or two systems? The third hypothesis of aging is that because all the organ systems in a complex organism are interdependent, a decline in the function of any one of them eventually impairs the function of the others.

For example, a decline in the secretion of growth hormone by the endocrine system could well tip the balance away from growth and replacement and toward a net loss of cells or organ function. Impaired function of the cardiovascular system would affect nutrient delivery to and waste removal from every cell in the body. A critical system in terms of aging may turn out to be the immune system. Indeed, the elderly have a higher incidence of cancer than the young, which may indicate that the immune system is not recognizing damaged cells as efficiently as it once did.

Useful gene section of DNA

Telomere section of DNA

Newly forming strand   Old (template) strand

Telomere

Figure 21.18 Telomeres and aging. Short terminal segments of DNA called telomeres are removed each time the DNA is replicated. As a result, the new strand will be shorter than the original strand. When all telomeres are gone, further cell divisions erode the useful genes.

MJ's Human Biology Blog

Caloric Restriction and Longevity

It has been known for some time that severe caloric restriction can retard the aging process and prolong the life of a variety of species, from worms to mice. But skeptics of the idea that caloric restriction might also slow the aging process in humans point out that the metabolisms of worms and mice are quite different from that of humans.

Now a new study reports that caloric restriction slows the aging process in primates, too. Macaque monkeys that have been on a calorie-restricted diet (by 30%) for the past 20 years are living longer and are healthier than their age-matched control counterparts. Excluding animals that died of non-age-related causes (accidents, for example), 50% of the animals on a normal diet have died of age-related causes, compared to only 20% in the restricted-diet group. The calorie-restricted animals also have fewer age-associated diseases such as diabetes, cancer, and cardiovascular disease. The study is still ongoing. ■

Reference: Coleman, R.J. et al. (2009). Caloric Restriction Delays Disease Onset and Mortality in Rhesus Monkeys. *Science* 325: 201–204.

It is also possible that aging is due to a specific type of tissue change that affects many organ systems at once. One hallmark of aging is that a certain type of protein cross-linking increases with age. This cross-linking could contribute to multiple aging processes, ranging from decreased elasticity of tendons and ligaments to functional changes in lungs, heart, and blood vessels. Of course, this hypothesis still leaves unanswered the question of what initiates the aging process at the cellular level.

 Quick Check What might be one negative consequence of increasing the amount of telomerase in cells? (Hint: What cells often have high levels of telomerase?) ■

Body systems age at different rates

What actually happens during aging is fairly well documented, even if we don't know exactly why. Next we describe important developments in each organ system.

Musculoskeletal system and skin Bone and muscle mass begin to decrease in early adulthood. Bone loss is more marked in women than in men, especially after menopause. The decline

in muscle mass is a combination of a loss of muscle cells and a decrease in the diameter of remaining cells. By age 80 the average person has lost 30% of his or her muscle mass, and strength declines accordingly. Often the joints become stiff and painful because they secrete less joint fluid. Ligaments and tendons become less elastic.

The skin becomes thinner, less elastic, and wrinkled. There are fewer sweat glands, so it becomes harder to adjust to warm temperatures. Pigmentation may change as well, with the appearance of pigmented blotches called liver spots.

Cardiovascular and respiratory systems Lung tissue becomes less elastic, decreasing lung capacity. The heart walls become slightly stiffer due to an increase in the ratio of collagen to muscle, with the result that an aging heart generates less force. Blood vessels also lose elasticity, so the systolic pressure and pulse pressure may rise in elderly people. Overall there is a reduced capacity for aerobic exercise as aging progresses.

Immune system The thymus decreases to only 10% of its maximum size by age 60, so there are fewer T cells. The activity, but not the number, of B cells declines as well. Wound healing slows, and susceptibility to infections increases—in the elderly, even the flu can become life-threatening. In addition the immune system fails to recognize "self" as well as it once did, leading to higher rates of autoimmune disorders.

Nervous and sensory systems Neurons in most areas of the brain do not divide in adults, so neurons that are lost throughout life generally are not replaced. The brain shrinks only slightly, however, because the neurons that remain tend to increase their connections with each other. Nevertheless, brain and sensory function do decline, in part because of degenerative changes that occur within the cell bodies of the remaining neurons. Virtually every neural activity is affected, including balance, motor skills, hearing, sight, smell, and taste. In addition to neural changes, sensory organs change as well. For example, hair cells in the inner ear stiffen, reducing hearing, and the lens of the eye stiffens, making it more difficult to focus on near objects. As a result of these many changes, elderly people become prone to injury. In addition, they may fail to eat properly because food seems less appealing.

Reproductive and endocrine systems Female reproductive capacity ceases when a woman reaches **menopause** (the cessation of ovulation and menstruation) in her late 40s or early 50s. Menopause occurs because the ovaries lose their responsiveness to LH and FSH and slowly reduce their secretion of estrogen. The onset of menopause may be gradual, with menstrual periods becoming irregular over several years.

The symptoms of menopause are variable. As a result of the decline in estrogen, some women experience "hot flashes" (periods of sweating and uncomfortable warmth). There is a tendency for vaginal lubrication to decline after menopause, making intercourse potentially less comfortable. Mood swings are not uncommon. However, female sexual

MJ's Human Biology Blog

Hormone Replacement Therapy Revisited

Back in the 1990s the federal government launched the Women's Health Initiative to investigate the health of older women. One of its goals was to determine the risks and benefits of hormone replacement therapy (HRT) in postmenopausal women. The HRT study was stopped early (in 2002) when it was discovered that HRT led to an increased risk of breast cancer, heart attacks, and strokes.

What about the women who were on HRT during the study—do they remain at higher risk even after discontinuing HRT? The good news is apparently most of the increased risk goes away with time. The first follow-up study indicates that three years after HRT is discontinued the increased risk of heart attacks disappears, and the increased risk of cancer declines significantly. Future reports will show whether any increased risk of cancer remains as time passes.

Current recommendations for HRT remain unchanged; women should consider HRT only if they have moderate to severe postmenopausal symptoms, and for the shortest time and lowest dose that is effective. ∎

Reference: Heiss, Gerardo et al. (2008). Health Risks and Benefits 3 Years After Stopping Randomized Treatment with Estrogen and Progestin. *JAMA* 299: 1036–1045, Mar. 5.

desire (libido) may be undiminished because the source of female androgens is the adrenal glands, not the ovaries. After menopause, women are at increased risk for cardiovascular disease and osteoporosis. Hormone replacement therapy with estrogen and progesterone is sometimes prescribed for postmenopausal women to reduce menopausal symptoms and other health problems, although the therapy itself adds an additional health risk.

Men exhibit a long, slow decline in the number of viable sperm as they age. In addition, after age 50 men may take longer to achieve an erection. Nevertheless, both sexes can and do remain sexually active into old age.

The rate of decline of various components of the endocrine system varies widely. Notably, the secretion of growth hormone decreases substantially. However, there is not an appreciable decline in renin, TSH, ACTH, or ADH.

Digestion and nutrition The digestive system continues to function well except that some vitamins are not absorbed as

efficiently. Nutritional requirements decline somewhat due to slower cellular metabolism, smaller muscle mass, and, often, less physical activity. Adequate nutrition can be a problem if the elderly fail to eat a balanced diet. Tooth loss varies widely among individuals, but people who have lost teeth may find it harder to eat fruit and other healthy but fibrous foods.

The liver's ability to detoxify and remove drugs from the body declines significantly. As a result, elderly people often require smaller doses of medications to achieve the appropriate therapeutic blood concentration.

Urinary system Kidney mass declines, and both blood flow and filtration rate may fall by as much as 50% by age 80. However, there is normally such a tremendous reserve of renal functional capacity that the decline is usually of little consequence. The elderly do lose some capacity to respond to extreme dehydration or overhydration, but otherwise the kidneys carry out their functions rather well.

The primary problem in men is usually prostate hypertrophy. The enlarged (hypertrophic) prostate presses against the ureter, making urination difficult. In women the most common problem is incontinence due to a weakened urethral sphincter.

✓ **Quick Check** Which of the above aging-related declines do you think could be minimized, or even temporarily reversed, with regular exercise? Which specific type of exercise might benefit specific age-related declines? ∎

Aging well

It would be easy to get depressed after reading the long list of declines in function that accompany aging. Bear in mind that not all of these developments occur to the same degree in everyone, nor do they all occur at the same age. The effects of aging are normal events in our long and exciting journey through life.

Will a healthy lifestyle that includes exercise and a proper diet slow the aging process? The answer is far from certain. However, regular exercise and healthy nutrition can certainly help us age better. Throughout human history, longevity has been determined primarily by the prevalence of disease, not by the "natural" length of a human life span (if indeed there is one). Exercise and a healthy diet improve cardiovascular and skeletomuscular fitness and reduce your risk of major killers such as cancer and cardiovascular disease. A healthy lifestyle also improves the quality of life, providing a heightened sense of well-being and increased energy and vigor (**Figure 21.19**).

With these benefits it really doesn't matter whether a healthy lifestyle can extend the "natural" human life span from 100 to, say, 105. If you can't live forever, you can at least age well and gracefully.

Figure 21.19 Aging well. Although it is not known for certain whether exercise and proper diet actually delay the aging process, they do improve performance and may provide an enhanced sense of wellness throughout life.

Recap Aging is a complex process that is still poorly understood. The number of times a cell can divide may have an upper limit. In addition, damage may accumulate in cells until they no longer function properly. All organs and organ systems decline in function with advancing age, though not necessarily at the same rate. Even if the aging process cannot be stopped, it is certainly possible to improve human health and wellness throughout life. Regular exercise and healthy nutrition are important factors in aging well. ∎

21.10 Death is the final transition

The final transition in the journey of life is death. In terms of human biology, it is as natural as life itself.

Death is defined as the cessation of life. The death of a complex organism such as a human being is not necessarily accompanied by the immediate death of all its organs or cells. Death occurs when an organ system that is essential for life in the short term fails to function, making it impossible to sustain the life of other systems.

The critical organ systems whose failures lead to death very quickly are the brain (because it controls respiration), the respiratory system (without oxygen, life cannot be sustained), and the cardiovascular system (without oxygen and nutrient delivery to all cells, life ends). Failure of other organs with more long-term functions, such as the kidneys or digestive system, does not necessarily kill immediately. You can live without either of these systems for days as long as their loss does not cause one of the critical systems to fail.

The challenge of defining the precise moment of death, then, is that all of the organ systems don't necessarily die at the same time. Death is a process that begins with the failure of certain life-critical organ systems and then proceeds through the death of other organ systems, organs, and cells until the very last of the body's 100 trillion cells is no longer alive. That could take hours, days, or (with the assistance of life-support systems) even years.

Although death may occur gradually, our society requires a practical way to define the moment at which an individual dies. We need a definition of death so the living can begin to grieve and make plans, and so health care workers can stop the medical treatments that sustain life. We need a legal definition so that the police and courts can make a distinction between assault and murder, and so that estates can be settled. A precise definition of death is especially important because organs that might still be alive can be transplanted into another person and save a life.

Legal and medical criteria for declaring a person dead vary according to state law. Basically, all such criteria attempt to define the moment when, to the best of our knowledge and abilities, the person no longer has even a remote possibility of sustaining life. In general they define death as either (1) *irreversible cessation of circulatory and respiratory functions,* or (2) *irreversible cessation of all functions of the entire brain, including the brain stem.* Note that the definitions focus on critical systems that we cannot live without for more than a few minutes.

The key, of course, is the word "irreversible." Usually it is easy to tell when a person has had no respiration or blood flow for 10 minutes or so, long enough to cause brain death. Brain death alone, especially in a person on cardiopulmonary life support, is harder to define. Medical professionals generally run down a checklist of criteria that must be met in order to declare a person's brain nonfunctional, and even then they usually repeat the tests 6–12 hours later before they are willing to declare irreversible brain death. Even for the medical profession, the final transition in life can be difficult to define.

Recap Death is the termination of life. Death starts with the failure of one or more critical organ systems and ends ultimately with the inability to maintain an internal environment consistent with cellular life. ∎

Chapter Summary

Fertilization begins when sperm and egg unite *p. 486*

- Sperm and egg meet (and fertilization occurs) in the upper third of the oviduct.
- Enzymes in the head of the sperm create a pathway through the corona radiata and the zona pellucida to the egg cell membrane.
- Entry of the sperm nucleus triggers the completion of meiosis II by the secondary oocyte. Thereafter the nuclei of sperm and egg fuse, creating a single diploid cell, the zygote.

Development: Cleavage, morphogenesis, differentiation, and growth *p. 489*

- Development begins by a process called *cleavage* and then proceeds with morphogenesis, differentiation, and growth.
- The periods of development prior to birth are known as pre-embryonic (the first two weeks), embryonic (weeks three through eight), and fetal (weeks nine to birth).

Pre-embryonic development: The first two weeks *p. 490*

- During the first two weeks of development, the single cell develops into a hollow ball with an embryonic disk in the center that will eventually become the embryo.
- At about one week the pre-embryo (now called a blastocyst) begins to burrow into the uterine wall, a process called implantation.

Embryonic development: Weeks three to eight *p. 491*

- Embryonic development is marked by the presence of three primary germ layers (ectoderm, mesoderm, and endoderm) in the embryonic disk.
- During development the embryo is completely surrounded by two membranes. The amnion contains amniotic fluid, and the chorion develops into fetal placental tissue. Two other membranes (the allantois and the yolk sac) have temporary functions only.
- By the fifth week the embryo begins to take on distinctly human features.
- By the eighth week the placenta and umbilical cord circulation are fully functional, and male or female gonads have begun to develop.

Gender development begins at six weeks *p. 494*

- Until six weeks an embryo is "sexually indifferent."
- The presence or absence of a Y chromosome determines whether the embryo will become phenotypically male or female.

Fetal development: Nine weeks to birth *p. 496*

- Months 3–9 are marked by rapid growth and development of the organ systems. By the fifth month the fetus begins to move.
- By the sixth month, life outside the womb is possible with good medical care.

Birth and the early postnatal period *p. 496*

- Birth occurs at about 38 weeks of development (nine months).
- The three phases of labor are dilation, expulsion, and afterbirth.
- Shortly after the newborn takes its first breath, the newborn's cardiovascular system undergoes substantial changes. Within hours or a few days the umbilical vessels regress and the ductus arteriosus and foramen ovale close, rerouting all blood through the pulmonary circulation.

From birth to adulthood *p. 500*

- Human neonates are relatively helpless because the nervous and muscular systems are not yet mature.
- There is a disproportionate increase in the development of the musculoskeletal and nervous system during infancy. The immune system remains relatively immature.
- Body shape changes in childhood. The brain achieves 95% of its adult size.
- Adolescence is marked by the last spurt of rapid growth and the attainment of sexual maturity.

Aging takes place over time *p. 501*

- Most cells have an internal mechanism that limits the number of times they can divide.
- Cumulative unrepaired cellular damage as a consequence of metabolic activity may limit the life of cells.
- Aging may affect all organ systems because a decline in one system will affect other systems.

Death is the final transition *p. 505*

- Death is a process. Death begins with the failure of one or more critical organ systems, leading to the failure of other organ systems and eventually to the death of all cells.

Terms You Should Know

amnion, *491*
amniotic fluid, *491*
blastocyst, *490*
cesarean delivery, *498*
chorion, *492*
colostrum, *499*
ectopic pregnancy, *490*
embryo, *491*
embryonic disk, *490*
endoderm, *491*
fertilization, *488*

fraternal twins, *488*
human chorionic gonadotropin (hCG), *492*
identical twins, *488*
menopause, *503*
miscarriage, *494*
morphogenesis, *489*
placenta, *493*
puberty, *501*
umbilical cord, *493*
zygote, *486*

Concept Review

Answers can be found at the Human Biology Place.
www.humanbiology.com

1. Why does only one sperm fertilize the egg, and why is this important?
2. What determines whether an embryo will develop into a male or a female?
3. Describe the function of the placenta and the hormones that it secretes.
4. Describe the stages of labor and delivery.
5. What causes the newborn to take its first breath immediately after birth?
6. Why aren't all of a child's vaccinations given in the first year?
7. What causes the onset of puberty (sexual maturation)?
8. Summarize the three main hypotheses of the causes of aging.
9. What has caused the increase in longevity of the U.S. population during the past 100 years?
10. Why is it sometimes so hard to define the moment when an individual dies?

Test Yourself

Answers can be found in Appendix A.

1. Which of the following lists the correct order of structures through which a sperm must pass to accomplish fertilization?
   a. zona pellucida, corona radiata, secondary oocyte plasma membrane
   b. zona pellucida, secondary ooctye plasma membrane, corona radiata
   c. ovum plasma membrane, corona radiata, zona pellucida
   d. corona radiata, zona pellucida, secondary oocyte plasma membrane

2. Which of the following will result in fraternal twins?
   a. the fertilization of one egg by two different sperm
   b. the division of a fertilized egg into two masses after differentiation has begun
   c. the fertilization of two eggs by two different sperm
   d. the division of a fertilized egg into two masses before differentiation has begun

3. Which of the following processes is the first to occur as a fertilized egg begins its journey down a uterine tube toward the uterus?
   a. differentiation   b. cleavage
   c. growth   d. all of these choices

4. A sperm with a defective acrosome would not be able to:
   a. swim   b. penetrate the egg
   c. penetrate through the cervical mucus
   d. provide the energy for the sperm to swim

5. Which of the following indicates the correct order of pre-embryonic developmental stages?
   a. morula, blastocyst, embryonic disk
   b. blastocyst, morula, embryonic disk
   c. trophoblast, blastocyst, morula embryonic disk
   d. embyronic disk, blastocyst, morula

6. Which of the following is a correct statement?
   a. The placenta is formed entirely from embryonic tissue.
   b. The umbilical cord brings maternal blood to the embryo.
   c. The embryonic/fetal blood and the maternal blood mix within the placenta.
   d. The placenta secretes hormones including HCG and progesterone.

7. Which of the following is mismatched?
   a. neural groove: spinal cord
   b. mesoderm: gastrointestinal tract lining
   c. urogenital groove: urethra
   d. placenta: chorion and endometrium

8. Which of the following events is correctly matched with the month of prenatal development?
   a. kidneys producing urine: month 3
   b. heart pumps to circulate blood: month 4
   c. gender development is completed: month 5
   d. bone marrow takes over blood cell production: month 7

9. Which fetal hormone plays a role in initiating labor?
   a. estrogen   b. ACTH
   c. progesterone   d. oxytocin

10. Which of the following would be detrimental to the health of a neonate?
    a. the ductus arteriosus stays open
    b. the foramen ovale closes
    c. the ductus venosus closes
    d. both (a) and (b)

11. _____ stimulates the production of milk while _____ causes ejection of milk from the breast.
    a. Estrogen…prolactin
    b. Prolactin…oxytocin
    c. Estrogen…oxytocin
    d. Oxytocin…prolactin

12. Possible activities of telomerase could be:
    a. stopping the aging process by rebuilding the telomeres
    b. contributing to cancer by allowing tumor cells to continue dividing indefinitely
    c. blocking cancer by restoring telomeres after cell divisions
    d. both (a) and (b)

13. The aging of which organ or organ system is least problematic during the normal aging process?
    a. bones   b. immune system
    c. nervous system   d. kidneys

14. All of the following are theories regarding causes of aging except:
    a. Cells run out of telomeres and stop dividing.
    b. Cortisol production increases and the stress response contributes to aging.
    c. Cells can no longer repair themselves properly, and damage accumulates.
    d. A decline in one critical body system induces parallel declines in others.

15. Which hormone plays a critical role both in childbirth as well as breast-feeding?
    a. oxytocin   b. prolactin
    c. estrogen   d. progesterone

Apply What You Know

Answers can be found at the Human Biology Place.
www.humanbiology.com

1. Why is it biologically impossible for humans and gorillas to mate and produce offspring?

2. Conjoined twins are twins that are attached to each other. They may even share limbs and organs. From what you know of early human development, how do you think conjoined twins come about?

3. Why are many sperm but only one egg produced in the process of fertilization? Would humans be able to reproduce if the cell numbers were inverted?

4. Each year it is estimated that over 40,000 newborn babies are impacted by the effects of alcohol consumed by the mother during pregnancy. Fetal alcohol syndrome is one of the leading causes of preventable birth defects. Why do you think alcohol is more likely to damage the fetus than the mother?

5. What exactly do home pregnancy tests test for as an indication that a woman is pregnant?

6. Babies born before a full nine months of gestation (premature babies) are often at a high risk of death. What is the earliest a baby can be born and stand a good chance of survival, and which organ system is generally the limiting one?

7. How is it possible that one's life could be prolonged without slowing the aging process?

22

Evolution and the Origins of Life

Skulls of human ancestors ranging from 3.0 million years ago (lower left) to 22,000 years ago (lower right).

Who Were the Flores People?

In 2004, archaeologists unearthed a female skeleton from Liang Bua Cave on the Indonesian island of Flores. They named the skeleton LB1, for Ling Bua #1. LB1 was barely 1 meter tall (about 3.3 feet) and weighed only about 25 kg (55 pounds). She was about 30 years old when she died.

Although similar in size and limb proportions to ancient hominids such as *Australopithecus afarensis*, LB1 bore an eerie resemblance to more recent members of the genus *Homo* in her upright walking posture and chewing ability. Her skull and brain were small, comparable to those of a chimpanzee. So archaeologists were

Liang Bua Cave, where the discovery of the Flores skeleton was made.

astounded when her bones were dated to only 18,000 years ago—a time when *Homo sapiens* were thought to be the only modern humans living on Earth. Subsequently, bone fragments and teeth of 14 other individuals were discovered in the same cave. The bones ranged in age from approximately 95,000 to 12,000 years, indicating a long and successful history in that area.

A New Human Species?

The discoverers proposed that the Flores people represent a new species descended from *Homo erectus*, which they named *Homo floresiensis*. They

LB1, the only nearly complete skeleton of *Homo floresiensis*.

clearly were advanced in their activities and social structure. For instance, they would have had to cross at least two water barriers from mainland Asia to reach Flores Island. Furthermore, they crafted an impressive variety of stone tools. Though most of the implements were relatively simple, some were complex tools comparable to those honed by prehistoric *Homo sapiens*. Many tools were found mingled with the bones of *Stegodon*, a 2,200-pound dwarf relative of modern elephants that, although small for an elephant, would have posed a significant challenge to hunters only 3 feet tall. This indicates that the Flores people engaged in group hunting activities that required planning and cooperation.

One hypothesis is that the brain of the Flores people became proportionally smaller only *after* the Flores people arrived on the island of Flores, not before. However, this runs counter to the observation that as human species become dwarfed from a larger ancestor, brain capacity doesn't necessarily decline. For example, there is not much difference in brain size between the Masai people of East Africa, who are over six feet tall, and the Bambuti people of the Congo, who are barely 4-1/2 feet tall. How, then, could the Flores people have descended from larger-brained ancestors and lost so much brain size?

A Unique Body Form

The LB1, the most complete Flores skeleton to date, has a number of unique physical characteristics. She had heavy brow ridges, a sloping forehead, and a small chin. Her arms were long but her lower extremities were short. But the real differences were in her hands and feet. Modern humans have a complex of five bones in the wrist that act as a shock absorber when we use the wrist forcefully, such as pounding with heavy tools or doing precision work. The shape of the wrist bones of LB1 more closely resemble those of an ape, or of *Homo habilis* who died out more than 1.6 million years ago.

The most striking feature of LB1 has to be her feet. Although her incredibly short big toes are in line with the rest of the toes, as they are in modern humans, the feet are primitive, lacking a proper arch. Most remarkable is the length of her feet—fully 70% as long as her thigh bone, compared to about 55% in modern humans. The feet of the Flores people were so long that they would have walked with a high-stepping gate and been poor runners. Taken together, the skull, hands, and feet of LB1 suggest that the Flores people probably diverged from the line that became modern humans sometime back about the time of *Homo erectus*, and then managed to survive until almost the present day.

Current thinking seems to be that the Flores people are sufficiently distinct to warrant their own Genus and species name— *Homo floresiensis*. Their features are primitive, indicating an early

Skulls of *Homo floresiensis* (left) and *Homo sapiens* (right).

divergence from other humans, and yet they managed to live on Earth with us until only 12,000 years ago. Their brains have probably always been small, and as a result they have shaken up anthropologists' thinking about the importance of brain size in human development.

Although the pendulum seems to be swinging in favor of calling the Flores people a separate species, don't expect the controversy to be over any time soon. So far all we have is one skull, a partial skeleton, and bone and teeth fragments from only about 14 individuals. It may be a while before we know for sure who the Flores people really were. In the meantime, it's likely to remain a contentious issue among anthropologists.

proposed that the Flores people became smaller than their *Homo erectus* ancestors as a result of an evolutionary adaptation to limited food supplies on the island, a phenomenon known as "Island dwarfism."

Almost immediately, anthropologists began arguing about whether the Flores people were, or were not, a new species. The most controversial aspect of LB1 was her brain size—about the size of a modern chimpanzee and only about a third the size of a modern human's. A large brain size, it has long been argued, is required for sophisticated tool-making and complex social organizations. And yet, the Flores people

Questions to consider

1 Do you think the Flores people should be classified as a separate species? Why or why not?

2 Do you think it's important that we try to discover the evolutionary history of humans? Who should finance anthropological research, and why?

The facts...

- Archaeologists recently discovered partial remains of up to 14 individuals of a small human species that became extinct only 12,000 years ago.

- Discoverers named this new species *Homo floresiensis*.

- The only skull found has a brain size comparable to that of a modern chimpanzee. However, these were intelligent people, as evidenced by their tools and their hunting abilities.

- The discovery challenges anthropologists' current beliefs about the relationship between brain size and human mental capacity and behavior.

>> The *theory of evolution* has been the best explanation for the complexity and diversity of life on Earth for over 150 years. Although the theory has undergone some modification over time, it has withstood all scientific examination and challenge.

>> Mutations coupled with natural selection cause slow evolutionary change in life-forms over time. Mass extinctions have dramatically affected the course of evolution at least five times during Earth's existence.

>> Life originated on Earth about 3.5 billion years ago, only a billion years after Earth was formed. The first life-forms probably originated when self-replicating molecules of RNA and then DNA became enclosed within a lipid-protein membrane.

>> Our human ancestors diverged from other primates about 4–5 million years ago. Chimpanzees are our closest living relatives.

>> Modern humans are classified as genus *Homo*, species *sapiens*. *Homo sapiens* originated in Africa around 140,000 to 100,000 years ago. Our two closest *Homo* relatives, *Homo neanderthalensis* and *Homo floresiensis*, became extinct only 28,000 and 13,000 years ago, respectively.

■ **Unpredictable and natural**. Evolution is affected by chance, *natural selection*, historical events, and changing environments.

Microevolution refers to evolution as a result of genetic changes that give rise to new species. **Macroevolution,** by contrast, describes large-scale evolutionary trends or changes that apply to whole groups of species, often as a result of changing environments or major historical events.

The theory that evolutionary processes shaped life on Earth is associated with Charles Darwin, a British naturalist of the mid-1800s. The heart of Darwin's hypothesis is that life arose only once, probably in the primordial sea long ago, and all life as we know it descended from that early life-form. Darwin actually used the word evolution sparingly, preferring instead "descent with change" or "descent with modification." Key to Darwin's hypothesis is the idea that descent with modification resulted from "natural selection," which we describe later in this chapter.

Scientists reserve the word "theory" for those few major concepts that offer the best explanation to fit a broad range of established facts. In that sense, scientists use the word "theory" very differently from opponents of evolution, who often say that evolution is "only a theory," as if the whole concept could be dismissed with a wave of the hand. Opponents of evolution also sometimes say "Darwin was wrong" by pointing to how specifics of his original explanation have had to be modified after extensive testing. Nevertheless, the *theory of evolution* remains the single most unifying concept in all of biology.

In this chapter we review the scientific evidence on which we base our understanding of evolution and discuss processes that affect evolutionary change. Then we pick up the intriguing question of life's origin on Earth. Finally, we trace the relatively recent evolution of human beings and consider the features that make us distinctively human.

↺ **Recap** Evolution is a process of descent over time with genetic modification. First proposed by Charles Darwin in the 1800s, the theory of evolution remains the single most unifying concept in biology. ■

22.1 Evidence for evolution comes from many sources

The evidence that humans and all other life-forms have evolved over time comes from several sources, including the fossil record, comparative anatomy and embryology, biochemistry, and biogeography.

The fossil record: Incomplete but valuable

Fossils, the preserved remnants of organisms, are one of our richest sources of information about life-forms that lived in the past. Soft tissues decay quickly, so most of the fossil record consists of bones, teeth, shells, and occasionally spores and seeds. Fossils of soft-tissue organisms such as jellyfish and bacteria do exist, but they are rare. In general, remnants of organisms are preserved only if they are covered soon after

ook around and you'll be impressed by the astounding variety of forms that life takes. From bacteria to beagles, hagfish to humans, living creatures come in a huge range of sizes, shapes, and colors. And yet, the evidence indicates that all living creatures on Earth are related. To the best of our knowledge, all of them descended from a single simple life-form that arose from the hot, steamy environment of Earth more than 3 billion years ago.

How did Earth's inhabitants come to be so different from each other? The diversity of life is thought to result from **evolution,** defined as an unpredictable and natural process of descent over time with genetic modification. The formal definition has three key elements.

■ **Descent over time**. As one generation of organisms gives rise to the next, populations of organisms undergo slow change that makes them different from their ancestors. Sometimes a population becomes two populations that are so different from each other that they no longer interbreed, giving rise to new species.

■ **Genetic modification.** The process of change depends on changes to the genes of the organisms.

a) **A body will decompose unless it is covered by sediment or volcanic ash.**

b) **Hard elements of a body may be preserved from decomposition if they are covered quickly.**

c) **Over time more layers of sediment, ash, or soil are deposited.** The hard elements of the organism become mineralized by the same minerals that comprise rocks.

d) **Occasionally erosion, uplifting of the earth's crust, or human excavation may expose fossils to the surface.**

e) **A fossilized human skeleton.**

Figure 22.1 **How fossils are formed.**

death with layers of sediment or volcanic ash (Figure 22.1). Over time the remains become mineralized with some of the same minerals that compose rocks, leaving a rocklike impression of the hard tissue of the organism, or a fossil.

Even though we have found fossils from over 200,000 species, there may be millions of species for which we will never have any record because they had no hard tissues. We have many fossils of ancient bony fishes, for example, but only a few fossils of jellyfish from the same time periods. We are also more likely to discover fossils on land (there may be many more, and possibly different, fossils deep under the sea) and to find fossils that were produced in great numbers. For example, we are more apt to find a seashell or spore from 100,000 years ago than to uncover the skeleton of an early human from the same time period. By its nature, then, the fossil record tends to be incomplete and heavily weighted toward fossils that are easy to find.

One of the most important pieces of information about a fossil is its age, which places it at a certain point in evolutionary history and shows how body structures have changed over time. Scientists know that sediment and volcanic ash are laid down layer upon layer, a process known

as stratification. Where fossils exist in the same location in different layers, it is usually possible to tell which one came first. This helps scientists establish relative relationships between fossils over time (Figure 22.2).

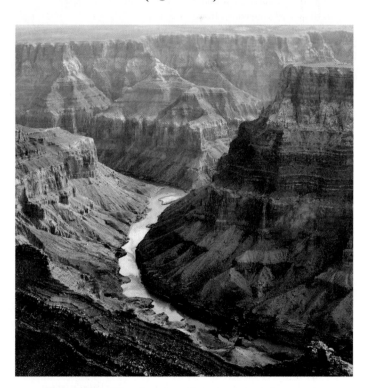

Figure 22.2 **Stratification of sedimentary deposits.** The rock layers of the Grand Canyon were formed by sequential deposits of sediments. The oldest layers at the bottom of the canyon are over 1.2 billion years old.

In addition, radiometric dating and other techniques can be used to date rocks and fossils. Radiometric dating takes advantage of the fact that some atoms, called radioactive isotopes, are inherently unstable. Radioactive isotopes radiate or give off particles until they reach a more stable state, sometimes becoming a different, more stable atom.

The isotope most often used to date the oldest fossils is radioactive potassium, which is slowly converted to argon, a gas. The half-life of radioactive potassium is 1.3 billion years, meaning that it takes 1.3 billion years for half of the radioactive potassium in a fossil to become argon. It would take another 1.3 billion years for half of the remaining radioactive potassium to become argon, and so on. The argon formed from the decay of radioactive potassium remains trapped in the fossil, so the ratio of radioactive potassium to argon in a fossil can be used to estimate the fossil's actual age, in billions of years. Radioactive potassium is of limited usefulness for dating most animal and plant fossils, however, because they are less than 600 million years old.

At the other end of the age spectrum, fossils newer than 50,000 years old are dated using carbon-14, the radioactive isotope of carbon, which decays with a half-life of only 5,700 years.

✅ **Quick Check** You find an interesting fossil of a tiny unicellular organism that is embedded in volcanic rock. Measurement of potassium-argon ratios shows that the volcanic rock contains 1/4 of the radioactive potassium that it originally had when it first solidified. Approximately how old is this rock (and the fossil that it contains)? ▪

Comparative anatomy and embryology provide more evidence

Additional evidence for evolution comes from comparing the anatomy of animals and the development of their embryos. When examining them in the context of evolution, scientists describe anatomical structures as homologous, analogous, or vestigial.

When different organisms share similar anatomical features, often it is because they evolved from a common ancestor. For example, the fins of some ancient fish and the legs of all living four-limbed vertebrates share the same basic arrangement of bones. This is true even though vertebrate forelimbs have

Answers to ✅ questions can be found in Appendix A.

undergone numerous modifications so animals can use them in a variety of specialized ways, such as flying, swimming, running, or grasping objects (**Figure 22.3**). The shared arrangement of bones is evidence that ancient groups of fish include the ancestors of all land vertebrates. **Homologous structures** are body structures that share a common origin. The degree to which homologous structures resemble each other can be used to infer the closeness of evolutionary relatedness.

In contrast, structures that serve the same function but do not arise from a common ancestor are called **analogous structures** (**Figure 22.4**). Good examples of analogous structures are the wings of birds and the wings of insects. Both types of wings evolved for flight. Nevertheless, they are not structurally similar, and birds and insects are not closely related.

Finally, **vestigial structures** serve little or no function at all. Vestigial structures may be homologous to body parts in other organisms, where they do still have an important function. The human coccyx (tailbone) is the vestigial remnant of a tail, suggesting that we share a common ancestry with other vertebrates with tails. The muscles that enable you to wiggle your ears are vestigial in humans. Other animals, such

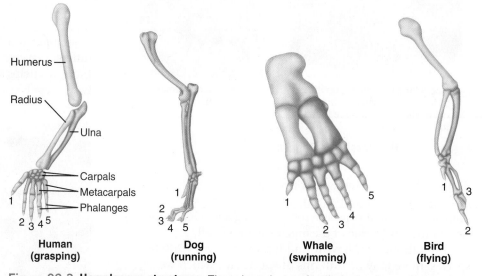

Figure 22.3 Homologous structures. These homologous forelimb structures share a common evolutionary origin. Numbers indicate digits. Notice how homologous bones (indicated by similar colors) have undergone modification in order to perform different functions in different vertebrates.

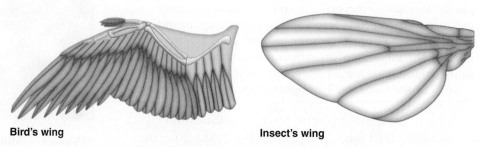

Figure 22.4 Analogous structures. The bird's wing and the insect's wing are both used for the same function, but they evolved from entirely different structures.

as dogs and deer, still use these muscles to rotate their ears toward a sound so that they can hear it better.

Compelling evidence for evolution also comes from comparisons of the embryos of animals, especially vertebrates. The vertebrates are a varied group: fishes, amphibians, reptiles, birds, and mammals. Yet despite their diversity, their early embryonic development follows the same general pattern (**Figure 22.5**). For example, all vertebrate embryos develop a primitive support structure called a notochord and a series of folds called somites that will become bone, muscle, and skin. But in addition, all vertebrate embryos develop a series of arches just below the head. In fishes and amphibians these are called

gill arches because they develop into gills. In humans the homologous *pharyngeal arches* become parts of the face, middle ear, and mouth. The most likely explanation for these common embryonic structures with completely different functions is common ancestry. Later in embryonic and fetal development the paths of morphogenesis and differentiation diverge, giving us the variety of vertebrates we know today. It is almost as if embryonic and fetal development represent an acceleration of the entire evolutionary history of the vertebrates.

 Quick Check Compare the dorsal fin (the stiff fin on the back) of a dolphin with that of a shark. Given that dolphins are mammals, do you think these are homologous or analogous structures? Explain. ▪

a) **Amphibian (frog).**

b) **Bird (chick).**

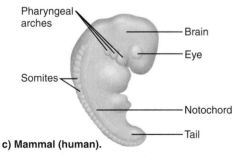

c) **Mammal (human).**

Figure 22.5 Early developmental similarity among vertebrate embryos. The embryos of all vertebrates tend to share certain developmental features, suggesting a common ancestry. In particular, all vertebrate embryos have either gill or pharyngeal arches. These arches ultimately have very different functions.

 In addition to the pharyngeal arches, human embryos have several other traits that disappear before birth. Three are visible in this diagram; can you find at least one of them? Propose an explanation for why we retain these traits as early embryos but lose them before birth.

Comparative biochemistry examines similarities between molecules

Proteins and genes also provide clues to the relationships among species. When two species possess identical or nearly identical biochemical molecules, common ancestry may be indicated. In principle, the two similar molecules are akin to homologous structures—you might call them "homologous molecules." They arise when evolutionary paths diverge, meaning that one species becomes two different species. Mutations occurring over long periods of time gradually modify the original molecule in each of the two new species. The greater the difference between two molecules serving the same function in two species, the earlier the two species probably diverged from a common ancestor.

For example, cytochrome *c*, an important protein in the metabolic pathways for energy exchange, is found in organisms ranging from yeast to humans. In humans, cytochrome *c* consists of 104 amino acids. A chimpanzee's cytochrome *c* is identical to that of a human, but they both differ by 1 amino acid from rhesus monkeys, 16 amino acids from chickens, and over 50 amino acids from yeast. From this data you could hypothesize that humans are more closely related to chimpanzees than to rhesus monkeys, and also that humans are more closely related to other primates than to chickens or yeast.

Similarities in other protein sequences and in DNA are also used as indicators of common ancestry. The degree to which DNA base pair sequences resemble each other helps scientists infer ancestry and relationships between humans and other species.

Biogeography: The impact of geographic barriers and continental drift on evolutionary processes

Biogeography is the study of the distribution of plants and animals around the world. Why do we find certain animals and plants in some locations on Earth and not in others? One explanation is that migrations are inhibited in regions

Figure 22.6 Geographic barriers. Animal and plant migrations are inhibited by barriers such as these mountains and desert sands in Death Valley, California.

isolated by barriers such as water, high mountain ranges, or deserts (Figure 22.6).

Where migrations of plants and animals were possible, the evidence indicates that closely related life-forms often evolved in one location and then spread to regions that were accessible to them at the time. Eventually their spread may have been blocked by physical barriers or by environments in which they could not survive, such as deserts or oceans. There are no snakes on the Hawaiian Islands, for example, because snakes evolved long before the volcanic islands existed, and the Pacific Ocean represents an effective barrier to snake migration. When we see how easily some plants and animals adapt (adjust to new conditions) when placed in new locations, we can appreciate the impact of geography on evolutionary processes and the distribution of life-forms.

Geologists tell us that Earth's continents are located on huge plates called tectonic plates, whose locations on the planet surface are not fixed. The plates slowly move over time, a process called **continental drift.** About 200 million years ago, all the continents were joined in one interconnected landmass called *Pangaea* (Figure 22.7). The continents drifted apart after the first animals and plants appeared. As a result, related groups of organisms that were isolated from each other evolved separately, but along almost parallel paths.

For instance, Australia separated from the other continents about 65 million years ago. Nevertheless, there are striking similarities between the unique marsupial (pouched) mammals of Australia and the predominant placental mammals (mammals that nourish their young with a true placenta) found on all other continents. These similarities exist because the two types of mammals evolved from common ancestors and also because evolution continued after separation, solving some of the same environmental challenges in the same way in both groups.

HBP **Web Animation** *Biogeography and Continental Movement* at www.humanbiology.com

Recap Comparisons of homologous anatomical structures, embryologic development, and DNA and protein structures can be used to estimate the closeness of relationships between different species. Fossils are the mineralized remains of the hard tissues of past life-forms. The fossil record provides compelling evidence for evolution. Radiometric dating can be used to date rocks and fossils. The slow drift of continents over the past 200 million years and physical barriers such as mountain ranges and deserts influence the evolution and distribution of life-forms. ■

HBP **Web Animation** *Principles of Evolution* at www.humanbiology.com

22.2 Natural selection contributes to evolution

Recall that an organism's DNA represents the set of instructions for all its life processes. If all creatures had an identical set of DNA they would all *be* identical in form and function. Differences in DNA account for the differences among species and even among individuals of a species.

Random mutations underlie evolution

A species' gene pool changes very slowly over time (measured in thousands to millions of years) as a result of random mutations to the genes of individual members of the species. Mutations are rare accidental events that produce a slightly different form of a gene, called an allele. Alleles that are not

a) 200 million years ago.

b) 90 million years ago.

c) Present.

Figure 22.7 The formation of Earth's continents from a common landmass. The continents were once connected in one giant landmass called Pangaea.

lethal may be passed on to future generations, to become part of the species' collective gene pool. Eventually these accumulating mutations may cause changes in the species' physical and functional traits. Ultimately, an accumulation of mutations may cause one species to diverge into two.

Without random mutations, there could be no evolution.

Natural selection encourages changes in the gene pool

Mutations alone do not amount to evolution. Living organisms interact with their local environments, and as a consequence of that interaction, some live long enough to reproduce and some do not. The frequency with which any allele is found in the gene pool of a species depends critically on whether that allele confers survival value on its holder.

Darwin understood this principle even though he did not fully comprehend the genetic basis for it. He referred to it as "survival of the fittest" by means of natural selection. By **natural selection,** Darwin meant that individuals with certain traits are more fit for their local environment and therefore are more likely to survive and reproduce. Mutations coupled with natural selection produce changes in the gene pool of a population.

 Quick Check A friend tells you that evolution occurs because of "good mutations." Explain what is wrong with this statement. ∎

Genetic drift and gene flow alter populations

A **population** is a group of individuals of the same species that occupy the same geographical area. Darwin was convinced that the fittest members of a population were always the ones to survive. We now know that there are several other contributors to evolution besides natural selection. Other factors that affect the gene pool of a population include genetic drift, gene flow, and selective hunting.

Genetic drift refers to random changes in allele frequency because of chance events. Genetic shift is more likely to occur in smaller populations than in large ones. One cause of genetic drift is the *bottleneck effect*, which occurs when a major catastrophe—such as a fire, a change in climate, or a new predator—wipes out most of a population without regard to any previous measure of fitness. When this happens the few genes left in the gene pool may no longer represent the original population, nor do they necessarily represent the most fit genes. An example of a species suffering from a severe bottleneck effect may be the cheetah. The 15,000 surviving cheetahs are all genetically similar, and their genomes include a few harmful alleles that affect fertility. It is not clear whether cheetahs will survive as a species.

Another cause of genetic drift, the *founder effect*, occurs when a few individuals leave the original group and begin a new population in a different location, or when some environmental change isolates a small population from a larger

one. Again, the new gene pool may not be a representative sample of the original population.

Differences in the gene pool of a particular population are also affected by movement of individuals into (immigration) or out of (emigration) the population. This geographical redistribution of alleles, called *gene flow*, tends to mix pools of genes that might not otherwise mix. The prevalence of international travel today has markedly increased the possibility of gene flow in the human population.

Certain human activities may also impact evolutionary processes. For example, the distinctly human practice of hunting big game animals for trophies specifically selects for larger, more fit males. How this practice might affect evolution is worth considering.

The slow process of genetic drift in higher organisms is quite different from what is known as *antigenic shift* in viruses. Antigenic shift is a very rapid and dramatic change in a virus, brought about when two or more viruses combine or exchange genetic material to form a new virus with entirely different properties. The possibility of antigenic shift is what makes viruses such as those causing bird flu or swine flu so potentially dangerous.

 Web Animation *Agents of Change* at www.humanbiology.com

 Quick Check Recently, Florida panthers have been reduced to an extremely small population size of just a few dozen individuals. The few Florida panthers that survive today often carry an allele which causes bent tails, although bent tails were rare in the past. Explain which evolutionary process is most likely to have caused the change in the frequency of bent tails, and state whether it is a random or nonrandom process. ∎

Mass extinctions eliminate many species

An **extinction** occurs when a life-form dies out completely. Throughout evolutionary history, major catastrophic events have sometimes wiped out whole groups of species regardless of fitness. In the last 530 million years (the only period for which we have a good fossil record) there have been at least five mass extinctions, each of which destroyed more than 50% of the species existing at that time.

The largest mass extinction, called the Triassic, occurred around 200–250 million years ago. It was accompanied by a period of high carbon dioxide concentration and substantial global warming, and it killed 70% of all land-dwelling species and 90% of all ocean-dwelling species. The most recent and best-known mass extinction, called the Cretaceous, claimed all of the dinosaurs about 65 million years ago. Some scientists attribute the Cretaceous mass extinction to a dramatic change in climate after a comet or asteroid collided with Earth near what is now Cancun, Mexico. It appears that we may now be entering a sixth mass extinction, this one due to human activity.

Evolutionary trees trace relationships between species

An **evolutionary tree** (also called a phylogenetic tree) illustrates evolutionary change and relationships among species. In an evolutionary tree each branch represents a point of divergence between two species (the creation of a new species), and the length of the branches represents time (Figure 22.8). If a species becomes extinct, the branch for that species ends short of the present time.

When conditions are right, many new species may develop in a relatively short time from a single ancestor. Such short bursts of evolutionary activity are called **adaptive radiation** and are shown as numerous branches from a single point on an evolutionary tree. Periods of adaptive radiation most commonly follow changes in the environment that create new habitats.

Populations of living creatures are constantly undergoing evolutionary change. Sometimes it is necessary to further subdivide species into subspecies that look slightly different but can still interbreed. At any given time some species are becoming extinct while others are in the process of diverging to become several new species. Evolutionary trees can be general, covering the entire history of all life-forms, or they can be detailed and focus on one species or one time period.

Recap Mutations coupled with natural selection provide much of the basis for evolution. In addition, the course of evolution can be affected by genetic drift, antigenic shift, gene flow, and mass extinctions. An evolutionary tree visually depicts relationships between species. ▪

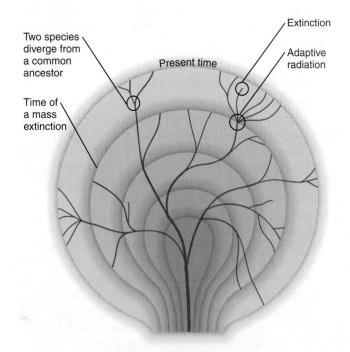

Figure 22.8 An evolutionary (phylogenetic) tree. Phylogenetic trees illustrate evolutionary change and relationships between species over time. Periods of mass extinctions are indicated by changes in color.

22.3 The young Earth was too hot for life

To the best of our knowledge, Earth formed about 4.6 billion years ago. At one time our sun was surrounded by a swirling cloud of gases, dust, and debris. As the outer region of the cloud cooled, it condensed into solid masses that today we know as the planets, one of which was Earth.

Four and a half billion years ago Earth was an inhospitable place. Its tremendous heat melted the interior, which condensed as a core of liquid nickel, iron, and other metals. The solid but thin outer crust was constantly changing as frequent cracks in the surface and numerous volcanoes released molten rock and hot gases. The early atmosphere consisted primarily of carbon dioxide (CO_2), water vapor (H_2O), hydrogen (H_2), nitrogen (N_2), methane (CH_4), and ammonia (NH_3), but no oxygen (O_2). There were no oceans yet because the tremendous heat vaporized all the water. Earth was not yet protected by an ozone layer, so it was constantly bombarded by ultraviolet radiation. The atmosphere crackled with electrical storms.

Over time Earth cooled enough for water vapor at higher elevations to condense as rain. Most of the rain immediately vaporized again as it struck hot rock. The hot, steamy cycle of condensation and vaporization continued for millions of years until the planet's surface cooled enough to permit water to remain liquid. Oceans began to form, but they were warm, small, and not very salty. The land mass remained steamy, hot, and subject to frequent volcanic eruptions and electrical storms. There was still no oxygen in the atmosphere.

Somehow, despite this harsh environment, living organisms appeared about 3.8 billion years ago, less than a billion years after Earth was formed. Now we turn to the question of how life began, certainly one of the most intriguing questions in all of biology.

Recap Earth was formed about 4.6 billion years ago and life appeared about 3.8 billion years ago. The environment at the time was hot, steamy, and devoid of oxygen. ▪

22.4 The first cells were able to live without oxygen

We know from the fossil record that by 3.5 billion years ago single-celled creatures resembling bacteria were present. Once life appeared, evolutionary processes over billions of years produced the varied life-forms that we know today, as well as many more that have long since become extinct.

Organic molecules formed from atmospheric gases

Current scientific evidence indicates that the first step on the path to life was the creation of simple organic molecules from atmospheric gases (Figure 22.9). In today's living organisms

a) The primitive Earth was hot and steamy with small, warm oceans.

b) Simple organic molecules formed from atmospheric gases and dissolved in the oceans. The energy for their formation was provided by electrical storms, heat, and intense ultraviolet radiation.

c) Self-replicating RNA probably formed on clay templates in shallow waters along the shorelines of oceans. RNA and other organic molecules became enclosed in a cell membrane and the first self-replicating cells were formed.

d) The development of photosynthesis created oxygen gas. The presence of oxygen permitted the later development of aerobic forms of metabolism.

Figure 22.9 The origin and early evolution of life.

 Outline an experiment that could test whether complex biological molecules could actually arise via the natural processes shown in this diagram. In your opinion, what is the greatest practical difficulty in replicating these processes in a laboratory experiment?

the formation of organic molecules requires enzymes, but of course enzymes had not yet evolved. In fact, it appears that it was only *because* the early environment was so inhospitable (by our standards) that organic molecules were able to form. Apparently the intense heat, ultraviolet radiation, and electrical discharges produced an extreme amount of energy, enough to combine molecules in the atmosphere into simple organic compounds, even without enzymes. These compounds (amino acids, simple sugars, fatty acids) dissolved in the sea, so that over time the oceans became a warm soup containing organic molecules and a little salt. The absence of oxygen was important, because oxygen is highly reactive and would have broken down the organic molecules as they formed.

✓ **Quick Check** Do you think a brand-new form of life could arise today on Earth's surface, i.e., a form of life newly derived from inorganic molecules and not from preexisting life? Why or why not? ∎

Self-replicating RNA and DNA formed

The mere presence of organic molecules does not create a living organism that can grow and reproduce. How did these compounds join into molecules that could reliably reproduce themselves?

The problem we face in our understanding of this process is this: the only *self-replicating* molecules we know of are primitive single-stranded RNA and our modern double-stranded DNA, neither of which is likely to have formed spontaneously in the sea. Our best guess is that single-stranded self-replicating RNA first formed on templates of clay, in mudflats along the ocean's edge. Indeed, experiments have shown that, under the right conditions of moisture and heat, thin layers of clay can promote the formation of fairly complex molecules. Alternatively, the dry conditions and intense heat near volcanoes may have provided just the right environment for RNA formation.

If RNA was the first self-replicating molecule, the more stable DNA molecule must have developed from RNA very early in evolution. Modern cells use DNA as their self-replicating molecule and RNA to direct the synthesis of proteins, including enzymes. Only viruses use single-stranded RNA as their genetic material, and consequently viruses do not replicate on their own.

The first living cells were anaerobic

At some point, self-replicating molecules and small organic molecules became enclosed within a lipid-protein membrane. We do not know exactly how this happened. However, experiments have demonstrated that, under the right conditions, certain mixtures of amino acids and lipids will spontaneously form hollow, water-filled spheres.

The first simple cells relied on anaerobic metabolism, meaning metabolism without oxygen. Their ability to synthesize compounds was limited, and so they relied on their

MJ's Human Biology Blog

Creating Synthetic Life

Scientists at the J. Craig Venter Institute in Rockville, MD, have succeeded in synthesizing the entire genome of the bacterium *Mycoplasma genitalium*, using only the four nucleotides of which all DNA is composed. It's an astonishing feat, considering that the *M. genitalium* genome is nearly 600,000 base pairs long. The next step, say the researchers, is to insert the synthetic DNA into a cell. If all goes well, the DNA will replicate itself and the cell will divide, becoming the first living, self-reproducing organism ever created synthetically in a laboratory.

The synthesis of the entire genome of an organism paves the way for experiments to determine the minimum number of genes required for life—the minimum operating system, so to speak. Looking further to the future, someday it may even be possible to create synthetic organisms for specific purposes, such as manufacturing medicines or cleaning up the environment. We'd better start thinking about how we want to regulate or control this new technology. ■

Reference: Gibson, Daniel G. et al. (2008). Complete Chemical Synthesis, Assembly, and Cloning of a *Mycoplasma genitalium* Genome. *Science* 319: 1215–1220.

immediate environment for the energy and raw materials they needed. When energy and raw materials were not available they died, or at least they did not replicate.

↩ **Recap** As best we can surmise, the sequence of events leading to self-replicating living cells was (1) the formation of organic molecules from atmospheric gases, (2) the formation of self-replicating RNA on templates of clay, and (3) the enclosure of RNA and organic molecules inside a cell membrane. All early cells used an anaerobic form of metabolism. ■

22.5 Photosynthetic organisms altered the course of evolution

Photosynthesis increased oxygen in the atmosphere

By about 3 billion years ago, certain cells had mutated and developed the ability to produce, internally, some of the molecules they needed. They drew on the abundant CO_2

and water in their environment and the energy available in sunlight to create complex organic molecules containing carbon. This process is called photosynthesis, and its by-product is oxygen. As a result, the oxygen concentration in the atmosphere began to rise. Note that oxygen appeared in Earth's atmosphere only after life began and as a direct consequence of life itself.

Aerobic organisms evolved

The availability of oxygen changed everything. Oxygen was toxic to anaerobic cells. In addition, because it is highly reactive, the oxygen began to break down the energy-containing molecules in the sea that the anaerobic organisms relied on for energy. Consequently, most of the anaerobic organisms that could not create their own organic molecules by photosynthesis died out.

In their place new cells evolved that could harness the reactive nature of oxygen. These cells extracted energy from the abundant organic molecules stored within the cell itself, using the now readily available oxygen. As you may recall, the ability to use oxygen to extract energy from organic molecules is called aerobic metabolism.

The rise of animals and our human ancestors

Even from the brief description above, you can see that evolutionary processes are remarkably efficient at shaping life-forms to fit their environments. At the risk of condensing several billion years into a few sentences so that we can move to the recent evolution of human beings, here are some highlights.

- DNA became enclosed within a nucleus about 1.7 billion years ago. Cells with nuclei, called eukaryotes, became the dominant cell type.
- The first multicellular organism, a type of seaweed, appeared approximately 1.3 billion years ago.
- Animals appeared about 600 million years ago.
- Dinosaurs became extinct about 65 million years ago.
- Our first distinctly human ancestors appeared only about 5 million years ago.

Figure 22.10 presents a time line of biological milestones in the history of Earth so far. And how will it all end? The best estimate is that about 8 billion years from now, the Earth will orbit closer and closer to an enlarged sun and be vaporized by the intense heat. If that is true, Earth is now about a third of the way through its period of existence.

↩ **Recap** Over time photosynthesis evolved, resulting in increased atmospheric concentrations of oxygen. Some organisms subsequently developed an aerobic form of metabolism. A cell nucleus formed to enclose the DNA in most cells, multicellular organisms arose, and animals appeared. Our first recognizably human ancestors emerged about 5 million years ago. ■

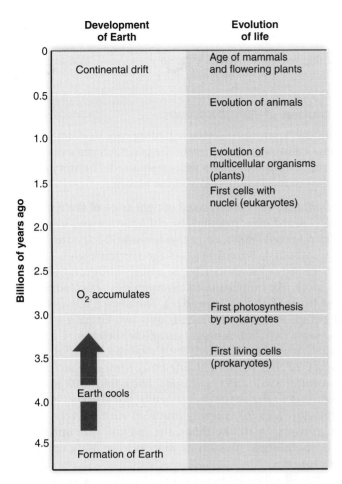

Figure 22.10 A time line of the evolution of life since Earth's formation. The entire span of human evolution lies within the horizontal line at the top of the figure.

22.6 Modern humans came from Africa

Humans are primates

Taxonomy, the branch of science that focuses on classifying and naming life-forms, seeks a precise way to describe the relationships between all life-forms. The eight levels of the taxonomic classification system are *domain, kingdom, phylum, class, order, family, genus,* and *species.* A **species** is defined as a group of organisms that under natural conditions tend to interbreed and produce fertile offspring.

Life-forms are generally referred to by their genus and species names. Names are italicized, with the genus name capitalized and listed first. Table 22.1 shows the eight taxonomic categories as they apply to humans, or ***Homo sapiens.***

Humans belong to the class *Mammalia.* **Mammals** are vertebrates that have hair during all or part of their life, and mammary glands that produce milk. Within the class Mammalia we belong to the order *Primata,* more commonly called **primates.** Primates are mammals with five digits on their hands, fairly flat fingernails and toenails rather than hooves or claws, and forward-facing eyes adapted for stereoscopic vision. They include the lemurs (sometimes called pre-monkeys), monkeys, apes, and humans. The primate hand has a thumb that opposes four fingers. All of these features apparently evolved as adaptations for an arboreal lifestyle, that is, life in trees. The fossil record indicates that all present-day primates share a common ancestor that lived about 60 million years ago (Figure 22.11 on the next page).

About 25 million years ago a subgroup of primates diverged from a common ancestor of the old-world monkeys to become the present-day apes (gibbons, orangutans, gorillas,

Table 22.1 The taxonomic classification system as it applies to modern humans

Taxonomic category	Scientific name	First appearance (bya, mya)*	Characteristics and examples of life forms in each category
Domain	Eukarya	1.7 bya	Organisms with a membrane-bound nucleus within its cells.
Kingdom	Animalia	600 mya	Many-celled organisms with eukaryotic cells and a complex anatomy. Includes all animals: sponges, worms, insects, fish, birds, mammals, and many more.
Phylum	Chordata	550 mya	Animal with a nerve cord and a backbone. Fish, birds, and mammals are examples.
Class	Mammalia	120 mya	Members of the phylum Chordata with hair during part of their life cycle and mammary glands. Includes mice, dogs, cows, whales, kangaroos, primates, and humans.
Order	Primata	60 mya	Members of the class Mammalia with five digits on the fore and hind limbs, flat fingernails, and stereoscopic vision. Lemurs, monkeys, apes, and humans are examples.
Family	Hominidae	5 mya	Primates that walk upright and that have slightly enlarged brains. Includes the genus *Australopithecus* and our *Homo* ancestors (now all extinct) and modern humans.
Genus	*Homo*	1.8 mya	Hominidae with an enlarged brain and a recognizable human skull and body form. Includes our extinct immediate *Homo* ancestors and modern humans.
Species	*sapiens*	0.1 mya	*Homo* with large brains and a complex cultural and social structure that includes spoken language. Modern humans.

*bya = billions of years ago; mya = millions of years ago.

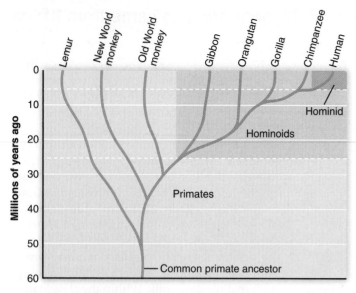

Figure 22.11 Evolution of modern primates from a common primate ancestor.

 Quick Check Suppose you are at a paleontological dig site in Africa, and you find a fossil skeleton that is about 60 million years old. What traits could you examine to assess whether it might be a primate? ▪

Evolution of *Homo sapiens*

Although scientists disagree regarding specific dates and the classification of certain fossils, the general framework of human evolution is fairly well established. The story of the origins of modern humans begins in Africa.

Ardipithecus ramidus: **Walked upright some of the time.** The fossil record indicates that the chimpanzee and human lines diverged from a common hominoid ancestor in Africa about 4–5 million years ago (review Figure 22.11). In the study of human origins, species of hominoids that are more like humans than chimpanzees are referred to as **hominids** (family Hominidae). The most likely first hominid, *Ardipithecus ramidus,* appeared about 4.4 million years ago in Africa. A nearly complete skeleton of a female *Ardipithecus ramidus* called "Ardi" was discovered in 1994 (Figure 22.13). Ardi had some body form and face features in common with later hominids. She could walk upright according to the shape of her pelvis, but she also had long opposable toes for grasping branches with her feet, and long fingers. In all likelihood she did not walk upright very well, spending as much time in trees as walking upright on the ground.

Not all scientists are convinced that Ardi is a true hominid, at least not yet. Part of the debate is over whether habitual upright posture and bipedal walking are the defining features of a hominid. The other problem is that evolution occurs slowly, so it's hard to tell when a transitional phase becomes final.

chimpanzees) and humans (Figure 22.12). **Hominoids** differ from the rest of the primates in that they are larger and have bigger brains. They also lack tails, and their social behaviors are more complex. Based on the number of differences in DNA, it appears that orangutans diverged first, followed by gorillas, chimpanzees, and humans. In other words, chimpanzees are our closest living animal relatives among the hominoids. However, this does not mean that we descended from chimpanzees. It means that chimpanzees and humans apparently evolved from a common hominoid ancestor.

a) Gibbon.

b) Orangutan.

c) Gorilla.

d) Chimpanzee.

Figure 22.12 The hominoids. Modern humans (not pictured) are also hominoids, of the family *Hominidae.* Chimpanzees are our closest living relatives.

 Is a gibbon more closely related to an Old World monkey or to a human? (Hint: Look at the time since the last common ancestor.)

a) The nearly complete skeleton of "Ardi", discovered in 1994.

b) An artist's depiction of Ardi.

Figure 22.13 *Ardipithecus ramidus.*

dominant males. Lucy, for example, was only about 1 meter tall. Their arms were long relative to their bodies, indicating that they still could travel from tree to tree (arboreal locomotion). But in terms of human evolution, the most important trait of *Australopithecus afarensis* was that they definitely walked upright (Figure 22.14).

Homo habilis:* The first toolmaker** Our first distinctly human ancestor was ***Homo habilis, which first appeared about 2.4 million years ago in Africa. Brain enlargement continued in *Homo habilis*, along with changes in teeth and facial features, a decline in sexual dimorphism, and conversion to a diet that included meat. As the first stone tools appear about this time, *Homo habilis* (which means "handy man") may have been the first toolmaker.

Homo erectus:* Out of Africa** The more human-looking ***Homo erectus (and its short-lived cousins *H. georgicus*, and *H. ergaster*) apparently arose from *Homo habilis* ancestors about 1.8 million years ago. *Homo erectus* migrated out of Africa to establish colonies as far away as Java and China. In the past, scientists thought that *Homo erectus* had been the direct ancestor of modern humans via populations that spread across Africa into Asia and Europe. However, *Homo erectus* died out in most locations by about 400,000 years ago. The very last *Homo erectus* lived in Java only 50,000 years ago, making them contemporary with modern humans.

***Australopithecus afarensis:* Most definitely walked upright.** The genus *Australopithecus* contains several related species, all of whom eventually became extinct. One of them, most likely an African hominid named *Australopithecus afarensis*, is thought to be the direct ancestor of modern humans. A nearly complete 3.2-million-year-old skeleton of an *Australopithecus afarensis*, named Lucy by her discoverers, was found in Ethiopia in 1974.

Australopithecus afarensis was still partially apelike in its anatomy and social structure, with just the beginnings of what we would call human features. Their brains were still relatively small relative to body size—more like chimpanzees than present-day humans. Their teeth had thick enamel, suggesting that they could grind hard foods; most likely they were vegetarian. The males were considerably larger than the females (*sexual dimorphism*), which suggests that *Australopithecus* had a social structure in which large, dominant males exerted control over small bands of females and less

Figure 22.14 Fossilized footprints of *Australopithecus afarensis.* Found by Mary Leakey in Tanzania in 1978, these 3.6-million-year-old footprints provided evidence of patterns of early hominid upright posture and locomotion. The prints appear to be of two adults and a child.

With *Homo erectus* came even more brain enlargement, a longer period of infant development, a continued decline in sexual dimorphism, and the continuation and expansion of toolmaking. By 1.7 million years ago, *Homo erectus* had developed specialized stone tools such as hand axes and cleavers. There were probably changes in social structure as well, including male–female pair bonds, shared responsibilities for offspring, and the formation of hunting and gathering groups that shared food.

Other species of *Homo* apparently coexisted with *Homo erectus* but later became extinct. Among them are *Homo heidelbergensis*, a species of archaic humans who created highly specialized stone tools, hunted cooperatively in groups, and may have had limited language skills. It is possible that some European descendants of *Homo heidelbergensis* became the Neanderthals, a controversial group of humans whose fossils were first discovered in the Neander Valley of Germany. Some researchers assign the Neanderthals to a separate species (*Homo neanderthalensis*), but this assignment remains controversial. The Neanderthals became extinct about 28,000 years ago.

Homo sapiens: Out of Africa again. The final phase of human development was marked by a substantial increase in brain size, the further development of spoken language, and the development of a physical structure that we would call thoroughly human. Although it is by no means certain, a current hypothesis is that *Homo sapiens* evolved in Africa from *Homo erectus* between 140,000 and 100,000 years ago. About 50,000 years ago a small band of *Homo sapiens* crossed

MJ's Human Biology Blog

http://www.humanbiologyblog.blogspot.com/

Share Report Abuse Next Blog» Create Blog Sign In

MJ's Human Biology Blog

Walking Like a Modern Human

Well-documented fossilized 3.8-million-year-old footprints show that our early ancestors walked with a decidedly ape-like primitive gait, with bent knees, short steps, and a predominant point of impact on the heels. It may have taken several million more years of evolution before changes in our ancestors' physical structure allowed them to walk with the long stride of modern humans. Analysis of 1.5-million-year-old footprints in Kenya indicates that the footprint-makers (most likely *Homo erectus*) had a long stride and the ability to push off with the big toe, as we do. The footprints also show that by 1.5 million years ago, foot anatomy was very much like ours. ■

Reference: Bennett, Matthew R. et. al. (2009). Early Hominin Foot Morphology Based on 1.5-Million-Year-Old Footprints from Ileret, Kenya. *Science* 323: 1197–1201.

the sea to what is now Yemen, on the Arabian peninsula. From there they migrated to all parts of the world, reaching the tip of South America about 10,000 years ago.

A key point is that wherever they went, these new modern humans from Africa replaced all other existing *Homo* species. Ultimately *Homo heidelbergensis* and the Neanderthals vanished completely. Why? We don't know. Although some scientists speculate that there was interspecies violence, there is no evidence of it to date. It is just as likely that they fell victim to disease or were out-competed for resources. Some researchers speculate that the Neanderthals interbred with *Homo sapiens*, but because they were so few in number compared to *Homo sapiens* their distinctive physical features eventually disappeared. Whatever the reason, the last Neanderthals disappeared about 28,000 years ago.

In what has been termed the greatest anthropological discovery in 50 years, archaeologists recently discovered evidence of what may have been yet another extinct human species, which they named *Homo floresiensis*. This latter group apparently existed as recently as 12,000 years ago, making them fully contemporary with modern humans. (See the CurrentIssue: Who Were The Flores People?)

Homo sapiens is the sole surviving human species (**Figure 22.15**). At a mere 140,000 years, modern *Homo sapiens* have not been around all that long. To put our existence in perspective, if the history of life on Earth were compressed into a 24-hour day, the entire history of *Homo sapiens* would represent less than the last 3 seconds.

✓ **Quick Check** In the early 20th century, before many hominid fossils had been found, most biologists thought that the first major human trait to evolve must have been an enlarged brain (much bigger than that of chimpanzees) and that upright walking must have evolved later. Explain whether this hypothesis turned out to be correct, and name the species in which these traits first appeared. ■

Differences within the human species

All humans belong to a single species. We are so similar that sometimes the sum of all humankind is called the human race. Nevertheless, there are certain distinct heritable physical differences between subpopulations of the human species that the dictionary defines as racial differences. How did these differences come about?

Throughout most of human history, the total human population existed mainly as smaller subpopulations that were geographically isolated from each other. Minor random mutations coupled with sexual reproduction within isolated subpopulations apparently led to slightly different human group phenotypes (referred to as *races*) that are still observable today. Human racial differences are a testament to the effectiveness of evolution over tens of thousands of years.

Some racial differences may have helped humans adapt more effectively to their environment. For example, in a climate with intense sunlight a dark skin color would have

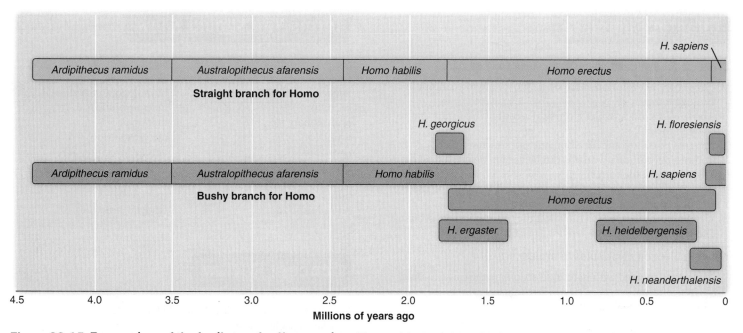

Figure 22.15 Two versions of the family tree for *Homo sapiens*. The straight branch version is the probable direct lineage. The bottom bushy branch shows our various extinct *Homo* relatives. The entire 4.5 million years shown in this figure correspond to a sliver of the zero time line at the top of Figure 22.10.

offered protection against excessive ultraviolet radiation. But in a northern climate where sunlight was less intense a light skin color would have been helpful because small amounts of ultraviolet radiation are required to activate vitamin D. Facial features and even blood types may also have had their special advantages at the time. Other differences may simply have been neutral, meaning that they were the result of mutations that did no harm and so were retained in the population in which they arose.

In the final analysis, racial differences are nothing more than differences in phenotypes between subgroups of our common species. But then, individual phenotypes are all very different, too (Figure 22.16). In fact, DNA analysis reveals that there are as many differences between individuals of the same race as there are between individuals of different races. Given that there are no biological barriers to reproduction and that global travel and communication have recently become commonplace, human racial differences may disappear over the next 10,000 to 100,000 years, assuming we survive that long as a species.

↺ Recap A taxonomic system of eight categories classifies the relationship of any species to all others. Modern humans descended from a primate ancestor that began to walk upright nearly 5 million years ago. Over time the brain enlarged, and toolmaking skills and spoken language developed. The current best hypothesis is that modern humans first appeared in Africa about 140,000 years ago. All human beings belong to the same species. Racial differences most likely arose as adaptations to different environments. ■

Figure 22.16 The human species. Genotypic and phenotypic differences between individuals of the same race are as great as the differences between races.

Chapter Summary

Evidence for evolution comes from many sources *p. 510*

- The theory of evolution is the single most unifying concept in biology today.
- The fossil record provides a physical record of life-forms that lived in the past.
- Patterns of embryonic development and structures with common evolutionary origins can be used to determine relationships between life-forms.
- Structural similarities between proteins and DNA can be used to define closeness of evolutionary relationships.
- Geographical separations can affect the distributions of evolving life-forms.

Natural selection contributes to evolution *p. 514*

- Mutations coupled with natural selection are the cause of most microevolution.
- The pool of alleles in a population is affected by chance events such as genetic drift and gene flow.
- At least five mass extinctions have altered the course of macroevolution.
- An evolutionary tree visually depicts relationships between species.

The young Earth was too hot for life *p. 516*

- Earth was formed about 4.6 billion years ago. At first, it was a harsh environment with constant cycles of condensation and vaporization.
- Life began about 3.8 billion years ago.

The first cells were able to live without oxygen *p. 516*

- Organic molecules formed from gases in Earth's atmosphere.
- The first self-replicating molecule was probably single-stranded RNA.
- The first cell was formed when RNA and organic molecules became enclosed in a membrane.
- The early atmosphere was devoid of oxygen, so the first living cells survived by anaerobic metabolism.

Photosynthetic organisms altered the course of evolution *p. 518*

- Oxygen appeared only after certain life-forms developed the process of photosynthesis as a way to produce organic molecules inside the cell.
- With oxygen available, other cells developed the process of aerobic metabolism as a way to extract energy from organic molecules.
- Eukaryotes became the dominant cell type, and multicellular organisms evolved.

Modern humans came from Africa *p. 519*

- Humans are a relatively recent development on Earth. Humans and modern chimpanzees evolved from a common ancestor that lived about 5 million years ago.
- *Australopithecus afarensis* and *Homo habilis* were early hominids in human evolution.
- Modern humans apparently first appeared in Africa about 140,000 years ago. From there they spread around the world.
- Heritable, observable, physical phenotype differences in human subpopulations are called racial differences. Racial differences are no more significant than differences between individuals. We all belong to the same human species.

Terms You Should Know

adaptive radiation, *516*
analogous structures, *512*
antigenic shift, *515*
evolution, *510*
evolutionary tree, *516*
extinction, *515*
fossil, *510*
gene flow, *515*
genetic drift, *515*
hominid, *520*
hominoid, *520*

Homo habilis, 521
Homo sapiens, 519
homologous structures, *512*
macroevolution, *510*
mammal, *519*
microevolution, *510*
natural selection, *515*
population, *515*
primate, *519*
species, *519*
vestigial structures, *512*

Concept Review

Answers can be found at the Human Biology Place.
www.humanbiology.com

1. Describe the three key elements of evolution.
2. Explain how a fossil is created.
3. List some of the main sources of evidence for evolution.
4. Describe how natural selection contributes to evolution.
5. Explain how genetic drift and gene flow affect populations.
6. Summarize the scientific theory of how life began on Earth.
7. Explain how rising atmospheric concentrations of oxygen affected life on Earth.
8. List the features that make Hominidae different from other families in the order Primata.
9. Describe the origins of *Homo sapiens*.
10. Explain why racial differences may disappear within the next 100,000 years.

Test Yourself

Answers can be found in Appendix A.

1. Which of the following statements about fossils is true?
   a. All body parts of an organism are equally likely to become fossilized.
   b. There are fossils of all species of organisms that have ever lived.
   c. Fossils in upper layers of rock are likely to be older than fossils in lower layers.
   d. Fossils are most likely to be found in sedimentary rock.

2. All the following would be considered homologous structures except:
   a. an elephant's foreleg
   b. a bat's wing
   c. a dolphin's flipper
   d. a butterfly's wing

3. Examination of the skeleton of whales shows bones similar to the human pelvic girdle. In whales, these bones would be considered:
   a. vestigial structures
   b. anachronisms
   c. analogous structures
   d. heterologous structures

4. Vestigial structures, homologous structures, and biochemical similarities all provide evidence of:
   a. descent from a common ancestor
   b. barriers to migration
   c. random mating
   d. sexual selection

5. Which of the following statements about mutations is true?
   a. Mutations are always detrimental to survival.
   b. Mutations decrease the size of the gene pool and result in more similarities than differences.
   c. Mutations provide the genetic variation upon which evolution works.
   d. While some mutations are random, many are caused by evolution.

6. Which of the following best describes evolutionary "fitness"?
   a. overall strength relative to other members of a population
   b. overall cardiovascular health relative to other members of a population
   c. ability to attract mates relative to other members of a population
   d. ability to survive and reproduce relative to other members of a population

7. Which of the following can affect the makeup of the gene pool within a population?
   a. natural selection
   b. mutation
   c. migration of individuals into or out of a population
   d. all of these choices

8. Genetic drift is most likely to affect:
   a. every population, regardless of size
   b. small populations
   c. large populations
   d. large and diverse communities

9. Which of the following would be described as genetic drift?
   a. the distribution of species resulting from the movement of the continents as they dispersed from Pangaea
   b. random mating that occurs in a large population
   c. the random changes that might occur in a gene pool as the result of a natural disaster that significantly reduced the size of a population
   d. the change in allele frequencies that result from sexual selection

10. Which of the following statements is true?
    a. Mass extinctions did not occur prior to the arrival of *Homo sapiens.*
    b. Increases in atmospheric CO_2 concentration and global warming have never occurred before.
    c. Environmental changes may create new habitats favoring the development of new species.
    d. Species can only become extinct; they can't diverge into more than one species.

11. In what order did events occur that led to life on Earth?
    (1) appearance of aerobic organisms, (2) appearance of photosynthetic organisms, (3) appearance and increase of O_2 in the atmosphere, (4) self-replicating RNA develops
    a. 4, 2, 3, 1   b. 2, 3, 1, 4
    c. 3, 1, 4, 2   d. 1, 2, 4, 3

12. Humans belong to all of the following taxonomic groups except:
    a. Primata   b. Mammalia
    c. Prokaryotae   d. Hominidae

13. Which of the following statements about the hominoids is false?
    a. Hominoids include gorillas, chimpanzees, and humans.
    b. Hominoids have larger brains than other primates.
    c. Hominoids lack tails.
    d. Hominoids typically are bipedal and walk upright.

14. Which of the following have coexisted with modern *Homo sapiens*?
    a. *Homo habilis*
    b. *Homo neanderthalensis*
    c. *Homo floresiensis*
    d. both (b) and (c)

15. Which of the following is **not** believed to be a direct ancestor of *Homo sapiens*?
    a. *Homo erectus*
    b. *Homo heidelbergensis*
    c. *Homo habilis*
    d. *Australopithecus afarensis*

Apply What You Know

Answers can be found at the Human Biology Place.
www.humanbiology.com

1. Because dogs are all one species, different breeds of dogs can mate and produce viable offspring. How, then, did different breeds of dogs with distinct characteristics evolve?

2. How would you expect an increase in gene flow in the human population to affect human phenotypes?

3. Critics of evolution often charge, correctly as it turns out, that the fossil record is incomplete and scientists cannot show how all living things evolved over time. Why don't we have a complete record?

4. Radiometric dating is a useful technique for identifying the age of fossils. Why does the suspected age of the fossil determine which isotope an investigator might use to try to date it?

5. There are species of fish that live in caves no longer connected to the surface. These fish have eye structures even though the fish are blind. Why do they have eyes even though the eyes don't function?

6. It is believed that humans evolved in Africa and then spread to all the other continents. If this is so, explain why the people of northern Europe do not have dark skin.

7. Throughout Earth's history many species, and perhaps even most of the species that ever existed, have become extinct. The causes are numerous: climate change, volcanic activity, the actions of oxygen-producing organisms, changes in carbon dioxide concentration, and even asteroid impacts. Have humans ever caused extinctions?

8. Of all the great evolutionary advances of life on Earth, which organisms have contributed the most significant ones: simple organisms or complex multicellular organisms such as plants and animals?

23

Ecosystems and Populations

A zebra and migrating wildebeest crossing a river in Kenya.

Waiting for the Next "Big One"

Nearly 700 years ago a mysterious bacterial disease nicknamed the Black Plague devastated the human population, killing roughly a third of the people of Europe and Asia. In just the past 100 years there have been four **pandemics**—disease outbreaks occurring over a wide geographical area and affecting a large part of the population. The worst of them was the deadly Spanish flu of 1918, which killed an estimated 20–40 million people.

The most recent pandemic was the swine flu of 2009. It didn't amount to much in the end, but it and the bird flu scare of a couple of years earlier

sensitized the public to the possibility of another "big one," a truly deadly pandemic that sweeps around the world, killing millions. Fortunately, such a pandemic is only likely to happen when a new disease-causing organism emerges that is both *virulent* (causing severe disease, even death) and *easily transmissible* (humans must catch it easily, usually from other humans.) The bird flu virus is highly virulent in humans, killing nearly 60% of its victims. But so far, human-to-human transmission of bird flu has not been documented. All of the rare human infections so far have come from direct close contact with infected birds. On the

other hand, swine flu is highly transmissible between humans. Swine flu proved not to be very virulent, however.

The next "big one" will come when a new virus emerges with the worst qualities of both bird flu and swine flu. No one can predict when or if it will happen—it's like trying to predict a California earthquake. But we do have a pretty good idea of how it will happen.

All flu viruses undergo a slow form of change over time called *genetic drift* as a result of normal reproductive processes. You can catch a seasonal flu each year because each year it's a little bit different. Less commonly, the change occurs by *genetic shift*—the wholesale re-assortment of the gene fragments of two very different viruses into a totally new viral combination. Genetic shift can occur when two different viruses meet in the same infected cell and recombine their genetic fragments. Genetic shift could occur literally overnight, and that's what keeps public health officials awake at night, worrying.

The genome of the deadly bird flu virus contains only eight separate gene segments. If the bird flu virus were to infect a human cell that was also infected with swine flu or the seasonal flu and genetic shift were to take place, one possible outcome might be a new virus with the virulence of the bird flu virus and the transmissibility of a seasonal flu virus. Every time a person becomes infected with the bird flu virus, there's a chance it could happen.

a) Bird flu virus (virulent).

b) Human flu virus (transmissible).

c) Novel virus (virulent *and* transmissible).

Segments of RNA from bird and human influenza viruses could recombine in humans to form a deadly new human strain.

Are We Prepared?

Should we worry about another "big one"? After all, the Black Plague occurred before we even knew about bacteria and viruses. The Spanish Flu preceded the development of antibiotics or vaccines. Aren't we better prepared now to handle a deadly outbreak? The short (and confusing) answer is "yes and no." In the case of bird flu there was a lack of preparedness in the areas of first outbreak, coupled with a reluctance of some countries to share information in the earliest stages when it might have been more easily contained. In the case of swine flu, the inability to produce adequate supplies of vaccines quickly enough would have had devastating consequences, had the swine flu virus actually been deadly.

Public health officials have taken notice of the deficiencies. The World Health Organization has outlined a strategic plan for preparing for pandemic influenza that includes reducing human flu exposure, strengthening early warning and detection systems worldwide, intensifying rapid containment efforts in the first areas of a deadly outbreak, and coordinating global scientific research and development. These efforts are meant to contain, control, and buy time for production of a vaccine, should an outbreak occur. A specific vaccine would probably not be ready for six months, but there are already several vaccines, most notably Tamiflu, that might prove of some use in an outbreak. Tamiflu would need to be available essentially worldwide on short notice, because it is most effective when given within 24 hours of onset of symptoms.

Public health efforts to try to prepare for the next "big one" are hampered by a lack of interest on the part of the public worldwide except when the danger seems personal and eminent. We tend to focus on the "dread disease of the month" but then lose interest quickly if it doesn't actually

materialize. For example, there were 208 articles using the phrase "bird flu" in the *New York Times* in the first half of 2006 (the peak of the bird flu scare). There were just two in the first half of 2010, and both of those articles were primarily about swine flu. Nevertheless, human infections and deaths from bird flu continue unabated (see graph).

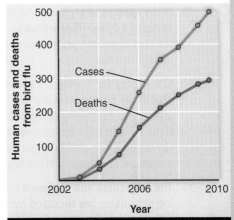

Cases and deaths from bird flu (cumulative since 2003) through June 2010. *Data from:* WHO/Map Production: Public Health Mapping and GIS Communicable Diseases (CDS) World Health Organization.

It's a small world now. Airline travel allows humans and disease-causing organisms to interact in ways and mingle across continents at speeds unimaginable 700 years ago, when the Black Plague devastated the human population. A new virus could be carried to every major city in the world within two or three days by its unknowing human hosts. The human species' best defense will be a coordinated worldwide strategy of surveillance, prevention, containment of the initial outbreak, and effective isolation and treatment of infected individuals. We need to plan for the worst and hope for the best.

Questions to consider

1 In the event of an initial outbreak of bird flu in another country, would you try to obtain Tamiflu for yourself and your family? Why or why not?

2 Would a regional outbreak of a deadly disease affect your interest in traveling to that region?

The facts...

- The next deadly pandemic is likely to be caused by a new strain of virus that is both *virulent* and *easily transmissible*.

- The new virus is likely to emerge rapidly as a result of *genetic shift*—the wholesale recombination of the gene fragments of two different viruses within an infected host cell.

- Effective preparation for the next deadly pandemic will require a worldwide coordinated effort.

>> **Under ideal conditions, a population of any species will grow** until growth is countered with *environmental resistance.*

>> **The human population has grown rapidly since about 1700.** Currently at 6.9 billion, it could increase another 30–50% by 2050. When it will reach a steady-state (or begin to decline) is unknown.

>> **Many species of organisms may live together in complex communities within an ecosystem.** Disruption of an ecosystem can have long-lasting consequences.

>> **The energy for fueling life's complex chemical processes comes from the sun.** Producer organisms (primarily plants and algae) make their own organic molecules; consumer organisms (primarily animals) get their organic molecules by consuming other organisms.

>> **The molecules and elements that compose living organisms are recycled over and over again between living organisms and Earth.** The speed of recycling can vary from minutes to millions of years depending on the nature of the molecule and its location within its particular geochemical cycle.

Throughout this book, we have focused on understanding enough basic biology to appreciate *human* biology. As you know by now, all living organisms have certain features in common. We all consist of one or more living cells, and we have the capacity to reproduce. We share a common origin and we have evolved over time. And, perhaps most importantly, we share the same planet. Our futures are inextricably linked.

Ecology is the study of the relationships between organisms—including humans—and their physical environment. In this chapter we begin by describing general principles of ecology and of population dynamics. We also examine the rapid growth of the human population and the possible effects of this growth on our world in the future. The dynamics of our population on Earth are no different, conceptually, from those of a bacterial population in a laboratory dish—it's just that our world is slightly bigger (and our generation times are longer).

23.1 Ecosystems: Communities interact with their environment

The basic unit of study in ecology is an *ecosystem*. *Eco-* implies a connection to the environment, and *system* indicates that an ecosystem is a functional unit. An **ecosystem** consists of a community of organisms *and* the physical environment in which they live. An ecosystem comprises all the living things, all the matter, and all the energy. In an ecosystem we consider not only who eats whom, but how nonliving matter is recycled and how energy flows into and out of the system. The components of an ecosystem form a hierarchy that includes populations, communities, and the physical environment in which they live.

As discussed in Chapter 22, a **population** is a group of individuals of the same species that occupy the same geographic area and interact with each other. A population of elk inhabits a mountain valley, and a population of frogs inhabits a small pond.

A **community** consists of the populations of all the species that occupy the same geographic area and therefore interact. Some organisms serve as food or shelter for others. Insects feed on decaying wood. Woodpeckers feed on insects living in trees and make nests in those same trees. Deer browse on shrubs, cougars hunt the deer, and so on. A community is the living part of an ecosystem.

Together, all of the ecosystems on Earth compose the **biosphere,** or planetary ecosystem (refer back to Figure 1.7 on p. 7). The biosphere is so complex that it is difficult to study as one entity. Studies of simpler ecosystems have, however, led to some interesting conclusions that may apply to the entire biosphere as well. Most importantly, it appears that ecosystems are delicately balanced self-sustaining units in which organisms interact with each other and their environment in complex ways. In the process, they can modify or control the physical and chemical nature of the ecosystem itself. Some ecologists suggest that disturbing the delicate balance of the biosphere could endanger the survival of the human species—and of all other species as well.

Recap An ecosystem consists of a community of organisms and their physical environment, and a community consists of all populations that interact within the same area. The biosphere comprises all ecosystems on Earth.

23.2 Populations: The dynamics of one species in an ecosystem

Characteristics of populations include habitat, range, size, actual growth rate, and potential capacity for growth.

Where a species lives: Habitat and range

A species' **habitat** is the type of location where it chooses to live (**Figure 23.1**). Typically this is determined by its tolerance for certain environmental conditions. For example, the habitat of the now rare northern spotted owl is mature evergreen forests, the habitat of the bison is grassland, and beavers prefer a habitat of mountain valleys with freshwater streams and small trees.

Each organism's habitat has certain chemical and physical characteristics that favor the organism's comfort and survival. Animals and plants that occupy the same or overlapping habitats form a community. Members of a community may benefit

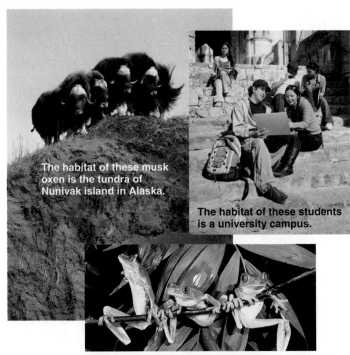

The habitat of these musk oxen is the tundra of Nunivak island in Alaska.

The habitat of these students is a university campus.

These red-eyed tree frogs inhabit the tree canopy of the Costa Rican rain forest.

Figure 23.1 Habitats. An organism's habitat is the place where it chooses to live.

each other in some way, or they may compete with each other for resources or even use each other for food and/or shelter.

Because the availability of ideal habitats varies, each species also has a **geographic range**—the area over which it may be found. An organism's range is limited by several factors: (1) competition for resources such as sunlight or nutrients, (2) intolerable conditions such as extreme temperatures or altitude, or (3) physical obstacles including mountain ranges, deserts, and bodies of water.

In the last few centuries the development of worldwide transportation has enabled some animals to expand their ranges beyond physical barriers. For example, innovations in navigation and shipbuilding allowed European humans to cross the Atlantic Ocean and expand their range into North and South America.

Population growth rate tends toward biotic potential

One of the most important topics in ecology is how populations change in size over time. The maximum rate of growth of any population under ideal conditions is called its *biotic potential*. Biotic potential is a function of certain characteristics of the species itself, including:

- The number of offspring produced by each reproducing member (female, if reproduction is sexual);
- The time span before the offspring reach reproductive maturity;
- The sex distribution if reproduction is sexual (ratio of males to females);
- The number of reproductive-age members of the population.

Typically, the biotic potential of any species follows an **exponential growth** curve with a characteristic J shape. For any species, the steepness of the growth curve, representing the biotic potential, is determined by the four factors above. The J shape reflects the fact that under ideal conditions, populations would double again and again over similar time periods. If you know the percent change in the population size per year you can calculate how long it will take to double in size by using the "Rule of 72": just divide 72 by the percent change per year. For example, a population that is increasing at a rate of 4% per year will double in 18 years (72/4 = 18); at 3% it would take 24 years. The rule of 72 works for any variable with a known % change per unit time. (If your savings account is earning 4% a year, your money will double in 18 years.)

✓ **Quick Check** If equal numbers of wolves and bears are introduced to a new area at the same time, the wolf population will often double in just a year, but the bear population can take a decade or more to double. Which species has greater biotic potential? Of the factors that affect biotic potential, which do you think are most likely to explain this difference in wolves compared to bears? ▪

Environmental resistance limits biotic potential

In the real world of finite resources and competition, an opposing force called **environmental resistance** limits any species' ability to consistently realize its biotic potential (**Figure 23.2**).

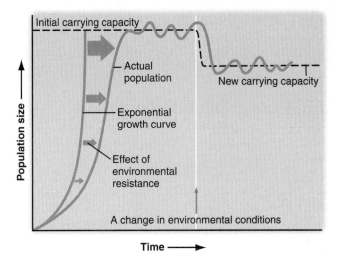

Figure 23.2 Effects of environmental resistance and carrying capacity on population size. As the population increases, the environmental resistance to continued ideal exponential growth increases. Eventually the population stabilizes near the carrying capacity. Changing environmental conditions can alter the carrying capacity.

✓ *What would this graph look like if this species had a lower biotic potential? Would a lower biotic potential affect the carrying capacity?*

Answers to ✓ **questions can be found in Appendix A.**

Environmental resistance consists of factors that kill organisms and/or prevent them from reproducing. Examples include limitations on nutrients, energy, and space; disease; predation by other organisms; and environmental toxins.

Even when conditions are ideal to begin with, eventually every growing population reaches a point where environmental resistance begins to increase. For bacteria in a petri dish or humans living on an island, it may be the food supply. For a population of mice it may be disease brought about by overcrowding. For a woodpecker it may be a lack of suitable nesting sites. Although we cannot always measure the specific environmental resistance, no population grows at its full biotic potential indefinitely.

Ultimately a population achieves an approximate balance between biotic potential and environmental resistance. At this point it stabilizes, reaching a steady-state. The population that the ecosystem can support indefinitely is called the **carrying capacity.** Population sizes may vary around their carrying capacity from time to time. In addition, the carrying capacity can change if environmental conditions change. A wet spring might increase the carrying capacity of a prairie ecosystem for certain grasses and the animals that feed on those grasses, and three years of drought might reduce the carrying capacity to below its average value. But in the end, every population tends to stabilize near its carrying capacity.

Recap Every species has a preferred habitat and a geographic range over which it is distributed. Under ideal conditions a population grows exponentially according to its biotic potential. Under normal circumstances, however, environmental resistance limits growth. The population of a particular species that an ecosystem can sustain indefinitely is called its carrying capacity. ▪

23.3 Human population growth

A graph of the global human population over just the past 4,000 years of human history shows that the human population is growing rapidly (**Figure 23.3**). How will this affect our world in the future?

For most of human history the global population remained relatively low and apparently stable, never exceeding 10 million people. People lived in widely dispersed small groups whose numbers were held in check by disease, famine, and a lack of tools and technology with which to control the environment or even to grow food. Apparently the environmental resistance was so high that our population was already at its carrying capacity.

About 4,000 years ago, the global human population began to rise slowly. The development of agriculture and the subsequent domestication of animals and plants for our own uses led to better conditions (a reduction in environmental resistance) and an increase in carrying capacity. Rapid growth began in the 1700s (barely 300 years ago) with the Industrial Revolution. Advances in communication, transportation,

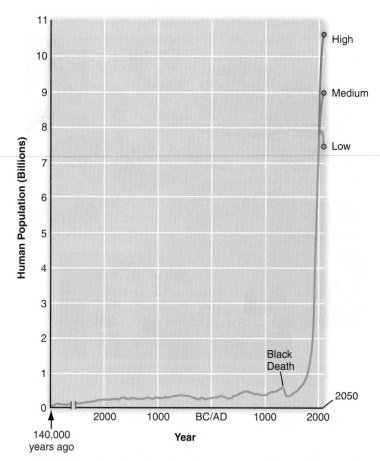

Figure 23.3 Past and projected growth of the human population. The human population currently is 6.9 billion. The dip in the population in the 1300s was caused by the Black Death, a bacterial plague that devastated the human population in Europe and Asia. The high, medium, and low world population projections are based on steady-state fertility rates of 2.6, 2.1, and 1.6 children per woman, respectively, by 2050.

Data after 1950 from: Population Division of the Department of Economic and Social Affairs at the United Nations Secretariat.

technologies for feeding and housing more people, and medical care, including the discovery of antibiotics and vaccines, have favored human reproduction and survival. As a result, the human population has entered a phase of explosive growth. Note that the human population curve in Figure 23.3 now resembles a J-shaped curve, typical of a population that is increasing according to its biotic potential, without any apparent environmental resistance.

Can we continue this trend? We do not know how high the human carrying capacity might be or how much it could be raised by future scientific discoveries. Some scientists believe that the concept of carrying capacity doesn't apply to humans, due to our ability to alter our environments to suit our needs. Humans may even have the capacity to limit our population growth voluntarily, before the Earth's human carrying capacity is actually reached.

A more pessimistic view is that we may be inadvertently lowering the human carrying capacity of Earth already, through environmental degradation or other human actions. How soon environmental degradation would limit the human population is unknown. But no matter how you choose to look at it, eventually the human population must stabilize at a point short of having a person standing on every square foot of Earth (Figure 23.4).

Zero Population Growth Has Not Yet Been Achieved

The **growth rate** of a population is calculated as the number of births per year minus the number of deaths per year, divided by the total population. Currently the human population is growing at a rate of about 1.2% per year. Although that may not sound like much, at this rate the human population would double in the next 60 years (review the Rule of 72).

Zero population growth is the point at which births equal deaths. To reach that point, of course, we would have to either decrease the birth rate or increase the death rate. Given that no one would advocate increasing the death rate on purpose, let's look at how we might reach a zero population growth through changing the birth rate.

The human **fertility rate** is the number of children born to each woman during her lifetime. Fertility experts agree that the average fertility rate required to achieve long-term zero population growth, called the **replacement fertility rate,** is about 2.1 children per woman. Those 2.1 children would exactly replace the equivalent of the two people (a woman and a man) who produced them. The replacement fertility rate is slightly higher than 2.0 because some children die before reaching their reproductive years.

The United Nations Population Division estimates that if replacement fertility rate is achieved by 2050, the world population would stabilize at about 9 billion. However, if fertility rates stabilize at lower or higher levels, the global population could range from a declining 7.4 billion to a rapidly increasing 10.6 billion (refer back to Figure 23.3). Note that the difference between the low and high projections is only one child per couple. (High, medium, and low population projections are based on fertility rates of 2.6, 2.1, and 1.6 children per couple.) Keep in mind that these graphs present global averages. Some areas, such as Africa and Asia, contribute to population growth more than others.

 Quick Check The fertility rate in the United Kingdom today is approximately 1.8 children per woman, and the population growth rate is 0.42%. Explain why this combination of numbers is surprising, and propose an explanation. ∎

Population age structure is linked to economic development

A nation's *age structure* describes the number of people in each age group within the population. As a nation develops, the age structure of its population undergoes a *demographic transition*, a progression of population changes that are tied to the region's industrial development and the economic well-being of its citizens. (*Demography* is the study of human populations.) The first stage of a demographic transition sees a shift from a society dominated by poor living conditions and a high death rate to one of rapidly improving economic conditions, declining death rate, and higher birth rate. In the final stages of transition the birth rate slowly declines until it equals the death rate, achieving zero population growth.

Demographic transitions have been under way around the world since the 1700s, with some countries further along in the transition than others. Demographers categorize nations with established industry-based economies as **more industrialized countries (MICs)** and those that are only just beginning to industrialize as **less industrialized countries (LICs).** The MICs, which include most countries of Europe, North America, northeastern Asia, and Australia, are in later stages of demographic transition than the LICs, which include most nations of Africa, Latin America, and Asia. The demographic patterns suggest that industrialization raises the human carrying capacity and allows the population to expand to a new steady state.

Figure 23.4 The human population. It is not known when, or even if, human overcrowding or environmental degradation will begin to limit Earth's human carrying capacity.

Figure 23.5 depicts the population age structures of the LICs versus the MICs. Males within various age groups are shown on the left of the figure and females are shown on the right. The population of the less industrialized countries has a pyramid-like age structure, with a large proportion of young people still below reproductive age. In a stable population the age structure is more bullet-shaped, with roughly the same number of people in each age group through the reproductive years.

Although it may seem counterintuitive, the population of an LIC is likely to continue to grow even after the replacement fertility rate of 2.1 children per woman has been achieved. This is because an LIC typically has more young people entering their reproductive years than are leaving. In this case, the country's population will continue to grow for several more decades simply because there are more women in their reproductive years.

The problem, therefore, is that the largest increases in population over the next 50 years are likely to occur in the very nations that are least able to provide for their citizens. To prevent poverty and starvation in these countries, we must seek solutions that promote economic development while at the same time slowing their population growth rate. Family planning and other social programs that encourage delaying childbirth or reducing the number of children per couple would certainly help. Even government action has

worked in some societies. In China, for example, it is against the law to have more than two children. In addition, substantial evidence supports the hypothesis that one of the most effective ways to reduce the birth rate is to improve the economic, social, and political standing of women in a society. Cultural norms and religious beliefs may conflict with this approach in some cultures, however.

Recap The human population began to rise rapidly about 300 years ago. For zero human population growth to be achieved, an average replacement fertility rate of 2.1 children per woman would need to be achieved and sustained worldwide. Some countries have achieved that goal; others have not. The largest increases in population over the next 50 years are likely to occur in the less industrialized countries, the very countries that are least able to provide for their citizens. ◼

23.4 Communities: Different species living together

Most species live in communities of many different species. Their relationships may be quite complex, ranging from intense competition to mutual benefit. Oftentimes they rely upon each other as a source of food, shelter, or protection.

Overlapping niches foster competition

An organism's *niche* is its role in the community—its functional relationship with all living and nonliving resources of its habitat. A woodpecker's habitat is woodland forests. Its niche is to rid trees of their insect infestations and serve as a food supply for hawks or raccoons, or ultimately for bacteria. A well-balanced ecosystem supports a wide variety of species, each with a different niche.

Although two species do not occupy exactly the same niche, niches can overlap enough that competition may occur between species for limited resources. Plants may compete for available sunlight or mineral nutrients, and animals often compete for similar food supplies (for example, both foxes and hawks prey on mice). Depending on their particular niches, one species may be able to out-compete another in a given location, a phenomenon known as *competitive exclusion*. As the human population expands into new habitats, humans are increasingly excluding other species.

Succession leads toward a climax community

Most natural communities undergo constant change. *Succession* is a natural sequence of change in terms of which organisms dominate in a community. Succession is determined by population growth rates, the niches occupied by various species, and the kinds of competition that exist between them.

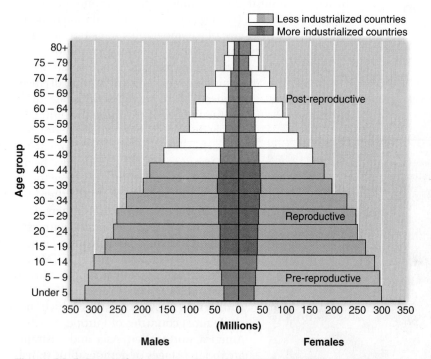

Legend:
- ☐ Less industrialized countries
- ▨ More industrialized countries

Age groups (top to bottom): 80+, 75–79, 70–74, 65–69, 60–64, 55–59, 50–54, 45–49, 40–44, 35–39, 30–34, 25–29, 20–24, 15–19, 10–14, 5–9, Under 5

Labels: Post-reproductive, Reproductive, Pre-reproductive

X-axis (Millions): 350 300 250 200 150 100 50 0 50 100 150 200 250 300 350
Males — Females

Figure 23.5 Population age structures in the LICs and MICs. The pyramid shape of the age structure of the LICs indicates that much of the population is younger than reproductive age. Therefore it is likely that their populations will continue to expand. The MICs have roughly the same number of people in each age group in and below the reproductive years, indicating a stable population in the future.

Source: U.S. Bureau of the Census, International Data Base.

If you have ever seen an abandoned field grow first to weeds, then to shrubs, and finally to trees, you have watched succession in action. Organisms that can reproduce and grow quickly under the prevailing conditions are the first to arrive in any area. These first organisms provide food and shelter for the next arrivals, which grow more slowly but occupy more specific niches. The number of species tends to increase as the succession advances and interrelationships between populations become ever more complex.

If the area suffers no major shocks, such as a fire, succession ends with the establishment of a **climax community**. Climax communities are the most efficient communities in terms of energy and nutrient utilization and also the most varied in terms of numbers of species. A climax community may take hundreds of years to develop, and retains its character for many hundreds to thousands of years.

The type of climax community found in any region is largely determined by the physical environment but can also be influenced by the organisms living in surrounding areas. In the Northeast United States, the typical climax community is dominated by large but slow-growing deciduous trees (broad-leaved trees such as oaks that lose their leaves each winter, then grow new leaves each spring). In the Midwest the climax community is prairie, in the Southwest it is semiarid desert, and in the mountainous West it is towering evergreen forests (**Figure 23.6**). Examples of climax communities in other parts of the world include the frozen tree-less Arctic tundra and tropical rainforests near the equator.

When climax communities are disturbed they do not recover easily. Consider the introduction of the giant reed, *Arundo donax*, to coastal stream communities in Southern California. The typical climax community in this area includes coast live oak and laurel sumac. However, since *A. donax* was brought to the United States in the 1800s for erosion control, its rapid growth characteristics have allowed it to supersede

Figure 23.6 A climax community. This old growth temperate rainforest in Washington is a climax community. The species present are those best adapted to the environment once the environment reaches a stable condition.

slower-growing native species until today it dominates these communities. The very complexity and slow growth characteristics of climax communities make them vulnerable.

 Quick Check An abandoned field in the Northeast United States will eventually become a forest. However, a similar abandoned field in the Midwest will remain a field (eventually becoming a prairie—a natural field). Why won't trees colonize the Midwestern field? More generally, why do different regions have different climax communities? ▪

Ecosystems: Communities and their physical environment

An ecosystem, as mentioned earlier, has both living and non-living components. The total living component of an ecosystem is called that ecosystem's **biomass.** The nonliving components consist of the *chemical elements* (the physical matter that makes up Earth) and the *energy* that drives all chemical reactions.

A constant supply of energy is essential for continued life because the transfer of nutrients and energy from one organism to the next is always an energy-absorbing process. Nearly all of Earth's energy comes from the sun, with only a tiny amount supplied by the residual heat of Earth itself (thermal vents and volcanoes). In contrast to energy, the chemical matter of Earth is recycled over and over again between Earth and the biomass. The next two sections discuss these two key concepts in more detail.

 Recap Species with similar niches may be in competition with each other. Communities go through progressive changes over time, called succession, which increases overall efficiency and species variety. Left undisturbed, eventually they achieve a climax community. Ecosystems consist of biomass, energy, and nonliving matter. ▪

23.5 Energy flows through ecosystems

Nearly all of the energy needed by living organisms comes from the sun. Energy in sunlight is used to create the complex organic molecules required by living organisms.

Producers capture and convert energy, consumers rely on stored energy

The flow of energy in an ecosystem demonstrates two key principles of physics, called the *laws of thermodynamics:*

- The *first law of thermodynamics* states that energy is neither created nor destroyed. However, energy can change form, and it can be stored.
- The *second law of thermodynamics* states that whenever energy changes form or is transferred, some energy is wasted (converted to non-useful forms). Generally this wasted energy is lost as heat.

Producer organisms can make their own organic molecules. **Consumer** organisms get their organic molecules (and energy) by eating other organisms or dead and decaying material. Energy flows in one direction through an ecosystem, from the sun to producers to consumers.

Most producers are capable of **photosynthesis** (Figure 23.7). A few producers that live entirely in the dark, such as bacteria located near deep-ocean hydrothermal vents, rely on a process called *chemosynthesis*. Producers are also sometimes called *autotrophs*, meaning "self-nutritive" or "self-growing." On land the producers are mainly green plants, whereas in aquatic ecosystems they are primarily species of algae.

The ability of producers such as plants to harness the energy of the sun depends on small organelles called *chloroplasts* within certain plant cells. Chloroplasts contain a green light-absorbing pigment called *chlorophyll* that gives plants their green color. In the first stage of photosynthesis, chlorophyll absorbs the energy of sunlight and then uses the energy to drive a chemical reaction in which water (H_2O) is disassembled and high-energy molecules of ATP and NADPH are produced. A by-product of the reaction is molecular oxygen (O_2), which is released to the atmosphere. In the second stage of photosynthesis, ATP and NADPH are used to fuel a biochemical cycle called the Calvin cycle, in which CO_2 from the atmosphere is used to produce high-energy carbohydrate molecules (sugars) consisting of multiple CH_2O units. Using the energy stored in the sugar and water and minerals obtained from the soil, plants are able to make all the other organic molecules they need, including proteins, lipids, and cellulose, a primary component of plant cell walls.

Notice how neatly the biochemical reactions of plants (producers) and animals (consumers) complement each other. Plants require CO_2 from the atmosphere for photosynthesis and produce O_2 as a by-product, whereas animals require O_2 for their metabolism and produce CO_2 as their waste product.

Most plants make more sugar than they need for their own use. The extra sugar is converted to starch, which is stored in the chloroplasts and in storage cells of roots, tubers, fruits, and seeds. On a global scale, it is estimated that photosynthesis accounts for the production of about 145 billion tons of carbohydrates per year, or nearly 23 tons for every person on Earth.

Animals, most bacteria, and fungi are *consumers* (also called *heterotrophs*). They cannot utilize the sun's energy to synthesize the molecules they need. Instead they must consume foods that already contain stored forms of energy. Consumers fall into four types depending on what they use as a food source.

- **Herbivores** feed on green plants. They are also called *primary* consumers because they eat only producers. Sheep, goats, and cattle are herbivores.
- **Carnivores** feed on other animals, and thus they are *secondary*, *tertiary* (third-level), and sometimes *quaternary* (fourth-level) consumers. The primary forms of stored energy in animals are fats and carbohydrates such as glycogen. Wolves and hawks are carnivores.
- **Omnivores** feed on both animals and plants. Humans, pigs, and bears are omnivores.
- **Decomposers** feed on *detritus,* the disintegrated matter of dead organisms. Bacteria, fungi, earthworms, and certain small arthropods (a group of organisms that includes insects and spiders) are all decomposers.

Figure 23.8 shows a pond ecosystem, where energy enters as sunlight and is converted to biomass by algae and aquatic plants (producers). Fish (consumers) eat the plant material, and when they die, bacterial or fungal decomposers

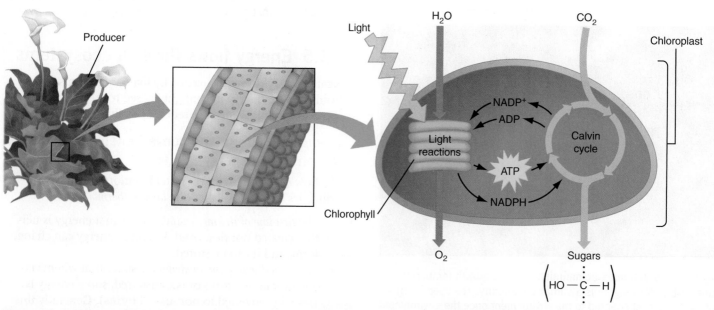

Figure 23.7 Photosynthesis. Producer organisms such as plants use the energy in sunlight, water, and CO_2 from the atmosphere to produce sugars. A by-product of photosynthesis is oxygen.

Figure 23.8 Energy flows and chemical cycling in a pond ecosystem. Energy flows in only one direction through ecosystems: from sunlight to producers and then to consumers. Most of the energy is eventually lost as heat. Chemicals are recycled continuously.

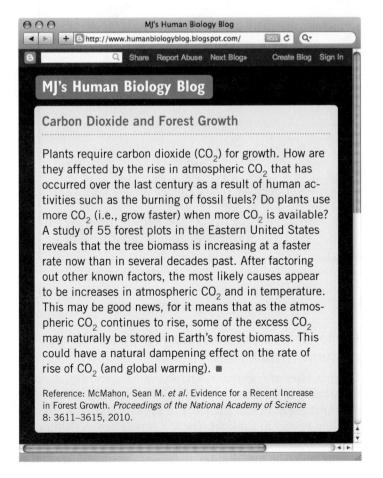

MJ's Human Biology Blog

Carbon Dioxide and Forest Growth

Plants require carbon dioxide (CO_2) for growth. How are they affected by the rise in atmospheric CO_2 that has occurred over the last century as a result of human activities such as the burning of fossil fuels? Do plants use more CO_2 (i.e., grow faster) when more CO_2 is available? A study of 55 forest plots in the Eastern United States reveals that the tree biomass is increasing at a faster rate now than in several decades past. After factoring out other known factors, the most likely causes appear to be increases in atmospheric CO_2 and in temperature. This may be good news, for it means that as the atmospheric CO_2 continues to rise, some of the excess CO_2 may naturally be stored in Earth's forest biomass. This could have a natural dampening effect on the rate of rise of CO_2 (and global warming). ■

Reference: McMahon, Sean M. *et al*. Evidence for a Recent Increase in Forest Growth. *Proceedings of the National Academy of Science* 8: 3611–3615, 2010.

(such as chytrids) return the nutrients to the water and soil for recycling. Heat is lost at each step.

✓ **Quick Check** In marine ecosystems, tiny arthropods called *krill* eat photosynthetic algae. Baleen whales feed on krill, and killer whales prey on baleen whales. When whales die and fall to the bottom of the ocean, their bodies are consumed by many organisms, including bacteria and fungi. Which of these species—algae, krill, baleen whales, killer whales, bacteria, and fungi—are producers and which are consumers? Which are herbivores, carnivores, or decomposers? ■

A food web: Interactions among producers and consumers

Feeding relationships between producers and consumers are sometimes described by a simple *food chain,* an example of which is when A (grass) is eaten by B (grasshopper), which is eaten by C (sparrow), which is eaten by D (fox). Although food chains are useful for presenting the relationships between types of consumers, they understate the degree of complexity in most ecosystems. Most organisms rely on more than one species for food and in turn may be eaten by more than one species.

A more accurate depiction of the complex balanced nature of the feeding relationships in an ecosystem is shown in a **food web** (Figure 23.9 on the next page). A particular consumer may eat different foods at different times depending on what is available. For example, as the population of a particular species of

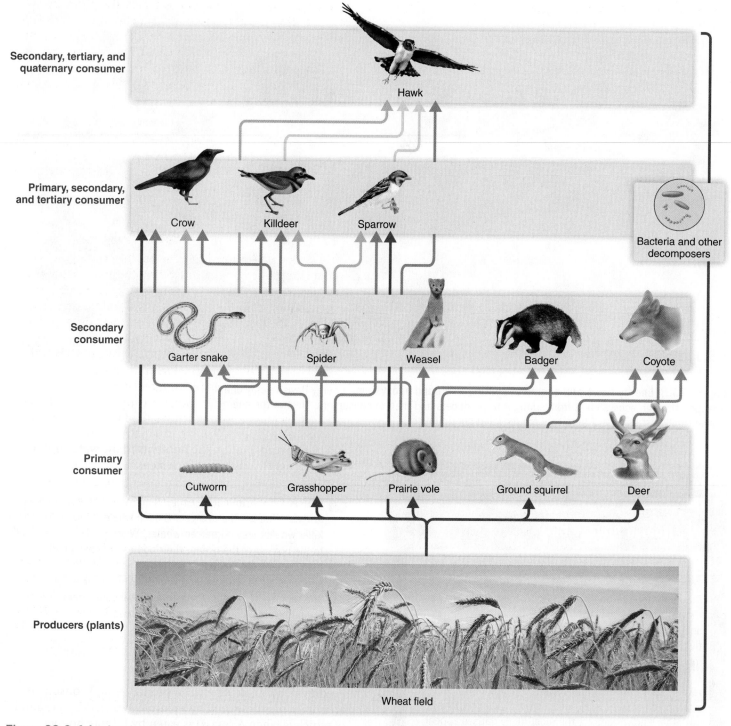

Figure 23.9 A food web. A food web shows the complexity of the feeding relationships among organisms in an ecosystem. Note that bacteria and other decomposers also play a role in the food web by feeding on dead organisms at all levels.

secondary consumer rises, it provides a more abundant food supply for a tertiary consumer. These tertiary consumers then rely less on other secondary consumer species whose populations are low at the moment. Consequently, no single population ever gets completely out of hand, and rarely does any species become extinct. In a stable ecosystem every population rises and falls periodically as part of the normal ebb and flow of life. These variations in a species population are always near the carrying capacity of the ecosystem for that particular species.

The lower levels of an ecological pyramid support consumer populations

The processes of energy conversion, transfer, and storage are remarkably inefficient at every level. For example, photosynthetic processes utilize and store in the biomass only 2% of the total energy in sunlight. The rest is reflected as light or heat energy. Only about 10% of the energy at one level of a food web can be found in the tissues of the consumers at the next level in the web. In other words, only about 10% of the stored energy available in plants is stored in herbivores, and only about 10% of that (or 1%) is stored in secondary consumers.

An **ecological pyramid** (Figure 23.10) depicts either the total amount of energy stored at each level of an ecosystem or the total biomass at each level. The total amount of energy represented by tertiary consumers at the top of the food chain is very small indeed. Ecological pyramids also generally indicate total biomass, as the amount of energy available usually defines how much biomass can be supported. To get a better image of the steepness of this pyramid, consider how many total pounds of large carnivores (wolves, bears, and eagles) there might be in a square mile of forest versus the total number of tons of grasses, shrubs, and trees.

✓ **Quick Check** One of the most famous questions of ecology is the simple question: Why are the largest carnivores so rare? Explain why lions, bears, and wolves are relatively uncommon in natural ecosystems—even when they are not being hunted by humans. ▪

Human activities disrupt ecological pyramids

Ecological pyramids illustrate that the population of consumers at any level depends critically on the populations of consumers directly below it. Tertiary consumers are especially vulnerable to ecosystem disruptions because the small amount of energy available to them depends on the energy transfers at all levels below them.

Humans can disrupt this balance rather easily with modern farming practices. When we create ecosystems with only one species of producer (mile after mile of wheat or corn), then harvest the entire crop, we exclude other species from their natural food web and their place on the ecological pyramid. We do this to feed more people, of course, and it works. But the short-term high yield comes with a long-term price in its effects on the field ecosystem and fertilizer requirements. In general, we must recognize that our actions affect other species on Earth.

Humans can be either primary or secondary consumers. We are secondary consumers when we eat meat and primary consumers when we eat plants. Eating meat is energetically expensive. When we feed grain to cattle and then eat the beef, we utilize only 10% of the energy that would otherwise be available to us in the grain. At some point, if we face hard choices about feeding the growing world population, scientists will surely point out that we can be more energy-efficient by eating plants rather than meat.

↩ **Recap** Producers are photosynthesizing organisms that use the sun's energy, available nutrients, and atmospheric carbon dioxide to produce energy-containing molecules. All other organisms, called consumers, obtain their energy from plants or other organisms. Energy is transferred in only one direction through the ecosystem: from the sun to producers to consumers. Energy transfer is inefficient, so each producer/consumer level in the ecological pyramid can support fewer organisms than the level below it. ▪

 Web Animation *Energy Flow and Food Webs* at www.humanbiology.com

50,000 units of energy from sunlight

49,000 units of energy lost (heat and light)

1 unit of energy in tertiary consumer

9 units of energy lost

10 units of energy in secondary consumer

90 units of energy lost

100 units of energy in primary consumer

900 units of energy lost

1,000 units of energy in producers

Figure 23.10 An ecological pyramid. This ecological pyramid depicts total energy stored in the biomass at each producer/consumer level. Because energy transfer is inefficient, only about 10% of the energy at one level can be found in the tissues of the consumer at the next highest level of the pyramid.

23.6 Chemical cycles recycle molecules in ecosystems

In contrast to the continuous input of energy from the sun, the total mass of nutrient matter that makes up Earth and its biomass remains completely fixed. With only a finite amount of each element available, the chemicals that compose living organisms are recycled over and over again between organisms and Earth itself. Such **biogeochemical cycles** include living organisms, geologic events, and even weather events.

As shown in **Figure 23.11**, every molecule and element in a living organism cycles between three different pools: biomass, the exchange pool, and reservoir. Some of the elements or molecules are found in the *exchange pool* (water, soil, and atmosphere), from which the primary producers draw their nutrients. Using the energy of the sun, the primary producers incorporate nutrients into the *biomass*. The biomass is the only biotic (living) compartment in biogeochemical cycles, hence, molecules in the biomass are exchanged rapidly as organisms consume each other.

When an organism dies and decomposes, its nutrients return to the exchange pool, where they again become available to primary producers. Coupled to the exchange pool is a large but hard-to-access pool of nutrients called the *reservoir*. Nutrients in reservoirs include minerals in solid rock and in sediments deposited at the bottom of the oceans. Other important reservoirs are carbon-containing compounds in **fossil fuels** (coal, oil, and gas) that were created millions of years ago from the remains of living organisms, and then covered by sediments. We turn now to the biogeochemical cycles for some of the most important molecules and elements of life—water, carbon, nitrogen, and phosphorus.

The water cycle is essential to other biogeochemical cycles

Water is probably more essential to life than any other molecule. It makes up about 60% of your body and covers approximately 75% of Earth's surface. However, freshwater—the

kind that humans and most other organisms need in order to live—actually accounts for less than 3% of our planet's total water supply (and two-thirds of that is trapped in glaciers and polar ice caps). The remaining 97% of the Earth's water is salty ocean water. Water is the universal solvent for many other elements—indeed, without the water cycle, all other biogeochemical cycles would cease.

Figure 23.12 illustrates the **water cycle**. Both the water cycle and the biomass are important to the biogeochemical cycles for minerals. Rain and surface water runoff slowly erode rocks and leach the minerals from them. Plants take up these minerals and incorporate them into the biomass. If there are no plants, the temperature is higher due to decreased evaporative cooling, and the minerals wash into the sea instead. When forests or grasslands are destroyed, nutrients lost from the soil may not be replenished from the reservoir pool for hundreds of years.

Water cycles between the atmosphere, ocean, and land. Although evaporation and precipitation occur primarily over the ocean, on land evaporation is mostly from leaf surfaces and soil. Overuse of water can deplete underground stores, compromising deep-rooting, formerly drought-tolerant vegetation and decreasing available water for irrigation. A consequence of this overuse is the loss of wetlands worldwide: for example, Greek Macedonia in Northern Greece has lost 95% of its marshland since 1930, primarily due to diversion of water for irrigation. In addition, the combination of pollutants from human activities and changes in landcover due to urban development can degrade streams, rivers, and even the ocean. For more about the impact of human activities on water, see Chapter 24.

Figure 23.11 The flow of nutrients in biogeochemical cycles. The three components of a biogeochemical cycle are the biomass, the exchange pool, and the reservoir. The chemicals that compose living organisms are constantly recycled among the three components. The relative magnitudes of nutrient flows between the three components and within the biomass are represented by differences in arrow thickness.

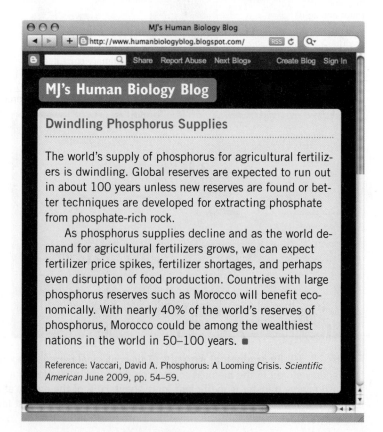

MJ's Human Biology Blog

Dwindling Phosphorus Supplies

The world's supply of phosphorus for agricultural fertilizers is dwindling. Global reserves are expected to run out in about 100 years unless new reserves are found or better techniques are developed for extracting phosphate from phosphate-rich rock.

As phosphorus supplies decline and as the world demand for agricultural fertilizers grows, we can expect fertilizer price spikes, fertilizer shortages, and perhaps even disruption of food production. Countries with large phosphorus reserves such as Morocco will benefit economically. With nearly 40% of the world's reserves of phosphorus, Morocco could be among the wealthiest nations in the world in 50–100 years. ■

Reference: Vaccari, David A. Phosphorus: A Looming Crisis. *Scientific American* June 2009, pp. 54–59.

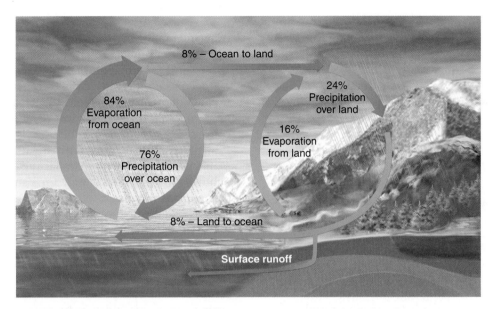

Figure 23.12 The water cycle. The amounts of water recycled relative to the total rates of evaporation and precipitation are indicated as percentages. Heat energy from the sun causes water to evaporate, primarily from oceans but also from leaf surfaces and soil on land. Most of the atmospheric water vapor precipitates as rain over the oceans, but approximately a quarter of it precipitates over land, replenishing surface water that was lost by evaporation and providing the extra water that becomes ponds, lakes, streams, and rivers.

 Web Animation *The Water Cycle* at www.humanbiology.com

✅ **Quick Check** The most productive areas for human agriculture are river floodplains; the most productive areas for marine fisheries are the estuaries where rivers meet the sea. Why do the most productive areas, both in land and sea, tend to occur near rivers? ▪

The carbon cycle: Organisms exchange CO$_2$ with the atmosphere

Carbon forms the backbone of organic molecules and the crystalline structure of bones and shells. Like the water cycle, the biogeochemical cycle for carbon (**Figure 23.13**) is called a *gaseous cycle* because carbon in living organisms is exchanged with atmospheric CO$_2$.

The carbon cycle is closely tied to photosynthesis by plants and aerobic respiration by both plants and animals. During photosynthesis, plants use the energy of sunlight to combine CO$_2$ with water, forming carbohydrates and releasing oxygen as a by-product. Both plants and animals then utilize aerobic respiration (metabolism in the presence of oxygen) to break down the carbohydrates, releasing the energy to make other complex molecules. In the process they produce CO$_2$ again. CO$_2$ is also released through the actions of decomposers such as fungi and bacteria.

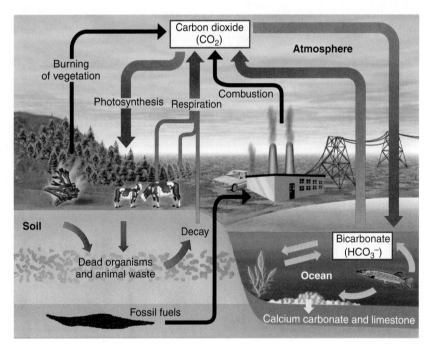

Figure 23.13 The carbon cycle. Carbon dioxide in the atmosphere and bicarbonate in the oceans constitute the exchange pool. The reservoir pool of carbon includes calcium carbonate and limestone from the shells of marine organisms and also fossil fuels formed several hundred million years ago from decomposed plant material. Fossil fuels represent a reservoir for carbon that would not normally be accessible to the exchange pool, but over the last several hundred years humans have tapped it for energy. The darker-colored arrows indicate the effect of human activities.

✅ *The amount of CO$_2$ in the Earth's atmosphere shows a marked annual cycle—a decrease every July and August followed by an increase every December and January. Why? Which arrow on this graph represents a process that changes on an annual cycle?*

In the absence of human interference, utilization and production of atmospheric CO_2 by the biomass would be in approximate balance. However, as shown in Figure 23.13, burning of vegetation and combustion of fossil fuels are tipping the balance toward increased CO_2 production, raising atmospheric CO_2 levels. The result may be global warming, discussed in Chapter 24.

 Web Animation *The Carbon Cycle and Global Warming* at www.humanbiology.com

Nitrogen: An essential component of nucleic acids and proteins

Nitrogen is an essential component of proteins and nucleic acids. By far the largest reservoir of nitrogen is the nearly 79% of the air that is nitrogen gas (N_2). However, the nitrogen in air is unavailable to living organisms because the two nitrogen atoms are held tightly together by a triple covalent bond. Consequently plant growth is more often limited by a shortage of useful nitrogen than a shortage of any other nutrient.

Figure 23.14 illustrates the nitrogen cycle. The process of converting nitrogen gas to ammonium (NH_4^+) is called **nitrogen fixation** because the nitrogen becomes trapped, or "fixed," in a form that ultimately can be used by plants. Nitrogen fixation is carried out by certain bacteria in the root nodules of plants such as peas, alfalfa, and clover and also by modern fertilizer factories. Most of the ammonium is converted to nitrate (NO_3^-) before it is used by plants. The formation of nitrate is called **nitrification.** Modern fertilizer plants use fossil fuels to manufacture nitrate from air, and lightning provides the energy for a small amount of nitrate formation as well.

Plants take up ammonium and nitrates from the soil and assimilate them into their proteins and nucleic acids in the course of their cellular metabolism. All other organisms must rely on plants or other animals for their usable nitrogen. When organisms die and decompose, the nitrogen compounds in their proteins and nucleic acids are converted again to ammonia, aided by bacteria. Most of the ammonia undergoes nitrification again and recycles through the biomass of living organisms. Similar nitrogen cycles occur on land and in the sea.

Finally, **denitrification** by certain denitrifying bacteria converts some nitrate back to atmospheric nitrogen gas. In a balanced ecosystem, denitrification should equal nitrogen fixation. The production of fertilizers by humans has tipped the balance in favor of nitrogen fixation. Excessive fertilizer runoff can pollute freshwater because the nitrates stimulate excessive growth of plants and algae.

 Web Animation *The Nitrogen Cycle* at www.humanbiology.com

Phosphorus: A sedimentary cycle

Phosphorus is an essential element of life that most commonly exists as phosphate ions (PO_4^{3-} and HPO_4^{2-}). Producer organisms use phosphate to make ATP for energy, phospholipids for cell membranes, and nucleotides for RNA and DNA. Consumer organisms that feed on the producers use phosphate for these purposes too, but they also incorporate phosphate into bones, teeth, and shells.

Figure 23.15 shows the phosphorus cycle. It is called a *sedimentary cycle* because, unlike carbon or nitrogen, phosphorus never enters the atmosphere. Phosphate ions deposited in sediments over the course of millions of years are brought to Earth's surface by the

Figure 23.14 The nitrogen cycle. The nitrogen in nitrogen gas is not readily available to living organisms. Once it is "fixed" as ammonium or nitrified to nitrate, the nitrogen can be recycled over and over through the biomass. Nitrogen fixation on land is carried out by bacteria in the root nodules of certain plants. Cycles of nitrogen utilization occur both on land and in the sea. Under normal circumstances, the amount of nitrogen gas returned to the atmosphere by denitrification approximately balances nitrogen fixation. Humans alter this natural balance through the large-scale commercial production of nitrate and ammonium fertilizers (darker blue arrows).

 Circle the three types of bacteria that catalyze key steps in the nitrogen cycle on land, and then, without looking at the figure, write down the type of nitrogen conversion that is performed by each one.

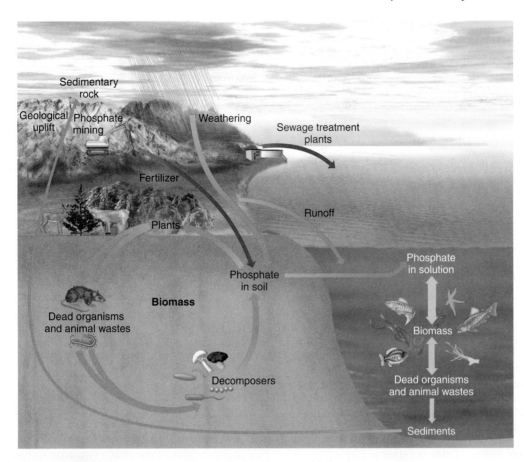

Figure 23.15 The phosphorus cycle. The reservoir pool for phosphorus is phosphate ions in sediments and sedimentary rocks. The slow weathering of rock makes "new" phosphate available to the exchange pool of phosphate in the soil. Some of this phosphate is taken up by plants and some runs off with the surface water to become the phosphorus supply for aquatic ecosystems. Eventually the phosphate ends up in sediments, where it may remain for millions of years until geologic upheaval brings the sedimentary rocks to the surface and a weathering process can begin again. Humans disturb the natural cycle by mining phosphate and applying it as fertilizer where it is in short supply. In urban areas, changes in land cover produce excess phosphate in runoff to streams and oceans. When organisms die the store of phosphate becomes available to producer organisms again.

upheaval of Earth's crust and the formation of new mountain ranges. The phosphates in sedimentary rocks are mined by humans and weathered by the forces of nature.

Without human intervention, phosphates from sedimentary rock are exchanged with water and soil, providing needed minerals for plant growth. With urbanization, an increase in covered surface areas (roads, parking lots, and buildings) leads to increased water runoff into aquatic ecosystems, with less water absorption by soil and plants. Excess phosphates leached from rock reservoirs and phosphate from fertilizers or from human or animal wastes may also enter aquatic ecosystems, causing excessive algae growth. When the algae die, decomposers using them for food take up so much oxygen for metabolism that other organisms may suffocate.

In our discussion, we have focused on several major nutrients of life. There are similar biogeochemical cycles for all other nutrients, including oxygen, calcium, potassium, and

sulfur. Pollutants and toxic substances such as heavy metals (mercury, cadmium, chromium, and lead), pesticides (DDT), and industrial chemicals (polychlorinated biphenyls or PCBs) also cycle between living organisms and Earth. Environmentalists need to understand these cycles to advise us on how to minimize the effects of pollution.

Recap The amount of matter in Earth is fixed. The chemicals and molecules that constitute all matter are recycled between the biomass, an exchange pool, and a reservoir pool. Most nutrients cycle rapidly within their biomass pool, less rapidly with their exchange pool, and rarely with their reservoir pool. Geochemical cycles can be described for virtually any chemical. Among the most important geochemical cycles for sustaining life are the cycles for water, carbon, nitrogen, and phosphorus. ■

Chapter Summary

Ecosystems: Communities interact with their environment *p. 528*

- The study of the relationships between organisms and their environment is called ecology.
- An ecosystem consists of a community of organisms and the physical environment in which they live.

Populations: The dynamics of one species in an ecosystem *p. 528*

- When there are no restrictions to its growth, a population grows exponentially according to its biotic potential.
- Within an ecosystem, the stable size of the population of a particular species ultimately is determined by carrying capacity.

Human population growth *p. 530*

- The human population has been growing rapidly since the Industrial Revolution, beginning in the 1700s.
- Currently there are wide differences in fertility rates between the more industrialized countries and the less industrialized countries.
- Even when (or if) the replacement fertility rate of 2.1 children per woman is reached, the world population will continue to grow for several more decades.
- Humans may be able to increase the magnitude of their carrying capacity temporarily using technology, but these activities may ultimately reduce that capacity.

Communities: Different species living together *p. 532*

- Species that occupy overlapping niches may be in competition for limited resources.
- Communities go through a natural sequence of changes called succession that ends with the establishment of a stable climax community.

Energy flows through ecosystems *p. 533*

- Because energy flows only one way through an ecosystem, a constant supply of energy (from the sun) is required.
- Producer organisms can utilize the energy of the sun through a process called photosynthesis. Consumer and decomposer organisms must rely on other organisms for energy.
- Less energy is available at each higher consumer level of a food web.

Chemical cycles recycle molecules in ecosystems *p. 538*

- Chemicals (nutrients) recycle in an ecosystem, sometimes quickly, sometimes slowly.
- Water is constantly being recycled from the oceans to land and back to the oceans.
- In the carbon cycle, plants use CO_2 available in the atmosphere and animals obtain organic carbon molecules by consuming plants or other animals.
- Nitrogen gas in the atmosphere must be converted to NH_4^+ or NO_3^- ("fixed") before it is useful to living organisms.
- The phosphorus cycle is a sedimentary cycle. Phosphorus from decaying organisms is available for reuse.

Terms You Should Know

biogeochemical cycle, *538*
biomass, *533*
biosphere, *528*
carnivore, *534*
carrying capacity, *530*
climax community, *533*
community, *528*
consumer, *534*
decomposer, *534*
denitrification, *540*
ecological pyramid, *537*
ecology, *528*
ecosystem, *528*
environmental resistance, *529*
fertility rate, *531*
food web, *535*

fossil fuels, *538*
geographic range, *529*
habitat, *528*
herbivore, *534*
less industrialized countries (LICs), *531*
more industrialized countries (MICs), *531*
nitrification, *540*
nitrogen fixation, *540*
omnivore, *534*
photosynthesis, *534*
population, *528*
producer, *534*
replacement fertility rate, *531*
zero population growth, *531*

Concept Review

Answers can be found at the Human Biology Place.
www.humanbiology.com

1. What are the components of an ecosystem?
2. Describe the concept of an ecosystem's carrying capacity.
3. Describe what happens to a country as it undergoes demographic transition. Give examples of nations in early and late stages of transition.
4. Define zero population growth and explain why it has not been achieved.
5. Describe the concept of a climax community and why such communities are sensitive to disruption.
6. Why is a constant supply of energy needed to sustain life?
7. Discuss the interactions between producers and consumers in an ecosystem.
8. Explain why the water cycle is essential to all other biogeochemical cycles.
9. Discuss the importance of the carbon cycle.
10. Discuss the phosphorus cycle and its importance.

Test Yourself

Answers can be found in Appendix A.

1. Your college campus, including all the humans and other living things that occupy it as well as its physical environment, such as the buildings, soil, water, and air, is a(n):
   a. population
   b. community
   c. ecosystem
   d. biosphere

2. Which of the following will affect the biotic potential of a species?
    a. the length of time it takes an individual to reach maturity
    b. the number of members of the population at reproductive age
    c. the number of offspring produced by each reproducing member
    d. all of the above

3. A population is growing at a rate of 2% per year. How many years will it take for this population to double in size?
    a. 16 years   b. 20 years
    c. 72 years   d. 36 years

4. Which of the following statements is false?
    a. The human population has grown steadily throughout human history.
    b. Human populations are now growing explosively.
    c. Human population growth can be represented by a J-shaped curve.
    d. Humans have been able to alter the carrying capacity of Earth.

5. Which statement about the human fertility rate is true?
    a. Fertility rates are the same in all human populations.
    b. Fertility rates can apply to both men and women.
    c. A sustained fertility rate of 2.1 would eventually lead to zero population growth.
    d. Death rates are factored in when calculating fertility rates.

6. The large intestine of humans is populated by a complex community of microorganisms which compete for nutrients and may prevent pathogenic microorganisms from becoming established. Which term would best describe this?
    a. competitive exclusion   b. selective pressure
    c. succession   d. environmental resistance

7. Which of the following characterizes climax communities?
    a. They contain a limited number of different species.
    b. They are generally stable and long-lived.
    c. They contain equal numbers of producers and consumers.
    d. They can be reestablished quickly following significant environmental change.

8. _____ flow(s) one-way through ecosystems while _____ is/are recycled.
    a. Biomass…elements   b. Energy…elements
    c. Elements…energy   d. Producers…consumers

9. All of the following terms can appropriately describe humans except:
    a. primary consumer   b. autotroph
    c. heterotroph   d. secondary consumer

10. A human maintaining a vegan diet (containing no animal products) would be a:
    a. producer   b. primary consumer
    c. secondary consumer   d. decomposer

11. All of the following processes are involved in the carbon cycle except:
    a. photosynthesis
    b. cell respiration
    c. evaporation
    d. decomposition

12. The shape of an ecological pyramid, with each successive layer being smaller than the one below it, can be explained most readily by:
    a. the first law of thermodynamics
    b. inefficient geochemical cycles
    c. the second law of thermodynamics
    d. industrialization

13. Plant growth is most likely to be limited by the available supply of usable:
    a. nitrogen
    b. carbon
    c. oxygen
    d. phosphorus

14. Which biogeochemical cycle does not include exchange with the atmosphere?
    a. nitrogen
    b. carbon
    c. phosphorus
    d. water

15. The biogeochemical cycles involve the cycling of key molecules and elements between a reservoir, an exchange pool, and:
    a. the biomass
    b. producers
    c. consumers
    d. decomposers

Apply What You Know

Answers can be found at the Human Biology Place.
www.humanbiology.com

1. Given absolutely no environmental resistance, how long would it take a population that was increasing by 8% a year to double?
2. Speculate as to what might eventually become the most important environmental resistance factor (or factors) determining the world's human carrying capacity.
3. Less industrialized countries tend to be poorer than more industrialized countries, and at the same time their populations are among the fastest growing. What sorts of interventions or assistance are likely to help less industrialized countries develop and prosper?
4. Both hawks and owls feed on small mammals to sustain their respective populations without causing competitive exclusion. How can this work when both species live in the same geographic area?
5. The introduction of the zebra mussel into the Great Lakes by oceangoing vessels has had a significant impact on the lake ecosystem. How could such a small organism have had such a large impact?
6. What would happen in an ecosystem if suddenly there were no decomposers?
7. The term "algae bloom" refers to rapid, excessive growth of algae. What might cause an algae bloom, and what would it cause?

24

Human Impacts, Biodiversity, and Environmental Issues

Air pollution over New York City.

Global Warming

Earth's surface temperature is slowly increasing, according to climatologists and Earth scientists. The phenomenon, known as global warming, is thought to be due to steadily rising atmospheric concentrations of greenhouse gases (primarily CO_2) brought about by the use of fossil fuels as a source of energy. In other words, "Earth's fever" is being caused by human activities.

Some people just don't want to hear about it because they can't see an easy solution. Others continue to deny that it is even happening. Still others argue that at the current rate of only about a 0.2° Celsius rise in temperature every ten years, it hardly seems like something to get too worried about.

To fully grasp the concept of global warming and why it may be a problem, you need to understand that the effect of global warming probably will *not* be that the summers feel hotter than normal. In fact, you are not likely to notice global warming at all. Global warming is a slow process; in the past 100 years the temperature has risen only about 0.8° Celsius (1.4° Fahrenheit). More disturbingly, the rate of rise has doubled in just the past 50 years. These small changes in temperature are expected to have effects that last for

hundreds, thousands, or even tens of thousands of years.

Current climate models suggest that if global warming continues, most of the polar ice cap may eventually melt. Rising sea levels will erode coastlines and flood low-lying coastal areas, including many of the world's major cities. The number and intensity of hurricanes, typhoons, and tornadoes could increase. Climate patterns may change; some regions of the world would probably experience more rain and frequent floods, and others would undergo prolonged draughts and become deserts. Crop failures and outbreaks of communicable diseases

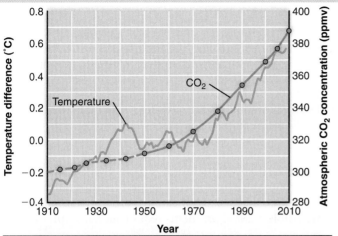

Global temperature differences (from a 1951–1980 base) and atmospheric CO_2 concentrations over the past 100 years. Data from NASA and the Carbon Dioxide Information Analysis Center.

may become more frequent. Fragile ecosystems could unravel, leading to a mass extinction of many species of organisms. In other words, life on Earth may change subtly, slowly, and for a very long time.

Is There a Global Warming Tipping Point?

Some Earth scientists and environmentalists talk of "tipping points" to try to convince the public and governments to do something about global warming before it is too late. A *tipping point* is loosely defined as a point at which patterns or relationships of the past no longer hold, and everything changes suddenly and dramatically. The concept of a tipping point in global warming would be that beyond some level of greenhouse gases that has not yet been reached, changes in atmospheric CO_2 concentrations would cause much larger increases in temperature than they do now, and global warming would spiral out of control. One tipping point scenario that has been proposed recently is that if the Arctic tundra begins to thaw due to global warming, vast quantities of methane (another greenhouse gas) might be released from the soil, causing a more rapid increase in global warming than is predicted by current models.

So far, the data of the last 100 years do not support the hypothesis that there is a tipping point in the progression of global warming. The relationship between the rise in atmospheric CO_2 concentrations and the rise in Earth's surface temperature has been consistent and fairly linear over that time. This is not proof that tipping points don't exist, however; it merely indicates that we haven't crossed one yet.

One of the dangers of talking about tipping points is the effect it might have on the public's perception of global warming. If proposed tipping points are passed and nothing dramatic happens, the public may become bored and lose interest. Then we might not work to solve the more fundamental problem—the slow, relentless warming of the planet.

What Can be Done?

Stopping global warming will require a commitment by all nations, not just a few. The solution seems obvious, but achieving it will be difficult. Simply put, if global warming is caused by rising concentrations of greenhouse gases, then we (the global society of humans) must halt the rise in atmospheric CO_2, and work to reverse it. To do that we must either reduce our dependency on fossil fuels for energy or find a way to keep the CO_2 that comes from burning fossil fuels out of the atmosphere.

A number of possible approaches have been discussed at recent international gatherings on the subject. One involves reducing emissions by trading "carbon credits." Each nation and company would be given a yearly allotment of carbon credits (representing the right to emit a certain amount of carbon into the atmosphere) based on an internationally agreed-upon formula. They could then trade carbon credits (buy additional credits or sell unused ones) on an open international market. The goal would be to cap worldwide CO_2 emissions by punishing countries and companies for emitting more than their share of CO_2 and rewarding countries and companies for reducing their CO_2 emissions. Another solution that individual countries can use is taxation; some countries already levy heavy taxes on companies for emitting CO_2 into the atmosphere.

Perhaps the most promising strategy is carbon sequestration; storing or burying CO_2 beneath the ocean floor or deep underground. Norway's Statoil company extracts the CO_2 from one of it's natural gas wells, liquefies it, and then pumps it back into a sandstone formation that is almost half a mile below the ocean floor. The company sequesters about 2,800 tons of CO_2 every day. It's expensive, but on the other hand if the company released the CO_2 into the atmosphere they would have to pay approximately \$175,000/day in Norwegian CO_2 tax.

Carbon credits, carbon taxation, and carbon sequestration are all potentially good ideas. But solving global warming will require more than good ideas; it will require a concerted effort over a long period of time by a coalition of most of the nations of the world. Whether or not that ever happens may depend on the perceived severity of the problem versus the costs of doing something about it. Right now we seem to be still searching for answers.

Questions to consider

1 Is it possible to protect world economies and still address global warming? Is it necessary? Explain your answer.

2 In what ways might your own lifestyle choices contribute to global warming? Would you be willing to support legislation that would affect your lifestyle choices in order to reduce global warming?

The facts...

- Unless action is taken soon, global warming is likely to have profound long-term effects on climate, weather, agriculture, and human health.

- Proponents of taking action argue that there may be "tipping points" after which global warming could spiral out of control.

- Approaches to addressing global warming include trading carbon credits, taxation of CO_2 released into the atmosphere, and sequestering (storing) CO_2 deep underground.

- Solving global warming will require cooperation among the nations of the world.

Key concepts

>> **Increased production of greenhouse gases is contributing to global warming.** Certain air pollutants produced by human activities also contribute to acid rain, cause poor air quality, and deplete Earth's protective ozone layer.

>> **Less than 1% of the water on Earth is freshwater.** Freshwater resources are declining in some regions of the world due to water pollution and inefficient use of water for agricultural purposes.

>> **Some human activities damage ecosystems and lead to a loss of biodiversity on Earth.** The consequences of loss of biodiversity are not well understood.

>> **Most of our energy comes from non-renewable resources (coal, oil, and gas).** A shift to renewable energy sources is slowly getting under way.

>> **Sustainability of life on Earth will require meeting our current needs without compromising the needs of future generations.** True sustainability will take into account the needs not only of humans, but also all other species.

---

thousands of chemicals or particles that have adverse effects on living organisms, collectively known as **pollution.** The major concerns regarding air pollution fall into four areas:

- Global warming
- Destruction of the ozone layer
- Acid precipitation
- Smog production

Figure 24.1 shows that for each of these issues there may be more than one contributing pollutant. Air pollution is difficult to remedy because it can be hard to determine who is responsible. Global problems require global solutions.

Excessive greenhouse gases lead to global warming

If you've ever been in a glass greenhouse on a sunny day you know that it permits sunlight to penetrate, but as sunlight is converted to heat energy, the greenhouse becomes warmer because the glass prevents the heat from escaping. Certain gases in the upper layer of the atmosphere (the stratosphere) act in much the same way for Earth; therefore they are called *greenhouse gases.*

The most significant greenhouse gas is water vapor, accounting for approximately 60% of the total. The remainder are human-made and consist of carbon dioxide (CO_2), methane (CH_4), nitrous oxide (N_2O), and small amounts of

---

The human capacity to shape the environment both creatively and destructively is causing global changes in biogeochemical cycles and affecting the populations of nearly all living species. Only humans live in houses that can be thousands of square feet per person. Only humans irrigate lands that otherwise could not grow crops, level entire mountains in search of minerals and fossil fuels, and spread herbicides, insecticides, and fertilizers over thousands of square miles. Our presence has altered air, water, and land quality not only in local ecosystems but globally as well. Along the way, we have favored the proliferation of some species and driven others to extinction.

This chapter examines how we *Homo sapiens* are altering our environment on a global scale. We will look at the causes and impacts of pollution on air, water, and land. We will discuss how our energy requirements affect the environment. Finally, we'll describe the benefits of biodiversity and look at steps we can take to build a sustainable future.

24.1 Pollutants impair air quality

The air we breathe is a mixture of primarily nitrogen (nearly 79%) and oxygen (nearly 21%), with trace amounts of carbon dioxide (0.03%). Air also contains trace amounts of

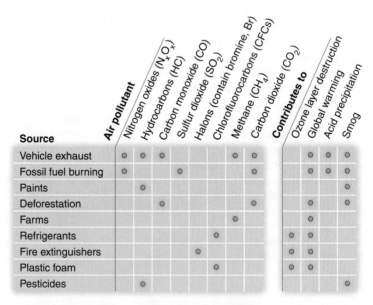

Figure 24.1 Major sources of air pollutants. A single activity (such as driving a car) may produce more than one type of air pollutant.

✓ *Which source of pollution releases the greatest number of different categories of pollutants? How many? Which environmental effect does it not contribute to, and why not?*

Answers to ✓ questions can be found in Appendix A.

two other air pollutants: the chlorofluorocarbons (CFCs), used primarily as refrigerants, and gases called halons that contain bromine, used in fire extinguishers. Together the greenhouse gases produce a **greenhouse effect,** allowing sunlight to come to Earth, but trapping most of the heat radiated from Earth on its way to space (**Figure 24.2**). This is a natural and normal phenomenon. Without the greenhouse effect, most of the sun's heat would radiate away from Earth and the average temperature would be well below freezing. In other words, to maintain a normal surface temperature on Earth, a certain amount of greenhouse gases is needed in the stratosphere.

Most of the CO_2 component of the greenhouse gases accumulated over millions of years, as a result of respiration. In modern times, however, human activities have increased the levels of greenhouse gases, especially CO_2. There is general agreement now among scientists that this excessive production of greenhouse gases is increasing the greenhouse effect and raising the average global temperature, a phenomenon known as **global warming.**

The main human activity that raises atmospheric CO_2 levels is the burning of fossil fuels for energy. The carbon in fossil fuels comes from decayed plant material that was buried underground by sedimentary processes over millions of years. When we burn fossil fuels we release the carbon into the air as CO_2.

The other main human activity that raises atmospheric CO_2 concentrations is **deforestation** (removing trees from large areas of land). Trees absorb CO_2 from the air during photosynthesis. A large growing tree will take up and store in its wood approximately 50 lb of CO_2 per year.

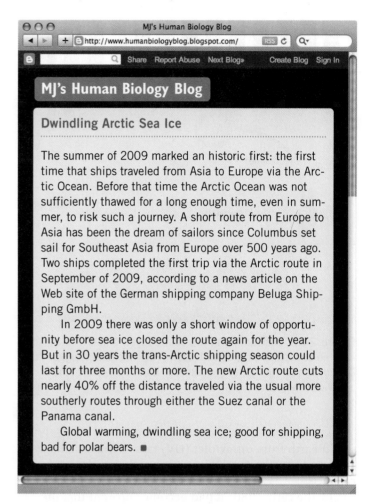

MJ's Human Biology Blog

Dwindling Arctic Sea Ice

The summer of 2009 marked an historic first: the first time that ships traveled from Asia to Europe via the Arctic Ocean. Before that time the Arctic Ocean was not sufficiently thawed for a long enough time, even in summer, to risk such a journey. A short route from Europe to Asia has been the dream of sailors since Columbus set sail for Southeast Asia from Europe over 500 years ago. Two ships completed the first trip via the Arctic route in September of 2009, according to a news article on the Web site of the German shipping company Beluga Shipping GmbH.

In 2009 there was only a short window of opportunity before sea ice closed the route again for the year. But in 30 years the trans-Arctic shipping season could last for three months or more. The new Arctic route cuts nearly 40% off the distance traveled via the usual more southerly routes through either the Suez canal or the Panama canal.

Global warming, dwindling sea ice; good for shipping, bad for polar bears. ▪

Deforestation is due to indiscriminate logging for wood, and to widespread burning of forests to create new cropland. Deforestation by burning is doubly damaging—not only are there fewer trees to absorb CO_2 from the atmosphere, but all of the carbon in the burned wood is immediately released back into the atmosphere.

✔ **Quick Check** One of the most active areas of global warming research includes an attempt to model weather to predict how much global warming might change the average amount of cloud cover on Earth. Why is cloud cover of such interest to researchers? ▪

CFCs deplete the ozone layer

Ozone (O_3) is found in two places in the atmosphere. In the atmospheric layer near the planet surface (the troposphere), ozone is an air pollutant formed by the reaction of oxygen with automobile exhaust and industrial pollution. Ozone is mildly toxic, causing plant damage and respiratory distress in animals, including humans.

Higher up in the atmosphere, ozone is actually very beneficial. It forms a thin layer in the stratosphere that helps

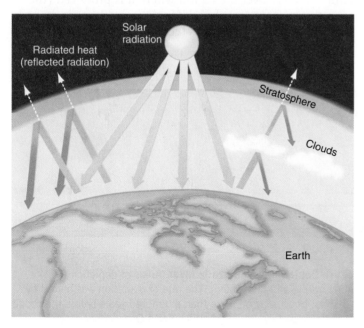

Figure 24.2 The greenhouse effect. A layer of stratospheric greenhouse gases consisting primarily of carbon dioxide (CO_2) allows sunlight to pass but traps most of the heat. An overabundance of greenhouse gases may contribute to global warming.

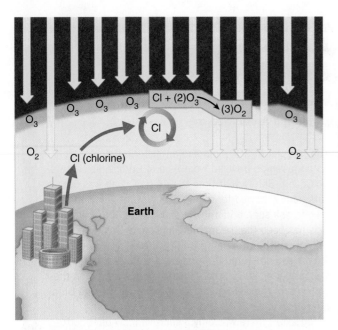

Figure 24.3 Destruction of the ozone layer. Chlorine released into the atmosphere reacts with ozone (O_3) to produce oxygen (O_2). Destruction of the ozone layer exposes Earth's surface to more ultraviolet radiation.

shield Earth from ultraviolet (UV) rays. UV rays damage DNA and contribute to skin cancer, and may also cause cataracts.

By the early 1980s it had become apparent that the stratospheric ozone layer was suffering significant damage due to chlorofluorocarbons (CFCs), a group of chemicals used in refrigerators, air conditioners, and aerosol sprays. CFCs released near ground level migrate slowly upward toward the stratosphere and decompose, releasing chlorine atoms. The chlorine atoms combine with ozone and destroy it, producing oxygen (**Figure 24.3**). Because the chlorine molecule can be used over and over in the reaction, a single chlorine atom may destroy as many as 10,000 ozone molecules.

By 1985 the stratospheric ozone layer had thinned noticeably. Holes began to develop in the ozone layer at certain times of the year, and the United Nations predicted that rates of skin cancer would begin to increase. Once the problem was recognized, quick action was taken. An international agreement to phase out the production of CFCs was signed by most industrial nations in 1987, and by the mid-1990s the size of the hole had stabilized. It remains at about the 1995 level today (**Figure 24.4**).

Models of the stratosphere predict that it may be 100–150 years before the ozone layer recovers completely. Nevertheless, quick action on the part of the

international community may have kept the problem from becoming even worse.

Pollutants produce acid precipitation

The major source of acid precipitation is sulfur dioxide, which is released into the air as a result of burning high-sulfur coal and oil for power. A secondary source is nitrogen oxide in automobile exhaust. Sulfur dioxide and nitrogen oxides combine with water vapor in the air, become sulfuric acid and nitric acid, and dissolve in raindrops, which fall as **acid precipitation.** Acid precipitation corrodes metal and stone and damages forests and aquatic ecosystems, particularly in the northeastern United States, southeastern Canada, and Europe. Some acid products also precipitate as solid particles of sulfate and nitrate salts.

Starting in the 1970s some coal-burning power plants have been required to install sulfur removal and capture systems known as "scrubbers" to reduce their emissions of sulfur dioxide. As a result, since 1985 the deposition of sulfur in rainwater has declined by about 33% across most of the U.S. Northeast. Recent regulatory actions taken by the Obama administration are expected to eliminate most of the remaining sulfur emissions by 2014.

Smog blankets industrial areas

Several pollutants in air, most notably nitrogen oxides and hydrocarbons (chains of carbons linked to hydrogens), react with each other in the presence of sunlight and water vapor. They form a hazy brown or gray layer of **smog** that tends to hover over the region where it is produced (the term comes from "smoke" + "fog"). Most smog is caused by the burning of fossil fuels (coal and oil) and by automobile exhausts.

1979   **1995**   **2009**

Figure 24.4 Depletion of the ozone layer. Areas of depletion are indicated by purple and blue colors. The rate of depletion was halted by the mid-1990s, but it may be more than a hundred years before the ozone layer recovers completely. Source: NASA's Earth Observatory.

 The country of New Zealand has the world's highest rate of death from skin cancer. Why? (Hint: There are two reasons.)

Smog also contains several chemicals that irritate the eyes and lungs and may lead to chronic respiratory illnesses such as asthma and emphysema. Depending on the source, smog may also contain small oil droplets and particles of wood or coal ash, asbestos, lead, dust, and animal waste. Smog can become especially troublesome during a **thermal inversion,** when a warm stagnant upper layer of air traps a cooler air mass containing smog beneath it.

Like acid precipitation, the problem of smog has been reduced due to widespread cleanup efforts and reduced emissions of pollutants. For example, smog was once a significant health problem in London and Pittsburgh, but both of these cities have succeeded in major efforts to reduce smog (**Figure 24.5**). The state of California has passed strict automobile antipollution laws. Many regions of the United States now require the use of cleaner-burning oxygenated automobile fuels containing ethanol or methanol during certain winter months, and the use of lead as a fuel additive has been significantly reduced as well. Other areas, including cities such as Cairo and Mexico City, have not yet solved their smog problems.

Recap Air quality and air pollution are a worldwide concern. Human activities are destroying the protective layer of ozone around Earth, producing acid precipitation and smog, and may be causing a slow global warming trend. Remedial efforts are effective, as shown by the reduction of smog in some cities and the application of new technologies for reducing acid precipitation. Global initiatives to slow global warming and reduce ozone layer destruction are already under way. ■

24.2 Pollution jeopardizes scarce water supplies

Human activities have three major detrimental effects on water quality and availability. First, humans use excessive amounts of water, depleting freshwater supplies. Second, the replacement of natural vegetation with buildings and roads, especially in urban areas, prevents rainwater from soaking in, causing runoff. Third, human activities sometimes pollute precious sources of water.

Water is scarce and unequally distributed

In one sense, water is a renewable resource because it evaporates continuously from the oceans and falls on land as rain or snow. However, all of the freshwater on the surface of the land and in **aquifers** (deep underground reservoirs) forms less than 1% of Earth's total water. Over 97% is salty ocean water and 2% is frozen in glaciers and polar ice caps.

In addition to general scarcity, water is not evenly distributed among human populations. Residents of the more industrialized countries use 10 to 100 times more water than people in less industrialized countries. Some desert and semiarid regions of the world have already reached their human carrying capacity as a result of water shortage. When we divert water for our purposes, we take it away from other species or limit their normal migration patterns. Water rights are a controversial subject in the western United States, where water is limited. Already, choices have to be made between irrigating crops, supplying growing cities, and encouraging the reproduction of Pacific salmon.

 Web Animation *Human Population Growth* at www.humanbiology.com

Urbanization increases storm water runoff

In urban areas, the shift from woodlands and fields to impermeable roads and buildings has caused major stormwater runoff problems. In older cities on the U.S. East Coast, stormwater combines with sewage, leading to combined sewage overflow (CSO), which overwhelms receiving streams and oceans. Twenty-eight billion gallons of CSO impact the New York harbor each year. This CSO is the major source of pathogens in the area, causing gastroenteritis, eye and ear infections, skin rashes, respiratory infections, and hepatitis in swimmers and kayakers. Furthermore, the use of pipes to transport stormwater runoff quickly from buildings and roads to streams leads to stream overflows during storms and insufficient water levels during dry periods. The result is erosion of stream beds and

a) Pittsburgh at the height of the steelmaking era, before smog cleanup efforts began.

b) Pittsburgh after programs were instituted to reduce smog.

Figure 24.5 Smog.

Figure 24.6 Urban stream erosion. Urbanization has adverse effects on streams due to excess and polluted runoff.

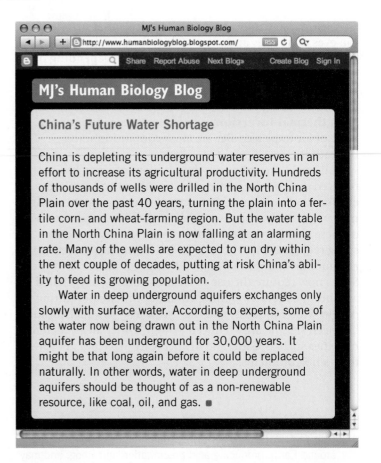

MJ's Human Biology Blog

China's Future Water Shortage

China is depleting its underground water reserves in an effort to increase its agricultural productivity. Hundreds of thousands of wells were drilled in the North China Plain over the past 40 years, turning the plain into a fertile corn- and wheat-farming region. But the water table in the North China Plain is now falling at an alarming rate. Many of the wells are expected to run dry within the next couple of decades, putting at risk China's ability to feed its growing population.

Water in deep underground aquifers exchanges only slowly with surface water. According to experts, some of the water now being drawn out in the North China Plain aquifer has been underground for 30,000 years. It might be that long again before it could be replaced naturally. In other words, water in deep underground aquifers should be thought of as a non-renewable resource, like coal, oil, and gas. ■

loss of aquatic life (Figure 24.6). Restoration of urban streams seeks to stabilize stream channels, reduce channel erosion, and restore aquatic wildlife populations.

Human activities pollute freshwater

Human activities tend to pollute what freshwater there is. Untreated sewage, chemicals from factories, the runoff of pesticides and fertilizers, and rubber and oil from city streets, all must go somewhere. Either they degrade chemically according to their natural cycle of decomposition or they end up in water or soil.

Some water pollutants are *organic nutrients* that arise from sewage treatment plants, food-packing plants, and paper mills. When these nutrients are degraded by bacteria, the rapid growth rate of bacteria can deplete the water of oxygen, threatening aquatic animals. Others are *inorganic nutrients* such as nitrate and phosphate fertilizers and sulfates in laundry detergents. These cause prolific growth of algae, which die and are also decomposed by bacteria.

Eutrophication refers to the rapid growth of plant life and the death of animal life in a shallow body of water as a result of excessive organic or inorganic nutrients (Figure 24.7). Eutrophication is part of the normal process that converts freshwater into marsh and then dry land, but human activities can accelerate the process.

A special problem is *toxic pollutants,* including polychlorinated biphenyls (PCBs), oil and gasoline, pesticides, herbicides, and heavy metals. Most of these cannot be degraded by biological decomposition, so they remain in the

Figure 24.7 Eutrophication. Water pollution accelerates the normal process of eutrophication, encouraging prolific growth of algae and plants at the expense of animal life.

✓ *Some rivers in North America have a "dead zone," with virtually no fish, at their mouth, where they join the sea. What causes the dead zone?*

environment for a long time. In addition, they tend to become more concentrated in the tissues of organisms higher up the food chain, a phenomenon known as **biological magnification** (Figure 24.8). Biological magnification

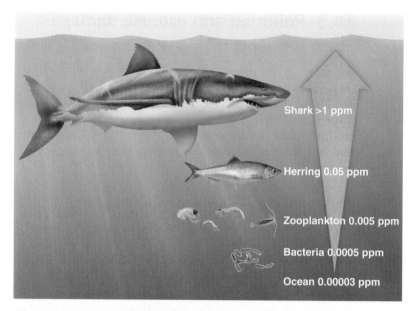

Shark >1 ppm

Herring 0.05 ppm

Zooplankton 0.005 ppm

Bacteria 0.0005 ppm

Ocean 0.00003 ppm

Figure 24.8 Biological magnification. The heavy metal mercury accumulates in body tissues as it moves up a food chain. Even though mercury may be present in only very trace amounts (about 0.00003 ppm or parts per million) in water, it becomes highly concentrated (more than 1 ppm) by the time it reaches top predators such as sharks and whales.

occurs because each animal in a food chain consumes many times its own weight in food throughout its lifetime.

A classic example of biological magnification is the heavy metal mercury. The primary sources of mercury in the environment are emissions from coal burning, gold production, smelters for nonferrous metal production, and cement production. Mercury released into the air or on land often ends up in aquatic ecosystems, where it accumulates by biological magnification in tertiary consumers such as sharks, tuna, and whales. Exposure to mercury can result in diverse symptoms including loss of coordination, decreased memory and intellect, and poor immune system function. Pregnant women and young children are especially at risk because mercury affects nervous system development. Many countries now have regulations governing the maximum allowable mercury concentrations in fish destined for human consumption.

Other important water pollutants worldwide include *disease-causing organisms* like those that cause typhoid fever and hepatitis, *sediments* from soil erosion that clog waterways and fill in lakes and shipping canals, excess *nitrogen fertilizers* used on land to grow crops, and even *heat pollution* from power plants. Heat pollution reduces the amount of oxygen that water can carry while at the same time increasing the oxygen demand by aquatic organisms whose activity level is temperature dependent. As a result, heat pollution may suffocate aquatic life.

✅ **Quick Check** Which of the following species do you think would have the very highest concentration of toxic pollutants: a swordfish (a short-lived top predator), a blue whale (a long-lived plankton feeder), a killer whale (a long-lived top predator), or a sardine (a short-lived plankton feeder)? ■

Groundwater pollution may impair human health

Many of the same pollutants that taint surface water also pollute groundwater. However, groundwater pollution poses two additional concerns. First, groundwater is often used as drinking water, so its pollutants can quickly affect human health. Second, groundwater is a slowly exchanging pool. Once it becomes polluted it may stay polluted for a long time.

The U.S. Environmental Protection Agency estimates that as many as 50% of all water systems and rural wells contain some type of pollutant. The most common are organic solvents such as carbon tetrachloride, pesticides, and fertilizers such as nitrates. We do not know the full effect of groundwater pollution because it is hard to separate from other possible causes of human disease. Public health officials suspect that some pollutants contribute to miscarriages, skin rashes, nervous disorders, and birth defects.

An issue of special concern regarding possible groundwater contamination is the disposal of radioactive wastes. Radioactive materials are used in nuclear power plants and in the diagnosis and treatment of human disease. Many radioactive wastes remain radioactive for thousands of years. They are usually stored in very dry environments in order to keep them from entering a water supply. Some radioactive wastes are so long-lived that for disposal they are incorporated into glass, which is then buried deep underground.

✅ **Quick Check** In 1997, New York City began purchasing thousands of wooded acres in the watershed that supplies its drinking water. Despite the high cost of buying this land (over a billion dollars), New York officials calculate that the city has *saved* money on its water supply. What might the city have had to spend money on if it had not bought the land? ■

Oil pollution damages oceans and shorelines

A massive oil spill in the Gulf of Mexico in 2010 highlighted the potentially huge environmental costs associated with dependency on fossil fuels. The spill began after a British Petroleum drilling facility named Deepwater Horizon was destroyed by an explosion and fire. Ultimately the Deepwater Horizon spill became the largest oil spill in U.S. history. Concerted efforts were made to burn the oil at sea, skim it off the surface, or disperse it with chemicals before it could reach shore. Despite these efforts, some of the oil ended up on pristine beaches and in fragile coastal estuaries and wetlands in the Gulf states (**Figure 24.9**). The full extent of the environmental and economic damage from just this one oil spill may not be known for decades.

In most years (2010 may have been an exception), several million tons of oil enter the world's oceans. About 50% comes from natural seepage; therefore, some oil pollution is a natural phenomenon. However, 30% of total oceanic oil pollution is caused by oil disposal on land that is washed to the sea in streams and rivers. The remaining 20% results from accidents at sea.

In general, when oil is spilled at sea about a quarter evaporates, nearly half eventually is degraded by bacteria, and the remaining quarter eventually settles to the ocean floor. In the short term, however, and especially if the spill is near shore, an oil spill can cause significant damage to marine and shoreline ecosystems. Before the oil dissipates it may coat living organisms, disrupting their ability to function and even choking and killing smaller organisms. Shoreline ecosystems may show signs of damage for years, including a loss of breeding grounds for shrimp and fish. Cleaning up an oil spill can sometimes save a shoreline ecosystem or help it recover more quickly, but it also shifts some of the pollution to land (if the oil is buried) or the air (if it is burned).

Recap Freshwater is scarce and distributed unevenly. Freshwater can become polluted by surface water runoff from urban areas and by fertilizers, pesticides, and herbicides in rural areas. Pollution of rivers eventually reaches the oceans. Oil spills at sea are an example of how human activities directly pollute oceans and shorelines. ■

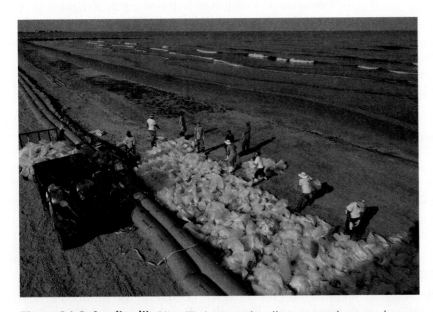

Figure 24.9 An oil spill. Oil spilled near a shoreline may wash up on shore.

24.3 Pollution and overuse damage the land

The main environmental issue with land may not be how we pollute it (though we do), but how much of it we use. Humans consume a lot, and we tend to alter the landscape to suit our own purposes. We dam river valleys to produce hydroelectric power, strip mountaintops to find coal, and cut down forests for lumber or to clear space for crops.

The economy of the United States alone accounts for the direct consumption of 22 tons of fuels, metals, minerals, and biomass (food and forest products) for every person every year. Add to this number the amount of earth moved to build roads and waterways and to find energy, plus the erosion of soil by agricultural and forestry activities, and the total use of natural resources amounts to nearly 88 tons per U.S. citizen each year. It has been estimated that human activities have already altered a third of Earth's land mass, including the removal of nearly half of its forest cover.

As the human population grows, people tend to migrate toward cities. Cities often expand into nearby farmland because it is more economical to build where it is relatively flat, even though only a fraction of Earth's surface will ever be suitable as farmland. Cities also require large quantities of water and power and generate waste and pollution in a relatively small area.

The problems of land use in rural areas are quite different but just as significant. More than half the people of the world live in rural poverty. Many rely almost entirely on their local environment to survive. It is not uncommon for impoverished rural communities to cut down all trees for fuel and shelter and overgraze communal lands with livestock. Stripping the biomass from fragile ecosystems leads to erosion and **desertification**—the transformation of marginal lands into near-desert conditions unsuitable for future agriculture (**Figure 24.10**). Every year an estimated 15 million more acres of once-productive land become desert. Because the very survival of the rural poor is linked so closely to the use of the available resources, it is unlikely that we can halt land degradation in these regions without first doing something about rural poverty.

Wars cause environmental damage that is often overlooked. In Iraq, the draining of the marshlands of the Euphrates/Tigris delta during the 1980s and 1990s resulted in the loss of valuable farmland. More recently the oil spill at the Jiyyeh power station during the Israeli-Hizbollah conflict resulted in damage to 150 km of Lebanese and Syrian coastline, polluting beaches and coastal water.

Figure 24.10 Desertification. Many poor residents of rural areas must depend on local resources to survive. Stripping the biomass from the ecosystem causes desertification, as once-productive land becomes eroded and barren. This photo was taken in Tanzania.

And finally, there is the issue of how we should dispose of our garbage. Landfills are one solution, but there may be other ways to dispose of our waste that have less environmental impact, including recycling as much as possible. Landfills are not necessarily an environmental problem—as long as they are well designed and do not contribute to ground or water pollution. Nevertheless, we could all seek ways to generate less garbage and recycle as much as possible.

Recap Land use and land pollution problems differ by region. In some areas humans alter the landscape in search of fossil fuels and minerals. Cities expand into productive land and place a burden on resources. In rural regions, deforestation and desertification damage ecosystems and limit their future productivity. ∎

24.4 Energy: Many options, many choices

Energy—we just can't get enough of it. We want inexpensive fuel for automobiles weighing thousands of pounds. We heat our homes and even the water in our swimming pools. We need energy to refine raw materials, manufacture the products we use, and process and cook our food.

Energy use is tied to consumption, so we have a choice in determining how much energy to use. We also make choices about the sources of energy—choices that can affect the environment.

First, there are *fossil fuels*—coal, oil, and gas. We've already examined the environmental costs of retrieving and transporting fossil fuels, and we know that burning them may contribute to global warming, acid rain, and smog. But we must also consider that fossil fuels are *non-renewable resources*. Their formation took place over millions of years, and we are on a path of consuming them all in a couple of centuries. Then what?

One way to reduce our dependence on fossil fuels would be to shift toward the use of renewable fuels or fuels with a seemingly infinite supply of energy. The possibilities include nuclear energy, biomass fuels, wind, water, geothermal energy, and solar power. Each has its advantages and its disadvantages.

Nuclear energy can supply a lot of power with the use of very little starting raw material. However, nuclear energy has proved to be relatively expensive because of the safeguards required to ensure that the core of the nuclear facility does not become overheated, suffer a meltdown, and release radioactivity into the environment. In addition, nuclear power plants generate wastes that remain radioactive for thousands of years. How to store radioactive wastes safely is a subject of current debate.

A growing trend is the production and use of *biomass fuels* (or just biofuels)—fuels made from plant materials. Many rural poor have always used biomass fuels such as wood and the dung of herbivores. These are renewable if they are not depleted too fast, but they pollute the air and are not easily transported. In more industrialized countries, ethanol for automobiles and biodiesel for trucks and buses are being produced on a commercial scale. Biofuels production can only take place where biofuel crops can be grown efficiently (e.g., corn in the United States and sugar cane in Brazil), so it is unlikely to be the fuel of choice of countries in cold climates. Biofuels production also removes valuable cropland from food production. Some countries are experimenting with processes that burn waste material, including municipal garbage, to extract energy.

Where flowing water is readily available, hydroelectric power plants can generate electricity. Hydroelectric power plants use the kinetic energy of water to turn turbines that drive electric generators. The environmental cost is the need to build dams that disrupt valley and river ecosystems. Some dams in the western United States are now being dismantled because of the damage they have done to the Pacific salmon population.

Other types of renewable energy sources include "wind farms" that harness wind to generate electricity (**Figure 24.11a**), and geothermal power that uses heat from underground sources deep within the Earth. Wind farms in particular have become popular lately, although not everyone wants one in their backyard. Wind and geothermal power sources have the advantage of providing renewable energy without the need for cooling water, as compared to fossil or nuclear power plants. Wind farms allow rural areas to retain jobs and to transmit energy to urban centers via utility grids. Geothermal energy is most

a) A wind farm. Wind farms are increasingly being used to generate power in areas that have sustained winds.

b) Photovoltaic solar panels. Photovoltaic panels can provide enough electricity to power a home.

c) A solar power plant. This power plant is in Australia.

Figure 24.11 **Alternatives to fossil fuel.**

plentiful near active volcanoes: 5% of California's energy and 25% of El Salvador's are from geothermal sources.

Ultimately we may have to rely on the one original and sustainable source of energy, *solar power*, just as plants do. Solar power can be used to generate electricity in two ways. One uses photovoltaic panels to convert light directly into electricity (Figure 24.11b). The panels absorb light units (photons) and release electrons as direct current. Photovoltaic panels are the method of choice where only a small amount of power is needed, such as for single homes. The other way is to concentrate the solar energy with powerful mirrors to create steam, which is then used to run turbines that generate electrical power on a commercial scale (Figure 24.11c).

At the moment, solar power technologies are less than 30% efficient at energy conversion, which makes them impractical to meet all our current energy demands. The efficiency of photovoltaic panels and solar power plants improves with each passing year, however. With the use of high-efficiency photovoltaic panels on residences and solar tracking mirrors at solar power plants, solar power could make a significant contribution to the world's energy needs, but right now they contribute less than 1%.

One way to reduce our energy consumption is to use passive solar principles in building construction and site selection. Examples include the use of south-facing windows, which can be shaded in the summer, and the use of building materials with high thermal mass. Materials with thermal mass can store heat in winter and prevent heat gain in summer.

Recap Humans consume a disproportionate amount of energy compared to other species. Most of the energy we use comes from non-renewable fossil fuels that contribute to pollution. Other possible sources include nuclear energy, biomass fuels, hydroelectric power, wind farms, geothermal energy, and solar energy. ∎

24.5 Humans are creating a biodiversity crisis

Biodiversity refers to species richness, the assortment of living organisms on Earth. One measure of biodiversity is the number of species: to date, we have identified approximately 1.75 million species, but many scientists estimate that the total could be closer to 10, 30, or even 80 million. Ultimately, biodiversity can be considered as the variety of all forms of life, from species all the way to ecosystems. In this sense, biodiversity is a measure of ecological "health."

Human impacts on the biosphere contribute to a reduction in biodiversity worldwide. Indeed, some scientists estimate that the global extinction rate is 50 times higher now than at any time during the past 100,000 years.

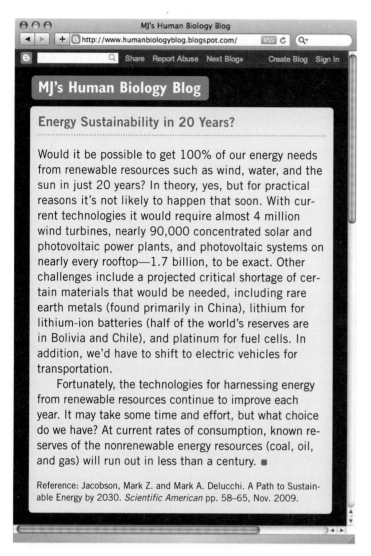

MJ's Human Biology Blog

Energy Sustainability in 20 Years?

Would it be possible to get 100% of our energy needs from renewable resources such as wind, water, and the sun in just 20 years? In theory, yes, but for practical reasons it's not likely to happen that soon. With current technologies it would require almost 4 million wind turbines, nearly 90,000 concentrated solar and photovoltaic power plants, and photovoltaic systems on nearly every rooftop—1.7 billion, to be exact. Other challenges include a projected critical shortage of certain materials that would be needed, including rare earth metals (found primarily in China), lithium for lithium-ion batteries (half of the world's reserves are in Bolivia and Chile), and platinum for fuel cells. In addition, we'd have to shift to electric vehicles for transportation.

Fortunately, the technologies for harnessing energy from renewable resources continue to improve each year. It may take some time and effort, but what choice do we have? At current rates of consumption, known reserves of the nonrenewable energy resources (coal, oil, and gas) will run out in less than a century. ▪

Reference: Jacobson, Mark Z. and Mark A. Delucchi. A Path to Sustainable Energy by 2030. *Scientific American* pp. 58–65, Nov. 2009.

farmland—they are fragile ecosystems with little topsoil and scant soil nutrients. Several years of intensive farming deplete the nutrients from the soil. To continue raising enough crops to survive, subsistence farmers must clear a fresh section of rain forest and start another destructive cycle. The clear-cut areas they leave behind are slow to recover, if they recover at all.

Commercial logging is also a factor. European and Middle Eastern logging companies harvest 800-year-old hardwood trees in Africa and ship them to other continents, where they are processed into furniture, floors, and buildings. In recent years the pace of logging has accelerated. The African nation of Ivory Coast once possessed 70 million acres of pristine tropical rain forest. Today only 5 to 7 million acres remain, and even these may be gone in 10 years.

Exploitation of scarce resources poses another threat. In Brazil, poachers sneak into Itatiaia National Park and cut down 100-year-old palm trees for the tender heart of palm, found only at the top of the tree. Hearts of palm are a popular gourmet product, and there is a growing market for them in the United States and other countries. Across Africa, illegal hunting of elephants for ivory and monkeys and antelopes for meat has soared. The United States has its own history of exploitation, including the complete extinction of the passenger pigeon and the near-extinction of the American bison in the late 1800s.

Clearly, habitat destruction and loss of biodiversity cannot be blamed on just one people, country, continent, or industry. After all, the logging industry is just responding to the demand from industrialized nations for paper, cardboard, and exotic woods. Subsistence farmers in the rain forests are just trying to feed themselves and their families (**Figure 24.12**). Halting the trend of habitat destruction and

Humans alter and destroy habitats

Pollution and overexploitation of natural resources are destroying whole habitats and driving many species of plants and animals toward extinction. The reasons are as varied as the places and the human cultures that inhabit them.

One cause of habitat destruction is farming. The widespread practice of planting just one type of crop over a wide geographic area (e.g., corn in the U.S. corn belt and wheat farther west) substantially alters the species variability within the local ecosystems. Intensive farming practices can also lead to soil erosion and loss of future productivity. The problem of erosion has been recognized and partially corrected with more modern farming practices, which may include crop rotation and "no-till" farming to reduce erosion.

In rain forest areas, forests are being cut down or burned to make way for subsistence farming. Despite their lush vegetation, rain forests do not make good

Figure 24.12 Deforestation. This rainforest in the Amazon region of Brazil is being cleared for subsistence farming.

loss of biodiversity will require recognition of the problems and the willingness to do something about them.

Urbanization is a major force for environmental change

The worldwide shift to urban living has widespread effects on ecosystems, including biodiversity. Although cities cover less than 3% of the Earth's surface, they are responsible for 78% of carbon emissions, 60% of residential water usage, and 76% of wood used for industrial purposes. The resource consumption and waste management requirements of a city typically expand well beyond the city limits. For example, the land area or "ecological footprint" required to serve London's resources and wastes is 120 times the size of the actual city.

Humans tend to plant specific trees, shrubs, and grasses in cities, resulting in fewer species overall, decreased native species, uneven plant coverage, and more invasive and exotic plants. Urbanization also shrinks the biodiversity of animal populations, with a shift to predominantly birds and insects. Furthermore, there is evidence that urbanization may reduce native species diversity on a regional or even global scale. For instance, urban sprawl in northern latitudes appears to reduce the populations of some migratory birds in southern latitudes.

✅ **Quick Check** Considering the examples presented so far, can you pick out two or three ecosystems that tend to be heavily affected by humans? In contrast, can you think of two or three other ecosystems that are relatively intact? How do you think this should impact decisions about land protection, such as which ecosystems to set aside? ▪

Biodiversity is healthy for humans too

Biodiversity is healthy for all organisms, including *Homo sapiens*, and large-scale disruptions to the biosphere threaten human populations, too. For example, we depend on plants for a healthy atmosphere. Through photosynthesis, plants consume carbon dioxide and produce oxygen, thereby removing CO_2 that might otherwise contribute to the greenhouse gases and producing oxygen for us to breathe.

We also depend on other species for clothing, shelter, food, and even medicine—25% of all prescription drugs sold in the United States are based on substances derived from fewer than 50 species of plants. It's estimated that there are 265,000 flowering plants in the world, and less than 1% have been tested so far for medicinal applications. By encouraging biodiversity, we maintain the potential to discover important new medicines in the future.

Biodiversity in the food supply may also be important. Nutrition studies provide strong evidence that eating a varied diet, particularly a wide range of fruits and vegetables, is essential for optimal health. However, our current agriculture practices limit our food choices. Very few of the many known

varieties of apples and tomatoes are typically available in U.S. supermarkets, and many leafy greens known to be high in antioxidants, such as arugula, watercress, and chard, can only be found in specialty stores or farmers markets.

↺ **Recap** Human impacts are leading to a global biodiversity crisis. Urbanization, pollution, habitat destruction, and overexploitation of natural resources are destroying habitats. Loss of biodiversity threatens human populations too, because humans depend on other species for oxygen, clothing, shelter, food, medicine, and other necessities. ▪

24.6 Sustainable development supports future economic needs

In a sustainable world, humans would be an integral part of ecosystems so that those ecosystems (and we humans) could exist on the planet far into the future. This would mean practicing **sustainable development,** defined by the 1987 World Commission on Environment and Development as "development that meets the needs of the present without compromising the ability of future generations to meet their economic needs."

Measuring sustainability and quality of life

How can we determine whether development is sustainable? Another way to think about this is to investigate the economics of sustainability. Economic progress is typically measured by the Gross Domestic Product (GDP), which is defined as the total market value of all goods and services produced within a country in a year. An alternative measure is the Genuine Progress Indicator (GPI). The GPI makes adjustments to the total market value of goods and services to account for environmental costs such as water and air pollution and loss of wetlands, as well as negative economic and social costs (crime, time spent commuting, loss of leisure time) associated with increased production. Quality of life is also taken into consideration by the GPI; it weighs the value of all goods and services produced against enjoyment of life, including such factors as family breakdown and loss of parklands.

Strategies to support sustainable development

Given what we know today, let's focus on strategies that might help us reach a sustainable world.

Consume less This approach isn't popular, but by consuming less you automatically use fewer resources, generate less garbage, and cause less pollution. If you use less, it leaves more for another person or even another species. *What* to consume less of may be a very personal choice, but unfortunately it's the big-ticket items that have the most impact. You could do a lot for the environment, for example, by living in a house that is no larger than your needs, keeping your car an extra year before replacing it (thus purchasing fewer cars in your lifetime), and living closer to work.

Recycle more This approach is more popular than consuming less. Recycling reusable materials such as paper, metal, and plastic reduces our consumption of raw materials. It also keeps the recycled substances from becoming either garbage or pollutants. We also need to consider not only how we recycle substances back into our manufacturing processes, but also how we recycle them back into the ecosystem. Some cities collect, compost, and then recycle leaves and organic material. Even landfills might be considered a slow form of recycling back into the ecosystem.

Encourage sustainable agriculture Current agriculture is based on an industrial model and is aptly referred to as "agribusiness" because it emphasizes high production. This approach has made food abundant and cheap but at the steep cost of degraded water and soil. In contrast, sustainable agriculture means developing systems to raise crops and live-stock following the principles found in nature and hence not depleting the Earth's resources. For example, shade-grown coffee involves coffee planted under shade trees. Larger trees and greater diversity of vegetation enhance the growth of coffee plants, while also allowing native plants and birds to flourish (**Figure 24.13**).

If you live in an area that still has small farms, consider participating in Consumer Supported Agriculture (CSA) and connecting directly with a local farm. CSA members make a commitment to support a local farm, usually on a quarterly basis, by purchasing a share of the farm's harvest.

Figure 24.14 A green roof. The roof of the California Academy of Sciences building in San Francisco is covered with native plants and recycled materials.

Support green roofs The replacement of conventional asphalt or tile on a flat roof with a living, vegetated roof system is a visually pleasing way to provide many benefits that support sustainability. Green roofs increase the energy efficiency of buildings, primarily due to evaporative cooling from the plant material and soil. Green roofs help restore the natural water cycle by filtering and retaining stormwater. Green roofs absorb airborne particulate pollutants, and also absorb harmful nitrogen oxides and carbon dioxide for photosynthesis.

One square meter of green roof absorbs the average amount of pollutants produced by an automobile driven 10,000 miles per year. Plants on green roofs typically consist of native species (for example, grasses in Scandinavia and succulents in California) and therefore require little irrigation or other maintenance (**Figure 24.14**).

Lower the worldwide fertility rate Fewer people on Earth translates to less resource consumption by humans and less environmental degradation. Fewer people to support would mean fewer cars and houses in North America, less deforestation in Brazil, fewer polluted rivers in India, less desertification in Africa, and less global warming. If the current fertility rate of the human species does not decline, eventually environmental degradation will reduce the human carrying capacity of Earth and the death rate will rise to equal the birth rate. We can already see this happening in certain areas of Africa where life expectancy is short and infant mortality is high.

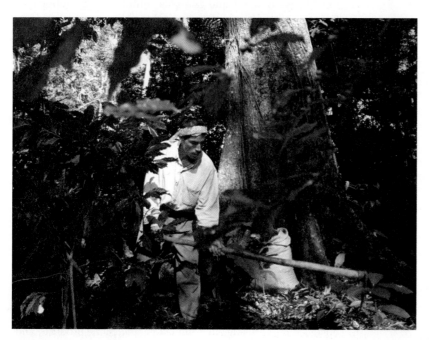

Figure 24.13 Shade-grown coffee, an example of sustainable agriculture.

 Web Animation *Age Structure and Population Growth* at www.humanbiology.com

Reduce rural world poverty People who are starving and without shelter think of their immediate needs first. It is virtually impossible to convey the long-term impact of desertification to someone who relies on a few goats and trees for day-to-day survival. More and more of the world will become uninhabitable due to deforestation and desertification unless the poor people in those areas are given a viable economic alternative to destruction of their own habitat.

Conserve energy in your home Opt for renewable energy at home, such as installing photovoltaic panels, and use energy-efficient appliances. Install low emittance (low-E) glass windows to reduce heat loss in winter and heat gain in summer. Switch incandescent bulbs to compact fluorescents to increase lighting efficiency.

Use environmentally preferable products Whenever possible, choose environmental products that are less harmful to the environment than their alternatives, such as paints that are low in volatile organic compounds (VOC), composite wood made from recycled materials, and carpets that are both low in VOC and high in recycled content.

Protect ecosystems that provide beneficial services From an economic point of view, it makes sense to protect our ecosystems, because if they are destroyed we will lose crucial sources of food, new medicines, and many other benefits. For example, forests provide climate control, raw materials, and food. Since a forest is a natural asset that provides valuable services, it is a form of "natural capital." Similarly, the ozone layer provides protection from damaging ultraviolet radiation, and wetlands can protect adjacent areas from floods, support biodiversity, and filter pollutants.

Will we be able to achieve a sustainable world? Given our history so far there is every reason for optimism. We humans have a remarkable capacity for making intelligent decisions and shaping our own environment. Already we can see cooperation on a global scale with the problem of global warming. More choices will be made in the future and you will have a part in them. Your decisions will help to shape humankind's future.

 Recap A sustainable world will require stabilizing the human population growth rate and the global environment. Strategies for doing this include consuming less, recycling more, supporting sustainable agriculture and green building practices, lowering the global fertility rate and poverty rate, utilizing renewable energy sources, and protecting ecosystems. Achieving this goal will test our capacity to make intelligent decisions on a worldwide scale. ■

Chapter Summary

Pollutants impair air quality *p. 546*

■ Air pollution leads to excessive production of greenhouse gases, especially CO_2, and may be contributing to global warming.

■ CFCs are thinning the ozone layer in the stratosphere. Although a number of nations are cooperating to reduce CFC emissions, ozone depletion remains a threat.

■ Acid precipitation damages forests and aquatic ecosystems.

■ Smog tends to hover over the area where it is produced and can cause chronic respiratory problems.

Pollution jeopardizes scarce water supplies *p. 549*

■ Water is scarce and often distributed unequally. Freshwater forms less than 1% of the world's total water supply.

■ Pollution of surface freshwater can contaminate groundwater and ultimately pollute the oceans.

■ Pollution can accelerate the process of eutrophication.

■ Because of biological magnification, toxic substances become more concentrated in organisms higher up the food chain.

Pollution and overuse damage the land *p. 552*

■ Humans alter the landscape for their own purposes.

■ Deforestation and desertification transform productive land into nonproductive land.

Energy: Many options, many choices *p. 553*

■ Humans use a great deal of energy compared to other species.

■ Most of the energy we consume comes from non-renewable fossil fuels—coal, oil, and gas—that contribute to pollution.

■ Other possible sources of energy include biomass fuels, nuclear power, hydroelectric power, wind farms, geothermal power, and solar power. Solar power is renewable, but current solar power technologies cannot fulfill all our energy demands.

Human impacts are creating a biodiversity crisis *p. 554*

■ Pollution, urbanization, habitat destruction, and over-exploitation of natural resources are damaging the biosphere and accelerating the global extinction rate. Human population growth, farming, logging, and exploitation of scarce resources are all factors.

■ Humans depend on other species for clothing, shelter, food, medicine, oxygen and a healthy atmosphere, and other necessities of life. Loss of biodiversity places human populations at risk, as well as other species.

Sustainable development supports future economic needs *p. 556*

■ Strategies for achieving a sustainable future include consuming less, recycling more, supporting sustainable agriculture, planting green roofs, lowering the worldwide fertility rate to the replacement fertility rate, reducing rural world poverty, conserving energy at home, using environmentally preferable products, and protecting ecosystems.

■ Substituting the GPI (genuine progress indicator) for the GDP (gross domestic product) provides a measure of progress that takes into account our quality of life issues.

■ Humans can shape the environment. It's up to us to shape it wisely.

Terms You Should Know

acid precipitation, *548*
aquifer, *549*
biodiversity, *554*
biological magnification, *550*
deforestation, *547*
desertification, *552*
eutrophication, *550*

global warming, *547*
greenhouse effect, *547*
ozone, *547*
pollution, *546*
smog, *548*
sustainable development, *556*
thermal inversion, *549*

Concept Review

Answers can be found at the Human Biology Place.
www.humanbiology.com

1. Describe the importance of maintaining the ozone layer that surrounds Earth.
2. Describe the effects of acid precipitation and smog.
3. Discuss how human activities pollute Earth's water supply.
4. Explain how pollution can increase the greenhouse effect, and how this might alter global temperatures.
5. Explain why tertiary consumers are more adversely affected by toxic pollutants than primary consumers.
6. Compare and contrast the various sources of energy that are available.
7. Describe the importance of biodiversity and impacts on human health.
8. Describe why sustainable development makes sense from an economic point of view.
9. List strategies that support sustainable development.

Test Yourself

Answers can be found in Appendix A.

1. Global warming is most likely due to the impact of human activity on the:
   a. carbon cycle
   b. nitrogen cycle
   c. oxygen cycle
   d. phosphorus cycle
2. Which problem is attributed primarily to the use of chlorofluorocarbons (CFC) in refrigerators and aerosol cans?
   a. greenhouse gases leading to global warming
   b. depletion of the ozone layer in the stratosphere
   c. production of air pollution leading to acid precipitation
   d. development of smog over heavily populated areas

3. Which of the following is the primary source of the acid in acid precipitation?
   a. ozone produced by lightning
   b. carbon monoxide in automobile exhaust
   c. CO_2 from the burning of coal and oil for power
   d. sulfur dioxide from the burning of coal and oil for power
4. Which of the following are associated with the burning of gasoline and diesel fuel in automobiles and trucks?
   a. production of acid precipitation
   b. production of smog
   c. production of greenhouse gases which contribute to global warming
   d. all of the above
5. Which of the statements about freshwater distribution and use is correct?
   a. Freshwater supplies are evenly distributed throughout the world.
   b. About 10% of the world's water is freshwater.
   c. The supply of freshwater can limit the carrying capacity of a region.
   d. Freshwater supplies generally are unaffected by the water cycle.
6. All of the following statements about eutrophication are true except:
   a. Eutrophication is beneficial as it encourages growth of CO_2-consuming water plants and algae.
   b. Eutrophication is the result of sewage pollution and fertilizer runoff.
   c. Eutrophication can eventually result in the conversion of a wetland into dry land.
   d. Eutrophication is a natural process that can be accelerated by human activities.
7. Which of the following would be most affected by biological magnification of toxins?
   a. tertiary consumer
   b. secondary consumer
   c. primary consumer
   d. producer
8. All of the following are water pollutants except:
   a. fertilizers
   b. ozone
   c. heat
   d. disease-causing bacteria
9. Which environmental issue is most closely associated with rural poverty-stricken regions where people rely almost exclusively on their local environment for survival?
   a. air pollution
   b. groundwater pollution
   c. desertification
   d. eutrophication
10. All of the following are examples of or sources of biomass fuels except:
    a. coal
    b. corn
    c. dung of herbivores
    d. sugar cane

11. The energy source raising the least concerns of pollution, environmental damage, or other risks is:
    a. nuclear energy
    b. hydroelectric energy
    c. solar energy
    d. biomass fuels

12. Which of the following exhibits the least biodiversity?
    a. a home vegetable garden
    b. a field of soybeans
    c. a tropical rain forest
    d. a northern pine forest

13. Which of the following contributes the most to the loss of biodiversity?
    a. water pollution
    b. air pollution
    c. habitat destruction
    d. mining and drilling for fossil fuels

14. Which of the following is likely to be a consequence of the burning of large areas of tropical rain forest to make way for croplands?
    a. a sustained increase in the economic well-being of the residents in the area
    b. an increase in the population of large game animals
    c. an increase in biodiversity
    d. an increase in atmospheric CO_2

15. All of the following are practices that promote sustainability except:
    a. planting large areas with the same crop
    b. reducing consumption
    c. recycling reusable materials
    d. reducing fertility rates worldwide

Apply What You Know

Answers can be found at the Human Biology Place.
www.humanbiology.com

1. Your uncle in Iowa builds a farm pond and stocks it with bass and bluegill sunfish. At first the pond water is fairly clear and the bass grow to large size, but within five years the pond is covered with a thick green scum and all the fish are dead. Explain to your uncle what might have happened to the pond.

2. Human warfare can have a devastating effect on an ecosystem. One of the most recent large-scale wars occurred in the early 1990s when a coalition of forces fought Saddam Hussein's takeover of the tiny country of Kuwait. What unsuccessful plan did Saddam's forces employ to hamper the invasion, and what impact did it have on the southern Iraqi and Kuwaiti landscape?

3. How can the seemingly simple choices a person makes in decorating a home on one continent affect the habitats of distant continents?

4. Most humans would agree that they wouldn't want to intentionally destroy something useful or valuable. If that is so, what is driving humans to destroy what most agree are valuable rain forests?

5. In the 1970s the Black Forest of Germany experienced a period of damaged trees. The number of trees dying in the forest increased substantially in a relatively short period of time. In addition, some lakes in the region became increasingly acidic to the point that fish were beginning to die. What could have caused these problems?

6. During the recent Olympic games in Beijing, China, the government temporarily closed down many factories and mandated that people drastically reduce the use of their vehicles until the close of the Olympic games. What was the government concerned about?

7. After Hurricane Katrina, many citizens of New Orleans were not able to return to their homes. Even those homes with only moderate water damage were declared condemned and ultimately torn down. Why were so many homes declared unfit and torn down rather than repaired?

8. Many different energy sources can be used to generate electricity. Which ones do you favor as the energy sources of the future, at least in your lifetime?

Glossary

A

ABO blood typing Technique for classifying blood according to the presence or absence of two protein markers, A and B.

acetylcholine (ACh) (AS-uh-teel-KOL-een) Neurotransmitter released by some nerve endings in both the peripheral and central nervous systems.

acid A substance that releases hydrogen ions when in solution. A proton donor.

acid rain Rain that has become acidic as the result of the airborne environmental pollutants sulfur dioxide and nitrogen oxide.

acromegaly (AK-roh-MEG-uh-lee) A condition of enlargement of the bones and muscles of the head, hands, face, and feet caused by an excess of growth hormone in an adult.

acrosome (AK-roh-sohm) A cap containing enzymes that covers most of the head of a sperm. Helps sperm penetrate an egg.

actin (AK-tin) A contractile protein of muscle. Forms the thin filaments in the myofibrils.

action potential A brief change in membrane potential of a nerve or muscle cell due to the movement of ions across the plasma membrane. Also called a nerve impulse.

active immunization The process of activation of the immune system in advance of the presence of a disease. Generally, active immunization is accomplished by administration of a vaccine.

active transport Process that transfers a substance into or out of a cell, usually against its concentration gradient. Requires a carrier and the expenditure of energy.

adaptation General term for the act or process of adjusting. In environmental biology the process of changing to better suit a particular set of environmental conditions. In sensory neurons, a decline in the frequency of nerve impulses even when a receptor is stimulated continuously and without change in stimulus strength.

adaptive radiation A short burst of evolutionary activity in which many new life-forms develop in a brief span of time from a single ancestor.

Addison's disease A condition caused by a failure of the adrenal cortex to secrete sufficient cortisol and aldosterone.

adipocyte (AH-di-po-syt) An adipose (fat) cell.

adipose tissue (AH-di-pohs) A connective tissue consisting chiefly of fat cells.

ADP (adenosine diphosphate) (uh-DEN-oh-seen dy-FOSS-fate) A nucleotide composed of adenosine with two phosphate groups. When ADP accepts another phosphate group it becomes the energy-storage molecule ATP.

adrenal cortex (ah-DREE-nul KOR-teks) External portion of the adrenal gland. Its primary hormones, aldosterone and cortisol, influence inflammation, metabolism, interstitial fluid volume, and other functions.

adrenal glands (ah-DREE-nul) Hormone-producing glands located atop the kidneys. Each consists of medulla and cortex areas.

adrenal medulla (ah-DREE-nul meh-DUL-uh) Interior portion of the adrenal gland. Its hormones, epinephrine and norepinephrine, influence carbohydrate metabolism and blood circulation.

adrenocorticotropic hormone (ACTH) (uh-DREE-noh-kort-ih-koh-TRO-pic) Anterior pituitary hormone that influences the activity of the adrenal cortex.

aerobic metabolism (air-OH-bik) Metabolic reactions within a cell that utilize oxygen in the process of producing ATP.

afferent Directed to or toward a point of reference. In the kidney, an afferent arteriole carries blood to each glomerulus.

agglutination (uh-glue-tin-AY-shun) The clumping together of (foreign) cells induced by cross-linking of antigen-antibody complexes.

agranular leukocyte A white blood cell that does not contain distinctive granules. Includes lymphocytes and monocytes.

AIDS (acquired immune deficiency syn-drome) Disease caused by human immunodeficiency virus (HIV). Symptoms include severe weight loss, night sweats, swollen lymph nodes, and opportunistic infections.

albumins (AL-byoo-mins) The most common group of proteins in blood plasma. Albumins have osmotic and transport functions.

aldosterone (al-DOS-ter-ohn or AL-do-STEER-ohn) Hormone produced by the adrenal cortex that regulates sodium reabsorption.

alleles (uh-LEELS) Genes coding for the same trait and found at the same locus on homologous chromosomes.

allergen (AL-ur-jen) Any substance that triggers an allergic response.

allergy (AL-ur-jee) Overzealous immune response to an otherwise harmless substance.

all-or-none principle The principle that muscle cells always contract completely each time they are stimulated by their motor neuron, and that they do not contract at all if they are not stimulated by their motor neuron.

alveolus (pl. alveoli) (al-VEE-oh-lus) One of the microscopic air sacs of the lungs.

Alzheimer's disease (AHLTZ-hy-merz) Degenerative brain disease resulting in progressive loss of motor control, memory, and intellectual functions.

amino acid (uh-MEE-no) Organic compound containing nitrogen, carbon, hydrogen, and oxygen. The 20 amino acids are the building blocks of protein.

amniocentesis A form of fetal testing in which a small sample of amniotic fluid is removed so that cells of the fetus can be examined for certain known diseases and genetic abnormalities.

amnionic fluid (am-nee-AHN-ik) A fluid similar to interstitial fluid that surrounds and protects the fetus during pregnancy. Amniotic fluid is in continuous exchange with maternal interstitial fluid.

ampulla (am-PULL-la) Base of a semicircular canal in the inner ear.

anabolic steroid Any one of over 100 synthetic and natural compounds, all of which are related to the male sex steroid hormone testosterone. Anabolic steroids increase muscle mass and contribute to the development and maintenance of masculine characteristics.

anabolism (ah-NAB-o-lizm) Energy-requiring building phase of metabolism, during which simpler substances are combined to form more complex substances.

analogous structures Structures that share a common function but not necessarily a common origin.

anaphase Third stage of mitosis, in which the two sets of daughter chromosomes move toward the poles of a cell.

androgen (AN-dro-jen) A hormone that controls male secondary sex characteristics, such as testosterone.

anemia (uh-NEE-mee-uh) A condition in which the blood's ability to carry oxygen is reduced because of a shortage of normal hemoglobin or too few red blood cells.

aneurysm (AN-yur-iz-um) A ballooning or bulging of the wall of an artery caused by dilation or weakening of the wall.

angina pectoris (AN-jin-uh PEK-tor-is) Sensation of severe pain and tightness in the chest caused by insufficient blood supply to the heart.

angiogram (AN-jee-oh-gram) An X-ray image of blood vessels.

angiotensin II (AN-jee-oh-TEN-sin TOO) A potent vasoconstrictor activated by renin. Also triggers the release of aldosterone from the adrenal gland.

anorexia nervosa (ah-nuh-REK-see-uh nur-VOH-suh) An eating disorder characterized by abnormally low body weight and unrealistic fear of becoming obese.

antibiotic (AN-tih-by-AH-tik) A substance that interferes with bacterial metabolism. Administered to cure bacterial diseases.

antibody A protein molecule released by a B cell or a plasma cell that binds to a specific antigen.

anticodon The three-base sequence in transfer RNA that pairs with a complementary sequence in messenger RNA.

antidiuretic hormone (ADH) (AN-tih-dy-yu-RET-ik) Hormone produced by the hypothalamus and released by the posterior pituitary. Stimulates the kidneys to reabsorb more water, reducing urine volume.

antigen (AN-tih-jen) A substance or part of a substance (living or nonliving) that the immune system recognizes as foreign. It activates the immune system and reacts with immune cells or their products, such as antibodies.

antigen-presenting cell (APC) A cell that displays antigen-MHC complexes on its surface. Triggers an immune response by lymphocytes.

antioxidant An agent that inhibits oxidation; specifically a substance that can neutralize the damaging effects of free radicals and other substances.

anus The outlet of the digestive tract.

aorta (ay-OR-tah) The main systemic artery. Arises from the left ventricle of the heart.

aortic bodies Receptors in the aorta that are sensitive primarily to the oxygen concentration of the blood. These receptors can also respond to H^+ and CO_2.

appendicular skeleton (ap-un-DIK-yuh-lur) The portion of the skeleton that forms the pectoral and pelvic girdles and the four extremities.

aquaporins Literally, water pores. Aquaporins are cell membrane proteins in certain specialized cells, especially the tubular cells in kidneys, that contain pores which permit the rapid diffusion of water.

appendix (uh-PEN-diks) A small finger-like pouch, with no known digestive function, that extends from the cecum of the large intestine.

aqueous humor (AY-kwee-us) Watery fluid between the cornea and the lens of the eye.

aquifer A deep underground reservoir of water.

Archaea (ar-KAY-uh) One of the three domains of life. The prokaryotes of Archaea have many of the features of eukaryotes, but they do not have true nuclei.

arrhythmia Irregular heart rhythm. Arrhythmias may be caused by defects in the cardiac conduction system or by drugs, stress, or cardiac disease.

arteriole (ar-TEER-ee-ohl) The smallest of the arterial blood vessels. Arterioles supply the capillaries, where nutrient and gas exchange takes place.

artery A blood vessel that conducts blood away from the heart and toward the capillaries.

arthritis Inflammation of the joints.

artificial insemination A technique to enhance fertility in which sperm are placed into the vagina or uterus with a syringe.

asthma (AZ-muh) A recurrent, chronic lung disorder characterized by spasmodic contraction of the bronchi, making breathing difficult.

astigmatism (uh-STIG-muh-tiz-um) An eye condition in which unequal curvatures in different parts of the lens (or cornea) lead to blurred vision.

atherosclerosis (ath-er-o-skler-OS-is) Buildup of fatty deposits on and within the inner walls of arteries. Atherosclerosis is a risk factor for cardiovascular disease.

atom Smallest particle of an element that exhibits the properties of that element. Atoms are composed of protons, neutrons, and electrons.

ATP (adenosine triphosphate) (uh-DEN-oh-seen try-FOSS-fate) A nucleotide composed of adenosine and three phosphate groups. ATP is an important carrier of energy for the cell because energy is released when one of the phosphate groups is removed.

atrial natriuretic hormone (ANH) (AY-tree-ul NAH-tree-u-RET-ik) A hormone secreted by the atria of the heart. Helps regulate blood pressure by causing the kidneys to excrete more salt.

atrium (pl. atria) (AY-tree-um) One of the two chambers of the heart that receive blood from veins and deliver it to the ventricles.

auditory canal Part of the outer ear. The auditory canal channels sounds to the eardrum.

auditory tube Hollow tube connecting the middle ear with the throat. The auditory tube is important for equalizing air pressure in the middle ear with atmospheric pressure.

autoimmune disorder A condition in which the immune system attacks the body's own tissues.

autonomic division Division of the peripheral nervous system composed of the sympathetic and parasympathetic divisions. The autonomic division carries signals from the central nervous system that control automatic functions of the body's internal organs. Sometimes referred to as the autonomic nervous system (ANS).

autosomes (AW-tow-sohms) Chromosomes 1 to 22. The term *autosome* applies to all chromosomes except the sex chromosomes.

autotroph An organism that uses carbon dioxide and energy from its physical environment to build its own large organic molecules.

AV (atrioventricular) bundle Bundle of specialized fibers that conduct impulses from the AV node to the right and left ventricles.

AV (atrioventricular) node Specialized mass of conducting cells located at the atrioventricular junction in the heart.

axial skeleton (AK-see-ul) The portion of the skeleton that forms the main axis of the body, consisting of the skull, ribs, sternum, and backbone.

axon (AK-sahn) An extension of a neuron that carries nerve impulses away from the nerve cell body. The conducting portion of a nerve cell.

B

B cell (B lymphocyte) (LIM-fo-syt) A white blood cell that matures in bone marrow and gives rise to antibody-producing plasma cells.

bacteria One of the three domains of life. Bacteria are prokaryotic microorganisms responsible for many human diseases.

balloon angioplasty Treatment to open a partially blocked blood vessel. The technique involves threading a small balloon into the blood vessel and then inflating the balloon, pressing the blocking material against the sides of the vessel.

baroreceptor Sensory receptor that is stimulated by increases in blood pressure.

base A substance capable of binding with hydrogen ions. A proton acceptor.

basement membrane Layer of nonliving extracellular material that anchors epithelial tissue to the underlying connective tissue.

basophil (BAY-so-fil) White blood cell whose granules stain well with a basic dye. Basophils release histamine and other substances during inflammation.

bile Greenish-yellow or brownish fluid that is secreted by the liver, stored in the gallbladder, and released into the small intestine. Bile is important in emulsifying fat.

biodiversity The variety of living species on Earth.

biogeochemical cycle A cycle in which the chemicals that compose living organisms are recycled between organisms and Earth itself.

biogeography Study of the distribution of plants and animals around the world.

biological magnification Process by which nonexcreted substances become more concentrated in organisms that are higher in the food chain.

biology The study of living things and life's processes.

biomass Total living component of an ecosystem.

biosphere (BY-oh-sfeer) The portion of the surface of Earth (land, water, and air) inhabited by living organisms.

biotechnology The technical application of biological knowledge for human purposes.

biotic potential (by-AHT-ic) Maximum growth rate of a population under ideal conditions.

birth control pill Oral contraceptive containing hormones that inhibit ovulation.

blastocyst (BLAS-toh-sist) Stage of early embryonic development. The product of cleavage.

blood A fluid connective tissue consisting of water, solutes, blood cells, and platelets that carries nutrients and waste products to and from all cells.

blood-brain barrier A functional barrier between the blood and the brain that in-hibits passage of certain substances from the blood into brain tissues.

blood doping The practice of increasing a person's red blood cell production above normal by artificial means, for the purpose of increasing oxygen-carrying capacity.

blood pressure The force exerted by blood against blood vessel walls, generated by the pumping action of the heart.

blood type Classification of blood based on the presence of surface antigens on red blood cells, and the presence of antibodies to surface antigens other than one's own.

BMR (basal metabolic rate) Rate at which energy is expended (heat produced) by the body per unit of time under resting (basal) conditions.

bone A connective tissue that forms the bony skeleton. Bone consists of a few living cells encased in a hard extracellular matrix of mineral salts.

bottleneck effect Evolutionary effect that occurs when a major catastrophe wipes out most of a population without regard to any previous measure of fitness.

botulism A form of poisoning caused by a bacterium, *Clostridium botulinum,* occasionally found in inadequately cooked or preserved foods.

Bowman's capsule Double-walled cup at the beginning of a renal tubule. Bowman's capsule, also called the glomerular capsule, encloses the glomerulus.

bronchiole (BRAHNG-kee-ohl) Small airways within the lungs that carry air to the alveoli. The walls of bronchioles are devoid of cartilage.

bronchitis (brahng-KI-tis) Inflammation of the bronchi.

bronchus (pl. bronchi) (BRAHNG-kus) Any one of the larger branching airways of the lungs. The walls of bronchi are reinforced with cartilage.

bulimia (byoo-LEE-mee-uh) Eating disorder involving episodes of binging and purging.

C

calcitonin (kal-suh-TOW-nin) Hormone released by the thyroid glands that promotes a decrease in blood calcium levels.

Calorie Amount of energy needed to raise the temperature of 1 kilogram of water 1° Celsius.

cancer A malignant, invasive disease in which cells become abnormal and divide uncontrollably. May spread throughout the body.

capillary (KAP-uh-lair-ee) The smallest type of blood vessel. Capillaries are the site of exchange between the blood and tissue cells.

carbohydrate (kar-boh-HY-drayt) Organic compound composed primarily of CH_2O groups. Includes starches, sugars, and cellulose.

carbonic anhydrase (kar-BAHN-ik an-HY-drase) Enzyme that facilitates the combination of carbon dioxide with water to form carbonic acid.

carcinogen (kar-SIN-o-jen) Cancer-causing agent.

cardiac cycle (KAR-dee-ak) Sequence of events encompassing one complete contraction and relaxation cycle of the atria and ventricles of the heart.

cardiac muscle Specialized muscle tissue of the heart.

cardiovascular system The organ system composed of the heart and blood vessels.

carnivore (KAR-nih-vor) Organism that feeds only on animals.

carrying capacity The maximum number of individuals in a population that a given environment can sustain indefinitely.

cartilage (KAR-til-ij) White, semiopaque, flexible connective tissue.

catabolism (kah-TAB-o-lizm) The process whereby larger molecules are broken down to smaller ones and energy is released.

catalyst (KAT-uh-list) A substance that changes the rate of a chemical reaction without itself being altered or consumed by the reaction.

cataracts (KAT-uh-rakts) Clouding of the eye's lens. Cataracts are often congenital or age-related.

cecum (SEE-kum) The blind-ended pouch at the beginning of the large intestine.

cell The smallest structure that shows all the characteristics of life. The cell is the fun-damental structural and functional unit of all living organisms.

cell body Region of a nerve cell that includes the nucleus and most of the cytoplasmic mass, and from which the dendrites and axons extend.

cell cycle A repeating series of events in which a eukaryotic cell grows, duplicates its DNA, and then undergoes nuclear and cytoplasmic division to become two cells. Includes interphase (growth and DNA synthesis) and mitosis (cell division).

cell doctrine Theory consisting of three basic principles: (1) All living things are composed of cells and cell products. (2) A single cell is the smallest unit that exhibits all the characteristics of life. (3) All cells come only from preexisting cells.

cellular respiration The collective term for all cellular metabolic processes that result in the production of ATP.

central canal In bone, the hollow central tube of an osteon that contains nerves and blood vessels.

central nervous system (CNS) The brain and spinal cord.

centriole (SEN-tree-ohl) Small organelle found near the nucleus of the cell, important in cell division.

centromere (SEN-troh-meer) Constricted region of a chromosome where sister chromatids are attached to each other and where they attach to the mitotic spindle during cell division.

cerebellum (ser-ah-BELL-um) Brain region most involved in producing smooth, coordinated skeletal muscle activity.

cerebral cortex (suh-REE-brul KOR-teks) The outer gray matter region of the cerebral hemispheres. Some regions of the cerebral cortex receive sensory information; others control motor responses.

cerebral hemispheres The large paired structures that together form most of the cerebrum of the brain.

cerebrospinal fluid (suh-REE-broh-SPY-nul) Plasmalike fluid that fills the cavities of the CNS and surrounds the CNS externally. Protects the brain and spinal cord.

cerebrum (suh-REE-brum) The cerebral hemispheres and nerve tracts that join them together. Involves higher mental functions.

cesarean delivery (C-section) Surgical delivery of a baby.

chemical bonds Attractive forces between atoms that cause atoms to be bound to each other.

chemoreceptor (KEE-moh-rih-sep-tur) Type of receptor sensitive to various chemicals.

chemotherapy The use of therapeutic drugs to selectively kill cancer cells.

chlamydia (kla-MID-ee-uh) Sexually transmitted disease involving infection by the bacterium *Chlamydia trachomatis.*

cholecystokinin (CCK) (kohl-uh-sis-tuh-KYN-in) An intestinal hormone that stimulates contraction of the gallbladder and the release of pancreatic juice.

cholesterol (ko-LES-ter-ahl) Steroid found in animal fats as well as in most body tissues. Made by the liver.

chondroblast A cartilage-forming cell. In the fetus, chondroblasts produce the hyaline cartilage that forms the rudimentary models of future bones.

chorion (KOR-ee-ahn) Outermost fetal membrane. Helps form the placenta.

chorionic villi (kor-ee-AHN-ik VIL-eye) Extensions of the fetal chorion that project into maternal tissues. Samples of chorionic villi are sometimes taken in early pregnancy to obtain fetal cells for diagnosis of fetal disease.

choroid (KOR-oyd) The vascular middle layer of the eye.

chromatin (KROH-muh-tin) Threadlike material in the nucleus composed of DNA and proteins.

chromosome (KROH-muh-som) Rodlike structure of tightly coiled chromatin. Visible in the nucleus during cell division.

chyme (kym) Semifluid mass consisting of partially digested food and gastric juice that is delivered from the stomach into the small intestine.

cilium (pl. cilia) (SIL-ee-um) A tiny hair-like projection on cell surfaces that moves in a wave-like manner.

cirrhosis (sir-ROH-sis) Chronic disease of the liver, characterized by an overgrowth of connective tissue (fibrosis).

citric acid cycle The metabolic pathway within mitochondria in which acetyl groups are completely disassembled into CO_2 and high-energy compounds. Also called the Krebs cycle.

cleavage An early phase of embryonic development consisting of rapid mitotic cell divisions without cell growth.

climax community The stable result of ecological succession in a given environment.

clitoris (KLIT-uh-ris) In the female, a sensitive erectile organ located in the vulva.

cloning Production of identical copies of a gene, a cell, or an organism.

cochlea (KOK-lee-uh) Snail-shaped chamber in the inner ear that houses the organ of Corti.

codominance Pattern of inheritance in which both alleles of a gene are equally expressed even though the phenotypes they specify are different.

codon (KOH-dahn) The three-base sequence on a messenger RNA molecule that provides the code for a specific amino acid in protein synthesis.

coenzyme A small molecule that assists an enzyme by transporting small molecular groups. Most vitamins are coenzymes.

collagen fiber Strong and slightly flexible type of connective tissue fiber, made of protein.

colon (KOH-lun) The region of the large intestine between the cecum and the rectum. The colon includes ascending, transverse, descending, and sigmoid portions.

colostrum (koh-LAHS-trum) Milky fluid secreted by the mammary glands shortly before and after delivery. Contains proteins and antibodies.

community Array of several different populations that coexist and interact within the same environment.

compact bone Type of dense bone tissue found on the outer surfaces and shafts of bones.

complement system A group of bloodborne proteins which, when activated, enhance the inflammatory and immune responses and may lead to lysis of pathogens.

complete dominance Pattern of genetic inheritance in which one allele completely masks or suppresses the expression of its complementary allele.

concussion An injury to the brain resulting from a physical blow that may result in a brief loss of consciousness.

cone cell One of the two types of photoreceptor cells in the retina of the eye. Cone cells provide for color vision.

congestive heart failure Progressive condition in which the pumping efficiency of the heart becomes impaired. The result is often an inability to adequately pump the venous blood being returned to the heart, venous congestion, and body fluid imbalances.

connective tissue A primary tissue; form and function vary extensively. Functions include support, energy storage, and protection.

consumer In ecosystems, an organism that cannot produce its own food and must feed on other organisms. Also called a heterotroph.

control center In control system theory, a structure that receives input from a sensor, compares it to a correct, internally set value of the controlled variable, and sends an output to an effector. The most important control center in a living organism is the brain.

control group In scientific experiments, a group of subjects that undergoes all the steps in the experiment except the one being tested. The control group is used to evaluate all possible factors that might influence the experiment other than the experimental treatment. Compare *experimental group*.

controlled variable Any physical or chemical property that might vary from time to time and that must be controlled in order to maintain homeostasis. Examples of controlled variables are blood pressure, body temperature, and the concentration of glucose in blood.

core temperature The body's internal temperature. In humans, normal core temperature is about 98.6°F (37°C).

cornea (KOR-nee-uh) The transparent anterior portion of the eyeball.

coronary artery (KOR-uh-nair-ee) One of the two arteries of the heart leading to capillaries that supply blood to cardiac muscle.

coronary artery bypass graft (CABG) Surgery to improve blood flow. A piece of blood vessel is removed from elsewhere in the body and grafted onto a blocked coronary artery to bypass the damaged region.

corpus luteum (KOR-pus LOOT-ee-um) Structure that develops from cells of a ruptured ovarian follicle. The corpus luteum secretes progesterone and estrogen.

cortex (KOR-teks) General term for the outer surface layer of an organ.

cortisol (KOR-tih-sahl) Glucocorticoid produced by the adrenal cortex, also called hydrocortisone.

covalent bond (koh-VAY-lent) Chemical bond created by electron sharing between atoms.

coxal bones The two large bones that connect the femur bones to the sacrum of the vertebral column. The coxal bones form our hips.

cranial nerve (KRAY-nee-ul) One of the 12 pairs of peripheral nerves that originate in the brain.

creatinine (kree-AT-uhn-een) A nitrogenous waste molecule excreted by the kidneys.

crossing-over Exchange of DNA segments between homologous chromosomes during prophase I of meiosis.

cystic fibrosis (SIS-tik fy-BRO-sys) Genetic disorder in which oversecretion of mucus clogs the respiratory passages. Cystic fibrosis can lead to fatal respiratory infections.

cytokines Signaling molecules secreted by helper T cells. Cytokines have a variety of functions, including promoting immune cell development, stimulating their activity, and attracting them to an area of infection.

cytokinesis (sy-tow-kih-NEE-sis) The division of cytoplasm that occurs after a cell nucleus has divided.

cytoplasm (SY-tow-plaz-um) The cellular material surrounding a cell nucleus and enclosed by the plasma membrane.

cytoskeleton A cell's internal "skeleton." The cytoplasm is a system of microtubules and other components that support cellular structures and provide the machinery to generate various cell movements.

cytotoxic T cell (sy-toh-TAHK-sik) A "killer T cell" that directly lyses foreign cells, cancer cells, or virus-infected body cells.

D

decomposer An organism that obtains energy by chemically breaking down the products, wastes, or remains of other organisms.

dehydration synthesis Process by which a larger molecule is synthesized by covalently bonding two smaller molecules together, with the subsequent removal of a molecule of water.

demographic transition A progression of changes followed by many nations, in which the society gradually moves from poor living conditions and a high death rate to improved economic conditions and a declining death rate.

denaturation (dee-nay-chur-AY-shun) Loss of a protein's normal shape and function produced by disruption of hydrogen bonds and other weak bonds.

dendrite (DEN-dryt) Branching neuron extension that serves as a receptive or input region. Dendrites conduct graded potentials toward the cell body.

denitrification (dee-nyt-ruh-fih-KAY-shun) Process in which nitrate is converted to nitrogen gas (N_2). Denitrification is part of the nitrogen cycle.

deoxyhemoglobin Hemoglobin that has given up its oxygen.

depolarization Loss of a state of polarity. In nerve and muscle cells, the loss or reduction of the inside-negative membrane potential.

dermis Layer of skin underlying the epidermis. Consists mostly of dense connective tissue.

desertification (deh-ZERT-uh-fuh-KAY-shun) Process by which marginal lands are converted to desertlike conditions.

detritus (dih-TRY-tus) Nonliving organic matter.

diabetes mellitus (dy-uh-BEE-teez MEL-ih-tus) Disease characterized by a high blood sugar concentration and glucose in the urine, caused by either deficient insulin production (Type 1 diabetes) or insufficient glucose uptake by cells (Type 2 diabetes).

dialysis (dy-AL-ih-sis) General term for several techniques that attempt to take the place of kidney function in patients whose kidneys have failed by letting the patient's blood exchange waste materials with artificial fluids. Two dialysis techniques are *continuous ambulatory peritoneal dialysis (CAPD)* and *hemodialysis.*

diaphragm (DY-uh-fram) A dome-shaped sheet of muscle that separates the thoracic cavity from the abdominal cavity. Also, a contraceptive device inserted into the vagina to cover the cervical opening.

diastole (dy-AS-toh-lee) Period of the cardiac cycle when a heart chamber is relaxed.

diastolic pressure (DY-uh-STAHL-ik) Lowest point of arterial blood pressure during a cardiac cycle. Diastolic pressure is the second of two numbers recorded by the health care provider, the first being systolic pressure.

differentiation Process by which a cell changes in form or function.

diffusion (dih-FYOO-shun) The movement of molecules from one region to another as the result of random motion. Net diffusion proceeds from a region of higher concentration to a region of lower concentration.

digestion Chemical or mechanical process of breaking down foodstuffs to substances that can be absorbed.

diploid The number of chromosomes in a body cell ($2n$), twice the chromosomal number (n) of a gamete. In humans, $2n = 46$.

disaccharide (dy-SAK-uh-ryd) Literally, a double sugar. A disaccharide consists of two monosaccharides linked together. Sucrose and lactose are disaccharides.

diuretic (dy-yoo-RET-ik) Any substance that enhances urinary output.

DNA (deoxyribonucleic acid) (dee-OX-ee-RY-bo-new-CLAY-ik) A nucleic acid found in all living cells. Carries the organism's hereditary information.

DNA ligases A class of enzymes that join fragments of DNA together.

DNA polymerase (poh-LIM-ur-ays) Enzyme that links separate nucleotides together as a new strand of DNA is formed.

dominant allele An allele that masks or suppresses the expression of its complementary allele.

Down syndrome A condition caused by inheriting three copies of chromosome 21.

duodenum (doo-oh-DEE-num) First part of the small intestine.

dysplasia (dis-PLAY-zhee-uh) Abnormal changes in the shape and/or organization of cells. Dysplasia may precede the development of cancer.

E

ECG (electrocardiogram) Graphic record of the electrical activity of the heart.

ecological pyramid Graph representing the biomass, energy content, or number of organisms of each level in a food web. An ecological pyramid includes both producer and consumer populations.

ecology (ee-KAHL-uh-jee) Study of the relationships between organisms and their physical environment.

ecosystem (EEK-oh-sis-tem) All living organisms, all matter, and all energy in a given environment.

ectoderm Embryonic germ layer. Ectoderm forms the epidermis of the skin and its derivatives, and nerve tissues.

ectopic pregnancy (ek-TOP-ik) A pregnancy that results from implantation outside the uterus. The most common form of ectopic pregnancy is a tubal pregnancy.

efferent Directed from or away from a point of reference. In the kidney, an efferent arteriole carries blood away from each glomerulus.

egg An ovum; a mature female gamete.

ejaculatory duct In the male, a short duct that carries sperm from the ductus deferens and seminal fluid from the seminal vesicle to the ureter.

electrocardiogram (ECG) Graphic record of the electrical activity of the heart.

electron Negatively charged subatomic particle with almost no weight. Electrons orbit the atom's nucleus.

electron transport system A series of electron- and energy-transfer molecules in the inner membrane of mitochondria that provide the energy for the active transport of hydrogen ions across the membrane. The electron transport system is essential to the ability of the mitochondria to produce ATP for the cell.

element One of a limited number of unique varieties of matter that composes substances of all kinds. Elements each consist of just one kind of atom.

embolus (EM-bo-lus) Material, such as a blood clot, that floats in the bloodstream and may block a blood vessel.

embryo (EM-bree-oh) An organism in an early stage of development. In humans, the embryonic period extends from the beginning of week 3 to the end of week 8.

embryonic disk A flattened disk of cells that develops in the blastocyst shortly after implantation. The embryonic disk will develop into the embryo.

emphysema (em-fih-SEE-muh) Chronic respiratory disorder involving damage to the bronchioles and alveoli. In emphysema, loss of elastic tissue in the bronchioles causes them to collapse during expiration, trapping air in the alveoli and eventually damaging them as well.

endocrine gland (EN-duh-krin) Ductless gland that secretes one or more hormones into the bloodstream.

endocrine system Body system that includes all of the hormone-secreting organs and glands. Along with the nervous system, the endocrine system is involved in coordination and control of body activities.

endocytosis (en-do-sy-TO-sis) Process by which fluids, extracellular particles, or even whole bacteria are taken into cells. In endocytosis, the materials become enclosed by a vesicle composed of cell membrane material and then internalized within the cell. Phagocytosis is an example of endocytosis.

endoderm (EN-do-derm) Embryonic germ layer. Endoderm forms the lining of the digestive tube and its associated structures.

endometrium (en-do-MEE-tree-um) Inner lining of the uterus. The endometrium becomes thickened and more vascular during the uterine cycle in preparation for pregnancy.

endoplasmic reticulum (ER) (en-do-PLAS-mik reh-TIK-yuh-lum) Membranous network of tubular or saclike channels in the cytoplasm of a cell. The endoplasmic reticulum is the site of most of the cell's production of proteins and other cell compounds.

energy The capacity to do work. Energy may be stored (potential energy) or in action (kinetic energy).

environmental resistance Factors in the environment that limit population growth in a particular geographic area.

enzyme (EN-zym) A protein that acts as a biological catalyst to speed up a chemical reaction.

eosinophil (ee-oh-SIN-oh-fil) Granular white blood cell whose granules stain readily with a red stain called eosin. Eosinophils attack parasites and function in allergic responses.

epidermis (ep-ih-DUR-mus) Outermost layer of the skin. The epidermis is composed of keratinized stratified squamous epithelium.

epididymis (ep-ih-DID-ih-mis) Portion of the male reproductive system in which sperm mature. The epididymis empties into the vas deferens.

epiglottis (ep-ih-GLAHT-is) Flaplike structure of elastic cartilage at the back of the throat that covers the opening of the larynx during swallowing.

epilepsy (EP-ih-lep-see) Condition involving abnormal electrical discharges of groups of brain neurons. Epilepsy may cause seizures.

epinephrine (ep-ih-NEF-rin) Primary hormone produced by the adrenal medulla, also called adrenaline.

epithelial tissue (ep-ih-THEE-lee-ul) A primary tissue that covers the body's surface, lines its internal cavities, and forms glands.

erythrocyte (red blood cell) Blood cell that transports oxygen from the lungs to the tissues and carbon dioxide from the tissues to the lungs.

erythropoietin (ih-rith-roh-PO-ih-tin) Hormone that stimulates production of red blood cells.

esophagus (ih-SOF-uh-gus) Muscular tube extending from the throat to the stomach. The esophagus is collapsed when it does not contain food.

essential amino acids Group of eight amino acids that the body cannot synthesize. Essential amino acids must be obtained from the diet.

estrogen (ES-tro-jen) Female sex hormone that stimulates the development of female secondary sex characteristics, helps oocytes mature, and affects the uterine lining during the menstrual cycle and pregnancy.

Eukarya One of the three domains of life. Eukarya includes all life-forms whose cells have nuclei.

eukaryote (yoo-KAIR-ih-ot) An organism composed of cells with nuclei and internal membrane-bound organelles. Animals, plants, protists, and fungi are all eukaryotes.

eutrophication (yoo-troh-fih-KAY-shun) Process by which mineral and organic nutrients accumulate in a body of water, leading to plant overgrowth and loss of oxygen. Eutrophication may be accelerated by pollution.

evolution An unpredictable and natural process of descent over time with genetic modification. Evolution is influenced by natural selection, chance, historical events, and changing environments.

exocrine gland Specialized epithelial tissue that secretes a product directly into a hollow organ or duct.

exocytosis (ex-oh-sy-TO-sis) Mechanism by which substances are moved from the cell interior to the extracellular space. In exocytosis, a membrane-bound secretory vesicle fuses with the plasma membrane and releases its contents to the exterior.

exon A nucleotide sequence of DNA that specifies a useful informational sequence. Compare *intron*.

experiment A procedure designed to test a working hypothesis. In a controlled experiment, all conditions of the environment are controlled except the specific factor dictated by the working hypothesis.

experimental group In a controlled experiment, the group of subjects that receives the experimental treatment. Compare *control group*.

exponential growth Growth pattern of a population that follows a characteristic J-curve. A population undergoing exponential growth will double repeatedly over similar time periods.

extinction The loss of a particular life-form completely.

F

facilitated transport Passive transport of a substance into or out of a cell along a concentration gradient. Facilitated transport requires a carrier protein.

FAD A transport molecule within mitochondria that can accept hydrogen ions and electrons liberated by the citric acid cycle, forming high-energy $FADH_2$.

farsightedness Condition in which the eyeball is too short, causing nearby objects to be focused behind the retina; also called hyperopia.

fatty acid Linear chain of carbon and hydrogen atoms (hydrocarbon chains) with an organic acid group at one end. Fatty acids are constituents of fats.

fertility rate Number of children that each female has during her lifetime.

fertilization Fusion of the nuclei of sperm and egg, thereby creating a zygote.

fever Body temperature that has risen above the normal set point (in humans, approximately 37°C, or 98.6°F). A fever often indicates illness.

fiber General term for a thin strand. A muscle fiber is a single muscle cell. A connective tissue fiber is a thin strand of extracellular material. In the digestive system, fiber refers to material that is indigestible but beneficial in the diet.

fibrin (FY-brin) Fibrous insoluble protein formed during blood clotting.

fibrinogen (fy-BRIN-oh-jen) A blood protein that is converted to fibrin during blood clotting.

fibroblast (FY-broh-blast) Young, actively mitotic cell that forms the fibers of connective tissue.

fimbriae (FYM-bree-ay) Fingerlike projections at the open end of an oviduct that help move an oocyte into the oviduct.

flagellum (pl. flagella) (flah-JEL-um) Long, whiplike extension of the plasma membrane used by sperm and some bacteria for propulsion.

follicle (FAHL-ih-kul) Ovarian structure consisting of a developing egg surrounded by one or more layers of granulosa cells.

follicle-stimulating hormone (FSH) Hormone produced by the anterior pituitary that stimulates ovarian follicle production in females and sperm production in males.

food web Hierarchy consisting of numerous food chains. A food web depicts interactions between producers and consumers in an ecosystem.

forebrain Anterior portion of the brain that includes the cerebrum, the thalamus, and the hypothalamus.

fossil Preserved remnant of an organism. The fossil record is one of our richest sources of information about life-forms that lived in the past.

fossil fuel A carbon-containing compound (coal, oil, or gas) that was formed millions of years ago by living organisms, then covered by sediments.

fovea centralis (FOH-vee-uh-sen-TRA-lis) A small region at the center of the macula of the retina that is responsible for the highest visual acuity. The fovea centralis consists of densely packed cone cells.

fraternal twins Twins resulting from the ovulation and fertilization of more than one oocyte in a particular cycle. Fraternal twins are as different as any two children by the same parents.

free radicals Highly reactive chemicals with unpaired electrons that can disrupt the structure of proteins, lipids, and nucleic acids.

fungi (FUN-jy) One of the five kingdoms of life. The fungi include yeasts, molds, and mushrooms.

G

gallbladder Organ located beneath the right lobe of the liver. The gallbladder stores and concentrates bile.

gallstones Hard crystals containing cholesterol, calcium, and bile salts that can obstruct the flow of bile from the gallbladder.

gamete (GAM-eet) Haploid sex or germ cell. Sperm and eggs are gametes.

ganglion (pl. ganglia) (GANG-lee-on)   Collection of nerve cell bodies outside the central nervous system.

gastric glands   Glands located beneath the stomach lining. Gastric glands secrete gastric juice.

gastric juice   Collective term for the hydrochloric acid, pepsinogen, and fluid secreted by gastric glands.

gastrin (GAS-trin)   Hormone secreted into the blood by cells located in the gastric glands. Gastrin regulates the secretion of gastric juice by stimulating production of hydrochloric acid.

gastrointestinal (GI) tract   A hollow tube that extends from the mouth to the anus. The GI tract includes mouth, pharynx, esophagus, stomach, small intestine, large intestine, rectum, and anus.

gel electrophoresis   A technique for separating fragments of DNA or protein according to size.

gene (JEEN)   The unit of heredity. Most genes encode for specific polypeptides, and each gene has a specific location on a particular chromosome.

gene flow   The physical movement of alleles caused by movement of individuals into (immigration) or out of (emigration) a population. Gene flow tends to mix pools of genes that might not otherwise mingle.

gene therapy   Technique in which defective genes are repaired or replaced with their normal counterparts.

genetic drift   Random changes in allele frequency because of chance events.

genetic engineering   The planned alteration of the genetic makeup of an organism by modifying, inserting, or deleting genes or groups of genes.

genital herpes (JEN-ih-tul HUR-peez)   A sexually transmitted disease that results from infection by the herpes simplex type 2 virus.

genital warts   A sexually transmitted disease caused by the human papillomavirus (HPV). Genital warts are a potential risk factor for cancers of the cervix or penis.

genome (JEE nohm)   The complete set of DNA in the chromosomes of a particular organism.

genotype   An individual's particular set of genes.

genus (JEE-nus)   Level of taxonomic classification. Modern humans belong to the genus *Homo*.

geographic range   The area of Earth over which a particular species is distributed.

GIFT (gamete intrafallopian transfer)   Technique to enhance fertility in which unfertilized eggs and sperm are placed directly into an oviduct (Fallopian tube) through a small incision in a woman's abdomen.

gigantism   Pituitary disorder caused by hypersecretion of growth hormone during childhood and adolescence.

gingivitis (jin-jih-VY-tis)   Condition of inflamed gum tissue (gingiva), often caused by tooth decay and poor oral hygiene.

gland   One or more epithelial cells that are specialized to secrete or excrete substances.

glaucoma (glaw-KO-muh)   Eye condition caused by blockage of the ducts that drain the aqueous humor in the eye, resulting in high intraocular pressures and compression of the optic nerve. Glaucoma can result in blindness unless detected early.

glial cells (GLEE-ul)   Cells in the nervous system that surround and protect neurons and supply them with nutrients. Glial cells do not generate action potentials.

global warming   Increase in average global temperature.

globulins   A group of proteins in blood plasma having transport and immune functions. Many antibodies are globulins.

glomerular filtration   The process of filtering protein-free plasma fluid from the glomerular capillaries into Bowman's capsule.

glomerulus (glow-MAIR-yoo-lus)   Cluster or tuft of capillaries inside the glomerular capsule in a kidney. The glomerulus is involved in the formation of the glomerular filtrate.

glottis (GLAHT-tis)   Opening between the vocal cords in the larynx.

glucagon (GLOO-kah-gon)   Hormone formed by alpha cells in the islets of Langerhans in the pancreas. Glucagon raises the glucose level of blood by stimulating the liver to break down glycogen.

glucocorticoids (gloo-kah-KOR-tih-koyds)   Adrenal cortex hormones that increase blood glucose levels and help the body resist long-term stressors. The principal glucocorti-coid is cortisol.

glycogen (GLY-ko-jen)   Main carbohydrate stored in animal cells. A polysaccharide.

glycolysis (gly-KOL-ih-sis)   Breakdown of glucose to pyruvic acid. Glycolysis is an anaerobic process.

goiter (GOY-tur)   An enlarged thyroid, caused by iodine deficiency or other factors.

Golgi apparatus (GOHL-jee)   Membranous system within a cell that packages proteins and lipids destined for export, packages enzymes into lysosomes for cellular use, and modifies proteins destined to become part of cellular membranes.

gonad (GOH-nad)   An organ that produces sex cells. The gonads are the testes of the male and the ovaries of the female.

gonadotropins (goh-nad-o-TRO-pinz)   Gonad-stimulating hormones produced by the anterior pituitary. The gonadotropins are FSH and LH.

gonorrhea (gahn-o-REE-uh)   Sexually transmitted disease caused by the bacterium *Neisseria gonorrhoeae*.

graded potential   A local change in membrane potential that varies directly with the strength of the stimulus. Graded potentials decline with distance from the stimulus.

granular leukocyte (GRAN-yoo-lur LOO-ko-syt)   A type of white blood cell with granules that stain readily.

Graves' disease   A disorder in which the thyroid gland produces too much thyroxine.

gray matter   Area of the central nervous system that contains cell bodies and unmyelinated fibers of neurons. The gray matter is gray because of the absence of myelin.

greenhouse effect   A natural phenomenon in which atmospheric gases allow sun-light to pass through but trap most of the heat that radiates outward from Earth's surface.

greenhouse gases   Atmospheric gases, notably CO_2 and methane, responsible for the greenhouse effect.

growth hormone   Hormone produced by the anterior pituitary that stimulates growth in general.

growth plate   Cartilage plate located near the ends of bone. Bones become longer during childhood and adolescence because new cartilage is continually being added to the outer surface of the growth plates.

growth rate   The percentage increase in a population per unit time. Growth rate is calculated as the number of births per year minus the number of deaths per year, divided by population number.

H

habitat   The type of location where an organism chooses to live. An organism's habitat typically is determined by its tolerance for certain environmental conditions.

haploid (HAP-loyd)   Half the diploid number of chromosomes; the number of chromosomes in a gamete.

heart attack   Condition characterized by the death of heart tissue caused by inadequate oxygen supply. The medical term for a heart attack is *myocardial infarction*.

helper T cell   Type of T lymphocyte that orchestrates cell-mediated immunity by direct contact with other immune cells and by releasing chemicals called lymphokines.

hematocrit (hee-MAHT-oh-krit)   The percentage of blood that consists of red blood cells.

hemoglobin (HEE-moh-gloh-bin)   Oxygen-transporting protein in red blood cells that gives the cells their characteristic red color.

hemophilia (hee-moh-FIL-ee-uh)   Collective term for several different hereditary bleeding disorders with similar symptoms. Hemophilia is caused by a deficiency of one or more clotting factors.

hemostasis (HEE-mo-STAY-sis) Stoppage of bleeding or of the circulation of blood to a part.

hepatic portal system (heh-PAT-ik) The system of blood vessels connecting the organs of the digestive tract to the liver.

hepatitis (hep-uh-TY-tis) General term for inflammation of the liver. Hepatitis is sometimes caused by viral infections.

herbivore (HUR-buh-vor) An organism that feeds on plants, utilizing the energy stored in them. Also called a primary consumer.

heterotroph (HET-ur-oh-trof) An organism that cannot utilize the sun's energy to synthesize the molecules it needs. Instead, it must consume foods that contain stored forms of energy. Also called a consumer.

heterozygous (HET-ur-oh-ZY-gus) Having different alleles at the same location (on a pair of homologous chromosomes).

hindbrain Region of brain connected to the spinal cord. The hindbrain comprises the medulla oblongata, cerebellum, pons, and part of the reticular formation.

hippocampus (hip-po-KAM-pus) In the brain, a region of the limbic system that plays a role in converting new information to long-term memories.

histamine (HIS-ta-meen) Substance produced by basophils and mast cells that causes vasodilation and increases vascular permeability.

HIV (human immunodeficiency virus) Virus that destroys helper T cells, thus depressing cell-mediated immunity. Symptomatic AIDS gradually appears when lymph nodes can no longer contain the virus.

homeostasis (ho-mee-oh-STAY-sis) State of body equilibrium characterized by a relatively constant and stable internal environment.

hominids (HAHM-ih-nids) Common name for members of the family *Hominidae*. Hominids include present-day humans *(Homo sapiens)* and all of their extinct ancestors of the genus *Australopithecus* and the genus *Homo*.

hominoids (HAHM-ih-noyds) A sub-group of primates that diverged from a common ancestor of old-world monkeys to become the great apes (gibbons, orangutans, gorillas, chimpanzees) and modern humans.

Homo erectus (HOH-moh ih-REK-tus) A species of *Homo* that diverged from *Homo ergaster* and spread across Africa into Asia and Europe. *Homo erectus* died out only 50,000 years ago.

Homo ergaster (HOH-moh er-GAS-tur) A species of *Homo* that arose from *Homo habilis* ancestors about 1.9 million years ago and migrated out of Africa to establish colonies as far away as Java and China.

Homo habilis (HOH-moh HAB-ih-lus) First distinctly human ancestor. *Homo habilis* showed brain enlargement, changes in teeth and facial features, a decline in sexual dimorphism, and conversion to a diet that included meat. *Homo habilis* may also have used tools.

Homo heidelbergensis (HOH-moh HY-del-bur-GEN-sus) Sometimes called "archaic" *Homo sapiens*, this species descended from *Homo ergaster* and eventually gave rise to modern humans, *Homo sapiens*.

Homo sapiens (HOH-moh SAY-pee-enz) Genus and species of modern humans, thought to have evolved about 140,000 to 100,000 years ago.

homologous chromosomes (ho-MAHL-uh-gus) Chromosomes that look identical under the microscope. Homologous chromosomes have the same genes in the same locations.

homologous structures Body structures that share a common origin.

homozygous (hoh-moh-ZY-gus) Having identical alleles at the same location (on a pair of homologous chromosomes).

hormone (HOR-mohn) A chemical messenger molecule secreted by an endocrine gland or cell into the bloodstream that has effects on specific target cells throughout the body.

human papillomavirus (HPV) (pap-ih-LOH-ma-vy-rus) Virus that causes genital warts. HPV is a risk factor for cancers of the cervix or penis.

hydrogen bond Weak bond that forms between a hydrogen atom with a partial positive charge and a nearby atom with a partial negative charge.

hydrolysis (hy-DRAHL-ih-sis) Process in which water is used to split a molecule into two smaller molecules.

hydrophilic A substance that is attracted to water. Most polar and charged compounds are hydrophilic, and therefore they dissolve easily in water.

hydrophobic A substance that is not attracted to water or is repelled by water. Nonpolar compounds such as oils are hydrophobic.

hyperopia (hy-per-OH-pee-uh) Condition in which the eyeball is too short, causing nearby objects to be focused behind the retina; also called farsightedness.

hyperplasia (hy-per-PLAY-zhee-uh) An increase in the number of cells in a tissue caused by an increase in the rate of cell division.

hyperpolarization Transient local change in the resting potential of a neuron that makes it even more negative than usual.

hypertension High blood pressure. Clinically hypertension is defined as a systolic blood pressure of 140 mm Hg or higher or a diastolic blood pressure of 90 mm Hg or higher.

hypertonic General term for above-normal tone or tension. In biology, a hypertonic solution is one with a higher solute concentration than plasma. Cells shrink when placed in a hypertonic solution.

hypotension Low blood pressure.

hypothalamus Region of the brain forming the floor of the third ventricle. The hypothalamus helps regulate the body's internal environment by secreting releasing factors that affect the secretion of the hormones of the anterior pituitary.

hypothesis An explanation or description of a proposed truth. A hypothesis is expressed as a testable statement about the natural world.

hypotonic General term for below-normal tone or tension. In biology, a hypotonic solution is one with a lower solute concentration than plasma. Cells swell when placed in a hypotonic solution.

I

identical twins Twins that arise when a single fertilized ovum splits in two shortly after fertilization. Identical twins have identical genotypes.

immune system A complex group of cells, proteins, and structures of the lymphatic system that work together to provide the immune response.

immunity Resistance to disease and pathogens.

immunization (im-yoo-nuh-ZAY-shun) A strategy for causing the body to develop immunity to a specific pathogen. Immunization may be active (the administraion of a vaccine) or passive (the administration of specific antibodies).

immunoglobulin (Ig) (im-yoo-noh-GLAHB-yoo-lin) Blood plasma proteins (gamma globulins) that play a crucial role in immunity. There are five classes.

immunosuppressive drugs Medications that block the immune response. Examples are corticosteroid drugs to suppress inflammation and cytotoxic medications to block activated lymphocytes.

immunotherapy Treatments that promote the general responsiveness of the immune system so that it can fight cancer more effectively.

incomplete dominance A pattern of genetic inheritance in which the heterozygous genotype results in a phenotype that is intermediate between the two homozygous phenotypes.

independent assortment Mendelian principle that genetic factors separate completely independently of each other during the formation of sperm and egg. Independent assortment applies fully only to genes located on different chromosomes.

inflammation (in-fluh-MAY-shun) A non-specific defensive response of the body to tissue injury characterized by dilation of blood vessels and an increase in vessel permeability. The presence of inflammation is indicated by redness, heat, swelling, and pain.

inheritance Characteristics or traits that are transmitted from parents to offspring via genes.

inner ear   Region of the ear consisting of the cochlea and the vestibular apparatus. The cochlea sorts sounds by tone and converts them into nerve impulses. The vestibular apparatus helps maintain balance.

insulin (IN-suh-lin)   A hormone secreted by the pancreas that enhances the uptake of glucose by cells, thus lowering blood glucose levels.

interferon (in-tur-FEER-ahn)   Chemical that is able to provide some protection against virus invasion of the body. Interferon inhibits viral reproduction.

interleukin (in-tur-LOO-kin)   A substance that promotes development of immune cells.

interneuron   A neuron within the central nervous system located between two other neurons. Interneurons receive input from one or more neurons and influence the functioning of other neurons.

interphase   One of two major periods in the cell life cycle. Includes the period from cell formation to cell division.

interstitial fluid (in-tur-STISH-ul)   Fluid between body cells.

intervertebral disk (in-tur-vur-TEE-brul)   A disk of fibrocartilage between vertebrae.

intron   A noncoding nucleotide sequence of DNA. Compare *exon.*

ion (EYE-ahn)   An atom or molecule with a positive or negative electric charge.

ionic bond (eye-AHN-ik)   Chemical bond formed by the attractive force between oppositely charged ions.

iris (EYE-ris)   A colored disk-shaped muscle that determines how much light enters the eye.

isotonic   General term for normal tone or tension. In biology, an isotonic solution is one with the same solute concentration as plasma. Cells maintain their normal cell volume in isotonic solutions.

isotopes   Different atomic forms of the same element, varying only in the number of neutrons they contain. The heavier forms tend to be radioactive.

IUD (intrauterine device)   Contraceptive device that is inserted into the uterus by a health care provider.

IVF (*in vitro* fertilization)   Fertilization outside the body. Eggs are harvested from a woman and fertilized, and the embryo is inserted into the woman's uterus through her cervix.

J

jaundice (JAWN-dis)   Yellowish color of the skin, mucous membranes, and some-times the whites of the eyes. Jaundice is often caused by liver malfunction and high circulating levels of bilirubin.

jejunum (jeh-JOO-num)   The part of the small intestine between the duodenum and the ileum.

joint   The junction or area of contact between two or more bones; also called an articulation.

juxtaglomerular apparatus   In the kidney, a region near the glomerulus where the afferent arteriole and the efferent arteriole make contact with the distal tubule. The hormone renin is secreted by specialized cells of the afferent arteriole of the juxtaglomerular apparatus.

K

karyotype (KAIR-ee-oh-typ)   The diploid chromosomal complement in any species. The human karyotype typically is shown as a composite display of the 22 pairs of autosomes arranged from longest to shortest, plus the sex chromosomes X and Y.

keratinocyte (kair-uh-TIN-oh-syt)   Type of cell in the epidermis that produces a tough, waterproof protein called keratin.

killer T cell   T lymphocyte that directly lyses foreign cells, cancer cells, or virus-infected body cells; also called a cytotoxic T cell.

kinetic energy   Energy in motion; energy actually doing work.

Klinefelter syndrome   A condition caused by an XXY genotype. People with Klinefelter syndrome show a tall male phenotype.

Krebs cycle   See *citric acid cycle.*

L

lactation   Production and secretion of milk.

lacteal (LAK-tee-ul)   Small lymphatic vessel in a villus of the small intestine that takes up lipids.

lactose intolerance (LAK-tohs)   A common disorder of digestion and absorption caused by insufficient quantities of the enzyme lactase, which digests the lactose in milk and dairy products.

large intestine   Portion of the digestive tract extending from the small intestine to the anus. Consists of the cecum, appendix, colon, rectum, and anal canal.

larynx (LAIR-inks)   Cartilaginous organ containing the vocal cords, located between the trachea and pharynx. The larynx is also called the voice box.

lens   A transparent, flexible, curved structure that focuses incoming light on the retina at the back of the eye.

leukemia (loo-KEE-mee-uh)   Cancer of the cells that form white blood cells, resulting in overproduction of abnormal white blood cells.

leukocyte (white blood cell)   One of several types of blood cells that are part of the body's defense system. Leukocytes are diverse in structure and specific function. They compose only about 1% of the volume of blood.

Leydig cell (LAY-dig)   Cells located between the seminiferous tubules of the testes. Leydig cells produce testosterone.

ligament   Dense fibrous connective tissue that connects bone to bone.

limbic system   Functional brain system involved in emotional responses.

lipase (LY-pace)   Any lipid-digesting enzyme.

lipid (LIP-id)   Organic compounds formed of carbon, hydrogen, and oxygen. Fats, oils, and cholesterol are lipids. Lipids are not very soluble in water.

lipoprotein (ly-poh-PRO-teen)   A compound containing both lipid and protein. Two medically important lipoproteins are the low-density lipoproteins (LDLs) and high-density lipoproteins (HDLs) that transport cholesterol.

liver   Large, lobed organ that overlies the stomach in the abdominal cavity. The liver produces bile to help digest fat, and serves other metabolic and regulatory functions.

lonevity   How long a person lives.

loop of Henle (HEN-lee)   Hairpin-shaped tubular structure that extends into the medulla of a nephron. The loop of Henle (also called the loop of the nephron) is important in the ability of the kidney to form either concentrated or dilute urine.

lumen (LOO-men)   Cavity inside a tube, blood vessel, or hollow organ.

lupus erythematosis (LOO-pus air-uh-them-uh-TOW-sis)   An autoimmune disorder in which the body attacks its own connective tissue.

luteinizing hormone (LH) (LOO-tee-in-eye-zing)   Anterior pituitary hormone that aids maturation of cells in the ovary and triggers ovulation. In males, LH causes the interstitial cells of the testis to produce testosterone.

lymph (LIMF)   Fluid derived from interstitial fluid transported by lymphatic vessels.

lymph nodes   Small masses of tissue and lymphatic vessels that contain macrophages and lymphocytes, which remove microor-ganisms, cellular debris, and abnormal cells from the lymph before it is returned to the cardiovascular system.

lymphatic system   System consisting of lymphatic vessels, lymph nodes, and other lymphoid organs and tissues. The lymphatic system returns excess interstitial fluid to the cardiovascular system and provides a site for immune surveillance.

lymphocyte (LIM-foh-syt)   One of several types of white blood cells (T cells and B cells) that participate in nonspecific and specific (immune) defense responses.

lymphoma   General term for cancers of the lymphoid tissues.

lysosome (LY-soh-sohm)   Organelle originating from the Golgi apparatus that contains strong digestive enzymes.

lysozyme (LY-soh-zym)   Enzyme in saliva and tears that kills certain kinds of bacteria.

M

macroevolution   Large-scale evolutionary trends or changes that apply to whole groups of species, often as a result of changing environments or major historical events.

macrophage (MAK-roh-fayj)   Large phagocytic white blood cell derived from a monocyte; it engulfs damaged tissue cells, bacteria, and other foreign debris. Macrophages are important as antigen presenters to T cells and B cells in the immune response.

macula (MAK-u-luh)   The central region of the retina, where photoreceptor density is the highest.

malignant tumor   A mass of cells that has metastasized to become invasive cancer.

mammal   Vertebrates that have hair during all or part of their lives, and mammary glands in the female that produce milk.

mammary gland   Milk-producing gland of the breast.

mast cell   Immune cell that detects foreign substances in the tissue spaces and releases histamine to initiate a local inflammatory response against them. Mast cells typically are clustered deep in an epithelium or along blood vessels.

matter   The material substance, composed of elements, that makes up the natural world. Matter takes up space and has weight.

mechanoreceptor (meh-KAN-oh-ree-sep-tur)   Receptor sensitive to mechanical pressure or distortion, such as that produced by touch, sound, or muscle contractions.

medulla (meh-DUL-uh)   General term for the inner zone of an organ. In the kidney, an inner pyramid-shaped zone of dense tissue that contains the loops of Henle. The term is also used for an area of the brain (medulla oblongata) and for the inner zone of the adrenal glands.

medulla oblongata (meh-DUL-uh ahb-long-GAH-tah)   Portion of the brain stem. Contains reflex centers for vital functions such as respiration and blood circulation.

meiosis (my-OH-sis)   Nuclear division process that reduces the chromosomal number by half. Meiosis results in the formation of four haploid (n) cells. It occurs only in certain reproductive organs.

melanocyte (meh-LAN-oh-syt)   Type of cell located near the base of the epidermis. Melanocytes produce a dark brown pigment called melanin that is largely responsible for skin color.

melanoma (mel-ah-NO-mah)   Least common but most dangerous form of skin cancer.

melatonin (mel-ah-TON-in)   A hormone secreted by the pineal gland that seems to synchronize our body's daily rhythms and may induce sleep. It may also play a role in determining the onset of puberty.

memory cell   Members of T cell and B cell clones that provide for immunologic memory.

meninges (meh-NIN-jeez)   Protective coverings of the central nervous system.

meningitis (meh-nin-JY-tis)   Inflammation of the meninges.

menopause (MEN-oh-pawz)   Period of a woman's life when, prompted by hormonal changes, ovulation and menstruation cease.

menstrual cycle (MEN-stroo-ul)   Pattern of changes in the ovaries and uterus. The menstrual cycle lasts about 28 days and is controlled by hormones.

menstruation (men-stroo-AY-shun)   Process in which the endometrial lining disintegrates, its small blood vessels rupture, and the tissue and blood are shed through the vagina.

mesoderm (MEEZ-oh-durm)   Primary germ layer that forms the skeleton and muscles of the body.

messenger RNA (mRNA)   Long nucleotide strand that complements the exact nucleotide sequence of genetically active DNA. Messenger RNA carries the genetic message to the cytoplasm.

metabolism   Sum total of the chemical reactions occurring in body cells.

metaphase   Second stage of mitosis, during which the chromosomes align themselves on one plane at the center of the cell.

metastasis (meh-TAS-tuh-sis)   The spread of cancer from one organ or location to another not directly connected to it.

MHC (major histocompatibility complex) proteins   A group of proteins on the cell surface that identify an individual's cells as "self" and normally signal the immune system to bypass the individual's cells.

microevolution   Evolution as a result of genetic changes that give rise to new species.

microvilli (sing. microvillus)   Tiny projections on the surface of some epithelial cells that function to increase a cell's surface area. Microvilli are often found on the luminal surface of cells involved in absorption of fluids, nutrients, and ions.

midbrain   Region of the brain stem. The midbrain is a coordination center for reflex responses to visual and auditory stimuli.

middle ear   Region of the ear consisting of an air-filled chamber within the temporal bone of the skull, bridged by the malleus, incus, and stapes. The middle ear amplifies sound.

mineral   An inorganic chemical compound found in nature. Some minerals must be in the diet because they are necessary for normal metabolic functioning.

mineralocorticoid (MIN-er-al-oh-KOR-tih-koyd)   Steroid hormone of the adrenal cortex that regulates salt and fluid balance.

mitochondrion (pl. mitochondria) (my-toh-KAHN-dree-ahn)   A cytoplasmic organelle responsible for ATP generation for cellular activities.

mitosis (my-TOH-sis)   Process of nuclear division during which duplicated chromosomes are distributed to two daughter nuclei, each with the same number of chromosomes as the parent cell. Mitosis consists of prophase, metaphase, anaphase, and telophase.

mitral valve (MY-trul)   Left atrioventricular valve in the heart, also called the bicuspid valve.

molecule   Particle consisting of two or more atoms joined together by chemical bonds.

Monera   One of the five kingdoms of life, comprising the bacteria.

monoclonal antibodies   Pure preparations of identical antibodies produced in the laboratory by a colony of genetically identical cells.

monocyte (MAHN-oh-syt)   Large agranular leukocyte that can differentiate into a macrophage that functions as a phagocyte.

mononucleosis (mahn-oh-nu-klee-OH-sis)   A contagious infection of lymphocytes in blood and lymph tissues, caused by the Epstein-Barr virus.

monosaccharide (mahn-oh-SAK-uh-ryd)   Simplest type of carbohydrate with only one sugar unit. Glucose is a monosaccharide.

morphogenesis (mor-foh-JEN-ih-sis)   Process involving dramatic changes in shape and form that an organism goes through during development.

motor neuron   A neuron in the peripheral nervous system that conducts nerve impulses from the central nervous system to body tissues and organs.

motor unit   A somatic motor neuron and all the muscle cells that it stimulates.

MRI (magnetic resonance imaging)   Technique that uses short bursts of a powerful magnetic field to produce cross-sectional images of body structures.

mRNA (messenger RNA)   Long nucleotide strand that complements the exact nucleotide sequence of genetically active DNA. Messenger RNA carries the genetic message to the cytoplasm.

mucosa (myoo-KO-suh)   The innermost tissue layer in the wall of the gastrointestinal tract.

mucus (MYOO-kus)   A sticky, thick fluid secreted by mucous glands and mucous membranes. Mucus keeps the surfaces of certain membranes moist.

multiple sclerosis (skler-OH-sis)   Progressive disorder of the central nervous system in which the myelin sheath around neuron axons disappears and is replaced by fibrous connective tissue.

muscle spindle   Encapsulated receptor in skeletal muscles that is sensitive to stretch.

muscle tension   The force exerted by a contracting muscle.

muscle tissue   Tissue consisting of cells that are specialized to shorten or contract, resulting in movement of some kind. There are three types of muscle tissue: skeletal, cardiac, and smooth.

muscular dystrophy (DIS-trof-ee)   A group of inherited muscle-destroying diseases.

muscularis   One of the four tissue layers of the gastrointestinal tract wall. The muscularis consists primarily of smooth muscle.

mutation   A change in the DNA base pair sequence of a cell.

mutator gene   A class of gene that normally is involved in DNA repair during DNA replication. Mutated mutator genes may contribute to cancer because errors in DNA replication may not be corrected.

myelin sheath (MY-ih-lin)   Fatty insulating sheath that surrounds the neuron axons of some types of neurons.

myocardial infarction (my-oh-KAR-dee-ul in-FAHRK-shun)   Condition characterized by death of heart tissue caused by inadequate oxygen supply; also called a heart attack.

myocardium (my-oh-KAR-dee-um)   Middle and largest layer of the heart wall, composed of cardiac muscle.

myopia (my-OH-pee-uh)   A condition in which the eyeball is too long, causing distant visual objects to be focused in front of the retina. Also called nearsightedness.

myosin (MY-oh-sin)   One of the principal contractile proteins found in muscle. Myosin is composed of thick filaments with cross-bridges.

N

NAD⁺   A transport molecule within mitochondria that can accept hydrogen ions and electrons liberated by the citric acid cycle, forming high-energy NADH.

natural killer (NK) cells   A type of lymphocyte that can lyse and kill cancer cells and virus-infected body cells before the immune system is activated.

natural selection   The process by which, according to Darwin, individuals with traits that make them more fit for their local environments tend to survive and reproduce. Most changes in allele frequency in a population are the result of mutations coupled with natural selection.

negative feedback   A homeostatic control mechanism in which a change in a controlled variable triggers a series of events that ultimately opposes (negates) the initial change, returning the controlled variable to its normal value, or set point.

neonate (NEE-oh-nayt)   A newborn baby during its first 28 days after birth.

neoplasm (NEE-oh-plazm)   Abnormal mass of proliferating cells; also called a tumor. Neoplasms may be benign or malignant.

nephron (NEF-rahn)   Structural and functional unit of the kidney. A nephron consists of the renal tubule and the blood vessels that supply it.

nerve   Cablelike bundle of many neuron axons, all wrapped together in a protective connective tissue sheath. Nerves carry nerve impulses to and from the central nervous system.

nerve impulse   A self-propagating wave of depolarization, also called an action potential.

nervous tissue   Tissue consisting of cells that are specialized for generating and transmitting nerve impulses throughout the body. Nervous tissue forms a rapid communication network for the body.

neuroendocrine cell   A type of cell in the hypothalamus that essentially functions as both a nerve cell and an endocrine cell because it generates nerve impulses and releases a hormone into blood vessels.

neuroglial cells (noo-RAHG-lee-ul)   Cells that provide physical support and protection to neurons and help maintain healthy concentrations of important chemicals in the fluid surrounding them.

neuromuscular junction   Region where a motor neuron comes in close contact with a muscle cell. At the neuromuscular junction, neurotransmitter released by the motor neuron causes an electrical impulse to be generated in the muscle cell.

neuron (NYOOR-ahn, NOOR-ahn)   Cell of the nervous system specialized to generate and transmit nerve impulses.

neurotransmitter   A chemical released by a neuron that may stimulate or inhibit other neurons or effector cells.

neutron   Uncharged subatomic particle found in the atomic nucleus.

neutrophil (NOO-truh-fil)   Most abundant type of white blood cell. Neutrophils respond rapidly during an infection, surrounding and engulfing bacteria.

niche   The role of an organism in its community.

nitrification   The formation of nitrate (NO_3^-) by any of several means, including nitrifying bacteria, lightning, and nitrate-producing fertilizer factories.

nitrogen fixation   The process whereby certain bacteria convert N_2 gas to ammonium (NH_4^+). Nitrogen fixation is an important part of the nitrogen cycle.

node of Ranvier (RAHN-vee-ay)   A short unmyelinated gap between adjacent Schwann cells where the surface of a neuron axon is exposed.

nondisjunction   Failure of sister chromatids to separate during mitosis or failure of homologous pairs to separate during meiosis. Nondisjunction causes abnormal numbers of chromosomes in the resulting daughter cells.

nonsteroid hormone   A hormone that consists of, or at least is partly derived from, the amino acid building blocks of proteins. Nonsteroid hormones generally are lipid insoluble, so they tend to act by binding to a receptor on the cell surface rather than by entering the cell cytoplasm.

norepinephrine (NOR-ep-ih-NEF-rin)   A neurotransmitter and adrenal medullary hormone, associated with sympathetic nervous system activation.

nucleolus (pl. nucleoli) (noo-KLEE-oh-lus)   A dense spherical body in the cell nucleus involved with ribosomal subunit synthesis and storage.

nucleotide   A structural unit consisting of a nitrogen base, a five-carbon sugar, and one or more phosphate groups. Nucleotides are the structural units of DNA and RNA. ATP and ADP are also nucleotides.

nucleus   General term for a central or essential part. The nucleus of an atom consists of neutrons and protons and contains most of the atomic mass. The nucleus of a eukaryotic cell contains the cell's DNA.

O

obesity   The condition of weighing more than 20% above ideal body weight.

olfactory receptor cell   A modified sensory neuron, located in the upper part of the nasal passages, that detects odors.

oligosaccharide   A short string of monosaccharides (single sugars) linked together by dehydration synthesis.

omnivore (AHM-nih-vor)   An organism that can derive energy from either plants or animals. Humans, pigs, and bears are omnivores.

oncogene (AHNG-koh-jeen)   A mutated or damaged proto-oncogene that contributes to cancer.

oocyte   An immature egg in the female ovary or newly released from the ovary.

optic disk   The area of the retina where the axons of the optic nerve exit the eye. The optic disk has no photoreceptors, so its presence creates a "blind spot" in each eye.

optic nerve   One of the two cranial nerves that transmit nerve impulses from the retina to the brain.

organ   A part of the body formed of two or more tissues and adapted to carry out a specific function. The stomach and the heart are examples of organs.

organ of Corti   Collective term for the hair cells and tectorial membrane of the inner ear. The organ of Corti converts pressure waves to nerve impulses.

organelle (or-guh-NEL)   One of many small cellular structures that perform specific functions for the cell as a whole.

organic molecule   A molecule that contains carbon and other elements, held together by covalent bonds.

orgasm (OR-gazm)   A brief, intensely pleasurable reflex event consisting of rhythmic, involuntary muscular contractions.

origin   The end of a muscle that is attached to the bone that does not move during muscular contraction.

osmosis (ahz-MO-sis)   The net diffusion of water across a selectively permeable membrane, such as a cell membrane.

osteoblast (AHS-tee-oh-blast)   A boneforming cell.

osteoclast (AHS-tee-oh-klast)   A cell that resorbs or breaks down bone.

osteocyte (AHS-tee-oh-syt)   A mature bone cell.

osteodystrophy (AHS-tee-oh-DIS-troh-fee)   Progressive disorder in which the bones become weak because of an abnormal mineral composition of bone.

osteon (AHS-tee-ahn)   A cylindrical structure in bone composed of layers of living bone cells and hard extracellular material, arranged like the layers of an onion. In the center is a central canal through which nerves and blood vessels pass.

osteoporosis (AHS-tee-oh-poh-ROH-sis)   Progressive disorder involving increased softening and thinning of the bone. Osteoporosis results from an imbalance in the rates of bone resorption and bone formation.

otoliths (OH-toh-lith)   Hard crystals of bonelike material embedded in gel in the inner ear. Otoliths contribute to sensations of head position and movement.

outer ear   Region of the ear consisting of the pinna (visible portion) and the auditory canal. The outer ear channels sound waves to the eardrum.

ovarian cycle (oh-VAIR-ee-un)   Monthly cycle of follicle development, ovulation, and corpus luteum formation in an ovary.

ovaries (OH-vair-eez)   The two female sex organs (gonads) in which ova (eggs) are produced.

oviduct (OH-vih-dukt)   Tube that leads from an ovary to the uterus, also called the Fallopian·tube or uterine tube.

ovulation (AHV-yoo-LAY-shun)   Ejection of an immature egg (oocyte) from the ovary.

ovum (pl. ova) (OH-vum)   Female gamete, also called an egg.

oxidative phosphorylation   The process of producing ATP from ADP plus a phosphate group (P_i), using the energy derived from the electron transport system within mitochondria. The process is called *oxidative* because it requires oxygen, and *phosphorylation* because a phosphate group is added (to ADP).

oxygen debt   The amount of oxygen required after exercise to oxidize the lactic acid formed by anaerobic metabolism during exercise.

oxyhemoglobin (ahk-see-HEE-moh-gloh-bin)   Oxygen-bound form of hemoglobin.

oxytocin (ahk-sih-TOH-sin)   Hormone secreted from the posterior pituitary by neuroendocrine cells of the hypothalamus. Oxytocin stimulates contraction of the uterus during childbirth and the ejection of milk during nursing.

ozone   In the stratosphere, a thin layer of O_3 gas that shields Earth from ultraviolet rays. Also an environmental pollutant near Earth's surface.

P

Pacinian corpuscle (puh-SIN-ee-un)   Type of encapsulated mechanoreceptor located in the dermis that responds to deep pressure or high-frequency vibration.

pancreas (PANG-kree-us)   Organ located behind the stomach. The pancreas secretes digestive enzymes and bicarbonate into the small intestine and the hormones insulin and glucagon into the bloodstream.

pancreatic amylase (pang-kree-AT-ik AM-ih-lays)   Pancreatic enzyme that continues the digestion of carbohydrates begun by salivary amylase.

parasympathetic division (par-uh-sym-puh-THET-ik)   The division of the autonomic nervous system that generally promotes activities associated with the resting state, such as digestion and elimination.

parathyroid glands (par-uh-THY-royd)   Small endocrine glands located on the thyroid gland that produce parathyroid hormone.

parathyroid hormone (PTH)   Hormone released by the parathyroid glands that regulates blood calcium levels.

Parkinson's disease   Progressive disorder of the nervous system involving death and degeneration of dopamine-containing nerve cells in the midbrain. Symptoms of Parkinson's disease include persistent tremor and rigid movements.

partial pressure   The pressure exerted by one particular gas in a mixture of gases.

passive immunization   The process of fighting an existing or anticipated infection by the administration of specific antibodies against the infective agent. The antibodies usually are isolated from a human or animal with immunity to the infective agent.

passive transport   Membrane transport processes that do not require cellular energy. Diffusion is a form of passive transport.

pathogen (PATH-uh-jen)   Disease-causing microorganism, such as a bacterium or virus.

pectoral girdle (PEK-tur-ul)   Portion of the skeleton that attaches the upper limbs to the axial skeleton, composed of the clavicle and scapula.

pelvic girdle   Portion of the skeleton that supports the weight of the upper body and attaches the lower limbs to the axial skeleton, composed of the two coxal bones and the sacrum and coccyx of the vertebral column.

pelvic inflammatory disease (PID)   Infection of the internal female reproductive organs. PID can result in scarring that blocks the oviducts.

pepsin (PEP-sin)   Protein-digesting enzyme secreted by the gastric glands of the stomach.

peptic ulcer   An open, sometimes bleeding sore that forms in the inner lining of the stomach, the esophagus, or the upper part of the small intestine.

perforin (PUR-for-in)   A chemical released by secretory vesicles of a cytotoxic T cell that perforates the membrane of an abnormal or foreign cell and kills it.

pericardium (pair-ih-KAR-dee-um)   Tough fibrous sac that encloses and protects the heart and prevents it from overfilling.

peripheral nervous system (PNS)   Portion of the nervous system that lies outside of the brain and spinal cord.

peristalsis (pair-ih-STAHL-sis)   Progressive wavelike contractions of muscle in a tubular structure. Peristalsis moves food through the digestive tract and urine through the ureters.

peritubular capillaries (pair-ih-TOO-byoo-lur)   Capillaries that form a network surrounding the proximal and distal tubules in the cortex of a nephron.

peroxisome (per-OX-ih-sohm)   Membrane-bound vesicle in the cell cytoplasm containing powerful enzymes that detoxify harmful or toxic substances.

PET (positron emission tomography)   Technique that employs radioactive substances to create three-dimensional images showing the metabolic activity of body structures.

pH   The measure of the relative acidity or alkalinity of a solution. Any pH below 7 is acidic and any pH above 7 is basic.

phagocyte (FAG-uh-syt)   A white blood cell that destroys foreign cells through the process of phagocytosis.

phagocytosis (fag-uh-sy-TOH-sis)   Process by which phagocytes surround, engulf, and destroy foreign cells.

pharynx (FAIR-inks)   The region of the digestive and respiratory systems that extends from behind the nasal cavities to the esoph-agus, also called the throat. Both air and food pass through the pharynx.

phenotype (FEE-noh-typ)   The observable physical and functional traits of an organism.

phospholipid   A modified lipid that has a polar (water-soluble) region containing a phosphate group. Phospholipids are the main structural components of the cell membrane.

photodynamic therapy A therapy for cancer. First, the patient takes light-sensitive drugs that are drawn into the rapidly dividing cancer cells. Then laser light is focused on the tumor, triggering a series of chemical changes in the light-sensitive drugs that kill malignant cells.

photopigment Any one of several proteins located in the rods and cones of the eye that changes its shape when exposed to energy in the form of light. The change in shape causes the rod or cone to alter the amount of neurotransmitter it releases.

photoreceptor A specialized receptor cell that responds to light energy.

photosynthesis The process by which plants capture the energy in sunlight and convert it into chemical energy for their own use.

pineal gland (PY-nee-ul) Endocrine gland thought to be involved in setting the biological clock and influencing reproductive function. The pineal gland secretes melatonin.

pituitary gland (pih-TOO-ih-tair-ee) Neuroendocrine gland located just below the hypothalamus of the brain. Together, the anterior and posterior lobes of the pituitary gland secrete eight hormones.

pituitary portal system A system of small blood vessels between the hypothalamus and anterior lobe of the pituitary gland.

placebo (pluh-SEE-bow) A false treatment in a controlled experiment, given or performed to minimize the possibility of bias by suggestion.

placenta (plah-SEN-tuh) Temporary organ formed from both fetal and maternal tissues that provides nutrients and oxygen to the developing fetus, removes fetal metabolic wastes, and produces the hormones of pregnancy.

plasma (PLAZ-muh) The fluid component of blood in which the formed elements are suspended.

plasma cell A cell derived from a B cell lymphocyte specifically to mass-produce and release antibodies.

plasma membrane Membrane surrounding the cell, consisting of a phospholipid bilayer with embedded cholesterol and proteins. The plasma membrane regulates the passage of substances into and out of the cell.

plasmid (PLAZ-mid) A small circular molecule of DNA found in the cytoplasm of some bacteria. Plasmids are used in recombinant DNA technology.

platelets (PLAYT-lets) Small cell fragments that are derived from certain cells in the bone marrow. Platelets are important in blood clotting.

pleural membranes (PLUR-ul) Membranes that line the thoracic cavity and cover the external surface of the lungs.

pneumonia (noo-MOHN-yuh) Respiratory condition involving inflammation and infection of the lungs. Pneumonia is usually caused by viruses or bacteria.

polar body Small nonfunctional cell with almost no cytoplasm that is formed when the primary oocyte completes stage I of meiosis.

polygenic inheritance (pahl-ee-JEN-ik) Pattern of inheritance in which a phenotypic trait depends on many genes. An example is eye color, controlled by three or more genes.

polymerase chain reaction (PCR) (poh-LIM-ur-ays) A technique for making multiple identical copies of DNA in a test tube.

polypeptide A chain of 3–100 amino acids (*poly* means "many"). Chains of more than 100 amino acids are generally referred to as proteins.

polysaccharide A chain of monosaccha-rides. Starch and glycogen are polysaccha-rides.

population A group of individuals of the same species that occupy the same geo-graphic area and interact with each other.

positive feedback A control system in which a change in the controlled variable sets in motion a series of events that ampli-fies the original change. The process of childbirth once labor has started is an example of positive feedback.

potential energy Stored or inactive energy.

precapillary sphincter (SFINK-ter) Band of smooth muscle that controls blood flow into individual capillaries.

pre-embryo An organism during the first two weeks after fertilization. The term preembryo is used because many of the cells that result from cleavage and morphogenesis in the first two weeks will eventually not be part of the embryo at all, but will instead constitute extra-embryonic membranes and parts of the placenta.

premenstrual syndrome (PMS) Recurring episodes of discomfort during the menstrual cycle, generally beginning after ovulation and lasting until menstruation. PMS can include symptoms such as food cravings, mood swings, anxiety, back and joint pain, water retention, and headaches.

presbyopia (prez-bee-OH-pee-uh) Eye condition in which an increase in stiffness of the lens results in an inability to focus on near objects. Presbyopia typically begins after age 40.

primary motor area Region of the frontal lobe of the brain that initiates motor activity.

primary somatosensory area Region in the parietal lobe of the brain that receives sensory input from the skin.

primate (PRY-mayt) An order of mammals with five digits on their hands, fairly flat fingernails and toenails rather than hooves or claws, and forward-facing eyes adapted for stereoscopic vision. Primates include lemurs, monkeys, apes, and humans.

primer A short double-stranded piece of DNA used to initiate DNA synthesis during DNA sequencing.

prion (PREE-on) An infectious mis-folded protein that replicates by causing a normal protein in the infected animal to misfold, producing another prion. Prions technically are not alive since they cannot reproduce on their own.

producer In ecosystems, an organism that makes its own organic molecules from inorganic compounds found in water, air, and soil, utilizing energy provided by the sun. Also called an autotroph. Green plants (on land) and algae (in water) are producers.

progesterone (proh-JES-tur-ohn) Hormone partly responsible for preparing the uterus for the fertilized ovum. Progesterone is secreted by the corpus luteum of the ovary and by the placenta.

prokaryote (proh-KAIR-ee-oht) A single-celled organism that lacks the nucleus and the membrane-bound organelles characteris-tic of eukaryotes. Bacteria are prokaryotes.

prolactin (proh-LAK-tin) Hormone secreted by the anterior pituitary gland that stimulates the mammary glands to produce milk.

promoter A unique base sequence that marks the beginning of a gene.

prophase The first stage of mitosis. During prophase the chromosomes condense and thicken, the pairs of centrioles migrate to opposite sides of the cell, and the mitotic spindle forms.

prostaglandins (prahs-tuh-GLAN-dins) A group of lipid-based chemicals synthesized by most tissue cells that act as local messengers. Some prostaglandins dilate blood vessels and others constrict them.

prostate gland (PRAHS-tayt) Accessory reproductive gland in males that produces approximately one-third of the semen volume, including fluids that activate sperm.

protein One or more polypeptide chains of more than 100 amino acids. The three-dimensional structure of a protein is deter-mined by the sequence of amino acids and by hydrogen bonds between different regions of the protein.

proteinuria (proh-tee-NYUR-ee-uh) Appearance of protein in urine. Persistent proteinuria may indicate glomerular damage.

prothrombin (proh-THRAHM-bin) A plasma protein that is converted to thrombin as part of the process of blood clot formation.

prothrombin activator Substance released by blood vessels and nearby platelets when blood vessels are damaged. Prothrombin activator activates the conversion of prothrombin into thrombin, facilitating the process of blood clotting.

Protista One of the five kingdoms of life, comprising unicellular and relatively simple multicellular eukaryotes such as protozoa, algae, and slime molds.

proton Positively charged subatomic particle in the nucleus of an atom.

proto-oncogene (proh-toh-AHNG-koh-jeen) Regulatory gene that promotes cell growth, differentiation, division, or adhesion. A proto-oncogene may contribute to cancer if it becomes mutated or damaged in such a way that it is turned on all the time.

proximal tubule Segment of a nephron that starts at Bowman's capsule and extends to the renal medulla.

puberty (PYOO-bur-tee) Period of life when reproductive maturity is achieved.

pulmonary circuit That part of the vascular system that takes deoxygenated blood to the lungs and returns oxygenated blood to the heart.

pulmonary embolism (EM-buh-liz-um) Condition that occurs when an embolus (blood clot) blocks an artery supplying blood to the lungs. A pulmonary embolism can cause sudden chest pain, shortness of breath, and even sudden death.

pulse Rhythmic expansion and recoil of arteries resulting from heart contraction that can be felt from the surface of the body.

Punnett square (PUN-et) A grid used for predicting patterns of inheritance and the probability that a particular genotype will be inherited.

pupil (PYOO-pul) Opening in the center of the iris through which light enters the eye.

Purkinje fibers (pur-KIN-jee) Modified cardiac muscle fibers that are part of the electrical conduction system of the heart.

pyloric sphincter (py-LOHR-ik SFINK-tur) A thickening of the circular layer of muscle at the distal end of the stomach that controls the rate at which the stomach empties.

pyrogen (PY-roh-jen) A fever-inducing agent. Macrophages and certain bacteria and viruses release pyrogens.

R

radiation High-energy waves or particles emitted by radioactive isotopes.

receptor In sensory physiology, a specialized cell or nerve ending that receives a sensory signal such as touch, smell, light, or a chemical, and ultimately generates nerve impulses in a sensory neuron.

receptor protein A protein molecule of the cell membrane whose function is to transmit information across the cell membrane. Typically, when a particular molecule in the interstitial fluid binds to the receptor protein, a specific series of events is triggered within the cell.

recessive allele An allele that does not manifest itself in the presence of a more dominant allele. Two recessive alleles must be present in order for the recessive phenotypic trait to be expressed.

recombinant DNA technology Field of applied science that explores applications of cutting, splicing, and creating DNA.

Recommended Daily Allowance (RDA) The U.S. National Research Council's current best estimate of how much of each vitamin and mineral we need to maintain good health.

red blood cell (RBC) Blood cell that transports oxygen from the lungs to the tissues, and carbon dioxide from the tissues to the lungs. Also called an erythrocyte.

reflex An involuntary, automatic response to a stimulus.

refractory period (rih-FRAK-tuh-ree) Period immediately following a nerve impulse during which a neuron is unable to conduct another nerve impulse.

regulatory gene A gene that codes for a repressor or activator regulatory protein.

REM (rapid eye movement) sleep Stage of sleep in which rapid eye movements occur. REM sleep is accompanied by an alert EEG pattern and dreaming.

renin (REE-nin) An enzyme secreted into the bloodstream by the kidneys that leads to an increase in blood pressure and the secretion of aldosterone.

replacement fertility rate The fertility rate required to achieve long-term zero population growth. Replacement fertility rate is estimated to be an average of 2.1 children per woman.

replication The process of copying DNA prior to cell division.

repolarization In a nerve or muscle cell, return of the membrane potential to the initial resting (polarized) state after depolarization.

reproductive system Organ system that functions to produce offspring.

residual volume The amount of air remaining in the lungs even after a forceful, maximal expiration. Normal residual volume is approximately 1200 ml.

respiratory center Groups of nerve cells in the medulla oblongata near the base of the brain that are responsible for the cyclic nature of respiration.

respiratory system Organ system responsible for gas exchange. The respiratory system includes the nose, pharynx, larynx, trachea, bronchi, and lungs.

resting potential The slight difference in voltage (electrical potential) between the inside and outside of a cell.

restriction enzyme An enzyme found in bacteria that can be used to cut DNA at specific nucleotide sequences called restriction sites.

reticular activating system (RAS) (reh-TIK-yoo-lur) A group of neurons in the reticular formation that controls levels of sleep and wakefulness.

reticular fibers Thin, interconnecting fibers of collagen found in fibrous connective tissue. Reticular fibers provide an internal structural framework for soft organs such as the liver, spleen, and lymph nodes.

reticular formation Collective term for neurons in the pons that work with the cerebellum to coordinate the skeletal muscle activity required to maintain posture, balance, and muscle tone. The reticular formation is also responsible for maintaining the level of wakefulness.

retina (RET-ih-nuh) Layers of tissue at the back and sides of the eye. The retina is composed primarily of photoreceptor cells, neurons, and a few blood vessels.

retrovirus (REHT-roh-vy-rus) A class of viruses that replicate by using their RNA to produce double-stranded DNA, which is inserted into the host cell's DNA. The host cell then produces RNA for new viruses. HIV is a retrovirus.

Rh factor A red blood cell surface antigen, first discovered in rhesus monkeys, that is a crucial consideration in blood transfusions.

rheumatoid arthritis (ROO-muh-toyd) A type of chronic arthritis involving inflammation of the synovial membrane that lines certain joints.

rhodopsin (rho-DOP-sin) The photopigment found in the rods of the eye.

ribosomal RNA (rRNA) The RNA component of a ribosome. Ribosomal RNA assists in protein synthesis.

ribosome (RY-boh-sohm) A cellular structure consisting of rRNA and protein at which amino acids are assembled into proteins. Some ribosomes float freely within the cytoplasm; others are attached to the endoplasmic reticulum.

RNA (ribonucleic acid) Nucleic acid that contains ribose and the bases A, G, C, and U. RNA carries out DNA's instructions for protein synthesis. Types of RNA include mRNA, rRNA, and tRNA.

RNA polymerase (poh-LIM-ur-ays) One of several enzymes that recognizes a promoter (the starting point of transcription). RNA polymerase attaches to the promoter, starts the DNA unwinding process, and assists in attaching the appropriate RNA segments to the growing chain.

rod cell One of the two types of photo-sensitive cells in the retina.

S

salivary amylase (SAL-ih-vair-ee AM-uh-layz) Enzyme in saliva that begins the process of digesting starch.

salivary glands Three pairs of glands that produce saliva to begin the process of digestion. The three pairs of salivary glands are the parotid, submandibular, and sublingual glands.

saltatory conduction Form of rapid transmission of a nerve impulse along a myelinated fiber, in which the nerve impulse leaps from node to node.

sarcomere (SAR-koh-mir)   The smallest contractile unit of a muscle myofibril. A sarcomere extends from one Z-line to the next.

sarcoplasmic reticulum (sar-koh-PLAZ-mik reh-TIK-yuh-lum)   Specialized endo-plasmic reticulum of muscle cells that surrounds the myofibrils and stores the calcium needed for the initiation of muscle contraction.

saturated fat   Type of fat with two hydrogen atoms for every carbon atom in the fatty acid tails. Saturated fats are found in meat and dairy products and in a few plant sources such as coconut and palm kernel oil.

Schwann cell   A type of supporting cell in the peripheral nervous system. Schwann cells form the myelin sheaths around myelinated neurons and are vital to peripheral nerve fiber regeneration.

science   The study of the natural world.

scientific method   The process of science; the way scientific knowledge is acquired.

sclera (SKLEAR-uh)   White, opaque outer layer of the eyeball.

scrotum (SKROH-tum)   External sac of skin enclosing the testes.

second messenger   Any intracellular messenger molecule generated by the binding of a chemical messenger (hormone or neurotransmitter) to a plasma membrane receptor on a cell's outer surface. A second messenger mediates intracellular responses to an extra-cellular (first) messenger.

secondary oocyte   The haploid female reproductive cell that is released from an ovary and begins traveling down the oviduct during the ovarian cycle. A secondary oocyte has completed stage I of meiosis but not stage II.

secretion   General term for the movement of a substance out of a cell of an endocrine or exocrine gland. Also, the substance secreted. The term is also used to denote the movement of any substance into the lumen of a tubule in the kidney.

segmentation   Type of gastrointestinal motility in which short sections of smooth muscle contract and relax in seemingly random fashion, mixing the contents of the intestinal lumen.

segregation   Mendelian principle that diploid organisms inherit two genes for each trait (on a pair of homologous chromo-somes), and that during meiosis the two genes are distributed to different gametes.

selective gene expression   The principle that, in any cell at any time, only a few of the genes are actually being expressed. Selective gene expression is responsible for differences in structure and function among different cells and for changes in cell function at different times.

selectively permeable   The quality of allowing certain substances to pass while restricting the movement of others. The plasma membrane of a cell is selectively permeable.

SEM (scanning electron microscope)   Microscope that bombards an object with beams of electrons to reveal what appears to be a three dimensional view of the object's surface.

semen (SEE-men)   Thick, whitish fluid mixture containing sperm and the secretions of the male accessory reproductive glands.

semicircular canals   Three fluid-filled canals in the vestibular apparatus of the inner ear that are important in sensing rotational movements of the head.

semilunar valves (sem-ee-LOO-nur)   Heart valves located between the ventricles and the two major arteries of the body (pulmonary and aortic) that prevent blood from returning to the ventricles after contraction.

seminal vesicles (SEM-ih-nul)   Male accessory reproductive structures that add fructose and prostaglandins to semen. Fructose serves as an energy source for sperm, and prostaglandins may induce muscle contractions in the female reproductive tract to help sperm travel more effectively.

seminiferous tubules (sem-ih-NIF-ur-us)   Highly convoluted tubes within the testes in which sperm are formed.

senescence (seh-NES-ens)   The progressive deterioration of multiple organs and organ systems over time; old age.

sensor   In control system theory, any structure that monitors the current value of the controlled variable and sends the information to a control center. In biology, also called a receptor. Most receptors send information in the form of either action potentials in nerves or hormones in the bloodstream.

sensory neuron   A neuron of the peripheral nervous system that is specialized to respond to a certain type of sensory stimulus, such as pressure or light. Sensory neurons conduct information about a stimulus to the central nervous system in the form of nerve impulses.

septicemia (sep-tih-SEE-mee-uh)   Systemic disease caused by the spread of microorganisms or their toxins in blood. Also called blood poisoning or toxemia.

septum (SEP-tum)   The muscular partition that separates the right and left sides of the heart.

serosa   The outermost tissue layer of the gastrointestinal tract wall. It consists of a thin layer of connective tissue that surrounds the other three layers and attaches the gastrointestinal tract to the walls of the body cavities.

Sertoli cells   Large cells that represent most of the bulk of seminiferous tubules in males. Also called *sustentacular cells,* they surround and nourish the developing sperm.

set point   The internally set (expected, normal) value of a controlled variable.

sex chromosomes   The chromosomes, X and Y, that determine gender (XXfemale, XY-male). The 23rd pair of chromosomes.

sex-influenced trait   A trait not inherited with a sex gene but influenced by the actions of the sex genes. An example is the phenotypic trait of male baldness.

sex-linked inheritance   Pattern of inheritance that depends on genes located on the sex chromosomes. Sex-linked inheritance is X-linked if the gene is located only on the X chromosome, and Y-linked if it is located only on the Y chromosome.

sickle-cell anemia   An inherited disorder in which the red blood cells assume a sickle shape when the oxygen concentration is low.

sinoatrial (SA) node (sy-noh-AY-tree-ul)   Group of specialized myocardial cells in the wall of the right atrium that initiates the heartbeat. The SA node is called the pacemaker of the heart.

sinus   Mucous-membrane–lined, air-filled cavity in certain cranial bones.

sister chromatids   The two identical chromosomes that remain attached at the centromere after replication. The centromere holds the sister chromatids together until they are physically pulled apart during the mitotic phase.

skeletal muscle   Muscle composed of cylindrical multinucleate cells with obvious striations. Skeletal muscles attach to the body's skeleton.

skeleton   The body's physical support system. The skeleton consists of bones and the various connective tissues that hold them together.

sliding filament mechanism   The mechanism of muscle contraction. Muscles contract when the thick and thin filaments slide past each other and sarcomeres shorten.

smog   Hazy brown or gray layer of atmospheric pollution that tends to hover over the region where it is produced. The burning of fossil fuels (coal and oil) and automobile exhaust are the primary culprits.

smooth muscle   Spindle-shaped muscle cells with one centrally located nucleus and no externally visible striations. Smooth muscle is found mainly in the walls of hollow organs.

sodium-potassium pump   An active transport protein of the plasma membrane that simultaneously transports three sodium ions (Na^+) out of the cell and two potassium ions (K^+) in. The sodiumpotassium pump is important for maintaining cell volume and for generating the resting membrane potential.

solute   Any substance dissolved in a liquid.

solvent   A liquid in which other substances dissolve.

somatic sensations   Sensations that arise from receptors located at many points throughout the body. The somatic sensations include temperature, touch, vibration, pressure, pain, and awareness of body movements and position.

somatostatin (soh-mah-toh-STAHT-in)   Hormone secreted from several locations in the body.

In the pancreas, somatostatin inhibits the secretion of glucagon and insulin.

species The smallest classification category of life. A species is a group of organisms that under natural conditions tend to breed within that group.

sperm Male reproductive cell (gamete).

sphincter (SFINK-tur) General term for a ring of muscle around a hollow tube or duct that functions to restrict the passage of materials.

sphygmomanometer (SFIG-moh-mah-NOM-uh-tur) Device used to measure blood pressure.

spinal cord The portion of the central nervous system that lies outside the brain, extending from the base of the brain to about the second lumbar vertebra. The spinal cord provides a conduction pathway to and from the brain.

spinal nerve Any one of the 31 paired nerves that arise from the spinal cord.

spinal reflex An involuntary muscle response that is mediated at the level of the spinal cord, with little or no involvement of the brain.

spleen Largest lymphoid organ. The spleen removes old and damaged red blood cells and helps fight infections.

spongy bone Type of bone tissue characterized by thin, hard interconnecting bony elements enclosing hollow spaces. Red blood cells are produced in the spaces between the bony elements.

starch The storage polysaccharide utilized by plants.

stem cells General term for any cells that have not yet differentiated. Stem cells in bone marrow are the source of all blood cells and platelets.

steroids (STEER-oydz) Group of lipids having four interconnected rings that includes cholesterol and several hormones. Steroids are fat-soluble molecules.

stomach Organ of the gastrointestinal tract where food is initially stored and where chemical breakdown of proteins begins.

stretch reflex Type of spinal reflex, important in maintaining upright posture and coordinating movement. Stretch receptors in a skeletal muscle stimulate sensory nerves when stretched, causing the sensory nerves to transmit nerve impulses to the spinal cord.

stroke Condition in which brain tissue is deprived of a blood supply, most often caused by blockage of a cerebral blood vessel. Also called a cerebrovascular accident.

stroke volume Volume of blood pumped out of a ventricle during one ventricular contraction.

structural gene A gene that codes for an enzyme or a structural protein.

submucosa The middle layer of the gastrointestinal tract wall. The submucosa is a layer of connective tissue that contains blood vessels, lymph vessels, and nerves.

substrate In chemistry and biology, the starting material of a chemical reaction.

succession A natural sequence of change in terms of which organisms dominate in a community.

summation Accumulation of effects, especially those of muscular or neural activity.

suppressor T cell Regulatory T lymphocyte that suppresses the immune response.

surfactant (sur-FAK-tant) Secretion produced by certain cells of the alveoli that reduces the surface tension of water molecules, preventing the collapse of the alveoli after each expiration.

sympathetic division The division of the autonomic nervous system that helps the body cope with stressors (danger, excitement, etc.) and with situations requiring high mental or physical activity.

synapse (SIN-aps) Functional junction or point of close contact between two neurons or between a neuron and an effector cell.

synaptic cleft (sih-NAP-tik) Fluid-filled space at a synapse.

synaptic transmission The process of transmitting information from a neuron to its target across a synapse.

synovial joint (sih-NO-vee-ul) A movable joint having a thin fluid-filled cavity between the bones.

syphilis (SIF-uh-lis) Sexually transmitted disease caused by infection with the bacterium *Treponema pallidum.*

systemic circuit The system of blood vessels that transport blood to all cells of the body except those served by the pulmonary circuit.

systole (SIS-toh-lee) Period when either the ventricles or the atria of the heart are contracting.

systolic pressure (sis-TOL-ik) The highest pressure reached in the arterial blood vessels during the cardiac cycle, and the first of two blood pressure numbers recorded by the health care provider. Systolic pressure is achieved during ventricular systole.

T

T cells (T lymphocytes) Lymphocytes responsible for cell-mediated immunity, which depends on the actions of several types of T cells.

taste bud A cluster of taste receptor cells and supporting cells that respond to dissolved food chemicals in the mouth. The human tongue has about 10,000 taste buds.

taxonomy (tak-SAHN-uh-mee) The science of classifying and naming life-forms.

tectorial membrane (tek-TOR-ee-ul) The gelatinous noncellular material in which the mechanoreceptor hair cells of the inner ear are embedded. With the hair cells, the tectorial membrane forms the organ of Corti, which converts pressure waves to nerve impulses.

telomerase (tel-OH-mer-ays) An enzyme produced primarily by cancer cells and reproductive cells that builds new telomeres to replace telomeres lost during cell division.

telomeres (TEEL-oh-meer) Short segments of DNA at the end of each DNA molecule that do not encode for a protein. In most cells, telomeres are removed every time a cell divides until there are no telomeres left.

telophase In cell division, the last phase of mitosis in which the two new sets of chromosomes arrive at opposite poles of the cell, new nuclear membranes form around the chromosomes, and the chromosomes uncoil and are no longer visible under a microscope.

TEM (transmission electron microscope) Microscope that bombards a very thin sample with a beam of electrons, some of which pass through the sample. A TEM provides images that are two-dimensional but highly magnified.

tendon A cord of dense fibrous connective tissue attaching muscle to bone.

teratogenic (TAIR-uh-toh-JEN-ik) Capable of producing abnormal development of a fetus. Teratogenic substances are one cause of birth defects.

testis (pl. testes) The male primary sex organ (gonad); produces sperm.

testosterone (tes-TAHS-teh-rohn) Male sex hormone produced primarily by the testes. (The adrenal cortex produces a small amount of testosterone in both sexes.) Testosterone promotes the development of secondary sexual characteristics and is necessary for normal sperm production.

tetanus (TET-uh-nus) In a muscle, a sustained maximal muscle contraction resulting from high-frequency stimulation. In medicine, an infectious disease also called lockjaw that is caused by an anaerobic bacterium.

thalamus (THAL-uh-mus) A region of the forebrain that receives and processes sensory information and relays it to the cerebrum.

theory A main hypothesis that has been extensively tested over time and that explains a broad range of scientific facts with a high degree of reliability. An example is the theory of evolution.

thermal inversion A situation in which a warm stagnant upper layer of air traps a cooler air mass beneath it. Thermal inversions can cause air pollutants to accumulate close to the ground by preventing them from dissipating.

thermoreceptor Receptor sensitive to temperature changes.

threshold In an excitable cell such as a nerve or muscle, the membrane voltage that must be reached to trigger an action potential, or nerve impulse.

thrombin Enzyme that facilitates the conversion of fibrinogen into long threads of fibrin. Thrombin promotes blood clotting.

thymus gland Endocrine gland that contributes to immune responsiveness. T cells mature in the thymus.

thyroid gland An endocrine gland, located in the neck, that produces the hormones thyroxine, triiodothyronine, and calcitonin.

thyroid-stimulating hormone (TSH) Hormone produced by the anterior pituitary. TSH regulates secretion of thyroxine and triiodothyronine.

thyroxine (thy-RAHK-sin) A hormone containing four iodine molecules that is secreted by the thyroid gland. Thyroxine and a closely related three-iodine hormone (triiodothyronine) accelerate cellular metabolism in most body tissues.

tidal volume Volume of air inhaled and exhaled in a single breath. Normal tidal volume is approximately 500 milliliters, or about a pint.

tissue A group of similar cells (and their intercellular substance) specialized to perform a specific function. Primary tissue types of the body are epithelial, connective, muscle, and nervous tissue.

tonsils Masses of lymphatic tissue near the entrance to the throat. Lymphocytes in the tonsils remove microorganisms that enter in food or in the air.

trachea (TRAY-kee-uh) Tube extending from the larynx to the bronchi that is the passageway for air to the lungs. Also called the windpipe.

transcription The production of a single strand of RNA from a segment (representing a gene) of one of the two strands of DNA. The base sequence of the RNA is complementary to that of the single strand of DNA.

***trans* fats** Vegetable oils that have been turned into solids by partial hydrogenation of their fatty acid tails. *Trans* fats increase the shelf life and stability of foods, but they also increase the risk of heart disease.

transfer RNA (tRNA) Short-chain RNA molecule that transfers amino acids to the ribosome.

transfusion reaction The adverse reaction that occurs in a recipient when a donor's red blood cells are attacked by the recipient's antibodies. During a transfusion reaction the donated cells may clump together (agglutinate), blocking blood flow in blood vessels.

transgenic organism A living organism that has had foreign genes inserted into it. Transgenic organisms are sometimes created to produce substances useful to humans, including human proteins.

translation The process by which the genetic code of mRNA is used to string together the appropriate amino acids to produce a specific protein.

translocation A change in chromosome location that occurs when a piece of a chromosome breaks off but reattaches at another site, either on the same chromosome or another chromosome. Translocations can result in subtle changes in gene expression and ability to function.

transport protein A protein molecule of the cell membrane whose function is to transport one or more molecules across the cell membrane.

tricuspid valve (try-KUS-pid) The right atrioventricular valve of the heart.

triglyceride A neutral fat molecule consisting of a molecule of glycerol and three fatty acid tails. Triglycerides are the body's most concentrated source of energy fuel.

triiodothyronine (try-i-oh-doh-THY-roh-neen) Hormone secreted by the thyroid gland, similar to thyroxine except that it contains three atoms of iodine instead of four. Like thyroxine, triiodothyronine accelerates cellular metabolism in most body tissues.

triplet code The genetic code of mRNA and tRNA in which three successive bases encode for one of the 20 amino acids.

trisomy-X syndrome (TRY-soh-mee) A condition caused by having three X chromosomes. Trisomy-X syndrome individuals are female phenotype, sometimes showing a tendency toward mild mental retardation.

trophoblast Outer sphere of cells of the blastocyst.

tubal ligation Procedure for female sterilization in which each oviduct is cut and tied.

tuberculosis Infectious disease of the lungs caused by the bacterium *Mycobacterium tuberculosis*.

tubular reabsorption The movement of fluid and solutes (primarily nutrients) from the renal tubules into the blood.

tubular secretion The movement of solutes (primarily undesirable substances such as drugs, metabolic wastes, and excess ions) from the blood into the renal tubules.

tumor A mass of cells derived from a single cell that began to divide at an abnormally high rate. The cells of a benign tumor remain at the site of origin, whereas those of a malignant tumor travel to distant sites (metastasize).

twitch A complete cycle of contraction and relaxation in a muscle cell.

tympanic membrane (tim-PAN-ik) Membrane between the outer ear and inner ear that receives sound waves and transmits the vibrations to the bones of the middle ear. The eardrum.

U

ultrasound Fetal imaging technique in which high-frequency (ultrasound) sound waves are aimed at the mother's uterus. The sound waves rebound when they hit hard tissues in the fetus. These reflecting waves are analyzed and converted to an image by a computer.

umbilical cord (um-BIL-uh-kul) Cablelike structure connecting the fetus to the mother, through which arteries and veins pass.

unsaturated fat Type of fat with fewer than two hydrogen atoms for every carbon atom in their fatty acid tails. Examples include plant oils—such as olive, safflower, canola, and corn oil—and fish oil.

urea (yoo-REE-uh) Main nitrogen-containing waste excreted in urine.

ureter (YUR-uh-tur) One of two tubes that transport urine from the kidneys to the urinary bladder.

urethra (yoo-REE-thruh) Tube through which urine passes from the bladder to the outside of the body.

uterine cycle (YOO-tur-in) A series of changes that occur in the uterus in preparation for the arrival of a fertilized egg. A complete cycle generally takes about 28 days.

uterus (YOO-tur-us) Hollow, thick-walled organ that receives, retains, and nourishes the fertilized egg. The uterus is the site of embryonic/fetal development.

V

vaccine Antigens prepared in such a way that when injected or taken orally they induce active immunity without causing disease.

vagina (vah-JY-nuh) Hollow muscular female organ extending from the cervix to the body exterior. The vagina functions as an organ of sexual intercourse and as the birth canal.

varicose veins (VAIR-ih-kohs) Permanently dilated veins, often caused by improperly functioning venous valves. Varicose veins can appear anywhere but are most common in the legs and feet.

vasa recta Long, straight capillaries that perfuse the inner (medullary) portion of the kidney, supplying the loops of Henle and collecting ducts.

vasectomy (vuh-SEK-tuh-mee) Procedure for male surgical sterilization in which the vas deferens (ductus deferens) is cut and tied in two places.

vasoconstriction A narrowing of blood vessel diameter caused by contraction of the smooth muscle in the vessel wall.

vasodilation A widening of blood vessel diameter caused by relaxation of the smooth muscle in the vessel wall.

vein A thin-walled blood vessel that returns blood toward the heart from the venules.

ventricle (VEN-trih-kul) A general term for a cavity. In the heart, one of the two chambers that receives blood from an atrium and pumps it into an artery. In the central nervous system, one of four hollow cavities in the brain.

ventricular fibrillation (ven-TRIK-yoo-lur fib-rih-LAY-shun) A type of abnormal heart rhythm involving rapid irregular ventricular contractions.

venule (VEEN-yool, VEN-yool) A small blood vessel that transports blood from capillaries to a vein.

vertebral column (VUR-tuh-brul) Structure of the axial skeleton formed of a number of individual bones (vertebrae) and two composite bones (sacrum and coccyx). Also called the spine or backbone.

vertebrate (VUR-tuh-brayt) Any animal with a backbone composed of vertebrae.

vesicle (VES-ih-kul) A small membrane-bound, fluid-filled sac that encloses and contains certain substances within a cell.

vestibular apparatus (ves-TIB-yoo-lur) A system of fluid-filled canals and chambers in the inner ear. The vestibular apparatus consists of three semicircular canals for sensing rotational movement of the head, and the vestibule for sensing static (non-moving) position and linear acceleration and deceleration.

vestigial structure (ves-TIJ-ee-ul) A structure that may have had a function in some ancestor but which no longer serves any function. The human coccyx (tailbone) is a vestigial structure.

villus (pl. villi) (VIL-us) Fingerlike projection of the small intestinal mucosa that functions to increase the surface area for absorption.

virus A noncellular infectious agent that can replicate only within a host cell. Viruses consist only of one or more pieces of genetic material (DNA or RNA) surrounded by a protein coat.

vital capacity The maximum volume of air that can be expelled from the lungs by forcible expiration after the deepest inspiration.

vitamin Any one of more than a dozen organic compounds that the body requires in minute amounts but generally does not synthesize.

vitreous humor (VIT-ree-us) The clear fluid within the main chamber of the eye, between the lens and the retina.

vocal cords Two folds of connective tissue that extend across the airway. Audible sounds are heard when the vocal cords vibrate as air passes by.

vulva Female external genitalia that surround the opening of the vagina.

W

white blood cell (WBC) One of several types of blood cells that are part of the body's defense system. WBCs are diverse in structure and specific function. They compose only about 1% of the volume of blood. Also called a leukocyte.

X

X chromosome One of the two human sex chromosomes. An embryo that inherits two X chromosomes will develop into a female.

Y chromosome One of the two human sex chromosomes. An embryo that inherits a Y chromosome will develop into a male.

yolk sac Extraembryonic membrane that serves as the first source for red blood cells for the fetus and as the source for primordial germ cells.

Z

zero population growth The condition that exists in a population when the birth rate equals the death rate.

ZIFT (zygote intrafallopian transfer) (ZY-goht IN-trah-fuh-LO-pee-un) A technique for enhancing fertility in which an egg is fertilized outside the body, then inserted into an oviduct through a small incision in the abdomen.

zona pellucida (ZO-nuh peh-LOOS-ih-duh) A noncellular coating around an oocyte.

zygote (ZY-goht) The diploid cell formed by the union of the ovum and a sperm. The zygote is the product of fertilization.

Answers to Figure Check, Quick Check, and Test Yourself Questions

Chapter 1 **Check Figure 1.8, p. 12:** Inductive reasoning takes a specific example and develops a generalization from it; deductive reasoning does the reverse, using a general statement to develop specific predictions. The scientific method uses both because science involves reasoning back and forth between a general hypothesis and specific experimental predictions. **Check Figure 1.9, p. 13:** Worse. The two groups of subjects would have different blood pressure even *before* receiving the drug, and this will make it very difficult to interpret the results. For example, the experimental group might respond to the drug, but their blood pressure, even after being lowered, may still be higher than the other group; this would make it appear as if the drug did not work at all. Alternatively, people with high blood pressure might simply respond *differently* to the drug than do people with normal blood pressure, making their results impossible to compare. Random assignment is a much better experimental design because it ensures that the only difference between the two groups is whether or not they receive the drug. **Check Figure 1.10, p. 17:** Standard error bars tell us how confident we can be about our estimate of the average value of the whole population. The large standard error of the last bar, compared to the first one, means that we have much lower confidence in our estimate of average freshman enrollment in large universities as compared to small ones. (You may have noticed that the size of the standard error bars closely matches the amount of "scatter" in first graph. This is not coincidence—high variation in data causes larger standard error bars.)

Quick Check, p. 7: You could test its molecular composition to look for molecules associated with life (such as proteins, DNA, carbohydrates, etc.); see whether it uses energy to transform raw materials into other compounds; look under a microscope for the presence of cells; and/or observe it to see whether it ever reproduces or responds to environmental changes. **Quick Check, p. 8:** The cell without a nucleus is a prokaryote that might be either in the domain Bacteria or the domain Archaea. The cell with a nucleus is a eukaryote; it is in the domain Eukarya, kingdom Protista. Eukaryotes are more closely related to human cells because both have a nucleus. **Quick Check, p. 9:** Promising areas to study would be the skull (to estimate the size of the brain); the hands (to look for opposable thumbs); and the hips, knees and feet (to look for signs of bipedalism). (The fourth trait of humans, complex language, is difficult to assess from skeletal material.) **Quick Check, first question on p. 13:** No. A scientific hypothesis is any statement about the natural world that can lead to testable deductions. This statement is about the natural world, but it doesn't make any testable deductions—because if sasquatches leave no evidence of their existence, there is no way to test whether they exist! **Quick Check, second question on p. 13:** Your prediction should specify the type and extent of exercise, and the exact nature of "sleeping better," in as much detail as possible. For example, one prediction could be: "People who jog for at least forty minutes per day should fall asleep more rapidly at night than people who do not do any exercise." (Many other predictions are possible.) **Quick Check, p. 14:** There

are several problems, including: There was no control group (no group of people who did not jog, and no measurement of how well the subjects would have slept without jogging); it was not a blind study, so the "power of suggestion" might well cause the subjects to overestimate how well they really slept; the amount of sleep was not objectively measured; a larger group of people should be studied. **Quick Check, p. 17:** The fact that one company has more personal testimonials than the other should carry no weight at all, since it is possible (even probable) that the second company simply surveyed more people—or paid more actors! These personal stories are anecdotes that give us literally no information about the likelihood that the medications will work for other people, nor about whether either supplement might be superior to the other (or, in fact, whether they're any better than a placebo).

Test Yourself, p. 22: 1. b; 2. d; 3. a; 4. b; 5. c; 6. d; 7. d; 8. a; 9. d; 10. b; 11. d; 12. d; 13. b; 14. d; 15. c

Chapter 2 **Check Figure 2.5, p. 29:** Your diagram should show one carbon atom in the center, surrounded by four hydrogen atoms. Each hydrogen atom should be sharing a pair of electrons (i.e., forming a single covalent bond) with the carbon atom. **Check Figure 2.15, p. 38:** Starch and cellulose. They are both made from glucose subunits. We can digest starch, but not cellulose, because of the way the glucose subunits are attached together. **Check Figure 2.17, p. 40:** The phospholipid head is polar—meaning it is partially charged—and that means the head is electrically attracted toward the polar water molecules outside the membrane. The phospholipid tails, in contrast, are nonpolar and are not attracted toward water molecules.

Quick Check, p. 27: An atomic number of 7 means a nitrogen atom always has 7 protons. Since in most atoms the number of protons equals the number of neutrons, it is a safe bet that a typical nitrogen atom also has 7 neutrons. If it's electrically neutral, it must also have 7 electrons—2 in the first shell and 5 in the second shell. **Quick Check, p. 30:** Carbon dioxide's structure is O=C=O, a carbon atom in the center connected by double bonds to each of the two oxygen atoms on either side. (We can deduce this by determining, from the number of electrons, the number of bonds that each atom needs to fill its outermost electron shell. Oxygen will form two bonds, and carbon will form four.) **Quick Check, p. 33:** Vinegar is polar and hydrophilic, like water. We can deduce this from knowing that olive oil is nonpolar and is hydrophobic and that the oil and water aren't mixing. **Quick Check, p. 35:** Since pH increased, she must have added a base. Before adding the base, the concentration of hydrogen ions was 10^{-3} moles /liter (0.001 moles/liter). After, it was 10^{-5} moles/liter (0.00001 moles/liter). **Quick Check, p. 38:** Oligosaccharides contain slightly more carbon (relative to hydrogen and oxygen) than monosaccharides do, but the ratio of hydrogen to oxygen is the same—2:1. This is because a molecule of water (H_2O) is removed in the process of forming the oligosaccharide. **Quick Check, p. 39:** Canola oil has more double bonds. Triglycerides that are liquid at room temperature typically have more double bonds in their fatty acid tails than triglycerides that are solid. The double bonds cause the fatty acid tails to bend, preventing the molecules from packing closely together.   **Quick Check, p. 42:** It's probably a lipid. Some proteins are insoluble in water, but proteins always contain nitrogen. Carbohydrates would contain a substantial amount of oxygen as well as carbon and hydrogen.

Test Yourself, p. 47: 1. a; **2.** c; **3.** b; **4.** a; **5.** b; **6.** d; **7.** c; **8.** a; **9.** d; **10.** a; **11.** e; **12.** a; **13.** d; **14.** c; **15.** d

Chapter 3   **Check Figure 3.7, p. 58:** There will be a net movement of water from left to right, both because of diffusion (because initially the *concentration* of glucose will be higher on the right side) and also because of osmotic pressure, which will also be driving water to the right. Water will continue to move to the right until the height of the water columns is the same.   **Check Figure 3.13, p. 63:** Her extracellular fluid will probably be hypertonic—that is, it will have more solutes (and less water) than her cells. If it is extremely hypertonic, her red blood cells will start to shrivel. **Check Figure 3.28, p. 74:** The ATP synthase is "powered" entirely by the high concentration gradient of H^+ ions on one side of the membrane as compared to the other side. NADH and $FADH_2$ never interact with ATP synthase directly; they just cause more H^+ ions to be pumped across the membrane, making the concentration gradient stronger.

Quick Check, p. 58: Na^+, Cl^- and water molecules cannot cross a lipid bilayer on their own. Though they are small molecules, they also are charged (the ions) or polar (water). Charged or polar molecules are not soluble in lipids, and thus, they cannot diffuse through lipid bilayers.   **Quick Check, p. 60:** The fructose is moving "downhill," from high to low concentration, which is the natural direction that it would diffuse on its own. Therefore, the membrane protein is most likely allowing facilitated diffusion (a form of passive transport) and should not require any ATP. **Quick Check, p. 61:** The cell will tend to swell. A living cell always tends to accumulate a greater concentration of solutes inside than are present outside, which would (if uncorrected) result in a continual influx of water into the cell. Sodium-potassium pumps reduce this problem by removing sodium from the cell, thus reducing the concentration of solutes inside the cell.   **Quick Check, p. 65:** Cells whose primary function is to secrete proteins typically have abundant ribosomes and rough ER, but don't have much smooth ER.   **Quick Check, p. 67:** It's probably a vesicle. The nucleus and mitochondrion would have a double membrane, whereas ribosomes have no membranes at all. It might contain products recently produced by the Golgi apparatus, or it might be a peroxisome or a lysosome.   **Quick Check, p. 74:** Yes. Without ATP synthase, oxidative phosphorylation does not occur—that is, the electron transport chain cannot generate any ATPs—but four ATP molecules can still be produced from each glucose molecule by substrate-level phosphorylation (2 from glycolysis, and 2 more from the 2 acetyl CoA molecules that enter the citric acid cycle). **Quick Check, p. 76:** The heart is a muscle, and the "muscle wasting" that occurs during protein catabolism affects the heart, too. During starvation, muscle protein is broken down as a source of fuel, causing muscles to literally get weaker. Eventually the heart becomes so weak that it cannot pump blood effectively.

Test Yourself, p. 78: 1. c; **2.** b; **3.** e; **4.** a; **5.** c; **6.** b; **7.** d; **8.** e; **9.** c; **10.** b; **11.** a; **12.** e; **13.** b; **14.** a; **15.** c

Chapter 4   **Check Figure 4.2, p. 84:** Gap junctions. The tiny pores in gap junctions allow exchange of ions, molecules, and other raw materials between cells; the other two types of junctions do not allow exchange between cells.   **Check Figure 4.9, p. 93:** He has a blister on the bottom of his right foot.   **Check Figure 4.13, p. 98:** Your diagram should look like the diagrams in Figure 4.13. The "sensors" are probably in the brain, along with the control center. If fat stores increase above the set point, hunger decreases and less food is eaten, causing fat stores to return to the set point. If fat stores decrease, hunger increases and more food is eaten, again causing fat stores to return to the set point. What isn't known precisely is what the "sensors" are actually sensing.

Quick Check, p. 84: Simple columnar epithelium. Columnar epithelium contains the goblet cells that can secrete mucus. A single cell layer (i.e., simple epithelium) would also absorb the food molecules more efficiently than a stratified layer.   **Quick Check, p. 87:** Ehler-Danlos syndrome (EDS) is a disorder of dense connective tissue caused by defective collagen. The connective tissue still has some normal elastin, but it has very little collagen, resulting in weak, stretchy ligaments that allow bones to dislocate easily and weak, stretchy skin. **Quick Check, p. 89:** Skeletal muscle does not have gap junctions. If it did, then whenever one skeletal muscle cell started to contract, all of the other cells in the muscle would contract too, causing a forceful, simultaneous contraction by the entire muscle. There would be no way to do a gentle or a slow contraction.   **Quick Check, p. 93:** A serous membrane. Each pleural cavity contains a lung. The lubrication provided by a serous membrane reduces friction, allowing the lung to smoothly expand and contract without sticking to the inside of the pleural cavity.   **Quick Check, p. 94:** If your keratinocytes began dividing more rapidly, your epidermis would simply become thicker (and might start flaking, as occurs in dandruff and psoriasis). Your skin should still be able to perform all of its major functions.   **Quick Check, p. 98:** Yes, this is a negative feedback system, because an elevation in the refrigerator's temperature will cause a response that brings the temperature back down (i.e., negating the original change). The controlled variable is internal temperature; the control center is a thermostat; the sensor is a thermometer (wired to the thermostat); the set point is whatever temperature you have set the thermostat to; and the effector is the refrigerator's compressor and refrigeration coils.   **Quick Check, p. 99:** During a fever, the negative feedback system continues to work but the set point is shifted to a higher-than-normal temperature. The "chills" and shivering represent the body's attempt to warm up to the new set point. When a fever "breaks," the set point returns to normal; this causes sweating to lower the temperature.

Test Yourself, p. 100: 1. a; **2.** a; **3.** d; **4.** b; **5.** b; **6.** d; **7.** d; **8.** e; **9.** a; **10.** c; **11.** d; **12.** e; **13.** c; **14.** d; **15.** c

Chapter 5   **Check Figure 5.3, p. 107:** Osteoblasts are most active on the underside of the epiphyseal growth plate, where they are rapidly converting producing hydroxyapatite (calcium phosphate crystals).   **Check Figure 5.5, p. 110:** Breastbone: sternum; Collarbone: clavicle; Shoulder blade: scapula; Hip bone: coxal bone; Thighbone: femur; Shinbone: tibia   **Check Figure 5.9, p. 112:** The ribs and the sternum assist in respiration, and to do this they must expand the size of the cavity they surround. The flexible cartilage allows the ribs to move up and slightly away from the sternum.   **Check Figure 5.12, p. 115:** Tendons attach bone to muscle; ligaments attach bone to bone.   **Check Figure 5.13, p. 116:** Hinge joints generally only allow flexion and extension. Ball-and-socket joints can usually produce all of the motions shown in a, b and c. Fibrous joints are not synovial joints, and cannot move at all.

Quick Check, p. 105: Osteocytes not in direct contact with a blood vessel die from lack of oxygen and food and an accumulation of waste products. Osteocytes that are not near blood vessels rely on their gap junctions to pass materials to and from other osteocytes. **Quick Check, first question on p. 109:** Bicycling causes compressive stress on the leg bones, but the arms and vertebral column have relatively little compressive stress placed on them. Over time this can cause lightening of the bones of the upper body.   **Quick Check, second question on p. 109:** His PTH levels would be high and his calcitonin levels would be low. This would stimulate his osteoclasts to break down bone in order to release the much-needed calcium into the bloodstream.   **Quick Check, p. 112:** Recall that one of the functions of the vertebral column is to serve as a site of attachment for limbs. Unlike most mammals, humans walk upright on two legs, and we therefore need strong site of attachment for the pelvis and legs. Fusing the sacral bones together results in an especially strong point of attachment.   **Quick Check, p. 117:** Bones generally heal more rapidly than ligaments. This is because living bone has a rich blood supply and many cells, while ligaments have a poor blood supply and few cells.   **Quick Check, p. 118:** A drug that increases chondroblast activity is most likely to help arthritis. This is because chondroblasts are the cells that make new cartilage, and arthritis is due to a wearing down of cartilage in the joints.

Test Yourself, p. 120: 1. b; **2.** c; **3.** a; **4.** b; **5.** d; **6.** c; **7.** b; **8.** c; **9.** d; **10.** d; **11.** a; **12.** d; **13.** b; **14.** a; **15.** b

Chapter 6   **Check Figure 6.6, p. 129:** The muscle would still produce an electrical impulse, but no Ca^{2+} would be released from the sarcoplasmic reticulum. The troponin-tropomyosin complex would continue to block myosin from binding to actin, and the muscle would not contract.   **Check Figure 6.8, p. 130:** The calcium must be taken back into the sarcoplasmic reticulum, allowing the troponin-tropomyosin complex to move back over the myosin binding sites. Additionally, a new molecule of ATP must bind to each myosin head before it can "let go" of its current binding site on the actin filaments.   **Check Figure 6.10, p. 135:** The tetanic contraction shown here is being produced by a single motor unit; recruitment involves contraction of additional motor units. They are similar in that they will both produce increased muscle tension.

Quick Check, p. 127: Muscle cells are almost completely packed with myofibrils, which in turn consist almost entirely of two types of protein molecules (actin and myosin). So, foods comprised primarily of muscle tissue are mostly protein.   **Quick Check, p. 130:** This person would probably be completely unable to contract any skeletal muscles (that is, the muscles would always be relaxed and the person would be paralyzed). This is because the troponin-tropomyosin complex would always block myosin from binding to actin; calcium ions would be unable to move the troponin-tropomyosin complex out of the way. In fact, he/she would

probably not live for long.   **Quick Check, p. 131:** Caffeine allows acetylcholine molecules to remain longer in the motor junction. The acetylcholine would stimulate the muscle cell for longer than normal, resulting in a prolonged series of contractions.   **Quick Check, p. 135:** The "all-or-none" principle means that all muscle cells in *one* motor unit contract simultaneously when stimulated by their motor neuron. Adjacent motor units controlled by other motor neurons may not contract.   **Quick Check, p. 136:** Marathons. The dark red color indicates that her muscle cells have an unusually high concentration of myoglobin and/or blood vessels. In other words, she has a lot of slow-twitch muscle cells in her leg muscles, and she will be well-suited for long-distance endurance events.   **Quick Check, p. 138:** The pacemaker cells could still beat on their own, but they would no longer be able to coordinate the beat of the rest of the heart. Each heart cell would revert to its own intrinsic rhythm of beating, and the heart as a whole would become uncoordinated. (This is known as fibrillation.) **Quick Check, p. 139:** The drug that blocks acetylcholine would be the better choice. This would allow the muscles to relax despite the nerves' continued release of acetylcholine. The most obvious problem is that, if the dose of this drug is too strong, it might also paralyze the patient's respiratory muscles.

Test Yourself, p. 141: 1. b; **2.** c; **3.** a; **4.** c; **5.** c; **6.** a; **7.** b; **8.** b; **9.** c; **10.** d; **11.** b; **12.** d; **13.** b; **14.** c; **15.** d

Chapter 7   **Check Figure 7.6, p. 149:** Not enough red blood cells will be made, blood oxygen levels will fall below normal, and the kidney will continue secreting high levels of erythropoietin. However, the bone marrow will be unable to respond—blood oxygen levels will simply continue to be low, and the kidney will continue to secrete very high levels of erythropoietin. **Check Figure 7.13, p. 156:** Native Americans are most likely to accept blood transfusions from each other without reactions, because 91% of the people in this population share the same blood type (type O). However, it would still be risky to try this.   **Check Figure 7.14, p. 157:** Your sketches should show that the Rh-positive blood will agglutinate when mixed with Rh antibodies; the Rh-negative blood will not.

Quick Check, p. 146: The artificial plasma is missing globulins. Without globulins, a patient would be unable to transport fats normally and would be more vulnerable to infections. The artificial plasma is also missing clotting proteins, which means the blood would be unable to clot normally—patients might suffer severe blood loss from just a small cut.   **Quick Check, first question on p. 147:** Both of these features will make hemoglobin less likely to release oxygen to body tissues. Hemoglobin releases oxygen most readily under conditions of high body temperature and low (acidic) pH.   **Quick Check, second question on p. 147:** This sample is probably from a man—the hematocrit is 44%, which is in the normal range for men but is slightly high for women.   **Quick Check, p. 151:** These are symptoms of inflammation, and they are most likely caused by basophils, a type of agranular leukocyte that is activated by injured body tissues. The inflammatory process helps bring in nutrients, cells and chemicals needed for tissue repair.   **Quick Check, p. 155:** He can donate blood to people of any blood type—A, B, AB, or O. The fact that he does not have A or B antigens means that his red blood cells will not be attacked by the recipient's antibodies, no matter what their blood type.   **Quick Check, p. 156:** No. First, the Rh positive woman does not have any Rh antibodies (because the Rh antigen is a normal "self" protein for her); and second, the Rh-negative fetus doesn't have any Rh antigens anyway.   **Quick Check, p. 158:** Erythropoietin levels will be high, at least in these four types of anemia. In anemia, the blood oxygen carrying capacity is below normal, and this will stimulate the kidneys to release more erythropoietin. Only when anemia is caused by kidney disease is erythropoietin low.

Test Yourself, p. 160: 1. d; **2.** a; **3.** c; **4.** d; **5.** a; **6.** b; **7.** b; **8.** c; **9.** d; **10.** b; **11.** a; **12.** b; **13.** c; **14.** d; **15.** a

Chapter 8   Check Figure 8.4, p. 167: Water is drawn back into the capillary because there is a high concentration of dissolved molecules (especially proteins) in the capillary, and a low concentration in the interstitial fluid. Water automatically moves to the area that has more dissolved molecules—that is, it moves into the capillary. **Check Figure 8.11, p. 175:** The AV and semilunar valves open and close passively due to changes in blood pressure. If blood pressure is higher "downstream" of the valve than "upstream", the valve automatically shuts.   **Check Figure 8.16, p. 178:** The noise is heard only when the cuff pressure is between systolic and diastolic pressure because only then does the artery keep opening and shutting with every cardiac cycle. This repeated opening and shutting is what creates the audible tapping noise. When cuff pressure is above systolic pressure the artery stays closed; when cuff pressure is below diastolic pressure the artery stays open.

Quick Check, p. 167: The walls of arteries and arterioles are too thick for gases and nutrients to be able to diffuse across them. Arteries must have thick muscular walls to withstand the high pressure of blood pumping out of the heart. Even the thinner walls of arterioles are still too thick for effective diffusion.   **Quick Check, p. 168:** If the lymphatic system in a limb is completely blocked, the excess fluid that leaves the capillaries every day cannot be drained from the limb. In that situation the limb will become very swollen—sometimes to several times its normal size.   **Quick Check, p. 172:** The pulmonary arteries are the only arteries in the body that carry deoxygenated blood. They are considered arteries because, by definition, arteries are thick-walled blood vessels that carry blood at high pressure away from the heart.   **Quick Check, p. 173:** The problem is that the systemic circuit is completely isolated from the pulmonary circuit. Blood in the systemic circuit goes to the body over and over again, without every going to the lungs—and without ever picking up any oxygen! (This defect must be corrected with surgery or the baby will die soon after birth.)   **Quick Check, p. 176:** The atria contract silently—no valves close when they contract, and the blood flows smoothly, so there is no noise. The ventricle contraction, however, makes the AV valves snap shut, which produce the first heart sound (the "lub" of the "lub-dub").   **Quick Check, p. 180:** Systolic blood pressure is 141 mm Hg, and diastolic blood pressure is 95 mm Hg. This indicates this patient may have Stage 1 Hypertension, but since one test is not consider sufficient to diagnose hypertension accurately, the patient should be tested again on several different days.   **Quick Check, p. 181:** Initially, the blood pressure will drop, causing the baroreceptors to send neural signals to the brain, which in turn will increase sympathetic nervous activity and decrease parasympathetic nervous activity. The increased sympathetic nervous activity will cause the heart to beat faster (both by direct nerve input and also via epinephrine), and the sympathetic nerves activation will cause blood vessels to constrict. Blood pressure will be restored to normal or close to it within minutes.

Test Yourself, p. 186: 1. d; **2.** c; **3.** a; **4.** a; **5.** c; **6.** b; **7.** c; **8.** b; **9.** b; **10.** a; **11.** d; **12.** b; **13.** d; **14.** c; **15.** b

Chapter 9   Check Figure 9.12, p. 204: If this macrophage encounters a T cell whose receptors can bind to that particular antigen fragment, the T cell will be activated and will divide to produce more T cells. If the original T cell has CD4+ receptors, it will divide to produce helper T cells and memory cells; if it has CD8+ receptors, it will produce cytotoxic and suppressor T cells.   **Check Figure 9.13, p. 204:** Clonal expansion has to occur because typically only a very few T cells (sometimes just one) have a receptor that can bind to that antigen. With so few T cells able to recognize the antigen, very few cytokines will be produced unless clonal expansion occurs to produce many more identical T cells that all can recognize the same antigen.   **Check Figure 9.16, p. 207:** A person lacking memory cells will not be able to mount a normal secondary immune response. Their first response will be normal (exactly like the green line), but their second response will be just like a first response (exactly like the green line again)—much smaller and slower than a normal secondary response.

Quick Check, p. 192: It is most likely a virus. A bacterium would have a cell membrane (and usually a cell wall); a prion would consist only of protein without any nucleic acid.   **Quick Check, p. 199:** Nasal congestion is due to histamine's vasodilation effect, which makes capillary walls "leaky" and causes swelling in the nasal tissues as part of the inflammation response to the common cold. When antihistamines prevent histamine from working, the capillaries in the nose return to their normal "non-leaky" state and the swelling diminishes.   **Quick Check, first question on p. 201:** During pregnancy, the mother's immune system must somehow tolerate the presence of a fetus and a placenta that are genetically different from the mother, and that therefore have different (non-self) MHC proteins. (It is still not understood why the mother's immune system does not attack the placenta during pregnancy.)   **Quick Check, second question on p. 201:** The child should still be able to produce at least some antibodies, which are made by B cells. B cells arise from the bone marrow, not the thymus.   **Quick Check, p. 203:** A memory cell can make only a single type of antibody that has unique amino acid sequences in the variable regions of the light and heavy chains. Because of the unique variable regions, these antibodies will (most likely) only bind to a certain antigen on the surface of the hepatitis virus, and will not be able to bind to the different antigens that occur on other pathogens.   **Quick Check, p. 208:** This was passive immunization that provided only temporary protection. It was essentially a gamma-globulin shot from another species.   **Quick Check, p. 210:** After each exposure, B cells produce more IgE antibodies, which bind to mast cells and basophils that were not previously involved in the allergy. On the next exposure, these mast cells and basophils will also react, causing a greater release of histamine, which causes worse symptoms.

Test Yourself, p. 217: 1. c; **2.** a; **3.** b; **4.** c; **5.** d; **6.** c; **7.** c; **8.** b; **9.** c; **10.** c; **11.** c; **12.** a; **13.** d; **14.** c; **15.** c

Chapter 10   Check Figure 10.3, p. 222: Nose, nasal cavity, pharynx, trachea, bronchus, bronchiole, alveolus, and finally the alveolar wall and the capillary wall. Gas exchange occurs only between the alveolus and the capillary.   **Check Figure 10.10, p. 229:** During exercise, tidal volume increases dramatically. Vital capacity remains the same (that is, the lungs do not suddenly grow

larger; exercising just means that more of their vital capacity is being used).   **Check Figure 10.11, p. 231:** PO_2 decreases continuously: from 104 mm Hg in the alveolar air, to 100 mm Hg in the pulmonary capillaries leaving the lungs, to 40 mm Hg in the interstitial fluid. These numbers must decrease in this manner in order for oxygen to diffuse in the right direction—from the alveoli, to the blood, to the interstitial fluid.

Quick Check, p. 222: The upper respiratory tract is directly connected to the ears via the two auditory tubes. The microorganisms that cause respiratory tract infections can often spread up these tubes to the middle ears, where they can then cause ear infections.   **Quick Check, first question on p. 230:** If the total atmospheric pressure at the summit of Everest is 260 mm Hg, the partial pressure of oxygen is 54.6 mm Hg (260 mm Hg $\times$ 0.21).   **Quick Check, second question on p. 230:** Oxygen will diffuse from the venous blood in the capillaries (P_{O_2} = 40 mm Hg) into the alveolar air (P_{O_2} = 35 mm Hg)—that is, the passengers will actually start losing oxygen from their blood. This occurs because gases always diffuse from areas of high partial pressure to areas of low partial pressure.   **Quick Check, p. 232:** These people most likely do not have enough hemoglobin in their red blood cells (recall that iron is a necessary component of hemoglobin). The small amount of O_2 that is dissolved in the blood plasma is not enough to sustain life; we also need the much greater amount of O_2 that binds directly to hemoglobin.   **Quick Check, p. 235:** A low concentration of H^+ ions in blood will result in low concentration of H^+ ions in cerebrospinal fluid as well. This will be detected by the receptor cells in the medulla oblongata, which will trigger an decrease in respiratory rate—resulting in less exhalation of CO_2 and causing $H+$ concentrations to rise back toward normal.   **Quick Check, 236:** Bronchitis is the most likely cause. The association with smoking suggests bronchitis or emphysema, but emphysema usually does not cause coughing; the coughing of asthma does not usually produce yellowish phlegm; and cystic fibrosis is a lifelong condition that would have been present since childhood.   **Quick Check, 238:** Pneumonia and tuberculosis can both cause these symptoms. Possible diagnostic tools include a chest x-ray, a tuberculin skin test, or a throat culture to look for *Mycobacterium tuberculosis*.

Test Yourself, p. 240: 1. a; **2.** c; **3.** a; **4.** c; **5.** d; **6.** c; **7.** d; **8.** b; **9.** a; **10.** d; **11.** c; **12.** b; **13.** c; **14.** a; **15.** b

Chapter 11   **Check Figure 11.4, p. 247:** The new action potential will look exactly the same as the first one, with the same height and the same duration. Once initiated, action potentials are "all or none."   **Check Figure 11.5, p. 248:** The green line represents *permeability* to K^+. When the green line goes up, that means K^+ channels have opened so that the membrane is now very permeable to K^+. K^+ ions then rush out of the cell, causing the interior of the cell to become *less positive* relative to the outside—which means the cell's membrane potential (blue line) falls.   **Check Figure 11.16, p. 262:** This can happen if the occipital lobe is damaged. The occipital lobe is at the back of the head, and its major function is processing of visual information from the eyes.

Quick Check, p. 247: The unusual pumps are producing a net movement of two (rather than the usual one) positive ions outward for every pumping cycle. This should make the resting potential more negative than the resting potential of a normal neuron.

Quick Check, p. 249: No. The neuron will still open its voltage-gated Na^+ channels—but there will be no Na^+ ions to enter the cell. Without $Na+$ ions entering the cell, the neuron will not depolarize and will not produce an action potential.   **Quick Check, p. 252:** The presynaptic neuron will still be able to produce action potentials (which requires only Na^+ and K^+), but will not be able to release neurotransmitter (which requires Ca^{++} to enter the neuron from the synaptic cleft).   **Quick Check, p. 255:** Both are produced by the spinal cord without requiring the brain, and both trigger contraction of a muscle. Differences include: flexor reflexes are triggered by a pain-detecting sensory neuron, while stretch reflexes are triggered by stretch receptors in a muscle; flexor reflexes cause a second compensating reflex in another limb, while stretch reflexes generally do not involve another limb.   **Quick Check, p. 257:** This would reduce the activity of the sympathetic division, but would not affect the parasympathetic or somatic divisions. Likely symptoms would include reduced heart rate and blood pressure, pupillary contraction, increased blood flow to the digestive tract, etc.—all the symptoms of *reduction* of sympathetic nervous activity.   **Quick Check, p. 259:** Antibodies cannot cross the blood-brain barrier, because they are too large and too polar to cross the lipid bilayer of the capillary cell membranes in the brain.   **Quick Check, p. 261:** Cessation of activity of the medulla oblongata function would cause death within minutes, because the medulla oblongata is required for respiration. Cessation of activity of the cerebellum would severely disrupt the ability to move normally, but if the person simply lies still, he would probably survive.

Test Yourself, p. 272: 1. a; **2.** d; **3.** c; **4.** c; **5.** a; **6.** d; **7.** c; **8.** b; **9.** a; **10.** d; **11.** b; **12.** c; **13.** a; **14.** b; **15.** c

Chapter 12   **Check Figure 12.8, p. 284:** The new graph should have more waves per unit time (high frequency, i.e. high pitch) than the "high" sounds illustrated in the original figure. The height of each wave (the "loudness") should be intermediate compared to the "loud" and "soft" sounds in the original figure.   **Check Figure 12.10, p. 286:** A person with this condition would probably hear a very high-pitched sound, such as a high-pitched hum or buzz. This is because activity of hair cells at the beginning of the cochlea normally generates impulses in sensory neurons that are interpreted by the brain as high-pitched sounds.   **Check Figure 12.23, p. 296:** Yes. Though she lacks blue cones, she has normal green cones and red cones, which will allow her to distinguish the colors green and red. (She will be unable to perceive the color blue, however—blue will probably appear to her as a shade of gray.)

Quick Check, p. 277: The person will most likely perceive action potentials from this neuron as some sort of light—even though the neuron was actually detecting sound waves. This is because the brain interprets sensory information according to the specific brain area being stimulated, and in this case the visual area of the brain would be stimulated.   **Quick Check, p. 280:** Though your friend might briefly touch his toes during the "downward" part of the bounce, the muscle spindles will immediately trigger a stretch reflex that will cause the hamstrings to contract. After the bounce, the hamstrings will probably be tighter than before—so this is not a good method for stretching muscles.   **Quick Check, p. 282:** The tastants in foods must dissolve in saliva (or water) before they can bind to the chemoreceptor cells in the taste buds. The hemoreceptors must be stimulated for the brain to perceive any sensations of taste.

Quick Check, p. 285: Our sense of hearing would be much less sensitive—that is, all sounds would seem much softer, and we would probably be unable to detect very faint sounds. This is because the functions of the outer and middle ears are to channel and amplify sounds. **Quick Check, p. 288:** When you suddenly stop spinning, the fluid in the semicircular canals keeps moving under its own inertia for a few moments, flowing around the canals on its own. This will bend at least one of the cupulas, and the hair cells within it. The brain interprets this as if the head is moving in the other direction. **Quick Check, p. 291:** The person would be able to see blurry images, but not sharp images. This is because most of the focusing of the light is done by the cornea. The lens is necessary only to do the second stage of focusing, bringing the blurry images produced by the cornea into sharp focus. **Quick Check, p. 294:** Animals that are active only at night have poor or no color vision—they see the world in shades of grey. This is because their retinas contain mostly rods, which provide sharp vision in very dim light, and very few (if any) cones, which are required for color vision.

Test Yourself, p. 298: 1. d; **2.** c; **3.** d; **4.** a; **5.** c; **6.** d; **7.** a; **8.** b; **9.** c; **10.** d; **11.** c; **12.** c; **13.** b; **14.** b; **15.** d

Chapter 13

Check Figure 13.3, p. 305: In this example, 100,000 molecules of final product will be produced by a single hormone molecule. Because of this amplifying effect, hormones can be present at very low concentrations in the blood but still have a significant impact on the body. **Check Figure 13.15, p. 316:** Your diagram should show alpha and beta cells of the pancreas on the right side of the diagram, responding to a rise in blood glucose by reducing glucagon (alpha cells) and increasing insulin (beta cells). On the left side should be three major effectors: the liver, muscle, and fat tissue. Liver and muscle cells both respond by converting the excess glucose into glycogen; fat cells respond by converting some of the excess glucose to fat.

Quick Check, p. 305: Steroid hormones generally don't use second messengers, because a second messenger is only needed for hormones that can't enter the cell. A second messenger is a molecule that relays the hormone's message into the interior of the cell. Steroid hormones simply enter the cell themselves. **Quick Check, p. 307:** The hypothalamus would detect that the blood has too much water (i.e., low solute concentration) and would secrete less ADH, causing a decline in the ADH concentration in the blood. The reduction in ADH would cause the kidney to reabsorb less water from the urine, so that more water would be excreted in the urine. **Quick Check, p. 309:** She will have trouble giving birth, because the mutant oxytocin will be unable to stimulate labor. Once the baby is born (presumably by Caesarian section or after a shot of artificial oxytocin), the mother will be able to produce milk normally (prolactin's role), but the milk will not be ejected from the mammary glands (oxytocin's role), so the baby will have trouble suckling. **Quick Check, p. 312:** Responding to a stressful situation usually requires energy (ATP). For example, healing a physical injury requires ATP (to rebuild the damaged tissues), and stressful emotions such as fear often indicate that we might need to run or fight, which also requires ATP. Cortisol helps raise blood glucose levels so that body tissues can use the glucose to produce more ATP. Put simply, cortisol gives you energy to do whatever needs doing. **Quick Check, 315:** Thyroxine levels usually drop sharply during starvation. (They also plummet during

crash diets!) This represents the body's attempt to save energy by its reducing metabolic rate. **Quick Check, p. 318:** People who are exposed only to dim office lights during the day often have elevated melatonin levels, so they must not have a normal 24-hour sleep-wakefulness cycle. High daytime levels of melatonin are thought to be one of the underlying causes of Seasonal Affective Disorder, a syndrome of low energy and poor mood that is associated with wintertime. **Quick Check, p. 320:** It is almost always easier to treat failure to produce a hormone, because you can simply give the patient an injection of the missing hormone if it's available. Failure to respond to a hormone is a much more difficult condition to treat because neither giving the hormone or blocking its action does very much of anything.

Test Yourself, p. 322: 1. c; **2.** b; **3.** a; **4.** d; **5.** a; **6.** d; **7.** a; **8.** d; **9.** c; **10.** c; **11.** b; **12.** a; **13.** d; **14.** a; **15.** d

Chapter 14

Check Figure 14.20, p. 349: Obesity is most prevalent in southeastern states and the Appalachian mountains, especially areas around Mississippi and West Virginia. These areas are notable for a preponderance of fried, high-fat foods in people's diets. (Many other cultural factors may also contribute, including a lack of exercise.)

Quick Check, p. 333: Pepsinogen would be produced, but in the absence of an acidic environment it would not be converted to pepsin. The stomach could still perform its food storage and regulation-of-delivery functions, and could still mechanically break up large pieces of food, but it would not be able to digest proteins effectively (because of the lack of pepsin) nor would bacteria in the stomach be killed as readily. **Quick Check, p. 336:** Liver is generally a very nutritious food because it contains not only glucose stores and fat stores, but also iron and many vitamins. But it can be dangerous if the animal has been exposed to environmental toxins, which tend to accumulate in the liver. In certain species, the liver's vitamin concentrations may be dangerously high. **Quick Check, p. 339:** No—the duodenum releases the hormone cholecystokinin (CCK) in response to the arrival of proteins and fats, but does not release any hormones in response to the arrival of carbohydrates. Carbohydrates are partially broken down in the mouth by the enzyme salivary amylase, and then are further broken down by enzymes from the pancreas and the small intestine. No hormones are involved. **Quick Check, first question on p. 342:** Cold-water fish must use unsaturated fats so that their fat tissue will stay soft and pliable. If cold-water fish stored lipids as saturated fats, the saturated fats would be in solid form, making the fish too stiff to swim. Warm-water fish, in contrast, are free to use saturated fats—and they do. **Quick Check, second question on p. 342:** Yes, eggs and milk both contain complete proteins. Unhatched bird embryos and baby mammals must assemble every protein in their bodies solely from the amino acids contained in the egg (for birds) and in milk (for mammals) **Quick Check, p. 343:** Vitamin B12. B12 is the only vitamin that is obtained only from animal products, and vegans do not eat any animal products. For this reason, many vegans take vitamin supplements and/or eat tofu that has been fortified with vitamin B12 **Quick Check, p. 346:** The woman who is lifting weights will probably lose more body fat. In addition to the 200 Calories burned during the exercise itself, the extra muscle that she develops will also increase her BMR throughout the rest of the day. **Quick Check, p. 347:** Peptic ulcers tend to occur only in the stomach and first portion of the duodenum because of the high acidity in those regions. The acid is neutralized by mid-duodenum by bicarbonate delivered from the pancreas.

Test Yourself, p. 352: 1. b; **2.** c; **3.** c; **4.** a; **5.** d; **6.** b; **7.** d; **8.** d; **9.** b; **10.** c; **11.** a; **12.** d; **13.** c; **14.** c; **15.** b

Chapter 15   **Check Figure 15.6, p. 362:** Water, ions, glucose, and amino acids are reabsorbed; creatinine and most urea are not. This makes sense because creatinine and urea are waste products that are not needed by the body.   **Check Figure 15.11, p. 367:** No. Concentrated urine is made by reabsorbing water via *passive* diffusion—no ATPs required. There is already a high solute concentration in the medulla that will cause water to move in the right direction on its own; the collecting tubules must merely become permeable to water.   **Check Figure 15.12, p. 368:** Desert animals need to conserve water. A longer loop of Henle is simply a longer countercurrent exchange system, which causes the fluid at the bottom of the loop to become even more concentrated. This hyperconcentrated interstitial fluid will draw more water out of the collecting duct, producing extremely concentrated urine—and saving water!   **Check Figure 15.15, p. 371:** ADH increases after all of the other events have occurred. It increases because the last step of the renin-angiotensin cycle, absorbing sodium, causes the blood solute levels to increase. The hypothalamus detects this and releases ADH in response.

Quick Check, p. 359: The doctor is checking to see if the bladder infection might have spread up the ureters to the kidneys. The kidneys are located at the back of the abdominal cavity, so kidney infections can cause back pain.   **Quick Check, p. 366:** Yes. The distal tubule will still be able to reabsorb sodium ions (Na^+) and chloride ions (Cl^-) by active transport, leaving water in the urine. (However, the kidney would be unable to produce concentrated urine.)   **Quick Check, p. 369:** The pizza has resulted in too much sodium and chloride in your blood, i.e., a high concentration of solutes. Your kidneys must get rid of the excess sodium while losing as little water as possible. ADH will increase (to conserve water) and aldosterone will decrease (to allow more sodium to be excreted). Your kidney will produce very concentrated urine.   **Quick Check, first question on p. 371:** The kidneys will sense the drop in blood pressure and will secrete renin, which will result in an increase in angiotensin II, which will stimulate aldosterone secretion from the adrenal gland. Meanwhile, the heart will decrease secretion of ANH. The final effects will include constriction of arterioles (via angiotensin II) and an increase in sodium reabsorption (via increased aldosterone, and also decreased ANH).   **Quick Check, second question on p. 371:** The kidney should excrete fewer H+ ions than usual, retaining H+ in the blood to bring the pH back down to normal. (This is, in fact, exactly what the kidneys do, and it is why most people begin to feel better after about three days at high altitude.)

Test Yourself, p. 375: 1. d; **2.** c; **3.** b.; **4.** b; **5.** a; **6.** c; **7.** a; **8.** d; **9.** d; **10.** c; **11.** a; **12.** c; **13.** a; **14.** b; **15.** d

Chapter 16   **Check Figure 16.2, p. 381:** A spermatogonium. A spermatogonium gives rise to millions of sperm, while each primary spermatocyte gives rise to just four. This is beause the spermatogonium continually buds off new primary spermatocytes.   **Check Figure 16.7, p. 387:** The corpus lutem is source of this progesterone. Progesterone levels rise when the corpus luteum

forms from the remnants of the follicle just after ovulation. Progesterone levels fall when the corpus luteum degenerates approximately twelve days later—unless pregnancy occurs.   **Check Figure 16.8, p. 388:** In both sexes, GnRH from the hypothalamus triggers release of LH and FSH from the anterior pituitary. FSH then stimulates development of the egg (and its follicle) or the sperm, while LH stimulates production of at least one steroid: estrogen and progesterone in women, testosterone in men. The steroid(s) then further supports development of the egg and sperm, and also exerts negative feedback on the anterior pituitary and hypothalamus—until ovulation in women, that is.   **Check Figure 16.14, p. 398:** One hypothesis is that because chlamydia's symptoms are very mild at first, many people do not realize that they have been infected.   **Quick Check, p. 382:** All oocytes pass from the ovary into the oviduct where fertilization normally occurs. Embryos can sometimes escape into the abdominal cavity because the oviduct has an open end.   **Quick Check, p. 386:** She will still have all of her primary oocytes, because those are all produced before birth. She won't be able to ovulate, though, because LH is the hormone that triggers ovulation.   **Quick Check, 391:** The woman will be sterile immediately (assuming she did not ovulate right before the operation), but the man will not. This is because maturing sperm are stored in the ductus deferens, so some viable sperm are usually present in the duct past the site of the vasectomy. It can take several weeks to get rid of these sperm.   **Quick Check, p. 392:** Pregnancy. In fact, most hormonal birth control methods are specifically designed to mimic the hormonal conditions of pregnancy. This makes sense because pregnancy is the one physiological state during which ovulation is completely prevented for a period of several months.   **Quick Check, p. 395:** Progesterone enables the uterus to thicken and maintain its inner lining, the endometrium—along with the embryo that is inside the endometrium. If progesterone is not high enough, the uterus will automatically lose its lining, just as it does during menstruation. A series of several miscarriages may indicate that the woman's progesterone is too low to maintain the uterine lining. If that's the case, then administering progesterone may help.   **Quick Check, p. 400:** Women generally are more susceptible to STDs than men are. The vagina is lined by a warm, moist mucus membrane. The outer surface of the penis is covered with dry skin, which is a much less hospitable environment for microbes.

Test Yourself, p. 402: 1. d; **2.** c; **3.** a; **4.** c; **5.** d; **6.** a; **7.** d; **8.** b; **9.** d; **10.** b; **11.** c; **12.** a; **13.** d; **14.** b; **15.** d

Chapter 17   **Check Figure 17.6, p. 411:** The promoter is most likely in the very bottommost loop of the DNA shown in part A—or possibly even further "down," off the figure entirely. The promoter is where transcription first began, and since transcription is already underway and the entire transcription complex is moving upward, the original starting point is now at the bottom of the figure.   **Check Figure 17.7, p. 412:** The mRNA will have the sequence AUG CCC GUA UAA. The peptide will be only three amino acids long, with a sequence of met-pro-val; that is, methionine, then proline, then valine. (The "Stop" codon UAA stops translation without adding a final amino acid.)   **Check Figure 17.8, p. 413:** Initiation will occur normally. Elongation will proceed normally until the ribosome meets the new stop codon. At that point, termination will occur—too early! A protein will be produced, but it will be an abnormally short protein that is missing its second half.

Quick Check, p. 408: A chromosome is a single long molecule of DNA along with its associated histone proteins. A gene is a short part of a chromosome that contains the instructions for making one or more proteins. Chromatin is simply a mixture of DNA with histone proteins, i.e., chromosomes are made of chromatin. Finally, a chromatid is one copy of a recently duplicated chromosome that is still attached to the other "sister" copy.   **Quick Check, p. 409:** Several thousand. Replication occurs simultaneously at several thousand places in the nucleus. Since DNA polymerase is the enzyme that forms new DNA, each of these replication sites must involve at least one DNA polymerase molecule.   **Quick Check, p. 411:** If RNA polymerase can attach more easily, then transcription is much more likely to happen, producing mRNAs. The gene is now much more likely to be expressed, meaning that whatever protein it codes for will probably be produced.   **Quick Check, first question on p. 415:** This chromosome will probably complete prophase and metaphase normally, but as anaphase begins, the chromosome will still be stuck in the center of the cell with both of its chromatids still attached together. The chromatids will be unable to separate. Mitosis may stall completely, or possibly both chromatids may be tugged to the same side of the cell (most likely with devastating consequences for both daughter cells). **Quick Check, second question on p. 415:** During development, skeletal muscle cells perform mitosis normally but then don't undergo cytokinesis. Many rounds of a cell cycle without cytokinesis will produce a very large cell containing many identical nuclei. **Quick Check, p. 417:** A chimpanzee cell just before meiosis will have 48 chromosomes, each with two chromatids. After meiosis I, each cell will have 24 chromosomes, each chromosome still with two chromatids. After meiosis II, each cell will still have 24 chromosomes, but now each chromosome consists just of one copy. (Technically, the moment the chromatids separate from each other they graduate to the status of "chromosomes" and are no longer called chromatids.)   **Quick Check, p. 421:** One strategy would be to collect eggs from the female, fertilize them with the male's sperm, split each embryo into 8 identical embryos, and then implant all eight embryos in surrogate females. The main problem (besides the likely technical difficulties) is that the surrogate mothers would have to be of a different species! This might make the pregnancies fail, and even if calves are born, they would have to be raised by mothers of another species and might end up with abnormal behavior. However, if a closely related, but non-endangered, antelope species could be used, this strategy might work. An additional concern for any future reproduction of the species would be that all the offspring would be identical.

Test Yourself, p. 424: 1. a; **2.** a; **3.** b; **4.** d; **5.** d; **6.** b; **7.** c; **8.** d; **9.** c; **10.** a; **11.** b; **12.** a; **13.** c; **14.** a; **15.** b

Chapter 18   **Check Figure 18.1, p. 428:** Most doctors would say that benign tumors need not be removed if they are not pressing on nearby tissues. Since benign tumors rarely progress to become malignant tumors, and since any surgical procedure inevitably has some risks, the risks of removal (of a tumor that is *known* to be benign) are often greater than the possible benefits. **Check Figure 18.7, p. 435:** The 70-year-old living in 2005 was an astounding *four and a half times more likely* to have lung cancer than a 70-year-old living in 1950. The 30-year-old in 2005, on the other hand, was much less likely (about 2/3 less) to have lung cancer than a 30-year-old living in 1950.

Quick Check, p. 431: The mutated gene is now an oncogene. Before it mutated, it was a proto-oncogene, because its normal function is to promote controlled cell growth.   **Quick Check, p. 433:** The elevated risk of lung cancer is due to radioactive radon gas that is emitted naturally by the rocks and soil around basements in some areas. Because radon is a gas, it primarily affects the lungs.   **Quick Check, first question on p. 436:** Radiation is a cancer risk. During cancer treatment, radiation may sometimes pass through tissues near the initial cancer, increasing the risk of other cancers in those tissues.   **Quick Check, second question on p. 436:** These three symptoms are all due to a shutdown of tissues that normally undergo unusually rapid cell division. Chemotherapy inhibits cell division in the bone marrow, which normally produces many new red blood cell and white blood cells every day. This causes the anemia and heightened infection risk. Hair follicles are also inhibited because they are a part of the skin, which under normal conditions grows very rapidly.   **Quick Check, p. 437:** Avastin slows healing of surgical incisions, because it blocks angiogenesis, the growth of new blood vessels. When a tissue grows or undergoes repair blood vessels are needed to supply the new tissue—including the new tissue made during wound healing.   **Quick Check, p. 438:** The natural ability of melanocytes to migrate makes them much more likely to metastasize, spreading from their source to other regions or organs of the body. Though they usually do not move after embryonic life, their ability to move seems to become re-activated if and when they become cancerous.   **Quick Check, p. 440:** A higher percentage of men with breast cancer die from the disease than women (26% vs. 19%). The most likely explanation is that men are not examined regularly for breast cancer, as women are. Consequently, the disease is generally not diagnosed as early in men. By the time the cancer is discovered, it's more likely to be too late.   **Quick Check, p. 441:** Cigarette smoke contains many carcinogens that don't just stay in the airborne smoke in the lungs; they dissolve into the blood plasma and circulate throughout the body. The leading hypothesis for the increase in bladder cancer is that many of these carcinogens are excreted in the urine, so the cells lining the bladder are exposed to high concentrations of these carcinogens while the bladder is holding urine.

Test Yourself, p. 444: 1. d; **2.** d; **3.** c; **4.** d; **5.** a; **6.** d; **7.** c; **8.** b; **9.** b; **10.** a; **11.** c; **12.** d; **13.** d; **14.** b; **15.** b

Chapter 19   **Check Figure 19.2, p. 449:** Each cell in a Punnett square has an equal probability of occurring. In this case, because there are four cells, each cell has a 25% probability of occurring. Therefore, the baby has a 25% chance of having the *AA* genotype and a 50% chance of having *Aa* (50% = 25 + 25, because there are two different cells with this genotype).   **Check Figure 19.7, p. 453:** The ratio is 9:3:3:1, which is to say, 9 out of 16 offspring have both dominant phenotypes (widow's peaks and free earlobes), 3 out of 16 offspring have widow's peaks and attached earlobes, another 3 out of 16 offspring have no widow's peaks and free earlobes, and 1 out 16 offspring has both recessive phenotypes (no widow's peak, attached earlobes). If you observed a similar 9:3:3:1 ratio in a natural mating that involved two traits, you could conclude that the parents were heterozygous for both traits. **Check Figure 19.10, p. 455:** Zero. The baby must inherit either the B or the A allele from his mother, and thus cannot be blood type O. **Check Figure 19.16, p. 460:** The daughter must have inherited the hemophilia allele from *both* parents. This can only occur if the woman was a carrier and she married a hemophiliac man.

Quick Check, p. 452: The man and the woman both have dark hair. The child will have a 25% probability of having red hair. **Quick Check, p. 458:** One experimental approach would be to raise many genetically unrelated young finches in an identical environment and see if they all end up with the same feather color. Another approach would be to take red finches or pink finches and raise them in different environments to see if the different environments somehow elicit differences in feather color. (It turns out that in this species, feather color is strongly affected by diet as well as by genes.) **Quick Check, p. 460:** No. Sons always inherit their father's Y chromosome, but never receive their father's X chromosome. (If they received their father's X chromosome, they would be daughters.) **Quick Check, first question on p. 462:** No. Down syndrome does not involve any gene mutations. There are extra copies of some genes (the ones on chromosome 21), but each copy is a normal copy. **Quick Check, second question on p. 462:** It is the presence or absence of the Y chromosome that primarily determines gender. The number of X's has only a minor effect, if any. That is, females are females not because they have two X's, but because they lack a Y. **Quick Check, p. 463:** The baby must be a homozygous recessive (because only homozygous recessives have PKU), and therefore must have inherited a PKU allele from each parent. Since neither parent has the disorder, they can't be homozygous recessives themselves, they must be heterozygotes. The probability that a child of two heterozygous parents will be homozygous recessive is 25%.

Test Yourself, p. 466: 1. b; **2.** a; **3.** c; **4.** a; **5.** b; **6.** c; **7.** d; **8.** a; **9.** d; **10.** c; **11.** d; **12.** d; **13.** d; **14.** b; **15.** a

Chapter 20   **Check Figure 20.1, p. 471:** Many copies of just a single green A* attached to the primer will be produced in the first step, but nothing else. This is because the only nucleotides available are the modified fluorescent ones that always terminate the growing strand. So the very first nucleotide used (the A*) will always be fluorescent *and* will always terminate the strand. The laser results in the third step will show only a single green peak. **Check Figure 20.3, p. 473:** The restriction enzyme would be unable to cut the human DNA into fragments. At the end of the procedure, the bacteria would have plasmids, but the plasmids would not contain any human DNA. **Check Figure 20.8, p. 476:** Yes, it's possible. Natural processes certainly do alter genomes through the process of random mutations. The result is evolutionary change over time. In addition, naturally occurring viruses sometimes transfer genes between species. So, a plant such as the one on the left in Figure 20.8 could indeed evolve through natural processes. The major differences between natural processes and genetic engineering are that natural processes are much slower than genetic engineering, and the direction evolution takes is random.

Quick Check, first question on p. 473: The human gene most likely contains the palindromic sequence that is the target of the restriction enzyme. That is, the restriction enzyme is cutting at a location *within* the gene, as well as at a site outside the gene. You could try a different restriction enzyme that cuts a different palindromic sequence. You could also try again a few more times with the same restriction enzyme, because with a bit of luck the gene fragments might end up in the right order in the same plasmid. **Quick Check, second question on p. 473:** You must know the sequences of the very beginning and the very end of each piece of the gene, so that you can make primers with

complementary sequences. To get the DNA copying started, the primers must be able to bind to both ends of the DNA, and they will only bind if they have complementary sequences. **Quick Check, p. 476:** Because transgenic bacteria cannot perform protein modification, they may not produce a functional version of the human protein of interest. It may take extensive trial and error to find the right combination of plasmid and bacteria. **Quick Check, p. 477:** The major problem is that plant cells do not take up plasmids as readily as bacteria do, so a variety of methods may have to be tried to coax embryonic plant cells into taking up the desired DNA. Plants also grow much more slowly than bacteria, so once a plant actually does take up the DNA, it will take longer to produce plant clones. **Quick Check, p. 479:** In theory this is possible by using genetic engineering techniques on a fertilized human egg that was produced by *in-vitro* fertilization. For example, the desired DNA could be microinjected directly into the fertilized human egg. However, recall that this technique fails in 90% of attempts, meaning that 90% of the embryos would either not take up the gene or might even die. It might work eventually—but many human embryos would probably be sacrificed in the process. And, even if it did work, the infant might be born with serious birth defects. At this time, attempting to perform gene therapy on a fertilized egg is just not practical. **Quick Check, first question on p. 480:** Cells of the immune system attack viruses—including the retrovirus vectors that are carrying the valuable DNA! Immune suppression gives the retroviruses a better chance to deliver their payload successfully without being attacked. **Quick Check, second question on p. 480:** The added gene sometimes lands in an unfortunate place in the genome, and in several patients, it apparently inserted next to or inside of one of the genes that controls cell division. Even if this happens in only a single one of the patient's cells, a single cell with a disabled cell-division gene could become a cancer.

Test Yourself, p. 482: 1. b; **2.** d; **3.** a; **4.** c; **5.** b; **6.** c; **7.** a; **8.** d; **9.** d; **10.** c; **11.** d; **12.** a; **13.** b; **14.** d; **15.** c

Chapter 21   **Check Figure 21.5, p. 490:** Cleavage has no period of cell growth between the rounds of cell division. The result is that individual cells get smaller and smaller—something that does not occur in normal cell division in adults. **Check Figure 21.6, p. 491:** The trophoblast secretes proteolytic enzymes that digest a path into the uterine tissue. **Check Figure 21.12, p. 495:** The XXY embryo will become male phenotype (though he might develop enlarged breasts at puberty as a result of the extra X chromosome.) It is the presence or absence of the Y that determines gender, not the number of Xs. **Check Figure 21.15, p. 499:** If either the ductus arteriosus or the foramen ovale passageway did not close properly, some of the baby's blood would bypass the lungs and return directly to the body, without picking up any oxygen! Depending how much blood is bypassing the lungs, a baby with either condition is likely to become oxygen deprived. Such conditions are correctable by surgery, if necessary, to close these passageways.

Quick Check, p. 488: The zona pellucida normally becomes impenetrable after the first sperm has entered, thus preventing other sperm from entering. If there is no zona pellucida present, multiple sperm could fertilize the same egg (a fatal condition). **Quick Check, p. 491:** Ectoderm. A defect in ectodermal function will often cause defects in many different ectoderm-derived organs, such as the peripheral nerves and the skin.

Quick Check, p. 493: The amnion cradles the embryo in a protective bag of fluid; the allantois helps form the blood vessels of the umbilical cord; the yolk sac forms part of the digestive tract, some blood cells, and the germ cells; and the chorion produces hCG and (together with maternal tissue) forms the placenta. **Quick Check, 499:** Without sufficient surfactants, the lungs will not inflate easily. Infants will have difficulty breathing and can become severely oxygen-deprived. Treatments include spraying a mist containing surfactant (often derived from cow lungs) into the infant's lungs. **Quick Check, p. 503:** If telomerase enables cells to divide indefinitely, it could increase the risk of cancer. The fact that cancer cells often have high levels of telomerase is certainly cause for caution. **Quick Check, p. 504:** Regular exercise can slow or temporarily reverse several of the aging-related declines in the musculoskeletal system, cardiovascular, and respiratory systems. For example, weight-training exercise increases muscle mass and bone density; endurance exercises (walking, hiking, etc.) increase heart function; flexibility exercises (yoga, etc.) increase the flexibility and range of motion of ligaments and tendons.

Test Yourself, p. 507: 1. d; **2.** c; **3.** b; **4.** b; **5.** a; **6.** d; **7.** b; **8.** a; **9.** b; **10.** a; **11.** b; **12.** d; **13.** d; **14.** b; **15.** a

Chapter 22 **Check Figure 22.5, p. 513:** In addition to the pharyngeal gill arches, the tail, the notochord, and the somites also disappear before birth in humans. All these structures were inherited from our distant ancestors and appear to be "programmed into" our embryological development. Their persistence during development indicates that they may still play important roles in cuing the embryonic cells to give rise to other, more recently evolved, structures. After the latter structures have formed, the former structures seem to be unnecessary. **Check Figure 22.9, p. 517:** The simplest approach would be simply to replicate, in a sealed flask, the conditions of the atmosphere, the sea, and the surface (e.g., presence of clay), leave the flask for a while and see whether the proposed molecules eventually appear. The flask would need to replicate the heat, electrical discharges, and UV radiation of ancient Earth. One of the many difficulties with such experiments is time: Some of the proposed steps may have taken millions of years to occur on ancient Earth. Typically, a scientific experiment can only run for a decade or two. Nevertheless, fascinating progress has been made at producing some of the molecules shown in panel b, and even some of the ones in panel c. **Check Figure 22.12, p. 520:** Gibbons split off from the Old World monkey lineage about 30 million years ago. More recently, about 25 million years ago, gibbons split off from the lineage that led to humans. Hence, a gibbon is more closely related to a human than to an Old World monkey. Note that the horizontal distance on the x-axis is not informative. It is the time from the last common ancestor that is the key for judging relationships.

Quick Check, p. 512: If 1/4 of the original potassium is still present, that means two half-lives have passed since the rock solidified. So the fossil is 2×1.3 billion years = 2.6 billion years old. **Quick Check, p. 513:** These are analogous structures because they share a similar function but are not descended from a similar structure in a common ancestor. Dolphins descended from a mammalian ancestor that had no dorsal fin, and hence the dolphins must have evolved their dorsal fin "from scratch," independently of the dorsal fins that evolved in sharks and their relatives. **Quick Check, first question on p. 515:** Mutations occur spontaneously and randomly—they are neither good nor bad. It just turns out that some mutations

increase an organism's chances of survival, others decrease it. Those that decrease the chance of survival tend to be eliminated from the gene pool over time by natural selection. **Quick Check, second question on p. 515:** The bent tails of Florida panthers are probably due to genetic drift, caused in this case by a population bottleneck. Genetic drift is a random process, so the bent-tail allele has become more common simply by chance, not because it is a superior allele. (Hunting probably played a role in that it helped reduce population size, but in this case hunting did not "select" for bent tails; that is, hunters were not specifically targeting cats with straight tails.) **Quick Check, p. 517:** Probably not. The same aspects that make today's atmosphere so hospitable to us also make it inhospitable for the formation of new kinds of life. The atmosphere today does not have enough energy (heat, UV radiation, and so forth) to produce complex macromolecules from simple precursors, and the abundant oxygen in today's atmosphere would quickly degrade any complex molecules that did manage to arise. (In addition, the abundant bacteria and other microorganisms that are present today would probably devour any fragile new life-forms.) **Quick Check, p. 520:** You could look for forward-facing eyes on the skull, opposable thumbs, the presence of five digits on each foot, and the presence of flat nails on the digits. There are other traits that could be used as well, such as the details of the teeth. **Quick Check, p. 522:** One of the great surprises of human paleontology is that this hypothesis has been completely disproven. Upright walking preceded the enlargement of the brain by several million years. Upright walking appears to a limited extent in *Ardipithecus*, with well-developed bipedal walking in *Australopithecus*, but the brain was still chimpanzee-sized in both these species. A substantial increase in brain size (along with the ability to make stone tools) did not begin until *Homo habilis*.

Test Yourself, p. 524: 1. d; **2.** d; **3.** a; **4.** a; **5.** c; **6.** d; **7.** d; **8.** b; **9.** c; **10.** c; **11.** a; **12.** c; **13.** d; **14.** d; **15.** b

Chapter 23 **Check Figure 23.2, p. 529:** If this species had lower biotic potential, both the orange and the green lines would have more gradual slopes. The species would take more time to reach carrying capacity. However, the carrying capacity itself would not necessarily be affected. Carrying capacity is determined by external environmental factors (such as food availability, predation, etc.), not by the species' biotic potential. **Check Figure 23.13, p. 539:** The annual cycle in atmospheric CO_2 occurs because of a strong cycle in the rate of photosynthesis. The vast forests of the large continents remove an enormous amount of CO_2 from the atmosphere every summer, but then cease to do so during the northern winter. The southern continents are much smaller and less forested, so they have a much less noticeable effect. **Check Figure 23.14, p. 540:** Nitrogen-fixing bacteria convert N_2 to ammonium (NH_4^+); nitrifying bacteria convert ammonium to nitrate (NO_3^-); and finally, denitrifying bacteria convert nitrate back to N_2.

Quick Check, p. 529: Wolves have a greater biotic potential than bears. This is because a female wolf produces many more young per year than a female bear. Wolves also start breeding at a younger age. Female wolves also generally breed every year, but female bears only breed every two or three years. **Quick Check, p. 531:** The fertility rate is well below the replacement rate, yet the population is growing anyway (i.e., the growth rate is positive). The United Kingdom's population is increasing because of immigration. Many other industrialized nations show a similar pattern.   **Quick Check, p. 533:** The climax communities in these two regions are different because of different environmental factors in the regions, particularly the climate. Some trees probably do colonize the Midwestern field, but they either die or they cannot reproduce. (Specifically, prairies occur where there is not enough rainfall to support tree growth.) **Quick Check, p. 535:** Algae are the producers; all the other species are consumers. Among the consumers, the krill are herbivores, the baleen whales and killer whales are both carnivores, and the bacteria and fungi are decomposers. In this simplified scenario, the baleen whales are secondary consumer and the killer whales are tertiary consumers. (In real life, killer whales function both as tertiary and quaternary consumers, because they also eat other tertiary consumers.) **Quick Check, p. 537:** Large carnivores are generally upper-level consumers—that is, they are secondary, tertiary, or quaternary consumers that prey on other animals. They are rarer than their prey because of the inefficiency of energy conversion between levels of an ecological pyramid. For example, it takes many acres of western rangeland to feed just one elk, and about twenty elk per year to feed just one wolf.   **Quick Check, p. 539:** Rivers contain minerals that have been eroded out of rocks in hills and mountains, and rivers deposit these valuable minerals in floodplains and estuaries. Hence plants and marine algae can grow abundantly in these areas, forming the base for a vibrant, productive ecosystem. In this way, the water cycle affects the productivity of ecosystems.

Test Yourself: 1. c; **2.** d; **3.** d; **4.** a; **5.** c; **6.** a; **7.** b; **8.** b; **9.** b; **10.** b; **11.** c; **12.** c; **13.** a; **14.** c; **15.** a

Chapter 24   **Check Figure 24.1, p. 546:** Vehicle exhaust is the number one "bad guy," producing five separate categories of pollutants that contribute to three of the four serious atmospheric problems. The only bright spot is that vehicle exhaust does not contribute to ozone destruction. Ozone destruction is caused by a single, very specific class of chemical (chlorine-containing gases, such as CFCs) not present in vehicle exhaust.   **Check Figure 24.4, p. 548:** Notice that the "ozone hole" is thinnest around the South Pole. New Zealand lies in a very southerly latitude close to this ozone hole, and thus receives much higher levels of ultraviolet radiation than most other countries. In addition, New Zealand has a higher percentage of light-skinned people than other nations at this latitude, and light-skinned people are especially vulnerable to ultraviolet radiation.   **Check Figure 24.7, p. 550:** Dead zones at the mouth of a river generally are due to eutrophication caused by fertilizer runoff into the river from agricultural areas. The fertilizer causes a rapid overgrowth of algae. Abundant bacteria which feed on the algae use up most of the oxygen in the water. Fish cannot survive under low oxygen conditions, so they either move away to other coastal areas or they die.

Quick Check, p. 547: Any change in the amount of cloud cover could have a large effect on global warming. This is because clouds are composed of water vapor, and water vapor is itself a greenhouse gas. **Quick Check, first question on p. 551:** Killer whales have the highest contamination load. (In fact, many killer whales are so highly contaminated that if a killer whale carcass washes up on shore, it must be handled as "hazardous waste.") Next would come swordfish, then blue whales, and last of all sardines. The most important factor affecting contamination load is the trophic level at which the animal feeds. Usually, the next most important factor is the animal's age, because contamination builds up over time.   **Quick Check, second question on p. 551:** Because almost the entire watershed is protected, New York City can prevent agricultural and industrial pollution into its water, and so the city's water is unusually clean. The financial benefit is that the city does not have to build a $6 billion water treatment plant, which would have been necessary if the watershed had not been protected.   **Quick Check, p. 556:** Human activity tends to target grassland ecosystems (for agriculture), forests (for logging), and riversides and coastlines (for cities, towns, and also near-shore fisheries). Least affected are high mountains, tundra, and taiga (northern coniferous forest). Thus, most biologists would say set aside ecosystems of grasslands, forests, and riverside/coastal areas.

Test Yourself: 1. a; **2.** b; **3.** d; **4.** d; **5.** c; **6.** a; **7.** a; **8.** b; **9.** c; **10.** a; **11.** c; **12.** b; **13.** c; **14.** d; **15.** a

Credits

Chapter 1
Opener: Gary I. Rothstein/epa/Corbis; **p. 3:** WENN/Newscom; **Fig. 1.1 clockwise from top left:** NASA; Image Quest Marine; Attila Kisbenedek/AFP/Getty Images; BIOS (A. Compost)/Photolibrary; **Fig. 1.2a–b:** Oliver Meckes/Photo Researchers, Inc.; **Fig. 1.3 left:** Philip Schermeister/Getty images; **Fig. 1.3 right:** SPL/Photo Researchers, Inc.; **Fig. 1.3 bottom:** Rick Wilking/Reuters/Corbis; **Fig. 1.6:** John Warburton-Lee/DanitaDelimont.com; **Fig. 1.9:** Digital Vision/Getty Images; **p. 15:** Scott Camazine/Alamy; **Fig. 1.12a:** Corbis; **Fig. 1.12b:** Philippe Psaila/Photo Researchers, Inc.; **Fig. 1.12c:** NASA; **Fig. 1.13:** Michael Rosenfeld/Stone/Getty Images Inc.

Chapter 2
Opener: Perennou Nuridsany/Photo Researchers, Inc.; **p. 24:** Kristin Piljay; **p. 25:** Cordelia Molloy/Photo Researchers, Inc.; **Fig. 2.1a:** Cordelia Molloy/Photo Researchers, Inc.; **Fig. 2.2b:** Richard Megna/Fundamental Photographs, NYC; **Fig. 2.4a:** All Sport Photography, USA/Getty Images, Inc.; **Fig. 2.4b:** Sports Illustrated/Getty Images; **Fig. 2.9:** Novastock Novastock/Photolibrary; **Fig. 2.11a:** Henry Chaplin/iStockphoto; **Fig. 2.11b:** James Driscoll/iStockphoto; **Fig. 2.15:** Don W. Fawcett/Photo Researchers, Inc.

Chapter 3
Opener: Dr. Torsten wittmann/SPL/Photo Researchers; **p. 50:** David McNew/Getty Images, Inc.; **p. 51:** Ron Sachs/Sygma/Corbis; **Fig. 3.1a:** Don W. Fawcett/Photo Researchers, Inc.; **Fig 3.1b:** Photo Researchers, Inc.; **Fig 3.2a:** Thomas Deerinck, NCMIR/Photo Researchers, Inc.; **Fig. 3.2b:** David McCarthy/Photo Researchers, Inc.; **Fig. 3.2c:** Ed Reschke/Peter Arnold/Photolibrary; **Fig. 3.3a:** Michael Abbey/Photo Researchers, Inc.; **Fig. 3.3b:** Biophoto Associates/Science Source/Photo Researchers; **Fig. 3.3c:** Oliver Meckes/Photo Researchers; **Fig. 3.6:** Erik Schweitzer, M.D., Ph.D; **Fig. 3.10:** CNRI/Photo Researchers, Inc.; **Fig. 3.13a–c:** David M. Phillips/Photo Researchers, Inc.; **Fig. 3.15:** Don W. Fawcett/Photo Researchers, Inc.; **Fig. 3.17:** Biology Media/Photo Researchers; **Fig. 3.19:** Bill Longcore/Photo Researchers; **Fig. 3.29:** Rick Rickman/Corbis.

Chapter 4
Opener: Steve Gschmeissner/Photo Researchers; **p. 80:** Liu Jin/Getty Images; **p. 81:** David M. Phillips/Photo Researchers, Inc.; **Fig. 4.4a–b:** Ed Reschke/Peter Arnold/Photolibrary; **Fig. 4.5a:** Ed Reschke/Peter Arnold/Photolibrary; **Fig. 4.5b:** Allen Bell; **Fig. 4.6a:** Eric Graves/Photo Researchers; **Fig. 4.6b:** Ed Reschke/Peter Arnold/Photolibrary; **Fig. 4.6c:** Allen Bell; **Fig. 4.7:** Ed Reschke/Peter Arnold/Photolibrary; **p. 96 left:** Stockbyte/Getty Images; **p. 96 right:** Keith Brofsky/Photolibrary.

Chapter 5
Opener: GustoImages/SPL/Photo Researchers; **p. 102:** Corbis; **p. 103 left:** AP Photo/Noah Berger; **p. 103 right:** Kevin Beebe/Newscom; **Fig. 5.1:**

Gene Cox/Photo Researchers; **p. 118:** Carolyn A. McKeone/Photo Researchers, Inc.; **Fig. 5.14a–b:** Reproduced from the J Bone Miner Res 1986; 1:15–21 with permission of the American Society for Bone and Mineral Research; **Fig. 5.15:** Larry Mulvehill/Science Source/Photo Researchers.

Chapter 6
Opener: John Burcham/National Geographic Creative/Getty Images; **p. 122 left:** Scott Barbour/Getty Images; **p. 122 right:** Hiroko Masuike/Getty Images; **p. 123 left:** Hiroko Masuike/Getty Images; **p. 123 right:** Dilip Vishwana/Getty Images; **Fig. 6.4:** Eric V. Grave/Photo Researchers, Inc.; **Fig. 6.5c:** Dr Rosalind King/Photo Researchers, Inc.; **Fig. 6.6d:** Don W. Fawcett/Photo Researchers, Inc.; **p. 133:** Ron Chapple/Getty Images; **Fig. 6.9:** Astrid & Hanns-Frieder Michler/Photo Researchers, Inc.; **Fig. 6.11a:** Andersen Ross/Getty Images; **Fig. 6.11b:** Lothar Schulz/Getty Images.

Chapter 7
Opener: Juergen Berger/SPL/Photo Researchers, Inc.; **p. 142:** Jaclyn Albanese; **p. 143 left:** Tissuepix/Photo Researchers, Inc.; **p. 143 right:** Burger/Photo Researchers, Inc.; **Fig. 7.2:** Burger/Photo Researchers, Inc.; **Fig. 7.3:** Susumu Nishinaga/SPL/Photo Researchers, Inc.; **Fig. 7.7:** Kari Lounatmaa/SPL/Photo Researchers, Inc.; **Fig. 7.9:** Steve Gschmeissner/Photo Researchers, Inc.; **p. 153:** LENEE/age footstock; **Fig. 7.10:** St. Mary's Hospital Medical School/SPL/Photo Researchers. Inc.; **Fig. 7.14:** Jack Scanlon, Holyoke Community College.

Chapter 8
Opener: SPL/Photo Researchers; **p. 163 left:** Popperfoto/Getty Images; **p. 163 right:** B Busco/Getty Images; **Table 8.1:** National Institutes of Health, National Heart, Lung, and Blood Institute, March 2003; **Fig 8.3a:** Matthew A. Boegehold, Ph.D.; **Fig. 8.3b:** Lennart Nilsson/Albert Bonniers Forlag AB; **Fig. 8.6:** A. & F. Michler/Photolibrary; **p. 173 left:** Biophoto Associates/Photo Researchers, Inc.; **Fig. 8.12a:** SPL/Photo Researchers, Inc.; **Fig. 8.12b:** Nathan Benn/Alamy; **Fig. 8.14:** Mark Hindley/Photolibrary; **Fig. 8.18:** Zephyr/Photo Researchers, Inc.; **Fig. 8.20:** Jupiter Images.

Chapter 9
Opener: SPL/Photo Researchers, Inc.; **p. 189:** Reuters/Corbis; **Fig. 9.1a:** S. Lowry/University of Ulster/Getty Images; **Fig. 9.1b:** Scimat/Photo Researchers, Inc.; **Fig. 9.1c:** S. Lowry/University of Ulster/Stone/Getty Images; **Fig. 9.4:** Steve Gschmeissner/Photo Researchers; **Fig. 9.5:** Tim Vernon, LTH NHS Trust/Photo Researchers; **Fig. 9.6:** SPL/Photo Researchers, Inc.; **Fig. 9.8:** Photo Lennart Nilsson/Albert Bonniers Forlag AB. **Fig. 9.15:** Steve Gschmeissner/Photo Researchers, Inc.; **Fig. 9.19 top:** SIU/Photolibrary; **Fig. 9.19 bottom:** CNRI/Science Photo Library/Photo Researchers; **Fig. 9.23a:** Centers for Disease Control and Prevention, HIV/AIDS Surveillance Report, 2007.

Chapter 10
Opener: Prof. P.M. Motta/Dept. of Anatomy/University "La Sapienza", Rome/Photo Researchers, Inc.; **p. 218:** Lucas Jackson/Reuters/CORBIS. **p. 219 left:** Olaf Doering/Alamy; **p. 219 right:** Gunter Marx/Alamy; **Fig. 10.4:** Chet Childs/Custom Medical Stock Photo. **Fig. 10.6a:** Charles Daghlian/Photo Researchers, Inc.; **Fig. 10.6b:** Andrew P. Evan; **Fig. 10.8:** David M. Phillips/Photo Researchers, Inc.; **Fig. 10.10:** L. Steinmark/Custom Medical Stock Photo; **p. 235:** AP Photo/Dale Sparks; **Fig. 10.14a:** Biophoto Associates/Photo Researchers, Inc.; **Fig. 10.14b:** John Burbidge/Science Photo Library/Photo Researchers;

Chapter 11
Opener: John Burbidge/Science Photo Library/Photo Researchers; **p. 243 left:** Morry Gash/AP Wide World Photos; **p. 243 right:** Mark J. Terrill/AP Wide World Photos; **Fig. 11.7:** Don W. Fawcett/Photo Researchers, Inc.; **Fig. 11.8 top:** Omikron/Photo Researchers, Inc.; **Fig. 11.8 bottom:** CNRI/Science Photo Library/Photo Researchers; **Fig. 11.11:** Jose Luis Pelaez/Corbis; **Fig. 11.14:** Manfred Kage/Peter Arnold; **Fig. 11.15:** Science Pictures Limited/Photo Researchers, Inc.; **p. 268:** KJ/Icon SMI/Corbis; **Fig. 11.20:** Substance Abuse and Mental Health Services Administration, 2009. Overview and Findings from the 2008 National Survey on Drug Use and Health; **Fig. 11.21a–b:** Photo Researchers.

Chapter 12
Opener: SPL/Photo Researchers, Inc.; **p. 275 left:** Sarah Phipps/New York Times/Redux; **p. 275 right:** Sarah Phipps/New York Times/Redux; **Fig. 12.5b:** Sarah Phipps/New York Times/Redux; **Fig. 12.5c:** Astrid & Hanns-Frieder Michler/Science Photo Library/Photo Researchers; **Fig. 12.7:** Corbis; **Fig. 12.10:** Photo Researchers; **Fig. 12.12:** Richard J. Mount, Auditory Science Laboratory, The Hospital for Sick Children, Toronto, Canada; **Fig. 12.13:** Photo Lennart Nilsson/Albert Bonniers Forlag AB; **p. 293:** Pascal Goetgheluck/Photo Researchers, Inc.; **Fig. 12.19:** Ralph C. Eagle/Photo Researchers; **Fig. 12.20a–b:** Scanning electron micrographs by Robert Preston and Joseph E. Hawkins, Kresge Hearing Research Institute, University of Michigan; **Fig. 12-21a:** Sue Ford/SPL/Photo Researchers; **Fig. 12-21b:** Western Opthamalic Hospital/SPL/Photo Researchers; **Fig. 12-22a–c:** National Eye Institute, National Institutes of Health.

Chapter 13
Opener: Quest/Photo Researchers, Inc.; **p. 300:** Bubbles Photolibrary/Alamy; **Fig. 13.8:** Science Photo Library/Photo Researchers; **Fig. 13.9:** Reproduced by permission from American Journal of Medicine 20:133, 1956. Copyright © 1956 by Elsevier Science Ltd.; **Fig. 13.14:** Bettmann/ Corbis; **Fig. 13-16a–b:** Courtesy of Dr. Charles B. Wilson, Neurological Surgery, University of California Medical Center, San Francisco.

Chapter 14
Opener: Steve Gschmeissner/Photo Researchers; **p. 325:** Daniel Berehulak/Getty Images; **Fig. 14.4:** David Scharf/Peter Arnold/Photolibrary; **Fig. 14.7:** Steve Gschmeissner/Photo Researchers, Inc.; **Fig. 14.9:** Steve Gschmeissner/Photo Researchers, Inc.; **Fig. 14.19:** SPL/Photo Researchers, Inc.; **Fig. 14.20:** Behavioral Risk Factor Surveillance System, Centers for Disease Control and Prevention; **Fig. 14.21:** Sipa Press.

Chapter 15
Opener: Steve Gschmeissner/Photo Researchers; **p. 355:** James A. Finley/AP Worldwide Photos; **Fig. 15.7a:** Steve Gschmeissner/Photo Researchers, Inc.; **Fig. 15.7b:** Dr. William Kriz, Ruprecgt-Karls-Universitat Heidelberg, Institut fur Anatomie und Zellbiologie; **Fig. 15.8:** Steve Gschmeisser/Photo Researchers, Inc.; **Fig. 15.17:** Medicimage/Photolibrary.

Chapter 16
Opener: Medicimage/Photolibrary; **p. 377:** Getty Images; **Fig. 16.2:** Ed Reschke/Peter Arnold/Photolibrary; **p. 389:** Saturn Stills/Photo Researchers; **Table 16.3:** Adapted from Birth Control Guide, FDA Office of Women's Health, 2007; **Fig. 16.9:** Photo Researchers; **Fig. 16.11a:** Getty Images; **Fig. 16.11b:** NuvaRing Inc., Organon USA Inc.; **Fig. 16.11c:** Garry Watson/Photo Researchers, Inc.; **Fig. 16.11d:** Dmitry V. Petrenko/Shutterstock; **Fig. 16.11e:** SIU/Photo Researchers; **Fig. 16.11f:** Scott Camazine/Sue Trainor/Photo Researchers; **Fig. 16.11g:** Gary Parker/Photo Researchers, Inc.; **Fig. 16.11h:** Cordelia Molloy/Photo Researchers; **Fig. 16.13:** Richard G. Rawlins, Ph.D./Custom Medical Stock Photo; **Fig. 16.15:** Data and graph style are from the Centers for Disease Control and Prevention (CDC) report, "Trends in Reportable Sexually Transmitted Diseases in the United States, 2007"; **Fig. 16.16a:** Science Source/SPL/Photo Researchers; **Fig. 16.16b:** Science Source/SPL/Photo Researchers; **Fig. 16.16c:** Biophoto Associates/Photo Researchers, Inc.; **Fig. 16.17:** Photo Researchers.

Chapter 17
Opener: Dr. Alexey Khodjakov/Photo Researchers; **p. 404:** Gusto/Photo Researchers, Inc.; **p. 405:** Seoul National University/Getty Images; **Fig. 17.2:** Biophoto Associates/Photo Researchers; **Fig. 17.9:** Michael Abbey/Photo Researchers; **Fig. 17.10:** Don W. Fawcett/Photo Researchers, Inc.

Chapter 18
Opener: Du Cane Medical Imaging Ltd./Photo Researchers, Inc.; **p. 427:** UNIMEDIA IMAGES/J.Lingo/Newscom; **Fig. 18.3:** Dr. P. Marazzi/Photo Researchers; **Fig. 18.4:** Richard R. Hansen/Photo Researchers; **Fig. 18.5:** Myrleen Ferguson Cate/PhotoEdit; **Fig. 18.7:** modernmechanix.com; **Fig. 18.8:** Chris Bjornberg/Photo Researchers; **Fig. 18.9:** Simon Fraser/SPL/Photo Researchers; **Fig. 18.10:** National Cancer Institute/Photo Researchers, Inc.; **Fig. 18.11a:** James Stevenson/Photo Researchers; **Fig. 18.11b:** CNRI/Photo Researchers; **Fig. 18.11c:** Dr. P. Marazzi/Photo Researchers; **Fig. 18.11d:** Biophoto Associates/Photo Researchers, Inc.; **Fig. 18.12:** David M. Martin, M.D./Photo Researchers; **Fig. 18.13:** Branislav Senic/iStockphoto.

Chapter 19
Opener: Andrew Syred/Photo Researchers; **p. 447:** Saturn Stills/Photo Researchers, Inc.; **Fig. 19.4 left:** Corbis; **Fig. 19.4 right:** image100/Jupiter Images; **Fig. 19.5 left:** Jennifer Graylock/AP Wide World Photos; **Fig. 19.5 right:** Chris Jackson/Getty Images; **Fig. 19.6a:** SPL/Photo Researchers; **Fig. 19.6b:** Paul Bricknell/Dorling Kindersley; **p. 452:** MENDIL/agefotostock; **Fig. 19.9 top:** Comstock Images/Jupiter Images; **Fig. 19.9 middle:** Photos.com; **Fig. 19.9 bottom:** Comstock Images/Jupiter Images; **Fig. 19.11:** Oliver Meckes and Nicole Ottawa/Photo Researchers; **Fig. 19.12-1:** Filippova Olga/iStockphoto; **Fig. 19.12-2:** Bo Valentino/Shutterstock; **Fig. 19.12-3:** Paul Maguire/Shutterstock; **Fig. 19.12-4:** Zhang Xiao Qiu/iStockphoto; **Fig. 19.14:** Leonard Lessin/Photo Researchers, Inc.; **Fig. 19.18:** Richard Hutchings/Photo Researchers; **Fig. 19.19a–b:** Courtesy of Dr. Nadler/Children's Memorial Hospital, Chicago, Illinois.

Chapter 20
Opener: VStock/Alamy; **p. 469 left:** AGStockUSA/Scott Sinklier; **p. 469 right:** Robert Brook/Photo Researchers; **Fig. 20.1:** Tek Image/Photo Researchers; **Fig. 20.6:** Maximillan Stock Ltd/SPL/Photo Researchers; **Fig. 20.7:** Michael L. Simpson; **Fig. 20.8:** AP Photo/John Gress; **Fig. 20.9:** Courtesy Syngenta; **Fig. 20.11:** Genzyme Transgenics Corp.; **Fig. 20.13:** AP Wide World Photos.

Chapter 21
Opener: Joe McBride/Getty Images; **p. 484:** Schindler Family/AP Wide World Photos; **p. 485:** David Kadlubowski/Corbis; **Fig. 21.4a:** Barbara Penoya/Photodisc/Getty Images; **Fig. 21.4b:** Barbara Penoya/Getty Images; **Fig. 21.10:** Photo Lennart Nilsson/Albert Bonniers Forlag AB. A Child is Born, Dell Publishing Company; **Fig. 21.11:** Photo Lennart Nilsson/Albert Bonniers Forlag AB. A Child is Born, Dell Publishing Company; **Fig. 21.13:** Photo Lennart Nilsson/Albert Bonniers Forlag AB. A Child is Born, Dell Publishing Company; **Fig. 21.17a:** The Image Works; **Fig. 21.17b:** Matthew Fern/Pool/Corbis; **Fig. 21.19:** Matthew Fern/Pool/Corbis.

Chapter 22
Opener: Pascal Goetgheluck/Photo Researchers, Inc.; **p. 508:** Ho New/Reuters; **p. 509 left:** Dr. William L. Jungers; **p. 509 right:** Peter Brown; **Fig. 22.1:** Pasquale Sorrentino/Photo Researchers, Inc.; **Fig. 22.2:** Shutterstock; **Fig. 22.6:** Jeffrey T. Kreulen/Shutterstock; **Fig. 22.12a:** porschelegend/fotolia.com; **Fig. 22.12b:** Philip Date/fotolia.com; **Fig. 22.12c:** Rod Williams/Nature Picture Library; **Fig. 22.12d:** John Giustina/Taxi/Getty Images; **Fig. 22.13:** T. White 2009, From "Ardipithecus ramidus and the Paleobiology of Early Hominids." T. White et al. Science. 2009 Oct 2; 326(5949): 75-86; **Fig. 22.14:** John Reader/Science Photo Library/Photo Researchers; **Fig. 22.18 top row left to right:** Photos.com; Comstock Complete; Andi Berger/Shutterstock; Photos.com; Photos.com; **second row left to right:** Robert Kneschke/iStockphoto; Comstock Complete; Phil Date/Shutterstock; Comstock Complete; Photos.com; **third row left to right:** Comstock Complete; Comstock Complete; Comstock Complete; Photos.com; Photos.com; **bottom row left to right:** Photos.com; Photos.com; Comstock Complete; Yuri Arcurs/Shutterstock; Comstock Complete.

Chapter 23
Opener: Steve Bloom Images/Alamy; **Fig. 23.1 top:** Joseph Van Os/Getty Images; **Fig. 23.1 middle:** Jeremy Woodhouse/Getty Images; **Fig. 23.1 bottom:** R. Andrew Odum/Peter Arnold; **Fig. 23.4:** Paul Carstairs/Alamy; **Fig. 23.5:** U.S. Bureau of the Census, International Data Base. www.census.gov/ipc/prod/wp98/wp98.pdf; **Fig. 23.6:** Greg Vaughn/Alamy; **Fig. 23.9:** MaxPhoto/Shutterstock; **Fig. 23.10:** MaxPhoto/Shutterstock.

Chapter 24
Opener: Novastock/Photo Researchers; **Fig. 24.CO1:** National Aeronautics and Space Administration and Carbon Dioxide Information Analysis Center; **Fig. 24.4:** NASA; **Fig. 24.5a:** Bettmann/Corbis; **Fig. 24.5b:** Joseph Sohm/VisionsofAmericacom/Photolibrary; **Fig. 24.6:** André Maslennikov/Photolibrary; **Fig. 24.7:** Michael P. Gadomski/Photo Researchers, Inc.; **Fig. 24.9:** x99/ZUMA Press/ Newscom; **Fig. 24.10:** Frans Lanting/ DanitaDelimont.com; **Fig. 24.11a:** bev/fotolia.com; **Fig. 24.11b:** Klein and Hubert/BIOS/Peter Arnold; **Fig. 24.11c:** SasPartout/Shutterstock; **Fig. 24.12:** Russell Illig/Photolibrary; **Fig. 24.13:** Russell Gordon/Danita Delimont/Alamy; **Fig. 24.14:** Charmagne Leung/California Academy of Sciences.

Index

A

ABCD rule for melanoma, 438
Abduction, 116
ABO blood group system, 154, 155
   compatibility, cross-matching for, 157
Abortions, 11
   elective abortion, 394
   gender of fetus and, 376–377
   stem cells and, 51
Abscesses, 198
Abscissa, 17
Absolute refractory period, 249
Abstinence, 215, 391
   periodic abstinence, 393
Acceleration, sensing, 288
Accessory glands, 379–380
Accommodation of eye, 201
ACE inhibitors, 370
Acetylcholine (ACh). *See* ACh (acetylcholine)
ACh (acetylcholine), 128
   actions of, 252
   Alzheimer's disease and, 269
   in somatic division of PNS, 257
Achilles tendon, 125
   PRP (platelet-rich plasma) therapy for injuries to, 152
Acid-base balance, 371, 372
Acid precipitation, 548
Acid reflux, 331
Acid-secreting cells, 332
Acidic solutions, 34
Acids, 34. *See also* pH
   acid-base balance, 371, 372
Acromegaly, 310–311
Acrosome, 380–381, 486–488
ACTH (adrenocorticotropic hormone), 303, 307
   cortisol secretion and, 312
   fetal pituitary gland secreting, 497
   functions of, 308–309
Actin, 127
Actinic keratosis, 429
Action potentials
   brain and, 260
   cranial nerves carrying, 254
   light and, 292
   membrane voltage and, 247–249
   muscle cells and, 134
   neurons initiating, 246–249
   as self-propagating, 249
   spinal cord and, 259–260
Active immunization, 207
Active transport, 59–60
Acu-Gen Biolab, Inc., 376
Acute leukemia, 158
Acute pain, 280
Acute renal failure, 372–373
Adam's apple, 223
Adaptation, 278, 281, 294
Adaptive radiation, 516
Addiction to psychoactive drugs, 266
Addison's disease, 320
Adduction, 116
Adductor longus, 125
Adenine, 44, 408
Adenoids, 194, 195–196
Adenosine deaminase (ADA), 480
Adenosine diphosphate (ADP). *See* ADP
Adenosine triphosphate (ATP). *See* ATP
ADH (antidiuretic hormone), 303
   concentrated urine, production of, 367
   diabetes insipidus and, 310
   functions of, 306–307
   hyposecretion of, 310
   water balance and, 368–369

Adhesion junctions, 84
Adipocytes, 87
Adipose tissue, 85, 87
   triglycerides stored in, 39
Adolescence, 501
ADP (adenosine diphosphate), 46
   active transport and, 59–60
Adrenal cortex
   ACTH (adrenocorticotropic hormone) and, 308–309
   Addison's disease, 320
   functions of, 312–313
   hormones of, 318
Adrenal glands, 303. *See also* Adrenal cortex;
      Adrenal medulla
   functions of, 312–313
   hormones produced by, 303
Adrenal medulla, 312–313, 318
Adrenaline. *See* Epinephrine
Adrenocorticotropic hormone. *See* ACTH
Advance directives, 485
Advanced LipoDissolve Center, 81
Aerobic metabolism, 220
   of glucose, 132
Aerobic organisms, 518
Aerobic training, 136–137
Afferent arterioles, 361
Aflatoxin, 433
Africa
   Ardithecus ramidus in, 520
   Ebola virus, 193
   HIV/AIDS in, 188–189, 214
   humans and, 519–520
Afterbirth, 497, 498
Age and aging
   atherosclerosis and, 173
   bone repair and, 109
   caloric restriction and, 502, 503
   causes of, 502–503
   cell damage and, 502
   congestive heart failure and, 184
   diverticulosis and, 348
   free radicals and, 11
   healthful aging, 504–505
   as hypertension risk factor, 179
   lens of eye and, 291
   longevity and, 502
   prostate cancer and, 440
Age-related macular degeneration (AMD), 296
Agglutination of red blood cells, 154
Aggression and anabolic steroids, 123
Agranular leukocytes, 149, 151
Agriculture. *See* Farming
AIDS. *See* HIV/AIDS
Airline travel and pandemics, 527
Alanine, 41
Albanese, Jaclyn, 142–143
Albumins in blood plasma, 145, 146
Alcohol
   blood-brain barrier and, 259
   cancer and, 433, 443
   cell differentiation and, 420
   cerebellum, effects on, 261
   as diuretic, 369
   as hypertension risk factor, 179
   liver and, 336
   placenta, crossing of, 493
   pre-embryonic development and, 490
   stomach absorbing, 333
Aldosterone, 303, 313, 318
   Addison's disease, 320
   angiotensin II and, 370
   renin-angiotensin system and, 369–370
   salt balance and, 369

Alendronate, 119
Alkaline solutions, 34
All-or-none principle, 134–135
Alleles, 448–449. *See also* Dominant alleles;
      Recessive alleles
   for baldness, 460
   codominance, 454–456
   evolution and mutation of, 14–515
   gene flow, 515
   incomplete dominance, 454, 455
   linked alleles, 458
   mutations of, 449
   Punnett squares, 449
   two-trait crosses, 452–454
Allergies, 209–210, 319, 347
Alonso-Coello, P., 119
Alpha cells secreting gucagon, 311
Alpha (α) helix, 42–43
Alpha interferon, 200
ALS (amyotrophic lateral sclerosis)
   breathing and, 233
   creatine phosphate and, 131
   myelin sheath and, 251
Alveoli, 221, 222
   atelectasis, 238
   emphysema and, 236
   first inflation of, 498
   gas exchange in, 226
Alzheimer's disease, 269–270
   prions and, 192
   stem cells and, 50
Ambien CR, 264
American Cancer Society, 437, 441, 447
American Heart Association, 173, 179
Amino acids, 40–44
   absorption of, 337–338
   codons specifying, 411
   foods with, 343
   with nonpolar R groups, 41
   nutrition and, 342–343
   with uncharged polar R groups, 41
Amino group, 40, 42
Ammonia, 357
Ammonium, 371, 540
Amniocentesis, 462, 489
Amnion, 491–492
Amniotic fluid, 491–492
Ampulla, 289
Amylase, 327
Amyotrophic lateral sclerosis (ALS). *See* ALS
Anabolic steroids, 122–123
Anabolism, 69–70
Anaerobic cells, 517–518
Anaerobic pathways, 76–77
Analogous structures, 512
Anaphase of cell division, 404, 414–415
Anaphylactic shock, 210
Andriole, G. R., 441
Androgens, 123
Androstenedione, 122–123
Anecdotal evidence, 17
Anemia, 159. *See also* Sickle-cell anemia
   beta-thalassemia and, 479
   chemotherapy and, 436
   hematocrit and, 147
   leukemia and, 158
   percutaneous umbilical blood sampling (PUBS) and, 489
   vitamin deficiencies and, 344
Anesthetics and blood-brain barrier, 259
Aneurysms, 165, 173
Angina, 182–183
Angiogenesis inhibition and cancer, 437

Angiotensin-converting enzyme (ACE), 370
Angiotensin I, 370
Angiotensin II, 370
Angiotensinogen, 370
Animalia, 7
Animals, 7. *See also* Chimpanzees; Monkeys
  apes, 4, 8, 519–520
  biogeography and, 513–514
  evolution of, 518
  gorillas, 519–520
  research and, 20–21
  transgenic animals, 477–478
Ankle sprains, treating, 118
Anorexia nervosa, 349–350
ANS (autonomic nervous system), 313
  parasympathetic division of, 255–257
  sympathetic division of, 255–257
Antagonistic muscles, 125–126
Anterior cruciate ligaments, 115
Anterior of body, 93
Anti-angiogenic drugs, 437
Anti-latent therapy for HIV, 193
Antibiotics and bacteria, 15, 191, 209
Antibodies
  antigens, attack on, 201
  autoimmune disorders and, 210–211
  B lymphocytes producing, 201–203
  blood types and, 154
  classes of, 203
  in colostrum, 499
  monoclonal antibodies, 208–209
  Rh factor and, 156
  structure of, 203
Antibody-mediated immunity, 201–203
Antidiuretic hormone. *See* ADH (antidiuretic hormone)
Antigen-presenting cells (APCs), 203–204
Antigenic shift, 515
Antigens
  autoimmune disorders and, 211
  blood types and, 154
  immune system and, 201
  monoclonal antibodies for, 208–209
Antiography, 182
Antioxidants, 433
Antiretroviral drugs (ARVs) and AIDS, 189
Antithrombin III, 478
Antrum, 385
Anus, 327, 337
Aorta, 170–171, 172, 174
  baroreceptors of, 180
  in systemic cycle, 172–173
Aortic bodies, 234
Apes, 4, 8, 519–520. *See also* Chimpanzees
Apoptosis, 480–481
Appendicitis, 337
Appendicular skeleton, 110, 112–114
Appendix, 327, 337
Aquaporins, 364
Aqueous humor, 289–290
Aquifers, 549
Arachnoid, 258
Archaea, 7, 8
Ardi, 520–521
Ardithecus ramidus, 520–521
Areola, 383
Arginine, 41, 342
Argon, radiometric dating and, 512
Arrhythmias, 177, 184
Arsenic, 343
Arteries, 164–166. *See also* Aorta; Blood pressure;
    Coronary arteries
  atherosclerosis, 173
  baroreceptors of, 180
  of embryonic circulation, 493
  major arteries, 172
  structures of, 165
  in systemic cycle, 172–173
Arterioles
  afferent arterioles, 361
  efferent arterioles, 361
  functions of, 166

structures of, 165
  in systemic cycle, 172–173
Arthritis. *See also* Rheumatoid arthritis
  osteoarthritis, 117–118
Arthropods, 534
Articulations, 111–112
Artificial insemination, 395
Artificial reproductive technologies (ART), 395–396
Arundo donax, 533
Asbestos, 432, 438
Ascending colon, 337
Ascorbic acid, 344
Asparagine, 41
Aspartame, PKU and, 463
Aspartic acid, 41
Aspirin, 333, 347
Asthma, 236
  smog and, 549
Astigmatism, 292, 293
Asymmetry of melanoma, 438
Atelectasis, 238, 243
Atherosclerosis, 173
Athletics
  anabolic steroids and, 122–123
  blood enhances and, 123
  creatine phosphate and, 131
  gene doping and, 123
  stretching and injuries, 137
Atmosphere of young Earth, 516–517
Atomic number, 27
Atoms, 9, 11, 25–26
  covalent bonds among, 29
  ionic bonds, 30
  isotopes, 27–28
  periodic table of elements, 27
  structure of, 28
ATP (adenosine triphosphate), 46
  active transport and, 59–60
  anaerobic pathways, 76–77
  bacteria using, 190–191
  catabolism and, 70
  cellular respiration and, 75
  electron transport system producing, 71, 72
  fast-twitch fibers and, 135–136
  for muscle contraction and relaxation, 131–132
  oxygen debt, 132
  phosphates and, 540
  photosynthesis and, 534
  slow-twitch fibers and, 135–136
  thyroxine and, 314
  tubular reabsorption and, 364
ATP synthase, 72
Atria, 170–171
Atrial natriuretic hormone, 303, 370–371
Atrial natriuretic hormone (ANH), 318
Atrial systole, 174–175
Atrioventricular (AV) bundle, 176
Atrioventricular (AV) node, 176
Atrioventricular (AV) valves, 171
Auditory canal, 289
Auditory tube, 284–285, 289
Auditory tubes, 222
AUG codon, 411
Australia, 469, 514
Australopithecus afarensis, 508, 521
Autism, 3, 8
Autoimmune disorders, 210–211. *See also* Rheumatoid arthritis
Automatic functions, hindbrain and, 260–261
Autonomic division of PNS, 255
Autonomic nervous system. *See* ANS
Autosomes, 415, 448
Axial skeleton, 109–112
Axon bulbs, 245–246, 251
Axon hillocks, 245–246
Axon membrane, 248
Axon terminals, 245–246
Axons, 89, 245–246. *See also* Nerves
  myelin sheath and, 250
AZT and HIV/AIDS, 215

B

B lymphocytes, 151
  antibody-mediated immunity and, 201
  functions of, 206
  rheumatoid arthritis and, 211
Bacteria, 7, 8
  antibiotics and, 15, 191, 209
  botulism, 237–238
  cancer and, 431–432
  in colon, 337
  complement proteins and, 200
  as decomposers, 534
  food poisoning, 347
  magnets for removing, 158
  of mouth, 330
  Mycoplasma genitalium, 518
  neutrophils and, 150
  as pathogens, 190–191
  as prokaryotes, 52–53
  resident bacteria, 197
  STDs, bacterial, 398–399
  on tooth surface, 330
  transgenic bacteria, 475–476
  urinary tract infections (UTIs) and, 372
Balance, 287–289, 295
Baldness, alleles and, 460
Ball-and-socket joints, 115–116
Balloon angioplasty, 162–163, 183
Baroreceptors, 180
Basal cell carcinoma, 431, 438
Basal metabolic rate (BMR), 345–346
Base, 34, 44
Basement membrane, 84
Basic solutions, 34
Basilar membrane, 285, 289
Basophils, 150
  allergies and, 210
  histamine, release of, 198–199
Behavior, 11, 465
Beluga Shipping GmbH, 547
Benign tumors, 428, 430
Bennett, Matthew R., 522
Benzene, 432
Beta amyloid, 269
Beta (β) sheet, 42–43
Beta-thalassemia, 479
Bicarbonate, 327
  in blood plasma, 145, 146
  as buffer, 35
  carbon dioxide transported as, 232–233
  filtration of, 364
  kidneys and reabsorption of, 371
  in saliva, 330
Biceps brachii, 125
Bicuspid valve, 171
Bile, 327, 336, 338–339, 348
Bile salts, 336, 339
Bilirubin, 148–149, 336
Binge eating, 350
Biodiversity, 554–556
Biogeochemical cycles, 538–539
  water cycle and, 538
Biogeography and evolution, 513–514
Biological magnification, 550–551
Biology, 4, 11
Biomass, 533, 538
Biomass fuels (biofuels), 553
Biosphere, 9, 11, 528
Biotechnology, 470
Biotic potential, 529–530
Biotin, 344
Bipedalism, 8
Biphosphonates, 119
Bird flu, 515, 526
Birth. *See* Childbirth
Birth control, 390–394. *See also* specific types
  abstinence, 391
  failure rates, 390, 393
  future of, 394
Birth defects, 405
Bison, 6

Bitter taste, 278, 282
Black Plague, 193, 526–527
  population DIP (digital imaging processing)
    chip and, 530
Bladder. *See* Urinary bladder
Blastocysts, 490
Blind experiments, 14
Blind spot, 290
Blindness, 296, 344
Blood. *See also* ABO blood group system; Blood-brain
    barrier; Blood plasma; Blood types; Glucose;
    Hemostasis; Red blood cells; White blood cells
  buffer pairs in, 35
  cholesterol in, 40
  components of, 144–151
  as connective tissue, 85, 87
  cord blood, banking, 142–143
  cytotoxic T cells in, 205
  diseases and disorders of, 157–159
  donating blood, 153
  of fetus, 498–499
  functions of, 144–145, 146
  hematocrit, 147
  hepatic portal system, 336
  Hepatitis C and, 348
  HIV/AIDS transmission in, 213
  lymphatic system, blood volume and, 167–168
  magnets for treating infections, 158
  of newborns, 498–499
  nitric oxide and blood flow, 319
  pH of, 34–35
  production of blood cells, 148
  prostaglandins and blood flow, 319
  PTH (parathyroid hormone) and blood calcium,
    315–316
  pulmonary circuit and, 171–172
  Rh factor, 156
  solute concentration, controlling, 369
  stem cells and, 147–148
Blood-brain barrier
  CNS (central nervous system), protection of, 258–259
  psychoactive drugs crossing, 266
Blood clots, formation of, 152–154
Blood clotting, 151. *See also* Hemostasis
  hemophilia and, 153–154
  vitamin K and, 344
Blood doping, 149
Blood enhancers, 123
Blood plasma, 87, 145, 146. *See also* Plasma proteins
  bicarbonate, transport of carbon dioxide as, 232–233
  blood poisoning and, 157
Blood poisoning, 157
Blood pressure, 177–180. *See also* Hypertension
  constant arterial blood pressure, 180
  diuretics and, 369
  exercise and, 182
  glomerular filtration and, 363
  hypotension, 180
  importance of, 180
  local requirements and, 181–182
  measurement of, 178–179
  renal failure and, 372
  renin-angiotensin system and, 370
Blood sugar. *See* Glucose
Blood transfusions
  blood typing and, 155
  compatibility, cross-matching for, 157
  cord blood, 142–143
  donating blood, 153
  transfusion reaction, 155, 209
Blood types, 154–157. *See also* ABO blood group system
  alleles for, 448
  codominance and, 454–456
  compatibility, cross-matching for, 157
  Rh factor and, 156
  tissue rejection and, 209
Blood vessels, 95–96. *See also* Arteries; Capillaries;
    Hemostasis; Veins
  ankle sprains and, 118
  embolisms, 184
  of embryonic circulation, 493

epinephrine and, 313
  functions of, 164–169
  nephrons, vessels serving, 360–361
  norepinephrine and, 313
  smoking and, 96
Blood volume
  atrial natriuretic hormone and, 370–371
  diuretics and, 369
  renin-angiotensin system and, 369–370
BMI (body mass index), 324–325, 326
  determining, 347
  health and, 346
BMR (basal metabolic rate), 345–346
Body cavities, 92–93
Body mass index. *See* BMI (body mass index)
Body position/direction, 93
Body temperature
  muscles and, 124
  negative feedback and, 98–99
  scrotum regulating, 379
  water and, 33
Bone marrow
  cell division in, 418
  compatibility issues, 143
  in fetal development, 496
  polycythemia, 147
  transplants, 143–144
Bones, 85, 87, 104
  and aging, 503
  black market in bones, 102–103
  calcitonin and bone growth, 315
  changes in, 108–109
  CNS (central nervous system), protection of,
    258–259
  embryo, development in, 106–107
  fractures, repair of, 109
  functions of, 104
  hormones and, 107, 109
  living cells in, 104–105
  remodeling and repair of, 108–109
  tooth and, 330
  vitamin K and, 344
Boniva Injection, 119
Books, scientific information in, 16
Borders of melanoma, 438
Bottleneck effect, 515
Botulism, 191, 237–238
Bovine growth hormone (bGH), 477
Brain, 257–260. *See also* Blood-brain barrier; Brain
    injuries; Strokes
  abscesses of, 198
  in adolescence, 501
  concussions, 267
  death and, 505
  encephalitis, 267
  of Flores people, 508–509
  functions of, 260–263
  human evolution and, 521–522
  memory and, 265–266
  MRI (magnetic resonance imaging) of, 435
  prions and, 192
  psychoactive drugs and, 266–267
  size of, 8–9
  sleep and, 263–265
  Tay-Sachs disease and, 463–464
  tumors, 268–269
  ventricles of, 258
Brain death, 505
Brain injuries, 242–243
  concussions, 267
  medically induced coma and, 243
  right to die and, 484–485
BRCA genes, 426–427, 440
  patents for isolating, 431
Breast cancer, 426–427, 437, 439–440
  HRT (hormone replacement therapy) and, 504
  screening for, 439
Breast feeding
  benefits of, 18
  HIV/AIDS transmission in milk, 213
  mammary glands and, 383

of newborns, 499–500
  oxytocin and, 307–308
Breasts
  mammary glands, 383
  mammograms of, 435
  mastitis, 198
  self-examination, 440
  smoking and sagging, 227
  voluntary removal of, 426–427
Breathing, 220
  conscious control over, 235
  homeostasis, monitoring for, 233–235
  nervous system regulating, 233–236
  process of, 227–229
  respiratory system and, 233
Brechbuhl, Julien, 276
Broad-spectrum antibiotics, 209
Bronchi, 221, 222
  asthma, 236
  bronchitis, 225, 236
  functions of, 224–225
Bronchioles, 222, 224–225
Bronchitis, 225, 236
Brush border, 334
  of proximal tubular cells, 364
BSE (breast self-examination), 440
Bubonic plague. *See* Black Plague
Buffers, 35
Bulbourethral glands, 380
Bulimia, 349–350
Bursae, 115, 117
Bursitis, 117

C

Cadaver kidneys, transplant of, 354–355
Cadmium, 541
Caffeine, 35, 259, 369
Calcitonin, 109, 303, 318
  bone growth and, 315
Calcium, 343, 345
  in blood plasma, 145, 146
  of enamel, 330
  ions, 357
  in living organisms, 31–32
  muscle cells releasing, 128–130
  muscle contraction and, 130
  muscular dystrophy and, 130
  osteoporosis and, 118–119
  PTH (parathyroid hormone) and blood calcium,
    315–316
  rigor mortis and, 131
Callus, 109
Calories
  aging, caloric restriction and, 502, 503
  carbohydrates and, 341
  exercise and, 346
  from lipids, 342
  MyPyramid guidelines, 340
  from proteins, 342
  weight control and, 345
Calvin cycle, 534
Campylobacter jejuni, 191
Canal of Schlemm, 289–290, 295–296
Canaliculi, 105
Cancer, 428. *See also* specific types
  bacteria and, 431–432
  brain tumors and, 268–269
  common cancers, 437–443
  development of, 429, 430–434
  diagnosis and, 434–436
  diet and, 433
  dysplasia, 429
  enzyme tests for, 435–436
  factors leading to, 431–433
  free radicals and, 11
  gene therapy and, 480–481
  imaging, types of, 434–435
  immune system and, 433
  internal factors and, 433
  metastasis of, 430
  mortality rate and, 434

Cancer (*Continued*)
  prevention of, 443
  screening recommendations, 439
  telomerase and, 502
  treatments for, 436–437
  viruses and, 431–432
Cancerous tumors, 428, 430. *See also* Cancer
Candida albicans, 399–400
Capillaries, 164. *See also* Glomerular capillaries
  functions of, 166–167
  lymphatic capillaries, 167–168, 193–195
  pulmonary capillaries, 226–227
  pulmonary circuit and, 171–172
  structures of, 165
  in systemic cycle, 172–173
Capillary beds, 166
Carbaminohemoglobin, 232
Carbohydrates, 5, 37–39
  absorption of, 337–338
  digestion, enzymes of, 335
  nutrition and, 341
  sources of, 341
Carbon, 36–37
  in living organisms, 31–32
  structural diversity of, 36
Carbon-14 isotope, 28, 57, 512
Carbon credits, 545
Carbon cycle, 539–540
Carbon dioxide. *See also* Gas exchange
  bicarbonate, transportation as, 232–233
  in blood plasma, 145, 146
  carbon cycle, 539–540
  cell differentiation and, 420
  deforestation and, 547
  forest growth and, 535
  global warming and, 544–545
  as greenhouse gas, 546–547
  hemoglobin transporting, 148
  lipid bilayer, crossing through, 58
  monitoring levels of, 233–235
  photosynthesis and, 518, 534
  red blood cells transporting, 146–147
Carbon monoxide, 220, 235
Carbon sequestration, 545
Carbon taxation, 545
Carbon tetrachloride, 551
Carbonic anhydrase, 232
Carboxyl group (COOH), 39
  in amino acids, 40, 42
Carboxypeptidase, 335
Carcinogenesis, 431
Carcinogens, 431
  in cigarette smoke, 218
  examples of, 432
  fiber and, 39
Cardiac conduction system, 176
Cardiac cycle, 174–175
Cardiac embolisms, 184
Cardiac muscle, 88–89, 124
  activation of, 137–138
  cardiac conduction system and, 176
  defining characteristics of, 139
  heart attacks and, 183
  repair system, 169
  speed of contraction, 138
  sustainability of contraction, 138
Cardiac output, 180–182
Cardiac pacemaker, 176
Cardiac veins, 172, 174
Cardiovascular center, 180, 261
Cardiovascular disease. *See* Heart disease
Cardiovascular system, 164. *See also* Blood
    vessels; Heart
  aging and, 503
  death and, 505
  of fetus, 498–499
  of newborns, 498–499
  regulation of, 180–182
  schematic representation of, 171
Cardioversion, 177
Carnivores, 534, 536

Carotid artery, 172
Carotid bodies, 234
Carpal bones, 113
Carpal tunnel syndrome, 113, 447
Carriers
  of hemophilia, 460
  of Tay-Sachs disease, 463–464
Carrying capacity, 530
Carter, Rosalynn, 3
Cartilage, 85, 87, 104
  function of, 106
  osteoarthritis and, 117–118
Cartilaginous joints, 114
Catabolism, 70
Catalysts, 43
Cataracts, 295
Causation *vs.* correlation, 18–19
CCK (cholecystokinin), 303, 318, 339
CDC (Centers for Disease Control)
  AIDS epidemic and, 214
  Gardasil vaccinations, 399
  on Type 2 diabetes, 300
Celiac disease, 347
Cell body, 89, 245–246
Cell cycle, 406–407, 409, 418–419
Cell differentiation, 489–490
  in early development, 419–420
  embryonic development and, 493–494
  environmental factors influencing, 419–420
  in later development, 420
  undifferentiated cells, re-creating, 422
Cell doctrine, 52
Cell junctions, 84
Cell membrane. *See* Plasma membrane
Cell phones
  DWD (driving while distracted), 274–275
  radiation from, 269
Cell products, 52
Cell reproduction, 406. *See also* Cytokinesis; Mitosis
  aging and, 502
  cell cycle, 406–407
  interphase, 406–407
  meiosis, 415–416
  regulation of, 418–419
  variations in rates of, 418
Cells, 5–6, 9, 11, 52. *See also* B lymphocytes; Muscle
    cells; Nucleus; Plasma cells; Plasma membrane;
    T lymphocytes; Target cells
  age of cells, determining, 57
  aging and, 502–503
  anaerobic cells, 517–518
  in bone development, 108
  centrioles, 69
  cilia, 69
  cytoskeleton supporting, 68–69
  energy and, 69–73
  first cells, 516–518
  flagella, 69
  glucose and, 70–71
  hormones affecting, 302–304
  interstitial fluid and, 97
  isotonic extracellular fluid, 63
  junctions between, 84
  matter and, 69–73
  neuroglial cells, 249–251
  Schwann cells, 250
  size of, 54–55
  stem cells, 50–51
  structure and function of, 53–54
Cellular respiration, 70, 72–73, 220
  anaerobic pathways, 76–77
  ATP (adenosine triphosphate) generated in, 75
Cellulose, 38–39
Celsentri, 215
Centers for Disease Control (CDC). *See* CDC
Central nervous system. *See* CNS
Centrioles, 64, 69, 404
Centromeres, 407, 408–409
CER (Comparative Effectiveness Research), 162–163
CERA, testing for, 131

Cerebellum, 260–261
Cerebral cortex, 262
Cerebral embolisms, 184
Cerebrospinal fluid, 258–259
Cerebrovascular accidents. *See* Strokes
Cerebrum, 260–262
Cerezme, 107
Cervical cancer, 439, 442
Cervical caps, 393
Cervical plug, 386–387
Cervical vertebrae, 111–112
Cervix, 383
Cesarean delivery, 498
CFCs (chlorofluorocarbons), 547–548
Chakravarty, Eliza F., 117
Channels, diffusion through, 59
Chemical bonds, 29–30
Chemical cycles, 538–541
Chemical elements, 533
Chemicals
  as carcinogens, 432
  cell differentiation and, 420
Chemistry, 26
Chemoreceptors, 277, 282–283
Chemotherapy, 436
Chewing, 328, 330
Chicken pox, 192
Childbirth, 496–498
  delivery of infants, 496–498
  HIV/AIDS transmission and, 213
  oxytocin and, 307–308
  positive feedback system and, 99
  stages of, 497
Childhood, 501
  protein deficiencies in, 342
Chimpanzees, 5, 8, 9, 519
China, 378, 532, 550
Chlamydia, 397, 398–399
Chlamydia trachomatis, 398
Chloride, 30, 343, 345
  in blood plasma, 145, 146
  filtration of, 364
  ions, 357
  tubular reabsorption and, 364
Chlorine, 32
Chlorofluorocarbons. *See* CFCs (chlorofluorocarbons)
Chlorophyll, 7, 534
Chloroplasts, 534
Choking, 220, 224
Cholecystokinin (CCK), 303, 318, 339
Cholesterol, 40, 342. *See also* HDL; LDL
  atherosclerosis and, 173
  in bile, 336
  familial hypocholesterolemia, 454, 455
  gallstones and, 348
  heart disease and, 40, 184
  in plasma membrane, 56
Chondroblasts, 106–107, 108
Chondrocytes, 87
Chondroitin, 24
Chordae tendineae, 171
Chorion, 492–493
Chorionic villi sampling, 462, 489
Choroid, 289–290
Christakis, N. A., 329
Chromaffin cells, 313
Chromatids, 404, 407–409
Chromatin, 407
Chromium, 343, 541
Chromosomes, 407–408. *See also* X chromosomes;
    Y chromosomes
  deletions, 463
  of diploid cells, 415
  Down syndrome and, 461–462
  homologous chromosomes, 448
  karyotypes, 458–459
  nondisjunction of, 461–463
  structure of, 407
  translocations of, 463
Chrondoblasts, 87
Chronic leukemia, 158

Chronic renal failure, 373
Chylomicrons, 338
Chyme, 332–333, 339
Chymotrypsin, 334–335
Cialis, 389
Cigarette smoking. *See* Smoking
Cilia, 69, 196, 222
Ciliary muscle, 289–290
Circadian cycle, 317
Circular layer of stomach wall, 332
Circulatory system, 91, 144. *See also* Blood
Circumduction, 116
Cirrhosis, 336, 348
Cisgenic plants, 468
Cities, land use and, 552
Citric acid, 72
Citric acid cycle, 71–72
 diagram of, 73
Classes, 519
Clavicles, 112–113
Clear-cutting, 555
Cleavage, 489–490
Climate, extinction and, 515
Climax communities, 532–533
Clitoris, 383
Clonal expansion of helper T cells, 204–205
Cloning, 11
 biological issues, 405
 cell differentiation and, 420
 DNA cloning, 471–475
 embryo splitting, 420–421
 humans, 404–405
 political and legal issues, 405
 polymerase chain reaction (PCR), 473–474
 reproductive cloning, 420–422
 somatic cell nuclear transfer, 421–422
 therapeutic cloning, 405, 422
Clostridium botulinum, 237–238
Clot-dissolving drugs, 184
Clotting factor VIII, 153
Clotting proteins in blood plasma, 145, 146
CNS (central nervous system), 244–245. *See also* Brain; Sensory mechanisms; Spinal cord
 incoming impulses, interpretation of, 277
 multiple sclerosis (MS) and, 250–251
 nerves carrying signals to and from, 253–254
 protections for, 258–259
 sensory neurons and, 254
Coagulation. *See* Blood clotting
Cobalt, 343
Cocaine, 267
 blood-brain barrier and, 259
 cell differentiation and, 420
 placenta, crossing of, 493
 tolerance and, 266
 tubular secretion and, 365
Coccygeal vertebrae, 111–112
Coccyx, 512
Cochlea, 285, 289
 function of, 286
 hair cells in, 277
 structure of, 286
Cochlear duct, 285, 286, 289
Cochlear implants, 295
Codominance, 454–456
Codons, 411
Coenzymes, 71
Coffee. *See also* Caffeine
 shade-grown, 557
Colds, 191–193, 237
Coleman, R. J., 503
Collagen
 in arteries, 164
 in cartilage, 87
 in ligaments, 106
Collagen fibers, 85–86
Collecting duct, 360
Colon, 337
 diverticulosis and, 348
 fiber and, 343
 polyps, 348

Colon cancer, 437, 440–441
 fiber deficiency and, 343
 screening for, 439
Colonoscopy, 348, 439, 447
Color blindness, 296
Color of melanoma, 438
Color vision, 294
Colostrum, 499
Columnar epithelium, 83
Comas, medically induced, 242–243
Combined sewage overflow (CSO), 549–550
Common iliac artery, 172
Common iliac vein, 172
Communities, 9, 11, 528, 532–533
 climax communities, 532–533
 ecosystems, 533
 succession and, 532–533
Compact bone, 104
Comparative anatomy and evolution, 512
Comparative biochemistry and evolution, 513
Comparative Effectiveness Research (CER), 162–163
Competition and niches, 532
Complement system, 198, 199–200
Complete proteins, 342–343
Complex carbohydrates, 341
Concentrated urine, production of, 366–367
Concentration gradient, 57–58
Conclusions *vs.* facts, 18
Concussions, 267
Condoms, 392, 393. *See also* Female condoms
 failure rate for, 390, 393
 HIV/AIDS and, 189, 215
Conduction deafness, 294–295
Congestive heart failure, 184
 diuretics and, 369
 lung function and, 238
Connective tissue, 85–88
 microchimerism and, 85
 specialized connective tissues, 87–88
Constant arterial blood pressure, 180
Constipation, 338, 343
Consumer organisms, 534
Consumer Supported Agriculture (CSA), 557
Consumers
 ecological pyramid supporting, 536–537
 food webs and, 535–536
 and sustainable development, 556–557
Contact lenses, 292
Continental drift, 514
Continuous ambulatory peritoneal dialysis (CAPD), 373
Continuous birth control pills, 391
Contraception. *See* Birth control
Contractions
 of muscles, 124
 of stomach, 333
Contreras, Ruben, 242–243
Control center, 98
Control group, 13–14
Controlled experiments, 13–14
Controlled variable, 13–14, 97–98
Convergence and presynaptic neurons, 253
Cooke, Alistair, 102–103
Copper, 345
Copy number variations, 458
Cord blood, banking, 142–143
Cornea, 289–290
 light and, 290–291
Coronary arteries, 162–163, 172–174
Coronary artery bypass graft (CABG), 162–163, 183
Corpus callosum, 260, 262
Corpus luteum, 385–387
Correlation *vs.* causation, 18–19
Cortex, 312
 of kidneys, 358
Cortisol, 303, 312, 318
 Addison's disease, 320
 Cushing's syndrome, 321
Cough reflex, 224
Coughing, 228
Countercurrent exchange mechanism, 367, 368

Covalent bonds, 29–30, 36
Coxal bones, 113–114
Cranberry juice, 25
Cranial bones, 109–110
Cranial nerves, 254
Creatine phosphate
 ATP (adenosine triphosphate) production and, 131
 muscles and, 131–132
Creatinine, filtration of, 364
Cretaceous extinction, 515
Cretinism, 320
Creutzfeldt-Jakob disease, 192
Cri-du-chat syndrome, 463
Critical thinking, 10, 16–19
Cross-bridges, 130
Cross-matching for blood type, 158
Cross-pollination, GE plants and, 469
Crossed extensor reflex, 254
Crossing-over process, 417
Crown of tooth, 330
Crowning, 498
Cruciate ligaments, 115
CT colonoscopy, 439
CT (computerized tomography) scans, 165
Cuboidal epithelium, 83
Cupula, 287–288, 289
Cushing's syndrome, 321
Cutaneous membranes, 92
Cyclic AMP, 305
Cyclins, 418
Cysteine, 41
Cystic fibrosis, 236, 237
 gene therapy and, 480
 recessive alleles and, 451, 452, 463
Cystic Fibrosis Foundation registry, 237
Cytochrome *c*, 513
Cytokines, 204–205, 206
Cytokinesis, 406–407, 413
 phases of, 414
 process of, 415
Cytoplasm, 52–53, 480
Cytosine, 44, 408
Cytoskeleton, 64, 68–69
Cytosol, 52–53, 64
Cytotoxic T cells, 205, 206, 210–211

D

Danger, sensing, 276
Danilova, Olga, 149
Darwin, Charles, 510, 515
Data, 16–17
Deafness, 294–295
Death, 505. *See also* Mortality
 rigor mortis, 131
Decibels, 284
Decomposers, 7, 534
Deductive reasoning, 12
 predictions and, 13
Deepwater Horizon oil spill, 551–552
Defecation, 196, 337, 357
Deforestation, 547, 555
Dehydration and urine, 366–367
Dehydration synthesis, 36–37
Dehydroepiandrosterone. *See* DHEA
DeKosky, Steven T., 270
Deletions of chromosomes, 463
Delivery of infants, 496–498
Delta cells secreting somatostatin, 311
Deltoid, 125
Delucchi, Mark A., 555
Dementia. *See also* Alzheimer's disease
 ginkgo and, 270
Demographics and age, 531–532
Denaturation of protein structure, 42
Dendrites, 89, 245–246
Denitrification, 540
Dense connective tissue, 85–86
Dental abscesses, 198
Dental caries, 330
Dental dams, 215
Dentin, 330

Deoxycholate, 81
Deoxyhemoglobin, 147
Deoxyribose, 38, 44
Department of Health and Human Services. *See* DHHS
Dependent variables, 17
Depo-Provera, 391
Depolarization, 248
Dermatitis, 344
Dermicidin, 95, 196
Dermis, 93–96
Descending colon, 337
Desertification, 552–553
Designer drugs, 123
DeSilva, Ashanthi, 480
Detritus, 534
Dextrose, 341
DHEA (dehydroepiandrosterone), 122–123
  adrenal gland producing, 317
DHHS (Department of Health and Human Services)
  kidneys for transplantation, allocation of, 355
  MyPyramid, 340–341
Diabetes, 319–320. *See also* Insulin
  as atherosclerosis risk factor, 173
  as autoimmune disorder, 211
  cloning of, 404
  cord blood and, 143
  heart disease and, 185
  as hypertension risk factor, 179
  microchimerism and, 85
  renal failure and, 373
  stem cells and, 50
  Type II diabetes, 300–301, 320
Diabetes insipidus, 310
Dialysis, 300, 373
Diameter of melanoma, 438
Diamonds, 35
Diaphragm, 112, 221, 227
Diaphragms (contraceptive), 392, 393
  failure rate for, 390
Diaphysis, 104–105
Diarrhea, 197, 338, 347
Diastole period, 174–175
Diastolic pressure, 177–179
Dickey-Wicker Amendment of 1996, 51
Diesel exhaust, 432
Diet. *See also* Weight control
  breast cancer and, 440
  cancer and, 433, 440, 443
  eating disorders, 349–350
Dietary Guidelines for Americans, 340–341
Dietary supplements, 24–25
  consumers, reactions of, 34
Dietary Supplements Health and Education Act of 1994, 25
Differentiation. *See* Cell differentiation
Diffusion, 57–59
Digestion, 328
  endocrine system regulating, 318, 339–340
  nervous system regulating, 339–340
  small intestine and, 333–334
Digestive acid, 196
Digestive enzymes
  pancreas producing, 334–335
  reabsorption of, 338–339
Digestive system, 90, 92, 326
  accessory organs of, 327, 334–336
  aging and, 504
  bulimia and, 350
  diseases and disorders of, 347–349
  endocrine functions of, 318, 339–340
  functions of, 327–329
  motility, types of, 328–329
  organs of, 327
  processes of, 328
  walls, layers of, 328
Digital rectal examination, 439, 441
Dilation phase of labor, 497
Dilute urine, production of, 366
Diploid cells, 380, 415
Disaccharides, 38

Disease-causing organisms, 551
Diseases and disorders. *See also* specific types
  bacteria causing, 191
  of blood, 157–159
  of cardiovascular system, 182–185
  of digestive system, 347–349
  of endocrine system, 319–321
  of gastrointestinal (GI) tract, 347–348
  with microchimerism link, 85
  of muscular system, 139–140
  of nervous system, 267–270
  of overweight, 324
  pandemics, 526–527
  of pituitary gland, 310–311
  prions causing, 192
  of respiratory system, 236–238
  of sensory mechanisms, 294–296
  of skeletal system, 117–119
  tissue products transmitting, 103
  of urinary system, 372–373
  urine, diagnosis and, 357
  viruses causing, 192
  X-linked diseases, 459–460
Distal point of reference, 93
Distal tubule, 360
Distracted drivers, 274–275
Disulfide bonds, 42–43
Diuretics, 369
Divergence and presynaptic neurons, 253
Diverticulitis, 348
Diverticulosis, 343, 347–348
DNA (deoxyribonucleic acid), 4, 44–45, 407–408. *See also* Cell reproduction; DNA replication; Genes and genetics; Genetic testing; Meiosis; Mitosis; Recombinant DNA technology
  aging and, 502
  bacterial DNA, 209
  carbon-14 isotope and, 57
  cell cycle, structure and, 409
  cervical cancer, test for, 442
  of chimpanzees, 5
  cloning of, 471–475
  double helical structure of, 44–45
  evolution and, 517
  free radicals and, 433
  gender of fetus, determining, 376
  gene therapy and, 478
  growth and, 6
  HIV/AIDS and, 212
  hydrogen bonds and, 31
  identical twins, DNA of, 458
  mitochondrial DNA, 407
  mutations and, 409–410
  nucleotides composing, 44
  nucleus and, 63
  palindromic double-stranded DNA, 471–472
  polymerase chain reaction (PCR), 473–474
  primers for sequencing, 470–471
  racial differences/similarities, 523
  repair mechanisms, 410
  steroid hormones and, 304
  synthetic life and, 518
  transcription, 408, 410–411
  transgenic organisms, creation of, 475–478
  translation, 408, 412–413
  viruses and, 193
DNA fingerprinting, 474–475
DNA ligases, 472
DNA polymerases, 408, 470–471
DNA repair enzymes, 410
DNA replication, 407–409, 502
DNA sequencing, 470–471
DNADirect.com, 446
Dolly the sheep, 404–405
Domains, 519
Dominant alleles, 451–452, 464
Donating blood, 153
Dopamine, 252, 270
Doping
  blood doping, 149
  gene doping, 123

  growth hormone doping, 306
  testosterone doping, 137
Dorsal root, 254
Double-blind experiments, 14
Double bonds, 30
Down syndrome, 461–462, 489
Doyle, Linda, 274–275
Driving while distracted (DWD), 274–275
Drug allergies, 209
Drugs, 25
  antiretroviral drugs (ARVs) and AIDS, 189
  for cancer treatment, 436–437
  cell differentiation and, 420
  fertility-enhancing drugs, 396
  genetically engineered drug, approval of, 478
  hemostasis and, 154
  HIV/AIDS and, 214–215
  as hypertension risk factor, 179
  illicit drug use, 266–267
  for osteoporosis, 119
  performance-enhancing drugs, 122–123
  placenta, crossing of, 493
  psychoactive drugs and brain, 266–267
  renal failure and, 372
  sleeping pills, 264
  tubular secretion and, 365
  Viagra, 389
Duchenne muscular dystrophy, 139
Ductus (vas) deferens, 379, 380
Duodenum, 334
Dura mater, 258
Durable power of attorney for health care, 485
Dwarfism, 310, 509
DWD (driving while distracted), 274–275
Dysmenorrhea, 387
Dystrophin, 139

E

Eardrum, 284, 289
Earlobes, alleles for, 451–452
Ears. *See also* Hair cells; Hearing; Inner ear; Middle ear
  diseases and disorders of, 294–295
  outer ear, 289
  structure of, 285, 289
  wiggling, muscles for, 512–513
Earth
  formation of, 516–517
  global warming, 544–545
  timeline of evolution of, 519
Earthworms, 534
Earwax, 196
Eating disorders, 349–350
Ebola virus, 193
ECG (electrocardiogram), 176–177
Echinacea purpurea, 24
Ecological pyramid, 536–537
Ecology, 528
Economics and population age structure, 531–532
Ecosystems, 9, 11, 528, 533
  ecological pyramid and, 537
  energy and, 533–537
  protecting, 558
Ecstasy, 267
Ectoderm, 491, 492
Ectopic pregnancy, 490
Edwards syndrome, 461
EEG (electroencephalogram), 243
  sleep, brain and, 263–265
Effectors, 98
Efferent arterioles, 361
Eggs, 382. *See also* Cell differentiation; Meiosis
  as cells, 55
  fertilization of, 390, 486–489
  as haploid cells, 415
  mutations in, 410
  somatic cell nuclear transfer, 405
  sperm contacting, 390
  sperm fertilizing, 379
Ehlers-Danlos Syndrome (EDS), 87
Ejaculation, 379, 389–390
Ejaculatory duct, 379, 380

Elastic cartilage, 106
Elastic fibers, 85–86
Elastin fibers, 85–86, 96
Elective abortion, 394
Electrocardiogram (ECG), 176–177
Electrolytes. *See* Ions
Electron transport system, 71–72, 74
Electrons, 26–27
    energy and, 29
    transport system, 71–72, 74
Electropherograms, 474–475
Elements, 4–5, 26–28, 31–32
Eli Lilly & Company insulin products, 320
Elimination process, 328
Elongation step of translation, 412–413
Embolisms, 184
Embryo splitting, 420–421
Embryonic development, 491–494
Embryonic disk, 490
Embryonic stem cells (ES cells), 51, 422
Embryos. *See also* Cloning; Prenatal development
    bone development in, 106–107
    embryonic development, 492–494
    pre-embryonic development, 490–491
    uterus and, 382–383
    vertebrate embryos, similarity between, 513
    in vitro fertilization (IVF) and, 380
Emergency contraception, 394
Emotions
    forebrain and, 261–263
    genetic testing and, 447
    limbic system and, 261, 265
    psychoactive drugs and, 266
Emphysema, 236
    congestive heart failure and, 184
    lung volume and, 228–229
    smog and, 549
    smoking and, 225
Employment discrimination and genetic testing, 447
Enamel of tooth, 330
Encephalitis, 192, 213, 267
End-stage renal disease (ESRD), 373
Endocarditis, 170
Endocardium, 170
Endocrine glands, 82, 83, 302, 305–306
Endocrine system, 90. *See also* Hormones
    aging and, 503–504
    characteristics of, 302–304
    components of, 303
    digestion, regulation of, 318, 339–340
    functions of, 302–304
    heart, endocrine functions of, 318
    kidneys, endocrine functions of, 318
    nervous system and, 304
    placenta, 493
Endocytosis, 60, 192
Endocytotic vesicles, 66
Endoderm, 491, 492
Endometriosis, 395
Endometrium, 383
    cancer, 439
    tissue samples, 439
Endoplasmic reticulum, 64–65
Endorphins, 252
Endothelium, 164–165
Energy, 28–29, 553–554. *See also* Mitochondria; Potential energy
    aerobic metabolism of glucose, 132
    cells and, 69–73
    citric acid cycle, 71–72
    communities and, 533
    conserving, 558
    ecological pyramid and, 536–537
    ecosystems and, 533–537
    fats and, 67, 75–76
    glucose and, 70–71
    glycogen and, 67
    life and, 5
    for macromolecules, 36
    for muscle contraction and relaxation, 131–132
    polysaccharides storing, 38–39

proteins and, 75–76
    sustainability, 555
    thermodynamics, laws of, 533–534
    weight control and, 345–347
Energy drinks, 24, 35
Environmental factors. *See also* Energy; Global warming; Pollutants and pollution
    adaptive radiation and, 516
    biodiversity crisis, 554–556
    breast cancer and, 440
    and cell differentiation, 419–420
    exploitation of environment, 555
    genetically engineered plants and, 469
    land pollution, 552–553
    phenotypes and, 457–458
    urbanization, 556
Environmental resistance, 529–530
Enzymes, 42
    cancer, tests for, 435–436
    of digestion, 335
    DNA repair enzymes, 410
    functions of, 43–44
    promoter enzyme, 410
    restriction enzymes, 471–472
Eosin, 150
Eosinophils, 150, 198
EPA (Environmental Protection Agency), 551
Epicardium, 170
Epidermis, 93–94
    diagram of, 95
    smoking and, 96
Epididymis, 379, 380
Epiglottis, 221, 223, 331
Epilepsy, 269
Epinephrine, 303, 318
    adrenal medulla producing, 313
    BMR (basal metabolic rate) and, 346
    cardiac output and, 181
Epiphyseal plate, 106–107
Epiphysis, 104–105
Episiotomy, 498
Epithelial cells, 164, 196, 418
Epithelial tissues, 82–85
EPO (erythropoietin). *See* Erythropoietin
Epstein-Barr virus, 157, 432
Erectile dysfunction, 389
Erection, 379
Erythrocytes. *See* Red blood cells
Erythropoietin, 123, 303, 318
    blood doping, 149
    cloning of, 404
    kidneys producing, 371–372
    red blood cell production and, 149
ES cells (embryonic stem cells), 422
Escherichia coli, 191
Esophageal sphincter, 331
Esophagus, 327, 331
Essential amino acids, 342–343
Essential fatty acids, 342
Estriol, 317, 382
Estrogen, 303, 318
    and birth process, 497
    in blood plasma, 145, 146
    bone loss and, 118
    mammary glands and, 383
    in ovarian cycle, 385
    ovaries producing, 317, 382–383
    placenta secreting, 493
    progesterone producing, 317
Estrone, 317, 382
Ethics
    fetus, gender of, 376–377
    of genetic testing, 447
    of organ gifting, 355
Ethmoid bone, 110
Ethnic discrimination, genetic testing and, 447
Eukarya, 7, 8
Eukaryotes, 7, 52–53
Eukaryotic cells, 191
European Union, 469
Eustachian tubes, 222, 284–285

Eutrophication, 550
Everett, Kevin, 268
Every Child By Two, 3
Evolution, 7, 510
    adaptive radiation, 516
    of aerobic organisms, 518
    biogeography and, 513–514
    comparative anatomy and, 512
    DNA (deoxyribonucleic acid), 517
    evidence for, 510–514
    extinction and, 515
    genetic drift and, 515
    of *Homo sapiens*, 520–523
    natural selection and, 514–516
    photosynthesis and, 518
    of primates, 520
    RNA (ribonucleic acid) and, 517
    timeline of, 519
Evolutionary trees, 516
Excitable muscle cells, 124
Excitatory neurotransmitters, 252
Excitement phase of sexual response, 389
Exercise. *See also* Athletics
    aerobic training, 136–137
    aging well and, 504–505
    BMR (basal metabolic rate) and, 346
    bone mass and, 108
    calories and, 346
    cancer and, 443
    cardiac output and, 182
    delayed onset soreness, 133
    heart disease and, 185
    osteoporosis and, 119
    oxygen debt, 132
    proteinuria and, 364
    strength training, 136–137
    Type 2 diabetes and, 301
Exocrine glands, 82, 83, 302
Exocytosis, 60
Exons, 411
Experimental group, 13–14
Experiments
    findings, publishing of, 14
    with humans, 20
    in scientific method, 13–14
Expiration, 227–228
Expiratory reserve volume, 228
Exploitation of environment, 555
Exponential growth, 529
Expulsion phase of labor, 497–498
Extension, 116
External anal sphincter, 337
External nose, 222
External oblique, 125
External respiration, 220, 230
Extinction, 515
Eyeglasses, 292
Eyes. *See also* Lens; Light; Vision
    accommodation and, 291
    diseases and disorders of, 295–296
    light, regulation of, 290–291
    shape of eyeball and focus, 291–292
    structure of, 289–290

F

Face, 109–110, 310–311
Facilitated transport, 59
Facts, conclusions *vs.*, 18
Faden, R. R., 380
$FADH_2$, 72
    in citric acid cycle, 73
    in electron transport system, 74
    in oxidative phosphorylation, 74
Familial hypocholesterolemia, 454, 455
Families, 519
    cancer, family history and, 443
Fanconi's anemia, 377
Farming
    habitat destruction and, 555
    sustainable agriculture, 557
    and technology, 19

Farsightedness, 292, 293
Fascia, 126–127, 140
Fascicles, 126–127
Fasciitis, 140
Fast pain, 280
Fast-twitch fibers, 135–136
Fat cells, 88
Fat genes, 348
Fat-soluble vitamins, 338, 343, 344
Fatigue, muscle, 132
Fats, 5, 37, 39–40
 absorption of, 338
 energy and, 67, 75–76
 LipoDissolve, 80–81
 metabolic pathways for, 76
 nutrition and, 342
Fatty acids, 39, 342, 338
FDA (Food and Drug Administration)
 biphosphonate medications, 119
 genetically engineered drug, approval of, 478
 glucose monitoring devices and, 312
 on herbal supplements, 25
 on Lipodissolve, 81
Fecal occult blood test, 439
Feces. See also Constipation; Defecation; Diarrhea
 diverticulosis and, 348
 fiber and, 343
 in sigmoid colon, 337
Female condoms, 392, 393
 failure rate for, 390
Female reproductive system, 382–384. See also
  Menstrual cycle
Femoral artery, 172
Femoral vein, 172
Femur, 114
Fertility. See also Infertility
 enhancing fertility, 395–396
 worldwide fertility, lowering, 557
Fertility rate, 531
Fertilization, 390, 486–489
Fertilizers
 nitrogen fixation and, 540
 water pollution and, 551
 in water systems, 551
Fetal alcohol syndrome, 420
Fetal pituitary gland, 496–497
Fetuses
 cardiovascular system of, 498–499
 development of, 496
 gender, discovery of, 376–377
 genetic testing of, 446–447
 microchimerism, 85
 research, use in, 11
Fever, 198, 200
Fiber
 benefits of, 343
 cellulose and, 38–39
 diverticulosis and, 348
Fibrin, 152–153
Fibrinogen, 152
Fibroblasts, 86, 95, 419
Fibrocartilage, 106
Fibrous connective tissue, 85–86
Fibrous joints, 114
Fibula, 114
Fight-or-flight response, 256–257, 313
Fingernails, 196
Fingerprinting, DNA, 474–475
First law of thermodynamics, 533–534
First messengers, 305
Fisher, Barbara Loe, 3
Five-kingdom system, 7
Flagella, 69
Flat bones, 109
Flexible sigmoidoscopy, 439
Flexor reflex, 254–255
Floating ribs, 112
Flores people, 508–509, 522
Flow cytometer, 377
Fluoride, 345
FocusDriven, 275

Folic acid, 344
Follicle-stimulating hormone. See FSH
Follicles, reproductive, 385
Follicles of hair, 95
Fombonne, Eric, 8
Fontanels, 115
Food allergies, 209, 347
Food and Drug Administration. See FDA
Food poisoning, 191, 347
Food webs, 535–536
Foods. See also Diet; Digestive system; Nutrition
 BMR (basal metabolic rate) and, 346
 functional foods, 24–25
 genetically engineered foods, 469
 genetically modified foods, vaccines in, 208
 golden rice, 472, 476
 mouth processing, 329–331
 pancreas, responses of, 311
 trans fats in, 335
 Type 2 diabetes and, 300–301
 vaccines in genetically modified foods,
  207–208
Foramen magnum, 110
Force generation, 133–135
Forebrain, 260, 261–263
Forest growth, 535
Fossil fuels, 538, 553
 carbon dioxide and, 547
Fossil record, 510–512
Fossils, 510–514
 of Australopithecus afarensis, 521
 formation of, 511
 of primates, 519
Founder effect, 515
Fovea centralis, 289–290
Fowler, J. H., 329
Fox, Michael J., 50, 51
Fractures, repair of, 109
Fragile X syndrome, 446
Fraternal twins, 488–489
Free radicals, 11, 433
Friends and obesity, 329
Frontal bone, 109–110
Frontal lobe, 262
Frontal plane, 93
Fructose, 38, 341
FSH (follicle-stimulating hormone), 303, 307
 in adolescence, 501
 functions of, 309
 inhibin affecting, 382
 menstrual cycle and, 388
 in ovarian cycle, 385
 testosterone and, 381–382
FTC (Federal Trade Commission), 269
Functional foods, 24–25
Fungi, 7, 534
Furosemide, 369

G

G_0 phase, 407
G_1 phase, 406
G_2 phase, 406
Gallbladder, 327
 functions of, 336
 gallstones and, 348
 location of, 335
Gallstones, 348
Gamete intrafallopian transfer (GIFT), 396
Gametes, 380. See also Eggs; Meiosis; Sperm
 fertilization and, 486–489
 mutations in, 410
Gamma-aminobutyric acid, 252
Gamma globulins, 203
 in blood plasma, 146
 as passive immunization, 208
Gamma interferon, 205
Ganglia, 255
Gap junctions, 84, 105, 138
Garbage, disposal of, 553
Gardasil, 399
Gas exchange, 226, 229–233

 atelectasis, 238
 external respiration, 230
 of fetus, 498–499
 hemoglobin in, 230–232
 internal respiration, 230
 partial pressure gradient, 229–230, 231
 pneumothorax, 238
Gaseous cycle, 539
Gasoline and water pollution, 550
Gastric juice, 332–333
 reabsorption of, 338–339
Gastric pits, 332
Gastrin, 303, 318, 339
Gastrocnemius, 125
Gastrointestinal (GI) tract, 327. See also Digestive system
 disorders of, 347–348
 hepatic portal system, 336
Gaucher disease, 107
GDP (Gross Domestic Product), 556
GE (genetically engineered) plants, 468–469
Gel electrophoresis, 470–471
Gender
 aging and, 503–504
 BMR (basal metabolic rate) and, 346
 confusion, 386
 eating disorders and, 348–349
 of fetus, 376–377
 as hypertension risk factor, 179
 preference for, 378
 prenatal development of, 494–495
 sex-linked inheritance and, 459
Gene doping, 123
Gene flow, 515
Gene pharming, 478
Gene therapy, 478–481
 cancer and, 480–481
 cystic fibrosis and, 480
 SCID (severe combined immunodeficiency disease)
  and, 480
 vectors and, 479–480
Generalizations in scientific method, 12
Generation Rescue, 3
Genes and genetics, 408, 448. See also Cancer; Genetic
  disorders; Genetic inheritance; Genetic testing
 bitter taste and, 278
 BMR (basal metabolic rate) and, 346
 breast cancer genes, 426–427
 celiac disease and, 347
 color blindness, 296
 cystic fibrosis and, 237
 fat genes, 348
 as hypertension risk factor, 179
 Mendel, Gregor and, 449–451
 patents for, 431
 proteins, codes for, 465
 testosterone doping and, 137
Genetic disorders, 11
 cloning and, 405
 miscarriages and, 494
 preimplantation genetic diagnosis (PGD) and, 377
 recessive alleles and, 463–464
 X-linked diseases, 459–460
Genetic drift, 515
Genetic engineering, 11, 470
 cloning and, 404–405
 golden rice, 472, 476
 of plants, 468–469
 transgenic organisms, creation of, 475–478
 for vaccines, 475
 vaccines in genetically modified foods, 208
Genetic Information Nondiscrimination Act
  (GINA), 447
Genetic inheritance, 448
 codominance, 454–456
 incomplete dominance, 454, 455
 law of independent assortment, 453–454
 law of segregation and, 450–451
 linked alleles and, 458
 patterns of, 449–454
 polygenic inheritance, 456–457
 Punnett squares and, 449

sex-linked inheritance, 458–461
two-trait crosses, 452–454
Genetic shift, 527
Genetic testing, 446–447, 464
benefits of, 446–447
for cystic fibrosis, 452
ethics of, 447
mutated genes, 435
Genital herpes, 397, 399
Genital warts, 192, 200, 397, 399
Genitalia in prenatal development, 495
Genomes, 448
of bird flu virus, 527
synthesis of, 518
Genotypes, 448–449
human beings, differences among, 523
phenotypes and, 457–458
Genuine Progress Indicator (GPI), 556
Genus, 8, 519
Genzyme, 107
Geographic range, 529
Geothermal power, 553–554
Germ layers, 491
German measles, 420
GH. See Growth hormone
Gibbons, 519–520
Gibson, Daniel G., 518
Gigantism, 310
Gingiva, 330
Gingivitis, 330
Ginkgo biloba, 24
and dementia, 270
Glands, 82
Glandular epithesia, 82
Glaucoma, 295–296
Glese, Jeanna, 242–243
Glial cells, 89
Global warming, 11, 544–545
greenhouse gases and, 546–547
Globulins in blood plasma, 145, 146
Glomerula filtration, 362–364
Glomerular capillaries, 361, 363–364
Glomerular capsules, 360–361
Glomerulus, 360
Glottis, 223
Glucagon, 303, 318
Glucocorticoids, 312, 318
Glucosamine, 24
Glucose, 38. See also Diabetes
in blood plasma, 145, 146
cortisol secretion and, 312
Cushing's syndrome, 321
energy and, 70–75
epinephrine and, 313
filtration of, 364
glycolysis, 71
insulin and, 311
monitoring devices for, 312
norepinephrine and, 313
pancreas and, 311–312
Glucose tolerance test (GTT), 301
Glutamate, 252
Glutamic acid, 41
Glutamine, 41
Gluten intolerance, 347
Gluteus maximus, 125
Glyceraldehyde-3-phosphate (PGAL), 71
Glycine, 41
Glycogen, 38, 67, 76, 132
Glycolysis, 71
anaerobic pathways for, 76–77
Glycoproteins, 38, 69, 385
GnRH (gonadotropin-releasing hormone), 381–382
in adolescence, 501
in ovarian cycle, 385
Goblet cells, 83
Goiters, 315
Golden rice, 472, 476
Golgi apparatus, 64–65
Gonadotropin-releasing hormone. See GnRH
Gonadotropins, 309

Gonorrhea, 397
Goodall, Jane, 5
Gorillas, 519–520
Graafian follicle, 385
Graded potentials, 247, 252
Granular leukocytes, 149, 150–151
Granulosa cells, 385, 486
Granzyme, 205
Graphite, 35
Graphs
reading graphs, 17
split axis in, 17, 18
Grasslands, 536
Graves' disease, 320
Gray matter, 260
Great saphenous vein, 172
Green roofs, 557
Greenhouse effect, 547
Greenhouse gases, 546–547
Greenpeace, 472
Ground substance, 86, 104
Groundwater pollution, 551
Growth, 6, 489–490
Growth factors, 319, 430
Growth hormone, 303, 307
in blood plasma, 145, 146
cloning of, 404
disorders of, 310–311
doping with, 306
functions of, 309–310
Growth inhibitors, 430
Growth plates, 106–107
GTC Biotherapeutics, 478
Guanine, 44, 408
Gucagon, 311
Guthrie, Arlo, 464
Guthrie, Woody, 464

H
Habitats, 528–529
destruction of, 555
Hair, 196
baldness, alleles and, 460
chemotherapy and, 436
in dermis, 95
Hair cells, 277, 285, 286, 289
damage to, 294–295
as mechanoreceptors, 287
Hamstring group, 125
Haploid cells, 380, 415
Haploid ovum, 488
Haversian systems, 105
Hay fever, 209
HCG (human chorionic gonadotropin), 386
chorion producing, 492–493
placenta secreting, 493
HDL (high-density lipoprotein), 173
in blood plasma, 146
Head position, sensing, 288
Health
BMR (basal metabolic rate) and, 346
weight and, 346–347
Health care proxy appointment, 485
Health insurance
pre-existing conditions and, 446–447
Type 2 diabetes and, 301
Hearing. See also Hair cells
deafness, 294–295
inner ear and, 285–287
mechanoreceptors and, 284–287
midbrain and, 261
middle ear and, 284–285
outer ear and, 284
Hearing aids, 295
Heart, 164, 169. See also Cardiac muscle; Congestive
heart failure; Heart attacks; Heart disease
arrhythmias, 177
cardiac conduction system, 176
cardiac cycle, 174–175
chambers of, 170–171
coronary angiogram, 182

ECG (electrocardiogram) of, 176–177
in embryonic development, 493–494
endocrine functions of, 318
of fetus, 498–499
functions of, 169–177
hormones produced by, 303
laboratory, creation in, 174
as organ, 89
pulmonary circuit and, 171–172
repair system for, 169
sounds of, 175–176
systemic cycle, 172–173
Heart attacks, 183
atherosclerosis and, 173
cardiac repair system and, 169
congestive heart failure and, 184
HRT (hormone replacement therapy) and, 504
referred pain and, 281
stress and, 185
Heart disease, 182–185. See also Congestive heart failure;
Heart attacks
angina, 182–183
CER (Comparative Effectiveness Research), 162–163
cholesterol and, 40, 184
cord blood and, 143
exercise and, 185
reducing risk of, 184–185
Heart murmurs, 175–176
Heart rate, 180
Heart sounds, 175–176
Heartburn, 331
Hearts of palm, 555
Heat
muscles generating, 124
pollution, 551
Heavy antibody chains, 203
Heavy metals
biological magnification, 551
cycles involving, 541
water pollution and, 550
Height and polygenic inheritance, 457
Heiss, Gerardo, 504
Helicobacter pylori, 196, 347
Helper T cells, 204–205, 206
HIV/AIDS and, 212
Hematocrit, 147
Hematoma, 109, 267
Hemochromatosis, 447
Hemodialysis, 373
Hemoglobin, 147
carbaminohemoglobin, 232
carbon monoxide and, 235
in gas exchange, 230–232
phagocytosis and, 148–149
Hemolytic anemia, 158
Hemolytic disease of the newborn (HDN), 156
Hemophilia, 153–154
percutaneous umbilical blood sampling (PUBS)
and, 489
as X-linked disease, 459–460
Hemorrhagic anemia, 158
Hemorrhoids, 343
Hemostasis, 151–154
blood clots, formation of, 152–154
platelets in, 152
stages of, 151
Hepatic portal system, 336
Hepatitis, 348
monoclonal antibodies of screening for, 209
viruses causing, 192
Hepatitis A, 348
Hepatitis B, 348, 397, 399
alpha interferon and, 200
genetically modified foods, vaccines in, 207
viruses and, 432
Hepatitis C, 205, 348, 432
Herbal remedies, 25
Herbicides and water pollution, 550
Herbivores, 534
Herd immunity, 3
Heritable diseases. See Genetic disorders

Heritable susceptibility, 431
Herniated disks, 112
Heroin and cell differentiation, 420
Herpes simplex, 399
Hertz (Hz), 284
Heterotrophs, 534
Heterozygous persons, 448
Hiatal hernia, 331
High blood pressure. *See* Hypertension
Hill, Christopher, 274–275
Hindbrain, 260–261
Hinge joints, 114–115, 116
Histadine, 342
Histamine, 319
    allergies and, 210
    in basophils, 150
    inflammation and, 198–199
Histidine, 41
Histones, 407
HIV/AIDS, 211–215, 397
    in Africa, 188–189, 214
    cell differentiation and, 420
    cures for, 193
    development of AIDS, 213
    as global health issue, 214
    helper T cells and, 205, 212
    hemophilia and, 154
    monoclonal antibodies of screening for, 209
    opportunistic viruses and, 213
    placenta, crossing of, 493
    risky behaviors and, 214–215
    safer sex guidelines, 215
    transmission of, 93, 212–213
    treatments for, 215
    in United States, 214
    vaccines for, 215
    viruses and, 192, 399, 432
HLA-A antigen, 143
HLA-B antigen, 143
HLA-DR antigen, 143
Hodgkin's disease, 441
Hogan, Hulk, 123
Home pregnancy tests, 386
Homeostasis, 6, 97
    blood calcium concentration and, 316
    of bone structure, 108–109
    breathing, monitoring, 233–235
    enzymes and, 43–44
    hormones and, 302
    hypertension and, 181–182
    kidneys maintaining, 368–372
    multicellular organisms maintaining, 97–99
    negative feedback and, 97–98
    parasympathetic nervous system and, 257
    red blood cell production and, 149
    sympathetic nervous system and, 257
    urinary system and, 356–357
Hominids, 520
Hominoids, 520
Homo erectus, 508–509, 521–522
Homo floresiensis, 508–509, 522
Homo habilis, 509, 521
Homo heidelbergensis, 522
Homo neanderthalensis, 522
Homo sapiens, 8, 508, 519, 522
    evolution of, 520–523
    family tree, versions of, 523
Homologous chromosomes, 448
Homologous structures, 512
Homosexuality and HIV/AIDS, 214–215
Homozygous persons, 448
Hormonal contraceptives, 390
Hormone replacement therapy (HRT), 504
Hormones, 82
    birth control methods, 390, 391–392
    in blood plasma, 145, 146
    bone development and, 107, 109
    cardiac output and, 180–181
    cell cycle and, 419
    cloning of, 404–405
    hypersecretion of, 310

hyposecretion of, 310
    menstrual cycle and, 388
    in negative feedback loops, 305–306
    nonsteroid hormones, 304–306
    pelvic girdle and, 113
    production of, 302–304
    steroid hormones, 304–306
    transgenic bacteria and, 475
Hot flashes, 503
HPV (human papillomavirus), 399. *See also*
        Genital warts
    cervical cancer and, 442
    viruses and, 432
HRT (hormone replacement therapy), 504
Hubble Space Telescope, 5
Human beings. *See also Homo sapiens*
    biological organization for, 9–10
    defining features of, 8–9
    differences among, 522–523
    evolution of, 518
    gene pool, 449
    taxonomic classification of, 519
Human chorionic gonadotropin. *See* hCG
Human development. *See also* Prenatal
        development
    adolescence, 501
    aging, 501–505
    childhood, 501
    death, 505
    neonatal period, 500
    stages of, 500
Human Genome Project, 464, 478
Human mutation rate, 410
Human papillomavirus (HPV). *See* HPV
Humerus, 113
Huntington disease, 192, 447, 463, 464
Hvistendahl, Mara, 378
Hyaline cartilage, 106, 114
Hybridoma cells, 208
Hydrocarbons, 432, 549–549
Hydrochloric acid, 332
Hydroelectric power, 553
Hydrogen
    in blood, 35
    in blood plasma, 145, 146
    carbon dioxide transport and, 232–233
    covalent bonds of, 29
    importance of, 33–35
    ions, 357
    in living organisms, 31–32
    molecules of, 28
    monitoring levels of, 233–235
Hydrogen bonds, 30–31
Hydrolysis and dehydration synthesis, 37
Hydrophilic molecules, 33
Hydrophobic molecules, 33
Hydroxide ions, 34
Hydroxyapatite, 107
Hymen, 383
Hyperextension, 116
Hyperopia, 292, 293
Hyperplasia, 428
Hypersecretion of hormone, 310
Hypertension, 179–180
    as atherosclerosis risk factor, 173
    congestive heart failure and, 184
    diuretics and, 369
    homeostatic controls and, 181–182
    risk factors for, 179
Hyperthyroidism, 320
Hypertonic solution, cells in, 63
Hypodermis, 93
Hyposecretion of hormones, 310
Hypotension, 180
Hypothalamus, 303, 306–311
    functions of, 261
    hormones produced by, 303
    temperature sensors in, 98–99
    water balance and, 368–369
Hypothesis, 12–15
Hypothyroidism, 320

Hypotonic solution, cells in, 63
Hysteroscopy, 391

I
Identical twins, 488–489
IgA, 203
IgD, 203
IgE, 203, 210
IgG, 203, 210
IgM, 203
Ileum, 334
Iliac region, 113–114
Immune memory, 206–207
Immune response, 201–206
Immune system, 190. *See also* HIV/AIDS;
        Lymphatic system
    aging and, 503
    allergies and, 209–210
    antigens and, 201
    asthma and, 236
    autoimmune disorders, 210–211
    cancer and, 433
    immune memory, 206–207
    of infants, 500–501
    microchimerism and, 85
    Rh factor and, 156
    severe combined immunodeficiency disease (SCID),
        211–212
    tissue rejection, 209
Immunity, 3, 206–207
Immunizations, 2–3
    active immunization, 207–208
    autism and, 3
    basis of, 207
    passive immunization, 208
    pathogens and, 207–208
Immunoglobulin (Ig), 203
Immunoglobulins, 206
Immunosuppressive therapy, 209
Immunotherapy as cancer treatment, 436–437
Implanon, 392
Implantation of blastocyst, 490–491
Implantation process, 383
In situ cancer, 429–430
In vitro fertilization (IVF), 51, 380, 395–396, 405
Incomplete dominance, 454–456
Incus, 289
Independent variables, 17
Induced pluripotent stem cells (iPS cells), 422
Inductive reasoning, 12
Industrial Revolution, 4, 530
Infant respiratory distress syndrome, 226
Infants, 500–501. *See also* Newborns
    delivery of, 496–498
    fontanels, 115
Infections
    chemotherapy and, 436
    of nervous system, 267–268
    renal failure and, 372
    urinary tract infections (UTIs), 372
Inferior point of reference, 93
Inferior vena cava, 172, 174
Infertility. *See also* Reproductive technologies
    causes of, 394–395
    drugs, fertility enhancing, 396
    enhancing fertility, 395–396
    treatments, 380
Inflammation
    histamine and, 319
    pathogens and, 198–199
    in rheumatoid arthritis, 211
Influenza, 193, 237
    bird flu, 515, 526
    monoclonal antibodies of screening for, 209
    pandemics and, 526–527
    Spanish flu, 526–527
    swine flu, 515, 526
Informed choice, 20–21
Inheritance. *See* Genes and genetics;
        Genetic inheritance
Inhibin, 382

Inhibiting hormones, 303
Inhibitory neurotransmitters, 252
Initiation step of translation, 412–413
Injected contraceptives, 390
Inner ear
　balance and, 287–289
　and hearing, 285–287
　Ménière's syndrome, 295
　structure and functions of, 189
Inorganic nutrients, water pollution and, 550
Insertion of muscle, 126
Inspiration, 227–228
Inspiratory reserve volume, 228
Insulin, 303, 318
　beta cells secreting, 311
　in blood plasma, 145, 146
　inhalable form of, 227
　inhaled insulin, 320
　transgenic bacteria for creating, 475
　Type 2 diabetes and, 300–301
Integumentary system, 90, 93–97. See also Skin
Intercalated discs, 138
Intercostal muscle, 221
Interferons, 198, 200, 205
Interleukins, 480
Internal anal sphincter, 337
Internal environment of organism, 97
Internal respiration, 220, 230
International Olympics Committee, 131
Interneurons, 245–246, 254
Interphase of cell cycle, 406–407
Interstitial fluid, 97
Intervertebral disk, 112
Intestinal enzymes, 335
Intestinal gas, 337, 347
Intestines. See also Large intestine; Small intestine
　hormones produced by, 303
Intrauterine devices. See IUDs (intrauterine devices)
Intrinsic factor, 333
Introns, 411
Involuntary muscle, 124, 137–138
Iodine, 315, 345
Ion channels, 304
Ionic bonds, 30
Ions, 30
　in bile, 336
　in blood plasma, 145, 146
　ingestion of, 357
　ion channels, 304
　ionic bonds, 30
　water and, 32–33
IPS cells (induced pluripotent stem cells), 422
Iris, 289–290
Iron, 32, 158, 345
Irregular bones, 109
Island dwarfism, 509
Islets of Langerhans, 311
Isolated systolic hypertension, 179
Isoleucine, 41, 342
Isometric contractions, 133
Isotonic contractions, 133
Isotonic extracellular fluid, 63
Isotopes, 27–28, 512
IUDs (intrauterine devices), 390, 392–393

J
J. Craig Venter Institute, 518
Jacob syndrome, 462
Jacobson, Mark Z., 555
Jagger, Mick, 501
Jejunum, 334
Jiyyeh power station oil spill, 552–553
Johnson, Ben, 123
Joint capsule, 114–115
Joint receptors, 277
Joints, 114–117, 211
Jones, Marion, 122, 123
Jones, R. K., 393
Jugular vein, 172
Juxtaglomerular apparatus, 370

K
Kaposi's sarcoma, 213, 432
Karyotypes, 458–459
Keratin, 196
Keratinocytes, 93–94, 96
Kidney cancer, 437, 442
Kidney stones, 372
Kidneys, 358. See also Urine
　acid-base balance and, 371, 372
　acute renal failure, 372–373
　aging and, 504
　buying and selling of, 359
　cancer of, 437, 442
　cells of, 53
　concentrated urine, production of, 366–367
　dilute urine, production of, 366
　donors, locating, 354–355
　endocrine functions of, 318
　erythropoietin, production of, 149, 371–372
　in fetal development, 496
　homeostasis and, 368–372
　hormones produced by, 303
　nitrogenous wastes, regulation of, 357
　pH and, 371
　red blood cells, regulation of, 371–372
　sale of, 355
　transplants, 354–355, 373
　Type 2 diabetes and, 301
　vitamin D, activation of, 372
　water balance and, 357, 368–369
Killer T cells, 205
Kinberly, M. M., 312
Kinetic energy, 28–29
Kingdoms, 519
Kinsty, L., 88
Klinefelter syndrome, 462
Knee-jerk reflex. See Patellar reflex
Knees. See also Patella
　running and, 117
Krebs, Hans, 71–72
Krebs cycle, 71–72, 73

L
L-dopa, 270
Labeling
　of dietary supplements, 25
　of functional foods, 25
Labia majora, 383
Labia minora, 383
Labor and delivery, 496–498
Lacrimal bones, 110
Lactation, 383, 499–500. See also Breast feeding
Lacteal, 338
Lactic acid, 77
Lactobacillus bacteria, 197
Lactose, 38, 347
Lacunae, 87, 105
LaHood, Ray, 275
Lalueza-Fox, Charles, 278
Landfills, 553
Language
　brain size and, 8–9
　capacity for, 9
Large intestine, 327
　diverticulosis, 347–348
　endocrine system regulating, 339
　functions of, 337
　motility of, 339
　nervous system regulating, 339
　water, absorption of, 338
Larynx, 221, 222
LASIK surgery, 292, 293
Lasix, 369
Latissimus dorsi, 125
Law of independent assortment, 453–454
Law of segregation, 450–451
Laxatives, 350
Lazutina, Larissa, 149
LDL (low-density lipoprotein), 173
　in blood plasma, 146
　saturated fats and, 342

Lead, 420, 541
Leakey, Mary, 521
Lecithin, 336
Left atrium, 170–171
Left coronary artery, 174
Left ventricle, 170–171
Legal issues
　of cloning, 405
　of death, 505
　gene patents, 431
　of genetic testing, 447
　right to die, 484–485
Legs, 113–114
Lemurs, 519
Lens, 289–291, 295
Less industrialized countries (LICs), 531–532
Leucine, 41, 342
Leukemia, 158–159, 437, 442
　alpha interferon and, 200
　cord blood and, 142–143
　spleen, effects on, 195
　stem cells and, 50
　translocations and, 463
　viruses and, 432
Leukocytes. See White blood cells
Levitra, 389
LH (luteinizing hormone), 303, 307
　in adolescence, 501
　corpus luteum formation and, 386
　functions of, 309
　menstrual cycle and, 388
　in ovarian cycle, 385
　and ovulation, 385
　testosterone and, 381–382
Liang Bua Cave, Flores, 508–509
Libido, aging and, 504
Lice, pubic, 397, 400
Life, characteristics of, 4–7
Life span, 502
Life-support systems, 505
Lifecare Advance Directive, 485
LifeSharers, 355
Lifestyles
　aging well and, 504–505
　as atherosclerosis risk factor, 173
　as hypertension risk factor, 179
　Type 2 diabetes and, 300–301
Ligaments, 104, 115–117
　function of, 106
　PRP (platelet-rich plasma) therapy for, 152
　sprains, 117
Light
　action potentials and, 292
　eyes regulating, 290–291
　ultraviolet light as carcinogen, 432, 433
Light antibody chains, 203
Light microscope (LM), 54
Limbic system, 261, 265
Linear acceleration, sensing, 288
Linkage mapping, 458
Linked alleles, 458
Linoleic acid, 342
Linolenic acid, 342
Lipases, 335, 336
Lipid bilayer, 56, 58
Lipids. See Fats
LipoDissolve, 80–81, 87
Lipoproteins, 173. See also HDL; LDL
　in blood plasma, 146
Liver, 327. See also Bilirubin
　abscesses of, 198
　aging and, 504
　cell division in, 418
　cholesterol, synthesis of, 342
　digestive functions of, 336
　in fetal development, 496
　hepatic portal system, 336
　hepatitis, 348
　location of, 335

Liver cancer
  Hepatitis C and, 348
  viruses and, 432
Living wills, 485
Local blood flow, 181–182
Locus of autosomes, 448
Logging, 547, 555
Long bones, 109
Long-term memory, 265
  Alzheimer's disease and, 269
Longevity, 502–503
Longitudinal layer of stomach wall, 332
Loop of Henle, 360–361
  concentrated urine, production of, 367
  dilute urine, production of, 366
  reabsorption by, 364
Loose connective tissue, 85–86
Lou Gehrig's Disease. See ALS
Lower respiratory tract, 221. See also Lungs
  functions of, 222–227
Lucy, 521
Lumbar vertebrae, 111–112
Lumen, 164–165, 168, 327
Lunelle, 391
Lunesta, 264
Lung cancer, 238, 437, 438–439
  secondhand smoke and, 219
  smoking and, 225, 435
Lungs, 221, 222. See also Gas exchange
  angiotensin I and, 370
  congestive heart failure and, 238
  functions of, 225–227
  lobes of, 226
  pneumonia, 237
  pulmonary circuit, 171–172
  volumes, 228–229
  water evaporation from, 357
Lupus erythematosus, 85, 211
Luteinizing hormone. See LH (luteinizing hormone)
Lybrel, 391
Lyerly, A. D., 380
Lymph, 167, 193–195, 205
Lymph nodes, 194, 195
Lymphatic capillaries, 167–168, 193–195
Lymphatic system, 91, 144, 193–196
  blood volume, maintenance of, 167–168
  functions of, 193–195
  structures of, 194
Lymphatic tissue, 205
Lymphatic vessels, 193–195
Lymphocytes
  antibodies produced by, 154
  in lymph nodes, 195
  monocytes, 151
  mononucleosis, 157–158
  multiple myeloma and, 159
  specific defenses and, 201
Lymphoma, 437, 441. See also Non-Hodgkin's lymphoma
Lysine, 41, 342, 343
Lysosomes, 66–67
Lysozyme, 196, 327, 330

M
Macroevolution, 510
Macromolecules, 36–37
Macrophages, 148
  as antigen-presenting cell (APC), 204
  in lymph, 194
  in lymph nodes, 195
  nitric oxide and, 319
  pathogens, defense to, 197–198
Macula, 289–290
Macular degeneration, 296
Mad cow disease, 192
Magazines. See Journals
Magnesium, 32, 145–146, 345
Magnetic resonance imaging. See MRI
Magnets
  blood infections, treating, 158
  as cancer treatment, 436
Maharishi University of Management, 185

Maize, genetically engineered, 469
Major histocompatibility complex (MHC) proteins, 201
Malaria, 456
Male birth control pills, 394
Male reproductive system, 378–382
Maleus, 289
Malignant tumors, 428, 430. See also Cancer
Malnutrition, 348
Mammals, 8, 519
Mammary glands, 8, 307–308, 383
Mammograms, 435, 439, 440
Mandated choice, 365
Mandible, 110
Manganese, 343
Mannitol, 243
Maraviroc, 215
March of Dimes, 464
Marijuana, 266–267
  cell differentiation and, 420
  tubular secretion and, 365
Marshall, Eliot, 431
Masai people, 509
Mass extinctions, 515
Mass of protons and neutrons, 27
Mast cells, 198–199, 210, 319
Mastitis, 198
Mastromarino, Michael, 103
MatchingDonors.com, 355
Maternal twins, 488–489
Matrix, 85
Matter, 26–28, 30, 69–73
Maxilla bones, 110
McCarthy, Jenny, 3
McCloy, Randal, 242–243
McDonalds, 335
McMahon, Sean M., 535
Mda-7 protein, 480
MDMA, 267
Measles, 2, 3
  genetically modified foods, vaccines in, 207
  rubella, 420
  vaccines and, 207
Mechanoreceptors, 276–277
  hair cells as, 287
  hearing and, 284–287
  limb position, indication of, 279–280
  muscle length, indication of, 279–280
  tension, indication of, 279–280
  touch and, 279
Medically induced coma, 242–243
Medicare/Medicaid, 301
  CER (Comparative Effectiveness Research) and, 163
Medications. See Drugs
Medulla. See also Adrenal medulla, 312
  of kidneys, 358
Medulla oblongata, 180–181, 260
  automatic functions, control of, 261
  breathing and, 234
Meiosis
  crossing-over process, 417
  nondisjunction during, 461
  process of, 415–417
  sex differences in, 417
  stages of, 416
Meiosis I, 415–417
Meiosis II, 415–417
Meissner's corpuscles, 277, 279
Melanin, 94
Melanocytes, 93–94, 96
Melanoma, 428, 437, 438
  ultraviolet light and, 433
Melatonin, 303, 317, 318
Membrane potential, 247–248
Membrane voltage, 247–249
Membranes. See also Plasma membrane
  nuclear membrane, 63
  tissue membranes, 92–93
Memory, 265–266
  Alzheimer's disease and, 269
  limbic system and, 261
Memory B cells, 206

Memory T cells, 205, 206
Mendel, Gregor, 449–451, 453
Ménière's syndrome, 295
Meninges, 258–259
Meningitis, 258–259, 267
  HIV/AIDS and, 213
Menisci, 115
Menopause, 118–119, 504
Menstrual cycle, 384–388. See also Menopause
  hormones, cyclic changes in, 388
  ovarian cycle, 384–386
  regulation of, 388
  uterine cycle, 386–387
Menstrual phase, 386–387
Menstruation, 386–387, 350
Mental retardation
  cri-du-chat syndrome, 463
  fragile X syndrome, 446
  Jacob syndrome and, 462
  Klinefelter syndrome and, 462
  phenylketonuria (PKU) and, 463
  Trisomy-X syndrome and, 462
Mercury
  biological magnification and, 551
  cell differentiation and, 420
  cycles involving, 541
Merkel disks, 277, 279
Mesoderm, 491, 492
Messenger RNA. See mRNA (messenger RNA)
Metabolic pathways, 69–70, 76
Metabolism, 5, 69, 181–182
Metabolites, 357
Metabolomics, 357
Metacarpal bones, 113
Metaphase, 414–415
Metastasis, 430
Metatarsal bones, 114
Methamphetamine, 266
Methane
  global warming and, 545
  as greenhouse gas, 546
Methionine, 41, 342, 343
Micelles, 338
Microchimerism, 85
Microevolution, 510
Microscopes, 54
Microtubules, 404
Microvilli, 55, 334
Micturition reflex, 367–368
Midbrain, 260, 261
Middle ear, 289
  and hearing, 284–285
  otitis media, 295
Midsagittal plane, 93
Mifeprex, 394
Mifepristone, 394
Milk
  lactation, 383, 499–500
  lactose intolerance, 347
Milk ejection, 499–500
Milk-producing lobules, 383
Milkweed, 6
Miller, Greg, 192
Mineralocorticoids, 313, 318
Minerals
  absorption of, 338–339
  list of, 345
  nutrition and, 343
Mirena, 392–393
Miscarriages, 395, 489, 494
Mitochondria, 67–68
  DNA (deoxyribonucleic acid) in, 407
Mitosis, 406–407
  daughter cells in, 413–415
  diploid cells from, 415
  of helper T cells, 204
  nondisjunction during, 461
  phases of, 414–415
Mitotic phase of cell cycle, 406–407
Mitotic spindle, 404, 414
Mitral valve, 171

Mm Hgs for blood pressure measurement, 178
Mode of transmission of pathogens, 192–193
Molecular treatments for cancer, 437
Molecules, 5, 9, 11, 28–32
  chemical bonds and, 29–30
  comparative biochemistry, 513
  hydrogen bonds, 30–31
  hydrophilic molecules, 33
  hydrophobic molecules, 33
  macromolecules, 36–37
  nonsteroid hormones and, 305
  organic molecules, 35–37, 516–517
  plasma membrane, crossing, 57–63
  polar molecules, 30
  producer organisms making, 534
  self-replicating molecules, 517
Moles, 34, 428
Molybdenum, 343
Monera, 7
Monkeys, 4, 8, 519–520
Monoclonal antibodies, 208–209
Monocytes, 148
  as agranular leukocytes, 151
  spleen, storage in, 150
Monoglycerides, 338
Mononucleosis, 157–158, 195
Monosaccharides, 37–38
More industrialized countries (MICs), 531–532
"Morning after" pills, 392, 393–394, 390
Morphogenesis, 489–490, 513
Mortality
  BMI (body mass index) and, 325, 326
  cancer and, 434
  fertility rate and, 557
Morula, 490
Motion sickness, 288
Motor neurons, 128, 134, 138, 245–246
  myelin and, 250
Motor units, 134
Mouth, 221, 327, 329–331
Movement
  acceleration, sensing, 288
  balance, inner ear and, 287–289
  head position, sensing, 288
  hindbrain and, 260–261
  rotational movement, sensing, 288
  stretch reflex and, 255
MRI (magnetic resonance imaging), 434–435
  of brain, 435
mRNA (messenger RNA), 408
  genetic code of, 412
  transcription process, 410–411
  translation, 412–413
MRSA (methicillin-resistant *Staphylococcus aureus*), 15
Mucin, 330
Mucosa, 328, 332
Mucous membranes, 92
Mucus
  bulbourethral gland producing, 380
  cells, mucus-secreting, 332
  gastric juice and, 333
  in nasal cavity, 222
  pathogens, defense to, 196
  smell and, 283
  smoking and, 224–225
  of stomach, 332
Muhammad Ali, 50, 51
Mühlegg, Johann, 149
Multicellular organisms, 5
  homeostasis in, 6
  tissues in, 82
Multiple myeloma, 159
Multiple sclerosis (MS)
  as autoimmune disorder, 211
  myelin sheath and, 250–251
Mumps, 2
Murmurs, 175–176
Muscle cells, 126–127
  all-or-non principle, 134–135
  ATP (adenosine triphosphate) in, 131
  cell division in, 418

contraction and relaxation of, 128–132
  glycogen cells, reliance on, 67
  structure of, 53
Muscle cramps, 140
Muscle fatigue, 132
Muscle fibers, 88, 126–127
Muscle spindles, 277, 279–280, 281
Muscle tension, 134–135
Muscle tissue, 88
Muscle tone, 134
Muscles, 88–89. *See also* Cardiac muscle; Smooth muscle
  all-or-none principle, 134–135
  calcium and contraction, 130
  contraction and relaxation of, 128–130
  defining characteristics of, 139
  delayed onset soreness, 133
  energy for contraction and relaxation, 131–132
  fast-twitch fibers, 135–136
  force, nerve activation and, 133–134
  isometric contractions, 133
  isotonic contractions, 133
  joints, stabilization of, 115–117
  movement, production of, 124–128
  myofibrils, 127–128
  myograms, 134–135
  nerves activating, 128
  rigor mortis, 131
  sliding filament mechanism, 128–130
  slow-twitch fibers, 135–136
  speed of contraction, 138
  stimulus-twitch relationship, 134–135
  structure of, 127
  summary of activation of, 135
  sustainability of contraction, 138
  veins, skeletal muscles and, 168
Muscular dystrophy, 139
Muscular system, 90, 122–141
  diseases and disorders of, 139–140
Muscularis
  of gastrointestinal (GI) tract, 328
  of stomach, 332
Mutations, 409–410
  of alleles, 449
  of BRCA genes, 426–427
  evolution and, 514–515
  between generations, 410
  genetic testing for, 435
  and pathogens, 190
Mutator genes, 430–431
Mycobacterium, 6
Mycobacterium tuberculosis, 237
Mycoplasma genitalium, 518
Myelin sheath, 250, 251
Myelinated neurons, 250
Myocardial infarction. *See* Heart attacks
Myocardium, 170
Myofibrils, 127–128
Myoglobin, 136
Myograms, 134–135
Myometrium, 383
Myopia, 291–292, 293
Myosin, 127
  of cardiac muscle, 138
  muscle contraction and, 130
  in sarcomeres, 127
MyPyramid, 340–341
Myriad Genetics, 431
Myxedema, 320

N

Na⁺–K⁺ pump. *See* Sodium-potassium pump
NAD⁺ (nicotinamide adenine dinucleotide), 71
NADH, 72
  in citric acid cycle, 73
  in electron transport system, 74
  in oxidative phosphorylation, 74
NADPH and photosynthesis, 534
Nasal bones, 110
Nasal cavity, 221, 222
Nasal passages, 194
National Cancer Act, 434

National Cancer Institute, 427
National Cattleman's Beef Association, 340–341
National Dairy Council and MyPyramid, 340–341
National Highway Traffic Safety Administration (NHTSA), 274–275
National Institute on Drug Abuse, 123
National Institutes of Health (NIH). *See* NIH
National Marrow Donor Program, 143
National Newborn Screening and Genetics Resource Center, 464
National Vaccine Information Center, 3
Natural killer (NK) cells, 198, 199
Natural selection, 510, 514–516
Nature, 16
Nausea and chemotherapy, 436
Neanderthals, 522
Nearsightedness, 291–292, 293
Negative feedback, 97–98
  blood solute concentration, controlling, 369
  core body temperature and, 98–99
  hormones and, 305–306
  hypertension and, 181–182
  and menstrual cycle, 388
  red blood cell production and, 149
  testosterone and, 381–382
  of thyroxine secretion, 314–315
Neoplasm, 428
Nephrons, 360–361
  blood vessels serving, 360–361
  tubular regions of, 360–361
Nerve cells
  cell division in, 418
  structure of, 53
Nerve deafness, 294–295
Nerve tracts, 259–260
Nerves
  cardiac output and, 180–181
  CNS (central nervous system), signals to and from, 253–254
  force, nerve activation and, 133–134
  micturition reflex and, 367
  muscles, activation of, 128
  of tooth, 330
Nervous system, 90. *See also* Brain; Neurons
  aging and, 503
  breathing, regulation of, 233–236
  characteristics of, 244
  digestion, regulation of, 339–340
  endocrine system and, 304
  in fetal development, 496
  infections, 267–268
  parts of, 244–245
  trauma and, 267
Nervous tissue, 89
Neurodegenerative disorders, 192
Neuroendocrine reflex, 308
Neuroglial cells, 249–251
Neuromuscular junction, 128
Neurons, 89. *See also* Action potentials; Motor neurons; Nerves; Neurotransmitters; Resting potential; Sensory neurons
  action potentials, initiation of, 246–249
  functions of, 245–246
  in G₀ phase, 407
  information transfer and, 251–253
  myelinated neurons, 250
  neuroglial cells supporting, 249–251
  postsynaptic neurons, 253
Neurotransmitters, 251–253. *See also* ACh (acetylcholine)
  in autonomic division of PNS, 257
  excitatory or inhibitory effects of, 252
  psychoactive drugs and, 266
  rods and cones releasing, 293
  in somatic division of PNS, 257
Neutral fats, 39
Neutral solutions, 34
Neutrons, 26–27
Neutrophils, 150
  pathogens, destruction of, 198
New Zealand, GE plants in, 469

Newborns, 498–499
  cardiovascular system of, 498–499
  first breath, 498
  lactation and, 499–500
  neonatal period, 500
Newspapers, scientific information in, 16
Newton, Terry, 306
Niacin, 344
Niches, overlapping, 532
Nickel, 343
Nicotine. *See also* Smoking
  blood-brain barrier and, 259
  as psychoactive drug, 266
Night blindness, 344
NIH (National Institutes of Health)
  dietary supplements, sales trends for, 34
  on secondhand smoke, 218–219
  stem cell research and, 51
Nipple, 383
Nitric oxide, 319
Nitrification, 540
Nitrogen
  kidneys regulating wastes, 357
  in living organisms, 31–32
  symbol for, 27
Nitrogen cycle, 540
Nitrogen fixation, 540
Nitrogen oxides
  acid precipitation and, 548
  and smog, 548–549
Nitrous oxide, 546
Nixon, Richard M., 434
Nodes of Ranvier, 250
  myelin sheath and, 250
Noise and hearing, 284
Non-Hodgkin's lymphoma, 441
  HIV/AIDS and, 213
  viruses and, 432
Nondisjunction, 461–463
  of sex chromosomes, 462
Nonmelanomas, 437, 438
Nonself markers, 201
Nonspecific defenses, 197–200
Nonsteroid hormones, 304–306
Norepinephrine, 303, 318
  actions of, 252
  adrenal medulla producing, 313
  in autonomic division of PNS, 257
  BMR (basal metabolic rate) and, 346
  cardiac output and, 181
North China Plain, aquifers in, 550
Norwalk virus, 208
Nose
  external nose, 222
  nasal bones, 110
  nasal cavity, 221, 222
Notochord, 513
Nova, 16
NSAIDs (non-steroidal anti-inflammatory drugs), 347
Nuclear energy, 553
Nuclear membrane, 63
Nuclear pores, 63
Nuclear reaction, 26
Nuclear weapons, 57
Nucleic acids, 5, 37, 44–45. *See also* DNA; RNA
Nucleolus, 63
Nucleotides, 44
Nucleus
  diagram of, 64
  of eukaryotes, 52–53
  role of, 63
Nursing. *See* Breast feeding
Nutraceuticals, 24
Nutrients, 326
  absorption of, 337–339
  in biogeochemical cycles, 538
  in blood plasma, 145, 146
  digestive system and, 327–329
  small intestine absorbing, 333–334
  stomach and, 333
  use of, 339–340

Nutrition, 340–345
  aging and, 504
  amino acids and, 342–343
  biodiversity and, 556
  carbohydrates, 341
  fiber and, 343
  lipids and, 342
  malnutrition, 348
  MyPyramid, 340–341
  proteins and, 342
  vitamins and, 343
NuvaRing, 392

O

Obama, Barack, 405
  stem cell research and, 51
Obesity
  as atherosclerosis risk factor, 173
  cancer and, 433
  epidemic of, 348–349
  friends and, 329
  as hypertension risk factor, 179
  overnutrition and, 348
  as risk factor, 324
  Type 2 diabetes and, 301
Oblique layer of stomach wall, 332
Observation in scientific method, 12
Occipital bone, 110
Occipital lobe, 262
Oil pollution, 551–552
  Jiyyeh power station oil spill, 552–553
Oils, 39, 342
Olfactory bulb, 283
Olfactory receptors, 277, 282–283
Oligodendrocytes, 250
Oligosaccharides, 38
Omnivores, 534
Oncogenes, 430
Oocytes, 382
  ovarian cycle and, 384–386
Opportunistic viruses, 213
Opposable thumbs, 8, 9, 113
Oprah Winfrey Show, 3
Optic disk, 289–290
Optic nerve, 289–290
Oral contraceptives, 391–392
  failure rate for, 390
  as hypertension risk factor, 179
Orangutans, 519–520
Orders, 519
Ordinates, 17
Organ of Corti, 285, 289
Organ Procurement Organizations (OPOs), 355
Organ systems, 9, 11, 89–92
Organelles, 52–53. *See also* Mitochondria
Organic metabolism pathways, 340
Organic molecules, 35–37, 516–517
Organic nutrients, water pollution and, 550
Organisms, 9, 11. *See also* Multicellular organisms;
    Unicellular organisms
  categorizing, 7
Organs, 9, 11, 89–93. *See also* Transplants
Orgasm
  in females, 390
  in males, 389–390
  phase of sexual response, 389
Origin of muscle, 126
Orpana, H. M., 326
Ortho Evra, 391
Osmosis, 58
Osmotic pressure, 58
Osteoarthritis, 117–118
  running and, 117
Osteoblasts, 106–107, 108
  hormones regulating, 109
Osteoclasts, 108–109
  hormones regulating, 109
Osteocytes, 105, 108
  cell division in, 418
  in G$_0$ phase, 407
Osteoids, 106–107

Osteons, 105
Osteopenia, 119
Osteoporosis, 108–109, 118–119
  anorexia and, 350
  milk consumption and, 340
  pre-osteoporosis, treatment of, 119
Otitis media, 295
Otoliths, 289
  head position, sensing, 288
Ott, Harald C., 174
Outer ear, 289
  and hearing, 284
Ova. *See* Eggs
Oval window, 284, 287
Ovarian cancer, 426–427
Ovarian cycle, 384–386
Ovaries, 382–383
  estrogen and, 317
  in fetal development, 496
  hormones of, 303, 318
  progesterone and, 317
  and puberty, 317
  voluntary removal of, 426–427
Overnutrition, 348
Overweight. *See also* Obesity
  BMI (body mass index) and, 324–325
  as health risk, 326
Oviducts, 382
Ovulation, 385–386
  LH (luteinizing hormone) and, 309
  in uterine phase, 386–387
Oxidative phosphorylation, 72
Oxygen, 343. *See also* Gas exchange
  aerobic metabolism and, 220
  anaerobic pathways, 76–77
  in blood plasma, 145, 146
  cell differentiation and, 420
  covalent bonds of, 29–30
  heart requiring, 172
  hematocrit and, 147
  hemoglobin binding to, 147
  hydrogen bonds and, 30–31
  lipid bilayer, crossing through, 58
  in living organisms, 31–32
  monitoring levels of, 233–235
  need for, 220
  photosynthesis and, 518
  red blood cells transporting, 146–147
  symbol for, 27
  in young Earth, 516–517
Oxygen debt, 132
Oxyhemoglobin, 147, 230–232
Oxytocin, 303, 307
  and birth process, 497
  in colostrum, 499–500
  functions of, 307
  lactation and, 383
Ozone layer, 11
  CFCs (chlorofluorocarbons) and, 547–548
  satellite map of, 19

P

P waves, 176–177
Pacemaker cells, 137–138
Pacinian corpuscles, 277, 279, 280
Pain
  angina, 182–183
  inflammation and, 199
  referred pain, 281
Pain receptors, 276–277, 280–281
Pain relievers, 267
Palatine bones, 110
Palindromic double-stranded DNA, 471–472
Pancreas, 327. *See also* Diabetes
  endocrine functions of, 311–312
  food, response to, 311
  functions of, 334–335
  hormones of, 303, 318
  location of, 335
Pancreatic amylase, 335
Pancreatic juice, 338–339

Pandemics, 526–527
Pangaea, 514
Pantothenic acid, 344
Pap smears, 439, 442
Papillae, 282
Papillary muscles, 171
Parasites, 150
Parasympathetic nervous system, 244, 255–257
cardiac output and, 181
Parathyroid glands
functions of, 313–316
hormones of, 303, 318
Parathyroid hormone. See PTH
Parietal bones, 109–110
Parietal lobe, 262
Parkinson's disease, 270
prions and, 192
stem cells and, 50–51, 55, 270
therapeutic cloning for, 422
Parotid gland, 330–331
Partial pressure gradient, 229–230, 231
Passive immunization, 208
Passive transport, 57–58
with concentration gradient, 58
facilitated transport, 59
forms of, 59
Patau syndrome, 461
Patella, 114
muscles and, 116
Patellar reflex, 255
muscle spindles and, 280
stretch reflex, 255
Patents for genes, 431
Pathogens, 190–193
bacteria as, 190–191
complement system, 198, 199–200
defecation ridding body of, 196
earwax as barrier, 196
fever and, 198, 200
immunization and, 207–208
inflammation and, 198–199
interferons and, 198, 200
medical assistance and, 207
monoclonal antibodies, 208–209
mucus as barrier, 196
natural killer (NK) cells and, 198, 199
nonspecific defenses, 197–200
saliva as barrier, 196
skin as defense, 196
specific defense mechanisms, 201–206
tears as barrier, 196
transmissibility of, 192–193
urination ridding body of, 196
vomiting ridding body of, 196
white blood cells and, 197–198
PCBs (polychlorinated biphenyls)
cell differentiation and, 420
cycles involving, 541
water pollution and, 550
PCDC (phosphotidylchone dioxycholate), 81
Pea plant experiments, 450–451
Pectoral girdle, 112–113
Pectoralis major, 125
Peer-reviewed journals, 14, 16
Peerbooms, Joost C., 152
Peet, Amanda, 3
Pellagra, 344
Pelvic examination, 439
Pelvic girdle, 113–114
Pelvic inflammatory disease (PID), 394–395
Pelvis, 113–114
Penicillin, 365
Penis, 379, 380
erectile dysfunction, 389
in prenatal development, 495
Pepsin, 332, 335
Pepsinogen-secreting cells, 332
Peptic ulcers, 333, 347
Helicobacter pylori and, 196
Perceptions, 276
Percussion therapy for cystic fibrosis, 452

Percutaneous umbilical blood sampling (PUBS), 489
Perforin, 205
Performance-enhancing drugs, 122–123
Pericardial cavity, 169–170
Pericarditis, 170
Pericardium, 169–170
Periodic abstinence, 393
Periodic table of elements, 26, 27
Periodontal membrane, 330
Periodontist, 330
Periosteum, 104–105, 106–107
Peripheral nervous system. See PNS
Peripheral vascular disease, 173
Peristalsis, 328–329
gravity and, 331
of stomach, 333
Peritubular capillaries, 361
Pernicious anemia, 158
Peroxisomes, 66–67
in animal cell, 64
free radicals and, 433
Persistent vegetative state, 485
Perspiration
body temperature and, 33
pH of, 196
Pertussis, 2
Pesticides
as carcinogens, 432
cycles involving, 541
tubular secretion and, 365
water pollution and, 550, 551
PET (positron-emission tomography), 434
P53, 430–431
Pfizer's Exubera, 320
pH, 34–35
buffers and changes in, 35
gastric juice and, 332
kidneys and, 371
of skin, 196
of vagina, 379–380
Phagocytes, 197–198
Phagocytosis, 148–149, 197–198
monocytes and, 151
Phalanges, 113
Pharmaceutical drugs. See Drugs
Pharyngeal arches, 493, 513
Pharynx, 221, 327
functions of, 222, 331
Phenotypes, 449
effects of, 456
environmental factors and, 457–458
genotypes, 457–458
human beings, differences among, 523
polygenic inheritance, 456–457
sex-influenced phenotypes, 460
two-trait crosses and, 453–454
Phenylalanine, 41, 342
codons specifying, 411
phenylketonuria (PKU) and, 463
Phenylketonuria (PKU), 463
Phospholipids, 39–40, 342, 540
in plasma membrane, 56
Phosphorus, 343, 345
in living organisms, 31–32
phosphorus cycle, 540–541
supplies of, 538
Phosphorus cycle, 540–541
Phosphorylation, oxidative, 72
Phosphotidylcholine, 81
Photodynamic therapy as cancer treatment, 436–437
Photopigments, 293–294
Photoreceptors, 277, 289. See also Rods and cones
Photosynthesis, 7, 220, 518
carbon cycle and, 539–540
producer organisms and, 534
Photovoltaic panels, 554
Phylogenetic trees, 516
Phylum, 519
Pia mater, 258

Pineal gland
functions of, 317
hormones of, 303, 318
Pinna, 289
Pitch of sound, 284
Pituitary dwarfism, 310
Pituitary gland, 303, 306–311
anterior lobe, 306, 308–310
disorders of, 310–311
fetal pituitary gland, 496–497
hormones of, 303, 307
posterior lobe, 306–308
water balance and, 368
Placebos in experiments, 14
Placenta, 383, 492, 493
afterbirth, 498
cord blood from, 143–144
as endocrine organ, 493
Plan B, 394
Planes of symmetry, 93
Plantae, 7
Plantar fasciitis, 140
Plants, 7
biogeography and, 513–514
genetically engineered, 468–469
nitrogen cycle an, 540
photosynthesis and, 534
transgenic plants, 476–477
Plaques, atherosclerotic, 173
Plasma. See Blood plasma
Plasma cells, 151
functions of, 206
multiple myeloma, 159
Plasma membrane, 52, 55–57
active transport, 59–60
in animal cell, 64
cytoskeleton supporting, 69
endocytosis, 60
of eukaryotes, 52–53
exocytosis, 60
information transfer across, 61
as lipid bilayer, 56
molecules crossing, 57–63
passive transport, 57–59
phospholipids and, 39–40
of prokaryotes, 52–53
selective permeability of, 58
viruses and, 192
Plasma proteins, 42, 56, 146
complement system and, 198, 199–200
Plasmids, 472
Plateau phase of sexual response, 389
Platelet-derived growth factor, 419
Platelets, 145, 146, 151
blood vessel ruptures and, 152
production of, 148
stem cells and, 147–148
thrombocytopenia, 159
Pleural cavity, 225
Pleural membranes, 221, 225
Pleurisy, 225
PMS (premenstrual syndrome), 387
Pneumonia, 237
bacteria causing, 191
HIV/AIDS and, 213
pleurisy and, 225
Pneumothorax, 238
PNS (peripheral nervous system), 244–245
autonomic division of, 255
functions of, 253–257
somatic division, functions of, 254–255
Poaching, 555
Podocytes, 363
Poison ivy, 209
Poisoning. See also Food poisoning
botulism, 237–238
carbon monoxide poisoning, 235
Polar bears, 547
Polar body, 385
Polar ice cap, 544–545

Polar molecules, 30
  water and, 33
Polio, 2, 3
  vaccine, 207
Political issues of cloning, 405
Pollutants and pollution, 11, 546
  acid precipitation and, 548
  air quality and, 546–549
  cell differentiation and, 420
  cycles involving, 541
  emphysema and, 236
  groundwater pollution, 551
  land, damage to, 552–553
  oil pollution, 551–552
  smog, 548–549
  of water, 549–552
Polychlorinated biphenyls. See PCBs
Polycythemia, 147
Polydactyly, 451
Polygenic inheritance, 456–457
Polymerase chain reaction (PCR), 473–474
Polypeptides, 42
Polyps, 440–441
  of colon, 348
Polysaccharides, 38–39
Pons, 260
  functions of, 261
Populations, 9, 11, 515, 528
  age structure, 531–532
  biotic potential and, 520
  carrying capacity, 530
  evolutionary trees and, 516
  habitat of, 528–529
  of human beings, 522–523, 530
  of less industrialized countries (LICs), 532
  zero population growth, 531
Portista, 7
Positive feedback, 99
  and birth process, 497
  in menstrual cycle, 388
Posterior cruciate ligaments, 115
Posterior of body, 93
Postganglionic neurons, 255
Postsynaptic membrane, 251
Postsynaptic neurons, 253
Posture
  muscles and, 124
  stretch reflex and, 255
Potassium, 343, 345. See also Sodium-potassium pump
  aldosterone and, 313
  in blood plasma, 145, 146
  filtration of, 364
  ions, 357
  in living organisms, 32
  radioactive potassium, 512
  symbol for, 27
Potential energy, 28–29
  matter and, 30
Poverty, 558
Pre-embryonic development, 490–491
Pre-osteoporosis, treatment of, 119
Precapillary sphincters, 166
  local requirements and blood flows, 181–182
Prediabetics, 320
Predictions in scientific method, 13
Preganglionic neurons, 255
Pregnancy. See also Abortions; Childbirth; Miscarriages; Prenatal development
  delivery of infants, 496–498
  ectopic pregnancy, 490
  HIV/AIDS transmission and, 213
  home pregnancy tests, 386
  monoclonal antibodies for pregnancy tests, 209
  prenatal diagnostic techniques, 489
  protein deficiencies in, 342
  Rh factor and, 156
  trimesters of, 490
Preimplantation genetic diagnosis (PGD), 377
Premature infants, 226
Premenstrual syndrome, 387
Prenatal development

embryonic development, 491–494
  extra-embryonic membranes, 491–492
  fetal development, 496
  of gender, 494–495
  pre-embryonic development, 490–491
  testing, prenatal, 489
Prenatal diagnosis, 489
Preparatory step, 71–72
Presbyopia, 291
  LASIK surgery for, 293
Prescription medications. See Drugs
Presumed consent laws, 355, 363
Presynaptic membrane, 251
Preven, 393–394
Primary consumers, 534
Primary immune response, 206
Primary motor area, 262–263
Primary oocytes, 385
Primary phase of syphilis, 398
Primary somatosensory area, 262–263
Primary spermatocytes, 380–381
Primary structure of protein, 42–43
Primary transcript, 410–411
Primates, 4, 8
  evolution of, 520
  humans as, 519–520
Primers, 470
Primitive streak, 493–494
Prions, 192
Privacy, genetic testing and, 447
PRL. See Prolactin
Process, science as, 11–15
Producer organisms, 534
Producers and food webs, 535–536
Products, 43
  of metabolic pathways, 69–70
Progesterone, 303, 318
  and birth process, 497
  LH (luteinizing hormone) and, 309
  mammary glands and, 383
  ovaries secreting, 382–383
  placenta secreting, 493
Prokaryotes, 7, 52–53
Prokaryotic cells, 191
Prolactin, 303, 307
  functions of, 309
Prolactin-inhibiting hormone, 309
Prolactin-releasing hormone, 309
Proliferative phase, 386–387
Proline, 41
Promoter enzyme, 410
Pronation, 116
Prophase, 414–415
Prostaglandins
  and birth process, 497
  functions of, 319
Prostate cancer, 437, 440, 441
  monoclonal antibodies of screening for, 209
  sarcosine as marker, 357
  screening for, 439
Prostate gland, 379–380
Proteases, 334–335
Proteins, 5, 37, 40–44. See also Amino acids; Plasma proteins
  absorption of, 337–338
  complete proteins, 342–343
  digestion, enzymes of, 335
  energy and, 75–76
  function and structure of, 42
  gene codes for, 465
  glycoproteins, 69
  major histocompatibility complex (MHC) proteins, 201
  metabolic pathways for, 76
  nutrition and, 342
  prions and, 192
  receptor proteins, 61
  self markers, 201
  synthesis of, 42
  transport protein, 59
  in urine, 364
Proteinuria, 364

Prothrombin, 154
  activator, 152
Protists, 7
Proto-oncogenes, 430–431
Protons, 26–27
Proximal point of reference, 93
Proximal tubules, 360
  reabsorption in, 364–365
PRP (platelet-rich plasma) therapy, 152
PSA (prostate-specific antigen) test, 439, 440, 441
Pseudomonas fluorescens bacteria, 476
Psychoactive drugs, 266–267
Psychological dependence, 266
Psychological impact of genetic testing, 447
PTH (parathyroid hormone), 109, 303, 318
  blood calcium levels and, 315–316
  vitamin D, activation of, 372
Puberty, 501
  FSH (follicle-stimulating hormone) and, 309
  growth plate, stimulation of, 107
  hormones and, 306
  LH (luteinizing hormone) and, 309
  ovaries and, 317
  testosterone and, 317
Pubic lice, 397, 400
Pubic symphysis, 113–114
Pulled muscles, 140
Pulmonary capillaries, 226–227
Pulmonary circuit, 171–172
Pulmonary embolisms, 184
Pulmonary surfactants, 296, 498
Pulmonary trunk, 174
Pulp cavity, 330
Pulse, 175
Punnett squares, 449
  for two-trait crosses, 453
Pupil, 289–290
  light and, 290–291
Purkinje fibers, 176
Pus, 198
Pyloric sphincter, 332
Pyruvate, 71
  anaerobic pathways and, 76–77

Q
QRS complex, 176–177
Quadriceps group, 125
Quaternary consumers, 534
Quaternary structure of protein, 42–43

R
R. J. Reynolds, 223
Rabies, 267
  viruses causing, 192
Rabin, Roni C., 264
Race
  as hypertension risk factor, 179
  lupus erythematosus and, 211
Racial differences, 522–523
Radiation
  as cancer treatment, 436
  as carcinogen, 432–433
  cell differentiation and, 420
  from cell phones, 269
  and lung cancer, 438
  radioisotopes emitting, 28
Radioactive potassium, 512
Radioactive wastes, 11
  in groundwater, 551
Radioisotopes, 28
Radiometric dating, 512
Radius, 113
Radon, 432
Rain forests, 19, 555
Range of population, 528–529
RBCs. See Red blood cells
RDA (Recommended Dietary Allowance), 343
Reactants, 43
Receptor proteins, 61
Receptor site, 61
Receptors, 276–278

adaptation of, 278
classification of, 276–277
nonsteroid hormones binding to, 304–305
Recessive alleles, 451–452
genetic disorders and, 463–464
phenylketonuria (PKU), 463
Tay-Sachs disease and, 463–464
Recombinant DNA technology, 470, 471–473
restriction enzymes and, 472
Recruitment, 134
Rectal cancer, 437, 440–441
screening for, 439
Rectum, 327, 337
Rectus abdominus, 125
Recycling, 557
Red blood cells, 87, 145, 146
in ABO blood group system, 155
carbon dioxide, transportation of, 146–147
kidneys regulating production of, 371–372
life span of, 148–149
negative feedback control and, 149
oxygen, transportation of, 146–147
phagocytosis, 148–149
Red bone marrow, 104–105
Red Bull energy drink, 24
Red-green color blindness, 296
Red pulp, 194, 195
Referred pain, 281
Reflexes. *See also* Patellar reflex
micturition reflex, 367–368
spinal reflexes, 254–255
Relative refractory period, 249
Releasing hormones, 303
REM (rapid eye movement) sleep, 265
Renal arteries, 358
Renal artery, 172
Renal failure, 372–373
Renal pelvis, 358
Renal veins, 172, 358
Renin, 303, 318, 370
Renin-angiotensin system, 318
aldosterone secretion and, 369–370
Repetitive stress syndrome, 113
Replacement fertility rate, 531
Replication. *See* DNA replication
Repolarization, 248
Reproduction, 6
Reproductive cloning, 420–422
Reproductive systems, 91, 378. *See also* Male
reproductive system
in adolescence, 501
aging and, 503–504
female reproductive system, 382–384
gonadotropins and, 309
Reproductive technologies
artificial reproductive technologies (ART), 395–396
preimplantation genetic diagnosis (PGD), 377
in vitro fertilization (IVF), 380
Research
stem cells and, 50–51, 55
transgenic animals for, 477
Resident bacteria, 197
Residual volume, 228
Resolution phase of sexual response, 389
Respiration, 220
Respiratory center
breathing rhythm and, 233
in medulla oblongata, 261
Respiratory cycle, 227–228
Respiratory system, 91, 220. *See also* Gas exchange
aging and, 503
diseases and disorders of, 236–238
microorganisms affecting, 236–238
structures of, 220–227
vocalization and, 220
Rest and repose effect, 257
Resting potential
action potential and, 248
graded potentials and, 247
reestablishment of, 248
sodium-potassium pump and, 246–247

Restriction enzymes, 471–472
Reticular activating system (RAS)
sleep, control of, 263
Reticular connective tissue, 85–86, 87
Reticular formation, 261
Retina, 289–290. *See also* Rods and cones
age-related macular degeneration (AMD), 296
detachment of, 295
structure of, 292
Retinal detachment, 295
Retroviruses, 212. *See also* HIV/AIDS
vectors of, 479–480
Rh factor, 156
percutaneous umbilical blood sampling (PUBS) and, 489
Rheumatoid arthritis, 118
as autoimmune disorder, 211
microchimerism and, 85
Rhodopsin, 293–294
RhoGAM, 156
Rhythm method, 393
Rib cage, 112
Riboflavin, 344
Ribose, 38, 45
Ribosomal RNA (rRNA), 412–413
Ribosomes
in animal cell, 64
role of, 63, 65
translation and, 412–413
Ribs, 112, 221
RICE (Rest, Ice, Compression, Elevation), 118
Rice, golden, 472, 476
Richman, Douglas D., 193
Right atrium, 170–171
Right coronary artery, 174
Right lymphatic duct, 193–195
Right to die, 484–485
Right ventricle, 170–171
Rigor mortis, 131
Risedronate, 119
RNA (ribonucleic acid), 44–45, 408. *See also* mRNA
DNA (deoxyribonucleic acid) compared, 45
evolution and, 517
HIV/AIDS and, 212
ribosomal RNA (rRNA), 412
transcription process, 410–411
translation, 412–413
tRNA (transfer RNA), 412–413
viruses and, 193
RNA polymerase, 410
Roach, J. C., 410
Rods and cones, 292
color blindness and, 296
color vision and, 294
dim light, vision in, 293–294
light and, 294
visual acuity and, 294
Roid rage, 123
Roofs, green, 557
Root canal, 330
Root of hair, 95
Rosenbaum, Levy-Izhak, 359
Rotation and joints, 116
Rotational movement, sensing, 288
Rough endoplasmic reticulum, 65
in animal cell, 64
Round window, 285, 287, 289
RRNA (ribosomal rRNA), 412–413
RU-486, 394
Rubella, 420
Ruffini endings, 277, 279
Rule of 72, 529
Running
and knees, 117
stretching and injuries, 137

S

S phase, 406
Sabin vaccine, 207
Saccule, 289
hair cells in, 277
head position, sensing, 288

Sacral vertebrae, 111–112
Safer sex guidelines, 215
Safety
of functional foods, 25
of LipoDissolve, 81
Sago Mine explosion, 235
Saliva, 327
functions of, 330
pathogens, defense to, 196
Salivary amylase, 330, 335
Salivary glands, 327, 330–331
Salt
aldosterone and salt balance, 369
cancer and, 433
taste, salty, 282
water and, 32–33
Saltatory conduction, 250
Salty taste, 282
Sankaranarayanan, R., 442
Sarcomeres, 127
delayed onset soreness and, 133
sliding filament mechanism, 128–130
Sarcoplasmic reticulum, 128
Sarcosine, 357
Sartorius, 125
Satellite map, 19
Saturated fats, 39, 342
Scanning electron microscope (SEM), 54
Scapulas, 112–113
Scatter plots, 17
Schiavo, Terri, 484–485
Schuchat, Anne, 3
Schulz, Jenny Jakobsson, 137
Schwann cells, 250
Schwarzenegger, Arnold, 123
SCID (severe combined immunodeficiency disease), 480
Science, 4
limits of, 20
as process, 11–15
shared responsibility concept, 20
society, role in, 19–21
Science, 16
Scientific evidence, 17
Scientific method, 11–15
Sclera, 289–290
Scleroderma, 85
Scrotum, 379, 380
Scrubbers for sulfur dioxide, 548
Scurvy, 344
Seaonale, 391
Sebaceous glands, 95
Second law of thermodynamics, 533–534
Second messengers, 305
Secondary consumers, 534
Secondary immune response, 206
Secondary oocytes, 385
Secondary phase of syphilis, 398
Secondary spermatocytes, 380–381
Secondary structure of protein, 42–43
Secondhand smoke, 218–219
Secretin, 303, 318, 339
Secretory phase, 387–388
Secretory vesicles, 66
in animal cells, 64
Sedimentary cycle, 540–541
Sedimentary deposits, 511
Sediments, 551
Segmentation, 329
Seizures in epilepsy, 269
Selective permeability of plasma membrane, 58
Selenium, 343
Self markers, 201
tissue rejection and, 209
Self-replicating RNA/DNA, 517
Semen, 379–380
HIV/AIDS transmission in, 213
Semenya, Caster, 386
Semicircular canals, 289
balance and, 287–288
hair cells in, 277

Semilunar valves, 171
Seminal fluid, 379
Seminal vesicles, 379, 380
Seminiferous tubules, 379
Senescence, 501–502
Sensations, 276
  somatic sensations, 278–281
Sensors, 97–98
Sensory mechanisms.
  aging and, 503
  diseases and disorders of, 294–296
  receptors, 276–278
Sensory nerve endings, 96
Sensory neurons, 245–246
  functions of, 254
  reflexes and, 254
Septicemia, 157
Septum, 170–171
Septuplets, 488
Serine, 41
Serosa
  of gastrointestinal (GI) tract, 328
  of stomach, 332
Serotonin, 252
Serous membranes, 92
Serratus anterior, 125
Sertoli cells, 380–381
  inhibin, secretion of, 382
Serum albumin, 477
Set point, 98
17ß-estradiol, 317, 382
Severe combined immunodeficiency disease (SCID), 211–212
Sewage, combined sewage overflow (CSO), 549–550
Sex education and HIV/AIDS, 189
Sex-influenced phenotypes, 460
Sex-linked inheritance, 458–461
Sextuplets, 488
Sexual response
  female sexual response, 390
  male sexual response, 389–390
  oxytocin and, 308
Sexually transmitted diseases. See STDs
Shaft of hair, 95
Shared responsibility concept, 20
Sharon, Ariel, 242
Sharp pain, 280
Shells, 26–27
Shivering, 124
Short bones, 109
Short-term memory, 265
  Alzheimer's disease and, 269
Sickle-cell anemia, 447
  codominance and, 455–456
  gene therapy for, 479
  hemolytic anemia and, 158
  percutaneous umbilical blood sampling (PUBS) and, 489
Sickle-cell trait, 455–456
Sigmoid colon, 327, 337
Sigmoidoscopy, 439
Sildenafil, 389
Silent receptors, 277
Silicon, 343
Simple columnar epithelium, 83
Simple cuboidal epithelium, 83
Simple epithelium, 83–84
Simple squamous epithelium, 83
Single-celled organisms. See Unicellular organisms
Sinoatrial (SA) node, 176
Sinuses, 110–111, 222
Sinusitis, 111
Skeletal muscle pump, 168
Skeletal muscles. See Muscles
Skeletal system, 90, 102–121
  aging and, 503
  diseases and disorders of, 117
  in fetal development, 496
Skeleton, 109–110
  appendicular skeleton, 112–114
  axial skeleton, 109–112

Skepticism, 16
Skin, 93–97. See also Touch
  adaptation of receptors in, 278
  cell division in skin cells, 418
  dermis, 93–94
  epidermis, 93–94
  functions of, 93
  pathogens, defense from, 196
  sensory receptors in, 279
  smoking and, 96, 227
  suntans and, 96
Skin cancer, 437, 438
  factors leading to, 431–432
  melanoma, 428
  ozone layer and, 548
Skin patch contraceptives, 391–392
  failure rate for, 390
Skull, 109–111
Sleep
  brain and, 263–265
  infants and, 500
  melatonin and, 317
  stages of, 264–267
Sleeping pills, 264
Sliding filament mechanism, 128–130
Slow pain, 280–281
Slow-twitch fibers, 135–136
Small intestine, 327
  endocrine system regulating, 339
  fats, absorption of, 338
  functions of, 333–334
  nervous system regulating, 339
  water, absorption of, 333–334, 338
Smell
  chemoreceptors and, 282–283
  receptors, 277
Smith, Jennifer, 275
Smog, 548–549
Smokeless tobacco, 223, 432
Smoker's cough, 224–225
Smoking
  as atherosclerosis risk factor, 173
  breast sagging and, 227
  cancer and, 432, 435, 438–439, 441, 443
  carbon monoxide exposure and, 235
  cell differentiation and, 420
  emphysema and, 236
  heart disease and, 184
  as hypertension risk factor, 179
  lung cancer and, 238, 435, 438–439
  pre-embryonic development and, 490
  respiratory tract and, 224–225
  secondhand smoke, 218–219
  and skin, 96
  urinary bladder cancer and, 441
Smooth endoplasmic reticulum, 65
  in animal cell, 64
Smooth muscle, 88–89, 124
  activation of, 137–138
  defining characteristics of, 139
  and hair follicle, 95
  speed of contraction, 138
  sustainability of contraction, 138
Snakes, migration of, 514
Sneezing, 196
  exhaling and, 228
Snus, 223
Society and science, 19–21
Sodium, 343, 345. See also Salt; Sodium-potassium pump
  aldosterone and, 313
  in blood plasma, 145, 146
  filtration of, 364
  as hypertension risk factor, 179
  ionic bonds and, 30
  ions, 357
  in living organisms, 32
  molecules of, 28
  symbol for, 27
  tubular reabsorption and, 364
Sodium bicarbonate, 335
Sodium chloride (NaCL). See Salt

Sodium-potassium pump, 60, 61–62
  cell volume and, 61–62
  resting potential and, 246–247
Soft paternalism, 219
Solar power, 554
Solutes, 32
Solvents, 32–33
Somatic cell nuclear transfer, 405, 421–422
Somatic cells, 410
Somatic division of PNS, 254–255
Somatic sensations, 278–281
Somatostatin, 303, 318
  actions of, 252
  delta cells secreting, 311
Somites, 513
Soreness after exercise, 133
Sound. See Hearing; Vocalization
Sound waves, 284–287
Sour taste, 282
Sources of scientific information, 15–16
Spanish flu, 526–527
Special senses, 278
Specialized connective tissues, 87–88
Species, 8, 519. See also Human beings
Specific connective tissues, 85
Specific defense mechanisms, 201–206
Speech, 223–224
Sperm, 378–379. See also Cell differentiation; Meiosis
  accessory glands and, 379–380
  artificial insemination, 395
  cell divisions and, 380–381
  eggs, contact with, 390
  fertilization and, 486–489
  FSH (follicle-stimulating hormone) and, 309
  as haploid cells, 415
  head of, 380–381
  infertility and, 394
  mutations in, 410
  tail of, 380–381
  testes producing, 379
Sperm sorting, 377
Spermatids, 380–381
Spermatogonia, 380
  cell division in, 418
Spermicidal agents, 215
Spermicide foam, 392
  failure rate for, 390
Spermicides, 393
SPF (sun protection factor), 96
Sphenoid bone, 110
Sphingolipids, 463
Sphygmomanometer, 178
Spider veins, 96
Spinal cord, 111–112, 257–260
  functions of, 259–260
  spinal nerves and, 254
Spinal cord injuries, 267–268
  repairing, 268
  stem cells and, 50
Spinal nerves, 254
Spinal reflexes, 254–255
Spirometer, 228–229
Spleen, 194
  functions of, 195
  monocytes, storage of, 150
  swelling of, 195
Split axis in graphs, 17, 18
Spongy bone, 104–105
Sports. See Athletics
Sprains, 117
  ankle sprains, treating, 118
Sprue, 347
Squamous cell cancer, 438
Squamous epithelium, 83
Sreekumar, Arun, 357
SRY (Sex-determining Region Y), 495
Stapes, 289
Staphylococcus aureus, 6
  antibiotics and, 15
Starch, 38
Statistics, value of, 16–17

Statoil company, 545
STDs (sexually transmitted diseases), 396–400.
    bacterial STDs, 398–399
    protecting against, 400
    viral STDs, 399
Stegodon, 509
Stem cells, 50–51
    blood from, 87, 147–148
    contoversies regarding, 51
    cord blood, 143–144
    embryonic stem cells (ES cells), 422
    induced pluripotent stem cells (iPS cells), 422
    Parkinson's disease and, 270
    platelets and, 147–148
    in red bone marrow, 104
    in vitro fertilization (IVF) and, 380
    white blood cells and, 149
Sterilization surgery, 391
    failure rate for, 390
Sternum, 112
Steroids, 40, 304–306
    anabolic steroids, 122–123
Stimulus
    adaptation of receptors to, 278
    incoming impulses, interpretation of, 277
    receptors and, 276–278
    stimulus-twitch relationship, 134–135
Stomach, 327. *See also* Peptic ulcers
    contractions of, 333
    endocrine system regulating, 339
    functions of, 332–333
    hormones produced by, 303
    nervous system regulating, 339
    wall of, 332
Stratification sedimentary deposits, 511
Stratified epithelium, 83–84
Strength training, 136–137
Streptococcus, 191
Stress
    BMR (basal metabolic rate) and, 346
    glomerular filtration and, 363
    heart attacks and, 185
    as hypertension risk factor, 179
Stretch reflex, 255
Stretching and sports injuries, 137
Striated muscle, 138–139
Stroke volume, 180
Strokes, 184
    atherosclerosis and, 173
    cerebral embolisms and, 184
    HRT (hormone replacement therapy) and, 504
Structural genes, 430
Sub-Saharan Africa, HIV/AIDS in, 188–189, 214
Subclavian artery, 172
Subclavian vein, 172
Subdural hematoma, 267
Sublingual gland, 330–331
Submandibular gland, 330–331
Submucosa
    of gastrointestinal (GI) tract, 328
    of stomach, 332
Substrate
    of metabolic pathways, 69–70
    phosphorylation, substrate-level, 71
Succession, 532–533
Sucrose as oligosaccharide, 38
Sugar. *See also* Glucose
    carbohydrates and, 341
    plants making, 534
Suicide, 235
Sulfur, 345
Sulfur dioxide, 548
Summation, 135, 247
Sunlight
    cancer and sun exposure, 96, 433, 443
    energy from, 5
Superior point of reference, 93
Superior vena cava, 172, 174
Supination, 116
Suppressive therapy for HIV, 193
Surfactant, 226, 498

Surgery
    as cancer treatment, 436
    Cesarean delivery, 498
    sterilization surgery, 390, 391
Surrogate motherhood, 396
Survival of the fittest, 515
Sustainable development, 556–558
    consumption and, 556–557
Swallowing, 331
Swallowing reflex, 331
Sweat. *See* Perspiration
Sweat glands, 95
Sweet taste, 282
Swine flu, 515, 526
Swirski, Filip K., 150
Syler, René, 426–427
Sympathetic nervous system, 244, 255–257
    cardiac output and, 181
Synapses, 251
Synaptic cleft, 251
Synergistic muscles, 125–126
Synovial fluid, 114–115
Synovial joints, 114–115
Synovial membranes, 92
Synthetic life, 518
Syphilis, 397, 399
    bacteria causing, 191
    cell differentiation and, 420
Systemic circuit, 171–172
Systemic cycle, 172–173
Systemic response, 210
Systole period, 174–175
Systolic pressure, 177–179

T

T lymphocytes, 151, 206. *See also* Cytotoxic T cells;
        Helper T cells
    cell-mediated immunity and, 201, 203–206
    cytotoxic T cells, 205, 206
    functions of, 206
    memory T cells, 205
    SCID (severe combined immunodeficiency disease)
        and, 480
    specific defenses and, 201
    thymus gland and, 195
T tubules, 128
T waves, 176–177
Tailbone, 512
Tamiflu, 527
Tanning beds, 96
Target cells, 302
    for ADH (antidiuretic hormone), 306–307
    nonsteroid hormones and, 304–305
    steroid hormones entering, 304
Tarsal bones, 114
Tastants, 282
Taste
    bitter taste, 278
    chemoreceptors and, 282
    receptors, 277
Taste buds, 282
Taxonomic classification of human beings, 519
Tay-Sachs disease, 447, 463–464
Tears, 196
Technology, 19
Tectorial membrane, 285, 286, 289
Teeth
    abscesses, 198
    anatomy of, 330
    chewing and, 330
    dental caries, 330
Telangiectasis, 96
Television, scientific information on, 16
Telomerase
    aging and, 502
    cancer, testing for, 435–436
Telomeres and aging, 502
Telophase, 414
Temperature. *See also* Body temperature; Global warming
    tendons, mechanoreceptors in, 280
    thermoreceptors and, 276–277

    upper respiratory tract and, 222
    of young Earth, 516–517
Temporal bones, 110
Temporal lobe, 262
Tendinitis, 117
Tendon receptors, 277, 281
Tendons, 115–117
    mechanoreceptors in, 280
    muscles, attaching, 126
    PRP (platelet-rich plasma) therapy for, 152
    tendinitis, 117
Tennis elbow, 117
Teriparatide, 119
Termination step of translation, 412–413
Tertiary consumers, 534
Tertiary phase of syphilis, 398
Tertiary structure of protein, 42–43
Testes, 380
    hormones of, 303, 318
    sperm production in, 379
    testosterone production and, 317
Testis-determining factor, 495
Testosterone, 303, 318
    baldness and, 460
    in blood plasma, 145, 146
    doping, 137
    LH (luteinizing hormone) and, 309
    male reproductive capacity and, 380–381
    in prenatal development, 495
    testes producing, 317
Tetanic contraction, 135
Tetanus, 139
Thalamus, 260
    functions of, 261
Theory, hypothesis as, 15
Therapeutic cloning, 405, 422
Thermal inversion, 549
Thermodynamics laws of, 533–534
Thermoreceptors, 276–277, 280
THG (tetrahydrogestrinone), 123
Thiamine, 344
Thimerosal, 8
Thoracic duct, 193–195
Thoracic vertebrae, 111–112
Thought, forebrain and, 261–263
Three-domain system, 7
Threonine, 41, 342
Threshold, 247
Throat. *See* Pharynx
Thrombin, 154
Thrombocytopenia, 159
Thumbs, 113
    opposable, 8, 9
Thymine, 44, 408
Thymopoietin, 303, 317, 318
Thymosin, 303, 317
Thymus gland, 194
    functions of, 195, 317
    hormones of, 303, 318
    lymphocytes and, 194
Thyosin, 318
Thyroid gland. *See also* TSH
    cellular metabolism and, 314–315
    functions of, 313–316
    hormones of, 303, 318
    hyperthyroidism, 320
    hypothyroidism, 320
Thyroid-stimulating hormone. *See* TSH
Thyroxine, 303, 314, 318
    Graves' disease, 320
    hypothyroidism, 320
Tibia, 114
Tibialis anterior, 125
Tidal volume, 228–229
Tight cell junctions, 84
Tilburt, Jon, 34
Tipping points, 545
Tissue membranes, 92–93
Tissue plasminogen activator (tPA), 475
Tissue rejection, 209
    stem cells and, 50–51

Tissues, 9, 82
  black market in, 102–103
  in embryonic development, 491
  epithelial tissues, 82–85
Tobacco. *See also* Smoking
  as carcinogen, 432
  smokeless tobacco, 223
Tolerance for psychoactive drugs, 266
Tone of sound, 284
Tongue, 194
  functions of, 330
  swallowing and, 331
  taste and, 282
Tonicity, 63
Tonsillitis, 195
  bacteria causing, 191
Tonsils, 194
  functions of, 195–196
Touch
  mechanoreceptors and, 279
  receptors, 278–279
Toxemia, 157
Toxic pollutants in water, 550
Toxic shock syndrome, 191
Trabeculae, 104–105
Trace minerals, 343
Trachea, 194, 221, 222
  functions of, 224
Trans fats, 335, 342
Transcendental meditation, 185
Transcription, 408, 410–411
Transfer RNA (tRNA), 412–413
Transfusion reaction, 155, 209
Transfusions. *See* Blood transfusions
Transgenic plants, 468
Translation, 408, 412–413
Translocation of chromosomes, 463
Transmissibility of disease, 526
Transmission electron microscope (TEM), 54
Transplantation of organs, 11
Transplants
  cord blood transplants, 143
  encouraging donations, 365
  kidney transplants, 354–355, 373
  tissue rejection and, 209
  "tourism," 359
Transport protein, 59
Transverse plane, 93
Trapezius, 125
Trauma and nervous system, 267
Travis, John, 306
Treponema pallidum, 398
Triassic extinction, 515
Triceps brachi, 125
Trichomonas vaginalis, 400
Trichomoniasis, 397, 400
Tricuspid valve, 171
Triglycerides, 39, 338
  absorption of, 338
Triiodothyronine, 303, 314, 318
Triphosphate, 46
Triplet code, 411
Triplets, 488
Trisomy 13, 461, 462
Trisomy 18, 461, 462
Trisomy 21, 461–462
Trisomy-X syndrome, 462
tRNA (transfer RNA), 412–413
Trophoblasts, 490, 492
Tropomyosin, 130
Troponin, 130
Troponin-tropomyosin protein complex, 130
Trypsin, 334–335
Tryptophan, 41, 342
  foods low in, 343
TSH (thyroid-stimulating hormone), 303, 307
  functions of, 309
  iodine deficiency and, 315
  thyroxine and, 314–315
Tubal ligation, 390, 391

Tuberculosis (TB), 237
  bacteria causing, 191
  HIV/AIDS and, 213
Tubular reabsorption, 362, 364–365
Tubular secretion, 362, 365
Tumor suppresser genes, 430–431
Tumors, 428. *See also* Cancer
  brain tumors, 268–269
  imaging of, 434–435
Turner syndrome, 462
Twins
  DNA (deoxyribonucleic acid) of, 458
  fraternal twins, 488–489
  identical twins, 488–489
Twitches, 134
Two-trait crosses, 452–454
Tympanic canal, 285, 286, 289
Tympanic membrane, 284
Type 1 diabetes. *See* Diabetes
Type 2 diabetes, 300–301, 320
Type A blood, 155
Type AB blood, 155
Type B blood, 155
Type O blood, 155
Tyrosine, 41

U

Ulcers. *See* Peptic ulcers
Ulna, 113
Ultraviolet (UV) light, 548
  as carcinogen, 432, 433
  and skin, 96
Umami, 282
Umbilical cord, 492, 493
  cord blood from, 143–144
  percutaneous umbilical blood sampling (PUBS) and, 489
UNAIDS, information from, 214
Undifferentiated cells, re-creating, 422
Unencapsulated dendritic endings, 279
Unicellular organisms, 5, 82, 516
  homeostasis in, 6
United Nations
  on AIDS in Africa, 188
  Population Division, 531
  UNAIDS, information from, 214
United Network for Organ Sharing (UNOS), 355
Universal blood donors, 157
Universal blood recipients, 157
Unsaturated fats, 39, 342
Upper respiratory infections (URIs), 237
Upper respiratory tract, 221
  functions of, 222
Urban stream erosion, 549–550
Urbanization, 556
  and water, 549–550
Urea, 357
  in blood plasma, 145, 146
  filtration of, 364
  lipid bilayer, crossing through, 58
Ureters, 358
Urethra, 359
Urethral sphincter, 367
Urinary bladder, 358–359
  cancer, 437, 441
Urinary system, 91. *See also* Kidneys; Urine
  aging and, 504
  diseases and disorders of, 372–373
  homeostasis and, 356–357
  organs of, 358–359
Urinary tract infections (UTIs), 372
Urine
  concentrated urine, production of, 366–367
  countercurrent exchange mechanism, 367, 368
  dilute urine, production of, 366
  disease diagnosis and, 357
  formation of, 362–365
  glomerula filtration and, 362–364
  kidney stones blocking flow, 372
  nephrons producing, 360–361
  pathogens, ridding body of, 196
  production of, 356–357

tubular reabsorption and, 362, 364–365
tubular secretion and, 362, 365
US Preventive Services Task Force (USPSTF), 440
USA Track & Field (USATF), 137
USDA (U.S. Department of Agriculture)
  MyPyramid, 340–341
  sweeteners, consumption of, 341
Uterine cancer, 437, 442
Uterine cycle, 386–387
Uterus, 382–383
  oxytocin and, 307–308
Utricle, 289
  hair cells in, 277
  head position, sensing, 288
UUC codon, 411
UUU codon, 411
UV light. *See* Ultraviolet (UV) light

V

Vaccari, David A., 538
Vaccines, 207. *See also* Immunizations
  for birth control, 394
  for flu pandemics, 527
  Gardasil, 399
  in genetically modified foods, 207–208
  for HIV/AIDS, 215
  transgenic bacteria for, 475
  transgenic plans for, 476
Vagina, 383–384
  acidity of, 379–380
  birth and, 497–498
  in prenatal development, 495
Vaginal contraceptive ring, 390
Vaginal secretions
  HIV/AIDS transmission in, 213
  pathogens, defense to, 196
Valine, 41, 342
Valves
  in heart, 171, 175–176
  in veins, 168
Vanadium, 343
Variables
  in controlled experiments, 13–14
  in graphs, 17
Varicose veins, 168
Vasa recta, 361
  countercurrent exchange mechanism, 367, 368
Vasectomy, 390, 391
Vasoconstriction, 166
Vasodilation, 166, 180
  inflammation and, 198–199
vCreutzfeldt-Jakob disease (vCJD), 192
Vectors, gene transfers with, 479–480
Vegetable oils, 342
Vegetarians, 342–343
Veins, 164
  cardiac veins, 172
  of embryonic circulation, 493
  functions of, 168–169
  major veins, 172
  pressures of breathing and blood and, 169
  pulmonary circuit and, 171–172
  skeletal muscles and, 168
  structures of, 165
  valves in, 168
  varicose veins, 168
Ventilation, 220
Ventral root, 254
Ventricles, 170–171
  for cerebrospinal fluid, 258
Ventricular fibrillation, 177
Ventricular systole, 174–175
Venules, 168
  in systemic cycle, 172–173
Vertebrae, 111–112
Vertebral column, 111–112
Vertebrates, 8
Vesicles, 60
  endocytotic vesicles, 66
  lysosomes, 66–67
  peroxisomes, 66–67

role of, 66–67
secretory vesicles, 66
shipping/storing cellular products, 66
Vestibular apparatus, 285, 287
balance and, 287–288
Vestibular canal, 285, 286, 289
Vestibule, 289. *See also* Saccule; Utricle
balance and, 287
head position, sensing, 288
otoliths in, 288
Vestigial structures, 512–513
Viagra, 3819
Vibration
in inner ear, 286–287
mechanoreceptors and, 279
of tympanic membrane, 284
Villi, 334
Virulence
of disease, 526
of pathogens, 192–193
Viruses, 191–192
alpha interferon and, 200
antigenic shift in, 515
cancer and, 431–432
colds and, 237
Ebola virus, 193
genetic shift and, 527
hepatitis and, 348
influenza and, 237
memory cells and, 206–207
opportunistic viruses, 213
STDs (sexually transmitted diseases), viral, 399
vectors and, 479–480
Vision, 289–294. *See also* Blindness; Rods and cones
abnormal vision, 291–292
midbrain and, 261
receptors, 277
Vital capacity, 228–229
Vitale, Susan, 292
Vitamin A, 344
as antioxidant, 433
golden rice and, 472
Vitamin B$_1$, 344
Vitamin B$_2$, 344
Vitamin B$_6$, 343, 344
Vitamin B$_9$, 344
Vitamin B$_{12}$, 344
pernicious anemia and, 158
Vitamin C, 344
as antioxidant, 433
Vitamin D, 343, 344
kidneys activating, 372
osteoporosis and, 118–119
PTH (parathyroid hormone) and, 315–316
skin and, 96
Vitamin E, 344
as antioxidant, 433
Vitamin K, 343, 344
bacteria releasing, 337
Vitamins
absorption of, 338–339
list of, 344
nutrition and, 343

Vitreous humor, 289–290
Vocal cords, 223
Vocalization
larynx and, 223
production of sound, 223–224
respiratory system and, 220
Vogel, Gretchen, 422
Voice box. *See* Larynx
Volatile organic compounds (VOCs), 558
Voluntary breast/ovary removal, 426–427
Voluntary muscle, 88, 124
Vomer bone, 110
Vomiting
bulimia and, 350
chemotherapy and, 436
pathogens, ridding body of, 196
Vulva, 383

W
Walking, humans and, 521–522
Wang, Jue, 185
Wars, environmental damage and, 552–553
Wastes. *See also* Urine
in blood plasma, 145, 146
Water, 32–33. *See also* Urine
ADH (antidiuretic hormone) and, 306–307, 368–369
aldosterone and, 313
as biological solvent, 32–33
in blood plasma, 145, 146
body temperature and, 33
diffusion of, 58
filtration of, 364
gain and loss, sources of, 357
as greenhouse gas, 546
human activities polluting, 550–551
hydrogen bonds and, 30–31
kidneys regulating, 357, 368–369
lipids and, 39
molecules of, 28
osmosis, absorption by, 338
photosynthesis and, 518, 534
pollution and, 549–552
scarcity of, 549
small intestine absorbing, 333–334, 338
urbanization and, 549–550
Water cycle, 538–539
Water rights, 549
Water-soluble vitamins, 343
absorption of, 338
list of, 344
WBCs. *See* White blood cells
Weight. *See also* Obesity; Overweight
cancer and, 443
eating disorders and, 348–349
energy balance and, 346
fat cells and, 88
healthy weight, 346–347
heart disease and, 185
mass and, 27
Weight control, 345–347
BMR (basal metabolic rate) and, 345–346
eating disorders and, 350
friends and, 329

Well water, pollutants in, 551
White blood cells, 87, 145, 146, 149–151. *See also* Agranular leukocytes; Granular leukocytes
life span of, 149–150
pathogens, destruction of, 197–198
White matter, 260
White pulp, 194, 195
WHO (World Health Organization)
on BMI (body mass index), 326
kidneys, buying and selling of, 359
on obesity, 348–349
on pandemic influenza outbreaks, 527
on Type 2 diabetes, 300
Whooping cough, 2
vaccines and, 207
Widow's peak, allele for, 450–451
Willed Body Programs, 103
Wilmut, Ian, 404
Wind farms-554, 553
Windpipe. *See* Trachea
Withdrawal method for contraception, 393
Withdrawal reflex, 254–255
Working hypothesis, 13
World Anti-Doping Agency, 137
World Commission on Environment and Development, 1987, 556
World Food Day, 472
World Health Organization. *See* WHO

X
X chromosomes, 459
gender of fetus, determining, 376–377
nondisjunction of, 462
sex-linked inheritance, 458–461
XO individuals, 462
XXX individuals, 462
XXY individuals, 462
XYY individuals, 462

Y
Y chromosomes, 459
gender development and, 376–377, 495
nondisjunction of, 462
sex-linked inheritance, 458–461
Yeast infections, 397, 399–400
Yellow bone marrow, 104
Yolk sac, 492
Yung, Chong W., 158

Z
Z-line, 127–128
Zero population growth, 531
Zinc, 345
Zona pellucida, 385, 486
fertilization of egg and, 390
Zygomatic bones, 110
Zygote intrafallopian transfer (ZIFT), 396
Zygotes, 486